CASE STUDIES

W9-CDA-072

ESSAYS

BIOLOGY
LIFE ON EARTH

ELEVENTH EDITION

BIOLOGY
LIFE ON EARTH

Teresa Audesirk
UNIVERSITY OF COLORADO DENVER

Gerald Audesirk
UNIVERSITY OF COLORADO DENVER

Bruce E. Byers
UNIVERSITY OF MASSACHUSETTS AMHERST

PEARSON

Editor-in-Chief: Beth Wilbur
Senior Acquisitions Editor: Star Burruto Mackenzie
Project Manager: Mae Lum
Program Manager: Leata Holloway
Development Editors: Erin Schnair
    and Kari Hopperstead
Editorial Assistant: Maja Sidzinska
Executive Development Manager: Ginnie Simione Jutson
Program Management Team Lead: Michael Early
Project Management Team Lead: David Zielonka
Production Management: Cenveo® Publisher Services
Copyeditor: Joanna Dinsmore
Proofreaders: Pete Shanks and Juliana Lewis
Indexer: Liz Schlembach

Compositor: Cenveo® Publisher Services
Design Manager: Marilyn Perry
Cover and Interior Designer: Elise Lansdon
Illustrators: Imagineering Art
Rights & Permissions Project Manager: Donna Kalal
Rights & Permissions Management: Candice Velez,
    QBS Learning
Photo Researcher: Kristin Piljay, Wanderlust Photos
Manufacturing Buyer: Stacey Weinberger
Executive Marketing Manager: Lauren Harp
Front Cover Photo Credit: John E. Marriott/All Canada
    Photos/Corbis
Back Cover Photo Credit: Markus Varesvuo/Nature
    Picture Library

Library of Congress Cataloging-in-Publication Data
Names: Audesirk, Teresa, author. | Audesirk, Gerald, author. | Byers, Bruce E.,
    author.
Title: Biology: Life on Earth / Teresa Audesirk, University of Colorado,
    Denver; Gerald Audesirk, University of Colorado, Denver; Bruce E. Byers,
    University of Massachusetts.
Description: Eleventh Edition. | Hoboken, New Jersey: Pearson, [2017]
Identifiers: LCCN 2015040314
Subjects: LCSH: Biology.
Classification: LCC QH308.2 .A93 2017 | DDC 570—dc23 LC record available at http://lccn.loc.gov/2015040314

3 17

www.pearsonhighered.com

ISBN 10: 0-134-16829-1; ISBN 13: 978-0-134-16829-6 (Student edition)
ISBN 10: 0-134-14295-0; ISBN 13: 978-0-134-142-951 (Books a la Carte
    edition)

ABOUT THE AUTHORS

TERRY AND GERRY AUDESIRK grew up in New Jersey, where they met as undergraduates, Gerry at Rutgers University and Terry at Bucknell University. After marrying in 1970, they moved to California, where Terry earned her doctorate in marine ecology at the University of Southern California and Gerry earned his doctorate in neurobiology at the California Institute of Technology. As postdoctoral students at the University of Washington's marine laboratories, they worked together on the neural bases of behavior, using a marine mollusk as a model system.

They are now emeritus professors of biology at the University of Colorado Denver, where they taught introductory biology and neurobiology from 1982 through 2006. In their research, funded primarily by the National Institutes of Health, they investigated the mechanisms by which neurons are harmed by low levels of environmental pollutants and protected by estrogen.

Terry and Gerry are long-time members of many conservation organizations and share a deep appreciation of nature and of the outdoors. They enjoy hiking in the Rockies, walking and horseback riding near their home outside Steamboat Springs, and singing in the community chorus. Keeping up with the amazing and endless stream of new discoveries in biology provides them with a continuing source of fascination and stimulation. They are delighted that their daughter Heather has become a teacher and is inspiring a new generation of students with her love of chemistry.

BRUCE E. BYERS is a Midwesterner transplanted to the hills of western Massachusetts, where he is a professor in the biology department at the University of Massachusetts Amherst. He has been a member of the faculty at UMass (where he also completed his doctoral degree) since 1993. Bruce teaches courses in evolution, ornithology, and animal behavior, and does research on the function and evolution of bird vocalizations.

With love to Jack, Lori, and Heather and in loving memory of Eve and Joe
— T. A. & G. A.

In memory of Bob Byers, a biologist at heart.
—B. E. B.

ABOUT THE COVER A young boreal owl (*Aegolius funereus*) peers out of a cavity. Boreal owls take their name from the boreal forest, the vast, northern coniferous forest in which they live. The owls inhabit boreal forest in Scandinavia, Siberia, Canada, and Alaska, as well as mountain forests a bit further south. Boreal owls hunt at night, using their keen hearing to find the mice, voles, and other small mammals that make up most of their diet. The owls do not build nests. Instead, a female lays her eggs in a cavity in a tree, often one excavated and abandoned by a woodpecker. About a month later, the eggs hatch. For another month or so, the young owls remain in the cavity, subsisting on food brought by their parents. Eventually, though, a young owl ventures to the mouth of the cavity and prepares to take flight. It will live out its life in a corner of the boreal forest, which is also home to endangered species such as the Amur tiger and the Siberian crane. Unfortunately, the boreal forest biome is threatened by widespread logging and a warming climate.

BRIEF CONTENTS

DETAILED CONTENTS

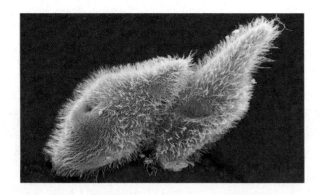

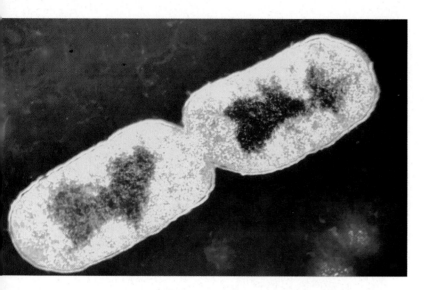

PREFACE

THE CASE FOR SCIENTIFIC LITERACY

Climate change, biofuels versus food and forests, bioengineering, stem cells in medicine, potential flu pandemics, the plight of polar bears and pandas, human population growth and sustainability: these are just some of the very real, urgent, and interrelated concerns sweeping our increasingly connected human societies. The Internet places a wealth of information—and a flood of misinformation—at our fingertips. Never have scientifically literate students been more important to humanity's future. As educators, we feel humbled before this massive challenge. As authors, we feel hopeful that the Eleventh Edition of *Biology: Life on Earth* will help lead introductory biology students along paths to understanding.

Scientific literacy requires a foundation of factual knowledge that provides a solid and accurate cognitive framework into which new information can be integrated. But more importantly, it endows people with the mental tools to separate the wealth of data from the morass of misinformation. Scientifically literate citizens are better able to evaluate facts and to make informed choices in both their personal lives and the political arena.

This Eleventh Edition of *Biology: Life on Earth* continues our tradition of:

- Helping instructors present biological information in a way that will foster scientific literacy among their students.
- Helping to inspire students with a sense of wonder about the natural world, fostering an attitude of inquiry and a keen appreciation for the knowledge gained through science.
- Helping students to recognize the importance of what they are learning to their future roles in our rapidly-changing world.

WHAT'S NEW IN THIS EDITION?

Each new edition gives the authors a fresh opportunity to ponder: "What can we do better?" With extensive help from reviewers, development editors, and our coauthors, we've answered this question with the following changes organized around three major goals:

Highlight an Inquiry-Driven Approach to Learning

- **Probing questions at the end of the extensively revised "Case Study Continued" segments** help students anticipate what they will learn.
- **Three unique question types in essays and figure captions** encourage students to think critically about the content: "Think Critically" questions focus on solving

problems, thinking about scientific data, or evaluating a hypothesis; "Evaluate This" questions ask students to interpret or draw conclusions from a hypothetical scenario; and "Consider This" questions invite students to form an opinion or pose an argument for or against an issue, based on valid scientific information. Answers to "Think Critically" and "Evaluate This" questions are included in the back of the book; hints for "Consider This" questions are included on MasteringBiology.
- **New multiple choice questions at the end of every chapter** address students' recall and comprehension and help them prepare for tests.

Create Connections for Students

- **"Health Watch" essays often include an "Evaluate This" question,** encouraging students to connect health topics to practical, real-world examples.
- **"Threads of Life" themes** in pertinent chapters weave together what may otherwise appear to be unrelated fields within the uniquely diverse science of biology. These threads—identified in our list of changes by chapter below—are the unifying theme of *Evolution*, the exploding science of *Biotechnology*, our increasing recognition of the impacts of *Climate Change*, and our emerging understanding of the importance of *Microbiomes* throughout the living world.
- **Dozens of entirely new and revised figures** illustrate concepts more clearly and engagingly than ever before. For example, negative feedback cycles are now illustrated in a consistent manner that allows students to instantly recognize the chain of events and relate it to negative feedback events in other chapters.

Encourage Critical Thinking

- **New "How Do We Know That?" essays** show students the process of science in a simple way, emphasizing the process and method to what scientists do. Essays go into the details of experiments, highlighting exciting technology and data. "How Do We Know That?" features include "Think Critically" or "Consider This" questions, encouraging students to analyze data or engage with the topics presented in the essay.
- **"Earth Watch" essays include more data.** Students will find more examples of real scientific data in the form of graphs and tables; the data are accompanied by "Think Critically" questions that challenge students to interpret the data, fostering increased understanding of how science is communicated.

In addition, **mitosis and meiosis are now covered in separate chapters** (Chapters 9 and 10, respectively), so students gain a stronger foundational understanding of some of the toughest topics in biology.

BIOLOGY: LIFE ON EARTH, ELEVENTH EDITION

... Is Organized Clearly and Uniformly

Navigational aids help students explore each chapter. An important goal of this organization is to present biology as a hierarchy of closely interrelated concepts rather than as a compendium of independent topics.

- Major sections are introduced as broad questions that stimulate students to think about the material to follow; subheadings are statements that summarize their specific content.
- A "Summary of Key Concepts" section ends each chapter, providing a concise, efficient review of the chapter's major topics.

... Engages and Motivates Students

Scientific literacy cannot be imposed on students—they must actively participate in acquiring the necessary information and skills. To be inspired to accomplish this, they must first recognize that biology is about their own lives. For example, we help students acquire a basic understanding and appreciation of how their own bodies function by including information about diet and weight, cancer, and lower back pain.

We fervently hope that students who use this text will come to see their world through keener eyes. For example, they will perceive forests, fields, and ponds as vibrant and interconnected ecosystems brimming with diverse life-forms rather than as mundane features of their everyday surroundings. If we have done our job, students will also gain the interest, insight, and information they need to look at how humanity has intervened in the natural world. If they ask the question, "Is this activity sustainable?" and then use their new knowledge and critical thinking skills to seek some answers, we can be optimistic about the future.

In support of these goals, the Eleventh Edition has updated features that make Biology more engaging and accessible.

- **Case Studies** Each chapter opens with an attention-grabbing "Case Study" that highlights topics of emerging relevance in today's world. Case Studies, including "Unstable Atoms Unleashed" (Chapter 2), "New Parts for Human Bodies" (Chapter 4), and "Unwelcome Dinner Guests" (Chapter 20), are based on news events, personal interest stories, or particularly fascinating biological topics. "Case Study Continued" segments weave the topic throughout the chapter, whereas "Case Study Revisited" completes the chapter, exploring the topic further in light of the information presented.

- **Boxed Essays** Four categories of essays enliven this text. "Earth Watch" essays explore pressing environmental issues; "Health Watch" essays cover important or intriguing medical topics; "How Do We Know That?" essays explain how scientific knowledge is acquired; and "In Greater Depth" essays make this text versatile for in-depth levels of instruction.

- **"Have You Ever Wondered" Questions** These popular features continue to demystify common and intriguing questions, showing the application of biology in the real world.

- **End-of-Chapter Questions** The questions that conclude each chapter allow students to review the material in different formats—multiple choice, fill-in-the-blank, and essay—that help them to study and test what they have learned. Answers to the multiple choice and fill-in-the-blank questions are included in the back of the book. Answers or hints for the essay questions are included on MasteringBiology.

- **Key Terms and a Complete Glossary** Boldfaced key terms are defined clearly within the text as they are introduced. These terms are also listed at the end of each chapter, providing users with a quick reference to the chapter's important vocabulary. The glossary, carefully written by the authors, provides exceptionally complete definitions for all key terms, as well as for many other important biological terms.

... Is a Comprehensive Learning Package

The Eleventh Edition of *Biology: Life on Earth* is a complete learning package, providing updated and innovative teaching aids for instructors and learning aids for students.

CHAPTER-BY-CHAPTER SUMMARY OF IMPORTANT CHANGES

Following the revision of chapters in response to reviews by instructors and experts, the text and artwork were carefully reviewed by each of the other two authors and the development editors. The coauthors provided valuable insights to one another, integrating the chapters more thoroughly, improving consistency between chapters, and explaining complex concepts more clearly. Our development editors brought trained eyes for order and detail to our work, helping us make the writing even more student-friendly. Following this intense scrutiny, each initial revision underwent a second, sometimes extensive revision. Specific changes include the following:

- **Chapter 1: An Introduction to Life on Earth** includes an entirely updated Case Study to reflect the recent Ebola epidemic. A new "Have You Ever Wondered: Why Scientists Study Obscure Organisms?" highlights unforeseen benefits that have emerged from investigating different organisms. Our *Evolution* "Thread of Life" is emphasized throughout and *Climate Change* is noted in the context of evolution.

UNIT 1 The Life of the Cell

- **Chapter 2: Atoms, Molecules, and Life** offers improved coverage of the unique properties of water. The essay "How Do We Know That? Radioactive Revelations" includes new PET images. The essay "Health Watch: Free Radicals—Friends and Foes?" incorporates new findings on antioxidant supplements. Figures 2-1, 2-2, 2-3, 2-4, 2-5, and 2-6 have been revised for greater clarity and consistency.

- **Chapter 3: Biological Molecules** now covers lipids last, because they are distinct in their structural diversity and in not forming polymers. The discussion of protein structure and intrinsically disordered proteins has been extensively revised. The "Health Watch" essay on trans fats and cholesterol has been extensively updated and rewritten, as has the "Have You Ever Wondered" essay on hair structure. Figures 3-1 and 3-3 and Table 3-2 have undergone major revisions.

- **Chapter 4: Cell Structure and Function** features an entirely new Case Study supporting our *Biotechnology* thread. There is new art for relative sizes as well as enhanced coverage and new art of the extracellular matrix and cytoskeleton (Figures 4-1, 4-6, and 4-7, respectively). Prokaryotic cells are now covered before eukaryotic cells. A new "Earth Watch" essay discusses the environmental impact of raising livestock and the culturing of cow muscle in the lab. "Have You Ever Wondered" has been revised and introduces our *Microbiome* thread.

- **Chapter 5: Cell Membrane Structure and Function** includes upgraded figures of the plasma membrane (Figure 5-1), phospholipids (Figure 5-2), membrane receptors (Figure 5-3), osmosis (Figure 5-6), and surface/volume relationship (Figure 5-13). Added micrographs illustrate cell junctions (Figure 5-14). The "How Do We Know That?" essay on aquaporins has been updated and now includes a data figure. Membrane fluidity has now been incorporated into a "Health Watch" essay, and there is a new "Have You Ever Wondered" essay describing how antibiotics destroy bacteria and supporting our *Evolution* thread.

- **Chapter 6: Energy Flow in the Life of a Cell** includes an updated Case Study, as well as revised art of coupled reactions (Figure 6-7), feedback inhibition (Figure 6-12), and regenerative braking (Figure E6-1). There are new images for entropy (Figure 6-3), activation energy (Figure 6-5b), and food preservation (Figure 6-14). Our explanation of the second law of thermodynamics now uses the phrase "isolated system." The section on solar energy incorporates the *Climate Change* thread. The revised "Health Watch" essay on lactose intolerance supports our *Evolution* thread and a revised "Have You Ever Wondered" about glowing plants supports our *Biotechnology* thread.

- **Chapter 7: Capturing Solar Energy: Photosynthesis** has a revised and updated Case Study, a new overview figure (Figure 7-1), and a chloroplast micrograph added to the figure illustrating photosynthetic structures (Figure 7-3). Figures describing energy transfer in the light reactions (Figure 7-7) and the C_4 and CAM pathways (Figures E7-1 and E7-2) have been significantly improved. The section The Calvin Cycle Captures Carbon Dioxide incorporates the *Biotechnology* thread. The "Earth Watch" essay on biofuels has been updated and supports our *Climate Change* thread.

- **Chapter 8: Harvesting Energy: Glycolysis and Cellular Respiration** features an entirely new Case Study on on the use of mitochondrial DNA in the identification of King Richard III of England. The essay "Health Watch: How Can You Get Fat by Eating Sugar?" has new art showing the conversion of sugar to fat. A micrograph of the mitochondrion has been added to Figure 8-4; the electron transport chain in Figure 8-6 has been redrawn; a new Figure 8-8 illustrates energy extraction from foods; and a new Table 8-1 summarizes glucose breakdown.

UNIT 2 Inheritance

- **Chapter 9: Cellular Reproduction** now covers only mitotic cell division and the control of the cell cycle; meiotic cell division and its importance in sexual reproduction are discussed in Chapter 10. Chapter 9 opens with a new Case Study describing the potential of stem cell therapy for healing injuries. Figure 9-2 illustrates the two important properties of stem cells: self-renewal and the ability of their daughter cells to differentiate into multiple cell types. Cloning is briefly introduced as a technology-based form of asexual reproduction continuing the *Evolution* thread.

- **Chapter 10: Meiosis: The Basis of Sexual Reproduction** begins with a new Case Study, which illustrates how the genetic variability produced by meiosis can be strikingly visible in everyday life. Descriptions of disorders such as Down syndrome and Turner syndrome have been moved into this chapter. A new "How Do We Know That?" essay describes hypotheses and experiments that explore selective forces that may favor the evolution of sexual reproduction, continuing the *Evolution* thread.

- **Chapter 11: Patterns of Inheritance** now includes photos in Figure 11-21, showing how the world looks to color-deficient people—highly accurate images, as verified by the color-deficient author. The "Have You Ever Wondered" essay on the inheritance of body size in dogs includes new information.

- **Chapter 12: DNA: The Molecule of Heredity** now features a streamlined description of the seminal Hershey-Chase experiment in "How Do We Know That? DNA Is the Hereditary Molecule."

- **Chapter 13: Gene Expression and Regulation** contains a revised and updated "Health Watch" essay on epigenetic control of gene expression.

- **Chapter 14: Biotechnology** begins with a new Case Study. The entire chapter has been updated with current information, including recently developed methods for using single-nucleotide polymorphisms to provide information on physical characteristics of both living and ancient humans; possible applications of biotechnology in environmental bioengineering; and using DNA microarrays to diagnose both inherited disorders and infectious diseases. The "How Do We Know That?" essay on prenatal genetic screening asks the students to use their knowledge of forensic DNA and prenatal testing in a simulated paternity case.

UNIT 3 Evolution and Diversity of Life

- **Chapter 15: Principles of Evolution** includes a largely new "How Do We Know That?" essay describing some of the evidence that led Darwin to formulate his theory. The section on evidence of natural selection in the wild includes a new example. "Earth Watch: People Promote High-Speed Evolution" supports our *Climate Change* thread.

- **Chapter 16: How Populations Evolve** includes a revised explanation of how population size affects genetic drift, with a new accompanying figure (Figure 16-5). The "In Greater Depth" essay includes a new figure to aid visualization of the Hardy–Weinberg principle. The section on mutation has been updated to reflect the latest research on mutation rates. A new "Health Watch" essay describes a Darwinian approach to thinking about cancer.

- **Chapter 17: The Origin of Species** presents a new Case Study about the discovery of new species. New, data-based graphics have been added to "Earth Watch: Why Preserve Biodiversity?" and "How Do We Know That? Seeking the Secrets of the Sea."

- **Chapter 18: The History of Life** includes a new Case Study about how our newfound ability to recover and sequence ancient (fossil) DNA provides insight into evolutionary history. We include updated information on fossils found since the previous edition. All dates have been updated to reflect the latest Geological Society revisions of the geological time scale. The human evolution section now contains information about *Homo floresiensis*. There is a new photo of a protist with an algal endosymbiont (Figure 18-6); new photos of early hominin tools (Figure 18-15); and a new artist's conception of a Carboniferous landscape (Figure 18-8).

- **Chapter 19: Systematics: Seeking Order Amid Diversity** includes a new "Have You Ever Wondered" essay about using systematics to estimate how long ago humans began to wear clothing. The account of current views on taxonomic ranks has been streamlined. Text and figures in "In Greater Depth: Phylogenetic Trees" have been revised for increased clarity.

- **Chapter 20: The Diversity of Prokaryotes and Viruses** presents a revised section on prokaryotic systematics that now includes descriptions of some specific clades. A new Table 20-1 summarizes the differences between Archaea and Bacteria. The chapter includes new descriptions of photosynthetic and subterranean bacteria. "Health Watch: Is Your Body's Ecosystem Healthy?" supports our *Microbiome* thread.

- **Chapter 21: The Diversity of Protists** includes a new "Health Watch" essay about diseases caused by protists. The sections on brown algae and red algae now include information on foods derived from those organisms. The description of chlorophytes has been revised to reflect improved understanding of the group's phylogeny, and the section also supports our *Biotechnology* thread. The chapter contains new photos of a parabasalid (Figure 21-3), a dinoflagellate (Figure 21-8), and chlorophytes (Figure 21-19).

- **Chapter 22: The Diversity of Plants** includes a new essay, "Health Watch: Green Lifesaver," about an important antimalarial derived from a plant, highlighting our *Biotechnology* thread. A new figure (Figure 22-3) illustrates some key adaptations for life on land.

- **Chapter 23: The Diversity of Fungi** contains a new essay, "Earth Watch: Killer in the Caves," which describes a fungal disease that threatens bat populations. The chapter contains new information on an airborne fungal disease of humans, the dangers of toxic mushrooms, and fungi known only from DNA sequences. A new segment on genetically engineered resistance to chestnut blight supports our *Biotechnology* thread.

- **Chapter 24: Animal Diversity I: Invertebrates** includes a new "Earth Watch" essay about coral reef bleaching. "How Do We Know That? The Search for a Sea Monster" focuses on the most recent expedition to search for giant squids. All species counts are updated to reflect the latest numbers from the Catalogue of Life.

- **Chapter 25: Animal Diversity II: Vertebrates** contains a new "Have You Ever Wondered" about shark attacks. The chapter contains new information about hagfish slime and new information about snake digestive physiology. "Earth Watch: Frogs in Peril" has been updated with new information and a new graph. All species counts are updated to reflect the latest numbers from the Catalogue of Life.

UNIT 4 Behavior and Ecology

- **Chapter 26: Animal Behavior** has been extensively revised and updated, including new material and many new figures.

- **Chapter 27: Population Growth and Regulation** opens with a new Case Study on the crash and subsequent regrowth of populations of northern elephant seals. Figure 27-1, illustrating exponential growth, has been revised. Section 27.3 offers a new discussion of life history strategies and their evolution, which also supports our *Evolution* thread. The chapter has been updated with current statistics and figures related to the growth of the human population.

- **Chapter 28: Community Interactions** begins with a new Case Study about endangered Channel Island foxes. Section 28.1 has been expanded to describe the different types of community interactions. Section 28.3 has been extensively revised to describe consumer–prey interactions as a general category that includes all situations in which one organism (the consumer) feeds on another (the prey), and encompasses predation (including herbivory) and parasitism. A new "Have You Ever Wondered" essay explains why rattlesnakes rattle. A new "Health Watch" essay explores how coevolution between parasites and their hosts can produce a range of outcomes, supporting our *Microbiome* thread.

- **Chapter 29: Energy Flow and Nutrient Cycling in Ecosystems** includes updated information on atmospheric carbon dioxide and supports our *Climate Change* thread. A new "How Do We Know That?" essay explores the ways in which scientists monitor Earth's conditions. The "Health Watch" essay on biological magnification includes a new figure.

- **Chapter 30: Earth's Diverse Ecosystems** provides a clear explanation of why global average temperature decreases with latitude, including a new illustration in Figure 30-2a. Descriptions of monsoons and the El Nino/Southern Oscillation have been added to Section 30.2.

- **Chapter 31: Conserving Earth's Biodiversity** opens with a new Case Study of the effects of extirpating, and then reintroducing, wolves in Yellowstone National Park. The description of ecosystem services is now organized into the four categories used by the *Millennium Ecosystem Assessment* and The Economics of Ecosystems and Biodiversity (TEEB). There are new images of rain-forest destruction (Figure 31-4) and wildlife corridors (Figure 31-8).

ACKNOWLEDGMENTS

Biology: Life on Earth enters its 11th edition invigorated by the oversight of the excellent team at Pearson. Beth Wilbur, our Editor-in-Chief, continues to oversee the huge enterprise with the warmth and competence that makes her such an excellent leader. Ginnie Simione Jutson, Executive Development Manager, and Leata Holloway, Program Manager, coordinated this complex and multifaceted endeavor. Senior Acquisitions Editor Star Burruto Mackenzie did a great job of helping us form a revision plan that even further expanded the text's appeal and its ability to convey fascinating information in a user-friendly manner. She listened and responded helpfully to our questions and suggestions—all while traveling extensively to share her enthusiasm for the text and its extensive ancillary resources with educators across the country. Mae Lum, as Project Manager, has done a marvelous job of keeping everything—especially the authors—on track and on schedule, not to mention helping us through the complexities of a rigorously upgraded permissions process. Erin Schnair carefully reviewed every word of the manuscript, making sure the sometimes extensive revisions and rearrangements flowed smoothly into the existing text. Her attention to detail and thoughtful suggestions have contributed significantly to the text's organization and clarity. Our outstanding copyeditor, Joanna Dinsmore, not only negotiated the intricacies of grammar and formatting, but also caught inconsistencies that we had overlooked. Erin and Joanna also looked carefully at the art, checking each piece for consistency with the text and helping us with instructions to the artists. As production advanced, Kari Hopperstead contributed her first-rate formatting skills to meld images and text into an integrated whole. The book boasts a large number of excellent new photos, tracked down with skill and persistence by Kristin Piljay. Kristin was always cheerfully responsive to our requests for still more photos when nothing in the first batch would do.

We are grateful to Imagineering Art, under the direction of Project Manager Wynne Au-Yeung, for deciphering our art instructions and patiently making new adjustments to already outstanding figures. We owe our beautifully redesigned text and delightful new cover to Elise Lansdon.

The production of this text would not have been possible without the considerable efforts of Norine Strang, Senior Project Manager at Cenveo Publisher Services. Norine brought the art, photos, and manuscript together into a seamless and beautiful whole, graciously handling last-minute changes. We thank Lauren Harp, Executive Marketing Manager, for making sure the finished product reached your desk.

In her role as Manufacturing Buyer, Stacey Weinberger's expertise has served us well. The ancillaries are an endeavor fully as important as the text itself. Mae Lum skillfully coordinated the enormous effort of producing a truly outstanding package that complements and supports the text, while Eddie Lee took the lead on the Instructor Resource DVD. Finally, thanks to Chloé Veylit for developing the outstanding MasteringBiology Web site that accompanies this text.

We are extremely fortunate to be working with the Pearson team. This Eleventh Edition of *Biology: Life on Earth* reflects their exceptional abilities and dedication.

With gratitude,

TERRY AUDESIRK, GERRY AUDESIRK, AND BRUCE BYERS

Eleventh Edition Reviewers

Aekam Barot,
Lake Michigan College

Mark Belk,
Brigham Young University

Karen Bledsoe,
Western Oregon University

Christine Bozarth,
Northern Virginia Community College

Britt Canada,
Western Texas College

Reggie Cobb,
Nash Community College

Rachel Davenport,
Texas State University, San Marcos

Diane Day,
Clayton State University

Lewis Deaton,
University of Louisiana at Lafayette

Peter Ekechukwu,
Horry-Georgetown Technical College

Janet Gaston,
Troy University

Mijitaba Hamissou,
Jacksonville State University

Karen Hanson,
Carroll Community College

Brian Ingram,
Jacksonville State University

Karen Kendall-Fite,
Columbia State Community College

Neil Kirkpatrick,
Moraine Valley Community College

Damaris-Lois Lang,
Hostos Community College

Tiffany McFalls-Smith,
Elizabethtown Community and Technical College

Mark Meade,
Jacksonville State University

Samantha Parks,
Georgia State University

Indiren Pillay,
Georgia College

John Plunket,
Horry-Georgetown Technical College

Cameron Russell,
Tidewater Community College

Roger Sauterer,
Jacksonville State University

Terry Sellers,
Spartanburg Methodist College

David Serrano,
Broward College

Philip Snider,
Gadsden State Community College

Judy Staveley,
Carroll Community College

Katelynn Woodhams,
Lake Michigan College

Min Zhong,
Auburn University

Deborah Zies,
University of Mary Washington

Previous Edition Reviewers

Mike Aaron,
Shelton State Community College

Kammy Algiers,
Ventura College

W. Sylvester Allred,
Northern Arizona University

Judith Keller Amand,
Delaware County Community College

William Anderson,
Abraham Baldwin Agriculture College

Steve Arch,
Reed College

George C. Argyros,
Northeastern University

Kerri Lynn Armstrong,
Community College of Philadelphia

Ana Arnizaut-Vilella,
Mississippi University for Women

Dan Aruscavage,
State University of New York, Potsdam

G. D. Aumann,
University of Houston

Vernon Avila,
San Diego State University

J. Wesley Bahorik,
Kutztown University of Pennsylvania

Peter S. Baletsa,
Northwestern University

Isaac Barjis
New York City College of Technology

John Barone,
Columbus State University

Bill Barstow,
University of Georgia–Athens

Mike Barton,
Centre College

Erin Baumgartner,
Western Oregon University

Michael C. Bell,
Richland College

Colleen Belk,
University of Minnesota, Duluth

Robert Benard,
American International College

Heather Bennett,
Illinois College

Gerald Bergtrom,
University of Wisconsin

Arlene Billock,
University of Southwestern Louisiana

Brenda C. Blackwelder,
Central Piedmont Community College

Melissa Blamires,
Salt Lake Community College

Karen E. Bledsoe,
Western Oregon University

Bruno Borsari,
Winona State University

Raymond Bower,
University of Arkansas

Robert Boyd,
Auburn University

Michael Boyle,
Seattle Central Community College

Marilyn Brady,
Centennial College of Applied Arts and Technology

David Brown,
Marietta College

Virginia Buckner,
Johnson County Community College

Arthur L. Buikema, Jr.,
Virginia Polytechnic Institute

Diep Burbridge,
Long Beach City College

Jamie Burchill,
Troy University

J. Gregory Burg,
University of Kansas

William F. Burke,
University of Hawaii

Robert Burkholter,
Louisiana State University

Matthew R. Burnham,
Jones County Junior College

Kathleen Burt-Utley,
University of New Orleans

Linda Butler,
University of Texas–Austin

W. Barkley Butler,
Indiana University of Pennsylvania

Jerry Button,
Portland Community College

Bruce E. Byers,
University of Massachusetts Amherst

Anne Casper,
Eastern Michigan University

Sara Chambers,
Long Island University

Judy A. Chappell,
Luzerne County Community College

Nora L. Chee,
Chaminade University

Joseph P. Chinnici,
Virginia Commonwealth University

Dan Chiras,
University of Colorado–Denver

Nicole A. Cintas,
Northern Virginia Community College

Bob Coburn,
Middlesex Community College

Joseph Coelho,
Culver Stockton College

Martin Cohen,
University of Hartford

Mary Colavito,
Santa Monica College

Jay L. Comeaux,
Louisiana State University

Walter J. Conley,
State University of New York at Potsdam

Mary U. Connell,
Appalachian State University

Art Conway,
Randolph-Macon College

Jerry Cook,
Sam Houston State University

Sharon A. Coolican,
Cayuga Community College

Clifton Cooper,
Linn-Benton Community College

Joyce Corban,
Wright State University

Brian E. Corner,
Augsburg College

Ethel Cornforth,
San Jacinto College–South

David J. Cotter,
Georgia College

Lee Couch,
  Albuquerque Technical Vocational Institute
Donald C. Cox,
  Miami University of Ohio
Patricia B. Cox,
  University of Tennessee
Peter Crowcroft,
  University of Texas–Austin
Carol Crowder,
  North Harris Montgomery College
Mitchell B. Cruzan,
  Portland State University
Donald E. Culwell,
  University of Central Arkansas
Peter Cumbie,
  Winthrop University
Robert A. Cunningham,
  Erie Community College, North
Karen Dalton,
  Community College of Baltimore County–Catonsville Campus
Lydia Daniels,
  University of Pittsburgh
David H. Davis,
  Asheville-Buncombe Technical Community College
Jerry Davis,
  University of Wisconsin, LaCrosse
Douglas M. Deardon,
  University of Minnesota
Lewis Deaton,
  University of Louisiana–Lafayette
Fred Delcomyn,
  University of Illinois–Urbana
Joe Demasi,
  Massachusetts College
David M. Demers,
  University of Hartford
Kimberly Demnicki,
  Thomas Nelson Community College
Lorren Denney,
  Southwest Missouri State University
Katherine J. Denniston,
  Towson State University
Charles F. Denny,
  University of South Carolina–Sumter
Jean DeSaix,
  University of North Carolina–Chapel Hill
Ed DeWalt,
  Louisiana State University
Daniel F. Doak,
  University of California–Santa Cruz
Christy Donmoyer,
  Winthrop University
Matthew M. Douglas,
  University of Kansas
Ronald J. Downey,
  Ohio University

Ernest Dubrul,
  University of Toledo
Michael Dufresne,
  University of Windsor
Susan A. Dunford,
  University of Cincinnati
Mary Durant,
  North Harris College
Ronald Edwards,
  University of Florida
Rosemarie Elizondo,
  Reedley College
George Ellmore,
  Tufts University
Joanne T. Ellzey,
  University of Texas–El Paso
Wayne Elmore,
  Marshall University
Thomas Emmel,
  University of Florida
Carl Estrella,
  Merced College
Nancy Eyster-Smith,
  Bentley College
Gerald Farr,
  Texas State University
Rita Farrar,
  Louisiana State University
Marianne Feaver,
  North Carolina State University
Susannah Feldman,
  Towson University
Linnea Fletcher,
  Austin Community College–Northridge
Doug Florian,
  Trident Technical College
Charles V. Foltz,
  Rhode Island College
Dennis Forsythe,
  The Citadel
Douglas Fratianne,
  Ohio State University
Scott Freeman,
  University of Washington
Donald P. French,
  Oklahoma State University
Harvey Friedman,
  University of Missouri–St. Louis
Don Fritsch,
  Virginia Commonwealth University
Teresa Lane Fulcher,
  Pellissippi State Technical Community College
Michael Gaines,
  University of Kansas
Cynthia Galloway,
  Texas A&M University–Kingsville
Irja Galvan,
  Western Oregon University
Gail E. Gasparich,
  Towson University
Janet Gaston,
  Troy University

Farooka Gauhari,
  University of Nebraska–Omaha
John Geiser,
  Western Michigan University
Sandra Gibbons,
  Moraine Valley Community College
George W. Gilchrist,
  University of Washington
David Glenn-Lewin,
  Iowa State University
Elmer Gless,
  Montana College of Mineral Sciences
Charles W. Good,
  Ohio State University–Lima
Joan-Beth Gow,
  Anna Maria College
Mary Rose Grant,
  St. Louis University
Anjali Gray,
  Lourdes College
Margaret Green,
  Broward Community College
Ida Greidanus,
  Passaic Community College
Mary Ruth Griffin,
  University of Charleston
Wendy Grillo,
  North Carolina Central University
David Grise,
  Southwest Texas State University
Martha Groom,
  University of Washington
Lonnie J. Guralnick,
  Western Oregon University
Martin E. Hahn,
  William Paterson College
Madeline Hall,
  Cleveland State University
Georgia Ann Hammond,
  Radford University
Blanche C. Haning,
  North Carolina State University
Richard Hanke,
  Rose State College
Helen B. Hanten,
  University of Minnesota
Rebecca Hare,
  Cleveland County Community College
John P. Harley,
  Eastern Kentucky University
Robert Hatherill,
  Del Mar College
William Hayes,
  Delta State University
Kathleen Hecht,
  Nassau Community College
Stephen Hedman,
  University of Minnesota
Jean Helgeson,
  Collins County Community College
Alexander Henderson,
  Millersville University

Wiley Henderson,
  Alabama A&M University
Timothy L. Henry,
  University of Texas–Arlington
James Hewlett,
  Finger Lakes Community College
Alison G. Hoffman,
  University of Tennessee–Chattanooga
Kelly Hogan,
  University of North Carolina–Chapel Hill
Leland N. Holland,
  Paso-Hernando Community College
Laura Mays Hoopes,
  Occidental College
Dale R. Horeth,
  Tidewater Community College
Harriette Howard-Lee Block,
  Prairie View A&M University
Adam Hrincevich,
  Louisiana State University
Michael D. Hudgins,
  Alabama State University
David Huffman,
  Southwest Texas State University
Joel Humphrey,
  Cayuga Community College
Donald A. Ingold,
  East Texas State University
Jon W. Jacklet,
  State University of New York–Albany
Kesmic Jackson,
  Georgia State University
Rebecca M. Jessen,
  Bowling Green State University
J. Kelly Johnson,
  University of Kansas
James Johnson,
  Central Washington University
Kristy Y. Johnson,
  The Citadel
Ross Johnson,
  Chicago State University
Florence Juillerat,
  Indiana University–Purdue University at Indianapolis
Thomas W. Jurik,
  Iowa State University
Ragupathy Kannan,
  University of Arkansas, Fort Smith
A. J. Karpoff,
  University of Louisville
L. Kavaljian,
  California State University
Joe Keen,
  Patrick Henry Community College
Jeff Kenton,
  Iowa State University
Hendrick J. Ketellapper,
  University of California, Davis
Jeffrey Kiggins,
  Blue Ridge Community College

Michael Koban,
Morgan State University

Aaron Krochmal,
*University of Houston–
Downtown*

Harry Kurtz,
Sam Houston State University

Kate Lajtha,
Oregon State University

Tom Langen,
Clarkson University

Patrick Larkin,
Santa Fe College

Stephen Lebsack,
*Linn-Benton Community
College*

Patricia Lee-Robinson,
*Chaminade University of
Honolulu*

David E. Lemke,
Texas State University

William H. Leonard,
Clemson University

Edward Levri,
*Indiana University of
Pennsylvania*

Graeme Lindbeck,
University of Central Florida

Jerri K. Lindsey,
*Tarrant County Junior
College–Northeast*

Mary Lipscomb,
*Virginia Polytechnic Institute
and State University*

Richard W. Lo Pinto,
*Fairleigh Dickinson
University*

Jonathan Lochamy,
Georgia Perimeter College

Jason L. Locklin,
Temple College

John Logue,
*University of South Carolina–
Sumter*

Paul Lonquich,
*California State University
Northridge*

William Lowen,
Suffolk Community College

Ann S. Lumsden,
Florida State University

Steele R. Lunt,
University of Nebraska–Omaha

Fordyce Lux,
*Metropolitan State College of
Denver*

Daniel D. Magoulick,
*The University of Central
Arkansas*

Bernard Majdi,
Waycross College

Cindy Malone,
*California State University–
Northridge*

Paul Mangum,
Midland College

Richard Manning,
*Southwest Texas State
University*

Mark Manteuffel,
*St. Louis Community
College*

Barry Markillie,
*Cape Fear Community
College*

Ken Marr,
*Green River Community
College*

Kathleen A. Marrs,
*Indiana University–Purdue
University Indianapolis*

Michael Martin,
University of Michigan

Linda Martin-Morris,
University of Washington

Kenneth A. Mason,
University of Kansas

Daniel Matusiak,
*St. Charles Community
College*

Margaret May,
*Virginia Commonwealth
University*

D. J. McWhinnie,
De Paul University

Gary L. Meeker,
*California State University,
Sacramento*

Thoyd Melton,
*North Carolina State
University*

Joseph R. Mendelson III,
Utah State University

Karen E. Messley,
Rock Valley College

Timothy Metz,
Campbell University

Steven Mezik,
*Herkimer County Community
College*

Glendon R. Miller,
Wichita State University

Hugh Miller,
*East Tennessee State
University*

Neil Miller,
Memphis State University

Jeanne Minnerath,
*St. Mary's University of
Minnesota*

Christine Minor,
Clemson University

Jeanne Mitchell,
Truman State University

Lee Mitchell,
*Mt. Hood Community
College*

Jack E. Mobley,
*University of Central
Arkansas*

John W. Moon,
Harding University

Nicole Moore,
Austin Peay University

Richard Mortenson,
Albion College

Gisele Muller-Parker,
*Western Washington
University*

James Mulrooney,
*Central Connecticut State
University*

Kathleen Murray,
University of Maine

Liz Nash,
*California State University,
Long Beach*

Robert Neill,
University of Texas

Russell Nemecek,
*Columbia College,
Hancock*

Harry Nickla,
Creighton University

Daniel Nickrent,
*Southern Illinois
University*

Jane Noble-Harvey,
University of Delaware

Murad Odeh,
South Texas College

David J. O'Neill,
*Community College of
Baltimore County–Dundalk
Campus*

James T. Oris,
Miami University of Ohio

Marcy Osgood,
University of Michigan

C. O. Patterson,
Texas A&M University

Fred Peabody,
*University of South
Dakota*

Charlotte Pedersen,
Southern Utah University

Harry Peery,
*Tompkins-Cortland
Community College*

Luis J. Pelicot,
*City University of New York,
Hostos*

Rhoda E. Perozzi,
*Virginia Commonwealth
University*

Gary B. Peterson,
*South Dakota State
University*

Bill Pfitsch,
Hamilton College

Ronald Pfohl,
Miami University of Ohio

Larry Pilgrim,
Tyler Junior College

Therese Poole,
Georgia State University

Robert Kyle Pope,
*Indiana University South
Bend*

Bernard Possident,
Skidmore College

Ina Pour-el,
DMACC–Boone Campus

Elsa C. Price,
*Wallace State Community
College*

Marvin Price,
Cedar Valley College

Kelli Prior,
*Finger Lakes Community
College*

Jennifer J. Quinlan,
Drexel University

James A. Raines,
*North Harris
College*

Paul Ramp,
*Pellissippi State Technical
College*

Robert N. Reed,
Southern Utah University

Wenda Ribeiro,
*Thomas Nelson Community
College*

Elizabeth Rich,
Drexel University

Mark Richter,
University of Kansas

Robert Robbins,
Michigan State University

Jennifer Roberts,
Lewis University

Frank Romano,
Jacksonville State University

Chris Romero,
*Front Range Community
College*

David Rosen,
Lee College

Paul Rosenbloom,
Southwest Texas State University

Amanda Rosenzweig,
Delgado Community College

K. Ross,
University of Delaware

Mary Lou Rottman,
University of Colorado–Denver

Albert Ruesink,
Indiana University

Cameron Russell,
Tidewater Community College

Connie Russell,
Angelo State University

Marla Ruth,
Jones County Junior College

Christopher F. Sacchi,
Kutztown University

Eduardo Salazar,
Temple College

Doug Schelhaas,
University of Mary

Brian Schmaefsky,
Kingwood College

Alan Schoenherr,
Fullerton College

Brian W. Schwartz,
*Columbus State
University*

Edna Seaman,
*University of Massachusetts,
Boston*

Tim Sellers,
Keuka College

Patricia Shields,
George Mason University

Marilyn Shopper,
*Johnson County Community
College*

Jack Shurley,
Idaho State University

Bill Simcik,
Lonestar College

Rick L. Simonson,
*University of Nebraska,
Kearney*

Howard Singer,
*New Jersey City
University*

Anu Singh-Cundy,
Western Washington University

Linda Simpson,
University of North Carolina–Charlotte

Steven Skarda,
Linn-Benton Community College

Russel V. Skavaril,
Ohio State University

John Smarelli,
Loyola University

Mark Smith,
Chaffey College

Dale Smoak,
Piedmont Technical College

Jay Snaric,
St. Louis Community College

Phillip J. Snider,
University of Houston

Shari Snitovsky,
Skyline College

Gary Sojka,
Bucknell University

John Sollinger,
Southern Oregon University

Sally Sommers Smith,
Boston University

Jim Sorenson,
Radford University

Anna Bess Sorin,
University of Memphis

Mary Spratt,
University of Missouri, Kansas City

Bruce Stallsmith,
University of Alabama–Huntsville

Anthony Stancampiano,
Oklahoma City University

Theresa Stanley,
Gordon College

Benjamin Stark,
Illinois Institute of Technology

William Stark,
Saint Louis University

Barbara Stebbins-Boaz,
Willamette University

Mary-Pat Stein,
California State University, Northridge

Kathleen M. Steinert,
Bellevue Community College

Barbara Stotler,
Southern Illinois University

Nathaniel J. Stricker,
Ohio State University

Martha Sugermeyer,
Tidewater Community College

Gerald Summers,
University of Missouri–Columbia

Marshall Sundberg,
Louisiana State University

Bill Surver,
Clemson University

Eldon Sutton,
University of Texas–Austin

Peter Svensson,
West Valley College

Dan Tallman,
Northern State University

Jose G. Tello,
Long Island University

Julienne Thomas-Hall,
Kennedy King College

David Thorndill,
Essex Community College

William Thwaites,
San Diego State University

Professor Peter Tobiessen,
Union College

Richard Tolman,
Brigham Young University

Sylvia Torti,
University of Utah

Dennis Trelka,
Washington and Jefferson College

Richard C. Tsou,
Gordon College

Sharon Tucker,
University of Delaware

Gail Turner,
Virginia Commonwealth University

Glyn Turnipseed,
Arkansas Technical University

Lloyd W. Turtinen,
University of Wisconsin, Eau Claire

Robert Tyser,
University of Wisconsin, La Crosse

Robin W. Tyser,
University of Wisconsin, La Crosse

Kristin Uthus,
Virginia Commonwealth University

Rani Vajravelu,
University of Central Florida

Jim Van Brunt,
Rogue Community College

F. Daniel Vogt,
State University of New York–Plattsburgh

Nancy Wade,
Old Dominion University

Susan M. Wadkowski,
Lakeland Community College

Jyoti R. Wagle,
Houston Community College–Central

Jerry G. Walls,
Louisiana State University, Alexandria

Holly Walters,
Cape Fear Community College

Winfred Watkins,
McLennan Community College

Lisa Weasel,
Portland State University

Janice Webster,
Ivy Tech Community College

Michael Weis,
University of Windsor

DeLoris Wenzel,
University of Georgia

Jerry Wermuth,
Purdue University–Calumet

Diana Wheat,
Linn-Benton Community College

Richard Whittington,
Pellissippi State Technical Community College

Jacob Wiebers,
Purdue University

Roger K. Wiebusch,
Columbia College

Carolyn Wilczynski,
Binghamton University

Lawrence R. Williams,
University of Houston

P. Kelly Williams,
University of Dayton

Roberta Williams,
University of Nevada–Las Vegas

Emily Willingham,
University of Texas–Austin

Sandra Winicur,
Indiana University–South Bend

Bill Wischusen,
Louisiana State University

Michelle Withers,
Louisiana State University

Chris Wolfe,
North Virginia Community College

Stacy Wolfe,
Art Institutes International

Colleen Wong,
Wilbur Wright College

Wade Worthen,
Furman University

Robin Wright,
University of Washington

Taek H. You,
Campbell University

Brenda L. Young,
Daemen College

Cal Young,
Fullerton College

Tim Young,
Mercer University

Marty Zahn,
Thomas Nelson Community College

Izanne Zorin,
Northern Virginia Community College–Alexandria

Michelle Zurawski,
Moraine Valley Community College

Hallmark Case Studies place biology in a real-world context

A **Case Study** describing a true and relevant event or phenomenon runs throughout each chapter, tying biological concepts to the real world.

All chapters open with a **Case Study,** a true yet extraordinary story that relates to the science presented in the chapter. The **Eleventh Edition** explores several **new** Case Study topics including the Ebola epidemic (Chapter 1), DNA Identification (Chapter 8), and Biotechnology (Chapter 14).

NEW! Chapter 9 now covers only mitotic cell division and the control of the cell cycle. Meiotic cell division and its importance in sexual reproduction are discussed in Chapter 10.

10 MEIOSIS: THE BASIS OF SEXUAL REPRODUCTION

The Giddings family is a rainbow of colors.

156

CASE STUDY

The Rainbow Connection

FIRST CAME JACOB, WHO HAS BLUE EYES like his mom, Tess, but curly brown hair and olive skin. Next came Savannah, who looks a lot like Jacob, though her hair is perhaps more dark blond than brown. Amiah, however, was truly a surprise when she was born—she has very pale skin, with straight, sandy-brown hair. Zion, the youngest child, has dark skin, black curly hair, and brown eyes, similar to his father, Chris. Even in today's multicultural England, a family like that is unusual.

Tess and Chris Giddings are as surprised as everyone else by their rainbow family. In fact, when Amiah was born, she had low blood sugar and needed to be checked out by a specialist right away. She was whisked away so fast that the hospital staff hadn't put an ID wristband on her yet. When she was returned to her parents a little while later, they were astounded at how white her skin was. They asked the inevitable question: Was she switched with another baby by mistake? Just to be sure, the Giddings agreed to a DNA test. The results showed that Tess and Chris were indeed Amiah's parents. When Zion was born a few years later, Chris burst out, "Oh my God, he's black!" To which the astounded midwife could only reply, "You do know you're a black man, don't you?"

How could one couple have such a diverse family? As we will see in this chapter, sexual reproduction can mix inherited characteristics from the parents into a remarkable variety of different offspring. How does sexual reproduction produce genetic diversity? And why would natural selection favor seemingly random shuffling of traits?

CASE STUDY \ CONTINUED

The Rainbow Connection

The genetic variability of the Giddings children started out as mutations that occurred thousands of years ago. Take hair color: Our distant ancestors probably all had dark hair, its color controlled by multiple genes located on several different chromosomes. The alleles that produced Tess's blond hair originated as mutations in genes that control the amount and type of hair pigment. Tess probably inherited only "pale hair" alleles of all of these genes, so for any given hair color gene, she has the same pale hair allele on both homologous chromosomes. Chris, on the other hand, inherited both dark and pale hair alleles for at least some of the genes, so his homologues have different alleles. As we will see in Chapter 11, in many cases one allele (in this case, the dark hair allele) overrides the effects of the other allele (the pale hair allele), so Chris has black hair. What combinations of alleles might have been packaged in Tess's eggs and Chris's sperm, which would combine to produce their diverse children?

Every chapter contains **Case Study Continued** sections that appear when you are well into the chapter. These sections expand on the **Chapter Opening Case Studies** and connect to biological concepts you will have learned.

CASE STUDY \ REVISITED

The Rainbow Connection

Many people are astounded by the diversity of the Giddings children. Basic biology, however, easily explains how such diversity arises. Most genes have multiple alleles, meiotic cell division separates homologous chromosomes—and the alleles they carry—into different sperm and eggs, and the sperm and eggs unite at random. From a biological perspective, perhaps the more interesting question is this: Why do alleles for dark pigmentation occur most frequently in people whose ancestors bodily functions. Folate deficiency can cause anemia and other disorders in adults and serious nervous system abnormalities in developing fetuses.

Ultraviolet rays in sunlight stimulate the synthesis of vitamin D, but they break down folate. In the fierce sunlight of equatorial regions, dark skin still allows for plenty of vitamin D production, while protecting against too much depletion of folate. In northern Europe, with far weaker sunlight and often cloudy skies, paler skin boosts vitamin D production, while folate levels remain adequate.

The selective advantage of blond hair in northern Europe is more uncertain. Some of the same genes contribute to hair and skin color, so selection for pale skin may have selected for pale hair as well. Another hypothesis is that the first few people with blond hair were very conspicuous in a population lived in equatorial regions, and alleles for pale pigmentation in people of northern European ancestry?

Natural selection probably favored different skin colors because of the differing amount of sunlight in equatorial versus northern regions and the importance of vitamin D and vitamin B$_9$ (folate) in human health. Vitamin D is needed for many physiological functions, including the absorption of calcium and other minerals by the digestive tract. Folate is also essential for many of otherwise dark-haired people. Novel appearance, within limits, is often attractive to members of the opposite sex. Some anthropologists have speculated that, a few thousand years ago, high-status men (proficient hunters or chieftains of small tribes, for example) preferentially chose blond-haired women as mates. Therefore, blond women produced more offspring than dark-haired women did. The result is that more than half the people in parts of Scandinavia have blond hair.

CONSIDER THIS Ultraviolet rays in sunlight cause skin cancer. In today's world, people of all skin colors, but especially pale-skinned people, are often urged to stay out of the sun and get their vitamin D from food or supplements. In the past, do you think that the risk of skin cancer selected against pale-skinned people, partially counterbalancing selection in favor of pale skin for vitamin D production?

A **Case Study Revisited** section wraps up the narrative of each chapter by connecting the biological themes described throughout the chapter with the everyday science brought out in the Case Study. The accompanying **Consider This** question allows further reflection on how the biology in the Case Study can be applied to a new situation.

NEW! Three-pronged taxonomy of questions in each chapter

Each chapter is organized around a consistent framework of questions that encourage students to look forward, look back, or dig deeper.

The section **headings** and **case study** sections give a preview of questions that will be addressed in the chapter.

CHECK YOUR LEARNING

Can you ...

- explain why people might be opposed to the use of genetically modified organisms in agriculture?
- envision circumstances in which it would be ethical to modify the genome of a human fertilized egg?

The **Check Your Learning** and **End of Chapter questions** ask students to look back, recall, and reinforce their comprehension of biology concepts.

CONSIDER THIS Genetic engineering is used both in food crops and in medicine. Golden Rice and almost all the corn and soybeans grown in the United States contain genes from other species. The he[...] by inserting a gene from th[...] The antibodies in ZMapp, [...] Ebola therapy, are part mo[...] scientifically important diff[...] engineering for food or for[...]

EVALUATE THIS In January 2012, the Pittsburgh Steelers football team played against the Denver Broncos in the "Mile-High City" (Denver's altitude is a mile above sea level). Steelers head coach Mike Tomlin did not allow safety Ryan Clark to play, because Clark has sickle-cell trait. What can happen when someone with sickle-cell trait exercises at high [...] right call in bench-

Applying the Concepts

1. As you may know, many insects have evolved resistance to common pesticides. Do you think that insects might evolve resistance to *Bt* crops? If this is a risk, do you think that *Bt* crops should be planted anyway? Why or why not?

2. All children born with X-linked SCID are boys. Can you e[...]

THINK CRITICALLY There are many other applications in which DNA barcoding might be useful. For example, how might ecologists use DNA barcoding to find out what species are present in a rain forest, or what kinds of animals a predator eats?

Thinking Through the Concepts

Multiple Choice

1. Which of the following is *not* true of a single nucleotide polymorphism?
   a. It is usually caused by a translocation mutation.
   b. It is usually caused by a nucleotide substitution mutation.
   c. It may change the phenotype of an organism.
   d. It is inherited from parent to offspring.

2. Imagine you are looking at a DNA profile that shows an STR pattern of a mother's DNA and her child's DNA. Will all of the bands of the child's DNA match those of the mother?
   a. Yes, because the mother's DNA and her child's DNA are identical.
   b. Yes, because the child developed from her mother's egg.
   c. No, because half of the child's DNA is inherited from its father.
   d. No, because the child's DNA is a random sampling of its mother's.

3. Which of the following is *not* a commonly used method of modifying the DNA of an organism?
   a. crossbreeding two plants of the same species
   b. crossbreeding two plants of different species
   c. the polymerase chain reaction
   d. genetic engineering

4. A restriction enzyme
   a. cuts DNA at a specific nucleotide sequence.
   b. cuts DNA at a random nucleotide sequence.
   c. splices pieces of DNA together at a specific nucleotide sequence.
   d. splices pieces of DNA together without regard to the nucleotide sequence.

5. DNA cloning is
   a. making multiple genetically identical cells.
   b. making multiple copies of a piece of DNA.
   c. inserting DNA into a cell.
   d. changing the nucleotide sequence of a strand of DNA.

Consider This, Think Critically, Evaluate This and **Applying the Concepts** ask students to dig deeper, reflect, and think critically about the chapter material.

NEW! Think Critically questions challenge readers to apply their knowledge to information presented in a photo, figure, graph, or table.

NEW! Evaluate This questions present a brief, realistic health care scenario and ask the reader to evaluate information before forming an opinion or making a decision.

Improved Figures and Photos appear throughout the text and include easy-to-follow process diagrams with labeled steps and a clearer use of color for distinguishing different structures.

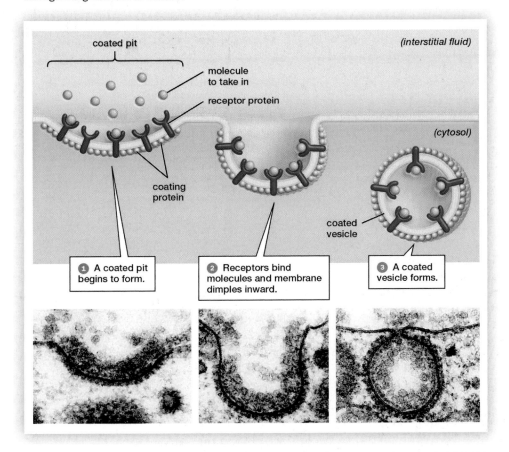

coated pit

(interstitial fluid)

molecule to take in

receptor protein

(cytosol)

coating protein

coated vesicle

1 A coated pit begins to form.

2 Receptors bind molecules and membrane dimples inward.

3 A coated vesicle forms.

NEW! How Do We Know That? Essays explore the process of scientific discovery, experimental design, and exciting new biotechnology techniques, explaining how scientists know what they know about biology.

---

548 UNIT 4 Behavior and Ecology

HOW DO WE KNOW THAT? | Monitoring Earth's Health

Carbon dioxide concentrations in the atmosphere are increasing; Earth is getting warmer; oceans are acidifying; glaciers are retreating; Arctic sea ice is decreasing. You may wonder—how do we know all this?

Estimating some conditions on Earth is fairly straightforward. For example, atmospheric CO_2 is measured at hundreds of stations in dozens of countries, including Mauna Loa in Hawaii (see Fig. 29-14a). Estimates of CO_2 concentrations in the distant past are obtained by analyzing gas bubbles trapped in ancient Antarctic ice.

In some places on Earth, people began keeping accurate temperature records well over a century ago. Now, air temperatures are measured at about 1,500 locations, on both land and sea, each day. Sophisticated computational methods compensate for the uneven distribution of weather stations (more in England than in the Arctic or Sahara Desert) and produce global average temperatures. Ancient temperatures can be estimated by "natural proxies"—natural phenomena that vary with temperature and leave long-lasting records. For example, isotopes of oxygen in air trapped in bubbles inside ice vary with the air temperature at the time the bubble formed. Ice cores collected from glaciers in Antarctica or Greenland can therefore be used to estimate "paleotemperatures." Chemical measurements of corals and mollusk shells, and even some types of sediments and fossils, also provide estimates of paleotemperatures.

However, some measurements of Earth's environment wouldn't have been possible even 20 or 40 years ago. Many involve data collected by satellites. For example, measuring areas of forest is a simple, if tedious, matter of carefully examining satellite photos. Other measurements are much more sophisticated. Accurate estimates of Arctic sea ice started in 1979, with the launch of satellites that measure microwave

radiation emitted from Earth's surface. Ice emits more microwave radiation than liquid water does, so the satellites can easily distinguish the two. Satellite data show that the extent of Arctic sea ice has declined about 13% per decade since 1979 (**FIG. E29-2**). Many other features of Earth have distinctive "signature wavelengths" that satellites can detect, from sulfur dioxide emitted by power plants to chlorophyll in the oceans (**FIG. E29-3**).

▲ FIGURE E29-2 **Changes in Arctic sea ice** Satellite measurements of Arctic sea ice began in 1979. By 2014, the area covered by ice at the end of the summer (September) had declined by more than a third.

CHAPTER 29 Energy Flow and Nutrient Cycling in Ecosystems **549**

Chlorophyll *a* Concentration (mg/m³)

0.01   0.1   1.0   10   60

▲ FIGURE E29-3 **Ocean chlorophyll** Satellite measurements of chlorophyll show which areas of the ocean have the greatest amount of phytoplankton. Purple/blue represent low chlorophyll concentrations, green/yellow intermediate amounts, and orange/red the highest concentrations.

Perhaps the most amazing measurements come from NASA's GRACE satellites—the Gravity Recovery and Climate Experiment. A satellite's orbiting speed is determined, in part, by the force of gravity exerted on it. Water and ice are heavy. Large volumes of ice on the land increase local gravity, tugging ever-so-slightly on the satellites, which then measure the extra gravitational pull. GRACE has found that land ice sheets in Antarctica and Greenland have declined dramatically over the past decade. Antarctica is losing about 150 billion tons of ice per year; Greenland is losing about 260 billion tons. GRACE can even measure water underground: the combination of prolonged drought and groundwater pumping for agriculture in California's Central Valley has greatly depleted the aquifers underlying the Valley (**FIG. E29-4**).

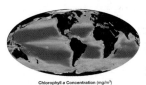

California Drying
Cumulative water storage changes from NASA GRACE (2002–2014)

June 2002   June 2008   June 2014

▲ FIGURE E29-4 **Changes in gravity show depletion of water in California's aquifers** Underground aquifers in California's Central Valley are losing about 4 trillion gallons of water each year. The transition from green to red in these false-color images shows water lost between 2002 and 2014.

THINK CRITICALLY People tend to be much more attuned to what's happening right now and less aware of long-term trends. Every time there's a blast of cold weather in winter or hot weather in summer, opinion polls show lesser or greater concern about global warming. Climatologists, however, take a very long view and look for trends in climate data. Using a ruler, estimate trend lines for the data in Figures 29-14 and E29-2. What do the trend lines predict about the future of atmospheric CO_2 concentrations, global temperatures, and Arctic sea ice? If these trends persist, will the Arctic become ice-free in late summer? If so, in what year? When will CO_2 concentrations double from preindustrial levels and reach 560 parts per million? Is it reasonable to extrapolate straight (linear) trend lines into the future? Why or why not?

Improve your grade—and your learning—with MasteringBiology®

Mastering is the most effective and widely used online homework and assessment system for the sciences. It delivers self-paced tutorials that focus on your course objectives, provide individualized coaching, and respond to your progress. Mastering motivates you to learn outside of class and arrive prepared for lecture or lab.

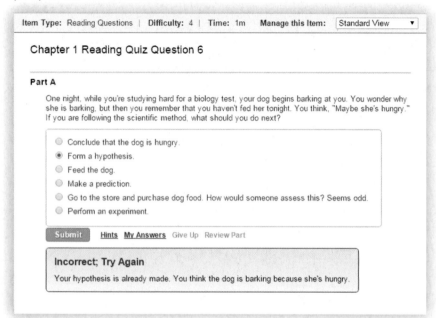

Reading Quizzes keep you on track with reading assignments. The quizzes require only 5–8 minutes for you to complete and make it possible for your instructor to understand your misconceptions before you arrive for class.

EXPANDED! Building Vocabulary Exercises help you learn the meaning of common prefixes, suffixes, and word roots, and then ask you to apply your knowledge to learn unfamiliar biology terms.

NEW! Working with Data activities ask students to analyze and apply their knowledge of biology to a graph or a set of data.

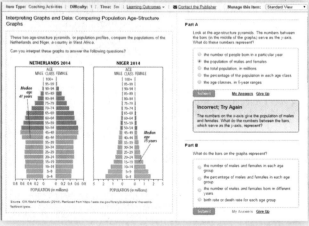

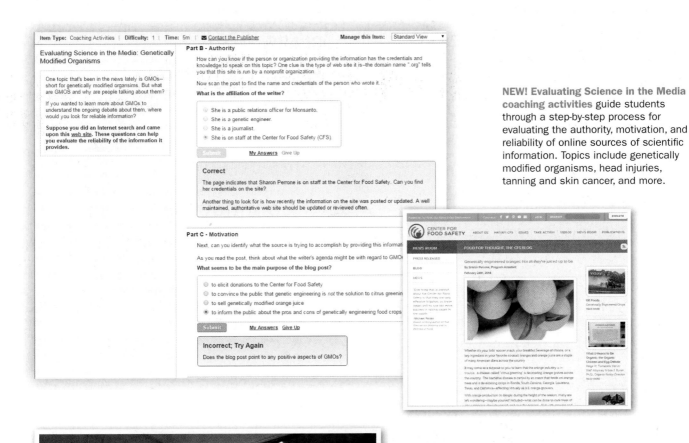

NEW! Evaluating Science in the Media coaching activities guide students through a step-by-step process for evaluating the authority, motivation, and reliability of online sources of scientific information. Topics include genetically modified organisms, head injuries, tanning and skin cancer, and more.

NEW! Everyday Biology Video activities briefly explore interesting and relevant biology topics that relate to concepts students learn about in class. These 20 videos, produced by the BBC, can be assigned in MasteringBiology with assessment questions.

MasteringBiology®

is an online homework, tutorial, and assessment program that helps you quickly master biology concepts and skills. Self-paced tutorials provide immediate wrong-answer feedback and hints to help keep you on track to succeed in the course.

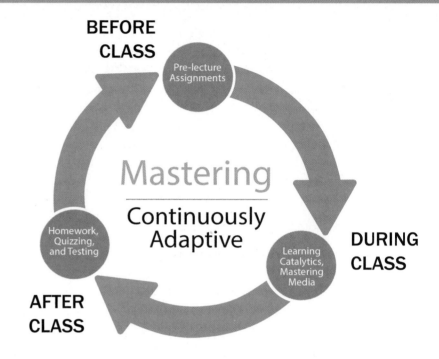

BEFORE CLASS

Pre-lecture Assignments

Mastering
Continuously Adaptive

Learning Catalytics, Mastering Media

DURING CLASS

Homework, Quizzing, and Testing

AFTER CLASS

BEFORE CLASS

NEW! Dynamic Study Modules help you acquire, retain, and recall information faster and more efficiently than ever before. These convenient practice questions and detailed review explanations can be accessed using a smartphone, tablet, or computer.

NEW! eText 2.0 allows students to access the text anytime, anywhere using a smartphone, tablet, or computer. The new eText is fully accessible and ready to use with screen-readers, re-sizable type, and more.

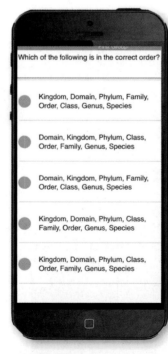

DYNAMIC STUDY MODULES

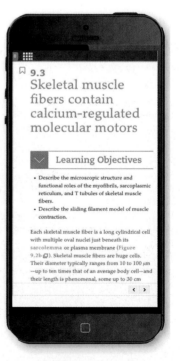

ETEXT 2.0

DURING CLASS

NEW!

Learning Catalytics is an assessment and classroom activity system that works with any web-enabled device and facilitates collaboration with your classmates. Your MasteringBiology subscription with eText includes access to Learning Catalytics.

AFTER CLASS

A wide range of question types and activities are available for homework assignments, including the following NEW assignment options for the Eleventh Edition:

- **EXPANDED! Building Vocabulary activities** help you learn the meaning of common prefixes, suffixes, and word roots, and then ask you to apply your knowledge to learn unfamiliar biology terms.
- **NEW! Working with Data questions** require you to analyze and apply your knowledge of biology to a graph or set of data.
- **NEW! Evaluating Science in the Media** challenge you to evaluate various information from websites, articles, and videos.

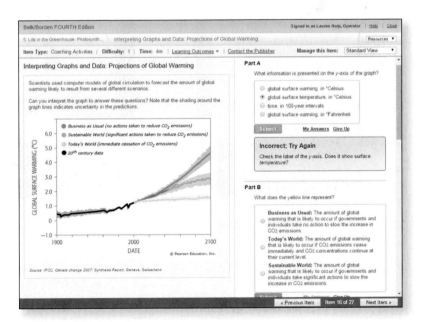

For Instructors: New Resources for Flipped Classrooms and More

New resources save valuable time both during course prep and during class.

NEW!

Learning Catalytics is a "bring your own device" assessment and classroom activity system that expands the possibilities for student engagement. Using Learning Catalytics, instructors can deliver a wide range of auto-gradable or open-ended questions that test content knowledge and build critical thinking skills. Eighteen different answer types provide great flexibility, including:

SKETCH/DIRECTION

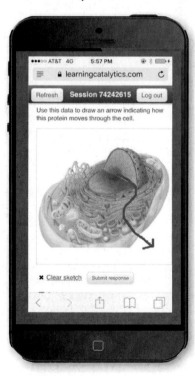

MANY CHOICE

REGION

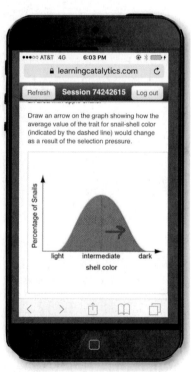

MasteringBiology®

MasteringBiology users may select from Pearson's library of Learning Catalytics questions, including questions developed specifically for *Biology: Life on Earth* 11e.

1 AN INTRODUCTION TO LIFE ON EARTH

The Ebola virus (inset) is so infectious and deadly that caregivers must protect themselves using isolation suits.

The Boundaries of Life

IN A SMALL VILLAGE in Guinea, a huge, hollow tree housed thousands of bats. The tree was a magnet for local children, who loved to play inside it and catch the bats. Scientists hypothesize that this is where two-year-old Emile Ouamouno, the first victim of the recent massive Ebola epidemic, may have become infected. Emile died in December 2013, followed by his mother and siblings. This set off a chain of transmission that has since killed more than 10,500 people, roughly half of those who became infected. The Ebola virus (see the inset photo) can lurk in rain-forest animals including certain types of bats, porcupines, chimpanzees, gorillas, and antelope—all of which are consumed in parts of Africa.

The threat of Ebola virus disease ("Ebola") strikes fear in anyone familiar with its symptoms, which often begin with fever, headache, joint and muscle aches, and stomach pains and progress to vomiting, bloody diarrhea, and organ failure. Internal hemorrhaging can leave victims bleeding from nearly every orifice. Death usually occurs within 7 to 16 days after

the onset of symptoms, and there is no cure; the death rate ranges from 25% to 90%. Ebola is so contagious that caregivers wear "moon suits" to avoid contact with any body fluids from their patients.

Ebola is one of many diseases caused by viruses. Although some viral diseases, such as smallpox and polio, have been largely eradicated, others, like the common cold and influenza (flu), continue to make us miserable. Most alarming are the contagious and deadly viruses that have emerged in recent history. AIDS (caused by the human immunodeficiency virus, HIV) was first documented in 1981 in San Francisco, and Ebola was first identified in 1976 (and named after Africa's Ebola River, where one of the first outbreaks occurred). New types of flu virus emerge regularly; a few of these cause a very high mortality rate and raise fears of a widespread epidemic.

No matter how you measure it, viruses are enormously successful. Although many consist only of a small amount of genetic material surrounded by protein, viruses infect every known form of life and are the most abundant biological entity on the planet. Viruses can rapidly increase in number and spread among organisms they infect. Yet in spite of these lifelike qualities, not all scientists agree about whether to classify viruses as living organisms or as inert parasitic biological particles. The basis for this argument may surprise you: There is no universally accepted scientific definition of life. What is life, anyway?

1.1 WHAT IS LIFE?

The word **biology** comes from the Greek roots "bio" meaning "life" and "logy" meaning "the study of" (see Appendix I for more word roots). But what is life? If you look up "life" in a dictionary, you will find definitions such as "the quality that distinguishes a vital and functioning being from a dead body," but you won't discover what that "quality" is. Life is intangible and defies simple definition, even by biologists. However, most agree that living things, or **organisms,** all share certain characteristics that, taken together, define life:

- Organisms acquire and use materials and energy.
- Organisms actively maintain organized complexity.
- Organisms sense and respond to stimuli.
- Organisms grow.
- Organisms reproduce.
- Organisms, collectively, evolve.

Nonliving objects may possess some of these attributes. Crystals can grow, and a desk lamp acquires energy from electricity and converts it to heat and light, but only living things can do them all.

The **cell** is the basic unit of life. A plasma membrane separates each cell from its surroundings, enclosing a huge variety of structures and chemicals in a fluid environment. The plasma membranes of many types of cells, including those of microorganisms and plants, are enclosed in a protective cell wall (**FIG. 1-1**). Although the most abundant organisms on Earth are unicellular (exist as single cells), the qualities of life are more easily visualized in **multicellular** organisms such as the water flea in **FIGURE 1-2**, an animal smaller than this letter "o." In the sections below, we introduce the characteristics of life.

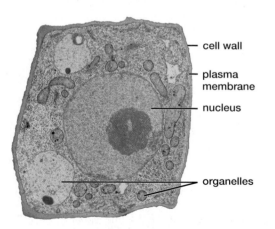

cell wall

plasma membrane

nucleus

organelles

▲ **FIGURE 1-1 The cell is the smallest unit of life** This artificially colored micrograph of a plant cell (a eukaryotic cell) shows a supporting cell wall (blue) that surrounds plant cells. Just inside the cell wall, the plasma membrane (found in all cells) has control over which substances enter and leave. Cells also contain several types of specialized organelles, including the nucleus, suspended within a fluid environment (orange).

▶ **FIGURE 1-2 Properties of life** The water flea uses energy from photosynthetic organisms that it consumes (green material in its gut) to maintain its amazing complexity. Eyes and antennae respond to stimuli. This adult female is reproducing, and she herself has grown from an egg like those she now carries. All the adaptations that allow this water flea to survive, grow, and reproduce have been molded by evolution.

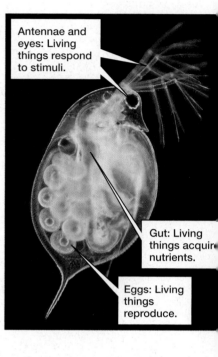

Antennae and eyes: Living things respond to stimuli.

Gut: Living things acquire nutrients.

Eggs: Living things reproduce.

Organisms Acquire and Use Materials and Energy

Organisms obtain the materials that make up their bodies—such as minerals, water, and other simple chemical building blocks—from the air, water, soil, and, in some cases, the bodies of other living things. Because life neither creates nor destroys matter, materials are continuously exchanged and recycled among organisms and their nonliving surroundings (**FIG. 1-3**).

Organisms use energy continuously to remain alive. For example, energy is needed to move and to construct the complex molecules that make up an organism's body. Essentially all the energy that sustains life comes from sunlight. Some organisms capture solar energy directly through a process called **photosynthesis.** Photosynthetic organisms (plants and many single-celled organisms) trap and store the sun's energy for their own use. The energy stored in their bodies also powers all nonphotosynthetic organisms. So energy flows in a one-way path from the sun to photosynthetic organisms to all other forms of life (see Fig. 1-3). Some energy is lost as heat at each transfer from one organism to another, making less energy available with each transfer.

Organisms Actively Maintain Organized Complexity

For both the books and papers on your desk and the fragile and dynamic intricacy of a cell, organization tends to disintegrate unless energy is used to maintain it (see Chapter 6). Living things, representing the ultimate in organized complexity, continuously use energy to maintain themselves.

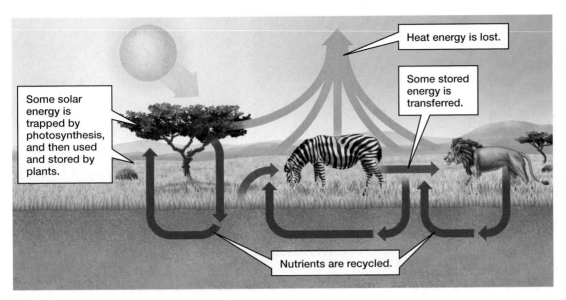

◀ **FIGURE 1-3** **The flow of energy and the recycling of nonliving nutrients**

Heat energy is lost.

Some stored energy is transferred.

Some solar energy is trapped by photosynthesis, and then used and stored by plants.

Nutrients are recycled.

THINK CRITICALLY
Describe the source of the energy stored in the meat and the bun of a hamburger, and explain how the energy got from the source to the two foodstuffs.

◀ **FIGURE 1-4** **Organisms maintain relatively constant internal conditions** Evaporative cooling by water, both from sweat and from a bottle, helps this athlete maintain his body temperature during vigorous exercise.

The ability of an organism to maintain its internal environment within the limits required to sustain life is called **homeostasis.** To maintain homeostasis, cell membranes constantly pump specific substances in and others out. People and other mammals use both physiological and behavioral mechanisms to maintain the narrow temperature range that allows life-sustaining reactions to occur in their cells (**FIG. 1-4**). Life, then, requires very precise internal conditions maintained by a continuous expenditure of energy.

Organisms Sense and Respond to Stimuli

To obtain energy and nutrients, organisms must sense and respond to stimuli in their environments. Animals use specialized cells to detect light, temperature, sound, gravity, touch, chemicals, and many other stimuli from their external and internal surroundings. For example, when your brain detects a low level of sugar in your blood (an internal stimulus), it causes your mouth to water at the smell of food (an external stimulus). Plants, fungi, and single-celled organisms use very different mechanisms that are equally effective for their needs (**FIG. 1-5**). Even many bacteria, the smallest and simplest life-forms, can move toward favorable conditions and away from harmful substances.

Organisms Grow

At some time in its life, every organism grows. The water flea in Figure 1-2 grew from the size of one of the eggs you see in its body. Single-celled organisms such as bacteria grow to about double their original size, copy their genetic material, and then divide in half to reproduce. Animals and plants use a similar process to produce more cells within their bodies, repeating the sequence until growth stops. Individual cells can also contribute to the growth of an organism by increasing in size, as occurs in muscle and fat cells in animals and in food storage cells in plants.

Organisms Reproduce

Organisms reproduce in a variety of ways (**FIG. 1-6**). These include dividing in half, producing seeds, bearing live young, and producing eggs (see Fig. 1-2). The end result is always the same: new versions of the parent organisms that inherit the instructions for producing and maintaining their particular form of life. These instructions—copied in every cell and passed on to descendants—are carried in the unique structure

◀ **FIGURE 1-5**
Bending toward the light Plants perceive and often bend toward light, which provides them with the energy they need to survive.

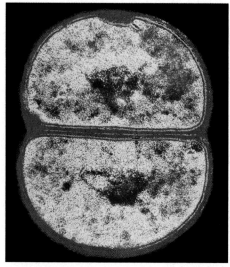
(a) Dividing *Streptococcus* bacterium

(b) Dandelion producing seeds

(c) Panda with its baby

▲ **FIGURE 1-6** **Organisms reproduce**

of the hereditary molecule **deoxyribonucleic acid (DNA)** (**FIG. 1-7**; see Chapter 12). The complete set of DNA molecules contained in each cell provides a detailed instruction manual for life, much like an architectural blueprint provides instructions for constructing a building.

▲ **FIGURE 1-7** **DNA** As James Watson, the codiscoverer of the structure of DNA, stated: "A structure this pretty just had to exist."

Organisms, Collectively, Have the Capacity to Evolve

A simple definition of **evolution** is the change in DNA that occurs in a population over time. Through the course of generations, changes in DNA within any **population** (a group of the same type of organism inhabiting the same area) are inevitable. In the words of biologist Theodosius Dobzhansky, "Nothing in biology makes sense except in the light of evolution." The next section provides a brief introduction to evolution—the unifying concept of biology.

CHECK YOUR LEARNING
Can you ...
- explain the characteristics that define life?
- explain why these characteristics are necessary to sustain life?
- describe how reproduction allows evolution to occur?

CASE STUDY \ **CONTINUED**
The Boundaries of Life

Are viruses alive? Viruses release their genetic material inside cells and then hijack the infected cell's energy supplies and biochemical machinery, turning the cell into a kind of factory that churns out many copies of viral parts. These parts assemble into an army of virus particles. The newly formed viruses then emerge from the host cell, often rupturing it in the process. Some types of viruses, including HIV and the Ebola virus, acquire an outer envelope made of the infected cell's plasma membrane as they emerge. Viruses do not obtain or use their own energy or materials, maintain themselves, or grow. Therefore, viruses do not meet our criteria for life. They do, however, possess a few characteristics of life: Viruses respond to stimuli by binding to specific sites on the cells they attack, and some scientists consider viral replication a form of reproduction. Viruses also evolve, often with stunning speed. How does evolution occur in viruses and other biological entities?

1.2 WHAT IS EVOLUTION?

Evolution is genetic change in a population over time. Cumulative changes over vast stretches of time explain the amazing diversity of organisms that now share this planet. The scientific theory of evolution was formulated in the mid-1800s by two English naturalists, Charles Darwin and Alfred Russel Wallace. Since that time, it has been supported by fossils, geological studies, radioactive dating of rocks, genetics, molecular biology, biochemistry, and breeding experiments. Evolution not only explains the enormous diversity of life, but also accounts for the remarkable similarities among different types of organisms. For example, people share many features with chimpanzees, and the sequence of our DNA is nearly identical to that of chimpanzees. This similarity is

◀ **FIGURE 1-8 Chimpanzees and people are closely related**

strong evidence that people and chimps descended from a common ancestor, but the obvious differences (**FIG. 1-8**) reflect the differences in our evolutionary paths.

Three Natural Processes Underlie Evolution

Evolution is an automatic and inevitable outcome of three natural occurrences: (1) differences among members of a population, (2) inheritance of these differences by offspring, and (3) natural selection, the process by which individuals that inherit certain characteristics tend to survive and reproduce better than other individuals. Let's take a closer look at these three factors.

Mutations Are the Source of Differences in DNA

Look around at your classmates and notice how different they are, or observe how dogs differ in size, in shape, and in the color, length, and texture of their coats. Although some of this variation (particularly among your classmates) is due to differences in environment and lifestyle, much of it results from differences in genes. **Genes,** which are specific segments of DNA, are the basic units of heredity. Before a cell divides, all of its DNA is copied, allowing its genes to be passed along to both resulting cells. Just as you would make mistakes if you tried to copy a blueprint by hand, cells make some errors as they copy their DNA. Changes in genes, such as those caused by these random copying errors, are called **mutations.** Mutations can also result from damage to DNA, caused, for example, by ultraviolet rays from sunlight, radiation released from a damaged nuclear power plant, or toxic chemicals from cigarette smoke. Just as changes to a blueprint will cause changes in the structure built from it, so may a new cell with altered DNA differ from its parent cell.

Some Mutations Are Inherited

Mutations that occur in sperm or egg cells may result in transmission of altered DNA from parent to offspring. Each cell in the offspring will carry the inherited mutation. Most mutations to genes are either harmful or neutral. For example, genetic diseases such as hemophilia, sickle-cell anemia, and cystic fibrosis are caused by harmful mutations. Other mutations have no observable effect or change the organism in a way that

is *neutral,* neither harmful nor beneficial. Almost all of the inherited variability among traits—such as human eye color—is caused by neutral mutations that occurred in the distant past and have been passed along harmlessly through generations. On rare occasions, however, an inherited mutation changes a gene in a way that helps offspring to survive and reproduce more successfully than those lacking the mutation. These infrequent events provide the raw material for evolution.

Some Inherited Mutations Help Individuals Survive and Reproduce

The most important process in evolution is natural selection, which acts on the natural variability in traits. **Natural selection** is the process by which organisms with certain inherited traits survive and reproduce better than others in a given environment. As a result, the advantageous inherited traits become increasingly common in the population as generations pass. Because these traits are caused by differences in genes, the genetic makeup of the population as a whole will change over time; that is, the population will evolve. Consider a likely scenario of natural selection. Imagine that ancient beavers had short front teeth like most other mammals. If a mutation caused one beaver's offspring to grow longer front teeth, these offspring would have gnawed down trees more efficiently, built bigger dams and lodges, and eaten more bark than beavers that lacked the mutation. These long-toothed beavers would have been better able to survive and would have raised more offspring that would inherit the genes for longer front teeth. Over time, long-toothed beavers would have become increasingly common; after many generations, all beavers would have long front teeth.

Structures, physiological processes, or behaviors that help an organism survive and reproduce in a particular environment are called **adaptations.** Most of the features that we admire so much in other life-forms, such as the fleet, agile limbs of deer, the broad wings of eagles, and the mighty trunks of redwood trees, are adaptations. Adaptations help organisms escape predators, capture prey, reach the sunlight, or accomplish other feats that help ensure their survival and reproduction. The huge array of adaptations found in living things today was molded by natural selection acting on random mutations.

But how did life's diversity, including deer, eagles, redwoods, and people, all arise from the first single-celled life that appeared billions of years ago? Natural selection is not uniform; a trait that is adaptive in one environment may not be helpful (or may even be a hindrance) in a different setting. After Darwin observed different but closely related organisms on clusters of islands, he hypothesized that different forms of life may evolve if a population becomes fragmented and groups of individuals are subjected to different environments. For example, a violent storm may carry some individuals from the mainland to an offshore island. The mainland and the island populations will initially consist of the same **species** (organisms of the same type that can interbreed). But if the island's environment differs from that of

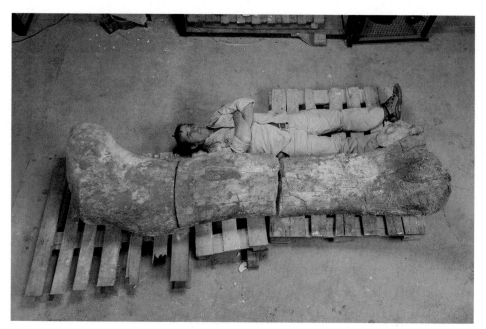

◀ **FIGURE 1-9 A fossil from a newly discovered dinosaur, *Titanosaurus*** The most widely accepted hypothesis for the extinction of dinosaurs about 65 million years ago is a massive meteorite strike that rapidly and radically altered their environment. This thigh bone, estimated to be 95 million years old, is from a plant-eating giant with an estimated length of 130 feet (40 meters) and a weight of about 176,000 pounds (80 metric tons).

THINK CRITICALLY The largest dinosaurs were plant-eaters. Based on Figure 1-3, can you suggest a reason why?

the mainland, the newcomers will be subjected to different forces of natural selection; as a result, they will evolve different adaptations. These differences may eventually become great enough that the two populations can no longer interbreed; a new species will have evolved.

What helps an organism survive today may become a liability in the future. If environments change—for example, as global climate change occurs—the traits that best adapt organisms to their environments will change as well. In the case of global climate change, if a random mutation helps an organism survive and reproduce in a warmer climate, the mutation will be favored by natural selection and will become more common in the population with each new generation.

If mutations that help an organism to adapt do not occur, a changing environment may doom a species to **extinction**—the complete elimination of this form of life. Dinosaurs flourished for 100 million years, but because they did not evolve fast enough to adapt to rapidly changing conditions, they became extinct (**FIG. 1-9**). In recent decades, human activities such as burning fossil fuels and converting tropical forests to farmland have drastically accelerated the rate of environmental change. Mutations that better adapt organisms to these altered environments are quite rare, and consequently the rate of extinction has increased dramatically.

CHECK YOUR LEARNING

Can you ...

- explain what mutations are, how they occur, what allows them to be inherited, and what general types of changes mutations can produce?
- explain how natural processes lead inevitably to evolution?
- describe how a new species can be produced by natural selection?

CASE STUDY \ **CONTINUED**

The Boundaries of Life

One lifelike property of viruses is their capacity to evolve. Through evolution, viruses sometimes become more infectious or more deadly, or they may gain the ability to infect new hosts. Certain types of viruses, including Ebola, HIV, and flu, are very sloppy in copying their genetic material and mutate about 1,000 times as often as the average animal cell. One consequence is that viruses such as flu evolve rapidly; flu shots must immunize you against different types of flu every year. Likewise, more than 200 different viruses can cause symptoms of the "common cold," explaining why you keep getting new colds throughout life. HIV in an infected person can produce up to 10 billion new viruses daily, with 10 million of these carrying a random mutation. Inevitably, some of these mutations will produce resistance to an antiviral drug. Therefore, antiviral drugs act as agents of natural selection that promote the survival and successful replication of drug-resistant viruses. For this reason, HIV victims are given "cocktails" of three or four different drugs; resistance to all of them would require multiple specific mutations to occur in the same virus, an enormously unlikely event.

1.3 HOW DO SCIENTISTS STUDY LIFE?

The science of biology encompasses many different areas of inquiry, each requiring different types of specialized knowledge. In fact, biology is not a single field, but many—linked by the amazing complexity of life.

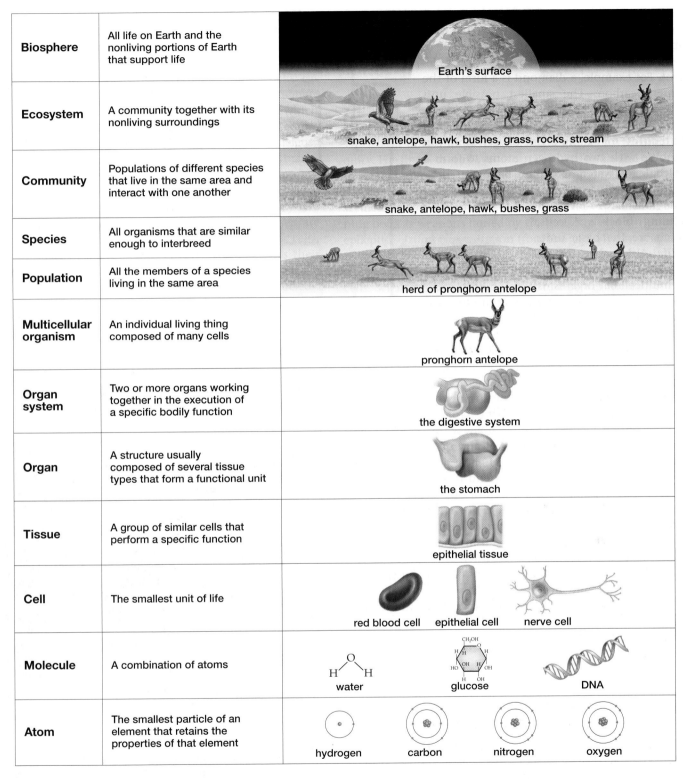

Biosphere	All life on Earth and the nonliving portions of Earth that support life	
Ecosystem	A community together with its nonliving surroundings	
Community	Populations of different species that live in the same area and interact with one another	
Species	All organisms that are similar enough to interbreed	
Population	All the members of a species living in the same area	
Multicellular organism	An individual living thing composed of many cells	
Organ system	Two or more organs working together in the execution of a specific bodily function	
Organ	A structure usually composed of several tissue types that form a functional unit	
Tissue	A group of similar cells that perform a specific function	
Cell	The smallest unit of life	
Molecule	A combination of atoms	
Atom	The smallest particle of an element that retains the properties of that element	

▲ **FIGURE 1-10 Levels of biological organization** Each level provides building blocks for the one above it, which has new properties that emerge from the interplay of the levels below.

THINK CRITICALLY What current, ongoing environmental change is likely to affect the entire biosphere?

Life May Be Studied at Different Levels

Let's look at the levels of organization that comprise life on Earth (FIG. 1-10). Biologists conduct research at nearly every level, from complex biological molecules such as DNA to entire ecosystems (for example, how forest ecosystems may be altered by climate change).

Each level of organization provides a foundation for the one above it, and each higher level has new, more inclusive

important date, so you rush to your car, turn the ignition key, and make the *observation* that the car won't start. Your *question*, "Why won't the car start?" leads to a *hypothesis*: The battery is dead. This leads to the *prediction* that a jump-start will solve the problem. You *experiment* by attaching jumper cables from your roommate's car battery to your own. The result? Your car starts immediately, leading to the *conclusion* that your experiment supported your hypothesis about the dead car battery.

Biologists Test Hypotheses Using Controlled Experiments

In controlled experiments, two types of situations are established. One is a baseline, or **control,** situation, in which all possible factors are held constant. The other is the experimental situation, where one factor, the **variable,** is manipulated to test the hypothesis that this variable is the cause of an observation. Often, the manipulation inadvertently changes more than one factor. In the preceding car example, jump-starting the car might have both delivered a charge to the battery and knocked some corrosion off the battery terminal that was preventing the battery from delivering power—your battery might actually have been fully charged. In real experiments, scientists must control for all the possible effects of any manipulation they perform, so frequently more than one control is needed.

Valid scientific experiments must be repeatable by the researcher and by other scientists. To help ensure this, a researcher performs multiple repetitions of an experiment, setting up several replications of each control group and an equal number of experimental groups. Data from control and experimental situations are often compared using *statistics,* mathematical formulas that can help interpret and draw conclusions from various types of numerical measurements. Statistics can determine the likelihood that the difference between control and experimental groups arose by random chance. If statistical tests show chance to be sufficiently unlikely, the difference between the groups is described as *statistically significant.*

Science must also be communicated, or it is useless. Good scientists publish their results, explaining their methods in detail so others can repeat and build on their experiments. Francesco Redi recognized this in the 1600s when he carefully recorded the methods of his classic controlled experiment testing the hypothesis that flies caused maggots to appear on rotting meat (see "How Do We Know That? Controlled Experiments Provide Reliable Data" on page 12).

Experimentation using variables and controls is powerful, but it is important to recognize its limitations. In particular, scientists can seldom be sure that they have controlled for *all* possible variables or performed all the manipulations that could possibly refute their hypothesis. Therefore, science mandates that conclusions are always subject to revision if new experiments or observations contradict them.

Fruit flies, bacteria from hot springs, sea jellies, Gila monsters, burdock burrs—why study these obscure forms of life? In fact, research on these organisms, and a host of others, has improved people's lives.

Fruit flies, for example, have been used for over 100 years to study how genes influence traits. Their genes are similar enough to ours that many human genetic diseases can be investigated to some extent in these flies—a pair of which can produce several hundred genetically identical offspring in a few weeks. An obscure bacterium from a hot spring in Yellowstone National Park is the source of a protein crucial to a process that rapidly copies DNA. Thanks to this discovery, the amount of DNA in a few skin cells left at a crime scene can now generate a sample large enough

Gila monster

to be compared to the DNA of a suspect. A fluorescent green protein discovered in a sea jelly can be attached to a gene, protein, or virus, making it glow and allowing researchers to monitor its activity. A protein found in the Gila monster's venomous saliva was approved in 2005 as a drug to help diabetics maintain more constant blood sugar levels. And what did microscopic examination of a burr lead to? The inspiration for Velcro.

Some people criticize governments for funding research into topics that seem obscure, like what makes a jellyfish glow. But no one can predict where such studies will lead; even lines of research that appear to be dead ends can provide unexpected and valuable insights.

Scientific Theories Have Been Thoroughly Tested

Scientists use the word "theory" in a way that differs from its everyday usage. If Dr. Watson asked Sherlock Holmes, "Do you have a theory as to the perpetrator of this foul deed?" in scientific terms, he would be asking Holmes for a hypothesis—a proposed explanation based on clues that provide incomplete evidence. A **scientific theory,** in contrast, is a general and reliable explanation of important natural phenomena that has been developed through extensive and reproducible observations and experiments. In short, a scientific theory is best described as a **natural law,** a basic principle derived from the study of nature that has never been disproven by scientific inquiry. For example, scientific theories such as the atomic theory (that all matter is composed of atoms) and the theory of gravitation (that objects exert attraction for one another) are fundamental to the science of physics. Likewise, the **cell theory** (that all living organisms are composed of cells) and the theory of evolution are fundamental to the study of biology.

Scientists describe fundamental principles as "theories" rather than "facts" because even scientific theories can potentially be disproved, or falsified. If compelling evidence arises that renders a scientific theory invalid, that theory must be modified or discarded. A modern example of the need to modify basic principles in the light of new scientific evidence is the discovery of prions, which are infectious proteins (see Chapter 3). Before the early 1980s, all known infectious disease agents copied themselves using instructions from genetic material. Then in 1982, neurologist Stanley Prusiner published evidence that scrapie (an infectious disease of sheep that causes brain degeneration) is actually triggered and transmitted by a protein and has no genetic material. Infectious proteins were unknown to science, and Prusiner's results were met with widespread disbelief. It took nearly two decades of further research to convince most of the scientific community that a protein alone could act as an infectious disease agent. Prions are now known to cause mad cow disease and two fatal human brain disorders. Stanley Prusiner was awarded the Nobel Prize in Physiology or Medicine for his pioneering work. Science is based on the premise that even basic scientific principles can be modified in light of new data. By accepting prions as infectious proteins, scientists maintained the integrity of the scientific process while expanding our understanding of how diseases can occur.

Ongoing scientific inquiry continuously tests scientific theories. This is a major difference between scientific principles and faith-based doctrines (such as creationism), which are impossible to prove or disprove and thus fall outside the scope of science.

Scientific Theories Involve Both Inductive and Deductive Reasoning

Scientific theories arise through **inductive reasoning,** the process of creating a broad generalization based on many observations that support it and none that contradict it. For example, the cell theory arises from the observation that all organisms that possess the characteristics of life are composed of one or more cells and that nothing that is not composed of cells shares all of these attributes.

Once a scientific theory has been formulated, it can be used to support deductive reasoning. In science, **deductive reasoning** starts with a well-supported generalization and uses it to generate hypotheses about how a specific experiment or observation will turn out. For example, based on the cell theory, if a scientist discovers a new entity that exhibits all the characteristics of life, she can confidently hypothesize that it will be composed of cells. Of course, the new organism must then be carefully examined to confirm its cellular structure.

Science Is a Human Endeavor

Scientists are people, driven by the pride, fears, and ambition common to humanity. Accidents, lucky guesses, controversies with competing scientists, and, of course, the intellectual curiosity of individual scientists all contribute to scientific advances. Even mistakes can play a role. Let's consider an actual case.

Microbiologists often study pure cultures—a single type of bacterium grown in sterile, covered dishes free from contamination by other bacteria and molds. At the first sign of contamination, a culture is usually thrown out, often with mutterings about sloppy technique. In the late 1920s, however, Scottish bacteriologist Alexander Fleming turned a ruined bacterial culture into one of the greatest medical advances in history.

One of Fleming's cultures became contaminated with a mold (a type of fungus) called *Penicillium*. But instead of discarding the dish, Fleming observed that no bacteria were growing near the mold (**FIG. 1-12**). He asked the question "Why aren't bacteria growing in this region?" Fleming then formulated the hypothesis that *Penicillium* releases a substance that kills bacteria, and he predicted that a solution in which the mold had grown would contain this substance and kill bacteria. To test this hypothesis, Fleming performed an experiment. He grew *Penicillium* in a liquid nutrient broth, and then filtered out the mold and poured some of the mold-free broth on a plate with a pure bacterial culture. Sure enough, something in the liquid killed the bacteria, supporting his hypothesis. This (and more experiments that confirmed his results) led to the conclusion that *Penicillium* secretes a substance that kills bacteria. Further research into these mold extracts resulted in the production of the first antibiotic—penicillin.

Fleming's experiments are a classic example of the scientific method, but they would never have happened without the combination of a mistake, a chance observation, and the curiosity to explore it. The outcome has saved millions of

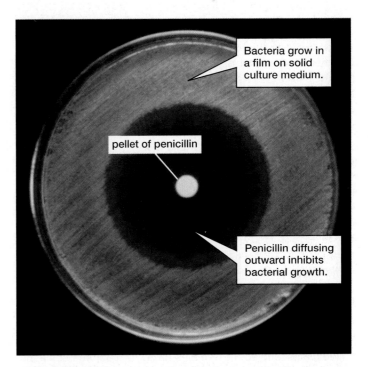

Bacteria grow in a film on solid culture medium.

pellet of penicillin

Penicillin diffusing outward inhibits bacterial growth.

▲ **FIGURE 1-12 Penicillin kills bacteria** Alexander Fleming observed similar inhibition of bacterial growth around colonies of *Penicillium* mold.

HOW DO WE KNOW THAT?

Controlled Experiments Provide Reliable Data

A classic experiment by the Italian physician Francesco Redi (1621–1697) beautifully demonstrates the scientific method and helps to illustrate the basic scientific principle that all events can be traced to natural causes. Redi investigated why maggots (fly larvae) appear on spoiled meat. In Redi's time, refrigeration was unknown, and meat was stored in the open. Many people of that time believed that the appearance of maggots on meat was evidence of **spontaneous generation,** the emergence of life from nonliving matter.

Redi *observed* that flies swarm around fresh meat and that maggots appear on meat left out for a few days. He *questioned* where the maggots came from. He then formed a testable *hypothesis:* Flies produce maggots. This led to the *prediction* that keeping flies off the meat would prevent maggots from appearing. In his *experiment,* Redi wanted to test one variable—the access of flies to the meat. Therefore, he placed similar pieces of meat in each of two clean jars. He left one jar open (the control jar) and covered the other with gauze to keep out flies (the experimental jar). He did his best to keep all the other conditions the same (for example, the type of jar, the type of meat, and the temperature). After a few days, he observed maggots on the meat in the open jar but saw none on the meat in the covered jar. Redi *concluded* that his hypothesis was correct and that maggots are produced by flies, not by the nonliving meat (**FIG. E1-1**). Only through this and other controlled experiments could the age-old belief in spontaneous generation be laid to rest.

Today, more than 300 years later, the scientific method is still used. Consider the experiments of Malte Andersson, who investigated the mating choices of female widowbirds. Andersson *observed* that male, but not female, widowbirds have extravagantly long tails, which they display while flying across African grasslands. Andersson asked the *question:* Why do male birds have such long tails? His *hypothesis* was that females prefer to mate with long-tailed males, and so these males have more offspring, who inherit their genes for long tails. Andersson *predicted* that if

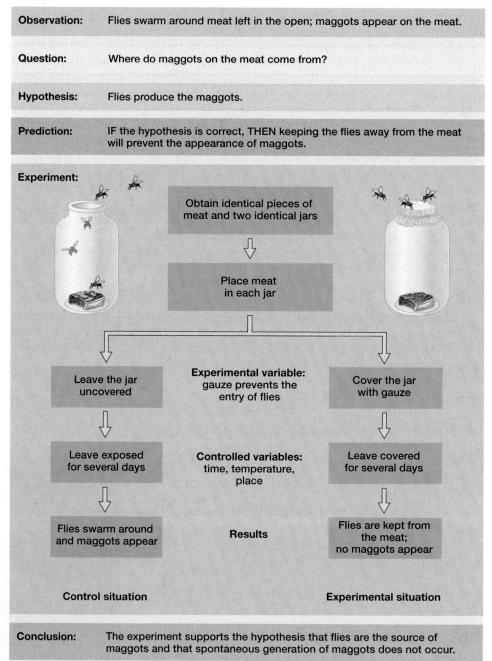

Observation:	Flies swarm around meat left in the open; maggots appear on the meat.
Question:	Where do maggots on the meat come from?
Hypothesis:	Flies produce the maggots.
Prediction:	IF the hypothesis is correct, THEN keeping the flies away from the meat will prevent the appearance of maggots.

Experiment:

Obtain identical pieces of meat and two identical jars

Place meat in each jar

Leave the jar uncovered	**Experimental variable:** gauze prevents the entry of flies	Cover the jar with gauze
Leave exposed for several days	**Controlled variables:** time, temperature, place	Leave covered for several days
Flies swarm around and maggots appear	**Results**	Flies are kept from the meat; no maggots appear

Control situation   **Experimental situation**

Conclusion:	The experiment supports the hypothesis that flies are the source of maggots and that spontaneous generation of maggots does not occur.

▲ **FIGURE E1-1** **The experiment of Francesco Redi illustrates the scientific method**

his hypothesis were true, more females would build nests on the territories of males with artificially lengthened tails than on the territories of males with artificially shortened tails. To test this, he captured some males, trimmed their tails to about half their original length, and released them (*experimental* group 1). He

took another group of males and glued on the tail feathers that he had removed from the first group, creating exceptionally long tails (*experimental* group 2). Then, in *control* group 1, he cut the tail feathers but then glued them back in place (to control for the effects of capturing the birds and manipulating their feathers). In *control* group 2, he simply captured and released a group of male birds to control for behavioral changes caused by the stress of being caught and handled. Later, Andersson counted the number of nests that females had built on each male's territory, which indicated how many females had mated with that

male. He found that males with lengthened tails had the most nests on their territories, males with shortened tails had the fewest, and control males (with normal-length tails, either untouched or cut and glued together) had an intermediate number (**FIG. E1-2**). Andersson *concluded* that his results supported the

hypothesis that female widowbirds prefer to mate with long-tailed males.

THINK CRITICALLY Did Redi's experiment (Fig. E1-1) convincingly demonstrate that flies produce maggots? What kind of follow-up experiment would help confirm the source of maggots?

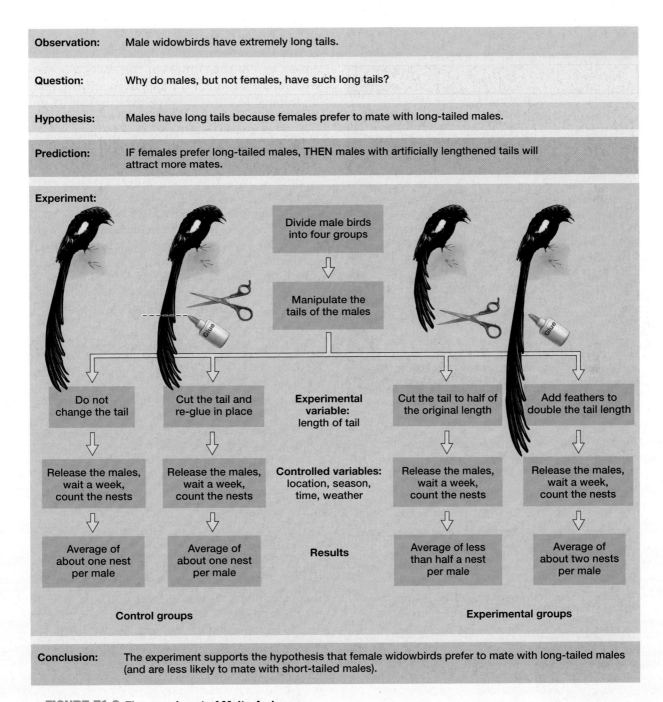

Observation: Male widowbirds have extremely long tails.

Question: Why do males, but not females, have such long tails?

Hypothesis: Males have long tails because females prefer to mate with long-tailed males.

Prediction: IF females prefer long-tailed males, THEN males with artificially lengthened tails will attract more mates.

Experiment:

Divide male birds into four groups

Manipulate the tails of the males

Do not change the tail

Cut the tail and re-glue in place

Experimental variable: length of tail

Cut the tail to half of the original length

Add feathers to double the tail length

Release the males, wait a week, count the nests

Release the males, wait a week, count the nests

Controlled variables: location, season, time, weather

Release the males, wait a week, count the nests

Release the males, wait a week, count the nests

Average of about one nest per male

Average of about one nest per male

Results

Average of less than half a nest per male

Average of about two nests per male

Control groups

Experimental groups

Conclusion: The experiment supports the hypothesis that female widowbirds prefer to mate with long-tailed males (and are less likely to mate with short-tailed males).

▲ **FIGURE E1-2** The experiment of Malte Andersson

lives. As French microbiologist Louis Pasteur said, "Chance favors the prepared mind."

Knowledge of Biology Illuminates Life

Some people regard science as a dehumanizing activity, thinking that too deep an understanding of the world robs us of wonder and awe. Nothing could be further from the truth. For example, let's look at lupine flowers. Their two

Pollen is forced onto the bee's abdomen.

▲ **FIGURE 1-13 Adaptations in lupine flowers** Understanding life helps people notice and appreciate the small wonders at their feet. (Inset) A lupine flower deposits pollen on a foraging bee's abdomen.

lower petals form a tube surrounding both male and female reproductive parts (**FIG. 1-13**). In young flowers, the weight of a bee on this tube forces pollen (carrying sperm) out of the tube onto the bee's abdomen. In older lupine flowers that are ready to be fertilized, the female part grows and emerges through the end of the tube. When a pollen-dusted bee visits, it deposits some pollen on the female organ, allowing the lupine to produce the seeds of its next generation.

Do these insights detract from our appreciation of lupines? Far from it. There is added delight in watching and understanding the intertwined form and function of bee and flower that resulted as these organisms evolved together. Soon after learning the lupine's pollination mechanism, two of the authors of this text crouched beside a wild lupine to watch it happen. An elderly man passing by stopped to ask what they were looking at so intently. He listened with interest as they explained about what happened when a bee landed on the lupine's petals and immediately went to observe another patch of lupines where bees were foraging. He, too, felt the heightened sense of appreciation and wonder that comes with understanding.

Throughout this text, we try to convey that biology is not just another set of facts to memorize. It is a pathway to understanding yourself and the life around you. It is also important to recognize that biology is not a completed work, but an ongoing exploration. As Alan Alda, best known for playing "Hawkeye" in the TV show *M*A*S*H,* stated: "With every door into nature we nudge open, 100 new doors become visible."

CHECK YOUR LEARNING

Can you ...

- describe the principles underlying science?
- outline the scientific method?
- explain why controls are crucial in biological studies?
- explain why fundamental scientific principles are called theories?
- distinguish between inductive and deductive reasoning?

CASE STUDY \ REVISITED

The Boundaries of Life

If viruses aren't a form of life, what are they? A virus by itself is an inert particle that doesn't approach the complexity of a cell. The simplest virus, such as that causing smallpox (**FIG. 1-14**), consists of a protein coat that surrounds genetic material. The uncompli-

protein coat

genetic material

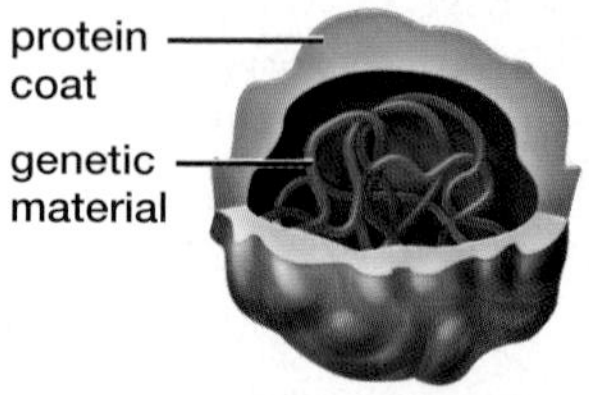

▲ **FIGURE 1-14 A smallpox virus**

cated structure of viruses, coupled with amazing advances in biotechnology, has allowed researchers to synthesize viruses in the laboratory. They have accomplished this using the blueprint contained in viral genetic material and readily purchased chemicals. The first virus to be synthesized was the small, simple poliovirus. This feat was accomplished in 2002 by Eckard Wimmer and coworkers at Stony Brook University, who titled their work "The Test-Tube Synthesis of a Chemical Called Poliovirus."

Did these researchers create life in the laboratory? A few scientists would say "yes," defining life by its ability to copy itself and to evolve. Wimmer himself describes viruses as entities that switch between a nonliving phase outside the cell and a living phase inside. Although most scientists agree that viruses aren't alive and support the definition of life presented in this text, the controversy continues. As virologist Luis Villarreal puts it, "Viruses are parasites that skirt the boundaries between life and inert matter."

CONSIDER THIS When Wimmer and coworkers announced that they had synthesized the poliovirus, they created considerable controversy. Some people feared that deadly and highly contagious viruses might be synthesized by bioterrorists. The researchers responded that they were merely applying current knowledge and techniques to demonstrate the principle that viruses are basically chemical entities that can be synthesized in the laboratory. Do you think scientists should synthesize viruses or other agents that can cause infectious disease? What are the implications of forbidding such research?

CHAPTER REVIEW

 Go to **MasteringBiology** for practice quizzes, activities, eText, videos, current events, and more.

Answers to Think Critically, Evaluate This, Multiple Choice, and Fill-in-the-Blank questions can be found in the Answers section at the back of the book.

Summary of Key Concepts

1.1 What Is Life?

Organisms acquire and use materials and energy. Materials are obtained from other organisms or the nonliving environment and are repeatedly recycled. Energy must be continuously captured from sunlight by photosynthetic organisms, whose bodies supply energy to all other organisms. Organisms also actively maintain organized complexity, perceive and respond to stimuli, grow, reproduce, and, collectively, evolve.

1.2 What Is Evolution?

Evolution is the scientific theory that modern organisms descended, with changes, from earlier organisms. Evolution occurs as a consequence of (1) genetic differences, originally arising as mutations, among members of a population; (2) inheritance of these differences by offspring; and (3) natural selection of the differences that produce the best adaptations to the organisms' environment.

1.3 How Do Scientists Study Life?

Scientists identify a hierarchy of levels of organization, each more encompassing than those beneath (see Fig. 1-10). Biologists categorize organisms into three domains: Archaea, Bacteria, and Eukarya. Members of Archaea and Bacteria consist of single prokaryotic cells, but fundamental molecular differences distinguish them. Members of Eukarya are composed of one or more eukaryotic cells. Organisms are assigned scientific names that identify each as a unique species within a specific genus.

1.4 What Is Science?

Science is based on three principles: (1) all events can be traced to natural causes that can be investigated; (2) the laws of nature are unchanging; and (3) scientific findings are independent of values except honesty in reporting data. Knowledge in biology is acquired through the scientific method, in which an observation leads to a question that leads to a hypothesis. The hypothesis generates a prediction that is then tested by controlled experiments or precise observation. The experimental results, which must be repeatable, lead to a conclusion that either supports or refutes the hypothesis. A scientific theory is a general explanation of natural phenomena developed through extensive and reproducible experiments and observations.

Key Terms

adaptation *5*
atom *8*
binomial system *9*
biology *2*
biosphere *8*
cell *2*
cell theory *10*
community *8*
conclusion *9*
control *10*
deductive reasoning *11*
deoxyribonucleic acid (DNA) *4*
domain *8*
ecosystem *8*
element *8*
eukaryotic *9*
evolution *4*
experiment *9*
extinction *6*
gene *5*
homeostasis *3*
hypothesis *9*
inductive reasoning *11*
molecule *8*
multicellular *2*

mutation *5*
natural law *10*
natural selection *5*
nucleus *9*
observation *9*
organ *8*
organ system *8*
organelle *9*
organism *2*
photosynthesis *2*
plasma membrane *8*
population *4*
prediction *9*
prokaryotic *9*
question *9*
science *9*
scientific method *9*
scientific theory *10*
species *5*
spontaneous generation *12*
tissue *8*
unicellular *2*
variable *10*

Thinking Through the Concepts

Multiple Choice

1. Evolution is
   a. a belief.
   b. a scientific theory.
   c. a hypothesis.
   d. never observed in the modern world.

2. Which of the following is *not* true of science?
   a. Science is based on the premise that all events can be traced to natural causes.
   b. Important science can be based on chance observations.
   c. A hypothesis is basically a wild guess.
   d. Scientific theories can potentially be disproved.

3. Which of the following does *not* apply to mutations?
   a. They occur to cause adaptive changes in response to the environment.
   b. They are usually either harmful or neutral.
   c. They are only inherited if they occur in a sperm or egg cell.
   d. They often occur when DNA is copied.

4. Viruses
   a. have DNA confined in a nucleus.
   b. are relatively rare compared to living organisms.
   c. do not evolve.
   d. may be surrounded by plasma membrane from their host cell.

5. Which one of the following is True?
   a. The presence of a cell nucleus distinguishes Bacteria from Archaea.
   b. All cells are surrounded by a plasma membrane.
   c. All members of Eukarya are multicellular.
   d. Viruses are the simplest cells.

Fill-in-the-Blank

1. Organisms respond to _____. Organisms acquire and use _____ and _____ from the environment. Organisms are composed of cells whose structure is both _____ and _____. Collectively, organisms _____ over time.

2. The smallest particle of an element that retains all the properties of that element is a(n) _____. The smallest unit of life is the _____. Cells of a specific type within multicellular organisms combine to form _____. A(n) _____ consists of all of the same type of organism within a defined area. A(n) _____ consists of all the interacting populations within the same area. A(n) _____ consists of the community and its nonliving surroundings.

3. A(n) _____ is a general explanation of natural phenomena supported by extensive, reproducible tests and observations. In contrast, a(n) _____ is a proposed explanation for observed events. To answer specific questions about life, biologists use a general process called the _____.

4. An important scientific theory that explains why organisms are at once so similar and so diverse is the theory of _____. This theory explains life's diversity as having originated primarily through the process of _____.

5. The molecule that guides the construction and operation of an organism's body is called (complete term) _____, abbreviated as _____. This large molecule contains discrete segments with specific instructions; these segments are called _____.

Review Questions

1. What properties are shared by all forms of life?

2. Why do organisms require energy? Where does the energy come from?

3. Define *evolution*, and explain the three natural occurrences that make evolution inevitable.

4. What are the three domains of life?

5. What are some differences between prokaryotic and eukaryotic cells? In which domain(s) is each found?

6. What basic principles underlie scientific inquiry?

7. What is the difference between a scientific theory and a hypothesis? Why do scientists refer to basic scientific principles as "theories" rather than "facts"?

8. What factors did Redi control for in his open jar of meat? What factors did Andersson control for?

9. Explain the differences between inductive and deductive reasoning. Which of these processes generates scientific theories?

10. List the steps in the scientific method with a brief description of each step.

Applying the Concepts

1. What misunderstanding causes some people to dismiss evolution as "just a theory"?

2. How would this textbook's definition of life need to be changed to allow viruses to qualify as life-forms? For prions to be considered alive?

3. Review Alexander Fleming's experiment that led to the discovery of penicillin. What would be an appropriate control for the experiment in which Fleming applied filtered medium from a *Penicillium* culture to plates of bacteria?

4. Explain an instance in which your own understanding of a phenomenon enhances your appreciation of it.

5. In using the scientific method to help start your car, if jump-starting didn't work, what hypothesis would you test next?

UNIT 1
The Life
of the Cell

Single cells can be complex, independent organisms such as this freshwater protist of the genus *Dendrocometes*. A rounded attachment region anchors the cell firmly to the gills of freshwater fish or crustaceans. Tentacles, resembling microscopic antlers, snare food as water passes over them.

"Any living cell carries with it the experiences of a billion years of experimentation by its ancestors." — MAX DELBRÜCK

2 ATOMS, MOLECULES, AND LIFE

The aftermath of explosions at the Fukushima nuclear power plant in Japan.

CASE STUDY

Unstable Atoms Unleashed

AN EARTHQUAKE OF EPIC MAGNITUDE—9.0 on the Richter scale—shook the northeast coast of Japan on March 11, 2011. It was the most violent earthquake in Japan's history, and one of the most powerful ever recorded worldwide. Soon after, a tsunami caused by the quake slammed into the Fukushima Daiichi nuclear power plant on Japan's eastern coast. Towering waves nearly 50 feet high flooded the plant and knocked out its main electrical power supply and backup generators, which caused its cooling system to fail.

The cores of nuclear reactors like those in the Fukushima plant contain thousands of fuel rods consisting of zirconium metal tubes filled with uranium fuel. Two thick steel containment vessels surround the nuclear core, and water is pumped continuously around the vessels to absorb the intense heat generated by the nuclear reactions within them. The heated water produces steam, which then expands, driving turbines that generate electricity.

When the power loss shut down the plant's water pumps, operators used firefighting equipment to inject seawater into the inner containment vessel in a desperate attempt to cool it. But their efforts failed; heat and pressure cracked the inner containment vessel, allowing water and steam to escape. The core temperature rose to over 1,800°F (about 1,000°C), melting the zirconium tubes and releasing the radioactive fuel into the inner vessel.

Because of the incredibly high temperatures, the zirconium reacted with the steam to generate hydrogen gas. As the pressure of the steam and hydrogen gas increased, it threatened to rupture the outer containment vessel. To prevent this, plant operators vented the mixture—which also contained radioactive elements from the melted fuel rods—into the atmosphere. As the hot hydrogen gas encountered oxygen in the atmosphere, the two combined explosively, destroying parts of the buildings housing the containment vessels (see the photo above). Despite venting, the intense heat and the pressure it generated eventually caused the outer containment structure to leak and disgorge contaminated water into the ocean for months following the disaster. Officials evacuated tens of thousands of people living within 12 miles of the plant, and many nearby villages remain uninhabitable.

Why were people evacuated from their homes when radioactive gases were released into the atmosphere? What are atoms composed of? How do the atoms of radioactive elements differ from non-radioactive elements?

2.1 WHAT ARE ATOMS?

If you write "atom" with a pencil, you are forming the letters with graphite, a form of carbon. Now imagine cutting up the carbon into finer and finer particles, until all you have left is a substance split into its basic subunits: individual carbon atoms, each with the structure unique to carbon. A carbon atom is so small that 100 million of them placed in a row would span less than half an inch (1 centimeter).

Atoms Are the Basic Structural Units of Elements

Carbon is an example of an **element**—a substance that can neither be separated into simpler substances nor converted into a different substance by ordinary **chemical reactions** (processes that form or break bonds between atoms). Elements, both alone and combined with other elements, form all matter. An **atom** is the smallest unit of an element, and each atom retains all the chemical properties of that element.

Ninety-two different elements occur in nature. Each is given an abbreviation, its *atomic symbol,* based on its name (sometimes in Latin; e.g., lead is Pb, for *plumbum*). Most elements are present in only small quantities in the biosphere, and relatively few are essential to life on Earth. **TABLE 2-1** lists the most common elements in living things.

TABLE 2-1	Common Elements in Living Organisms		
Element	**Atomic Number[1]**	**Mass Number[2]**	**% by Weight in the Human Body**
Oxygen (O)	8	16	65.0
Carbon (C)	6	12	18.5
Hydrogen (H)	1	1	9.5
Nitrogen (N)	7	14	3.0
Calcium (Ca)	20	40	1.5
Phosphorus (P)	15	31	1.0
Potassium (K)	19	39	0.35
Sulfur (S)	16	32	0.25
Sodium (Na)	11	23	0.15
Chlorine (Cl)	17	35	0.15
Magnesium (Mg)	12	24	0.05
Iron (Fe)	26	56	Trace
Fluorine (F)	9	19	Trace
Zinc (Zn)	30	65	Trace

[1]Atomic number: number of protons in the atomic nucleus.
[2]Mass number: total number of protons and neutrons.

TABLE 2-2	Mass and Charge of Subatomic Particles	
Subatomic Particle	**Mass (in atomic mass units)**	**Charge**
Neutron (n)	1	0
Proton (p^+)	1	+1
Electron (e^-)	0.00055	−1

Atoms Are Composed of Still Smaller Particles

Atoms are composed of *subatomic particles:* **neutrons** (n), which have no charge; **protons** (p^+), each of which carries a single positive charge; and **electrons** (e^-), each of which carries a single negative charge. An atom as a whole is uncharged, or *neutral,* because it contains equal numbers of protons and electrons, whose positive and negative charges electrically balance each other. Subatomic particles are assigned their own unit of mass, measured in *atomic mass units.* As you can see in **TABLE 2-2**, each proton and neutron has a mass unit of 1, while the mass of an electron is negligible compared to these larger particles. The **mass number** of an atom is the total number (which equals the total mass) of the protons and neutrons in its nucleus.

Protons and neutrons cluster together in the center of each atom, forming its **atomic nucleus.** An atom's tiny electrons are in continuous rapid motion around its nucleus within a large, three-dimensional space, as illustrated by the two simplest atoms, hydrogen and helium, in **FIG. 2-1**. These *orbital models* of

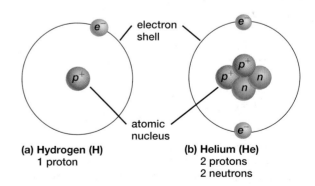

(a) Hydrogen (H)
1 proton

(b) Helium (He)
2 protons
2 neutrons

▲ **FIGURE 2-1 Atomic models** Orbital models of **(a)** hydrogen (the only atom with no neutrons) and **(b)** helium. In these simplified models, the electrons (pale blue) are represented as miniature planets, orbiting around a nucleus that contains protons (brown) and neutrons (olive green).

THINK CRITICALLY What is the mass number of hydrogen? Of helium?

atomic structure are extremely simplified to make atoms easy to imagine. Atoms are never drawn to scale; if they were, and if this dot • were the nucleus, the electrons would be somewhere in the next room (or outside)—roughly 30 feet away!

Elements Are Defined by Their Atomic Numbers

The number of protons in the nucleus—called the **atomic number**—is the feature that defines each element, making it distinct from all others. For example, every hydrogen atom has one proton, every carbon atom has six, and every oxygen atom has eight, giving these atoms atomic numbers of 1, 6, and 8, respectively. The **periodic table** in Appendix II organizes the elements according to their atomic numbers (rows) and their general chemical properties (columns).

Isotopes Are Atoms of the Same Element with Different Numbers of Neutrons

Although every atom of an element has the same number of protons, the atoms of that element may have different numbers of neutrons. Atoms of the same element with different numbers of neutrons are called **isotopes.** Isotopes can be distinguished from one another because each has a different mass number, which is written as a superscript preceding the atomic symbol.

Some Isotopes Are Radioactive

Most isotopes are stable; their nuclei do not change spontaneously. A few, however, are **radioactive,** meaning that their nuclei spontaneously break apart, or decay. Radioactive decay always emits energy and often subatomic particles as well. The decay of radioactive nuclei may form different elements. For example, nearly all carbon exists as stable ^{12}C. But a radioactive isotope called carbon-14 (^{14}C; 6 protons + 8 neutrons; 1 in every trillion carbon atoms) is produced continuously by atmospheric reactions involving cosmic rays. ^{14}C atoms disintegrate spontaneously at a slow, predictable rate. When one decays, energy is released and a neutron is converted to a proton, producing a stable nitrogen atom (^{14}N; 7 protons + 7 neutrons).

Radioactive Isotopes Are important in Scientific Research and Medicine

Scientists often make use of radioactive isotopes. For example, archeologists take advantage of the fact that after an organism dies, the ratio of ^{14}C to ^{12}C in its body declines predictably as the ^{14}C decays. By measuring this ratio in artifacts such as mummies, ancient trees, skeletons, or tools made of wood or bone, researchers can accurately assess the age of artifacts up to about 50,000 years old.

In laboratory research, scientists often expose organisms to radioactive isotopes and trace the isotopes' movements during physiological processes. For example, experiments with radioactively labeled DNA and protein allowed scientists to conclude that DNA is the genetic material of cells (described in Chapter 12).

Modern medicine also makes extensive use of radioactive isotopes. For example, radiation therapy is frequently used to treat cancer. DNA can be destroyed by radiation, so rapidly dividing cancer cells (which require intact DNA to copy themselves) are particularly vulnerable. A radioactive isotope may be introduced into the bloodstream or implanted in the body near the cancer, or radiation may be directed into the tumor by an external device. The radiation that kills cancer cells can also cause mutations in the DNA of healthy cells. This slightly increases the chance that the patient will develop cancer again in the future, but most patients consider this a risk worth taking. You'll learn more uses for radioactive isotopes in "How Do We Know That? Radioactive Revelations" on page 22.

CASE STUDY | CONTINUED
Unstable Atoms Unleashed

Because exposure to radioactivity can cause cancer, Japanese authorities have performed regular cancer screenings on hundreds of thousands of children exposed to radioactivity by the Fukushima power plant disaster. Fortunately, recent surveys have found no evidence of increased cancer rates.

But months after the meltdown, engineers at the Fukushima power plant—using specialized cameras located outside the plant—discovered hot spots of radiation so intense that a person exposed for an hour would be dead within a few weeks. How could death come so fast? Extremely high doses of radiation damage DNA and other biological molecules so badly that cells—particularly those that divide rapidly—can no longer function. Skin cells are destroyed. Cells lining the stomach and intestine break down, causing nausea and vomiting. Bone marrow, where blood cells and platelets are produced, is destroyed. Lack of white blood cells allows infections to flourish, and the loss of platelets crucial for blood clotting leads to internal bleeding.

Fortunately, radioactive substances such as those released by the Fukushima disaster are rare in nature. Why do most elements remain stable?

Electrons Are Responsible for the Interactions Among Atoms

Nuclei and electrons play complementary roles in atoms. Nuclei (unless they are radioactive) provide stability; they remain unchanged during ordinary chemical reactions. Electrons, in contrast, are dynamic; they can capture and release energy, and as we describe later, they form the bonds that link atoms together into molecules.

Electrons Occupy Shells of Increasing Energy

Electrons occupy **electron shells,** complex three-dimensional regions around the nucleus. For simplicity, we will depict these shells as increasingly large, concentric

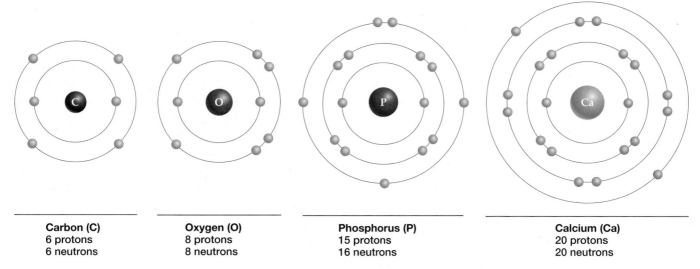

Carbon (C)	Oxygen (O)	Phosphorus (P)	Calcium (Ca)
6 protons	8 protons	15 protons	20 protons
6 neutrons	8 neutrons	16 neutrons	20 neutrons

▲ **FIGURE 2-2 Electron shells in atoms** Most biologically important atoms have two or more shells of electrons. The shell closest to the nucleus can hold two electrons; the next three shells can each contain eight electrons.

THINK CRITICALLY Why do atoms with unfilled outer electron shells tend to react with one another?

rings around the nucleus where electrons travel like planets orbiting the sun (**FIG. 2-2**). Each shell has a specific energy associated with it. The farther away from the nucleus, the greater the amount of energy stored in the electrons occupying the shell.

Electrons Can Capture and Release Energy

When an atom is excited by energy, such as heat or light, this energy can cause an electron to jump from a lower-energy electron shell to a higher-energy shell. Soon afterward, the electron spontaneously falls back into its original electron shell and releases its extra energy, some in the form of heat, and often in the form of light as well (**FIG. 2-3**).

We make use of the ability of electrons to capture and release energy every time we switch on a light bulb. Although incandescent bulbs are rapidly becoming obsolete, they are the easiest type to understand. Electricity flows through a thin wire, heating it to around 4,500°F (about 2,500°C) for a 100-watt bulb. The heat energy bumps some electrons in the wire into higher-energy electron shells. As the electrons drop back down into their original shells, they emit some of the energy as light. Unfortunately, more than 90% of the energy absorbed by the wire is re-emitted as heat rather than light, making an incandescent bulb an extremely inefficient light source.

As Atomic Number Increases, Electrons Fill Shells Increasingly Distant from the Nucleus

Each electron shell can hold a specific number of electrons; the shell nearest the nucleus can hold only two, and more

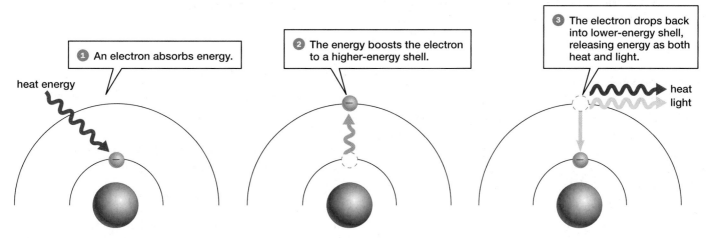

① An electron absorbs energy.

② The energy boosts the electron to a higher-energy shell.

③ The electron drops back into lower-energy shell, releasing energy as both heat and light.

heat energy

heat
light

▲ **FIGURE 2-3 Energy capture and release in an incandescent bulb.**

THINK CRITICALLY What causes the coals of a campfire to glow?

HOW DO WE KNOW THAT? | Radioactive Revelations

How do doctors know the location and size of a cancerous brain tumor? Or how brain activity diminishes with Alzheimer's disease? Or what brain regions are active when a person performs a math problem? These and many more questions can be investigated using positron emission tomography (PET) scans. To perform PET scans, sugar molecules tagged with a radioactive isotope are injected into a patient's bloodstream. More metabolically active regions of the body use more sugar for energy, accumulating larger amounts of radioactivity. To identify these regions, the person's body is moved through a ring of detectors that respond to the energetic particles (positrons) emitted as the isotope decays. A powerful computer then uses these data

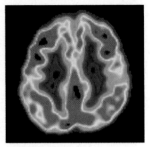

(a) Alzheimer's patient   **(b) Healthy brain**

▲ **FIGURE E2-2 PET reveals differences in brain function** Brain activity is rainbow color-coded, with red indicating the highest activity and blue the lowest; black areas are fluid-filled.

to calculate precisely where the decays occurred and generates a color-coded map of the frequency of decays within each "slice" of body passing through the detector ring (**FIG. E2-1a**). PET can be used to study the working brain, because regions activated by a specific mental task—such as a math problem—will have increased energy needs and will "light up" as they accumulate more radioactive glucose. Cancerous tumors show up in PET scans as "hot spots" because their rapid cell division uses large amounts of glucose (**FIG. E2-1b**). PET also reveals that the brain of an Alzheimer's patient is far less active than that of a healthy individual (**FIG. E2-2**).

THINK CRITICALLY In addition to lower brain activity, what other problem has occurred in the Alzheimer's victim's brain as shown by the images in Fig. E2-2?

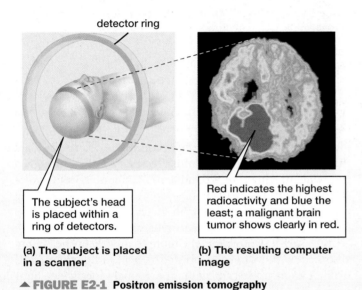

detector ring

The subject's head is placed within a ring of detectors.

Red indicates the highest radioactivity and blue the least; a malignant brain tumor shows clearly in red.

(a) The subject is placed in a scanner

(b) The resulting computer image

▲ **FIGURE E2-1 Positron emission tomography**

---

distant shells can hold eight or more. Electrons always fill the lowest-energy shell (the shell nearest the nucleus) first, and then fill higher-energy shells. Elements with increasingly large numbers of protons in their nuclei require more electrons to balance these protons, so their electrons will occupy shells at increasing distances from the nucleus. For example, the two electrons in helium (He) occupy the first electron shell (see Fig. 2-1b). A carbon atom (C) with six electrons will have two electrons filling its first shell and four occupying its second shell, which can contain a total of eight electrons (see Fig. 2-2).

CHECK YOUR LEARNING
Can you ...
- define *element* and *atom*?
- name and describe the *subatomic particles* that make up the atom?
- explain atomic number and mass number?
- explain radioactivity and its dangers and benefits?
- describe electron shells?

2.2 HOW DO ATOMS INTERACT TO FORM MOLECULES?

Most forms of matter that we encounter in our daily lives consist of atoms of the same or different elements linked together to form **molecules.** Simple examples are oxygen gas (O_2; two oxygen atoms) and water (H_2O; two hydrogen atoms and one oxygen atom). How and why do molecules form?

Atoms Form Molecules by Filling Vacancies in Their Outer Electron Shells

In most elements, the electrons needed to balance the protons fill one or more inner shells, but they do not completely fill the outer shell. Atoms generally behave according to two basic principles:

- An atom whose outermost electron shell is completely full will not react with other atoms. Such an atom (e.g., helium in Fig. 2-1b) is extremely stable and is described as *inert.*

TABLE 2-3	Common Types of Bonds in Biological Molecules	
Type	**Type of Interaction**	**Example**
Ionic bond	An electron is transferred between atoms, creating positive and negative ions that attract one another.	Occurs between the sodium (Na^+) and chloride (Cl^-) ions of table salt (NaCl)
Covalent bond	Electrons are shared between atoms.	
Nonpolar	Electrons are shared equally between atoms.	Occurs between the two hydrogen atoms in hydrogen gas (H_2)
Polar	Electrons are shared unequally between atoms.	Occurs between hydrogen and oxygen atoms of a water molecule (H_2O)
Hydrogen bond	Attractions occur between polar molecules in which hydrogen is bonded to oxygen or nitrogen. The slightly positive hydrogen attracts the slightly negative oxygen or nitrogen of a nearby polar molecule.	Occurs between water molecules, where slightly positive charges on hydrogen atoms attract slightly negative charges on oxygen atoms of nearby molecules

- An atom whose outermost electron shell is only partially full will react readily with other atoms. Such an atom (e.g., hydrogen in Fig. 2-1a) is described as *reactive*.

Among atoms and molecules with unfilled outer shells, some—called **free radicals**—are so reactive that they can tear other molecules apart. Free radicals are produced in large numbers in the body by reactions that make energy available to cells. Although these reactions are essential for life, over time, the stress that free radicals place on living cells may contribute to aging and eventual death. Learn more in "Health Watch: Free Radicals—Friends and Foes?" on page 25.

Chemical Bonds Hold Atoms Together in Molecules

Reactive atoms form **chemical bonds,** which are attractive forces that hold atoms together in molecules. Bonds are formed when atoms acquire, lose, or share electrons to gain stability. There are three major types of bonds: *ionic bonds, covalent bonds,* and *hydrogen bonds* (**TABLE 2-3**).

Ionic Bonds Form Among Ions

Atoms, including those that are reactive, have equal numbers of protons and electrons. The equal number of protons and electrons gives atoms an overall neutral charge, but that does not make them stable. An atom with an almost empty outermost electron shell can become more stable by losing electrons and completely emptying the outer shell; this gives it a positive charge. An atom with a nearly full outer shell can become more stable by gaining electrons and filling the shell completely, giving it a negative charge.

When an atom becomes stable by losing or gaining one (or a few) electrons and thus acquiring an overall positive or negative charge, it becomes an **ion** (**FIG. 2-4**). Ions with

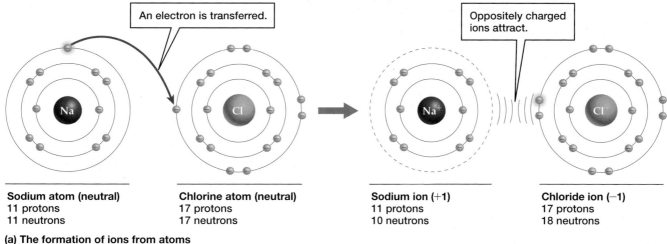

An electron is transferred.

Oppositely charged ions attract.

Sodium atom (neutral)
11 protons
11 neutrons

Chlorine atom (neutral)
17 protons
17 neutrons

Sodium ion (+1)
11 protons
10 neutrons

Chloride ion (−1)
17 protons
18 neutrons

(a) The formation of ions from atoms

(b) An ionic molecule: NaCl

◀ **FIGURE 2-4 Ions and ionic bonds (a)** Stable ions form when sodium loses an electron (Na^+) and chlorine gains an electron (Cl^-). **(b)** Sodium and chloride ions nestle closely together in cubic crystals of table salt (NaCl).

opposite charges attract one another, and the electrical attraction between positively and negatively charged ions forms **ionic bonds.** For example, the white crystals in your salt shaker are sodium and chloride ions linked by ionic bonds. Sodium (Na) has only one electron in its outermost electron shell, so it can become stable by losing this electron, forming the ion Na$^+$. Chlorine (Cl) has seven electrons in its outer shell, which can hold eight electrons. So chlorine can become stable by gaining an electron (in this case from sodium), forming the ion Cl$^-$ (**FIG. 2-4a**) and producing an ionic bond between Na$^+$ and Cl$^-$. These ionic bonds result in crystals composed of a repeating, orderly array of Na$^+$ and Cl$^-$ (**FIG. 2-4b**). As we describe later, water is attracted to ions and can break ionic bonds, as occurs when water dissolves salt. Because biological molecules function in a watery environment, most are held tightly together by covalent bonds.

Covalent Bonds Form When Atoms Share Electrons

Atoms with partially full outermost electron shells can become stable by sharing electrons with one another, filling both of their outer shells and forming **covalent bonds.** The atoms in most biological molecules, such as proteins, sugars, and fats, are joined by covalent bonds (**TABLE 2-4**).

TABLE 2-4	Electrons and Bonds in Atoms Common in Biological Molecules		
Atom	Capacity of Outer Electron Shell	Electrons in Outer Shell	Number of Covalent Bonds Usually Formed
Hydrogen (H)	2	1	1
Carbon (C)	8	4	4
Nitrogen (N)	8	5	3
Oxygen (O)	8	6	2
Sulfur (S)	8	6	2

Covalent Bonds May Produce Nonpolar or Polar Molecules

In all covalent bonds between atoms of the same element, and in covalent bonds between some pairs of atoms of different elements, the participating atoms share electrons equally or nearly equally. This creates **nonpolar covalent bonds** in which there is no charge on any part of the molecule. For example, two hydrogen atoms can become more stable if they share their outer electrons, allowing each to behave almost as if it had two electrons in its outer shell (**FIG. 2-5**). This reaction forms hydrogen gas (H$_2$).

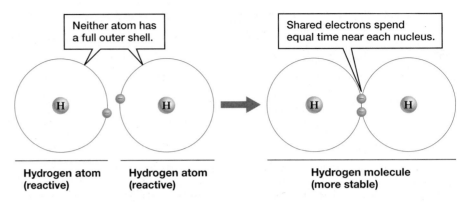

Neither atom has a full outer shell.

Shared electrons spend equal time near each nucleus.

Hydrogen atom (reactive)   Hydrogen atom (reactive)   Hydrogen molecule (more stable)

▲ **FIGURE 2-5 Nonpolar covalent bonds** A hydrogen molecule (H$_2$) is formed when an electron from each of two hydrogen atoms is shared equally, forming a single nonpolar covalent bond.

Because the two H nuclei are identical, their electrons spend equal time near each, and so neither end, or *pole,* of the molecule is charged. Other examples of nonpolar molecules include oxygen gas (O$_2$), nitrogen gas (N$_2$), carbon dioxide (CO$_2$), and certain biological molecules such as oils and fats (described in Chapter 3). In each of these molecules, the nuclei exert a roughly equal pull on the shared electrons.

Some covalently bonded atoms share electrons unequally, because the nucleus of one molecule attracts the electrons more strongly than the nucleus of the other. Unequally shared electrons produce **polar covalent bonds** in molecules, which are then described as *polar molecules.* Although the molecule as a whole is electrically neutral, a polar molecule has charged poles. In water (H$_2$O), for example, each hydrogen atom shares an electron with the single oxygen atom (**FIG. 2-6**). The oxygen nucleus exerts a stronger attraction on the electrons than does either hydrogen nucleus. By attracting electrons, the oxygen pole of a water molecule becomes slightly negative, leaving each hydrogen atom slightly positive.

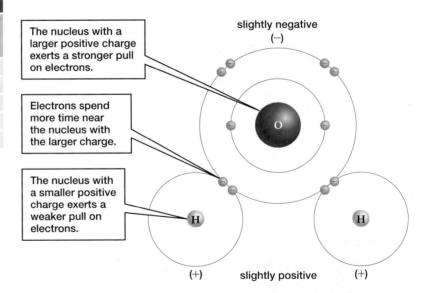

The nucleus with a larger positive charge exerts a stronger pull on electrons.

Electrons spend more time near the nucleus with the larger charge.

The nucleus with a smaller positive charge exerts a weaker pull on electrons.

slightly negative (−)

(+)   slightly positive   (+)

▲ **FIGURE 2-6 Polar covalent bonds** Oxygen (O) needs two electrons to fill its outer shell, allowing it to form covalent bonds with two hydrogen atoms (H), which produces water (H$_2$O). The oxygen atom exerts a greater pull on the electrons than do the hydrogen atoms, producing a slight negative charge near the oxygen and a slight positive charge near the two hydrogens.

Health WATCH

Free Radicals—Friends and Foes?

Any atom with a partially full outer electron shell will be reactive, but this reactivity increases dramatically if the unfilled shell also contains an uneven number of electrons. Like people, electrons like to pair up. Free radicals are molecules containing atoms with unpaired electrons in their outer shells. They react vigorously with other molecules, capturing or releasing electrons to achieve a more stable arrangement. Such reactions may result in damage to crucial biological molecules, including proteins and DNA.

Our bodies continuously produce oxygen-containing free radicals, such as hydrogen peroxide (H_2O_2), as a by-product of reactions that generate cellular energy. Free radicals are also formed when our cells are bombarded by sunlight, X-rays, radioactive isotopes, and various toxic chemicals in the environment. Our bodies counteract free radicals by generating **antioxidants,** molecules that react with free radicals and render them harmless. We also obtain antioxidants in our diets, since they occur naturally in many plant-derived foods. But when free radicals are produced that overwhelm the body's ability to counteract them, the resulting *oxidative stress* can injure cells. For example, free radicals generated by ultraviolet rays can damage DNA and promote skin cancer. Oxidative stress contributes to cardiovascular disease, lung disorders such as asthma, and neurological diseases, including Alzheimer's. The most visible signs of aging—graying hair from free radical damage to hair follicles and wrinkles that occur when ultraviolet rays from sunlight damage proteins in skin—are a result of oxidative stress (**FIG. E2-3**).

Strong evidence suggests that diets high in antioxidant-containing fruits and vegetables are associated with a lower incidence of cardiovascular disease and some cancers. Many people have concluded that a shortcut to health can be provided by antioxidant supplements such as vitamins C and E and beta-carotene (found in many fruits and vegetables). But an analysis that combined numerous studies using large groups of people showed no health benefits from taking these supplements and suggested that beta-carotene and vitamin E supplements may in some cases have adverse health effects. Why might this be?

During the course of evolution, organisms have continuously been exposed to free radicals and have evolved ways to both use them constructively and deactivate them. Free radicals are involved in regulating blood pressure, wound healing, and defense against disease-causing microbes. Growing evidence supports the hypothesis that health requires a complex balance of free radicals

and antioxidants and that the high doses of purified antioxidants in supplements can upset this delicate equilibrium. Fruits and vegetables contain natural antioxidants at levels generally well below those found in supplements, so it is most prudent to obtain antioxidants from a diet rich in fruits and vegetables (unless a medical condition requires using supplements).

What types of fruits and vegetables should you eat? While it's a stretch to put chocolate in the fruit or vegetable category, cocoa (a powder made from the cacao bean; **FIG. E2-4**) is especially rich in natural plant molecules called *flavonols*. Flavonols—also found in green tea, cranberries, apples, onions, kale, and other plant foods—possess antioxidant and other beneficial properties. Although it sounds almost too good to be true, controlled studies have reported beneficial effects of consuming dark chocolate (with a high percentage of cocoa) on risk factors for cardiovascular disease including high blood pressure. Several large studies on human populations have also found a correlation between higher consumption of chocolate and other flavonol-rich foods and a reduced incidence of cardiovascular disease, including high blood pressure, strokes, and heart attacks (correlation studies provide suggestive evidence, but are far less rigorous than controlled studies).

EVALUATE THIS At a physical exam, Thomas, a sedentary individual, was warned of dangerously high blood pressure. After reading that chocolate is good for cardiovascular health, he immediately stocked up on 6-oz dark chocolate bars and resolved to add one daily to his regular diet. Predict some results of his next annual physical exam and provide him with some good health advice.

◀ **FIGURE E2-3 Free-radical damage** Free radicals interfere with the production of hair pigment and damage proteins that give skin its elasticity. The results? Gray or white hair and wrinkled skin.

▲ **FIGURE E2-4 Chocolate** This substance comes from cacao beans found inside pods (inset) that grow from the trunks of trees native to South America.

Hydrogen Bonds Are Attractive Forces Between Certain Polar Molecules

Biological molecules, including sugars, proteins, and nucleic acids, often have many polar covalent bonds between either hydrogen and oxygen or hydrogen and nitrogen. In these cases, the hydrogen is slightly positive and the oxygen or nitrogen is slightly negative. As you know, opposite charges attract. A **hydrogen bond** is the attraction between the slightly positive and slightly negative regions of polar molecules. Hydrogen bonding occurs in water molecules between their slightly positive hydrogen poles and slightly negative oxygen poles, linking water molecules into a loosely connected, ever-changing network (**FIG. 2-7**). As you will see shortly, hydrogen bonds among water molecules give it several unusual properties that make water crucial for life as we know it.

CHECK YOUR LEARNING

Can you ...

- explain what makes an atom reactive?
- define *molecules* and *chemical bonds*?
- describe and provide examples of ionic, covalent, and hydrogen bonds?

2.3 WHY IS WATER SO IMPORTANT TO LIFE?

As naturalist Loren Eiseley eloquently stated, "If there is magic on this planet, it is contained in water." Water has many special properties that all result from the polarity of its molecules and the hydrogen bonds that form among them. What makes water unique?

Water Molecules Attract One Another

Hydrogen bonds interconnect water molecules. But, like square dancers continually moving from one partner to

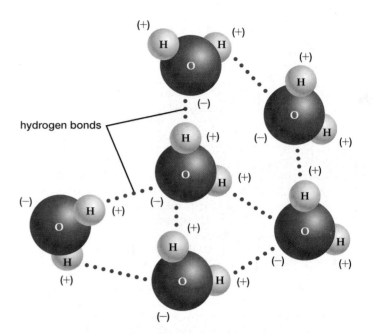

▲ **FIGURE 2-7 Hydrogen bonds in water** The slight charges on opposite poles of water molecules (shown in parentheses) produce hydrogen bonds (dotted lines) between the oxygen and hydrogen atoms in adjacent water molecules. Each water molecule can form up to four hydrogen bonds. In liquid water, these bonds constantly break as new ones form.

the next, joining and releasing hands as they go, hydrogen bonds in liquid water constantly break and re-form, allowing water to flow. Hydrogen bonds among water molecules cause **cohesion,** the tendency for molecules of a single type to stick together; this causes water to form droplets (**FIG. 2-8a**). Cohesion also produces **surface tension,** the tendency for a water surface to resist being broken. Surface water molecules have nothing above them to bond with, so they are attracted more strongly to one another and to the water molecules

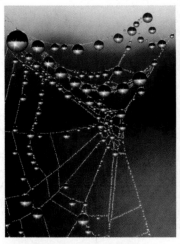

(a) Cohesion and adhesion   **(b) Cohesion causes surface tension**   **(c) Capillary action**

▲ **FIGURE 2-8 Water molecules have cohesive and adhesive properties (a)** Cohesion causes water to form droplets; adhesion sticks them to spiderweb silk. **(b)** This basilisk lizard uses surface tension as it races across the water's surface to escape a predator. **(c)** Cohesion and adhesion work together in capillary action, which draws water into narrow spaces among charged surfaces.

beneath them. For example, water brimming slightly above the top of a glass just before it overflows is caused by surface tension. Surface tension in a pond can support water insects and even a running basilisk lizard (**FIG. 2-8b**).

Water also exhibits **adhesion,** the tendency for different surfaces to cling to one another. Water adheres to substances whose molecules contain charged regions; these substances include glass, the cellulose in wood and paper, and the silk of spiderwebs (see Fig. 2-8a).

When water is attracted onto a surface by adhesion and then draws more water molecules along by cohesion, this produces **capillary action.** During capillary action, water moves spontaneously into very narrow spaces, such as those between the cellulose fibers of a paper towel (**FIG. 2-8c**).

The strong cohesion among polar water molecules plays a crucial role in the life of land plants. How does nutrient-laden water absorbed by a plant's roots ever reach its leaves, especially if the plant is a 300-foot-tall redwood tree? Water fills tubes that connect the roots, stem, and leaves. As water molecules continuously evaporate from leaves, each pulls the water molecule below it to the leaf's surface, much like a chain being dragged up from the top. The hydrogen bonds that link water molecules are stronger than the downward pull of gravity, so the water chain doesn't break. In addition, water adheres to the walls of the conducting tubes, which are composed of cellulose and are microscopically narrow. These properties allow capillary action to contribute to the transport of water from roots to leaves. Without the cohesion of water, there could be no large land plants, and life on Earth would be radically different.

Water Interacts with Many Other Molecules

A **solvent** is a substance that completely surrounds and disperses the individual atoms or molecules of another substance. When this occurs, the solvent is said to **dissolve** the substance it disperses. A solvent that contains one or more dissolved substances is called a **solution.** The positive and negative poles of water are attracted to charges on other polar molecules and ions, making water an excellent solvent. Polar molecules and ions are described as **hydrophilic** (Gk. *hydro,* water, and *phylos,* loving) because they are attracted to (and dissolve in) water.

A crystal of table salt, for example, is held together by ionic bonds between positively charged sodium ions (Na^+) and negatively charged chloride ions (Cl^-; see Fig. 2-4b). When a salt crystal is dropped into water, the positively charged hydrogen poles of water molecules are attracted to the Cl^-, and the negatively charged oxygen poles are attracted to the Na^+. As water molecules surround the ions, shielding them from interacting with each other, the ions separate from the crystal and drift away in the water—thus, the salt dissolves (**FIG. 2-9**).

Gases such as oxygen and carbon dioxide also dissolve in water even though they are nonpolar. How? These molecules are so small that they fit into the spaces between water

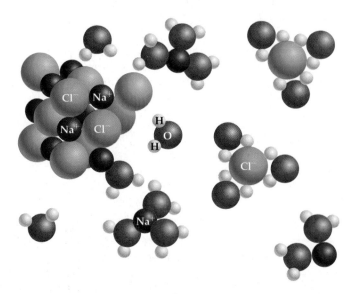

▲ **FIGURE 2-9 Water as a solvent** When a salt crystal (NaCl) is dropped into water, the water surrounds the sodium and chloride ions, with positive poles of water molecules facing the Cl^-, and the negative poles facing the Na^+. The ions disperse as the surrounding water molecules isolate them from one another, dissolving the salt crystal.

THINK CRITICALLY If you placed a salt crystal in a nonpolar liquid (like oil), would it dissolve?

molecules without disrupting the hydrogen bonds. The ability of water to dissolve oxygen allows fish to flourish, even when swimming under a layer of ice.

Larger molecules with nonpolar covalent bonds, such as fats and oils, are **hydrophobic** (Gk. *phobic,* fearing) and do not dissolve in water. Nevertheless, water has an important effect on such molecules. By sticking together, water molecules exclude oil molecules. The nonpolar oil molecules are forced together into drops, surrounded by water molecules that form hydrogen bonds with one another but not with the oil (**FIG. 2-10**).

◀ **FIGURE 2-10 Oil and water don't mix** Yellow oil poured into water remains in discrete droplets as it rises to the surface. Oil floats because it is less dense than water.

THINK CRITICALLY Predict how a drop of water on an oil-coated surface would differ in shape from a drop on a clean glass surface.

HAVE YOU EVER WONDERED...

Why It Hurts So Much to Do a Belly Flop?

The slap of a belly flop provides firsthand experience of the power of cohesion among water molecules. Because of the hydrogen bonds that interconnect its molecules, the water surface resists being broken. When you suddenly force a large number of water molecules apart with your belly, the result can be a bit painful. Do you think that belly-flopping into a pool of (nonpolar) vegetable oil would hurt as much? If you encountered a deep pool filled with vegetable oil, could you float or swim in it?

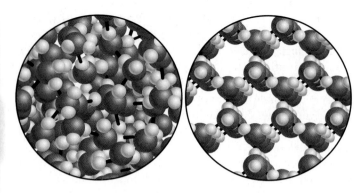

▲ **FIGURE 2-11** Liquid water (left) and ice (right)

THINK CRITICALLY How do these configurations explain why ice floats?

Water Moderates the Effects of Temperature Changes

The hydrogen bonds linking water molecules (see Fig. 2-7) allow water to moderate temperature changes.

It Takes a Lot of Energy to Heat Water

The energy required to heat 1 gram of a substance by 1°C is called its **specific heat.** The specific heat of water is far higher than that of any other common substance. Why? At any temperature above absolute zero (−459°F or −273°C), all molecules are in constant motion. The warmer the substance, the faster its molecules move. But forcing water molecules to speed up requires breaking their hydrogen bonds more frequently. Breaking the bonds consumes a considerable amount of energy, and so less energy is available to raise the water's temperature. In contrast, when substances lack hydrogen bonds, a greater proportion of added energy is available to raise their temperatures. For example, a given amount of energy would increase the temperature of granite rock (which lacks hydrogen bonds) about five times as much as it would the same weight of water.

Because of its high specific heat, water moderates temperature changes. One reason you can sit on hot sand in the hot sun without instantly overheating is that your body is about 60% water, which must absorb considerable heat to change its temperature.

It Takes a Lot of Energy to Evaporate Water

Overheating still poses a real threat, however, because the molecules in our bodies function only within a narrow range of temperatures. We use another property of water when we perspire to maintain our body temperature in hot conditions. Water has an extremely high **heat of vaporization,** which is the amount of heat needed to cause a substance to evaporate (change from a liquid to a vapor). Because of the polar nature of water molecules, water must absorb enough energy to break the hydrogen bonds that interconnect its molecules before the molecules can move fast enough to escape and evaporate into the air. Water in sweat absorbs a great deal of body heat, and it cools us as its fastest-moving molecules vaporize.

Water Forms an Unusual Solid: Ice

Even solid water is unusual. Most liquids become denser when they solidify, but ice is actually less dense than liquid water. When water freezes, each molecule forms stable hydrogen bonds with four other water molecules, creating an open, hexagonal (six-sided) arrangement (**FIG. 2-11**). This keeps the water molecules farther apart than their average distance in liquid water. Thus, ice is less dense than liquid water, which is why icebergs and ice cubes float.

This property of water is crucial to the distribution of aquatic life. When a pond or lake starts to freeze in winter, the floating ice forms an insulating layer that delays the freezing of the rest of the water. This insulation allows fish and other aquatic organisms to survive in the liquid water below (**FIG. 2-12**). If ice were to sink, many ponds and lakes around the world would freeze solid from the bottom up during the winter, killing most of their inhabitants. The ocean floor at higher latitudes would be covered with extremely thick layers of ice that would never melt.

Water-Based Solutions Can Be Acidic, Basic, or Neutral

At any given time, a tiny fraction of water molecules (H_2O) will have split into hydroxide ions (OH^-) and hydrogen ions (H^+) (**FIG. 2-13**). Pure water contains equal concentrations of each.

CASE STUDY **CONTINUED**

Unstable Atoms Unleashed

The high specific heat of water makes it an ideal coolant for nuclear power plants. Compared to nonpolar liquids like alcohol, a great deal of heat is required to raise the temperature of water. The tsunami that hit the Fukushima power plant disrupted the electrical supply running the pumps that kept water circulating over the fuel rods. Without enough water to absorb the excess heat, the metal tubes surrounding the fuel rods melted.

◀FIGURE 2-12 Ice floats
The floating ice insulates the water beneath it, helping to protect bodies of water from freezing solid and allowing fish and other organisms to survive beneath it.

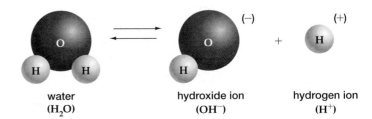

▲ **FIGURE 2-13 Some water is always ionized**

When ion-forming substances that release OH^- or H^+ are added to water, the solution no longer has equal concentrations of OH^- and H^+. If the concentration of H^+ exceeds the concentration of OH^-, the solution is **acidic.** An **acid** is a substance that releases hydrogen ions when it dissolves in water. For example, when hydrochloric acid (HCl) is added to pure water, almost all of the HCl molecules separate into H^+ and Cl^-. Therefore, the concentration of H^+ exceeds the concentration of OH^-, and the resulting solution is acidic. Acidic substances—think lemon juice (containing citric acid) or vinegar (acetic acid)—taste sour because sour receptors on your tongue respond to excess H^+.

If the concentration of OH^- is greater than the concentration of H^+, the solution is **basic.** A **base** is a substance that combines with hydrogen ions, reducing their number. If, for instance, sodium hydroxide (NaOH) is added to water, the NaOH molecules separate into Na^+ and OH^-. Some OH^- ions combine with H^+ to produce H_2O, reducing the number of H^+ ions and creating a basic solution. Bases are used in many cleaning solutions. Bases are also in many antacids like Tums to neutralize heartburn caused by excess hydrochloric acid in the stomach.

The **pH scale** of 0 to 14 measures how acidic or basic a solution is (**FIG. 2-14**). Neutral pH (equal concentrations of H^+ and OH^-) is 7. Pure water has a pH of 7, acids have a pH below 7, and bases have a pH above 7. Each unit on the pH scale represents a tenfold change in the concentration of H^+. Thus, the concentration of H^+ is 10,000 times greater in a soft drink with a pH of 3 than in water (pH 7).

A **buffer** is a molecule that tends to maintain a solution at a nearly constant pH by accepting or releasing H^+ in response to small changes in H^+ concentration. In the presence of excess H^+, a buffer combines with the H^+, reducing its concentration. In the presence of excess OH^-, buffers release H^+, which

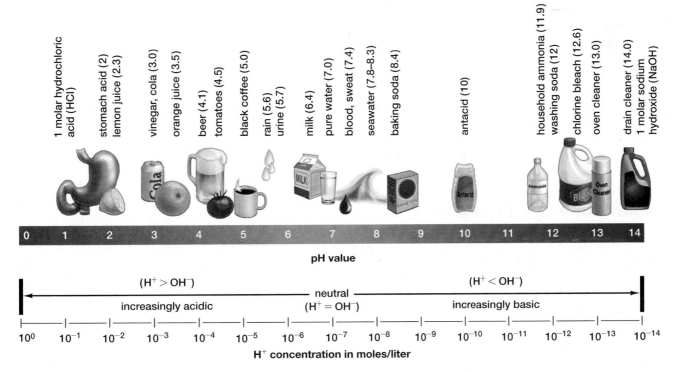

▲ **FIGURE 2-14 The pH scale** The pH scale reflects the concentration of hydrogen ions in a solution. Notice that pH (upper scale; 0–14) is the negative log of the H^+ concentration (lower scale). Each unit on the scale represents a tenfold change. Lemon juice, for example, is about 10 times more acidic than orange juice.

combines with the OH⁻ to form H_2O. Humans and other mammals maintain a pH in body fluids that is just slightly basic (about 7.4). If your blood became as acidic as 7.0 or as basic as 7.8, you would likely die because even small changes in pH cause drastic changes in both the structure and function of biological molecules. Nevertheless, living cells seethe with chemical reactions that take up or release H^+. The pH of body fluids remains remarkably constant because it is controlled by several different buffers.

CHECK YOUR LEARNING

Can you ...

- describe the unique properties of water and the importance of these properties to life?
- explain how polar covalent and hydrogen bonds contribute to the unique properties of water?
- explain the concept of pH and how acids, bases, and buffers affect solutions?

CASE STUDY \ REVISITED
Unstable Atoms Unleashed

Scientists believe that the isotopes of uranium were forged in the explosion of a star and became incorporated into Earth as our solar system formed. Today, the radioactive form of this rare element is mined and concentrated to help satisfy humanity's unquenchable desire for energy.

The chain reaction that generates heat in nuclear power plants begins when neutrons are released from radioactive uranium. These bombard other uranium atoms and cause them to split, in a self-sustaining chain reaction. When the tsunami struck the Fukushima plant, neutron-absorbing rods were immediately lowered around the fuel, halting the chain reaction. But the breakdown of uranium generates additional radioactive isotopes, and these continued to spontaneously decay and generate heat. This caused the disastrous breach that released these isotopes into the environment. One isotope of particular concern is radioactive iodine.

Iodine enters the body in food and water. It becomes concentrated in the thyroid gland, which uses iodine to synthesize thyroid hormone. Unfortunately, the thyroid gland does not distinguish between radioactive and non-radioactive iodine. Children exposed

to radioactive iodine are at increased risk for thyroid cancer, which may occur decades after exposure. To help protect them, Japanese authorities distributed iodine tablets to children near the failed reactor. This non-radioactive iodine saturates the thyroid, which then does not take up the radioactive form. Only time will reveal the full health effects of the fallout from Fukushima.

CONSIDER THIS The Fukushima disaster led to a reassessment of safety precautions in nuclear power plants and a worldwide dialogue about the dangers of nuclear power, which also generates waste that remains radioactive for thousands of years. How can societies evaluate and compare the safety of nuclear power versus the safety of burning fossil fuels, from which humanity currently gets most of its energy? How can one compare the possibility of events that might cause a nuclear disaster—such as a magnitude 9 earthquake or the accidental escape of radioactive waste—with the certainty of continued carbon dioxide emissions and global climate change resulting from fossil fuel use? To what extent should societies invest in renewable energy, including wind and solar power?

CHAPTER REVIEW

Go to **MasteringBiology** for practice quizzes, activities, eText, videos, current events, and more.

Answers to Think Critically, Evaluate This, Multiple Choice, and Fill-in-the-Blank questions can be found in the Answers section at the back of the book.

Summary of Key Concepts

2.1 What Are Atoms?

An element is a substance that can neither be broken down nor converted to different substances by ordinary chemical reactions. The smallest particle of an element is the atom, which is composed of positively charged protons, uncharged neutrons, and negatively charged electrons. All atoms of a given element have the same unique number of protons. Neutrons and protons cluster to form atomic nuclei. Electrons orbit the nucleus within regions called electron shells. Shells at increasing distances from the nucleus contain electrons with increasing amounts of energy. Each shell can contain a fixed maximum number of electrons. An atom is most stable when its outermost shell is full. Isotopes are atoms of the same element with different numbers of neutrons. Nuclei of

radioactive isotopes spontaneously break down, forming new elements and releasing energy and often subatomic particles.

2.2 How Do Atoms Interact to Form Molecules?

Atoms gain stability by filling or emptying their outer electron shells. They do this by acquiring, losing, or sharing electrons during chemical reactions. This produces attractive forces called chemical bonds, which link atoms to form molecules.

There are three types of bonds: ionic, covalent, and hydrogen. Ions are atoms that have lost or gained electrons. Ionic bonds link negatively and positively charged ions in crystals. Covalent bonds form when atoms fill their outer electron shells by sharing electrons. In a nonpolar covalent bond, the two atoms share electrons equally. In a polar covalent bond, one atom attracts electrons more strongly than the other atom does, giving the molecule slightly positive and negative poles. Polar covalent bonds allow hydrogen bonding, the attraction between the slightly positive hydrogen of one molecule and the slightly negative regions of other polar molecules.

2.3 Why Is Water So Important to Life?

Water's unique properties allowed life as we know it to evolve. Water is polar and dissolves polar substances and ions. Water forces nonpolar substances, such as oil, to form clumps. Water molecules cohere to each other using hydrogen bonds, producing surface tension. Water also adheres to other polar surfaces. Water's extremely high specific heat and high heat of vaporization function in some animals to maintain relatively stable body temperatures despite large outside temperature fluctuations. Water is unusual in being less dense in its frozen than in its liquid state.

Pure water contains equal numbers of H^+ and OH^- (pH 7), but dissolved substances can make solutions acidic (more H^+ than OH^-) or basic (more OH^- than H^+). Buffers help maintain a constant pH.

Key Terms

acid *29*	heat of vaporization *28*
acidic *29*	hydrogen bond *26*
adhesion *27*	hydrophilic *27*
antioxidant *25*	hydrophobic *27*
atom *19*	ion *23*
atomic nucleus *19*	ionic bond *24*
atomic number *20*	isotope *20*
base *29*	mass number *19*
basic *29*	molecule *22*
buffer *29*	neutron *19*
capillary action *27*	nonpolar covalent bond *24*
chemical bond *23*	periodic table *20*
chemical reaction *19*	pH scale *29*
cohesion *26*	polar covalent bond *24*
covalent bond *24*	proton *19*
dissolve *27*	radioactive *20*
electron *19*	solution *27*
electron shell *20*	solvent *27*
element *19*	specific heat *28*
free radical *23*	surface tension *26*

Thinking Through the Concepts

Multiple Choice

1. Which of the following is False?
   a. An element is defined by its atomic number.
   b. Ninety-two elements occur naturally.
   c. An atom consists of subatomic particles.
   d. Electron shells increase in energy closer to the nucleus.

2. The mass number of an element is equal to
   a. the mass of its atom's protons and neutrons.
   b. the mass of its atom's neutrons.
   c. the mass of its atom's protons.
   d. its atomic number.

3. Isotopes are defined as atoms of
   a. the same element with different numbers of protons.
   b. radioactive elements.
   c. stable elements.
   d. the same element with different numbers of neutrons.

4. Molecules
   a. always consist of different elements bonded together.
   b. may be polar or nonpolar.
   c. may be held together entirely by hydrogen bonds.
   d. cannot be held together by ionic bonds.

5. Covalent bonds
   a. link water molecules in ice.
   b. create only nonpolar molecules.
   c. bind sodium and chlorine in table salt.
   d. link hydrogen to oxygen in water.

Fill-in-the-Blank

1. An atom consists of an atomic nucleus composed of positively charged _____ and uncharged _____. Orbiting around the nucleus are _____ that occupy discrete spaces called _____.

2. An atom that has lost or gained one or more electrons is called a(n) _____. If an atom loses an electron it takes on a(n) _____ charge. Atoms with opposite charges attract one another, forming _____ bonds.

3. Atoms of the same element that differ in the number of neutrons in their nuclei are called _____. Some of these atoms spontaneously break apart, and in this process, they sometimes become different _____. Atoms that behave this way are described as being _____.

4. An atom with an outermost electron shell that is either completely full or empty is described as _____. Atoms with partially full outer electron shells are _____. Covalent bonds are formed when atoms _____ electrons, filling their outer shells.

5. Water is described as _____ because each water molecule has slightly negative and positive poles. This property allows water molecules to form _____ bonds with one another. The bonds between water molecules give water a high _____ that produces surface tension.

Review Questions

1. Based on Table 2-1, how many neutrons are there in oxygen? In hydrogen? In nitrogen?

2. Distinguish between atoms and molecules and among protons, neutrons, and electrons.

3. Compare and contrast covalent bonds and ionic bonds.

4. Explain how polar covalent bonds allow hydrogen bonds to form, and provide an example.

5. Why can water absorb a great amount of heat with little increase in its temperature? What is this property called?

6. Describe how water dissolves a salt.

7. Define *pH scale, acid, base,* and *buffer.* How do buffers reduce changes in pH when hydrogen ions or hydroxide ions are added to a solution? Why is this phenomenon important in organisms?

Applying the Concepts

1. Detergents help clean by dispersing fats and oils in water so that they can be rinsed away. What general chemical structures (for example, polar or nonpolar parts) must a soap or detergent have, and why?

2. What do people mean when they say, "It's not the heat, it's the humidity"? Why does high humidity make a hot day less bearable?

3. Artificial ice cubes ("whiskey stones") made of granite can be cooled in your freezer and used in drinks. Would these cool drinks more or less effectively than an equal weight of ice cubes from the same freezer? Explain.

3

BIOLOGICAL MOLECULES

Puzzling Proteins

"YOU KNOW, LISA, I think something is wrong with me," Charlene Singh told her sister. The vibrant 22-year-old scholarship winner had begun to lose her memory and experience mood swings. During the next 3 years, her symptoms worsened. Singh's hands shook, she was subject to uncontrollable episodes of biting and striking people, and she became unable to walk or swallow. Ultimately, Charlene Singh became the first U.S. resident to die of the human form of mad cow disease, which she had almost certainly contracted more than 10 years earlier while living in England.

It was not until the mid-1990s that health officials recognized that mad cow disease, or BSE (bovine spongiform encephalitis), could spread to people who ate meat from infected cattle. Although millions of people may have eaten tainted beef, fewer than 200 people worldwide have contracted the human version of BSE, called vCJD (variant Creutzfeldt-Jakob disease; CJD is a human genetic disorder with similar symptoms). For those infected, however, the disease is always fatal, riddling the brains of people and cows with microscopic holes that give the brain a spongy appearance.

Where did mad cow disease come from? One hypothesis is that it was derived from a disease in sheep called scrapie, whose symptoms are almost identical to those of BSE. Scrapie was named after the tendency of infected sheep to scrape off their wool. They also lose weight and coordination and may become nervous or aggressive, and their brains become spongy. A mutated form of scrapie may have become capable of infecting cattle, perhaps in the early 1980s. At that time, cattle feed often included parts from sheep, some of which may have harbored scrapie. BSE was first identified in British cattle in 1986, and the use of sheep, cow, and goat parts in cattle feed was banned in 1988. In 1996 British beef exports were temporarily halted after experts confirmed that the disease could spread to people who ate infected meat. As a precautionary measure at that time, more than 4.5 million cattle in Britain were slaughtered and their bodies were burned—a tragedy for British farmers.

Why is mad cow disease particularly fascinating to scientists? In the early 1980s, Dr. Stanley Prusiner, a researcher at the

Friends don't eat friends. Mad cow disease may have emerged as a result of cows eating feed containing protein from the remains of sheep infected with scrapie.

University of California–San Francisco, provided evidence that a protein caused scrapie and that this protein could transmit the disease to experimental animals. He dubbed the infectious proteins "prions" (pronounced PREE-ons). No entity lacking genetic material (DNA or RNA) had ever been shown to be contagious before, and Prusiner's results were initially met with skepticism. But after other scientists confirmed them, his findings expanded our understanding of the importance of proteins.

What are proteins? How do they differ from DNA and RNA? How can a protein infect another organism, increase in number, and produce a fatal disease? Are BSE and vCJD still a threat?

AT A GLANCE

3.1 WHY IS CARBON SO IMPORTANT IN BIOLOGICAL MOLECULES?

You have probably seen fruits and vegetables in the supermarket labeled as "organic," meaning that they were grown without synthetic fertilizers or pesticides. But in chemistry, the word **organic** describes molecules that always contain carbon and usually contain oxygen and hydrogen. Many organic molecules are synthesized by organisms, hence the name organic. In contrast, **inorganic** molecules lack carbon atoms (examples are water and salt). Inorganic molecules, such as those that make up Earth's rocks and metal deposits, are far less diverse and generally much simpler than organic molecules.

Life is characterized by an amazing variety of **biological molecules,** which we define as all molecules produced by living things. Nearly all of these are based on the carbon atom. Biological molecules interact in dazzlingly complex ways that are governed by the chemical properties that arise from their structures. As molecules within cells interact with one another, their structures and chemical properties change. Collectively, these precisely orchestrated changes give cells the ability to acquire and use nutrients, eliminate wastes, move and grow, and reproduce. This complexity is made possible by the versatile carbon atom.

The Bonding Properties of Carbon Are Key to the Complexity of Organic Molecules

As described in Chapter 2, atoms whose outermost electron shells are only partially filled tend to react with one another, gaining stability by filling their shells and forming covalent bonds. Depending on the number of vacancies in their outer shells, two atoms can share two, four, or six electrons—forming a single, double, or triple covalent bond (**FIG. 3-1**). The bonding patterns in the four most common types of atoms found in biological molecules are shown in **FIGURE 3-2**. Covalent bonds are represented by solid lines drawn between atomic symbols.

The bonding versatility of the carbon atom is key to the tremendous variety of biological molecules that make life on Earth possible. The carbon atom (C) has four electrons in its outermost shell, which can accommodate eight electrons. Carbons readily form single or double bonds with each other,

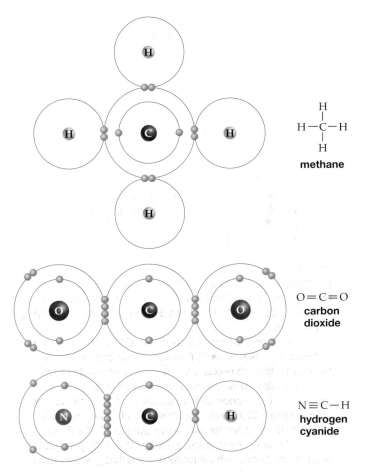

▲ **FIGURE 3-1 Covalent bonding by carbon atoms** Carbon must form four covalent bonds to fill its outer electron shell and become stable. It can do this by forming single, double, or triple covalent bonds. In these examples, carbon forms methane (CH_4), carbon dioxide (CO_2), and hydrogen cyanide (HCN).

THINK CRITICALLY Which of these is/are polar molecules? (You may need to refer back to Chapter 2.)

▶ **FIGURE 3-2 Bonding patterns** The bonding patterns of the four most common atoms in biological molecules. Each line indicates a covalent bond.

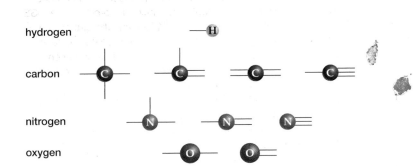

but can also bond with two, three, or four other atoms (see Fig. 3-1). Additional diversity arises from the range of complex shapes that organic molecules can assume, including branched chains, rings, sheets, and helices.

Functional Groups Attach to the Carbon Backbone of Organic Molecules

Functional groups are commonly occurring atoms or groups of atoms that are attached to the carbon backbone of organic molecules. Functional groups are less stable than the carbon backbone and are more likely to participate in chemical reactions. The functional groups in biological molecules endow them with unique properties and tendencies to react with other molecules. **TABLE 3-1** describes seven functional groups that are important in biological molecules.

CHECK YOUR LEARNING

Can you ...

- define *organic* molecules and explain why carbon is so important to life?
- explain why functional groups are important in biological molecules?
- name and describe the properties of seven functional groups?

3.2 HOW ARE LARGE BIOLOGICAL MOLECULES SYNTHESIZED?

Although a complex molecule could be made by laboriously attaching one atom after another, the machinery of life works far more efficiently by preassembling molecular subunits and hooking them together. Just as trains are made by joining a series of train cars, small organic molecules (for example, sugars or amino acids) are joined to form large biological molecules (for example, starches or proteins). The individual subunits are called **monomers** (Gk. *mono,* one); chains of monomers are called **polymers** (Gk. *poly,* many).

Biological Polymers Are Formed by the Removal of Water and Broken Down by the Addition of Water

The subunits of large biological molecules are usually joined by a chemical reaction called **dehydration synthesis,** literally meaning "removing water to put together." In dehydration synthesis, a hydrogen ion (H^+) is removed from one subunit and a hydroxyl ion (OH^-) is removed from a second subunit, leaving openings in the

TABLE 3-1	Important Functional Groups in Biological Molecules		
Group	**Structure**	**Properties**	**Found In**
Hydroxyl	—O—H	Polar; involved in dehydration and hydrolysis reactions; forms hydrogen bonds	Sugars, polysaccharides, nucleic acids, alcohols, some amino acids, steroids
Carbonyl	C=O	Polar; makes parts of molecules hydrophilic (water soluble)	Sugars (linear forms), steroid hormones, peptides and proteins, some vitamins
Carboxyl (ionized form)	C(=O)O⁻	Polar and acidic; the negatively charged oxygen may bond H^+, forming carboxylic acid (—COOH); involved in peptide bonds	Amino acids, fatty acids, carboxylic acids (such as acetic and citric acids)
Amino	N with H, H	Polar and basic; may become ionized by binding a third H^+; involved in peptide bonds	Amino acids, nucleic acids, some hormones
Sulfhydryl	S—H	Nonpolar; forms disulfide bonds in proteins	Cysteine (an amino acid), many proteins
Phosphate (ionized form)	O—P(O⁻)(O⁻)—O⁻	Polar and acidic; links nucleotides in nucleic acids; forms high-energy bonds in ATP (ionized form occurs in cells)	Phospholipids, nucleotides, nucleic acids
Methyl	C with H, H, H	Nonpolar; may be attached to nucleotides in DNA (methylation), changing gene expression	Steroids, methylated nucleotides in DNA

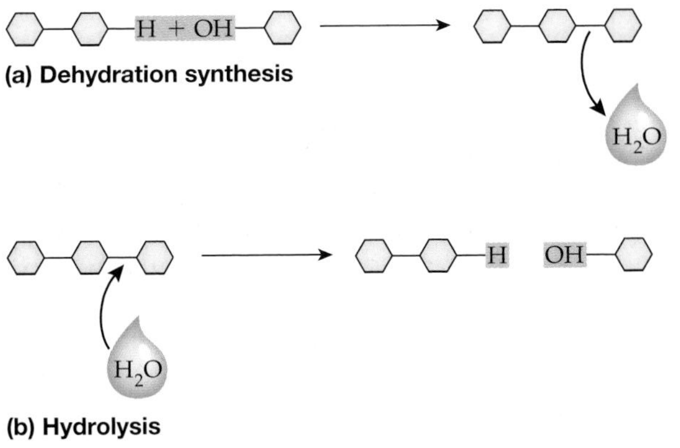

(a) Dehydration synthesis

(b) Hydrolysis

▲ **FIGURE 3-3 Dehydration synthesis and hydrolysis** Biological polymers are formed by **(a)** linking monomer subunits in a reaction that removes H_2O from polar functional groups. These polymers may be broken apart by adding the atoms in H_2O **(b)**, which recreates the subunits.

outer electron shells of atoms in the two subunits. These openings are filled when the subunits share electrons, creating a covalent bond that links them. The hydrogen ion and the hydroxyl ion combine to form a molecule of water (H_2O), as shown in **FIGURE 3-3a**.

The reverse reaction is hydrolysis (Gk. *hydro*, water, and *lysis*, break apart). **Hydrolysis** breaks apart the molecule into its original subunits, with water donating a hydrogen ion to one subunit and a hydroxyl ion to the other (**FIG. 3-3b**). Digestive enzymes use hydrolysis to break down food. For example, enzymes in our saliva and small intestines promote hydrolysis of starch, which consists of a chain of sugar molecules, into individual sugar molecules that can be absorbed into the body.

Although there is a tremendous diversity of biological molecules, nearly all fall into one of four general categories: carbohydrates, proteins, nucleic acids, and lipids (**TABLE 3-2**).

CHECK YOUR LEARNING

Can you ...
- name and describe the reactions that create and break apart biological polymers?

TABLE 3-2	Four Principal Classes of Biological Molecules			
Name and General Structure	**Types**	**Example(s)**	**Typical Function**	
Carbohydrates: Molecules composed primarily of C, H, and O in the ratio $(CH_2O)_n$, where *n* is the number of C's in the molecule's backbone. *Subunit:* Monosaccharide	**Monosaccharides:** Simple sugars	Glucose, fructose	Short-term energy storage in plants	
	Disaccharides: Two linked monosaccharides	Sucrose		
	Polysaccharides: Polymers of monosaccharides	Starch, glycogen	Long-term energy storage in plants and animals, respectively	
		Cellulose, chitin	Structural support in plants and arthropods, respectively	
Proteins: Molecules with one or more chains of amino acids. Proteins have up to four levels of structure. *Subunit:* Amino acid	**Peptides:** Short chains of amino acids	Insulin, oxytocin	Hormones involved in blood sugar regulation and reproduction, respectively	
	Polypeptides: Long chains (polymers) of amino acids	Hemoglobin	Oxygen transport	
		Keratin	Structural component of hair	
Nucleic acids: Molecules composed of polymers of nucleotides, each consisting of a simple sugar, an N-containing base, and a phosphate group. *Subunit:* Nucleotide	**Deoxyribonucleic acid (DNA):** A polymer of nucleotides whose simple sugar is deoxyribose	DNA	Codes for genetic information	
	Ribonucleic acids (RNA): Polymers of nucleotides whose simple sugar is ribose	Messenger RNA, transfer RNA, ribosomal RNA	Work together to form proteins from amino acids based on nucleotide sequences in DNA	
Lipids: Diverse group of molecules containing non-polar (hydrophobic) regions that make them insoluble in water. *Subunit:* No consistent subunit; not polymers	**Fats, oils, and waxes:** Contain one or more fatty acids, hydrophobic chains of carbon atoms that terminate in a carboxylic acid functional group	Animal fats, vegetable oils	Long-term energy storage in animals and plants, respectively	
		Beeswax	Structural component of bee hives	
	Phospholipids: Contain two fatty acids (hydrophobic) and two hydrophilic functional groups, one of which is phosphate	Lecithin	Structural component of cell membranes	
	Steroids: Contain four rings of carbon atoms, with different functional groups attached	Cholesterol	Component of cell membranes	
		Testosterone, estrogen	Male and female sex hormones, respectively	

3.3 WHAT ARE CARBOHYDRATES?

Carbohydrate molecules are composed of carbon, hydrogen, and oxygen in the approximate ratio of 1:2:1. This ratio explains the origin of the word "carbohydrate," which literally means "carbon plus water." All carbohydrates are either small, water-soluble **sugars** or polymers of sugar, such as starch. If a carbohydrate consists of just one sugar molecule, it is called a **monosaccharide** (Gk. *mono,* one, and *saccharum,* sugar). When two monosaccharides are linked, they form a **disaccharide.** For example, sucrose, or table sugar, is a disaccharide composed of fructose and glucose. If you've stirred sugar into coffee, you know that sugar dissolves in water. Sugar molecules are hydrophilic; their hydroxyl functional groups are polar and form hydrogen bonds with polar water molecules (**FIG. 3-4**).

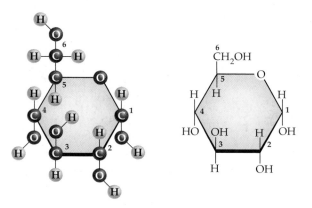

▲ **FIGURE 3-5 Depictions of chemical structures** The molecule glucose ($C_6H_{12}O_6$) can be drawn as (left) a ball-and-stick model showing each atom or (right) a simplified version in which each unlabeled joint is a carbon atom. The carbon atoms are numbered for reference. The space-filling structure of glucose is shown in Figure 3-4.

▼ **FIGURE 3-4 Sugar dissolving in water** Glucose dissolves as the polar hydroxyl groups of each sugar molecule form hydrogen bonds with nearby water molecules.

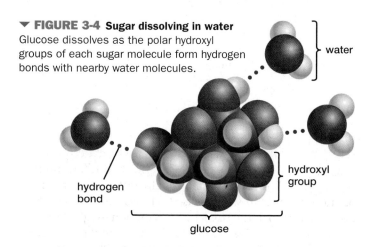

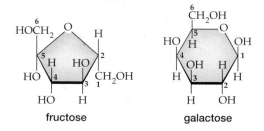

fructose   galactose

▲ **FIGURE 3-6 Some six-carbon monosaccharides**

Different Monosaccharides Have Slightly Different Structures

Monosaccharides have a backbone of three to seven carbon atoms. Most of these carbon atoms have both a hydrogen (—H) and a hydroxyl group (—OH) attached to them; therefore, carbohydrates generally have the approximate chemical formula $(CH_2O)_n$, where n is the number of carbons in the backbone. When a sugar molecule is dissolved in water, such as inside a cell, its carbon backbone usually forms a ring.

Glucose is the most common monosaccharide in organisms and the primary energy source of cells. Glucose has six carbons, so its chemical formula is $C_6H_{12}O_6$. Figure 3-4 and **FIGURE 3-5** show various ways of depicting the chemical structure of glucose; keep in mind that any unlabeled "joint" in a ring or chain actually represents a carbon atom.

Many organisms synthesize other monosaccharides that have the same chemical formula as glucose but slightly different structures. For example, some plants store energy in *fructose* (L. *fruct,* fruit), which we consume in fruits, juices, honey, corn syrup, and soft drinks. *Galactose* is secreted by mammals in their milk as an energy source for their young (**FIG. 3-6**). Fructose and galactose must be converted to glucose before cells can use them as a source of energy.

Other common monosaccharides, such as ribose and deoxyribose (found in the nucleic acids of RNA and DNA, respectively), have five carbons. Notice in **FIGURE 3-7** that deoxyribose has one fewer oxygen atom than ribose because one of the hydroxyl groups in ribose is replaced by a hydrogen atom in deoxyribose.

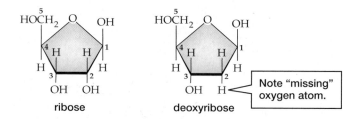

ribose   deoxyribose

Note "missing" oxygen atom.

▲ **FIGURE 3-7 Some five-carbon monosaccharides**

Disaccharides Consist of Two Monosaccharides Linked by Dehydration Synthesis

Monosaccharides can be linked by dehydration synthesis to form disaccharides or polysaccharides (**FIG. 3-8**). Disaccharides are often used for short-term energy storage in plants. When energy is required, the disaccharides are broken apart by hydrolysis into their monosaccharide

glucose   fructose   sucrose

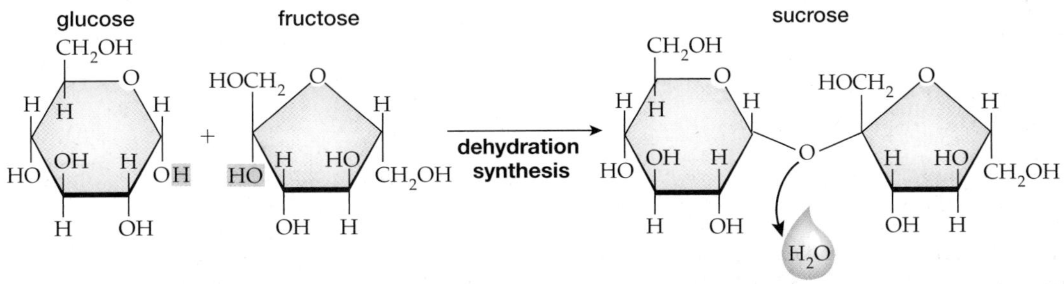

dehydration synthesis

H_2O

▲ **FIGURE 3-8 Synthesis of a disaccharide** Sucrose is synthesized by a dehydration reaction in which a hydrogen is removed from glucose and a hydroxyl group is removed from fructose. This forms water and leaves the two monosaccharide rings joined by single bonds to the remaining oxygen atom.

THINK CRITICALLY Describe hydrolysis of this molecule.

subunits (see Fig. 3-3) and converted to glucose, which is broken down further to release energy stored in its chemical bonds. Perhaps you had toast and coffee with cream and sugar for breakfast. You stirred *sucrose* (glucose plus fructose) into your coffee and then added cream containing *lactose* (glucose plus galactose). The disaccharide *maltose* (glucose plus glucose) is rare in nature, but it is formed when enzymes in your digestive tract hydrolyze starch, such as the wheat starch in your toast. Other digestive enzymes then hydrolyze each maltose into two glucose molecules that cells can absorb and break down to liberate energy.

If you've ever added Splenda or Equal to your coffee instead of sugar, you know that these artificial sweeteners contain few calories. How is this possible? We discuss artificial sweeteners and artificial fats in "Health Watch: Fake Foods" on page 38.

Polysaccharides Are Chains of Monosaccharides

A polymer of many monosaccharides is called a **polysaccharide.** Most polysaccharides do not dissolve in water at body temperatures because the polar hydroxyl groups of their sugars have been lost during dehydration synthesis, which links the monomers together and releases water (see Fig. 3-8). Despite their lack of solubility, polysaccharides can be hydrolyzed under the right conditions. For example, if you take a bite of a bagel and chew it for a minute or so, you may notice that it gradually tastes sweeter. This is because enzymes in saliva cause hydrolysis of the **starch** (a polysaccharide) in the bagel into its component glucose molecules, which dissolve in your saliva and stimulate receptors on your tongue that respond to sweetness. Plants often use starch (**FIG. 3-9**) as an energy-storage molecule. Starch, which

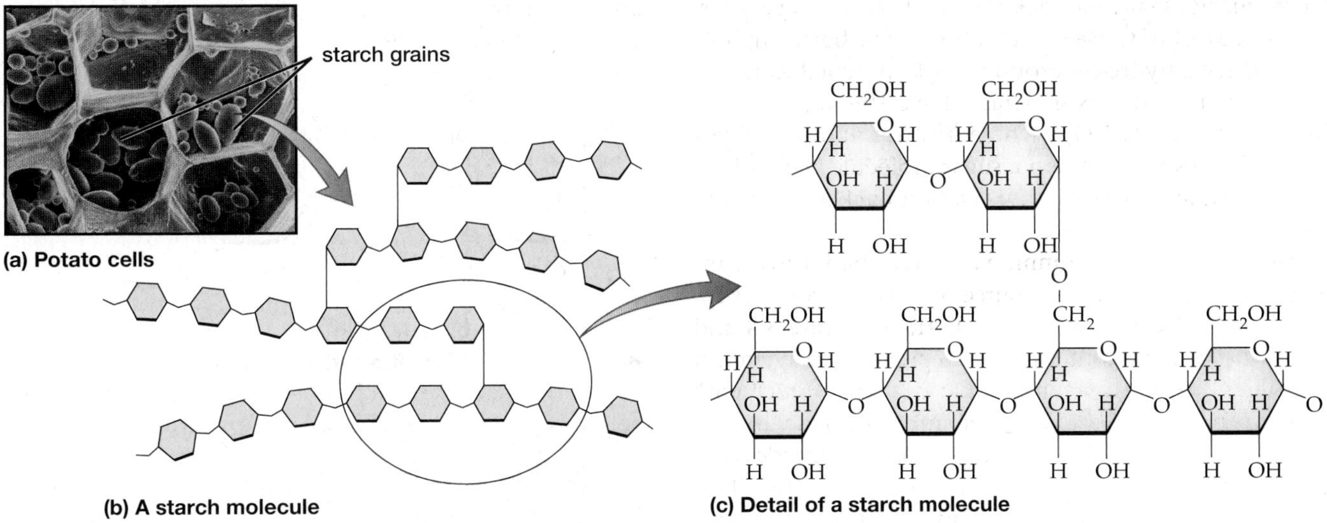

starch grains

(a) Potato cells

(b) A starch molecule

(c) Detail of a starch molecule

▲ **FIGURE 3-9 Starch structure and function (a)** Starch grains inside potato cells store energy that will allow the potato to generate new plants in the spring. **(b)** A section of a single starch molecule. Starches consist of branched chains of up to half a million glucose subunits. **(c)** The precise structure of the circled portion of the starch molecule in (b). Notice the linkage between the individual glucose subunits for comparison with cellulose (Fig. 3-10).

Health WATCH

Fake Foods

People evolved to enjoy sweets and fats because they are high in the calories we need to survive. But in societies blessed with an overabundance of food, obesity is a serious health problem. In response, food scientists have modified biological molecules to make them noncaloric (**FIG. E3-1**). How are these fake food molecules made?

The artificial oil olestra, used in fat-free potato chips, tastes and feels similar to oil. The olestra molecule has a core of sucrose, but with six to eight fatty acids attached to its carbon atoms, which prevent digestive enzymes from breaking it down. Fat-soluble vitamins (A, D, E, and K) dissolve in olestra, so foods containing olestra are supplemented with small amounts of these vitamins to compensate. After its initial debut on grocery shelves, olestra became unpopular with consumers and has been largely phased out.

The sweetener sucralose (Splenda) is a modified sucrose molecule in which three hydroxyl groups are replaced with chlorine atoms. Aspartame (Equal, NutraSweet) is a combination of the amino acids aspartic acid and phenylalanine (see Figs. 3-14a, b). Both taste hundreds of times as sweet as the same amount of sugar.

▲ **FIGURE E3-1 Artificial foods** Products made with artificial sweeteners and olestra are marketed to people trying to control their weight.

High-fructose corn syrup is a fake food often substituted for sucrose in soft drinks because it is easier for manufacturers to work with. Although it does not occur in nature, high-fructose corn syrup very closely resembles sucrose (50% fructose + 50% glucose) but with fructose increased to 55%.

Are fake foods bad for you? Olestra interferes with the absorption of carotenoids (pigments with antioxidant properties) from fruits and vegetables, a possible drawback to eating it frequently. Products containing aspartame carry a warning for people with the rare genetic disorder phenylketonuria, which prevents them from metabolizing phenylalanine. There is no good scientific evidence that high-fructose corn syrup is more detrimental to health than the equivalent amount of sucrose, despite numerous popular press reports.

But our understanding of the health effects of fake foods is limited. Controlled studies of human nutrition are notoriously difficult to design and interpret correctly. Hundreds of recent nutritional studies have reached conclusions that differ considerably and often seem contradictory, due to differences in study groups, sample sizes, methods, controls, and interpretation of the results. The ways in which the human body processes food are staggeringly complex and differ among individuals. Although studies using laboratory animals can be far better controlled, their results may not directly apply to people. Nonetheless, the popular press often pounces on conclusions that can be stated most dramatically in headlines, often omitting important caveats by the investigators. The bottom line? Get your health information from reliable sources and be aware that even expert opinions in this field often change. There is no magic bullet for weight loss, and everything you eat has consequences for your body. The safest approach is to use moderation in consuming natural sugars and fats as well as the fake foods that might replace them.

EVALUATE THIS A 19-year-old 6′ 2″ male weighing 297 pounds comes to his doctor's office with high blood sugar, a symptom of adult-onset diabetes (often associated with obesity), and claims to have an insatiable sweet tooth. He proposes to switch to sugar-free cakes and donuts. What advice should the doctor give him?

---

is commonly formed in roots and seeds, consists of branched chains of up to half a million glucose subunits. **Glycogen,** a short-term energy-storage molecule in animals (including people), is also a chain of glucose subunits but is much more highly branched than starch. Glycogen is stored primarily in the liver and muscles.

Many organisms use polysaccharides as structural materials. One of the most important structural polysaccharides is **cellulose,** which makes up most of the walls of the living cells of plants, the fluffy white bolls of cotton plants, and about half the dry weight of tree trunks (**FIG. 3-10**). Scientists estimate that plants synthesize about a trillion tons

of cellulose each year, making it the most abundant organic molecule on Earth.

Cellulose, like starch, is a polymer of glucose, but in cellulose, every other glucose is "upside down," as you will see when you compare Figure 3-9c with Figure 3-10d. Although most animals easily digest starch, no vertebrates synthesize an enzyme that can attack the bonds between glucose molecules in cellulose. A few animals, such as cows and termites, harbor cellulose-digesting microbes in their digestive tracts and can benefit from the glucose subunits that the microbes release. In humans, cellulose fibers pass intact through the digestive system; cellulose supplies no

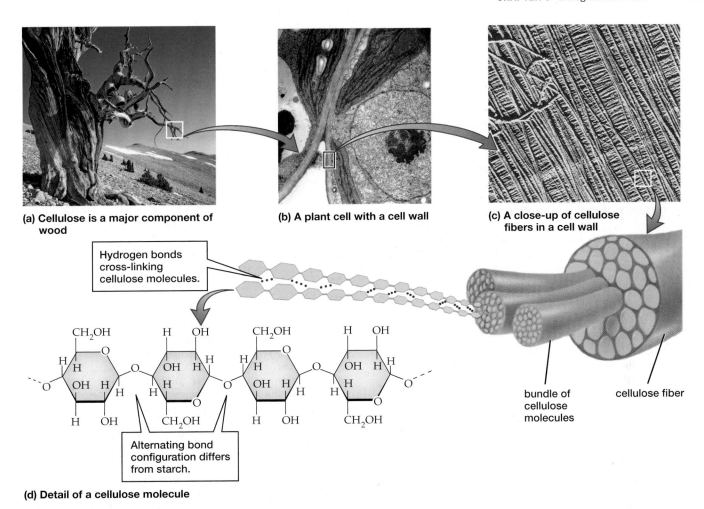

(a) Cellulose is a major component of wood

(b) A plant cell with a cell wall

(c) A close-up of cellulose fibers in a cell wall

Hydrogen bonds cross-linking cellulose molecules.

bundle of cellulose molecules

cellulose fiber

Alternating bond configuration differs from starch.

(d) Detail of a cellulose molecule

▲ **FIGURE 3-10 Cellulose structure and function (a)** Wood in this 3,000-year-old bristlecone pine is primarily cellulose. **(b)** Cellulose forms the cell wall that surrounds each plant cell. **(c)** Plant cell walls often consist of cellulose fibers in layers that run at right angles to each other to resist tearing in both directions. **(d)** Cellulose is composed of up to 10,000 glucose subunits. Compare this structure with Figure 3-9c and notice that every other glucose molecule in cellulose is "upside down."

nutrients but provides roughage with several digestive benefits.

Another supportive polysaccharide is **chitin,** which makes up the outer coverings (exoskeletons) of insects, crabs, and spiders. Chitin also stiffens the cell walls of many fungi, including mushrooms. Chitin is similar to cellulose, except the glucose subunits bear a nitrogen-containing functional group (**FIG. 3-11**).

Carbohydrates may also form parts of larger molecules; for example, the plasma membrane that surrounds each cell

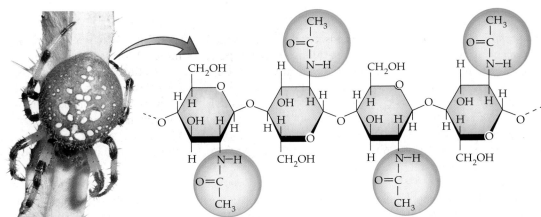

▶ **FIGURE 3-11 Chitin structure and function** Chitin has the same glucose bonding configuration as cellulose, but the glucose subunits have a nitrogen-containing functional group replacing one of the hydroxyls. Chitin supports the otherwise soft bodies of arthropods (including spiders such as this one, insects, and crabs and their relatives) as well as most fungi.

is studded with proteins to which carbohydrates are attached. Nucleic acids (discussed later) also contain sugar molecules.

CHECK YOUR LEARNING

Can you ...

- describe the major types of carbohydrates?
- provide examples of each type of carbohydrate and explain how organisms use them?

3.4 WHAT ARE PROTEINS?

Like starches, **proteins** are biological polymers synthesized by linking simple subunits. Scientists estimate that the human body has between 250,000 and 1 million different proteins (**TABLE 3-3**). Most cells contain hundreds of different **enzymes,** which are proteins that promote specific chemical reactions. Other proteins are structural. *Keratin*, for example, forms hair, horns, nails, scales, and feathers (**FIG. 3-12**). Silk proteins are secreted by silk moths and spiders to make cocoons and webs, respectively. Nutritional proteins, such as albumin in egg white and casein in milk, provide amino acids to developing animals. The protein hemoglobin transports oxygen in the blood. Actin and myosin in muscle are contractile proteins that allow animal bodies to move. Some proteins are hormones (insulin and growth hormone, for example), others are antibodies (which help fight disease and infection), and a few are toxins (such as rattlesnake venom).

Proteins Are Formed from Chains of Amino Acids

The subunits of proteins are **amino acids.** There are 20 different amino acids commonly found in proteins, all of which have the same basic structure. A central carbon is bonded to a hydrogen atom and to three functional groups: a nitrogen-containing amino group ($-NH_2$), a

TABLE 3-3	Functions of Proteins
Function	**Example(s)**
Structural	Keratin (forms hair, nails, scales, feathers, and horns); silk (forms webs and cocoons)
Movement	Actin and myosin (found in muscle cells; allow contraction)
Defense	Antibodies (found in the bloodstream; fight disease organisms; some neutralize venoms); venoms (found in venomous animals; deter predators and disable prey)
Storage	Albumin (in egg white; provides nutrition for an embryo)
Signaling	Insulin (secreted by the pancreas; promotes glucose uptake into cells)
Catalyzing reactions	Amylase (found in saliva and the small intestine; digests carbohydrates)

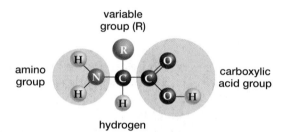

▲ FIGURE 3-13 Amino acid structure

carboxylic acid group ($-COOH$), and an "R" group that varies among different amino acids (**FIG. 3-13**). The R group gives each amino acid distinctive properties (**FIG. 3-14**). Some amino acids are hydrophilic and water soluble because their R groups are polar. Others are hydrophobic, with nonpolar R groups that are insoluble in water. The amino acid cysteine (Fig. 3-14c) is unique in having a sulfur-containing (sulfhydryl) R group that can form covalent **disulfide bonds** with the sulfur of another cysteine molecule. These disulfide bonds play important roles in proteins, as described later.

(a) Hair

(b) Horn

(c) Silk

▲ **FIGURE 3-12 Structural proteins** Keratin is a common structural protein. It is the predominant protein found in **(a)** hair, **(b)** horn, and **(c)** the silk of a spiderweb.

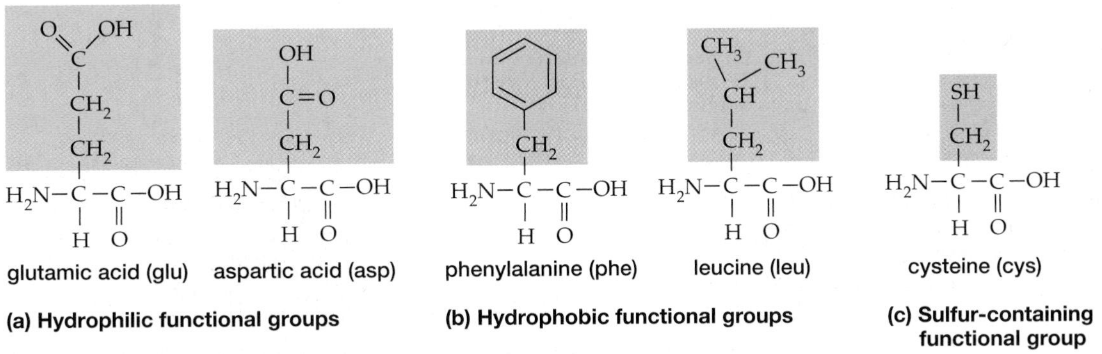

glutamic acid (glu)   aspartic acid (asp)

phenylalanine (phe)   leucine (leu)

cysteine (cys)

(a) Hydrophilic functional groups

(b) Hydrophobic functional groups

(c) Sulfur-containing functional group

▲ **FIGURE 3-14 Amino acid diversity** The diversity of amino acids is caused by the different R functional groups (green backgrounds), which may be **(a)** hydrophilic or **(b)** hydrophobic. **(c)** The R group of cysteine has a sulfur atom that can form covalent bonds with the sulfur in other cysteines.

THINK CRITICALLY Look up the rest of the amino acids and, based on their structures, identify three others that have hydrophobic functional groups.

Like polysaccharides and lipids, proteins are formed by dehydration synthesis. In proteins, the nitrogen in the amino group of one amino acid is joined to the carbon in the carboxylic acid group of another amino acid by a single covalent bond, and water is liberated (**FIG. 3-15**). This process forms a **peptide bond,** and the resulting chain of two amino acids is called a **peptide,** a term used for relatively short chains of amino acids (up to 50 or so). Additional amino acids are added, one by one, until a long *polypeptide* chain is completed. Polypeptide chains in cells can be up to thousands of amino acids in length. A protein consists of one or more polypeptide chains.

A Protein Can Have up to Four Levels of Structure

Interactions among amino acids and their R groups cause twists, folds, and interconnections that give proteins their three-dimensional structure. Up to four levels of well-defined protein structure are possible: primary, secondary, tertiary, and quaternary. The first three structures

are characteristic of many proteins, and the fourth level occurs in proteins such as hemoglobin that include two or more polypeptide chains (**FIG. 3-16**). The sequence of amino acids in a protein is called its **primary structure** and is specified by genetic instructions in a cell's DNA (see Fig. 3-16a). The positions of specific amino acids in the sequence allow hydrogen bonds to form at particular sites within the polypeptide, causing it to assume a **secondary structure,** which is most commonly either a helix or a pleated sheet. These hydrogen bonds do not involve the R groups, but form between the slightly negative C=O (carbonyl) group in one amino acid and the slightly positive N—H group from an amino acid farther along the peptide chain (see the carbonyl group in Table 3-1 and on the right side of Figure 3-15). When such hydrogen bonds form between every fourth amino acid, they create the coils of the spring-like **helix** found in the polypeptide subunits of the hemoglobin molecule (see Fig. 3-16b). The keratin protein of hair also forms helices. Other proteins, such as silk, contain polypeptide chains that repeatedly fold back upon themselves, anchored by hydrogen bonds in

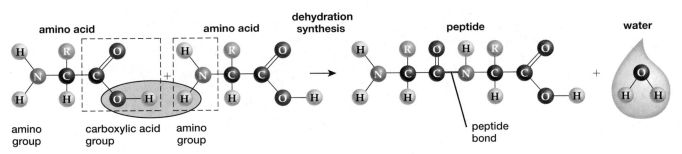

▲ **FIGURE 3-15 Protein synthesis** A dehydration reaction forms a peptide bond between the carbon of the carboxylic acid group of one amino acid and the nitrogen of the amino group of a second amino acid.

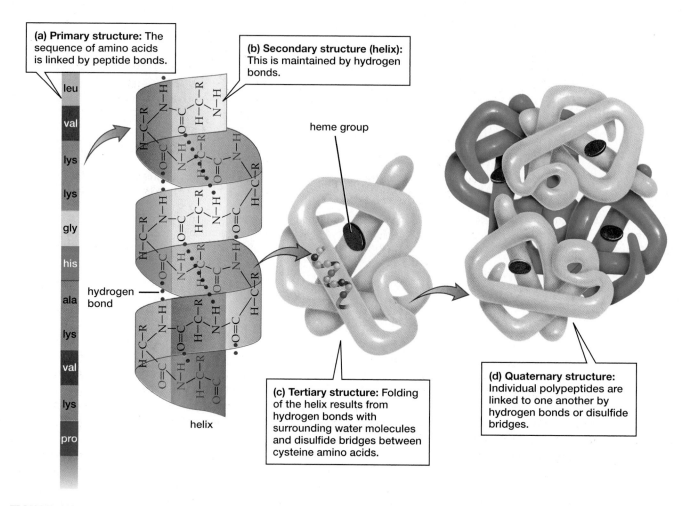

(a) Primary structure: The sequence of amino acids is linked by peptide bonds.

(b) Secondary structure (helix): This is maintained by hydrogen bonds.

heme group

hydrogen bond

helix

leu
val
lys
lys
gly
his
ala
lys
val
lys
pro

(c) Tertiary structure: Folding of the helix results from hydrogen bonds with surrounding water molecules and disulfide bridges between cysteine amino acids.

(d) Quaternary structure: Individual polypeptides are linked to one another by hydrogen bonds or disulfide bridges.

▲ **FIGURE 3-16 The four levels of protein structure** Hemoglobin is the oxygen-carrying protein in red blood cells. Red disks represent the iron-containing heme group that binds oxygen.

THINK CRITICALLY Why do many proteins, when heated excessively, lose their ability to function?

a **pleated sheet** arrangement (**FIG. 3-17a**). Silk proteins also contain disordered segments with no defined secondary structure that link their pleated sheets and allow silk to stretch (**FIG. 3-17b**).

CASE STUDY **CONTINUED**
Puzzling Proteins

Infectious prions such as those that cause mad cow disease are abnormally folded versions of a protein that is found throughout the body. The secondary structure of the normal prion protein is primarily helical. Infectious prions, however, fold into pleated sheets. The pleated sheets are so stable they are unaffected by the enzymes that break down normal prion protein. As a result, infectious prions accumulate destructively in the brain.

Helices and pleated sheets are the two major secondary structures of proteins. What do protein tertiary and quaternary structures look like?

Each protein also contorts into a **tertiary structure** (see Fig. 3-16c) determined by the protein's primary and secondary structure and also by its environment. For example, a protein in the watery interior of a cell folds in a roughly spherical way that exposes its hydrophilic amino acids to the water and buries its hydrophobic amino acids deeper within the molecule. These somewhat water-soluble *globular proteins* include hemoglobin, enzymes, and albumin in eggs. Tertiary structure also forms fibrous proteins such as the keratin in hair, hoofs, horns, and fingernails. Fibrous proteins are insoluble in water and contain many hydrophobic amino acids. In keratin, tertiary structure consists of paired helical strands held together by disulfide bonds that form between cysteines of each helical polypeptide (see Fig. E3-2 on page 44). The more disulfide bonds, the stiffer the keratin; for example, keratin in fingernails has more cysteine and more disulfide bonds than keratin in hair.

A fourth level of protein organization, called **quaternary structure,** occurs in proteins that contain individual polypeptides linked by hydrogen bonds, disulfide bonds, or attractions between oppositely charged portions of

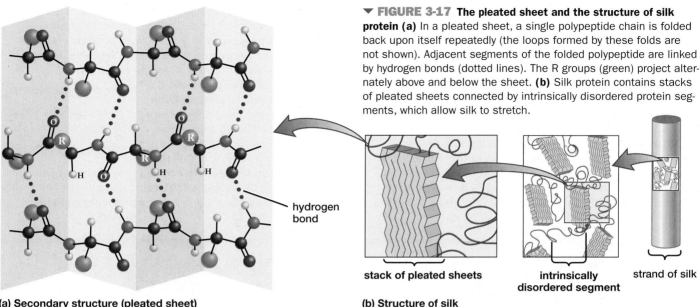

protein (a) In a pleated sheet, a single polypeptide chain is folded back upon itself repeatedly (the loops formed by these folds are not shown). Adjacent segments of the folded polypeptide are linked by hydrogen bonds (dotted lines). The R groups (green) project alternately above and below the sheet. **(b)** Silk protein contains stacks of pleated sheets connected by intrinsically disordered protein segments, which allow silk to stretch.

hydrogen bond

stack of pleated sheets

intrinsically disordered segment

strand of silk

(a) Secondary structure (pleated sheet)

(b) Structure of silk

different amino acids. Hemoglobin, for example, consists of four polypeptide chains held together by hydrogen bonds (see Fig. 3-16d). Each of the four polypeptides holds an iron-containing organic molecule called a heme group that can bind one molecule of oxygen. In keratin, quaternary structure links paired helical subunits into groups of eight.

Protein Function Is Determined by Protein Structure

The highly organized structures of many proteins are essential for their ability to function. Various enzymes, for example, fold in ways that only allow very specific molecules to interact with them, fitting together like a lock and key, as described in Chapter 6. Primary protein structure specifies the location of amino acids bearing specific R groups. In hemoglobin, for example, specific R groups must occur in precisely the right places to hold the heme group that binds oxygen. Interactions of hydrophilic and hydrophobic R groups with their watery environment are important in determining whether and how a protein will fold and what other molecules it can react with. A mutation that replaces a hydrophilic with a hydrophobic amino acid can sometimes cause significant distortion of the protein. For example, this type of mutation in hemoglobin causes the genetic disorder sickle-cell anemia (described in Chapters 11 and 13).

Many proteins, notably enzymes, hemoglobin, and structural proteins such as keratin, rely on a precise three-dimensional structure to perform their functions in the body. But there has recently been an explosion of interest and research into proteins or segments of protein with very flexible structures. These **intrinsically disordered proteins** have a precisely ordered primary structure dominated by hydro-philic amino acids that float rather freely in their watery surroundings. As a result, they lack stable secondary and tertiary structure. Intrinsically disordered proteins are flexible and versatile; some can fold in various ways that allow them to interact with several different molecules, like master keys that fit several locks. But unlike rigid keys, intrinsically disordered proteins change their configuration as they bind different targets. For example, the protein p53 (see Chapter 9) has disordered segments that can bind multiple target molecules. This allows the p53 protein to regulate such diverse processes as cell division and the repair of defective DNA molecules.

A protein is described as **denatured** when its normal three-dimensional structure is destroyed, leaving its primary structure intact. For example, egg white consists of albumin protein, which is normally transparent and relatively fluid. But the heat of a frying pan rips its hydrogen bonds apart, destroying the albumin's secondary and tertiary structure and causing it to become opaque, white, and solid (**FIG. 3-18**). Keratin in hair is denatured by a permanent wave, as described in "Have You Ever Wondered: Why a Perm Is (Temporarily) Permanent?" on page 44. Bacteria and viruses can be destroyed by denaturing their proteins using heat, ultraviolet light, or solutions that are highly salty or acidic. Water is sometimes sterilized with ultraviolet light, and dill pickles are preserved from bacterial attack by their salty, acidic brine.

▲ **FIGURE 3-18 Heat denatures albumin**

HAVE YOU EVER
WONDERED...

Whether your hair is naturally straight or curly is determined by the properties of your hair follicles and the shape of the hair shafts they produce. The follicles of straight hair are round in cross section, whereas the follicles of curly hair are flatter. But genes are not destiny! You can create curls chemically.

Each hair consists of bundles within bundles of keratin (FIG. E3-2). Keratin's helical, spring-like secondary structure is created by hydrogen bonds, which are easily disrupted by the attraction of polar water molecules. So if you let wet straight hair dry on curlers, as the water disappears, new hydrogen bonds form among the keratin molecules in different places because of the distortion caused by the curler. But if it's rainy (or even humid), you can say goodbye to your hydrogen-bonded curls as the moisture disrupts them and your hair reverts to its natural straightness.

How is a permanent wave created? Keratin has lots of cysteine amino acids, and a perm alters the locations of the strong covalent disulfide bonds between cysteines. First, the hair is soaked in a solution that breaks the natural disulfide bonds linking adjacent keratin molecules. Set on curlers, the hair is then saturated with a different solution that causes disulfide bonds to re-form. The curlers force these bonds to form in new locations, and the strong disulfide bonds permanently maintain the curl (see Fig. E3-2). Genetically straight hair has been transformed into artificially curly hair—until new hair grows in.

> *Why a Perm Is (Temporarily) Permanent?*

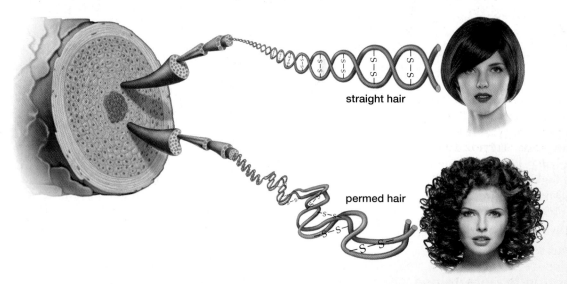

straight hair

permed hair

▲ **FIGURE E3-2 A permanent distortion** A perm changes the location of covalent disulfide bonds between adjacent keratin molecules throughout the hair shaft, making it curl.

CHECK YOUR LEARNING
Can you ...
- describe protein subunits and how proteins are synthesized?
- explain the four levels of protein structure and why a protein's three-dimensional structure is important?
- list several functions of proteins and provide examples of proteins that perform each function?
- describe the properties of intrinsically disordered proteins?

3.5 WHAT ARE NUCLEOTIDES AND NUCLEIC ACIDS?

A **nucleotide** is a molecule with three parts: a five-carbon sugar, a phosphate functional group, and a nitrogen-containing **base.** The sugar may be either ribose or deoxyribose (see Fig. 3-7). The bases are composed of carbon and nitrogen atoms linked together in either a single ring (in the bases thymine, uracil, and cytosine) or double rings (in the bases adenine and guanine). A deoxyribose nucleotide with the adenine base is illustrated in FIGURE 3-19. Nucleotides may function as energy-carrier molecules, intracellular messenger molecules, or subunits of polymers called nucleic acids.

Some Nucleotides Act As Energy Carriers or Intracellular Messengers

Adenosine triphosphate (ATP) is a ribose nucleotide with three phosphate functional groups (FIG. 3-20). This molecule is formed in cells by reactions that release energy, such as the reaction that breaks down a sugar molecule. ATP stores energy in bonds between its phosphate groups and releases energy when the bond linking the last phosphate to the ATP molecule is broken. This energy is then available to drive energy-demanding reactions, such as linking amino acids to form proteins.

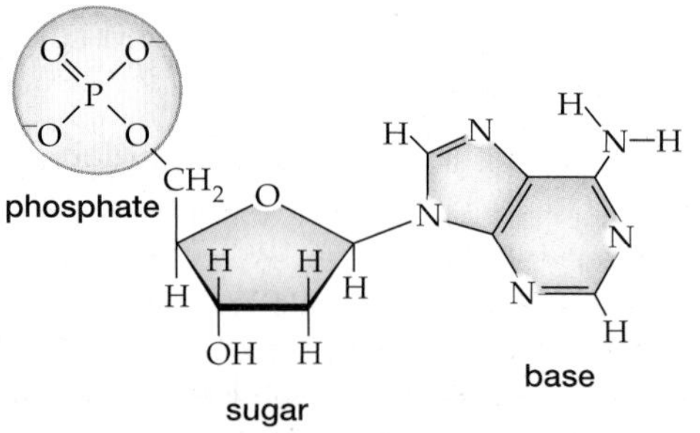

phosphate

sugar

base

▲ **FIGURE 3-19 A deoxyribose nucleotide**

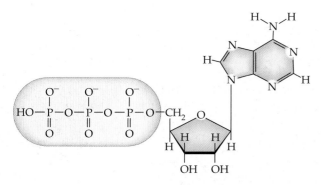

▲ **FIGURE 3-20 The energy-carrier molecule adenosine triphosphate (ATP)**

The ribose nucleotide cyclic adenosine monophosphate (cAMP) acts as a messenger molecule in cells. Many hormones exert their effects by stimulating cAMP to form within cells, where it initiates a series of biochemical reactions (see Chapter 38).

Other nucleotides (such as NAD$^+$ and FAD) are known as *electron carriers* because they transport energy in the form of high-energy electrons. Their energy and electrons are used in ATP synthesis, for example, when cells break down sugar (see Chapter 8).

DNA and RNA, the Molecules of Heredity, Are Nucleic Acids

Single nucleotides (monomers) may be strung together in long chains by dehydration synthesis, forming polymers called **nucleic acids.** In nucleic acids, an oxygen atom in the phosphate functional group of one nucleotide is covalently bonded to the sugar of the next. The polymer of deoxyribose nucleotides, called **deoxyribonucleic acid (DNA),** can contain millions of nucleotides. A DNA molecule consists of two strands of nucleotides entwined in the form of a double helix and linked by hydrogen bonds (**FIG. 3-21**). DNA forms the genetic material of all cells. Its sequence of nucleotides, like the letters of a biological alphabet, spells out the genetic information needed to construct the proteins of each organism. Single-stranded chains of ribose nucleotides, called **ribonucleic acid (RNA),** are copied from the DNA and direct the synthesis of proteins (see Chapters 11 and 12).

Puzzling Proteins

All cells use DNA as a blueprint for producing more cells, and viruses use either DNA or RNA. Before the discovery of prions, however, no infectious agent had ever been discovered that completely lacked genetic material composed of nucleic acids. Scientists were extremely skeptical of the hypothesis that prion proteins could reproduce themselves, until repeated studies found no trace of genetic material associated with prions.

In addition to lacking genetic material, prions also lack another component that all other infectious agents possess: a surrounding membrane. What kinds of molecules are involved in the construction of membranes?

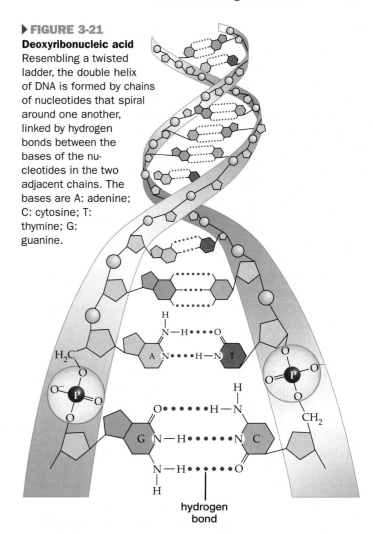

▶ **FIGURE 3-21**
Deoxyribonucleic acid
Resembling a twisted ladder, the double helix of DNA is formed by chains of nucleotides that spiral around one another, linked by hydrogen bonds between the bases of the nucleotides in the two adjacent chains. The bases are A: adenine; C: cytosine; T: thymine; G: guanine.

CHECK YOUR LEARNING

Can you ...

- describe the general structure of nucleotides?
- list three different functions of nucleotides?
- explain how nucleic acids are synthesized?
- give two examples of nucleic acids and their functions?

3.6 WHAT ARE LIPIDS?

Lipids are a diverse group of molecules that contain regions composed almost entirely of hydrogen and carbon, with nonpolar carbon–carbon and carbon–hydrogen bonds. These regions are hydrophobic, which makes lipids insoluble in water. Unlike carbohydrates, proteins, and nucleic acids, lipids are not formed by linking monomer subunits into polymers. Lipids can store energy, provide waterproof coatings on plants, form major components of cell membranes, or function as hormones. Lipids fall into three major groups: (1) oils, fats, and waxes, (2) phospholipids, and (3) steroids.

Oils, Fats, and Waxes Contain Only Carbon, Hydrogen, and Oxygen

Oils, fats, and waxes are built from only three types of atoms: carbon, hydrogen, and oxygen. Each contains one or more **fatty acids,** long chains of carbon and hydrogen with a carboxylic acid functional group (—COOH) at one end. Fats and oils are formed by dehydration synthesis linking three fatty acid subunits to one molecule of *glycerol,* a three-carbon molecule (**FIG. 3-22**). This structure gives fats and oils their chemical name: **triglycerides.**

Fats and **oils** are used primarily as energy-storage molecules; they contain more than twice as many calories per gram as do carbohydrates and proteins. Fats (such as butter and lard) are produced primarily by animals, whereas oils (found in corn, canola, olives, and avocados) are produced primarily by plants (**FIG. 3-23**). The difference between fats, which are solid at room temperature, and oils, which are liquid at room temperature, lies in the structure of their fatty acid subunits.

In fats, the carbons of fatty acids are linked entirely by single bonds, and the fatty acids are described as **saturated,** because they contain as many hydrogen atoms as possible. Saturated fatty acid chains are straight and can pack closely together, thus forming a solid at room temperature (**FIG. 3-24a**). If, however, some of the carbons are linked by double bonds and the fatty acids consequently contain fewer hydrogens, the fatty acids are **unsaturated.** The double

(a) Fat is stored prior to hibernation.   (b) Avocado flesh is rich in oil.

▲ **FIGURE 3-23 Energy storage (a)** A grizzly bear stores fat to provide both insulation and energy as he prepares to hibernate. If he stored the same amount of energy in carbohydrates, he would probably be unable to walk. **(b)** Oily avocado flesh likely originally evolved to entice enormous seed-dispersing mammals (such as giant ground sloths, extinct for about 10,000 years), which would swallow the seeds and excrete them intact.

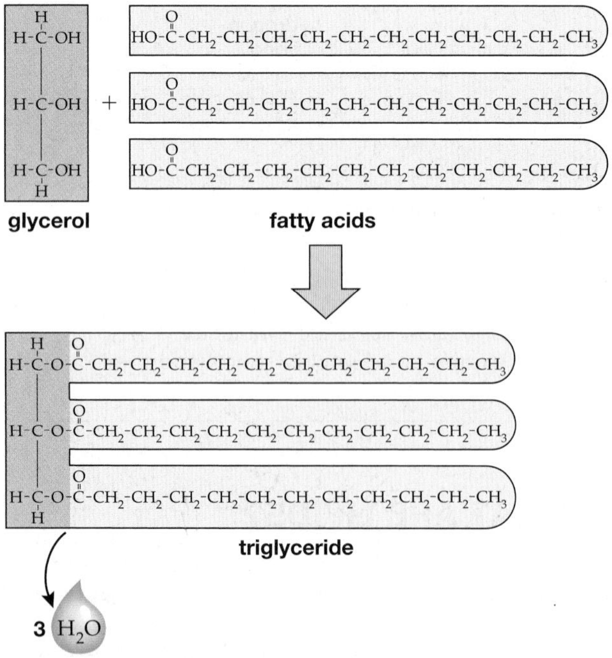

bonds produce kinks in the fatty acid chains, which prevent oil molecules from packing closely together (**FIG. 3-24b**). The commercial process of hydrogenation—which breaks some of the double bonds and adds hydrogens to the carbons—can convert liquid oils to solids, but with health consequences (see "Health Watch: Cholesterol, *Trans* Fats, and Your Heart" on page 48).

▲ **FIGURE 3-22 Synthesis of a triglyceride** Dehydration synthesis links a single glycerol molecule with three fatty acids to form a triglyceride and three water molecules.

THINK CRITICALLY What kind of reaction breaks this molecule apart?

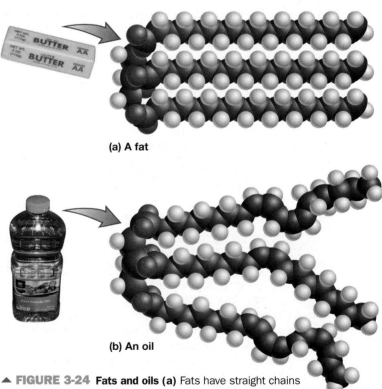

(a) A fat

(b) An oil

▲ **FIGURE 3-24 Fats and oils (a)** Fats have straight chains of carbon atoms in their fatty acid tails. **(b)** The fatty acid tails of oils have double bonds between some of their carbon atoms, creating kinks in the chains. Oils are liquid at room temperature because their kinky tails keep the molecules farther apart.

▲ FIGURE 3-25 **Waxes** Waxes are highly saturated lipids that remain solid at outdoor temperatures. Bees form wax into the hexagons of this honeycomb.

Although **waxes** are chemically similar to fats, humans and most other animals do not have the appropriate enzymes to break them down. Waxes are highly saturated and are solid at outdoor temperatures. They form a water repellent coating on the leaves and stems of land plants, and birds distribute waxy secretions over their feathers, causing them to shed water. Honey bees use waxes to build intricate honeycomb structures, where they store honey and lay their eggs (**FIG. 3-25**).

Phospholipids Have Water-Soluble "Heads" and Water-Insoluble "Tails"

The plasma membrane that surrounds each cell contains several types of **phospholipids.** A phospholipid resembles an oil with one of its three fatty acids replaced by a phosphate group. The phosphate is linked to one of several polar functional groups that typically contain nitrogen (**FIG. 3-26**). A

phospholipid molecule has two dissimilar ends. At one end are the two nonpolar fatty acid "tails," which are insoluble in water. At the other end is the phosphate–nitrogen "head," which is polar and water soluble. These properties of phospholipids are crucial to the structure and function of cell membranes (see Chapter 5).

Steroids Contain Four Fused Carbon Rings

All **steroids** are composed of four rings of carbon atoms. As shown in **FIGURE 3-27**, the rings share one or more sides, with various functional groups protruding from them. One steroid, *cholesterol*, is a vital component of the membranes

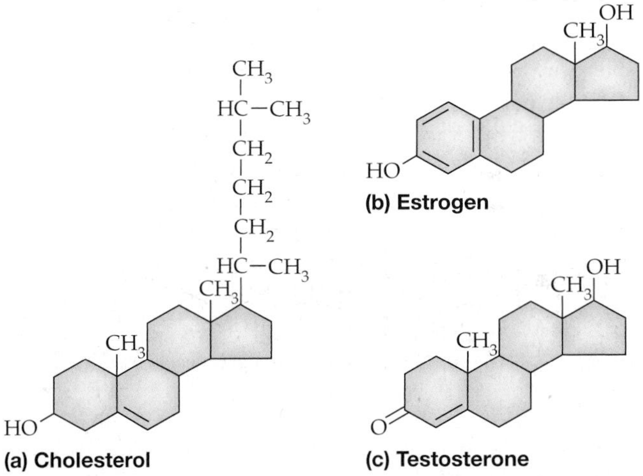

(a) Cholesterol

(b) Estrogen

(c) Testosterone

▲ FIGURE 3-27 **Steroids** All steroids have a similar, nonpolar molecular structure with four fused carbon rings. Differences in steroid function result from differences in functional groups attached to the rings. **(a)** Cholesterol, the molecule from which other steroids are synthesized; **(b)** the female sex hormone estrogen (estradiol); **(c)** the male sex hormone testosterone. Note the similarities in structure between the sex hormones.

THINK CRITICALLY Why are steroid hormones able to diffuse through cell membranes to exert their effects?

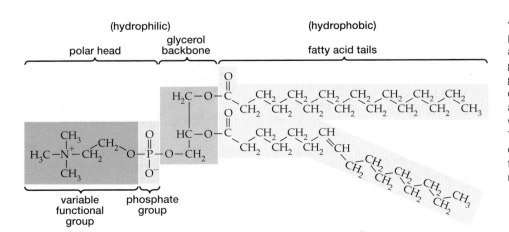

◀ FIGURE 3-26 **Phospholipids** Phospholipids have two hydrophobic fatty acid tails attached to the three-carbon glycerol backbone. The third position on glycerol is occupied by a polar head (left) consisting of a phosphate group to which a second (usually nitrogen-containing) variable functional group is attached. The phosphate group bears a negative charge, and the nitrogen-containing functional group bears a positive charge, making the head hydrophilic.

Cholesterol, *Trans* Fats, and Your Heart

Why are so many foods advertised as "cholesterol free" or "low in cholesterol"? Cholesterol is crucial to life, so what makes it bad?

Cholesterol molecules are insoluble and are transported in the bloodstream as microscopic particles surrounded by hydrophilic phospholipids and proteins. These *lipoprotein* (lipid plus protein) particles differ considerably in their relative amounts of cholesterol and protein. Those with less cholesterol and more protein are described as *high-density lipoprotein* (*HDL*), because proteins are denser than lipids. Those with more cholesterol and less protein are *low-density lipoprotein* (*LDL*).

Blood tests can identify the levels of HDL and LDL present in the body. Why is this important to know? Because elevated HDL is associated with a reduced risk of heart disease and stroke, whereas elevated LDL is a risk factor for cardiovascular disease. LDL deposits cholesterol in artery walls, where it participates in the formation of complex fatty deposits called *plaques* (**FIG. E3-3**). Blood clots may form around the plaques. If a clot breaks loose and blocks an artery supplying blood to the heart muscle or brain, it can cause a heart attack or a stroke. In contrast to the damaging effects of LDL, HDL particles can absorb cholesterol from plaque accumulating in artery walls and transport it to the liver, which uses cholesterol to synthesize bile. The bile is secreted into the small intestine to aid in fat digestion.

Most cholesterol is synthesized by the body. Saturated fats (such as those in dairy and red meat) stimulate the liver to churn out more LDL cholesterol. Foods containing cholesterol, which include egg yolks, sausages, bacon, whole milk, and butter, typically only contribute about 15% to 20% of our blood cholesterol, but diets in which unsaturated fats (found in fish, nuts, and most vegetable oils) replace saturated fats are associated with a decreased risk of heart disease. Lifestyle also contributes: Exercise tends to increase HDL, whereas obesity and smoking increase LDL cholesterol levels.

The worst dietary fat is **trans fat,** made artificially when hydrogen atoms are added to oil in a configuration that causes the kinky fatty acid tails to straighten and the oil to become solid at room temperature. Because they are very stable, *trans* fats extend the shelf life and help retain the flavor of processed foods such as margarine, cookies, crackers, and fried foods. *Trans* fat, also called "partially hydrogenated oil," simultaneously decreases HDL and increases LDL and so places consumers at a higher risk of heart disease. Since 2006, when the danger was clearly recognized and new laws required that the *trans* fat content of foods be specified, most manufacturers and fast food chains have greatly reduced or eliminated *trans* fats from their products. In 2015, based on the scientific evidence, the FDA concluded that *trans* fats are "not generally recognized as safe" and gave food manufacturers three years to eliminate them.

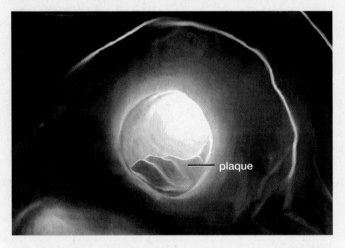

▲ **FIGURE E3-3 Plaque** A plaque deposit (rippled structure) partially blocks this carotid artery.

EVALUATE THIS An obese 55-year-old woman consults her physician about minor chest pains during exercise. Explain the physician's preliminary diagnosis, list the questions she should ask her patient, describe the tests she would perform, and provide the advice she should give.

---

of animal cells. It makes up about 2% of the human brain, where it is an important component of the lipid-rich membranes that insulate nerve cells. Cholesterol is also used by cells to synthesize other steroids, such as the female and male sex hormones *estrogen* and *testosterone*. Too much of the wrong form of cholesterol is linked to cardiovascular disease, as explained in "Health Watch: Cholesterol, *Trans* Fats, and Your Heart."

CHECK YOUR LEARNING

Can you ...

- compare and contrast the structure and synthesis of fats and oils?
- describe the functions of fats, oils, and waxes?
- provide two reasons why cholesterol is important in the body?

Puzzling Proteins

Stanley Prusiner and his associates coined the term "prion" to refer to the misfolded version of a normal protein called PrP^c, found on cell membranes. But how do prions replicate themselves? Researchers have discovered that prions interact with normal helical PrP^c proteins, forcing segments of them to change into the pleated sheet configuration of the infectious form. These new "prion converts" then go on to transform other normal PrP^c proteins in an ever-expanding chain reaction. The chain reaction apparently occurs slowly enough that, as in Charlene Singh's case, it can be a decade or more after infection before disease symptoms occur. Fortunately, both vCJD and BSE have been nearly eradicated worldwide. However, there is evidence that the gene coding for PrP^c can mutate in extremely rare cases, causing cows that have not been exposed to BSE prions to develop the disease. Careful surveillance of cattle continues.

Recognition of the prion as the disease agent in vCJD has focused attention on the role of the normal PrP^c protein. It is found in cell membranes throughout the body and in relatively high levels in the brain, a prime target of infectious prions. Preliminary evidence suggests that PrP^c has diverse functions, which include protecting cells from oxidative stress, contrib-

uting to the growth of neurons, and helping to maintain the new connections between neurons that form when an animal learns.

In 1997, when accepting the Nobel Prize for his prion research, Prusiner predicted that our understanding of this novel disease process could lead to insights into the prevention and treatment of other neurodegenerative disorders. Researchers are now exploring the hypothesis that diseases including Alzheimer's, Parkinson's, and amyotrophic lateral sclerosis (ALS) may arise from misfolded proteins that propagate and accumulate within the nervous system.

CONSIDER THIS A disorder called CWD (chronic wasting disease) of deer and elk, first identified in the late 1960s, has now been reported in at least 20 U.S. states. Like scrapie and BSE, CWD is a fatal brain disorder caused by prions. The disease spreads among animals by contact with saliva, urine, and feces, which may contain prions. Prions have also been found in the muscles of deer with CWD. There is no evidence that CWD can be transmitted to people or domestic livestock. If you were a hunter in an affected region, would you continue to hunt? Would you eat deer or elk meat? Explain why or why not.

CHAPTER REVIEW

 Go to **MasteringBiology** for practice quizzes, activities, eText, videos, current events, and more.

Answers to Think Critically, Evaluate This, Multiple Choice, and Fill-in-the-Blank questions can be found in the Answers section at the back of the book.

Summary of Key Concepts

3.1 Why Is Carbon So Important in Biological Molecules?

Organic molecules have a carbon backbone. They are so diverse because the carbon atom is able to form bonds with up to four other molecules. This allows organic molecules to form complex shapes, including branched chains, helices, pleated sheets, and rings. The presence of functional groups produces further diversity among biological molecules (see Table 3-1).

3.2 How Are Large Biological Molecules Synthesized?

Most large biological molecules are polymers synthesized by linking many smaller monomer subunits using dehydration

synthesis. Hydrolysis reactions break these polymers apart. The most important organic molecules fall into four classes: carbohydrates, lipids, proteins, and nucleotides/nucleic acids (see Table 3-2).

3.3 What Are Carbohydrates?

Carbohydrates include sugars, starches, cellulose, and chitin. Sugars include monosaccharides and disaccharides. They are used for temporary energy storage and the construction of other molecules. Starches and glycogen are polysaccharides that provide longer-term energy storage in plants and animals, respectively. Cellulose forms the cell walls of plants, and chitin strengthens the exoskeletons of many invertebrates and the cell walls of fungi.

3.4 What Are Proteins?

Proteins consist of one or more amino acid chains called polypeptides with up to four levels of structure. Primary structure is the sequence of amino acids; secondary structure consists of helices or pleated sheets. These may fold to produce tertiary structure.

Proteins with two or more linked polypeptides have quaternary structure. Some proteins or parts of proteins are disordered and lack a stable secondary or tertiary structure. The function of a protein is determined by its shape and by how its amino acids interact with their surroundings and with each other. See Table 3-3 for protein functions and examples.

3.5 What Are Nucleotides and Nucleic Acids?

A nucleotide is composed of a phosphate group, a five-carbon sugar (ribose or deoxyribose), and a nitrogen-containing base. Molecules formed from single nucleotides include energy-carrier molecules (ATP) and messenger molecules (cyclic AMP). Nucleic acids are chains of nucleotides. DNA carries the hereditary blueprint, and RNA is copied from DNA and directs the synthesis of proteins.

3.6 What Are Lipids?

Lipids are nonpolar, water-insoluble molecules. Oils, fats, waxes, and phospholipids all contain fatty acids, which are chains of carbon and hydrogen atoms with a carboxylic acid group at the end. Steroids all have four fused rings of carbon atoms with functional groups attached. Lipids are used for energy storage (oils and fats), as waterproofing for the outside of many plants and animals (waxes), as the principal component of cellular membranes (phospholipids and cholesterol), and as hormones (steroids).

Key Terms

adenosine triphosphate
  (ATP)   *44*
amino acid   *40*
base   *44*
biological molecules   *33*
carbohydrate   *36*
cellulose   *38*
chitin   *39*
dehydration synthesis   *34*
denatured   *43*
deoxyribonucleic
  acid (DNA)   *45*
disaccharide   *36*
disulfide bond   *40*
enzyme   *40*
fat   *46*
fatty acid   *46*
functional group   *34*
glucose   *36*
glycogen   *38*
helix   *41*
hydrolysis   *35*
inorganic   *33*
intrinsically disordered
  protein   *43*
lipid   *45*
monomer   *34*

monosaccharide   *36*
nucleic acid   *45*
nucleotide   *44*
oil   *46*
organic   *33*
peptide   *41*
peptide bond   *41*
phospholipid   *47*
pleated sheet   *42*
polymer   *34*
polysaccharide   *37*
primary structure   *41*
protein   *40*
quaternary structure   *42*
ribonucleic acid
  (RNA)   *45*
saturated   *46*
secondary
  structure   *41*
starch   *37*
steroid   *47*
sugar   *36*
tertiary structure   *42*
trans fat   *48*
triglyceride   *46*
unsaturated   *46*
wax   *47*

Thinking Through the Concepts

Multiple Choice

1. Polar molecules
   a. dissolve in lipids.
   b. are hydrophobic.
   c. form covalent bonds.
   d. form ionic bonds.

2. Which match is correct?
   a. monosaccharide–sucrose
   b. polysaccharide–maltose
   c. disaccharide–lactose
   d. disaccharide–glycogen

3. Which of the following statements is False?
   a. In starch breakdown, water is formed.
   b. In chitin, glucoses are linked as in cellulose.
   c. Peptide bonds determine primary peptide structure.
   d. Disulfide bridges are formed by covalent bonds.

4. Which of the following is *not* composed of repeating subunits?
   a. starch
   b. protein
   c. nucleic acid
   d. lipid

5. Which of the following statements is False?
   a. Carbohydrates are the most efficient energy-storage molecules by weight.
   b. Intrinsically disordered protein segments have mostly hydrophilic amino acids.
   c. Nucleotides may act as energy-carrier molecules.
   d. Very acidic or salty solutions may denature proteins.

Fill-in-the-Blank

1. In organic molecules made of chains of subunits, each subunit is called a(n) _____ , and the chains are called _____ . Carbohydrates consisting of long chains of sugars are called _____ . These sugar chains can be broken down by _____ reactions. Three types of carbohydrates consisting of long glucose chains are _____ , _____ , and _____ .

2. Fill in the following with the specific bond(s): Maintain(s) the helical structure of many proteins: _____ ; link(s) polypeptide chains and can cause proteins to bend: _____ and _____ ; join(s) the two strands of the double helix of DNA: _____ ; link(s) amino acids to form the primary structure of proteins: _____ .

3. Proteins are synthesized by a reaction called _____ synthesis, which releases _____ . Subunits of proteins are called _____ . The sequence of protein subunits is called the _____ structure of the protein. Two regular configurations of secondary protein structure are _____ and _____ . When a protein's secondary or higher-order structure is destroyed, the protein is said to be _____ .

4. A nucleotide consists of three parts: _____, _____, and _____. A nucleotide that acts as an energy carrier is _____. The four bases found in deoxyribose nucleotides are _____, _____, _____, and _____. Two important nucleic acids are _____ and _____. The functional group that joins nucleotides in nucleic acids is _____.

5. Fill in the following with the appropriate type of lipid: Unsaturated, liquid at room temperature: _____; bees use to make honeycombs: _____; stores energy in animals: _____; sex hormones are synthesized from these: _____; the LDL form of this contributes to heart disease: _____; a major component of cell membranes that has polar heads: _____.

Review Questions

1. What does the term "organic" mean to a chemist?

2. List the four principal classes of biological molecules and give an example of each.

3. What roles do nucleotides play in living organisms?

4. How are fats and oils similar? How do they differ, and how do their differences explain whether they are solid or liquid at room temperature?

5. Describe and compare dehydration synthesis and hydrolysis. Give an example of a substance formed by each chemical reaction, and describe the specific reaction in each instance.

6. Describe the synthesis of a protein from amino acids. Then describe the primary, secondary, tertiary, and quaternary structures of a protein.

7. Where in nature do we find cellulose? Where do we find chitin? In what way(s) are these two polymers similar? How are they different?

Applying the Concepts

1. Based on their structure, sketch and explain how phospholipids would organize themselves in water.

2. Compare the way fat and carbohydrates interact with water, and explain why this interaction gives fat an extra advantage for weight-efficient energy storage.

3. In an alternate universe where people could digest cellulose molecules, how might this affect our way of life?

4 CELL STRUCTURE AND FUNCTION

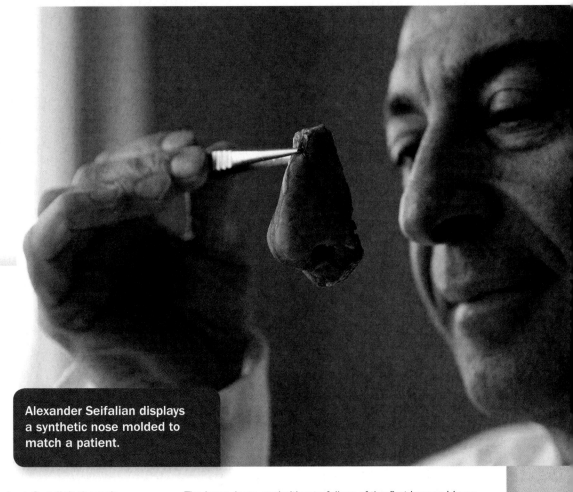

Alexander Seifalian displays a synthetic nose molded to match a patient.

New Parts for Human Bodies

ANDEMARIAM BEYENE, A STUDENT FROM AFRICA, was pursuing a postgraduate degree at the University of Iceland when he developed a tracheal tumor that persisted despite surgery and radiation therapy. No donor windpipe was available, so Beyene agreed to become the first person to receive an artificial body part embedded with his own cells.

An international team overseen by a Swedish thoracic surgeon collaborated to create Beyene's artificial trachea. In England, researchers used a CT scan (a series of X-rays combined by computer to generate a three-dimensional image) to create an exact glass replica of Beyene's trachea. The replica formed a mold for a plastic scaffolding material with microscopic pores that could be infiltrated with human cells. In Sweden, the plastic trachea scaffold was placed in a temperature-controlled "bioreactor" that had been specially designed in the United States to seed cells onto artificial scaffolds. Doctors isolated stem cells (which are capable of forming a variety of adult cell types) from Beyene's bone marrow and placed them in a nutrient broth in the bioreactor. The plastic trachea scaffold was rotated within the bioreactor, allowing the stem cells to attach. Doctors then replaced Beyene's cancerous trachea with the newly-constructed bio-artificial organ. The new trachea bought time for Beyene. He resumed his studies and graduated in 2012 with a Master's degree. But in 2014, the bioartificial trachea loosened and Beyene died.

The hope, hype, and ultimate failure of the first human biosynthetic organ underscores both the immense possibilities and the practical challenges facing bioengineered human body parts. At the Royal Free Hospital in London, Professor Alexander Seifalian and his team have been working to grow noses, ears, and blood vessels in an endeavor that may eventually transform thousands of lives. The body parts with the greatest immediate promise are the simple ones, which don't move or perform complex functions. When a British businessman lost his nose to cancer, the group molded him a new one, slanted slightly to the left, just like the original. The structure was seeded with the patient's stem cells and incubated in a nutrient solution formulated to stimulate the stem cells to develop into cartilage. It was then implanted under the skin of his forearm to acquire blood vessels, nerves, and a coating of skin in preparation for transfer to his face.

Bioengineering simple human organs demonstrates our rapidly-expanding ability to manipulate cells, the fundamental units of life. What structures make up cells? What new bioengineering techniques involving human or animal cells are being developed and tested?

AT A GLANCE

4.1 WHAT IS THE CELL THEORY?

Because cells are so small, no one had ever seen them until the first microscope was invented in the mid-1600s (see "How Do We Know That? The Search for the Cell" on page 54). But seeing cells was only the first step toward understanding their importance. In 1838, the German botanist Matthias Schleiden concluded that cells and substances produced by cells form the basic structure of plants and that plant growth occurs by adding new cells. In 1839, German biologist Theodor Schwann (Schleiden's friend and collaborator) drew similar conclusions about animal cells. The work of Schleiden and Schwann provided a unifying theory of cells as the fundamental units of life. In 1855, the German physician Rudolf Virchow completed the **cell theory**—a fundamental concept of biology—by concluding that all cells come from previously existing cells.

The cell theory consists of three principles:

1. Every organism is made up of one or more cells.
2. The smallest organisms are single cells, and cells are the functional units of multicellular organisms.
3. All cells arise from preexisting cells.

CHECK YOUR LEARNING

Can you ...

• trace the historical development of the cell theory?
• list the three principles of the cell theory?

4.2 WHAT ARE THE BASIC ATTRIBUTES OF CELLS?

All living things, from microscopic bacteria to a giant sequoia tree, are composed of cells. Cells perform an enormous variety of functions, including obtaining energy and nutrients, synthesizing biological molecules, eliminating wastes, interacting with other cells, and reproducing.

Most cells range in size from about 1 to 100 micrometers (μm; millionths of a meter) in diameter (**FIG. 4-1**). Why are most cells so small? The answer lies in the need for cells to exchange nutrients and wastes with their external environment through the plasma membrane. Many nutrients and wastes move into, through, and out of cells by **diffusion,** the process by which molecules dissolved in fluids disperse

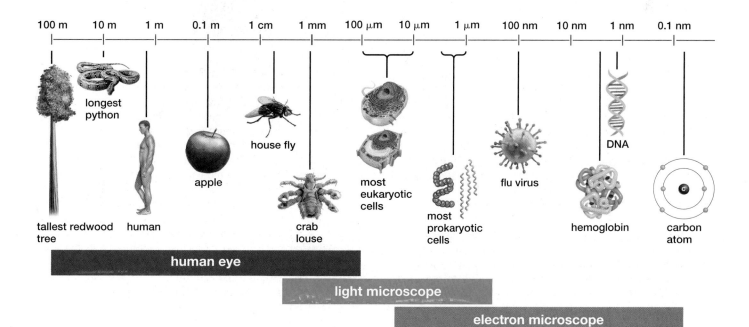

▲ **FIGURE 4-1 Relative sizes** Dimensions encountered in biology range from about 100 meters (the height of the tallest redwood trees) to a few nanometers (the diameter of many large molecules).

Although cells form the basis of life, they're so small that it wasn't until we could actually see them that we realized they existed. In 1665, the English scientist and inventor Robert Hooke aimed his primitive light microscope at an "exceeding thin ... piece of Cork" and saw "a great many little Boxes" (**FIG. E4-1a**). Hooke called the boxes "cells," because he thought they resembled the tiny rooms (called cells) occupied by monks in a monastery. Cork comes from the dry outer bark of the cork oak, and we now know that he was looking at the nonliving cell walls that surround all plant cells. Hooke wrote that in the living oak and other plants, "These cells [are] fill'd with juices."

In the 1670s, Dutch microscopist Anton van Leeuwenhoek constructed his own simple light microscopes and observed a previously unknown living world (**FIG. E4-1b**). Although van Leeuwenhoek's microscopes appear much more primitive than Hooke's, their superior lenses provided clearer images and

higher magnification, down to almost 1 micron (1 μm; see Fig. 4-1). A self-taught amateur scientist, van Leeuwenhoek's descriptions of myriad "animalcules" (mostly single-celled organisms) in rain, pond, and well water were greeted with amazement. Over the years, he described an enormous range of microscopic specimens, including blood cells, sperm cells, and the eggs of aphids and fleas, helping overturn the belief that these insects emerged spontaneously from dust or grain. Observing white matter scraped from his teeth, van Leeuwenhoek saw swarms of cells that we now recognize as bacteria. Disturbed by these animalcules in his mouth, he tried to kill them with vinegar and hot coffee—but with little success.

Since the pioneering efforts of early microscopists, biologists, physicists, and engineers have collaborated to develop a variety of advanced microscopes to view the cell and its components. *Light microscopes* use lenses made of

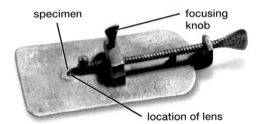

specimen   focusing knob

location of lens

(b) van Leeuwenhoek's light microscope

(c) Electron microscope

(a) Robert Hooke's light microscope and his drawing of cork cells

▲ **FIGURE E4-1** **Microscopes yesterday and today (a)** Robert Hooke saw the walls of cork cells through his elegant light microscope, and drew them with great skill. **(b)** Hooke and van Leeuwenhoek were contemporaries. Hooke admitted that van Leeuwenhoek's microscopes produced better images, but described these extremely simple microscopes as "offensive to my eye." **(c)** This modern machine is both a transmission electron microscope (TEM) and a scanning electron microscope (SEM).

glass or quartz to bend, focus, and transmit light rays that have passed through or bounced off a specimen. The light microscope produces images depending on how the specimen is illuminated and how it has been stained. Fluorescent stains targeted to specific molecules and viewed under specific wavelengths of light are now revolutionizing our view of cells. The resolving power (the smallest structure distinguishable under ideal conditions) of modern light microscopes is about 200 nanometers (nm; see Fig. 4-1). This is sufficient to see most prokaryotic cells, some structures inside eukaryotic cells, and living cells such as a swimming *Paramecium* (**FIG. E4-2a**).

Electron microscopes (**FIG. E4-1c**) use beams of electrons focused by magnetic fields rather than light focused by lenses. *Transmission electron microscopes* pass electrons through a thin specimen and can reveal the details of interior cell structure (**FIG. E4-2b**). Some modern transmission electron microscopes can resolve structures as small as 0.05 nanometer, allowing scientists to see molecules such

as DNA and even individual carbon atoms (seen here forming a six-carbon ring).

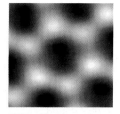

TEM of carbon atoms

Scanning electron microscopes bounce electrons off specimens that are dry and hard (such as shells) or that have been covered with an ultrathin coating of metal such as gold. Scanning electron microscopes can be used to view the three-dimensional surface details of structures that range in size from entire small insects down to cells and their components, with a maximum resolution of about 1.5 nanometers (**FIGS. E4-2c, d**).

THINK CRITICALLY Based on the images in Fig. E4-2, what advantages are there in visualizing microscopic structures using each of these techniques?

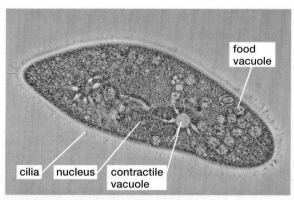

(a) Light micrograph *(Paramecium)*

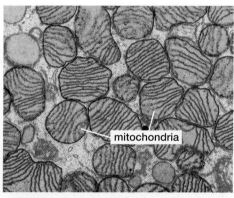

(b) Transmission electron micrograph

(c) Scanning electron micrograph *(Paramecia)*

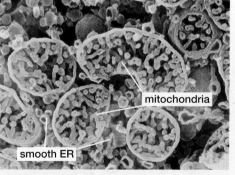

(d) Scanning electron micrograph

▲ **FIGURE E4-2 A comparison of microscope images (a)** A living *Paramecium* (a single-celled freshwater protist) photographed through a light microscope. **(b)** A transmission electron micrograph (TEM) showing mitochondria. **(c)** A scanning electron micrograph (SEM) of two *Paramecia*. **(d)** An SEM showing mitochondria and smooth endoplasmic reticulum. All colors in electron micrographs (SEMs or TEMs) have been added artificially.

from regions where their concentration is higher to regions where their concentration is lower (see Chapter 5). Diffusion is a relatively slow process, so to meet the constant metabolic demands of cells, even their innermost parts must remain close to the external environment. Thus, cells maintain a very small diameter, whether they are round or elongated.

All Cells Share Common Features

All cells arose from a common ancestor that evolved about 3.5 billion years ago. Modern cells include the simple prokaryotic cells of bacteria and archaea and the complex eukaryotic cells of protists, fungi, plants, and animals. All cells, whether simple or complex, share some important features.

The Plasma Membrane Encloses the Cell and Allows Interactions Between the Cell and Its Environment

Each cell is surrounded by an extremely thin, rather fluid membrane called the **plasma membrane** (FIG. 4-2). The plasma membrane, like all membranes in and around cells, contains proteins embedded in a double layer, or *bilayer*, of phospholipids interspersed with cholesterol molecules.

The phospholipid and protein components of cellular membranes play very different roles. The phospholipid bilayer helps isolate the cell from its surroundings, allowing the cell to maintain essential differences in the concentrations of materials inside and out. In contrast, the huge variety of proteins within the bilayer facilitate communication between the cell and its environment. For example, *channel proteins* allow passage of specific molecules or ions into or out of the cell (see Fig. 4-2). *Glycoproteins*, which have short carbohydrate chains extending outside the cell, both facilitate interactions between cells and respond to external signaling molecules that promote chemical reactions within the cell (described in Chapter 5).

All Cells Contain Cytoplasm

The **cytoplasm** consists of all the fluid and structures that lie inside the plasma membrane but outside of the nucleus (see Figs. 4-4 and 4-5). The fluid portion of the cytoplasm in both prokaryotic and eukaryotic cells, called the **cytosol,** contains water, salts, and an assortment of organic molecules, including proteins, lipids, carbohydrates, sugars, amino acids, and nucleotides (described in Chapter 3). Most of the cell's metabolic activities—the biochemical reactions that support life—occur in the cell cytoplasm.

The **cytoskeleton** consists of a variety of protein filaments within the cytoplasm. These provide support, transport structures within the cell, aid in cell division, and allow cells to move and change shape (see Figs. 4-2 and 4-7).

All Cells Use DNA As a Hereditary Blueprint and RNA to Copy the Blueprint and Guide Construction of Cell Parts

The genetic material in all cells is **deoxyribonucleic acid (DNA),** an inherited blueprint consisting of segments called genes. Genes store the instructions for making all the parts of the cell and for producing new cells (see Chapter 12). **Ribonucleic acid (RNA),** which is chemically similar to DNA, copies the genes of DNA and helps construct proteins based on this genetic blueprint. The construction of protein from RNA in all cells occurs on **ribosomes,** cellular workbenches composed of a specialized type of RNA called *ribosomal RNA.*

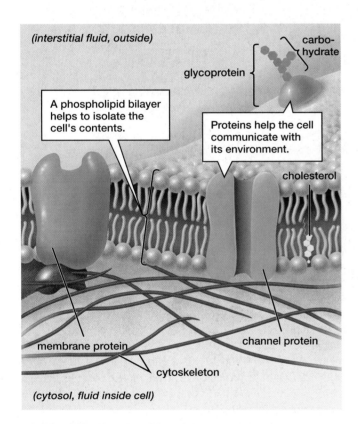

(interstitial fluid, outside)

carbo-hydrate

glycoprotein

A phospholipid bilayer helps to isolate the cell's contents.

Proteins help the cell communicate with its environment.

cholesterol

membrane protein

channel protein

cytoskeleton

(cytosol, fluid inside cell)

▲ **FIGURE 4-2 The plasma membrane** The plasma membrane encloses the cell in a double layer of phospholipids associated with a variety of proteins. The membrane is supported by the cytoskeleton.

There Are Two Basic Types of Cells: Prokaryotic and Eukaryotic

All forms of life are composed of one of two types of cells. **Prokaryotic** cells (Gk. *pro*, before, and *kary*, nucleus) form the bodies of **bacteria** and **archaea,** the simplest forms of life. **Eukaryotic** cells (Gk. *eu*, true) are far more complex and make up the bodies of animals, plants, fungi, and protists. As their names suggest, one striking difference between prokaryotic and eukaryotic cells is that the genetic material of eukaryotic cells is contained within a membrane-enclosed nucleus. **TABLE 4-1** summarizes the principal features of prokaryotic and eukaryotic cells.

TABLE 4-1	Functions and Distribution of Cell Structures			
Structure	Function	Prokaryotes	Eukaryotes: Plants	Eukaryotes: Animals
Cell Surface				
Extracellular matrix	Surrounds cells, providing biochemical and structural support	Absent	Present	Present
Cilia	Move the cell through fluid or move fluid past the cell surface	Absent	Absent (in most)	Present
Flagella	Move the cell through fluid	Present[1]	Absent (in most)	Present
Plasma membrane	Isolates the cell contents from the environment; regulates movement of materials into and out of the cell; allows communication with other cells	Present	Present	Present
Organization of Genetic Material				
Genetic material	Encodes the information needed to construct the cell and to control cellular activity	DNA	DNA	DNA
Chromosomes	Contain and control the use of DNA	Single, circular	Many, linear	Many, linear
Nucleus[2]	Contains chromosomes and nucleoli	Absent	Present	Present
Nuclear envelope	Encloses the nucleus; regulates movement of materials into and out of the nucleus	Absent	Present	Present
Nucleolus	Synthesizes ribosomes	Absent	Present	Present
Cytoplasmic Structures				
Ribosomes	Provide sites for protein synthesis	Present	Present	Present
Mitochondria[2]	Produce energy by aerobic metabolism	Absent	Present	Present
Chloroplasts[2]	Perform photosynthesis	Absent	Present	Absent
Endoplasmic reticulum[2]	Synthesizes membrane components, proteins, and lipids	Absent	Present	Present
Golgi apparatus[2]	Modifies, sorts, and packages proteins and lipids	Absent	Present	Present
Lysosomes[2]	Contain digestive enzymes; digest food and worn-out organelles	Absent	Absent (in most)	Present
Plastids[2]	Store food, pigments	Absent	Present	Absent
Central vacuole[2]	Contains water and wastes; provides turgor pressure to support the cell	Absent	Present	Absent
Other vesicles and vacuoles[2]	Transport secretory products; contain food obtained through phagocytosis	Absent	Present	Present
Cytoskeleton	Gives shape and support to the cell; positions and moves cell parts	Present	Present	Present
Centrioles	Produce the basal bodies of cilia and flagella	Absent	Absent (in most)	Present

[1]Some prokaryotes have structures called flagella, which lack microtubules and move in a fundamentally different way than do eukaryotic flagella.
[2]Indicates organelles, which are surrounded by membranes and found only in eukaryotic cells.

CASE STUDY CONTINUED
New Parts for Human Bodies

Why was Beyene's bioartificial trachea considered a scientific breakthrough? One reason is that the patient's own cells were used to grow the new body part, so his immune system was unlikely to reject the cells. The plasma membranes of all cells bear surface molecules called glycoproteins that are unique to the individual and allow the person's immune system to recognize the cells as "self." Cells from any other person (except an identical twin), however, bear different glycoproteins. The immune system will identify the different cells as foreign and attack them, which can cause rejection of a transplanted organ. To prevent organ rejection, physicians must find donor cells that match the patient's as closely as possible. But even then, the patient must take drugs that suppress the immune system, which increases vulnerability to cancers and infections that the immune system would normally target and destroy. The cells most likely to cause problems for immune-suppressed patients are prokaryotic. What are the features of these simple cells?

CHECK YOUR LEARNING
Can you ...
- describe the structure and features shared by all cells?
- distinguish prokaryotic from eukaryotic cells?

4.3 WHAT ARE THE MAJOR FEATURES OF PROKARYOTIC CELLS?

Prokaryotic cells have a relatively simple internal structure and are generally less than 5 micrometers in diameter (in comparison, eukaryotic cells range from 10 to 100 micrometers in diameter). Prokaryotes also lack the complex internal membrane-enclosed structures that are the most prominent features of eukaryotic cells.

Single prokaryotic cells make up two of life's domains: Archaea and Bacteria. Many Archaea inhabit extreme environments, such as hot springs and cow stomachs, but Archaea are increasingly being discovered in more familiar

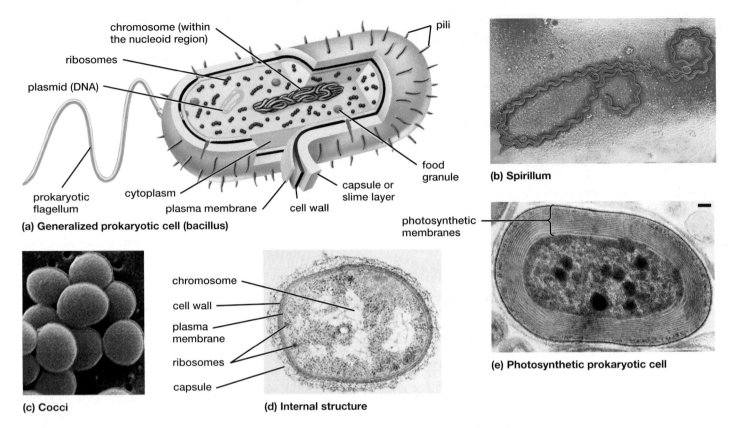

▲ **FIGURE 4-3** **Prokaryotic cells** Prokaryotes come in different shapes, including **(a)** rod-shaped bacilli, **(b)** spiral-shaped spirilla, and **(c)** spherical cocci. Internal structures are revealed in the TEMs in **(d)** and **(e)**. Some photosynthetic bacteria have internal membranes where photosynthesis occurs, as shown in **(e)**.

locales, such as the soil and oceans. None are known to cause disease. In this chapter, we focus on the more familiar bacteria as representative prokaryotic cells (**FIG. 4-3**).

Prokaryotic Cells Have Specialized Surface Features

Nearly all prokaryotic cells are surrounded by a **cell wall,** which is a relatively stiff coating that the cell secretes around itself to provide protection and help maintain its shape. The cell walls of bacteria are composed of *peptidoglycan* (a unique molecule consisting of short peptides that link chains of sugar molecules which have amino functional groups). Bacteria include rod-shaped *bacilli*, spherical *cocci*, and spiral-shaped *spirilla* (see Figs. 4-3a, b, c).

Many bacteria secrete polysaccharide coatings called *capsules* and *slime layers* outside their cell walls (see Fig. 4-3a). In bacteria such as those that cause tooth decay, diarrhea, pneumonia, or urinary tract infections, capsules and slime layers help them adhere to specific host tissues, such as the surface of a tooth or the lining of the small intestine, lungs, or bladder. Capsules and slime layers allow some bacteria to form surface films (such as those that may coat unbrushed teeth or unwashed toilet bowls). They also protect the bacteria and help keep them moist.

Pili (singular, pilus; meaning "hairs") are surface proteins that project from the cell walls of many bacteria (see Fig. 4-3a). There are two types of pili: attachment pili and sex pili. Short,

abundant *attachment pili* may work on their own or with capsules and slime layers to help bacteria adhere to structures. For example, various types of *Streptococcus* bacteria (which can cause strep throat, skin infections, pneumonia, and toxic shock syndrome) use pili to help them infect their victims. Many bacteria form *sex pili*, which are few in number and quite long. A sex pilus from one bacterium binds to a nearby bacterium of the same type and draws them together. The two bacteria form a short bridge that links their cytoplasm and allows them to transfer small rings of DNA called **plasmids.**

Some bacteria and archaea possess **flagella** (singular, flagellum; "whip"), which extend from the cell surface and rotate to propel these cells through a fluid environment (see Fig. 4-3a). Prokaryotic flagella differ from those of eukaryotic cells, which are described later in this chapter.

Prokaryotic Cells Have Specialized Cytoplasmic Structures

The cytoplasm of a typical prokaryotic cell contains several specialized structures. A distinctive region called the **nucleoid** (meaning "like a nucleus"; see Fig. 4-3a) contains a single circular chromosome that consists of a long, coiled strand of DNA that carries essential genetic information. Unlike the nucleus of a eukaryotic cell, the nucleoid is not separated from the cytoplasm by a membrane. Most prokaryotic cells also contain plasmids outside the nucleoid. Plasmids usually carry genes that give the cell special properties; for example, some disease-causing bacteria possess plasmids that encode proteins that inactivate antibiotics, making the bacteria much more difficult to kill. Bacterial cytoplasm also includes ribosomes, where proteins are synthesized, as well as food granules that store energy-rich molecules, such as glycogen.

Although prokaryotic cells lack internal structures surrounded by membranes, some bacteria use internal membranes to organize enzymes. These enzymes facilitate biochemical processes requiring several reactions, and they are situated in a specific sequence along the membrane that corresponds to the sequence in which the reactions must occur. For example, photosynthetic bacteria possess extensive internal membranes where light-capturing proteins and enzymes are embedded, allowing the bacteria to harness the energy of sunlight to synthesize high-energy molecules (Fig. 4-3e).

Prokaryotes also contain an extensive cytoskeleton that includes some proteins that resemble those of the eukaryotic cytoskeleton (see Fig. 4-7) and others that are unique. The prokaryotic and eukaryotic cytoskeletons serve many similar functions; for example, both are essential for cell division and contribute to regulating the shape of the cell.

CHECK YOUR LEARNING

Can you ...

- describe the structure and function of the major features of prokaryotic cells?
- describe the internal features of bacteria, including how some bacteria utilize internal membranes?

4.4 WHAT ARE THE MAJOR FEATURES OF EUKARYOTIC CELLS?

Eukaryotic cells make up the bodies of organisms in the domain Eukarya: animals, plants, protists, and fungi. As you might imagine, these cells are extremely diverse. The cells that form the bodies of unicellular protists can perform all the activities necessary for independent life. Within the body of any multicellular organism, cells are specialized to perform a variety of functions. Here, we focus on plant and animal cells.

Unlike prokaryotic cells, eukaryotic cells (**FIG. 4-4**) have **organelles** ("little organs"), membrane-enclosed structures specialized for a specific function (see Table 4-1). Organelles contribute to the complexity of eukaryotic cells. Figure 4-4

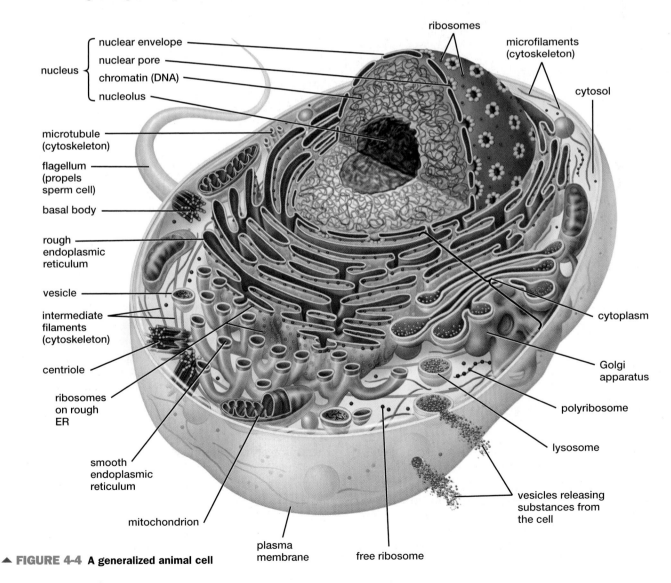

▲ **FIGURE 4-4 A generalized animal cell**

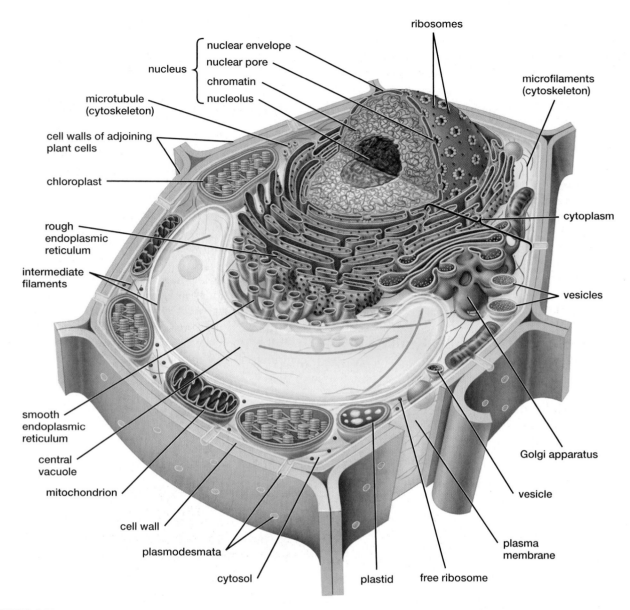

▲ **FIGURE 4-5** **A generalized plant cell**

illustrates a generalized animal cell, and **FIGURE 4-5** illustrates a generalized plant cell, each with some distinctive structures. Animal cells have centrioles, lysosomes, cilia, and flagella, which are not found in the most common plant cells, and plant cells have cell walls, central vacuoles, and plastids (including chloroplasts), which are absent in animal cells.

Extracellular Structures Surround Animal and Plant Cells

The plasma membrane, which is only about two molecules thick and has the consistency of viscous oil, would be torn apart without reinforcing structures. The reinforcing structure for animal cells is the complex **extracellular matrix (ECM),** secreted by the cell. The ECM includes an array of supporting and adhesive proteins embedded in a gel composed of polysaccharides that are linked together by proteins (**FIG. 4-6a**). The

ECM (which differs among cell types) provides both structural and biochemical support, including proteins called growth factors, which promote cell survival and growth. The ECM attaches adjacent cells, transmits molecular signals between cells, and guides cells as they migrate and differentiate during development. It anchors cells within tissues and provides a supporting framework within tissues; for example, a stiff extracellular matrix forms the scaffolding for bone and cartilage (**FIG. 4-6b**).

The extracellular matrix of plant cells is the cell wall, which protects and supports each cell. Plant cell walls, composed mainly of overlapping cellulose fibers, are porous and allow oxygen, carbon dioxide, and water with its dissolved substances to flow through them. Cell walls attach adjacent plant cells to one another and are perforated by plasma membrane-lined openings called *plasmodesmata* that connect the cytoplasm of adjacent cells (see Fig. 4-5).

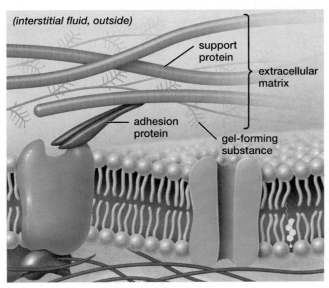

(interstitial fluid, outside)

support protein

extracellular matrix

adhesion protein

gel-forming substance

(a) The extracellular matrix

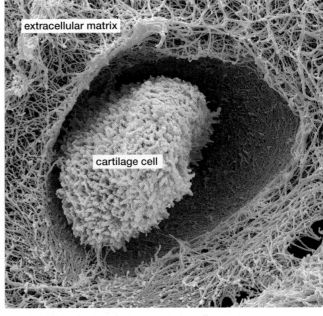

extracellular matrix

cartilage cell

(b) Extracellular matrix of a cartilage cell

▲ **FIGURE 4-6 The extracellular matrix (a)** Extracellular proteins perform a variety of functions.
(b) An SEM of a cartilage cell surrounded by its extracellular matrix.

CASE STUDY ⟍ **CONTINUED**

New Parts for Human Bodies

Besides tracheas and noses, researchers are working on growing bioartificial muscles as well. In the past, major muscle injuries could mean amputation and a prosthetic limb, because muscles have limited ability to regenerate, and scar tissue forms and interferes with their function. But Stephen Badylak and colleagues at the McGowen Institute for Regenerative Medicine are investigating the use of the ECM to help muscles heal and even regenerate.

After 28-year-old Marine Ron Strang's quadriceps muscle was almost ripped from his leg by a roadside bomb in Afghanistan, he volunteered for a new bioartificial muscle treatment developed by Badylak. Badylak used ECM from pig bladders with the cells removed (which prevents tissue rejection) to recreate Strang's muscle. Badylak's team then cut away scar tissue from the Marine's thigh muscle and placed the pig matrix in the resulting cavity. There, its unique combination of natural scaffolding proteins and growth factors recruited muscle stem cells and worked a major transformation. After 6 months, the pig matrix was broken down and replaced by healthy human tissue, and Strang went from hobbling to hiking and riding a bike.

Strang's treatment worked in part because the ECM helps support tissues and facilitates communication between cells. Which structures provide support and facilitate communication within a cell?

The Cytoskeleton Provides Shape, Support, and Movement

The cytoskeleton is a dynamic network of protein fibers within the cytoplasm (**FIG. 4-7**). Cytoskeletal proteins come in three major types: thin **microfilaments** (composed of actin protein), medium-sized **intermediate filaments** (composed of various proteins), and thick **microtubules** (composed of tubulin protein). Cytoskeletal proteins provide the cell with both internal support and the ability to change shape and divide, directed by signals from the ECM. The cytoskeleton is important in regulating the following properties of cells:

- *Cell Shape* Cytoskeletal proteins can alter the shapes of cells using energy released from ATP, either by changing their length (by adding or removing subunits) or by sliding past one another. In animal cells, a scaffolding of intermediate filaments supports the cell, helps determine its shape, and links cells to one another and to the ECM. An array of microfilaments concentrated just inside the plasma membrane provides additional support and also connects with the surrounding ECM.

- *Cell Movement* Cell movement can occur in animal cells as microtubules and microfilaments extend by adding subunits at one end and releasing subunits at the other end. Microtubules and intermediate filaments may be associated with *motor proteins,* which are specialized to release energy stored in ATP and use it to generate molecular movement. Another form of movement is generated as motor proteins cause actin microfilaments

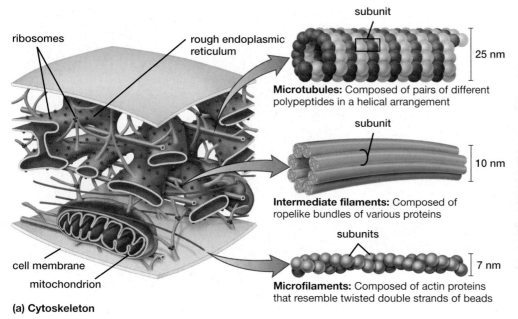

Microtubules: Composed of pairs of different polypeptides in a helical arrangement

subunit

25 nm

Intermediate filaments: Composed of ropelike bundles of various proteins

subunit

10 nm

Microfilaments: Composed of actin proteins that resemble twisted double strands of beads

subunits

7 nm

ribosomes

rough endoplasmic reticulum

cell membrane

mitochondrion

(a) Cytoskeleton

microfilaments (red)

DNA in nucleus (blue)

microtubules (green)

(b) Light micrograph showing the cytoskeleton

▲ **FIGURE 4-7** **The eukaryotic cytoskeleton (a)** Three types of protein strands form the cytoskeleton. **(b)** In this light micrograph, cells treated with fluorescent stains reveal microtubules, microfilaments, and nuclei.

to slide past one another; a well-known example occurs during the contraction of muscle cells. Muscle cells must contract to increase in size, as scientists discovered when they attempted to grow muscle protein in the laboratory for possible human consumption. Learn more in "Earth Watch: Would You Like Fries with Your Cultured Cow Cells?" on page 65.

- *Organelle Movement* Motor proteins use microfilaments and microtubules as "railroad tracks" to transport organelles within the cell.
- *Cell Division* Microtubules guide chromosome movements, and microfilaments in animal cells pinch the dividing cell into two daughter cells. (Cell division is covered in Chapter 9.)

Cilia and Flagella May Move Cells Through Fluid or Move Fluid Past Cells

Both **cilia** (singular, cilium; "eyelash") and eukaryotic **flagella** are beating hair-like structures covered by plasma membrane that extend outward from some cell surfaces. They are supported and moved by microtubules of the cytoskeleton. Each cilium or flagellum contains a ring of nine fused pairs of microtubules surrounding an unfused pair (**FIG. 4-8**). Cilia and flagella beat almost continuously, powered by motor proteins that extend like tiny arms and attach neighboring pairs of microtubules (see Fig. 4-8a). These sidearms use ATP energy to slide the microtubules past one another, causing the cilium or flagellum to bend.

In general, cilia are shorter and more numerous than flagella. Cilia beat in unison to produce a force on the surrounding fluid that is similar to that created by oars on a rowboat. A flagellum, in contrast, rotates in a corkscrew motion that propels a cell through fluid, acting somewhat like the propeller on a motorboat. Cells with flagella usually have only one or two of them.

HAVE YOU EVER WONDERED...

Over the years, scientists have wondered how many cells are in the human body. They don't yet agree, but 10 trillion seems a reasonable estimate. There is a consensus, however, that there are at least 10 times as many prokaryotic cells associated with the body, residing in a community called the microbiome. We each host a unique community consisting of about 3 pounds (1.4 kilograms) of prokaryotic life, which includes roughly 100 different types of bacteria. These cells colonize the nose, skin, vagina, and the digestive tract from mouth to anus.

How Many Cells Form the Human Body?

Because the digestive tract is a tube open to the outside at both ends, our microbiome occupies a unique niche that is simultaneously integral to—yet outside of—our bodies. With recent advances allowing identification of microorganisms by their unique DNA sequences, scientists are increasingly studying our relationships with our microbial residents. Our gut microbiome helps digest food and synthesize vitamins, and it allows the immune system to develop properly. Even though our bacterial populations changes in response to food intake and states of disease and health, one thing is clear: We would not be ourselves without them.

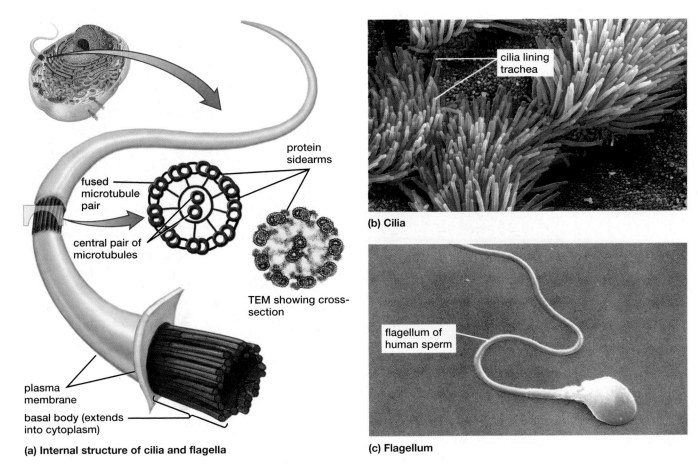

protein sidearms

fused microtubule pair

central pair of microtubules

TEM showing cross-section

plasma membrane

basal body (extends into cytoplasm)

(a) Internal structure of cilia and flagella

cilia lining trachea

(b) Cilia

flagellum of human sperm

(c) Flagellum

▲ **FIGURE 4-8 Cilia and flagella (a)** These structures are filled with microtubules produced by the basal body. **(b)** Cilia, shown in this SEM, line the trachea and sweep out debris. **(c)** A human sperm cell, shown in this SEM, uses its flagellum to swim to the egg.

THINK CRITICALLY What problems would arise if the trachea were lined with flagella instead of cilia?

Protists use cilia or flagella to swim through water; the *Paramecium* in Figures E4-2a, c (see "How Do We Know That? The Search for the Cell" on page 54) uses cilia. In animals, cilia usually move fluids past a surface. Ciliated cells line such diverse structures as the gills of oysters (where they circulate water rich in food and oxygen), the female reproductive tract of vertebrates (where cilia transport the egg cell to the uterus), and the respiratory tracts of most land vertebrates (where cilia convey mucus that carries debris and microorganisms out of the air passages; see Fig. 4-8b). Flagella propel the sperm cells of nearly all animals (see Fig. 4-8c).

Each cilium or flagellum arises from a **basal body** just beneath the plasma membrane. Basal bodies are produced by **centrioles,** and, like centrioles, they differ from the outer portion of flagella and microtubules in having fused triplets and no central pair of microtubules (see Fig. 4-8a). A single pair of centrioles is found in animal cells (see Fig. 4-4), and these play a role in organizing cytoskeletal proteins during cell division (described in Chapter 9).

The Nucleus, Containing DNA, Is the Control Center of the Eukaryotic Cell

A cell's DNA stores all the information needed to construct the cell and direct the countless chemical reactions necessary for life and reproduction. A cell uses only a portion of the instructions in DNA at any given time, depending on the cell's stage of development, its environment, and its function in a multicellular body. In eukaryotic cells, DNA is housed within the nucleus. The **nucleus** is a large organelle with three major parts: the nuclear envelope, chromatin, and the nucleolus (**FIG. 4-9**).

The Nuclear Envelope Allows Selective Exchange of Materials

The nucleus is isolated from the rest of the cell by a double membrane, the **nuclear envelope,** which is perforated by protein-lined nuclear pores. Water, ions, and small molecules can pass freely through the pores, but the passage of large molecules—particularly proteins, parts of ribosomes, and RNA—is regulated by gatekeeper proteins called the **nuclear pore complex** (see Fig. 4-9) that line each nuclear pore. Ribosomes stud the outer

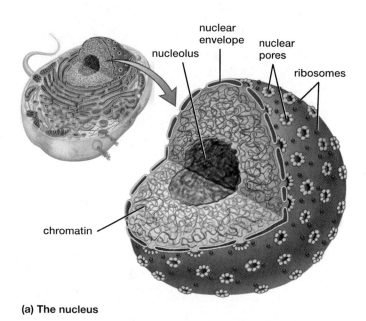

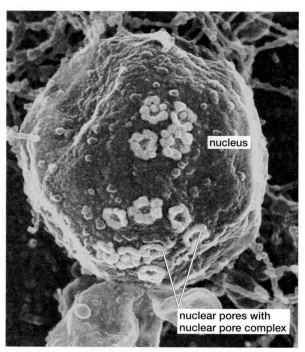

(a) The nucleus

(b) Nucleus of a yeast cell

▲ **FIGURE 4-9** **The nucleus (a)** The nucleus is bounded by a double outer membrane perforated by pores. **(b)** SEM of the nucleus of a yeast cell.

nuclear membrane, which is continuous with membranes of the rough endoplasmic reticulum, described later.

Chromatin Consists of Strands of DNA Associated with Proteins

Early observers of the nucleus noted that it was darkly colored by the stains used in light microscopy and named the nuclear material **chromatin** (meaning "colored substance"). Biologists have since learned that chromatin consists of **chromosomes** (literally, "colored bodies") made of DNA molecules and their associated proteins. When a cell is not dividing, the chromosomes are extended into extremely long strands that are so thin that they cannot be distinguished from one another with a light microscope. During cell division, the individual chromosomes become condensed and are easily visible with a light microscope (**FIG. 4-10**).

The genes of DNA, consisting of specific sequences of nucleotides, provide a molecular blueprint for the synthesis of proteins and ribosomes. Some proteins form structural components of the cell, others regulate the movement of materials through cell membranes, and still others are enzymes that promote chemical reactions within the cell.

Proteins are synthesized in the cytoplasm, but DNA is confined to the nucleus. This means that copies of the genetic code for proteins must be ferried from the nucleus into the cytoplasm. To accomplish this, the genetic information is copied from DNA in the nucleus into molecules of *messenger RNA* (mRNA). The mRNA then moves through the nuclear pores into the cytosol. In the cytosol, the sequence of nucleotides in mRNA is used to direct protein synthesis, a process that

occurs on ribosomes (see Fig. 4-11). (Protein synthesis is described in Chapter 13.)

The Nucleolus Is the Site of Ribosome Assembly

Eukaryotic nuclei contain at least one **nucleolus** (plural, **nucleoli**; meaning "little nuclei") (see Fig. 4-9). The nucleolus is the site of ribosome synthesis. It consists of *ribosomal RNA* (rRNA), parts of chromosomes that carry genes coding for rRNA, proteins, and ribosomes in various stages of synthesis.

Ribosomes are small particles composed of ribosomal RNA combined with proteins. A ribosome serves as a kind

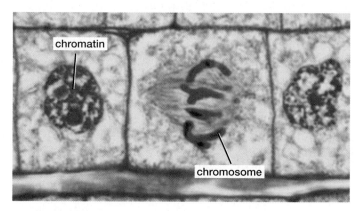

▲ **FIGURE 4-10** **Chromosomes** Chromosomes, seen in a light micrograph of a dividing cell (center) in an onion root tip. Chromatin is visible in adjacent cells.

THINK CRITICALLY Why do the chromosomes in chromatin condense in dividing cells?

Earth WATCH

Would You Like Fries with Your Cultured Cow Cells?

What do you get when you combine 20,000 paper-thin yellowish-pink strips of cells, some lab-grown fat cells, beet coloring, egg powder, bread crumbs, and a dash of salt? These unlikely ingredients make up the world's first lab-grown hamburger (FIG. E4-3). To create it, cow muscle stem cells were allowed to multiply in a nutrient broth. The cells were then seeded into strips of gel and stimulated repeatedly by pulses of electricity. This caused their actin-based filaments to contract and the cells to "bulk up," much as human muscle cells do when exercised. The resulting artificial burger made up of 20,000 cells cost roughly $425,000 to produce, and its flavor was found to be somewhat lacking by fast-food aficionados. So what was the point?

Demand for meat is growing, fuelled partly by an expanding population, but also by increasing incomes and appetite for meat (FIG. E4-4). This is particularly true in China, whose meat consumption between 1971 and 2011 increased at 10 times the rate of its population growth (from 841 million to 1.3 billion people. The Food and Agriculture Organization of the UN estimates that world meat production in 2050 will be 500 million tons (compared to about 350 million tons in in 2016).

To accommodate our increasing demand for meat, we are stripping Earth of its natural ecosystems and altering its climate. Grazing and growing food for livestock already require about 30% of Earth's total land (compared to about 6% used for growing crops directly for human consumption), and meat production accounts for roughly 18% of human-caused greenhouse gas emissions. Cattle have by far the greatest environmental impact among meat-producing livestock; raising beef cattle requires about three times as much land per pound of protein as does raising chicken or pork. Increasing beef production occurs primarily at the expense of rain forest, which is cleared to provide low-quality land for cattle grazing.

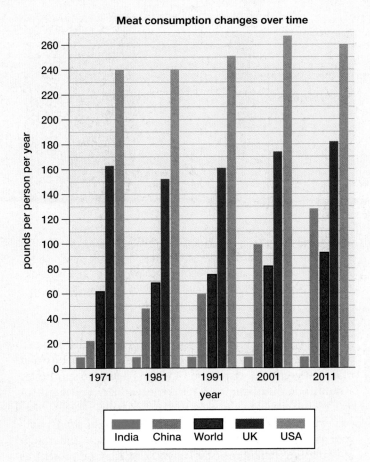

Meat consumption changes over time

India China World UK USA

▲ **FIGURE E4-4 Changes in meat consumption in selected countries**
Source: FAOSTAT (Food and Agricultural Organization of the United Nations), Food Supply

▲ **FIGURE E4-3 Will hamburger of the future be grown in a lab?**

As livestock compete for Earth's limited resources, beef is likely to become an expensive luxury in the relatively near future. Clearly, our present course is unsustainable. Scientists growing "test-tube beef" argue that if their techniques could be refined and scaled up, meat would require almost no killing of animals and use 99% less land. It would also greatly reduce greenhouse gas emissions and energy and water use.

THINK CRITICALLY Using Fig. E4-4, plot the changes in each country over the 40-year period shown and use a ruler to create a trend line to predict the meat consumption per person in each country in the years 2020, 2030, 2040, and 2050 if the current trends continue. Would the ranking of the countries change over this period? Now look up the current population of each of these countries and determine which is the largest total meat consumer. Was this true in 1980?

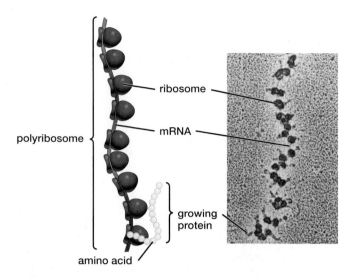

polyribosome

ribosome

mRNA

growing protein

amino acid

▲ **FIGURE 4-11** **A polyribosome** Ribosomes strung along a messenger RNA molecule form a polyribosome. In the TEM (right), individual ribosomes are synthesizing multiple copies of a protein, visible as strands projecting from some of the ribosomes.

of workbench for the synthesis of proteins within the cell cytoplasm. Just as a workbench can be used to construct many different objects, a ribosome can be used to synthesize a multitude of different proteins (depending on the mRNA to which it attaches). In electron micrographs of cells, ribosomes appear as dark granules; they may appear singly, may stud the membranes of the nuclear envelope and rough endoplasmic reticulum (see Fig. 4-4), or may be present as *polyribosomes* (GK. *poly,* many) strung along strands of mRNA within the cytoplasm (**FIG. 4-11**).

Eukaryotic Cytoplasm Contains Membranes That Compartmentalize the Cell

All eukaryotic cells contain internal membranes that create loosely connected compartments within the cytoplasm. These membranes, collectively called the **endomembrane system,** segregate molecules from the surrounding cytosol and ensure that biochemical processes occur in an orderly fashion. The endomembrane system encloses regions within which an enormous variety of molecules are synthesized, broken down, and transported for use inside the cell or export outside the cell. This system of intracellular membranes includes the nuclear envelope (described earlier), vesicles, the endoplasmic reticulum, the Golgi apparatus, and lysosomes.

Vesicles Bud from the Endomembrane System and the Plasma Membrane

Vesicles are temporary sacs that bud from parts of the endomembrane system and from the plasma membrane to ferry biological molecules throughout the cell. The fluid property of membranes permits vesicles to fuse with and release their contents into different endomembrane compartments for processing. Vesicles may also fuse with the plasma membrane, exporting

their contents outside the cell, a process called *exocytosis* (Gk. *exo,* outside). Conversely, the plasma membrane may extend and surround material just outside the cell and then fuse and pinch off to form a vesicle inside the cell, a process called *endocytosis* (Gk. *endo,* inside). As they move about the cells, vesicles not only transport their cargo but also transport their membranes, which become integrated into the membranes that they fuse with.

The vesicles are transported within the cell by motor proteins running along tracks of microtubules. How do the vesicles know where to go? Proteins embedded in vesicle membranes contain specific sequences of amino acids that serve as "mailing labels," providing the address for delivery of the vesicle and its payload. Membrane proteins and proteins exported from the cell are synthesized in the rough endoplasmic reticulum, described in the next section.

The Endoplasmic Reticulum Forms Membrane-Enclosed Channels Within the Cytoplasm

The **endoplasmic reticulum (ER)** (endoplasmic, "inside the cytoplasm," and reticulum, "network") is a labyrinth of narrow channels that form interconnected sacs and tubules throughout the cytosol. The ER typically makes up at least 50% of the total cellular membrane (**FIG. 4-12**). This organelle plays a major role in synthesizing, modifying, and transporting biological molecules throughout the cell. Some of these molecules are incorporated into the ER membranes; others are processed within the ER channels and tubules. The ER has both rough and smooth membranes, which are continuous with one another.

Rough Endoplasmic Reticulum Rough ER emerges from the ribosome-covered outer nuclear membrane (see Fig. 4-4). Ribosomes studding the outer surface make it appear rough under the electron microscope. These ribosomes are the most important sites of protein synthesis in the cell. As they are synthesized, some proteins on ER ribosomes are inserted into the ER membrane. Some remain there, whereas others become part of vesicle membranes budded from the ER. Proteins destined to be secreted from the cell or used in lysosomes are inserted into the interior of the ER, where they are chemically modified and folded into their proper three-dimensional structures (see Chapter 3). Eventually, the proteins accumulate in pockets of ER membrane that pinch off as vesicles and travel to the Golgi apparatus. Proteins produced by the rough ER for export differ with cell type; they include digestive system enzymes, infection-fighting antibodies, and proteins that form the extracellular matrix. Proteins that remain in the cell include the digestive enzymes within lysosomes (described later) and plasma membrane proteins.

Enzymes produced for the synthesis of membrane phospholipids are located on the outer surfaces of ER membranes. Phospholipids become incorporated into the ER membrane as they are formed, along with membrane proteins synthesized in the rough ER. Thus, the ER produces new membrane that, through vesicle fusion, becomes distributed throughout the endomembrane system.

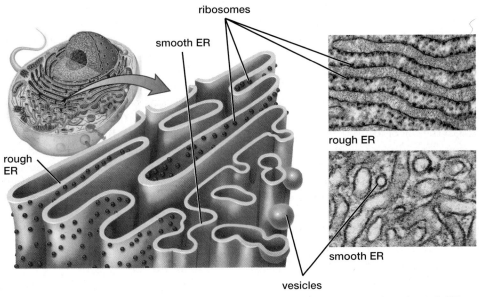

(a) Endoplasmic reticulum may be rough or smooth

(b) Smooth and rough ER

◀ **FIGURE 4-12 Endoplasmic reticulum (a)** Ribosomes (black dots) stud the outside of the rough ER membrane. Rough ER is continuous with the outer nuclear envelope. Smooth ER is less flattened and more cylindrical than rough ER and may be continuous with rough ER. **(b)** TEMs of rough and smooth ER with vesicles.

Smooth Endoplasmic Reticulum Smooth ER, which lacks ribosomes, is also involved in the synthesis of cell membrane phospholipids. It is scarce in most cell types, but abundant and specialized in others. For example, smooth ER packs the cells of vertebrate reproductive organs that synthesize steroid sex hormones. Membranes of smooth ER of liver cells have a variety of enzymes embedded within them. Some participate in converting stored glycogen into glucose to provide energy. Others promote the synthesis of the lipid portion of lipoproteins. Finally, smooth ER enzymes break down metabolic wastes such as ammonia, drugs such as alcohol, and poisons such as certain pesticides. In muscle cells, smooth ER is specialized to store calcium ions, which play a central role in muscle contraction.

The Golgi Apparatus Modifies, Sorts, and Packages Important Molecules

Named for the Italian physician and cell biologist Camillo Golgi, who discovered it in 1898, the **Golgi apparatus** (or simply **Golgi**) is a specialized set of membranes resembling a stack of flattened and interconnected sacs (**FIG. 4-13**). The compartments of the Golgi act like the finishing rooms of a factory, where final touches are added to products to be packaged and exported. Vesicles from the rough ER fuse with the receiving side of the Golgi apparatus, adding their membranes to the Golgi and emptying their contents into the Golgi sacs. Within the Golgi compartments, some of the proteins synthesized in the rough ER are modified further; many are tagged with molecules that specify their destinations in the cell. Finally, vesicles bud off from the "shipping" face of the Golgi, carrying away finished products for use in the cell or export out of the cell.

The Golgi apparatus performs the following functions:

- The Golgi modifies some molecules; an important role of the Golgi is to add carbohydrates to proteins to make glycoproteins. Some of these carbohydrates

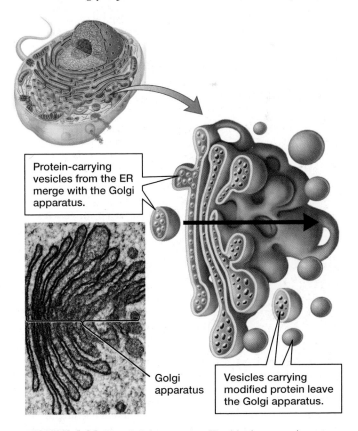

Protein-carrying vesicles from the ER merge with the Golgi apparatus.

Golgi apparatus

Vesicles carrying modified protein leave the Golgi apparatus.

▲ **FIGURE 4-13 The Golgi apparatus** The black arrow shows the direction of movement of materials through the Golgi as they are modified and sorted. Vesicles bud from the face of the Golgi opposite the ER.

act as "mailing labels" that specify the proteins' destination.

- The Golgi separates various proteins received from the ER according to their destinations. For example, the Golgi apparatus separates the digestive enzymes that are bound for lysosomes from the protein hormones that the cell will secrete.
- The Golgi packages the finished molecules into vesicles that are then transported to other parts of the cell or to the plasma membrane for export.

The Endomembrane System Synthesizes, Modifies, and Transports Proteins to Be Secreted

To understand how some of the components of the endomembrane system work together, let's look at the manufacture and export of antibodies (**FIG. 4-14**). Antibodies, produced by white blood cells, are glycoproteins that bind to foreign invaders (such as disease-causing bacteria) and help destroy them. Antibody proteins are synthesized on ribosomes of the rough ER and released into the ER channels ❶, where they are packaged into vesicles formed from ER membrane. These vesicles travel to the Golgi ❷, where their membranes fuse with the Golgi membranes and release the antibodies inside. Within the Golgi, carbohydrates are attached to the antibodies (transforming them into glycoproteins) ❸, which are then repackaged into vesicles formed from Golgi membrane ❹. The vesicle containing the completed antibodies travels to the plasma membrane and fuses with it, releasing the antibodies outside the cell by exocytosis ❺. From there, they will make their way into the bloodstream to help defend the body against infection.

Lysosomes Serve as the Cell's Digestive System

Lysosomes are membrane-bound sacs that digest food particles ranging from individual proteins to microorganisms such as bacteria (**FIG. 4-15**). Lysosomes contain dozens of different enzymes. These enzymes use hydrolysis to break down almost all large biological molecules including carbohydrates, lipids, proteins, and nucleic acids. The enzymes of lysosomes require an acidic environment (pH 5) to function effectively, so they are almost nonfunctional at the cytosolic pH of about 7.2 that exists in the ER compartment where they are manufactured ❶. Lysosomal enzymes are transported to the Golgi in vesicles that bud from the ER ❷. In the Golgi, a carbohydrate "mailing label" is added to the enzymes ❸; this tag directs them into specific Golgi vesicles that will travel to lysosomes ❹. Lysosomal membranes expend energy to pump hydrogen ions inside, creating an acidic environment (about pH 5) that allows the enzymes to perform optimally. The lysosomal membrane is chemically modified to resist the action of the enzymes it encloses.

Many cells of animals and protists "eat" by endocytosis—that is, by engulfing particles from just outside the cell ❺. The plasma membrane with its enclosed food then pinches off inside the cytosol and forms a large vesicle called a **food vacuole**. Lysosomes merge with these food vacuoles ❻, and the lysosomal enzymes digest the food into small molecules such as monosaccharides, fatty acids, and amino acids. Lysosomes also digest worn-out or defective organelles within the cell, breaking them down into their component molecules. All of these small molecules are released into the cytosol through the lysosomal membrane, where they are used in the cell's metabolic processes.

Vacuoles Serve Many Functions, Including Water Regulation, Storage, and Support

Some types of vacuoles, such as food vacuoles, are temporary structures. Other vacuoles, however, persist for the

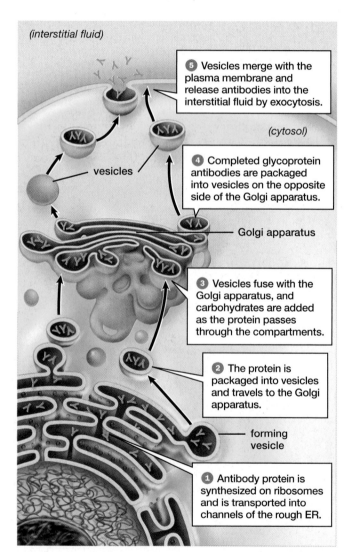

(interstitial fluid)

❺ Vesicles merge with the plasma membrane and release antibodies into the interstitial fluid by exocytosis.

(cytosol)

vesicles

❹ Completed glycoprotein antibodies are packaged into vesicles on the opposite side of the Golgi apparatus.

Golgi apparatus

❸ Vesicles fuse with the Golgi apparatus, and carbohydrates are added as the protein passes through the compartments.

❷ The protein is packaged into vesicles and travels to the Golgi apparatus.

forming vesicle

❶ Antibody protein is synthesized on ribosomes and is transported into channels of the rough ER.

▲ **FIGURE 4-14 A protein is manufactured and exported through the endomembrane system** The formation of an antibody is an example of the process of protein manufacture and export.

THINK CRITICALLY Why is it advantageous for all cellular membranes to have a fundamentally similar composition?

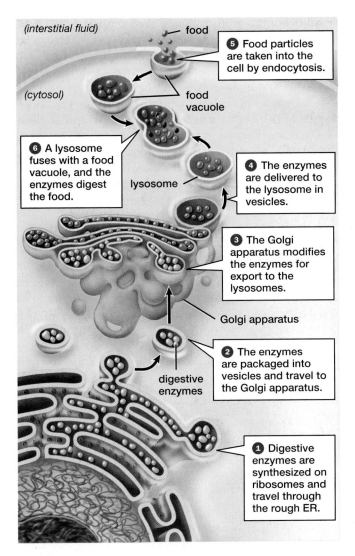

▲ **FIGURE 4-15 Lysosomes and food vacuoles are formed by the endomembrane system**

THINK CRITICALLY Why is it important for lysosomal enzymes to be inactive at pH 7.2?

lifetime of a cell. In the following sections, we describe the permanent vacuoles found in some freshwater protists and in plant cells.

Freshwater Protists Have Contractile Vacuoles

Freshwater protists such as *Paramecium* possess **contractile vacuoles** composed of collecting ducts, a central reservoir, and a tube leading to a pore in the plasma membrane (**FIG. 4-16**). Fresh water constantly leaks into the cell through the plasma membrane and then into contractile vacuoles. This influx of water would soon burst the fragile organism if it did not use cellular energy to draw water from its cytosol into collecting ducts. The water then drains into the central reservoir. When the reservoir is full, the contractile vacuole contracts, squirting the water out through a pore in the plasma membrane.

Plant Cells Have Central Vacuoles

A large **central vacuole** occupies three-quarters or more of the volume of most mature plant cells (see Fig. 4-5) and serves several functions. Its membrane helps to regulate the ion content of the cytosol and secretes wastes and toxic substances into the water that fills the central vacuole. Some plants store substances in central vacuoles that deter animals from munching on their otherwise tasty leaves. Vacuoles may also store sugars and amino acids not immediately needed by the cell. Blue or purple pigments stored in central vacuoles are responsible for the colors of many flowers.

Central vacuoles also provide support for plant cells. Dissolved substances cause water to move by osmosis into the vacuole. The resulting water pressure, called *turgor pressure,* within the vacuole pushes the fluid portion of the cytoplasm up against the cell wall with considerable force. Cell walls are usually somewhat flexible, so both the overall shape and the rigidity of the cell depend on turgor pressure within the cell. Turgor pressure thus provides support for the non-woody parts of plants (see Fig. 5-7).

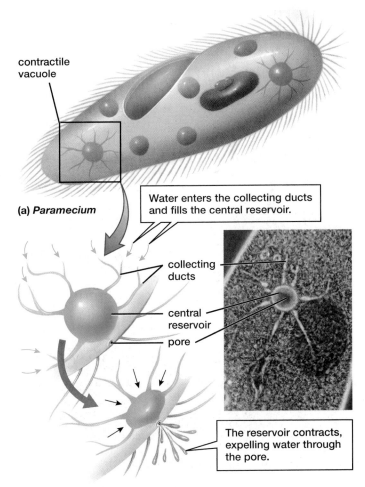

(a) *Paramecium*

(b) Contractile vacuole

▲ **FIGURE 4-16 A contractile vacuole (a)** *Paramecium* lives in freshwater ponds and lakes. **(b)** (Left) The vacuole collects and expels water. (Right) The contractile vacuole seen under a light microscope using fluorescent dyes.

Mitochondria Extract Energy from Food Molecules and Chloroplasts Capture Solar Energy

Both mitochondria and chloroplasts are complex organelles with a unique origin. Nearly all biologists accept the **endosymbiont hypothesis** (see Chapter 18) that both mitochondria and chloroplasts evolved from prokaryotic bacteria. Roughly 1.7 billion years ago, these prokaryotes took up residence within other prokaryotic cells, a process called endosymbiosis (Gk. *symbiosis,* living together). Both mitochondria and chloroplasts are surrounded by a double membrane; the outer membrane may have come from the original host cell and the inner membrane from the guest cell. Mitochondria and chloroplasts resemble each other, and prokaryotic cells, in several ways. They are both the size of a typical prokaryotic cell (1 to 5 micrometers in diameter). Both are surrounded by double membranes. Both have assemblies of enzymes that synthesize ATP, as would have been needed by an independent cell. Finally, both possess their own DNA and ribosomes that more closely resemble prokaryotic than eukaryotic DNA and ribosomes.

Mitochondria Use Energy Stored in Food Molecules to Produce ATP

All eukaryotic cells contain **mitochondria** (singular, mitochondrion), organelles that are sometimes called the "powerhouses" of the cell because they extract energy from food molecules and store it in the high-energy bonds of ATP.

Mitochondria possess a pair of membranes (**FIG. 4-17**). The outer membrane is smooth, whereas the inner membrane forms deep folds called cristae (singular, crista; meaning "crest"). The mitochondrial membranes enclose two fluid-filled spaces: the intermembrane compartment lies between the two membranes, and the matrix fills the space within the inner membrane. Some of the reactions that break down high-energy molecules occur in the fluid of the matrix; the rest are conducted by a series of enzymes attached to the membranes of the cristae. (The role of mitochondria in energy production is described in Chapter 8.)

Chloroplasts Are the Sites of Photosynthesis

Photosynthesis, which captures sunlight and provides the energy to power life, occurs in the chloroplasts found in the cells of plants and some protists. **Chloroplasts** are a type of plastid (described below), surrounded by a double membrane (**FIG. 4-18**). The inner membrane of the chloroplast encloses a fluid called the *stroma.* Within the stroma are interconnected stacks of hollow, membranous sacs. An individual sac is called a *thylakoid,* and a stack of thylakoids is a *granum* (plural, grana).

The thylakoid membranes contain the pigment molecule **chlorophyll** (which gives plants their green color). During photosynthesis, chlorophyll captures the energy of sunlight and transfers it to other molecules in the thylakoid membranes. These molecules transfer the energy to ATP and other energy carriers. The energy

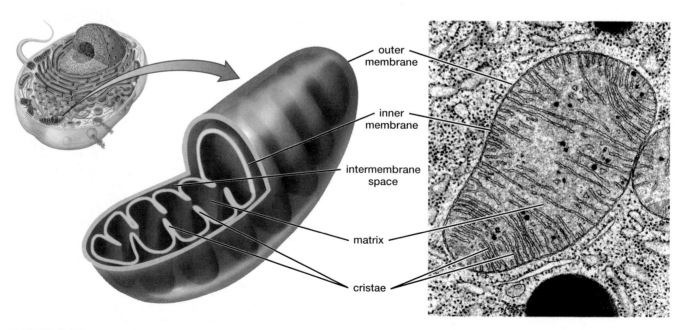

outer membrane

inner membrane

intermembrane space

matrix

cristae

▲ **FIGURE 4-17 A mitochondrion** (Left) Mitochondrial membranes enclose two fluid compartments. The inner membrane forms deep folds called cristae. (Right) A TEM showing mitochondrial structures.

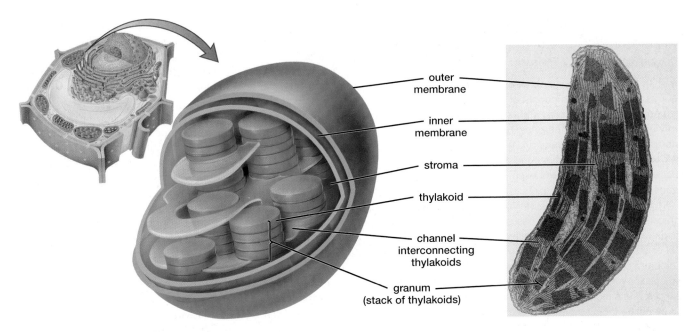

▲ **FIGURE 4-18 The chloroplast is a complex plastid** (Left) A chloroplast is surrounded by a double membrane that encloses the fluid stroma. Within the stroma are stacks of thylakoid sacs called grana. (Right) A TEM showing chloroplast structures.

carriers diffuse into the stroma, where their energy is used to drive the synthesis of sugar from carbon dioxide and water. The sugar stores energy that powers nearly all life on Earth.

Plants Use Some Plastids for Storage

Chloroplasts are highly specialized **plastids,** organelles surrounded by double membranes and used for the synthesis and/or storage of pigments or food molecules. Plastids are found only in plants and photosynthetic protists (**FIG. 4-19**); all are thought to have originated from prokaryotic cells. Some storage plastids are packed with pigments that give ripe fruits or flower petals their yellow, orange, or red colors. In plants that continue growing from one year to the next, plastids store food produced during the growing season, usually in the form of starch granules. For example, potato cells are stuffed with starch-filled plastids, food for the next spring's growth (see Fig. 4-19, upper right).

CHECK YOUR LEARNING

Can you ...

- define organelles?
- list the structures found in animal but not plant cells, and vice versa?
- describe the structure and function of each major structure found in eukaryotic cells?

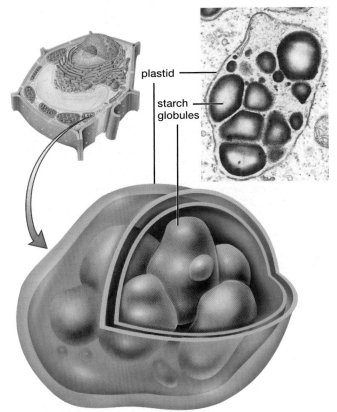

▲ **FIGURE 4-19 A simple storage plastid** Plastids are surrounded by a double outer membrane. This potato plastid stores starch, also seen in the upper right TEM.

New Parts for Human Bodies

Rapid advances in bioengineered tissues and organs requires the coordinated efforts of biochemists, biomedical engineers, cell biologists, and physicians—experts who rarely communicated in the past. But now teams of scientists from various disciplines are working together to grow not only windpipes, but also bone, cartilage, heart valves, bladders, blood vessels, and even small, partially functioning organs.

Recent progress in understanding how cells function within the extracellular matrix they create has set the stage for constructing organs with no synthetic parts. Scientists now recognize that the ECM exists in a dynamic partnership with the cells that secrete it; different cell types secrete unique matrices that support their own specific needs. Even when isolated from their cells, these matrices retain molecular cues that attract appropriate stem cells and stimulate them to differentiate into functioning cells typical of the original organ. This knowledge has generated hope that entire organs from animals such as pigs can be stripped of cells, leaving the matrix intact. The ECM might then serve as both a physical and a biochemical scaffold upon which to recreate the organ, using stem cells taken from the person needing the transplant to avoid tissue rejection.

One challenge is to infuse cells throughout the entire three-dimensional scaffolding of a major organ, such as the heart or liver. Stem cells will need to infiltrate deeply into the ECM and produce not only cells unique to the organ, but also an extensive network of blood vessels to keep it alive and functioning. Scientists have made significant progress by perfusing detergent solution through the natural blood supply of animal—and recently—entire human livers. This procedure washes away all the liver and blood vessel cells, but leaves the three-dimensional channels of the blood vessel network intact within the ECM. Researchers injected suspensions of immature human liver and blood vessel cells into the entry vessel, which conducted the cells throughout the vessel network. After a week in a nutrient-filled bioreactor, the human blood vessel cells had formed a lining within the blood vessel channels, and the liver cells had multiplied within the rest of the scaffold. There is a long road to travel before these one-inch-diameter liver-like tissues are ready to help people whose livers have failed, but the rapidly expanding field of bioengineering raises the prospect that bioartificial body parts will help people live longer, healthier lives in the coming decades.

CONSIDER THIS What advantages do bioengineered organs have over donor transplants? If you had a failing organ and an experimental bioengineered organ was just starting human trials, would you volunteer to be a recipient?

CHAPTER REVIEW

 Go to **MasteringBiology** for practice quizzes, activities, eText, videos, current events, and more.

Answers to Think Critically, Evaluate This, Multiple Choice, and Fill-in-the-Blank questions can be found in the Answers section at the back of the book.

Summary of Key Concepts

4.1 What Is the Cell Theory?

The cell theory states that every living organism consists of one or more cells, the smallest organisms are single cells, cells are the functional units of multicellular organisms, and all cells arise from preexisting cells.

4.2 What Are the Basic Attributes of Cells?

Cells are small because they must exchange materials with their surroundings by diffusion, a slow process that requires the interior of the cell to be close to the plasma membrane. All cells are surrounded by a plasma membrane that regulates the interchange of materials between the cell and its environment. All cells use DNA as a genetic blueprint and RNA to direct protein synthesis based on DNA. There are two fundamentally different types of cells. Prokaryotic cells lack membrane-enclosed organelles. Eukaryotic cells are generally larger than prokaryotic cells and have a variety of organelles, including a nucleus. See Table 4-1 for a comparison of prokaryotic and eukaryotic cells.

4.3 What Are the Major Features of Prokaryotic Cells?

All members of the domains Archaea and Bacteria consist of prokaryotic cells. Prokaryotic cells are generally smaller than eukaryotic cells and have a simpler internal structure that lacks membrane-enclosed organelles. Some prokaryotes have flagella. Most bacteria are surrounded by a cell wall made of peptidoglycan. Some bacteria, including many that cause disease, attach to surfaces using external capsules or slime layers and/or hair-like protein strands called attachment pili. Sex pili draw bacteria together to allow transfer of plasmids, small rings of DNA that confer special features such as antibiotic resistance. Most bacterial DNA is in a single chromosome in the nucleoid region. Bacterial cytoplasm includes ribosomes and a cytoskeleton. Photosynthetic bacteria may have internal membranes where the reactions of photosynthesis occur.

4.4 What Are the Major Features of Eukaryotic Cells?

Eukaryotic cells have a variety of membrane-enclosed structures called organelles, some of which differ between plant and animal cells (see Table 4-1). Both secrete an extracellular matrix. In animal cells, the ECM consists of proteins and polysaccharides that provide structural and biochemical support. In plants cells, the ECM is the supportive and porous cell wall composed primarily of cellulose.

Eukaryotic cells have an internal cytoskeleton of protein filaments that transports and anchors organelles, and that, in animal cells, shapes the cells, aids in cell division, and allows certain cells to move. Some eukaryotic cells have cilia or flagella, extensions of the plasma membrane that contain microtubules in a characteristic pattern. These structures move fluids past the cell or move the cell through a fluid.

Genetic material (DNA) is contained within the nucleus, surrounded by the double membrane of the nuclear envelope. Pores in the nuclear envelope regulate the movement of molecules between nucleus and cytoplasm. The genetic material is organized into strands called chromosomes, which consist of DNA and proteins. The nucleolus within the nucleus is the site of ribosome synthesis. Ribosomes, composed of rRNA and protein, are the sites of protein synthesis within the cytoplasm.

The endomembrane system within eukaryotic cells includes the nuclear envelope, endoplasmic reticulum (ER), Golgi apparatus, and lysosomes and other vesicles. The ER forms a series of interconnected membranous compartments and is a major site of new membrane production. Rough ER, continuous with the outer nuclear envelope, bears ribosomes where proteins are manufactured. These proteins are modified, folded, and transported within the ER channels. Smooth ER, which lacks ribosomes, manufactures lipids such as steroid hormones, detoxifies drugs and metabolic wastes, and stores calcium. The Golgi apparatus is a series of flattened membranous sacs. The Golgi processes and modifies materials that are synthesized in the rough ER. Substances modified in the Golgi are sorted and packaged into vesicles for transport elsewhere in the cell. Lysosomes are specialized sacs of membrane containing digestive enzymes. These merge with food vacuoles and break down food particles. Lysosomes also digest defective organelles.

Some freshwater protists have contractile vacuoles that collect and expel excess water. Plants use central vacuoles to support their cells and may also use these vacuoles to store nutrients, pigments, wastes, and toxic materials.

All eukaryotic cells contain mitochondria, which extract energy from food molecules and store it in the high-energy bonds of ATP. Cells of plants and photosynthetic protists contain plastids. Plastids include chloroplasts, in which photosynthesis captures solar energy and stores it in sugar molecules. Other plastids store pigments or starch. The endosymbiont hypothesis states that mitochondria and chloroplasts (as well as other plastids) originated from prokaryotic cells.

Key Terms

archaea *56*
bacteria *56*
basal body *63*
cell theory *53*
cell wall *58*
central vacuole *69*
centriole *63*
chlorophyll *70*
chloroplast *70*
chromatin *64*
chromosome *64*
cilium *62*
contractile vacuole *69*
cytoplasm *56*
cytoskeleton *56*
cytosol *56*
deoxyribonucleic acid
  (DNA) *56*
diffusion *53*
endomembrane system *66*
endoplasmic reticulum (ER) *66*
endosymbiont hypothesis *70*
eukaryotic *56*
extracellular matrix *60*
flagellum *58*
food vacuole *68*
Golgi apparatus *67*
intermediate filament *61*
lysosome *68*
microfilament *61*
microtubule *61*
mitochondrion *70*
nuclear envelope *63*
nuclear pore complex *63*
nucleoid *58*
nucleolus *64*
nucleus *63*
organelle *59*
pilus *58*
plasma membrane *56*
plasmid *58*
plastid *71*
prokaryotic *56*
ribonucleic acid (RNA) *56*
ribosome *56*
vesicle *66*

Thinking Through the Concepts

Multiple Choice

1. Which of the following is/are found only in prokaryotic cells?
 a. plasmids
 b. a cytoskeleton
 c. mitochondria
 d. ribosomes

2. Which of the following is *not* a function of the cytoskeleton?
 a. organelle movement
 b. extracellular support
 c. cell movement
 d. maintenance of cell shape

3. Which of the following statements is False?
 a. Cilia and flagella can move cells through fluids.
 b. Cilia and flagella are supported and moved by microfilaments.
 c. Cilia are shorter and more numerous than flagella.
 d. Flagella propel human sperm.

4. Which of the following is *not* a location of ribosomes?
 a. on the nuclear membrane
 b. free in the cytoplasm
 c. strung along messenger RNA
 d. inside the rough ER

5. Which of the following statements is False?
 a. Photosynthesis occurs in plastids.
 b. Chloroplasts extract energy from food storage molecules.
 c. Mitochondria likely originated from prokaryotic cells.
 d. Plant cell walls are a type of extracellular matrix.

Fill-in-the-Blank

1. The plasma membrane is composed of two major types of molecules, _____ and _____.
Which type of molecule is responsible for each of the following functions? Isolation from the surroundings: _____; interactions with other cells: _____.

2. The three types of cytoskeleton fibers are _____, _____ , and _____ . Which of these supports cilia? _____ . Moves organelles? _____ . Allows muscle contraction? _____ . Provides a supporting internal framework for the cell? _____ .

3. After each description, fill in the appropriate term: "workbenches" of the cell: _____; comes in rough and smooth forms: _____; site of ribosome production: _____; a stack of flattened membranous sacs: _____; outermost layer of plant cells: _____; ferries blueprints for protein production from the nucleus to the cytoplasm: _____ .

4. Antibody proteins are synthesized on ribosomes associated with the _____ . The antibody proteins are packaged into membranous sacs called _____ and are then transported to the _____ . There, what type of molecule is added to the protein? _____ After the antibody is completed, it is packaged into vesicles that fuse with the _____ .

5. After each description, fill in the appropriate structure: "powerhouses" of the cell: _____; capture solar energy: _____; structure outside of animal cells: _____; region of prokaryotic cell containing DNA: _____; propel fluid past cells: _____; consists of the cytosol and the organelles within it: _____ .

6. Two organelles that are believed to have evolved from prokaryotic cells are _____ and _____ . Evidence for this hypothesis is that both have _____ membranes, both have groups of enzymes that synthesize _____ , and their _____ is similar to that of prokaryotic cells.

7. What structure in bacterial cells is composed of peptidoglycan? _____ What structure in prokaryotic cells serves a similar function to the nucleus in eukaryotic cells? _____ Short segments of DNA that confer special features such as antibiotic resistance on bacteria are called _____ . Bacterial structures called _____ pull bacterial cells together so they can transfer DNA.

Review Questions

1. What are the three principles of the cell theory?

2. Which cytoplasmic structures are common to both plant and animal cells, and which are found in one type but not the other?

3. Name the proteins of the eukaryotic cytoskeleton; describe their relative sizes and major functions.

4. Describe the nucleus and the function of each of its components, including the nuclear envelope, chromatin, chromosomes, DNA, and the nucleolus.

5. What are the functions of mitochondria and chloroplasts? Why do scientists believe that these organelles arose from prokaryotic cells? What is this hypothesis called?

6. What is the function of ribosomes? Where in the cell are they found? Are they limited to eukaryotic cells?

7. Describe the structure and function of the endoplasmic reticulum (smooth and rough) and the Golgi apparatus and how they work together.

8. How are lysosomes formed? What is their function?

9. Diagram the structure of eukaryotic cilia and flagella. Describe how each moves and what their movement accomplishes.

10. List the structures of bacterial cells that have the same name and function as some eukaryotic structures, but a different molecular composition.

Applying the Concepts

1. If samples of muscle tissue were taken from the legs of a world-class marathon runner and a sedentary individual, which would you expect to have a higher density of mitochondria? Why?

2. One of the functions of the cytoskeleton in animal cells is to give shape to the cell. Plant cells have a fairly rigid cell wall surrounding the plasma membrane. Does this mean that a cytoskeleton is unnecessary for a plant cell? Explain.

3. What problems would an enormous round cell encounter? What adaptations might help a very large cell survive?

5 CELL MEMBRANE STRUCTURE AND FUNCTION

A diamondback rattlesnake prepares to strike.

Vicious Venoms

THIRTEEN-YEAR-OLD JUSTIN SCHWARTZ was enjoying his 3-week stay at a summer camp near Yosemite National Park. But that all changed when, after hiking 4.5 miles, Justin rested on some sunny rocks, hands hanging loosely at his sides. Suddenly, he felt a piercing pain in his left palm. A 5-foot rattlesnake—probably feeling threatened by Justin's dangling arm—had struck without warning.

His campmates stared in alarm as the snake slithered into the undergrowth, but Justin focused on his hand, where his palm was swelling and the pain was becoming agonizing. He suddenly felt weak and dizzy. As counselors and campmates spent the next 4 hours carrying him down the trail, pain and discoloration spread up Justin's arm, and his hand felt as if it were going to burst. A helicopter whisked him to a hospital, where he fell unconscious. A day later, he regained consciousness at the University of California Davis Medical Center. There, Justin spent more than a month undergoing 10 surgeries. These relieved the enormous pressure from swelling in his arm,

removed dead muscle tissue, and began the long process of repairing the extensive damage to his hand and arm.

Diane Kiehl's ordeal began as she dressed for an informal Memorial Day celebration with her family, pulling on blue jeans that she had tossed on the bathroom floor the previous night. Feeling a sting on her right thigh, she ripped off the jeans and watched with irritation as a long-legged spider crawled out. Living in an old house in the Kansas countryside, Diane had grown accustomed to spiders—which are often harmless—but this was an exception: a brown recluse. The two small puncture wounds seemed merely a minor annoyance until the next day, when an extensive, itchy rash appeared at the site. By the third day, intermittent pain pierced like a knife through her thigh. A physician gave her painkillers, steroids to reduce the swelling, and antibiotics to combat bacteria introduced by the spider's mouthparts. The next 10 days were a nightmare of pain from the growing sore, now covered with oozing blisters and underlain with clotting blood. It took 4 months for the lesion to heal. Even a year later, Diane sometimes felt pain in the large scar that remained.

How do rattlesnake and brown recluse spider venoms cause leaky blood vessels, disintegrating skin and tissue, and sometimes life-threatening symptoms throughout the body? What do venoms have to do with cell membranes?

5.1 HOW IS THE STRUCTURE OF THE CELL MEMBRANE RELATED TO ITS FUNCTION?

All cells, as well as organelles within eukaryotic cells, are surrounded by membranes. All the membranes of a cell have a similar basic structure: proteins suspended in a double layer, or bilayer (Gk. *bi*, double), of phospholipids (**FIG. 5-1**). Beyond this basic structure, membranes differ from one tissue type to another. The structures of their proteins and phospholipids can change dynamically in response to the environment and the cell's changing needs.

Cell membranes perform several crucial functions:

- They isolate the contents of membrane-enclosed organelles from the surrounding cytosol and the contents of the cell from the surrounding interstitial fluid.
- They regulate the exchange of substances between the cell and the interstitial fluid or between membrane-enclosed organelles and the surrounding cytosol.
- They allow communication among the cells of multicellular organisms.
- They create attachments within and between cells.
- They regulate many biochemical reactions.

These are formidable tasks for a structure so thin that 10,000 membranes stacked atop one another would scarcely equal the thickness of a book's page.

Membranes Are "Fluid Mosaics" in Which Proteins Move Within Layers of Lipids

Before the 1970s, cell biologists knew that cell membranes consist primarily of proteins and a double layer of lipids, but they

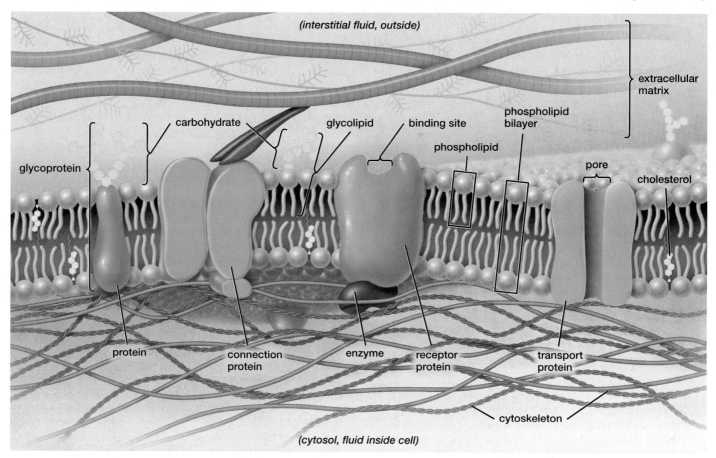

(interstitial fluid, outside)

extracellular matrix

glycoprotein
carbohydrate
glycolipid
binding site
phospholipid bilayer
phospholipid
pore
cholesterol

protein
connection protein
enzyme
receptor protein
transport protein

cytoskeleton

(cytosol, fluid inside cell)

▲ **FIGURE 5-1 The plasma membrane** The plasma membrane is a bilayer of phospholipids interspersed with cholesterol molecules and embedded with proteins (blue). Membrane proteins include recognition, connection, receptor, and transport proteins, as well as enzymes. There are also many glycoproteins and glycolipids with attached carbohydrates.

did not know how these molecules were arranged within the membrane. In 1972, S. J. Singer and G. L. Nicolson proposed the **fluid mosaic model** of cell membranes, which forms the basis for our understanding of membrane structure and function. A **fluid** is any substance whose molecules can flow past one another; fluids include gases, liquids, and cell membranes. According to the fluid mosaic model, the cell membrane consists of a fluid formed by the bilayer of phospholipids, with a variety of different embedded proteins forming a sort of "mosaic" patchwork within this fluid (see Fig. 5-1).

The Fluid Phospholipid Bilayer Helps to Isolate the Cell's Contents

A phospholipid consists of two very different parts: a "head" that is polar and hydrophilic (attracted to water) and a pair of fatty acid "tails" that are nonpolar and hydrophobic (not attracted to water). Cell membranes contain a variety of phospholipids with structures similar to those shown in FIGURE 5-2. Most phospholipids help isolate the cell from its surroundings, but some have other functions, such as identifying cells. For example, *glycolipids* (lipids with carbohydrates attached) on red blood cell membranes identify blood as type A, B, AB, or O. Other glycolipids help identify a cell as belonging to a specific individual.

Phospholipids in membranes arrange themselves in a particular way (see Fig. 5-1) due to the fact that all cells are immersed in watery solutions. Single-celled organisms may live in marine or freshwater environments, and water saturates the cellulose walls that surround plant cells. The outer surfaces of animal plasma membranes are bathed in watery **interstitial fluid,** a weakly salty liquid resembling blood, but without its cells or large proteins. Inside the plasma membrane, the *cytosol* (the fluid portion of the cytoplasm) is mostly water. In these watery surroundings, phospholipids spontaneously arrange themselves into a double layer called a **phospholipid bilayer.** Hydrogen bonds between water and the hydrophilic phospholipid heads cause the heads to face the water on either side. The hydrophobic phospholipid tails cluster together within the bilayer.

The components of cell membranes are in constant motion. To understand why, it is important to be aware that, at any temperature above absolute zero (−459.4°F or −273°C), atoms, molecules, and ions are in constant random motion. As the temperature increases, their rate of motion increases; at temperatures that support life, these particles move rapidly indeed. The phospholipid molecules in a membrane are not chemically bonded, so at body temperature their constant random motion causes them to shift about freely, although they rarely flip between the two layers of membrane.

The flexibility of the bilayer is crucial for membrane function. If plasma membranes were stiff, your cells would break open and die as you moved about. The fluid nature of membranes also allows membranes to merge with one another. For example, vesicles from the Golgi expel their contents to the outside of the cell by merging with the plasma membrane (see Chapter 4). To maintain flexibility, most membranes have a consistency similar to room-temperature olive oil. But olive oil becomes a solid if you refrigerate it, so what happens to membranes when an organism gets cold? To find out, see "Health Watch: Membrane Fluidity, Phospholipids, and Fumbling Fingers."

All animal cell membranes contain cholesterol (see Fig. 5-1), which is especially abundant in the plasma membrane. Interactions between cholesterol and phospholipids help to stabilize the membrane, making it less fluid at higher temperatures and less solid at lower temperatures. A high cholesterol content reduces the permeability of the membrane to hydrophilic substances and small molecules that would otherwise diffuse through it. Reducing permeability allows the cell to exert greater control over which substances enter and leave it.

Some biological molecules, including fat-soluble vitamins and steroid hormones such as estrogen and testosterone (see Chapter 3), are hydrophobic and can diffuse directly through the phospholipid bilayer. However, most molecules used by cells, including salts, amino acids, and sugars, are hydrophilic. Because these molecules are polar and water soluble, they cannot move through the nonpolar, hydrophobic fatty acid tails of the phospholipid bilayer. Movement of these substances in or out of the cell relies on the mosaic of proteins within the membrane, which we will discuss next.

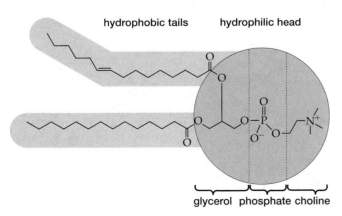

▲ **FIGURE 5-2 A phospholipid** Phosphatidylcholine (shown here) is abundant in cell membranes. The double bond in one of the fatty acid tails causes the tail to bend.

Health WATCH

Membrane Fluidity, Phospholipids, and Fumbling Fingers

Cell membranes need to maintain optimal fluidity to allow the many embedded proteins in the membrane to function. Temperature has a significant impact on membrane fluidity. As temperature increases, phospholipid molecules jiggle around more vigorously and maintain greater distance from one another, which makes the membrane more fluid. As temperatures cool, the molecules pack together more tightly, causing the membrane to stiffen.

Many organisms do not maintain constant body temperatures, so they must possess mechanisms for maintaining optimal membrane fluidity in the face of seasonal changes. Scientists studying plants, fish, frogs, snails, and mammals have found that these organisms can modify the composition of their cell membranes. As the temperature falls, they incorporate phospholipids containing more unsaturated fatty acids into their membranes. When the temperature rises, the organisms restore the saturated phospholipids. Why? How fluid a membrane is at a given temperature is strongly influenced by the relative amounts of saturated and unsaturated fatty acid tails in membrane phospholipids. Saturated fatty acids (with no double-bonded carbons) are straight and can pack tightly together, forming a relatively stiff membrane. Unsaturated fatty acids have one or more double-bonded carbons, each introducing a kink into the tail (see Fig. 5-2). This structure forces the phospholipids farther apart, making the membrane more fluid (**FIG. E5-1**). Adjusting the fatty acids in the phospholipids helps organisms maintain optimal membrane fluidity during seasonal temperature changes.

But when you get caught out in the cold, your membranes don't have time to adjust their phospholipids. If your core temperature suddenly begins to fall, your body will reduce blood flow to your hands and feet, conserving warmth for vital organs such as your heart and brain. As your hands get colder, it becomes harder to control your fingers, and your sense of touch will diminish (**FIG. E5-2**). What is happening?

Cooling causes the nerves that control muscles and carry sensations to conduct nerve impulses more slowly, which makes it difficult to coordinate delicate hand movements like zipping a coat or lighting a fire. Scientists do not yet know exactly why this happens, but there is evidence supporting the hypothesis that as membranes stiffen in the cold, the functioning of embedded proteins—including the ion channels responsible for transmitting nerve impulses—is hampered. If your hands get extremely cold (but your tissues have not frozen, which eliminates sensation), you are likely to feel excruciating "cold pain." The pain will persist even though you will be practically unable to move your hands and you will have lost your sense of touch.

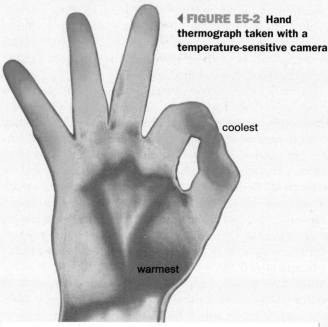

◀ **FIGURE E5-2 Hand thermograph taken with a temperature-sensitive camera**

coolest

warmest

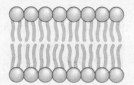

more saturated fatty acids
less fluidity

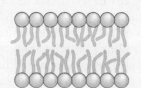

more unsaturated fatty acids
greater fluidity

▲ **FIGURE E5-1 Tail kinks in phospholipids increase membrane fluidity**

THINK CRITICALLY Researchers have recently discovered ion channels in pain-generating skin nerve cells whose ability to conduct ions is actually enhanced by cold, making the nerve cell more active. Form a hypothesis as to why these receptors evolved.

A Variety of Proteins Form a Mosaic Within the Membrane

Thousands of different membrane proteins are embedded within or attached to the surfaces of the phospholipid bilayer of cell membranes. Many membrane proteins bear carbohydrate groups that project from the outer membrane surface (see Fig. 5-1). These proteins are called **glycoproteins** (Gk. *glyco,* sweet; referring to the carbohydrate sugar subunits). Membrane proteins may be grouped into five major categories based on their function: enzymes, recognition proteins, receptor proteins, transport proteins, and connection proteins.

Enzymes

Proteins called **enzymes** promote chemical reactions that synthesize or break apart biological molecules (see Chapter 6). A variety of enzymes are associated with cell membranes, performing different functions in the plasma membrane and in various organelles. For example, enzymes involved in ATP synthesis are embedded in the inner mitochondrial membrane. Plasma membrane enzymes help synthesize the supportive extracellular matrix that fills spaces between animal cells. In cells lining the small intestine, plasma membrane enzymes complete the breakdown of carbohydrates and proteins as these nutrients are taken into the cells.

Recognition Proteins

Recognition proteins are glycoproteins that serve as identification tags. The cells of each individual organism bear distinctive glycoproteins that identify the cells as "self." Immune cells ignore self-cells and attack invading cells, such as bacteria, that have different recognition proteins on their membranes. For successful organ transplants, the most important recognition glycoproteins of the donor must match those of the recipient so the organ won't be attacked by the recipient's immune system.

Transport Proteins

Transport proteins span the phospholipid bilayer and regulate the movement of hydrophilic molecules across the membrane. Some transport proteins form pores (channels) that can be opened or closed to allow specific substances to pass across the membrane. Other transport proteins bind substances and conduct them through the membrane, sometimes using cellular energy. Transport proteins are described later in this chapter.

Receptor Proteins

Most cells bear dozens of types of **receptor proteins** (some of which are glycoproteins; see Fig. 5-1) that span their plasma membranes. Each has a binding site specific for a messenger molecule, such as a particular hormone or neurotransmitter (a nerve cell messenger). When the appropriate messenger molecule binds, it activates the receptor protein.

Some membrane proteins act as both receptors and ion channels. One of the ways they do this is through direct receptor action. When a messenger molecule binds the receptor, it immediately and directly causes an ion channel within the same protein to open (**FIG. 5-3a**). The best known example of such direct receptor action occurs in a protein in the membranes of skeletal muscle cells (which move the body). This protein binds the neurotransmitter acetylcholine, which opens an ion channel in the same protein and allows ions to flow that stimulate the muscle cell to contract. Other receptor proteins produce their effects through an indirect action. Upon binding a messenger molecule—which can be a neurotransmitter or a hormone—the receptor protein changes its shape and biochemical activity. This starts a series of reactions within the cell that produce effects (such as opening many ion channels) at different sites (**FIG. 5-3b**). Most neurotransmitters and hormones act in this manner.

Connection Proteins

A diverse group of **connection proteins** anchors cell membranes in various ways. Some connection proteins help maintain cell shape by linking the plasma membrane to the cell's cytoskeleton. Other connection proteins span the plasma membrane, linking the cytoskeleton inside the cell with the extracellular matrix outside, which helps anchor the cell in place within a tissue (see Fig. 5-1). Connection proteins also link adjacent cells, as described later.

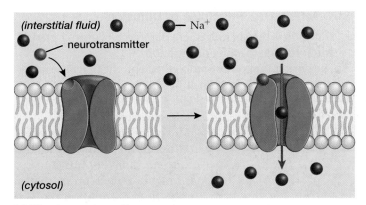

(a) Direct receptor action

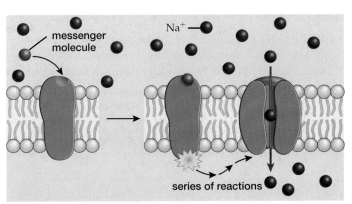

(b) Indirect receptor action

▲ **FIGURE 5-3 Receptor protein activation (a)** A neurotransmitter binds a receptor site on a membrane channel protein, causing the channel to open and allow a flow of ions. **(b)** A messenger molecule (such as a hormone or neurotransmitter) binds a membrane receptor, which stimulates a series of reactions inside the cell that cause channel proteins to open.

CHECK YOUR LEARNING

Can you ...

- describe the components, structure, and functions of cell membranes?
- diagram and describe the fluid mosaic model of cell membranes?
- explain how the different components of cell membranes contribute to their functions?

CASE STUDY \ CONTINUED

Vicious Venoms

Most snake venoms are nasty cocktails of toxins. In addition to breaking down phospholipids, some rattlesnake venoms contain toxins that bind to and inhibit the activity of receptor proteins for acetylcholine. Inhibiting the action of acetylcholine on heart muscle cells increases heart rate, which speeds the travel of venom through the body. Krait snakes, native to Asia, produce a venom protein that binds to the direct-acting membrane channel receptors for acetylcholine, including those on skeletal muscle cells. Once bound, the protein remains firmly attached, blocking acetylcholine from binding. This prevents muscles from contracting; an untreated krait bite victim often dies when the skeletal muscles that control breathing are paralyzed.

The ability of substances to move across membranes is crucial not only to controlling heart rate and breathing, but also to all other aspects of staying alive. How do proteins control the movement of substances across membranes?

5.2 HOW DO SUBSTANCES MOVE ACROSS MEMBRANES?

Some substances, especially individual molecules and ions, can move across membranes by diffusing through the phospholipid bilayer or traveling through specialized transport proteins. To provide some background on how membrane transport works, we begin our study with a few definitions:

- A **solute** is a substance that can be dissolved (dispersed into individual atoms, molecules, or ions) in a **solvent,** which is a fluid (usually a liquid) capable of dissolving the solute. Water, in which all of the cell's chemical processes occur, dissolves so many different solutes that it is sometimes called the "universal solvent."
- **Concentration** is the amount of solute in a given volume of solvent.
- A **gradient** is a difference in certain properties—such as temperature, pressure, electrical charge, or concentration—between two adjacent regions. Energy must be expended to create a gradient. Gradients decrease over time unless an impenetrable barrier separates the adjacent regions or energy is supplied to maintain them.

Molecules in Fluids Diffuse in Response to Gradients

Recall that atoms, molecules, and ions are in constant random motion. As a result of this motion, molecules and ions in solution are continuously bombarding one another and the structures surrounding them. Over time, random movements of solutes produce a net movement from regions of high concentration to regions of low concentration, a process called **diffusion.** The greater the concentration gradient, the more rapidly diffusion occurs. If nothing opposes this diffusion (such as electrical charge, pressure differences, or physical barriers), then the random movement of molecules will eventually cause the substance to become evenly dispersed throughout the fluid. In an analogy to gravity, molecules moving from regions of high concentration to regions of low concentration are described as moving "down" their concentration gradients.

To watch diffusion in action, place a drop of food coloring in a glass of water (**FIG. 5-4 ❶**). Random motion propels dye molecules both into and out of the dye droplet, but there is a net transfer of dye into the water and of water into the dye, down their respective concentration gradients ❷. The net movement of dye will continue until it is uniformly dispersed in the water ❸. If you compare the diffusion of a dye in hot water to that in cold water, you will see that heat increases the diffusion rate by causing molecules to move faster.

SUMMING UP: Principles of Diffusion

- Diffusion is the net movement of molecules down a gradient from high to low concentration.
- The greater the concentration gradient, the faster the rate of diffusion.
  - The higher the temperature, the faster the rate of diffusion.
  - If no other processes intervene, diffusion will continue until the concentrations become equal throughout the solution, that is, until the concentration gradient is eliminated.

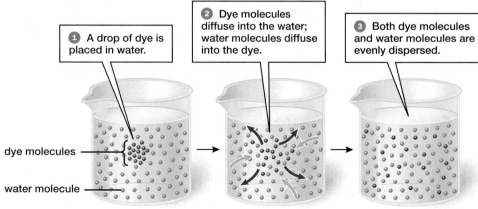

❶ A drop of dye is placed in water.

❷ Dye molecules diffuse into the water; water molecules diffuse into the dye.

❸ Both dye molecules and water molecules are evenly dispersed.

dye molecules

water molecule

◀ **FIGURE 5-4 Diffusion of a dye in water**

TABLE 5-1	Transport Across Membranes
Passive Transport	Diffusion of substances across a membrane down a gradient of concentration, pressure, or electrical charge; does not require cellular energy
Simple diffusion	Diffusion of water, dissolved gases, or lipid-soluble molecules through the phospholipid bilayer of a membrane
Facilitated diffusion	Diffusion of water, ions, or water-soluble molecules through a membrane via a channel or carrier protein
Osmosis	Diffusion of water across a selectively permeable membrane from a region of higher free water concentration to a region of lower free water concentration
Energy-Requiring Transport	Movement of substances through membranes using cellular energy, usually supplied by ATP
Active transport	Movement of individual small molecules or ions against their concentration gradients through membrane-spanning proteins
Endocytosis	Movement of fluids, specific molecules, or particles into a cell; occurs as the plasma membrane engulfs the substance in a membranous sac that pinches off and enters the cytosol
Exocytosis	Movement of particles or large molecules out of a cell; occurs as a membrane within the cell encloses the material, moves to the cell surface, and fuses with the plasma membrane, allowing its contents to diffuse out

Movement Through Membranes Occurs by Passive Transport and Energy-Requiring Transport

To stay alive, cells must generate and maintain concentration gradients, or the differences in solute concentrations across their membranes. Plasma membranes are described as **selectively permeable** because their proteins allow only specific ions or molecules to pass through, or permeate. The selective permeability of the plasma membrane creates a barrier that helps maintain the cell's concentration gradients.

The plasma membrane permits substances to move through it in two different ways: passive transport and energy-requiring transport (**TABLE 5-1**). **Passive transport** involves diffusion of substances across cell membranes down their concentration gradients, whereas **energy-requiring transport** requires that the cell expend energy to move substances across membranes. Energy-requiring transport occurs when transporting substances against concentration gradients, or when moving particles or fluid droplets into or out of the cell.

Passive Transport Includes Simple Diffusion, Facilitated Diffusion, and Osmosis

Diffusion can occur within a fluid or across a membrane that is permeable to the diffusing substance. Many molecules cross plasma membranes by diffusion, driven by concentration differences between the cytosol and the surrounding interstitial fluid.

Some Molecules Move Across Membranes by Simple Diffusion

Some molecules diffuse directly through the phospholipid bilayer of cell membranes, a process called **simple diffusion** (**FIG. 5-5a**). Very small molecules with no net charge, such

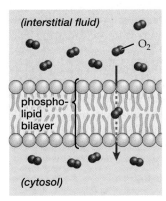

(a) Simple diffusion through the phospholipid bilayer

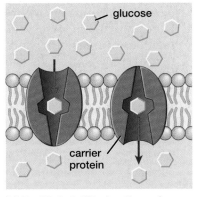

(b) Facilitated diffusion through carrier proteins

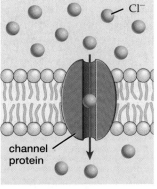

(c) Facilitated diffusion through channel proteins

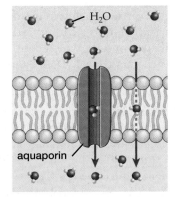

(d) Osmosis through aquaporins or the phospholipid bilayer

▲ **FIGURE 5-5 Types of diffusion through the plasma membrane (a)** Small, uncharged, or lipid-soluble molecules diffuse directly through the phospholipid bilayer. Here, oxygen molecules diffuse down their concentration gradient (red arrow). **(b)** Carrier proteins have binding sites for specific molecules, such as glucose. Binding causes the carrier to change shape and shuttle the molecule across the membrane down its concentration gradient. **(c)** Facilitated diffusion through specific channel proteins allows ions, such as chloride, to cross membranes. **(d)** Osmosis is the diffusion of water. Water molecules can pass through the phospholipid bilayer by simple diffusion, or they move far more rapidly by facilitated diffusion through aquaporins.

as water, oxygen, and carbon dioxide, can travel across cell membranes by simple diffusion, as can lipid-soluble molecules, including alcohol, certain vitamins, and steroid hormones. The rate of simple diffusion is increased by larger concentration gradients, higher temperatures, smaller molecular sizes, and greater solubility in lipids.

How can water—a polar molecule—diffuse directly through the hydrophobic (literally "water-fearing") phospholipid bilayer? Water molecules are so small, some stray into the thicket of phospholipid tails, and the random movement of these water molecules carries them through the membrane. Because simple diffusion of water through the phospholipid bilayer is relatively slow, many cell types have specific transport proteins for water, described later.

Some Molecules Cross Membranes by Facilitated Diffusion Using Membrane Transport Proteins

The phospholipid bilayers of cell membranes are quite impermeable to most polar molecules, for example, sugars, whose size and lack of lipid solubility keeps them out. Ions, such as K^+, Na^+, Cl^-, and Ca^{2+}, are also excluded even though they are small, because their charges cause polar water molecules to cluster around them, forming an aggregation that is too large to move directly through the phospholipid bilayer. Therefore, ions and polar molecules must use specific transport proteins to move through cell membranes, a process called **facilitated diffusion.** Two types of proteins allow facilitated diffusion: carrier proteins and channel proteins. The cell does not expend energy when using these transport proteins, which facilitate diffusion down a pre-existing concentration gradient either into or out of the cell.

Carrier proteins span the cell membrane and have regions that loosely bind certain ions or specific molecules such as sugars or small proteins. This binding of ions or molecules causes the carrier proteins to change shape and transfer the bound molecules across the membrane. For example, glucose carrier proteins in plasma membranes allow this sugar to diffuse down its concentration gradient from the interstitial fluid into cells, which continuously use up glucose to meet their energy needs (**FIG. 5-5b**).

Channel proteins form pores through the cell membrane. For example, the cell membranes of mitochondria and chloroplasts have pores that allow passage of many different water-soluble substances. In contrast, ion channel proteins (ion channels) are small and highly selective (**FIG. 5-5c**). Many ion channels help preserve concentration gradients by remaining closed unless they are opened by specific stimuli, such as a neurotransmitter binding to a receptor protein. Ion channel proteins are selective because their interior diameters limit the size of the ions that can pass through and because specific amino acids that line the pore have weak electrical charges that attract specific ions and repel others. For example, slight negative charges on some amino acids inside Na^+ channels attract Na^+ but repel Cl^-.

Many cells have specialized water channel proteins called **aquaporins** (literally, "water pores"; **FIG. 5-5d**). The narrow aquaporin channels are selective for water molecules, which are extremely small. Some amino acids lining the aquaporin channel protein have slight positive charges that attract the negative poles of water but repel positive ions. To learn more about aquaporins, see "How Do We Know That? The Discovery of Aquaporins."

Osmosis Is the Diffusion of Water Across Selectively Permeable Membranes

Osmosis is the diffusion of water across a membrane that is selectively permeable to water in response to gradients of concentration, pressure, or temperature. Here, we will focus on osmosis from a region of higher water concentration to a region of lower water concentration.

What do we mean by a "high water concentration" or a "low water concentration"? Water containing no solutes has the highest possible water concentration, causing more water molecules to collide with—and so move through—a water-permeable membrane. Any solute reduces the water concentration by replacing some of the water molecules in a given volume of the solution. In addition to displacing water molecules, polar solutes and ions also attract and weakly bind water molecules around them, so the water molecules aren't as free to move across a water-permeable membrane. For these reasons, the higher the concentration of solute, the lower the concentration of water. Osmosis (like other forms of diffusion) will cause a net movement of water molecules from the solution with a higher water concentration (a lower solute concentration) into the solution with a lower water concentration (a higher solute concentration). For example, water will move by osmosis from a solution with less dissolved sugar into a solution with more dissolved sugar.

Solutions with equal concentrations of solute—and thus equal concentrations of water—are described as being **isotonic** to one another (Gk. *iso*, same). When isotonic solutions are separated by a water-permeable membrane, water moves equally through the membrane in both directions, so there is no net movement of water. When comparing two solutions with different concentrations of a solute, the solution with the greater concentration of solute is described as being **hypertonic** (Gk. *hyper*, greater than) to the less concentrated solution. The more dilute solution is described as **hypotonic** (Gk. *hypo*, below). Water tends to move through water-permeable membranes from hypotonic solutions into hypertonic solutions. This movement will continue until the water concentrations (and thus the solute concentrations) on both sides of the membrane are equal.

SUMMING UP: Principles of Osmosis

- Osmosis is the movement of water through a selectively water-permeable membrane by simple diffusion or by facilitated diffusion through aquaporins.
- Water moves down its concentration gradient from a higher concentration of free water molecules to a lower concentration of free water molecules.

HOW DO WE KNOW THAT?

The Discovery of Aquaporins

Sometimes a chance observation leads to a scientific breakthrough. Scientists had long observed that osmosis directly through the phospholipid bilayer is much too slow to account for water movement across certain cell membranes, such as those of red blood cells (see Fig. 5-6). But attempts to identify selective transport proteins for water repeatedly failed. Then, in the mid-1980s, Peter Agre (**FIG. E5-3**), working at the Johns Hopkins School of Medicine in Maryland, was attempting to determine the structure of a glycoprotein on red blood cells. The glycoprotein he isolated was contaminated, however, with large quantities of an unknown protein. Instead of discarding the mystery protein, Agre and his coworkers collaborated with researchers at other universities to determine its structure and function.

To test their hypothesis that the protein was involved with water transport, they performed an experiment using frog egg cells, whose membranes are nearly impermeable to water. Agre's team predicted that if the proteins were water channels, inserting the mystery protein into the egg cells would cause them to swell when they were placed in a hypotonic solution. The researchers injected frog eggs with messenger RNA that coded for the unidentified protein, causing the eggs to synthesize the protein and insert it into their plasma membranes. Control eggs were injected with an equal quantity of water. Three days later, the eggs appeared identical—that is, until they were placed in a hypotonic solution. The control eggs swelled only slightly, whereas those with the inserted protein swelled rapidly and burst (**FIG. E5-4**). Further studies revealed that only water could move through this channel protein. Agre reached the conclusion that these were water channels and named them aquaporins. In 2000, Agre's group and other research teams reported the three-dimensional structure of aquaporin. Billions of water molecules can move through an aquaporin in single file every second, while larger molecules and small positively charged ions (such as hydrogen ions) are excluded.

Many subtypes of aquaporin proteins have now been identified, and these water channels have been found in all forms of life that have been investigated. For example, the membrane of the central vacuole of plant cells is rich in aquaporins, allowing it to fill rapidly when water is available (see Fig. 5-7). Kidney cells insert aquaporins into their plasma membranes when the body becomes dehydrated and water needs to be conserved. Aquaporins have also been implicated in many pathological conditions including brain swelling, glaucoma, and cancer; researchers are working to design therapeutic drugs for these disorders that will block or facilitate water movement through these channels.

In 2003, Peter Agre shared the Nobel Prize in Chemistry for his discovery. In his Nobel lecture, he shared an insight that is fundamental to scientific advances today: "In science, one should use all available resources to solve difficult problems. One of our most powerful resources

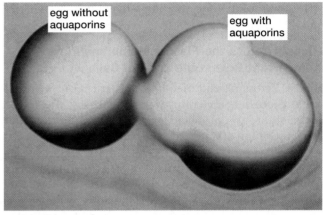

(a) Frog eggs

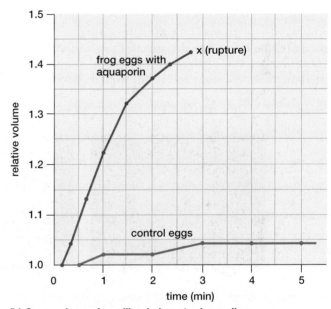

(b) Comparison of swelling in hypotonic medium

▲ **FIGURE E5-4 Investigating aquaporins (a)** The frog egg on the right with aquaporins inserted into its plasma membrane burst after immersion in a hypotonic solution. The normal frog egg on the left swelled only slightly. **(b)** When transferred from a normal to a hypotonic solution, the average volume of frog eggs increased rapidly until they burst, while control eggs swelled very slightly (the relative volume of 1 is the size of the egg just prior to transfer).

is the insight of our colleagues." As a result of collaboration, careful observation, persistence, and perhaps a bit of what he described modestly as the "scientific approach known as sheer blind luck," Agre and his team identified the elusive transport protein for water.

THINK CRITICALLY Based on Figure E5-4b, graph what likely would have happened if eggs with and without aquaporins had been immersed in a concentrated salt solution, extending the y-axis if needed. Explain your reasoning.

◀ **FIGURE E5-3 Peter Agre**

- Dissolved substances, called solutes, reduce the concentration of free water molecules in a solution.
- When comparing two solutions, the one with a higher solute concentration is hypertonic, and the solution with the lower solute concentration is hypotonic.

Osmosis Across the Plasma Membrane Plays an Important Role in the Lives of Cells

Osmosis across plasma membranes is crucial to many biological processes, including water uptake by the roots of plants, absorption of dietary water from the intestine, and the reabsorption of water into the bloodstream that occurs in kidneys.

In animals, the interstitial fluid that surrounds cells is isotonic to the cell cytosol. Although the concentrations of specific solutes are rarely the same both inside and outside of cells, the total concentrations of water and solutes inside and outside are equal. As a result, there is no overall tendency for water to enter or leave animal cells. To demonstrate the importance of maintaining isotonic conditions between fragile cells and their surrounding interstitial fluid, we can observe cells placed in solutions with different solute concentrations, illustrated in **FIGURE 5-6**.

Red blood cells have an abundance of aquaporins in their plasma membranes, making them very water permeable. Placed in an isotonic salt solution, they retain their normal size. But if the salt solution is hypertonic to the cytosol of the blood cells, water leaves by osmosis, causing the cells to shrivel. Immersion in a hypotonic salt solution, in contrast, causes the cells to swell (and eventually burst) as water diffuses in.

Freshwater organisms must continuously expend energy to counteract osmosis because their cells are hypertonic to the surrounding water. For example, protists such as *Paramecium* use cellular energy to pump salts from the cytosol into their contractile vacuoles. Water follows by osmosis and is squirted out through a pore in the plasma membrane (see Fig. 4-16).

Nearly every living plant cell is supported by water that enters through osmosis. Most plant cells have a large central vacuole enclosed by a membrane that is rich in aquaporins. Dissolved substances stored in the vacuole make its contents hypertonic to the surrounding cytosol, which in turn is usually hypertonic to the interstitial fluid that bathes the cells. Water therefore flows through the cell wall into the cytosol and then into the central vacuole by osmosis. This produces **turgor pressure,** which inflates

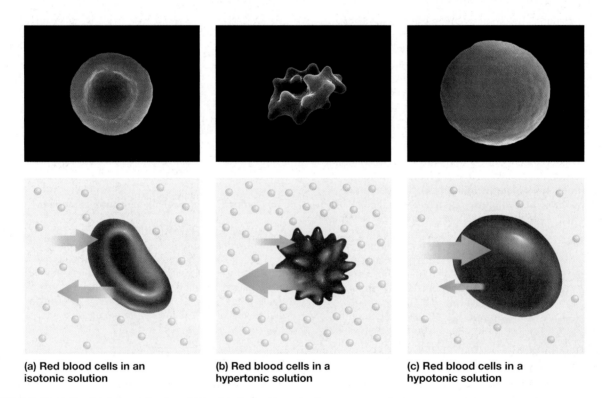

(a) Red blood cells in an isotonic solution

(b) Red blood cells in a hypertonic solution

(c) Red blood cells in a hypotonic solution

▲ **FIGURE 5-6 The effects of osmosis on red blood cells** Red blood cell plasma membranes are rich in aquaporins, so water flows readily in or out along its concentration gradient. **(a)** Cells immersed in an isotonic solution retain their normal dimpled shape. **(b)** Cells in a hypertonic solution shrivel as more water moves out than flows in. **(c)** Cells in a hypotonic solution expand.

THINK CRITICALLY A student pours some distilled water into a sample of blood. Later, she looks at the blood under a microscope and sees no blood cells at all. What happened?

the cell, forcing the cytosol within its plasma membrane against the cell wall (FIG. 5-7a). If you forget to water a houseplant, the cytosol and cell vacuole lose water, causing the cell to shrink away from its cell wall. Like a leaky balloon, the plant droops as its cells lose turgor pressure (FIG. 5-7b). Now you know why grocery stores are always spraying their leafy produce: to keep it looking perky and fresh with full central vacuoles.

Energy-Requiring Transport Includes Active Transport, Endocytosis, and Exocytosis

Many cellular activities rely on energy-requiring transport. Active transport, endocytosis, and exocytosis are crucial to maintaining concentration gradients, acquiring food, excreting wastes, and (in multicellular organisms) communicating with other cells.

Cells Maintain Concentration Gradients Using Active Transport

By building gradients and then allowing the gradients to run down under specific circumstances, cells generate ATP and respond to stimuli. For example, concentration gradients of various ions provide the energy to form ATP in mitochondria and chloroplasts (see Chapters 7 and 8), power the electrical signals of neurons, and trigger the contraction of muscles. But gradients cannot form spontaneously—they require active transport across a membrane.

During **active transport,** membrane proteins use cellular energy to move molecules or ions across a plasma membrane *against* their concentration gradients, which means that the substances are transported from areas of lower concentration to areas of higher concentration. For example, every cell must use active transport to acquire some nutrients that are less concentrated in the environment than in the cell's cytoplasm. In addition, substances such as sodium and calcium ions are actively transported to maintain them at much lower concentrations in the cytosol than in the interstitial fluid. Nerve cells maintain large ion concentration gradients because generating their electrical signals requires rapid, passive flow of ions when channels are opened. After these ions diffuse into or out of the cell, their concentration gradients must be restored by active transport.

Active transport proteins span the width of the membrane and have two binding regions (FIG. 5-8). One of these loosely binds with a specific molecule or ion, such as a calcium ion (Ca^{2+}); the second region, on the inside of the membrane,

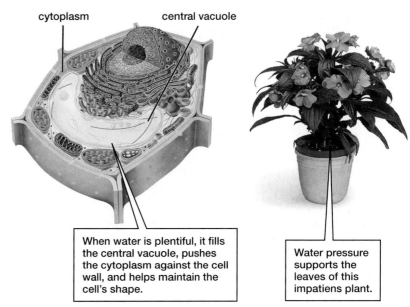

When water is plentiful, it fills the central vacuole, pushes the cytoplasm against the cell wall, and helps maintain the cell's shape.

Water pressure supports the leaves of this impatiens plant.

(a) Turgor pressure provides support

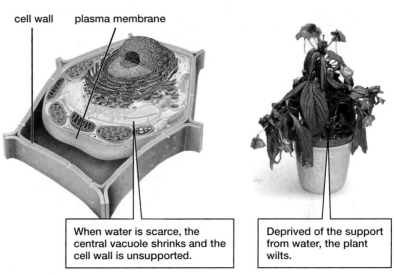

When water is scarce, the central vacuole shrinks and the cell wall is unsupported.

Deprived of the support from water, the plant wilts.

(b) Loss of turgor pressure causes the plant to wilt

▲ **FIGURE 5-7 Turgor pressure in plant cells** Aquaporins allow water to move rapidly in and out of the central vacuoles of plant cells. **(a)** The cell and the plant are supported by turgor pressure. **(b)** The cell and the plant have lost turgor pressure and support due to dehydration.

THINK CRITICALLY If a plant cell is placed in water containing no solutes, will the cell eventually burst? Explain.

binds ATP ❶. The ATP donates energy to the protein, causing the protein to change shape and move the calcium ion across the membrane ❷. The energy for active transport comes from breaking the high-energy bond that links the last of the three phosphate groups in ATP. As it loses a phosphate group, releasing its stored energy, ATP becomes ADP (adenosine diphosphate) plus a free phosphate ❸. Active transport proteins are often called *pumps* because, like pumping water into an elevated storage tank, they use energy to move ions or molecules "uphill" against a concentration gradient.

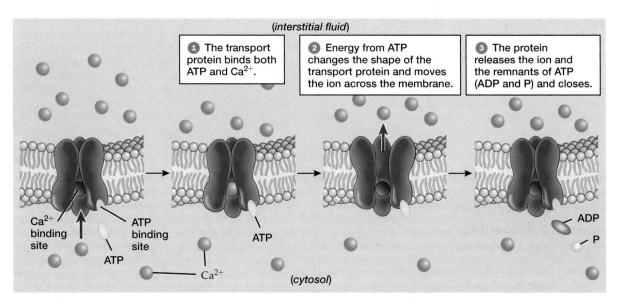

① The transport protein binds both ATP and Ca^{2+}.

② Energy from ATP changes the shape of the transport protein and moves the ion across the membrane.

③ The protein releases the ion and the remnants of ATP (ADP and P) and closes.

(interstitial fluid)

Ca^{2+} binding site

ATP binding site

ATP

Ca^{2+}

ATP

(cytosol)

ADP

P

▲ **FIGURE 5-8 Active transport** Cellular energy is used to move molecules across the plasma membrane against a concentration gradient. The active transport protein has an ATP binding site and a binding site for the transported substance, such as these calcium ions. When ATP donates its energy, it loses its third phosphate group and becomes ADP plus a free phosphate.

THINK CRITICALLY Would a cell ever use active transport to move water across its membrane? Explain.

HAVE YOU EVER WONDERED...

Why Bacteria Die When You Take Antibiotics?

Penicillin and related antibiotics fight bacterial infections by interfering with cell wall synthesis in newly forming bacterial cells. Under normal conditions, a bacterium uses active transport to maintain an internal environment that is hypertonic to its surroundings. This allows the bacterial cytoplasm to maintain turgor pressure against its tough cell wall, much as in a plant cell. In the absence of this confining wall, osmosis into the hyperosmotic cytosol of the bacterium causes it to swell and rupture its fragile plasma membrane. We are fortunate to have medical antibiotics to help us fight bacterial infections, but unfortunately bacteria evolve rapidly, and antibiotics select for resistant strains. Resistance to antibiotics can occur as a result of several different adaptations. These include bacterial enzymes that break down the antibiotic, bacterial membranes that block entry of the antibiotic, and membrane pumps that expel the antibiotic. As a result, hospitals are now battling some bacterial strains that cause serious infections but resist nearly all antibiotics.

Endocytosis Allows Cells to Engulf Particles or Fluids

A cell may need to acquire materials from its extracellular environment that are too large to move directly through the membrane. These materials are engulfed by the plasma membrane and are transported within the cell inside vesicles. This energy-requiring process is called **endocytosis** (Gk. *endo,* inside). Here, we describe three forms of endocytosis based on the size and type of material acquired and the method of acquisition: pinocytosis, receptor-mediated endocytosis, and phagocytosis.

In **pinocytosis** (Gk. *pino,* drink; **FIG. 5-9**), a very small patch of plasma membrane dimples inward as it surrounds

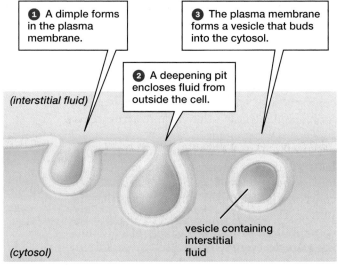

① A dimple forms in the plasma membrane.

② A deepening pit encloses fluid from outside the cell.

③ The plasma membrane forms a vesicle that buds into the cytosol.

(interstitial fluid)

vesicle containing interstitial fluid

(cytosol)

(a) Pinocytosis

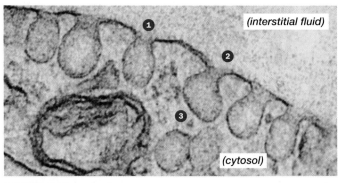

(interstitial fluid)

(cytosol)

(b) TEM of pinocytosis

▲ **FIGURE 5-9 Pinocytosis**

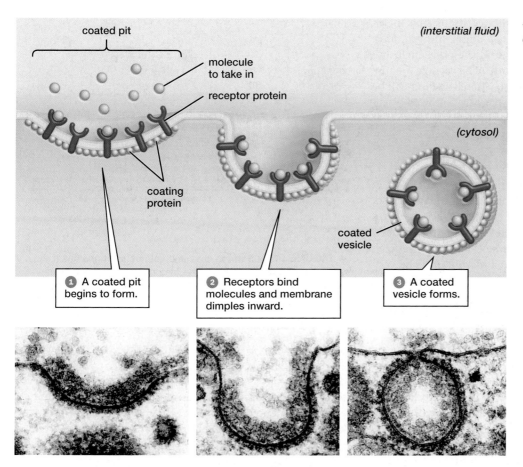

coated pit

(interstitial fluid)

molecule
to take in

receptor protein

coating
protein

(cytosol)

coated
vesicle

1 A coated pit
begins to form.

2 Receptors bind
molecules and membrane
dimples inward.

3 A coated
vesicle forms.

◀ **FIGURE 5-10** **Receptor-mediated endocytosis**

interstitial fluid, and then the membrane buds off into the cytosol as a tiny vesicle. Pinocytosis moves a droplet of interstitial fluid, contained within the dimpling patch of membrane, into the cell. Therefore, the cell acquires materials in the same concentration as in the interstitial fluid. Virus particles in interstitial fluid may be taken up into cells by pinocytosis, resulting in infection. Countering this, some immune system cells patrol for viruses by routinely taking in large quantities of interstitial fluid via pinocytosis. After ingesting viruses, they alert other immune cells to produce antibodies that will destroy the viruses.

Cells use **receptor-mediated endocytosis** to selectively take up specific molecules or complexes of molecules that cannot move through channels or diffuse through the plasma membrane (**FIG. 5-10**). Receptor-mediated endocytosis occurs in thickened depressions called *coated pits*. The coating material consists of proteins on the inside surface of the plasma membrane that assist in forming the pit. Receptor proteins for a specific substance project from the plasma membrane. These receptors bind the molecules to be transported. Then the depression deepens into a pocket that pinches off, forming a coated vesicle that carries the molecules into the cytosol. Molecules moved by receptor-mediated endocytosis include most protein hormones and packets of lipoprotein containing cholesterol.

Phagocytosis (Gk. *phago*, eat) moves large particles—sometimes whole microorganisms—into the cell (**FIG. 5-11a**). When the predatory protist *Amoeba*, for example, senses a

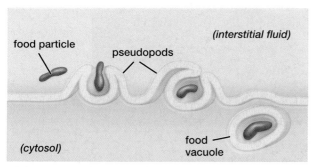

food particle

pseudopods

(interstitial fluid)

(cytosol)

food
vacuole

(a) Phagocytosis

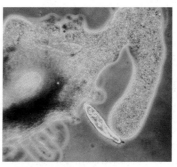

(b) An *Amoeba* engulfs a *Paramecium*

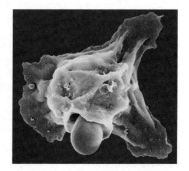

(c) A white blood cell engulfs a disease-causing fungal cell

▲ **FIGURE 5-11** **Phagocytosis (a)** The mechanism of phagocytosis. **(b)** Amoebas use phagocytosis to feed, and **(c)** white blood cells use phagocytosis to engulf disease-causing microorganisms.

Paramecium, the *Amoeba* extends parts of its plasma membrane, forming pseudopods (Gk. *pseudo,* false, and *pod,* foot; FIG. 5-11b). The pseudopods fuse around the prey, enclosing it inside a vesicle called a **food vacuole.** The vacuole will fuse with a lysosome (described in Chapter 4) where the food will be digested. White blood cells use phagocytosis, followed by digestion, to engulf and destroy invading bacteria, a drama that occurs continuously within your body (FIG. 5-11c).

Exocytosis Moves Material Out of the Cell

Cells also use energy to dispose of undigested particles or to secrete substances such as hormones into the interstitial fluid, a process called **exocytosis** (Gk. *exo,* outside; FIG. 5-12). During exocytosis, a membrane-enclosed vesicle carrying material to be expelled moves to the cell surface, where the vesicle's membrane fuses with the cell's plasma membrane. The vesicle's contents then diffuse into the fluid outside the cell.

Exchange of Materials Across Membranes Influences Cell Size and Shape

As you learned in Chapter 4, most cells are too small to be seen with the naked eye; they range from about 1 to 100 micrometers (millionths of a meter) in diameter. Why are cells so small? To acquire nutrients and eliminate wastes, all parts of a cell rely on the slow process of diffusion, so the cell must be small enough that no part of it is too far removed from the surrounding fluid.

Assuming that a cell is roughly spherical, the larger its diameter, the farther its innermost contents are from the plasma membrane. In a hypothetical giant cell 8.5 inches (20 centimeters) in diameter, oxygen molecules would take more than 200 days to diffuse to the center of the cell, but the cell would be long dead. In addition, all cellular wastes and nutrients must diffuse through the cell's plasma membrane. As a hypothetical spherical cell enlarges, its volume of cytoplasm

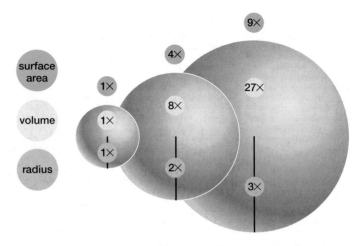

▲ **FIGURE 5-13 Surface area and volume relationships** If the radius of a sphere increases by a factor of 3, then the volume increases by a factor of 27 but the surface area only increases by a factor of 9.

(where all its metabolic reactions occur) increases more rapidly than does its surface area (through which it must exchange nutrients and wastes; FIG. 5-13).

These constraints limit the size of most cells. However, some cells, such as nerve and muscle cells, have an extremely elongated shape that increases their membrane surface area, keeping the ratio of surface area to volume relatively high.

CHECK YOUR LEARNING

Can you ...

- explain simple diffusion, facilitated diffusion, and osmosis?
- describe active transport, endocytosis, and exocytosis?
- explain how the need to exchange materials across membranes influences cell size and shape?

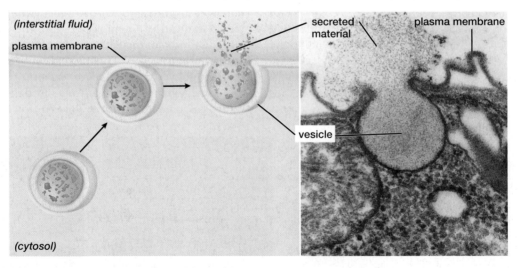

▲ **FIGURE 5-12 Exocytosis** Exocytosis is functionally the reverse of endocytosis. Material enclosed in a vesicle from inside the cell is transported to the cell surface. It then fuses with the plasma membrane, releasing its contents into the surrounding fluid.

THINK CRITICALLY How does exocytosis differ from diffusion of materials out of a cell?

5.3 HOW DO SPECIALIZED JUNCTIONS ALLOW CELLS TO CONNECT AND COMMUNICATE?

In multicellular organisms, some specialized structures on plasma membranes hold cells together, whereas others provide avenues through which cells communicate with neighboring cells. Here we discuss four major types of cell-to-cell connections: adhesive junctions, tight junctions, gap junctions, and plasmodesmata. The first three types of junctions are found only in animal cells; plasmodesmata are restricted to plant cells.

Adhesive Junctions Attach Cells Together

Adhesive junctions are specialized groups of proteins that link cells to one another within tissues. Adhesive junctions function by connecting the cytoskeleton to the inner plasma membrane and extending through the plasma membrane to extracellular linking proteins that join to the plasma membranes of adjacent cells. There are several types of adhesive junctions; here we focus on *desmosomes* (FIG. 5-14a). Desmosomes join cells in tissues that are repeatedly stretched, such as those found in skin, intestines, and the heart. These strong adhesive junctions prevent forces on the tissues from pulling them apart. In a desmosome, anchoring proteins lie on the inner side of the membranes of adjacent cells. The anchoring proteins are attached to intermediate filaments of the cytoskeleton that extend into the cytoplasm. Linking proteins join to the anchoring proteins and span the narrow space between the adjacent cells, linking them firmly together.

Tight Junctions Make Cell Attachments Leakproof

Tight junctions are formed by proteins that span the plasma membranes at corresponding sites on adjacent cells (FIG. 5-14b), joining the cells almost as if their adjacent membranes had been stitched together. Interlocking tight junction proteins create barriers that prevent nearly all substances from passing between the linked cells. For example,

▶ FIGURE 5-14 **Links between cells (a)** In desmosomes, anchoring proteins on adjacent plasma membranes are bound together by linking proteins. Intermediate filaments of the cytoskeleton inside each cell strengthen the connection. (Right) A transmission electron micrograph of a desmosome. **(b)** Tight junction proteins of adjacent cells fuse to one another, forming a stitch-like pattern. (Right) A scanning electron micrograph of a membrane whose bilayer has been split reveals the pattern of tight junction proteins. **(c)** Gap junctions consist of protein channels interconnecting the cytosol of adjacent cells to allow small molecules and ions to pass through. (Right) An atomic force micrograph looking down on connexons on one of the two membranes they connect. **(d)** Plasmodesmata connect the plasma membranes and cytosol of adjacent plant cells and allow large molecules to move between them. (Right) A transmission electron micrograph showing a cross-section of plasmodesmata connecting adjacent plant cells. SEM in part (b) from Claude, P., and Goodenough, D. 1973. "Fracture Faces of Zonulae Occludentes from 'Tight' and 'Leaky' Epithelia." *Journal of Cell Biology* 58:390–400.

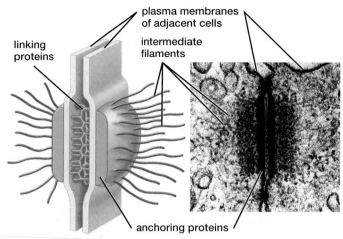

plasma membranes of adjacent cells
linking proteins
intermediate filaments
anchoring proteins

(a) Adhesive junction (desmosome)

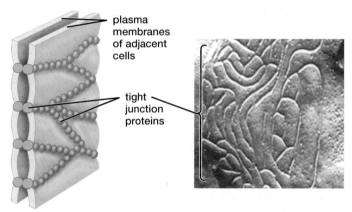

plasma membranes of adjacent cells
tight junction proteins

(b) Tight junctions

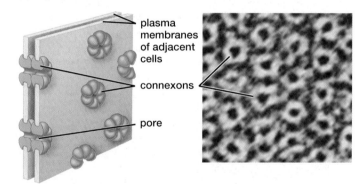

plasma membranes of adjacent cells
connexons
pore

(c) Gap junctions

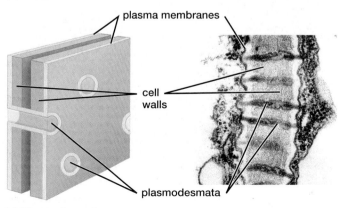

plasma membranes
cell walls
plasmodesmata

(d) Plasmodesmata

tight junctions in the bladder prevent cellular wastes in urine from leaking back into the blood. Tight junctions between cells lining the digestive tract protect the rest of the body from the acids, digestive enzymes, and bacteria found in its various compartments.

Gap Junctions and Plasmodesmata Allow Direct Communication Between Cells

The cells of many tissues in the animal body are interconnected by **gap junctions** (FIG. 5-14c), clusters of channels ranging in number from a few to thousands. The channels are formed by six-sided tubes of protein called *connexons* that span the plasma membrane. Connexons line up so that their central pores link the cytosol of adjacent cells. The small size of the pore allows small water-soluble molecules—including sugars, various ions, amino acids, and small messenger molecules such as cAMP—to pass between cells, but excludes organelles and large molecules such as proteins. Gap junctions coordinate the metabolic activities of many cells. They allow electrical signals to pass extremely rapidly among certain groups of nerve cells, and they synchronize contraction of heart muscle and of smooth muscles, such as in the walls of the digestive tract, bladder, and uterus.

Plasmodesmata are channels that link nearly all adjacent plant cells and allow movement of large molecules between them (FIG. 5-14d). These openings, which are lined with plasma membrane and filled with cytosol, make the membranes and the cytosol of adjacent cells continuous with one another. Many plant cells have thousands of plasmodesmata, allowing water, nutrients, and hormones to pass freely from one cell to another. These connections among plant cells serve a function somewhat similar to the gap junctions of animal cells in coordinating metabolic activities among groups of cells.

CHECK YOUR LEARNING

Can you ...

- describe the major types of junctions between cells?
- explain how these junctions function and provide an example of where each is found?

CASE STUDY \ **REVISITED**

Vicious Venoms

The "witches' brews" of rattlesnake and brown recluse spider venoms contain phospholipases that destroy the tissue around the bite (FIG. 5-15). When phospholipases attack the membranes of capillary cells, these tiny blood vessels rupture and release blood into the tissue surrounding the wound. In extreme cases, capillary damage can lead to internal bleeding. By attacking the membranes of red blood cells, rattlesnake venoms can cause anemia (an inadequate number of oxygen-carrying red blood cells). Rattlesnake phospholipases also attack muscle cell membranes; this attack caused extensive damage to muscles in Justin Schwartz's forearm. Justin required large quantities of antivenin, which contains specialized proteins that bind and neutralize the snake venom proteins. Unfortunately, no antivenin is available for brown recluse bites, and treatment generally consists of preventing infection, controlling pain and swelling, and waiting—sometimes for months—for the wound to heal.

Although both snake and spider bites can have serious consequences, very few of the spiders and snakes found in the Americas are dangerous to people. The best defense is to learn

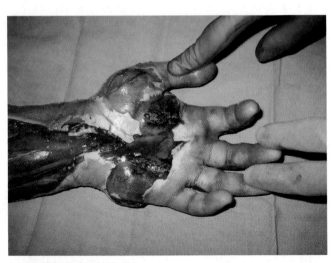

(a) Justin's rattlesnake bite

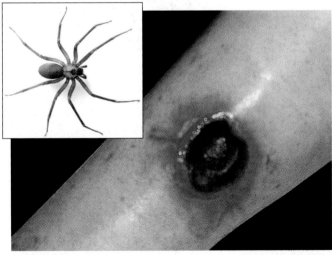

(b) Brown recluse spider bite

▲ **FIGURE 5-15 Phospholipases in venoms can destroy cells (a)** Justin Schwartz's hand 36 hours after the rattlesnake bite. **(b)** A brown recluse spider bite. (Inset) A brown recluse spider.

which venomous animals live in your area and where they prefer to hang out. If your activities bring you to such places, wear protective clothing—and always look before you reach! Knowledge can help us coexist with spiders and snakes, avoid their bites, and keep our cell membranes intact.

CHAPTER REVIEW

MB Go to **MasteringBiology** for practice quizzes, activities, eText, videos, current events, and more.

Answers to Think Critically, Evaluate This, Multiple Choice, and Fill-in-the-Blank questions can be found in the Answers section at the back of the book.

Summary of Key Concepts

5.1 How Is the Structure of the Cell Membrane Related to Its Function?

The cell membrane consists of a bilayer of phospholipids in which a variety of proteins are embedded, often described as a fluid mosaic. The plasma membrane isolates the cytoplasm from the external environment, regulates the flow of materials into and out of the cell, allows communication between cells, allows attachments within and between cells, and regulates many biochemical reactions. There are five major types of membrane proteins: (1) enzymes, which promote chemical reactions; (2) recognition proteins, which label the cell; (3) transport proteins, which regulate the movement of most water-soluble substances through the membrane; (4) receptor proteins, which bind molecules and trigger changes within the cell; and (5) connection proteins, which anchor the plasma membrane to the cytoskeleton and extracellular matrix or bind cells to one another.

5.2 How Do Substances Move Across Membranes?

Diffusion is the movement of particles from regions of higher concentration to regions of lower concentration. In simple diffusion, water, dissolved gases, and lipid-soluble molecules diffuse through the phospholipid bilayer. During facilitated diffusion, carrier proteins or channel proteins allow water and water-soluble molecules to cross the membrane down their concentration gradients without expending cellular energy. Osmosis is the diffusion of water across a selectively permeable membrane down its concentration gradient through the phospholipid bilayer or through aquaporins.

Energy-requiring transport includes active transport, in which carrier proteins use cellular energy (ATP) to drive the movement of molecules across the plasma membrane against concentration gradients. Interstitial fluid, large molecules, and food particles may be acquired by endocytosis, which includes pinocytosis, receptor-mediated endocytosis, and phagocytosis. The secretion of substances and the excretion of particulate cellular wastes are accomplished by exocytosis.

Cells exchange materials between the cytoplasm and the external environment principally by the slow process of diffusion through the plasma membrane. This requires that no part of the cell be too far from the plasma membrane, limiting the diameter of cells.

5.3 How Do Specialized Junctions Allow Cells to Connect and Communicate?

Animal cell junctions include (1) adhesive junctions, which attach adjacent cells within tissues and prevent tissues from tearing apart during ordinary movements; (2) tight junctions, which leak-proof the spaces between adjacent cells; (3) gap junctions, which connect the cytosol of adjacent animal cells; and (4) plasmodesmata, which connect adjacent plant cells.

Key Terms

active transport   85
adhesive junction   89
aquaporin   82
carrier protein   82
channel protein   82
concentration   80
connection protein   79
diffusion   80
endocytosis   86
energy-requiring transport   81
enzyme   79
exocytosis   88
facilitated diffusion   82
fluid   77
fluid mosaic model   77
food vacuole   88
gap junction   90
glycoprotein   78
gradient   80
hypertonic   82
hypotonic   82
interstitial fluid   77
isotonic   82
osmosis   82
passive transport   81
phagocytosis   87
phospholipid bilayer   77
pinocytosis   86
plasmodesmata   90
receptor protein   79
receptor-mediated endocytosis   87
recognition protein   79
selectively permeable   81
simple diffusion   81
solute   80
solvent   80
tight junction   89
transport protein   79
turgor pressure   84

Thinking Through the Concepts

Multiple Choice

1. Animal cells are surrounded by _____ fluid. This fluid is _____ to the cytosol.
a. phospholipid; isotonic
b. plasma; hypertonic
c. interstitial; isotonic
d. interstitial; hypotonic

2. Which of the following *cannot* enter a cell by simple diffusion?
   a. water
   b. sugar
   c. estrogen
   d. oxygen

3. Glycoproteins are important for
   a. binding hormones.
   b. cell recognition by the immune system.
   c. forming ion channels.
   d. active transport through the membrane.

4. Diffusion
   a. is always facilitated.
   b. occurs by active transport.
   c. is increased when temperature increases.
   d. requires aquaporin proteins.

5. Which of the following is *not* true of endocytosis?
   a. It is a form of passive transport.
   b. It includes pinocytosis.
   c. It can occur in coated pits.
   d. It is used by *Amoeba* to feed on *Paramecium*.

Fill-in-the-Blank

1. Membranes consist of a bilayer of _____ . The five major categories of protein within the bilayer are _____ , _____ , _____ , _____ , and _____ proteins.

2. A membrane that is permeable to some substances but not to others is described as being _____ . The movement of a substance through a membrane down its concentration gradient is called _____ . When applied to water, this process is called _____ . Channels that are specific for water are called _____ . The process that moves substances through a membrane against their concentration gradient is called _____ .

3. Facilitated diffusion involves either _____ proteins or _____ proteins. Diffusion directly through the phospholipid bilayer is called _____ diffusion, and molecules that take this route must be soluble in _____ or be very small and have no net electrical charge.

4. The three types of cell attachment structures in animal cells are _____ , _____ , and _____ . The structures that interconnect plant cells are called _____ .

5. After each molecule, place the two-word term that most specifically describes the *process* by which it moves through a plasma membrane. Carbon dioxide: _____ ; ethyl alcohol: _____ ; a sodium ion: _____ ; glucose: _____ .

6. The general process by which fluids or particles are transported out of cells is called _____ . Does this process require energy? _____ The substances to be expelled are transported within the cell in membrane-enclosed sacs called _____ .

Review Questions

1. Describe and diagram the structure of a plasma membrane. What are the two principal types of molecules in plasma membranes, and what is the general function of each?

2. Sketch the configuration that 10 phospholipid molecules would assume if placed in water. Explain why they arrange themselves this way.

3. What are the five categories of proteins commonly found in plasma membranes, and what is the function of each one?

4. Define *diffusion* and *osmosis*. Explain how osmosis helps plant leaves remain firm. What is the term for water pressure inside plant cells?

5. Define *hypotonic*, *hypertonic*, and *isotonic*. What would be the fate of an animal cell immersed in each of these three types of solution?

6. Describe the following types of transport processes in cells: simple diffusion, facilitated diffusion, active transport, pinocytosis, receptor-mediated endocytosis, phagocytosis, and exocytosis.

7. Name the protein that allows facilitated diffusion of water. What experiment demonstrated the function of this protein?

8. Imagine a container of glucose solution, divided into two compartments (A and B) by a membrane that is permeable to both water and glucose. If some glucose is added to compartment A, how will the contents of compartment B change? Explain.

9. Name four types of cell-to-cell junctions, and describe the function of each. Which are found in plants and which in animals?

Applying the Concepts

1. Different cells have different plasma membranes. The plasma membrane of a *Paramecium,* for example, is only about 1% as permeable to water as the plasma membrane of a human red blood cell. Hypothesize why this is the case. Is *Paramecium* likely to have aquaporins in its plasma membrane? Explain your answer.

2. Predict and sketch the configuration that ten phopholipid molecules would assume if they were completely submerged in vegetable oil. Explain your prediction.

6

ENERGY FLOW IN THE LIFE OF A CELL

The bodies of these runners in the New York Marathon convert stored energy to the energy of movement and heat. Their pounding footsteps shake the Verrazano Narrows Bridge.

Energy Unleashed

PICTURE THE NEW YORK CITY MARATHON, where well over 50,000 people from countries throughout the world gather to run 26.2 miles. All participate in a personal odyssey and a testimony to persistence, endurance, and the ability of the human body to utilize energy. On average, each runner expends roughly 3,000 Calories before reaching the finish line. Once finished, they douse their overheated bodies with water and replenish their energy stores with celebratory meals. Finally, subways, trains, cars, buses, and airplanes—burning vast quantities

of fossil fuels and releasing enormous amounts of heat—transport the runners home.

Training for a marathon takes months, especially for someone unaccustomed to running long distances. During training, several important physiological changes occur that prepare the body to expend the tremendous amount of energy necessary for the race. Muscle mitochondria, with their glucose-metabolizing enzymes, increase in number. The cells of muscles that move the body increase their ability to store glycogen, a polymer of glucose. Capillaries around muscles proliferate to supply the extra oxygen needed to break down glucose in the mitochondria.

What exactly is energy? Do our bodies use energy according to the same principles that govern its use in the engines of cars and airplanes? Why do our bodies generate heat, and why do we give off more heat when exercising than when studying or watching TV?

6.1 WHAT IS ENERGY?

Energy is the capacity to do work. **Work,** in turn, is the transfer of energy to an object that causes the object to move. It is obvious that marathoners are working; their chests heave, their arms pump, and their legs stride, moving their bodies relentlessly forward for 26.2 miles. This muscular work is powered by the energy available in the bonds of molecules. The molecules that provide this energy—including glucose, glycogen, and fat—are stored in the cells of the runners' bodies. Within each cell, specialized molecules such as ATP accept, briefly store, and transfer energy from the reactions that release energy to those that demand it, such as muscle contraction.

There are two fundamental types of energy: potential energy and kinetic energy, each of which takes several forms. **Potential energy** is stored energy, including the elastic energy stored in a compressed spring or a drawn bow and the gravitational energy stored in water behind a dam or a roller-coaster car about to begin its downward plunge (**FIG. 6-1**). Potential energy also includes **chemical energy,** which is energy stored in, for example, batteries, the biological molecules that power marathon runners, and the fossil fuels used by vehicles to transport the runners. **Kinetic energy** is the energy of movement. It includes radiant energy (such as waves of light, X-rays, and other forms of *electromagnetic radiation*), heat or thermal energy (the motion of molecules or atoms), electrical energy (electricity; the flow of charged particles), and any motion of larger objects, such as the plummeting roller-coaster car or running marathoners. Under the right conditions, kinetic energy can be transformed into potential energy, and vice versa. For example, the roller-coaster car converts the kinetic energy of its downward plunge into gravitational potential energy as it coasts to the top of the next rise. At a molecular level, during photosynthesis, the kinetic energy of light is captured and transformed into the potential energy of chemical bonds (see Chapter 7). To understand energy flow and change, we need to know more about its properties.

The Laws of Thermodynamics Describe the Basic Properties of Energy

The laws of thermodynamics describe some basic properties of energy. The **first law of thermodynamics** states that energy can be neither created nor destroyed by ordinary processes. (Nuclear reactions, in which matter is converted into

▲ **FIGURE 6-1 Converting potential energy to kinetic energy** Roller coasters convert gravitational potential energy to kinetic energy as they plummet downhill.

THINK CRITICALLY Could one design a roller coaster that didn't use any motors to pull the cars uphill after they were released from a high point?

energy, are the exception.) This means that within an **isolated system**—a space where neither mass nor energy can enter or leave—the total amount of energy before and after any process will be unchanged. For this reason, the first law of thermodynamics is often called the **law of conservation of energy.** An isolated system is a theoretical concept, but for practical purposes, you can visualize energy transformations occurring in an enormous, perfectly sealed and insulated chamber.

To illustrate the law of conservation of energy, consider a gasoline-powered car. Before you turn the ignition key, the energy in the car is all potential energy, stored in the chemical bonds of its fuel. As you drive, only about 20% of this potential energy is converted into the kinetic energy of motion. But if energy is neither created nor destroyed, what happens to the other 80% of the energy? The burning fuel also heats

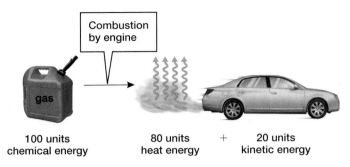

▲ FIGURE 6-2 All energy conversions result in a loss of useful energy

100 units
chemical energy

80 units
heat energy

+

20 units
kinetic energy

up the engine, the exhaust system, and the air around the car, while friction from the tires heats the road. So, as the first law dictates, the total amount of energy remains the same, although it has changed in form—about 20% of it converted to kinetic energy and 80% of it to heat (**FIG. 6-2**).

The **second law of thermodynamics** states that when energy is converted from one form to another, the amount of useful energy decreases. In other words, all ordinary (non-nuclear) processes cause energy to be converted from more useful into less useful forms, such as the heat in our combustion example, which increased the random movement of molecules in the car, the air, and the road. One way of reducing the chemical energy lost as heat while driving is discussed in "Earth Watch: Step on the Brakes and Recharge Your Battery."

Now consider the human body. Whether running or reading, your body "burns" food to release the chemical energy stored in its molecules. Your body warmth results from the heat given off, which is radiated to your surroundings. This heat is not available to power muscle contraction or to help brain cells interpret written words. Thus, the second law tells us that no energy conversion process, including those that occur in the body, is 100% efficient; some energy is lost to the environment—almost always in the form of heat—which cannot be used to power muscles or brain activity.

The second law of thermodynamics also tells us something about the organization of matter. Useful energy tends to be stored in highly ordered matter, such as in the bonds of complex molecules. As a result, whenever energy is used within an isolated system, there is an overall loss of organization as complex molecules are broken apart into simpler ones. The loss of organization also occurs as we perform activities of daily life: Dirty dishes accumulate, clothes collect in confusion, the bed gets rumpled, and books and papers pile up (**FIG. 6-3**). This randomness and disorder can only be reversed by adding energy to the system through energy-demanding cleaning and organizing efforts. You'll see a bit later where this energy comes from.

This tendency toward the loss of complexity, orderliness, and useful energy is called **entropy.** At the molecular level, we see the same principle at work. Let's look at what happens when we burn glucose sugar.

$$C_6H_{12}O_6 + 6\,O_2 \longrightarrow 6\,CO_2 + 6\,H_2O + \text{heat energy}$$

glucose   oxygen   carbon   water
dioxide

▲ FIGURE 6-3 Entropy at work

Although you'll find the same number and types of atoms on both sides of the equation, if you count the molecules, you will see that there is an overall increase in simple product molecules (carbon dioxide and water) as the single molecule of sugar is broken down and combined with oxygen. The heat energy that is released causes the product molecules to move about randomly and more rapidly. To counteract entropy—for example, to synthesize glucose from carbon dioxide and water—energy must be infused into the system from an outside source, ultimately the sun. When the eminent Yale scientist G. Evelyn Hutchinson stated, "Disorder spreads through the universe, and life alone battles against it," he was making an eloquent reference to entropy and the second law of thermodynamics.

CASE STUDY \ **CONTINUED**

Energy Unleashed

Much like a car's engine, the marathoner's muscles are only about 20% efficient in converting chemical energy into movement; much of the other 80% is lost as heat. Sweating helps to prevent overheating because the water in sweat absorbs large amounts of heat as it evaporates. But even while sitting at the computer and doing other non-sweaty activities, we still burn energy, just to stay alive. Where does this energy come from?

Living Things Use Solar Energy to Maintain Life

If you think about the second law of thermodynamics, you may wonder how life can exist at all. If chemical reactions, including those inside living cells, cause the amount

Earth WATCH — Step on the Brakes and Recharge Your Battery

As carbon dioxide levels in the atmosphere continue to rise, fueling global climate change, it is increasingly urgent that we reduce our impact by improving auto fuel economy.

In a typical car powered only by an internal combustion engine, whenever you step on the brake, brake pads are forced against brake discs. The resulting friction that stops the car converts the kinetic energy of forward motion almost entirely into waste heat. Fortunately, engineers have devised a way to capture and use some of this squandered energy. Called regenerative braking, this technology is used in hybrid or all-electric cars, which are partly or entirely driven by battery-powered electric motors. As you drive, potential chemical energy stored in a large battery is converted to kinetic energy by the car's electric motor, which drives the wheels. Stepping on the brake flips a switch that reverses this process, forcing the turning wheels to drive the electric motor in the opposite direction. This reversal converts the electric motor into a generator of electricity (**FIG. E6-1**), and the resistance to reversing the electric motor helps slow the car. This electrical energy, derived from the kinetic energy of the car's forward motion, is transmitted back to the battery, where it is stored as chemical energy. This chemical energy can be used to propel the car forward when you start up again. Of course, the second law of thermodynamics tells us that each of these energy conversions will generate some heat, but regenerative braking wastes 30% to 50% less energy than conventional friction braking. Regenerative braking allows cars to travel much farther on less energy.

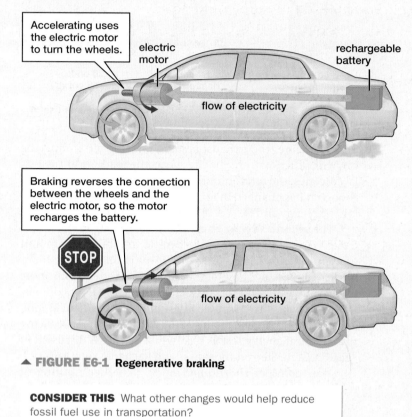

▲ FIGURE E6-1 Regenerative braking

CONSIDER THIS What other changes would help reduce fossil fuel use in transportation?

---

of usable energy to decrease, and if matter tends toward increasing randomness and disorder, how can organisms maintain the amazingly organized complexity of life? Where does useful energy originate, and where does all the waste heat go? The answer is that cells, the bodies of organisms, and Earth itself are not isolated systems; they receive useful solar energy released by nuclear reactions in the sun, 93 million miles away. As they generate light and other forms of electromagnetic energy, these nuclear reactions liberate an almost unimaginable amount of heat; the temperature of the sun's core is estimated to be 27 million °F (16 million °C).

Living things "battle against disorder" by using a continuous influx of sunlight to synthesize complex molecules and maintain their intricate bodies. All solar energy enters the biosphere through photosynthetic organisms such as plants and algae. The chemical equation for photosynthesis is:

$$6\,CO_2 + 6\,H_2O + \text{light energy} \longrightarrow C_6H_{12}O_6 + 6\,O_2$$

Notice that photosynthesis reverses the reaction that breaks down glucose; the energy of sunlight is captured and stored in the chemical bonds of glucose. Thus, the highly ordered (and therefore low-entropy) systems that characterize life do not violate the second law of thermodynamics.

Such systems are produced and maintained through a continuous influx of useful energy from the sun. So when you organize your desk or make your bed, your muscles are (indirectly) using solar energy that was originally trapped by photosynthesis.

What happens to the solar energy that is not trapped in living things? Most of it is converted to heat. Some heat remains in the atmosphere and maintains Earth within a temperature range that sustains life. But humans are also changing the composition of the atmosphere. Since the 1800s, we have increased atmospheric CO_2 levels by about 30% by burning fossil fuels to use their stored chemical energy. Because CO_2 traps heat in the atmosphere somewhat like the glass of a greenhouse or the closed windows of a car in the sun, we are changing Earth's climate.

CHECK YOUR LEARNING
Can you ...
- define energy and work?
- define potential energy and kinetic energy and provide three examples of each?
- state and explain the first and second laws of thermodynamics?

6.2 HOW IS ENERGY TRANSFORMED DURING CHEMICAL REACTIONS?

A **chemical reaction** is a process that breaks and forms chemical bonds. Chemical reactions convert one combination of molecules, the **reactants,** into different molecules, the **products,** that contain the same numbers and types of atoms. Since all chemical reactions transfer energy, all release some heat. A reaction is **exergonic** if there is an overall release of heat, that is, if the products contain less energy than the original reactants, as in burning sugar (**FIG. 6-4a**). A reaction is **endergonic** if it requires a net input of energy, that is, if the products contain more energy than the reactants. Endergonic reactions require an overall influx of energy from an outside source (**FIG. 6-4b**).

Exergonic Reactions Release Energy

Sugar can be ignited by heat, as any cook can tell you. As it burns, sugar undergoes the same overall reaction as it does in our bodies and in most other forms of life. Organisms combine sugar with oxygen, producing carbon dioxide and water, while generating both stored chemical energy (ATP) and releasing heat. The total energy in the reactant molecules (glucose and oxygen) is much higher than in the product molecules (carbon dioxide and water), so burning sugar is an exergonic reaction. It may be helpful to think of exergonic reactions as running "downhill," from a higher energy state to a lower energy state, like a rock rolling down a hill to rest at the bottom.

Endergonic Reactions Require a Net Input of Energy

Many reactions in living things are endergonic, requiring a net input of energy and yielding products that contain more energy than the reactants. The synthesis of large biological molecules is endergonic. For example, the proteins in a muscle cell contain more energy than the individual amino acids that were linked together to synthesize them. How do organisms power endergonic reactions? They use high-energy

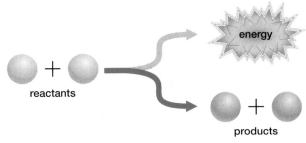
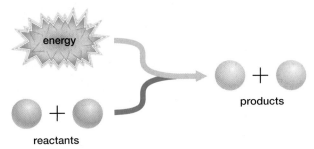

(a) An exergonic reaction

(b) An endergonic reaction

▲ **FIGURE 6-4 Exergonic and endergonic reactions (a)** In an exergonic reaction, the products contain less energy than the reactants. **(b)** In an endergonic reaction, the products contain more energy than the reactants.

THINK CRITICALLY Is glucose breakdown endergonic or exergonic? What about photosynthesis?

molecules synthesized using solar energy that was captured during photosynthesis. We can think of endergonic reactions as "uphill" reactions because they require a net input of energy, just as pushing a rock up a hill requires effort.

All Chemical Reactions Require Activation Energy to Begin

All chemical reactions require **activation energy** to get started (**FIG. 6-5a**). Think of a rock sitting at the top of a hill; it will remain there indefinitely unless a push starts it rolling down. Like the rolling rock, many chemical reactions continue spontaneously if enough activation energy is supplied to start them. We see this in wood burning in a campfire or in a marshmallow ignited by the fire's flames (**FIG. 6-5b**). After the sugar in a marshmallow is ignited as it reacts with oxygen from the air, the

high / energy content of molecules / low

Activation energy required to start the reaction

energy level of reactants

reactants

energy level of products

products

progress of reaction

(a) An exergonic reaction

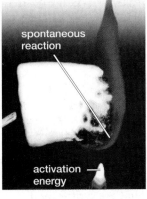

spontaneous reaction

activation energy

(b) A flame ignites the sugar in a marshmallow

◀ **FIGURE 6-5 Activation energy in an exergonic reaction (a)** After the activation hump is overcome, there will be a net release of energy. **(b)** Heat released by the burning sugar will allow the reaction to continue spontaneously.

reaction releases enough heat to sustain itself, and the marshmallow burns spontaneously.

Why is activation energy required for chemical reactions? Shells of negatively charged electrons surround all atoms (see Chapter 2). These negative charges repel one another and tend to keep the atoms separated. Activation energy is required to overcome this repulsion and force the atoms close enough together to react and form new chemical bonds.

Activation energy can be provided by the kinetic energy of moving molecules. Atoms and molecules are in constant motion. If they are moving fast enough, collisions between reactive molecules force their electron shells to mingle and react. Because molecules move faster as the temperature increases, most chemical reactions occur more readily at high temperatures; this is why a flame's heat can ignite a marshmallow.

CHECK YOUR LEARNING

Can you ...

- describe how energy is captured and released by chemical reactions?
- explain exergonic and endergonic reactions and provide examples of each?
- explain activation energy?

CASE STUDY \ CONTINUED
Energy Unleashed

Marathoners rely on glycogen stored in their muscles and liver for much of the energy to power their run. Glycogen consists of long, branched chains of glucose molecules. When energy is needed, glucose molecules are cleaved from the chain and then broken down into carbon dioxide and water. This exergonic reaction generates the chemical energy in ATP that will power muscle contraction. The carbon dioxide is exhaled as the runners breathe rapidly to supply their muscles with adequate oxygen. The water generated by glucose breakdown (and a lot more that the runners drink during the race) is lost as cooling sweat.

The glucose molecules in muscle and other cells do not need to be ignited and do not literally burn as they are broken down, so how do cells provide activation energy and control the release of chemical energy to allow it to do work?

6.3 HOW IS ENERGY TRANSPORTED WITHIN CELLS?

Most organisms are powered by the chemical energy supplied by the exergonic breakdown of glucose. But the chemical energy stored in glucose must first be transferred to energy-carrier molecules, such as ATP. **Energy-carrier molecules** are high-energy molecules that are synthesized at the site of an exergonic reaction, where they capture and temporarily store some of the released chemical energy. Just as rechargeable flashlight batteries can store electrical

energy that is later released as light, energy-carrier molecules are charged up by exergonic reactions and then release their energy to drive endergonic reactions. They can then be recharged, as described later. Energy-carrier molecules capture and transfer energy only within cells; they cannot ferry energy through cell membranes, nor are they used for long-term energy storage.

ATP and Electron Carriers Transport Energy Within Cells

Many exergonic reactions in cells, such as breaking down sugars and fats, produce ATP, the most common energy-carrier molecule in the body. ATP (adenosine triphosphate) is a nucleotide composed of the nitrogen-containing base adenine, the sugar ribose, and three phosphate groups. Because ATP provides energy to drive a wide variety of endergonic reactions, it is sometimes called the "energy currency" of cells. It is produced when energy from exergonic reactions is used to combine the lower-energy molecules of inorganic phosphate (HPO_4^{2-}, also designated P_i) with ADP (adenosine diphosphate) (**FIG. 6-6a**). Because it requires a net input of energy, ATP synthesis is endergonic.

ATP diffuses throughout the cell, carrying energy to sites where endergonic reactions occur. There its energy is liberated as it is broken down, regenerating ADP and P_i (**FIG. 6-6b**). The life span of an ATP molecule in a cell is very short; each molecule is recycled roughly 1,400 times every day. A marathon runner may use a pound (0.45 kilogram) of ATP molecules every minute, so if ATP were not almost instantly recycled, marathons would never happen. In contrast to ATP, far more stable molecules such as starch in plants and

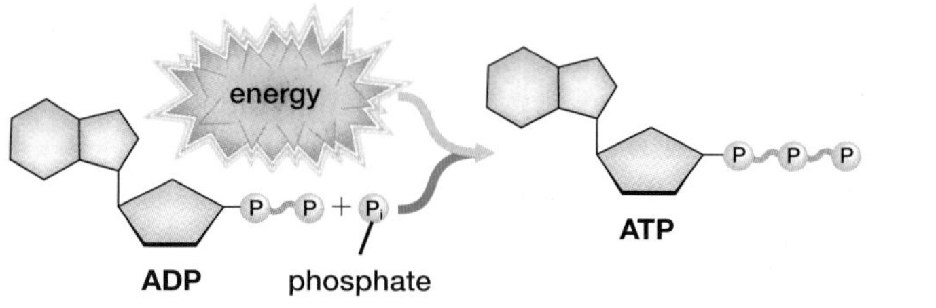

(a) ATP synthesis: Energy is stored in ATP

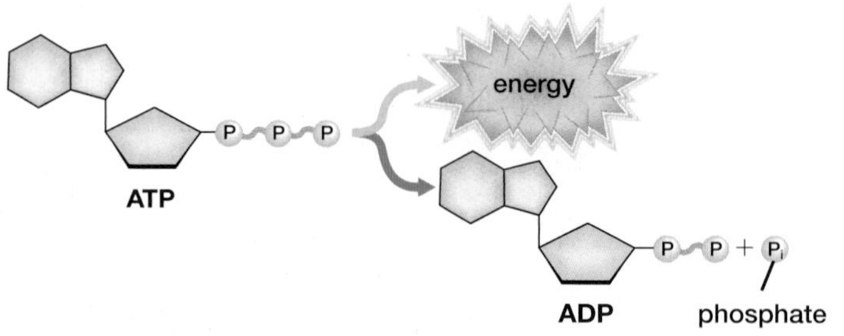

(b) ATP breakdown: Energy is released

▲ **FIGURE 6-6 The interconversion of ADP and ATP (a)** Energy is captured when a phosphate group (P_i) is added to adenosine diphosphate (ADP) to make adenosine triphosphate (ATP). **(b)** Energy to power cellular work is released when ATP is broken down into ADP and P_i.

glycogen and fat in animals can store energy for hours, days, or—in the case of fat—years.

ATP is not the only energy-carrier molecule within cells. In some exergonic reactions, including both glucose breakdown and the light-capturing stage of photosynthesis, some energy is transferred to electrons. These energetic electrons, along with hydrogen ions (H^+; present in the cytosol of cells), are captured by molecules called **electron carriers.** The loaded electron carriers donate their high-energy electrons to other molecules, which are often involved in pathways that generate ATP. Common electron carriers include NADH (nucleotide nicotinamide adenine dinucleotide) and its relative, $FADH_2$ (flavin adenine dinucleotide). You will learn more about electron carriers in Chapters 7 and 8.

Coupled Reactions Link Exergonic with Endergonic Reactions

In a **coupled reaction,** an exergonic reaction provides the energy needed to drive an endergonic reaction (**FIG. 6-7**), using ATP or electron carriers as intermediaries. During photosynthesis, for example, plants use sunlight (from exergonic reactions in the sun's core) to drive the endergonic synthesis of high-energy glucose molecules from lower-energy reactants. Nearly all organisms use the energy released by exergonic reactions (such as the breakdown of glucose) to drive endergonic reactions (such as the synthesis of proteins from amino acids). Because some energy is lost every time it is transformed, in coupled reactions the energy released by exergonic reactions always exceeds the energy needed to drive the endergonic reactions. Thus, the coupled reaction overall is exergonic.

The exergonic and endergonic portions of coupled reactions often occur in different places within a cell, so energy is transferred by energy-carrier molecules such as ATP. In its role as an intermediary in coupled reactions, ATP is constantly being synthesized to capture the energy released during exergonic reactions and then broken down to power endergonic reactions.

CHECK YOUR LEARNING
Can you ...
- name and describe two important energy-carrier molecules in cells?
- explain coupled reactions?

6.4 HOW DO ENZYMES PROMOTE BIOCHEMICAL REACTIONS?

Ignite sugar and it will go up in flames as it combines rapidly with oxygen, releasing carbon dioxide and water. The same overall reaction occurs in our cells, although not with uncontrolled blasts of heat. To capture energy in ATP, cells channel the release of energy produced by the breakdown of sugar in controlled steps. These steps are important because a single sugar molecule contains sufficient energy to produce dozens of ATP molecules.

Catalysts Reduce the Energy Required to Start a Reaction

In general, how likely a reaction is to occur is determined by its activation energy, that is, by how much energy is required to overcome the barrier created by repelling forces between atoms (see Fig. 6-5a). Some reactions, such as sugar dissolving in water, have low activation energies and occur rapidly at human body temperature (approximately 98.6°F, or 37°C). In contrast, you could store sugar at body temperature in the presence of oxygen for decades and it would remain virtually unchanged. Why? Although the reaction of sugar with oxygen to yield carbon dioxide and water is exergonic, this reaction has a high activation energy. The heat of a flame can overcome this activation energy barrier by increasing the rate of movement of sugar molecules and nearby oxygen

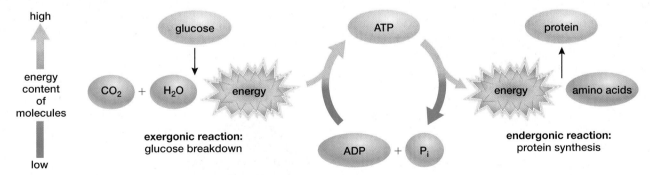

▲ **FIGURE 6-7 Coupled reactions within cells** Exergonic reactions, such as glucose breakdown, drive the endergonic reaction that synthesizes ATP from ADP and P_i. The ATP molecule carries its chemical energy to a part of the cell where energy is needed to drive an endergonic reaction, such as protein synthesis.

THINK CRITICALLY Why is the overall coupled reaction exergonic?

molecules, causing them to collide with sufficient force to react; the sugar burns as it releases energy. But, obviously, ignition doesn't start biological reactions in the body. Instead, life takes a different approach, lowering activation energy with catalysts.

Catalysts speed up the rate of a reaction by reducing its activation energy (**FIG. 6-8**); in the process catalysts themselves are neither used up nor permanently altered. Consider the catalytic converters on automobile exhaust systems. Incomplete combustion of gasoline generates poisonous carbon monoxide gas, which could reach dangerous levels in heavy traffic. Catalytic converters consist of a metallic catalyst that provides a surface upon which the carbon monoxide combines rapidly with atmospheric oxygen to produce carbon dioxide.

Enzymes Are Biological Catalysts

Inorganic catalysts speed up a number of different chemical reactions. But in cells, indiscriminately speeding up dozens of reactions would almost certainly be deadly. Instead, cells employ highly specific biological catalysts called **enzymes,** nearly all of which are proteins. Most enzymes catalyze one or a few types of chemical reactions involving specific molecules, leaving very similar molecules unchanged.

Both exergonic and endergonic reactions are catalyzed by enzymes. The synthesis of ATP from ADP and P_i, for example, is catalyzed by the enzyme ATP synthase. When energy is required to drive an endergonic reaction, ATP is broken down by an ATPase. As you read about enzymes, be aware that enzyme names are not consistent. In some cases, the suffix "ase" is added to what the enzyme does (ATP synthase), in other cases, "ase" is added to the molecule upon which the enzyme acts (e.g. ATPase), while some enzymes have their own unique names (e.g. pepsin).

The Structure of Enzymes Allows Them to Catalyze Specific Reactions

The function of an enzyme, like the function of any protein, is determined by its structure (see Chapter 3). Each enzyme's distinctive shape is determined by its amino acid sequence and the precise way in which the amino acid chain is twisted and folded. The three-dimensional structure of enzymes allows them to orient, distort, and reconfigure other molecules, causing these molecules to react, while the enzyme emerges unchanged.

Each enzyme has a pocket, called the **active site,** into which reactant molecules, called **substrates,** can enter. The shape of the active site, as well as the charges on the amino acids that form the active site, determine which molecules can enter. Consider the enzyme amylase, for example. Amylase breaks down starch molecules by hydrolysis, but leaves cellulose molecules intact, even though both starch and cellulose consist of chains of glucose molecules. Why? Because the bonding pattern between glucose molecules in starch allows the glucose chain to fit into the active

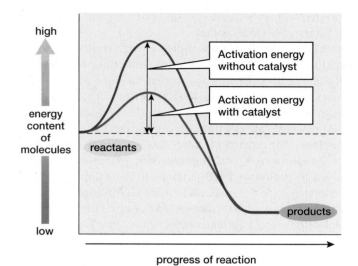

▲ **FIGURE 6-8 Catalysts lower activation energy** At any given temperature, a reaction is much more likely to proceed in the presence of a catalyst.

THINK CRITICALLY Can an enzyme catalyst make an endergonic reaction occur spontaneously at body temperature?

HAVE YOU EVER WONDERED...

If Plants Can Glow in the Dark?

You may have seen the almost magical glow of fireflies, but did you know that plants can be bioengineered to glow in the dark, too? The firefly's natural glow comes from specialized cells in their abdomens that are bioluminescent, meaning that they produce light from biological reactions. These cells are rich in both ATP and the fluorescent chemical luciferin (L. *lucifer*, light-bringer). Luciferin and ATP serve as substrates for the enzyme luciferase. In the presence of oxygen, luciferase catalyzes a reaction that modifies luciferin, using the energy from ATP to boost electrons briefly into a higher-energy electron shell. As they fall back into their original shell, the electrons emit their excess energy as light.

Plants don't naturally glow, but bioluminescence has been bioengineered into glowing *Arabidopsis* plants by the Glowing Plant team. First, the team ordered commercial luciferin and luciferase DNA synthesized from its computerized gene sequence. Then they implanted these genes into a special form of bacterial DNA that incorporates itself into plant cells. They used a "gene gun" to shoot microscopic particles coated with the modified bacterial DNA into masses of plant stem cells, which developed into plants. Choosing the brightest-glowing plants, the team harvested their seeds to grow new generations, whose seeds are now sold online to buyers who wish to brighten their abodes.

site of amylase, but the bonding pattern in cellulose does not. In the stomach, the enzyme pepsin breaks down proteins, attacking them at certain sites along their amino acid chains. Certain other protein-digesting enzymes (trypsin, for example) will break bonds only between specific amino acids. Therefore, digestive systems must manufacture several different enzymes that work together to completely break down dietary protein into its individual amino acids.

How does an enzyme catalyze a reaction? You can follow the events in **FIGURE 6-9**, which illustrates two substrate molecules combining into a single product. The shape and charge of the active site allow substrates to enter the enzyme only in specific orientations ❶. When appropriate substrates enter the active site, both the substrates and active site change shape slightly as weak chemical bonds form between specific amino acids in the active site and specific parts of the substrate ❷. This shape change distorts the original bonds within the substrate, making these bonds easier to break. The combination of substrate selectivity, substrate orientation, temporary bonds, and the distortion of existing bonds promotes the specific chemical reaction catalyzed by a particular enzyme. This holds true whether the enzyme is causing two molecules to react with one another or causing a single molecule to split into smaller products. When the reaction is complete, the product no longer fits properly into the active site and diffuses away ❸. The enzyme reverts to its original configuration, and it is then ready to accept another set of the same substrates. When substrate molecules are abundant, some fast-acting enzymes can catalyze tens of thousands of reactions per second, while others act far more slowly.

Enzymes, Like All Catalysts, Lower Activation Energy

The breakdown or synthesis of a molecule within a cell usually occurs in many small, discrete steps, each catalyzed by a different enzyme. Each of these enzymes lowers the activation energy for its particular reaction, allowing the reaction to occur readily at body temperature. Imagine how much easier it is to walk up a flight of stairs compared to scaling a cliff of the same height. In a similar manner, a series of reaction "stair steps"—each requiring a small amount of activation energy and each catalyzed by an enzyme that lowers activation energy—allows the overall reaction to surmount its high activation energy barrier and to proceed at body temperature.

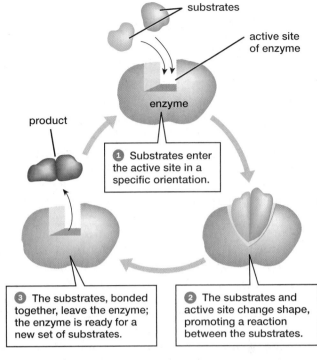

▲ **FIGURE 6-9 The cycle of enzyme–substrate interactions** This diagram shows two reactant substrate molecules combining to form a single product. Enzymes can also catalyze the breakdown of a single substrate into two product molecules.

CHECK YOUR LEARNING

Can you ...

- explain how catalysts reduce activation energy?
- explain how enzymes function as biological catalysts?

6.5 HOW ARE ENZYMES REGULATED?

The **metabolism** of a cell is the sum of all its chemical reactions. Many of these reactions, such as those that break down glucose into carbon dioxide and water, are linked in sequences called **metabolic pathways** (FIG. 6-10). In a metabolic pathway, a starting reactant molecule is converted, with the help of an enzyme, into a slightly different intermediate molecule, which is modified by yet another enzyme to form a second intermediate, and so on, until an end product is produced. Photosynthesis (see Chapter 7), for example, is a metabolic pathway, as is the breakdown of glucose (Chapter 8). Different

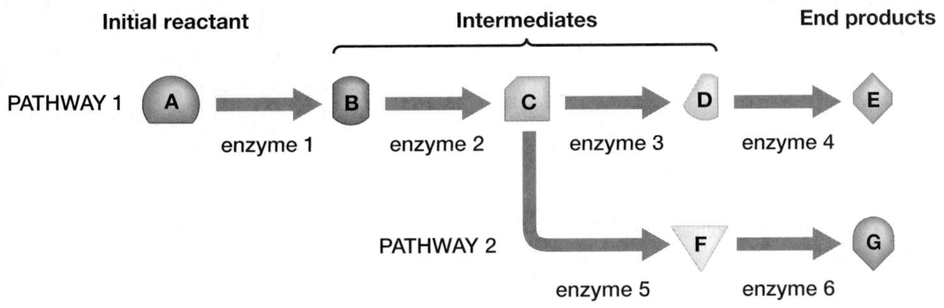

◀ **FIGURE 6-10 Simplified metabolic pathways** The initial reactant molecule (A) undergoes a series of reactions, each catalyzed by a specific enzyme. The product of each reaction serves as the reactant for the next reaction in the pathway. Metabolic pathways are commonly interconnected, so the product of a step in one pathway (C in pathway 1) often serves as a substrate for an enzyme in a different pathway (enzyme 5 in pathway 2).

metabolic pathways often use some of the same molecules; as a result, all the thousands of metabolic pathways within a cell are directly or indirectly interconnected.

Cells Regulate Metabolic Pathways by Controlling Enzyme Synthesis and Activity

In a test tube under constant, ideal conditions, the rate of a particular reaction will depend on how many substrate molecules diffuse into the active sites of enzyme molecules in a given time period. This, in turn, will be determined by the concentrations of the enzyme and substrate molecules. Generally, increasing the concentrations of either the enzyme or the substrate (or both) will increase the reaction rate, because it will boost the chances that the two types of molecules will meet. But living cells must precisely control the rate of reactions in their metabolic pathways, requiring a system far more complex than that in a test tube. Cells must keep the amounts of end products within narrow limits, even when the amounts of reactants (enzyme substrates) fluctuate considerably. For example, when glucose molecules flood into the bloodstream after a meal, it would not be desirable to metabolize them all at once, producing far more ATP than the cell needs. Instead, some of the glucose molecules should be stored as glycogen or fat for later use. To be effective, then, metabolic reactions within cells must be precisely regulated;

they must occur at the proper times and proceed at the proper rates. Cells regulate their metabolic pathways by controlling the type, quantity, and activity levels of the enzymes they produce.

Genes That Code for Enzymes May Be Turned On or Off

A very effective way for cells to regulate enzymes is to turn the genes that code for specific enzymes on or off depending on the cell's changing needs. Gene regulation may cause enzymes to be synthesized in larger quantities when more of their substrate is available. Larger concentrations of an enzyme make it more likely that substrate molecules will encounter the enzyme, speeding up the rate at which the reaction occurs. For example, glucose entering the bloodstream after a starchy meal triggers an elaborate series of metabolic adjustments. One of these causes the pancreas to release the hormone insulin. Insulin activates the gene that codes for the first enzyme in the metabolic pathway that breaks down glucose. Insulin also activates a gene that codes for fatty acid synthase, an enzyme that helps convert molecules liberated by glucose breakdown into fats that store energy for later use.

Some enzymes are synthesized only during specific stages in an organism's life. For example, the enzyme that allows organisms to digest lactose (milk sugar) is typically lost after the animal is weaned. A mutation can alter this type

Health WATCH   Lack of an Enzyme Leads to Lactose Intolerance

If you enjoy ice cream and pizza, it might be hard for you to imagine life without these treats. However, such dairy-containing treats cannot be enjoyed by much of the world's population. Although all young children normally produce lactase (the enzyme that breaks down lactose, or "milk sugar"), about 65% of people worldwide, including 30 to 50 million people in the United States, produce less of this enzyme as they progress through childhood, a condition called *lactose intolerance*. In the worst cases, people may experience abdominal pain, flatulence, nausea, and diarrhea after consuming milk products (**FIG. E6-2**).

Why do people stop synthesizing the enzyme for this nutritious food? From an evolutionary perspective, it makes sense not to continue expending energy to produce an enzyme that has no function. In our early ancestors (who had not yet domesticated livestock), lactase lost its function in very early childhood because, after weaning, these people no longer had access to milk—the main source of lactose. As a result, many modern adults cannot digest lactose because the gene that encodes lactase is regulated by being turned off after weaning.

Lactose intolerance is particularly prevalent in people of East Asian, West African, and Native American descent. Genetic studies have revealed that between 10,000 and 6,000 years ago, some people in northern Europe and the Middle East acquired mutations that allowed them to digest lactose throughout their lives. These mutations were advantageous and gradually spread because they provided better nutrition for members of agricultural societies, who could obtain milk as well as meat from their livestock. Their descendants today continue to enjoy milk, ice cream, and extra-cheese pizzas.

▲ **FIGURE E6-2 Risky behavior?** For the majority of the world's adults, drinking milk invites unpleasant consequences.

EVALUATE THIS A family brings their 8-year-old adopted child to a pediatric clinic because she has begun to suffer from diarrhea and stomach cramps after drinking milk. What would the pediatrician suspect was the cause? If tests confirm his suspicions, what approaches would he recommend to deal with the issue? Are there dairy products that would not cause the reaction? How might they work?

of regulation, as described in "Health Watch: Lack of an Enzyme Leads to Lactose Intolerance."

Some Enzymes Are Synthesized in Inactive Forms

Some enzymes are synthesized in an inactive form that is activated under the conditions found where the enzyme is needed. Examples include the protein-digesting enzymes pepsin and trypsin, which cells produce with the active site covered, preventing the enzyme from digesting and killing the cell that manufactures it. Acid conditions in the stomach cause a transformation of the inactive pepsin that exposes its active site and allows it to begin breaking down proteins from a meal. Trypsin (which helps to finish protein digestion) is released into the small intestine in an inactive form that is activated by a different enzyme secreted by intestinal cells.

Enzyme Activity May Be Controlled by Competitive or Noncompetitive Inhibition

After an enzyme has been synthesized and is in its active state, there are two additional ways in which the enzyme can be inhibited to control metabolic pathways: competitive inhibition and noncompetitive inhibition. In both cases, an inhibitor molecule binds temporarily to the enzyme. The higher the concentration of inhibitor molecules, the more likely they are to bind to enzymes.

We know that for an enzyme to catalyze a reaction, its substrate must bind to the enzyme's active site (**FIG. 6-11a**). In **competitive inhibition,** a substance that is not the enzyme's normal substrate can also bind to the active site of the enzyme, competing directly with the substrate for the active site (**FIG. 6-11b**). Usually, a competitive inhibitor molecule has structural similarities to the usual substrate that allow it to occupy the active site.

In **noncompetitive inhibition,** a molecule binds to a site on the enzyme that is distinct from the active site. This causes the active site to change shape and become unavailable, making the enzyme unable to catalyze the reaction (**FIG. 6-11c**).

Allosteric Regulation of Enzymes Is Important in Controlling Reaction Rates

The most important mechanism for adjusting the rate at which metabolic reactions occur to meet the needs of the cell is through **allosteric regulation** (Gk. *allosteric*, other shape). During allosteric regulation, the same enzyme is either activated or inhibited by molecules binding to an allosteric site on the enzyme; these sites are always distinct from the active site. Enzymes regulated in this manner are called *allosteric enzymes*. Allosteric enzymes switch easily and spontaneously between an active and an inactive configuration but can be stabilized in either form. *Allosteric inhibition* is a form of noncompetitive inhibition in which an allosteric inhibitor molecule binds an allosteric inhibiting site and stabilizes the enzyme in its inactive form (a process similar to that shown in Fig. 6-11c).

Allosteric activation occurs when an allosteric activator molecule binds to an allosteric activating site, stabilizing the

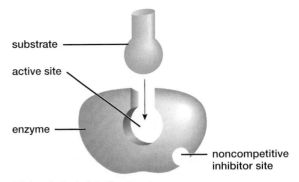

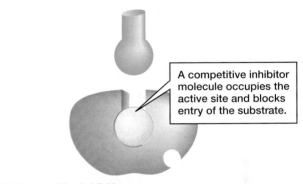

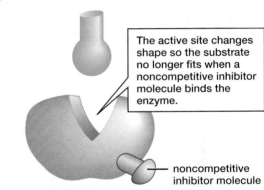

(a) A substrate binding to an enzyme

substrate
active site
enzyme
noncompetitive inhibitor site

(b) Competitive inhibition

A competitive inhibitor molecule occupies the active site and blocks entry of the substrate.

(c) Noncompetitive inhibition

The active site changes shape so the substrate no longer fits when a noncompetitive inhibitor molecule binds the enzyme.

noncompetitive inhibitor molecule

▲ **FIGURE 6-11 Competitive and noncompetitive enzyme inhibition (a)** The normal substrate fits into the enzyme's active site. **(b)** In competitive inhibition, a competitive inhibitor molecule that resembles the substrate enters and blocks the active site. **(c)** In noncompetitive inhibition, a molecule binds to a different site on the enzyme, distorting the active site so that the enzyme's substrate no longer fits.

enzyme in its active form. Allosteric activators and inhibitors bind briefly and reversibly to allosteric sites. As a result of this temporary binding, the number of enzyme molecules being activated (or inhibited) is proportional to the numbers of activator (or inhibitor) molecules that are present at any given time.

To see an example of allosteric regulation, let's look at ATP synthesis from its substrate molecule ADP. ADP is an allosteric activator and ATP an allosteric inhibitor of PFK (phosphofructokinase), an enzyme near the beginning of the metabolic pathway that breaks down glucose. ADP builds up in cells when a lot of ATP has been broken down. At high concentrations, ADP molecules are very likely to encounter the allosteric activator site on PFK, stabilizing PFK in its active state. The activated enzyme will cause an increase in ATP production, using up ADP. Then, as ADP levels fall and ATP levels increase, ATP will become much more likely to bind the allosteric inhibitor site on PFK, stabilizing PFK in its inactive state and causing ATP levels to fall and ADP to build up. This balancing act by allosteric regulation very precisely controls cellular ATP levels.

The inhibition of the enzyme PFK by ATP is an example of an important form of allosteric regulation called feedback inhibition. In **feedback inhibition,** the activity of an enzyme near the beginning of a metabolic pathway is inhibited by the end product, which acts as an allosteric inhibitor (**FIG. 6-12**). Feedback inhibition causes a metabolic pathway to stop producing its end product when the product concentration reaches an optimal level, much as a thermostat turns off a heater when a room becomes warm enough.

Poisons, Drugs, and Environmental Conditions Influence Enzyme Activity

Poisons and drugs that act on enzymes usually inhibit them, either competitively or noncompetitively. In addition, environmental conditions can denature enzymes, distorting the three-dimensional structure that is required for their function.

Some Poisons and Drugs Are Competitive or Noncompetitive Inhibitors of Enzymes

Competitive inhibitors of enzymes, including nerve gases such as sarin and certain insecticides such as malathion, permanently block the active site of the enzyme acetylcholinesterase, which breaks down acetylcholine (a substance that nerve cells release to activate muscles). This allows acetylcholine to build up and overstimulate muscles, causing paralysis. Death may ensue because victims become unable to breathe. Other poisons are noncompetitive inhibitors of enzymes; these include the heavy metals arsenic, mercury, and lead. The poison potassium cyanide causes rapid death

by noncompetitively inhibiting an enzyme that is crucial for the production of ATP.

Many drugs act as competitive inhibitors of enzymes. For example, the antibiotic penicillin destroys bacteria by competitively inhibiting an enzyme that is needed to synthesize bacterial cell walls. Both aspirin and ibuprofen (Advil) act as competitive inhibitors of an enzyme that catalyzes the synthesis of molecules that contribute to swelling, pain, and fever. Statin drugs (such as Lipitor) competitively inhibit an enzyme in the pathway that synthesizes cholesterol, thus reducing blood cholesterol levels. Many anticancer drugs block cancer cell proliferation by interfering with one or more of the numerous enzymes required to copy DNA, because each cell division requires the synthesis of new DNA. Unfortunately, these anticancer drugs also interfere with the growth of other rapidly dividing cells, such as those in hair follicles and the lining of the digestive tract. This explains why cancer chemotherapy may cause hair loss and nausea.

The Activity of Enzymes Is Influenced by Their Environment

The complex three-dimensional structures of enzymes are sensitive to environmental conditions. Recall that hydrogen bonds between polar amino acids are important in determining the three-dimensional structure of proteins (see Chapter 3). These bonds only occur within a narrow range of chemical and physical conditions, including the proper pH, temperature, and salt concentration. Thus, most enzymes have a very narrow range of conditions in which they function optimally. When conditions fall outside this range, the enzyme becomes **denatured,** meaning that it loses the exact three-dimensional structure required for it to function properly.

In humans, cellular enzymes generally work best at a pH around 7.4, the level maintained in and around our cells (**FIG. 6-13a**). For these enzymes, an acid pH alters

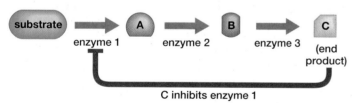

(a) Allosteric regulation by feedback inhibition

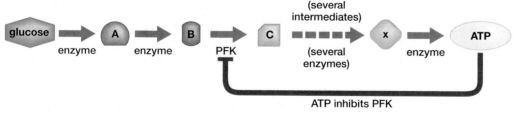

(b) Feedback inhibition by ATP on the enzyme PFK

◀ **FIGURE 6-12 Allosteric regulation by feedback inhibition (a)** In the general process of feedback inhibition, the end product of a metabolic pathway acts as an allosteric inhibitor, thus reducing the rate at which that end product is produced. **(b)** PFK inhibition is an example of feedback inhibition.

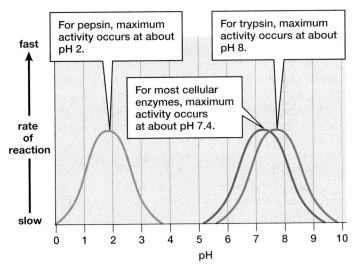

(a) Effect of pH on enzyme activity

For pepsin, maximum activity occurs at about pH 2.

For trypsin, maximum activity occurs at about pH 8.

For most cellular enzymes, maximum activity occurs at about pH 7.4.

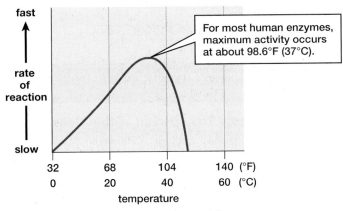

(b) Effect of temperature on enzyme activity

For most human enzymes, maximum activity occurs at about 98.6°F (37°C).

▲ **FIGURE 6-13 Human enzymes function best within narrow ranges of pH and temperature (a)** The digestive enzyme pepsin, released into the stomach, works best at an acidic pH. Trypsin, released into the small intestine, works best at a basic pH. Most enzymes within cells work best at the pH found in the blood, interstitial fluid, and cytosol (about 7.4). **(b)** The maximum activity of most human enzymes occurs at human body temperature.

the charges on amino acids by adding hydrogen ions to them, which in turn will change the enzyme's shape and compromise its ability to function. Enzymes that operate in the human digestive tract, however, may function outside of the pH range maintained within cells. The protein-digesting enzyme pepsin, for example, requires the acidic conditions of the stomach (pH around 2). In contrast, the protein-digesting enzyme trypsin, found in the small intestine where alkaline conditions prevail, works best at a pH close to 8.

Temperature affects the rate of enzyme-catalyzed reactions, which are slowed by lower temperatures and accelerated by moderately higher temperatures (**FIG. 6-13b**). Why?

Recall that molecular motion increases as temperature increases and decreases as the temperature falls. The rate of movement of molecules, in turn, influences how likely they are to encounter the active site of an enzyme. Cooling the body can drastically slow human metabolic reactions. Consider the real-life example of a young boy who fell through the ice on a lake, where he remained submerged for about 20 minutes before being rescued. At normal body temperature, the brain dies from lack of ATP after about 4 minutes without oxygen. But, fortunately, this child recovered because the icy water drastically reduced his need for oxygen by lowering his body temperature and thus slowing his metabolic rate.

In contrast, when temperatures rise too high, the hydrogen bonds that regulate protein shape may be broken apart by the excessive molecular motion, denaturing the protein. Think of the protein in egg white and how its appearance and texture are completely altered by cooking. Far lower temperatures than those required to fry an egg can still be too hot to allow enzymes to function properly. Excessive heat may be fatal; every summer, children in the United States die when left unattended in overheated cars.

Food remains fresh longer in the refrigerator or freezer because cooling slows the enzyme-catalyzed reactions that allow bacteria and fungi (which can spoil food) to grow and reproduce. Before the advent of refrigeration, meat was commonly preserved by using concentrated salt solutions, which kill most bacteria; think of bacon or salt pork. Salts dissociate into ions, which form bonds with amino acids in enzyme proteins. Too much salt interferes with the three-dimensional structure of enzymes, destroying their activity. Dill pickles are very well preserved in a vinegar-salt solution, which combines both highly salty and acidic conditions (**FIG. 6-14**). The enzymes of organisms that live in salty environments, as you might predict, have configurations that depend on a relatively high concentration of salt ions.

▶ **FIGURE 6-14 Preservation** Only the pickled cucumbers will be edible months from now.

CHECK YOUR LEARNING

Can you ...

- describe how cells regulate the rate at which metabolic reactions proceed?
- explain how poisons, drugs, and environmental conditions influence enzyme activity, and provide examples?

Energy Unleashed

During the course of a 26.2-mile race, a marathoner burns a great deal of glucose to provide enough ATP to power muscles through roughly 34,000 running steps. People store glucose molecules linked together in long branched chains of glycogen, primarily in the muscles and liver. Adults typically store about 3.5 ounces (100 grams) of glycogen in the liver and about 10 ounces (280 grams) in muscles. Highly trained distance athletes may store over 50% more glycogen than average in their livers, and the ability of their muscles to store glycogen may be more than double that of non-athletes.

Glycogen storage is crucial for marathon runners. During a marathon, a runner depletes essentially all of his or her body's stored glycogen, often about 90 minutes into the race. The body then begins converting fat into glucose. This is a much slower process, which can leave the muscles and brain starved for

glucose. Low blood glucose levels can cause extreme muscle fatigue, loss of motivation, and occasionally even hallucinations. Runners describe this sensation as "hitting the wall" or "bonking." To store the greatest possible amount of glycogen, endurance athletes practice carbo-loading by consuming large quantities of starches and sugars during the 3 days preceding the race. By packing their livers and muscles with glycogen before the race, and also by consuming energy drinks during the race, some runners manage to cross the finish line before they hit the wall.

THINK CRITICALLY When a runner's body temperature begins to rise, the body activates several mechanisms, including sweating and circulating more blood to the skin. How does this response both resemble and differ from feedback inhibition in enzymes?

CHAPTER REVIEW

Go to **MasteringBiology** for practice quizzes, activities, eText, videos, current events, and more.

Answers to Think Critically, Evaluate This, Multiple Choice, and Fill-in-the-Blank questions can be found in the Answers section at the back of the book.

Summary of Key Concepts

6.1 What Is Energy?

Energy is the capacity to do work. Potential energy is stored energy, and kinetic energy is the energy of movement. Potential energy and kinetic energy can be interconverted. The first law of thermodynamics states that in an isolated system, the total amount of energy remains constant, although the energy may change form. The second law of thermodynamics states that any use of energy causes a decrease in the quantity of useful energy and an increase in entropy (disorder or less useful forms of energy such as heat). The highly organized, low-entropy systems that characterize life do not violate the second law of thermodynamics because they are achieved through a continuous influx of usable energy from the sun, accompanied by an enormous increase in solar entropy.

6.2 How Is Energy Transformed During Chemical Reactions?

All chemical reactions involve making and breaking bonds, and all release some heat. In exergonic reactions, the reactant molecules have more energy than do the product molecules, so the overall reaction releases energy. In endergonic reactions, the reactants have less energy than do the products, so the reaction requires a net input of energy. Exergonic reactions can occur spontaneously, but all reactions, including exergonic ones, require an initial input of activation energy to overcome electrical repulsions between reactant molecules. Exergonic and endergonic reactions may be coupled such that the energy liberated by an exergonic reaction is stored in ATP, which can then drive an endergonic reaction.

6.3 How Is Energy Transported Within Cells?

Energy released by chemical reactions within a cell is captured and transported within the cell by unstable energy-carrier molecules, such as ATP and the electron carriers NADH and $FADH_2$. These molecules are the major means by which cells couple exergonic and endergonic reactions occurring at different places in the cell.

6.4 How Do Enzymes Promote Biochemical Reactions?

Enzymes are proteins that act as biological catalysts by lowering activation energy and allowing biochemical reactions to occur without a permanent change of the enzyme. Enzymes usually promote one or a few specific reactions. The reactants temporarily bind to the active site of the enzyme, causing less activation energy to be needed to produce the product. Cells control their metabolic reactions by regulating the synthesis and use of enzymes. Enzymes allow the breakdown of energy-rich molecules such as glucose in a series of small steps so that energy is released gradually and can be captured in ATP for use in endergonic reactions.

6.5 How Are Enzymes Regulated?

Cellular metabolism involves complex, interconnected sequences of reactions called metabolic pathways. Each reaction is catalyzed by an enzyme. Cells precisely control the amounts and activities of these enzymes. Enzyme action can be regulated by altering the rate of enzyme synthesis, activating previously inactive enzymes, competitive and noncompetitive inhibition, and allosteric regulation, which includes feedback inhibition. Many poisons and drugs act as enzyme inhibitors. Environmental conditions—including pH, salt concentration, and temperature—can promote or inhibit enzyme function by altering or preserving the enzyme's three-dimensional structure.

Key Terms

activation energy *97*	first law of
active site *100*	thermodynamics *94*
allosteric regulation *103*	isolated system *94*
catalyst *100*	kinetic energy *94*
chemical energy *94*	law of conservation of
chemical reaction *97*	energy *94*
competitive inhibition *103*	metabolic pathway *101*
coupled reaction *99*	metabolism *101*
denatured *104*	noncompetitive
electron carrier *99*	inhibition *103*
endergonic *97*	potential energy *94*
energy *94*	product *97*
energy-carrier molecule *98*	reactant *97*
entropy *95*	second law of
enzyme *100*	thermodynamics *95*
exergonic *97*	substrate *100*
feedback inhibition *104*	work *94*

Thinking Through the Concepts

Multiple Choice

1. Which of the following is True?
   a. Enzymes increase activation energy requirements.
   b. Activation energy is required to initiate exergonic reactions.
   c. Heat *cannot* supply activation energy.
   d. Stomach acid inactivates pepsin.

2. Which is *not* an example of an exergonic reaction?
   a. photosynthesis
   b. a nuclear reaction in the sun
   c. ATP → ADP + P$_i$
   d. glucose breakdown

3. Which of the following is True?
   a. ATP is a long-term energy storage molecule.
   b. ATP can carry energy from one cell to another.
   c. ADP inhibits glucose breakdown in cells.
   d. ATP can act as an allosteric regulator molecule.

4. Coupled reactions
   a. are endergonic overall.
   b. both synthesize and break down ATP.
   c. are catalyzed by the same enzyme.
   d. end with reactants that contain more energy than their products.

5. Which of the following is False?
   a. Allosteric inhibition is noncompetitive.
   b. Allosteric regulation can either stimulate or inhibit enzyme activity.
   c. Feedback inhibition is a form of allosteric regulation.
   d. Competitive inhibition is a form of allosteric enzyme regulation.

Fill-in-the-Blank

1. According to the first law of thermodynamics, energy can be neither _____ nor _____. Energy occurs in two major forms: _____, the energy of movement, and _____, or stored energy.

2. According to the second law of thermodynamics, when energy changes forms, some is always converted into _____ useful forms. This tendency is called _____.

3. Once started, some reactions release energy and are called _____ reactions. Others require a net input of energy and are called _____ reactions. Which type of reaction will continue spontaneously once it starts? _____. Which type of reaction allows the formation of complex biological molecules from simpler molecules? _____.

4. The abbreviation ATP stands for _____. The molecule is synthesized by cells from _____ and _____. This synthesis requires an input of _____, which is temporarily stored in ATP.

5. Enzymes are what type of biological molecule? _____ Enzymes promote reactions in cells by acting as biological _____ that lower the _____. Each enzyme possesses a region called a(n) _____ that binds specific biological molecules.

6. Some poisons and drugs act by _____ enzymes. When a drug is similar to the enzyme's substrate, it acts as a(n) _____ inhibitor.

Review Questions

1. Explain why organisms do not violate the second law of thermodynamics. What is the ultimate energy source for most forms of life on Earth?

2. Define *potential energy* and *kinetic energy* and provide two specific examples of each. Explain how one form of energy can be converted into another. Will some energy be lost during this conversion? If so, what form will it take?

3. Define *metabolism,* and explain how reactions can be coupled to one another.

4. What is activation energy? How do catalysts affect activation energy? How do catalysts affect the reaction rate?

5. Compare breaking down glucose in a cell to setting it on fire with a match. What is the source of activation energy in each case?

6. Compare the mechanisms of competitive and noncompetitive inhibition of enzymes.

7. Describe the structure and function of enzymes. How is enzyme activity regulated?

Applying the Concepts

1. While vacuuming, you show off by telling a friend that you are using electrical energy to create a lower-entropy state. She replies that you are taking advantage of increasing solar entropy. Explain this conversation.

2. Refute the following: "According to evolutionary theory, organisms have increased in complexity through time. However, an increase in complexity contradicts the second law of thermodynamics. Therefore, evolution is impossible."

3. Can a bear use all the energy contained in the body of the fish it eats? Explain, and based on your explanation, predict and further explain whether a forest would likely have more predators or more prey animals (by weight)?

7

CAPTURING SOLAR ENERGY: PHOTOSYNTHESIS

Did the Dinosaurs Die from Lack of Sunlight?

ABOUT 66 MILLION YEARS AGO, the Cretaceous-Tertiary (K-T) extinction event brought the Cretaceous period to a violent end, and life on Earth suffered a catastrophic blow. The fossil record indicates that a devastating mass extinction eliminated at least 80% of all forms of life—both marine and terrestrial—that are known to have existed at that time. The 160-million-year reign of the dinosaurs, including the massive *Triceratops* and its predator *Tyrannosaurus rex*, ended abruptly. It would be many millions of years before Earth became repopulated with a diversity of species even approaching that of the late Cretaceous.

In 1980, Luis Alvarez, a Nobel Prize–winning physicist, his geologist son Walter Alvarez, and nuclear chemists Helen Michel and Frank Asaro published what was then a very controversial hypothesis. They proposed that an invader from outer space—a massive asteroid—had brought the Cretaceous period to an abrupt and violent end. Their evidence consisted of a thin layer of clay deposited at the end of the Cretaceous period and found at sites throughout the world. Known as the "K-T boundary layer," this clay deposit contains from 30 to 160 times the iridium level typically found in Earth's crust. Iridium is a silvery-white metal that, although extremely rare on Earth, is abundant in certain types of asteroids. How large must an iridium-rich asteroid have been to create the K-T boundary layer encircling Earth? The Alvarez team calculated that this iridium-enriched space rock must have been at least 6 miles (10 kilometers) in diameter. As it crashed into Earth, its impact released the energy equivalent to 8 billion of the atomic bombs that destroyed Hiroshima and Nagasaki. The asteroid's impact blasted out a plume of pulverized rock, some of which reached the moon and beyond. Most of the asteroid's fragments and pellets of debris grew incandescent as they re-entered the atmosphere,

The end of the reign of the dinosaurs?

plummeting down in a fiery shower that ignited wildfires, possibly over most of Earth's surface. A shroud of dust and soot blocked the sun's rays, and the broiling heat gave way to a cooling darkness that enveloped Earth.

How could an asteroid impact have eliminated most known forms of life? In the months following the firestorms, one of the most damaging effects would have been darkness that disrupted photosynthesis, the most important biochemical pathway on Earth. What occurs during photosynthesis? What makes this process so important that interrupting it would wipe out much of Earth's biodiversity?

7.1 WHAT IS PHOTOSYNTHESIS?

Roughly 3.5 billion years ago, chance mutations allowed a prokaryotic (bacterial) cell to harness the energy of sunlight. Exploiting abundant water and sunshine, early photosynthetic cells proliferated and filled the shallow seas. The evolution of photosynthesis made life as we know it possible. This amazing process provides not only fuel for life but also the oxygen required to burn this fuel efficiently. **Photosynthesis** is the process by which light energy is captured and then stored as chemical energy in the bonds of organic molecules such as sugar (**FIG. 7-1**). In lakes and oceans, photosynthesis is performed primarily by photosynthetic protists and certain bacteria, and on land, mostly by plants. Collectively, these organisms incorporate close to 100 billion tons of carbon into their bodies annually. The carbon- and energy-rich molecules of photosynthetic organisms eventually become available to feed all other forms of life. Fundamentally similar reactions occur in all photosynthetic organisms; here we will concentrate on the most familiar of these: land plants.

Leaves and Chloroplasts Are Adaptations for Photosynthesis

The leaves of plants are beautifully adapted to the demands of photosynthesis. A leaf's flattened shape exposes a large surface area to the sun, and its thinness ensures that sunlight can penetrate to reach the light-trapping chloroplasts inside. Both the upper and lower surfaces of a leaf consist of a layer of transparent cells that form the **epidermis,** which protects

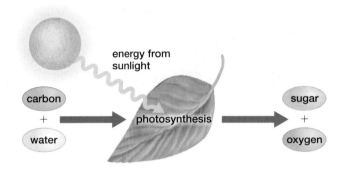

▲ **FIGURE 7-1 An overview of photosynthesis**

the inner parts of the leaf while allowing light to penetrate. The outer surface of the epidermis is covered by the **cuticle,** a transparent, waxy, waterproof covering that reduces the evaporation of water from the leaf.

A leaf obtains the carbon dioxide (CO_2) necessary for photosynthesis from the air, through adjustable pores in the epidermis called **stomata** (singular, stoma; **FIG. 7-2**). Inside the leaf are layers of cells collectively called **mesophyll** (Gk. *meso,* middle; **FIG. 7-3**) where most chloroplasts are located. Mesophyll cells in the leaf's center are loosely packed, allowing air to circulate around them and CO_2 and O_2 to be exchanged through their moist membranes. *Vascular bundles,* which form veins in the leaf (see Fig. 7-3b), supply water and minerals to the leaf's cells and carry the sugar molecules produced during photosynthesis to other parts of the plant. Surrounding the vascular bundles are **bundle sheath cells,** which lack chloroplasts in most plants.

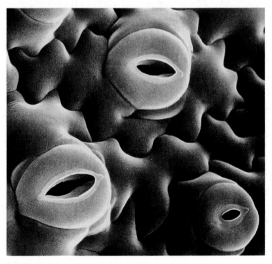

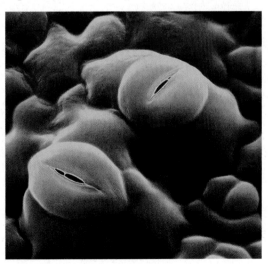

◀ **FIGURE 7-2 Stomata** **(a)** Open stomata allow CO_2 to diffuse in and O_2 to diffuse out. **(b)** Closed stomata reduce water loss by evaporation but prevent CO_2 from entering and O_2 from leaving.

(a) Stomata open

(b) Stomata closed

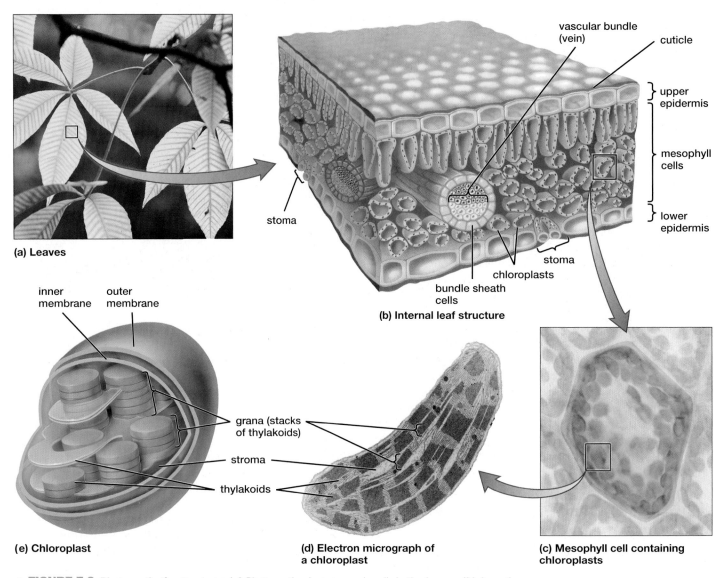

(a) Leaves

(b) Internal leaf structure

vascular bundle (vein)

cuticle

upper epidermis

mesophyll cells

lower epidermis

stoma

chloroplasts

bundle sheath cells

stoma

inner membrane

outer membrane

grana (stacks of thylakoids)

stroma

thylakoids

(e) Chloroplast

(d) Electron micrograph of a chloroplast

(c) Mesophyll cell containing chloroplasts

▲ **FIGURE 7-3** **Photosynthetic structures (a)** Photosynthesis occurs primarily in the leaves. **(b)** A section of a leaf. **(c)** A light micrograph of a single mesophyll cell, packed with chloroplasts. **(d)** A TEM of a single chloroplast, showing the stroma and thylakoids where photosynthesis occurs. **(e)** An illustrated chloroplast.

Photosynthesis in plants takes place within **chloroplasts,** most of which are contained within mesophyll cells. A single mesophyll cell often contains 40 to 50 chloroplasts (see Fig. 7-3c), and up to 500,000 of them may be packed into a 1 mm^2 area of leaf. Chloroplasts are organelles that consist of a double outer membrane enclosing a semifluid substance, the **stroma.** Embedded in the stroma are interconnected membrane-enclosed compartments called **thylakoids.** Thylakoids often form disk-shaped structures, which are arranged into stacks called **grana** (see Figs. 7-3d, e). Each of these sacs encloses a fluid-filled region called the thylakoid space.

Photosynthesis Consists of the Light Reactions and the Calvin Cycle

Starting with the simple molecules of carbon dioxide and water, photosynthesis converts the energy of sunlight into chemical energy stored in the bonds of glucose and releases oxygen as a by-product. The overall chemical reaction for photosynthesis is:

$$6\,CO_2 + 6\,H_2O + \text{light energy} \rightarrow C_6H_{12}O_6\,(\text{sugar}) + 6\,O_2$$

This straightforward equation obscures the fact that photosynthesis actually involves dozens of individual reactions, each catalyzed by a separate enzyme. These reactions occur in two distinct stages: the light reactions and the Calvin cycle. Each stage takes place in a different region of the chloroplast, but the two are connected by an important link: energy-carrier molecules.

In the **light reactions** (the "photo" part of photosynthesis), **chlorophyll** and other molecules embedded in the thylakoid membranes of the chloroplast capture sunlight energy and convert it into chemical energy. This chemical

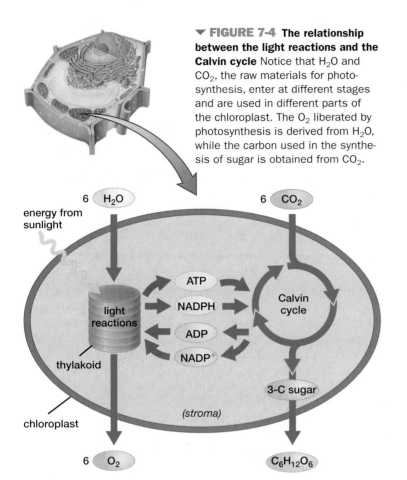

▼ **FIGURE 7-4 The relationship between the light reactions and the Calvin cycle** Notice that H_2O and CO_2, the raw materials for photosynthesis, enter at different stages and are used in different parts of the chloroplast. The O_2 liberated by photosynthesis is derived from H_2O, while the carbon used in the synthesis of sugar is obtained from CO_2.

CASE STUDY   CONTINUED

Did The Dinosaurs Die from Lack of Sunlight?

Over 2 billion years before the K-T extinction event, the first photosynthetic cells filled the seas and released what was then a deadly gas: oxygen. Oxygen accumulated in what had originally been an oxygen-free atmosphere, radically altering Earth's environment. This "Great Oxygenation Event" triggered a massive extinction. But unlike the K-T extinction, most of the casualties left no trace, because life had not evolved beyond single cells, which are rarely preserved in the fossil record. Fortunately, these simple life-forms reproduce rapidly, allowing their DNA to accumulate mutations over a relatively short time span. A serendipitous combination of random mutations allowed some early cells to not only survive exposure to oxygen, but also use it to their advantage. These organisms became the ancestors of nearly all modern forms of life. Eventually, plants invaded the land, and by the Cretaceous period, plants growing in luxuriant profusion provided sustenance for herbivorous giants such as the 12-ton *Triceratops*.

What chemical reactions allow plants to capture solar energy and store it in chemical bonds, releasing oxygen in the process?

energy is stored in the energy-carrier molecules ATP (adenosine triphosphate) and **NADPH (NADP+; nicotinamide adenine dinucleotide phosphate).** Water is split apart, and oxygen gas is released as a by-product.

The reactions of the **Calvin cycle** (the "synthesis" part of photosynthesis) can occur in either light or darkness. During these reactions, enzymes in the stroma surrounding the thylakoids combine CO_2 from the atmosphere and chemical energy from ATP and NADH. The end product is a three-carbon sugar that will be used to make glucose.

FIGURE 7-4 shows the locations at which the light reactions and the Calvin cycle occur and illustrates the interdependence of the two processes. In the following sections, we examine each stage of photosynthesis.

CHECK YOUR LEARNING

Can you ...

- explain why photosynthesis is important?
- diagram the structure of leaves and chloroplasts and explain how these structures function in photosynthesis?
- write out and explain the basic equation for photosynthesis?
- summarize the main events of the light reactions and the Calvin cycle and explain the relationship between these two processes?

7.2 THE LIGHT REACTIONS: HOW IS LIGHT ENERGY CONVERTED TO CHEMICAL ENERGY?

Recall that the light reactions capture the energy of sunlight, storing it as chemical energy in ATP and NADPH. The molecules that make these reactions possible, including light-capturing pigments and enzymes, are anchored in a precise array within the membranes of the thylakoids. As you read this section, notice how the thylakoid membranes and the spaces they enclose support the light reactions.

Light Is Captured by Pigments in Chloroplasts

The sun emits energy that spans a broad spectrum of electromagnetic radiation. The **electromagnetic spectrum** ranges from short-wavelength gamma rays, through ultraviolet, visible, and infrared light, to very long-wavelength radio waves (**FIG. 7-5**). Light and all other electromagnetic waves are composed of individual packets of energy called **photons.** The energy of a photon corresponds to its wavelength: Short-wavelength photons, such as gamma and X-rays, are very energetic, whereas long-wavelength photons, such as microwaves and radio waves, carry lower energies. Visible light consists of wavelengths with energies that are high enough to alter biological pigment molecules (light-absorbing molecules) such as chlorophyll, but not high enough to break the bonds of crucial molecules such as DNA.

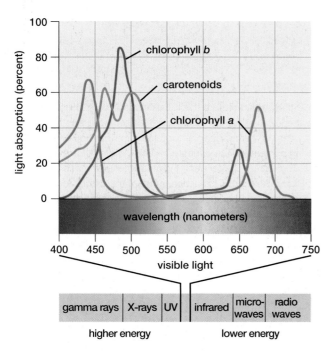

▲ **FIGURE 7-5 Light and chloroplast pigments** The rainbow colors that we perceive are a small part of the electromagnetic spectrum. Chlorophyll *a* and *b* (green and blue curves, respectively) strongly absorb violet, blue, and red light, reflecting a green or yellowish-green color to our eyes. Carotenoids (orange curve) absorb blue and green wavelengths.

THINK CRITICALLY You continuously monitor the photosynthetic oxygen production from the leaf of a plant illuminated by white light. How and why would oxygen production change if you placed filters in front of the light source that transmit (a) only red, (b) only infrared, and (c) only green light onto the leaf?

(It is no coincidence that these wavelengths, with just the right amount of energy, also stimulate the pigments in our eyes, allowing us to see.)

When a specific wavelength of light strikes an object such as a leaf, one of three events occurs: The light may be reflected (bounced back), transmitted (passed through), or absorbed (captured). Wavelengths of light that are reflected or transmitted can reach the eyes of an observer; these wavelengths are seen as the color of the object. Light energy that is absorbed can drive biological processes such as photosynthesis.

Chloroplasts contain a variety of pigment molecules that absorb different wavelengths of light. **Chlorophyll *a*,** the key light-capturing pigment molecule in chloroplasts, strongly absorbs violet, blue, and red light, but reflects green, thus giving green leaves their color (see Fig. 7-5). Chloroplasts also contain other molecules, collectively called **accessory pigments,** which absorb additional wavelengths of light energy and transfer their energy to chlorophyll *a.* Accessory pigments include chlorophyll *b*, a slightly different form of chlorophyll *a* that reflects yellow-green light and absorbs some of the blue and red-orange wavelengths of light that are missed by chlorophyll *a.* **Carotenoids** are

accessory pigments found in all chloroplasts. They absorb blue and green light and therefore appear mostly yellow or orange (see Fig. 7-5). Carotenoid accessory pigments include beta-carotene, which gives many vegetables and fruits (including carrots, squash, oranges, and cantaloupes) their orange colors. Interestingly, animals convert beta-carotene into vitamin A, which is used to synthesize the light-capturing pigment in our eyes. Thus, in a beautiful symmetry, the beta-carotene that captures light energy in plants is converted into a substance that captures light in animals as well.

Although carotenoids are present in leaves, their color is usually masked by the more abundant green chlorophyll. In temperate regions, as leaves begin to die in autumn, chlorophyll breaks down before carotenoids do, revealing these bright yellow and orange pigments as fall colors (**FIG. 7-6**).

The Light Reactions Occur in Association with the Thylakoid Membranes

The light reactions occur in and on the thylakoid membranes. These membranes contain many **photosystems,** each consisting of a cluster of chlorophyll and accessory

▲ **FIGURE 7-6 Loss of chlorophyll reveals carotenoid pigments** As winter approaches, chlorophyll in these aspen leaves breaks down, revealing yellow and orange carotenoid pigments.

pigment molecules surrounded by various proteins. There are two types of photosystems—photosystem II and photosystem I—that work together during the light reactions. The photosystems are named according to the order in which they were discovered, but the light reactions start with photosystem II and then proceed to photosystem I.

Adjacent to each photosystem is an **electron transport chain (ETC)** consisting of a series of electron-carrier molecules embedded in the thylakoid membrane. Electrons flow through the following pathway in the light reactions: photosystem II → electron transport chain → photosystem I → electron transport chain → NADPH.

You can think of the light reactions as a sort of arcade pinball game: Energy (sunlight) is introduced when a player pulls back and releases a knob. The energy is transferred from spring-driven pistons (chlorophyll molecules) to a ball (an electron), propelling it upward (into a higher-energy level). As the ball bounces back downhill, the energy it releases can be used to turn a wheel (generate ATP) and ring a bell (generate NADPH). With this overall scheme in mind, let's look more closely at the sequence of events in the light reactions.

Photosystem II and Its Electron Transport Chain Capture Light Energy, Create a Hydrogen Ion Gradient, and Split Water

As you read the following descriptions, refer to the numbered steps in **FIGURE 7-7**. The light reactions begin when photons of light are absorbed by pigment molecules clustered in photosystem II ❶. The energy hops from one pigment molecule to the next until it is funneled into the photosystem II reaction center ❷. The **reaction center** of each photosystem consists of a pair of specialized chlorophyll *a* molecules and a *primary electron acceptor* molecule embedded in a complex of proteins. When the energy from light reaches the reaction center, it boosts an electron from one of the reaction center

◀ **FIGURE 7-7 Energy transfer and the light reactions of photosynthesis** Light reactions occur in and immediately adjacent to the thylakoid membrane. The vertical axis indicates the relative energy levels of the molecules involved.

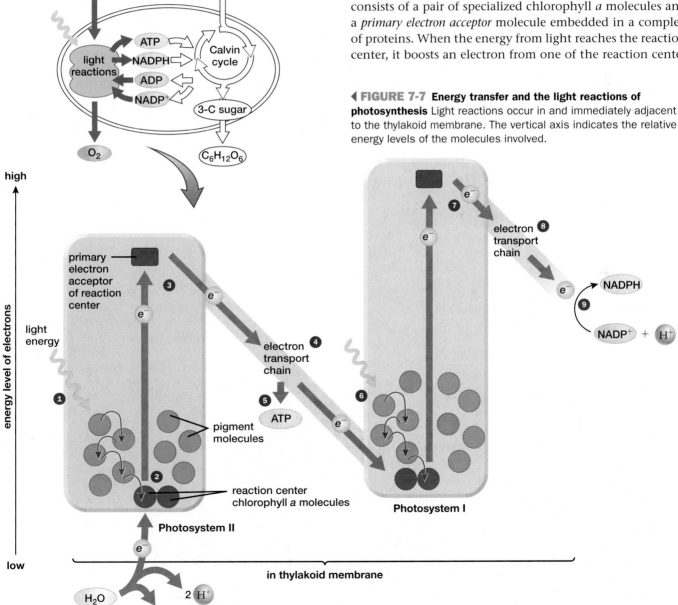

chlorophylls to the primary electron acceptor, which captures the energized electron ❸.

For photosynthesis to continue, the electrons that were boosted out of the reaction center of photosystem II must be replaced. The replacement electrons come from water (see ❷). Water molecules are split by an enzyme associated with photosystem II, liberating electrons that will replace those lost by the reaction center chlorophyll molecules. Splitting water also releases two hydrogen ions, and for every two water molecules split, one molecule of O_2 is produced.

Once the primary electron acceptor in photosystem II captures the electron, it passes the electron to the first molecule of the adjacent ETC in the thylakoid membrane ❹. The electron then travels from one electron carrier molecule to the next, releasing energy as it goes. Some of this energy is harnessed to pump H^+ across the thylakoid membrane and into the thylakoid space, where it contributes to the H^+ gradient that generates ATP (❺; to be discussed shortly). Finally, the energy-depleted electron leaves the ETC and enters the reaction center of photosystem I, where it replaces the electron ejected when light strikes photosystem I ❻.

Photosystem I and Its Electron Transport Chain Generate NADPH

Meanwhile, light has also been striking the pigment molecules of photosystem I. This light energy is passed to a chlorophyll *a* molecule in the reaction center ❻. Here, it energizes an electron that is absorbed by the primary electron acceptor of photosystem I ❼. (This energized electron is immediately replaced by an energy-depleted electron from the first electron transport chain.) From the primary electron acceptor of photosystem I, the energized electron is passed to a second ETC adjacent to photosystem I in the thylakoid membrane ❽. Here, the final electron carrier is an enzyme that catalyzes the synthesis of NADPH. To form NADPH, the enzyme combines $NADP^+$ and H^+ (both dissolved in the stroma) with two energetic electrons from the ETC ❾.

The Hydrogen Ion Gradient Generates ATP by Chemiosmosis

FIGURE 7-8 shows how electrons move through the thylakoid membrane and how their energy is used to create an H^+ gradient that drives ATP synthesis through a process called **chemiosmosis.** As an energized electron travels along the ETC associated with photosystem II, it releases energy in steps. Some of this energy is harnessed to pump H^+ across the thylakoid membrane and into the thylakoid space ❶. This creates a high concentration of H^+ inside the space ❷

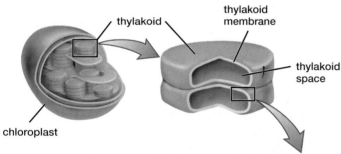

◀ **FIGURE 7-8 Events of the light reactions occur in and near the thylakoid membrane**

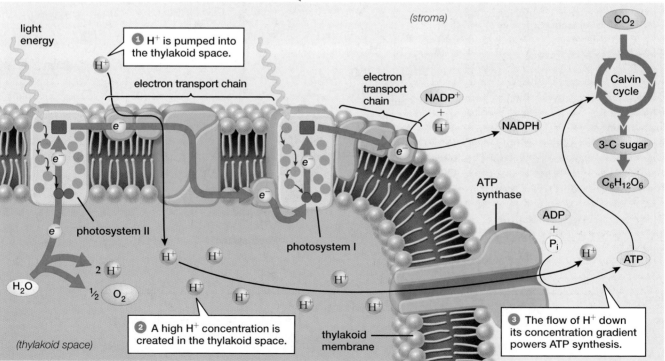

and a low concentration in the surrounding stroma. During chemiosmosis, H^+ flows back down its concentration gradient through a special type of channel called **ATP synthase** that spans the thylakoid membrane. ATP synthase produces ATP using ADP and phosphate dissolved in the stroma ❸. It takes the energy from about three H^+ passing through ATP synthase to synthesize one ATP molecule.

The H^+ gradient serves the same function as water stored behind a dam at a hydroelectric plant. When the stored water is released at the hydroelectric plant, it is channeled downward through turbines. Similarly, the hydrogen ions in the thylakoid space are funneled through ATP synthase channels. In the hydroelectric plant, turbines convert the energy of moving water into electrical energy. In an analogous way, ATP synthase converts the energy liberated by the flow of H^+ into chemical energy stored in the bonds of ATP.

SUMMING UP: Light Reactions

- Chlorophyll and carotenoid pigments of photosystem II absorb light that energizes and ejects an electron from a reaction center chlorophyll *a* molecule. The energized electron is captured by the primary electron acceptor molecule.
- The electron is passed from the primary electron acceptor to the adjacent ETC, where it moves from molecule to molecule, releasing energy with each transfer. Some of the energy is used to create a hydrogen ion gradient across the thylakoid membrane. This gradient is used to drive ATP synthesis by chemiosmosis.
- Enzymes associated with photosystem II split water. This releases electrons that replace those ejected from the reaction center chlorophylls, supplies H^+ that enhances the H^+ gradient for ATP production, and liberates O_2.
- Chlorophyll and carotenoid pigments in photosystem I absorb light that energizes and ejects an electron from a reaction center chlorophyll *a* molecule into the primary electron acceptor molecule. This electron is replaced by an energy-depleted electron from the ETC associated with photosystem II.
- The energized electron passes from the primary electron acceptor into the adjacent ETC, where it moves from molecule to molecule, releasing energy.
- The final molecule in this second ETC is an enzyme that synthesizes the energy-carrier NADPH from $NADP^+$ and H^+ for every two energized electrons that reach it.
- The overall products of the light reactions are the energy carriers NADPH and ATP; O_2 is released as a by-product.

CHECK YOUR LEARNING

Can you ...

- list the light-capturing molecules in chloroplasts and describe their functions?
- diagram and describe the molecules within the thylakoid membranes and explain how they capture and transfer light energy?
- explain how NADPH and ATP are generated?

CASE STUDY ⟍ **CONTINUED**

Did the Dinosaurs Die from Lack of Sunlight?

Air bubbles trapped in amber from the Cretaceous period have revealed that oxygen made up nearly 35% of the atmosphere at that time, compared to 21% today. Abundant oxygen would have intensified the conflagrations caused by the flaming re-entry of debris from the asteroid impact. Marine ecosystems, which relied on photosynthesizing microorganisms, would have collapsed rapidly in the twilight conditions.

A large portion of Earth's terrestrial vegetation was likely consumed by fire, and many of those land plants that survived the fires would have succumbed during the cold, dark "global winter" that began as the planet was encompassed by soot and dust. Most plant-eating animals that survived the initial blast would have soon starved, especially enormous ones like the 12-ton *Triceratops*, which needed to consume hundreds of pounds of vegetation daily. Predators such as *Tyrannosaurus*, which relied on plant-eaters for food, would have died soon afterward.

What reactions allow plants to store the high-energy molecules that they and most other forms of life still rely on today?

7.3 THE CALVIN CYCLE: HOW IS CHEMICAL ENERGY STORED IN SUGAR MOLECULES?

Our cells produce carbon dioxide as we burn sugar for energy (described in Chapter 8), but they can't form organic molecules by capturing (or *fixing*) the carbon atoms in CO_2. Although this feat can be accomplished by a few types of *chemosynthetic* bacteria that fix carbon using energy gained by breaking down inorganic molecules, nearly all carbon fixation is performed by photosynthetic organisms. The carbon is captured from atmospheric CO_2 during the Calvin cycle using energy from sunlight harnessed during the light reactions. The details of the Calvin cycle were discovered in the 1950s by chemists Melvin Calvin, Andrew Benson, and James Bassham. Using radioactive isotopes of carbon (see Chapter 2), they were able to follow carbon atoms as they moved from CO_2 through the various compounds of the cycle and, finally, into sugar molecules.

The Calvin Cycle Captures Carbon Dioxide

The ATP and NADPH synthesized during the light reactions are dissolved in the fluid stroma that surrounds the thylakoids. There, these energy carriers power the synthesis of the three-carbon sugar *glyceraldehyde-3-phosphate (G3P)* from CO_2 in the Calvin cycle. This metabolic pathway is described as a "cycle" because it begins and ends with the same five-carbon molecule, *ribulose bisphosphate (RuBP)*.

For simplicity, we illustrate the cycle starting and ending with three molecules of RuBP. Each "turn" of the cycle captures three molecules of CO_2 and produces one molecule of

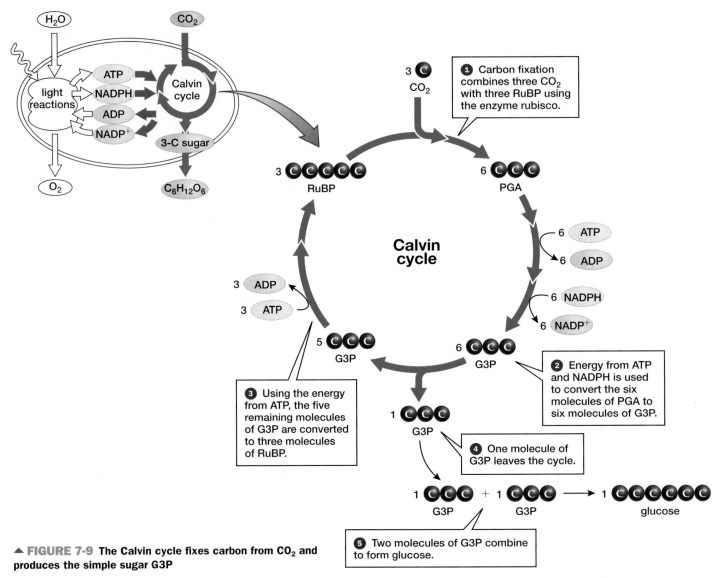

▲ **FIGURE 7-9 The Calvin cycle fixes carbon from CO$_2$ and produces the simple sugar G3P**

the simple sugar end product: G3P. The Calvin cycle is best understood if we divide it into three parts: (1) carbon fixation, (2) the synthesis of G3P, and (3) the regeneration of RuBP that allows the cycle to continue (**FIG. 7-9**).

During **carbon fixation,** carbon from CO$_2$ is incorporated into organic molecules. The enzyme **rubisco** combines three CO$_2$ molecules with three RuBP molecules to produce three unstable six-carbon molecules that immediately split in half, forming six molecules of *phosphoglyceric acid* (*PGA, a three-carbon molecule*) ❶. Because carbon fixation generates this three-carbon PGA molecule, the Calvin cycle is often referred to as the **C$_3$ pathway.**

The synthesis of the simple three-carbon sugar G3P occurs *via* a series of reactions using energy donated by ATP and NADPH. During these reactions, six three-carbon PGA molecules are rearranged to form six three-carbon G3P molecules ❷. Five of the six G3P molecules are used to regenerate three five-carbon RuBP molecules, using ATP generated during the light reactions ❸. The single remaining G3P molecule exits the Calvin cycle ❹.

Carbon fixation, the first step in the Calvin cycle, can be disrupted by O$_2$. The enzyme rubisco that fixes carbon is not completely selective, and it will allow O$_2$ instead of CO$_2$ to combine with RuBP. When O$_2$ replaces CO$_2$, the result is a wasteful process called **photorespiration,** which reduces the rate of carbon fixation by roughly 33%. If photorespiration could be avoided, plants could capture solar energy much more efficiently. Many researchers are working to genetically modify the enzyme rubisco to make it more selective for CO$_2$, with the hope of increasing the yields of crops such as wheat. Other researchers have already taken genes for a faster-acting version of rubisco from photosynthetic bacteria and inserted it into plants.

A small percentage of Earth's terrestrial plants have evolved biochemical pathways that consume a bit more energy but increase the efficiency of carbon fixation in hot, dry environments. These pathways—the **C$_4$ pathway** and the **crassulacean acid metabolism (CAM)** pathway—are explored in "In Greater Depth: Alternate Pathways Increase Carbon Fixation."

IN GREATER DEPTH | Alternate Pathways Increase Carbon Fixation

In hot, dry conditions, stomata remain closed much of the time to prevent water from evaporating. But this also prevents the exchange of gases, so as photosynthesis occurs, the concentration of CO_2 drops and the concentration of O_2 rises. The O_2 can bind to the active site of the enzyme rubisco and prevent CO_2 from binding, an example of competitive inhibition. The O_2 then combines with RuBP, causing photorespiration, which greatly reduces the rate of carbon fixation. Plants, particularly fragile seedlings, may die under these circumstances because they are unable to capture enough energy to meet their metabolic needs.

Rubisco is the most abundant protein on Earth and arguably one of the most important. It catalyzes the reaction by which carbon enters the biosphere, and all life is based on carbon. Why, then, is rubisco so unselective? When rubisco first evolved, Earth's atmosphere was high in CO_2 but contained little O_2, so there was no threat of competitive inhibition. As atmospheric O_2 increased, the chance mutations that would have prevented competitive inhibition never occurred. Instead, certain flowering plants evolved two different but closely related mechanisms that circumvent photorespiration: the C_4 pathway and crassulacean acid metabolism (CAM). Each uses the Calvin cycle, but each also involves several additional reactions and consumes more ATP than does the Calvin cycle alone. But in compensation for the loss of ATP, these plants conserve more water under hot, dry conditions.

C_4 Plants Capture Carbon and Synthesize Sugar in Different Cells

In typical plants, known as C_3 *plants* (because they use only the C_3 cycle, another name for the Calvin cycle), the chloroplasts in which the Calvin cycle occurs are located primarily in mesophyll cells. No chloroplasts are found in bundle sheath cells that surround the leaf's veins (see Fig. 7-3). In contrast, C_4 *plants* have chloroplasts in both mesophyll and bundle sheath cells. Such plants use an initial series of reactions, called the C_4 pathway, to selectively capture carbon in their mesophyll chloroplasts. The mesophyll chloroplasts lack Calvin cycle enzymes and use the enzyme *PEP carboxylase* to fix CO_2. Unlike rubisco, PEP carboxylase is highly selective for CO_2 over O_2. PEP carboxylase causes CO_2 to react with a three-carbon molecule called phosphoenolpyruvate (PEP). So in C_4 plants, carbon fixation produces the four-carbon molecule oxaloacetate, from which the C_4 pathway gets its name. Oxaloacetate is rapidly converted into another four-carbon molecule, malate, which diffuses from the mesophyll cells into bundle sheath cells. The malate acts as a shuttle for CO_2.

In C_4 plants, Calvin cycle enzymes (including rubisco) are present only in the chloroplasts of the bundle sheath cells. In the bundle sheath cells, malate is broken down, forming the three-carbon molecule pyruvate and releasing CO_2. This generates a high CO_2 concentration in the bundle sheath cells (up to 10 times higher than atmospheric CO_2). The resulting high CO_2 concentration allows rubisco to fix carbon with little competition from O_2, minimizing photorespiration. The pyruvate is then actively transported back into the mesophyll cells. Here, more ATP energy is used to convert pyruvate back into PEP, allowing the cycle to continue (**FIG. E7-1**). Plants using C_4 photosynthesis include crabgrass, corn, daisies, and some thistles.

CAM Plants Capture Carbon and Synthesize Sugar at Different Times

CAM plants also use the C_4 pathway, but in contrast to C_4 plants, CAM plants do not use different cell types

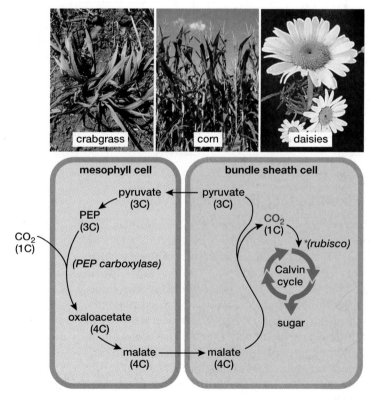

▲ **FIGURE E7-1** **The C_4 pathway** Both the C_4 and CAM pathways form the same molecules, fixing carbon into oxaloacetate using the selective enzyme PEP carboxylase, and then storing it in malate. In C_4 plants, atmospheric CO_2 is trapped in mesophyll cells and enters the Calvin cycle in bundle sheath cells.

THINK CRITICALLY Why do C_3 plants have an advantage over C_4 plants under cool, moist conditions?

to capture carbon and to synthesize sugar. Instead, they perform both activities in the same mesophyll cells, but at different times: Carbon fixation occurs at night, and sugar synthesis occurs during the day (**FIG. E7-2**).

The stomata of CAM plants open at night, when less water will evaporate because temperatures are cooler and humidity is higher. Carbon dioxide diffuses into the leaf and is captured in mesophyll cells using the C_4 pathway. The malate produced by the C_4 pathway is then shuttled into the central vacuole, where it is stored as malic acid until daytime. During the day, when stomata are closed to conserve water, the malic acid leaves the vacuole and re-enters the cytoplasm as malate. The malate is broken down, forming pyruvate (which will be converted to PEP) and releasing CO_2, which enters the Calvin cycle (via rubisco) to produce sugar. CAM plants include pineapples, succulents, and cacti.

C_4 plants and CAM Pathways Are Specialized Adaptations

Because C_4 and CAM plants greatly reduce competitive inhibition of rubisco by O_2, why don't all plants use these pathways? The trade-off is that both the C_4 and the CAM pathways consume more energy than does the Calvin cycle by itself; hence, these plants waste some of the solar energy that they capture. As a result, they only

have an advantage in warm, sunny, dry environments. This explains why a lush spring lawn of Kentucky bluegrass (a C_3 plant) may be taken over by spiky crabgrass (a C_4 plant) during a hot, dry summer.

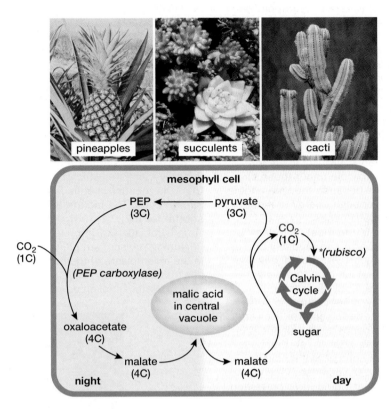

pineapples   succulents   cacti

▲ **FIGURE E7-2 The CAM pathway** As with the C_4 pathway, the CAM pathway fixes carbon into oxaloacetate using PEP carboxylase and stores it in malate. In CAM plants, both processes occur in mesophyll cell, but CO_2 capture occurs at night and CO_2 enters the Calvin cycle during the day.

Carbon Fixed During the Calvin Cycle Is Used to Synthesize Glucose

In reactions that occur outside of the Calvin cycle, two three-carbon G3P sugar molecules can be combined to form one six-carbon glucose molecule (see Fig. 7-9 ❺). Glucose can then be used to synthesize sucrose (table sugar), a disaccharide storage molecule consisting of a glucose linked to a fructose. Glucose molecules can also be linked together in long chains to form starch (another storage molecule) or cellulose (a major component of plant cell walls). Some plants convert glucose into lipids for storage. Glucose is also broken down during cellular respiration to provide energy for the plant's cells.

The storage products of photosynthesis are being eyed by Earth's growing and energy-hungry human population as a substitute for fossil fuels. These "biofuels" have the potential advantage of not adding additional CO_2 (a "greenhouse gas" that contributes to global climate change) to the atmosphere, but do they live up to their promise? We explore this question in "Earth Watch: Biofuels—Are Their Benefits Bogus?"

SUMMING UP: The Calvin Cycle

The Calvin cycle can be divided into three stages:

1. **Carbon fixation:** Three RuBP capture three CO_2, forming six PGA.
2. **G3P synthesis:** A series of reactions, driven by energy from ATP and NADPH (from the light reactions), produces six G3P. One G3P leaves the cycle and is available to form glucose.
3. **RuBP regeneration:** Three RuBP molecules are regenerated from the remaining five G3P using ATP energy, allowing the cycle to continue.

In a separate process outside the chloroplast, two G3P molecules produced by the Calvin cycle combine to form glucose.

CHECK YOUR LEARNING

Can you ...

- describe the function of the Calvin cycle and where it occurs?
- list the three stages of the Calvin cycle, including the molecules that enter the cycle and those that are formed at each stage?
- describe the fate of the simple sugar G3P generated by the Calvin cycle?

Earth WATCH

Biofuels—Are Their Benefits Bogus?

When you drive your car, turn up the thermostat, or flick on your desk lamp, you are actually unleashing the energy of prehistoric sunlight trapped by prehistoric photosynthetic organisms. This is because over hundreds of millions of years, heat and pressure converted the bodies of these organisms—with their stored solar energy and carbon captured from ancient atmospheric CO_2—into coal, oil, and natural gas. Without human intervention, these fossil fuels would have remained trapped deep underground.

A major contributor to global climate change is increased burning of fossil fuels by a growing human population. This combustion releases CO_2 into the atmosphere; the added carbon dioxide traps heat in the atmosphere that would other-wise radiate into space. Since we began using fossil fuels dur-ing the industrial revolution in the mid-1800s, humans have increased the CO_2 content of the atmosphere by about 38%. As a result, Earth is growing warmer, and many experts fear that a hotter future climate will place extraordinary stresses on Earth's inhabitants, ourselves included (see Chapter 29).

To reduce CO_2 emissions and reliance on imported oil, many governments are subsidizing and promoting the use of biofuels, especially ethanol and biodiesel. Ethanol is produced by fermenting plants rich in sugars, such as sugarcane and corn, to produce alcohol (fermentation is de-scribed in Chapter 8). Biodiesel fuel is made primarily from oil derived from plants such as soybeans, canola, or palms. Because the carbon stored in biofuels was removed from the modern atmosphere by photosynthesis, burning them seems to simply restore CO_2 that was recently present in the atmo-sphere. Is this a solution to global climate change?

The environmental and social benefits of burning fuels derived from food crops in our gas tanks are hotly debated. Over 35% of the U.S. corn crop is now feeding vehicles rather than animals and people; this transition has driven up corn prices worldwide (**FIG. E7-3**). Higher corn prices translate into increased food prices as cars compete with animal and human consumers. Another concern is that growing corn and

▲ **FIGURE E7-4 Cleared tropical forest** This aerial view shows the aftermath of clearing lush tropical forest, the former home of rare Sumatran tigers, elephants, leopards, orangutans, and a wealth of bird species. The cleared area will become a palm oil plantation for biofuels. Endangered orangutans (such as the ones pictured) are increasingly rendered homeless by deforestation and are often killed as they are forced closer to human settlements.

converting it into ethanol uses large quantities of fossil fuel, negating corn ethanol's advantages over burning gasoline. The environmental costs of using food crops as an alternative fuel source are also enormous. For example, Indonesia's luxurious tropical rain forests—home to orangutans, Sumatran tigers, and clouded leopards—are being destroyed to make room for oil palm plantations for biofuels, with at least 15 million acres (an area the size of West Virginia) cleared between 2000 and 2012, and over 2 million acres annually in more recent years (**FIG. E7-4**). In Brazil, soybean plantations for biofuels have replaced large expanses of rain forest. Ironically, clear-ing these forests for agriculture increases atmospheric CO_2 because rain forests trap far more carbon than the crops that replace them.

Biofuels would have far less environmental and social impact if they were not produced from food crops or by destroying Earth's dwin-dling rain forests. Algae show great promise as an alternative. Some algae produce starch that can be fermented into ethanol; others produce oil that can become biodiesel. Some of these microscopic photosynthesizers can potentially produce 60 times as much oil per acre as soybeans and 5 times as much as oil palm. Researchers are also attempting to cleave cel-lulose into its component sugars, which would allow ethanol to be generated from corn stalks, wood chips, or grasses. Commercial scale cellulosic biorefineries have recently opened in the United States; their long-term success re-mains uncertain. Although the benefits of most biofuels in wide use today may not justify their environmental costs, there is hope that this will change as we develop better technologies to harness the energy captured by photosynthesis.

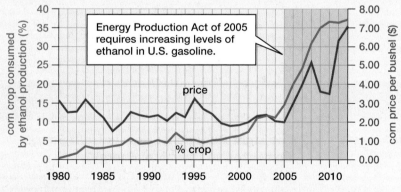

▲ **FIGURE E7-3 Corn prices have increased dramatically since corn ethanol has been added to gasoline**

THINK CRITICALLY The percentage of ethanol in gasoline has reached 10% and is not likely to increase in the near future. Sketch a possible scenario for corn prices and the percentage of corn used for ethanol by continuing the graph in Figure E7-3 to 2030, and provide a rationale to support your projection. Your scenario should assume a constant-sized corn harvest.

CASE STUDY \ REVISITED

Did the Dinosaurs Die from Lack of Sunlight?

Did an asteroid end the reign of dinosaurs? The Alvarez hypothesis was initially met with skepticism. If such a cataclysmic event had occurred, where was the crater? In 1991, scientists finally located it near the coastal town of Chicxulub on Mexico's Yucatán Peninsula. The crater, estimated at over 110 miles in diameter and 10 miles deep, was filled with debris and sedimentary rock laid down during the 66 million years since the impact. Ocean and dense vegetation hid most remaining traces from satellite images. The final identification of the Chicxulub crater was based on rock core samples, unusual gravitational patterns, and faint surface features.

Some paleontologists argue that the asteroid's impact may have exacerbated more gradual changes in climate, to which the dinosaurs (with the exception of those ancestral to modern-day birds) could not adapt. Such changes might have been caused by prolonged intense volcanic activity, such as occurred at a site in India at about the time of the K-T extinction. Volcanoes spew out soot and ash, and iridium is found in higher levels in lava from Earth's molten mantle than in its crust. So furious volcanism could significantly reduce the amount of sunlight for plant growth, spew climate-changing gases into the air, and also contribute to the iridium-rich K-T boundary layer.

In 2010, alternative hypotheses to the asteroid impact were dealt a blow when an expert group of 41 researchers published a review article in the journal *Science*. This publication analyzed the previous 20 years of research by paleontologists, geochemists, geophysicists, climatologists, and sedimentation experts dealing with the K-T extinction event. The conclusion: Land and ocean ecosystems were destroyed extremely rapidly, and evidence overwhelmingly supports the asteroid impact hypothesis first proposed by the Alvarez group 30 years earlier.

Recently, researchers applied high-precision radioactive dating techniques to the K-T boundary layer and to debris kicked up by the Chicxulub asteroid. These findings dramatically narrow the time frame for both occurrences, revealing with as much accuracy as is currently possible that the two coincided. The authors of this study stress that the asteroid impact may have been the "final straw," culminating a series of ecological perturbations that had already stressed existing ecosystems. The hothouse conditions of the late Cretaceous period had previously been interrupted by several rapid drops in temperature and sea level. Pulses of volcanic activity may have caused these cold snaps, leaving the biosphere particularly vulnerable to the devastating climate change associated with the asteroid impact.

CONSIDER THIS The K-T extinction event was the most recent of five major extinctions documented in the fossil record. The ultimate cause of any such event is a massive environmental change occurring on a timescale too rapid to allow species to adapt. In 1968, biologist Paul Ehrlich published the controversial book *The Population Bomb* to describe the impact of overpopulation on Earth's ecosystems. Since Ehrlich's publication, human numbers have more than doubled. Many scientists are now recognizing a sixth mass extinction caused entirely by people; current extinction rates are estimated to be from 100 to 1,000 times the extinction rate that would occur in the absence of human activity. We have modified roughly half of all Earth's land area, co-opting a significant percentage of all terrestrial photosynthesis to feed ourselves. We are changing Earth's climate at a rate at least 10 times that of previous natural cycles between warming and ice ages. The impact of humanity is collective, but the human population consists of individuals. What global policy changes and what individual choices can help us sustain the planet that sustains us?

CHAPTER REVIEW

Go to **MasteringBiology** for practice quizzes, activities, eText, videos, current events, and more.

Answers to Think Critically, Evaluate This, Multiple Choice, and Fill-in-the-Blank questions can be found in the Answers section at the back of the book.

Summary of Key Concepts

7.1 What Is Photosynthesis?

Photosynthesis is the process that captures the energy of sunlight and uses it to convert inorganic molecules of carbon dioxide and water into a high-energy sugar molecule, releasing oxygen as a by-product. In plants, photosynthesis takes place in the chloroplasts, using two major reaction sequences: the light reactions and the Calvin cycle.

7.2 The Light Reactions: How Is Light Energy Converted to Chemical Energy?

The light reactions occur in the thylakoids of chloroplasts. Light energizes electrons in chlorophyll molecules located in photosystems II and I. Energetic electrons jump to a primary electron acceptor and then move into adjacent electron transport chains. Energy lost as the electrons travel through the first ETC is used to pump hydrogen ions into the thylakoid space, creating an H^+ gradient across the thylakoid membrane. Hydrogen ions flow down this concentration gradient through ATP synthase channels in the membrane, driving ATP synthesis by chemiosmosis. For every two energized electrons that pass through the second ETC, one molecule of the energy-carrier NADPH is formed from $NADP^+$ and H^+. Electrons lost from photosystem II are replaced by electrons liberated by splitting water, which also generates H^+ and O_2. Energized electrons lost from photosystem I are replaced by energy-depleted enzymes from photosystem II.

7.3 The Calvin Cycle: How Is Chemical Energy Stored in Sugar Molecules?

The Calvin cycle, which occurs in the stroma of chloroplasts, uses energy from the ATP and NADPH generated during the light reactions to drive the synthesis of G3P. Two molecules of G3P may then be combined to form glucose. The Calvin cycle has three parts: (1) *Carbon fixation*: Carbon dioxide combines with ribulose bisphosphate (RuBP) to form phosphoglyceric acid (PGA). (2) *Synthesis of G3P*: PGA is converted to glyceraldehyde-3-phosphate (G3P), using energy from ATP and NADPH. (3) *Regeneration of RuBP*: Five molecules of G3P are used to regenerate three molecules of RuBP, using ATP energy. One molecule of G3P exits the cycle; this G3P may be used to synthesize glucose and other molecules.

Key Terms

accessory pigment *112*	electron transport
ATP synthase *115*	chain (ETC) *113*
bundle sheath cells *109*	epidermis *109*
C_3 pathway *116*	grana *110*
C_4 pathway *116*	light reactions *110*
Calvin cycle *111*	mesophyll *109*
carbon fixation *116*	NADPH (NADP+; nicotinamide
carotenoid *112*	adenine dinucleotide
chemiosmosis *114*	phosphate) *111*
chlorophyll *110*	photon *111*
chlorophyll *a* *112*	photorespiration *116*
chloroplast *110*	photosynthesis *109*
crassulacean acid	photosystem *112*
metabolism	reaction center *113*
(CAM) *116*	rubisco *116*
cuticle *109*	stoma (plural, stomata) *109*
electromagnetic	stroma *110*
spectrum *111*	thylakoid *110*

Thinking Through the Concepts

Multiple Choice

1. Which of the following is True?
   a. Photosynthesis evolved in an atmosphere with little or no oxygen.
   b. Photosynthesis occurs only in plants.
   c. Oxygen is necessary for photosynthesis.
   d. Carbon dioxide is necessary for photorespiration.

2. The Calvin cycle
   a. can only occur when light is present.
   b. is the part of photosynthesis where carbon is captured.
   c. produces ATP and NADPH.
   d. occurs in the thylakoids.

3. Which of the following is True?
   a. Chloroplasts are found primarily in leaf mesophyll cells.
   b. The space between the outer and inner membranes of chloroplasts is called the thylakoid space.
   c. Mesophyll cells in the leaf's center are packed tightly together.
   d. Vascular bundles of the leaf circulate CO_2 and O_2.

4. Carotenoids
   a. include chlorophylls *a* and *b*.
   b. serve as accessory pigments.
   c. are produced in the fall in temperate climates.
   d. absorb mostly yellow and orange light.

5. Which of the following is False?
   a. Photosystem II splits water.
   b. Photosystem II captures energy directly from light.
   c. Reaction centers contain chlorophyll *a* molecules.
   d. Photosystem I generates O_2.

Fill-in-the-Blank

1. Plant leaves contain pores called _____ that allow the plant to release _____ and take in _____. In hot dry weather, these pores are closed to prevent _____. Photosynthesis occurs in organelles called _____ that are concentrated within the _____ cells of most plant leaves.

2. Chlorophyll *a* captures wavelengths of light that correspond to the three colors _____, _____, and _____. What color does chlorophyll reflect? _____ Accessory pigments that reflect yellow and orange are called _____. These pigments are located in clusters called _____ in the _____ membrane of the chloroplast.

3. During the first stage of photosynthesis, light is captured and funneled into _____ chlorophyll *a* molecules. The energized electron is then passed into a(n) _____. From here, it is transferred through a series of molecules called the _____. Energy lost during these transfers is used to create a gradient of _____. The process that uses this gradient to generate ATP is called _____.

4. The oxygen produced as a by-product of photosynthesis is derived from _____, and the carbons used to make glucose are derived from _____. The biochemical pathway that captures atmospheric carbon is called the _____. The process of capturing carbon is called _____.

5. In plants, the enzyme that catalyzes carbon capture is _____, which binds _____ as well as CO_2. When it binds the "wrong" molecule, this enzyme causes _____ to occur. Two pathways that reduce this process are called the _____ and the _____.

6. Light reactions generate the energy-carrier molecules _____ and _____, which are then used in the _____ cycle. Carbon fixation combines carbon dioxide with the five-carbon molecule _____. Two molecules of _____ can be combined to produce the six-carbon sugar _____.

Review Questions

1. Explain what would happen to life if photosynthesis ceased. Why would this occur?

2. Write and then explain the equation for photosynthesis.

3. Draw a simplified diagram of a leaf cross-section and label it. Explain how a leaf's structure supports photosynthesis.

4. Draw a simplified diagram of a chloroplast and label it. Explain how the individual parts of the chloroplast support photosynthesis.

5. Explain photorespiration and why it occurs. Describe the two mechanisms that some groups of plants have evolved to reduce photorespiration.

6. Trace the flow of energy in chloroplasts from sunlight to ATP, including an explanation of chemiosmosis.

7. Summarize the events of the Calvin cycle. Where does it occur? What molecule is fixed? What is the product of the cycle? Where does the energy come from to drive the cycle? What molecule is regenerated?

Applying the Concepts

1. Suppose an experiment is performed in which plant I is supplied with normal carbon dioxide but with water that contains radioactive oxygen atoms. Plant II is supplied with normal water but with carbon dioxide that contains radioactive oxygen atoms. Each plant is allowed to perform photosynthesis, and the oxygen gas and sugars produced are tested for radioactivity. Which plant would you expect to produce radioactive sugars, and which plant would you expect to produce radioactive oxygen gas? Explain why.

2. If you were to measure the pH in the space surrounded by the thylakoid membrane in an actively photosynthesizing plant, would you expect it to be acidic, basic, or neutral? Explain your answer.

3. Assume you want to add an accessory pigment to the photosystems of a plant chloroplast to help it photosynthesize more efficiently. What wavelengths of light would the new pigment absorb? Describe the color of your new pigment. Would this be a useful project for genetic engineers? Explain.

8 HARVESTING ENERGY: GLYCOLYSIS AND CELLULAR RESPIRATION

The skeleton of King Richard III (inset) is revealed.

Raising a King

"A HORSE! A HORSE! MY KINGDOM FOR A HORSE!" shouts King Richard III in Shakespeare's *King Richard III*, moments after his horse is slain at the Battle of Bosworth Field in Leicester, England in 1485. The 32-year-old monarch had ruled for only 2 years before this final battle of the 30-year War of the Roses, and his short life was riddled with political intrigue and conspiracy. Now, thanks to the DNA in mitochondria, his skeleton has been identified, providing details of his final moments in battle, as well as a broad sketch of his appearance. For example, although he did not have the hunched back portrayed by Shakespeare, his skeleton does reveal a severe spinal deformity (see the chapter-opening photo).

No one knows what King Richard III was really shouting as he died, but we do know that his wounds were horrendous. His eight head injuries included a sword wound that pierced his skull, penetrating entirely through his brain, and mutilation from an axe-like weapon that hacked out a large chunk of his skull.

Wounds on his ribs and pelvis suggest that the defeated king's body was further mutilated after death. Franciscan friars apparently buried him hastily in the Greyfriars Church—naked, without a coffin, and in a shallow grave that was too short for his body.

Prior to the excavation in 2012, the Franciscan church and its associated monastery had not been seen since 1538, when they were leveled and newer buildings were erected on the site. The location of the church and monastery was subsequently forgotten, but a historian and archaeologists (aided by ground-penetrating radar) did some detective work that led to the hypothesis that the ruins were likely to be found beneath a Leicester parking lot. Despite this compelling hypothesis, experts still believed there was only a very remote possibility that Richard III's body was buried there, so the discovery of a skeleton disfigured by battle wounds and severe spinal curvature (consistent with historical records; see arrow in photo) was met with astonishment.

Painstaking investigations led by geneticist Turi King of the University of Leicester identified the king's body with near-certainty. King analyzed mitochondrial DNA from a tooth of the 529-year-old skeleton. A mitochondrion has several copies of a tiny circular loop of DNA, with each loop containing about 1/200,000 as much DNA as is found in the nucleus.

Mitochondria provide energy for every cell in the bodies of eukaryotic organisms. These complex organelles powered the muscles of the soldiers and their horses at the Battle of Bosworth, while simultaneously performing the mundane task of keeping their teeth alive. How do mitochondria work? Why do we begin dying in seconds if their function is blocked? And why do mitochondria have their own DNA?

8.1 HOW DO CELLS OBTAIN ENERGY?

Cells require a continuous supply of energy to power the multitude of metabolic reactions that are essential just to stay alive. In this chapter, we describe the cellular reactions that transfer energy from energy-storage molecules, particularly glucose, to energy-carrier molecules, such as ATP.

The second law of thermodynamics tells us that every time a spontaneous reaction occurs, the amount of useful energy in a system decreases and heat is produced (see Chapter 6). Cells are relatively efficient at capturing chemical energy during glucose breakdown when oxygen is available, storing about 40% of the chemical energy from glucose in ATP molecules, and releasing the rest as heat. (If 60% waste heat sounds high, compare this to the 80% of chemical energy released as heat by conventional engines burning gasoline.)

Photosynthesis Is the Ultimate Source of Cellular Energy

The energy utilized by life on Earth comes almost entirely from sunlight, captured during photosynthesis by plants and other photosynthetic organisms and stored in the chemical bonds of sugars and other organic molecules (see Chapter 7). Almost all organisms, including those that photosynthesize, use glycolysis and cellular respiration to break down these sugars and other organic molecules and capture some of the energy as ATP. **FIGURE 8-1** illustrates the interrelationship between photosynthesis and the breakdown of glucose. Glucose ($C_6H_{12}O_6$) breakdown begins with glycolysis in the cell cytosol, liberating small quantities of ATP. Then the end product of glycolysis is further broken down during cellular respiration in mitochondria, supplying far greater amounts of energy in ATP. In forming ATP during cellular respiration, cells use oxygen (originally released by photosynthetic organisms) and liberate both water and carbon dioxide—the raw materials for photosynthesis.

Photosynthesis

$$6\,CO_2 + 6\,H_2O + \text{light energy} \rightarrow C_6H_{12}O_6 + 6\,O_2$$

The chemical equation describing complete glucose breakdown is the reverse of glucose formation by photosynthesis.

Complete Glucose Breakdown

$$C_6H_{12}O_6 + 6\,O_2 \rightarrow 6\,CO_2 + 6\,H_2O + \text{ATP energy}$$

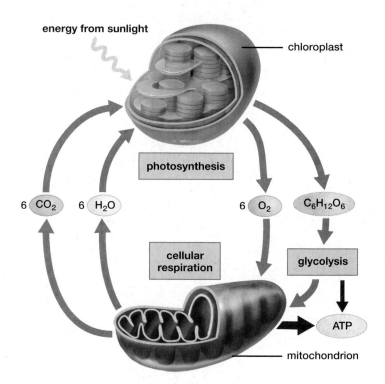

energy from sunlight

chloroplast

photosynthesis

$6\;CO_2$ $6\;H_2O$ $6\;O_2$ $C_6H_{12}O_6$

cellular respiration

glycolysis

ATP

mitochondrion

▲ **FIGURE 8-1 The interrelationship between photosynthesis and glucose breakdown** The products of each process are used by the other. The ultimate source of energy is sunlight, captured during photosynthesis and liberated during glycolysis and cellular respiration.

The only difference is in the forms of energy involved. The light energy stored in glucose during photosynthesis is released during glucose breakdown and used to generate ATP, with some lost as heat during each conversion.

All Cells Can Use Glucose As a Source of Energy

Few organisms store glucose in its simple form. Plants convert glucose to sucrose or starch for storage. Humans and many other animals store energy in molecules such as glycogen (a long chain of glucose molecules) and fat (see Chapter 3). Although most cells can use a variety of organic molecules to produce ATP, in this chapter, we focus on the breakdown of glucose, which all cells can use as an energy source. Glucose breakdown occurs via two major processes: It starts with *glycolysis* and proceeds to *cellular respiration* if oxygen is available. Some energy is captured in ATP during glycolysis and far more is captured during cellular respiration (see **FIG. 8-2**).

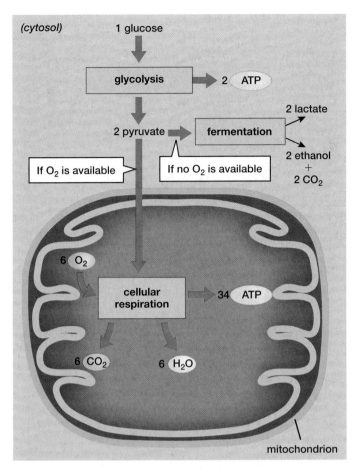

(cytosol)

▲ **FIGURE 8-2 A summary of glucose breakdown**

8.2 HOW DOES GLYCOLYSIS BEGIN BREAKING DOWN GLUCOSE?

Glycolysis (Gk. *glyco,* sweet, and *lysis,* to split apart) splits a six-carbon glucose molecule into two molecules of pyruvate. Glycolysis has an *energy investment stage* and an *energy harvesting stage,* each with several steps (**FIG. 8-3**). Extracting energy from glucose first requires an investment of energy from ATP. During a series of reactions that constitutes the energy investment stage, each of two ATP molecules donates a phosphate group and energy to glucose, forming an "energized" molecule of *fructose bisphosphate.* Fructose is a monosaccharide sugar similar to glucose; "bisphosphate" (L. *bis,* two) refers to the two phosphate groups acquired from the ATP molecules. Fructose bisphosphate is much more easily broken down than glucose because of the extra energy it has acquired from ATP.

Next, during the energy harvesting stage, fructose bisphosphate is converted into two three-carbon molecules of *glyceraldehyde-3-phosphate,* or *G3P.* Each G3P molecule, which retains one phosphate and some energy from ATP, then undergoes a series of reactions that convert the G3P to pyruvate. During these reactions, energy is stored when two high-energy electrons and a hydrogen ion (H^+) are added to the electron carrier **nicotinamide adenine dinucleotide (NAD⁺)** to produce **NADH.** Two molecules of NADH are produced for every glucose molecule broken down. Additional energy is captured in two ATP from each G3P, for a total of four ATP per glucose molecule. But because two ATP were used up to form fructose bisphosphate, there is a net gain of only two ATP per glucose molecule during glycolysis. For the details of glycolysis, see "In Greater Depth: Glycolysis" on page 126.

CHECK YOUR LEARNING

Can you ...

- explain how photosynthesis and glucose breakdown are related to one another using their overall chemical equations?
- summarize glucose breakdown in the presence and absence of oxygen?

CHECK YOUR LEARNING

Can you ...

- explain the energy investment and energy-harvesting phases of glycolysis?
- describe the two types of high-energy molecule produced by glucose breakdown?

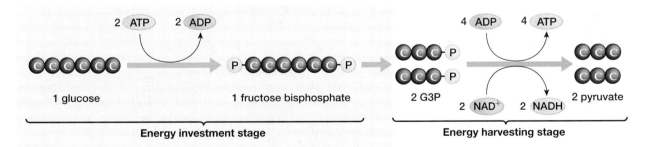

Energy investment stage

Energy harvesting stage

▲ **FIGURE 8-3 The essentials of glycolysis** In the energy investment stage, the energy of two ATP molecules is used to convert glucose into the fructose bisphosphate, which then breaks down into two molecules of G3P. In the energy harvesting stage, the two G3P molecules undergo a series of reactions that capture energy in four ATP and two NADH molecules. (Only carbon skeletons are shown.)

THINK CRITICALLY What is the net energy yield in ATP and NADH produced?

IN GREATER DEPTH | Glycolysis

Glycolysis is a series of enzyme-catalyzed reactions that break down a single molecule of glucose into two molecules of pyruvate. In **FIGURE E8-1**, we show only the carbon skeletons of molecules. Blue arrows represent enzyme-catalyzed reactions.

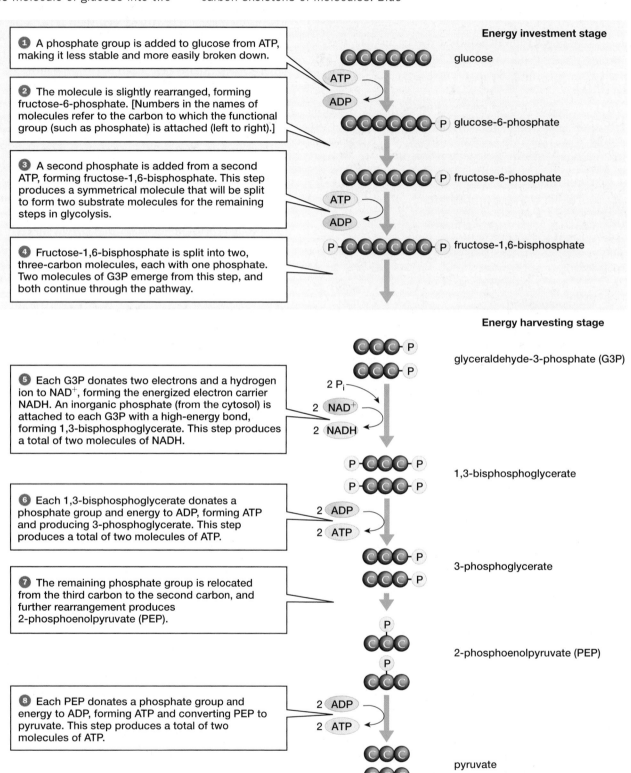

Energy investment stage

❶ A phosphate group is added to glucose from ATP, making it less stable and more easily broken down.

glucose

❷ The molecule is slightly rearranged, forming fructose-6-phosphate. [Numbers in the names of molecules refer to the carbon to which the functional group (such as phosphate) is attached (left to right).]

glucose-6-phosphate

❸ A second phosphate is added from a second ATP, forming fructose-1,6-bisphosphate. This step produces a symmetrical molecule that will be split to form two substrate molecules for the remaining steps in glycolysis.

fructose-6-phosphate

❹ Fructose-1,6-bisphosphate is split into two, three-carbon molecules, each with one phosphate. Two molecules of G3P emerge from this step, and both continue through the pathway.

fructose-1,6-bisphosphate

Energy harvesting stage

glyceraldehyde-3-phosphate (G3P)

❺ Each G3P donates two electrons and a hydrogen ion to NAD⁺, forming the energized electron carrier NADH. An inorganic phosphate (from the cytosol) is attached to each G3P with a high-energy bond, forming 1,3-bisphosphoglycerate. This step produces a total of two molecules of NADH.

2 P_i

2 NAD⁺

2 NADH

1,3-bisphosphoglycerate

❻ Each 1,3-bisphosphoglycerate donates a phosphate group and energy to ADP, forming ATP and producing 3-phosphoglycerate. This step produces a total of two molecules of ATP.

2 ADP

2 ATP

3-phosphoglycerate

❼ The remaining phosphate group is relocated from the third carbon to the second carbon, and further rearrangement produces 2-phosphoenolpyruvate (PEP).

2-phosphoenolpyruvate (PEP)

❽ Each PEP donates a phosphate group and energy to ADP, forming ATP and converting PEP to pyruvate. This step produces a total of two molecules of ATP.

2 ADP

2 ATP

pyruvate

▲ **FIGURE E8-1 Glycolysis**

8.3 HOW DOES CELLULAR RESPIRATION EXTRACT ENERGY FROM GLUCOSE?

In most organisms, if oxygen is available, the second process in glucose breakdown, called **cellular respiration,** occurs. Cellular respiration breaks down the two pyruvate molecules produced by glycolysis into six carbon dioxide molecules and six water molecules. During this process, the chemical energy from the two pyruvate molecules is used to produce 34 ATP.

In eukaryotic cells, cellular respiration occurs within **mitochondria,** organelles that are sometimes called the "powerhouses of the cell." A mitochondrion has two membranes. The inner membrane encloses a central compartment containing the fluid **matrix,** and the outer membrane surrounds the organelle, producing an **intermembrane space** between the two membranes. Each structure is crucial to the process of cellular respiration (**FIG. 8-4**).

In the following sections, we discuss the two major stages of cellular respiration: first, the formation of acetyl CoA and its breakdown via the Krebs cycle, and second, the transfer of electrons along the electron transport chain and the generation of ATP by chemiosmosis.

Cellular Respiration Stage 1: Acetyl CoA Is Formed and Travels Through the Krebs Cycle

Pyruvate, the end product of glycolysis, is synthesized in the cytosol. Before cellular respiration can occur, the pyruvate diffuses from the cytosol through the porous outer mitochondrial membrane. It is then actively transported through the inner mitochondrial membrane and into the matrix, where cellular respiration begins.

Two sets of reactions occur within the mitochondrial matrix during stage 1 of cellular respiration: the formation of acetyl CoA and the Krebs cycle (**FIG. 8-5**). Acetyl CoA consists of a two-carbon functional (acetyl) group attached to a molecule called *coenzyme A (CoA).* To generate acetyl CoA, pyruvate is split, releasing CO_2 and leaving behind an acetyl group. The acetyl group reacts with CoA, forming acetyl CoA. This reaction liberates and stores energy by transferring two high-energy electrons and a hydrogen ion to NAD^+, forming NADH.

The next set of reactions is known as the **Krebs cycle,** named after its discoverer, Hans Krebs, who won a Nobel Prize for this work in 1953. The Krebs cycle is also called the *citric acid cycle* because citrate (the dissolved, ionized form of citric acid) is the first molecule produced in the cycle. This metabolic pathway is called a cycle because it continuously regenerates the same substrate molecule with which it begins: oxaloacetate (see Fig. E8-2 on page 130).

With each pass around the Krebs cycle, the two carbon atoms that enter in the form of acetate are released as carbon dioxide, liberating energy. Some of this energy is captured

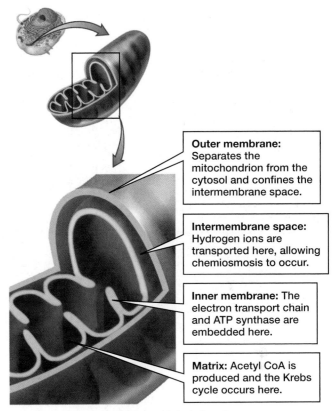

Outer membrane: Separates the mitochondrion from the cytosol and confines the intermembrane space.

Intermembrane space: Hydrogen ions are transported here, allowing chemiosmosis to occur.

Inner membrane: The electron transport chain and ATP synthase are embedded here.

Matrix: Acetyl CoA is produced and the Krebs cycle occurs here.

(a) Mitochondrial structures and their functions

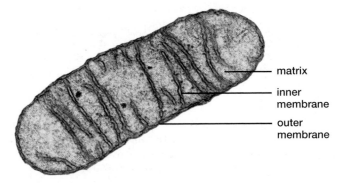

matrix

inner membrane

outer membrane

(b) TEM of a mitochondrion

▲ **FIGURE 8-4** The mitochondrion

during the Krebs cycle in high-energy electron carriers (described later) and some in ATP.

The breakdown of acetate begins when acetyl CoA is combined with the four-carbon molecule oxaloacetate, forming a six-carbon citrate molecule and releasing the catalyst CoA. CoA is not permanently altered during these reactions and is reused many times. As the Krebs cycle proceeds, enzymes within the mitochondrial matrix break down the acetyl group, releasing two CO_2 and regenerating the oxaloacetate molecule to continue the cycle. In your body, the CO_2 generated in cells during the stage 1 reactions diffuses into your blood, which carries the CO_2 to your lungs. This is why the air you breathe out contains more CO_2 than the air you breathe in.

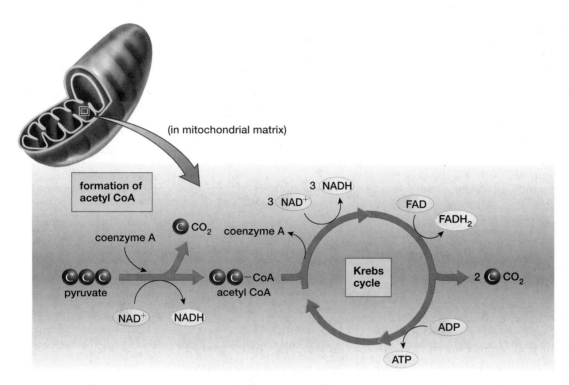

◀ **FIGURE 8-5** **Reactions in the mitochondrial matrix: acetyl CoA formation and the Krebs cycle**

As the Krebs cycle continues, chemical energy is captured in energy-carrier molecules. The breakdown of each acetyl group from acetyl CoA produces one ATP and three NADH. It also produces one **flavin adenine dinucleotide (FADH₂),** a high-energy electron carrier similar to NADH. During the Krebs cycle, FAD picks up two energetic electrons along with two H⁺, forming FADH₂. Remember that for each glucose molecule, two pyruvate molecules are formed during glycolysis, so the energy generated per glucose molecule is twice that generated for one pyruvate (for details, see "In Greater Depth: Acetyl CoA Production and the Krebs Cycle" on page 130.)

Cellular Respiration Stage 2: High-Energy Electrons Traverse the Electron Transport Chain and Chemiosmosis Generates ATP

By the end of stage 1, the cell has gained only four ATP from the original glucose molecule (a net of two during glycolysis and two during the Krebs cycle). However, the cell has also captured many high-energy electrons in a total of 10 NADH and two FADH₂ molecules for each glucose molecule.

In the second stage of cellular respiration, the high-energy electron carriers each release two high-energy electrons into an **electron transport chain (ETC),** a series of electron-transporting molecules, many copies of which are embedded in the inner mitochondrial membrane (**FIG. 8-6**). The depleted carriers are then available for recharging by glycolysis and the Krebs cycle.

The Electron Transport Chain Releases Energy in Steps

The ETCs in the mitochondrial membrane serve the same function as those embedded in the thylakoid membrane of chloroplasts (see Chapter 7). High-energy electrons jump from molecule to molecule along the ETC, releasing small amounts of energy at each step. The energy liberated in these stages is just the right amount to pump H⁺ across the inner membrane, from the matrix into the intermembrane space (although some is always lost as heat). This ion pumping produces a concentration gradient of H⁺, high in the intermembrane space and low in the matrix (see Fig. 8-6). Expending energy to create an H⁺ gradient is similar to charging a battery. This H⁺ battery will be discharged as ATP is generated by chemiosmosis, discussed later.

Finally, at the end of the electron transport chain, the energy-depleted electrons are transferred to oxygen, which acts as an electron acceptor. The energy-depleted electrons, oxygen, and hydrogen ions combine, forming water. One water molecule is produced for every two electrons that traverse the ETC (see Fig. 8-6).

Without oxygen to accept electrons, the ETC would become saturated with electrons and could not acquire more from NADH and FADH₂. With electrons unable to move through the ETC, H⁺ could not be pumped across the inner membrane. The H⁺ gradient would rapidly dissipate, and ATP synthesis by chemiosmosis would stop. Because of their high demand for energy from ATP, most eukaryotic cells die within minutes without a steady supply of oxygen to accept electrons.

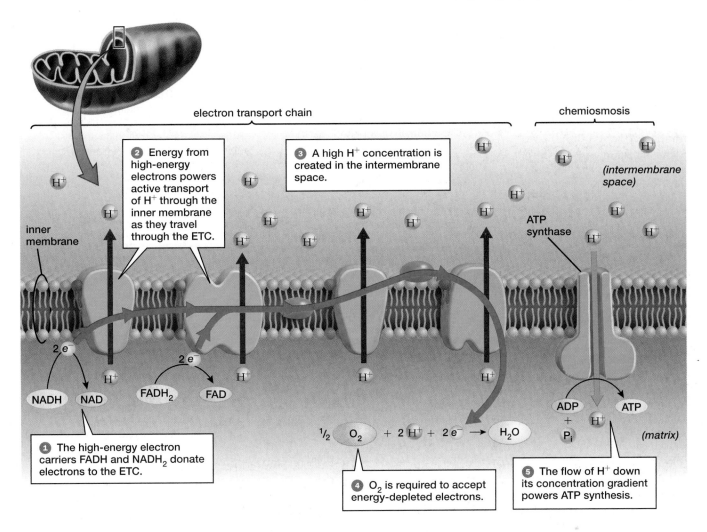

electron transport chain   chemiosmosis

2 Energy from high-energy electrons powers active transport of H⁺ through the inner membrane as they travel through the ETC.

3 A high H⁺ concentration is created in the intermembrane space.

(intermembrane space)

ATP synthase

inner membrane

2 e⁻

2 e⁻

NADH   NAD

FADH₂   FAD

1 The high-energy electron carriers FADH and NADH₂ donate electrons to the ETC.

$$\frac{1}{2}\ O_2\ +\ 2\ H^+\ +\ 2\ e^-\ \longrightarrow\ H_2O$$

ADP
+
P$_i$

ATP

H⁺

(matrix)

4 O₂ is required to accept energy-depleted electrons.

5 The flow of H⁺ down its concentration gradient powers ATP synthesis.

▲ **FIGURE 8-6 The electron transport chain and chemiosmosis** Many copies of the electron transport chain and ATP synthase are embedded in the inner mitochondrial membrane.

THINK CRITICALLY How would the rate of ATP production be affected by the absence of oxygen?

Your body obtains oxygen through the air you breathe, which enters your lungs and is transported in the bloodstream to every cell. Because of cellular respiration, the air you breathe out contains less oxygen than the air you breathe in.

HAVE YOU EVER
WONDERED...

Cyanide is a favorite poison in old murder mysteries, causing the hapless victim to die almost instantly. Cyanide exerts its lethal effects by blocking the last protein in the ETC: an enzyme that combines energy-depleted electrons with oxygen. If these electrons are not carried away by oxygen, they act like a plug in a pipeline. Additional high-energy electrons cannot travel through the ETC, so no more hydrogen can be pumped across the membrane, and ATP production by chemiosmosis stops abruptly. Because the energy demands of our cells are so great, blocking cellular respiration with cyanide can kill a person within a few minutes.

Why Cyanide Is So Deadly?

Chemiosmosis Captures Energy in ATP

Chemiosmosis is the process by which some of the energy stored in the concentration gradient of H⁺ is captured in ATP as H⁺ flows down its gradient. How is the energy captured? The inner membranes of mitochondria are permeable to H⁺ only at channels that are part of an ATP synthase enzyme. As hydrogen ions flow from the intermembrane space into the matrix through these ATP-synthesizing enzymes, ATP is formed from ADP and inorganic phosphate ions dissolved in the matrix.

ATP Is Transported out of the Mitochondrion

How does ATP escape from the mitochondrion to power reactions throughout the cell? Movement through the inner mitochondrial membrane is highly regulated, so a specialized carrier protein in the inner membrane selectively exchanges ATP for ADP. The protein does this by simultaneously exporting ATP from the matrix into the intermembrane space while

IN GREATER DEPTH | Acetyl CoA Production and the Krebs Cycle

Two sets of reactions occur in the mitochondrial matrix: (1) the formation of acetyl CoA from pyruvate and (2) the Krebs cycle (**FIG. E8-2**).

Formation of Acetyl CoA

Pyruvate is split to form an acetyl group and CO_2. The formation of CO_2 releases energy that is captured in 2 high-energy electrons and an H^+, converting NAD^+ to NADH. The acetyl group attaches to CoA, forming acetyl CoA, which enters the Krebs cycle.

The Krebs Cycle

Each acetyl CoA entering the Krebs cycle is broken down into 2 CO_2, releasing energy that is captured in 1 ATP, 3 NADH, and 1 $FADH_2$.

Total Energy Capture

Acetyl CoA formation and the Krebs cycle together produce 3 CO_2, 1 ATP, 4 NADH, and 1 $FADH_2$ from each pyruvate. Each glucose molecule produces 2 pyruvates, doubling the number of product molecules. The electron-carrier molecules NADH and $FADH_2$ will deliver their high-energy electrons to the electron transport chain (ETC). The ETC will store energy in an H^+ gradient that will be used to synthesize ATP by chemiosmosis.

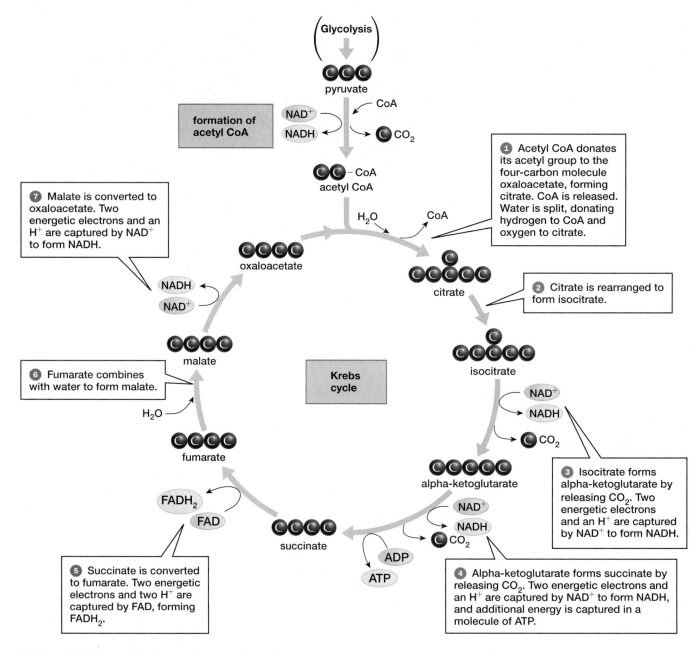

▲ **FIGURE E8-2** Acetyl CoA production and the Krebs cycle

▶ **FIGURE 8-7 A summary of the ATP harvest from glycolysis and cellular respiration** Cellular respiration provides the biggest ATP payoff. Nearly all of the ATP comes from high-energy electrons donated by NADH and FADH$_2$. As the electrons flow through the electron transport chain, they generate the H$^+$ gradient, which allows chemiosmosis to occur.

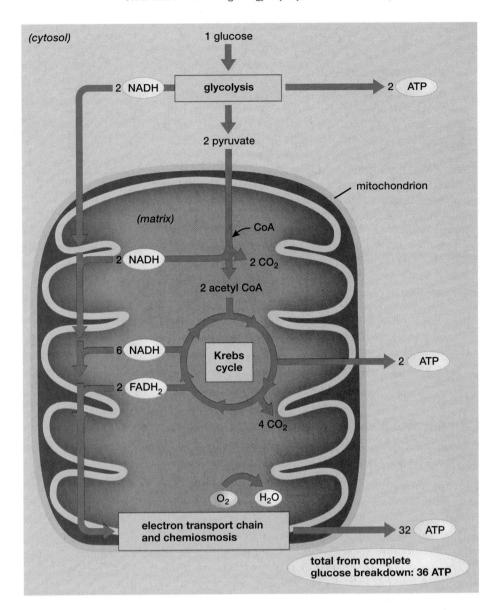

importing ADP from the intermembrane space into the matrix. The outer mitochondrial membrane, in contrast to the inner membrane, is perforated by large pores, through which ATP and ADP can diffuse freely along their concentration gradients. Thus, from the intermembrane space, ATP diffuses through the outer membrane to power reactions throughout the cell, while energy-depleted ADP diffuses into the intermembrane space. Without this continuous recycling, life would cease. Each day, a person produces, uses, and then regenerates the equivalent of roughly his or her body weight of ATP.

You now know why glycolysis followed by cellular respiration generates far more ATP than glycolysis alone. **FIGURE 8-7** and **TABLE 8-1** summarize the breakdown of one glucose molecule in a eukaryotic cell with oxygen present, showing the energy produced during each stage and the general locations where the pathways occur. In summary, the two ATPs formed during glycolysis are supplemented by two more formed during the Krebs cycle and an additional 32 via chemiosmosis, for a total of 36 ATP per glucose molecule.

TABLE 8-1	A Summary of Glucose Breakdown		
Stage of glucose breakdown	**Electron carriers**	**Net ATP produced**	**Location**
Glycolysis	Energy captured in 2 NADH	2 ATP	Cytosol
Cellular respiration stage 1: Acetyl CoA formation and the Krebs cycle	Energy captured in 8 NADH and 2 FADH$_2$	2 ATP	Mitochondrial matrix
Cellular respiration stage 2: Electron transport chain and chemiosmosis	Energy released from 10 NADH and 2 FADH$_2$	32 ATP	Inner mitochondrial membrane and intermembrane space
Fermentation	2 NAD regenerated	0 ATP	Cytosol
Total	0 NADH and 0 FADH$_2$	36 ATP	Cytosol and mitochondrion

CASE STUDY **CONTINUED**
Raising a King

The mitochondrial DNA (mtDNA) that identified Richard III serves a unique role in the human body. Although the human nucleus has about 20,000 genes, mtDNA has only 37. Twenty-four of these code for RNA that helps translate genes into proteins, and the remaining 13 genes code for proteins that are subunits of enzymes that participate in the ETC and chemiosmosis. Some contribute to the ETC enzymes that cause NADH and FADH$_2$ to release their high-energy electrons into the chain. Other mtDNA genes help to produce the final enzyme in the chain, which combines the energy-depleted electrons with oxygen, forming water. Some genes in mtDNA code for parts of the ATP synthase enzyme on the inner mitochondrial membrane. If mtDNA were to disappear, cellular respiration would come to a screeching halt!

We've seen how NADH and FADH$_2$ can gain high-energy electrons that originated in glucose. Can these electron carriers also obtain high-energy electrons from other molecules in our diets, such as fat or protein?

Cellular Respiration Can Extract Energy from a Variety of Foods

Glucose often enters the body as starch (a long chain of glucose molecules) or sucrose (table sugar; glucose linked to fructose), but the typical human diet also provides considerable energy in the form of fat and some from protein. This is possible because various intermediate molecules of cellular respiration can be formed by other metabolic pathways. These intermediates then enter cellular respiration at various stages and are broken down to produce ATP (**FIG. 8-8**). For example, some of the 20 amino acids from protein can be directly converted into pyruvate, and the others can be transformed through complex pathways into molecules of the Krebs cycle. To release the energy stored in fats, the long fatty acid tails (which comprise most of each fat molecule; see Chapter 3) are broken into two-carbon fragments and combined with CoA, producing acetyl CoA, which enters the Krebs cycle.

An excess of intermediate molecules from glucose breakdown can be converted to fat. So if you overeat, not only are the fats from your meal stored in your body, but excess sugar and starch are also used to synthesize body fat, as described in "Health Watch: How Can You Get Fat by Eating Sugar?"

CHECK YOUR LEARNING
Can you ...

• summarize the two major stages of cellular respiration?
• explain how ATP is generated by chemiosmosis?
• describe the role of oxygen in cellular respiration?

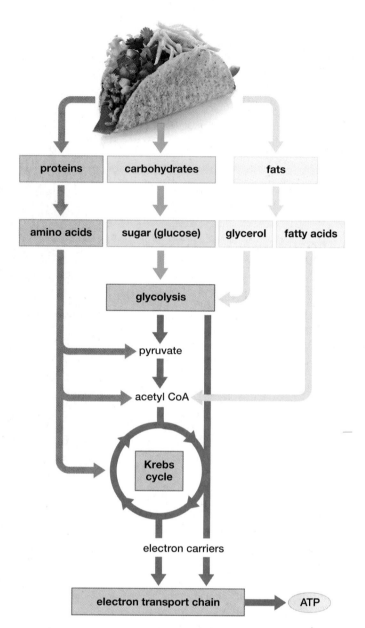

▲ **FIGURE 8-8 Proteins, carbohydrates, and fats are broken down and release ATP**

8.4 HOW DOES FERMENTATION ALLOW GLYCOLYSIS TO CONTINUE WHEN OXYGEN IS LACKING?

Glycolysis is employed by virtually every organism on Earth, providing evidence that this is one of the most ancient of all biochemical pathways. Under **aerobic** conditions—that is, when oxygen is available—cellular respiration usually follows. But scientists have concluded that the earliest forms of life appeared under the **anaerobic** (no oxygen) conditions that existed before photosynthesis evolved and enriched the air with oxygen. These pioneering life-forms relied entirely on glycolysis for energy production. Many

How Can You Get Fat by Eating Sugar?

From an evolutionary perspective, feeling hungry even if you are overweight and overeating when rich food is abundant are highly adaptive behaviors. During the famines common during early human history, heavier people were more likely to survive. It is only recently (from an evolutionary vantage point) that many people have had continuous access to high-calorie food. As a result, obesity is an expanding health problem. Why do we accumulate fat? Fats (triglycerides) are more difficult to break down and also store twice as much energy for their weight as do carbohydrates. Stockpiling energy with minimum weight was important to our prehistoric ancestors who needed to move quickly to catch prey or to avoid becoming prey themselves.

Acquiring fat by eating sugar and other carbohydrates is common among animals. How is fat made from sugar? As glucose is broken down during the Krebs cycle, acetyl CoA is formed. Excess acetyl CoA molecules are used as raw materials to synthesize the fatty acids that will be linked together to form a fat molecule (**FIG. E8-3**). Starches, such as those in bread, potatoes, or pasta, are actually long chains of glucose molecules, so you can see how eating excess starch can also make you fat.

To understand why storing fat rather than sugar can be advantageous, let's look at the ruby-throated hummingbird, which begins the summer weighing

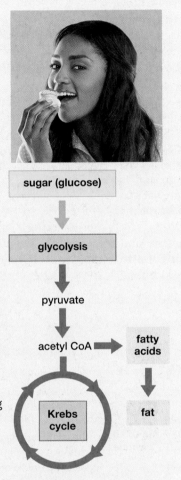

▲ **FIGURE E8-3 How sugar is converted to fat**

3 to 4 grams (in comparison, a nickel weighs 5 grams). In late summer, hummingbirds feed voraciously on the sugary nectar of flowers and nearly double their weight in stored fat. The energy from fat powers their migration from the eastern United States across the Gulf of Mexico and into Mexico or Central America for the winter. If hummingbirds stored sugar instead of fat, they would be too heavy to fly.

EVALUATE THIS Colin, a 45-year-old obese man, comes to you, his physician, complaining that he has been on a fat-free diet for months without losing weight. What do you hypothesize about Colin's weight issue? What questions would you ask him to develop your hypothesis? If the answers support your hypothesis, what dietary recommendations would you make?

---

microorganisms still thrive in places where oxygen is rare or absent, such as in the stomach and intestines of animals (including humans), deep in soil, or in bogs and marshes. Most of these rely on glycolysis, whose end product is pyruvate. In the absence of oxygen, this metabolic pathway continues through **fermentation,** the process by which pyruvate is converted either into lactate or into ethanol and CO_2, depending on the organism. Some microorganisms lack the enzymes for cellular respiration and are completely dependent on fermentation. Others, such as yeasts, are opportunists, using fermentation when oxygen is absent, but switching to more efficient cellular respiration when oxygen is available.

Fermentation is not limited to microorganisms. **Lactate fermentation,** which converts pyruvate to *lactic acid,* is a

temporary recourse in most vertebrates, especially during intense muscular activity. If you feel your muscles "burning" during vigorous exercise, they are probably fermenting pyruvate into lactic acid.

Compared to cellular respiration, fermentation is an extremely inefficient way to metabolize pyruvate; it produces no additional ATP. So what good is it? Glycolysis generates two ATP and two NADH for each molecule of glucose metabolized; without oxygen, there is no final acceptor for the electrons accumulated by NAD^+ to form NADH. Fermentation is required to convert the NADH produced during glycolysis back to NAD^+. If the supply of NAD^+ were to be exhausted—which would happen quickly without fermentation—glycolysis would stop, energy production would cease, and the organism would rapidly die.

Fermentation Produces Either Lactate or Alcohol and Carbon Dioxide

Muscles contracting so vigorously that blood cannot supply adequate oxygen for cellular respiration briefly use glycolysis to generate ATP. The pathway then regenerates NAD⁺ by using the electrons and hydrogen ions from NADH to convert pyruvate into lactate (the dissolved form of lactic acid; **FIG. 8-9**). Glycolysis uses a great deal of glucose relative to the meager 2 ATPs per glucose molecule it produces, but this far simpler pathway also generates ATP much faster than does cellular respiration. The ATPs can provide the energy needed for a final, brief burst to the finish line (**FIG. 8-10**) or for fighting, fleeing, or pursuing prey, when the ability to persist just a bit longer can make the difference between life and death. Most of the lactate generated by muscle cells during fermentation diffuses into the bloodstream and is carried to cells of the liver, which convert the lactate back to pyruvate and then back to glucose. Many microorganisms also utilize lactate fermentation, as described later.

Certain microorganisms, including some bacteria and all forms of yeast (single-celled fungi), engage in **alcoholic fermentation** under anaerobic conditions. During alcoholic

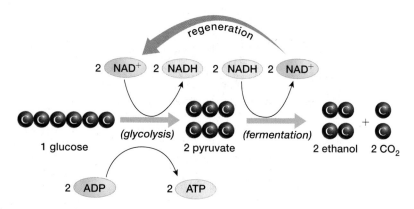

▲ **FIGURE 8-11** Glycolysis followed by alcoholic fermentation

THINK CRITICALLY What would happen if cells were prevented from producing lactic acid or alcohol after glycolysis?

fermentation, pyruvate is converted into ethanol and CO_2. This process converts NADH into NAD⁺, which is then available to accept more high-energy electrons during glycolysis (**FIG. 8-11**).

CASE STUDY   **CONTINUED**

Raising a King

Richard III included plenty of fermented products in his diet. How do we know? The ratio of isotopes of certain minerals and oxygen derived from food and stored in teeth and bones provides evidence of both the types of food consumed and where these foods originated. Researchers have been analyzing these isotopes to glean information about Richard III's life history and lifestyle. Based on oxygen isotope ratios in his bones, investigators hypothesize that during the last few years of his life, starting about when he became king, roughly one-quarter of his fluid intake consisted of imported wine. This beverage—fit for a king—was produced, of course, by alcoholic fermentation.

What other ancient and modern staples result from fermentation?

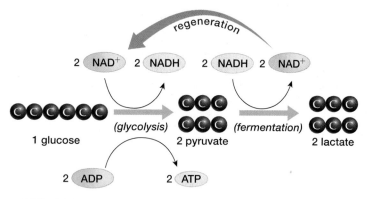

▲ **FIGURE 8-9** Glycolysis followed by lactic acid fermentation

▲ **FIGURE 8-10** Lactic acid fermentation in action

Fermentation Has Played a Long and Important Role in the Human Diet

The poet Omar Khayyam (1048–1122) described his vision of paradise on Earth as "A Jug of Wine, a Loaf of Bread—and Thou Beside Me" (**FIG. 8-12**). Historical evidence suggests that wine and beer, whose alcohol is produced by yeast, were being made roughly 7,000 years ago. Yeasts are opportunists; they engage in efficient cellular respiration if oxygen is available, but switch to alcoholic fermentation (producing alcohol and CO_2) if they run out of oxygen. To make beer or wine, sugars from mashed grain (beer) or grapes (wine) are fermented by specialized strains of yeast. Fermentation is carried out in

▲ **FIGURE 8-12 Some products of fermentation**

Fermentation also gives bread its airy texture. Enzymes in yeast cells break the starch in flour into its component glucose molecules. As the yeast cells rapidly grow and divide, they release CO_2, first during cellular respiration and later during alcoholic fermentation after the O_2 dissolved in the water used to make the dough is used up. The dough, made stretchy and resilient by kneading, traps the CO_2 gas, which expands in the heat of the oven. The alcohol evaporates as the bread is baked.

A variety of microorganisms rely primarily on glycolysis followed by lactate fermentation for energy. These include the lactic acid bacteria that assist in transforming milk into yogurt, sour cream, and cheese. These bacteria first split lactose (milk sugar, a disaccharide) into glucose and galactose; then these simple sugars enter glycolysis followed by fermentation that produces lactic acid. Lactic acid denatures milk protein, altering its three-dimensional structure and giving sour cream and yogurt their semisolid textures. Like all acids, lactic acid tastes sour and contributes to the distinctive tastes of these foods. Lactic acid bacteria are also used to begin the coagulation of milk during cheese production. In addition, lactate fermentation by salt-tolerant bacteria convert sugars in vegetables such as cucumbers and cabbage into lactic acid. The result: dill pickles and sauerkraut.

casks with valves that prevent air from entering (so cellular respiration can't occur) but that allow the CO_2 to escape (so the cask doesn't explode). To put the characteristic fizz in beer and champagne, fermentation is allowed to continue after the bottle is sealed, trapping CO_2 under pressure.

CHECK YOUR LEARNING
Can you ...

- explain the function of fermentation and the conditions under which it occurs?
- describe the two types of fermentation?
- list some examples of human uses of each type of fermentation?

CASE STUDY \ **REVISITED**

Raising a King

Richard III's skeleton yielded a wealth of information about his life, and ultimately it confirmed his identity because mtDNA is uniquely valuable for tracing the hereditary relationships of ancient remains. MtDNA originates from the mitochondria in the cytoplasm of the mother's egg cell; sperm mitochondria do not enter the egg when it is fertilized. As a result, mtDNA is passed directly from mother to child in an unbroken chain that can extend through thousands of generations on the mother's side of the family.

Over millennia, harmless mutations have accumulated in specific regions of the mtDNA that do not code for functional proteins. The original, ancient mutations persist while new mutations are gradually added to the noncoding regions. Scientists can sequence these regions and define distinct subgroups by their newer (and therefore less common and less

widespread) mutations. Each of these subgroups originated from the mutated mtDNA in an egg of one woman, whose female descendants formed a population that was originally localized to a specific geographic area; the unique mtDNA signature will remain common in that area even today. As a result, modern people with specific mutations can be traced to ancestors with the same genetic signature in their mtDNA and also to the specific areas of the world where the mutations first emerged.

Because there are many identical copies of mtDNA in each cell, modern techniques can reconstruct crucial mtDNA nucleotide sequences even from severely decomposed remains, such as the skeleton under the Leicester parking lot. Because even a trace of contaminating DNA would render the results useless, Turi King performed the analyses in two separate specially designed ultra-clean laboratories. In each lab, she extracted and sequenced mtDNA from inside a tooth of the skeleton, where it remained relatively well-preserved. The results from the two labs matched, independently verifying one another. The skeleton's mtDNA sequence is relatively rare, shared by only 1% to 2% of the population of the United Kingdom.

Meanwhile, genealogists had identified two living descendants of an unbroken maternal line from Cecily Neville, mother of Richard III. One remains anonymous; the other is Michael Ibsen, a Canadian-born carpenter living in London (**FIG. 8-13**). The five centuries and 18 generations that separate the descendants from Cecily Neville to Michael Ibsen have not altered their mitochondrial DNA, which matches that of the skeleton. The location of the skeleton in Greyfriars Church, its deformity and battle wounds, and, most importantly, the remarkable match of its mtDNA to the only known descendants, led Leicester University's lead archaeologist to state: ". . . beyond reasonable doubt, the individual exhumed at Greyfriars in September 2012 is indeed Richard III, the last Plantagenet king."

▲ **FIGURE 8-13** **Michael Ibsen provides cheek cells to Turi King, who used them to sequence his mitochondrial DNA**

EVALUATE THIS Jeremy has always had difficulty walking rapidly and for long distances. Shortly before Jeremy's wedding, genetic testing revealed that his problem was caused by a mtDNA mutation. Should Jeremy be concerned about a future daughter inheriting the faulty gene? What about a son?

CHAPTER REVIEW

 Go to **MasteringBiology** for practice quizzes, activities, eText, videos, current events, and more.

Answers to Think Critically, Evaluate This, Multiple Choice, and Fill-in-the-Blank questions can be found in the Answers section at the back of the book.

Summary of Key Concepts

8.1 How Do Cells Obtain Energy?

The ultimate source of energy for nearly all life is sunlight, captured during photosynthesis and stored in molecules such as glucose. Cells produce chemical energy by breaking down glucose and capturing some of the released energy as ATP. During glycolysis, glucose is broken down in the cytosol, forming pyruvate and generating two ATP and two NADH molecules, which are high-energy electron carriers. If oxygen is available, the NADH from glycolysis is captured in ATP, and pyruvate is broken down through cellular respiration in the mitochondria, generating a total of 34 additional molecules of ATP.

8.2 How Does Glycolysis Begin Breaking Down Glucose?

Figures 8-3 and E8-1 and Table 8-1 summarize glycolysis. During the energy investment stage of glycolysis, glucose is energized by adding energy-carrying phosphate groups from two ATP molecules, forming fructose bisphosphate. Then, during the energy-harvesting stage, a series of reactions breaks down the fructose bisphosphate into two molecules of pyruvate. This produces a net energy yield of two ATP molecules and two NADH molecules.

8.3 How Does Cellular Respiration Extract Energy from Glucose?

Cellular respiration, which requires O_2 and generates 17 times as much ATP as does glycolysis, is summarized in Figures 8-5, 8-6, and E8-2 and Table 8.1. Before cellular respiration begins, pyruvate is transported into the mitochondrial matrix. During stage 1 of cellular respiration, acetyl CoA is formed from pyruvate, releasing CO_2 and generating NADH. The acetyl CoA then enters the Krebs cycle, which releases the CoA for reuse and releases the remaining two carbons as CO_2. One ATP, three NADH, and one $FADH_2$ are also formed for each acetyl group that goes through the cycle. In the mitochondrial matrix, each molecule of glucose that originally entered glycolysis produces a total of two ATP, eight NADH, and two $FADH_2$ (see Fig. 8-7).

During stage 2 of cellular respiration, the NADH and $FADH_2$ deliver their high-energy electrons to the electron transport chain (ETC) within the inner mitochondrial membrane. As the electrons pass along the ETC, energy is released and used to pump hydrogen ions across the inner membrane from the matrix into the intermembrane space, creating a hydrogen ion gradient. At the end of the ETC, the depleted electrons combine with hydrogen ions and oxygen to form water. During chemiosmosis, the energy stored in the hydrogen ion gradient is used to produce ATP as the hydrogen ions diffuse down their concentration gradient across the inner membrane through ATP synthase channels. Chemiosmosis yields 32 ATP from the complete breakdown of a glucose molecule; all are generated by energy carried in $FADH_2$ and NADH (including two NADH from glycolysis). Two additional ATPs are formed directly during glycolysis, and two more during the Krebs cycle. So altogether, a single molecule of glucose provides a net yield of 36 ATP when it is broken down by glycolysis followed by cellular respiration.

8.4 How Does Fermentation Allow Glycolysis to Continue When Oxygen Is Lacking?

Glycolysis uses NAD^+ to produce NADH as glucose is broken down into pyruvate. For these reactions to continue, NAD^+ must be continuously recycled. Under anaerobic conditions, NADH cannot release its high-energy electrons to the electron transport chain because there is no oxygen to accept them. Fermentation regenerates NAD^+ from NADH by converting pyruvate to lactate (via lactic acid fermentation) or to ethanol and CO_2 (via alcoholic fermentation), allowing glycolysis to continue.

Key Terms

aerobic   *132*
alcoholic fermentation   *134*
anaerobic   *132*
cellular respiration   *127*
chemiosmosis   *129*
electron transport chain
   (ETC)   *128*
fermentation   *133*
flavin adenine dinucleotide
   (FAD or $FADH_2$)   *128*

glycolysis   *125*
intermembrane space   *127*
Krebs cycle   *127*
lactate fermentation   *133*
matrix   *127*
mitochondrion (plural,
   mitochondria)   *127*
nicotinamide adenine
   dinucleotide (NAD^+ or
   NADH)   *125*

Thinking Through the Concepts

Multiple Choice

1. Which of the following is True for one glucose molecule?
   a. Fermentation produces 2 ATP.
   b. Glycolysis followed by fermentation nets 4 ATP.
   c. Ethanol is one end product of glycolysis.
   d. The overall equation for photosynthesis is the reverse of that for aerobic glucose breakdown.

2. The portion of glucose breakdown that produces the most ATP is
   a. chemiosmosis.
   b. glycolysis.
   c. the Krebs cycle.
   d. fermentation.

3. ATP synthase enzymes are located in the
   a. cytosol.
   b. inner mitochondrial membrane.
   c. intermembrane space.
   d. mitochondrial matrix.

4. Fermentation
   a. regenerates NADH.
   b. follows cellular respiration when oxygen is lacking.
   c. generates additional ATP after glycolysis.
   d. uses pyruvate as its substrate.

5. Which of the following is produced in the intermembrane space of mitochondria?
   a. ATP
   b. a high concentration of H^+
   c. NADH and $FADH_2$
   d. acetyl CoA

Fill-in-the-Blank

1. The complete breakdown of glucose in the presence of oxygen occurs in two major stages: _____ and _____. The first of these stages occurs in the _____ of the cell, and the second stage occurs in organelles called _____. Conditions in which oxygen is present are described as _____.

2. Conditions in which oxygen is absent are described as _____. Some microorganisms break down glucose in the absence of oxygen using _____, which generates only _____ molecules of ATP. This process is followed by _____, in which no more ATP is produced, but the electron-carrier molecule _____ is regenerated so it can be used in further glucose breakdown.

3. Yeasts in bread dough and alcoholic beverages use a type of fermentation that generates _____ and _____. Muscles pushed to their limit use _____ fermentation. Which form of fermentation is used by microorganisms that produce yogurt, sour cream, and sauerkraut? _____

4. During cellular respiration, the electron transport chain pumps H^+ out of the mitochondrial _____ into the _____, producing a _____ of H^+. The ATP produced by cellular respiration is generated by a process called _____. ATP is generated as H^+ travels through membrane channels within _____.

5. The cyclic portion of cellular respiration is called the _____ cycle. The molecule that enters this cycle is _____. How many ATP molecules are generated by this cycle per molecule of glucose? _____ What types of high-energy electron-carrier molecules are generated during the cycle? _____ and _____.

Review Questions

1. Starting with glucose ($C_6H_{12}O_6$), write the overall equation for glucose breakdown in the presence of oxygen, compare this to the overall equation for photosynthesis, and explain how the energy components of the equations differ.

2. Draw and label a mitochondrion, and explain how each structure relates to its function.

3. What role do the following play in breaking down and harvesting energy from glucose: glycolysis, cellular respiration, chemiosmosis, fermentation, and the electron carriers NAD^+ and FAD?

4. Outline the two major stages of glycolysis. How many ATP molecules (overall) are generated per glucose molecule during glycolysis? Where in the cell does glycolysis occur?

5. What molecule is the end product of glycolysis? How are the carbons of this molecule used in stage 1 of cellular respiration? In what form is most of the energy from the Krebs cycle captured?

6. Describe the electron transport chain and the process of chemiosmosis.

7. Why is oxygen necessary for cellular respiration to occur?

8. Compare the structure of chloroplasts (described in Chapter 7) to that of mitochondria, and describe how the similarities in structure relate to similarities in function.

Applying the Concepts

1. Some species of bacteria use aerobic respiration, and other species use fermentation. In an oxygen-rich environment, would either type be at a competitive advantage? What about in an oxygen-poor environment?

2. Many microorganisms in lakes use cellular respiration to generate energy. Dumping large amounts of raw sewage into rivers or lakes typically leads to massive fish kills, even if the sewage itself is not toxic to fish. What kills the fish? How might you reduce fish mortality after raw sewage is accidentally released into a small pond?

3. Imagine a hypothetical situation in which a starving cell reaches the stage where every bit of its ATP has been depleted and converted to ADP plus phosphate. If at this point you place the cell in a solution containing glucose, will it recover and survive? Explain your answer based on what you know about glucose breakdown.

4. Some species of bacteria that live at the surface of sediment on the bottom of lakes are capable of using either glycolysis plus fermentation or cellular respiration to generate ATP. There is very little circulation of water in lakes during the summer. Predict and explain what will happen to the bottommost water of a deep lake as the summer progresses, and describe how this situation will affect the amount of energy production by bacteria.

UNIT 2
Inheritance

The striking similarities and amazing diversity of life on Earth are both based on inheritance: remarkable fidelity from generation to generation, accompanied by occasional mistakes that allow new functions and structures to emerge.

"A structure of astounding elegance, a ladder delicately twisting into a double helix, packing into one, efficient strand all the information to create a living being." —G. SANTIS, CYPRUS

9 CELLULAR REPRODUCTION

Healthy again following stem cell therapy for shoulder and elbow injuries, Bartolo Colón hurls another fastball.

Body, Heal Thyself

WITH A 95 MILES-PER-HOUR FASTBALL, Bartolo Colón was at the top of his game when he won the Cy Young Award as the best pitcher in the American League in 2005. But throwing that hard takes its toll on a pitcher's arm. Colón stretched and tore ligaments and tendons in his shoulder and elbow, which kept him on the bench for much of the next four years. Why didn't Colón's arm heal after all that time?

Ligaments and tendons consist mostly of specialized proteins organized in a precise, orderly arrangement that provides both strength and flexibility. If Colón was ever to throw as fast as he once did, his joints needed to rebuild the damaged tissues with new proteins of the correct types, amounts, and organization. How? When a joint is injured, broken blood vessels leak blood. Some blood cells, called platelets, release a number of proteins, collectively called growth factors, into the injured tissue. Ideally, growth factors attract various types of cells to the site of injury and stimulate cell division. Growth factors also cause cells to specialize and become the cell types needed to repair the ligaments and tendons, so they return to their original size, strength, and flexibility. Unfortunately, this process is slow and isn't always completely successful. It didn't work very well for Colón.

In the spring of 2010, physicians removed stem cells from Colón's bone marrow and fat and injected them into his shoulder and elbow. Stem cells are cells that, with the right stimuli, can multiply and produce populations of specialized cells that

form cartilage, ligaments, tendon, bone, or many other tissues. The hope was that the stem cells would repair Colón's damaged ligaments and tendons. Because Colón's own cells were used, there was no risk of rejection.

By late 2010, Colón was pitching again, in a Puerto Rican winter league. Meanwhile, the New York Yankees were looking for a good pitcher, and Colón was hoping to make a comeback in the major leagues. The Yankees worried that Colón might never return to top form, but signed him anyway. They were rewarded: In 2011, Colón was throwing his trademark fastballs once again, winning 8 games. In 2013, playing for the Oakland Athletics, Colón won 18 games and made the All-Star team. Before the start of the 2014 season, the 40-year-old Colón signed a two-year contract with the New York Mets for $20 million. It paid off for both Colón and the Mets—he won 15 games, making him the eighth winningest pitcher in the National League that year.

Did stem cells heal Colón's injuries? How do growth factors cause cells to divide and form new tissue? When cells divide, why are the offspring cells genetically identical to the cells they came from?

AT A GLANCE

9.1 WHAT ARE THE FUNCTIONS OF CELL DIVISION?

"All cells come from cells." This insight, first stated by the German physician Rudolf Virchow in the mid-1800s, captures the critical importance of cellular reproduction for all living organisms. Cells reproduce by **cell division,** in which a parent cell divides into two **daughter cells.** In typical cell division, each daughter cell receives a complete set of hereditary information, identical to that of the parent cell, and about half the parent cell's cytoplasm.

The hereditary information of all living cells is contained in **deoxyribonucleic acid (DNA).** DNA is a polymer composed of subunits called **nucleotides** (**FIG. 9-1a**; see also Chapter 3). Each nucleotide consists of a phosphate, a sugar (deoxyribose), and one of four bases—adenine (A), thymine (T), guanine (G), or cytosine (C). In all cells, DNA is packaged into **chromosomes.** The DNA in a chromosome consists of two long strands of nucleotides wound around each other, like a ladder twisted into a corkscrew shape. This structure is called a double helix (**FIG. 9-1b**). Each chromosome contains a double helix of DNA as well as proteins that organize the

DNA's three-dimensional structure and regulate its use. The units of inheritance, called **genes,** are segments of the DNA of a chromosome, ranging from a few hundred to many thousands of nucleotides in length. Like the letters of an alphabet spelling out very long sentences, the specific sequences of nucleotides in genes spell out the instructions for making the proteins of a cell.

Cell Division Is Required for Growth, Development, and Repair of Multicellular Organisms

Mitotic cell division, which produces two daughter cells that are genetically identical to the parent cell, is the most common form of cell division in eukaryotic cells (see Sections 9.4 and 9.6). As you grew and developed from a fertilized egg, mitotic cell division produced all the cells in your body. Even now that you have attained your adult size, mitotic cell division continues to be essential, replacing cells that are killed by everyday life, such as cells in your digestive tract that are destroyed by stomach acid and digestive enzymes, or skin cells that are worn away by rubbing on your clothes. Mitotic cell division is also required to repair injuries, such as the damage that throwing thousands of fastballs inflicted in Bartolo Colón's arm.

The daughter cells formed by cell division may grow and divide again, in a repeating pattern called the **cell cycle.** Many of the daughter cells **differentiate,** becoming specialized for specific functions, such as contraction (muscle cells) or fighting infections (white blood cells). Most multicellular eukaryotic

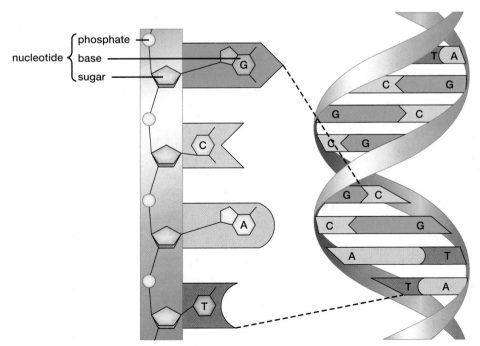

(a) A single strand of DNA

(b) The double helix

◄ **FIGURE 9-1 The structure of DNA**
(a) A nucleotide consists of a phosphate, a sugar, and one of four bases—adenine (A), thymine (T), guanine (G), or cytosine (C). A single strand of DNA consists of a long chain of nucleotides held together by chemical bonds between the phosphate of one nucleotide and the sugar of the next. **(b)** Two DNA strands twist around one another to form a double helix.

organisms typically have three categories of cells, based on their abilities to divide and differentiate:

- **Stem Cells**   Most of the cells formed by the first few cell divisions of a fertilized egg, and some cells in adult animals, including certain cells in the skin, intestines, fat, brain, ovaries, testes, bone marrow, and heart, are **stem cells** (FIG. 9-2). Stem cells have two important characteristics: self-renewal and potency. *Self-renewal* means that stem cells retain the capacity to divide, in some cases for the entire life of the organism. Usually, when a stem cell divides, one of its daughters remains a stem cell. Therefore, the number of stem cells remains about the same over time. The other daughter cell often undergoes several rounds of mitotic cell division, but the resulting cells eventually differentiate. *Potency* means that dividing stem cells produce daughter cells that can differentiate into a variety of specialized cell types. Some stem cells in early embryos can produce any of the specialized cell types of the entire body. Stem cells in adults are usually more limited and produce daughter cells that can differentiate into only a few cell types. The cellular environment, especially the specific "cocktail" of growth factors secreted by nearby cells, determines the type of differentiation that the daughter cells undergo.

  Plants also have stem cells, usually called meristem cells. Growing points in plants contain clusters of meristem cells, often at the tips of roots, stems, and branches. Cell division and differentiation of some of the daughter cells produce the various structures of the plant body.

- **Other Cells Capable of Dividing**   Some differentiated cells can divide, but their daughter cells typically differentiate into only one or two cell types. For example, if most of your

liver is seriously damaged, differentiated liver cells start dividing to replace the lost liver tissue; their daughter cells can only become more liver cells.

- **Permanently Differentiated Cells**   Some cells differentiate and never divide again. For example, most of the cells in your heart and brain cannot divide.

Cell Division Is Required for Sexual and Asexual Reproduction

Organisms reproduce by either or both of two fundamentally different processes: sexual reproduction and asexual reproduction. **Sexual reproduction** in most eukaryotic organisms occurs when offspring are produced by the fusion of **gametes** (sperm and eggs). To produce gametes, cells in the adult's reproductive system undergo a specialized type of cell division called meiotic cell division, which we will describe in Chapter 10.

Reproduction in which offspring are formed from a single parent, without having a sperm fertilize an egg, is called **asexual reproduction.** Asexual reproduction produces offspring that are genetically identical to the parent and to each other—they are **clones.** Bacteria (FIG. 9-3a) reproduce asexually by a type of cell division called prokaryotic fission (see Section 9.2). Many single-celled eukaryotic organisms, such as *Paramecium* (FIG. 9-3b), reproduce asexually by mitotic cell division. Some multicellular eukaryotes can also reproduce by asexual reproduction, using mitotic cell division, followed by differentiation of daughter cells, to produce new, genetically identical, miniature versions of the adult. For example, a *Hydra* reproduces by budding. First it grows a small replica of itself, called a bud, on its body (FIG. 9-3c). Eventually, the bud separates from its parent, forming a new *Hydra*. Many plants and fungi can reproduce both asexually and sexually. Aspen groves, for example, develop asexually from shoots growing up from the root system of a single parent tree (FIG. 9-3d). Although a grove looks like a cluster of separate trees, it is often a single individual whose multiple trunks are interconnected by a common root system. Aspen can also reproduce by seeds, which result from sexual reproduction.

Cloning Produces Genetically Identical Plants and Animals

Humans often assist asexual reproduction to produce clones of genetically identical, valuable plants and animals. Consider navel oranges, which don't produce seeds. Navel orange trees are propagated by cutting a piece of stem from an adult navel orange tree and grafting it onto the top of the root of a seedling of another type of orange tree. The cells of the aboveground, fruit-bearing parts of the grafted tree are clones of the original navel orange stem. All navel oranges originated from a single mutant bud of an orange tree discovered in Brazil in the early 1800s and propagated asexually ever since. Without cloning, there would be no navel oranges today.

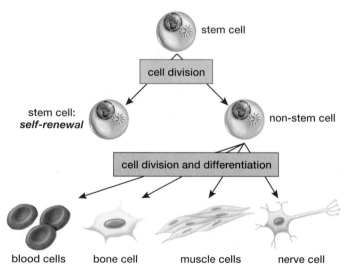

▲ **FIGURE 9-2 Stem cells** When a stem cell divides, one daughter cell remains a stem cell (self-renewal, middle left). The other daughter cell may divide a few times, but eventually differentiates into a specialized cell type (potency, bottom).

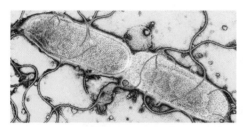

(a) Dividing bacteria

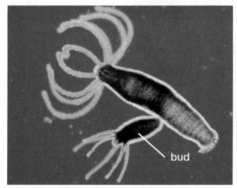

(b) Cell division in *Paramecium*

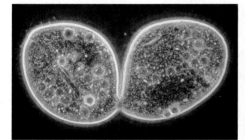

bud

(c) *Hydra* reproduces asexually by budding

The trees in this grove have already lost their leaves.

The trees in this grove have begun to change color.

The trees in this grove are still green.

(d) A grove of aspens often consists of genetically identical trees produced by asexual reproduction

◀ **FIGURE 9-3** **Cell division enables asexual reproduction (a)** Bacteria reproduce asexually by dividing in two. **(b)** In single-celled eukaryotic microorganisms, such as the freshwater protist *Paramecium*, cell division produces two new, independent organisms. **(c)** *Hydra*, a freshwater relative of the sea anemone, grows a miniature replica of itself (a bud) on its side. When fully developed, the bud breaks off and assumes independent life. **(d)** The trees in aspen groves are often genetically identical. Here, the timing of fall colors and leaf drop shows the genetic identity within a grove and the genetic difference between separate groves.

People have also cloned a variety of animals. The usual procedure is to obtain cells from an especially valuable animal, perhaps a racehorse or a particularly talented drug-sniffing dog (**FIG. 9-4**). Unfertilized eggs are collected from an unremarkable animal of the same species. The nucleus is removed from the unfertilized egg and replaced with a nucleus taken from a cell of the valuable animal. The egg cell is stimulated to divide a few times in culture, and then the resulting embryo is implanted into the uterus of a surrogate mother animal to complete development. Because mitotic cell division produces genetically identical daughter cells, the cloned animal will be genetically identical to the animal that provided the nucleus.

Cloning mammals is typically quite inefficient. Usually, only about 5% to 15% of the implanted embryos produce live offspring. Cloning mammals is also quite expensive—it would cost about $100,000 to clone your dog. In most cases, therefore, mammals are cloned for experimental

◀ **FIGURE 9-4** **Cloned drug-sniffer dogs** These yellow Labrador retrievers are genetically identical clones of an especially good sniffer dog, Chase. Although usually only 30% of candidate sniffer dogs successfully complete their training, all seven of Chase's clones passed with flying colors.

purposes; to reproduce individuals, usually livestock, that possess highly valuable, genetically determined traits; or for emotional reasons, such as attempting to replicate a beloved family pet.

CHECK YOUR LEARNING

Can you ...

- describe the types of cells found in a multicellular organism, distinguished by their ability to divide and differentiate?
- describe the functions of cell division in single-celled and multicellular eukaryotic organisms?

9.2 WHAT OCCURS DURING THE PROKARYOTIC CELL CYCLE?

The prokaryotic cell cycle consists of a relatively long period of growth, during which the cell replicates its DNA, followed by a type of cell division called **prokaryotic fission** (FIG. 9-5a). Prokaryotic fission is often called "binary fission." However, many biologists use the term binary fission to describe cell division in both prokaryotes and single-celled eukaryotes. To avoid confusion, we will use the term prokaryotic fission.

FIGURE 9-5b shows the process of prokaryotic fission. The DNA of a prokaryotic cell is contained in a single, circular chromosome about a millimeter or two in circumference. The prokaryotic chromosome is not contained in a membrane-bound nucleus (see Chapter 4). Instead, the chromosome is usually attached to the inside of the plasma membrane of the cell ➊. During the growth phase of the prokaryotic cell cycle, the DNA is replicated, producing two identical chromosomes that become attached to the plasma membrane at nearby, but separate, sites ➋. As the cell grows, new plasma membrane is added between the attachment sites of the chromosomes, pushing them apart ➌. When the cell has approximately doubled in size, the plasma membrane around the middle of the cell grows inward between the two attachment sites ➍. The plasma membrane then fuses along the equator of the cell, producing two daughter cells, each containing one of the chromosomes ➎. Because DNA replication yields two identical DNA molecules, the two daughter cells are genetically identical to one another (and to the parent cell that produced them).

CHECK YOUR LEARNING

Can you ...

- describe the prokaryotic cell cycle and the major events of prokaryotic fission?

▶ FIGURE 9-5 **The prokaryotic cell cycle (a)** The prokaryotic cell cycle consists of growth and DNA replication, followed by prokaryotic fission. **(b)** The process of prokaryotic fission.

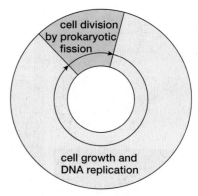

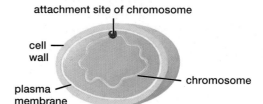

(a) The prokaryotic cell cycle

➊ The prokaryotic chromosome, a circular DNA double helix, is attached to the plasma membrane at one point.

➋ The DNA replicates and the resulting two chromosomes attach to the plasma membrane at nearby points.

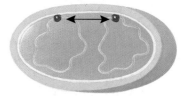

➌ New plasma membrane is added between the attachment points, pushing the two chromosomes farther apart.

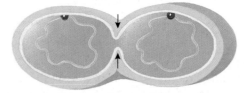

➍ The plasma membrane grows inward at the middle of the cell.

➎ The parent cell divides into two daughter cells.

(b) Prokaryotic fission

9.3 HOW IS THE DNA IN EUKARYOTIC CHROMOSOMES ORGANIZED?

Eukaryotic chromosomes differ from prokaryotic chromosomes in several respects. They are separated from the cytoplasm within a membrane-bound nucleus, and they are linear, instead of circular, as prokaryotic chromosomes are. Eukaryotic chromosomes also contain much more protein than prokaryotic chromosomes do, and their proteins are very different. Finally, eukaryotic chromosomes usually contain far more DNA than prokaryotic chromosomes do. Human chromosomes, for example, contain 10 to 50 times more DNA than the typical prokaryotic chromosome; depending on the chromosome, their length ranges from about 50 million to 250 million nucleotides. If the DNA in a human cell were completely relaxed and extended, each chromosome would be about 0.6 to 3.0 inches long (15 to 75 millimeters); a single human cell would contain about 6 feet (1.8 meters) of DNA. The number of chromosomes in eukaryotic organisms varies tremendously—the smallest number, 1, is found in the cells of male jack jumper ants, but most animals have dozens, and some plants have more than 1,200!

The complex events of eukaryotic cell division are largely an evolutionary solution to the problem of duplicating and parceling out a large number of long chromosomes. To understand eukaryotic cell division, we will begin by taking a closer look at the structure of the eukaryotic chromosome.

The Eukaryotic Chromosome Consists of a Linear DNA Double Helix Bound to Proteins

Fitting a huge amount of DNA into a nucleus only a few ten-thousandths of an inch in diameter is no trivial task. The eukaryotic cell solves this problem by wrapping the DNA around protein supports, greatly reducing its length (**FIG. 9-6**). For most of a cell's life, the DNA double helix in a chromosome is wound around proteins called histones ❶, ❷. Other proteins coil up the DNA/histone beads, much like a spring or Slinky toy ❸. These coils are attached in loops to protein "scaffolding" to complete the chromosome packaging as it occurs during most of the life of a cell. All of this winding, coiling, and looping condenses the DNA to about 1/1,000th of its extended length ❹, but even this enormous degree of compaction still leaves the chromosomes much too long to be sorted out and moved into daughter nuclei during cell division. However, as cell division begins, proteins fold up the chromosome, yielding about another 10-fold condensation ❺. The chromosome is now a compact structure less than 2 ten-thousandths of an inch long (about 4 micrometers).

Every chromosome has specialized regions that are crucial to its structure and function: two telomeres and one centromere. **Telomeres** ("end part" in Greek) are protective caps at each end of a chromosome (see Fig. 9-6). Without telomeres, genes located at the ends of the chromosomes would be lost during DNA replication. Telomeres also keep chromosomes from fusing with one another and

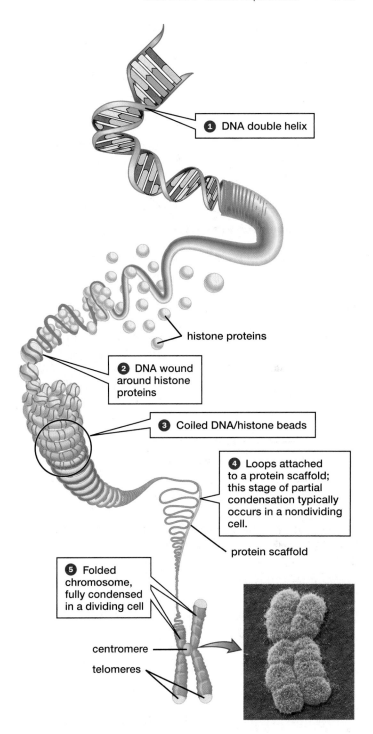

❶ DNA double helix

histone proteins

❷ DNA wound around histone proteins

❸ Coiled DNA/histone beads

❹ Loops attached to a protein scaffold; this stage of partial condensation typically occurs in a nondividing cell.

protein scaffold

❺ Folded chromosome, fully condensed in a dividing cell

centromere

telomeres

▲ **FIGURE 9-6 Chromosome structure** Proteins in a eukaryotic chromosome wrap, coil, and fold the DNA into a compact structure. The ends of the chromosome are protected by telomeres. The centromere will be the site of attachment of microtubules that move the chromosome during mitotic cell division. (Inset) The fuzzy edges visible in the scanning electron micrograph are loops of folded chromosome.

forming long, unwieldy structures that probably could not be distributed properly to the daughter cells during cell division. The second specialized region of a chromosome is its **centromere** ("central part"). As we will see, the centromere has two principal functions: (1) It temporarily holds two daughter DNA double helices together after DNA replication, and (2) it is the attachment site for microtubules that move the chromosomes during cell division.

CHECK YOUR LEARNING

Can you ...
- describe the structure of a eukaryotic chromosome?
- describe the functions of telomeres and centromeres?

9.4 WHAT OCCURS DURING THE EUKARYOTIC CELL CYCLE?

In the eukaryotic cell cycle, newly formed cells usually acquire nutrients from their environment, synthesize more cytoplasm and organelles, and grow larger. After a variable amount of time—depending on the organism, the type of cell, and the nutrients available—the cell may divide. Each daughter cell may then enter another cell cycle and divide again. Most cells, however, divide only if they receive chemical signals, such as growth factors, that cause them to enter another cell cycle (see Section 9.6). Other cells may differentiate and never divide again.

The Eukaryotic Cell Cycle Consists of Interphase and Mitotic Cell Division

The eukaryotic cell cycle is divided into two major phases: interphase and mitotic cell division (**FIG. 9-7**).

During Interphase, a Cell Grows in Size, Replicates Its DNA, and Often Differentiates

Most eukaryotic cells spend the majority of their time in **interphase,** the period between cell divisions. For example, some cells in human skin spend roughly 22 hours in interphase and only a couple of hours dividing. Interphase contains three subphases: G_1 (the first *g*rowth phase and the first *g*ap in DNA *s*ynthesis), S (when DNA *s*ynthesis occurs), and G_2 (the second *g*rowth phase and the second *g*ap in DNA *s*ynthesis).

A newly formed daughter cell enters the G_1 portion of interphase. During G_1, a cell carries out one or more of three activities. First, it almost always grows in size. Second, it often differentiates, developing the structures and biochemical pathways that allow it to perform a specialized function. For example, most nerve cells grow long strands, called axons, that allow them to connect with other cells, whereas liver cells produce bile, proteins that aid blood clotting, and enzymes that detoxify many poisonous materials. Third, the cell responds to internal and external signals that determine whether or not it will divide. If the cell is stimulated to divide, it must first duplicate its chromosomes, including making exact copies of the DNA of each chromosome. Duplicating the chromosomes occurs during the S phase. When the chromosomes have been duplicated, the cell proceeds to the G_2

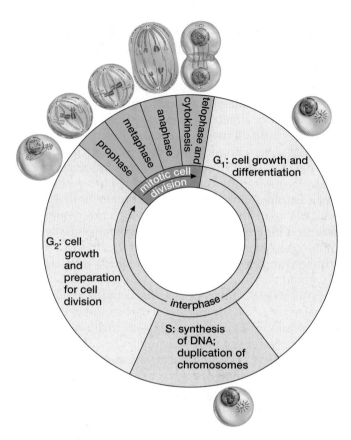

▲ **FIGURE 9-7** **The eukaryotic cell cycle** The eukaryotic cell cycle consists of interphase and mitotic cell division.

phase, during which it may grow some more and synthesize the proteins needed for cell division.

Many differentiated cells, such as liver cells, can be recalled from the differentiated state back into the dividing state, whereas others, such as most heart muscle and nerve cells, remain in the G_1 phase and never divide again.

Mitotic Cell Division Consists of Nuclear Division and Cytoplasmic Division

Mitotic cell division consists of two processes: mitosis and cytokinesis. **Mitosis** is the division of the nucleus. The word "mitosis" is derived from a Greek word meaning "thread," because, as the chromosomes condense and shorten, they become visible in a light microscope as thread-like structures. Mitosis produces two daughter nuclei, each containing one copy of each of the chromosomes that were present in the parent nucleus. **Cytokinesis** (from Greek words meaning "cell movement") is the division of the cytoplasm. Cytokinesis places about half the cytoplasm, half the organelles (such as mitochondria, ribosomes, and Golgi apparatus), and one of the newly formed nuclei into each of two daughter cells. Thus, mitotic cell division typically produces daughter cells that are physically similar and genetically identical to each other and to the parent cell.

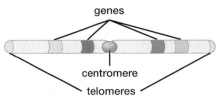

(a) A eukaryotic chromosome (one DNA double helix) before DNA replication

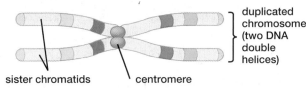

(b) A eukaryotic chromosome after DNA replication

(c) Separated sister chromatids become independent chromosomes

▲ **FIGURE 9-8 A eukaryotic chromosome during cell division** **(a)** Before DNA replication. **(b)** After DNA replication, the two sister chromatids are held together at the centromere. **(c)** The sister chromatids separate during cell division to become two independent, genetically identical chromosomes.

CHECK YOUR LEARNING

Can you ...

- describe the events of the eukaryotic cell cycle?
- explain the difference between mitotic cell division and mitosis?

Ligaments and tendons have a limited capacity for self-repair. They tend to have a meager blood supply and contain only a small number of specialized cells that produce proteins, such as collagen and elastin, that provide flexibility and strength. In Bartolo Colón's case, the hope was that the stem cells injected into his shoulder and elbow would progress rapidly through the cell cycle, producing large populations of specialized daughter cells that would regenerate his ligaments and tendons. How would mitotic cell division ensure that the daughter cells contained accurate copies of all of Colón's chromosomes, including the genes that specify all of the proteins needed to repair his arm?

9.5 HOW DOES MITOTIC CELL DIVISION PRODUCE GENETICALLY IDENTICAL DAUGHTER CELLS?

Remember that a chromosome consists of genes, two telomeres, and one centromere (**FIG. 9-8a**). All of a cell's chromosomes are copied during the S phase of interphase, before mitotic cell division starts. Each resulting **duplicated chromosome** consists of two identical DNA double helices (and their associated proteins), called sister **chromatids,** which are attached to each other at the centromere (**FIG. 9-8b**). During mitotic cell division, the two sister chromatids separate, each becoming an independent chromosome that is delivered to one of the two daughter cells (**FIG. 9-8c**).

For convenience, biologists divide mitosis into four phases, based on the appearance and behavior of the chromosomes: prophase, metaphase, anaphase, and telophase (**FIG. 9-9**). However, these phases are not really discrete events; they instead form a continuum, with each phase merging into the next.

During Prophase, the Chromosomes Condense, the Spindle Forms, the Nuclear Envelope Breaks Down, and the Chromosomes Are Captured by Spindle Microtubules

The first phase of mitosis is called **prophase** (meaning "the stage before" in Greek). During prophase, four major

events occur: (1) The duplicated chromosomes condense (see Fig. 9-6), (2) the spindle microtubules form, (3) the nuclear envelope breaks down, and (4) the chromosomes are captured by the spindle microtubules (**FIGS. 9-9b, c**). Chromosome condensation also causes the nucleolus to disappear. The nucleolus consists of partially assembled ribosomes and the genes that code for the RNA component of the ribosomes (see Chapter 4). These genes are located on several different chromosomes. As the chromosomes condense, they separate from one another and ribosome synthesis ceases, so the nucleolus fades away.

As the duplicated chromosomes condense, the **spindle** begins to form. The spindle is composed of microtubules, called **spindle microtubules** (see Fig. 9-9c). In all eukaryotic cells, the movement of chromosomes during mitosis depends on the spindle microtubules. In animal cells, the spindle microtubules originate from a region that contains a pair of microtubule-containing structures called **centrioles.** The cells of plants, fungi, many algae, and even some mutant fruit flies do not contain centrioles. Nevertheless, these cells form functional spindles during mitotic cell division, showing that centrioles are not required for spindle formation.

In animal cells, a new pair of centrioles forms during interphase near the previously existing pair. During prophase, the two centriole pairs migrate to opposite sides of the nucleus (see Fig. 9-9b). The area of cytoplasm around each centriole pair, called the spindle pole, controls the formation of

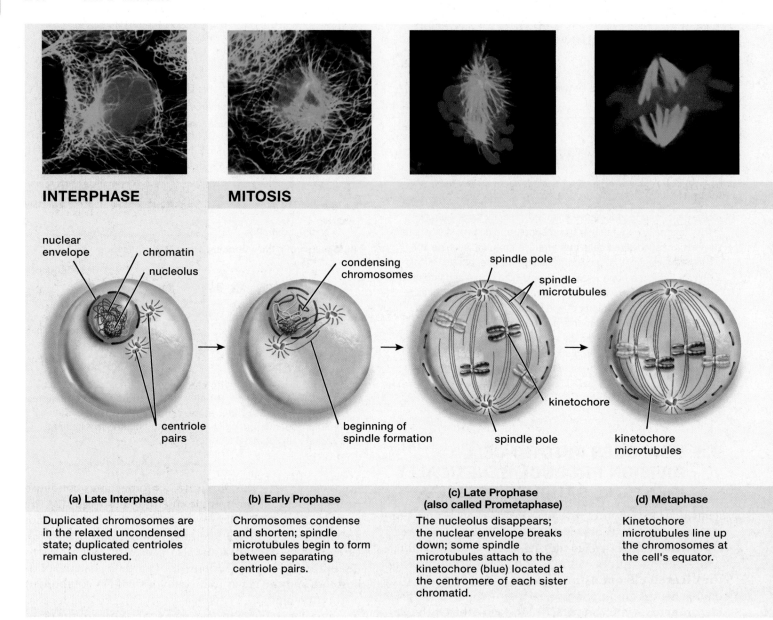

INTERPHASE

MITOSIS

nuclear
envelope   chromatin
                  nucleolus

condensing
chromosomes

spindle pole

spindle
microtubules

centriole
pairs

beginning of
spindle formation

kinetochore

spindle pole

kinetochore
microtubules

(a) Late Interphase	(b) Early Prophase	(c) Late Prophase (also called Prometaphase)	(d) Metaphase
Duplicated chromosomes are in the relaxed uncondensed state; duplicated centrioles remain clustered.	Chromosomes condense and shorten; spindle microtubules begin to form between separating centriole pairs.	The nucleolus disappears; the nuclear envelope breaks down; some spindle microtubules attach to the kinetochore (blue) located at the centromere of each sister chromatid.	Kinetochore microtubules line up the chromosomes at the cell's equator.

▲ **FIGURE 9-9** **Mitotic cell division in an animal cell**

THINK CRITICALLY What would the consequences be if one set of sister chromatids failed to separate at anaphase?

the spindle microtubules. These microtubules radiate inward toward the nucleus and outward toward the plasma membrane (see Fig. 9-9c). (To visualize this, picture the cell as a globe. The spindle poles are roughly where the north and south poles would be, and the spindle microtubules correspond to the lines of longitude. As on a globe, the equator of the cell cuts across the middle, halfway between the poles.) Because one pair of centrioles is located at each spindle pole, each daughter cell will receive a pair of centrioles when the cell divides.

As the spindle microtubules form around the nucleus, the nuclear envelope disintegrates, releasing the duplicated chromosomes. Each sister chromatid in a duplicated

chromosome has a protein-containing structure, called a **kinetochore,** located at its centromere. The kinetochores of the two sister chromatids are arranged back-to-back, facing away from one another. The kinetochore of one sister chromatid binds to the ends of spindle microtubules leading to one pole of the cell, while the kinetochore of the other sister chromatid binds to spindle microtubules leading to the opposite pole (see Fig. 9-9c). The microtubules that bind to kinetochores are called kinetochore microtubules to distinguish them from polar microtubules, which do not bind to a kinetochore. When the sister chromatids separate later in mitosis, the newly independent chromosomes will move along the kinetochore microtubules to opposite poles.

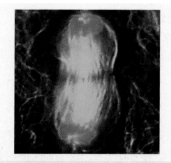

INTERPHASE

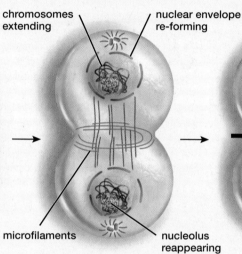
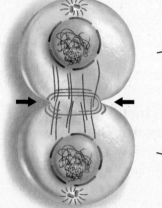
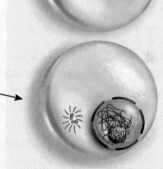

polar
microtubules

chromosomes
extending

nuclear envelope
re-forming

microfilaments

nucleolus
reappearing

(e) Anaphase	(f) Telophase	(g) Cytokinesis	(h) Interphase of daughter cells
Sister chromatids separate and move to opposite poles of the cell; polar microtubules push the poles apart.	One set of chromosomes reaches each pole and begins to decondense; nuclear envelopes start to form; nucleoli begin to reappear; spindle microtubules begin to disappear; microfilaments form rings around the equator.	The ring of microfilaments contracts, dividing the cell in two; each daughter cell receives one nucleus and about half of the cytoplasm.	Spindles disappear, intact nuclear envelopes form, and the chromosomes extend completely.

Polar microtubules do not attach to the chromosomes; rather, they have free ends that overlap along the cell's equator. As we will see, the polar microtubules push the two spindle poles apart later in mitosis.

During Metaphase, the Chromosomes Line Up Along the Equator of the Cell

At the end of prophase, the two kinetochores of each duplicated chromosome are connected to kinetochore microtubules leading to opposite poles of the cell. As a result, each duplicated chromosome is connected to both spindle poles. During **metaphase** (the "middle stage"), the two kinetochores on a duplicated chromosome pull toward opposite poles of the cell. During this molecular "tug-of-war," the microtubules lengthen or shorten until each chromosome lines up along the equator of the cell, with one kinetochore facing each pole (FIG. 9-9d).

During Anaphase, Sister Chromatids Separate and Are Pulled to Opposite Poles of the Cell

At the beginning of **anaphase** (FIG. 9-9e), the sister chromatids separate, becoming independent daughter chromosomes. This separation allows each kinetochore to move its chromosome poleward, while simultaneously nibbling off the end of the attached microtubule, thereby shortening it (a mechanism appropriately called "Pac-Man" movement). One of the two daughter chromosomes derived from each parental chromosome moves to each pole of the cell. Because the daughter chromosomes are identical copies of the parental chromosomes, each cluster of chromosomes that forms at opposite poles of the cell contains one copy of every chromosome that was in the parent cell.

At about the same time that the daughter chromosomes begin to move toward the poles, polar microtubules radiating from each pole grab one another where they overlap at the equator. These polar microtubules then simultaneously lengthen and push on one another, forcing the poles of the cell apart (see Fig. 9-9e).

During Telophase, a Nuclear Envelope Forms Around Each Group of Chromosomes

When the chromosomes reach the poles, **telophase** (the "end stage") begins (**FIG. 9-9f**). The spindle microtubules disintegrate and a nuclear envelope forms around each group of chromosomes. The chromosomes revert to their extended state, and nucleoli begin to re-form. In most cells, cytokinesis occurs during telophase, isolating each daughter nucleus in its own daughter cell (**FIG. 9-9g**). However, mitosis sometimes occurs without cytokinesis, producing cells with multiple nuclei.

During Cytokinesis, the Cytoplasm Is Divided Between Two Daughter Cells

Cytokinesis differs considerably between animal cells and plant cells. In animal cells, microfilaments attached to the plasma membrane assemble into a ring around the equator of the cell, usually late in anaphase or early in telophase (see Fig. 9-9f). The ring contracts and constricts the cell's equator, much like pulling the drawstring on sweatpants tightens the waist (see Fig. 9-9g). Eventually the "waist" of the parent cell constricts completely, dividing the cytoplasm into two new daughter cells (**FIG. 9-9h**).

Cytokinesis in plant cells is quite different, perhaps because their stiff cell walls make it impossible to divide one cell into two by pinching at the middle. Instead, carbohydrate-filled sacs called vesicles bud off the Golgi apparatus and line up along the equator of the cell between the two nuclei (**FIG. 9-10**). The vesicles fuse, producing a structure called the **cell plate**, which is shaped like a flattened sac, surrounded by membrane and filled with carbohydrates. When enough vesicles have fused, the edges of the cell plate merge with the plasma membrane around the circumference of the cell. The membranes on the two sides of the cell plate become new plasma membranes between the two daughter cells. The carbohydrates formerly contained in the vesicles remain between the plasma membranes as the beginning of the new cell wall.

CHECK YOUR LEARNING

Can you ...
- describe the steps of mitotic cell division?
- describe the usual outcome of mitotic cell division?
- explain how cytokinesis differs in plant and animal cells?

CASE STUDY \ **CONTINUED**

Body, Heal Thyself

The precision of mitotic cell division is essential for repairing damaged tissues like those in Bartolo Colón's pitching arm. Imagine what might happen if DNA synthesis during interphase did not copy all of the genes accurately, or if mitotic cell division sent random numbers and types of chromosomes into the daughter cells. Some of the daughter cells might not contain all the genes needed to form the cell types that are required to repair damaged tissues. Other daughter cells might have genetic changes that stimulate unrestrained cell division and cause cancer. In cancer cells, the cell cycle spins out of control, but under normal circumstances cell division is precisely regulated. How does the body usually control the cell cycle?

9.6 HOW IS THE CELL CYCLE CONTROLLED?

Cell division is regulated by a diverse array of molecules, not all of which have been identified and studied. Nevertheless, some general principles apply to cell cycle control in most eukaryotic cells.

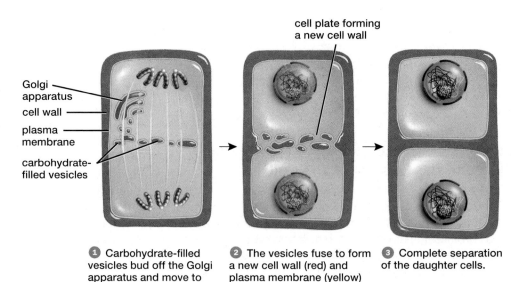

Golgi apparatus
cell wall
plasma membrane
carbohydrate-filled vesicles

cell plate forming a new cell wall

❶ Carbohydrate-filled vesicles bud off the Golgi apparatus and move to the equator of the cell.

❷ The vesicles fuse to form a new cell wall (red) and plasma membrane (yellow) between the daughter cells.

❸ Complete separation of the daughter cells.

▲ **FIGURE 9-10** Cytokinesis in a plant cell

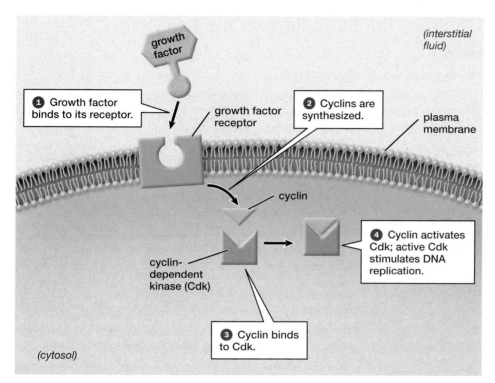

(interstitial fluid)

1 Growth factor binds to its receptor.

growth factor

growth factor receptor

2 Cyclins are synthesized.

plasma membrane

cyclin

4 Cyclin activates Cdk; active Cdk stimulates DNA replication.

cyclin-dependent kinase (Cdk)

3 Cyclin binds to Cdk.

(cytosol)

◀ **FIGURE 9-11 Growth factors stimulate cell division** Progress through the cell cycle is under the overall control of cyclin and cyclin-dependent kinases (Cdks). In most cases, growth factors stimulate synthesis of cyclin proteins, which activate Cdks, starting a cascade of events that lead to DNA replication and cell division.

THINK CRITICALLY What would happen if a cell suffered a mutation that turned a growth factor receptor "on" all the time so that it activated the intracellular cascade even without growth factors present?

The Activities of Specific Proteins Drive the Cell Cycle

During development, after an injury, or to compensate for normal wear and tear, many cells in the body release hormone-like molecules called **growth factors**. Most growth factors stimulate cell division by controlling the synthesis of intracellular proteins collectively called cyclins, which in turn regulate the activity of enzymes called cyclin-dependent kinases. The proteins are named "cyclins" because they help to govern the cell cycle. Cyclin-dependent kinases (Cdks) get their name from two features: A "kinase" is an enzyme that adds a phosphate group to another protein, stimulating or inhibiting the activity of the target protein. "Cyclin dependent" means that the kinase is active only when it binds cyclin.

As an example, let's see how growth factors, cyclins, and Cdks stimulate cell division to heal a cut in your skin (**FIG. 9-11**). Platelets (blood cells that are involved in clotting) accumulate at the wound site and release several types of growth factors. These growth factors bind to receptors on the surfaces of cells in damaged areas of the skin **1**, stimulating the cells to synthesize cyclin proteins **2**. Cyclins bind to specific Cdks **3**, forming cyclin–Cdk complexes that promote the manufacture and activity of the proteins required for DNA synthesis **4**. The cells enter the S phase of the cell cycle and replicate their DNA. After DNA replication is complete, other Cdks become activated during G_2 and mitosis, causing chromosome condensation, breakdown of the nuclear envelope, formation of the spindle, and attachment of the chromosomes to the spindle microtubules. Finally, still other Cdks stimulate processes that allow the sister chromatids to separate into individual chromosomes and move to opposite poles during anaphase.

HAVE YOU EVER

WONDERED...

The saliva of dogs, like the saliva of most mammals (including humans), contains enzymes, antibacterial compounds, and growth factors. When a dog licks a wound, it not only cleans out some of the dirt and kills some of the bacteria that may have entered, but also leaves growth factors behind. The growth factors speed up the synthesis of cyclins, thereby stimulating the division of cells that regenerate the skin, helping to heal the wound more rapidly.

Why Dogs Lick Their Wounds?

Checkpoints Regulate Progress Through the Cell Cycle

Unregulated cell division can be dangerous. If a cell contains mutations in its DNA or if its daughter cells receive too many or too few chromosomes, the daughter cells may die. If they survive, they may become cancerous. To prevent this, the eukaryotic cell cycle has three major **checkpoints,** where proteins in the cell determine whether the cell has successfully completed a specific phase of the cycle:

- **G_1 to S** Is the cell's DNA intact and suitable for replication?
- **G_2 to Mitosis** Has the DNA been completely and accurately replicated?
- **Metaphase to Anaphase** Are all the chromosomes attached to the spindle and aligned properly at the equator of the cell?

The checkpoint proteins usually regulate the production of cyclins or the activity of Cdks, or both, thereby regulating

Cancer—Running the Stop Signs at the Cell Cycle Checkpoints

A cancer is a cluster of cells that multiply without control and can invade other parts of the body. The ultimate causes of most cancers are **mutations**—damage to DNA from a variety of sources, including mistakes during replication, infection by certain viruses, exposure to ultraviolet light, or chemicals in the environment (such as pesticides, industrial products, and chemicals naturally produced by plants or fungi). In most cases, a mutation is quickly fixed by enzymes that repair DNA, or the defective cell is killed. Occasionally, however, a renegade cell survives and multiplies.

The cell cycle is regulated by two interacting processes: responses to growth factors that start or speed up the cell cycle, and checkpoints that stop the cell cycle if problems, such as mutations in DNA or misalignment of chromosomes, have occurred. Cancers develop when mutated cells evade these controls.

Responses to Growth Factors

Most cells divide only when stimulated by growth factors. Many cancerous cells have mutated genes, collectively called *oncogenes* (literally, "to cause cancer"), that promote uncontrolled cell division. Some oncogenes overproduce growth factor receptors or produce receptors that are permanently activated, even in the absence of growth factors (FIG. E9-1). Mutations in cyclin genes may cause cyclins to be synthesized at a high rate, again independently of growth factors. The result: an abnormally large supply of activated Cdks and other molecules that stimulate cell division. Like a driver who hits the accelerator instead of the brake while approaching a stop sign, a cell with these mutations is likely to barge right through the checkpoints and multiply without control.

Evading the Checkpoint Stop Signs

Cells, however, have ways of enforcing the checkpoint stop signs. All cells contain a variety of proteins collectively called *tumor suppressors*. These proteins prevent uncontrolled cell division and block the production of daughter cells that have mutated DNA. For example, a tumor suppressor called p53 monitors the integrity of a cell's DNA. Healthy cells, with intact DNA, contain little p53. However, p53 levels rapidly increase in cells with damaged DNA. The p53 protein activates intracellular processes that inhibit Cdks and block DNA synthesis, halting the cell cycle at the checkpoint between the G₁ and S phases. The p53 protein also stimulates the synthesis of DNA repair enzymes. After the DNA has been repaired, p53 levels decline, Cdks become active, and the cell enters the S phase. If the DNA cannot be repaired, p53 triggers a special form of cell death called apoptosis, in which the cell cuts up its DNA and effectively commits suicide. Thus, p53 acts as a checkpoint enforcer, much like the tire-spiking strips that police sometimes use to prevent criminals from driving through roadblocks. Most cells with dangerous mutations cannot plow through the G₁ to S checkpoint, so they cannot continue through the cell cycle.

But what if the gene encoding the p53 protein is mutated, causing the production of defective p53? Then, even if a cell's DNA is damaged, the cell skips through the G₁ to S checkpoint. Not surprisingly, about half of all can-

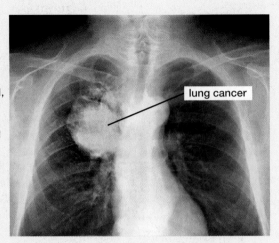

▶ **FIGURE E9-1**
A colorized X-ray of advanced lung cancer In women and in people who have never smoked, about 40% to 50% of lung cancers seem to be caused by too many receptors for growth factors or by mutated receptors that are active even in the absence of growth factors.

lung cancer

cers—including breast, lung, brain, pancreas, bladder, stomach, and colon cancers—have mutations in the p53 gene.

From Mutated Cells to Metastasizing Tumors

All of us probably have some cells with mutations in oncogenes or tumor suppressor genes, or both. Usually, a single mutation in one of these gene families will cause cells to multiply faster than usual and form a *benign tumor*—a cluster of cells that has multiplied independently of its surroundings, forming a distinct patch or lump. Benign tumors are common; moles, birthmarks, and some types of warts are benign tumors. "Benign" means that the tumor is not cancerous, or at least not yet. It grows slowly, if at all, and it doesn't spread, or *metastasize*, to other parts of the body. Some benign tumors, however, can become cancerous, or *malignant*, over time.

A *malignant tumor* is a lump of cells that grows rapidly and often metastasizes. All malignant tumors are cancers, but some cancers, such as leukemia, do not form discrete tumors. Tumors may become metastatic by several mechanisms. Generally, cells in the tumor accumulate mutations over time. Some mutations promote the growth of blood vessels in the tumor, nourishing the cancerous cells and helping the tumor to grow larger. Other mutations allow some of the cells to break away, invade the blood vessels, and spread throughout the body. Finally, some of the cells emerge from the circulatory system and invade other parts of the body. Once a cancer metastasizes and tumors begin to grow in multiple sites in the body, the cancer is extremely difficult to treat.

EVALUATE THIS Yesterday, when Daniel was showering after a basketball game in the gym, one of his friends asked, "Have you always had that big brown thing on your back?" Looking in the mirror, Daniel saw a large, dark brown, irregularly shaped mole. He checked in with a physician at the health center. She told him, "It's probably just a large mole, but we should do a biopsy to find out for sure." What genetic differences would you expect the pathology lab to find between a malignant tumor and an ordinary mole?

progression from one phase of the cell cycle to the next. In most cases, if the checkpoint proteins are activated, for example, by mutated DNA or misaligned chromosomes, they stop the cell cycle until the defect is repaired. If the defect is not repaired, the defective cells usually either destroy themselves or are killed by the immune system. When checkpoint control malfunctions, the result may be cancer, as we explore in "Health Watch: Cancer—Running the Stop Signs at the Cell Cycle Checkpoints."

CHECK YOUR LEARNING
Can you ...
- describe the interactions among growth factors, cyclins, and cyclin-dependent kinases that control the eukaryotic cell cycle?
- explain how a cell protects against producing defective daughter cells?

CASE STUDY \ REVISITED
Body, Heal Thyself

Bartolo Colón's physicians wanted to give Colón's arm every possible chance to heal rapidly and completely. In any wound, platelets leak from nearby blood vessels and deliver growth factors that stimulate cell division and promote healing. However, the limited blood supply of ligaments and tendons may not provide enough platelets, and hence enough growth factors, to allow full healing. To correct this deficit, Colón's physicians administered platelet-rich plasma (PRP) therapy a few weeks after his stem cell injection. Some of Colón's blood was removed, the platelets were concentrated into a small volume, and the resulting PRP was injected into the wound.

Bartolo Colón's saga sounds like a fairy tale come true: Injured, aging pitcher receives stem cell and PRP therapy and returns to stardom. But did stem cell and PRP therapy really help Colón? The truth is, no one really knows. Although there are several reports of spectacular results on individuals such as Colón, maybe he would have healed anyway. Or maybe he just happened to have an injury that stem cells and PRP worked for, and most other people would not be so lucky. Perhaps there will be long-term problems, such as migration of some injected stem cells to other locations in his body, that Colón won't discover for 20 years or more.

There have been very few clinical trials of PRP therapies in humans. Research in dogs and horses has found that arthritic or injured joints improved following PRP therapy, but the studies often had small sample sizes, used different methodologies, or were not designed as clinical trials. Finally, not all the studies found significant improvement in PRP-treated animals compared to the controls. PRP therapy is now an almost routine part of some joint surgeries in both humans and animals, but physician and patient confidence may be based as much on hope as on evidence.

Stem cells are even more of an unknown. Stem cells taken from bone marrow are routinely used as treatments for certain cancers of the blood and immune system, but clinical trials of other stem cell therapies are just beginning. Although researchers can't be sure that they will work, the range of possible applications is breathtaking: not just joint injuries, but multiple sclerosis, Parkinson's disease, amyotrophic lateral sclerosis (ALS, or Lou Gehrig's disease), and certain types of blindness.

CONSIDER THIS Colón's miraculous recovery and similar stories may give the impression that soon a "weekend warrior" with torn knee ligaments will be able to hobble into a clinic, have some bone marrow and blood removed, and a few hours later have stem cells and PRP injected into the injured knee. Just a few weeks later, the would-be athlete will be back on the basketball court or furiously pedaling a bicycle up steep hills. The U.S. Food and Drug Administration agrees that stem cells offer great promise, but also cautions against hasty overenthusiasm. Search the Internet for information about PRP and stem cell therapies (be sure that you use authoritative sites such as the FDA or the National Institutes of Health). What are the likely benefits, and what are the potential risks? Would you be willing to try PRP or stem cell therapies, knowing that they haven't yet been clinically proven to be either safe or effective?

CHAPTER REVIEW

 Go to **MasteringBiology** for practice quizzes, activities, eText, videos, current events, and more.

Answers to Think Critically, Evaluate This, Multiple Choice, and Fill-in-the-Blank questions can be found in the Answers section at the back of the book.

Summary of Key Concepts

9.1 What Are the Functions of Cell Division?

Growth of multicellular eukaryotic organisms and replacement of cells that die during an organism's life occur through cell division and differentiation of the daughter cells. Asexual reproduction also occurs through cell division.

9.2 What Occurs During the Prokaryotic Cell Cycle?

A prokaryotic cell contains a single, circular chromosome. The prokaryotic cell cycle consists of growth, replication of the DNA, and division of the cell by prokaryotic fission. The two resulting daughter cells are genetically identical to one another and to the parent cell.

9.3 How Is the DNA in Eukaryotic Chromosomes Organized?

Each chromosome in a eukaryotic cell consists of a single DNA double helix and proteins that organize the DNA and regulate its use. Genes are segments of DNA found at specific locations on a chromosome. During cell division, the chromosomes are duplicated and condense into short, thick structures.

9.4 What Occurs During the Eukaryotic Cell Cycle?

The eukaryotic cell cycle consists of interphase and mitotic cell division. During interphase, the cell grows and duplicates its chromosomes. Interphase is divided into G_1 (growth phase 1), S (DNA synthesis), and G_2 (growth phase 2). During G_1, cells may differentiate to perform a specific function. Some differentiated cells can re-enter the dividing state; other cells remain differentiated for the life of the organism and never divide again. Eukaryotic cells divide by mitotic cell division.

9.5 How Does Mitotic Cell Division Produce Genetically Identical Daughter Cells?

A cell's chromosomes are duplicated during interphase, prior to mitotic cell division. A duplicated chromosome consists of two identical sister chromatids that remain attached to one another at the centromere during the early stages of mitotic cell division. Mitosis (nuclear division) consists of four phases, usually accompanied by cytokinesis (cytoplasmic division) during the last phase (see Fig. 9-9):

- *Prophase* The chromosomes condense and their kinetochores attach to kinetochore microtubules that form at this time.
- *Metaphase* Kinetochore microtubules move the chromosomes to the equator of the cell.
- *Anaphase* The two chromatids of each duplicated chromosome separate and become independent chromosomes. The kinetochore microtubules move the chromosomes to opposite poles of the cell. Meanwhile, polar microtubules force the cell to elongate.
- *Telophase* The chromosomes decondense, and nuclear envelopes re-form around each new daughter nucleus.
- *Cytokinesis* Cytokinesis usually occurs at the end of telophase and divides the cytoplasm into approximately equal halves, each containing a nucleus. In animal cells, a ring of microfilaments pinches the plasma membrane in along the equator. In plant cells, new plasma membrane forms along the equator by the fusion of vesicles produced by the Golgi apparatus.

9.6 How Is the Cell Cycle Controlled?

Complex interactions among many proteins, particularly cyclins and cyclin-dependent protein kinases, drive the cell cycle. There are three major checkpoints where progress through the cell cycle is regulated: between G_1 and S, between G_2 and mitosis, and between metaphase and anaphase. These checkpoints ensure that the DNA is intact and replicated accurately and that the chromosomes are properly arranged for mitosis before the cell divides.

Key Terms

anaphase 149	gene 141
asexual reproduction 142	growth factor 151
cell cycle 141	interphase 146
cell division 141	kinetochore 148
cell plate 150	metaphase 149
centriole 147	mitosis 146
centromere 145	mitotic cell division 141
checkpoint 151	mutation 152
chromatid 147	nucleotide 141
chromosome 141	prokaryotic fission 144
clone 142	prophase 147
cytokinesis 146	sexual reproduction 142
daughter cell 141	spindle 147
deoxyribonucleic	spindle
acid (DNA) 141	microtubule 147
differentiate 141	stem cell 142
duplicated chromosome 147	telomere 145
gamete 142	telophase 150

Thinking Through the Concepts

Multiple Choice

1. A cell that remains capable of dividing throughout the life of an organism, and that produces daughter cells that can mature into any of several different cell types is a
   a. cancerous cell.
   b. differentiated cell.
   c. stem cell.
   d. gamete.

2. The chromosomes of a cell are lined up along the equator during
   a. prophase.
   b. metaphase.
   c. anaphase.
   d. telophase.

3. The chromosomes first attach to the spindle during
   a. prophase.
   b. metaphase.
   c. anaphase.
   d. telophase.

4. How does prokaryotic fission differ from eukaryotic cell division?
   a. Prokaryotic cells do not have chromosomes.
   b. Daughter cells are not genetically identical to the parent cells.
   c. Prokaryotic cell division does not require replication of DNA.
   d. Prokaryotic cells do not form spindles during cell division.

5. Which of the following is NOT true of mitotic cell division?
   a. The daughter cells are genetically identical.
   b. Chromosomes are moved to opposite poles of the cell.
   c. Mitotic cell division is required for asexual reproduction.
   d. Mitotic cell division is the mechanism by which bacterial cells divide.

Fill-in-the-Blank

1. The genetic material of all living organisms is _____, which is contained in chromosomes.
2. Prokaryotic cells divide by a process called _____.
3. Growth and development of eukaryotic organisms occur through _____ cell division and _____ of the resulting daughter cells. _____ cells in multicellular eukaryotes remain capable of dividing throughout the life of the organism; their daughter cells can differentiate into a variety of cell types.
4. Eukaryotic cells are often stimulated to divide by hormone-like molecules called _____. _____ monitor progress through the cell cycle. Two categories of genes that, when mutated, often allow unregulated cell division are _____ and _____.
5. The four phases of mitosis are _____, _____, _____, and _____. Division of the cytoplasm into two cells, called _____, usually occurs during which phase? _____
6. Chromosomes attach to spindle microtubules at structures called _____. Some spindle microtubules, called _____ microtubules, do not bind to chromosomes, but have free ends that overlap along the equator of the cell. These microtubules push the poles of the cell apart.

Review Questions

1. Diagram and describe the eukaryotic cell cycle. Name the phases, and briefly describe the events that occur during each.
2. Define *mitosis* and *cytokinesis*. What changes in cell structure would result if cytokinesis did not occur after mitosis?
3. Diagram the stages of mitosis. How does mitosis ensure that each daughter nucleus receives a full set of chromosomes?
4. Define the following terms: *centromere, telomere, kinetochore, chromatid,* and *spindle.*
5. Describe and compare the process of cytokinesis in animal cells and in plant cells.
6. How is the cell cycle controlled? Why is it important to regulate progression through the cell cycle?
7. Diagram and describe the prokaryotic cell cycle.

Applying the Concepts

1. Most nerve cells in the adult human central nervous system, as well as heart muscle cells, do not divide. In contrast, cells lining the inside of the small intestine divide frequently. Discuss this difference in terms of why damage to the nervous system and heart muscle cells (for example, that caused by a stroke or heart attack) is so dangerous. What do you think might happen to tissues such as the intestinal lining if a disorder blocked mitotic cell division in all cells of the body?
2. Cancer cells divide out of control. Side effects of the cancer treatments chemotherapy and radiation therapy include loss of hair and of the intestinal lining, the latter producing severe nausea. What can you infer about the mechanisms of these treatments? What would you look for in an improved cancer therapy?

MEIOSIS: THE BASIS OF SEXUAL REPRODUCTION

The Giddings family is a rainbow of colors.

The Rainbow Connection

FIRST CAME JACOB, WHO HAS BLUE EYES like his mom, Tess, but curly brown hair and olive skin. Next came Savannah, who looks a lot like Jacob, though her hair is perhaps more dark blond than brown. Amiah, however, was truly a surprise when she was born—she has very pale skin, with straight, sandy-brown hair. Zion, the youngest child, has dark skin, black curly hair, and brown eyes, similar to his father, Chris. Even in today's multicultural England, a family like that is unusual.

Tess and Chris Giddings are as surprised as everyone else by their rainbow family. In fact, when Amiah was born, she had low blood sugar and needed to be checked out by a specialist right away. She was whisked away so fast that the hospital

staff hadn't put an ID wristband on her yet. When she was returned to her parents a little while later, they were astounded at how white her skin was. They asked the inevitable question: Was she switched with another baby by mistake? Just to be sure, the Giddings agreed to a DNA test. The results showed that Tess and Chris were indeed Amiah's parents. When Zion was born a few years later, Chris burst out, "Oh my God, he's black!" To which the astounded midwife could only reply, "You do know you're a black man, don't you?"

How could one couple have such a diverse family? As we will see in this chapter, sexual reproduction can mix inherited characteristics from the parents into a remarkable variety of different offspring. How does sexual reproduction produce genetic diversity? And why would natural selection favor seemingly random shuffling of traits?

AT A GLANCE

10.1 HOW DOES SEXUAL REPRODUCTION PRODUCE GENETIC VARIABILITY?

There are two fundamentally different methods by which organisms reproduce: asexual reproduction and sexual reproduction. Asexual reproduction uses mitotic cell division to create offspring that are genetically identical to the parent organism, whether single-celled, such as *Paramecium* or *Amoeba,* or multicellular, such as *Hydra* or aspen trees. In contrast, **sexual reproduction,** in which offspring are produced through the union of **gametes** (sperm and egg), creates offspring that are different from one another and from either parent. The production of gametes requires a specialized form of cell division, called meiotic cell division, which we will explore in this chapter.

Asexual reproduction was the original method of reproducing, billions of years ago, and many modern organisms, including bacteria, fungi, many plants and protists, and some animals such as flatworms, sea anemones, and *Hydra,* reproduce asexually, at least some of the time. Therefore, asexual reproduction can be a successful evolutionary strategy under some circumstances. Why did sexual reproduction evolve, and why do two such different methods of reproduction persist today, even among multicellular organisms? It has long been assumed that there must be evolutionary advantages to both sexual reproduction and asexual reproduction, perhaps in different organisms and at different times, depending on the environment. Recently, research has provided support for the hypothesis that the evolutionary advantage to sexual reproduction is the continual generation of genetic variability, as we explore in "How Do We Know That? The Evolution of Sexual Reproduction" on page 164.

Genetic Variability Originates as Mutations in DNA

The hereditary information of all living cells resides in molecules of deoxyribonucleic acid, or DNA, packaged into one or more chromosomes. Each unit of inheritance, called a **gene,** consists of a sequence of nucleotides at a specific place, or **locus** (plural, loci), on a chromosome. A eukaryotic chromosome typically contains a few dozen to a few thousand genes. All the members of a species have extremely similar, but usually not identical, nucleotide sequences in their genes. The slightly different nucleotide sequences of a gene are called **alleles.** If we could survey all the members of a given species, we might find a few, several dozen, or even hundreds of alleles of each gene. As alleles interact with different factors in the environment, such as nutrition or exercise, they may produce differences in structure and function, such as height, weight, or muscle strength.

Where do alleles come from? Alleles are the result of **mutations,** which can occur when a cell makes a mistake copying its DNA prior to cell division (see Chapter 12) or when a ray of ultraviolet light from the sun or certain chemicals in the environment cause changes in a cell's DNA. When a mutation happens in the cells that produce sperm or eggs, it may be passed down from generation to generation. A given mutation might have happened yesterday, or it may have occurred hundreds or thousands of years ago and been inherited ever since.

Sexual Reproduction Generates Genetic Variability Between the Members of a Species

Different members of a species usually have different combinations of alleles, and consequently have different traits. For example, you may have some classmates who are tall with straight blond hair, others who are tall with curly black hair, and still others who are short with straight brown hair. To understand how sexual reproduction generates allele combinations of such remarkable variety, we'll begin by looking at the numbers and types of chromosomes found in the cells of eukaryotic organisms.

Eukaryotic Chromosomes Usually Occur in Pairs Containing Similar Genetic Information

The complete set of chromosomes from a single cell is its **karyotype** (FIG. 10-1). For most eukaryotic organisms, a karyotype consists of pairs of chromosomes. Humans have 23 pairs, for a total of 46 chromosomes per cell. The two chromosomes that make up a pair are called **homologous chromosomes,** or **homologues,** from Greek words that mean "to say the same thing," because homologous chromosomes contain the genes that control the same inherited characteristics. Despite their name, the two homologues in

one duplicated
chromosome

sister
chromatids

a pair of
homologous
chromosomes

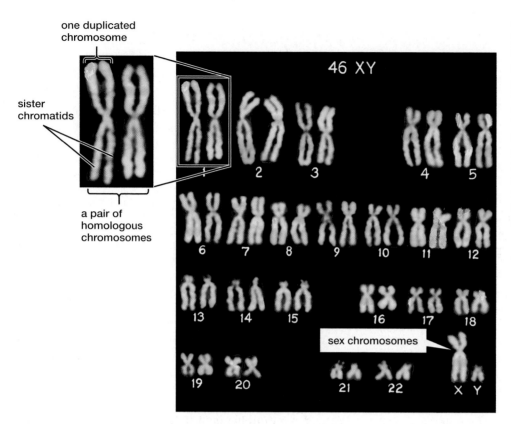

46 XY

sex chromosomes

◀ **FIGURE 10-1 The karyotype of a human male** Staining and photographing the entire set of duplicated, condensed chromosomes from a single cell produces a karyotype. Pictures of the individual chromosomes are cut out and arranged in descending order of size. The chromosome pairs (homologues) are usually similar in size and have similar genetic material. Chromosomes 1 through 22 are the autosomes; the X and Y chromosomes are the sex chromosomes. If this were a female karyotype, it would have two X chromosomes and no Y chromosome.

a pair seldom say exactly the "same thing": Although homologous chromosomes contain the same genes, a pair of homologues may have the same alleles of some genes and different alleles of other genes (**FIG. 10-2**).

Cells with pairs of homologous chromosomes are called **diploid,** meaning "double." One homologue of each pair, which we will call the maternal homologue, is inherited from the mother, and the other, called the paternal homologue, is inherited from the father. Pairs of chromosomes with nearly identical DNA sequences and that are found in diploid cells of both sexes are called **autosomes.** People have 22 pairs of autosomes. In addition to autosomes, humans and almost all other mammals have two **sex chromosomes:** either two X chromosomes (in females) or an X and a Y chromosome (in males). Although X and Y chromosomes are quite different in size (see Fig. 10-1) and in genetic composition, small portions of the X and Y chromosomes are homologous to each other.

Not all cells are diploid: If a cell contains only one member of each pair of homologues, it is **haploid.** As we will see in Section 10.2, the sperm and eggs produced by diploid organisms contain only one member of each pair of homologous chromosomes and so are haploid. Some organisms, such as the bread mold *Neurospora,* have haploid cells for most of their life cycle.

In biological shorthand, the number of different types of chromosomes in a species is called the *haploid number* and is designated *n.* For humans, $n = 23$ because we have 23 different types of chromosomes (22 autosomes plus one sex chromosome). Diploid cells contain $2n$ chromosomes.

Other organisms may have more than two copies of each homologous chromosome in each cell and are **polyploid.** Many plants, for example, have more than two copies of each homologue, with four (tetraploid; $4n$), six (hexaploid; $6n$), or even more copies per cell. Many common flowers, including some daylilies, orchids, lilies, and phlox, are tetraploid; most wheat is either tetraploid or hexaploid.

gene 1   gene 2

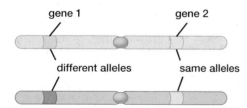

different alleles   same alleles

▲ **FIGURE 10-2 Homologous chromosomes are usually not identical** Homologous chromosomes have the same genes at the same locations. The homologues may have the same allele of some genes (right) and different alleles of other genes (left).

CHECK YOUR LEARNING

Can you ...

- describe the relationships between genes, mutations, and alleles?
- define the terms *homologous chromosome, autosome,* and *sex chromosome*?
- explain the differences between diploid, haploid, and polyploid cells?

CASE STUDY   CONTINUED

The Rainbow Connection

The genetic variability of the Giddings children started out as mutations that occurred thousands of years ago. Take hair color: Our distant ancestors probably all had dark hair, its color controlled by multiple genes located on several different chromosomes. The alleles that produced Tess's blond hair originated as mutations in genes that control the amount and type of hair pigment. Tess probably inherited only "pale hair" alleles of all of these genes, so for any given hair color gene, she has the same pale hair allele on both homologous chromosomes. Chris, on the other hand, inherited both dark and pale hair alleles for at least some of the genes, so his homologues have different alleles. As we will see in Chapter 11, in many cases one allele (in this case, the dark hair allele) overrides the effects of the other allele (the pale hair allele), so Chris has black hair. What combinations of alleles might have been packaged in Tess's eggs and Chris's sperm, which would combine to produce their diverse children?

10.2 HOW DOES MEIOTIC CELL DIVISION PRODUCE GENETICALLY VARIABLE, HAPLOID CELLS?

Sexual reproduction starts with genetically similar, but not identical, pairs of homologues and produces offspring through two steps (**FIG. 10-3**): ❶ During **meiotic cell division,** a diploid cell gives rise to haploid daughter cells containing a single member of each pair of homologues. The haploid cells, or their descendants produced by mitotic cell division, become gametes. In animals, the haploid cells produced by meiotic cell division differentiate into sperm or eggs. ❷ Fertilization of an egg by a sperm restores the diploid number of chromosomes in the offspring.

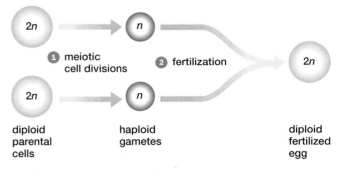

▲ **FIGURE 10-3 Meiotic cell division is essential for sexual reproduction** In sexual reproduction, specialized diploid reproductive cells of the parents (2n) undergo meiosis to produce haploid cells (n). In animals, these cells become gametes (sperm or eggs). When an egg is fertilized by a sperm, the resulting fertilized egg, or zygote, is diploid once again (2n).

Meiotic cell division consists of **meiosis,** a specialized type of nuclear division in which a diploid nucleus divides twice, producing four haploid nuclei, and cytokinesis, which packages the four nuclei into separate cells. (Fittingly, "meiosis" comes from a Greek word meaning "to diminish.") Although many of the structures and events of meiotic cell division are similar to those of mitotic cell division, there are several important differences. A crucial difference involves DNA replication: In mitotic cell division, the parent cell undergoes one round of DNA replication followed by one nuclear division. In meiotic cell division, there are two nuclear divisions; the DNA is replicated before the first division (**FIG. 10-4a**), but it is *not replicated again* between the first and second divisions.

The first division of meiosis (called **meiosis I**) separates the pairs of homologous chromosomes and sends one homologue from each pair into each of two daughter nuclei, which are therefore haploid. Each chromosome, however, still consists of two chromatids (**FIG. 10-4b**). The second division (called **meiosis II**) separates the chromatids into independent chromosomes and parcels one chromosome into each of two daughter

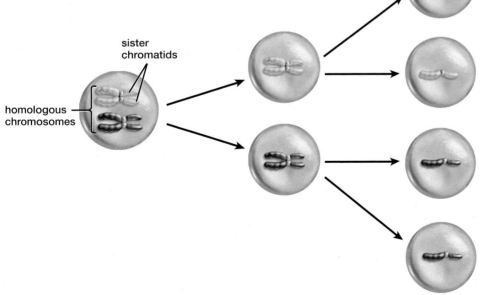

(a) Duplicated homologues prior to meiosis (diploid)

(b) After meiosis I (haploid)

(c) After meiosis II (haploid)

sister chromatids

homologous chromosomes

◀ **FIGURE 10-4 Meiosis halves the number of chromosomes (a)** Both members of a pair of homologous chromosomes are duplicated prior to meiosis. **(b)** During meiosis I, each daughter cell receives one member of each pair of homologues. **(c)** During meiosis II, sister chromatids separate into independent chromosomes, and each daughter cell receives one of these chromosomes. Maternal chromosomes are colored violet; paternal chromosomes are colored yellow.

MEIOSIS I

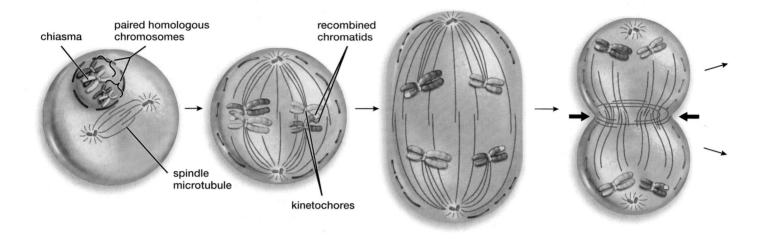

(a) Prophase I	(b) Metaphase I	(c) Anaphase I	(d) Telophase I
Duplicated chromosomes condense. Homologous chromosomes pair up and chiasmata occur as chromatids of homologues exchange parts by crossing over. The nuclear envelope disintegrates, and spindle microtubules form.	Paired homologous chromosomes line up along the equator of the cell. One homologue of each pair faces each pole of the cell and attaches to the spindle microtubules via the kinetochore (blue).	Homologues separate, one member of each pair going to each pole of the cell. Sister chromatids do not separate.	Spindle microtubules disappear. Two clusters of chromosomes have formed, each containing one member of each pair of homologues. The daughter nuclei are therefore haploid. Cytokinesis commonly occurs at this stage. There is little or no interphase between meiosis I and meiosis II.

▲ **FIGURE 10-5 Meiotic cell division** In meiotic cell division, the homologous chromosomes of a diploid cell are separated, producing four haploid daughter cells. Two pairs of homologous chromosomes are shown.

THINK CRITICALLY What would be the consequences for the resulting gametes and offspring if one pair of homologues failed to separate at anaphase I?

nuclei. Therefore, at the end of meiosis, there are four haploid daughter nuclei, each with one copy of each homologous chromosome. Because each nucleus is usually enclosed in a separate cell, meiotic cell division typically produces four haploid cells from a single diploid parent cell (**FIG. 10-4c**).

Meiosis I Separates Homologous Chromosomes into Two Haploid Daughter Nuclei

The phases of meiosis have the same names as similar phases in mitosis, followed by I or II to distinguish the two nuclear divisions that occur in meiosis (**FIG. 10-5**). When meiosis I begins, the chromosomes have already been duplicated during

interphase, and the sister chromatids of each chromosome are attached to each other at the centromere.

During Prophase I, Homologous Chromosomes Pair Up and Exchange DNA

In mitosis, homologous chromosomes move independently of each other. In contrast, during prophase I of meiosis, the duplicated homologous chromosomes line up side by side and their chromatids exchange segments of DNA (**FIG. 10-5a** and **FIG. 10-6**). This process begins when proteins bind the maternal and paternal homologues together so that they align precisely along their entire length. Enzymes then cut through the DNA of both homologues and graft the cut ends together, often exchanging part of a chromatid of the maternal

MEIOSIS II

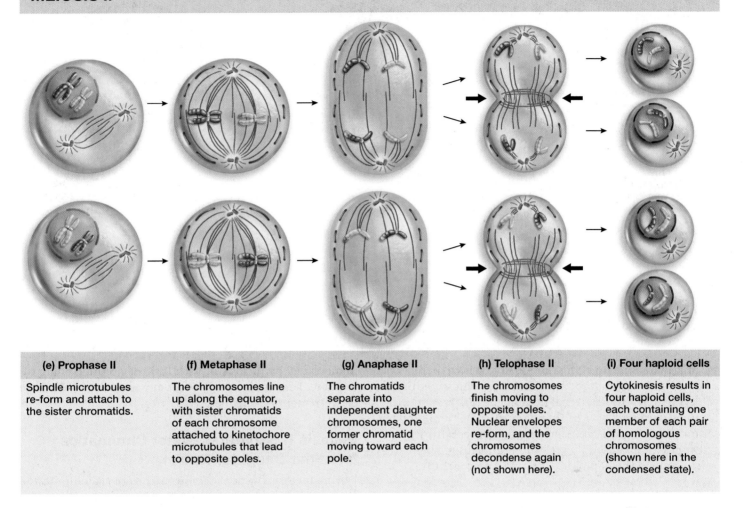

(e) Prophase II	(f) Metaphase II	(g) Anaphase II	(h) Telophase II	(i) Four haploid cells
Spindle microtubules re-form and attach to the sister chromatids.	The chromosomes line up along the equator, with sister chromatids of each chromosome attached to kinetochore microtubules that lead to opposite poles.	The chromatids separate into independent daughter chromosomes, one former chromatid moving toward each pole.	The chromosomes finish moving to opposite poles. Nuclear envelopes re-form, and the chromosomes decondense again (not shown here).	Cytokinesis results in four haploid cells, each containing one member of each pair of homologous chromosomes (shown here in the condensed state).

homologue for part of a chromatid of the paternal homologue. The binding proteins and enzymes then depart, leaving crosses, or **chiasmata** (singular, chiasma), where chromatids of the maternal and paternal chromosomes have exchanged parts (see Fig. 10-6). In human cells, each pair of homologues usually forms two or three chiasmata during prophase I. The mutual exchange of DNA between maternal and paternal chromosomes at chiasmata is called **crossing over.** Even after the exchange of DNA, the arms of the homologues remain temporarily entangled at the chiasmata. This keeps the two homologues together until they are pulled apart during anaphase I.

As in prophase of mitosis, the spindle microtubules begin to assemble outside the nucleus during prophase I. Near the

end of prophase I, the nuclear envelope breaks down and spindle microtubules invade the nuclear region, capturing the chromosomes by attaching to their kinetochores.

During Metaphase I, Paired Homologous Chromosomes Line Up at the Equator of the Cell

During metaphase I, interactions between the kinetochores and the spindle microtubules move the paired homologues to the equator of the cell (**FIG. 10-5b**). Unlike in metaphase of mitosis, in which *individual* duplicated chromosomes line up along the equator, in metaphase I of meiosis, *homologous pairs* of duplicated chromosomes, held together

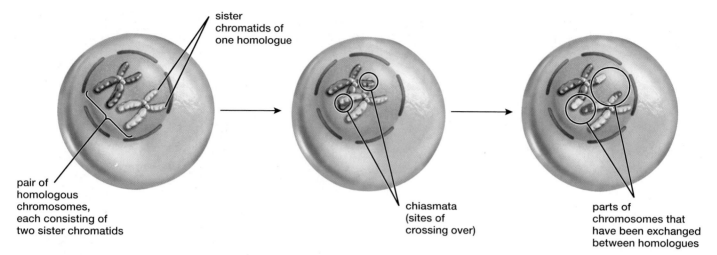

sister
chromatids of
one homologue

pair of
homologous
chromosomes,
each consisting of
two sister chromatids

chiasmata
(sites of
crossing over)

parts of
chromosomes that
have been exchanged
between homologues

▲ **FIGURE 10-6 Crossing over** Nonsister chromatids of different members of a homologous pair of chromo-somes exchange DNA at chiasmata.

THINK CRITICALLY What would be the genetic consequences for the gametes and offspring if crossing over occurred between two nonhomologous chromosomes?

by chiasmata, line up along the equator. Which member of a pair of homologous chromosomes faces which pole of the cell is random—the maternal homologue may face "north" for some pairs and "south" for other pairs. This randomness (also called *independent assortment*), together with genetic recombination caused by crossing over, causes genetic diversity among the haploid cells produced by meiosis.

During Anaphase I, Homologous Chromosomes Separate

Anaphase in meiosis I differs considerably from anaphase in mitosis. In anaphase of mitosis, the sister chromatids separate and move to opposite poles. In contrast, in anaphase I of meiosis, the sister chromatids of each duplicated homologue remain attached to each other and move to the same pole. However, the chiasmata joining the two homologues untangle, allowing the homologues to separate and move to opposite poles (FIG. 10-5c). At the end of anaphase I, the cluster of chromosomes at each pole contains one member of each pair of homologous chromosomes. Therefore, each cluster contains the haploid number of chromosomes (although each chromosome is still duplicated and consists of sister chromatids attached at the centromere).

During Telophase I, Two Haploid Clusters of Duplicated Chromosomes Form

Telophase I in meiosis is similar to telophase in mitosis. In telophase I, the spindle microtubules disappear. Cytokinesis commonly occurs during telophase I (FIG. 10-5d). Nuclear envelopes may re-form. Telophase I is usually followed

immediately by meiosis II, with little or no intervening interphase. Remember that the chromosomes do not replicate between meiosis I and meiosis II.

Meiosis II Separates Sister Chromatids into Four Daughter Nuclei

During meiosis II, the sister chromatids of each duplicated chromosome separate in a process that is virtually identical to mitosis in a haploid cell. During prophase II, the spindle microtubules re-form (FIG. 10-5e) and the kinetochores of the sister chromatids of each duplicated chromosome attach to spindle microtubules extending to opposite poles of the cell. During metaphase II, the duplicated chromosomes line up at the cell's equator (FIG. 10-5f). During anaphase II, the sister chromatids separate and move to opposite poles (FIG. 10-5g). Telophase II and cytokinesis conclude meiosis II as nuclear envelopes re-form, the chromosomes decondense into their extended state, and the cytoplasm divides (FIG. 10-5h). Both daughter cells produced in meiosis I usually undergo meiosis II, producing a total of four haploid cells from the original diploid parental cell (FIG. 10-5i).

TABLE 10-1 compares mitotic and meiotic cell division, pointing out similarities and differences between the two.

CHECK YOUR LEARNING
Can you ...
- describe the steps and outcome of meiotic cell division?
- explain the results of crossing over?

TABLE 10-1 A Comparison of Mitotic and Meiotic Cell Division in Animal Cells

Feature	Mitotic Cell Division	Meiotic Cell Division
Cells in which it occurs	Body cells	Gamete-producing cells
Final chromosome number	Diploid—2n; two copies of each type of chromosome (homologous pairs)	Haploid—1n; one member of each homologous pair
Number of daughter cells	Two, identical to the parent cell and to each other	Four, containing recombined chromosomes due to crossing over
Number of cell divisions per DNA replication	One	Two
Function in animals	Development, growth, repair, and maintenance of tissues; asexual reproduction	Gamete production for sexual reproduction

MITOSIS

no stages comparable to meiosis I

interphase   prophase   metaphase   anaphase   telophase   two diploid cells

MEIOSIS

Recombination occurs. | Homologues pair. | Sister chromatids remain attached.

interphase   prophase   metaphase   anaphase   telophase   prophase   metaphase   anaphase   telophase   four haploid cells

MEIOSIS I   MEIOSIS II

In these diagrams, comparable phases are aligned. In both mitosis and meiosis, chromosomes are duplicated during interphase. Meiosis I, with the pairing of homologous chromosomes, formation of chiasmata, exchange of chromosome parts, and separation of homologues to form haploid daughter nuclei, has no counterpart in mitosis. Meiosis II, however, is virtually identical to mitosis in a haploid cell.

CASE STUDY   CONTINUED

The Rainbow Connection

As Tess and Chris Giddings produced eggs and sperm, meiosis separated their homologous chromosomes. Let's assume that Tess has only "pale hair" alleles for all of the genes that might contribute to hair color, but that Chris has alleles for both dark and pale hair. During egg formation in Tess's ovaries, crossing over and separating the homologues wouldn't make any difference for the hair color genes, and all of her eggs would contain only pale hair alleles. For Chris, on the other hand, crossing over and separating the homologues would matter a lot. Some of his sperm might receive a dark hair allele for one gene, but a pale hair allele for another gene. Other sperm would have different combinations of dark and pale hair alleles, including some sperm with all pale hair alleles and others with all dark hair alleles. Can this diversity of sperm and eggs explain the diversity of the Giddings children?

HOW DO WE KNOW THAT?

The Evolution of Sexual Reproduction

Asexual reproduction has some distinct advantages over sexual reproduction. Asexual reproduction is much more efficient because it does not require energy to seek out and court mates (with the possibility of failing) or to produce huge numbers of sperm so that a few of them might fertilize eggs. In addition, because asexual reproduction is based on mitosis, which generates genetically identical cells, an asexually reproducing organism passes on all of its genes to all of its offspring. In contrast, a sexually reproducing organism passes on only half of its genes to any given offspring. Therefore, all other factors being equal, an asexually reproducing organism passes twice as many of its genes to the next generation as a sexually reproducing organism does—the genetic equivalent of having twice as many offspring. Finally, if an asexually reproducing organism is well adapted to its environment, then so are all of its offspring, whereas the genetic variability created by sexual reproduction might break up a good combination of alleles.

Not surprisingly, then, some very successful organisms routinely reproduce asexually. For example, many of the grasses and weeds in a suburban lawn can reproduce by sprouting new plants from their stems or roots. Some, like Kentucky bluegrass and dandelions, even bear flowers that can produce seeds without being fertilized. Nevertheless, almost all eukaryotic organisms reproduce sexually (even bluegrass and dandelions reproduce sexually some of the time). How might natural selection favor sexual reproduction, despite its significant costs?

No one knows for sure, partly because it's difficult to design experiments to test possible hypotheses. For example, all mammals and birds reproduce exclusively by sexual reproduction, so scientists cannot compare the reproductive success of sexual versus asexual populations of these animals under varying conditions. Despite this difficulty, a handful of inventive experiments indicate that sexual reproduction may be favored in certain situations:

- **Variable Environments**   If the environment is stable, and a population of organisms is already well adapted, then asexual reproduction, by producing identical, well-adapted offspring, will be favored. But if the environment is variable, then sexual reproduction is often favored, as new combinations of traits may promote the success of some offspring, even though others, with different combinations, might die young or fail to reproduce in their turn: The especially successful offspring more than compensate for the unsuccessful ones. Experiments using yeasts (single-celled fungi) and rotifers (tiny freshwater animals) support this hypothesis. Yeasts and rotifers can reproduce either sexually or asexually; in both, variable environments favor sexual reproduction.
- **Parasites**   Many organisms are plagued by parasites. If the parasites evolve to become more efficient at infecting and disabling their host organisms, the host population will decline. Sexual reproduction may help to foil the parasites by constantly changing the defenses of the hosts. This mechanism was demonstrated in New Zealand mud snails, which can reproduce either sexually or asexually (**FIG. E10-1**). In just a few years, originally successful,

▲ **FIGURE E10-1   The selective advantages of sexual reproduction** Populations of the New Zealand mud snail are controlled in their native habitat by tiny parasitic worms that multiply until they practically fill up the inside of the snail, displacing its reproductive organs. In much of Europe and the western United States, the snail is a rapidly spreading invasive species, because its parasites aren't found in the new habitats.

asexually reproducing populations of snails become heavily infested by parasitic worms, which effectively castrates female snails. Sexual reproduction generates genetically variable, continually changing populations of snails, some of which resist infection.

- **Accumulation of Harmful Mutations**   Harmful mutations that appear in an asexually reproducing population can never be removed. Over time, the genome accumulates more and more harmful mutations, and fitness declines. In a sexually reproducing population, however, meiosis shuffles chromosomes, and even parts of chromosomes, which are then recombined when a sperm fertilizes an egg (see Sections 10.2 and 10.3). In this way, sexual reproduction can reduce the number of harmful mutations in some lucky offspring, who then survive and reproduce successfully. Experiments with yeasts support this hypothesis.

Will biologists ever conclusively prove why natural selection drove the evolution of sexual reproduction and maintains it today in so many species? Perhaps not. What is clear, however, is that sexual reproduction has been a powerful force in the evolution of many species, even when the survival of individual organisms is put at risk: Peacocks display glorious, but unwieldy, tails; some female spiders digest their own bodies to feed their young; male elk and deer grow elaborate antlers and fight for mates, sometimes suffering serious wounds as a result; and some male spiders and insects are routinely killed and eaten by their mates. Without sexual reproduction, life on Earth would be very different, and a lot less interesting.

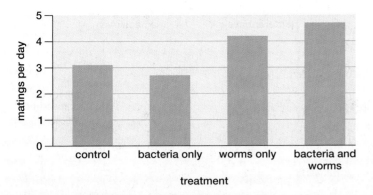

(a) The number of pairs of snails mating per day

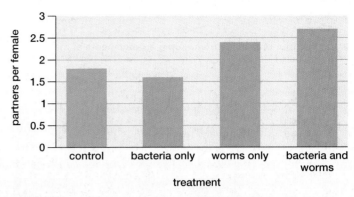

(b) The number of different male snails mating with each female snail

▲ **FIGURE E10-2 The effects of parasitism on mating in New Zealand mud snails**
Modified from Soper, D. M., et al., 2014. *Biology Letters* 10:20131091.

THINK CRITICALLY In many freshwater lakes, duck feces are a major source of both bacteria and the eggs of parasitic worms. Researcher Curt Lively and his colleagues collected duck feces and treated the feces in one of four ways: (1) sterilizing the feces by heating, which kills both bacteria and worm eggs (called the control condition); (2) sterilizing the feces by heating and then replacing the bacteria (leaving bacteria only); (3) killing the bacteria with bleach, which does not kill worm eggs (leaving worms only); and (4) no treatment (leaving bacteria and worms). They added one of the four types of fecal samples to aquaria housing parasite-free New Zealand mud snails and observed mating (**FIG. E10-2**). From these data, what can you conclude about the role of worm parasitism in snail mating? How would parasitism affect genetic variability in mud snails?

HAVE YOU EVER
WONDERED...

A mule is a cross between a horse and a donkey. A horse has 64 chromosomes ($n = 32$) and a donkey has 62 ($n = 31$), so a mule has a total of 63 chromosomes. An odd number of chromosomes cannot all pair up during meiosis I. In addition, many horse chromosomes are not homologous

Why Mules Are Sterile?

to donkey chromosomes, so many of a mule's chromosomes do not have a homologue to match up with. Therefore, in almost all cases, gametes resulting from meiosis in a mule receive neither a full set of horse chromosomes nor a full set of donkey chromosomes. The nearly random number and parentage of the chromosomes in mule gametes means that crucial genes are almost always missing, so whether a mule mates with another mule, a horse, or a donkey, the resulting fertilized eggs cannot develop. Very rarely, however, a mule does reproduce: There is one report of a female mule that apparently produced an egg with *all* horse chromosomes and *no* donkey chromosomes, and had a foal sired by a male donkey. Appropriately, the foal was named Blue Moon.

10.3 HOW DO MEIOSIS AND UNION OF GAMETES PRODUCE GENETICALLY VARIABLE OFFSPRING?

Mutations occurring randomly over millions of years provide the original sources of genetic variability: new alleles. However, mutations in gametes, or in precursor cells that produce gametes, are very rare events. Therefore, the genetic variability that occurs from one generation to the next results almost entirely from meiosis and sexual reproduction.

Shuffling the Homologues Creates Novel Combinations of Chromosomes

One major source of genetic diversity is the random distribution of maternal and paternal homologues to the daughter nuclei during meiosis I. Remember that at metaphase I the paired homologues line up at the cell's equator. In each pair of homologues, the maternal chromosome faces one pole and the paternal chromosome faces the opposite pole, but which homologue faces which pole is random and is not

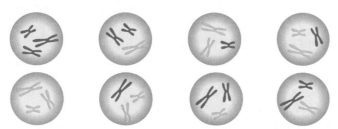

(a) The four possible chromosome arrangements at metaphase of meiosis I

(b) The eight possible sets of chromosomes after meiosis I

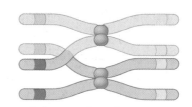

(c) The eight possible types of gametes after meiosis II

▲ **FIGURE 10-7 Random separation of homologous pairs of chromosomes produces genetic variability** For clarity, the chromosomes are depicted as large, medium, and small.

affected by the orientation of the homologues of other chromosome pairs.

Let's consider meiosis in mosquitoes, which have three pairs of homologous chromosomes ($n = 3$, $2n = 6$). At metaphase I, the chromosomes can align in four possible configurations (**FIG. 10-7a**). Therefore, anaphase I can yield eight possible sets of chromosomes ($2^3 = 8$; **FIG. 10-7b**). At the conclusion of meiotic cell division, a mosquito can thus produce gametes with any one of eight unique sets of chromosomes (**FIG. 10-7c**). In a human, meiosis randomly shuffles 23 pairs of homologous chromosomes and can theoretically produce gametes with any one of more than 8 million (2^{23}) different combinations of maternal and paternal chromosomes.

Crossing Over Creates Chromosomes with Novel Combinations of Genes

Recall that the two members of a pair of homologous chromosomes may have different alleles of some genes (see Fig. 10-2). If they do, then crossing over creates genetic **recombination**: the formation of chromosomes with combinations of alleles that differ from those of either parent (**FIG. 10-8**). Chromosomes are very long—human chromosomes range from about 50 million to 250 million nucleotides in length—and crossing over can occur almost anywhere along the chromosome. Therefore, even in a single person, gamete production can yield a tremendous number of genetically unique, recombined chromosomes.

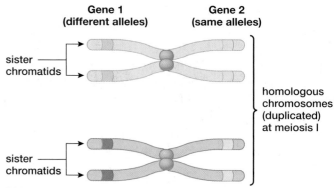

(a) Duplicated chromosomes in prophase of meiosis I

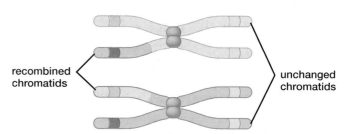

(b) Crossing over during prophase I

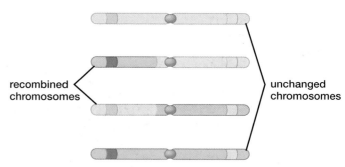

(c) Homologous chromosomes separate at anaphase I

(d) Unchanged and recombined chromosomes after meiosis II

▲ **FIGURE 10-8 Crossing over recombines alleles on homologous chromosomes (a)** During prophase of meiosis I, duplicated homologous chromosomes pair up. **(b)** Nonsister chromatids of the two homologues exchange parts by crossing over. **(c)** When the homologous chromosomes separate during anaphase of meiosis I, one chromatid of each of the homologues now contains a piece of DNA from a chromatid of the other homologue. **(d)** After meiosis II, two chromosomes are unchanged and two chromosomes show genetic recombination, with allele arrangements that did not occur in the parental chromosomes.

Fusion of Gametes Adds Further Genetic Variability to the Offspring

At fertilization, two gametes, each containing a unique combination of alleles, fuse to form a diploid offspring. As we have seen, if we ignore crossing over, a single person can produce gametes with any of 8 million chromosome combinations. Therefore, the chances that your parents could produce another child who is genetically identical to you are about $1/8{,}000{,}000 \times 1/8{,}000{,}000$, or about 1 in 64 trillion! When we factor in the almost endless variability produced by crossing over, we can confidently say that (unless you are an identical twin) there never has been, and never will be, anyone just like you.

CHECK YOUR LEARNING

Can you ...

- explain how meiosis and sexual reproduction generate genetic variability in populations?

10.4 WHEN DO MITOTIC AND MEIOTIC CELL DIVISION OCCUR IN THE LIFE CYCLES OF EUKARYOTES?

The life cycles of almost all eukaryotic organisms share a common pattern. First, two haploid cells fuse during the process of fertilization, bringing together genes from different parental organisms and endowing the resulting diploid cell with new gene combinations. Second, at some point in the life cycle, meiotic cell division occurs, re-creating haploid cells. Third, mitotic cell division of either haploid or diploid cells, or both, results in the growth of multicellular bodies or in asexual reproduction.

The seemingly vast differences between the life cycles of, say, ferns and humans are caused by variations in two aspects: (1) the points in the life cycle at which mitotic and meiotic cell division occur and (2) the relative proportions of the life cycle spent in the diploid and haploid states. We will name eukaryotic life cycles according to the relative dominance of diploid and haploid stages.

In Diploid Life Cycles, the Majority of the Cycle Is Spent as Diploid Cells

In most animals, virtually the entire life cycle is spent in the diploid state (**FIG. 10-9**). Diploid adults produce short-lived haploid gametes by meiotic cell division. Sperm and egg fuse to form a diploid fertilized egg, called a **zygote.** Development of the zygote to the adult organism results from mitotic cell division and differentiation of diploid cells.

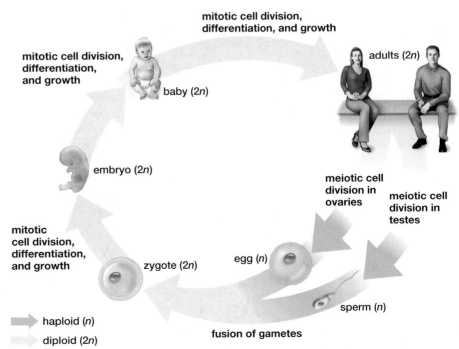

▶ **FIGURE 10-9 The human life cycle**
Through meiotic cell division, the two sexes produce gametes—sperm in males and eggs in females—that fuse to form a diploid zygote. Mitotic cell division and differentiation of the daughter cells produce an embryo, child, and ultimately a sexually mature adult. The haploid stages last only a few hours to a few days; the diploid stages may survive for a century.

mitotic cell division, differentiation, and growth

mitotic cell division, differentiation, and growth

adults (2n)

baby (2n)

embryo (2n)

meiotic cell division in ovaries

meiotic cell division in testes

mitotic cell division, differentiation, and growth

zygote (2n)

egg (n)

sperm (n)

fusion of gametes

→ haploid (n)

→ diploid (2n)

▶ FIGURE 10-10 **The life cycle of the single-celled alga *Chlamydomonas*** *Chlamydomonas* reproduces asexually by mitotic cell division of haploid cells. When nutrients are scarce, specialized haploid reproductive cells (usually from genetically different populations) fuse to form a diploid cell. Meiotic cell division then immediately produces four haploid cells, usually with different genetic compositions than either of the parental strains.

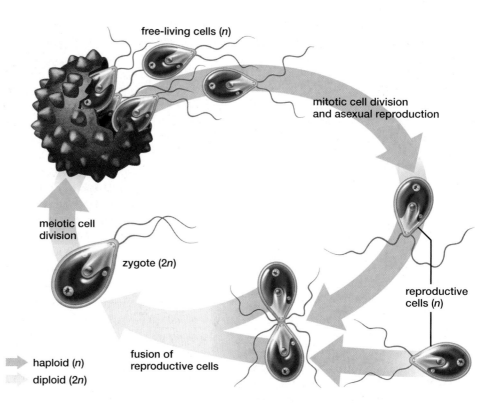

In Haploid Life Cycles, the Majority of the Cycle Is Spent as Haploid Cells

Some eukaryotes, such as many fungi and single-celled algae, spend most of their life cycles in the haploid state (FIG. 10-10). Asexual reproduction by mitotic cell division produces a population of identical, haploid cells. Under certain environmental conditions, some differentiate into reproductive cells. Two haploid reproductive cells, usually from genetically different strains, fuse to form a diploid zygote. The zygote immediately undergoes meiotic cell division, producing haploid cells again. In organisms with haploid life cycles, mitotic cell division never occurs in diploid cells.

In Alternation of Generations Life Cycles, There Are Both Diploid and Haploid Multicellular Stages

The life cycle of plants is called *alternation of generations,* because it alternates between multicellular diploid and multicellular haploid stages. In the typical pattern (FIG. 10-11), specialized cells of a multicellular diploid adult stage (the diploid generation) undergo meiotic cell division, producing haploid cells called spores. The spores undergo many rounds of mitotic cell division and their daughter cells

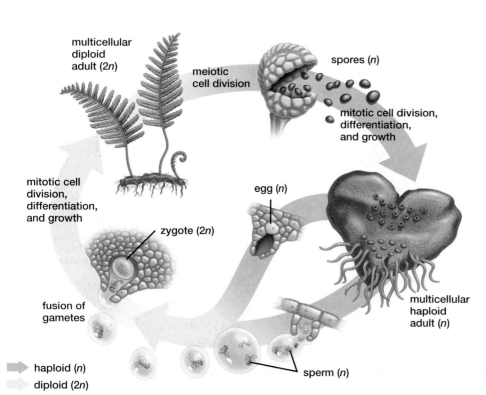

◀ FIGURE 10-11 **Alternation of generations** In plants, such as this fern, specialized cells in the multicellular diploid adult stage undergo meiotic cell division to produce haploid spores. The spores undergo mitotic cell division and differentiation of the daughter cells to produce a multicellular haploid adult stage. Sometime later, perhaps after many weeks, some of these haploid cells differentiate into sperm and eggs. These fuse to form a diploid zygote. Mitotic cell division and differentiation once again give rise to a multicellular diploid adult stage.

differentiate, producing a multicellular haploid adult stage (the haploid generation). At some point, certain haploid cells differentiate into haploid gametes. Two gametes then fuse to form a diploid zygote. The zygote grows by mitotic cell division into another multicellular diploid adult stage.

In some plants, such as ferns, both the haploid and diploid stages are free-living, independent plants. Flowering plants, however, have reduced haploid stages, represented only by the pollen grain and a small cluster of cells in the ovary of the flower (see Chapters 22 and 45).

CHECK YOUR LEARNING

Can you ...

- compare and contrast the three main types of eukaryotic life cycles, and give examples of organisms that exhibit each type?

10.5 HOW DO ERRORS IN MEIOSIS CAUSE HUMAN GENETIC DISORDERS?

As we have seen, the intricate mechanisms of meiotic cell division are essential to sexual reproduction and producing genetic diversity. However, this elaborate dance of the chromosomes comes with a cost: There are occasional stumbles, resulting in gametes that have too many or too few chromosomes. Such errors in meiosis, called **nondisjunction,** can affect the number of sex chromosomes or autosomes in a gamete (**FIG. 10-12**). In humans, most embryos that arise from the fusion of gametes with abnormal chromosome numbers spontaneously abort, accounting for 20% to 50% of all miscarriages. However, some embryos with abnormal numbers of chromosomes survive to birth or beyond.

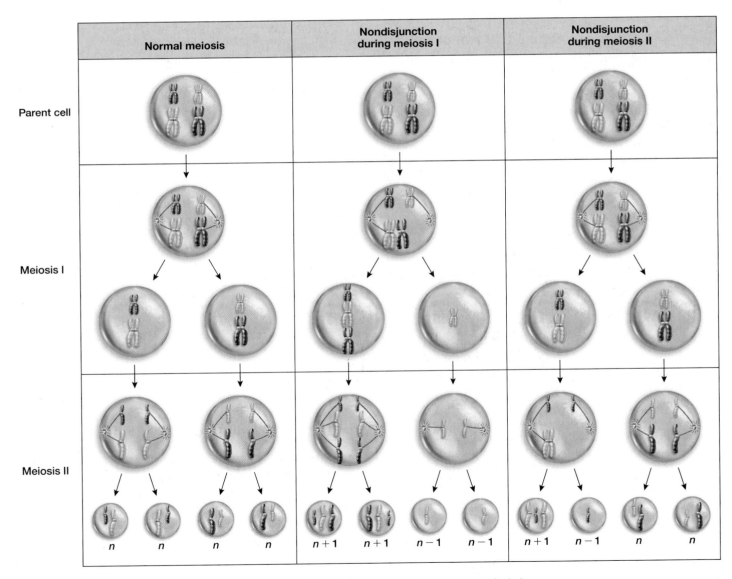

▲ **FIGURE 10-12 Nondisjunction during meiosis** Nondisjunction may occur either during meiosis I or meiosis II, resulting in gametes with too many ($n + 1$) or too few ($n - 1$) chromosomes.

TABLE 10-2	Effects of Nondisjunction of the Sex Chromosomes During Meiosis			
Nondisjunction in the Father				
Sex Chromosomes of Defective Sperm	**Sex Chromosomes of Normal Egg**	**Sex Chromosomes of Offspring**	**Characteristics of Offspring**	
O (none)	X	XO	Female—Turner syndrome	
XX	X	XXX	Female—Trisomy X	
XY	X	XXY	Male—Klinefelter syndrome	
YY	X	XYY	Male—Jacob syndrome	
Nondisjunction in the Mother				
Sex Chromosomes of Normal Sperm	**Sex Chromosomes of Defective Egg**	**Sex Chromosomes of Offspring**	**Characteristics of Offspring**	
X	O (none)	XO	Female—Turner syndrome	
Y	O (none)	YO	Dies as early embryo	
X	XX	XXX	Female—Trisomy X	
Y	XX	XXY	Male—Klinefelter syndrome	

Some Disorders Are Caused by Abnormal Numbers of Sex Chromosomes

In humans and other mammals, sperm normally contain either an X or a Y chromosome, and all eggs contain an X chromosome. Nondisjunction of sex chromosomes in males produces sperm with either no sex chromosome (often called "O" sperm) or two sex chromosomes (XX, YY, or XY). Nondisjunction of the sex chromosomes in females produces O or XX eggs. When normal gametes fuse with these defective sperm or eggs, the zygotes have normal numbers of autosomes but abnormal numbers of sex chromosomes (TABLE 10-2). The most common abnormalities are XO, XXX, XXY, and XYY. Genes on the X chromosome are essential to survival, so any embryo without at least one X chromosome spontaneously aborts very early in development.

Turner Syndrome (XO)

About 1 in every 2,500 female babies has only one X chromosome, a condition known as **Turner syndrome** (also called monosomy X, meaning "having one X chromosome"). The ovaries of girls with Turner syndrome usually degenerate before birth, and the girls do not undergo puberty. Treatment with estrogen can promote the development of secondary sexual characteristics, such as enlarged breasts. However, because most women with Turner syndrome do not have functioning ovaries and therefore cannot produce eggs, hormone treatment does not make it possible for them to bear children. Other common characteristics of women with Turner syndrome include short stature, folds of skin around the neck, and increased risk of cardiovascular disease, kidney defects, and hearing loss.

Trisomy X (XXX)

About 1 in every 1,000 women has three X chromosomes, a condition known as **trisomy X,** or triple X. Most of these women have no detectable differences from XX women, except for a tendency to be taller and to have a higher incidence of learning disabilities. Unlike women with Turner syndrome, most trisomy X women are fertile and, interestingly enough, almost always bear XX and XY children. Some unknown mechanism must operate during meiosis to prevent an extra X chromosome from being included in their eggs.

Klinefelter Syndrome (XXY)

About 1 in every 500 to 1,000 males is born with two X chromosomes and one Y chromosome. Men with **Klinefelter syndrome** usually have small testes that do not produce as much testosterone as the testes of XY men typically do. At puberty, some show mixed secondary sexual characteristics, such as partial breast development, broadening of the hips, and thin beards. XXY men may be infertile because of low sperm count, but they are not impotent. Klinefelter syndrome is usually diagnosed when an XXY man and his female partner seek medical help because they are unable to have children.

Jacob Syndrome (XYY)

Jacob syndrome occurs in about 1 male in every 1,000. Y chromosomes contain few active genes, and in most men with Jacob syndrome, having an extra Y chromosome doesn't change function or appearance very much. The most common effect is that XYY males tend to be taller than average. There may also be a slightly increased likelihood of learning disabilities.

Some Disorders Are Caused by Abnormal Numbers of Autosomes

Nondisjunction of the autosomes produces eggs or sperm that are missing an autosome or that have two copies of an autosome. Fusion with a normal gamete (bearing one copy of each autosome) leads to an embryo with either one or three copies

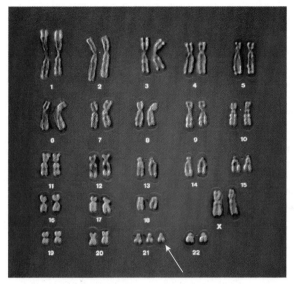 

(a) Karyotype showing three copies of chromosome 21   **(b) Girl with Down syndrome and her older sister**

▲ **FIGURE 10-13 Trisomy 21, or Down syndrome (a)** This karyotype of a Down syndrome child reveals three copies of chromosome 21 (arrow). **(b)** Down syndrome is almost always caused by nondisjunction and seldom runs in families. The older girl on the left received a single copy of chromosome 21 from each of her parents; her younger sister received two copies from one of the parents.

of the affected autosome. Embryos that have only one copy of any of the autosomes almost always abort so early in development that the woman never knows she was pregnant. Embryos with three copies of an autosome (trisomy) also usually spontaneously abort. However, a small fraction of embryos with three copies of chromosomes 13, 18, or 21 survive to birth. In the case of trisomy 21, the child may live into adulthood.

Trisomy 21 (Down Syndrome)

An extra copy of chromosome 21, a condition called **trisomy 21,** or **Down syndrome,** occurs in about 1 of every 700 births, although this rate varies tremendously with the age of the parents (see below). Children with Down syndrome often show several distinctive physical characteristics, including weak muscle tone, a small mouth held partially open because it cannot accommodate the tongue, and distinctively shaped eyes (**FIG. 10-13**). More serious problems include varying degrees of mental impairment, low resistance to infectious diseases, and heart defects.

The frequency of nondisjunction increases with the age of the parents, especially the mother. Down syndrome occurs in only about 0.05% of children born to 20-year-old women, but in more than 3% of children born to women over 45 years of age. Nondisjunction in sperm accounts for about 10% of the cases of Down syndrome, and there is a small increase with increasing age of the father. Trisomy 21 can be diagnosed before birth by examining the chromosomes of fetal cells and, with less certainty, by biochemical tests and ultrasound examination of the fetus (see "Health Watch: Prenatal Genetic Screening" in Chapter 14).

CHECK YOUR LEARNING
Can you ...
- explain how nondisjunction causes offspring to have too many or too few chromosomes?
- describe some of the human genetic disorders that are caused by nondisjunction?

CASE STUDY \ REVISITED

The Rainbow Connection

Many people are astounded by the diversity of the Giddings children. Basic biology, however, easily explains how such diversity arises. Most genes have multiple alleles, meiotic cell division separates homologous chromosomes—and the alleles they carry—into different sperm and eggs, and the sperm and eggs unite at random. From a biological perspective, perhaps the more interesting question is this: Why do alleles for dark pigmentation occur most frequently in people whose ancestors

lived in equatorial regions, and alleles for pale pigmentation in people of northern European ancestry?

Natural selection probably favored different skin colors because of the differing amount of sunlight in equatorial versus northern regions and the importance of vitamin D and vitamin B_9 (folate) in human health. Vitamin D is needed for many physiological functions, including the absorption of calcium and other minerals by the digestive tract. Folate is also essential for many bodily functions. Folate deficiency can cause anemia and other disorders in adults and serious nervous system abnormalities in developing fetuses.

Ultraviolet rays in sunlight stimulate the synthesis of vitamin D, but they break down folate. In the fierce sunlight of equatorial regions, dark skin still allows for plenty of vitamin D production, while protecting against too much depletion of folate. In northern Europe, with far weaker sunlight and often cloudy skies, paler skin boosts vitamin D production, while folate levels remain adequate.

The selective advantage of blond hair in northern Europe is more uncertain. Some of the same genes contribute to hair and skin color, so selection for pale skin may have selected for pale hair as well. Another hypothesis is that the first few people with blond hair were very conspicuous in a population of otherwise dark-haired people. Novel appearance, within limits, is often attractive to members of the opposite sex. Some anthropologists have speculated that, a few thousand years ago, high-status men (proficient hunters or chieftains of small tribes, for example) preferentially chose blond-haired women as mates. Therefore, blond women produced more offspring than dark-haired women did. The result is that more than half the people in parts of Scandinavia have blond hair.

CONSIDER THIS Ultraviolet rays in sunlight cause skin cancer. In today's world, people of all skin colors, but especially pale-skinned people, are often urged to stay out of the sun and get their vitamin D from food or supplements. In the past, do you think that the risk of skin cancer selected against pale-skinned people, partially counterbalancing selection in favor of pale skin for vitamin D production?

CHAPTER REVIEW

 Go to **MasteringBiology** for practice quizzes, activities, eText, videos, current events, and more.

Answers to Think Critically, Evaluate This, Multiple Choice, and Fill-in-the-Blank questions can be found in the Answers section at the back of the book.

Summary of Key Concepts

10.1 How Does Sexual Reproduction Produce Genetic Variability?

Eukaryotic cells typically contain pairs of chromosomes, called homologues, that carry the same genes with similar, although usually not identical, nucleotide sequences. These slightly different nucleotide sequences of a gene are called alleles. Cells containing paired homologous chromosomes are called diploid. Cells with only a single copy of each type of chromosome are called haploid. Cells with three or more copies of each type of chromosome are called polyploid.

10.2 How Does Meiotic Cell Division Produce Genetically Variable, Haploid Cells?

Meiotic cell division (meiosis followed by cytokinesis) separates homologous chromosomes and produces haploid cells with only one homologue from each pair. During interphase before meiosis, chromosomes are duplicated. The cell then undergoes two specialized divisions—meiosis I and meiosis II—to produce four haploid daughter cells (see Fig. 10-5).

Meiosis I

During prophase I, homologous duplicated chromosomes, each consisting of two chromatids, pair up and exchange parts by crossing over. During metaphase I, homologues move together as pairs to the cell's equator, one member of each pair facing opposite poles of the cell. Homologous chromosomes separate during anaphase I, and two nuclei form during telophase I. Cytokinesis also usually occurs during telophase I. Each daughter nucleus receives only one member of each pair of homologues and, therefore, is haploid. The sister chromatids of each chromosome remain attached to each other throughout meiosis I.

Meiosis II

Meiosis II resembles mitosis in a haploid cell. The duplicated chromosomes move to the cell's equator during metaphase II. The two chromatids of each chromosome separate and move to opposite poles of the cell during anaphase II. This second division produces four haploid nuclei. Cytokinesis normally occurs during or shortly after telophase II, producing four haploid cells.

10.3 How Do Meiosis and Union of Gametes Produce Genetically Variable Offspring?

The random shuffling of homologous maternal and paternal chromosomes during meiosis I creates new chromosome combinations. Crossing over creates chromosomes with allele combinations that may never have occurred before on single chromosomes. Because of the separation of homologues and crossing over, a parent probably never produces any gametes that are completely identical. The fusion of two genetically unique gametes adds further genetic variability to the offspring.

10.4 When Do Mitotic and Meiotic Cell Division Occur in the Life Cycles of Eukaryotes?

Most eukaryotic life cycles have three parts: (1) Sexual reproduction combines haploid gametes to form a diploid cell. (2) At some point in the life cycle, diploid cells undergo meiotic cell division to produce haploid cells. (3) Mitosis of a haploid cell, a diploid cell, or both, results in the growth of multicellular bodies. When these stages occur, and what proportion of the life cycle is occupied by each stage, varies greatly among different species.

10.5 How Do Errors in Meiosis Cause Human Genetic Disorders?

Errors in meiosis can result in gametes with abnormal numbers of sex chromosomes or autosomes. Many people with abnormal numbers of sex chromosomes have distinguishing physical characteristics and some have difficulty reproducing. Abnormal numbers of autosomes typically lead to spontaneous abortion early in pregnancy. In rare instances, the fetus may survive to birth, but mental or physical deficiencies always occur. The likelihood of abnormal numbers of chromosomes increases with increasing age of the mother and, to a lesser extent, the father.

Key Terms

allele *157*	meiosis *159*
autosome *158*	meiosis I *159*
chiasma (plural,	meiosis II *159*
chiasmata) *161*	meiotic cell
crossing over *161*	division *159*
diploid *158*	mutation *157*
Down syndrome *171*	nondisjunction *169*
gamete *157*	polyploid *158*
gene *157*	recombination *166*
haploid *158*	sex chromosome *158*
homologous chromosome *157*	sexual
homologue *157*	reproduction *157*
Jacob syndrome *170*	trisomy 21 *171*
karyotype *157*	trisomy X *170*
Klinefelter syndrome *170*	Turner syndrome *170*
locus (plural, loci) *157*	zygote *167*

Thinking Through the Concepts

Multiple Choice

1. Pairs of chromosomes with almost identical genetic composition and that are found in cells of both males and females are called
   a. sex chromosomes.
   b. autosomes.
   c. polyploid.
   d. chromatids.

2. A cell with three or more copies of each homologous chromosome is called
   a. a gamete.
   b. haploid.
   c. trisomy X.
   d. polyploid.

3. During crossing over,
   a. chromatids of homologous chromosomes exchange parts.
   b. mutations occur with higher than average frequency.
   c. chromatids of nonhomologous chromosomes exchange parts.
   d. nondisjunction occurs.

4. Which of the following does *not* contribute to genetic variability?
   a. accurate replication of DNA
   b. crossing over
   c. random alignment of homologous chromosomes during metaphase I of meiosis
   d. union of sperm and egg

5. Haploid nuclei are first formed at what stage of meiosis?
   a. metaphase I
   b. telophase I
   c. metaphase II
   d. telophase II

Fill-in-the-Blank

1. Meiotic cell division produces _____ (how many) haploid daughter cells from each diploid parental cell. In animals, the haploid daughter cells produced by meiotic cell division become _____.

2. During _____ of meiosis I, homologous chromosomes form structures called _____. These structures are the sites of what event? _____

3. Three processes that promote genetic variability of offspring during sexual reproduction are _____, _____, and _____.

4. Plants have a life cycle called _____. In this type of life cycle, some cells in the multicellular diploid stage undergo _____ cell division to form spores, which divide by _____ cell division to form a multicellular haploid stage.

5. Women with _____ syndrome have a single X chromosome. They typically _____ (do/do not) undergo puberty and _____ (can/cannot) bear children. Men with _____ syndrome typically have reduced male secondary sexual characteristics. Their sex chromosomes are _____ (list the number of X and Y chromosomes).

Review Questions

1. Diagram the events of meiosis. At which stage do homologous chromosomes separate?

2. Describe crossing over. At which stage of meiosis does it occur? Name two functions of chiasmata.

3. In what ways are mitosis and meiosis similar? In what ways are they different?

4. Diagram and describe the three main types of eukaryotic life cycles. When do meiotic cell division and mitotic cell division occur in each?

5. Describe how meiosis provides for genetic variability. If an animal had a haploid number of two (no sex chromosomes), how many genetically different gametes could it produce? (Assume no crossing over.) What if it had a haploid number of five?

6. Define *nondisjunction,* and describe common syndromes caused by nondisjunction of sex chromosomes and autosomes.

Applying the Concepts

1. Many plants can reproduce sexually or asexually. Strawberries, for example, can reproduce asexually by sending out horizontal stems called runners that root and form new plants, or they can reproduce sexually by flowering and producing fruit and seeds. Describe some advantages and disadvantages of each type of reproduction in wild plants. Include in your discussion the important aspects of the environments in which runners and seeds are likely to find themselves.

11

PATTERNS OF INHERITANCE

Sudden Death on the Court

FLO HYMAN, 6 feet, 5 inches tall, graceful and athletic, was probably the best woman volleyball player of her time. Captain of the American women's volleyball team that won the silver medal in the 1984 Olympics, Hyman later joined a professional Japanese squad. In 1986, she was taken out of a game for a short breather and died while sitting quietly on the bench. Hyman was only 31 years old. How could this happen to someone so young and fit?

Hyman had a rare genetic disorder called Marfan syndrome. People with Marfan syndrome are typically tall and slender, with long limbs and large hands and feet. For some people with Marfan syndrome, these characteristics contribute to fame and fortune. Unfortunately, Marfan syndrome can also be deadly.

Hyman died from a ruptured aorta, the massive artery that carries blood from the heart to most of the body. Why did Hyman's aorta burst? What does a weak aorta have in common with tallness and large hands? Marfan syndrome is caused by a mutation in the gene that encodes for a protein called fibrillin. Normal fibrillin forms long fibers that give strength and elasticity to tendons that attach muscles to bones, ligaments that fasten bones to other bones in joints, and the walls of arteries. Fibrillin also traps certain growth factors, preventing them from stimulating excessive cell division in cells that produce connective tissue, including bone, cartilage, ligaments, and tendons. Defective fibrillin cannot trap these growth factors, so the arms, legs, hands, and feet

Olympic volleyball silver medalist Flo Hyman was struck down by Marfan syndrome at the height of her career.

of people with Marfan syndrome tend to become unusually long. The combination of defective fibrillin and high concentrations of growth factors weakens bone, ligaments, tendons, and artery walls.

Diploid organisms, including people, generally have two copies of each gene, one on each homologous chromosome. One defective copy of the fibrillin gene is enough to cause Marfan syndrome. What does this tell us about the inheritance of Marfan syndrome? Are all inherited diseases caused by a single defective copy of a gene? To find out, we must go back in time and visit the garden of Gregor Mendel.

11.1 WHAT IS THE PHYSICAL BASIS OF INHERITANCE?

Inheritance is the process by which the traits of organisms are passed to their offspring. We will begin our exploration of inheritance with a brief review of the structures that form its physical basis. In this chapter, we will confine our discussion to diploid organisms, including most plants and animals, that reproduce sexually by the fusion of haploid gametes.

Genes Are Sequences of Nucleotides at Specific Locations on Chromosomes

A chromosome consists of a double helix of DNA, packaged with a variety of proteins (see Figs. 9-1 and 9-6). Segments of DNA ranging from a few hundred to many thousands of nucleotides in length are the units of inheritance—the **genes**—that encode the information needed to produce proteins, cells, and entire organisms. Therefore, genes are parts of chromosomes (**FIG. 11-1**). A gene's physical location on a chromosome is called its **locus** (plural, loci). The chromosomes of diploid organisms occur in pairs called homologues. Both members of a pair of homologues carry the same genes, located at the same loci. However, the nucleotide sequences of a given gene may differ in different members of a species, or even on the two homologues of a single individual. These different versions of a gene at a given locus are called **alleles** (see Fig. 11-1). To understand the relationship between genes and alleles, it may be helpful to think of genes as very long sentences, written in an alphabet of nucleotides instead of letters. The alleles of a gene are like slightly different spellings of individual words in different copies of the same nucleotide sentence.

Mutations Are the Source of Alleles

The alleles on your chromosomes were almost all inherited from your parents. But where did these alleles come from in the first place? All alleles originally arose as **mutations**—changes in the sequence of nucleotides in the DNA of a gene. If a mutation occurs in a cell that becomes a sperm or egg, it can be passed on from parent to offspring. Most of the alleles in an organism's DNA first appeared as mutations in the reproductive cells of the organism's ancestors, perhaps hundreds or even millions of years ago, and have been inherited, generation after generation, ever since. A few alleles, which we will call "new mutations," may have occurred in the reproductive cells of the organism's own parents, but this is rare.

An Organism's Two Alleles May Be the Same or Different

Because a diploid organism has pairs of homologous chromosomes, and both members of a pair contain the same gene loci, the organism has two copies of

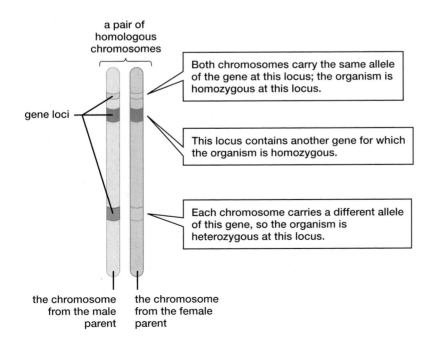

▲ **FIGURE 11-1 The relationships among genes, alleles, and chromosomes** Each homologous chromosome carries the same set of genes. Each gene is located at the same position, or locus, on its chromosome. Differences in nucleotide sequences at the same gene locus produce different alleles of the gene. Diploid organisms have two alleles of each gene, one on each homologue. The alleles on the two homologues may be the same or different.

each gene. If both homologues have the same allele at a given gene locus, the organism is said to be **homozygous** at that locus. (*Homozygous* comes from Greek words meaning "same pair.") The chromosomes shown in Figure 11-1 are homozygous at two loci. If two homologous chromosomes have different alleles at a locus, the organism is **heterozygous** ("different pair") at that locus. The chromosomes in Figure 11-1 are heterozygous at one locus.

CHECK YOUR LEARNING

Can you ...

- describe the relationships among chromosomes, DNA, genes, mutations, and alleles?
- explain what it means for an organism to be heterozygous or homozygous for a gene?

11.2 HOW WERE THE PRINCIPLES OF INHERITANCE DISCOVERED?

In the mid-1800s, experiments by an Austrian monk, Gregor Mendel (**FIG. 11-2**), revealed many important principles of inheritance. Although Mendel worked long before DNA, chromosomes, or meiosis had been discovered, his research revealed essential facts about genes and alleles and how they are inherited during sexual reproduction. Because his experiments are elegant examples of science in action, let's follow Mendel's paths of discovery.

Doing It Right: The Secrets of Mendel's Success

There are three key steps to any successful experiment in biology: choosing a suitable "system" to work on (the system could be as diverse as an enzyme, a metabolic pathway, an organism, or an ecosystem), designing and performing the experiment correctly, and analyzing the data properly. Mendel was the first geneticist to complete all three steps.

Mendel chose the edible pea for his experiments (**FIG. 11-3**). The male reproductive structures of a flower,

◀ **FIGURE 11-2 Gregor Mendel**

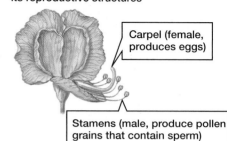

> Carpel (female, produces eggs)

> Stamens (male, produce pollen grains that contain sperm)

▲ **FIGURE 11-3 Flowers of the edible pea** In the intact pea flower (left), the lower petals enclose the reproductive structures—the stamens (male) and carpel (female). Pollen normally cannot enter the flower from outside, so peas usually self-pollinate and, hence, self-fertilize. If the flower is opened (right), it can be cross-pollinated by hand.

called stamens, produce pollen. Each pollen grain contains sperm. Pollination allows a sperm to fertilize an egg, which is located within the ovary of the flower's female reproductive structure, called the carpel. In pea flowers, the petals enclose all of the reproductive structures, preventing another flower's pollen from entering. Therefore, the eggs in a pea flower must be fertilized by sperm from the pollen of the same flower. When an organism's sperm fertilize its own eggs, the process is called **self-fertilization.**

Mendel, however, often wanted to mate two different pea plants to see what characteristics their offspring would inherit. To do this, he opened a pea flower and removed its stamens, preventing self-fertilization. Then he dusted the sticky tip of the carpel with pollen from the flower of another plant. When sperm from one organism fertilize eggs from a different organism, the process is called **cross-fertilization.**

Mendel's experimental design was simple, but brilliant. He studied traits with unmistakably different forms, such as white versus purple flowers. He also began by studying only one trait at a time. Earlier researchers had generally tried to study inheritance by simultaneously considering all of the features of entire organisms, including traits that differed only slightly among organisms. Not surprisingly, the investigators were often confused rather than enlightened.

To help interpret his results, Mendel followed the inheritance of traits for several generations, counting the numbers of offspring with each type of trait. When he analyzed these numbers, the basic patterns of inheritance became clear. Today, quantifying experimental results and applying statistical analysis are essential tools in virtually every field of biology. In Mendel's time, numerical analysis was an innovation.

CHECK YOUR LEARNING

Can you ...

- distinguish between self-fertilization and cross-fertilization?
- explain the important features of Mendel's experimental design?

11.3 HOW ARE SINGLE TRAITS INHERITED?

True-breeding organisms possess a trait, such as purple flowers, that is inherited unchanged by all offspring produced by self-fertilization. In his first set of experiments, Mendel cross-fertilized pea plants that were true-breeding for different forms of a single trait. The offspring of parents that differ in at least one genetically determined trait are called **hybrids.** To determine the traits of the offspring, Mendel saved the hybrid seeds and grew them the following year.

In one of these experiments, Mendel cross-fertilized true-breeding, white-flowered plants with true-breeding, purple-flowered plants. This was the parental generation, denoted by the letter P. When he grew the hybrid seeds, he found that all the first-generation offspring (the "first filial," or F_1 generation) produced purple flowers (**FIG. 11-4**). What had happened to the white color? The flowers of the F_1 hybrids were just as purple as their true-breeding purple parent. The white color of their true-breeding white parent seemed to have disappeared.

Mendel then allowed the F_1 flowers to self-fertilize, collected the seeds, and planted them the next spring. In the second (F_2) generation, Mendel counted 705 plants with purple flowers and 224 plants with white flowers. These numbers are approximately three-fourths purple flowers and one-fourth white flowers, or a ratio of about 3 purple to 1 white (**FIG. 11-5**). This result showed that the capacity to produce white flowers had not disappeared in the F_1 hybrids, but had only been hidden.

Mendel allowed the F_2 plants to self-fertilize and produce a third (F_3) generation. He found that all the white-flowered F_2 plants produced white-flowered offspring; that is, they were true-breeding. In contrast, when purple-flowered F_2 plants self-fertilized, their offspring were of two types. About one-third were true-breeding for purple, but the other two-thirds were hybrids that produced both purple- and white-flowered offspring, again in the ratio of 3 purple to 1 white. Therefore, the F_2 generation included one-quarter true-breeding white plants, one-quarter true-breeding purple, and one-half hybrid purple.

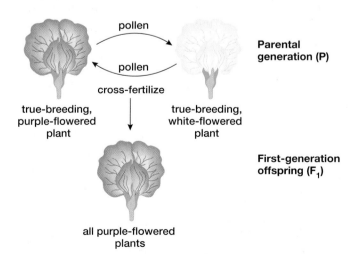

First-generation offspring (F_1)

▲ **FIGURE 11-4 Cross of pea plants true-breeding for white or purple flowers** All of the offspring bear purple flowers.

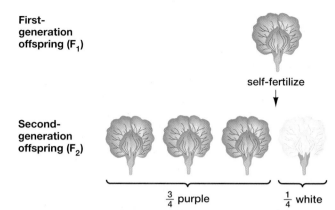

▲ **FIGURE 11-5 Self-fertilization of F_1 pea plants with purple flowers** Three-quarters of the offspring bear purple flowers and one-quarter bear white flowers.

The Inheritance of Dominant and Recessive Alleles on Homologous Chromosomes Explains the Results of Mendel's Crosses

Mendel's results, supplemented by modern knowledge of genes and chromosomes, allow us to develop a five-part hypothesis to explain the inheritance of single traits:

- Each trait is determined by pairs of discrete physical units called genes. Each organism has two alleles for each gene, one on each homologous chromosome. True-breeding, white-flowered peas have different alleles of the flower-color gene than true-breeding, purple-flowered peas do.
- True-breeding organisms have two copies of the same allele for a given gene and are therefore homozygous for that gene. All of the gametes from a homozygous individual receive the same allele for that gene (**FIG. 11-6a**). Hybrid

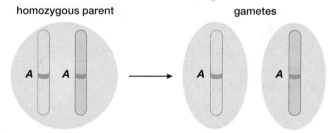

(a) Gametes produced by a homozygous parent

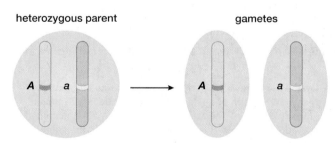

(b) Gametes produced by a heterozygous parent

▲ **FIGURE 11-6 The distribution of alleles in gametes (a)** All of the gametes produced by homozygous organisms contain the same allele. **(b)** Half of the gametes produced by heterozygous organisms contain one allele, and half of the gametes contain the other allele.

organisms have two different alleles for a given gene and so are heterozygous for that gene. Half of a heterozygote's gametes will contain one allele for that gene and half will contain the other allele (**FIG. 11-6b**).

- When two different alleles are present in an organism, one—the **dominant** allele—may mask the expression of the other—the **recessive** allele. The recessive allele, however, is still present. In the edible pea, the allele for purple flowers is dominant, and the allele for white flowers is recessive.
- Homologous chromosomes separate, or segregate, from each other during meiosis, thus separating the alleles they carry. This is known as Mendel's **law of segregation:** Each gamete receives only one allele of each pair of genes. When a sperm fertilizes an egg, the resulting offspring receives one allele from the father (in his sperm) and one from the mother (in her egg).
- Because homologous chromosomes separate randomly during meiosis, the distribution of alleles into the gametes is also random.

Let's see how this hypothesis explains the results of Mendel's experiments with flower color (**FIG. 11-7**). We will use letters to represent the different alleles, assigning the uppercase letter P to the dominant allele for purple flower color and the lowercase letter p to the recessive allele for white flower color. A homozygous purple-flowered plant has two alleles for purple flower color (PP); a homozygous white-flowered plant has two alleles for white flower color (pp). Therefore, all the sperm and eggs produced by a PP plant carry the P allele, and all the sperm and eggs of a pp plant carry the p allele (**FIG. 11-7a**).

The cross-fertilized F_1 offspring were produced when P sperm fertilized p eggs or when p sperm fertilized P eggs. In both cases, the F_1 offspring were Pp. Because P is dominant over p, all of the offspring were purple (**FIG. 11-7b**).

For the F_2 generation, Mendel allowed the heterozygous F_1 plants to self-fertilize. A heterozygous plant produces equal numbers of P and p sperm and equal numbers of P and p eggs. When a Pp plant self-fertilizes, each type of sperm has an equal chance of fertilizing each type of egg (**FIG. 11-7c**). Therefore, the F_2 generation contained three types of offspring: PP, Pp, and pp. The three types occurred in the approximate proportions of one-quarter PP (homozygous purple), one-half Pp (heterozygous purple), and one-quarter pp (homozygous white).

Two organisms that look alike may actually have different combinations of alleles. The combination of alleles carried

▶ **FIGURE 11-7 Segregation of alleles and fusion of gametes predict the distribution of alleles and traits in the inheritance of flower color in peas (a)** The parental generation: All of the gametes of homozygous PP parents contain the P allele; all of the gametes of homozygous pp parents contain the p allele. **(b)** The F_1 generation: Fusion of gametes containing the P allele with gametes containing the p allele produces only Pp offspring. (Note that Pp is the same genotype as pP.) **(c)** The F_2 generation: Half of the gametes of heterozygous Pp parents contain the P allele and half contain the p allele. Fusion of these gametes produces PP, Pp, and pp offspring.

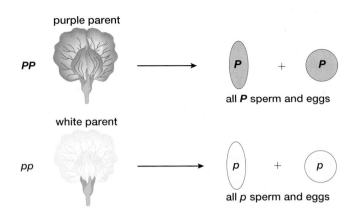

(a) Gametes produced by homozygous parents

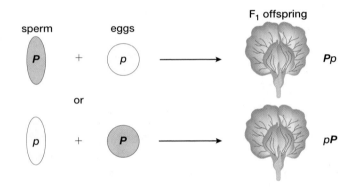

(b) Fusion of gametes produces F₁ offspring

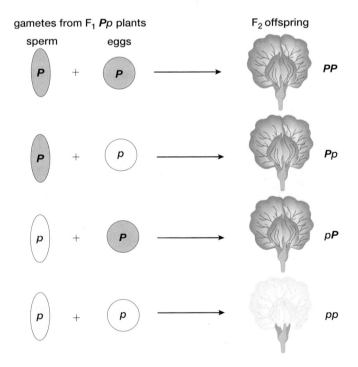

(c) Fusion of gametes from the F₁ generation produces F₂ offspring

by an organism (for example, *PP* or *Pp*) is its **genotype.** The organism's traits, including its outward appearance, behavior, digestive enzymes, blood type, or any other observable or measurable feature, make up its **phenotype.** As we have seen, plants with either the *PP* or the *Pp* genotype have the phenotype of purple flowers. Therefore, the F_2 generation of Mendel's peas consisted of three genotypes (one-quarter *PP*, one-half *Pp*, and one-quarter *pp*), but only two phenotypes (three-quarters purple and one-quarter white).

"Genetic Bookkeeping" Can Predict Genotypes and Phenotypes of Offspring

The **Punnett square method,** named after R. C. Punnett, a famous geneticist of the early 1900s, is a convenient way to predict the genotypes and phenotypes of offspring. **FIGURE 11-8a** shows how to use a Punnett square to determine the expected proportions of offspring that arise from breeding two organisms that are heterozygous for a single trait. **FIGURE 11-8b** shows how to calculate the proportions of offspring using the probabilities that each type of sperm will fertilize each type of egg.

▶ **FIGURE 11-8 Determining the outcome of a single-trait cross**
(a) The Punnett square allows you to predict both genotypes and phenotypes of specific crosses; here we use it for a cross between pea plants that are heterozygous for a single trait—flower color.

1. Assign letters to the different alleles; use uppercase for dominant alleles and lowercase for recessive alleles.

2. Determine all the types of genetically different gametes that can be produced by the male and female parents.

3. Draw the Punnett square, with the columns labeled with all possible genotypes of the eggs and the rows labeled with all possible genotypes of the sperm. (We also show the fractions of each genotype.)

4. Fill in the genotype of the offspring in each box by combining the genotype of the sperm in its row with the genotype of the egg in its column. (Multiply the fraction of sperm of each type in the row headers by the fraction of eggs of each type in the column headers.)

5. Count the number of offspring with each genotype. Note that *Pp* is the same genotype as *pP*.

6. Convert the number of offspring of each genotype to a fraction of the total number of offspring. In this example, out of four fertilizations, only one is predicted to produce the *pp* genotype, so one-quarter of the total number of offspring produced by this cross is predicted to be white. To determine phenotypic fractions, add the fractions of genotypes that would produce a given phenotype. For example, purple flowers are produced by $\frac{1}{4}$ *PP* + $\frac{1}{4}$ *Pp* + $\frac{1}{4}$ *pP*, for a total of three-quarters of the offspring.

(b) Probabilities may also be used to predict the outcome of a single-trait cross. Determine the fractions of eggs and sperm of each genotype and multiply these fractions together to calculate the fraction of offspring of each genotype. When two genotypes produce the same phenotype (e.g., *Pp* and *pP*), add the fractions of each genotype to determine the phenotypic fraction.

THINK CRITICALLY If you crossed a heterozygous *Pp* plant with a homozygous recessive *pp* plant, what would be the expected ratio of offspring? How does this differ from the offspring of a *PP* × *pp* cross? Try working this out before you read further in the text.

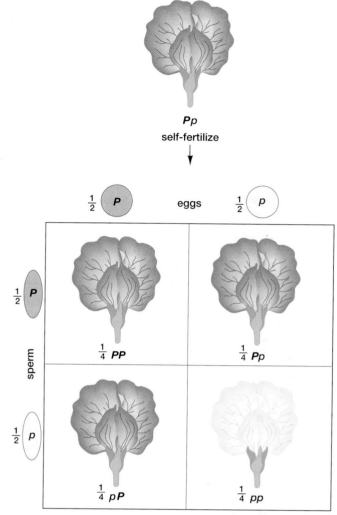

(a) Punnett square of a single-trait cross

sperm	eggs	offspring genotypes	genotypic ratio (1:2:1)	phenotypic ratio (3:1)
$\frac{1}{2}$ **P** ×	$\frac{1}{2}$ **P** =	$\frac{1}{4}$ **PP**	$\frac{1}{4}$ **PP**	
$\frac{1}{2}$ **P** ×	$\frac{1}{2}$ **p** =	$\frac{1}{4}$ **Pp**		
			$\frac{1}{2}$ **Pp**	$\frac{3}{4}$ purple
$\frac{1}{2}$ **p** ×	$\frac{1}{2}$ **P** =	$\frac{1}{4}$ **pP**		
$\frac{1}{2}$ **p** ×	$\frac{1}{2}$ **p** =	$\frac{1}{4}$ **pp**	$\frac{1}{4}$ **pp**	$\frac{1}{4}$ white

(b) Using probabilities to determine the offspring of a single-trait cross

As you use these genetic bookkeeping techniques, keep in mind that in a real experiment, the actual offspring will not occur in exactly the predicted proportions. Why not? Let's consider a familiar example. Each time a baby is conceived, it has an equal chance of being a boy or a girl. However, many families with two children do not have one girl and one boy. The 1:1 ratio of girls to boys occurs only if we average the sexes of the children in many families.

Mendel's Hypothesis Can Be Used to Predict the Outcome of New Types of Single-Trait Crosses

You have probably recognized that Mendel used the scientific method: He made an observation and used it to formulate a hypothesis. But does Mendel's hypothesis accurately predict the results of further experiments? Based on the hypothesis that heterozygous F_1 plants have one allele for purple flowers and one for white (that is, they have the Pp genotype), Mendel predicted the outcome of cross-fertilizing Pp plants with

homozygous recessive white plants (pp): There should be equal numbers of Pp (purple) and pp (white) offspring. This is indeed what he found.

This type of experiment has practical uses for breeders of domestic plants and animals, who may want to know if an organism with a desirable, dominant trait will pass that trait on to all of its offspring or only to some of them. Cross-fertilization of an organism with a dominant phenotype (in this case, a purple flower) but an unknown genotype with a homozygous recessive organism (a white flower) is called a **test cross,** because it tests whether the organism with the dominant phenotype is homozygous or heterozygous (**FIG. 11-9**). When crossed with a homozygous recessive (pp), a homozygous dominant (PP) produces all phenotypically dominant offspring, whereas a heterozygous dominant (Pp) yields offspring with both dominant and recessive phenotypes in a 1:1 ratio.

CHECK YOUR LEARNING

Can you ...

- describe the pattern of inheritance of a trait controlled by a single gene with two alleles, one dominant and one recessive?
- distinguish between genotype and phenotype?
- calculate the proportions of offspring with each genotype and phenotype that would be produced by mating parents with various combinations of the two alleles?

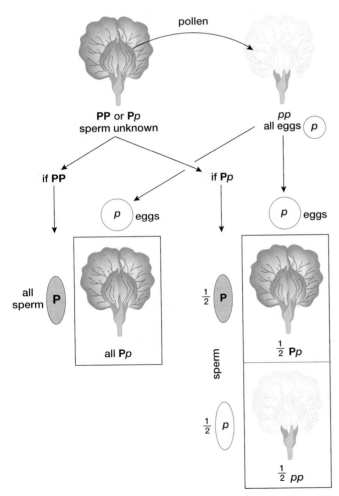

▲ **FIGURE 11-9 Punnett square of a test cross** An organism with a dominant phenotype may be either homozygous or heterozygous. Crossing such an organism with a homozygous recessive organism can determine whether the dominant organism is homozygous (left) or heterozygous (right).

CASE STUDY **CONTINUED**

Sudden Death on the Court

Many traits, in humans and other organisms, are inherited in a simple Mendelian fashion. Marfan syndrome, for example, is inherited as a dominant trait, which means that a single defective fibrillin allele is enough to cause the disorder. Flo Hyman inherited her defective allele from her father. Are all genetically determined traits inherited according to the straightforward patterns worked out by Gregor Mendel? We'll return to this question in Section 11.5.

11.4 HOW ARE MULTIPLE TRAITS INHERITED?

Mendel turned next to the inheritance of multiple traits (**FIG. 11-10**). He cross-fertilized plants that differed in two traits—for example, seed color (yellow or green) and seed shape (smooth or wrinkled). From earlier crosses of plants with these traits, Mendel already knew that the smooth allele of the seed shape gene (S) is dominant to the wrinkled allele (s) and that the yellow allele of the seed color gene (Y) is dominant to the green allele (y). He crossed a true-breeding plant that produced smooth, yellow seeds ($SSYY$) with a true-breeding plant that produced wrinkled, green seeds

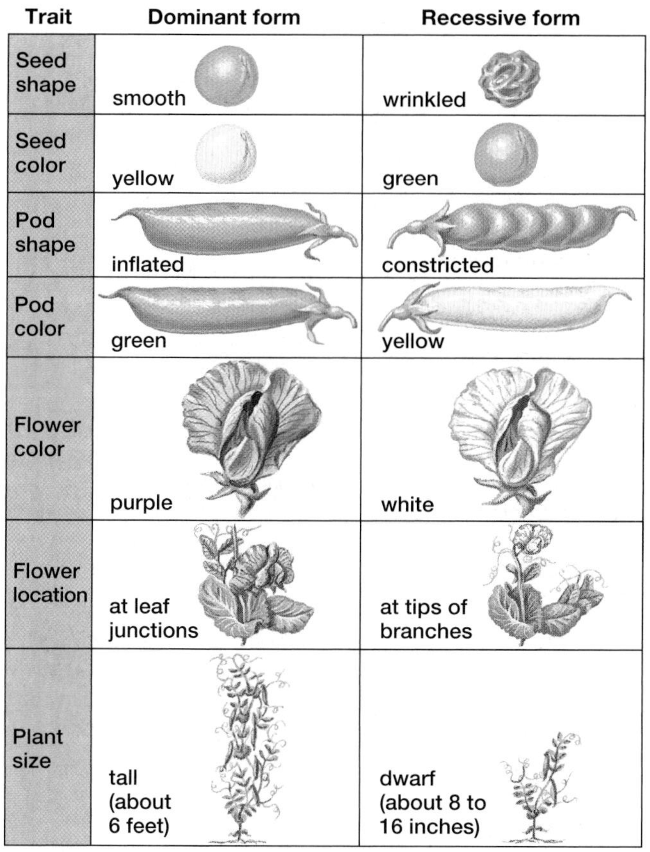

Trait	Dominant form	Recessive form
Seed shape	smooth	wrinkled
Seed color	yellow	green
Pod shape	inflated	constricted
Pod color	green	yellow
Flower color	purple	white
Flower location	at leaf junctions	at tips of branches
Plant size	tall (about 6 feet)	dwarf (about 8 to 16 inches)

▲ **FIGURE 11-10** **Traits of pea plants studied by Gregor Mendel**

(*ssyy*). The *SSYY* plant can produce only *SY* gametes, and the *ssyy* plant can produce only *sy* gametes. Therefore, all the F₁ offspring were heterozygotes: genotypically *SsYy* with the phenotype of smooth, yellow seeds.

Mendel allowed these heterozygous F₁ plants to self-fertilize. The F₂ generation consisted of 315 plants with smooth, yellow seeds; 101 with wrinkled, yellow seeds; 108 with smooth, green seeds; and 32 with wrinkled, green seeds—a ratio of about 9:3:3:1. The offspring produced from other crosses of plants that were heterozygous for two traits also had phenotypic ratios of about 9:3:3:1.

Mendel Hypothesized That Traits Are Inherited Independently

Mendel realized that these results could be explained if the genes for seed color and seed shape were inherited independently of each other and did not influence each other during gamete formation. If this hypothesis is correct, then for each trait, three-quarters of the offspring should show the dominant phenotype and one-quarter should show the recessive phenotype. This result is just what Mendel observed. He found 423 plants with smooth seeds (of either color) and 133 with wrinkled seeds (of either color), a ratio of about 3:1; 416 plants produced yellow seeds (of either shape) and 140 produced green seeds (of either shape), also about a 3:1 ratio. **FIGURE 11-11** shows how a Punnett square or probability

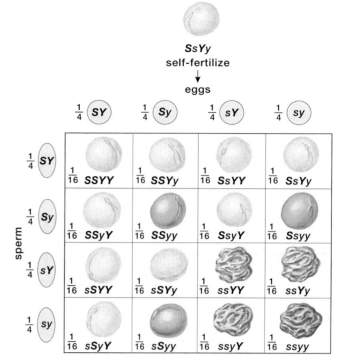

(a) Punnett square of a two-trait cross

seed shape	seed color	phenotypic ratio (9:3:3:1)
$\frac{3}{4}$ smooth ×	$\frac{3}{4}$ yellow =	$\frac{9}{16}$ smooth yellow
$\frac{3}{4}$ smooth ×	$\frac{1}{4}$ green =	$\frac{3}{16}$ smooth green
$\frac{1}{4}$ wrinkled ×	$\frac{3}{4}$ yellow =	$\frac{3}{16}$ wrinkled yellow
$\frac{1}{4}$ wrinkled ×	$\frac{1}{4}$ green =	$\frac{1}{16}$ wrinkled green

(b) Using probabilities to determine the offspring of a two-trait cross

▲ **FIGURE 11-11** **Predicting genotypes and phenotypes for a cross between parents that are heterozygous for two traits** In pea seeds, yellow color (*Y*) is dominant to green (*y*), and smooth shape (*S*) is dominant to wrinkled (*s*). **(a)** In this cross, an individual heterozygous for both traits (*SsYy*) self-fertilizes. In a cross involving two independent genes, there will be equal numbers of gametes with all of the possible combinations of alleles of the two genes—*SY*, *Sy*, *sY*, and *sy*. Place these gamete combinations as the labels for the rows and columns in the Punnett square and then calculate the offspring as explained in Figure 11-8. Note that the Punnett square predicts both the frequencies of combinations of traits ($\frac{9}{16}$ smooth, yellow; $\frac{3}{16}$ smooth, green; $\frac{3}{16}$ wrinkled, yellow; and $\frac{1}{16}$ wrinkled, green) and the frequencies of individual traits ($\frac{3}{4}$ yellow, $\frac{1}{4}$ green, $\frac{3}{4}$ smooth, and $\frac{1}{4}$ wrinkled). **(b)** The probability of two independent events is the product (multiplication) of their individual probabilities. For example, to find the probability of tossing two coins and having both come up heads, multiply the probabilities of each coin coming up heads ($\frac{1}{2} \times \frac{1}{2} = \frac{1}{4}$). Seed shape is independent of seed color. Therefore, multiplying the individual probabilities of the genotypes or phenotypes for each trait produces the predicted frequencies for the combined genotypes or phenotypes of the offspring. These frequencies are identical to those generated by the Punnett square.

THINK CRITICALLY Can the genotype of a plant bearing smooth, yellow seeds be revealed by a test cross with a plant bearing wrinkled, green seeds?

▶**FIGURE 11-12 Independent assortment of alleles** Chromosome movements during meiosis produce independent assortment of alleles, shown here for two genes. Each combination of alleles is equally likely to occur, producing gametes in the predicted proportions $\frac{1}{4}$ SY, $\frac{1}{4}$ sy, $\frac{1}{4}$ Sy, and $\frac{1}{4}$ sY.

THINK CRITICALLY If the genes for seed color and seed shape were on the same chromosome rather than on different chromosomes, would their alleles assort independently? Why or why not?

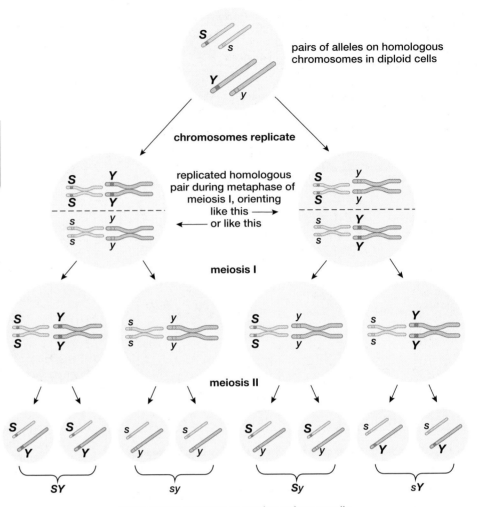

pairs of alleles on homologous chromosomes in diploid cells

chromosomes replicate

replicated homologous pair during metaphase of meiosis I, orienting like this ⟶ ⟵ or like this

meiosis I

meiosis II

SY   sy   Sy   sY

independent assortment produces four equally likely allele combinations during meiosis

calculation can be used to estimate the proportions of genotypes and phenotypes of the offspring of a cross between organisms that are heterozygous for two traits.

The independent inheritance of two or more traits is called the **law of independent assortment.** Multiple traits are inherited independently if the alleles of the gene controlling any given trait are distributed to gametes independently of the alleles for the genes controlling all the other traits. Independent assortment will occur when the traits being studied are controlled by genes on different pairs of homologous chromosomes. Why? During meiosis, paired homologous chromosomes line up at metaphase I. Which homologue faces which pole of the cell is random, and the orientation of one homologous pair does not influence other pairs (see Chapter 10). Therefore, when the homologues separate during anaphase I, which homologue of pair 1 moves "north" does not affect which homologue of pair 2 moves "north," and so on. The result is that the alleles of genes on different chromosomes are distributed, or assorted, independently of one another (**FIG. 11-12**).

CHECK YOUR LEARNING

Can you ...

- describe the pattern of simultaneous inheritance of two traits if each of the traits is controlled by a separate gene with only two alleles, one dominant and one recessive?
- explain the law of independent assortment?
- calculate the frequencies of the genotypes and phenotypes of the offspring that would be produced by mating organisms with various combinations of the two alleles of each gene, assuming independent assortment of the two genes?

11.5 DO THE MENDELIAN RULES OF INHERITANCE APPLY TO ALL TRAITS?

In our discussion thus far, we have assumed that each trait is completely controlled by a single gene, that there are only two possible alleles of each gene, and that one allele is completely dominant to the other. Most traits, however, are influenced in more varied and subtle ways.

In Incomplete Dominance, the Phenotype of Heterozygotes Is Intermediate Between the Phenotypes of the Homozygotes

When one allele is completely dominant over a second allele, heterozygotes with one dominant allele have the same phenotype as homozygotes with two dominant alleles (see Figs. 11-8 and 11-9). However, in some cases the heterozygous phenotype is intermediate between the two homozygous phenotypes, a pattern of inheritance called **incomplete dominance.** For example, the golden palomino is regarded as one of the most beautifully colored

▶ **FIGURE 11-13 Incomplete dominance**
The inheritance of palomino coat color in horses is an example of incomplete dominance. Palominos are heterozygotes with one chestnut allele (C_1) and one cremello allele (C_2). Foals produced by breeding palominos may have chestnut, palomino, or cremello coat colors, in the approximate ratio of $\frac{1}{4}$ chestnut: $\frac{1}{2}$ palomino: $\frac{1}{4}$ cremello.

THINK CRITICALLY What is the only breeding combination that will ensure a palomino foal?

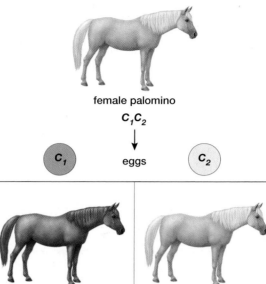

female palomino
C_1C_2

eggs

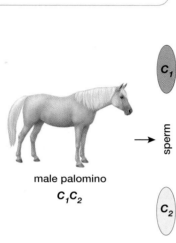

male palomino
C_1C_2

sperm

chestnut
C_1C_1

palomino
C_1C_2

palomino
C_1C_2

cremello
C_2C_2

horses. Palominos are heterozygous for two incompletely dominant alleles we will call chestnut (C_1) and cremello (C_2). Horses with reddish-brown chestnut coats are homozygous for the C_1 allele, and cremellos, with pale creamy coats, are homozygous for the C_2 allele. Because palominos are heterozygotes (C_1C_2), they do not breed true; a cross between palominos can produce chestnut, palomino, or cremello foals, with probabilities of one-quarter chestnut (C_1C_1), one-half palomino (C_1C_2), and one-quarter cremello (C_2C_2; **FIG. 11-13**).

A Single Gene May Have Multiple Alleles

Recall that alleles originate as mutations, which may then be inherited from generation to generation. Over thousands of generations and millions of organisms of a given species, many different mutations may occur in the same gene, resulting in multiple alleles of the gene. Although an individual organism can have at most two different alleles of a gene (one on each of two homologous chromosomes), if we examined the genes of all the members of a species, we might find dozens, even hundreds, of different alleles for some genes. Which of these alleles an offspring inherits, of course, depends on which alleles were present in its parents.

Human blood types are a familiar example of multiple alleles of a single gene. The blood types A, B, AB, and O arise

as a result of three different alleles of a gene (we will designate the alleles *A*, *B*, and *o*). This gene codes for an enzyme that adds sugar molecules to the ends of glycoproteins that protrude from the surfaces of red blood cells. Alleles *A* and *B* code for enzymes that add different sugars to the glycoproteins (we'll call the resulting molecules type A and type B glycoproteins, respectively). Allele *o* codes for a nonfunctional enzyme that doesn't add any sugar molecules.

A person may have one of six genotypes: *AA, BB, AB, Ao, Bo,* or *oo.* Alleles *A* and *B* are dominant to *o.* Therefore, people with genotypes *AA* or *Ao* make only type A glycoproteins and have type A blood. Those with genotypes *BB* or *Bo* synthesize only type B glycoproteins and have type B blood. Homozygous recessive *oo* individuals lack both types of glycoproteins and have type O blood. In people with type AB blood, both enzymes are present, so their red blood cells have both A and B glycoproteins. When a heterozygote expresses the phenotypes of both of the homozygotes (in this case, both A and B glycoproteins), the pattern of inheritance is called **codominance,** and the alleles are said to be codominant to one another.

The fact that people have different blood types affects the safety of blood transfusions. The human immune system produces proteins called antibodies, which bind to complex molecules that are not produced by a person's own body (if they did bind to "self" molecules, your immune system would destroy the cells of your body). In their usual role in defending against disease, antibodies bind to molecules on the surfaces of invading bacteria or viruses and help to destroy them. However, certain antibodies complicate blood transfusions. These antibodies will bind to "foreign" glycoproteins on red blood cells—that is, glycoproteins bearing sugars that are different from the sugars on a person's own red blood cells. If people are given transfusions of the wrong blood type, their antibodies bind to the foreign glycoproteins, which causes the red blood cells in the transfused blood to clump together and rupture. The resulting clumps and fragments can clog small blood vessels and damage vital organs such as the brain, heart, lungs, or kidneys.

TABLE 11-1	Human Blood Group Characteristics					
Blood Type	Genotype	Red Blood Cells	Has Plasma Antibodies to:	Can Receive Blood from:	Can Donate Blood to:	Frequency in the U.S.
A	AA or Ao	A glycoprotein	B glycoprotein	A or O (no blood with B glycoprotein)	A or AB	42%
B	BB or Bo	B glycoprotein	A glycoprotein	B or O (no blood with A glycoprotein)	B or AB	10%
AB	AB	Both A and B glycoproteins	Neither A nor B glycoprotein	AB, A, B, O (universal recipient)	AB	4%
O	oo	Neither A nor B glycoprotein	Both A and B glycoproteins	O (no blood with A or B glycoprotein)	O, AB, A, B (universal donor)	44%

Therefore, blood type must be carefully matched before a blood transfusion.

TABLE 11-1 summarizes human blood types and safe transfusions. Obviously, a person can donate blood to anyone with the same blood type. In addition, type O blood, with red blood cells that lack any sugars, can be safely transfused to all other blood types, because type O red blood cells are not attacked by the antibodies found in A, B, or AB blood. (The antibodies in the donor's blood become too diluted by the much larger volume of the recipient's blood to cause problems.) People with type O blood are called "universal donors." But type O blood contains antibodies to both A and B glycoproteins, so type O individuals can receive transfusions only of type O blood. Type AB blood doesn't contain antibodies against any type of red blood cells, so a person with type AB blood can receive blood from people with any other blood type; thus, they are called "universal recipients."

Single Genes Typically Have Multiple Effects on Phenotype

Single genes often have multiple phenotypic effects, a phenomenon called **pleiotropy**. For example, a mutation in a single gene in a lab mouse produced a nude mouse (**FIG. 11-14**). Researchers rapidly discovered that nude mice not only are hairless but also lack a thymus gland and have virtually no immune response, and females do not develop functional mammary glands, so they can't nurse their pups.

▲ **FIGURE 11-14 Nude mice**

CASE STUDY CONTINUED

Sudden Death on the Court

In Marfan syndrome, a single defective fibrillin allele causes increased height, long limbs, large hands and feet, weak walls in the aorta, and often dislocated lenses in one or both eyes—a striking example of pleiotropy in humans. However, the types and severity of symptoms vary, even among family members who carry the same defective fibrillin allele. This variability suggests that environmental factors or the actions of other genes may affect the Marfan phenotype. Are most traits significantly influenced by the environment and by the alleles of other genes that an individual inherits?

▲ **FIGURE 11-15 Skin color in humans** Polygenic inheritance and variable amounts of suntan produce a continuous gradation of skin colors.

Many Traits Are Influenced by Several Genes

Your class probably contains people of varied heights, skin colors, and body builds—variation that cannot be divided into convenient, easily defined phenotypes. Traits such as these are influenced by interactions among two or more genes, a process called **polygenic inheritance.** As you might imagine, the more genes that contribute to a single trait, the greater the number of possible phenotypes and the finer the gradations among them.

For example, human skin color is affected by at least ten different genes (**FIG. 11-15**). Some genes have extremely large effects: People who are homozygous for a recessive allele of one particular gene lack pigmentation in skin, eyes, and hair (see Section 11.8). Other genes have small effects, with various alleles causing slightly darker or slightly lighter skin. At least 400 genes contribute to human height; not surprisingly, variation in height is continuous, with no discrete increments.

The Environment Influences the Expression of Genes

An organism is not just the sum of its genes. In addition to its genotype, the environment in which an organism lives also profoundly influences its phenotype. Fur color in Siamese cats vividly illustrates environmental effects on gene action. All Siamese cats are born with pale fur, but within the first few weeks, the ears, nose, paws, and tail turn dark (**FIG. 11-16**). One of a Siamese cat's genes codes for an enzyme that produces dark

HAVE YOU EVER WONDERED ...

Why Dogs Vary So Much in Size?

Dogs evolved from wolves. Although all wolves are about the same size, dogs vary in size more than any other mammal—from huge Great Danes and Irish wolfhounds to minuscule toy breeds such as Chihuahuas and Pomeranians. Researchers have identified six genes that account for most of the size difference between breeds. Toy breeds are usually homozygous for "small" alleles of most of these genes. All known wolves, along with most large dogs such as Danes and wolfhounds, are homozygous for the "large" alleles of all six. Medium-sized dogs tend to be heterozygous for about half the genes. These patterns suggest that polygenic inheritance with incomplete dominance between two or more alleles of each gene controls size in dogs. Why do only dogs, and not wolves, have small alleles? Small alleles could arise as mutations in dogs or wolves. However, once the mutations occurred, people who preferred small dogs selectively bred small dogs to one another, often keeping the smallest of each litter, and thereby unwittingly selected for the small alleles of these genes. Human protection prevented natural selection from weeding out the small alleles. In contrast, small alleles that might arise in wolves are quickly eliminated by natural selection—just imagine the fate of a Chihuahua-sized wolf in the wild!

◀ **FIGURE 11-16 Environmental influence on phenotype** The distribution of dark fur in the Siamese cat is an interaction between genotype and environment, producing a particular phenotype. Newborn Siamese kittens have pale fur everywhere on their bodies. In an adult Siamese, the allele for dark fur is expressed only in the cooler areas (nose, ears, paws, and tail).

fur. This enzyme is synthesized in pigment cells everywhere on the cat's body. So why aren't Siamese cats completely black? Because the enzyme that produces dark pigment is inactive at temperatures above about 93°F (34°C). While inside their mother's uterus, unborn kittens are warm all over, so newborn Siamese kittens have pale fur on their entire bodies. After they are born, the ears, nose, paws, and tail become cooler than the rest of the body, so dark pigment is produced in those areas.

Most environmental influences are more complicated and subtle than this. For example, exposure to sunlight significantly affects skin color. When combined with complex polygenic inheritance, the result is virtually continuous variation in phenotype (see Fig. 11-15). Human height is strongly influenced by nutrition, which not only contributes to a continuously variable phenotype, but also has caused average heights to change profoundly over time: In many countries, average height increased by about 4 inches over the last 150 years, as improved nutrition allowed more people to achieve their full genetic potential.

CHECK YOUR LEARNING

Can you ...

- describe the patterns of inheritance of traits showing incomplete dominance, codominance, and multiple alleles?
- explain how polygenic inheritance and environmental influences combine to produce nearly continuous variation in many phenotypes?

11.6 HOW ARE GENES LOCATED ON THE SAME CHROMOSOME INHERITED?

Every chromosome contains many genes, up to several thousand in a really large chromosome. This fact has important implications for inheritance.

Genes on the Same Chromosome Tend to Be Inherited Together

Chromosomes, not individual genes, assort independently during meiosis I. Therefore, genes located on *different chromosomes* assort independently into gametes. In contrast, genes on the *same chromosome* tend to be inherited together, a phenomenon called **gene linkage.** One of the first pairs of linked genes to be discovered was found in the sweet pea, a different species from Mendel's edible pea. In sweet peas, the gene for flower color (purple versus red) and the gene for pollen grain shape (round versus long) are carried on the same chromosome (**FIG. 11-17**). Thus, the alleles for these genes usually assort together into gametes during meiosis and are inherited together.

Consider a heterozygous sweet pea plant with purple flowers and long pollen. Let's assume that the dominant purple allele of the flower-color gene and the dominant long allele of the pollen-shape gene are located on one homologous chromosome (Fig. 11-17, top) and that the recessive red allele of the flower-color gene and the recessive round allele

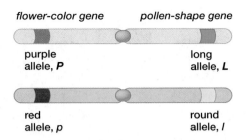

▲ **FIGURE 11-17 Linked genes on homologous chromosomes in the sweet pea** The genes for flower color and pollen shape are on the same chromosome, so they tend to be inherited together.

of the pollen-shape gene are located on the other homologue (Fig. 11-17, bottom). Therefore, the gametes produced by this plant are likely to have either purple and long alleles or red and round alleles. This pattern of inheritance does not conform to the law of independent assortment because the alleles for flower color and pollen shape do not segregate independently of one another, but tend to stay together during meiosis.

Crossing Over Creates New Combinations of Linked Alleles

However, genes on the same chromosome do not always stay together. If you cross-fertilized two sweet peas with the chromosomes shown in Figure 11-17, you might expect that all of the offspring would have either purple flowers with long pollen grains or red flowers with round pollen grains. (Try working this out with a Punnett square.) In reality, you would usually find a few offspring with purple flowers and round pollen and a few with red flowers and long pollen, as if, sometimes, the genes for flower color and pollen shape became unlinked. How can this happen?

During prophase I of meiosis, homologous chromosomes sometimes exchange parts, a process called crossing over (see Chapter 10, Fig. 10-8). In most chromosomes, at least one exchange between each homologous pair occurs during meiotic cell division. The exchange of corresponding segments of DNA during crossing over produces **genetic recombination:** new combinations of alleles of the genes that are located on homologous chromosomes. Then, when homologues separate at anaphase I, the haploid daughter cells will receive chromosomes with different sets of alleles than the chromosomes of the parent cell had.

Let's look at the sweet pea chromosomes during meiosis. During prophase I, the duplicated, homologous chromosomes pair up (**FIG. 11-18a**). Each homologue will have one or more regions where crossing over occurs. Imagine that crossing over exchanges the alleles for flower color between nonsister chromatids of the two homologues (**FIG. 11-18b**). At anaphase I, the separated homologues will each now have one chromatid bearing a piece of DNA from a chromatid of the other homologue (**FIG. 11-18c**). During meiosis II, four types of chromosomes will be distributed, one to each of the four daughter cells: two unchanged chromosomes and two recombined chromosomes (**FIG. 11-18d**).

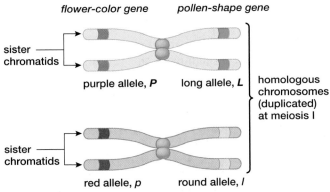

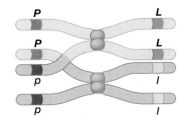

(a) Duplicated chromosomes in prophase of meiosis I

(b) Crossing over during prophase I

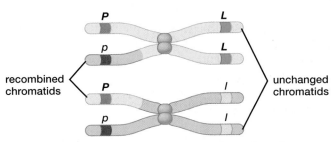

(c) Homologous chromosomes separate at anaphase I

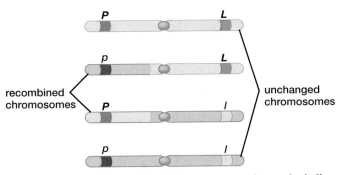

(d) Unchanged and recombined chromosomes after meiosis II

▲ **FIGURE 11-18 Crossing over recombines alleles on homologous chromosomes (a)** During prophase of meiosis I, duplicated homologous chromosomes pair up. **(b)** Nonsister chromatids of the two homologues exchange parts by crossing over. **(c)** When the homologous chromosomes separate during anaphase of meiosis I, one chromatid of each of the homologues now contains a piece of DNA from a chromatid of the other homologue. **(d)** After meiosis II, two of the haploid daughter cells receive unchanged chromosomes, and two receive recombined chromosomes. The recombined chromosomes contain allele arrangements that did not occur in the original parental chromosomes.

Therefore, some gametes will be produced with each of four configurations: *PL* and *pl* (the same configurations as on the original parental chromosomes) and *Pl* and *pL* (new configurations on the recombined chromosomes). If a sperm with a *Pl* chromosome fertilizes an egg with a *pl* chromosome, the offspring plant will have purple flowers (*Pp*) and round pollen (*ll*). If a sperm with a *pL* chromosome fertilizes an egg with a *pl* chromosome, then the offspring will have red flowers (*pp*) and long pollen (*Ll*).

The farther apart the genes are on a chromosome, the more likely it is that crossing over will occur between them. Think of a pair of homologous chromosomes as two long strings, each with a red stripe at one end, a blue stripe very close to the red one, and a yellow stripe at the opposite end. If you throw the strings on the floor so that one lands on top of the other, the strings will almost always cross between the blue and yellow stripes, but will very seldom cross between the red and blue stripes. Similarly, two genes close together on a chromosome are strongly linked and will rarely be separated by a crossover. However, if two genes are very far apart, crossing over between the genes occurs so often that they seem to be independently assorted, just as if they were on different chromosomes. When Gregor Mendel discovered independent assortment, he was not only clever and careful, he was also lucky. The seven traits that he studied were controlled by genes on only four different chromosomes. He observed independent assortment because the genes that were on the same chromosomes were far apart.

CHECK YOUR LEARNING

Can you ...

- describe how the patterns of inheritance differ between traits controlled by genes on a single chromosome and traits controlled by genes on different chromosomes?

11.7 HOW ARE SEX AND SEX-LINKED TRAITS INHERITED?

In many animals, an individual's sex is determined by its **sex chromosomes.** In mammals, females have two identical sex chromosomes, called **X chromosomes,** whereas males have one X chromosome and one **Y chromosome** (**FIG. 11-19**). Despite their huge differences in size and genetic composition, the X and Y chromosomes act like homologues: They pair up during prophase of meiosis I and separate during anaphase I. The other chromosomes, which occur in homologous pairs with identical appearance in males and females, are called **autosomes.**

In Mammals, the Sex of an Offspring Is Determined by the Sex Chromosome in the Sperm

During sperm formation, the sex chromosomes segregate, and each sperm receives either an X or a Y chromosome (plus one member of each pair of autosomes). The sex

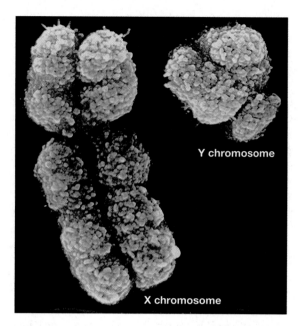

▲ **FIGURE 11-19 Human sex chromosomes** The Y chromosome (right), which carries relatively few genes, is much smaller than the X chromosome (left).

Image courtesy of Indigo® Instruments: http://www.indigo.com.

chromosomes also segregate during egg formation, but because females have two X chromosomes, every egg receives one X chromosome (and one member of each pair of autosomes). Thus, a male offspring is produced if an egg is fertilized by a Y-bearing sperm, and a female offspring is produced if an egg is fertilized by an X-bearing sperm (**FIG. 11-20**).

▶ **FIGURE 11-20 Sex determination in mammals** Male offspring receive their Y chromosome from their father; female offspring receive the father's X chromosome (labeled X_m). Both male and female offspring receive an X chromosome (either X_1 or X_2) from their mother.

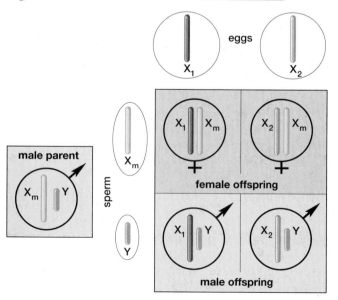

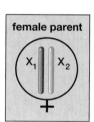

Sex-Linked Genes Are Found Only on the X or Only on the Y Chromosome

Genes that are located only on sex chromosomes are referred to as **sex-linked.** In mammals, the Y chromosome carries relatively few genes. The human Y chromosome contains several dozen genes, many of which play a role in male reproduction. The most well-known Y-linked gene is the sex-determining gene, called *SRY*. During embryonic life, the action of *SRY* sets in motion the entire male developmental pathway. Under normal conditions, *SRY* causes the male sex to be 100% linked to the Y chromosome.

In contrast to the small Y chromosome, the human X chromosome contains more than 1,000 genes, most of which have no counterpart on the Y chromosome. Most of the genes on the X chromosome determine traits that are important in both sexes, such as color vision, blood clotting abilities, and the presence of specific structural proteins in muscles. Because they have two X chromosomes, females can be either homozygous or heterozygous for genes on the X chromosome, and dominant versus recessive relationships among alleles will be expressed. Males, in contrast, fully express all the alleles they have on their single X chromosome, regardless of whether those alleles would be dominant or recessive in females.

Let's look at a familiar example: red-green color deficiency, more commonly—though usually incorrectly—called color blindness (**FIG. 11-21**). Color deficiency is caused by recessive alleles of either of two genes located on the X chromosome. The normal, dominant alleles of these genes (we will call them both *C*) encode proteins that allow one set of color-vision cells in the eye, called cones, to be most sensitive to red light and another set to be most sensitive to green light. There are several defective recessive alleles of these genes (we will call them all *c*). Certain extremely defective alleles encode proteins that make both sets of cones equally sensitive to red and green light. Therefore, the affected person cannot distinguish red from green and is truly red-green color-blind. The more common, moderately defective alleles, however, produce cones that respond differently to red and green light, just not as differently as normal red and green cones do. Men with these moderately defective alleles are color-deficient: Fire engines still look red and grass still looks green, but many "reddish" or "greenish" colors cannot be distinguished from one another (**FIG. 11-21a**).

How is color deficiency inherited? A man can have the genotype *CY* or *cY*, meaning that he has a color-vision allele *C* or *c* on his X chromosome and no color-vision gene on his Y chromosome. He will have normal color vision if his X chromosome bears the *C* allele or be color-deficient if it bears the *c* allele. A woman may be *CC*, *Cc*, or *cc*. Women with *CC* or *Cc* genotypes will have normal color vision; only women with *cc* genotypes will be color-deficient. Roughly 7% of men have defective color vision. Among women, about 93% are homozygous normal *CC*, 7% are heterozygous normal *Cc*, and less than 0.5% are homozygous color-deficient *cc*.

(a) Normal color vision (left); simulation of red-green color deficiency (right)

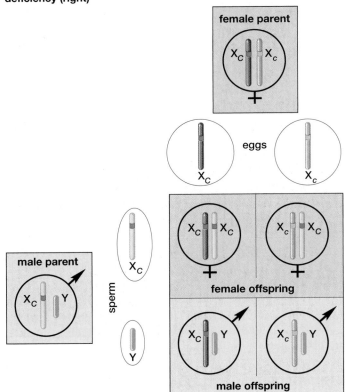

(b) Expected children of a man with normal color vision (CY), and a heterozygous woman (Cc)

◀ **FIGURE 11-21 Sex-linked inheritance of red-green color deficiency (a)** These photographs show people with normal color vision what the world looks like through the eyes of a person with red-green color deficiency. To one of the authors of this textbook (GA), the left and right photos of each pair look almost the same. **(b)** A Punnett square shows the inheritance of color deficiency from a heterozygous woman (*Cc*) to her sons.

A color-deficient man (*c*Y) can pass his defective *c* allele only to his daughters, because only his daughters inherit his X chromosome. Usually, however, his daughters will have normal color vision, because they also inherit a normal *C* allele from their mother, who is very likely homozygous normal *CC*. The sons of a heterozygous woman (*Cc*) have a 50% chance of inheriting her defective allele (**FIG. 11-21b**). Sons who receive the defective allele are color-deficient (*c*Y), whereas sons who inherit the functional allele have normal color vision (*C*Y).

CHECK YOUR LEARNING
Can you ...
- explain why sperm determine the sex of offspring in mammals?
- explain why most sex-linked traits are controlled by genes on the X chromosome?
- describe the pattern of inheritance of sex-linked traits?

11.8 HOW ARE HUMAN GENETIC DISORDERS INHERITED?

Many human diseases are influenced by genetics to a greater or lesser degree. Because experimental crosses with people are out of the question, human geneticists search medical, historical, and family records to study past crosses. Records extending across several generations can be arranged in the form of family **pedigrees,** diagrams that show the genetic relationships among a set of related individuals (**FIG. 11-22**).

Careful analysis of human pedigrees, combined with molecular genetic technology, has produced great strides in understanding human genetic diseases. For instance, geneticists now know the genes responsible for dozens of inherited diseases, including sickle-cell anemia, hemophilia, muscular dystrophy, Marfan syndrome, and cystic fibrosis. Research in molecular genetics has increased our ability to predict genetic diseases and in some cases even to cure them (see Chapter 14).

Disorders arising from abnormal numbers of chromosomes, which are caused by errors in meiosis, were discussed in Chapter 10. Here, we will focus on disorders caused by defective alleles of a single gene. However, just as common traits such as height and skin color are often influenced by several genes (see Section 11.5), multiple genes, interacting with complex environmental factors, may predispose people to develop health problems such as Parkinson's and Alzheimer's diseases, cancer, and schizophrenia.

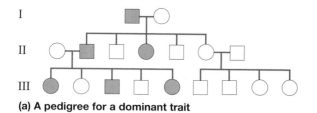

(a) A pedigree for a dominant trait

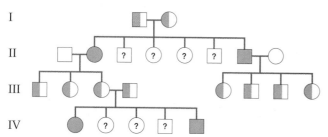

(b) A pedigree for a recessive trait

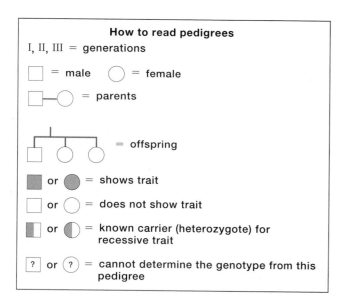

▲ **FIGURE 11-22 Family pedigrees (a)** A pedigree for a dominant trait. Note that any offspring showing a dominant trait must have at least one parent with the trait. **(b)** A pedigree for a recessive trait. Any individual showing a recessive trait must be homozygous recessive. If that person's parents did not show the trait, then both parents must be heterozygotes (carriers). Note that the genotype cannot be determined for some offspring, who may be either carriers or homozygous dominants.

Some Human Genetic Disorders Are Caused by Recessive Alleles

The human body depends on the actions of thousands of enzymes and other proteins. A mutation in an allele of the gene coding for one of these proteins can impair or destroy its function. However, the presence of one normal allele may generate enough functional protein to enable heterozygotes to have the same phenotype as homozygotes with two normal alleles. In these cases, a mutant allele encoding a nonfunctional

protein is recessive to a normal allele encoding a functional protein, and an abnormal phenotype occurs only in people who inherit two copies of the mutant allele.

A **carrier** for a genetic disorder is a person who is heterozygous, with one normal, dominant allele and one defective, recessive allele. Carriers are phenotypically healthy but can pass on defective alleles to their offspring. In all likelihood, we all carry some recessive alleles that would cause serious genetic disorders in homozygotes. Every time we have a child, there is a 50:50 chance that we will pass on the defective allele. This is usually harmless, because an unrelated man and woman will usually have defective alleles of different genes, and their children will develop a genetic disorder only if they are homozygous for a defective allele of the same gene. Related couples, however (especially first cousins or closer), have inherited some of their genes from recent common ancestors and so are more likely to carry a defective allele of the same gene. If a man and woman are both heterozygous for a defective recessive allele of the same gene, they have a 1 in 4 chance of having a child with the genetic disorder (see Fig. 11-22).

Albinism Results from a Defect in Melanin Production

An enzyme called tyrosinase is needed to produce melanin—the dark pigment in skin, hair, and the iris of the eye. Normal melanin production will occur if a person has either one or two functional tyrosinase alleles. However, if a person is homozygous for an allele that encodes defective tyrosinase, **albinism** occurs (**FIG. 11-23**). Albinism in humans and other mammals results in very pale skin and hair.

Sickle-Cell Anemia Is Caused by a Defective Allele for Hemoglobin Synthesis

Red blood cells are packed with hemoglobin proteins, which transport oxygen and give the cells their red color. Anemia is

(a) Human **(b) Wallaby**

▲ **FIGURE 11-23 Albinism (a)** Albinism occurs in most vertebrates, including people. This boy's irises are extremely pale, so his eyes are very sensitive to bright light. **(b)** The albino wallaby in the foreground is safe in a zoo, but in the wild, its bright white fur would make it very conspicuous to predators.

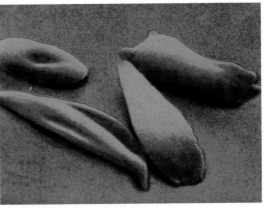

◄ **FIGURE 11-24 Sickle-cell anemia (a)** Normal red blood cells are disk shaped with indented centers. **(b)** When blood oxygen is low, the red blood cells in a person with sickle-cell anemia become long, slender, and curved, resembling a sickle.

(a) Normal red blood cells **(b) Sickled red blood cells**

a generic term given to a number of diseases, all characterized by a low red blood cell count or below-normal hemoglobin in the blood. **Sickle-cell anemia** is an inherited form of anemia that results from a mutation in the hemoglobin gene. A change in a single nucleotide places an incorrect amino acid at a crucial position in the hemoglobin protein (see Section 13.4 in Chapter 13). When people with sickle-cell anemia exercise or move to high altitude, oxygen concentrations in their blood drop, and the sickle-cell hemoglobin proteins inside their red blood cells stick together. The resulting clumps of hemoglobin force red blood cells out of their usual flexible, disk shapes (**FIG. 11-24a**) into long, stiff sickle shapes (**FIG. 11-24b**). The sickled cells are fragile and easily damaged. Anemia occurs because the sickled red blood cells are destroyed before their usual life span is completed.

The sickle shape also causes other complications. Sickle cells jam up in capillaries, causing blood clots. Tissues downstream of the clot do not receive enough oxygen. Paralyzing strokes can result if blocks occur in blood vessels in the brain.

People homozygous for the sickle-cell allele synthesize only defective hemoglobin. Consequently, many of their red blood cells become sickled, and they suffer from sickle-cell anemia. Although heterozygotes produce about half normal and half abnormal hemoglobin, they have very few sickled red blood cells and seldom show any symptoms. Because only people who are homozygous for the sickle-cell allele typically show any symptoms, sickle-cell anemia is usually considered to be a recessive disorder. However, during exceptionally strenuous exercise, some heterozygotes may experience life-threatening complications, as we explore in "Health Watch: The Sickle-Cell Allele and Athletics."

About 5% to 25% of sub-Saharan Africans and 8% of African Americans are heterozygous for sickle-cell anemia, but the allele is very rare in Caucasians. Why? Shouldn't natural selection work to eliminate the sickle-cell allele in both African and Caucasian populations? The difference arises because heterozygotes have some resistance to the parasite that causes malaria, which is common in Africa and other places with warm, humid climates, but not in colder

regions such as most of Europe. This "heterozygote advantage" explains the higher prevalence of the sickle-cell allele in people of African origin.

Some Human Genetic Disorders Are Caused by Incompletely Dominant Alleles

In some cases, the amount of functional protein produced by one normal allele is not enough to compensate for a defective allele, so the defective allele is incompletely dominant to the normal allele. For example, incomplete dominance explains the variable severity of familial hypercholesteremia, a disease in which an affected person cannot clear low-density lipoprotein (LDL, the "bad" cholesterol) from the bloodstream. The resulting high cholesterol levels cause hardening of the arteries. People who are homozygous for the defective allele have extremely high cholesterol levels and develop heart disease at a very young age, often suffering serious heart attacks in childhood. Male heterozygotes usually have heart attacks in their 40s or 50s, female heterozygotes about a decade later.

Some Human Genetic Disorders Are Caused by Dominant Alleles

Some serious genetic disorders, such as Huntington disease, are caused by dominant alleles. Just as a pea plant needs only one dominant allele for purple color to bear purple flowers (see Figs. 11-7 and 11-8), so too a person needs to have only one defective dominant allele in order to suffer from these disorders. Therefore, everyone who inherits a dominant genetic disorder must have at least one parent with the disease (see Fig. 11-22a). In rare cases, a dominant allele that causes a genetic disorder may result not from an allele passed down generation after generation, but from a mutation in the egg or sperm of a parent who is otherwise unaffected. In this case, neither parent would have the disease.

How can a defective allele be dominant to the normal, functional allele? Some defective dominant alleles encode an abnormal protein that interferes with the function of the

Health WATCH

The Sickle-Cell Allele and Athletics

Sickle-cell anemia is considered to be a recessive trait because only homozygous recessive people usually show any symptoms. At the molecular level, though, half the hemoglobin proteins in a heterozygote are defective. Does this really have no effect at all?

For the vast majority of heterozygotes (often described as having "sickle-cell trait"), there indeed are no health effects. However, a very small number of heterozygotes may experience serious medical problems during extreme exercise. Consider Devard and Devaughn Darling, identical twin brothers, who shared all their genes, including one copy of the sickle-cell allele.

The Darling brothers starred in multiple sports in high school. Both were probable starters for the Florida State University football team when the unthinkable happened one day during practice: Devaughn collapsed and died. No one could prove that Devaughn's death was caused by the combination of strenuous workouts and the sickle-cell trait, but suspicions ran high. The university decided that it didn't want to risk Devard suffering the same fate and barred Devard from playing football. Devard, however, transferred to Washington State University and played football for the Cougars for 2 years. He then played for five seasons in the National Football League (FIG. E11-1).

The Darling brothers epitomize the rare, but real, dilemmas facing athletes with sickle-cell trait. Devard's football career and the accomplishments of many other heterozygotes show that having sickle-cell trait does not preclude strenuous athletics. Although the National Collegiate Athletic Association requires sickle-cell screening of all Division I and II athletes, the Association agrees that "Student-athletes with sickle-cell trait should not be excluded from athletics participation." However, Devaughn's tragic death underscores the need to take appropriate precautions. Dehydration during extreme exercise, especially in hot weather, is probably the most important risk to heterozygotes, so the NCAA recommends that athletes "stay well hydrated at all times." These and other simple precautions have helped the U.S. Army to eliminate excess

▲ FIGURE E11-1 Devard Darling runs to daylight for the Kansas City Chiefs Devard's identical twin Devaughn died during football practice in college, probably from complications of sickle-cell trait.

deaths caused by sickle-cell trait during basic training. In fact, the Army no longer even screens for sickle-cell trait. Medically appropriate and humane training procedures—realizing, for example, that failing to "tough it out" in the face of serious physical distress is not a sign of mental weakness——help all athletes, not only those with sickle-cell trait.

EVALUATE THIS In January 2012, the Pittsburgh Steelers football team played against the Denver Broncos in the "Mile-High City" (Denver's altitude is a mile above sea level). Steelers head coach Mike Tomlin did not allow safety Ryan Clark to play, because Clark has sickle-cell trait. What can happen when someone with sickle-cell trait exercises at high elevation? Do you think Tomlin made the right call in benching Clark? Explain your reasoning.

normal one. Other dominant alleles may encode proteins that carry out new, toxic reactions. Still other dominant alleles may encode a protein that is overactive, performing its function at inappropriate times and places in the body.

Huntington Disease Is Caused by a Defective Protein That Kills Cells in Specific Brain Regions

Huntington disease is a dominant disorder that causes a slow, progressive deterioration of parts of the brain, resulting in loss of coordination, flailing movements, personality disturbances, and eventual death. The symptoms of Huntington disease typically do not appear until 30 to 50 years of age. Therefore, before they experience their first symptoms, many Huntington victims pass the allele to their children. Geneticists isolated the Huntington gene in 1993 and, a few years

later, identified the gene's product, a protein they named "huntingtin." Normal huntingtin affects gene transcription, cytoskeleton function, and the movement of organelles within brain cells. Mutant huntingtin is cut up into toxic fragments inside cells, ultimately killing them.

Some Human Genetic Disorders Are Sex-Linked

As we described earlier, the X chromosome contains many genes that have no counterpart on the Y chromosome. Because men have only one X chromosome, they have only one allele for each of these genes. Therefore, men show the phenotypes produced by these single alleles, even if the alleles are recessive and would be masked by dominant alleles in women.

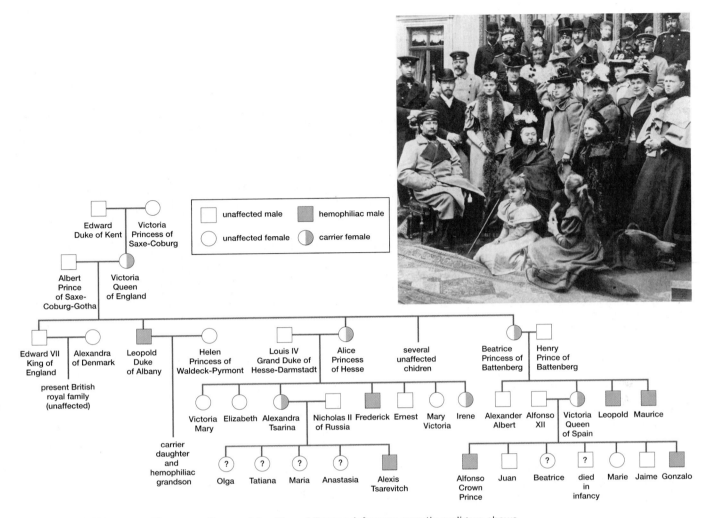

▲ **FIGURE 11-25 Hemophilia among the royal families of Europe** A famous genetic pedigree shows the transmission of sex-linked hemophilia from Queen Victoria of England (seated center front, with cane, in 1885) to her offspring and eventually to virtually every royal house in Europe, because of the extensive intermarriage of her children to the royalty of other European nations. Because Victoria's ancestors were free of hemophilia, the hemophilia allele must have arisen as a mutation either in Victoria herself or in one of her parents (or as a result of marital infidelity).

THINK CRITICALLY Why is it not possible that a mutation in Victoria's husband, Albert, was the original source of hemophilia in this family pedigree?

A son receives his X chromosome from his mother and passes it only to his daughters. Thus, X-linked disorders caused by recessive alleles have a unique pattern of inheritance. Such disorders appear far more frequently in males and typically skip generations: An affected male passes the trait to a phenotypically normal, carrier daughter, who in turn bears some affected sons. The most familiar genetic defects due to recessive alleles of X-chromosome genes are red-green color-vision deficiency (see Fig. 11-21), hemophilia, and muscular dystrophy.

Hemophilia is caused by a recessive allele on the X chromosome that results in a deficiency in one of the proteins needed for blood clotting. People with hemophilia bruise easily and may bleed extensively from minor injuries. They often have anemia due to blood loss. Nevertheless,

even before modern treatment with clotting factors, some hemophiliac males survived to pass on their defective allele to their daughters, who in turn could pass it to their sons (**FIG. 11-25**). We describe muscular dystrophy, a fatal degeneration of the muscles in young boys, in "Health Watch: Muscular Dystrophy."

CHECK YOUR LEARNING

Can you ...

- use pedigrees to determine the pattern of inheritance of a trait?
- describe why some genetic disorders might be dominant, incompletely dominant, or recessive, and give examples of each?

Health WATCH

Muscular Dystrophy

When weightlifter Tatiana Kashirina of Russia set a new world record in the "snatch" at the 2012 London Olympics, she lifted almost 333 pounds (151 kilograms), about 50% more than her own body weight (FIG. E11-2). How could her muscles withstand the stress? Muscle cells are firmly tied together by a very long protein called dystrophin. The almost 3,700 amino acids of dystrophin form a supple yet strong rod that connects the cytoskeleton inside a muscle cell to proteins in its plasma membrane, which in turn attach to supporting proteins in the extracellular matrix surrounding each muscle cell. When a muscle contracts, its cells remain intact because the forces are evenly distributed throughout each cell and to the extracellular matrix.

Unfortunately, about 1 in 3,500 boys makes faulty dystrophin proteins and suffers from **muscular dystrophy,** which literally means "degeneration of the muscles." Duchenne muscular dystrophy is the most devastating form of the disease; Becker muscular dystrophy is a less severe form. Muscular dystrophy may be caused by more than 1,000 different defective alleles of the dystrophin gene. The lack of functional dystrophin means that ordinary muscle contraction tears the muscle cells, which die and are replaced by fat and connective tissue (FIG. E11-3). By the age of 7 or 8, boys with Duchenne muscular dystrophy can no longer walk. Death usually occurs in the early 20s from heart and respiratory problems.

Girls almost never have Duchenne muscular dystrophy because the dystrophin gene is on the X chromosome, and muscular dystrophy alleles are recessive. Therefore, a boy will suffer muscular dystrophy if he has a defective dystrophin allele on his single X chromosome, but a girl, with two X chromosomes, would need two defective copies to suffer the disorder. This virtually never happens, because a girl would have to inherit one defective dystrophin allele from her mother, on one of her X chromosomes, and one from her father, on his X chromosome. Because they suffer early disability and death, boys with Duchenne muscular dystrophy almost never have children.

◄ FIGURE E11-2
Tatiana Kashirina sets a world record in the snatch.

If affected boys virtually never reproduce, shouldn't natural selection have almost completely eradicated defective dystrophin alleles? Actually, natural selection does rapidly eliminate these alleles. However, the dystrophin gene is enormous—about 2.4 *million* nucleotides long, compared to about 28 *thousand* nucleotides for the average human gene. Why does this matter? Remember, alleles arise as mutations in DNA. The longer the gene, the greater the chances for a mutation to occur: Because the dystrophin gene is almost a hundred times longer than the average gene, its mutation rate is also about a hundred times higher. As a result, about one-third of the boys with muscular dystrophy receive a new mutation that occurred in a reproductive cell of their mother, and two-thirds inherit a pre-existing mutation. The new mutations counterbalance natural selection, resulting in the steady incidence of about 1 in 3,500 boys.

Right now, there are no cures, although treatments are available that slow muscle degeneration, prolong life, and make the affected boys more comfortable. However, clinical trials have shown that a novel molecular technique can trick the muscles of about 13% of the boys with muscular dystrophy into making partially functional dystrophin from a faulty dystrophin allele. Perhaps most promising, studies in mice have found that utrophin, a different, naturally occurring muscle protein, may be able to partially substitute for dystrophin. In 2014, a small clinical trial showed that boys treated with an experimental drug that increases utrophin synthesis had less muscle damage than untreated boys did. If further trials confirm these results, this new drug may greatly improve the health and lifespan of all boys with muscular dystrophy.

▲ FIGURE E11-3 **The effects of muscular dystrophy** The micrograph on the left shows a normal muscle, with little space between the cells. A dystrophic muscle (right) has fewer and more irregular muscle cells, with spaces between the cells filled with fat and connective tissue.

EVALUATE THIS A mother of a young boy is devastated to find that her son has Duchenne muscular dystrophy. She takes a DNA test and discovers that she is a carrier for a defective dystrophin allele. If she decides to have another child, what is the likelihood that the second child will have the disorder? The woman has two sisters. What is the likelihood that they are also carriers?

Sudden Death on the Court

Marfan syndrome caused Flo Hyman's death, but it need not be fatal if detected in time. In 2014, Baylor University basketball star Isaiah Austin (**FIG. 11-26**) decided to play professional ball after his sophomore year in college. Luckily for Austin, the National Basketball Association extensively screens all players for health problems before they are eligible for the draft. NBA physicians diagnosed Austin with Marfan syndrome, and found that he has an enlarged aorta, probably with weak walls. If Austin had continued to play college basketball instead of trying to turn pro, he may well have suffered Flo Hyman's fate. Austin cannot play competitive sports; in fact, he should not exercise strenuously at all, because exercise increases blood pressure, which may put too much stress on his aorta and cause it to rupture. However, with careful monitoring and perhaps drugs to keep his blood pressure down, he should be able to live a normal life span.

CONSIDER THIS In some genetic disorders, including Duchenne muscular dystrophy, cystic fibrosis, sickle-cell anemia, and most cases of Marfan syndrome, defective alleles can be detected in both adults and embryos. If you and your spouse knew that you carried alleles for a serious genetic disorder, would you seek prenatal diagnosis of an embryo? What would you do if your embryo were destined to be born with Marfan syndrome? Duchenne muscular dystrophy?

◀ **FIGURE 11-26**

Isaiah Austin Because he has Marfan syndrome, the exertion and increased blood pressure of a slam dunk could have ruptured Austin's aorta.

CHAPTER REVIEW

Go to **MasteringBiology** for practice quizzes, activities, eText, videos, current events, and more.

Answers to Think Critically, Evaluate This, Multiple Choice, and Fill-in-the-Blank questions can be found in the Answers section at the back of the book.

Summary of Key Concepts

11.1 What Is the Physical Basis of Inheritance?

The units of inheritance are genes, which are segments of DNA found at specific locations (loci) on chromosomes. Genes may exist in two or more alternative forms, called alleles. When both homologous chromosomes carry the same allele at a given locus, the organism is homozygous for that gene. When the two homologous chromosomes have different alleles at a given locus, the organism is heterozygous for that gene.

11.2 How Were the Principles of Inheritance Discovered?

Gregor Mendel deduced many principles of inheritance in the mid-1800s, before the discovery of DNA, genes, chromosomes, or meiosis. He did this by choosing an appropriate experimental subject, designing his experiments carefully, following progeny for several generations, and analyzing his data statistically.

11.3 How Are Single Traits Inherited?

A trait is an observable or measurable feature of an organism's phenotype, such as flower color or blood type. Each parent provides its offspring with one allele of every gene, so the offspring inherits a pair of alleles for every gene. The combination of alleles in the offspring determines its phenotype. Dominant alleles mask the expression of recessive alleles. The masking of recessive alleles can result in organisms with the same phenotype but different genotypes. Organisms with two dominant alleles (homozygous dominant) have the same phenotype as do organisms with one dominant and one recessive allele (heterozygous). Because each allele segregates randomly during meiosis, we can predict the relative proportions of offspring with a particular trait, using Punnett squares or probability.

11.4 How Are Multiple Traits Inherited?

If the genes for two traits are located on separate chromosomes, their alleles assort independently of one another into the egg or sperm; that is, the distribution of alleles of one gene into the gametes does not affect the distribution of the alleles of the other gene. Thus, breeding two organisms that are heterozygous at two loci on separate chromosomes produces offspring with nine different genotypes. For typical dominant and recessive alleles, the offspring will display only four different phenotypes.

11.5 Do the Mendelian Rules of Inheritance Apply to All Traits?

Not all inheritance follows the simple dominant-recessive pattern. In incomplete dominance, heterozygotes have a phenotype that is intermediate between the two homozygous phenotypes. If we examine the genes of many members of a given species, we find that many genes have more than two alleles. Codominance results when two alleles of a single gene independently contribute to the observed phenotype. Pleiotropy occurs when a single gene has effects on several, seemingly unrelated, aspects of an organism's phenotype. In polygenic inheritance, several different genes contribute to the phenotype. The environment influences the phenotypic expression of virtually all traits.

11.6 How Are Genes Located on the Same Chromosome Inherited?

Genes on the same chromosome tend to be inherited together. However, crossing over will result in some recombination of alleles on each chromosome. Crossing over will occur more often the farther apart on a chromosome the genes are located.

11.7 How Are Sex and Sex-Linked Traits Inherited?

In many animals, sex is determined by sex chromosomes, often designated X and Y. In mammals, females have two X chromosomes; males have one X and one Y chromosome. Male sperm contain either an X or a Y chromosome, whereas a female's egg cells always have an X chromosome. Therefore, sex is determined by the sex chromosome in the sperm that fertilizes an egg.

Sex-linked genes are found on the X or Y chromosome. In mammals, the Y chromosome has many fewer genes than the X chromosome, so most sex-linked genes are found on the X chromosome. Because males have only one copy of X chromosome genes, recessive traits on the X chromosome are more likely to be phenotypically expressed in males.

11.8 How Are Human Genetic Disorders Inherited?

Molecular genetic techniques and analysis of family pedigrees are used to determine the mode of inheritance of human traits. Some genetic disorders are inherited as recessive traits; therefore, only homozygous recessive persons show symptoms of the disease. Heterozygotes are called carriers; they carry the recessive allele but do not express the trait. Some disorders are inherited as incompletely dominant traits. Heterozygotes, with only one defective allele, show some symptoms of the disorder, while people who are homozygous for the defective allele have a more severe disorder. Other disorders are inherited as simple dominant traits. In such cases, only one copy of the dominant allele is needed to cause full disease symptoms. Some human genetic disorders are sex-linked.

Key Terms

albinism   *190*
allele   *175*
autosome   *187*
carrier   *190*
codominance   *183*
cross-fertilization   *176*

dominant   *178*
gene   *175*
gene linkage   *186*
genetic recombination   *186*
genotype   *179*
hemophilia   *193*
heterozygous   *176*
homozygous   *176*
Huntington disease   *192*
hybrid   *177*
incomplete dominance   *182*
inheritance   *175*
law of independent assortment   *182*
law of segregation   *178*
locus (plural, loci)   *175*
muscular dystrophy   *194*

mutation   *175*
pedigree   *189*
phenotype   *179*
pleiotropy   *184*
polygenic inheritance   *185*
Punnett square method   *179*
recessive   *178*
self-fertilization   *176*
sex chromosome   *187*
sex-linked   *188*
sickle-cell anemia   *191*
test cross   *180*
true-breeding   *177*
X chromosome   *187*
Y chromosome   *187*

Thinking Through the Concepts

Multiple Choice

1. The physical position of a gene on a chromosome is its _____; slightly different forms of a gene are called _____.
  a. locus; alleles
  b. locus; polygenic
  c. chiasma; alleles
  d. trait; hybrids

2. If an organism has two different alleles (call the alleles *a* and *b*) of a gene,
  a. its phenotype will be the same as an organism with two identical alleles of this gene.
  b. all of its gametes will contain both the *a* allele and the *b* allele.
  c. it is homozygous for that gene.
  d. it is heterozygous for that gene.

3. Independent assortment means that
  a. two genes tend to be inherited together.
  b. which allele of a gene is included in a gamete has no effect on which allele of a second gene is included in the same gamete.
  c. which allele of a gene is included in a gamete determines which allele of a second gene is included in the same gamete.
  d. homologous chromosomes do not separate during meiosis.

4. If a gene is located on the X chromosome of a mammal, it is
  a. expressed only in females.
  b. expressed only in males.
  c. sex-linked, with females more likely to show recessive traits.
  d. sex-linked, with males more likely to show recessive traits.

5. A test cross is used to determine
  a. the genotype of an organism with a phenotypically dominant trait.
  b. the genotype of an organism with a phenotypically recessive trait.

c. the genotype of an organism showing pleiotropic effects of a gene.
d. if a trait is inherited polygenically.

Fill-in-the-Blank

1. An organism is described as *Rr,* with red coloring. *Rr* is the organism's _____, while red color is its _____. This organism would be _____ (homozygous/heterozygous) for this color gene.
2. The inheritance of multiple traits depends on the locations of the genes that control the traits. If the genes are on different chromosomes, then the traits are inherited _____ (as a group/independently). If the genes are located close together on a single chromosome, then the traits tend to be inherited _____ (as a group/independently). Genes on the same chromosome are said to be _____.
3. In mammals, males have _____ (XX/XY/YY) sex chromosomes and females have _____ (XX/XY/YY) sex chromosomes. The sex of offspring depends on which chromosome is present in the _____ (sperm/egg).
4. Genes that are present on one sex chromosome but not the other are called _____.
5. When the phenotype of heterozygotes is intermediate between the phenotypes of the two homozygotes, this pattern of inheritance is called _____. When heterozygotes express phenotypes of both homozygotes (not intermediate, but showing both traits), this is called _____. In _____, many genes, usually with similar effects on phenotype, control the inheritance of a trait.

Review Questions

1. Define the following terms: *gene, allele, dominant, recessive, true-breeding, homozygous, heterozygous, cross-fertilization,* and *self-fertilization.*
2. Explain why genes located on the same chromosome are said to be linked. Why do alleles of linked genes sometimes separate during meiosis?
3. Define *polygenic inheritance.* Why does polygenic inheritance sometimes allow parents to produce offspring that are notably different in skin color than either parent?
4. What is sex linkage? In mammals, which sex would be most likely to show recessive sex-linked traits?
5. What is the difference between a phenotype and a genotype? Does knowledge of an organism's phenotype always allow you to determine the genotype? What type of experiment would you perform to determine the genotype of a phenotypically dominant individual?
6. In the pedigree of part (a) of Figure 11-22, do you think that the individuals showing the trait are homozygous or heterozygous? How can you tell from the pedigree?

Applying the Concepts

1. Sometimes the term *gene* is used rather casually. Compare the terms *allele* and *gene.*
2. In an alternate universe, all the genes in all species have only two alleles, one dominant and one recessive. Would every trait have only two phenotypes? Would all members of a species that are dominant for a given gene have exactly the same phenotype? Explain your reasoning.

Genetics Problems

1. In certain cattle, hair color can be red (homozygous R_1R_1), white (homozygous R_2R_2), or roan (a mixture of red and white hairs, heterozygous R_1R_2).
   a. When a red bull is mated to a white cow, what genotypes and phenotypes of offspring could be obtained?
   b. If one of the offspring bulls in part (a) were mated to a white cow, what genotypes and phenotypes of offspring could be produced? In what proportion?

2. In the edible pea, tall (*T*) is dominant to short (*t*), and green pods (*G*) are dominant to yellow pods (*g*). List the types of gametes and offspring that would be produced in the following crosses:
   a. *TtGg* × *TtGg*
   b. *TtGg* × *TTGG*
   c. *TtGg* × *Ttgg*

3. In tomatoes, round fruit (*R*) is dominant to long fruit (*r*), and smooth skin (*S*) is dominant to fuzzy skin (*s*). A true-breeding round, smooth tomato (*RRSS*) was crossbred with a true-breeding long, fuzzy tomato (*rrss*). All the F_1 offspring were round and smooth (*RrSs*). When these F_1 plants were bred, the following F_2 generation was obtained:

   Round, smooth: 43
   Long, fuzzy: 13

   Are the genes for skin texture and fruit shape likely to be on the same chromosome or on different chromosomes? Explain your answer.

4. In the tomatoes of Problem 3, an F_1 offspring (*RrSs*) was mated with a homozygous recessive (*rrss*). The following offspring were obtained:

   Round, smooth: 583   Long, fuzzy: 602
   Round, fuzzy: 21   Long, smooth: 16

   What is the most likely explanation for this distribution of phenotypes?

5. In humans, hair color is controlled by two interacting genes. The same pigment, melanin, is present in both brown-haired and blond-haired people, but brown hair has much more of it. Brown hair (*B*) is dominant to blond (*b*). Whether any melanin can be synthesized depends on another gene. The dominant form of this second gene (*M*) allows melanin synthesis; the recessive form (*m*) prevents melanin synthesis. Homozygous recessives (*mm*) are albino. What will be the expected proportions of phenotypes in the children of the following parents?
   a. BBMM × *BbMm*
   b. BbMm × *BbMm*
   c. BbMm × *bbmm*

6. In humans, one of the genes determining color vision is located on the X chromosome. The dominant form (*C*) produces normal color vision; red-green color deficiency (*c*) is recessive. If a man with normal color vision marries a color-deficient woman, what is the probability of them having a color-deficient son? A color-deficient daughter?

7. In the couple described in Problem 6, the woman gives birth to a color-deficient but otherwise normal daughter. The husband files for a divorce on the grounds of adultery. Will his case stand up in court? Explain your answer.

12 DNA: THE MOLECULE OF HEREDITY

Ordinary bull or incredible hulk? A tiny change in DNA makes all the difference.

Muscles, Mutations, and Myostatin

NO, THE BULL in the top photo hasn't been pumping iron—he's a Belgian Blue, which always have bulging muscles. What makes a Belgian Blue look like a bodybuilder compared to an ordinary bull, such as the Hereford in the bottom photo, which just looks bulky and fat? It's all in their genes.

When a mammal develops, its cells divide many times, a process that is controlled by proteins synthesized from the instructions contained in its genes. Eventually, most cells stop dividing and become specialized for a specific function. Muscle cells are no exception. When you were very young, cells destined to form your muscles multiplied, fused together to form long, relatively thick cells with numerous nuclei, and synthesized the specialized proteins that enable muscles to contract. A protein called myostatin puts the brakes on

muscle development. "Myostatin" literally means "to make muscles stay the same," and that is exactly what it does. As muscles develop, myostatin slows down—and eventually stops—the multiplication of pre-muscle cells. Myostatin also regulates the ultimate size of muscle cells.

Belgian Blues have more, and larger, muscle cells than ordinary cattle do because they don't produce normal myostatin. Why not? As you know, genes are made of deoxyribonucleic acid (DNA). A Belgian Blue has a change, or mutation, in the DNA of its myostatin gene, making it slightly different from the DNA of the myostatin gene in most other cattle. As a result, a Belgian Blue produces defective myostatin. Their pre-muscle cells multiply more than normal, and the cells become extra-large as they differentiate, producing remarkably buff cattle.

How does DNA encode the instructions for traits such as muscle size, flower color, and sex? How are these instructions passed from generation to generation? And why do the instructions sometimes change?

AT A GLANCE

12.1 HOW DID SCIENTISTS DISCOVER THAT GENES ARE MADE OF DNA?

By the early 1900s, scientists had learned that genetic information exists in discrete units that they called genes, and that genes are parts of chromosomes. Chromosomes are composed only of protein and DNA, so one of these must be the molecule of heredity. But which one?

In the late 1920s, Frederick Griffith, a British researcher, attempted to develop a vaccine to prevent bacterial pneumonia.

Some vaccines consist of a weakened strain of bacteria, which can't cause illness. Injecting a weakened, but still living, strain into an animal may stimulate immunity against disease-causing (virulent) strains. Other vaccines use virulent bacteria that have been killed by exposure to heat or chemicals.

Griffith experimented with two strains of the bacterium *Streptococcus pneumoniae*. One strain, named R, did not cause pneumonia when injected into mice (**FIG. 12-1a**), but injecting mice with another strain, called S, caused pneumonia, killing the mice in a day or two (**FIG. 12-1b**). As expected,

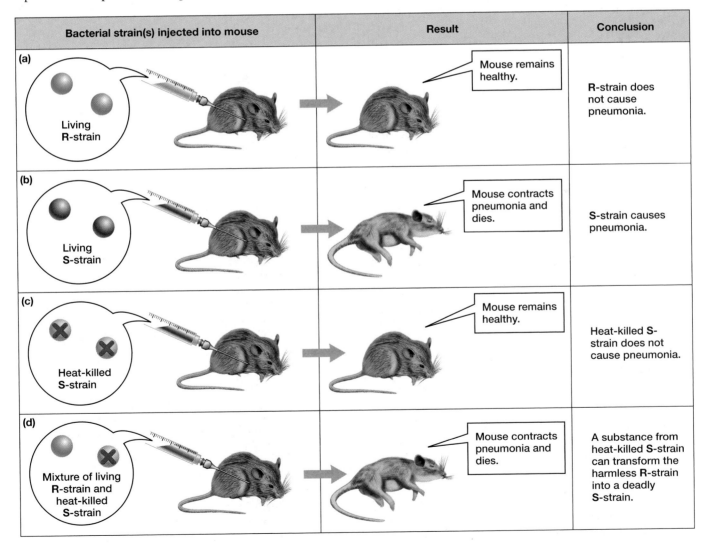

▲ **FIGURE 12-1 Transformation in bacteria** Griffith's discovery that bacteria can be transformed from harmless to deadly laid the groundwork for the discovery that genes are composed of DNA.

when the S-strain was heat-killed before being injected into the mice, it did not cause disease (**FIG. 12-1c**). Unfortunately, neither the live R-strain nor the heat-killed S-strain provided immunity against live S-strain bacteria.

Griffith also tried injecting a mixture of living R-strain bacteria and heat-killed S-strain bacteria (**FIG. 12-1d**). Because neither caused pneumonia on its own, he expected the mice to remain healthy. To his surprise, they sickened and died. When he autopsied the mice, he recovered living S-strain bacteria from them. How did the mice acquire living S-strain bacteria? Griffith hypothesized that some substance in the heat-killed S-strain changed the living, harmless R-strain bacteria into the deadly S-strain, a process he called transformation. These transformed bacteria could cause pneumonia.

Griffith never discovered an effective pneumonia vaccine, so in that sense his experiments were a failure (in fact, an effective vaccine against *Streptococcus pneumoniae* was not developed until the late 1970s). However, Griffith's experiments marked a turning point in our understanding of genetics because other researchers suspected that the substance that causes transformation might be the long-sought molecule of heredity.

The Transforming Molecule Is DNA

In 1944, Oswald Avery, Colin MacLeod, and Maclyn McCarty discovered that the transforming molecule is DNA. They isolated DNA from S-strain bacteria, mixed it with live R-strain bacteria, and produced live S-strain bacteria. They treated some samples with protein-destroying enzymes and other samples with DNA-destroying enzymes. Protein-destroying enzymes did not prevent transformation, but DNA-destroying enzymes did. Therefore, they concluded that transformation must be caused by DNA, and not by protein contaminating the DNA.

This discovery helps us to interpret the results of Griffith's experiments. Heating S-strain cells killed them but did not completely destroy their DNA. When heat-killed S-strain bacteria were mixed with living R-strain bacteria, fragments of DNA from the dead S-strain cells entered into some of the R-strain cells and became incorporated into the chromosome of the R-strain bacteria (**FIG. 12-2**). Some of these DNA fragments contained the genes needed to cause pneumonia, transforming a harmless R-strain cell into a virulent S-strain cell. Thus, Avery, MacLeod, and McCarty concluded that DNA is the molecule of heredity.

Over the next decade, evidence continued to accumulate that DNA is the genetic material. For example, before dividing, a eukaryotic cell duplicates its chromosomes (see Chapter 9) and exactly doubles its DNA content, but not its protein content—just what would be expected if genes are made of DNA, and not protein. Nevertheless, not everyone was convinced, until Alfred Hershey and Martha Chase showed that DNA is the genetic material of bacteriophages (viruses that infect bacteria), as we describe in "How

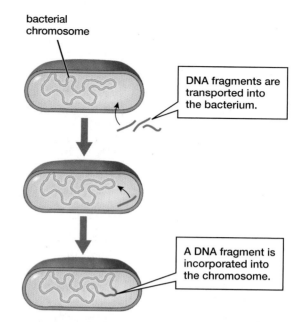

▲ **FIGURE 12-2 The molecular mechanism of transformation** Transformation may occur when a living bacterium takes up pieces of DNA from its environment and incorporates those fragments into its chromosome.

Do We Know That? DNA Is the Hereditary Molecule" on page 202.

CHECK YOUR LEARNING

Can you …

- describe the experiments of Griffith; Avery, MacLeod, and McCarty; and Hershey and Chase?
- explain why these experiments showed that DNA is the hereditary molecule?

12.2 WHAT IS THE STRUCTURE OF DNA?

Knowing that genes are made of DNA still does not answer critical questions about inheritance: How does DNA encode genetic information? How is DNA replicated so that a cell can pass its hereditary information to its daughter cells? The secrets of DNA function and replication are found in the three-dimensional structure of the DNA molecule.

DNA Is Composed of Four Nucleotides

DNA consists of long chains made of subunits called **nucleotides.** Each nucleotide consists of three parts: a phosphate group, a sugar called deoxyribose, and one of four nitrogen-containing **bases.** The bases in DNA are **adenine (A), guanine (G), thymine (T),** and **cytosine (C)** (**FIG. 12-3**). Adenine and guanine both consist of fused five- and six-member rings of carbon and nitrogen atoms, with different functional

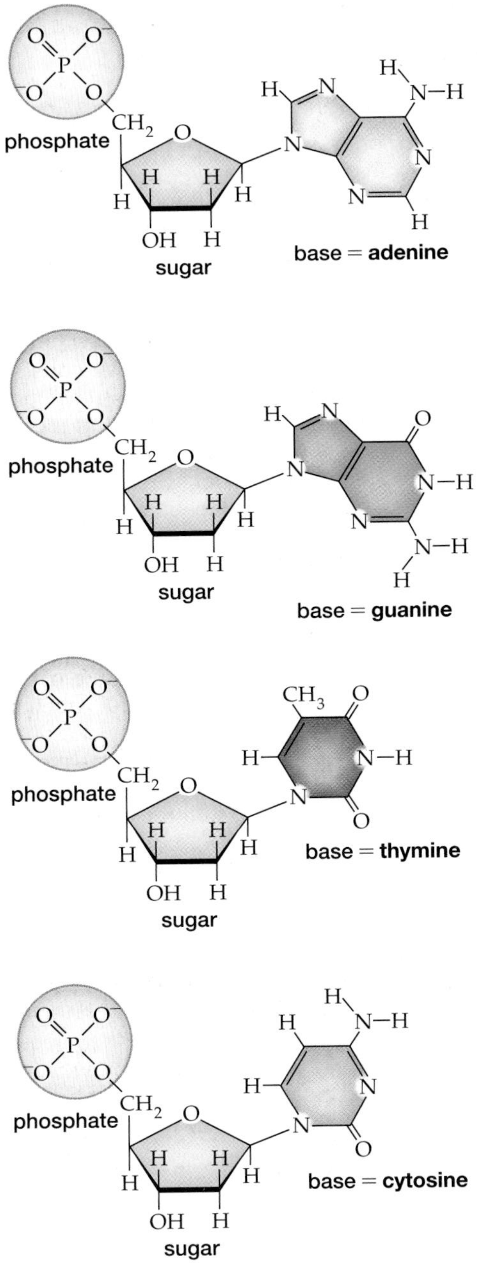

phosphate

sugar

base = **adenine**

phosphate

sugar

base = **guanine**

phosphate

sugar

base = **thymine**

phosphate

sugar

base = **cytosine**

▲ **FIGURE 12-3** DNA nucleotides

DNA Is a Double Helix of Two Nucleotide Strands

In the late 1940s, several scientists began to investigate the structure of DNA. British researchers Maurice Wilkins and Rosalind Franklin used a technique called X-ray diffraction to study the DNA molecule (**FIG. 12-4**). Although X-ray diffraction patterns do not provide a direct picture of molecules, they do provide considerable information about molecular shape and structure. Wilkins and Franklin made several deductions from their experiments. First, a molecule of DNA is long and thin, with a uniform width of 2 nanometers (2 billionths of a meter).

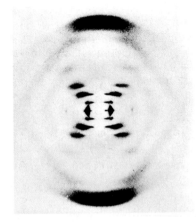

▲ **FIGURE 12-4 X-ray diffraction image of DNA** The crossing pattern of dark spots is characteristic of helical molecules such as DNA. Measurements of various aspects of the pattern indicate the dimensions of the DNA helix; for example, the distance between the dark spots corresponds to the distance between turns of the helix.

Second, DNA is helical, twisted like a spiral staircase. Third, DNA is a double helix; that is, two strands of nucleotides coil around one another. Fourth, DNA consists of repeating subunits. And fifth, the phosphates are probably on the outside of the helix.

Given enough time, Franklin and Wilkins would probably have deduced the correct structure of DNA. However, they were scooped by two young scientists, James Watson and Francis Crick (**FIG. 12-5**). Wilkins shared the X-ray diffraction

▲ **FIGURE 12-5 James Watson (left) and Francis Crick with their model of DNA**

groups attached to the six-member ring. Thymine and cytosine consist of a single six-member ring of carbon and nitrogen atoms, again with different functional groups attached to the ring.

In the 1940s, biochemist Erwin Chargaff analyzed the amounts of the four bases in DNA from organisms as diverse as bacteria, sea urchins, fish, and humans. He found a curious consistency: Although the proportions of each base differ from species to species, for any given species, there are always equal amounts of adenine and thymine and equal amounts of guanine and cytosine. However, it would be almost another decade before anyone figured out why this consistency, called "Chargaff's rule," holds true.

HOW DO WE KNOW THAT?

DNA Is the Hereditary Molecule

Avery, MacLeod, and McCarty showed that the transforming molecule in bacteria was DNA. Did that mean that DNA was the long-sought molecule of heredity? Some weren't so sure, until Alfred Hershey and Martha Chase convinced virtually all the remaining skeptics in a marvelous set of experiments in 1952.

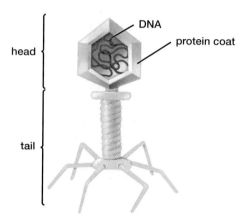

head

tail

DNA

protein coat

(a) Structure of a bacteriophage

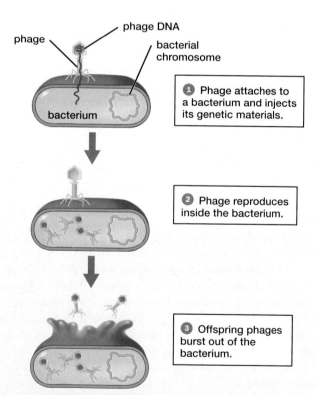

phage

phage DNA

bacterial chromosome

bacterium

❶ Phage attaches to a bacterium and injects its genetic materials.

❷ Phage reproduces inside the bacterium.

❸ Offspring phages burst out of the bacterium.

(b) Bacteriophage reproduction

▲ **FIGURE E12-1 Bacteriophages (a)** Many bacteriophages have complex structures, including a head containing genetic material, tail fibers that attach to the surface of a bacterium, and an elaborate apparatus for injecting their genetic material into the bacterium. **(b)** A bacteriophage reproduces inside a bacterium.

Hershey and Chase studied a type of virus, called a **bacteriophage** ("phage" for short), that infects bacteria (**FIG. E12-1**). When a phage encounters a bacterium, it attaches to the bacterial cell wall and injects its genetic material into the bacterium ❶. The outer coat of the phage remains outside. The bacterium cannot distinguish phage genes from its own genes, so it "reads" the phage genes and uses that information to produce more phages ❷. Finally, the bacterium bursts, freeing the new phages ❸.

Most phages are chemically very simple, consisting only of DNA and protein. Therefore, one of these two molecules must be the phage genetic material. DNA and protein both contain carbon, oxygen, hydrogen, and nitrogen. DNA also contains phosphorus but not sulfur, whereas proteins contain sulfur but not phosphorus. Hershey and Chase used these differences in the composition of DNA and protein to deduce that DNA is the hereditary molecule of bacteriophages (**FIG. E12-2**).

Hershey and Chase forced one culture of phages to synthesize DNA using radioactive phosphorus, thereby labeling the phage DNA. They forced another culture of phages to synthesize protein using radioactive sulfur, labeling the phage protein ❶. Bacteria were infected by one of these two labeled phage cultures ❷. Then the bacteria were whirled in a blender to shake the phage coats off the bacteria ❸, followed by centrifugation to separate the phage coats from the bacteria ❹.

Hershey and Chase found that, if bacteria were infected by phages containing radioactively labeled protein, the resulting phage coats were radioactive but the bacteria were not. If bacteria were infected by phages containing radioactive DNA, the bacteria became radioactive but the phage coats were not ❺. Therefore, the substance injected by the phages into the bacteria was DNA, not protein. Further, the infected bacteria produced new phages, even after the protein coats were removed, showing that the injected DNA, not the protein in the coat, was the genetic material. In the words of James Watson, this experiment provided "powerful new proof that DNA is the primary genetic material."

THINK CRITICALLY Some viruses, such as the tobacco mosaic virus (TMV), consist of a protein coat surrounding ribonucleic acid (RNA) instead of DNA. A few years after the Hershey-Chase experiments, Heinz Fraenkel-Conrat and several colleagues separated TMV of two different strains (normal and HR) into their protein and RNA components. They then mixed the protein from strain HR with RNA from the normal strain, and vice versa. Hybrid viruses (either HR protein coats with normal RNA or normal protein coats with HR RNA) spontaneously assembled in these mixtures. They then allowed the hybrid viruses to infect tobacco plants and produce new viruses. If RNA is the genetic material of TMV, predict the type of protein coats formed by the offspring of hybrid viruses.

Observations:
1. Bacteriophage viruses consist of only DNA and protein.
2. Bacteriophages inject their genetic material into bacteria, forcing the bacteria to synthesize more phages.
3. The outer coat of bacteriophages stays outside of the bacteria.
4. DNA contains phosphorus but not sulfur.
   • DNA can be "labeled" with radioactive phosphorus.
5. Protein contains sulfur but not phosphorus.
   • Protein can be "labeled" with radioactive sulfur.

Question: Is DNA or protein the genetic material of bacteriophages?

Hypothesis: DNA is the genetic material.

Prediction:
1. If bacteria are infected with bacteriophages containing radioactively labeled DNA, the bacteria will be radioactive.
2. If bacteria are infected with bacteriophages containing radioactively labeled protein, the bacteria will not be radioactive.

Experiment:

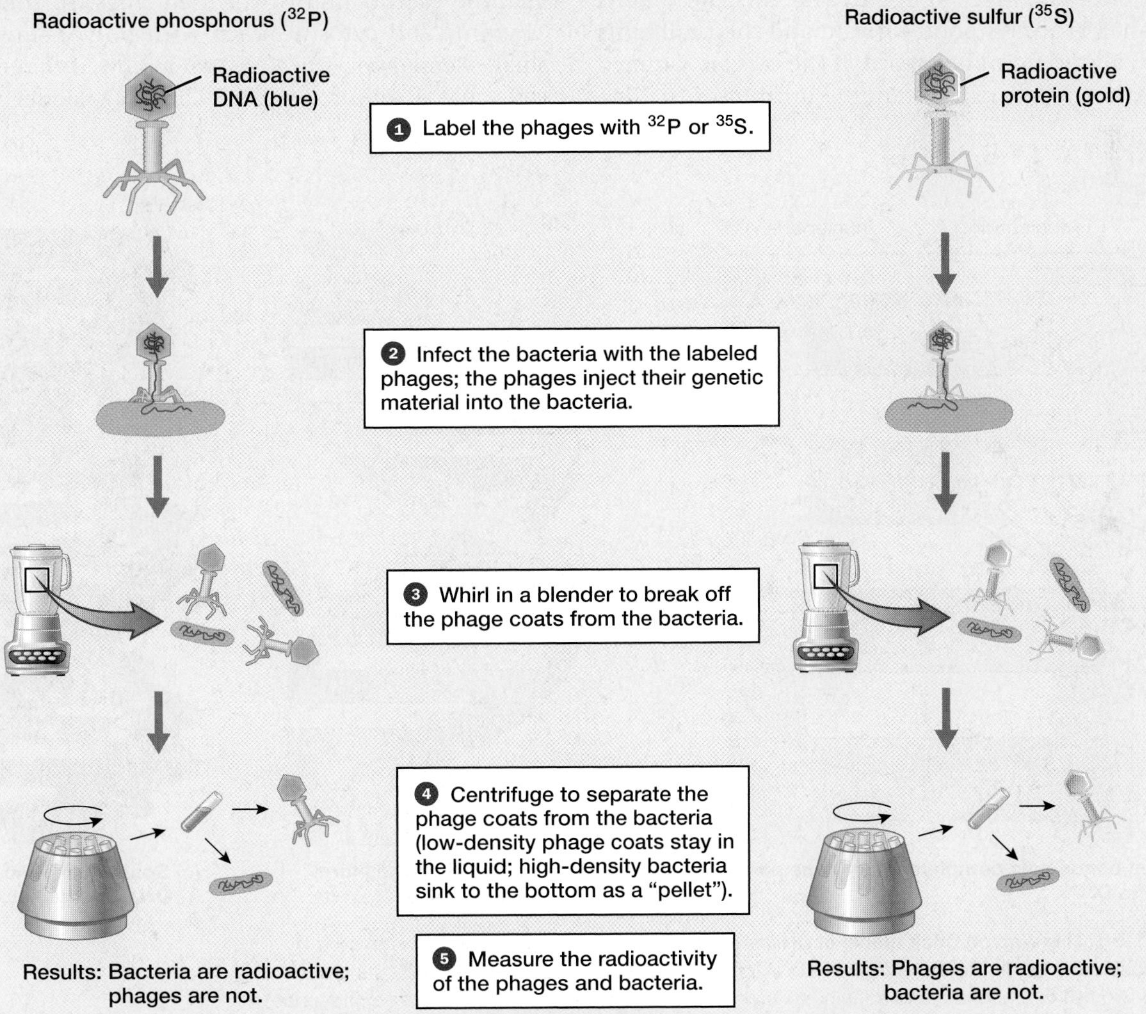

Radioactive phosphorus (^{32}P)

Radioactive sulfur (^{35}S)

Radioactive DNA (blue)

Radioactive protein (gold)

❶ Label the phages with ^{32}P or ^{35}S.

❷ Infect the bacteria with the labeled phages; the phages inject their genetic material into the bacteria.

❸ Whirl in a blender to break off the phage coats from the bacteria.

❹ Centrifuge to separate the phage coats from the bacteria (low-density phage coats stay in the liquid; high-density bacteria sink to the bottom as a "pellet").

❺ Measure the radioactivity of the phages and bacteria.

Results: Bacteria are radioactive; phages are not.

Results: Phages are radioactive; bacteria are not.

Conclusion: Infected bacteria contain radioactive phosphorus but not radioactive sulfur, supporting the hypothesis that the genetic material of bacteriophages is DNA, not protein.

▲ **FIGURE E12-2 The Hershey-Chase experiment** By radioactively labeling either the DNA or the protein of bacteriophages, Hershey and Chase tested whether the genetic material of phages is DNA (left side of the experiment) or protein (right side).

data with them, so they knew the general size and shape of a DNA molecule. Using an understanding of how complex organic molecules bond together and an intuition that "important biological objects come in pairs," as Watson put it, Crick and Watson offered a detailed molecular model for the structure of DNA.

Watson and Crick proposed that a single strand of DNA is a polymer consisting of many nucleotide subunits. The phosphate group of one nucleotide is bonded to the sugar of the next nucleotide in the strand, thus producing a **sugar-phosphate backbone** of alternating, covalently bonded sugars and phosphates (**FIG. 12-6**). The bases of the nucleotides stick out from this sugar-phosphate backbone.

All of the nucleotides in a single DNA strand are oriented in the same direction. Therefore, the two ends of a DNA strand differ; one end has a "free" or unbonded sugar, and the other end has a "free" or unbonded phosphate (**FIG. 12-6a**). Picture a long line of cars stopped on a crowded one-way street at night; the cars' headlights (free phosphates) always point forward and their taillights (free sugars) always point backward. If the cars are jammed tightly together, a pedestrian standing in front of the line

of cars will see only the headlights on the first car; a pedestrian at the back of the line will see only the taillights of the last car.

Hydrogen Bonds Between Complementary Bases Hold Two DNA Strands Together in a Double Helix

Watson and Crick's crucial insight was that the DNA in a chromosome of a living organism consists of two strands, assembled like a ladder made out of similar, but not identical, nucleotide modules. The sugar-phosphate backbones of the two strands form the two "uprights" of the DNA ladder. The protruding bases of each strand attach to one another with hydrogen bonds, forming the "rungs" of the ladder (see Fig. 12-6a). Now look at the sizes of the bases: Adenine and guanine each contain two fused rings, so they are large. Thymine and cytosine, each with only a single ring, are small. Remember, the X-ray data showed that a DNA molecule has a uniform width. The DNA ladder will have a

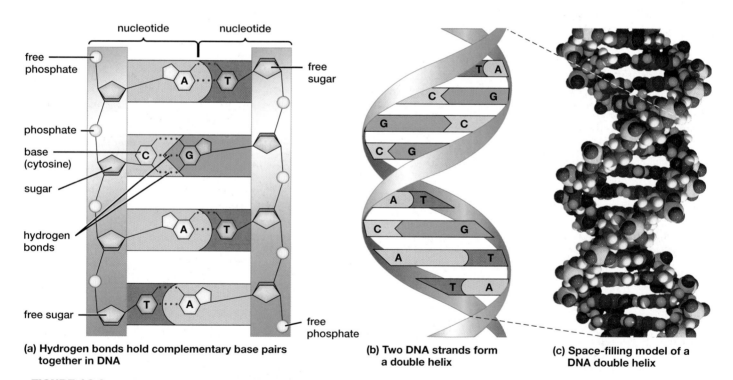

(a) Hydrogen bonds hold complementary base pairs together in DNA

(b) Two DNA strands form a double helix

(c) Space-filling model of a DNA double helix

▲ **FIGURE 12-6 The Watson-Crick model of DNA structure (a)** Hydrogen bonding between complementary base pairs holds the two strands of DNA together. Three hydrogen bonds hold guanine to cytosine, and two hydrogen bonds hold adenine to thymine. Note that each strand has a free phosphate on one end and a free sugar on the opposite end, but the two strands run in opposite directions. **(b)** Strands of DNA wind about each other in a double helix, like a twisted ladder, with the sugar-phosphate backbone forming the uprights and the complementary base pairs forming the rungs. **(c)** A space-filling model of DNA structure.

THINK CRITICALLY Which do you think would be more difficult to break apart: an A–T base pair or a C–G base pair?

uniform width only if each rung consists of one small and one large base. Which base pairs plug together to form a rung?

Take a close look at the pairs of bases in the rungs of Figure 12-6a. Adenine can form hydrogen bonds only with thymine, and guanine can form hydrogen bonds only with cytosine. These A–T and G–C pairs are called **complementary base pairs.** Every rung of the DNA ladder is made of complementary base pairs. Therefore, the base sequence of one DNA strand tells you the base sequence of the other strand. For example, if one strand reads A-T-T-C-C, the other strand must read T-A-A-G-G.

Complementary base pairs explain "Chargaff's rule"— that the DNA of a given species contains equal amounts of adenine and thymine and equal amounts of cytosine and guanine. Because an A in one DNA strand always pairs with a T in the other strand, the amount of A always equals the amount of T. Similarly, because a G in one strand always pairs with a C in the other DNA strand, the amount of G always equals the amount of C.

Finally, as the X-ray data showed, the DNA ladder isn't straight: The two strands are wound about each other to form a **double helix,** like a ladder twisted lengthwise into the shape of a spiral staircase (FIG. 12-6b). Further, the two strands in a DNA double helix are antiparallel to one another; that is, they are oriented in opposite directions. In Figure 12-6a, note that the left-hand DNA strand has a free phosphate group at the top and a free sugar on the bottom; the ends are reversed on the right-hand DNA strand. Again imagine an evening traffic jam, this time on a crowded two-lane highway. A pedestrian on an overpass would see only the headlights of cars in one lane and only the taillights of cars in the other lane.

The structure of DNA was solved. On March 7, 1953, at the Eagle Pub in Cambridge, England, Francis Crick proclaimed to the lunchtime crowd, "We have discovered the secret of life." This claim was not far from the truth. Although further data would be needed to confirm the details, within just a few years, the discovery of the double helix revolutionized much of biology, including genetics, evolutionary biology, and medicine. The revolution continues today.

CHECK YOUR LEARNING
Can you ...
- describe the four nucleotides found in DNA, how individual DNA strands are constructed, and the three-dimensional structure of DNA?

12.3 HOW DOES DNA ENCODE GENETIC INFORMATION?

Look again at the structure of DNA shown in Figure 12-6. Can you see why many scientists had trouble believing that DNA could be the carrier of genetic information? Consider the many characteristics of just one organism. How can the color of a bird's feathers, the size and shape of its beak, and its ability to sing all be determined by a molecule made from only four different nucleotides?

Genetic Information Is Encoded in the Sequence of Nucleotides

The answer is that it's not the *number* of different nucleotides but their *sequence* that's important. Within a DNA strand, the four nucleotides can be arranged in any order, and each unique sequence of nucleotides represents a unique set of genetic instructions. An analogy might help: You don't need a lot of different letters to make up a language. English has 26 letters, but Hawaiian has only 12, and the binary language of computers uses only two "letters" (0 and 1, or "off" and "on"). Nevertheless, all three languages can spell out millions of different sentences. A stretch of DNA that is just 10 nucleotides long can form more than a million different sequences of the four nucleotides. Because an organism has millions (in bacteria) to billions (in plants or animals) of nucleotides, DNA can encode a staggering amount of information. As we will describe in Chapter 13, the DNA sequence of most genes encodes the information needed to synthesize a protein.

To make sense, the letters of a language must be in the correct order. Similarly, a gene must have the right nucleotides in the right sequence. Just as "friend" and "fiend" mean different things, and "fliend" doesn't mean anything, different sequences of nucleotides in DNA may encode very different pieces of information or no information at all. The resulting proteins might be fully functional, partially functional, or nonfunctional.

CHECK YOUR LEARNING
Can you ...
- explain how DNA encodes hereditary information?

CASE STUDY \ CONTINUED
Muscles, Mutations, and Myostatin

The sequence of nucleotides in a gene determines the function of the protein that it encodes. The myostatin gene of Herefords and most other breeds of cattle has a nucleotide sequence that differs from the sequence in the Belgian Blue myostatin gene. The Hereford gene codes for a protein that limits muscle size; the Belgian Blue gene, however, codes for a completely nonfunctional myostatin protein, so their muscles become oversized. Both Hereford and Belgian Blue cattle breed true—their offspring have the same nucleotide sequence in their myostatin genes as their parents do, which occurs because DNA replication, from cell to cell and from parent to offspring, almost always produces exactly the same nucleotide sequences, time after time. How do cells replicate their DNA so precisely?

HAVE YOU EVER WONDERED...

Face it—you'll never run like Usain Bolt. How much of his fantastic ability is genetic? A few genes are known to make significant contributions to athletic performance. For example, myostatin mutations can boost strength and speed. Different alleles of a gene called *ACTN3* seem to favor sprinting and power sports over distance running and other endurance sports. However, at least 240 genes contribute to human athletic performance, and the effects of most individual genes (including *ACTN3*) are small. In all likelihood, super-athletes like Bolt won the "genetic lottery" and inherited scores of alleles that each boost his performance just a little but add up to unsurpassed athleticism.

How Much Genes Influence Athletic Prowess?

12.4 HOW DOES DNA REPLICATION ENSURE GENETIC CONSTANCY DURING CELL DIVISION?

In the 1850s, Austrian pathologist Rudolf Virchow realized that "all cells come from cells." All the trillions of cells of your body are the offspring of other cells, going all the way back to when you were a fertilized egg. Moreover, almost every cell of your body contains identical genetic information—the same genetic information that was present in that fertilized egg. When cells reproduce by mitotic cell division, each daughter cell receives a nearly perfect copy of the parent cell's genetic information. Therefore, before cell division, the parent cell must synthesize two exact copies of its DNA. A process called **DNA replication** produces these two identical DNA double helices.

DNA Replication Produces Two DNA Double Helices, Each with One Original Strand and One New Strand

In their paper describing DNA structure, Watson and Crick included one of the greatest understatements in all of science: "It has not escaped our notice that the specific [base] pairing we have postulated immediately suggests a possible copying mechanism for the genetic material." In fact, base pairing is the foundation of DNA replication. Because an adenine on one strand must pair with a thymine on the other strand, and a cytosine must pair with a guanine, the base sequence of each strand contains all the information needed to replicate the other strand.

Conceptually, DNA replication is quite simple (**FIG. 12-7**). The essential ingredients are the parental DNA strands ❶, **free nucleotides** (not yet part of a DNA strand) that were previously synthesized in the cytoplasm and imported into the nucleus, and a variety of enzymes that

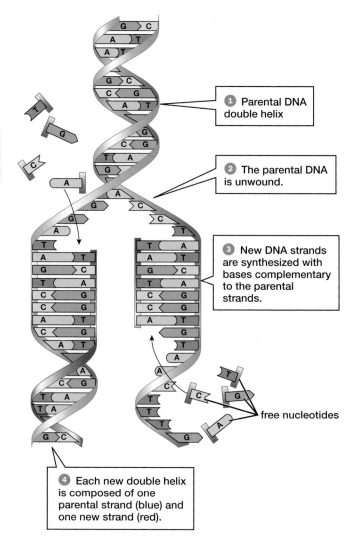

❶ Parental DNA double helix

❷ The parental DNA is unwound.

❸ New DNA strands are synthesized with bases complementary to the parental strands.

free nucleotides

❹ Each new double helix is composed of one parental strand (blue) and one new strand (red).

▲ **FIGURE 12-7 Basic features of DNA replication** During replication, the two strands of the parental DNA double helix separate. Free nucleotides that are complementary to those in each strand are joined to make new daughter strands. Each parental strand and its new daughter strand then form a new double helix.

unwind the parental DNA double helix and synthesize new DNA strands.

First, enzymes called **DNA helicases** (meaning "enzymes that break the DNA helix") pull apart the parental double helix, so that the bases of the two DNA strands are no longer bonded to one another ❷. Second, enzymes called **DNA polymerases** ("enzymes that synthesize a DNA polymer") move along each separated parental DNA strand, matching bases on the parental strands with complementary free nucleotides ❸. For example, DNA polymerase pairs an exposed adenine in the parental strand with a free thymine. DNA polymerase also connects these free nucleotides with one another to form two new DNA strands, one new strand complementary to each parental strand. Thus, if a parental DNA strand reads T–A–G, DNA polymerase will synthesize a new strand with the complementary

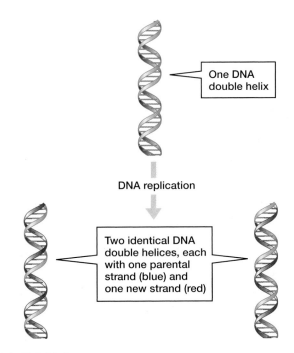

▲ **FIGURE 12-8 Semiconservative replication of DNA**

In the figure:
- One DNA double helix
- DNA replication
- Two identical DNA double helices, each with one parental strand (blue) and one new strand (red)

sequence A–T–C. For more information on how DNA is replicated, refer to "In Greater Depth: DNA Structure and Replication" on page 208.

When replication is complete, each parental DNA strand and its newly synthesized, complementary daughter DNA strand wind together to form new double helices ❹. In making each new double helix, DNA replication uses, or conserves, one parental DNA strand and synthesizes one new strand, so the process is called **semiconservative replication** (FIG. 12-8).

If no mistakes have been made, the base sequences of both new DNA double helices are identical to the base sequence of the parental DNA double helix and, of course, to each other.

CHECK YOUR LEARNING

Can you ...

- describe the process of DNA replication, including the enzymes involved and the actions that they perform?
- explain why DNA replication is called "semiconservative"?

CASE STUDY **CONTINUED**

Muscles, Mutations, and Myostatin

"Double-muscled" cattle were first reported in the early 1800s. Sometime in the late 1700s or early 1800s, a mutation must have occurred in the myostatin gene of the Belgian Blue ancestor, changing the nucleotide sequence of the gene. If DNA replication is so precise, how do such mutations happen?

12.5 WHAT ARE MUTATIONS, AND HOW DO THEY OCCUR?

The nucleotide sequence of DNA is preserved, with great precision, from cell division to cell division, and from generation to generation. However, changes in the nucleotide sequence sometimes do occur: These are **mutations,** and they are the source of all genetic variation. Mutations are often harmful, much as randomly changing words in the middle of Shakespeare's *Hamlet* would probably interrupt the flow of the play. If a mutation is really damaging, a cell or organism inheriting it may quickly die. Other mutations have no effect on the organism or, in very rare instances, are even beneficial. Mutations that are advantageous, at least in certain environments, will be favored by natural selection, and are the basis for the evolution of life on Earth (see Unit 3).

Accurate Replication, Proofreading, and DNA Repair Produce Almost Error-Free DNA

The specificity of hydrogen bonding between complementary base pairs makes DNA replication highly accurate. DNA polymerase incorporates incorrect bases about once in every 10 thousand to 1 million base pairs. However, completed DNA strands contain only about one mistake in every 100 million to 10 billion base pairs (in humans, usually less than one per chromosome per replication). This phenomenally low error rate is the result of DNA repair enzymes that proofread each daughter strand during and after its synthesis. For example, some forms of DNA polymerase recognize a base pairing mistake as it is made. These types of DNA polymerase pause, fix the mistake, and then continue synthesizing more DNA. Other changes in the DNA base sequence that may occur during the life of a cell are also usually fixed by DNA repair enzymes.

Toxic Chemicals, Radiation, or Occasional Mistakes During DNA Replication May Cause Mutations

Despite the amazing accuracy of DNA replication, no organism has error-free DNA. Occasionally, mistakes made during normal DNA replication are not repaired. DNA may also be damaged by toxic chemicals (such as free radicals formed during normal cellular metabolism, some components of cigarette smoke, and toxins produced by some molds) and some types of radiation (such as ultraviolet rays in sunlight). Toxic chemicals and radiation increase the likelihood of base-pairing errors during replication. Some damage DNA between replications. Although most changes in DNA sequence are fixed by repair enzymes, those that remain are mutations.

IN GREATER DEPTH | DNA Structure and Replication

DNA Structure

To fully understand DNA replication, we must return to the structure of DNA. Biochemists keep track of the atoms in a complex molecule by numbering them. In nucleotides (FIG. E12-3), the atoms that form the "corners" of the base are numbered 1 through 6 for the single rings of cytosine and thymine, or 1 through 9 for the double rings of adenine and guanine. The carbon atoms of the sugar are numbered 1′ (1-prime) through 5′ (5-prime). The prime symbol (′) is used to distinguish atoms in the sugar from atoms in the base.

The sugar of a nucleotide has two "ends" that can be involved in synthesizing the sugar-phosphate backbone of a DNA strand: a 3′ end, which has a free –OH (hydroxyl) group attached to the 3′ carbon of the sugar, and a 5′ end, which has a phosphate group attached to the 5′ carbon. When a DNA strand is synthesized, the phosphate of one nucleotide bonds with the hydroxyl group on the sugar of the next nucleotide (FIG. E12-4).

This still leaves a free hydroxyl group on the 3′ end of one nucleotide and a free phosphate group on the 5′ end of the other nucleotide. No matter how many nucleotides are joined, there is always a free hydroxyl on the 3′ end of the strand and a free phosphate on the 5′ end.

The sugar-phosphate backbones of the two strands of a double helix are antiparallel—they run in opposite directions. Therefore, at one end of the double helix, one strand has a sugar with a free hydroxyl (the 3′ end) and the other strand has a free phosphate (the 5′ end). On the other end of the double helix, the positions of the free sugar and phosphates are reversed (FIG. E12-5).

DNA Replication

DNA replication involves three major events (FIG. E12-6 on page 210). First, the DNA double helix is unwound and the two strands are separated, allowing the nucleotide sequence to be read. Then new DNA strands with nucleotide sequences complementary to the two original strands are synthesized. In eukaryotic cells, these new DNA strands are synthesized in short

pieces, so the third step in DNA replication is to stitch the pieces together to form a continuous new strand of DNA. Each step is carried out by a distinct set of enzymes.

DNA Helicase Unwinds and Separates the Parental DNA Strands

Acting in concert with several other enzymes, DNA helicase breaks the hydrogen bonds between complementary base pairs that hold the two parental DNA strands together. This unwinds a segment of the parental double helix and separates the two strands, forming a replication bubble **1**, **2**. Each replication bubble contains a replication fork at each end, where

the two parental DNA strands are just beginning to be separated. Within the replication bubble, the bases of the parental DNA strands are no longer bonded to one another.

DNA Polymerase Synthesizes New DNA Strands

Replication bubbles are essential because they allow a second enzyme, DNA polymerase, to bind to the separated DNA strands. At each replication fork, a complex of DNA polymerase and other proteins binds to each parental strand **3**. DNA polymerase recognizes an unpaired base in the parental strand and matches it up with a complementary base in a free

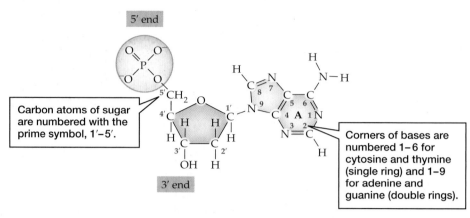

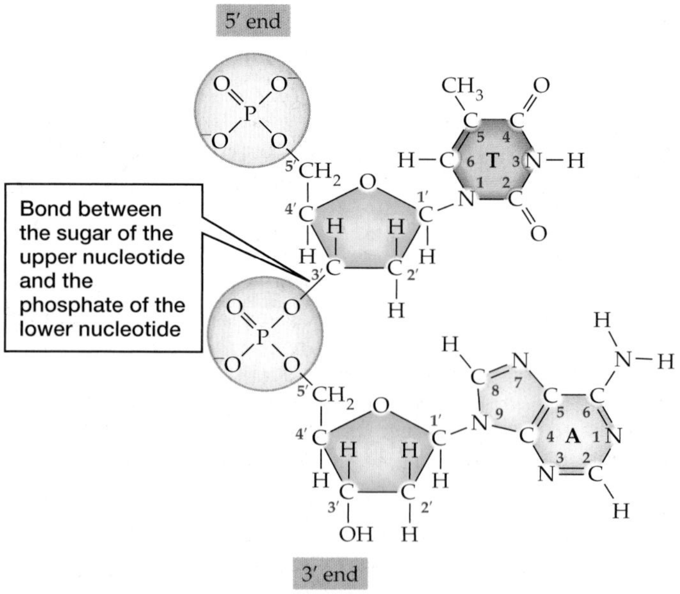

▲ **FIGURE E12-3 Numbering of carbon atoms in a nucleotide**

Carbon atoms of sugar are numbered with the prime symbol, 1′–5′.

Corners of bases are numbered 1–6 for cytosine and thymine (single ring) and 1–9 for adenine and guanine (double rings).

Bond between the sugar of the upper nucleotide and the phosphate of the lower nucleotide

▲ **FIGURE E12-4 Numbering of carbon atoms in a dinucleotide**

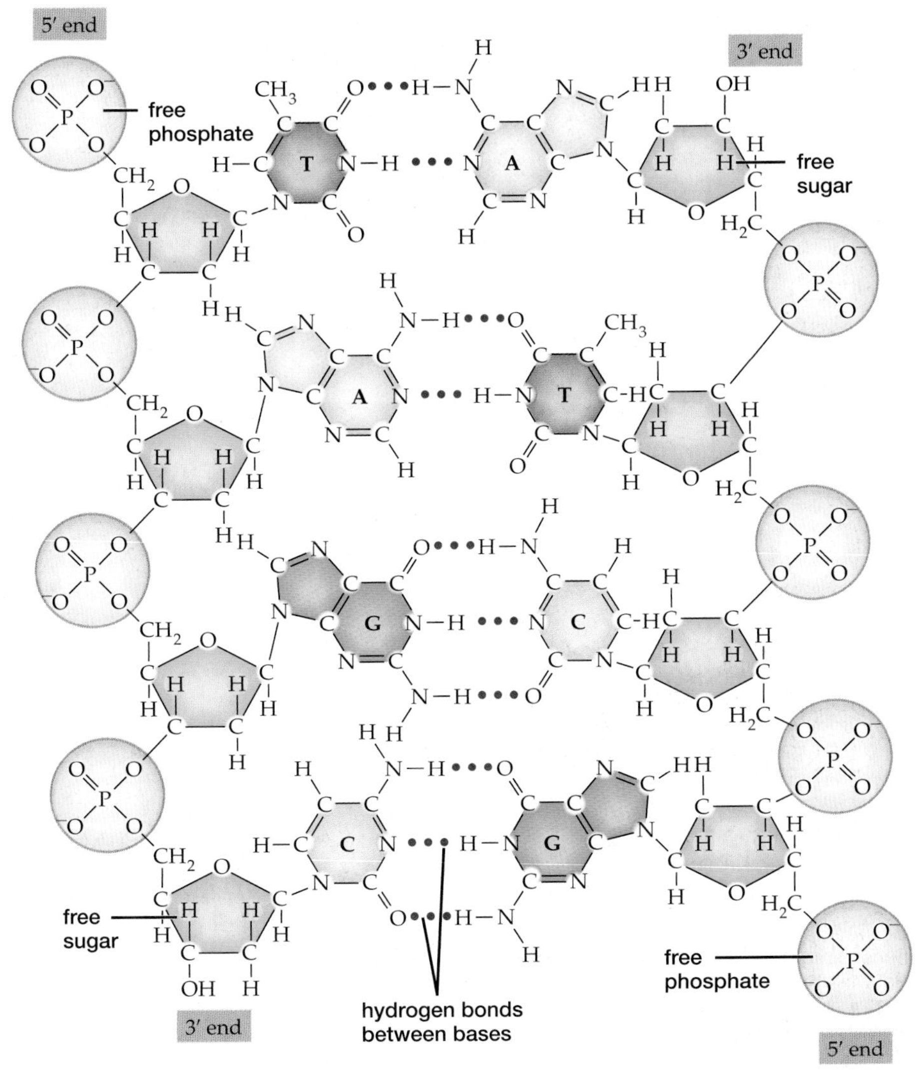

▲ **FIGURE E12-5 The two strands of a DNA double helix are antiparallel**

to the 3′ end of the daughter strand. Because the two strands of the parental DNA double helix are oriented in opposite directions, the DNA polymerase molecules move in opposite directions on the two parental strands (see step ❸).

DNA helicase and DNA polymerase work together ❹. A DNA helicase binds to the double helix and moves along, unwinding the double helix and separating the strands. Because the two DNA strands run in opposite directions, as a DNA helicase enzyme moves toward the 5′ end of one parental strand, it is simultaneously moving toward the 3′ end of the other parental strand. Now visualize two DNA polymerases landing on the separated strands of DNA. One DNA polymerase (call it polymerase #1) can follow behind the helicase toward the 5′ end of the parental strand and can synthesize a continuous daughter DNA strand until it runs into another replication bubble. This continuous daughter DNA strand is called the leading strand. On the other parental strand, however, DNA polymerase #2 moves *away from* the helicase: In step ❸, note that the helicase moves to the left, whereas DNA polymerase #2 moves to the right. Therefore, DNA synthesis on this strand will be discontinuous: DNA polymerase #2 will synthesize a short new DNA strand, called the lagging strand, but meanwhile, the helicase continues to move to the left, unwinding more of the double helix ❹, ❺. Additional DNA polymerases (#3, #4, and so on) land on this strand and synthesize more short lagging strands.

DNA Ligase Joins Segments of DNA

Multiple DNA polymerases synthesize pieces of DNA of varying lengths. Each chromosome may form hundreds of replication bubbles. Within each bubble, there will be one leading strand and dozens to thousands of lagging strands. Therefore, a cell might synthesize millions of pieces of DNA while replicating a single chromosome. How are all of these pieces sewn together? This is the job of the third major enzyme, **DNA ligase** ("an enzyme that ties DNA together"; see step ❺). Many DNA ligase enzymes stitch the fragments of DNA together until each daughter strand consists of one long, continuous DNA polymer.

nucleotide. Then DNA polymerase bonds the phosphate of the incoming free nucleotide (the 5′ end) to the sugar of the most recently added nucleotide (the 3′ end) of the growing daughter strand. In this way, DNA polymerase synthesizes the sugar-phosphate backbone of the daughter strand.

Why make replication bubbles, rather than simply starting at one end of a double helix and copying the DNA in one continuous piece all the way to the other end? Recall that eukaryotic chromosomes are very long: Human chromosomes range from about 50 million nucleotides in the relatively tiny Y chromosome to about 250 million nucleotides in chromosome 1. Eukaryotic DNA is copied at a rate of

about 50 nucleotides per second, so it would take about 12 to 58 days to copy a human chromosome in one continuous piece. To replicate an entire chromosome in a reasonable time, many DNA helicase enzymes open up many replication bubbles simultaneously, allowing many DNA polymerase enzymes to copy the strands in fairly small pieces all at the same time. Each individual bubble enlarges as DNA replication progresses, and the bubbles merge when they contact one another.

DNA polymerase always moves away from the 3′ end of a parental DNA strand (the end with the free hydroxyl group of the sugar) toward the 5′ end (with a free phosphate group). New nucleotides are always added

(continued)

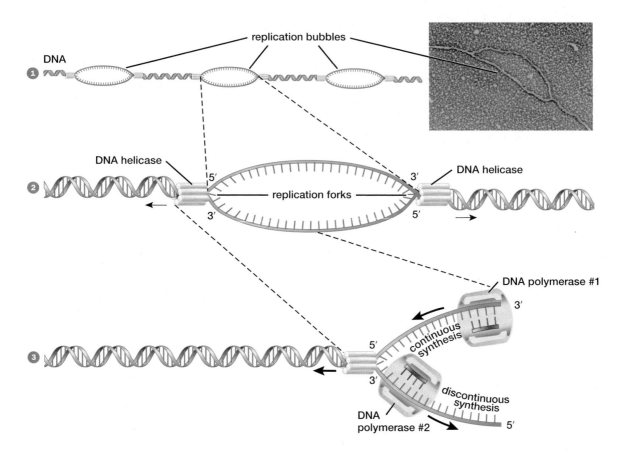

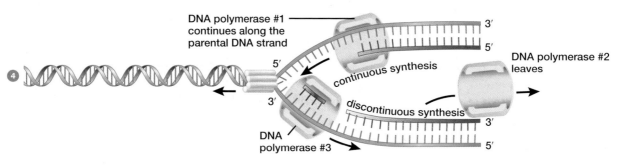

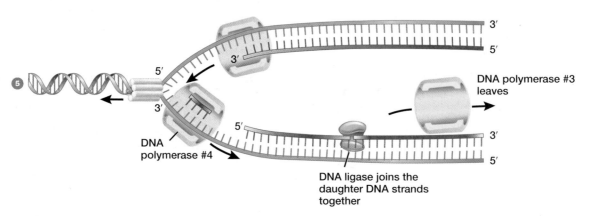

▲ **FIGURE E12-6 DNA replication**

THINK CRITICALLY During DNA replication, why doesn't DNA polymerase move away from the replication fork on both strands?

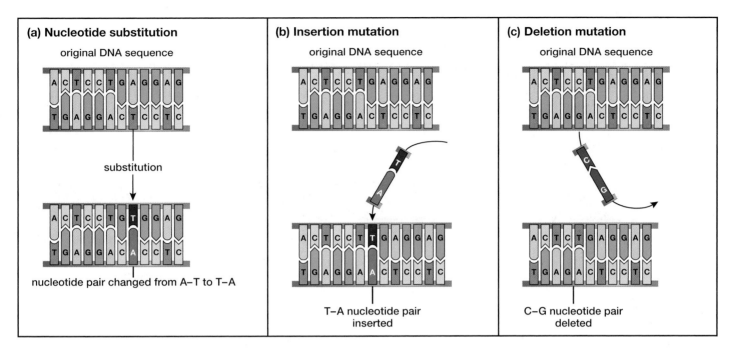

(a) Nucleotide substitution

original DNA sequence

substitution

nucleotide pair changed from A–T to T–A

(b) Insertion mutation

original DNA sequence

T–A nucleotide pair
inserted

(c) Deletion mutation

original DNA sequence

C–G nucleotide pair
deleted

▲ **FIGURE 12-9 Mutations involving only one or a few pairs of nucleotides**
(a) Nucleotide substitution. **(b)** Insertion mutation. **(c)** Deletion mutation.
The original DNA bases are in pale colors with black letters; mutations
are in dark colors with white letters.

Mutations Range from Changes in Single Nucleotide Pairs to Movements of Large Pieces of Chromosomes

If a pair of bases is mismatched during replication, repair enzymes usually recognize the mismatch, cut out the incorrect nucleotide, and replace it with a nucleotide containing the complementary base. Sometimes, however, the enzymes replace the parental nucleotide instead of the incorrect daughter nucleotide. Although the resulting base pair is complementary, it is different from the original pair; there has been a **nucleotide substitution mutation** (**FIG. 12-9a**). Because the incorrect base pair is complementary, accurate DNA replication during future cell divisions will perpetuate the mutation: It has become a permanent part of the chromosome and will be inherited by all the cell's descendants. An **insertion mutation** occurs when one or more nucleotide pairs are inserted into the DNA double helix (**FIG. 12-9b**). A **deletion mutation** occurs when one or more nucleotide pairs are removed from the double helix (**FIG. 12-9c**). Both insertion and deletion mutations have correctly base-paired DNA, so these mutations will also be permanent.

Pieces of chromosomes ranging in size from a single nucleotide pair to massive pieces of DNA are occasionally rearranged. An **inversion** occurs when a piece of DNA is cut out of a chromosome, turned around, and reinserted into the gap (**FIG. 12-10a**). A **translocation** results when a chunk of DNA, sometimes very large, is removed from one chromosome and attached to a different one (**FIG. 12-10b**). As with insertions and deletions, the DNA resulting from inversions and translocations has correct, complementary base pairs.

As we will describe in Chapter 13, different mutations can have very different consequences for the protein encoded by the mutated gene, ranging from no effect at all, through slightly altered function, to complete loss of function.

CHECK YOUR LEARNING
Can you ...
- explain what mutations are and how they occur?
- explain why mutations are rare?
- describe the different types of mutations?

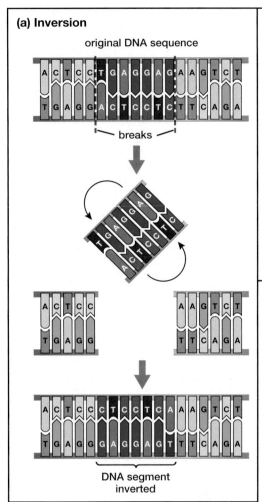

(a) Inversion

original DNA sequence

← breaks →

DNA segment inverted

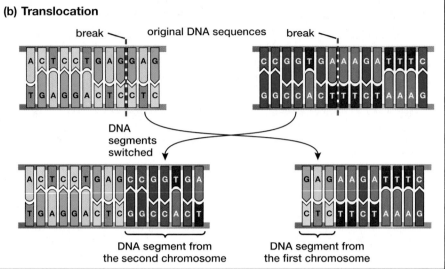

(b) Translocation

break — original DNA sequences — break

DNA segments switched

DNA segment from the second chromosome

DNA segment from the first chromosome

▲ **FIGURE 12-10 Mutations that rearrange pieces of chromosomes (a)** Inversion mutation. **(b)** Translocation of pieces of DNA between two different chromosomes. In part **(a)**, bases in the unchanged part of the chromosome are in pale colors with black letters; bases in the part of the chromosome that is inverted are in dark colors with white letters. In part **(b)**, the DNA bases of one chromosome are in pale colors with black letters, and the DNA bases of the second chromosome are in dark colors with white letters.

CASE STUDY \ REVISITED

Muscles, Mutations, and Myostatin

Belgian Blue cattle are homozygous for a deletion mutation in their myostatin gene. As a result, their cells stop synthesizing the myostatin protein about halfway through. Other animals may also have mutated myostatin. For example, "bully" whippet dogs have a deletion mutation, different from the one in Belgian Blue cattle, that also produces short, nonfunctional myostatin and a huge increase in muscle size (**FIG. 12-11**). Piedmontese, another breed of "double-muscled" cattle, have a substitution mutation. Although a full-length myostatin protein is synthesized, it doesn't fold into the correct three-dimensional structure and is completely inactive, so Piedmontese cattle have essentially the same phenotype as Belgian Blues. Some horses can inherit a different substitution mutation, which creates an allele that encodes myostatin with slightly altered function. Thoroughbred racehorses with this mutation tend to be good sprinters; those with the original nucleotide sequence tend to be better at long distances.

▲ **FIGURE 12-11 Myostatin mutation in whippets** "Bully" whippets have nonfunctional myostatin, resulting in enormous muscles.

Humans have myostatin, too. A few people inherit defective myostatin alleles from their parents, resulting in a very rare condition called myostatin-related muscle hypertrophy. In some cases, an insertion mutation causes the synthesis of a short, nonfunctional myostatin protein. As in whippets, the functional and defective human myostatin alleles are incompletely dominant to one another. Homozygotes for the defective myosin allele have far greater muscle bulk and strength than people who are homozygous for the functional allele; heterozygotes have an intermediate increase in muscle size and strength.

Myostatin mutations reveal an important feature of the language of DNA: The nucleotide words must be spelled just right, or at least really close (as in the horse mutation), for the resulting proteins to function. In contrast, any one of an enormous number of possible mistakes will render the proteins useless.

CONSIDER THIS Mutations, even those that produce completely nonfunctional proteins, may be neutral, harmful, or beneficial to an organism. Into which category do myostatin mutations fall? It seems to depend on the species. In people, there seem to be no harmful effects of myostatin-related muscle hypertrophy in either homozygotes or heterozygotes. Homozygous Belgian Blue cattle, however, are born so muscular, and consequently so large, that they usually must be delivered by cesarean section. Whippets that are homozygous for the defective "bully" allele have lots of muscle but often suffer from cramps in the shoulder and thigh, and they are not fast runners. Homozygous normal whippets are skinny and quite fast, sometimes fast enough to race. However, the majority of successful racing whippets are heterozygous, with intermediate muscling and phenomenal speed. If whippets were wild dogs that chased down their prey, how do you think that natural selection might operate on inheritance of defective myostatin alleles?

CHAPTER REVIEW

 Go to **MasteringBiology** for practice quizzes, activities, eText, videos, current events, and more.

Answers to Think Critically, Evaluate This, Multiple Choice, and Fill-in-the-Blank questions can be found in the Answers section at the back of the book.

Summary of Key Concepts

12.1 How Did Scientists Discover That Genes Are Made of DNA?

Studies by Griffith showed that genes can be transferred from one bacterial strain into another. This transfer could transform the bacterial strain from harmless to deadly. Avery, MacLeod, and McCarty showed that DNA was the molecule that could transform bacteria. Hershey and Chase found that DNA is the hereditary material of bacteriophage viruses. Thus, genes must be made of DNA.

12.2 What Is the Structure of DNA?

DNA consists of nucleotides that are linked into long strands. Each nucleotide consists of a phosphate group, the five-carbon sugar deoxyribose, and a nitrogen-containing base. Four types of bases occur in DNA: adenine, guanine, thymine, and cytosine. The sugar of one nucleotide is linked to the phosphate of the next nucleotide, forming a sugar-phosphate backbone for each strand. The bases stick out from this backbone. Two nucleotide strands wind together to form a DNA double helix, which resembles a twisted ladder. The sugar-phosphate backbones form the sides of the ladder. The bases of each strand pair up in the middle of the helix, held together by hydrogen bonds and forming the rungs of the ladder. Only complementary base pairs can bond together in the helix: Adenine bonds with thymine, and guanine bonds with cytosine.

12.3 How Does DNA Encode Genetic Information?

Genetic information is encoded as the sequence of nucleotides in a DNA molecule. Just as a language can form thousands of words

and complex sentences from a small number of letters, DNA can encode large amounts of information by varying the sequences and numbers of nucleotides in different genes. Because DNA molecules are usually millions of nucleotides long, DNA can encode huge amounts of information in its nucleotide sequence.

12.4 How Does DNA Replication Ensure Genetic Constancy During Cell Division?

When cells reproduce, they must replicate their DNA so that each daughter cell receives all the original genetic information. During DNA replication, enzymes unwind and separate part of the two parental DNA strands. Then DNA polymerase enzymes bind to each parental DNA strand. Free nucleotides form hydrogen bonds with complementary bases on the parental strands, and DNA polymerase links the free nucleotides to form new DNA strands. Replication is semiconservative because both new DNA double helices consist of one parental DNA strand and one newly synthesized, complementary daughter strand. The two new DNA double helices are duplicates of the parental DNA double helix.

12.5 What Are Mutations, and How Do They Occur?

Mutations are changes in the base sequence in DNA. DNA polymerase and other repair enzymes "proofread" the DNA, minimizing the number of mistakes during replication, but mistakes do occur. Other mutations occur as a result of radiation and damage from toxic chemicals. Mutations include substitutions, insertions, deletions, inversions, and translocations. Most mutations are harmful or neutral, but a few are beneficial and may be favored by natural selection.

Key Terms

adenine (A)   *200*   base   *200*
bacteriophage   *202*   complementary base pair   *205*

cytosine (C)   *200*
deletion mutation   *211*
DNA helicase   *206*
DNA ligase   *209*
DNA polymerase   *206*
DNA replication   *206*
double helix   *205*
free nucleotide   *206*
guanine (G)   *200*
insertion mutation   *211*

inversion   *211*
mutation   *207*
nucleotide   *200*
nucleotide substitution
   mutation   *211*
semiconservative
   replication   *207*
sugar-phosphate backbone   *204*
thymine (T)   *200*
translocation   *211*

Thinking Through the Concepts

Multiple Choice

1. If a parental DNA strand has the base sequence A-T-T-G-C-A-C-T, DNA polymerase would synthesize a new strand with the sequence
   a. A-T-T-G-C-A-C-T.
   b. T-A-A-C-G-T-G-A.
   c. C-G-G-T-A-C-A-G.
   d. The sequence of the new strand cannot be determined from the information given.

2. What happens at the conclusion of DNA replication?
   a. The daughter double helices each consist of one original DNA strand and one new DNA strand.
   b. One daughter double helix consists of the two original DNA strands and the other daughter double helix consists of two new DNA strands.
   c. Each resulting DNA strand consists of part of one of the original DNA strands and part of a new DNA strand.
   d. The resulting DNA daughter strands contain nucleotide sequences that were not present in the parental DNA strands.

3. An insertion mutation occurs when
   a. a nucleotide is replaced by a different nucleotide.
   b. one or more nucleotide pairs are added in the middle of DNA.
   c. one or more nucleotides are removed from the middle of DNA.
   d. a piece of DNA is removed from one chromosome and attached to a different chromosome.

4. The "rungs" of the DNA double helix consist of
   a. any combination of bases.
   b. any combination of one double-ring base and one single-ring base.
   c. specific combinations of double-ring bases.
   d. specific combinations of single-ring and double-ring bases.

5. The "rungs" of the DNA double helix are held together by
   a. ionic bonds.
   b. hydrogen bonds.
   c. covalent bonds.
   d. the force of the backbones on the outside of the helix pushing them together.

Fill-in-the-Blank

1. DNA consists of subunits called _____.
   Each subunit consists of three parts: _____, _____, and _____.

2. The subunits of DNA are assembled by linking the _____ of one nucleotide to the _____ of the next. As it is found in chromosomes, two DNA polymers are wound together into a structure called a(n) _____.

3. The "base pairing rule" in DNA is that adenine pairs with _____, and guanine pairs with _____. Bases that can form pairs in DNA are called _____.

4. When DNA is replicated, two new DNA double helices are formed, each consisting of one parental strand and one new, daughter strand. For this reason, DNA replication is called _____.

5. The DNA double helix is unwound by an enzyme called _____. Daughter DNA strands are synthesized by the enzyme _____. In eukaryotic cells, the daughter DNA strands are synthesized in pieces; these pieces are joined by the enzyme _____.

6. Sometimes mistakes are made during DNA replication. If uncorrected, these mistakes are called _____. When a single nucleotide is changed, this is called a(n) _____.

Review Questions

1. Describe the experimental evidence that DNA is the hereditary material of bacteriophages.

2. Draw the general structure of a nucleotide. Which parts are identical in all nucleotides, and which can vary? Name the four types of nitrogen-containing bases found in DNA.

3. Describe the structure of DNA. Where are the bases, sugars, and phosphates in the structure? Which bases are complementary to one another? How are they held together in the double helix of DNA?

4. How is information encoded in the DNA molecule?

5. Describe the process of DNA replication.

6. How do mutations occur? Describe the principal types of mutations.

Applying the Concepts

1. In an alternate universe, although proteins are still constructed of combinations of 20 different amino acids, DNA is constructed of six different nucleotides, not four as on Earth. Would you expect organisms in this universe to have more precise genetic instructions or more different genes than life on Earth? Would you expect the length of a typical gene to be the same, shorter, or longer than that of a typical gene on Earth?

2. Genetic information is encoded in the sequence of nucleotides in DNA. Let's suppose that the nucleotide sequence on one strand of a double helix encodes the information needed to synthesize a hemoglobin molecule. Do you think that the sequence of nucleotides on the other strand of the double helix also encodes useful information? Why or why not? Why do you think DNA is double-stranded?

13

GENE EXPRESSION AND REGULATION

Alice Martineau, shown here in a portrait painted by her brother Luke, hoped that "... people will realize when they hear the music, I am a singer-songwriter who just happens to be ill."

Cystic Fibrosis

IF ALL YOU knew was her music, you'd think Alice Martineau had it made—a young, pretty singer-songwriter under contract with a major recording label. However, like about 70,000 other people worldwide, Martineau had cystic fibrosis. This recessive genetic disorder is caused by defective alleles of a gene that encodes a crucially important protein called CFTR (the CF in the name of the protein stands for "cystic fibrosis"). Cystic fibrosis occurs when a person is homozygous for defective *CFTR* alleles. Before modern medical care, most people with

cystic fibrosis died by age 4 or 5; even now, their average life span is only 35 to 40 years. Martineau died when she was 30.

The CFTR protein is found in many parts of the body, including the pancreas, intestines, and sweat glands, but probably its most essential role is in the cells lining the airways of the lungs. Normally, because of the action of the CFTR protein, the airways are covered with a film of thin, watery mucus, which traps bacteria and debris. The bacteria-laden mucus is then swept out of the lungs by cilia on the cells of the airways.

The CFTR protein forms channels that allow chloride to move across plasma membranes down its concentration gradient. CFTR also regulates some channels that allow sodium ions to move across plasma membranes. In the lungs, chloride moves through CFTR channels out of the airway cells into the mucus. At the same time, CFTR inhibits the movement of sodium ions from the mucus back into the airway cells. The resulting high concentration of sodium chloride in the mucus causes water to move into the mucus by osmosis, resulting in a thin liquid that the cilia can move very easily. However, people with cystic fibrosis produce defective CFTR proteins. As a result, chloride does not move from the cells into the mucus, and extra sodium is reabsorbed from the mucus into the cells. With more sodium chloride in the cells and less in the mucus, water moves by osmosis out of the mucus and into the cells. The mucus becomes so thick that the cilia can't move it out of the lungs, leaving the airways clogged. Bacteria multiply in the mucus, causing chronic lung infections.

In this chapter, we examine the processes by which the instructions in genes are translated into proteins. How do changes in those instructions—mutations—alter the structure and function of proteins such as CFTR?

215

13.1 HOW IS THE INFORMATION IN DNA USED IN A CELL?

Information itself doesn't do anything. For example, a blueprint may provide all the information needed to build a house, but unless that information is translated into action by construction workers, no house will be built. Likewise, although the base sequence of DNA, the molecular blueprint of every cell, contains an incredible amount of information, DNA cannot carry out any actions on its own. So how does DNA determine whether you have black, blond, or red hair or whether you have normal lung function or cystic fibrosis?

Although DNA is the hereditary molecule of all cells, proteins are a cell's "molecular workers." Proteins form many cellular structures, such as the cytoskeleton and ion channels in the plasma membrane. The enzymes that catalyze chemical reactions within a cell are also proteins. Therefore, to build and operate a cell, information must flow from DNA to protein.

DNA Provides Instructions for Protein Synthesis via RNA Intermediaries

DNA directs protein synthesis through intermediary molecules of **ribonucleic acid,** or **RNA.** RNA is structurally similar to DNA but differs in three respects: (1) Instead of the deoxyribose sugar found in DNA, the backbone of RNA contains the sugar ribose (the "R" in RNA); (2) RNA is usually single-stranded instead of double-stranded; and (3) RNA has the base uracil instead of the base thymine (**TABLE 13-1**).

DNA codes for the synthesis of many types of RNA, three of which play specific roles in protein synthesis: messenger RNA, transfer RNA, and ribosomal RNA (**FIG. 13-1**). There are several other types of RNA, including RNA used as the genetic material in some viruses, such as HIV; enzymatic RNA molecules, called ribozymes, that catalyze certain chemical reactions; and "regulatory" RNA, which we will discuss later in this chapter. Here we will introduce the roles of messenger RNA, transfer RNA, and ribosomal RNA.

Messenger RNA Carries the Code for Protein Synthesis from DNA to Ribosomes

The DNA of a eukaryotic cell is stored in the nucleus, like a valuable document in a library, whereas **messenger RNA (mRNA),** like a molecular photocopy, carries the information to ribosomes in the cytoplasm, where it will be used to direct protein synthesis (**FIG. 13-1a**). As we will see shortly, groups of three bases in mRNA, called **codons,** specify which amino acids will be incorporated into a protein.

TABLE 13-1	A Comparison of DNA and RNA		
	DNA	**RNA**	
Strands	Two	One	
Sugar	Deoxyribose	Ribose	
Types of bases	Adenine (A), thymine (T) cytosine (C), guanine (G)	Adenine (A), uracil (U) cytosine (C), guanine (G)	
Base pairs	DNA–DNA	RNA–DNA	RNA–RNA
	A–T	A–T	A–U
	T–A	U–A	U–A
	C–G	C–G	C–G
	G–C	G–C	G–C
Function	Contains genes; the sequence of bases in most genes determines the amino acid sequence of a protein	Messenger RNA (mRNA): carries the code for a protein-coding gene from DNA to ribosomes Transfer RNA (tRNA): carries amino acids to the ribosomes Ribosomal RNA (rRNA): combines with proteins to form ribosomes, the structures that link amino acids to form a protein	

▶ **FIGURE 13-1 Cells synthesize three major types of RNA that are required for protein synthesis**

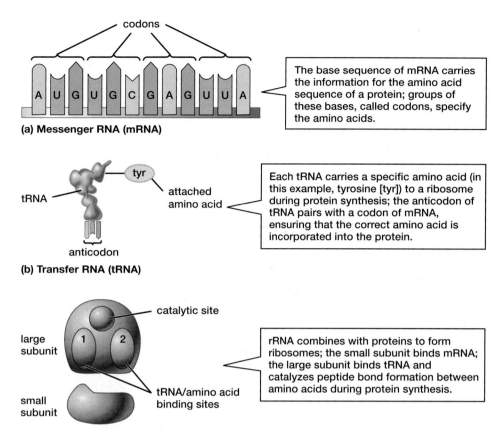

(a) Messenger RNA (mRNA)

The base sequence of mRNA carries the information for the amino acid sequence of a protein; groups of these bases, called codons, specify the amino acids.

(b) Transfer RNA (tRNA)

Each tRNA carries a specific amino acid (in this example, tyrosine [tyr]) to a ribosome during protein synthesis; the anticodon of tRNA pairs with a codon of mRNA, ensuring that the correct amino acid is incorporated into the protein.

(c) Ribosome: contains ribosomal RNA (rRNA)

rRNA combines with proteins to form ribosomes; the small subunit binds mRNA; the large subunit binds tRNA and catalyzes peptide bond formation between amino acids during protein synthesis.

Transfer RNA Carries Amino Acids to the Ribosomes

Transfer RNA (tRNA) delivers amino acids to a ribosome, where they will be incorporated into a protein. Every cell synthesizes at least one type of tRNA for each of the 20 amino acids used in proteins. Twenty enzymes in the cytoplasm, one for each amino acid, recognize the different tRNA molecules and use the energy of ATP to attach the correct amino acid to one end of the tRNA molecule (**FIG. 13-1b**). These "loaded" tRNA molecules bring their amino acids to a ribosome. A group of three bases, called an **anticodon,** protrudes from each tRNA. Complementary base pairing between codons of mRNA and anticodons of tRNA specifies which amino acids are used during protein synthesis.

Ribosomal RNA and Proteins Form Ribosomes

Ribosomes, the cellular structures that synthesize proteins from the instructions in mRNA, are composed of **ribosomal**

RNA (rRNA) and dozens of proteins. Each ribosome consists of two subunits—one small and one large (**FIG. 13-1c**). The small subunit has binding sites for mRNA, a "start" tRNA, and several proteins that are essential for assembling the ribosome and beginning protein synthesis. The large subunit has binding sites for two tRNA molecules and a site that catalyzes the formation of the peptide bonds that join amino acids into proteins. During protein synthesis, the two subunits come together, clasping an mRNA molecule between them.

Overview: Genetic Information Is Transcribed into RNA and Then Translated into Protein

Information in DNA is used to direct the synthesis of proteins in two steps, called *transcription* and *translation* (**FIG. 13-2** and **TABLE 13-2**).

Process	Information for the Process	Product	Major Enzyme or Structure Involved in the Process	Type of Base Pairing Required
TABLE 13-2	**Transcription and Translation**			
Transcription (synthesis of RNA)	A segment of one DNA strand	One RNA molecule (e.g., mRNA, tRNA, or rRNA)	RNA polymerase	RNA with DNA: RNA bases pair with DNA bases as an RNA molecule is synthesized
Translation (synthesis of a protein)	mRNA	One protein molecule	Ribosome (also requires tRNA)	mRNA with tRNA: A codon in mRNA forms base pairs with an anticodon in tRNA

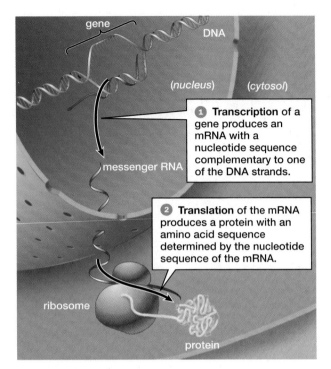

1. **Transcription** of a gene produces an mRNA with a nucleotide sequence complementary to one of the DNA strands.

2. **Translation** of the mRNA produces a protein with an amino acid sequence determined by the nucleotide sequence of the mRNA.

FIGURE 13-2 Genetic information flows from DNA to RNA to protein During transcription, the base sequence in a gene specifies the base sequence of a complementary RNA molecule. For protein-encoding genes, the product is an mRNA molecule that exits from the nucleus and enters the cytoplasm. During translation, the base sequence in an mRNA molecule specifies the amino acid sequence of a protein.

1. In **transcription,** the information contained in the DNA of a gene is copied into RNA. The base sequence of mRNA encodes the amino acid sequence of a protein. In eukaryotic cells, transcription occurs in the nucleus.

2. During protein synthesis, or **translation,** the mRNA base sequence is decoded. Messenger RNA binds to a ribosome, where base pairing between mRNA and tRNA (which brings amino acids to the ribosome) converts the base sequence of mRNA into the amino acid sequence of the protein. In eukaryotic cells, ribosomes are found in the cytoplasm, so translation occurs there as well.

It's easy to confuse the terms "transcription" and "translation." It may help to compare their common English meanings with their biological meanings. In English, to "transcribe" means to make a written copy of something, almost always in the same language. In an American courtroom, for example, verbal testimony is transcribed into a written copy, and both the testimony and the transcriptions are in English. In biology, transcription is the process of copying information from DNA to RNA using the common language of the bases found in their nucleotides. In contrast, the English meaning of "translation" is to convert words from one language to another language. In biology, translation means to convert information from the "base language" of RNA to the "amino acid language" of proteins.

The Genetic Code Uses Three Bases to Specify an Amino Acid

Before we examine transcription and translation in detail, let's see how geneticists deciphered the **genetic code—** the biological dictionary that spells out the rules for translating base sequences in DNA and mRNA into amino acid

sequences in proteins. DNA and RNA each have four different bases: DNA contains adenine (A), guanine (G), cytosine (C), and thymine (T); RNA also contains adenine, guanine, and cytosine, but uracil (U) replaces thymine (see Table 13-1). However, proteins are made of 20 different amino acids, so one base cannot directly translate into one amino acid. If a sequence of two bases codes for an amino acid, there would be 16 possible combinations (each of four possible first bases paired with each of four possible second bases, or $4 \times 4 = 16$). This still isn't enough to code for 20 amino acids. A three-base sequence, however, gives 64 possible combinations ($4 \times 4 \times 4 = 64$). Using this reasoning, physicist George Gamow hypothesized in 1954 that sets of three bases in mRNA, called *codons,* specify the amino acids. In 1961, Francis Crick and three coworkers demonstrated that this hypothesis is correct.

For a language to be understood, its users must know what the words mean, where words start and stop, and where sentences begin and end. To decipher the codons, which are the "words" of the genetic code, Marshall Nirenberg and Heinrich Matthaei ruptured bacteria, producing a cytoplasmic mixture that could synthesize proteins if mRNA was added. To this mixture, they added artificial mRNA that they synthesized to have a known sequence of nucleotides so they could see which amino acids were incorporated into protein. For example, they found that an mRNA strand composed entirely of uracil (UUUUUU ...) directed the mixture to synthesize a protein composed solely of the amino acid phenylalanine. Therefore, the triplet UUU must be the codon that translates into phenylalanine. Because the genetic code was deciphered using artificial mRNAs, it is usually written in terms of the base triplets in mRNA (rather than in DNA) that code for each amino acid (**TABLE 13-3**).

How does a cell recognize where individual codons start and stop and where the code for an entire protein starts and stops? Translation always begins with the codon AUG, appropriately known as the **start codon.** Because AUG also codes for the amino acid methionine, all proteins originally begin with methionine, although it may be removed after the protein is synthesized. Only the first AUG codon in an mRNA acts as a start codon; AUG codons that occur further on in the mRNA simply code for methionine. Three codons—UAG, UAA, and UGA—are **stop codons** and don't code for any amino acids. When the ribosome encounters a stop codon, it releases both the newly synthesized protein and the mRNA. Because all codons consist of three bases, and the beginning and end of a protein are specified by start and stop codons, respectively, then "spaces" between codon "words" are unnecessary. Why? Consider what would happen if English used only three-letter words: A sentence such as THEDOGSAWTHECAT would be perfectly understandable, even without spaces between the words.

TABLE 13-3	The Genetic Code (Codons of mRNA)							
		Second Base						
		U	C	A	G			
U	UUU	Phenylalanine (Phe)	UCU	Serine (Ser)	UAU	Tyrosine (Tyr)	UGU	Cysteine (Cys)	U
	UUC	Phenylalanine	UCC	Serine	UAC	Tyrosine	UGC	Cysteine	C
	UUA	Leucine (Leu)	UCA	Serine	UAA	Stop	UGA	Stop	A
	UUG	Leucine	UCG	Serine	UAG	Stop	UGG	Tryptophan (Trp)	G
C	CUU	Leucine	CCU	Proline (Pro)	CAU	Histidine (His)	CGU	Arginine (Arg)	U
	CUC	Leucine	CCC	Proline	CAC	Histidine	CGC	Arginine	C
	CUA	Leucine	CCA	Proline	CAA	Glutamine (Gln)	CGA	Arginine	A
	CUG	Leucine	CCG	Proline	CAG	Glutamine	CGG	Arginine	G
A	AUU	Isoleucine (Ile)	ACU	Threonine (Thr)	AAU	Asparagine (Asp)	AGU	Serine (Ser)	U
	AUC	Isoleucine	ACC	Threonine	AAC	Asparagine	AGC	Serine	C
	AUA	Isoleucine	ACA	Threonine	AAA	Lysine (Lys)	AGA	Arginine (Arg)	A
	AUG	Methionine (Met) Start	ACG	Threonine	AAG	Lysine	AGG	Arginine	G
G	GUU	Valine (Val)	GCU	Alanine (Ala)	GAU	Aspartic acid (Asp)	GGU	Glycine (Gly)	U
	GUC	Valine	GCC	Alanine	GAC	Aspartic acid	GGC	Glycine	C
	GUA	Valine	GCA	Alanine	GAA	Glutamic acid (Glu)	GGA	Glycine	A
	GUG	Valine	GCG	Alanine	GAG	Glutamic acid	GGG	Glycine	G

(Row labels U, C, A, G at left = First Base; column U, C, A, G at right = Third Base)

Because the genetic code has three stop codons, 61 triplets remain to specify only 20 amino acids. Therefore, several different codons may code for the same amino acid. For example, six codons—UUA, UUG, CUU, CUC, CUA, and CUG—code for leucine (see Table 13-3). However, each individual codon specifies one, and only one, amino acid.

Translating the codons of mRNA into proteins is the job of tRNA and ribosomes. Remember that tRNA transports amino acids to the ribosomes and that distinct tRNA molecules carry each different type of amino acid. Each of these unique tRNAs has three exposed bases, called an *anticodon*. The bases of an anticodon are complementary to the bases of a codon in mRNA. For example, the mRNA codon GUU forms base pairs with the anticodon CAA of a tRNA that has the amino acid valine attached to it. A ribosome will then incorporate valine into a growing protein chain.

CHECK YOUR LEARNING

Can you ...

- describe how information is encoded in DNA and RNA, and how this information flows from DNA to RNA to protein?
- explain the difference between transcription and translation and how each process is used to convert information in DNA to the amino acid sequence of a protein?

13.2 HOW IS THE INFORMATION IN A GENE TRANSCRIBED INTO RNA?

Transcription (**FIG. 13-3**) consists of three steps: (1) initiation, (2) elongation, and (3) termination. These three steps correspond to the three major parts of most genes in both eukaryotes and prokaryotes: (1) a promoter region at the beginning of the gene, where transcription is started, or initiated; (2) the "body" of the gene, where elongation of the RNA strand occurs; and (3) a termination signal at the end of the gene, where RNA synthesis stops, or terminates.

Transcription Begins When RNA Polymerase Binds to the Promoter of a Gene

The enzyme **RNA polymerase** catalyzes the synthesis of RNA. Near the beginning of every gene is a DNA sequence called the **promoter.** When RNA polymerase binds to the promoter of a gene, the DNA double helix at the beginning of the gene unwinds and transcription begins (**FIG. 13-3 ❶**).

In eukaryotic cells, a promoter consists of two main parts: (1) a short sequence of bases, often TATAAA, that binds RNA polymerase; and (2) one or more other sequences called *response elements,* so named because they allow a cell to respond to changing conditions. Proteins called transcription factors, which are activated in a cell in response to developmental or environmental changes, attach to a response element, enhancing or suppressing binding of RNA polymerase to the promoter and, consequently, enhancing or suppressing transcription of the gene. We will return to the topic of gene regulation in Section 13.5.

Elongation Generates a Growing Strand of RNA

After binding to the promoter, RNA polymerase travels down one of the DNA strands, called the **template strand,** synthesizing a single strand of RNA with bases complementary to those in the DNA (**FIG. 13-3 ❷**). Like DNA polymerase,

▶ **FIGURE 13-3 Transcription is the synthesis of RNA from instructions in DNA** A gene is a segment of a chromosome's DNA. One of the DNA strands that make up the double helix will serve as the template for the synthesis of an RNA molecule with bases complementary to the bases in the DNA strand.

THINK CRITICALLY If the other DNA strand of this molecule were the template strand, in which direction would the RNA polymerase travel?

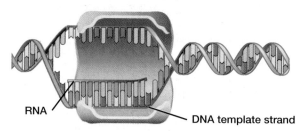

1 **Initiation:** RNA polymerase binds to the promoter region of DNA near the beginning of a gene, separating the double helix near the promoter.

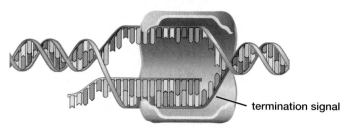

2 **Elongation:** RNA polymerase travels along the DNA template strand (blue), unwinding the DNA double helix and synthesizing RNA by catalyzing the addition of ribose nucleotides into an RNA molecule (red). The nucleotides in the RNA are complementary to the template strand of the DNA.

RNA polymerase always travels along the DNA template strand starting at the 3′ end of a gene and moving toward the 5′ end. Base pairing between RNA and DNA is the same as between two strands of DNA, except that uracil in RNA pairs with adenine in DNA (see Table 13-1).

After about 10 nucleotides have been added to the growing RNA chain, the first nucleotides of the RNA separate from the DNA template strand. This separation allows the two DNA strands to rewind into a double helix (**FIG. 13-3 ❸**). As the RNA molecule continues to elongate, one end drifts away from the DNA, while RNA polymerase keeps the other end attached to the template strand of the DNA. Sometimes multiple RNA polymerases land on the template strand of DNA, one after another, and transcribe dozens of strands of RNA in rapid succession (**FIG. 13-4**).

3 **Termination:** At the end of the gene, RNA polymerase encounters a DNA sequence called a termination signal. RNA polymerase detaches from the DNA and releases the RNA molecule.

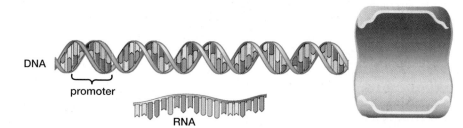

4 **Conclusion of transcription:** After termination, the DNA completely rewinds into a double helix. The RNA molecule is free to move from the nucleus to the cytoplasm for translation, and RNA polymerase may move to another gene and begin transcription once again.

Transcription Stops When RNA Polymerase Reaches the Termination Signal

RNA polymerase continues along the template strand of the gene until it reaches a sequence of DNA bases known as the *termination signal*. The termination signal causes RNA polymerase to release the completed RNA molecule and detach from the DNA (**FIG. 13-3 ❸, ❹**).

In Eukaryotes, a Precursor RNA Is Processed to Form mRNA

Although termination is the final step in transcription, most types of RNA molecules must be modified before they can carry out their functions. Here, we will describe how the RNA molecules transcribed from a gene are processed by eukaryotic cells to form active messenger RNAs.

finished mRNA through pores in the nuclear envelope to the cytoplasm, bind the mRNA to a ribosome, and protect the mRNA molecule from degradation by cellular enzymes. To produce a finished mRNA, enzymes in the nucleus cut this RNA molecule apart at the junctions between introns and exons, splice together the protein-coding exons, and discard the introns ❸. The finished mRNA molecule leaves the nucleus and enters the cytoplasm through pores in the nuclear envelope ❹. In the cytoplasm, the mRNA binds to ribosomes, which synthesize the protein specified by the mRNA base sequence.

Functions of Intron–Exon Gene Structure

Why do eukaryotic genes contain introns and exons? This gene structure appears to serve at least two functions. The first is to allow a cell to produce several different proteins from a single gene by splicing exons together in different ways. For example, a gene called *CT/CGRP* is transcribed in both the thyroid and the brain. In the thyroid, one splicing arrangement results in the synthesis of the hormone calcitonin, which helps regulate calcium concentrations in the blood. In the brain, a different splicing arrangement results in the synthesis of a protein used as a messenger for communication between nerve cells. Most vertebrate genes are spliced into two or more final mRNA molecules, although it is not known how many of these mRNAs are actually translated into functional proteins.

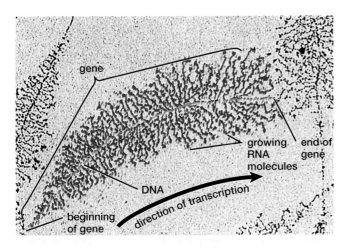

▲ **FIGURE 13-4 RNA transcription in action** This colorized electron micrograph shows the progress of RNA transcription in the egg of an African clawed toad. In each treelike structure, the central "trunk" is DNA and the "branches" are RNA molecules. A series of RNA polymerase enzymes (too small to be seen here) is traveling down the DNA, each synthesizing a strand of RNA. The beginning of the gene is on the left. The short RNA molecules on the left have just begun to be synthesized; the long RNA molecules on the right are almost finished.

THINK CRITICALLY Why do you think so many mRNA molecules are being transcribed from the same gene?

Most eukaryotic genes consist of two or more segments of DNA with nucleotide sequences that code for a protein, interrupted by sequences that are not translated into protein. The coding segments are called **exons,** because they are *ex*pressed in protein; the untranslated segments are called **introns,** because they are *intra*genic, meaning "within a gene" (**FIG. 13-5a**). In humans, the average gene contains eight or nine exons.

Transcription of a eukaryotic protein-coding gene produces a very long RNA strand, called a *precursor mRNA* or *pre-mRNA*, which starts before the first exon and ends after the last exon (**FIG. 13-5b** ❶). More nucleotides are added at the beginning and end of this pre-mRNA molecule, forming a "cap" and "tail" ❷. These nucleotides will help move the

▶ **FIGURE 13-5 Messenger RNA synthesis in eukaryotic cells (a)** Eukaryotic genes consist of exons (medium blue), which code for the amino acid sequence of a protein, and introns (dark blue), which do not. **(b)** Eukaryotic cells synthesize mRNA (red) in several steps.

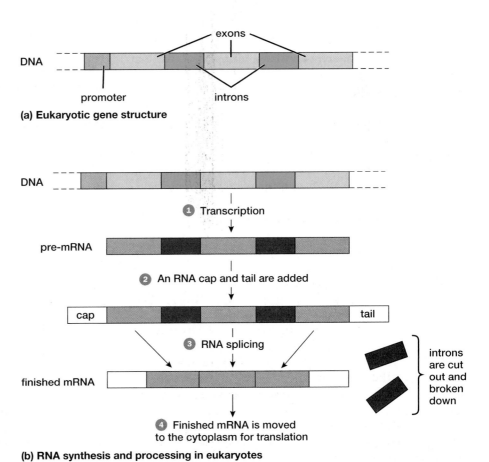

(a) Eukaryotic gene structure

(b) RNA synthesis and processing in eukaryotes

The second advantage is that fragmented genes may provide a quick and efficient way for eukaryotes to evolve new proteins with new functions. In a process called exon shuffling, exons may be moved intact from one gene to another. Most exon shuffling is harmful. But sometimes, exon shuffling produces new genes whose protein products enhance the survival and reproduction of the organism that carries them. These beneficial genes would be favored by natural selection.

CHECK YOUR LEARNING

Can you ...

- describe the process of transcription, explaining how DNA, RNA, and RNA polymerase interact to produce a strand of RNA?

13.3 HOW IS THE BASE SEQUENCE OF mRNA TRANSLATED INTO PROTEIN?

Prokaryotic and eukaryotic cells differ in the organization of their genes, how they produce a functional mRNA molecule from the instructions in their DNA, and the timing and location of translation.

In the prokaryotic genome, most or all of the genes for a complete metabolic pathway sit side by side on the chromosome (**FIG. 13-6a**). Most prokaryotic genes do not contain introns. Therefore, all the nucleotides in a prokaryotic gene usually code for the amino acids in a protein. Finally, prokaryotic mRNA can be directly translated into protein, without further processing.

Prokaryotic cells usually transcribe a single, long mRNA from a series of adjacent genes, each of which specifies a different protein in a metabolic pathway. Because prokaryotic cells do not have a nuclear membrane separating their DNA from the cytoplasm (see Fig. 4-3), transcription and translation usually occur at the same place and time. In most cases, as soon as the beginning of an mRNA molecule separates from the DNA during transcription, ribosomes attach to the mRNA and start translating its codons into protein (**FIG. 13-6b**).

Converting the genetic information in DNA to protein is much more complex in eukaryotes. For example, the DNA of eukaryotic cells is contained in the nucleus, whereas the ribosomes reside in the cytoplasm. The genes that encode the proteins needed for a metabolic pathway in eukaryotes are not clustered together as they are in prokaryotes, but may be dispersed among several chromosomes. And, as we have seen, the RNA molecules copied from protein-coding genes during transcription cannot be directly translated into protein, but must first be processed to produce functional mRNA.

Although the translation of mRNA into protein is quite similar in prokaryotic and eukaryotic cells, our discussion will focus on eukaryotic cells.

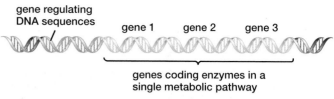

(a) Gene organization on a prokaryotic chromosome

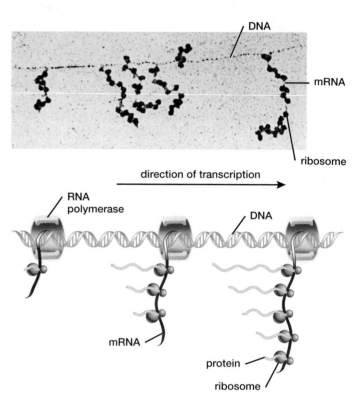

(b) Simultaneous transcription and translation in prokaryotes

▲ **FIGURE 13-6 Transcription and translation are coupled in prokaryotic cells (a)** In prokaryotes, many or all of the genes for a complete metabolic pathway lie side by side on the chromosome. **(b)** Transcription and translation are simultaneous in prokaryotes. In the electron micrograph, RNA polymerase (not visible at this magnification) travels from left to right on a strand of DNA. As it synthesizes an mRNA molecule, ribosomes bind to the mRNA and immediately begin synthesizing a protein (not visible). The diagram below the micrograph shows the key molecules involved.

During Translation, mRNA, tRNA, and Ribosomes Cooperate to Synthesize Proteins

Like transcription, translation has three steps: (1) initiation, (2) elongation of the protein chain, and (3) termination (**FIG. 13-7**).

Initiation: tRNA and mRNA Bind to a Ribosome

A "preinitiation complex"—composed of a small ribosomal subunit, a start (methionine) tRNA, and several other proteins ❶—binds to the beginning of an mRNA molecule. The preinitiation complex moves along the mRNA until it finds a start (AUG) codon, which forms base pairs with the UAC

Initiation:

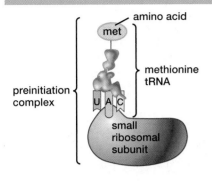

① A tRNA with an attached methionine amino acid binds to a small ribosomal subunit, forming a preinitiation complex.

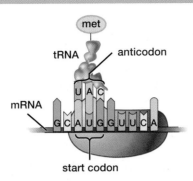

② The preinitiation complex binds to an mRNA molecule. The methionine (met) tRNA anticodon (UAC) base-pairs with the start codon (AUG) of the mRNA.

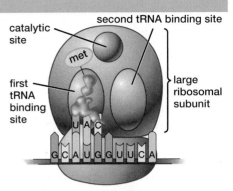

③ The large ribosomal subunit binds to the small subunit. The methionine tRNA binds to the first tRNA site on the large subunit.

Elongation:

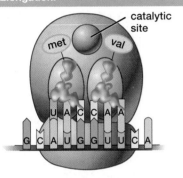

④ The second codon of mRNA (GUU) base-pairs with the anticodon (CAA) of a second tRNA carrying the amino acid valine (val). This tRNA binds to the second tRNA site on the large subunit.

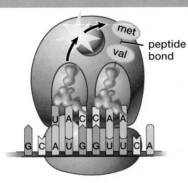

⑤ The catalytic site on the large subunit catalyzes the formation of a peptide bond linking the amino acids methionine and valine. The two amino acids are now attached to the tRNA in the second binding site.

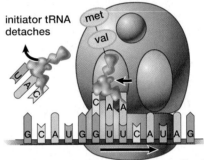

ribosome moves one codon to the right

⑥ The "empty" tRNA is released and the ribosome moves down the mRNA, one codon to the right. The tRNA that is attached to the two amino acids is now in the first tRNA binding site and the second tRNA binding site is empty.

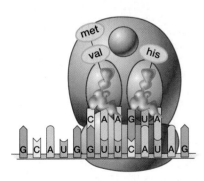

⑦ The third codon of mRNA (CAU) base-pairs with the anticodon (GUA) of a tRNA carrying the amino acid histidine (his). This tRNA enters the second tRNA binding site on the large subunit.

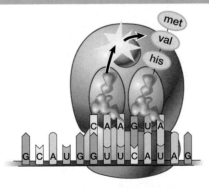

⑧ The catalytic site forms a peptide bond between valine and histidine, leaving the peptide attached to the tRNA in the second binding site. The tRNA in the first site leaves, and the ribosome moves one codon over on the mRNA.

Termination:

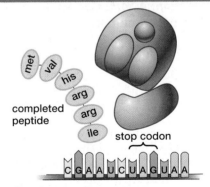

⑨ This process repeats until a stop codon is reached; the mRNA and the completed peptide are released from the ribosome, and the subunits separate.

▲ **FIGURE 13-7 Translation is the process of protein synthesis** Translation decodes the base sequence of an mRNA into the amino acid sequence of a protein.

THINK CRITICALLY Examine step 9. If mutations changed all of the guanine molecules visible in the mRNA sequence shown here to uracil, how would the translated peptide differ from the one shown?

anticodon of the methionine tRNA ❷. A large ribosomal subunit then attaches to the small subunit, sandwiching the mRNA between the two subunits and holding the methionine tRNA in the first tRNA binding site ❸. The ribosome is now ready to translate the mRNA.

Elongation: Amino Acids Are Added One at a Time to the Growing Protein Chain

A ribosome holds two mRNA codons aligned with the two tRNA binding sites of the large subunit. A second tRNA, with an anticodon complementary to the second codon of the mRNA, moves into the second tRNA binding site on the large subunit ❹. The catalytic site of the large subunit breaks the bond holding the first amino acid (methionine) to its tRNA and forms a peptide bond between this amino acid and the amino acid attached to the second tRNA ❺. Ribosomal RNA, and not one of the proteins of the large subunit, catalyzes the formation of the peptide bond. Because it is made of RNA, not protein, the catalytic site of a ribosome is called a ribozyme.

After the peptide bond is formed, the first tRNA is no longer attached to an amino acid, and the second tRNA carries a two-amino-acid chain. The ribosome releases the empty tRNA and shifts to the next codon on the mRNA molecule ❻. The tRNA holding the chain of amino acids also shifts, moving from the second to the first binding site of the ribosome. A new tRNA, with an anticodon complementary to the third codon of the mRNA, binds to the empty second site ❼. The catalytic site now joins the third amino acid to the growing protein chain ❽. The empty tRNA leaves the ribosome, the ribosome shifts to the next codon on the mRNA, and the process repeats, one codon at a time.

Termination: A Stop Codon Signals the End of Translation

When the ribosome reaches a stop codon in the mRNA, protein synthesis terminates. Stop codons do not bind to tRNA. Instead, the ribosome releases the finished protein chain and the mRNA ❾. The ribosome then disassembles into its large and small subunits.

SUMMING UP: Decoding the Sequence of Bases in DNA into the Sequence of Amino Acids in Protein

Let's summarize how a cell decodes the genetic information of DNA and synthesizes a protein (**FIG. 13-8**):

❶ With some exceptions, such as the genes for tRNA and rRNA, each gene codes for the amino acid sequence of a protein. The DNA of a gene consists of the template strand, which is transcribed into mRNA, and its complementary strand, which is not transcribed.

❷ Transcription produces an RNA molecule that is complementary to the template strand. In prokaryotes, this RNA is the messenger RNA that will be translated into protein.

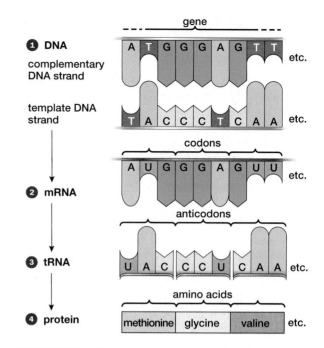

▲ **FIGURE 13-8 Complementary base pairing is required to decode genetic information**

In eukaryotes, this RNA molecule undergoes splicing to produce the final mRNA that will be translated. Sequences of three bases in mRNA, called codons, specify either the beginning of translation (the start codon, AUG), an amino acid, or the end of translation.

❸ Meanwhile, enzymes in the cytoplasm attach the appropriate amino acid to each tRNA, as determined by the tRNA's anticodon.

❹ The mRNA moves out of the nucleus to a ribosome in the cytoplasm. Transfer RNAs carry their attached amino acids to the ribosome. There, the bases in tRNA anticodons bind to complementary bases in mRNA codons. The ribosome catalyzes the formation of peptide bonds that join the amino acids to form a protein with the amino acid sequence specified by the sequence of bases in mRNA. When a stop codon is reached, the finished protein is released from the ribosome.

This decoding chain, from DNA bases to mRNA codons to tRNA anticodons to amino acids, results in the synthesis of a protein with an amino acid sequence determined by the base sequence of a gene.

CHECK YOUR LEARNING
Can you ...
- describe the process of translation?
- explain how the production of mRNA differs between prokaryotic and eukaryotic cells?
- describe how ribosomes, mRNA, and tRNA cooperate to produce a protein?

CASE STUDY   CONTINUED

Cystic Fibrosis

Some mutations in the *CFTR* gene result in a complete absence of correctly spliced mRNA molecules and cause severe cystic fibrosis. Other mutations seem to "confuse" the splicing machinery so that both correct and incorrect mRNA molecules are made. However, most mutations in the *CFTR* gene change codons in the exons of the gene. As you know, individual codons either specify an amino acid or stop translation. How do altered codons affect protein structure and function?

13.4 HOW DO MUTATIONS AFFECT PROTEIN STRUCTURE AND FUNCTION?

Mistakes during DNA replication, ultraviolet rays in sunlight, chemicals in cigarette smoke, and a host of other environmental factors may cause mutations—changes in the sequence of bases in DNA. The consequences for an organism's structure and function depend on how the mutation affects the protein encoded by the mutated gene.

The Effects of Mutations Depend on How They Alter the Codons of mRNA

Mutations may be categorized as inversions, translocations, deletions, insertions, and substitutions (see Figs. 12-9 and 12-10). These different types of mutations differ greatly in how they affect DNA and, consequently, their likelihood of producing significant alterations in protein structure and function.

Inversions and Translocations

Inversions are mutations that occur when a piece of DNA is cut out of a chromosome, flipped around, and reinserted in a reversed orientation. **Translocations** are mutations that occur when a piece of DNA is removed from one chromosome and attached to another. Inversions and translocations may be relatively benign if entire genes, including their promoters, are merely moved from one place to another. In these cases, the mRNA transcribed from the gene will contain all of the original codons. However, if a gene is split in two, it will no longer code for a complete, functional protein. For example, almost half the cases of severe hemophilia are caused by an inversion in the gene that encodes a protein required for blood clotting.

Deletions and Insertions

In a **deletion mutation,** one or more pairs of nucleotides are removed from a gene. In an **insertion mutation,** one or more pairs of nucleotides are inserted into a gene. If one or two pairs of nucleotides are removed or added, protein function is usually completely ruined. Why? Think back to

the genetic code: Three nucleotides encode a single amino acid. Therefore, deleting or inserting one or two nucleotides, or any number that isn't a multiple of three, changes all of the codons that follow the deletion or insertion. Consider this sentence, composed of all three-letter words: THE DOG SAW THE CAT SIT AND THE FOX RUN. Deleting or inserting a letter (deleting the first E, for example), changes all of the following words: THD OGS AWT HEC ATS ITA NDT HEF OXR UN. Most of the amino acids of a protein synthesized from an mRNA containing such a mutation will be incorrect, so the protein will be nonfunctional.

Deleting or inserting three pairs of nucleotides sometimes has only minor effects on the protein, regardless of whether the three nucleotide pairs that were deleted or inserted make up a single codon or overlap into two codons. Returning to our model sentence, let's suppose that we delete OGS. The sentence now reads: THE DAW THE CAT SIT AND THE FOX RUN, most of which still makes sense. If we add a new three-letter word, such as FAT, even in the middle of one of the original words, most of the sentence still makes sense, such as THE DOG SAF ATW THE CAT SIT AND THE FOX RUN.

Substitutions

In a **nucleotide substitution mutation,** a single base pair in DNA is changed. A substitution within a protein-coding gene can produce one of four possible outcomes. Let's consider substitutions that occur in the gene encoding beta-globin, one of the subunits of hemoglobin, the oxygen-carrying protein in red blood cells (**TABLE 13-4**). The other type of subunit in hemoglobin is called alpha-globin; a normal hemoglobin molecule consists of two alpha and two beta subunits. In the first three examples, we will consider the results of mutations that occur in the sixth codon of the beta-globin gene (CTC in DNA, GAG in mRNA), which specifies glutamic acid—a charged, hydrophilic, water-soluble amino acid (see Chapter 3). The fourth example is a mutation that changes the 17th codon to a stop codon.

- *The amino acid sequence of the protein may be unchanged.* Recall that many amino acids can be encoded by several different codons. If a substitution mutation changes the beta-globin DNA base sequence from CTC to CTT, this sequence still codes for glutamic acid. Therefore, the protein synthesized from the mutated gene remains unchanged.

- *The amino acid sequence may be altered, but protein function may be essentially unchanged.* Many proteins have regions in which the exact amino acid sequence is relatively unimportant. In beta-globin, the amino acids on the outside of the protein must be hydrophilic to keep the protein dissolved in the cytoplasm of red blood cells. Exactly *which* hydrophilic amino acids are on the outside doesn't matter much. Substitutions in which the resulting amino acid is the same as, or functionally equivalent to, the original amino acid are called **neutral mutations** because they do not detectably change the function of the encoded protein. There is little or no natural selection for or against a neutral mutation.

TABLE 13-4	Effects of Mutations in the Hemoglobin Gene					
	DNA (Template Strand)	mRNA	Amino Acid	Properties of Amino Acid	Functional Effect on Protein	Disease
Original codon 6	CTC	GAG	Glutamic acid	Hydrophilic	Normal protein function	None
Mutation 1	CT**T**	GA**A**	Glutamic acid	Hydrophilic	Neutral; normal protein function	None
Mutation 2	**G**TC	**C**AG	Glutamine	Hydrophilic	Neutral; normal protein function	None
Mutation 3	C**A**C	G**U**G	Valine	Hydrophobic	Loses water solubility; compromises protein function	Sickle-cell anemia
Original codon 17	TTC	AAG	Lysine	Hydrophilic	Normal protein function	None
Mutation 4	**A**TC	**U**AG	Stop codon	Ends translation after amino acid 16	Synthesizes only part of the protein; eliminates protein function	Beta-thalassemia

- *Protein function may be changed by an altered amino acid sequence.* A mutation from CTC to CAC replaces glutamic acid (hydrophilic) with valine (hydrophobic). Hydrophobic valines on the outside of the hemoglobin molecules cause them to clump together, distorting the shape of the red blood cells. This substitution is the genetic defect that causes sickle-cell anemia (see Chapter 11).
- *Protein function may be destroyed by a premature stop codon.* A particularly catastrophic mutation occasionally occurs in the 17th codon of the beta-globin gene (TTC in DNA, AAG in mRNA). This codon specifies the amino acid lysine. A mutation from TTC to ATC (UAG in mRNA) results in a stop codon, halting translation of beta-globin mRNA before the protein is completed. People who inherit this mutant allele from both parents do not synthesize any functional beta-globin protein; they manufacture hemoglobin consisting entirely of alpha-globin subunits. This "pure alpha" hemoglobin does not bind oxygen very well. People with this condition, called beta-thalassemia, require regular blood transfusions throughout life.

CHECK YOUR LEARNING

Can you ...

- describe the different types of mutations?
- explain why different mutations can have different effects on protein function?

13.5 HOW IS GENE EXPRESSION REGULATED?

The complete human genome contains about 20,000 genes that code for proteins and probably thousands of genes for "noncoding RNA," that is, genes whose final product is RNA, not protein. All of these genes are present in almost every body cell, but any individual cell expresses (transcribes and, if the gene product is a protein, translates) only a small fraction of them. Some genes are expressed in all cells because they encode proteins or RNA molecules that are essential for the life of any cell. For example, all cells need to synthesize proteins, so they all transcribe the genes for tRNA, rRNA, and

CASE STUDY **CONTINUED**

Cystic Fibrosis

There are more than 1,900 different defective alleles of the *CFTR* gene. The most common defective *CFTR* allele originated as a deletion mutation that removed three nucleotides—one codon. Losing this codon deletes a crucial amino acid from the CFTR protein, causing it to be misshapen. Normally, the CFTR protein is synthesized by ribosomes on rough endoplasmic reticulum (ER), enters the ER, and then is transported to the plasma membrane. The misshapen CFTR protein, however, is broken down within the ER and never reaches the plasma membrane. Four other common mutant *CFTR* alleles are substitutions that introduce a stop codon in the middle of the mRNA, so translation terminates partway through. Still other substitution mutations produce proteins that are completely synthesized and inserted into the plasma membrane, but do not form functional chloride channels.

Some *CFTR* alleles can produce functional chloride channels, but nevertheless cause cystic fibrosis. How can that be? These alleles affect gene expression, including how often a gene is transcribed and translated, and how the activity of the resulting protein is controlled, as we describe in Section 13.5.

ribosomal proteins. Other genes are expressed exclusively in certain types of cells, at certain times in an organism's life, or under specific environmental conditions. For example, even though every cell in your body contains the gene for the milk protein casein, that gene is expressed only in women, only in certain breast cells, and only when a woman is breast-feeding.

Some aspects of the regulation of gene expression in eukaryotes and prokaryotes are similar. In both, not all genes are transcribed and translated all the time. Further, controlling the rate of transcription of specific genes is an important mechanism of gene regulation in both. However, there are substantial differences as well, as we describe below.

In Prokaryotes, Gene Expression Is Primarily Regulated at the Level of Transcription

Bacterial DNA is often organized in packages called **operons,** in which the genes for related functions lie close to one another (FIG. 13-9a). An operon consists of four parts: (1) a **regulatory**

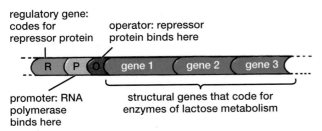

(a) **Structure of the lactose operon**

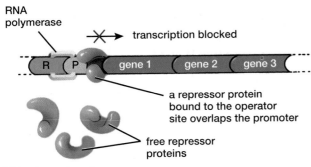

(b) **Lactose absent**

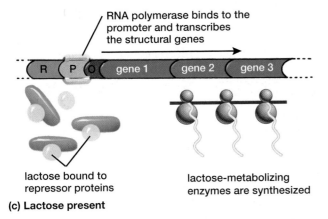

(c) **Lactose present**

▲ **FIGURE 13-9 Regulation of the lactose operon (a)** The lactose operon consists of a regulatory gene, a promoter, an operator, and three structural genes that code for enzymes necessary for lactose metabolism. **(b)** In the absence of lactose, repressor proteins bind to the operator of the lactose operon, preventing RNA polymerase from transcribing the structural genes. **(c)** When lactose is present, it binds to the repressor proteins, making the repressor proteins unable to bind to the operator. RNA polymerase binds to the promoter, moves past the unoccupied operator, and transcribes the structural genes.

gene, which controls the timing or rate of transcription of other genes; (2) a promoter, which RNA polymerase recognizes as the place to start transcription; (3) an **operator,** which governs the access of RNA polymerase to the promoter, and (4) the **structural genes,** which encode the related enzymes or other proteins. Operons are regulated as units; therefore, proteins that work together to perform a specific function may be synthesized simultaneously when the need arises.

Prokaryotic operons may be regulated in a variety of ways. Some operons encode enzymes that are needed by the

Bruises typically progress from purple to green to yellow. This sequence is visual evidence of the control of gene expression. If you bang your shin on a chair, blood vessels break and release red blood cells, which burst and spill their hemoglobin. Hemoglobin and its iron-containing heme group are dark bluish-purple in the deoxygenated state, so fresh bruises are purple. Heme, which is toxic to the liver, kidneys, brain, and blood vessels, stimulates transcription of the heme oxygenase gene. Heme oxygenase is an enzyme that converts heme to biliverdin, which is green. A second enzyme, which is always present because its gene is always expressed, converts biliverdin to bilirubin, which is yellow. The bruise finally disappears as bilirubin moves to the liver, which secretes it into the bile. You can follow the detoxification of heme by watching your bruise change color.

Why Bruises Turn Colors?

cell just about all the time, such as the enzymes that synthesize many amino acids. Such operons are usually transcribed continuously, unless the bacterium encounters a surplus of that particular amino acid. Other operons encode enzymes that are needed only occasionally, for instance, to digest a relatively rare food. They are transcribed only when the bacterium encounters that food.

Consider the common intestinal bacterium, *Escherichia coli* (*E. coli*). This bacterium must live on whatever types of nutrients its host eats, and it can synthesize many different enzymes to metabolize a wide variety of foods. The genes that code for these enzymes are transcribed only when the enzymes are needed. The enzymes that metabolize lactose, the principal sugar in milk, are a case in point. The **lactose operon** contains three structural genes, each coding for an enzyme that aids in lactose metabolism (see Fig. 13-9a).

The lactose operon is shut off, or repressed, unless activated by the presence of lactose. The regulatory gene of the lactose operon directs the synthesis of a **repressor protein.** When the repressor binds to the operator site, RNA polymerase cannot transcribe the structural genes. Consequently, the bacterium does not synthesize lactose-metabolizing enzymes (**FIG. 13-9b**).

When *E. coli* colonize the intestines of a newborn mammal, however, they find themselves bathed in lactose whenever their host nurses from its mother. Lactose molecules enter the bacteria and bind to the repressor proteins, changing their shape (**FIG. 13-9c**). The lactose–repressor complex cannot attach to the operator site. Therefore, RNA polymerase binds to the promoter of the lactose operon and transcribes the genes for lactose-metabolizing enzymes, allowing the bacteria to use lactose as an energy source. After the young mammal is weaned, it usually does not consume milk again. The

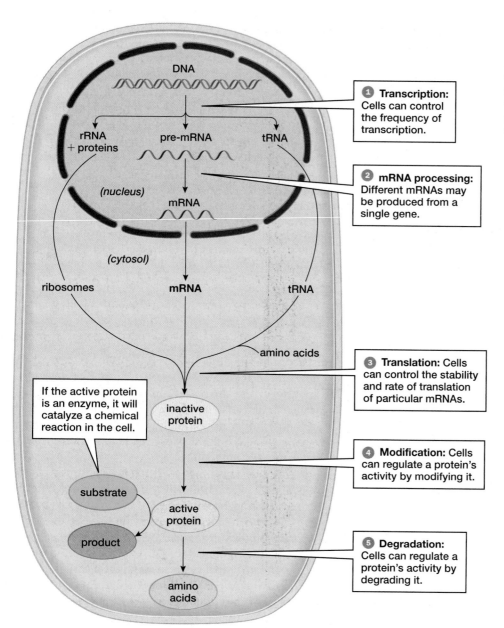

intestinal bacteria no longer encounter lactose, the repressor proteins bind to the operator, and the genes for lactose metabolism are shut off.

In Eukaryotes, Gene Expression Is Regulated at Many Levels

Gene expression in a eukaryotic cell is a multistep process, beginning with transcription of DNA and commonly ending with a protein performing a particular function. Regulation of gene expression can occur at any of these steps, as shown in **FIGURE 13-10**:

❶ *Cells can control the frequency at which a gene is transcribed.* The rate of transcription of specific genes differs among organisms, among cell types in a given

organism, within a given cell type at different stages in the organism's life, and within a cell or organism depending on environmental conditions. Some cases of cystic fibrosis are caused by mutations in the promoter site, so transcription of the gene into mRNA is slowed down or never even begins.

❷ *A single gene may be used to produce different mRNAs and proteins.* A single gene may produce more than one protein (as we described in Section 13.3), depending on how the pre-mRNA is spliced to form the finished mRNA that is translated into protein.

❸ *Cells can control the stability and translation of mRNAs.* Some mRNAs are long lasting and are translated into protein many times. Others are translated only a few times before they are degraded. In addition, certain small RNA molecules may block translation of some mRNAs or may target some mRNAs for destruction. Some cases of cystic fibrosis arise from mutations that cause *CFTR* mRNA to be degraded more rapidly than usual or that slow down the translation of the mRNA into CFTR protein.

④ *Cells may modify proteins to regulate their activity.* Many proteins, especially enzymes, may be modified after translation, thereby temporarily or permanently regulating their function. Adding or removing phosphate groups changes the activity of many enzymes, receptors, ion channels and other proteins, providing second-to-second control of the protein's activity. For example, adding a phosphate to the CFTR chloride channel protein opens the channel, allowing chloride ions to flow across the plasma membrane down their concentration gradient. Some *CFTR* mutations cause cystic fibrosis because the channel cannot be phosphorylated. Other proteins require permanent modification to activate them. The protein-digesting enzymes produced by cells in your stomach wall and pancreas, for instance, are initially synthesized in an inactive form, which prevents the enzymes from digesting the cells that produce them. After these inactive forms are secreted into the digestive tract, portions of the enzymes are snipped out to unveil the active site, allowing the enzymes to digest the proteins in food.

⑤ *Cells can control the rate at which proteins are degraded.* By preventing or speeding up a protein's degradation, a cell can rapidly adjust the amount of a particular protein it contains.

Let's examine some of the mechanisms by which cells control transcription and translation.

Regulatory Proteins Binding to a Gene's Promoter Alter Its Rate of Transcription

The promoter regions of virtually all genes contain several different response elements. Therefore, whether these genes are transcribed depends on which transcription factors are synthesized by the cell and whether those transcription factors are active. For example, when cells are exposed to free radicals (see Chapter 2), a transcription factor binds to antioxidant response elements in the promoters of several genes. As a result, the cell produces enzymes that break down free radicals to harmless substances.

Many transcription factors require activation before they can affect gene transcription. One of the best-known examples is the role that the female sex hormone, estrogen, plays in controlling egg production in birds. The gene for albumin, the major protein in egg white, is not transcribed in winter when birds are not breeding and estrogen levels are low. During the breeding season, the ovaries of female birds release estrogen, which enters cells in the oviduct and binds to a transcription factor. The complex of estrogen and its

transcription factor then attaches to an estrogen response element in the promoter of the albumin gene, making it easier for RNA polymerase to bind to the promoter and start transcribing mRNA. The mRNA is translated into large amounts of albumin. Similar activation of gene transcription by steroid hormones occurs in other animals, including humans. The importance of hormonal regulation of transcription during development is illustrated by genetic defects in which receptors for sex hormones are nonfunctional (see "Health Watch: Androgen Insensitivity Syndrome" on page 230).

Epigenetic Controls Alter Gene Transcription and Translation

Epigenetics (which means "in addition to genetics") is the study of how cells and organisms change gene expression and function without changing the base sequence of their DNA. There is disagreement about which processes should be considered to be epigenetic. In general, however, epigenetic control works in three ways: (1) modification of DNA; (2) modification of chromosomal proteins; and (3) changing transcription and translation through the actions of several types of RNA collectively called noncoding RNA. Many types of epigenetic controls can be inherited from parent to daughter cell during mitotic cell division. In organisms as diverse as bacteria, plants, and mice—and maybe even people—epigenetic tags may even be inherited from one generation to the next, as we explore in "Health Watch: The Strange World of Epigenetics" on page 231.

Epigenetic Modification of DNA May Suppress Transcription Certain enzymes in a cell add methyl groups ($-CH_3$) to cytosine bases in specific locations in the cell's DNA, a process called *methylation*. If a gene or its promoter has lots of methylated cytosines, the gene usually will not be transcribed into mRNA, and its instructions will not be used to make proteins. The number and location of methyl groups on DNA are important in normal development and in some diseases. In cancer cells, for example, growth factor genes (see Chapter 9) often have too few methyl groups. This can cause the genes to be transcribed at a very high level, producing high concentrations of growth factors that inappropriately stimulate cell division. If tumor suppressor genes have too many methyl groups, shutting down their transcription, the body is robbed of one of its most effective weapons against cancer. Defective epigenetic control has also been implicated in disorders as varied as heart disease, obesity, and infertility.

Epigenetic Modification of Histones May Enhance Transcription In eukaryotic chromosomes, DNA is wound around "spools" made of proteins called *histones* (see Chapter 9). When the DNA is tightly wound, RNA polymerase can't get to the promoters of genes, so transcription occurs slowly, if at all. However, when acetyl groups ($-COCH_3$) are added to histones, the DNA partially unwinds and RNA polymerase has better access to the promoters, making gene transcription easier.

Health WATCH

Androgen Insensitivity Syndrome

Sometime between 7 and 14 years of age, a girl usually goes through puberty: Her breasts swell, her hips widen, and she begins to menstruate. In rare instances, however, the years go by, but the girl never menstruates. If her physician performs a chromosome test, in some cases the results show that the girl's sex chromosomes are XY. The reason she has not begun to menstruate is that she lacks ovaries and a uterus but instead has immature testes inside her abdominal cavity. She has about the same concentrations of androgens (male sex hormones, such as testosterone) circulating in her blood as would be found in most boys her age. In fact, androgens have been present since early in her development. However, her cells cannot respond to them—a condition called androgen insensitivity syndrome.

The affected gene codes for a protein known as the androgen receptor. In typical males, androgens bind to the receptor proteins, stimulating the transcription of multiple genes that help to produce many male features, including the formation of a penis and the descent of the testes into sacs outside the body cavity. Androgen insensitivity and varying degrees of disruption of male sexual development may be caused by any of 400 recessive mutant alleles of the gene encoding the androgen receptor. Mutations that create a premature stop codon completely eliminate androgen receptor function.

The androgen receptor gene is on the X chromosome. A person who is genetically XY inherits a single allele for the androgen receptor. If this allele codes for nonfunctional androgen receptor proteins, then the person's cells will be unable to respond to testosterone, and male characteristics will not develop. In many respects, female development is the "default" option in humans, and without functional androgen receptors, the affected person's body will develop female characteristics. Thus, a mutation that changes the nucleotide sequence of a single gene, causing a single type of nonfunctional protein to be produced, can cause a person who is genetically XY to be female (**FIG. E13-1**).

▲ **FIGURE E13-1 Androgen insensitivity leads to female features** The cells of these women have both X and Y chromosomes. The women have testes that produce testosterone, but a mutation in their androgen receptor genes make their cells unable to respond to testosterone. (For more information on androgen insensitivity, visit http://aisdsd.org.)

EVALUATE THIS Envision yourself as a physician. A mother, father, and their daughter come to you because the daughter is 16 years old and hasn't had her first menstrual cycle, whereas all of her girlfriends started menstruating years ago. You do a karyotype and find that she is XY. Further molecular genetic testing reveals that she has a mutated androgen receptor allele on her X chromosome. The parents want to know how their daughter inherited the syndrome, why they don't have it, and, if they were to have other children, if they would be androgen insensitive. How would you explain, in terms understandable to a layperson, the inheritance of androgen insensitivity and the likelihood that the parents would have another child with androgen insensitivity syndrome? Include diagrams to help them understand.

Noncoding RNA May Alter Transcription or Translation

Protein-coding genes make up only a small percentage of human DNA. Does that mean that the rest of our DNA is pointless? Far from it. Recently, molecular biologists have found that some of this DNA is transcribed into hundreds, perhaps thousands, of distinct types of noncoding RNA that help to control gene expression.

Noncoding RNA May Regulate Transcription Some types of noncoding RNA affect gene transcription. Some inhibit the binding of RNA polymerase to specific gene promoters, thereby blocking transcription. Others stimulate or inhibit epigenetic changes to DNA or histones in specific locations on specific chromosomes. These noncoding RNAs may enhance or reduce transcription, depending on the exact nature of the epigenetic controls that are affected.

Perhaps the best-known noncoding RNA silences transcription in mammalian X chromosomes. As you know, male mammals have an X and a Y chromosome (XY), and females have two X chromosomes (XX). As a consequence, females have the capacity to synthesize mRNA from genes on their two X chromosomes, whereas males, with only one X chromosome, may produce only half as much. In 1961, Mary Lyon, an English geneticist, hypothesized that one of the two X chromosomes in females is inactivated in some way, so that its genes are not expressed. Subsequent research showed that she was correct. In female mammals, one of the X chromosomes is inactivated, and about 85% of its genes are not transcribed. Early in embryonic development (about

The Strange World of Epigenetics

Most of the controls over gene expression work for time periods ranging from a few seconds to a few days and then fade away. Epigenetic controls, however, often work for the life of an organism. Some may even be passed down from parent to offspring.

Epigenetic controls are important regulators of gene transcription and translation. For example, adding methyl groups to the promoter of the insulin gene turns off transcription. All of the cells of early embryos have methylated, silenced insulin genes. Later in development, methyl groups on the insulin gene are selectively removed in cells destined to become insulin-secreting cells of the pancreas. The other cells of the body contain methylated, silenced insulin genes.

Some cells, such as those in the intestinal lining, divide every day or two—thousands of divisions during a lifetime. Throughout all of these divisions, the insulin genes remain methylated. How? Recall that DNA replication is semiconservative (see Chapter 12). When an intestinal cell divides, each daughter cell receives one parental DNA strand with methyls on the insulin gene and one new DNA strand without methyls on the gene. However, an enzyme in the daughter cells adds the parental methyl pattern to the daughter DNA strand. The result: Intestinal cells have silenced insulin genes.

Coat color in mice provides a striking example of epigenetic control of gene expression (**FIG. E13-2**). In certain strains of mice, the offspring in a single litter of genetically identical mice can have fur that ranges from yellow to mousy brown. Studies have shown that DNA methylation of a single gene controls the color: the more methylation, the less the gene is expressed and the browner the fur. Mice with very little methylation of this gene have high gene expression and yellow fur. They also become obese and have a much higher risk of diabetes and cancer. Feeding pregnant mice a diet high in methionine, folate, soy protein, and vitamin B_{12}, which increases DNA methylation, produces litters in which all the pups are brown.

In the vast majority of cases, methyl patterns on DNA are erased during meiotic cell division or gamete development, so epigenetic changes do not pass from generation to generation. However, there are exceptions. Methyl groups may be added to certain clusters of genes in either the sperm or the egg, resulting in genomic imprinting, in which a given gene will be expressed only if it is inherited from either the father or the mother, respectively. For example, Angelman syndrome, a rare genetic disease characterized by seizures, speech defects, and motor disabilities, is the result of a deletion mutation in chromosome 15. Angelman syndrome occurs only when the mutation has been inherited from the mother. The normal, functional genes on the father's chromosome are silenced by methyl groups and cannot compensate for the mother's mutation. In mice, yellow versus brown fur can also be inherited across generations, principally from the mother.

Epigenetic changes that last for generations have been found in bacteria, protists, fungi, plants, and animals. Even behaviors can be inherited across generations: If mice are trained to associate a specific odor with receiving an electrical shock, their grandchildren can inherit both the memory and a slightly larger brain region that responds to this odor.

▲ **FIGURE E13-2 Epigenetic differences can cause phenotypic differences in genetically identical mice** The obese yellow mouse has far fewer methyl groups on a gene that controls fur color than the slim brown mouse does.

These findings lead to a provocative question: In people, can epigenetic changes caused by parents' life experiences or environment become part of the inheritance of their offspring?

For now, the answer seems to be "Maybe." No one can perform controlled, multigenerational experiments on people, so good data are hard to obtain. Evidence for multigenerational epigenetic inheritance in humans comes from "natural experiments" in which some major event affected a fairly large number of people. One such natural experiment has already occurred in a remote northern area of Sweden called Norrbotten. Until fairly recent times, Norrbotten was extremely isolated. Little food entered or left the region. If crop harvests were good, people stuffed themselves during the following winter; if harvests were bad, people starved. Researchers tracked birth and death records and correlated those with harvests during the 1800s. They found that the grandsons of boys who lived during the years of abundant harvests, and therefore probably overate, lived remarkably shorter lives—from 6 to 32 years shorter, depending on how the data were analyzed—than the grandsons of boys who suffered through winters of near starvation. Similar effects were found in girls.

In other natural experiments, people whose fathers were conceived during the Dutch famine of 1944–1945 at the end of World War II are more likely to be obese than the offspring of fathers who were not undernourished prenatally. A study in England found that when men start smoking at a very early age (before 11 years old), their sons tend to be overweight. These results are intriguing, but no one knows what genes were involved or whether epigenetic methyl groups on DNA caused the difference.

THINK CRITICALLY In some people with type 2 diabetes, the pancreas doesn't secrete enough insulin. If you could analyze the epigenetic tags on the insulin gene and its promoter, what might you expect to see? If you could add or remove epigenetic tags on the gene, what would you do to attempt to normalize insulin secretion in type 2 diabetics?

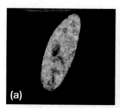 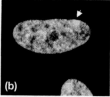 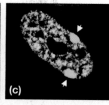

▲ **FIGURE 13-11 Barr bodies provide visible evidence of X-chromosome inactivation** In mammalian cells, only one X chromosome is active; additional X chromosomes are condensed into Barr bodies, visible as bright spots in these micrographs. **(a)** Nuclei from a male (XY) animal have no Barr bodies. **(b)** Nuclei from a female (XX) animal have one Barr body. **(c)** Nuclei from a female with trisomy X (see Chapter 10) have two Barr bodies.

the 16th day in humans), one X chromosome in each of a female's cells begins to produce large amounts of a noncoding RNA molecule called Xist. Xist RNA coats most of that X chromosome, condenses it into a tight mass, and prevents further transcription. The condensed X chromosome, called a **Barr body** after its discoverer, Murray Barr, forms a discrete spot in the nuclei of the cells of female mammals (**FIG. 13-11**).

Usually, large clusters of cells (each cluster descended from a single cell in the early embryo) have the same X chromosome inactivated. As a result, the bodies of female mammals consist of patches of cells in which one of the X chromosomes is fully active and patches of cells in which the other X chromosome is active. The results of this phenomenon are easily observed in tortoiseshell and calico cats (**FIG. 13-12**). The X chromosome of a cat contains a gene encoding an enzyme that produces fur pigment. There are

two common alleles of this gene: One produces orange fur and the other produces black fur. If one X chromosome in a female cat has the orange allele and the other X chromosome has the black allele, the cat will have patches of orange and black fur. These patches represent areas of skin that developed from cells in the early embryo in which different X chromosomes were inactivated. Calico coloring is almost exclusively found in female cats. Because male cats normally have only one X chromosome, a male cat may have black fur or orange fur, but not both.

MicroRNA and RNA Interference Regulate Translation The quantity of any particular protein that a cell synthesizes depends both on how much mRNA is made and on how rapidly and for how long mRNA is translated. Enter RNA interference. Some of the DNA of organisms as diverse as plants, roundworms, and people is transcribed into hundreds of different noncoding RNAs that are subsequently cut up into very short strands appropriately named **microRNA**. Each microRNA is complementary to part of a specific mRNA. These microRNA molecules interfere with translation of the mRNA (hence, the term "RNA interference"). In some cases, these small RNA strands base-pair with the complementary mRNA, forming a little section of double-stranded RNA that cannot be translated. In other cases, the short RNA strands combine with enzymes to cut up complementary mRNA, which also prevents translation.

It may seem strange that a cell would interfere with the translation of its own mRNA. However, RNA interference is important for the development of eukaryotic organisms. For example, in mammals microRNAs influence the development of the heart and brain, the secretion of insulin by the pancreas, and even learning and memory. Defects in microRNA production—either too much or too little of certain microRNAs—can lead to cancer or heart disease.

CHECK YOUR LEARNING

Can you ...

- describe the ways in which information flow from DNA to RNA to protein synthesis to protein function can be regulated?
- explain which controls over gene expression are likely to be very brief, which may be long lasting, and why they differ?

◀ **FIGURE 13-12 Inactivation of the X chromosome regulates gene expression** This female calico kitten carries a gene for orange fur on one X chromosome and a gene for black fur on her other X chromosome. Inactivation of different X chromosomes produces the black and orange patches. The white color is due to an entirely different gene that prevents pigment formation altogether.

Cystic Fibrosis

All of the defective alleles of the *CFTR* gene are recessive to the functional *CFTR* allele. People who are heterozygous, with one normal *CFTR* allele and one copy of any of the defective alleles, produce enough functional CFTR protein for adequate chloride transport. Therefore, they produce normal, watery secretions in their lungs and do not develop cystic fibrosis. Someone with two defective alleles will produce only proteins that don't work properly and will develop the disease.

Genetic diseases such as cystic fibrosis can't be "cured" in the way that an infection can be cured by killing offending bacteria or viruses. Typically, genetic diseases are treated by replacing the lost function, such as by giving insulin to diabetics, or by relieving the symptoms. In cystic fibrosis, the most common therapies relieve some of the symptoms. These treatments include antibiotics, medicines that open the airways, and physical therapy to drain the lungs.

What ultimately happens to a person with cystic fibrosis depends on how defective the mutant alleles are. Canadian triathlete Lisa Bentley, for example, has a relatively mild case of cystic fibrosis (**FIG. 13-13**). However, during a 9-hour triathlon, she produces copious amounts of very salty sweat. Why? One of the roles of the CFTR protein is to promote reabsorption of salt from sweat and transport the salt back into the blood. In cystic fibrosis, salt reabsorption fails, and people can lose life-threatening amounts of salt during exercise. Therefore, it's a constant challenge for Bentley to keep her body supplied with salt during a race. Nevertheless, Bentley has won 11 Ironman triathlons, including the Australian Ironman Triathlon five straight years. Bentley carefully controls her diet, especially her salt intake. Vigorous exercise helps to clear out her lungs. She also scrupulously avoids situations where she might be exposed to contagious diseases. Her cystic fibrosis hasn't kept her from becoming one of the finest athletes in the world.

▲ **FIGURE 13-13 Lisa Bentley, sometimes called the Iron Queen, wins another triathlon**

CONSIDER THIS About 5% of the cases of cystic fibrosis arise from a substitution mutation in which a full-length CFTR protein is synthesized and inserted into the plasma membrane, but fails to transport chloride. In 2012, the U.S. Food and Drug Administration approved a drug called ivacaftor to treat this form of cystic fibrosis. Ivacaftor binds to the CFTR protein and helps to open the chloride channel. As of 2014, ivacaftor treatment costs about $300,000 a year. How do you think that such treatments should be financed: by the patient, by health insurance, or by governments?

CHAPTER REVIEW

MB Go to **MasteringBiology** for practice quizzes, activities, eText, videos, current events, and more.

Answers to Think Critically, Evaluate This, Multiple Choice, and Fill-in-the-Blank questions can be found in the Answers section at the back of the book.

Summary of Key Concepts

13.1 How Is the Information in DNA Used in a Cell?

Genes are segments of DNA that can be transcribed into RNA and, for most genes, translated into protein. Transcription produces the three types of RNA needed for translation: mRNA, tRNA, and rRNA. Messenger RNA carries the genetic information of a gene from the nucleus to the cytoplasm, where ribosomes use the information to synthesize a protein. There are many different tRNAs. Each tRNA binds a specific amino acid and carries it to a ribosome for incorporation into a protein. Ribosomes are composed of rRNA and proteins, organized into large and small subunits. The genetic code consists of codons, sequences of three bases in mRNA that specify the start of translation (start codon), the amino acids in the protein chain, or the end of protein synthesis (stop codons).

13.2 How Is the Information in a Gene Transcribed into RNA?

Within an individual cell, only certain genes are transcribed. When the cell requires the product of a gene, RNA polymerase binds to the promoter region of the gene and synthesizes a single strand of RNA. This RNA is complementary to the template strand in the gene's DNA double helix. Cellular proteins called transcription factors may bind to DNA near the promoter and enhance or suppress transcription of a given gene.

13.3 How Is the Base Sequence of mRNA Translated into Protein?

In prokaryotic cells, all of the nucleotides of a protein-coding gene code for amino acids; therefore, the RNA transcribed from the gene is the mRNA that will be translated on a ribosome. In eukaryotic cells, protein-coding genes consist of two parts: exons, which are translated into the amino acids in a protein, and introns, which are not. The entire gene, including both introns and exons, is transcribed into a pre-mRNA molecule. The introns of the pre-mRNA are cut out and the exons are spliced together to produce a finished mRNA.

In eukaryotes, mRNA carries the genetic information from the nucleus to the cytoplasm, where ribosomes use this information to synthesize a protein. The two ribosomal subunits come together at the start codon of the mRNA molecule to form a complete protein-synthesizing assembly. Transfer RNAs deliver the appropriate amino acids to the ribosome for incorporation into the growing protein, which depends on base pairing between the anticodon of the tRNA and the codon of the mRNA. Two tRNAs, each carrying an amino acid, bind simultaneously to the ribosome; the large subunit catalyzes the formation of peptide bonds between the amino acids. As each new amino acid is attached, one tRNA detaches, and the ribosome moves over one codon, binding to another tRNA that carries the next amino acid specified by mRNA. Addition of amino acids to the growing protein continues until a stop codon is reached, causing the ribosome to disassemble and to release both the mRNA and the newly formed protein.

13.4 How Do Mutations Affect Protein Structure and Function?

A mutation is a change in the nucleotide sequence of a gene. Mutations can be caused by mistakes in base pairing during replication, by chemical agents, and by environmental factors such as radiation. Common types of mutations include inversions, translocations, deletions, insertions, and substitutions. Mutations vary in their effects on protein function. Neutral mutations produce codons that specify the same amino acid, or a very similar one, as the original codon; in these cases, protein function usually will not change significantly. Other mutations may substitute a functionally different amino acid or may encode a stop codon. These mutations may destroy protein function.

13.5 How Is Gene Expression Regulated?

The expression of a gene requires that it be transcribed and translated and that the resulting protein perform some action within the cell. Which genes are expressed in a cell at any given time is regulated by the function of the cell, the developmental stage of the organism, and the environment. Control of gene regulation can occur at many steps. The amount of mRNA synthesized from a particular gene can be regulated by increasing or decreasing the rate of its transcription, as well as by changing the stability of the mRNA itself. A single gene may be used to produce different proteins, depending on how the pre-mRNA is spliced into the final mRNA. Rates of translation of mRNAs can also be regulated. Regulation of transcription and translation affects how many protein molecules are produced from a particular gene. After they are synthesized, some proteins are cut up into smaller, functional proteins with different functions in distinct cell types. Other proteins

must be modified before they can function. Proteins also vary in how rapidly they are degraded in a cell. In epigenetic regulation, adding methyl groups to DNA often suppresses gene transcription, whereas adding acetyl groups to histones increases gene transcription. Noncoding RNA may suppress transcription, speed up mRNA degradation, or inhibit translation of mRNA.

Key Terms

anticodon 217
Barr body 232
codon 216
deletion mutation 225
epigenetics 229
exon 221
genetic code 218
insertion mutation 225
intron 221
inversion 225
lactose operon 227
messenger RNA (mRNA) 216
microRNA 232
neutral mutation 225
nucleotide substitution mutation 225
operator 227
operon 226
promoter 219
regulatory gene 226
repressor protein 227
ribonucleic acid (RNA) 216
ribosomal RNA (rRNA) 217
ribosome 217
RNA polymerase 219
start codon 218
stop codon 218
structural gene 227
template strand 219
transcription 218
transfer RNA (tRNA) 217
translation 218
translocation 225

Thinking Through the Concepts

Multiple Choice

1. The molecule that carries the genetic information of DNA from the nucleus to the cytoplasm to be used in protein synthesis is
   a. ribosomal RNA.
   b. messenger RNA.
   c. transfer RNA.
   d. microRNA.

2. Which of the following is *not* true of RNA?
   a. It contains the base thymine.
   b. It contains the sugar ribose.
   c. It contains the base adenine.
   d. It is transcribed from DNA.

3. A stop codon
   a. signals the end of protein synthesis on a ribosome.
   b. codes for the amino acid methionine.
   c. signals the end of RNA synthesis.
   d. marks the boundary between an exon and an intron.

4. If the template strand of DNA reads ATTCGTAG,
   a. its complementary strand reads UAAGCAUC.
   b. the base sequence of the RNA transcribed from this sequence is AUUCGUAG.
   c. the base sequence of the RNA transcribed from this sequence is UAAGCAUC.
   d. this segment of DNA contains the information for four codons.

5. Epigenetic modification of gene expression
    a. always inhibits gene transcription.
    b. always stimulates gene expression.
    c. is erased from the DNA following mitotic cell division.
    d. may sometimes be transmitted from generation to generation.

Fill-in-the-Blank

1. Synthesis of RNA from the instructions in DNA is called _____. Synthesis of a protein from the instructions in mRNA is called _____. Which structure in the cell is the site of protein synthesis? _____

2. The three types of RNA that are essential for protein synthesis are _____, _____, and _____. Another type of RNA, which can interfere with translation, is called _____.

3. The genetic code uses _____ (how many?) bases to code for a single amino acid. This sequence of bases in mRNA is called a(n) _____. The complementary sequence of bases in tRNA is called a(n) _____.

4. The enzyme _____ synthesizes RNA from the instructions in DNA. For any given gene, only one DNA strand, called the _____ strand, is transcribed. To begin transcribing a gene, this enzyme binds to a specific sequence of DNA bases located at the beginning of the gene. This DNA sequence is called the _____. Transcription ends when the enzyme encounters a DNA sequence at the end of the gene called the _____.

5. Translation begins with the _____ codon of mRNA and continues until a(n) _____ codon is reached. Individual amino acids are brought to the ribosome by _____. These amino acids are linked into protein by _____ bonds.

6. If a nucleotide is replaced by a different nucleotide, this is called a(n) _____ mutation. _____ mutations occur if nucleotides are added in the middle of a gene. _____ mutations occur if nucleotides are removed from the middle of a gene.

Review Questions

1. How does RNA differ from DNA?

2. Name the three types of RNA that are essential to protein synthesis. What is the function of each?

3. Define the following terms: *genetic code, codon,* and *anticodon.* What is the relationship among the bases in DNA, the codons of mRNA, and the anticodons of tRNA?

4. How is mRNA formed from a eukaryotic gene?

5. Diagram and describe protein synthesis.

6. Explain how complementary base pairing is involved in both transcription and translation.

7. Describe the principal mechanisms of regulating gene expression.

8. Define *mutation.* Describe four different effects of nucleotide substitution mutations on protein sequence and function.

Applying the Concepts

1. Many years ago, some researchers reported that they could transfer learning from one animal (a flatworm) to another by feeding trained animals to untrained animals. Further, they claimed that RNA was the active molecule of learning. Given your knowledge of the roles of RNA and protein in cells, do you think that a *specific* memory (for example, remembering the base sequences of codons of the genetic code) could be encoded by a *specific* molecule of RNA and that this RNA molecule could transfer that memory to another person? In other words, in the future, could you learn biology by popping an RNA pill? If so, how would this work? If not, can you propose a reasonable hypothesis for the results with flatworms? How would you test your hypothesis?

2. The defective *CFTR* alleles that cause cystic fibrosis vary tremendously in their severity. Some produce completely nonfunctional proteins, while others produce proteins that can transport chloride, just not as efficiently as the normal proteins do. Many people with cystic fibrosis are heterozygotes: They have two different defective *CFTR* alleles. How might the severity of cystic fibrosis symptoms depend on the allele combination in heterozygotes? As good as the better-functioning allele? As bad as the worse-functioning one? Somewhere in between? Why?

"This is my best birthday. Nothing can compare to this." Thomas Haynesworth, shown here with his sister, Sandra (far left) and his Mid-Atlantic Innocence Project attorney, Shawn Armbrust (between Sandra and Thomas), was released from prison on his 46th birthday.

Guilty or Innocent?

IMAGINE SPENDING WELL OVER HALF YOUR LIFE in prison for crimes you didn't commit. For 27 years, this nightmare was real life for Thomas Haynesworth.

It began in early 1984, when a young black man sexually assaulted five women in the East End neighborhood of Richmond, Virginia. Shortly thereafter, Haynesworth, then 18 years old, was walking to the grocery store. He was spotted by one of the women, who identified him as her assailant. The other four women subsequently picked him out of a photo lineup. Haynesworth was swiftly convicted of two rapes and one count of attempted robbery and kidnapping.

But the rapes didn't stop. Between April and December, at least 12 other women were raped, also by a young black man. Finally, Leon Davis was arrested on December 19, and the epidemic of rapes in the East End ceased. Davis was sentenced to prison for multiple life sentences.

By the time Davis was arrested, however, Haynesworth was already serving time. Although Davis's crimes were extremely similar to those for which Haynesworth was convicted, no one thought to revisit Haynesworth's case. At

last, in 2005, then-Governor Mark Warner ordered a review of any biological evidence remaining in thousands of case files dating from 1973 to 1988. In 2009, DNA preserved in one of the Haynesworth files was tested. It showed that Haynesworth was innocent; Davis had sexually assaulted the woman.

You might think that Haynesworth would immediately be set free; instead, he continued to languish in prison. However, the Mid-Atlantic Innocence Project, an organization based at the George Washington University Law School, and a member of the worldwide Innocence Network, took Haynesworth's case. Finally, on March 21, 2011, Haynesworth was released on parole (see photo above). On December 6, he was declared innocent of all the charges against him.

In this chapter, we'll investigate the techniques of biotechnology that now pervade so much of modern life. How do crime scene investigators decide that two DNA samples match? How can biotechnology diagnose inherited disorders? Should biotechnology be used to change the genetic makeup of crops, livestock, or even people?

AT A GLANCE

14.1 WHAT IS BIOTECHNOLOGY?

Biotechnology is the use, and especially the alteration, of organisms, cells, or biological molecules to produce food, biofuels, drugs, or other goods. Some aspects of biotechnology are ancient. People have used yeast to produce bread, beer, and wine for the past 10,000 years. Many plants and animals, including wheat, grapes, dogs, pigs, and cattle, were domesticated and selectively bred for desirable traits 6,000 to 15,000 years ago. For example, selective breeding rapidly transformed relatively slim wild boars, with long tusks and fierce temperaments, into much heavier, more placid domestic pigs.

Although selective breeding is still an important tool for improving livestock and crops, modern biotechnology also uses **genetic engineering** to isolate and manipulate the genes that control inherited characteristics. Genetically engineered cells or organisms have had genes deleted, added, or changed. In addition to its use in improving plants and animals for agriculture, genetic engineering can be used to study how cells and genes work; to combat disease; to produce valuable biological molecules, including hormones and vaccines; and maybe even to restore endangered species or resurrect extinct ones.

A key tool in modern biotechnology is **recombinant DNA,** which is DNA that contains genes or parts of genes from two or more organisms, usually of different species. Recombinant DNA can be produced in bacteria, viruses, or yeast and then transferred into other species. Organisms that contain DNA that has been modified or derived from other species through genetic engineering are called **transgenic** or **genetically modified organisms (GMOs).**

Modern biotechnology includes many methods of analyzing and manipulating DNA, whether or not the DNA is subsequently put into a cell. For example, determining the nucleotide sequence of DNA is crucial for fields as diverse as forensic science, medicine, and evolutionary biology.

In this chapter, we will provide an overview of the methods and applications of biotechnology and discuss the impacts of biotechnology on society. We will organize our discussion around five major themes: (1) recombinant DNA mechanisms found in nature, (2) biotechnology in criminal forensics, (3) production of transgenic plants and animals, (4) analysis of the genomes of humans and other organisms, and (5) applications of biotechnology in medicine.

CHECK YOUR LEARNING

Can you ...
- define biotechnology?
- describe applications of genetic engineering and recombinant DNA?

14.2 WHAT NATURAL PROCESSES RECOMBINE DNA BETWEEN ORGANISMS AND BETWEEN SPECIES?

The process of recombining DNA is not unique to modern laboratories. Many natural processes can transfer DNA from one organism to another, sometimes even to organisms of different species.

Sexual Reproduction Recombines DNA

Homologous chromosomes, inherited from an organism's two parents, exchange DNA by crossing over during meiosis I (see Chapter 10), thereby recombining DNA from two different organisms. When these chromosomes become packaged in sperm and eggs that unite to form zygotes, the resulting offspring contain the recombined chromosomes. In these cases, the recombined DNA almost always comes from members of a single species.

Transformation May Combine DNA from Different Bacterial Species

In **transformation,** bacteria pick up pieces of DNA from the environment (FIG. 14-1). The DNA may be part of the chromosome from another bacterium (FIG. 14-1a), sometimes from another species, or it may be tiny circular DNA molecules called **plasmids** (FIG. 14-1b). A single bacterium

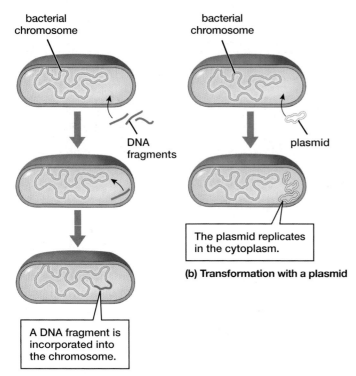

(b) Transformation with a plasmid

The plasmid replicates in the cytoplasm.

A DNA fragment is incorporated into the chromosome.

(a) Transformation with a DNA fragment

▲ **FIGURE 14-1** **Transformation in bacteria** Bacterial transformation occurs when living bacteria take up **(a)** fragments of chromosomes or **(b)** plasmids.

may contain dozens or even hundreds of copies of a plasmid. When the bacterium dies, its plasmids are released into the environment, where they may be picked up by other bacteria of the same or different species. In addition, living bacteria can often pass plasmids directly to other living bacteria. Plasmids may also move from certain bacteria to yeast or plants, transferring genes from a prokaryotic cell to a eukaryotic cell.

Although a bacterium's chromosome contains all the genes the cell needs for basic survival, genes carried by plasmids may permit the bacteria to thrive in novel environments. Some plasmids contain genes that allow bacteria to metabolize unusual energy sources, such as oil. Other plasmids carry genes that enable bacteria to grow in the presence of antibiotics. In environments where antibiotic use is high, particularly in hospitals, bacteria carrying antibiotic-resistance plas-

mids can quickly spread among patients and health care workers, making antibiotic-resistant infections a serious problem.

Viruses May Transfer DNA Between Species

Viruses, which are often little more than genetic material encased in a protein coat, can reproduce only inside cells (**FIG. 14-2**). A virus attaches to specific molecules on the surface of a suitable host cell ❶. Usually the virus then enters the cytoplasm of the host ❷, where it releases its genetic material ❸. The host cell replicates the viral genetic material (DNA or RNA) and synthesizes viral proteins ❹. The replicated genes and viral proteins assemble into new viruses inside the cell ❺. Eventually, the viruses are released and may infect other cells ❻.

Some viruses can transfer genes from one organism to another. In these instances, the viral DNA is inserted into one of the host cell's chromosomes (see Fig. 14-2 ❸). The viral DNA may remain there for days, months, or even years. Every time the cell divides, it replicates the viral DNA along with its own DNA. (Researchers believe that about 8% of the human genome consists of "fossil" viral genes, inserted into our ancestors' DNA thousands to millions of years ago.)

▼ **FIGURE 14-2** **The life cycle of a typical virus** In some cases, viral infections may transfer DNA from one host cell to another.

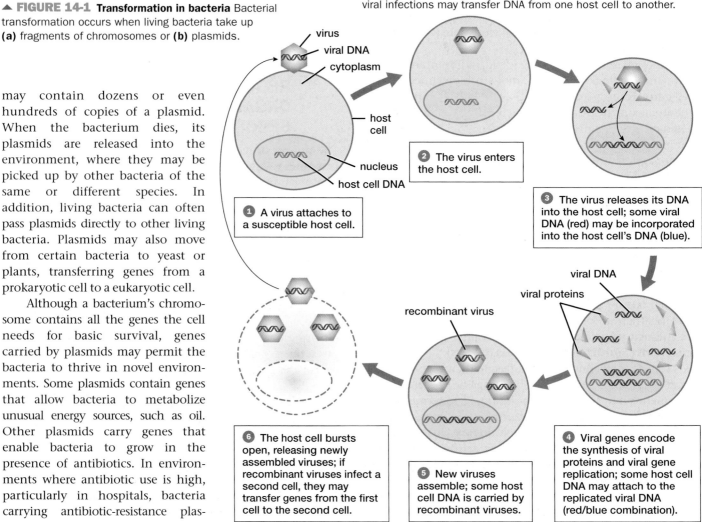

❶ A virus attaches to a susceptible host cell.

❷ The virus enters the host cell.

❸ The virus releases its DNA into the host cell; some viral DNA (red) may be incorporated into the host cell's DNA (blue).

❹ Viral genes encode the synthesis of viral proteins and viral gene replication; some host cell DNA may attach to the replicated viral DNA (red/blue combination).

❺ New viruses assemble; some host cell DNA is carried by recombinant viruses.

❻ The host cell bursts open, releasing newly assembled viruses; if recombinant viruses infect a second cell, they may transfer genes from the first cell to the second cell.

When new viruses are finally produced, some of the host cell's genes may be attached to the viral DNA. If these recombinant viruses infect another cell and insert their DNA into the new host cell's chromosomes, pieces of the previous host's DNA may also be inserted.

Most viruses infect and replicate only in the cells of specific bacterial, animal, or plant species. Therefore, most of the time, viruses move host DNA among different individuals of a single, or closely related, species. However, some viruses may infect species only distantly related to one another. For example, influenza infects birds, pigs, and humans; other viruses have jumped from bats to people, cats to dogs, horses to dogs, and dogs to seals. Gene transfer among viruses that infect multiple species can produce extremely lethal recombined viruses. This happened in 1957 and again in 1968, when recombination between bird and human flu viruses caused global epidemics that killed hundreds of thousands of people.

CHECK YOUR LEARNING

Can you ...

- describe natural processes that recombine DNA, including mechanisms that may combine DNA across species?

14.3 HOW IS BIOTECHNOLOGY USED IN FORENSIC SCIENCE?

Applications of DNA biotechnology vary, depending on the goals of the forensic scientists, biotechnology firms, pharmaceutical companies, physicians, and others who use it. We will begin by describing a few common methods of manipulating DNA and discussing their application to forensic DNA analysis.

The Polymerase Chain Reaction Amplifies DNA

Developed by Kary Mullis in 1986, the **polymerase chain reaction (PCR)** can be used to make billions, even trillions, of copies of selected pieces of DNA. PCR is so crucial to molecular biology that it earned Mullis a share in the Nobel Prize for Chemistry in 1993. PCR involves two major steps: (1) synthesizing two short pieces of DNA, called *primers,* that identify the DNA segment to be copied, often called the *target DNA,* and (2) running repetitive reactions to make multiple copies of the DNA. The nucleotide sequence of one primer is complementary to the beginning of the DNA segment on one strand of the double helix, and the sequence of the other primer is complementary to the beginning of the target DNA on the other strand. During the copying process, DNA polymerase recognizes the primers as the place where DNA replication should begin.

In PCR, the target DNA is mixed with primers, free nucleotides, and DNA polymerase in a small test tube. The

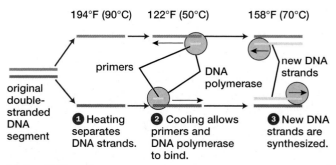

(a) One PCR cycle

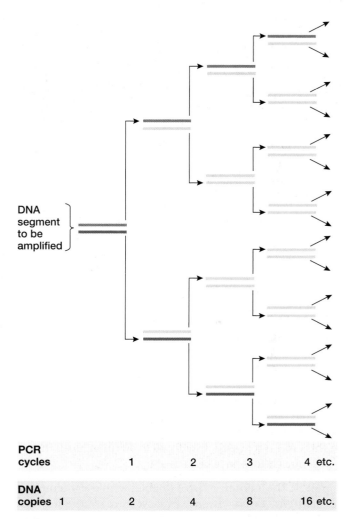

(b) Each PCR cycle doubles the number of copies of the DNA

▲ **FIGURE 14-3 PCR copies a specific DNA sequence (a)** The polymerase chain reaction consists of a cycle of heating, cooling, and warming that is typically repeated 30 to 40 times. **(b)** Each cycle doubles the amount of target DNA. After a little more than 30 cycles, a billion copies of the target DNA have been synthesized.

reaction mixture is then cycled through a series of temperature changes (**FIG. 14-3a**):

1. The test tube is heated to 194° to 203°F (90° to 95°C). High temperatures break the hydrogen bonds between complementary bases, separating the DNA into single strands ❶.

Eight side-by-side (tandem) repeats of the same four-nucleotide sequence

▲ **FIGURE 14-5 Short tandem repeats** This STR locus contains the sequence AGAT, repeated from 7 to 15 times in different alleles.

▲ **FIGURE 14-4 Thomas Brock surveys Mushroom Spring** Brock discovered the bacterium *Thermus aquaticus* in Mushroom Spring in Yellowstone National Park. The DNA polymerase from *T. aquaticus* functions best at the high temperatures required by PCR.

2. The temperature is lowered to about 122°F (50°C), which allows the two primers to form complementary base pairs with the beginning of the target DNA on each strand ❷.

3. The temperature is raised to 158° to 162°F (70° to 72°C). DNA polymerase uses the free nucleotides to make copies of the DNA segment bounded by the primers ❸. Most DNA polymerases do not function at temperatures much higher than 105°F (40°C). However, PCR uses a special DNA polymerase isolated from bacteria that live in hot springs (FIG. 14-4), which actually works best at these high temperatures.

4. This cycle is repeated, usually 30 to 40 times, until the free nucleotides have been used up.

In PCR, the amount of DNA doubles with every temperature cycle (FIG. 14-3b). Twenty PCR cycles make about a million copies, and a little over 30 cycles make a billion copies. Each cycle takes only a few minutes, so PCR can produce billions of copies of a DNA segment in an afternoon. The DNA is then available for forensics, cloning, making transgenic organisms, or many other purposes.

Differences in Short Tandem Repeats Are Used to Identify Individuals by Their DNA

In many criminal investigations, PCR is used to amplify the DNA so that there is enough to compare the DNA left at a crime scene with a suspect's DNA. How do crime labs compare DNA? After years of painstaking work, forensics experts have found that specific segments of DNA, called **short**

tandem repeats (STRs), can be used to identify people with astonishing accuracy. Think of STRs as very small genes (FIG. 14-5). STRs are *short* (about 20 to 250 nucleotides), *repeating* (consisting of the same sequence of 2 to 5 nucleotides repeated up to 50 times), and *tandem* (having all of the repetitions side by side). As with any gene, there may be alternative forms, or alleles. The alleles of any given STR simply have different numbers of repeats of the same short nucleotide sequence. To identify individuals from DNA samples, the U.S. Department of Justice established a standard set of 13 STR loci that have highly variable numbers of repeats in different people. Most crime labs also examine a gene that shows whether the DNA sample came from a man or a woman. European countries use an overlapping, but not identical, set of STRs. In 2014, biotech companies developed methods of analyzing as many as 24 STR loci at a time, so the U.S. and European STRs can all be tested simultaneously in a single sample.

Forensics labs use PCR primers that amplify only the STRs and the DNA immediately surrounding them. Because STR alleles vary in how many repeats they contain, they vary in size: An STR allele with more repeats is larger than one with fewer repeats. Therefore, a forensic lab needs to identify each STR in a DNA sample and then determine its size to find out which alleles occur in the sample.

Modern forensics labs use sophisticated and expensive machines to analyze STRs. Most of these machines are based on two methods that are used in molecular biology labs around the world: (1) separating DNA segments by size and (2) labeling specific DNA segments of interest.

CASE STUDY CONTINUED

Guilty or Innocent?

When biological evidence was found in one of the Haynesworth case files in 2009, the sample was 25 years old. Fortunately, DNA doesn't degrade very fast. Forensic lab technicians amplified the DNA with PCR so that they had enough material to analyze. How did the lab use STRs to determine that the semen profile matched Davis, and not Haynesworth?

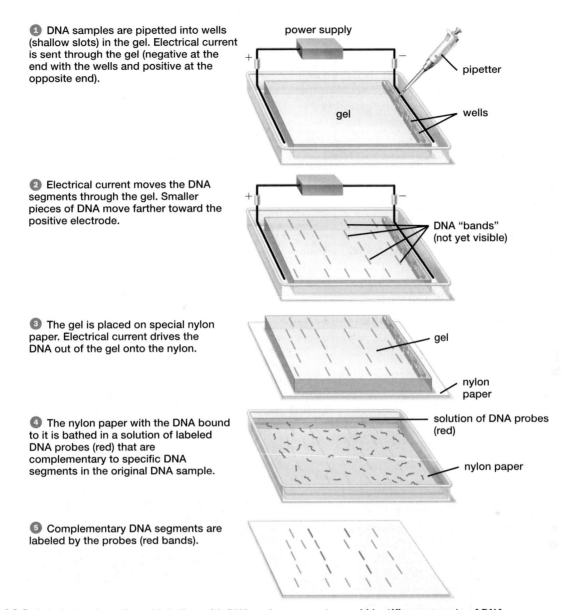

1 DNA samples are pipetted into wells (shallow slots) in the gel. Electrical current is sent through the gel (negative at the end with the wells and positive at the opposite end).

power supply

pipetter

gel

wells

2 Electrical current moves the DNA segments through the gel. Smaller pieces of DNA move farther toward the positive electrode.

DNA "bands" (not yet visible)

3 The gel is placed on special nylon paper. Electrical current drives the DNA out of the gel onto the nylon.

gel

nylon paper

4 The nylon paper with the DNA bound to it is bathed in a solution of labeled DNA probes (red) that are complementary to specific DNA segments in the original DNA sample.

solution of DNA probes (red)

nylon paper

5 Complementary DNA segments are labeled by the probes (red bands).

▲ **FIGURE 14-6 Gel electrophoresis and labeling with DNA probes separates and identifies segments of DNA**

Gel Electrophoresis Separates DNA Segments

A mixture of DNA fragments can be separated by a technique called **gel electrophoresis** (**FIG. 14-6**). First, a laboratory technician loads the DNA mixture into shallow grooves, or wells, in a slab of gel **1**. The gel consists of a meshwork of fibers with minuscule holes of various sizes between the fibers. The gel is put into a chamber with electrodes connected to each end. One electrode is made positive and the other negative; therefore, current will flow between the electrodes through the gel. How does this process separate pieces of DNA? Remember, the phosphate groups in the backbones of DNA are negatively charged. When electrical current flows through the gel, the negatively charged DNA fragments move toward the positively charged electrode. Smaller fragments slip through the holes in the gel more easily than larger fragments do, so they move more rapidly toward the positively

charged electrode. Eventually, the DNA fragments are separated by size, forming distinct bands on the gel **2**.

DNA Probes Are Used to Label Specific Nucleotide Sequences

Unfortunately, the DNA bands are invisible. There are several dyes that stain DNA, but these are often not very useful in forensics or medicine. Why not? Because the dyes stain all DNA molecules, regardless of their nucleotide sequence, and there may be many different DNA fragments of approximately the same size. For example, five or six different STRs might be mixed together in the same band. Therefore, researchers and lab technicians identify specific sequences of DNA the same way nature does—by base pairing.

When the gel has finished running, the technician treats it with chemicals that break apart the double helices into single DNA strands. These DNA strands are transferred out of the gel onto a piece of paper made of nylon ❸. Because the DNA samples are now single-stranded, pieces of synthetic DNA, called **DNA probes,** can base-pair with specific DNA fragments in the sample. DNA probes are short pieces of single-stranded DNA that are labeled, either by radioactivity or by attaching colored molecules to them. To locate a specific piece of DNA, the paper is bathed in a solution containing a DNA probe with a nucleotide sequence that is complementary to the nucleotide sequence of the target DNA ❹. The probe base-pairs with, and binds to, the target DNA, but not to any of the other DNA fragments on the paper. Any extra DNA probe is then washed off. The result: The DNA probe shows where the target DNA ran in the gel ❺. (Visualizing DNA fragments with radioactive or colored DNA probes is standard procedure in many research applications. In forensics labs, the STRs are directly labeled with colored molecules during PCR and are immediately visible in the gel, so DNA probes are not necessary.)

Unrelated People Almost Never Have Identical DNA Profiles

The locations of STRs that are run on gels produce a pattern called a **DNA profile** (FIG. 14-7). The positions of the bands on the gel are determined by the numbers of repeats of the short nucleotide sequence of each STR allele. What does a DNA profile tell us? As with any gene, every person has two alleles of each STR (see Chapter 11). The two alleles of a given STR might have the same number of repeats (the person would be homozygous for that STR) or a different number of repeats (the person would be heterozygous). For example, in the D16 STR samples shown on the right side of Figure 14-7, the first person's DNA has a single band at 12 repeats (this person is homozygous for the D16 STR), but the second person's DNA has two bands—at 13 and 12 repeats (this person is heterozygous for the D16 STR). If you look closely at all of the DNA samples in Figure 14-7, you will see that, although the DNA from some people had the same repeats for one of the STRs (for example, the second, fourth, and fifth samples for D16), no one's DNA had the same repeats for all four STRs.

Are 13 STRs enough to uniquely identify people, given the huge human population? Worldwide, different people may have as few as 3 to as many 50 repeats in a given STR. Although there are some complicating factors that forensics labs take into account, let's take a simple case: Assume that a crime lab analyzes five STRs, each with 10 possible numbers of repeats (for example, all people have either 6, 7, 8, 9, 10, 11, 12, 13, 14, or 15 repeats). Let's also assume that all of the numbers of repeats occur with equal probability in the human population, that is, 1 in 10, or 1/10. Finally, the STRs are independently assorted (see Chapter 11). Therefore, the probability of two unrelated people sharing the same number of repeats of all five STRs is simply the product (multiplication) of the separate probabilities, or $1/10 \times 1/10 \times 1/10 \times 1/10 \times 1/10 = 1$ chance in 100,000.

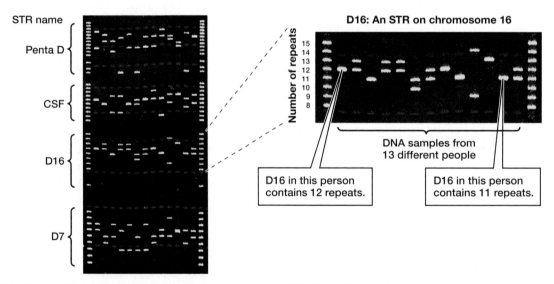

▲ **FIGURE 14-7 DNA profiling** The lengths of short tandem repeats of DNA form characteristic patterns on a gel. This gel displays four different STRs (Penta D, CSF, D16, and D7). The columns of evenly spaced yellow bands on the far left and far right sides of the gel show the number of repeats in the different STR alleles. DNA samples from 13 different people were run between these standards, resulting in one or two yellow bands in each vertical lane. The position of each band corresponds to the number of repeats in that STR allele (more repeats means more nucleotides, so the allele is larger). (Photo courtesy of Dr. Margaret Kline, National Institute of Standards and Technology.)

THINK CRITICALLY For any single person, a given STR always has either one or two bands. Why? Further, single bands are always about twice as bright as each band of a pair. For example, in the D16 STR on the right, the single bands of the first and third DNA samples are twice as bright as the pairs of bands of the second, fourth, and fifth samples. Why?

With 13 STRs, containing up to 50 repeats each, the chances of a random match are incredibly small. A perfect match of both alleles for all 13 STRs used in the United States means that there is far less than one chance in a trillion that the two DNA samples match purely by chance. Identical twins, of course, have the same DNA profile. In addition, for complicated statistical reasons, there are probably a few unrelated people in the world who have the same DNA profile. However, the odds that anyone who would be a likely suspect in a criminal case being misidentified are extremely low. Finally, a *mismatch* in DNA profiles is absolute proof that two samples did not come from the same source.

In the United States, anyone convicted of certain crimes (assault, burglary, attempted murder, etc.) must give a blood sample. Crime lab technicians then determine the criminal's DNA profile and code the results as the number of repeats in each STR. The profile is stored in computer files at a state agency, at the FBI, or both. (On TV crime shows, actors often refer to "CODIS," which stands for "Combined DNA Index System," a DNA profile database kept on FBI computers.) Because all U.S. forensic labs use the same 13 STRs, computers can easily determine if DNA left behind at another crime scene matches one of the profiles stored in the CODIS database. If the STRs match, then the odds are overwhelming that the crime scene DNA was left by the person with the matching profile. Whether or not there is a match, the crime scene DNA profile will remain in CODIS permanently. Sometimes, years later, a new DNA profile will match an archived crime scene profile, and a "cold case" will be solved.

CASE STUDY CONTINUED
Guilty or Innocent?

In the Haynesworth case, the DNA profiles of the semen sample and Haynesworth were not a match, so he could not have been the assailant. The DNA profile of Davis, which had been stored in CODIS, was a match, with only one chance in 6.5 billion that the semen sample did not come from him. Of course, when a crime is committed by a first-time offender, there won't be a DNA profile in CODIS to identify the perpetrator. In such cases, can biotechnology help?

Forensic DNA Phenotyping May Aid the Search for Criminals and Victims

What happens when the DNA left at a crime scene does not match a profile in CODIS? For now, it's back to traditional police work. But that may change, thanks to SNPs (pronounced "snips"). SNP stands for "single nucleotide polymorphism," an allele created by a nucleotide substitution mutation in a gene sometime in the distant past and passed down from generation to generation. In some cases, different SNPs produce clearly different phenotypes. This holds true for easily recognizable physical features in people, such as sex, height, and the colors of hair, eyes, and skin. Except for sex, these features are controlled by multiple genes (over 400 in the case

of height) and are strongly influenced by the environment (for example, exposure to sunlight darkens the skin, and poor nutrition in childhood causes reduced height).

Despite the complications of polygenic inheritance and environmental influences on phenotype, the right SNPs in forensic DNA phenotyping can provide a pretty good idea of what a person looks like. The HIrisPlex system is a good example. "HIris" is short for "hair and iris." HIrisPlex uses modified PCR and electrophoresis to determine the SNPs of 24 genes involved in determining hair and eye color. In test studies, HIrisPlex was 70 to 90% accurate for hair color and over 90% accurate for eye color. The Identitas Forensic Chip analyzes 200,000 SNPs and can determine sex, hair and eye color, and geographical ancestry, generally with 50 to 95% accuracy, depending on the characteristic. Sex determination, of course, has nearly 100% accuracy. Forensic DNA phenotyping is not yet accurate enough for the courtroom but would be useful for the detective who can tell his investigators that they are "probably looking for a white male, with blue eyes and brown hair." DNA phenotyping has also been used by paleoanthropologists to try to determine the phenotype of ancient humans, even Neanderthals, some of whom probably had red hair and pale skin (**FIG. 14-8**).

People aren't the only organisms that can be identified by their DNA sequences. An international group of government and private organizations is putting together the "Barcode of Life" to enable rapid DNA identification of all of the species of life on Earth. DNA barcodes have both serious and entertaining applications, as we explain in "Earth Watch: What's Really in That Sushi?" on page 244.

▲ **FIGURE 14-8 Using DNA to visualize ancient humans** Analysis of DNA isolated from Neanderthal bones suggests that some had red hair and pale skin. However, contrary to some reports in the popular press, modern humans did not inherit red hair through interbreeding with Neanderthals; their red-hair allele differs from ours.

What's Really in That Sushi?

Walk into a sushi bar and chances are that the most expensive item on the menu is tuna sushi. But is it really tuna? In 2008, two New York teenagers, Kate Stoeckle and Louisa Strauss, decided to find out (FIG. E14-1). Sounds difficult—after all, the fish are beheaded, cleaned, and skinned, and only a chunk of meat is presented to the diner—but biotechnology makes it simple, using DNA barcoding.

DNA barcoding sequences a small fragment of DNA from a gene found in the mitochondria of virtually all eukaryotic organisms—a fragment only 650 nucleotides long (FIG. E14-2). Although plants have mitochondria, the mitochondrial DNA barcode sequence doesn't differ very much between species of flowering plants, so a segment of chloroplast DNA is often used instead. Only extremely closely related species have the same nucleotide sequence in either of these particular pieces of DNA. Thus, DNA barcoding is a simple, inexpensive way to identify species.

Kate and Louisa visited restaurants and grocery stores and brought home samples of raw fish. They cut off little pieces from each sample, preserved them in alcohol, and sent them off to a lab at the University of Guelph in Canada for barcoding. Surprise! About a quarter of the sushi samples were imposters. And no surprise—the "mistakes" almost always labeled a cheap, readily available fish as a more expensive species. One specimen sold as red snapper was actually Acadian redfish, an endangered species. One "tuna" sushi turned out to be tilapia, a freshwater species often raised in fish farms. Some restaurants had mislabeled half their sushi.

DNA barcoding is useful for more than just checking up on your local sushi bar. The U.S. Food and Drug Administration uses barcodes to authenticate fish sold for food. Barcoding is often used to identify agricultural pests such as fruit flies and public health threats such as disease-carrying mosquitoes. The Federal Aviation Administration barcodes feathers to find out what kinds of birds collide with planes.

DNA barcoding can also help to stop illegal trafficking in endangered species, which is extremely lucrative (thought to be second only to illegal narcotics). Identifying the species of origin of meat, skin, feathers, and many other animal parts is often difficult, even for experts, but DNA barcoding can't be fooled. The day may come when barcoding not only verifies your sushi but also puts a stop to the exploitation of endangered species.

▲ FIGURE E14-1 **Kate Stoeckle and Louisa Strauss with their research subjects**

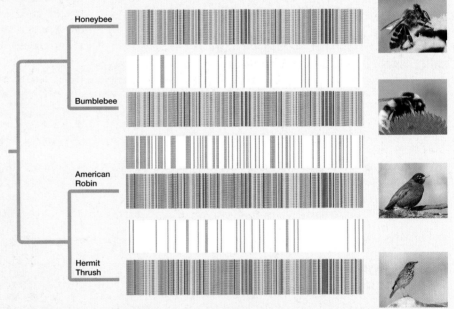

▲ FIGURE E14-2 **DNA barcoding** The different colors in the barcodes represent different bases in the DNA sequence of a fragment of a mitochondrial gene. Closely related organisms have more similar barcodes than distantly related organisms do, but every species has a unique barcode.

THINK CRITICALLY There are many other applications in which DNA barcoding might be useful. For example, how might ecologists use DNA barcoding to find out what species are present in a rain forest, or what kinds of animals a predator eats?

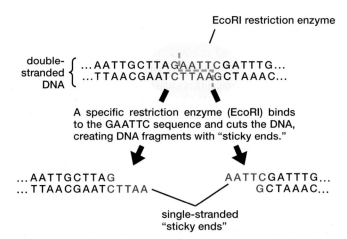

FIGURE 14-9 Restriction enzymes cut DNA at specific nucleotide sequences

THINK CRITICALLY Restriction enzymes are isolated from bacteria. Why would bacteria synthesize enzymes that cut up DNA? (*Hint:* Bacteria can be infected by viruses called bacteriophages; see Chapter 12.) Why wouldn't a bacterium's restriction enzymes destroy the DNA of its own chromosome?

CHECK YOUR LEARNING

Can you ...

- explain the uses of the polymerase chain reaction, gel electrophoresis, and DNA probes, and how they work?
- describe how DNA profiles are produced?
- explain why a DNA profile is usually unique to each individual person?

14.4 HOW IS BIOTECHNOLOGY USED TO MAKE GENETICALLY MODIFIED ORGANISMS?

Biotechnology has applications far beyond forensic science. Biotechnology can be used to identify, isolate, and modify genes; combine genes from different organisms; and move genes from one species to another. Let's see how some of these techniques can be used to make genetically modified organisms (GMOs).

There are three major steps to make a GMO: (1) Obtain the desired gene, (2) clone the gene, and (3) insert the gene into the cells of the host organism. Various technologies can be used for each step, often involving complex procedures. We will provide only a brief overview of the general processes.

The Desired Gene Is Isolated or Synthesized

Two common methods are used to obtain a gene. For a long time, the only practical method was to isolate the target gene from the organism that possessed it. Chromosomes can be isolated from cells of the gene donor and cut up with enzymes (see below). DNA fragments containing the desired gene can then be separated from the rest of the DNA by gel electrophoresis (see Fig. 14-6). Today, biotechnologists can often synthesize the gene—or a modified version of it—in the lab, using a DNA synthesizer.

The Gene Is Cloned

Once the gene has been obtained, it can be used to make transgenic organisms, shared with other scientists around the world, or used for medical treatments. It is useful, or even essential, to have a huge number of copies of the gene, far more than are usually made by PCR. The simplest way to generate lots of copies of a gene is to let living organisms do it, by **DNA cloning.** In DNA cloning, the gene is usually inserted into single-celled organisms, such as bacteria or yeasts, that multiply very rapidly, manufacturing copies of the gene as they do.

The most common method of DNA cloning is to insert the gene into a bacterial plasmid (see Fig. 14-1), which will be replicated when bacteria containing the plasmid multiply. Inserting the gene into a plasmid, rather than the bacterial chromosome, also allows it to be easily separated from the rest of the bacterial DNA. The target gene may be isolated from the plasmid, or the whole plasmid may be used to make transgenic organisms, including plants, animals, or other bacteria.

Genes are inserted into plasmids using **restriction enzymes,** each of which cuts DNA at a specific nucleotide sequence. There are hundreds of different restriction enzymes.

Many cut straight across the double helix of DNA. Others make a staggered cut, snipping the DNA in a different location on each of the two strands so that single-stranded sections hang off the ends of the DNA. These single-stranded regions are commonly called "sticky ends," because they can base-pair with, and thus stick to, other single-stranded pieces of DNA with complementary bases (**FIG. 14-9**). Restriction enzymes that make a staggered cut are used in DNA cloning.

To insert a gene into a plasmid, the same restriction enzyme is used to cut the DNA on both ends of the gene and to split open the circle of plasmid DNA (**FIG. 14-10 ❶**). As

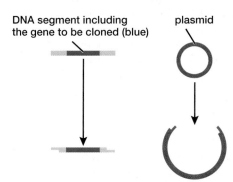

❶ The plasmid and the DNA segment containing the desired gene are cut with the same restriction enzyme.

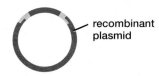

❷ The plasmids and the DNA segment containing the gene, both with the same complementary sticky ends, are mixed together; DNA ligase bonds the genes into the plasmids.

FIGURE 14-10 Inserting a gene into a plasmid for DNA cloning

a result, the ends of the DNA containing the gene and the opened-up plasmid both have complementary nucleotides in their sticky ends and can base-pair with each other. When the cut genes and plasmids are mixed together, some copies of the genes will be temporarily inserted between the cut ends of the plasmids, held together by their complementary sticky ends. Adding DNA ligase (see Chapter 12) permanently bonds the genes into the plasmids ❷.

Bacteria are then transformed with these recombinant plasmids. Under the right conditions, when the bacteria multiply, they replicate the plasmids, too. Huge vats of bacteria produce as many copies of the gene as are needed.

The Gene Is Inserted into a Host Organism

Now comes the hard part—**transfecting** the host organism. To provide a useful function, the gene must be inserted into the host and expressed in the appropriate cells, at the appropriate times, and at the desired level. Host organisms may be transfected by one of several different methods. In some cases, the recombinant plasmids or genes purified from them are inserted into harmless bacteria or viruses, called *vectors,* and then the host organism is infected with them. In the ideal case, the bacteria or viruses insert the new gene into the chromosomes of the host organism's cells, where it becomes a permanent part of the host's genome, and is replicated whenever the host's DNA is replicated. This is how some plants were transfected with genes for herbicide resistance and insect resistance (see Section 14.5).

A simpler method is to use a "gene gun." Microscopically small pellets of gold or tungsten are coated with DNA (either plasmids or purified genes) and then shot at cells or organisms. Ideally, the pellets penetrate individual cells without damaging them. Inside the cells, the DNA dissolves off the pellets into the cytoplasm, makes its way into the nucleus, and becomes incorporated into a chromosome. This process is literally "hit or miss," but is often quite effective for plants, cells in culture, and sometimes even whole animals (usually small ones, such as roundworms or fruit flies). Gene guns are often used when host organisms are easily available in large numbers, so that a low success rate doesn't really matter.

Various chemical treatments also can be used to transfect cultured animal cells and plant or fungal cells that have had their cell walls removed or disrupted. Typically, the DNA is incorporated into tiny lipid vesicles, which can fuse with the plasma membrane of the target cell and move the DNA to the target cell cytoplasm. Other transfection methods temporarily render the plasma membrane permeable so that DNA can enter the target cells.

Finally, plasmids or purified genes can be directly injected into animal cells, usually fertilized eggs (**FIG. 14-11**). Tiny glass pipettes are loaded with a suitable solution containing the DNA. The pipettes have tips that are sharp enough to impale a cell without damaging it. Pressure applied to the back of the pipette pushes some of the DNA into the cell.

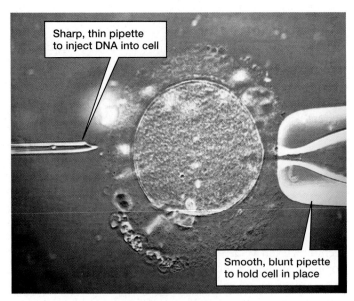

Sharp, thin pipette to inject DNA into cell

Smooth, blunt pipette to hold cell in place

▲ **FIGURE 14-11 Transfecting a fertilized egg by injecting foreign DNA** The large pipette on the right holds the egg stationary during the procedure. The small, sharp pipette on the left penetrates the egg and injects DNA.

CHECK YOUR LEARNING

Can you ...

- explain how genes are inserted into a plasmid, and why that is useful in making a genetically modified organism?
- describe the procedures used to transfect an organism with a foreign gene?

14.5 HOW ARE TRANSGENIC ORGANISMS USED?

Transgenic organisms are widely used in agriculture and biomedical research. Recently, genetically modified organisms have been developed to help control insect-borne diseases, clean up mine wastes, and repopulate American forests with endangered chestnut trees. Genetic engineers are also working to improve photosynthesis and develop algae that make biofuels cheaply and efficiently.

Many Crops Are Genetically Modified

The main goal of agriculture is to grow as much food as possible, as cheaply as possible, with minimal loss from pests such as insects and weeds. Many seed suppliers have turned to biotechnology to achieve these goals. Some people, however, feel that the risks of genetically modified food to human health or the environment are not worth the benefits. We will explore this controversy in Section 14.8.

According to the U.S. Department of Agriculture, 93% of the corn, 96% of the cotton, and 94% of the soybeans grown in the United States in 2014 were transgenic; that is, they contained genes from other species. Globally, 18 million farmers planted more than 430 million acres of land with transgenic crops in 2013.

TABLE 14-1	Genetically Engineered Crops with USDA Approval	
Genetically Engineered Trait	**Potential Advantage**	**Examples**
Resistance to herbicide	Application of herbicide kills weeds but not crop plants, producing higher crop yields	Beet, canola, corn, cotton, flax, potato, rice, soybean, tomato
Resistance to pests	Crop plants suffer less damage from insects, producing higher crop yields	Corn, cotton, potato, rice, soybean
Resistance to disease	Plants are less prone to infection by viruses, bacteria, or fungi, producing higher crop yields	Papaya, potato, squash
Sterile	Transgenic plants cannot cross with wild varieties, making them safer for the environment and more economically valuable for the seed companies that produce them	Chicory, corn
Altered oil content	Oils can be made healthier for human consumption or can be made similar to more expensive oils (such as palm or coconut)	Canola, soybean

Crops are most commonly modified to improve their resistance to insects, herbicides, or both (TABLE 14-1). Herbicide-resistant crops allow farmers to kill weeds without harming their crops. Less competition from weeds means more water, nutrients, and light for the crops and, hence, larger harvests. Many herbicides kill plants by inhibiting an enzyme that is used by plants, fungi, and some bacteria—but not animals—to synthesize specific amino acids. Without these amino acids, the plants die because they cannot synthesize proteins. Most herbicide-resistant transgenic crops have been given bacterial genes that encode enzymes that either rapidly metabolize the herbicides or that function even in the presence of the herbicide. In either case, the transgenic plants can synthesize normal amounts of amino acids and proteins.

The insect resistance of many crops has been enhanced by giving them a gene, called *Bt,* from the bacterium *Bacillus thuringiensis*. The protein encoded by the *Bt* gene damages the digestive tract of insects, but not mammals. Transgenic *Bt* crops often suffer far less damage from insects than regular crops do, so farmers can apply less pesticide to their fields (FIG. 14-12).

HAVE YOU EVER
WONDERED...

If the Food You Eat Has Been Genetically Modified?

Corn and soy products are found in an amazing variety of foods, in addition to the obvious ones like tortilla chips, soy sauce, and margarine. For example, corn syrup is an ingredient in foods as diverse as soda, ketchup, and bran flakes; soybean oil or protein is an important ingredient in cookies, cake mixes, and veggie burgers. Almost all of the corn and soy grown in the United States is genetically modified (GM), with the result that about 80% of the packaged foods in American supermarkets contain substances made from GM plants. Many countries, including those in the European Union, require labeling of GM foods, but the U.S. Food and Drug Administration does not, so in the United States you can't tell if a food contains GM ingredients by reading the label. Therefore, unless you are extremely motivated to avoid them, you probably eat GM foods.

Genetically Modified Plants May Be Used to Produce Medicines

The tools of biotechnology can also be used to insert medically useful genes into plants, producing medicines down on the "pharm." In 2012, replacement enzyme therapy for Gaucher's disease became the first treatment based on transgenic plants to be approved for clinical use by the U.S. Food and Drug Administration (FDA). Patients with Gaucher's disease fail to make an enzyme that breaks down a certain type of lipid. Without the enzyme, the lipid accumulates in the body, causing anemia, joint pain, neurological disorders, poor resistance to infectious diseases, and many other symptoms.

◀ FIGURE 14-12 *Bt* plants resist insect attack
Transgenic cotton plants expressing the *Bt* gene (right) resist attack by bollworms, which eat cotton seeds. The transgenic plants therefore produce far more cotton than nontransgenic plants do (left).

Plant-derived enzymes can substitute for the patient's missing enzymes.

Many researchers are working on engineering plants to make vaccines. To make a vaccine, a plant is given the genes needed to produce harmless proteins that are normally found in disease-causing bacteria or viruses. The proteins are then extracted and purified before injection, much like conventional vaccines. Plant-produced vaccines against hepatitis B, measles, rabies, tooth decay, flu, infant diarrhea, and other diseases are in various stages of animal or human trials, but none has yet been approved for use in humans.

Could people be vaccinated simply by eating suitable transgenic plants? It wouldn't be as easy as it sounds. First, the proteins would need to be modified to resist digestion in the stomach and intestine. Second, there is no simple way to control the dose: Too little and the user doesn't develop decent immunity; too much, and the vaccine proteins might be harmful.

Molecular biologists can also engineer plants to produce human antibodies that combat specific diseases. When a disease-causing microbe invades your body, it takes several days for your immune system to respond and produce enough antibodies to overcome the infection. Meanwhile, you feel terrible and might even die if the disease is serious enough. A direct injection of large quantities of the right antibodies might be able to cure the disease quickly enough to save your life. During the 2014 Ebola outbreak in West Africa, ZMapp, an experimental cocktail of three antibodies made in transgenic tobacco plants, may have helped to save the lives of some health professionals who became infected while caring for Ebola patients. In early 2015, clinical trials of ZMapp began in Liberia, with the hope that the antibody treatment will be ready before the next outbreak of Ebola.

Genetically Modified Animals May Be Useful for Agriculture, Medicine, and Industry

There are a number of ways to produce transgenic animals, including injecting the desired DNA into a fertilized egg. The egg is allowed to divide a few times in culture before being implanted into a surrogate mother. If the offspring are healthy and express the foreign gene, they are then bred with one another to produce homozygous transgenic animals. So far, it has proven difficult to create commercially valuable transgenic livestock, but several companies are working on it.

For example, biotechnology companies have made genetically modified sheep that produce more wool, cattle that produce more protein in their milk, and pigs that produce meat that has less fat or that has high concentrations of omega-3 fatty acids, which are thought to provide a variety of health benefits. The transgenic Enviropig, developed at the University of Guelph in Canada, metabolizes phosphate much more efficiently than regular pigs and so excretes less phosphate in its feces. Phosphate runoff from pig farms often enters streams and lakes, causing harmful algal blooms that kill aquatic animal life. This would be much less likely to occur with Enviropigs. Researchers in England and Scotland have developed transgenic chickens that cannot spread the H5N1 influenza virus, which causes avian flu. Worldwide, many millions of chickens have been killed to stop outbreaks of avian flu, so flu-resistant chickens could potentially be very valuable birds. There are even goats at Utah State University that secrete spider silk proteins in their milk. Spider silk is far stronger than steel or Kevlar®, the fiber usually used in bulletproof vests, so the hope is that lightweight, nearly impenetrable vests could be made using silk protein from these goats.

Biotechnologists are also developing animals that produce medicines, such as human antibodies or other essential proteins. For example, there are genetically modified sheep whose milk contains a protein, alpha-1-antitrypsin, that may prove valuable in treating cystic fibrosis and emphysema. Other GM sheep produce human clotting factors, which could be used to treat hemophilia. Livestock have also been engineered so that their milk contains erythropoietin (a hormone that stimulates red blood cell synthesis) or clot-busting proteins (to treat heart attacks caused by blood clots in the coronary arteries).

In addition, biomedical researchers have made a large number of transgenic animals, primarily mice, that carry genes associated with human diseases, such as Alzheimer's disease, Marfan syndrome, and cystic fibrosis. These animals are used to investigate the causes of disease and to develop possible treatments.

Genetically Modified Organisms May Be Used for Environmental Bioengineering

Transgenic organisms have been developed that could be used for environmental bioengineering: using biotechnology to remedy environmental mishaps. Environmental bioengineering may include such diverse activities as restoring rare or endangered species, mine cleanup, and reducing the incidence of insect-borne diseases. For example, when Europeans first came to America, about a quarter of the trees in the deciduous forests were American chestnuts—possibly 4 billion trees. Then around 1900 the chestnut blight fungus arrived, accidentally imported with Chinese chestnut trees. American chestnuts had no resistance to the blight and soon became very rare. For decades, horticulturists have been crossbreeding American and Chinese chestnuts, trying to get an almost-American variety that resists blight. Some of these hybrids are now being planted in restoration projects. Biotechnology offers a second path to blight resistance: making transgenic American chestnut trees with a gene from wheat that prevents the fungus from harming the trees. The best transgenic varieties resist blight even better than pure Chinese chestnuts do.

Many old mine sites are heavily contaminated with heavy metals, including mercury, lead, and cadmium. GM bacteria have been developed that thrive in high concentrations of some heavy metals. They also remove the metals from water and soil and store them in their cells. The hope is that, by sequestering heavy metals, the bacteria can be used

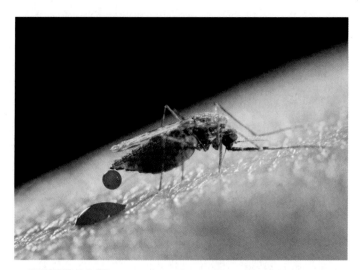

▲ **FIGURE 14-13 An *Anopheles* mosquito taking a meal of human blood** The World Health Organization estimates that *Anopheles* mosquitoes transmit malaria to almost 200 million people each year, causing about 600,000 deaths.

to clean up polluted streams and allow plants to grow in contaminated soils.

A serious health problem is the spread of infectious diseases by insects, particularly mosquitoes. The females of many mosquito species need a blood meal to produce eggs (**FIG. 14-13**). When they feed on people, some of these species transmit potentially deadly diseases, including malaria and dengue fever, both of which infect hundreds of millions of people each year, principally in warmer areas of Earth. Malaria is carried by several species of *Anopheles* mosquitoes. To help control malaria, researchers have engineered a bacterium commonly found in the *Anopheles* digestive tract so that the bacterium secretes a toxin that kills the malaria parasite, but is harmless to both mosquitoes and people. Another project has created mosquitoes with genetically modified immune systems that kill the malaria parasite.

A different approach seeks to wipe out the mosquitoes. Researchers have engineered *Anopheles* mosquitoes to carry a gene from slime molds encoding a protein that damages X chromosomes in mosquito sperm. Therefore, almost all viable sperm carry a Y chromosome and produce male offspring. GM male mosquitoes would be released into the wild, where they would mate with wild-type females. These matings would produce almost no female offspring. The male offspring would still carry the slime mold gene, so the process would continue. Over time, with very few females being born, the population should crash.

A British biotech firm, Oxitec, has genetically modified *Aedes aegypti,* the mosquito that carries dengue fever and chikungunya. When the GM males mate with wild-type females, the offspring inherit a lethal gene and die. In limited trials, releasing huge numbers of GM male mosquitoes reduces the mosquito population by 80 to 96%. Brazil has authorized the widespread release of Oxitec's GM mosquitoes in the state of Bahia to fight dengue fever.

CHECK YOUR LEARNING
Can you ...
- describe the advantages of genetically modified crops and animals in agriculture, and provide some examples?
- list some examples of how GM animals might be useful in medicine?
- describe how GM organisms might be used in environmental bioengineering?

14.6 HOW IS BIOTECHNOLOGY USED TO LEARN ABOUT THE GENOMES OF HUMANS AND OTHER ORGANISMS?

Genes influence virtually all the traits of human beings, including susceptibility to infectious diseases, mental disorders, heart disease, and diabetes. The Human Genome Project and ongoing research have determined that the human genome contains about 20,000 genes, comprising approximately 2% of our DNA. Some of the other 98% consists of promoters, regions that regulate how often genes are transcribed, and noncoding RNA, but it's not really known what most of our DNA does.

Improved understanding of our genome is having an enormous impact on medical practice. The World Health Organization estimates that more than 10,000 human diseases are caused, or made more likely to occur, by defective alleles. Although many of these diseases are extremely rare, some are common, devastating disorders. Defective alleles predispose tens of millions of people to develop conditions such as breast cancer, alcoholism, schizophrenia, heart disease, Alzheimer's disease, and many others. An increasing number of these defective alleles can be discovered through genetic testing. In 2013, actress Angelina Jolie brought the impact of genetic testing to the public's attention when she underwent a preventive double mastectomy after finding out that she has a defective allele of *BRCA1,* a tumor suppressor gene that is crucial to preventing breast and ovarian cancer (**FIG. 14-14**). In 2015, Jolie also had her ovaries removed. Some defective *BRCA1* alleles increase the lifetime risk

▲ **FIGURE 14-14 Angelina Jolie's preventive double mastectomy and ovariectomy publicized the risks of genetic predisposition to diseases**

of developing breast cancer from 12% to about 65% to 85% and the lifetime risk of ovarian cancer from a little over 1% to about 30% to 60%.

The Human Genome Project, along with companion projects that have sequenced the genomes of organisms as diverse as bacteria, fungi, mice, and chimpanzees, also helps us to appreciate our place in the evolution of life on Earth. For example, the DNA of humans and chimps is extremely similar; researchers hope that studying the relatively few differences may help us to understand why humans are so much more intelligent than chimps. Recently, researchers have deciphered the genomes of Neanderthals and Denisovans, another group of ancient hominins. Modern humans, depending on their origins, may have up to a few percent Neanderthal or Denisovan genes.

CHECK YOUR LEARNING
Can you ...
- explain why it is medically useful to understand the human genome?
- explain how knowledge of the genomes of humans and other organisms helps us to understand evolution?

14.7 HOW IS BIOTECHNOLOGY USED FOR MEDICAL DIAGNOSIS AND TREATMENT?

For over two decades, biotechnology has been used to diagnose some inherited disorders, even in fetuses (see "How Do We Know That? Prenatal Genetic Screening" on page 252). More recently, medical researchers have begun using biotechnology in an attempt to cure, or at least treat, genetic diseases.

DNA Technology Can Be Used to Diagnose Inherited Disorders

A person inherits a genetic disease when he or she inherits one or more defective alleles, which differ from normal, functional alleles because they have different nucleotide sequences. Many methods of diagnosing genetic disorders begin with PCR to make multiple copies of specific genes, and sometimes specific alleles.

Using PCR to Obtain Disease-Specific Alleles

Recall that PCR uses specific DNA primers that determine which DNA sequences are amplified. If the DNA sequences of the defective alleles responsible for a genetic disorder are known, medical testing companies can sometimes design primers that amplify only the defective alleles that cause a given disorder and not the normal alleles, making PCR itself a diagnostic tool.

Restriction Enzymes May Cut Different Alleles of a Gene at Different Locations

Sickle-cell anemia is an inherited form of anemia—not having enough red blood cells—caused by a nucleotide substitution mutation in which thymine replaces adenine near the beginning of the globin gene (see Chapters 11 and 13). A common diagnostic test for sickle-cell anemia relies on the fact that restriction enzymes cut DNA only at specific nucleotide sequences. To diagnose the presence of the sickle-cell allele, DNA is extracted from cells of a patient, a parent who might be a carrier of the allele, or even a fetus. PCR is used to amplify a section of DNA that includes the mutation site. A restriction enzyme called MstII can cut the normal sequence (CCTG**A**GGAG), but not the sickle-cell sequence (CCTG**T**GGAG). The result is that MstII cuts the normal globin allele in half, but the sickle-cell allele remains intact. Gel electrophoresis easily separates the intact sickle-cell allele from the pieces of the normal allele, which are smaller.

Different Alleles Bind to Different DNA Probes

Cystic fibrosis is a disease caused by a defect in a protein, called CFTR, that normally helps to move chloride ions across the plasma membranes of many cells, including those in the lungs, sweat glands, and intestines (see the case study in Chapter 13). There are more than 1,900 known *CFTR* alleles, all at the same locus, each encoding a different, defective CFTR protein. Fortunately, 32 alleles account for about 90% of the cases of cystic fibrosis—the other alleles are extremely rare. Although some expensive tests sequence the entire *CFTR* gene and can detect all defective alleles, rapid, relatively inexpensive tests focus on these 32 common alleles.

Each defective *CFTR* allele has a unique nucleotide sequence. Although there are many different technologies for detecting them, all are based on complementary base pairing: Under the right conditions, a DNA probe will bind to a target DNA strand only if the probe and target have perfectly complementary sequences. The simplest cystic fibrosis screening tests consist of an array of single-stranded DNA probes bound to a piece of specialized paper (**FIG. 14-15**). Each probe is complementary to one strand of a unique *CFTR* allele (**FIG. 14-15a**). A person's DNA is tested by cutting it into small pieces, separating the pieces into single strands, and labeling the strands with a colored molecule (**FIG. 14-15b**). The paper is then bathed in a solution containing the labeled DNA fragments. The person's DNA will bind only to a probe with a perfectly complementary nucleotide sequence, thereby showing which *CFTR* alleles the person possesses (**FIG. 14-15c**).

A similar technology for diagnosing genetic diseases, or potentially any genetic feature of interest, is the DNA microarray. A microarray is a glass or plastic slide spotted with hundreds to hundreds of thousands of DNA probes. Each probe can bind only a single allele of one specific

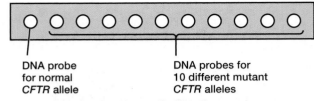

(a) Linear array of probes for cystic fibrosis

DNA probe for normal *CFTR* allele

DNA probes for 10 different mutant *CFTR* alleles

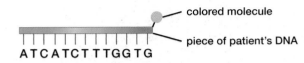

colored molecule

piece of patient's DNA

ATCATCTTTGGTG

(b) *CFTR* allele labeled with a colored molecule

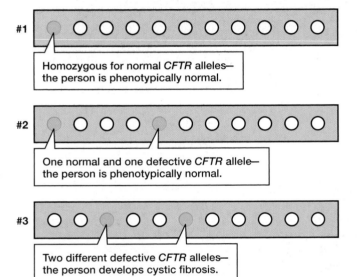

#1 Homozygous for normal *CFTR* alleles— the person is phenotypically normal.

#2 One normal and one defective *CFTR* allele— the person is phenotypically normal.

#3 Two different defective *CFTR* alleles— the person develops cystic fibrosis.

(c) Linear arrays with labeled DNA samples from three different people

▲ **FIGURE 14-15 A cystic fibrosis diagnostic array (a)** A typical diagnostic array for cystic fibrosis consists of special paper to which DNA probes complementary to the normal *CFTR* allele (far left spot) and several of the most common defective *CFTR* alleles (the other 10 spots) are attached. **(b)** DNA from a patient is cut into small pieces and separated into single strands, and the *CFTR* alleles are labeled with colored molecules. **(c)** The array is bathed in a solution of the patient's labeled DNA. The labeled DNA binds to different spots on the array, depending on which *CFTR* alleles the patient possesses.

gene; a large array can test for dozens of alleles of hundreds of genes. Several companies make arrays that test for specific disease alleles, such as defective *BRCA1* alleles. Ideally, large DNA microarrays could help to provide more effective, customized medical care, based on which alleles people have that might make them more or less susceptible to many diseases or respond more or less well to various therapies.

DNA Technology Can Be Used to Diagnose Infectious Diseases

DNA microarrays can also be used to diagnose diseases caused by bacteria or viruses. An infectious disease microarray is spotted with DNA probes specific for individual types of microbes. DNA extracted from the patient is labeled, and then binding of the extracted DNA to the microarray probes reveals which microbes infect the patient. One such microarray, the Virochip, although not yet in clinical practice, can identify more than 1,500 different viruses.

Sequencing the DNA of infectious bacteria or viruses is another way to diagnose an infection. For example, in 2014, physicians were baffled by the cause of the fevers, seizures, and brain inflammation that threatened the life of a teenager in Wisconsin. They enlisted the help of Joseph DeRisi and Charles Chiu of the University of California at San Francisco. They sequenced DNA from the patient's cerebrospinal fluid and then searched databases of bacterial DNA sequences, identifying the culprit, *Leptospira santarosai*, in less than two days at a cost of about $1,000. If similar procedures become commercialized, the cost should drop, making fast diagnosis of bacterial and viral infections practical for the clinic. In DeRisi's words, "This is one test to rule them all."

DNA Technology Can Help to Treat Disease

There are two principal applications of DNA technology for treating disease: (1) producing medicines using recombinant DNA techniques and (2) **gene therapy,** which seeks to cure diseases by inserting, deleting, or altering genes in a patient's cells.

Using Biotechnology to Produce Medicines

Thanks to recombinant DNA technology, several medically important proteins are now made in bacteria or cultured eukaryotic cells. The first human protein made by recombinant DNA technology was insulin. Prior to 1982, when recombinant human insulin was first licensed for use, the insulin needed by people with diabetes was extracted from the pancreases of cattle or pigs slaughtered for meat. Although the insulin from these animals is very similar to human insulin, the slight differences caused an allergic reaction in about 5% of people with diabetes. Recombinant human insulin does not cause allergic reactions.

Other human proteins, including growth hormone, clotting factors, antibodies, and some enzymes, are also being produced in transgenic bacteria or eukaryotic cells. Some of these proteins, such as human growth hormone and clotting factors, were formerly obtained from either human blood or human cadavers; these sources are expensive and sometimes dangerous. As you probably know, blood can be contaminated by the human immunodeficiency virus

HOW DO WE KNOW THAT?

Prenatal Genetic Screening

Thanks to modern biotechnology, physicians can now perform prenatal screening both for normal attributes, such as the sex and paternity of the embryo, and for genetic disorders, including cystic fibrosis, sickle-cell anemia, muscular dystrophy, and Down syndrome. Prenatal screening requires samples of fetal cells or chemicals produced by the fetus. Three techniques are commonly used to obtain samples for prenatal diagnosis: amniocentesis, chorionic villus sampling, and maternal blood collection.

Amniocentesis

The human fetus, like all animal embryos, develops in a watery environment. A waterproof membrane called the amnion surrounds the fetus and holds the fluid. As the fetus develops, it releases various chemicals (often in its urine) and sheds some of its cells into the amniotic fluid. When a fetus is 15 weeks or older, amniotic fluid can be collected by a procedure called **amniocentesis.**

First, the physician determines the position of the fetus by ultrasound scanning. High-frequency sound is broadcast into a pregnant woman's abdomen, and sophisticated instruments convert the echoes bouncing off the fetus into a real-time image (**FIG. E14-3**). Using the ultrasound image as a guide, the physician carefully inserts a sterilized needle through the abdominal wall, the uterus, and the amnion (being sure to avoid the fetus and placenta), and withdraws 10 to 20 milliliters of amniotic fluid (**FIG. E14-4**). Amniocentesis carries a slight risk of miscarriage, about 0.5% or less.

Chorionic Villus Sampling

The chorion is a membrane that is produced by the fetus and becomes part of the placenta. The chorion produces many small projections, called villi. In **chorionic villus sampling (CVS),** a physician inserts a small tube into the uterus through the mother's vagina and suctions off a few villi for analysis (see Fig. E14-4). The loss of a few villi does not harm the fetus. CVS has two major advantages over amniocentesis. First, it can be done much earlier in pregnancy—as early as the 8th week, but usually between the 10th and 12th

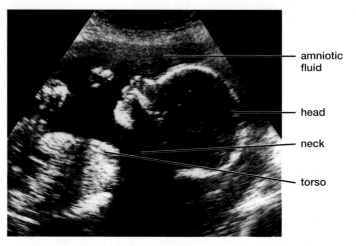

▲ **FIGURE E14-3** **A human fetus imaged with ultrasound**

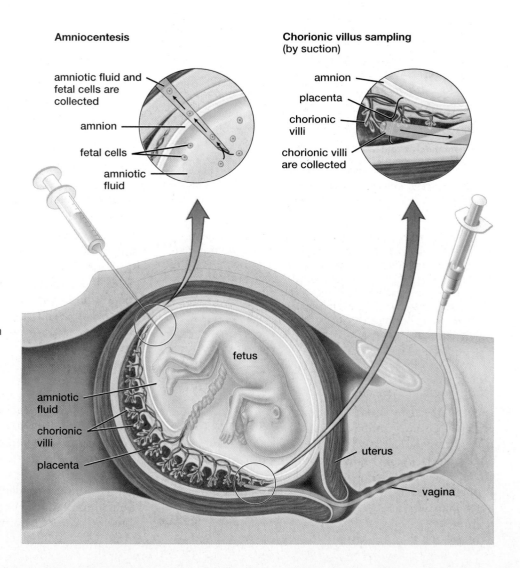

▶**FIGURE E14-4** **Prenatal sampling techniques** The two most common ways of obtaining samples for prenatal diagnosis are amniocentesis and chorionic villus sampling. (In reality, CVS is usually performed when the fetus is much younger than the one depicted in this illustration.)

weeks. This is especially important if the woman is contemplating a therapeutic abortion if her fetus has a major defect. Second, the sample contains far more fetal cells than can be obtained by amniocentesis. However, CVS appears to have a slightly greater risk of causing a miscarriage than amniocentesis does. Also, because the chorion is outside of the amniotic sac, CVS does not obtain a sample of the amniotic fluid, which is needed to diagnose certain disorders. Finally, in some cases chorionic cells have chromosomal abnormalities that are in fact not present in the fetus, which complicates karyotyping. For these reasons, CVS is less commonly performed than amniocentesis.

Maternal Blood Collection

A tiny number of fetal cells cross the placenta and enter the mother's bloodstream as early as the sixth week of pregnancy. Collecting a sample of the mother's blood is quick, easy, and poses no risk to the fetus. Separating fetal cells (perhaps as few as one per milliliter of blood) from the huge numbers of maternal cells is challenging, but it can be done. There is also fetal DNA floating free in the mother's blood. In addition, proteins and other chemicals produced by the fetus may enter the mother's bloodstream.

Analyzing the Samples

Information about the gestational stage, overall health, certain developmental disorders, and possible genetic abnormalities can be gleaned from chemicals in amniotic fluid or maternal plasma and from fetal chromosomes or DNA. Amniotic fluid and maternal blood are briefly centrifuged to separate the cells from the fluids. Biochemical analysis may be performed to measure the concentrations of hormones, enzymes, or other proteins in the fluids. For example, if the amniotic fluid contains high concentrations of an embryonic protein called alpha-fetoprotein, this indicates that the fetus may have nervous system disorders, such as spina bifida, in which the spinal cord is incomplete, or anencephaly, in which major portions of the brain fail to develop. Specific combinations of alpha-fetoprotein, estrogen, and other chemicals in maternal plasma indicate the likelihood of Down syndrome, spina bifida, or certain other disorders.

However, these biochemical screening tests do not provide completely definitive diagnoses. Therefore, if screening tests indicate that a disorder is present, then other tests, such as karyotyping, DNA analysis, or highly detailed ultrasound examination of the fetus, are employed to find out whether or not the fetus actually has one of these conditions.

Fetal cells are required for karyotyping. Amniotic fluid contains very small numbers of fetal cells, so to obtain enough cells for karyotyping or DNA analysis, the usual procedure is to grow the cells in culture for a week or two. The large number of fetal cells obtained by CVS means that karyotyping and DNA analyses can usually be performed without culturing the cells first. Karyotyping the fetal cells can show if there are too many or too few copies of the chromosomes and if any chromosomes show structural abnormalities. Down syndrome, for example, results from the presence of three copies of chromosome 21 (see Chapter 10).

Biotechnology techniques can be used to analyze fetal DNA for many defective alleles, such as those that cause sickle-cell anemia or cystic fibrosis. For Down syndrome, fetal DNA in maternal blood can now be examined as early as the 10th week of pregnancy, with about 99% accuracy. If the fetal DNA tests positive for Down syndrome, then amniocentesis and karyotyping are usually performed to confirm the diagnosis.

THINK CRITICALLY Fetal DNA in the mother's blood can be used for paternity testing, assuming, of course, that the mother's and presumed father's DNA are available for testing. Consider the following STR testing data, which includes the DNA profile from a mother, child, and two potential fathers:

| STR locus | Numbers of Repeats | | | |
	Mother	Child	Man 1	Man 2
TPOX	10	10	8, 10	6, 10
CSF	6, 8	6, 8	8, 10	8
D5S	9, 13	9, 12	7, 13	7, 12
D13S	9, 14	9, 11	10, 11	10, 11
D7S	8, 12	8, 9	7, 9	8, 9
D18S	15, 17	15	13, 15	15, 17

At which STR loci is the child homozygous? Heterozygous? Which man is a possible father for the child, and which cannot be the father? Why?

(HIV), which causes acquired immune deficiency syndrome (AIDS). Cadavers may also contain several hard-to-diagnose infectious diseases, such as Creutzfeldt-Jakob syndrome, in which an abnormal protein can be passed from the tissues of an infected cadaver to a patient and cause fatal brain degeneration (see the Chapter 3 case study). Engineered proteins grown in bacteria or other cultured cells avoid these dangers.

Treating Diseases with Gene Therapy

Gene therapies include "fixing" a defective allele, inactivating an allele that increases disease susceptibility, or adding a functional allele to substitute for a defective one. Almost all gene therapies are still in the experimental stage or in clinical trials and have not been approved for routine medical practice.

Gene Editing for AIDS

The human immunodeficiency virus (HIV) enters several kinds of immune cells, including helper T cells that play a crucial role in responses to infection. HIV kills helper T cells. When the body's supply of helper T cells becomes too low, the immune response falters, ordinarily trivial infections become life-threatening, and full-blown AIDS develops. However, a few people resist HIV infection. HIV binds to a receptor protein, called CCR5, found on the surface of susceptible immune cells. HIV then moves into the cells and begins its deadly infectious cycle. But a tiny number of people have a mutated *CCR5* gene and don't make the receptors, so they aren't infected with the usual strains of HIV.

Biotechnology offers the possibility of eliminating the CCR5 receptor in patients with AIDS and curing, or at least greatly alleviating, their disease. Through a process called gene editing, molecular biologists can manufacture specialized enzymes tailor-made to cut up specific genes, such as the one that encodes the CCR5 receptors. The treatment would work like this: Immune cells are removed from a patient, and the cells' *CCR5* genes are damaged with the enzyme. Although the cells try to repair the damaged DNA, about a quarter of them fail and can never make CCR5 receptors again. These CCR5-deleted cells are transfused back into the patient. In two small clinical trials of AIDS patients, the numbers of functioning immune cells were greatly increased in most of the patients receiving this treatment. The amount of HIV in several patients also decreased.

Gene Replacement for Severe Combined Immune Deficiency

Severe combined immune deficiency (SCID) is a rare disorder in which a child fails to develop an immune system. About 1 in 80,000 children is born with some form of SCID. Infections that would be trivial in a normal child become life-threatening. In some cases, a bone marrow transplant from a compatible donor can give the child functioning stem cells so that he or she can develop a working immune system. Most children with SCID, however, die before their first birthday.

Most forms of SCID are caused by defective recessive alleles of one of several genes. In one type of SCID, affected children are homozygous for a recessive defective allele that normally codes for an enzyme called adenosine deaminase (this condition is called ADA-SCID). In 1990, gene therapy was performed on four-year-old Ashanti DeSilva, who suffered from ADA-SCID. Some of her white blood cells were removed, genetically altered with a virus containing a functional version of her defective allele, and then returned to her bloodstream. The treatment was a partial success, but not a complete cure. Ashanti, now a healthy adult, continues to receive regular injections of a form of adenosine deaminase to boost her immune system. Recent clinical trials have used a different gene therapy to cure ADA-SCID. The researchers removed bone marrow stem cells from children with ADA-SCID, inserted a functional copy of the adenosine deaminase gene into the cells, and returned the repaired cells into the children. Because bone marrow stem cells continue to produce new white blood cells throughout life, the hope is that these children might be permanently cured. More than 40 children have been given gene therapy; as of 2014, all were healthy, and 70% seem to be fully cured.

A second type of SCID, called X-linked SCID, is caused by a defective recessive allele of a gene located on the X chromosome. More than 20 children have been given gene therapy to insert a functional copy of this gene into their bone marrow stem cells. Almost all appear to be cured, some for as long as 10 years after treatment. However, gene therapy for X-linked SCID is not without risks: Several children developed leukemia, apparently because the gene insertion turned on an oncogene (see Chapter 9). More recent methods seem to have reduced, and perhaps eliminated, this danger. None of the nine boys in a recent trial developed leukemia, and eight have developed functional immune systems.

Other Gene Therapies Several other disorders have been treated with varying success by gene therapy, mostly by adding an active allele to the appropriate cells of the patient. These include alleviating the symptoms of Parkinson's disease, partially restoring eyesight in patients with a type of inherited blindness, restoring blood clotting to hemophiliacs, curing beta-thalassemia (a type of anemia; see Chapter 13), and "training" the immune system to destroy some types of leukemia.

CHECK YOUR LEARNING

Can you ...
- explain how biotechnology is used to diagnose both inherited and infectious diseases?
- describe the procedures and advantages of gene therapy to treat inherited diseases?

14.8 WHAT ARE THE MAJOR ETHICAL ISSUES OF MODERN BIOTECHNOLOGY?

Modern biotechnology offers the promise—some would say the threat—of greatly changing our lives and the lives of many other organisms on Earth. Is humanity capable of handling the responsibility of biotechnology? Here we will explore two important issues: the use of genetically modified organisms in agriculture or environmental bioengineering and the prospects for genetically modifying human beings.

Should Genetically Modified Organisms Be Permitted?

The aims of traditional and modern agricultural biotechnology are the same: to modify the genetic makeup of living organisms to make them more useful. However, there are three significant differences. First, traditional biotechnology is slow; many generations of selective breeding are usually necessary to produce useful new strains of plants or animals. Genetic engineering, in contrast, can potentially introduce massive genetic changes in a single generation. Second, traditional biotechnology almost always recombines genetic material from the same, or a very closely related, species, whereas genetic engineering can recombine DNA from very different species. Finally, traditional biotechnology has no

Health WATCH

Golden Rice

Rice is the principal food for about two-thirds of the people on Earth. Rice provides carbohydrates and some protein, but is a poor source of many vitamins, including vitamin A. Unless people eat enough fruits and vegetables, they often lack sufficient vitamin A and may suffer from poor vision, immune system defects, and damage to their respiratory, digestive, and urinary tracts. According to the World Health Organization, about 250 million children suffer from vitamin A deficiency, principally in Asia, Africa, and Latin America. As a result, each year 250,000 to 500,000 children become blind; half of those children die. Vitamin A deficiency typically strikes the poor, because rice may be all they can afford to eat. Biotechnology offers a possible remedy: rice genetically engineered to contain beta-carotene, a pigment that makes daffodils yellow and that the human body easily converts into vitamin A.

Creating rice with high levels of beta-carotene wasn't simple. Molecular biologists Ingo Potrykus and Peter Beyer inserted three genes into the rice genome, two from daffodils and one from a bacterium. As a result, "Golden Rice" grains synthesize beta-carotene. Unfortunately, the original Golden Rice didn't make very much beta-carotene, so people would have had to eat enormous amounts to get enough vitamin A. However, Golden Rice 2, with genes from corn instead of daffodils, produces 23 times more beta-carotene than the original Golden Rice does and consequently is bright yellow (FIG. E14-5). One to two cups of cooked Golden Rice 2 would provide enough beta-carotene to equal the full recommended daily amount of vitamin A. Golden Rice 2 was given, free, to the Humanitarian Rice Board for experiments and planting in Southeast Asia.

However, Golden Rice faces other hurdles. Many people strongly resist large-scale planting of Golden Rice (or any transgenic crop). When the first field trials of Golden Rice began in the Philippines in 2008, some of the fields were destroyed by activists. Nevertheless, field trials continued, and Golden Rice could soon become available to farmers, pending approval by the Philippine Department of Agriculture.

Golden Rice cannot solve all the problems of malnutrition in poor people, of course. For one thing, poor people's diets are often deficient in many nutrients, not just vitamin A. To help solve that problem, the Bill and Melinda Gates Foundation is funding research to increase the levels of vitamin E, iron, and zinc in rice. Further, not all poor people eat mostly rice. In parts of Africa, sweet

▲ **FIGURE E14-5 Golden Rice** The high beta-carotene content of Golden Rice 2 gives it a bright yellow color. Normal rice lacks beta-carotene and is off-white.

potatoes are the main source of calories. Eating orange, instead of white, sweet potatoes, has dramatically increased vitamin A intake for many of these people. Finally, in many parts of the world, governments and humanitarian organizations have started vitamin A supplementation programs. In some parts of Africa and Asia, as many as 80% of the children receive large doses of vitamin A a few times when they are very young. Someday, the combination of these efforts may result in a world in which no children suffer blindness from the lack of a simple nutrient in their diets.

CONSIDER THIS Genetic engineering is used both in food crops and in medicine. Golden Rice and almost all the corn and soybeans grown in the United States contain genes from other species. The hepatitis B vaccine is produced by inserting a gene from the hepatitis virus into yeast. The antibodies in ZMapp, currently in clinical trials as an Ebola therapy, are part mouse and part human. Are there scientifically important differences in the use of genetic engineering for food or for medical purposes? Would you accept GMO products for medicine but not food? Defend your position.

way to directly manipulate the DNA sequence of genes themselves. Genetic engineering can produce new genes never before seen on Earth.

The best transgenic crops have clear advantages for farmers. Herbicide-resistant crops allow farmers to rid their fields of weeds, which reduce harvests by 10% or more, through the use of powerful herbicides at virtually any stage of crop growth. Insect-resistant crops decrease the need to apply pesticides, saving the cost of the pesticides themselves, as well as tractor fuel and labor. Therefore, transgenic crops may produce larger harvests at lower cost. These savings may be passed along to the consumer. Transgenic crops also have the potential to be more nutritious than standard crops (see "Health Watch: Golden Rice").

However, many people strenuously object to transgenic crops or livestock. The principal concerns are that GMOs may be harmful to human health or dangerous to the environment.

Are Foods from GMOs Dangerous to Eat?

In most cases, there is no reason to think that GMOs are dangerous to eat. For example, tests have shown that the protein encoded by the *Bt* gene is not toxic to mammals, so it should not be dangerous to human health. If growth-enhanced livestock are ever marketed, they will simply have more meat, composed of the same proteins that exist in nontransgenic animals, so they shouldn't be dangerous either. For example, a company called AquaBounty has produced transgenic Atlantic salmon containing extra genes for growth hormone. The fish grow faster than wild-type Atlantic salmon, but will have the same proteins in their flesh as wild salmon. The U.S. FDA has declared that AquaBounty salmon are "as safe as food from conventional Atlantic salmon."

On the other hand, some people might be allergic to genetically modified plants. In the 1990s, a gene from Brazil nuts was inserted into soybeans in an attempt to improve the balance of amino acids in soybean protein. It was soon discovered that people allergic to Brazil nuts would probably also be allergic to the transgenic soybeans. These transgenic soybean plants never made it to the farm. The FDA now requires all new transgenic crop plants to be tested for allergenic potential.

There have been other concerns raised about GMO foods, such as decreased nutritional value or increased levels of naturally occurring plant toxins, but so far these concerns have not been supported by convincing evidence. In late 2012, the board of directors of the American Association for the Advancement of Science (AAAS) wrote: ". . . contrary to popular misconceptions, GM crops are the most extensively tested crops ever added to our food supply." The AAAS board added: "Indeed, the science is quite clear: crop improvement by the modern molecular techniques of biotechnology is safe." Over the past 15 years, similar statements have been issued by the U.S. National Academy of Sciences, the World Health Organization, and many other health and scientific organizations.

Are GMOs Hazardous to the Environment?

The environmental effects of GMOs are more debatable. One clear positive effect of *Bt* crops is that farmers usually apply less insecticide to their fields. This should translate into less pollution of the environment and less harm to the farmers. For example, in India, farmers growing *Bt* cotton use less than half as much insecticide as farmers growing conventional cotton. This also reduces the incidence of pesticide poisoning by about a factor of 8. The United States Department of Agriculture finds that the increasing adoption of *Bt* corn has resulted in a parallel drop in pesticide application on cornfields, which fell by about 90% between 1995 and 2010. On the other hand, herbicide-resistant GMO crops have encouraged a great increase in the use of glyphosate herbicides. Widespread glyphosate use has resulted in the evolution of dozens of resistant weeds.

An undesirable side effect of growing GMO crops is that *Bt* or herbicide-resistance genes might spread outside a farmer's fields. Because these genes are incorporated into the genome of the transgenic crop, the genes will be in its pollen, too. A farmer cannot control where pollen from a transgenic crop will go. In 2006, researchers at the U.S. Environmental Protection Agency discovered herbicide-resistant grasses more than 2 miles away from a test plot in Oregon. Based on genetic analyses, the scientists concluded that some of the herbicide-resistance genes escaped in pollen (most grasses are wind pollinated) and some escaped in seeds (most grasses have very lightweight seeds). In 2010, researchers found that transgenic canola plants carrying genes for herbicide resistance are widespread in North Dakota, where over 90% of the canola in the United States is grown.

Does this matter? Many crops, including corn, canola, and sunflowers in America and wheat, barley, and oats in Eastern Europe and the Middle East, have wild relatives living nearby. Suppose these wild relatives interbred with transgenic crops and became resistant to herbicides or pests. Would the accidentally transgenic wild plants become significant weed problems? Would they displace other plants in the wild because they would be less likely to be eaten by insects? Even if transgenic crops have no close relatives in the wild, bacteria and viruses sometimes transfer genes among unrelated plant species. Could viruses spread unwanted genes into wild plant populations? No one knows the answers to these questions.

What about transgenic animals? Most domesticated animals, such as cattle or sheep, are relatively immobile. Further, most have few wild relatives with which they might exchange genes, so the dangers to natural ecosystems appear minimal. However, some transgenic animals, especially fish, have the potential to pose more significant threats because they can disperse rapidly and are nearly impossible to recapture. If transgenic fish were more aggressive, grew faster, or matured faster than wild fish, they might replace native populations. One possible way out of this dilemma, suggested by AquaBounty, is to sell only sterile transgenic fish to growers, so that any escapees would die without reproducing and thus have minimal impact on natural ecosystems. AquaBounty says that their sterilization procedure is 99.8% effective. Some argue that nothing short of 100% sterility is good enough to guarantee that there will be no harm to aquatic ecosystems.

Should the Genome of Humans Be Changed by Biotechnology?

Many of the ethical implications of human applications of biotechnology are fundamentally the same as those connected with other medical procedures. For example, for about the past 40 years, trisomy 21 (Down syndrome) could be diagnosed in embryos by counting the chromosomes in cells taken from amniotic fluid (see "How Do We Know That? Prenatal Genetic Screening" on page 252); this information is sometimes used as the basis for an abortion. Other ethical concerns, however, have arisen purely as a result of advances in biotechnology. For instance, should people be allowed to select, or even change, the genomes of their offspring?

Selecting offspring genomes can be a relatively straightforward part of in vitro fertilization (IVF). Shortly after an egg has been fertilized in vitro, and before it is implanted into the uterus, it divides a few times, forming an embryo. A cell can be removed from the early embryo, usually without harm. Karyotyping or even genome sequencing can then be performed, and only embryos with desired phenotypes would be implanted into the mother. Usually, physicians screen only for genetic disorders, but in principle the same procedures could be used to select for physical traits such as sex or eye color. Most countries regulate preimplantation genetic diagnosis and allow selection of embryos based only on the absence or presence of serious inherited disorders.

The same technologies used to insert genes into stem cells to cure SCID could be used to insert or change the genes of fertilized eggs (**FIG. 14-16**). Suppose it were possible to insert functional *CFTR* alleles into human eggs, thereby preventing cystic fibrosis. Would this be an ethical change to the human genome? How about increasing intelligence or reducing the likelihood of obesity? Or making bigger football players and more beautiful supermodels? If and when the technology is developed to cure genetic diseases, it will be difficult to prevent it from being used for nonmedical purposes. Who will determine which uses are appropriate and which are trivial vanity?

CHECK YOUR LEARNING

Can you ...

- explain why people might be opposed to the use of genetically modified organisms in agriculture?
- envision circumstances in which it would be ethical to modify the genome of a human fertilized egg?

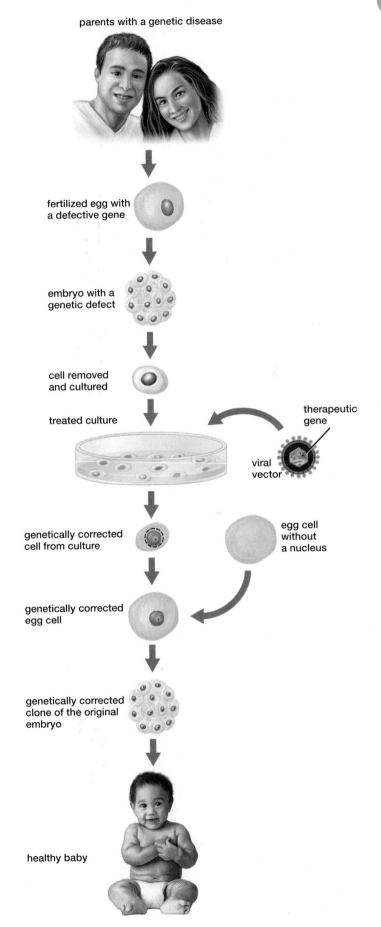

▶ **FIGURE 14-16 Using biotechnology to correct genetic defects in human embryos** In this hypothetical example, a couple who carry the alleles for a serious genetic disorder wish to have a child. The woman's eggs are fertilized in vitro by her partner's sperm. When an embryo containing a defective gene grows into a small cluster of cells, a single cell is removed from the embryo, and the defective allele in the cell is replaced using an appropriate vector, usually a disabled virus. The nucleus of another egg cell (taken from the same mother) is removed. The genetically repaired cell is then injected into the egg whose nucleus had been removed. The now repaired egg cell is allowed to divide a few times, and the resulting embryo is implanted in the woman's uterus for fetal development.

Guilty or Innocent?

▲ **FIGURE 14-17 Mary Jane Burton**

Former Virginia Governor Mark Warner credits his state's massive review of old cases mostly to one woman: lab technician Mary Jane Burton (**FIG. 14-17**). During her time at the state forensic lab, the standard practice was to return evidence to local authorities. Because of space constraints at courthouses and police departments across the state, old evidence was routinely destroyed a few years after a case was closed. But Burton kept bits of evidence taped to her case files—no one is really sure why. Burton's efforts have given new life, not only to Thomas Haynesworth, but to Marvin Anderson, Julius Ruffin, Arthur Whitfield, Philip Thurman, Victor Burnette, and Willie Davidson, as well. Burton, who died in 1999, did not live to see her legacy, but these men will never forget her.

DNA evidence cannot always clear the wrongly convicted: There was no longer any biological evidence in two of the rapes for which Thomas Haynesworth was accused. Mid-Atlantic Innocence Project attorney Shawn Armbrust and former Virginia Attorney General Ken Cuccinelli stepped up and persuaded the courts to issue a "writ of actual innocence" for Haynesworth, officially clearing him of all crimes. Cuccinelli also hired Haynesworth to work in the Attorney General's office mailroom, where he is now the supervisor. Haynesworth, a truly remarkable man, also volunteers for the Innocence Project, to help other people who may be accused and convicted of crimes they did not commit.

CONSIDER THIS Who are the heroes in these stories? There are the obvious ones, of course—Mary Jane Burton; the lawyers of the Innocence Project who have helped to free over 300 people wrongly convicted of crimes they did not commit; and, of course, innocent men who, like Thomas Haynesworth, have become gracious, productive members of society. But what about molecular biologist Kary Mullis, who discovered PCR? Or Thomas Brock, whose discovery of *Thermus aquaticus* in Yellowstone hot springs provided the source of heat-stable DNA polymerase that is so essential to PCR (see Fig. 14-4)? Or the hundreds of biologists, chemists, and mathematicians who developed procedures for gel electrophoresis, DNA labeling, and statistical analysis of sample matching?

Scientists often say that science is worthwhile for its own sake, and that it is difficult to predict which discoveries will lead to the greatest benefits for humanity. Nonscientists, when asked to pay the costs of scientific projects, are sometimes skeptical of such claims. How do you think that public support of science should be allocated? Fifty years ago, would you have voted to give Thomas Brock public funds to see what types of organisms lived in hot springs?

CHAPTER REVIEW

Go to **MasteringBiology** for practice quizzes, activities, eText, videos, current events, and more.

Answers to Think Critically, Evaluate This, Multiple Choice, and Fill-in-the-Blank questions can be found in the Answers section at the back of the book.

Summary of Key Concepts

14.1 What Is Biotechnology?

Biotechnology is the use, and especially the alteration, of organisms, cells, or biological molecules to produce food, biofuels, drugs, or other goods. Modern biotechnology uses genetic engineering, frequently combining DNA from different organisms, even different species. When DNA is transferred from one organism to another, the recipients are called transgenic or genetically modified organisms (GMOs). Applications of modern biotechnology include increasing our understanding of gene function, treating disease, improving agriculture, and solving crimes.

14.2 What Natural Processes Recombine DNA Between Organisms and Between Species?

DNA recombination occurs naturally through processes such as sexual reproduction; bacterial transformation, in which bacteria acquire DNA from plasmids or other bacteria; and viral infection, in which viruses incorporate fragments of DNA from their hosts and transfer the fragments to members of the same or other species.

14.3 How Is Biotechnology Used in Forensic Science?

Specific regions of very small quantities of DNA can be amplified by the polymerase chain reaction (PCR). The most common regions used in forensics are short tandem repeats (STRs). The pattern of STRs, called a DNA profile, can be used to match DNA found at a crime scene with DNA from suspects with extremely high accuracy. DNA phenotyping may make it possible to derive a general physical description of a person from DNA samples.

14.4 How Is Biotechnology Used to Make Genetically Modified Organisms?

There are three steps to making a genetically modified organism. First, the desired gene is obtained from another organism or, less commonly, synthesized. Second, the gene is cloned, often into a bacterial plasmid, to provide multiple copies of the gene. Third, the gene is inserted into a host organism, often through the action of bacteria or viruses, with gene guns or by injection into cells (especially fertilized eggs).

14.5 How Are Transgenic Organisms Used?

Many crop plants have been modified by the addition of genes that promote herbicide resistance or insect resistance. Plants may also be modified to produce human proteins, vaccines, or antibodies. Transgenic animals may be produced, with properties such as faster growth, increased production of valuable products such as milk, or the ability to produce human proteins, vaccines, or antibodies. Transgenic organisms may be useful to remediate contaminated areas or to decrease the population of disease vectors.

14.6 How Is Biotechnology Used to Learn About the Genomes of Humans and Other Organisms?

Techniques of biotechnology were used to discover the complete nucleotide sequence of the human genome. This knowledge is being used to discover medically important genes and to better understand the evolutionary relationships between humans and other organisms.

14.7 How Is Biotechnology Used for Medical Diagnosis and Treatment?

Inherited diseases are caused by defective alleles of crucial genes. Biotechnology, including PCR, gel electrophoresis, and DNA microarrays, may be used to diagnose genetic disorders such as sickle-cell anemia and cystic fibrosis. Genetic engineering may be used to insert functional alleles into normal cells, stem cells, or even eggs to correct genetic disorders. Biotechnology may be used to identify microbes that cause infectious diseases. Biotechnology is also widely used to produce medicines and vaccines.

14.8 What Are the Major Ethical Issues of Modern Biotechnology?

The use of genetically modified organisms in agriculture is controversial for two major reasons: food safety and potentially harmful effects on the environment. In general, GMOs contain proteins that are harmless to mammals, are readily digested, or are already found in similar foods. Environmental effects of GMOs are more difficult to predict. It is possible that foreign genes, such as those for pest or herbicide resistance, might be transferred to wild plants, with resulting damage to agriculture and/or disruption of ecosystems. If they escape, highly mobile transgenic animals might displace their wild relatives.

Genetically selecting or modifying human embryos is highly controversial. As technologies improve, society may be faced with decisions about the extent to which parents should be allowed to correct or enhance the genomes of their children.

Key Terms

amniocentesis *252*	plasmid *237*
biotechnology *237*	polymerase chain reaction
chorionic villus sampling	(PCR) *239*
(CVS) *252*	recombinant
DNA cloning *245*	DNA *237*
DNA probe *242*	restriction
DNA profile *242*	enzyme *245*
gel electrophoresis *241*	short tandem repeat
gene therapy *251*	(STR) *240*
genetic engineering *237*	transfect *246*
genetically modified	transformation *237*
organism (GMO) *237*	transgenic *237*

Thinking Through the Concepts

Multiple Choice

1. Which of the following is *not* true of a single nucleotide polymorphism?
   a. It is usually caused by a translocation mutation.
   b. It is usually caused by a nucleotide substitution mutation.
   c. It may change the phenotype of an organism.
   d. It is inherited from parent to offspring.

2. Imagine you are looking at a DNA profile that shows an STR pattern of a mother's DNA and her child's DNA. Will all of the bands of the child's DNA match those of the mother?
   a. Yes, because the mother's DNA and her child's DNA are identical.
   b. Yes, because the child developed from her mother's egg.
   c. No, because half of the child's DNA is inherited from its father.
   d. No, because the child's DNA is a random sampling of its mother's.

3. Which of the following is *not* a commonly used method of modifying the DNA of an organism?
   a. crossbreeding two plants of the same species
   b. crossbreeding two plants of different species
   c. the polymerase chain reaction
   d. genetic engineering

4. A restriction enzyme
   a. cuts DNA at a specific nucleotide sequence.
   b. cuts DNA at a random nucleotide sequence.
   c. splices pieces of DNA together at a specific nucleotide sequence.
   d. splices pieces of DNA together without regard to the nucleotide sequence.

5. DNA cloning is
   a. making multiple genetically identical cells.
   b. making multiple copies of a piece of DNA.
   c. inserting DNA into a cell.
   d. changing the nucleotide sequence of a strand of DNA.

Fill-in-the-Blank

1. _____ are organisms that contain DNA that has been modified (usually through use of recombinant DNA technology) or derived from other species.

2. _____ is the process whereby bacteria pick up DNA from their environment. This DNA may be part of a chromosome or it may be tiny circles of DNA called _____.

3. The _____ is a technique for multiplying DNA in the laboratory.

4. Matching DNA samples in forensics uses a specific set of small "genes" called _____. The alleles of these genes in different people vary in the _____ of the allele. The pattern of these alleles that a given person possesses is called his or her _____.

5. Pieces of DNA can be separated according to size by a process known as _____. The identity of a specific sample of DNA is usually determined by binding a synthetic piece of DNA called a(n) _____, which binds to the sample DNA by _____.

Review Questions

1. Describe three natural forms of genetic recombination, and discuss the similarities and differences between recombinant DNA technology and these natural forms of genetic recombination.

2. What is a plasmid? How are plasmids involved in bacterial transformation?

3. What is a restriction enzyme? How can restriction enzymes be used to splice a piece of human DNA into a plasmid?

4. Describe the polymerase chain reaction.

5. What is a short tandem repeat? How are short tandem repeats used in forensics?

6. How does gel electrophoresis separate pieces of DNA?

7. How are DNA probes used to identify specific nucleotide sequences of DNA? How are they used in the diagnosis of genetic disorders?

8. Describe several uses of genetic engineering in agriculture.

9. Describe several uses of genetic engineering in human medicine.

10. Describe amniocentesis and chorionic villus sampling, including the advantages and disadvantages of each. What are their medical uses?

Applying the Concepts

1. As you may know, many insects have evolved resistance to common pesticides. Do you think that insects might evolve resistance to *Bt* crops? If this is a risk, do you think that *Bt* crops should be planted anyway? Why or why not?

2. All children born with X-linked SCID are boys. Can you explain why?

UNIT 3
Evolution and Diversity of Life

All of Earth's species, including this strikingly colored chameleon, are linked by descent from a common ancestor.

"... from so simple a beginning endless forms most beautiful and most wonderful have been, and are being, evolved." —CHARLES DARWIN, in *On the Origin of Species*

15
PRINCIPLES OF EVOLUTION

This massive, earthbound ostrich has wings, a legacy of its evolutionary heritage.

CASE STUDY

What Good Are Wisdom Teeth and Ostrich Wings?

HAVE YOU HAD YOUR WISDOM TEETH REMOVED YET? If not, it's probably only a matter of time. Almost all of us will visit an oral surgeon to have our wisdom teeth extracted. There's just not enough room in our jaws for these rearmost molars, and removing them is the best way to prevent the pain, infections, and gum disease that can accompany the development of wisdom teeth. Removal is harmless because we don't really need wisdom teeth.

If you've already suffered through a wisdom tooth extraction, you may have found yourself wondering why we even have these extra molars. Biologists hypothesize that we have them because our apelike ancestors had them and we inherited them, even though we don't need them. Other living species, such as apes, also have these teeth, but with their ancestral function preserved. Even though in people these rearmost molars have lost their original function, the fact that they are present in apes as well reveals that we share an ancestor with apes.

Flightless birds also illustrate the connection between evolutionary ancestry and structures that do not perform their original function. Consider the ostrich, a bird that can grow to

8 feet tall and weigh 300 pounds (see the photo above). These massive creatures cannot fly. Nonetheless, they have wings, just as sparrows and ducks do. Why do ostriches have wings? Because the ancestor of all living birds had wings, and so do all of its descendants, even those that cannot fly. Many other organisms have, like ostriches and people, inherited hand-me-downs that no longer serve their original functions. What does this observation tell us about evolution? What other evidence shows us that evolution has occurred and reveals the mechanisms that cause evolution?

AT A GLANCE

15.1 HOW DID EVOLUTIONARY THOUGHT DEVELOP?

When you began studying biology, you may not have seen a connection between your wisdom teeth and an ostrich's wings. But the connection is there, provided by the concept that unites all of biology: **evolution,** or change over time in the characteristics of a population. (A **population** consists of all the individuals of one species in a particular area.)

Modern biology is based on our understanding that life has evolved, but early scientists did not recognize this fundamental principle. The main ideas of evolutionary biology became widely accepted only after the publication of Charles Darwin's work in the nineteenth century. Nonetheless, the intellectual foundation on which these ideas rest developed gradually over the centuries before Darwin's time.

Early Biological Thought Did Not Include the Concept of Evolution

Pre-Darwinian science, heavily influenced by theology, held that all organisms were created simultaneously by God and that each distinct life-form remained fixed and unchanging from the moment of its creation. This explanation of how life's diversity arose was elegantly expressed by the ancient Greek philosophers, especially Plato and Aristotle. Plato (427–347 B.C.) proposed that each object on Earth is merely a temporary reflection of its divinely inspired "ideal form." Plato's student Aristotle (384–322 B.C.) categorized all organisms into a linear hierarchy that he called the "Ladder of Nature" (FIG. 15-1).

These ideas formed the basis of the view that the form of each type of organism is permanently fixed. This view reigned unchallenged for more than 2,000 years. By the eighteenth century, however, several lines of newly emerging evidence began to undermine this static view of creation.

Exploration of New Lands Revealed a Staggering Diversity of Life

The Europeans who explored and colonized Africa, Asia, and the Americas were often accompanied by naturalists who observed and collected the plants and animals of these previously unknown (to Europeans) lands. By the 1700s, the accumulated observations and collections of the naturalists had begun to reveal the true scope of life's variety. The number of species, or different types of organisms, was much greater than anyone had suspected.

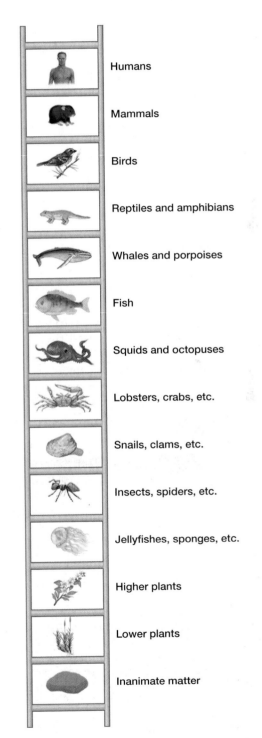

Humans

Mammals

Birds

Reptiles and amphibians

Whales and porpoises

Fish

Squids and octopuses

Lobsters, crabs, etc.

Snails, clams, etc.

Insects, spiders, etc.

Jellyfishes, sponges, etc.

Higher plants

Lower plants

Inanimate matter

▲ **FIGURE 15-1 Aristotle's "Ladder of Nature"** In Aristotle's view, fixed, unchanging species can be arranged in order of increasing closeness to perfection.

Stimulated by the new evidence of life's incredible diversity, some eighteenth-century naturalists began to take note of fascinating patterns. They noticed, for example, that each geographic area had its own distinctive set of species. In addition, the naturalists saw that some of the species in a given location closely resembled one another, yet differed in some characteristics. To some scientists of the day, the differences between the species of different geographic areas and the existence of clusters of similar species within areas seemed inconsistent with the idea that species were fixed and unchanging. (You may wish to refer to the timeline in **FIGURE 15-2** as you read the following account.)

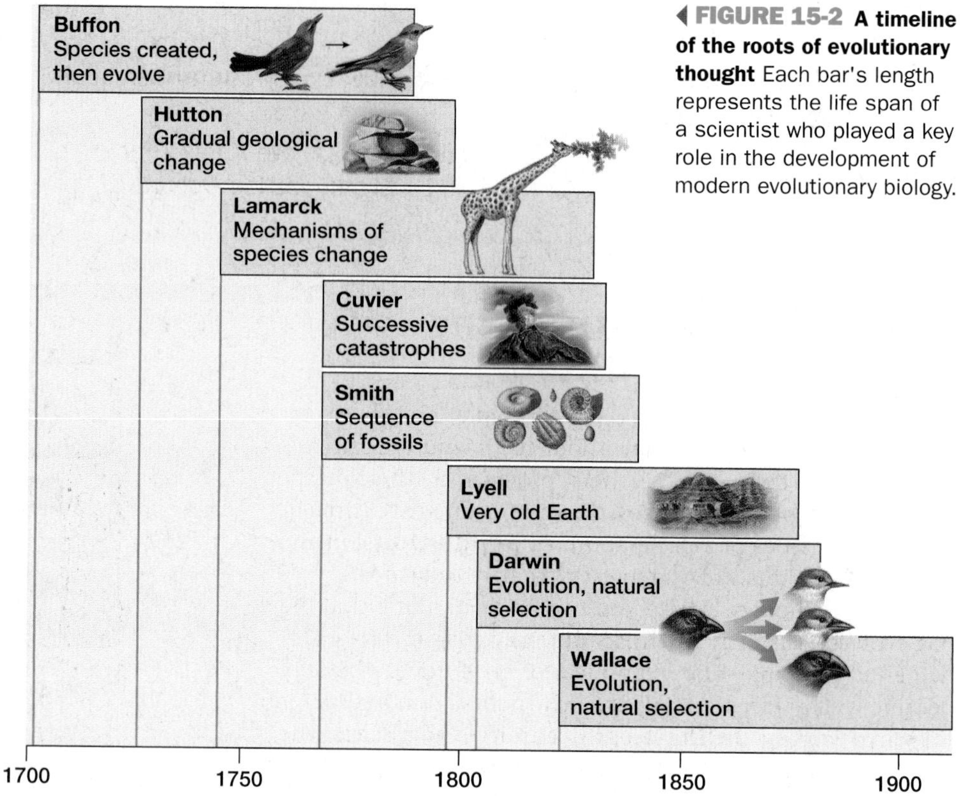

> ◀ **FIGURE 15-2 A timeline of the roots of evolutionary thought** Each bar's length represents the life span of a scientist who played a key role in the development of modern evolutionary biology.

Buffon
Species created, then evolve

Hutton
Gradual geological change

Lamarck
Mechanisms of species change

Cuvier
Successive catastrophes

Smith
Sequence of fossils

Lyell
Very old Earth

Darwin
Evolution, natural selection

Wallace
Evolution, natural selection

1700   1750   1800   1850   1900

A Few Scientists Speculated That Life Had Evolved

A few eighteenth-century scientists went so far as to speculate that species had, in fact, changed over time. For example, the French naturalist Georges Louis Leclerc (1707–1788), known by the title Comte de Buffon, suggested that the original creation provided a relatively small number of founding species, after which some might have "improved" or "degenerated," perhaps after moving to new geographic areas. That is, Buffon suggested that species had changed over time through natural processes.

Fossil Discoveries Showed That Life Has Changed over Time

As Buffon and his contemporaries pondered the implications of new biological discoveries, developments in geology cast further doubt on the idea of permanently fixed species. Especially important was the discovery, during excavations for roads, mines, and canals, of rock fragments that resembled parts of living organisms. People had known of such objects since the fifteenth century, but most thought they were ordinary rocks that wind, water, or people had worked into lifelike forms. As more and more organism-shaped rocks were discovered, however, it became obvious that they were **fossils,** the preserved remains or traces of organisms that had died long ago (**FIG. 15-3**). Many fossils are bones, wood, shells, or their

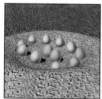
eggs in nest

fossilized feces (coprolites)

bones

footprint

skin impression

▲ **FIGURE 15-3 Types of fossils** Any preserved part or trace of an organism is a fossil.

impressions in mud that have been petrified, or converted to stone. Fossils also include other kinds of preserved traces, such as tracks, burrows, pollen grains, eggs, and feces.

By the beginning of the nineteenth century, some pioneering investigators realized that the distribution of fossils in rock was also significant. Many rocks occur in layers, with newer layers positioned over older layers. The British surveyor William Smith (1769–1839), who studied rock layers and the fossils embedded in them, recognized that certain fossils were always found in the same layers of rock. Further, the organization of fossils and rock layers was consistent across different areas: Fossil type A could always be found in a rock layer resting beneath a younger layer containing fossil type B, which in turn rested beneath a still-younger layer containing fossil type C, and so on.

Scientists of the period also discovered that fossil remains showed a remarkable progression. Most fossils found in the oldest layers were very different from modern organisms, and the resemblance to modern organisms gradually increased in progressively younger rocks (**FIG. 15-4**). Many of the fossils

youngest rocks

oldest rocks

(a) Trilobite   **(b) Seed ferns**   **(c)** *Allosaurus*

▲ **FIGURE 15-4** **Different fossils are found in different rock layers** Fossils provide strong support for the idea that today's organisms were not created all at once but arose over time by the process of evolution. If all species had been created simultaneously, we would not expect **(a)** the earliest trilobites to be found in older rock layers than **(b)** the earliest seed ferns, which in turn would not be expected in older layers than **(c)** dinosaurs, such as *Allosaurus*. Trilobites first appeared about 520 million years ago, seed ferns (which were not actually ferns but had fern-like foliage) about 380 million years ago, and dinosaurs about 230 million years ago.

were from plant or animal species that had gone *extinct*; that is, no members of the species still lived on Earth.

Putting all of these facts together, some scientists came to an inescapable conclusion: Different types of organisms had lived at different times in the past.

Some Scientists Devised Nonevolutionary Explanations for Fossils

Despite the growing fossil evidence, many scientists of the period did not accept the proposition that species changed and new ones arose over time. To account for extinct species while preserving the notion of a single creation by God, Georges Cuvier (1769–1832) advanced the idea of *catastrophism*. Cuvier, a French anatomist and paleontologist, hypothesized that a vast supply of species was created initially. Successive catastrophes (such as the Great Flood described in the Bible) produced layers of rock and destroyed many species, fossilizing some of their remains in the process. The organisms of the modern world, he speculated, are the species that survived the catastrophes.

Geology Provided Evidence That Earth Is Exceedingly Old

Cuvier's hypothesis of a world shaped by successive catastrophes was challenged by the work of the geologist Charles Lyell (1797–1875). Lyell, building on the earlier thinking of James Hutton (1726–1797), considered the forces of wind, water, and volcanoes and concluded that there was no need to invoke catastrophes to explain the findings of geology. Don't flooding rivers lay down layers of sediment? Don't lava flows produce layers of basalt? Shouldn't we conclude, then, that layers of rock are evidence of ordinary natural processes, occurring repeatedly over long periods of time? This concept, that Earth's present landscape was produced by past action of the same gradual geological processes that we observe today, is called *uniformitarianism*. Acceptance of uniformitarianism by scientists of the time had a profound impact, because the idea implies that Earth is very old.

Before the 1830 publication of Lyell's evidence in support of uniformitarianism, few scientists suspected that Earth could be more than a few thousand years old. Counting generations in the Old Testament, for example, yields a maximum age of 4,000 to 6,000 years. An Earth this young poses problems for the idea that life has evolved. For example, ancient writers such as Aristotle described wolves, deer, lions, and other organisms that were identical to those present in Europe more than 2,000 years later. If organisms had changed so little over that time, how could whole new species possibly have arisen if Earth was created only a couple of thousand years before Aristotle's time?

But if, as Lyell suggested, rock layers thousands of feet thick were produced by slow, natural processes, then Earth must be old indeed, many millions of years old. Lyell, in fact, concluded that Earth was eternal. Modern geologists estimate that Earth is about 4.5 billion years old (see "How Do We Know That? Discovering the Age of a Fossil" in chapter 18.

Lyell (and his intellectual predecessor Hutton) showed that there was enough time for evolution to occur. But what was the mechanism? What process could cause evolution?

Some Pre-Darwin Biologists Proposed Mechanisms for Evolution

One of the first scientists to propose a mechanism for evolution was the French biologist Jean Baptiste Lamarck (1744–1829). Lamarck was impressed by the sequences of organisms in rock layers. He observed that older fossils tend to be less like existing organisms than are more recent fossils. In 1809, Lamarck published a book in which he hypothesized that organisms evolved through the inheritance of acquired characteristics, a process in which the bodies of living organisms are modified through the use or disuse of parts, and these modifications are inherited by offspring. Why would bodies be modified? Lamarck proposed that all organisms possess an innate drive for perfection. For example, if ancestral giraffes tried to increase their feeding opportunities by stretching upward to reach leaves growing high up in trees, their necks became slightly longer as a result. Their offspring would inherit these longer necks and then stretch even farther to reach still higher leaves. Eventually, this process would produce modern giraffes with very long necks indeed.

Today, we understand how inheritance works and can see that Lamarck's proposed evolutionary process could not work. Acquired characteristics are not inherited. The fact that a prospective father pumps iron doesn't mean that his child will look like a champion bodybuilder. Remember, though, that in Lamarck's time the principles of inheritance had not yet been discovered. Gregor Mendel's pioneering work demonstrating inheritance in pea plants was not widely recognized until 1900 (see Chapter 11). In any case, Lamarck's insight that inheritance plays an important role in evolution had an important influence on the later biologists who discovered the key mechanism of evolution.

Darwin and Wallace Proposed a Mechanism of Evolution

By the mid-1800s, a growing number of biologists had concluded that present-day species had evolved from earlier ones. But how? In 1858, Charles Darwin (1809–1882) and Alfred Russel Wallace (1823–1913), working separately, provided convincing evidence that evolution was driven by a simple yet powerful process.

Although their social and educational backgrounds were very different, Darwin and Wallace were quite similar in some respects. Both had traveled extensively in the tropics and had studied the plants and animals living there. Both observed that some species differed in only a few features

(a) Large ground finch: beak suited to large seeds

(b) Small ground finch: beak suited to small seeds

(c) Warbler finch: beak suited to insects

(d) Vegetarian tree finch: beak suited to leaves

◀ **FIGURE 15-5 Darwin's finches, residents of the Galápagos Islands** Darwin studied a group of closely related species of finches on the Galápagos Islands. Each species specializes in eating a different type of food and has a beak of characteristic size and shape because past individuals whose beaks were best suited to exploit each local food source produced more offspring than did individuals with less effective beaks.

(**FIG. 15-5**). Both were familiar with the fossils that had been discovered, many of which showed a trend through time of increasing similarity to modern organisms. Finally, both were aware of the studies of Hutton and Lyell, who had proposed that Earth is extremely ancient. These facts suggested to both Darwin and Wallace that species change over time. Both men sought a mechanism that might cause such evolutionary change.

Of the two, Darwin was the first to propose a mechanism for evolution, which he sketched out in 1842 and described more fully in an essay in 1844. He sent the essay to a few colleagues, but did not submit it for publication, perhaps because he was fearful of the controversy that publication would cause. Some historians wonder if Darwin would ever have published his ideas had he not received, some 16 years after his initial draft, a paper by Wallace that outlined ideas remarkably similar to Darwin's own. Darwin realized that he could delay no longer.

In separate but similar papers that were presented to the Linnaean Society in London in 1858, Darwin and Wallace each described the same mechanism for evolution. Initially, their papers had little impact. The secretary of the society, in fact, wrote in his annual report that nothing very interesting happened that year. Fortunately, the next year, Darwin published his monumental book, *On the Origin of Species by Means of Natural Selection,* which attracted a great deal of attention to the new ideas about how species evolve. (To learn more

about Darwin's life, see "How Do We Know That? Charles Darwin and the Mockingbirds" on page 268.)

CHECK YOUR LEARNING

Can you ...
- identify some of the thinkers whose ideas set the stage for the development of the theory of evolution?
- describe the key ideas of those thinkers?
- define *evolution*?

15.2 HOW DOES NATURAL SELECTION WORK?

Darwin and Wallace proposed that life's huge variety arose by a process of descent with modification, in which individuals in each generation differ slightly from the members of the preceding generation. Over long stretches of time, these small differences accumulate to produce major transformations.

Darwin and Wallace's Theory Rests on Four Postulates

The chain of logic that led Darwin and Wallace to their proposed process of evolution turns out to be surprisingly

HOW DO WE KNOW THAT?

Charles Darwin and the Mockingbirds

Charles Darwin

How did mockingbirds provide key evidence for evolution? The story begins in 1831, when 22-year-old Charles Darwin secured a position as "gentleman companion" to Captain Robert Fitzroy of the HMS *Beagle*. The *Beagle* soon embarked on a 5-year surveying expedition along the coastline of South America and then around the world.

In addition to his duties as companion to the captain, Darwin served as the expedition's official naturalist, whose task was to observe and collect geological and biological specimens. The *Beagle* sailed to South America and made many stops along its coast. But perhaps the most significant stopover of the voyage was the month spent on the Galápagos Islands, off the northwestern coast of South America. There, along with many other fascinating plants and animals, Darwin found mockingbirds. The data he recorded about them played a crucial role in bringing him to conclude that species could change over time.

When the *Beagle* reached the Galápagos, the first island that Darwin visited was then called Chatham Island. He noticed that the mockingbirds there seemed different than the ones he had observed on the South American mainland. A short while later, Darwin visited Charles Island and was surprised to find mockingbirds that differed in many respects from the ones on nearby Chatham. And, traveling onward, he found mockingbirds on James Island that, to his eye, were different from any he had seen on the other islands.

Darwin was astonished and impressed by his discovery that different islands had different mockingbirds (**FIG. E15-1**). He was similarly struck by reports that the archipelago's gigantic tortoises also differed from island to island. Darwin began to wonder if the differences in the tortoises and mockingbirds arose after they had become isolated on separate islands. A few weeks after leaving the Galápagos, Darwin, pondering the variety and distribution of the animals, wrote that "such facts undermine the stability of Species." This journal entry represents the first tentative indication that Darwin had accepted the impermanence of species.

When the *Beagle* returned to London in 1836, Darwin asked various experts to examine the specimens he had collected during his journey. The ornithologist John Gould studied Darwin's bird specimens and judged that the roving naturalist had indeed collected three different species of mockingbird, each of which inhabited a different island or small set of islands. This conclusion, possible only because of Darwin's systematic collecting and careful documentation, proved to be one of the clinching pieces of evidence in Darwin's conversion to evolutionary thinking. A few months later, Darwin drew in his journal a small tree-like diagram, a representation of his emerging idea that species are linked by descent from a common ancestor (**FIG. E15-2**).

THINK CRITICALLY A recent study found that Galápagos mockingbirds on a given island are more genetically similar to mockingbirds on nearby islands than to mockingbirds on more distant islands. From this information, what can you conclude about the evolutionary history of Galápagos mockingbirds?

▲ **FIGURE E15-1** **Galápagos mockingbirds** Each island in the Galápagos contains a unique mockingbird species. Darwin and his contemporary John Gould described three species, but modern ornithologists recognize four. Clockwise from top left: the Galápagos, Hood, Charles, and Chatham mockingbirds.

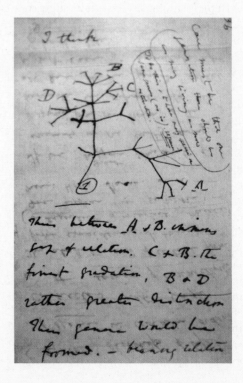

◀ **FIGURE E15-2**
A sketch from Darwin's notebook, 1837

simple and straightforward. It is based on four postulates about populations:

Postulate 1: Individual members of a population differ from one another in many respects.

Postulate 2: At least some of the differences among members of a population are due to characteristics that may be passed from parent to offspring.

Postulate 3: In each generation, some individuals in a population survive and reproduce successfully but others do not.

Postulate 4: The fate of individuals is not determined entirely by chance or luck. Instead, an individual's likelihood of survival and reproduction depends on its characteristics. Individuals with advantageous traits survive longest and leave the most offspring, a process known as **natural selection.**

Darwin and Wallace understood that if all four postulates were true, populations would inevitably change over time. If members of a population have different traits, and if the individuals that are best suited to their environment leave more offspring, and if those individuals pass their favorable traits to their offspring, then the favorable traits will become more common in subsequent generations. The characteristics of the population will change slightly with each generation. This process is evolution by natural selection.

Are the four postulates true? Darwin thought so, and devoted much of *On the Origin of Species* to describing supporting evidence. Let's briefly examine each postulate, in some cases with the advantage of knowledge that had not yet come to light during the lifetimes of Darwin and Wallace.

Postulate 1: Individuals in a Population Vary

The accuracy of postulate 1 is apparent to anyone who has glanced around a crowded room. People differ in size, eye color, skin color, and many other physical features. Similar variability is present in populations of other organisms, although it may be less obvious to the casual observer (**FIG. 15-6**).

Postulate 2: Traits Are Passed from Parent to Offspring

The principles of genetics had not yet been discovered when Darwin published *On the Origin of Species*. Therefore, although observation of people, pets, and farm animals seemed to show that offspring generally resemble their parents, Darwin and Wallace did not have scientific evidence in support of postulate 2. Mendel's later work, however, demonstrated conclusively that particular traits can be passed to offspring. Since Mendel's time, genetics researchers have produced a detailed picture of how inheritance works.

Postulate 3: Some Individuals Fail to Survive and Reproduce

Darwin's formulation of postulate 3 was heavily influenced by Thomas Malthus's *Essay on the Principle of Population*

(1798), which described the perils of unchecked growth of human populations. Darwin was keenly aware that organisms can produce far more offspring than are required merely to replace the parents. He calculated, for example, that a single pair of elephants would multiply to a population of 19 million in 750 years if each descendant had six offspring that lived to reproduce.

But we aren't overrun with elephants. The number of elephants, like the number of individuals in most natural populations, tends to remain relatively constant. Therefore, more organisms must be born than survive long enough to reproduce. In each generation, many individuals must die young. Even among those that survive, many must fail to reproduce, produce few offspring, or produce less-vigorous offspring that, in turn, fail to survive and reproduce. As you might expect, whenever biologists have measured reproduction in a population, they have found that some individuals have more offspring than others.

Postulate 4: Survival and Reproduction Are Not Determined by Chance

If unequal reproduction is the norm in populations, what determines which individuals leave the most offspring? A large amount of scientific evidence has shown that reproductive success depends on an individual's characteristics. For example, scientists found that larger male elephant seals in a California population have more offspring than smaller males

▲ **FIGURE 15-6 Variation in a population of snails** Although these snail shells are all from members of the same population, no two are exactly alike.

THINK CRITICALLY Is sexual reproduction required to generate the variability in structures and behaviors that is necessary for natural selection?

(because females are more likely to mate with large males). In a Colorado population of snapdragons, plants with white flowers have more offspring than plants with yellow flowers (because pollinators find white flowers more attractive). These results, and hundreds of other similar ones, show that in the competition to survive and reproduce, winners are for the most part determined not by chance but by the traits they possess.

Natural Selection Modifies Populations over Time

Observation and experiment suggest that the four postulates of Darwin and Wallace are sound. Logic suggests that the resulting consequence ought to be change over time in the characteristics of populations. In *On the Origin of Species*, Darwin proposed the following example: "Let us take the case of a wolf, which preys on various animals, securing [them] by . . . fleetness . . . . The swiftest and slimmest wolves would have the best chance of surviving, and so be preserved or selected . . . . Now if any slight innate change of habit or structure benefited an individual wolf, it would have the best chance of surviving and of leaving offspring. Some of its young would probably inherit the same habits or structure, and by the repetition of this process, a new variety might be formed." The same logic applies to the wolf's prey: The fastest or most alert or best camouflaged would be most likely to avoid predation and would pass these traits to its offspring.

Notice that natural selection acts on individuals. Eventually, however, the influence of natural selection on the fates of individuals has consequences for the population as a whole. Over generations, the population changes as the percentage of individuals inheriting favorable traits increases. An individual cannot evolve, but a population can.

CHECK YOUR LEARNING

Can you ...

- explain how natural selection works and how it affects populations?
- describe the logic, based on four postulates, by which Darwin and Wallace deduced that populations must evolve by natural selection?

15.3 HOW DO WE KNOW THAT EVOLUTION HAS OCCURRED?

Today, evolution is an accepted scientific theory. (A scientific theory is a general explanation of important natural phenomena, developed through extensive, reproducible observations; see Chapter 1.) An overwhelming body of evidence supports the conclusion that evolution has occurred. The key lines of evidence come from fossils, comparative anatomy (the study of how body structures differ among species), embryology (the study of developing organisms in the period from fertilization to birth or hatching), biochemistry, and genetics.

Fossils Provide Evidence of Evolutionary Change over Time

If many fossils are the remains of species ancestral to modern species, we might expect to find fossils in a progressive series that starts with an ancient organism, progresses through several intermediate stages, and culminates in a modern species. Such series have indeed been found. For example, fossils of the ancestors of modern whales illustrate stages in the evolution of an aquatic species from land-dwelling ancestors (**FIG. 15-7**). Series of fossil giraffes, elephants, horses, and mollusks also show the evolution of body structures over time. These fossil series suggest that new species evolved from, and replaced, previous species.

Comparative Anatomy Gives Evidence of Descent with Modification

Fossils provide snapshots of the past that allow biologists to trace evolutionary changes, but careful examination of today's organisms can also uncover evidence of evolution. Comparing the bodies of organisms of different species can reveal similarities that can be explained only by shared ancestry and differences that could result only from evolutionary change during descent from a common ancestor. In this way, the study of comparative anatomy has supplied strong evidence that different species are linked by a common evolutionary heritage.

Homologous Structures Provide Evidence of Common Ancestry

A body structure may be modified by evolution to serve different functions in different species. The forelimbs of birds and mammals, for example, are variously used for flying, swimming, running, and grasping objects. Despite this enormous diversity of function, the internal anatomy of all bird and mammal forelimbs is remarkably similar (**FIG. 15-8**). It seems inconceivable that the same bone arrangements would be used to serve such diverse functions if each animal had been created separately. Such similarity is exactly what we would expect, however, if bird and mammal forelimbs were derived from the forelimb of a common ancestor. Through natural selection, the ancestral forelimb has undergone different modifications in different kinds of animals. The resulting internally similar structures are called **homologous structures,** meaning that they have the same evolutionary origin despite any differences in current function or appearance.

Vestigial Structures Are Inherited from Ancestors

A **vestigial structure** no longer performs the function for which it evolved in a species' ancestors. Although vestigial structures are sometimes co-opted for new uses, they often seem to serve no function at all. Examples of functionless

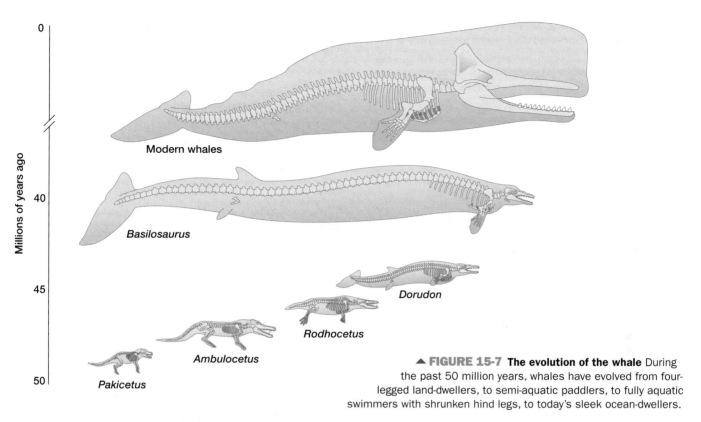

Millions of years ago

Modern whales

Basilosaurus

Dorudon

Rodhocetus

Ambulocetus

Pakicetus

▲ **FIGURE 15-7 The evolution of the whale** During the past 50 million years, whales have evolved from four-legged land-dwellers, to semi-aquatic paddlers, to fully aquatic swimmers with shrunken hind legs, to today's sleek ocean-dwellers.

THINK CRITICALLY The fossil history of some kinds of modern organisms, such as sharks and crocodiles, shows that their structure and appearance have changed very little over hundreds of millions of years. Is this lack of change evidence that such organisms have not evolved during that time?

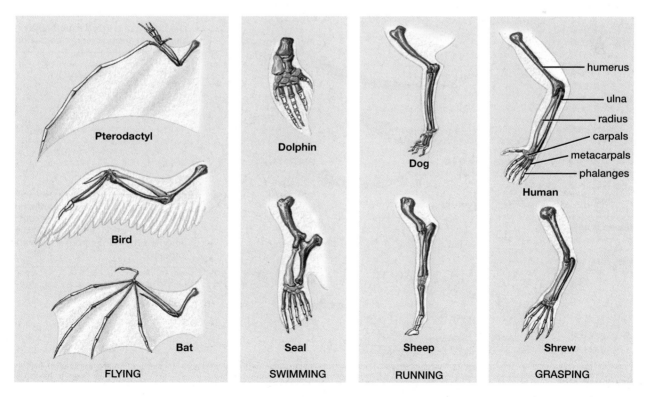

Pterodactyl

Bird

Bat

FLYING

Dolphin

Seal

SWIMMING

Dog

Sheep

RUNNING

humerus

ulna

radius

carpals

metacarpals

phalanges

Human

Shrew

GRASPING

▲ **FIGURE 15-8 Homologous structures** Despite wide differences in function, the forelimbs of all of these animals contain the same set of bones, inherited from a common ancestor. The different colors of the bones highlight the correspondences among the various species.

vestigial structures include molar teeth in vampire bats (which live on a diet of blood and, therefore, don't chew their food) and pelvic bones in whales and certain snakes (FIG. 15-9). Both of these vestigial structures are clearly homologous to structures that are found in—and used by—other vertebrates (animals with a backbone). The continued existence in organisms of structures for which they have no use is best explained as a sort of "evolutionary baggage." For example, the ancestral mammals from which whales evolved had four legs and a well-developed set of pelvic bones (see Figure 15-7). Whales do not have hind legs, yet they have small pelvic and leg bones embedded in their sides. During whale evolution, losing the hind legs provided an advantage, better streamlining the body for movement through water. The result is the modern whale with small, useless pelvic bones that persist because they have shrunk to the point that they no longer constitute a survival-reducing burden.

Some Anatomical Similarities Result from Evolution in Similar Environments

The study of comparative anatomy has demonstrated the shared ancestry of life by identifying a host of homologous structures that different species have inherited from common ancestors, but comparative anatomists have also identified many anatomical similarities that do not stem from common ancestry. Instead, these similarities arose through **convergent evolution**, in which natural selection

CASE STUDY **CONTINUED**
What Good Are Wisdom Teeth and Ostrich Wings?

Ostrich wings are vestigial because they are too rudimentary to perform the function for which they evolved in the species' flying ancestor. Nonetheless, the ostrich uses its wings for other purposes. For example, an ostrich may extend its wings to the side while running, to help maintain balance, and it may spread its wings as part of a threat display. These uses show that evolution by natural selection can sometimes repurpose vestigial structures that have lost the function for which they originally evolved. But whether a vestigial structure remains useless or acquires a new function, it is homologous to the version that retains its original function in other organisms and provides evidence of common ancestry. But are all similarities between different organisms the result of shared ancestry?

causes non-homologous structures that serve similar functions to resemble one another. For example, both birds and insects have wings, but this similarity did not arise from evolutionary modification of a structure that both birds and insects inherited from a common ancestor. Instead, the similarity arose from parallel modification of two different, non-homologous structures. Because natural selection

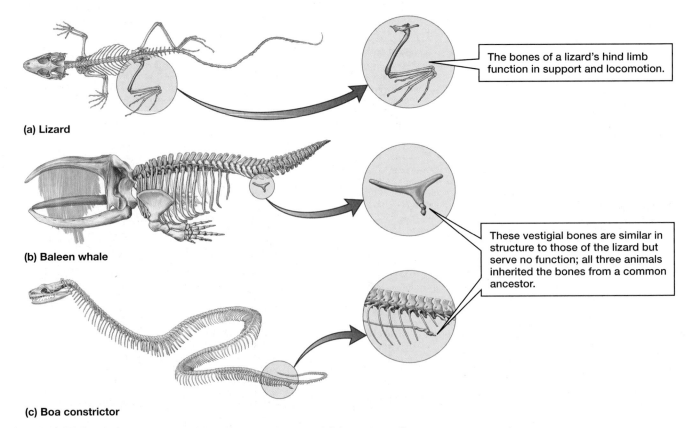

(a) Lizard

The bones of a lizard's hind limb function in support and locomotion.

(b) Baleen whale

These vestigial bones are similar in structure to those of the lizard but serve no function; all three animals inherited the bones from a common ancestor.

(c) Boa constrictor

▲ **FIGURE 15-9 Vestigial structures** Many organisms have vestigial structures that serve no apparent function. The **(a)** lizard, **(b)** baleen whale, and **(c)** boa constrictor all inherited hind limb bones from a common ancestor. These bones remain functional in the lizard but are vestigial in the whale and snake.

(a) Damselfly

(b) Swallow

▲ FIGURE 15-10 **Analogous structures** Convergent evolution can produce outwardly similar structures that differ anatomically, such as the wings of **(a)** insects and **(b)** birds.

THINK CRITICALLY Are a peacock's tail and a dog's tail homologous structures or analogous structures?

favored flight in both birds and insects, the two groups evolved wings of roughly similar appearance, but the similarity is superficial. Such outwardly similar but non-homologous structures are called **analogous structures** (FIG. 15-10). Analogous structures are typically very different in internal anatomy, because the parts are not derived from common ancestral structures.

Embryological Similarity Suggests Common Ancestry

Evidence of common ancestry is apparent in the striking similarity of embryos of different species (FIG. 15-11). For example, in their early embryonic stages, fish, turtles, chickens, mice, and humans all develop tails and gill slits (also called gill grooves).

Why are vertebrates that are so different as adults so similar at an early stage of development? The only plausible explanation is that all of these species descended from an ancestral vertebrate that possessed genes that directed the development of gills and tails. All of the descendants still have those genes. In fish, these genes are active throughout development, resulting in adults with fully developed tails and gills. In humans and chickens, these genes are active only during early developmental stages; the structures are lost or become inconspicuous before adulthood.

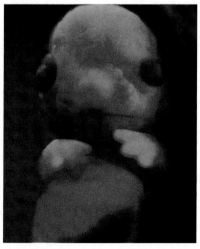

(a) Lemur

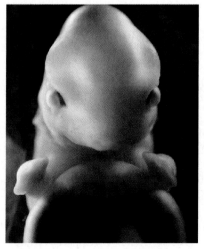

(b) Pig

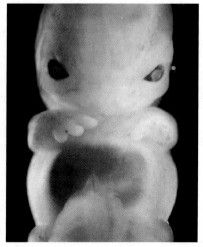

(c) Human

▲ FIGURE 15-11 **Embryological stages reveal evolutionary relationships** Early embryonic stages of a **(a)** lemur, **(b)** pig, and **(c)** human, showing strikingly similar anatomical features.

Between 70% and 85% of people will experience lower back pain at some point in life, and for many people, the condition is chronic. This state of affairs is an unfortunately painful consequence of the evolutionary process. We walk upright on two legs, but our distant ancestors walked on all fours. Thus, natural selection formed our vertically oriented spine by remodeling one whose normal orientation was parallel to the ground. Our spinal anatomy evolved some modifications in response to its new posture, but as is often the case with evolution, the changes involved some trade-offs. The arrangements of bone and muscle that permit our smooth, bipedal gait also generate vertical compression of the spine, and the resulting pressure can, and frequently does, cause painful damage to muscle and nerve tissues.

Why Backaches Are So Common?

Modern Biochemical and Genetic Analyses Reveal Relatedness Among Diverse Organisms

Biologists have been aware of anatomical and embryological similarities among organisms for centuries, but it took the emergence of modern technology to reveal similarities at the molecular level. Biochemical similarities among organisms provide perhaps the most striking evidence of their evolutionary relatedness. Just as relatedness is revealed by homologous anatomical structures, it is also revealed by homologous molecules.

Today's scientists have access to a powerful tool for revealing molecular homologies: DNA sequencing. It is now possible to quickly determine the sequence of nucleotides in a DNA molecule and to compare the DNA of different organisms. For example, consider the gene that encodes the protein cytochrome *c* (see Chapters 12 and 13 for information on DNA and how it encodes proteins). Cytochrome *c* is present in all plants and animals (and many single-celled organisms) and performs the same function in all of them. The sequence of nucleotides in the gene for cytochrome *c* is similar in these diverse species (**FIG. 15-12**). The widespread presence of the same complex protein, encoded by the same gene and performing the same function, is evidence that the common ancestor of plants and animals had cytochrome *c* in its cells. At the same time, though, the sequence of the cytochrome *c* gene differs slightly in different species, showing that variations arose during the independent evolution of Earth's multitude of plant and animal species.

Some biochemical similarities are so fundamental that they extend to all living cells. For example:

- All cells have DNA as the carrier of genetic information.
- All cells use RNA, ribosomes, and approximately the same genetic code to translate that genetic information into proteins.
- All cells use roughly the same set of 20 amino acids to build proteins.
- All cells use ATP as a cellular energy carrier.

The most plausible explanation for such widespread sharing of complex and specific biochemical traits is that the

◀ **FIGURE 15-12 Molecular similarity shows evolutionary relationships** The DNA sequences of the genes that code for cytochrome *c* in a human and a mouse. Of the 315 nucleotides in the gene, only 30 (shaded blue) differ between the two species.

traits are homologies. That is, they arose only once, in the common ancestor of all living things, from which all of today's organisms inherited them.

CHECK YOUR LEARNING

Can you ...

- describe the evidence that evolution has occurred?
- explain the difference between similarity due to homology and similarity due to convergent evolution?

CASE STUDY **CONTINUED**

What Good Are Wisdom Teeth and Ostrich Wings?

Just as anatomical homology can lead to vestigial structures such as human wisdom teeth and the wings of flightless birds, genetic homology can lead to vestigial DNA sequences. For example, most mammal species produce an enzyme, L-gulonolactone oxidase, that catalyzes the last step in the production of vitamin C. The species that produce the enzyme are able to do so because they all inherited the gene that encodes it from a common ancestor. Humans, however, do not produce L-gulonolactone oxidase, so we can't produce vitamin C ourselves and must consume it in our diets. But even though we don't produce the enzyme, our cells do contain a stretch of DNA with a sequence very similar to that of the enzyme-producing gene present in rats and most other mammals. The human version, though, does not encode the enzyme (or any protein). We inherited this stretch of DNA from an ancestor that we share with other mammal species, but in us, the sequence has undergone a change that rendered it nonfunctional. (The change probably did not confer a strong disadvantage, because our ancestors got sufficient vitamin C in their diets.) The nonfunctional sequence remains as a vestigial trait, evidence of our shared ancestry.

Vestigial traits are evidence of both shared ancestry and change in traits over time. What kinds of observations and experiments show that natural selection contributes to evolutionary change?

15.4 WHAT IS THE EVIDENCE THAT POPULATIONS EVOLVE BY NATURAL SELECTION?

We have seen that evidence of evolution comes from many sources. But what is the evidence that evolution occurs by the process of natural selection?

Controlled Breeding Modifies Organisms

One line of evidence supporting evolution by natural selection is **artificial selection,** the breeding of domestic plants and animals to produce specific desirable features. The various dog breeds provide a striking example of artificial selection (**FIG. 15-13**). Dogs descended from wolves, and even today, the two will readily crossbreed. With few exceptions,

(a) Gray wolf

(b) Diverse dogs

▲ **FIGURE 15-13 Dog diversity illustrates artificial selection** A comparison of **(a)** the ancestral dog (the gray wolf, *Canis lupus*) and **(b)** various breeds of dog. Artificial selection by humans has caused great divergence in the size and shape of dogs in only a few thousand years.

however, modern dogs do not closely resemble wolves. Some breeds are so different from one another that they would be considered separate species if they were found in the wild. Humans produced these radically different dogs in a few thousand years by doing nothing more than repeatedly selecting individuals with desirable traits for breeding. Therefore, it is quite plausible that natural selection could, by a comparable process acting over hundreds of millions of years, produce the spectrum of living organisms. Darwin was so impressed by the connection between artificial selection and natural selection that he devoted a chapter of *On the Origin of Species* to the topic.

Evolution by Natural Selection Occurs Today

Additional evidence of natural selection comes from scientific observation and experimentation. The logic of natural selection gives us no reason to believe that evolutionary change is limited to the past. After all, inherited variation and competition for access to resources are certainly not limited to the past. If Darwin and Wallace were correct that those conditions lead inevitably to evolution by natural selection, then researchers ought to be able to detect evolutionary change as it occurs. And they have. Next, we will consider some examples that give us a glimpse of natural selection at work.

Natural Selection Has Silenced Calling Crickets

Among the species found on the Hawaiian island of Kauai is the Polynesian field cricket (**FIG. 15-14a**). Researchers studying the crickets on Kauai in the 1990s found that males produced a loud call by rubbing together two structures on their wings: the scraper (a raised ridge) and the file (a modified vein studded with evenly spaced tiny teeth). The males called at night, and female crickets moved toward the calls to find and choose mates. Strangely, however, each year that the researchers returned to Kauai, they heard fewer and fewer cricket calls. By 2003, the silence was virtually complete; almost no crickets were calling. But the crickets had not disappeared; a nighttime searcher with a strong flashlight could still find a multitude of quietly active crickets. Why didn't the male crickets make any calls? Because the files on their wings were tiny and distorted, these "flatwing" males were not capable of calling (**FIG. 15-14b**). What had happened?

Some time during the 1990s, the crickets' environment had changed drastically with the arrival from North America of a fly species that had not previously been present on Kauai. The fly is a deadly parasite, with larvae that burrow into the bodies of crickets, eating them alive. How does a fly find a cricket to parasitize? By moving toward the sound of calling crickets. Thus, after the arrival of the flies, loudly calling crickets were less likely to survive than those that happened to call more softly or not at all. Natural selection favored quieter crickets, and within fewer than 20 generations, the silent males had almost completely replaced callers. As a result of natural selection, the structure of male cricket wings had changed significantly. There must have been a corresponding evolutionary change in female mating behavior, as females are now willing to mate with silent flatwing males.

Natural Selection Can Lead to Herbicide and Pesticide Resistance

Natural selection is also evident in weed species that have evolved resistance to the herbicides with which we try to control them. Successful agriculture depends on farmers' ability to kill weeds that compete with crop plants, but many members of weed species can no longer be killed by a formerly lethal dose of glyphosate, the active ingredient in Roundup®, the world's most widely used herbicide. How did these glyphosate-resistant "superweeds" arise? They arose because the herbicide has acted as an agent of natural selection. Consider, for example, giant ragweed, one of the highly destructive weed species that are now resistant to glyphosate in some places. When a field is sprayed with Roundup, almost all of the giant ragweed plants there are killed, because glyphosate inactivates an enzyme that is essential to the plants' survival. A few ragweed plants, however, survive. Researchers have discovered that some of these survivors carry a mutation that causes them to produce a tremendous amount of the enzyme that glyphosate attacks, more enzyme than the usual dose of glyphosate can destroy. In the face of repeated applications of Roundup, the formerly rare protective mutation has become common in many giant ragweed populations. (For additional examples of how humans influence evolution, see "Earth Watch: People Promote High-Speed Evolution.")

The evolution of glyphosate-resistant superweeds was a direct result of changes in agricultural practice. In the 1990s, the biotechnology company Monsanto began selling seeds that had been genetically engineered to produce crops that are not harmed by glyphosate. These "Roundup Ready" crops, which now account for the vast majority of soybean, corn, and cotton plantings in the United States and other countries, allow farmers to freely apply glyphosate to their fields without fear of harming crop plants. As a result, use of glyphosate has skyrocketed. Today, large-scale agriculture is

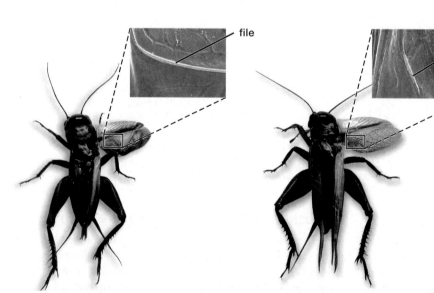

(a) Polynesian cricket, wing with file   **(b) Polynesian cricket, wing with lost file**

◀ **FIGURE 15-14 Crickets evolve to become silent when calls attract parasites** A few decades ago, Polynesian field crickets on the island of Kauai used their wings to produce loud calls. But when the environment changed to include parasitic flies that are attracted to cricket calls, natural selection favored crickets with wings than cannot produce sound. **(a)** Noise-producing "files" were present on male cricket wings before the flies appeared in the environment, but **(b)** the files all but disappeared within a few years of the flies' arrival.

Earth WATCH — People Promote High-Speed Evolution

You probably don't think of yourself as a major engine of evolution. Nonetheless, as you go about the routines of your daily life, you are contributing to what is perhaps today's most significant cause of rapid evolutionary change. Human activity has changed Earth's environments tremendously, and the biological logic of natural selection, spelled out so clearly by Darwin, tells us that environmental change leads inevitably to evolutionary change. Thus, by changing the environment, humans have become a major agent of natural selection.

Unfortunately, many of the evolutionary changes we have caused have turned out to be bad news for us. Our liberal use of pesticides has selected for resistant pests that frustrate efforts to protect our food supply. By overmedicating ourselves with antibiotics and other drugs, we have selected for resistant "supergerms" and diseases that are ever more difficult to treat (see Chapter 16). Heavy fishing in the world's oceans has favored smaller fish that can slip through nets more easily, thereby selecting for slow-growing fish that remain small even as mature adults. As a result, fish of many commercially important species are now so small that our ability to extract food from the sea is compromised.

Our use of pesticides, antibiotics, and fishing technology has caused evolutionary changes that threaten our health and welfare, but the scope of these changes may be dwarfed by those that will arise from human-caused modification of Earth's climate. Human activities, especially activities that use energy derived from fossil fuels, modify the climate by contributing to global warming. In coming years, species' evolution will be increasingly influenced by environmental changes associated with a warming climate, such as reduced ice and snow, longer, hotter summers, and shifts in the life cycles of other species that provide food or shelter.

There is growing evidence that climate change is already causing evolutionary change. Warming-related evolution has been found in a number of plant and animal populations. For example, in Finland, tawny owls have changed color in response to a warmer climate (**FIG. E15-3**). Owls of this species come in two varieties, gray or brown. In the past, most owls were gray, and the brown variety was rare. Today, however, about 40% of Finnish owls are brown, and researchers have shown that the increase in brown owls was probably caused by climate change. In particular, the researchers found that in snowy winters, gray owls survive much better than brown owls, perhaps because they are better camouflaged against the snow and therefore suffer less predation

▲ **FIGURE E15-3 Tawny owl populations have evolved in response to global climate change** The proportion of brown owls is increasing because brown owls survive better than gray ones over winters with less snow.

by eagles. In winters with less snow, brown owls are better camouflaged and more likely to survive. As temperatures have risen in recent decades, snowy winters have become increasingly rare, and the resulting reduced snow cover has favored the survival of brown owls. Thus, a brown owl is more likely than a gray one to survive the winter and, because feather color is inherited, produce brown offspring. The proportion of brown owls in the population has grown in response to natural selection associated with warming temperatures.

The available evidence suggests that global climate change will have an enormous evolutionary impact, potentially affecting the evolution of almost every species. How will these evolutionary changes affect us and the ecosystems on which we depend? This question is not readily answerable, because the path of evolution is not predictable. We can hope, however, that careful monitoring of evolving species and increased understanding of evolutionary processes will help us take appropriate steps to safeguard our health and well-being as Earth warms.

THINK CRITICALLY To reduce the incidence of pesticide resistance, farmers are advised to place fields that are free of pesticides or pesticide-containing crops beside fields in which pesticides are used as usual. Given your understanding of how evolution works, explain how this method would slow the evolution of pesticide resistance in insects.

---

extremely dependent on this single herbicide, even as its effectiveness declines steadily due to the evolution of resistant weeds.

Just as weeds have evolved to resist herbicides, many of the insects that attack crops have also evolved to resist the pesticides that farmers use to control them. Such resistance has been documented in more than 500 species of crop-damaging insects, and virtually every pesticide has fostered the evolution of resistance in at least one insect species. We pay a heavy price for this evolutionary phenomenon. The additional pesticides

that farmers apply in their attempts to control resistant insects cost almost $2 billion each year in the United States alone and add millions of tons of poisons to Earth's soil and water.

Experiments Can Demonstrate Natural Selection

In addition to observing natural selection in the wild, scientists have also devised numerous experiments that confirm the action of natural selection. For example, one group of evolutionary biologists released small groups of *Anolis sagrei* lizards onto 14 small Bahamian islands that were previously uninhabited by lizards. The original lizards came from a population on Staniel Cay, an island with tall vegetation, including plenty of trees. In contrast, the islands to which the lizards were introduced had few or no trees and were covered mainly with small shrubs and other low-growing plants.

The biologists returned to those islands 14 years later and found that the original small groups of lizards had given rise to thriving populations of hundreds of individuals. On all 14 of the experimental islands, lizards had legs that were shorter and thinner than those of lizards from the original source population on Staniel Cay. In just over a decade, it appeared, the lizard populations had changed in response to new environments.

Why had the new lizard populations evolved shorter, thinner legs? The researchers found that lizards with short, thin legs were slower but more agile than lizards with long, thick legs. They hypothesized that speed was especially important for escaping predators on the thick-branched trees of Staniel Cay, but that agility was more important on the thin-branched bushes of the experimental islands. Therefore, in the new environment, agile individuals with shorter, thinner legs were better able to survive and produce a greater number of offspring, so members of subsequent generations had shorter, thinner legs on average.

Selection Acts on Random Variation to Favor Traits That Work Best in Particular Environments

Two important points underlie the evolutionary changes just described:

- **The variations on which natural selection works are produced by chance mutations.** Silent wings in Hawaiian crickets, extra enzyme in giant ragweed plants, and shorter legs in Bahamian lizards were not *produced* by the parasitic flies, Roundup herbicide, or thinner branches. The mutations that produced each of these beneficial traits arose spontaneously.

- **Natural selection favors organisms that are best adapted to a particular environment.** Natural selection is not a process for producing ever-greater degrees of perfection. Natural selection does not select for the "best" in any absolute sense, but only for what is best in the context of a particular environment, which varies from place to place and which may change over time. A trait that is advantageous under one set of conditions may become disadvantageous if conditions change. For example, lizards with longer legs were better at escaping predators in the forested environment of Staniel Cay, but were worse at escaping predators in the shrubby environments of other islands.

CHECK YOUR LEARNING
Can you ...
- describe some observations and experiments that demonstrate that populations evolve by natural selection?

What Good Are Wisdom Teeth and Ostrich Wings?

Wisdom teeth are but one of many human anatomical structures that appear to no longer serve an important function (**FIG. 15-15**). Darwin himself noted many of these "useless, or nearly useless"

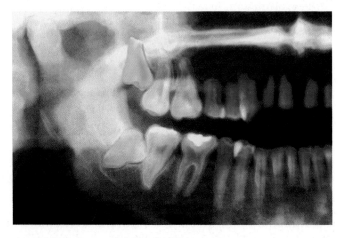

▲ **FIGURE 15-15 Wisdom teeth** Squeezed into a jaw that is too short to contain them, wisdom teeth often become impacted—unable to erupt through the surface of the gum. The leftmost upper and lower teeth in this X-ray image are impacted wisdom teeth.

traits in the very first chapter of *On the Origin of Species* and declared them to be prime evidence that humans had evolved from earlier species.

Body hair is another vestigial human trait. It seems to be an evolutionary relic of the fur that kept our distant ancestors warm (and that still warms our closest evolutionary relatives, the great apes). Not only do we retain useless body hair, we also still have the muscles that allow other mammals to puff up their fur for better insulation. In humans, these vestigial structures just give us goose bumps.

Though humans don't have and don't need a tail, we nonetheless have a tailbone. The tailbone consists of a few tiny vertebrae fused into a small structure at the base of the backbone, where a tail would be if we had one. People born without a tailbone or who have theirs surgically removed suffer no ill effects.

CONSIDER THIS Some advocates of the view that all organisms were created simultaneously by God argue that vestigial structures do not constitute evidence of evolution, because they show only that a divinely created structure can degenerate over time. According to this view, human tailbones are not evidence of evolution because they do not show that an adaptive improvement has occurred. Is this a valid argument?

CHAPTER REVIEW

 Go to MasteringBiology for practice quizzes, activities, eText, videos, current events, and more.

Answers to Think Critically, Evaluate This, Multiple Choice, and Fill-in-the-Blank questions can be found in the Answers section at the back of the book.

Summary of Key Concepts

15.1 How Did Evolutionary Thought Develop?

Historically, the most common explanation for the origin of species was the divine creation of each species in its present form, and species were believed to remain unchanged after their creation. This view was challenged by evidence from fossils, geology, and biological exploration. Since the middle of the nineteenth century, scientists have realized that species originate and evolve by the operation of natural processes that change the genetic makeup of populations.

15.2 How Does Natural Selection Work?

Charles Darwin and Alfred Russel Wallace independently proposed the theory of evolution by natural selection. Their theory expresses the logical consequences of four postulates about populations: (1) populations are variable, (2) the variable traits can be inherited, (3) there is unequal reproduction, and (4) differences in reproductive success depend on the traits of individuals. If these four postulates four true, then the characteristics of successful individuals will be "naturally selected" and become more common over time.

15.3 How Do We Know That Evolution Has Occurred?

Many lines of evidence indicate that evolution has occurred, including the following:

- Fossils of ancient species tend to be simpler in form than modern species. Sequences of fossils have been discovered that show a graded series of changes in form. Both of these observations would be expected if modern species evolved from older species.
- Species thought to be related through evolution from a common ancestor possess many similar anatomical structures.
- Stages in early embryological development are quite similar among very different types of vertebrates.
- Living cells share similarities in biochemical traits, such as the use of DNA as the carrier of genetic information.

15.4 What is the Evidence That Populations Evolve by Natural Selection?

Similarly, many lines of evidence indicate that natural selection is the chief mechanism driving changes in the characteristics of species over time, including the following:

- Inheritable traits have been changed rapidly in populations of domestic animals and plants by selectively breeding organisms with desired features (artificial selection). The immense variations in species produced in a few thousand years by artificial selection make it almost inevitable that much larger changes would be wrought by hundreds of millions of years of natural selection.

- Both natural and human activities can drastically change the environment over short periods of time. Inherited characteristics of species have been observed to change significantly in response to such environmental changes.

Key Terms

analogous structure *273*
artificial selection *275*
convergent evolution *272*
evolution *263*
fossil *264*
homologous structure *270*
natural selection *269*
population *263*
vestigial structure *270*

Thinking Through the Concepts

Multiple Choice

1. Whale skeletons contain nonfunctional pelvic bones
   a. as a result of convergent evolution.
   b. due to catastrophism.
   c. because whales evolved from ancestors that had hind legs.
   d. because the bones might be needed for a future adaptation.

2. Darwin was influenced by Malthus's thinking about _____ and by Lamarck's conclusion that _____.
   a. the formation of fossils; Earth has a complex history
   b. the geographic distribution of species; individuals compete to survive and reproduce
   c. the principles of geology; all organisms are related by common descent
   d. limits on population growth; species change over time

3. Natural selection is
   a. a preference for natural traits.
   b. the method by which domestic dog breeds originated.
   c. increased reproduction due to particular traits.
   d. the reason that mutations occur.

4. Which of the following is *not* required for evolution by natural selection to occur?
   a. Individuals in a population vary.
   b. Offspring inherit traits from their parents.
   c. Thousands of generations must pass.
   d. Some individuals have more offspring than others.

5. Which of the following is *not* evidence of evolution (i.e., that species are linked by shared ancestry and have changed over time)?
   a. Different species share homologous structures.
   b. Different species share analogous structures.
   c. The older a fossil is, the less likely it is to be similar to a living organism.
   d. The cells of all species contain DNA.

Fill-in-the-Blank

1. The flipper of a seal is homologous with the _____ of a bird, and both of these are homologous with the

_____ of a human. The wing of a bird and the wing of a butterfly are described as _____ structures that arose as a result of _____ evolution. Remnants of structures in animals that have no use for them, such as the small hind leg bones of whales, are described as _____ structures.

2. The finding that all organisms share the same genetic code provides evidence that all descended from a(n) _____. Further evidence is provided by the fact that all cells use roughly the same set of _____ to build proteins, and all cells use the molecule _____ as an energy carrier.

3. Georges Cuvier espoused a concept called _____ to explain layers of rock with embedded fossils. Charles Lyell, building on the work of James Hutton, proposed an alternative explanation called _____, which states that layers of rock and many other geological features can be explained by gradual processes that occurred in the past just as they do in the present. This concept provided important support for evolution because it required that Earth be extremely _____.

4. The process by which inherited characteristics of populations change over time is called _____. Variability among individuals is the result of chance changes called _____ that occur in the hereditary molecule _____.

5. The process by which individuals with traits that provide an advantage in their natural habitats are more successful at reproducing is called _____. People who breed animals or plants can produce large changes in their characteristics in a relatively short time, a process called _____.

6. Darwin's postulate 2 states that _____. The work of _____ provided the first experimental evidence for this postulate.

Review Questions

1. Natural selection acts on individuals, but only populations evolve. Explain why this statement is true.

2. Distinguish between catastrophism and uniformitarianism. How did these hypotheses contribute to the development of evolutionary theory?

3. Describe Lamarck's theory of inheritance of acquired characteristics. Why is it invalid?

4. What is natural selection? Describe how natural selection might have caused unequal reproduction among the ancestors of a fast-swimming predatory fish, such as the barracuda.

5. Describe how evolution occurs. In your description, include discussion of the reproductive potential of species, the stability of natural population sizes, variation among individuals of a species, inheritance, and natural selection.

6. What is convergent evolution? Give an example.

7. How do biochemistry and molecular genetics contribute to the evidence that evolution has occurred?

8. In what sense are humans currently acting as agents of selection on other species? Name some traits that are favored by the environmental changes humans cause.

Applying the Concepts

1. In discussions of untapped human potential, it is commonly said that the average person uses only 10% of his or her brain. Is this conclusion likely to be correct? Explain your answer in terms of natural selection.

2. Does evolution through natural selection produce "better" organisms in an absolute sense? Are we climbing the "Ladder of Nature"? Defend your answer.

16 HOW POPULATIONS EVOLVE

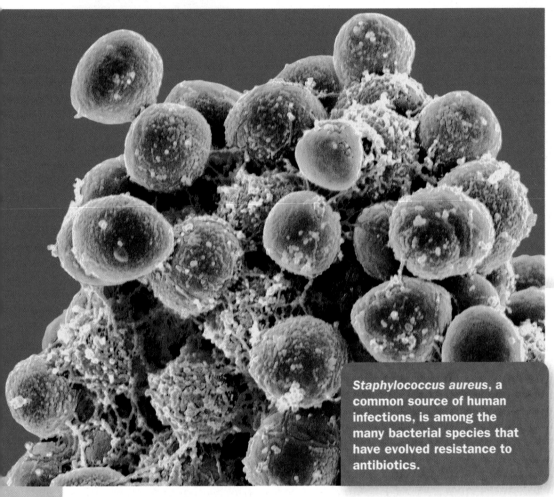

Staphylococcus aureus, a common source of human infections, is among the many bacterial species that have evolved resistance to antibiotics.

Evolution of a Menace

ON A FEBRUARY DAY NOT LONG AGO, a 20-year-old student arrived at the health center at Western Washington University. He had been bothered by a lingering cough for a couple of weeks, and when his symptoms worsened to include a fever and vomiting, he sought medical attention. The health center staff quickly determined that the student had pneumonia and began treatment. His condition deteriorated, however, and he was transferred to the local hospital. A few days later, he died.

Why couldn't doctors save a previously healthy young man from a normally curable disease? Because the victim's pneumonia was caused by methicillin-resistant *Staphylococcus aureus* (MRSA) bacteria. *Staphylococcus aureus*, sometimes referred to as "staph," is a common bacterium that can infect the skin, blood, or respiratory system. Many staph infections can be successfully treated with antibiotics, but MRSA bacteria are antibiotic resistant and cannot be killed by many of the most commonly used antibiotics. Until recently, MRSA infections occurred almost exclusively in

hospitals. Today, however, resistant staph is widespread, and more than half of MRSA infections occur outside of hospitals, in homes, schools, and workplaces.

In the United States, MRSA infections kill at least 11,000 people each year. And, unfortunately, *Staphylococcus* is by no means the only disease-causing bacterium that is becoming less vulnerable to antibiotics. For example, antibiotic resistance has also appeared in the bacteria that cause tuberculosis, a disease that kills almost 2 million people each year. In an increasing number of tuberculosis cases, the disease does not respond to any of the drugs commonly used to treat it. Multidrug resistance is also becoming more prevalent in the bacteria responsible for the widespread sexually transmitted disease gonorrhea. In addition, drug resistance is common in the bacteria that cause food poisoning, blood poisoning, dysentery, pneumonia, meningitis, and urinary tract infections. We are experiencing a global onslaught of resistant "supergerms," and are facing the specter of diseases that cannot be cured.

Many physicians and scientists believe that the most effective way to combat the rise of resistant diseases is to reduce the use of antibiotics. Why might such a strategy be effective? Because the upsurge of antibiotic resistance is a consequence of evolutionary change in populations of bacteria, and the agent of this change is natural selection imposed by antibiotic drugs. Can understanding the mechanisms by which populations evolve help us understand how the crisis of antibiotic resistance arose, and how we might resolve it?

AT A GLANCE

16.1 HOW ARE POPULATIONS, GENES, AND EVOLUTION RELATED?

If you live in an area with a seasonal climate and you own a dog or cat, you have probably noticed that your pet's fur gets thicker and heavier as winter approaches. Has the animal evolved? No. The changes that we see in an individual organism over the course of its lifetime are not evolutionary changes. Instead, evolutionary changes occur from generation to generation, causing descendants to be different from their ancestors.

Furthermore, we can't detect evolutionary change across generations by looking at a single set of parents and offspring. For example, if you observed that a 6-foot-tall man had an adult son who stood 5 feet tall, could you conclude that humans were evolving to become shorter? Obviously not. Rather, if you wanted to learn about evolutionary change in human height, you would begin by measuring many humans over many generations to see if the average height is changing with time. Evolution is a property not of individuals but of populations. A **population** is a group that includes all the members of a species living in a given area.

Recognizing that evolution is a population-level phenomenon was one of Darwin's key insights. But populations are composed of individuals, and the actions and fates of individuals determine which characteristics will be passed on to descendant populations. In this fashion, inheritance provides the link between the lives of individual organisms and the evolution of populations. We will therefore begin our discussion of the processes of evolution by reviewing some principles of genetics as they apply to individuals. We will then extend those principles to the genetics of populations.

Genes and the Environment Interact to Determine Traits

Each cell of every organism contains genetic information encoded in the DNA of its chromosomes. The combined DNA in an organism's set of chromosomes is its *genome*. A *gene* is a segment of DNA located at a particular place on a chromosome (see Chapter 11). The sequence of nucleotides in a gene encodes the sequence of amino acids in a protein, usually an enzyme that catalyzes a particular reaction in the cell. At a given gene's location, different members of a species may have slightly different nucleotide sequences, called *alleles*. Different alleles encode different forms of the same enzyme. For example, various alleles of the genes that influence eye color in humans generate enzymes that help produce eyes that are brown, or blue, or green, and so on.

In any population of organisms, there are usually two or more alleles of each gene. An individual of a diploid species whose alleles of a particular gene are both the same is *homozygous* for that gene, and an individual with different alleles for that gene is *heterozygous*. The specific alleles borne on an organism's chromosomes (its *genotype*) influence the development of its physical and behavioral traits (its *phenotype*) (**FIG. 16-1**).

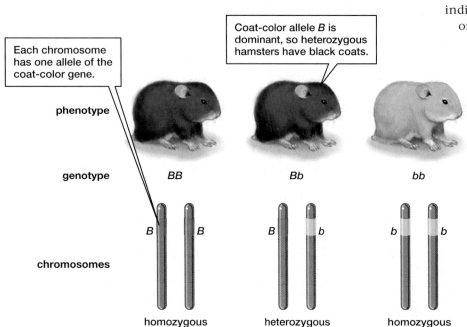

Each chromosome has one allele of the coat-color gene.

Coat-color allele *B* is dominant, so heterozygous hamsters have black coats.

phenotype

genotype   *BB*   *Bb*   *bb*

chromosomes   *B*   *B*   *B*   *b*   *b*   *b*

homozygous   heterozygous   homozygous

◀ **FIGURE 16-1 Alleles, genotype, and phenotype in individuals** An individual's particular combination of alleles is its genotype. The word "genotype" can refer to the alleles of a single gene (as shown here), to a set of genes, or to all of an organism's genes. An individual's phenotype is determined by its genotype and environment. "Phenotype" can refer to a single trait, a set of traits, or all of an organism's traits.

Let's illustrate these principles with an example. A black hamster's coat is colored black because a chemical reaction in its hair follicles produces a black pigment. When we say that a hamster has the allele for a black coat, we mean that a particular stretch of DNA on one of the hamster's chromosomes contains a sequence of nucleotides that codes for the enzyme that catalyzes a pigment-producing reaction that results in a black coat. A hamster with the allele for a brown coat has a different sequence of nucleotides at the corresponding chromosomal position. That different sequence codes for an enzyme that cannot produce black pigment. If a hamster is homozygous for the black allele (two black alleles) or is heterozygous (one black allele and one brown allele), its fur contains the pigment and is black. But if a hamster is homozygous for the brown allele, its hair follicles produce no black pigment and its coat is brown. Because the hamster's coat is black even when only one copy of the black allele is present, the black allele is considered *dominant* and the brown allele *recessive*.

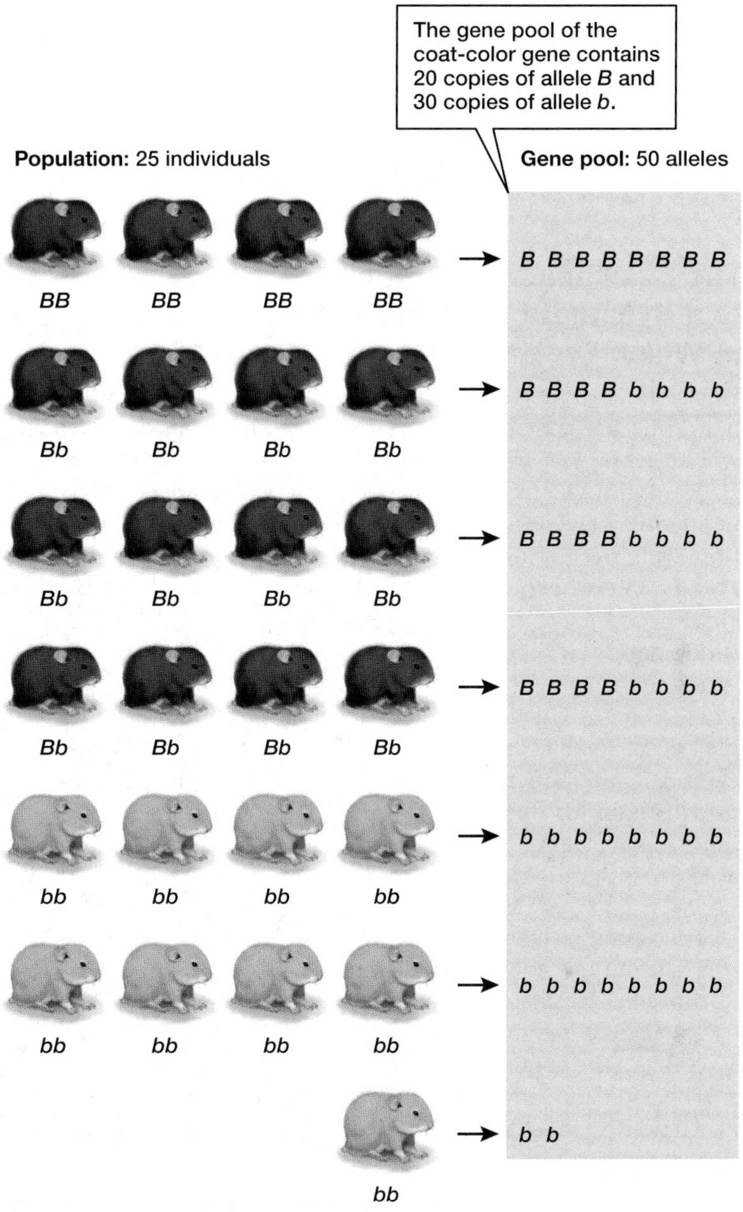

The Gene Pool Comprises All of the Alleles in a Population

Looking at evolution in terms of its effects on genes has proven to be a useful way to understand evolutionary processes. In particular, evolutionary biologists have made excellent use of the tools of a branch of genetics, called *population genetics*, that deals with the frequency, distribution, and inheritance of alleles in populations. To take advantage of this powerful aid to understanding evolution, you will need to learn a few of the basic concepts of population genetics.

Population genetics defines a **gene pool** as a set that contains all of the alleles of all of the genes from all of the individuals in a population. A gene pool is not an actual physical entity but is instead a concept that can help us understand the process of evolution. You can think of a gene pool as the contents of an imaginary bucket into which each member of a population has tossed one copy of its genotype. Thus, for any given allele, the gene pool receives one copy from each individual that is heterozygous for that allele (and therefore has only one copy of the allele in question), and two copies from each individual that is homozygous for that allele (and therefore has two copies).

Each particular gene can be considered to have its own gene pool, which comprises all of the alleles of that specific gene in a population (FIG. 16-2). If we count the number of copies of each allele present in a gene pool, we can determine the relative proportion of each allele in the gene pool. An allele's proportion in the gene pool is its **allele frequency.** For example, the gene pool of the population

The gene pool of the coat-color gene contains 20 copies of allele *B* and 30 copies of allele *b*.

Population: 25 individuals   Gene pool: 50 alleles

▲ FIGURE 16-2 **A gene pool** In diploid organisms, each individual in a population contributes two alleles of each gene to the gene pool.

of 25 hamsters portrayed in Figure 16-2 contains 50 alleles of the gene that controls coat color (because hamsters are diploid and each hamster thus has two copies of each gene). Twenty of those 50 alleles are of the type that codes for black coats, so the frequency of that allele in the population is $20/50 = 0.40$ (or 40%).

Evolution Is the Change of Allele Frequencies in a Population

A casual observer might define evolution on the basis of changes in the outward appearance or behaviors of the members of a population. However, many of the outward changes that we observe in the individuals that make up the population can also be viewed as the visible expression of underlying changes to the gene pool. A population geneticist, therefore, defines evolution as changes over time in the allele frequencies of a gene pool. In other words, evolution is change in the genetic makeup of populations over generations.

The Equilibrium Population Is a Hypothetical Population in Which Evolution Does Not Occur

It is easier to understand what causes populations to evolve if we first consider the characteristics of a population that would *not* evolve. In 1908, English mathematician Godfrey H. Hardy and German physician Wilhelm Weinberg independently developed a simple mathematical model of a non-evolving population. This model, now known as the **Hardy–Weinberg principle,** showed that under certain conditions, allele frequencies and genotype frequencies in a population will remain constant no matter how many generations pass. (For more information on how the model works, see "In Greater Depth: The Hardy–Weinberg Principle" on page 289.) In other words, this population will not evolve. Population geneticists use the term **equilibrium population** for this hypothetical non-evolving population in which allele frequencies do not change as long as the following conditions are met:

- There must be no mutation.
- There must be no **gene flow.** That is, there must be no movement of alleles into or out of the population (as would be caused, for example, by the movement of organisms into or out of the population).
- The population must be very large.
- All mating must be random, with no tendency for certain genotypes to mate with specific other genotypes.
- There must be no natural selection. That is, all genotypes must reproduce with equal success.

Under these conditions, allele frequencies in a population will remain the same indefinitely. If one or more of these conditions is violated, then allele frequencies may change: The population will evolve.

As you might expect, few natural populations are truly in equilibrium. What, then, is the importance of the Hardy–Weinberg principle? The Hardy–Weinberg conditions are useful starting points for studying the mechanisms of evolution. In the following sections, we will examine some of these conditions, show that natural populations typically fail to meet them, and illustrate the consequences of such failures. In this way, we can better understand both the inevitability of evolution and the processes that drive evolutionary change.

CHECK YOUR LEARNING
Can you …
- define *evolution* in terms of concepts from population genetics?
- define *equilibrium population* and describe the conditions under which a population is expected to remain at equilibrium?

16.2 WHAT CAUSES EVOLUTION?

Population genetics predicts that the Hardy–Weinberg equilibrium can be disturbed by deviations from any of its five conditions. Therefore, we can predict five major causes of evolutionary change: mutation, gene flow, small population size, nonrandom mating, and natural selection.

Mutations Are the Original Source of Genetic Variability

A population remains in evolutionary equilibrium only if there are no **mutations** (changes in DNA sequence). Most mutations occur during cell division, when a cell makes a copy of its DNA. Sometimes, errors occur during the copying process and the copied DNA does not match the original. Most such errors are quickly corrected by cellular systems that identify and repair DNA copying mistakes, but some changes in nucleotide sequence slip past the repair systems. An unrepaired mutation in a cell that gives rise to gametes (eggs or sperm) may be passed to offspring and enter the gene pool of a population.

Inherited Mutations Are Rare But Important

How significant are mutations in changing the gene pool of a population? For any given gene, only a tiny proportion of a population inherits a new mutation from the previous generation. For example, the best estimates of human mutation rates suggest that a mutation at any particular site (base pair) in the genome will appear in only about 1 out of every 80 million newborns. Therefore, mutation by itself generally causes only very small changes in the frequency of any particular allele.

Despite the rarity of inherited mutations at any particular location in the genome, the cumulative effect of mutations is essential to evolution. The genomes of most organisms contain a large number of DNA base pairs, so although the rate of mutation is low for any particular base pair, the sheer number of possibilities means that each new generation of a population is likely to include some mutations. For example, the diploid human genome contains about 6 billion base pairs. Thus, even though each base pair has, on average, only a 1 in 80 million chance of mutation, most newborns will probably inherit 70 or 80 mutations. These mutations are new alleles—new variations on which other evolutionary processes can work. As such, they are the

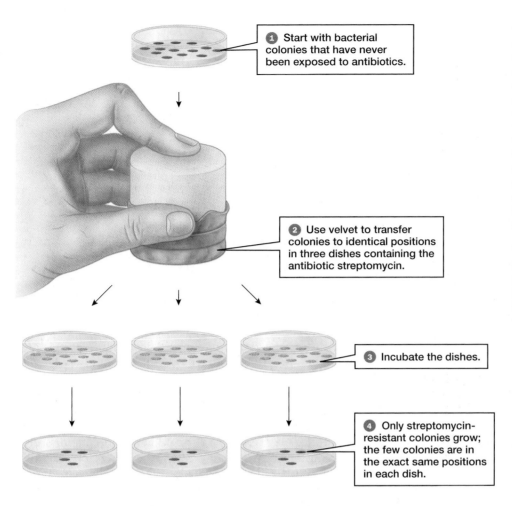

1 Start with bacterial colonies that have never been exposed to antibiotics.

2 Use velvet to transfer colonies to identical positions in three dishes containing the antibiotic streptomycin.

3 Incubate the dishes.

4 Only streptomycin-resistant colonies grow; the few colonies are in the exact same positions in each dish.

◀ FIGURE 16-3 Mutations occur spontaneously This experiment demonstrates that mutations occur spontaneously and not in response to environmental conditions. When bacterial colonies that have never been exposed to antibiotics are exposed to the antibiotic streptomycin, only a few colonies grow. The observation that these surviving colonies grow in the exact same positions in all dishes shows that the mutations for resistance to streptomycin were present in the original dish before exposure to streptomycin.

THINK CRITICALLY If it were true that mutations do occur in response to the presence of antibiotics, how would the result of this experiment have differed from the actual result?

foundation of evolutionary change. Without mutations, there would be no evolution.

Mutations Are Not Goal Directed

A mutation does not arise as a result of, or in anticipation of, the needs of an organism. A mutation simply happens and may produce a change in a structure or function of the organism. Whether that change is helpful or harmful or neutral, now or in the future, depends on environmental conditions over which the organism has little or no control (FIG. 16-3). The mutation merely provides a potential for evolutionary change. Other processes, especially natural selection, may act to spread the mutation through the population or to eliminate it.

Gene Flow Between Populations Changes Allele Frequencies

The movement of alleles between populations, known as *gene flow*, changes how alleles are distributed among populations. When individuals move from one population to another and interbreed at the new location, alleles are transferred from one gene pool to another. For example, baboons live in social groupings called troops, and some individuals—usually juvenile males—routinely leave their troop and move to new populations. If the departing baboons are fortunate, they join another troop and achieve sufficient social status to breed. In this way, the male offspring of one troop may carry alleles to the gene pools of other troops.

In some kinds of organisms, alleles move between populations only at certain stages of the life cycle. In flowering plants, for example, most gene flow is due to the movement of seeds and pollen. Pollen, which contains sperm cells, may be carried long distances by wind or by animal pollinators. If the pollen ultimately reaches the flowers of a different population of its species, it may fertilize eggs and add its collection of alleles to the local gene pool. Similarly, seeds may be borne by wind, water, or animals to distant locations where they can germinate to become part of a population far from their place of origin.

The main evolutionary effect of gene flow is to increase the genetic similarity of different populations of a species. Movement of alleles from one population to another tends to change the gene pool of the destination population so that it is more similar to the source population.

(a) Generation 1 frequency of *B* = 50%
frequency of *b* = 50%

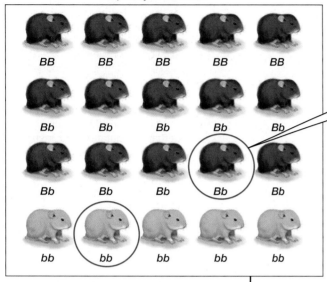

(b) Generation 2 frequency of *B* = 25%
frequency of *b* = 75%

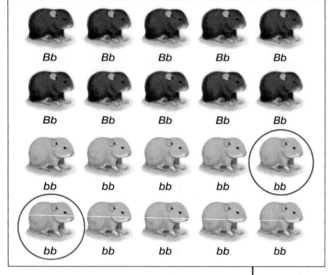

(c) Generation 3 frequency of *B* = 0%
frequency of *b* = 100%

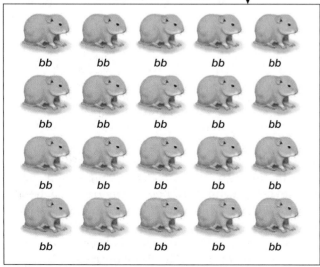

In each generation, only two randomly chosen individuals breed; their offspring form the entire next generation.

◀ **FIGURE 16-4 Genetic drift** If chance events prevent some members of a population from reproducing, allele frequencies can change randomly.

THINK CRITICALLY Explain how the distribution of genotypes in generation 2 is calculated.

Allele Frequencies May Change by Chance in Small Populations

Allele frequencies in populations can be changed by chance events other than mutations. For example, if bad luck prevents some members of a population from reproducing, their alleles will ultimately be removed from the gene pool, altering its makeup. Any unpredictable event, such as a flood or a fire, that arbitrarily cuts lives short or otherwise allows only a random subset of a population to reproduce can cause random changes in allele frequencies. The process by which chance events change allele frequencies is called **genetic drift.**

To see how genetic drift works, imagine a population of 20 hamsters in which the frequency of the black coat color allele *B* is 0.50 and the frequency of the brown coat color allele *b* is 0.50 (**FIG. 16-4a**). If all of the hamsters in the population were to interbreed to yield another population of 20 animals, the frequencies of the two alleles would not change in the next generation. But if we instead allow only two, randomly chosen hamsters (the ones circled in Figure 16-4, top) to breed and become the parents of the next generation of 20 animals, allele frequencies might be quite different in generation 2 (**FIG. 16-4b**; the frequency of *B* has decreased and the frequency of *b* has increased). And if breeding in the second generation were again restricted to two randomly chosen hamsters (circled in Figure 16-4b), allele frequencies

HAVE YOU EVER

WONDERED...

A flu vaccination stimulates your immune system to recognize and attack the viruses that cause influenza. The immune system recognizes the viruses by the proteins they contain, but these proteins change from year to year. Why? Because of genetic drift, which is especially rapid in flu viruses due to their high mutation rate. After a year of genetic drift in the flu virus population, the immune system of a previously vaccinated person can no longer recognize the virus. The person needs a fresh vaccination designed to protect against the evolved version of the virus.

Why You Need to Get a Flu Shot Every Year?

might change again in generation 3 (**FIG. 16-4c**). Allele frequencies will continue to change in random fashion as long as reproduction is restricted to a random subset of the population. Note that the changes caused by genetic drift can include the disappearance of an allele from the population, as illustrated by the disappearance of the *B* allele (and therefore the black coat phenotype) in generation 3 in Figure 16-4.

Population Size Matters

Genetic drift occurs more rapidly and has a greater effect in small populations than in large ones. If a population is large, chance events that randomly prevent some individuals from breeding are unlikely to significantly alter the population's genetic composition. The individuals that do breed will constitute a random sample large enough to ensure that its genetic composition is roughly the same as that of the source population. In a small population, however, a random sample of breeders may be quite small, and its genetic composition is therefore more likely to differ from that of the source population. An allele that occurs at low frequency in a small population (and is therefore present in only a few individuals) can be completely eliminated from the population if a chance event prevents its only carriers from breeding.

One way to test these predictions about genetic drift in large versus small populations is to write a computer program that simulates how the frequencies of the alleles would change over many generations in which only a random subset of the population breeds. **FIGURE 16-5** shows the results of simulations of three populations of the hypothetical hamsters we introduced in Figure 16-4. In these simulations, the program began with a population of a specified size in which the initial frequency of each allele was 50%. Then gametes were selected at random from the population's gene pool and combined to create a new generation with the same population size; the process was repeated for 100 generations.

FIGURE 16-5a shows the results of ten runs in which the simulated population size was large (2,000 individuals). Notice that the frequency of allele *B* remains close to its initial frequency, but nonetheless changes over time. **FIGURE 16-5b** shows ten simulations of a smaller population (40 individuals). In this case, the frequency of allele *B* is quite likely to diverge from its initial level, in some instances reaching a frequency of 100% (all hamsters are black) or 0% (all hamsters are brown). **FIGURE 16-5c** shows the fate of allele *B* in ten runs of a simulation of a very small population (4 individuals). Here, allele frequencies change very rapidly; in all ten runs, allele *B* reaches a frequency of 100% or 0% after no more than 20 generations. Overall, the smaller the population, the more dramatic the effects of genetic drift. In populations of all sizes, however, each run of the simulation had a different outcome, because genetic drift is a random process.

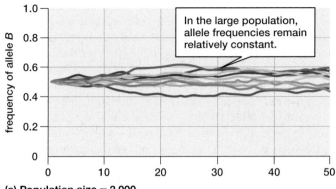

In the large population, allele frequencies remain relatively constant.

(a) Population size = 2,000

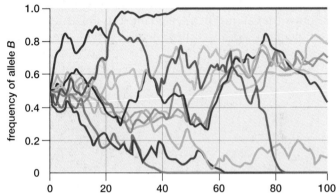

(b) Population size = 40

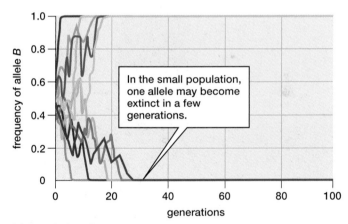

In the small population, one allele may become extinct in a few generations.

generations

(c) Population size = 4

▲ **FIGURE 16-5 The effect of population size on genetic drift** Each colored line represents one computer simulation of the change over time in the frequency of allele *B* in **(a)** a large population, **(b)** a smaller population, and **(c)** a very small population. Half of the alleles in each starting population were *B* (50%) and, in each generation, randomly chosen individuals reproduced.

A Population Bottleneck Can Cause Genetic Drift

Two causes of genetic drift, the population bottleneck and the founder effect, further illustrate the impact that small population size may have on the allele frequencies of a species. In a **population bottleneck,** a population's size is drastically reduced by an event such as a natural catastrophe or overhunting. After such a bottleneck, only a few individuals are available to contribute genes to the next generation. Population bottlenecks can rapidly change allele frequencies and can reduce genetic variability by eliminating alleles (FIG. 16-6a). Even if the population later increases, the genetic effects of the bottleneck may remain for hundreds or thousands of generations.

Loss of genetic variability due to bottlenecks has been documented in numerous species, including the northern elephant seal (FIG. 16-6b). The elephant seal was hunted almost to extinction in the 1800s; by the 1890s, only about 20 individuals survived. Dominant male elephant seals typically monopolize breeding, so a single male may have fathered all the offspring at this extreme bottleneck point. Since the late nineteenth century, elephant seals have increased in number to more than 200,000 individuals, but biochemical analysis shows that all northern elephant seals are genetically almost identical. With so little genetic variation, the elephant

seal has little potential to evolve in response to environmental changes (see "Earth Watch: The Perils of Shrinking Gene Pools" on page 291). Because of the limited genetic variation in this species, the species remains vulnerable to extinction regardless of how many elephant seals there are.

Isolated Founding Populations May Produce Bottlenecks

The **founder effect** occurs when isolated colonies are founded by a small number of organisms. A small flock of birds, for instance, that becomes lost during migration or is blown off course by a storm may settle on an isolated island. This founder group may, by chance, have allele frequencies that are very different from the frequencies of the parent population. If this is the case, the gene pool

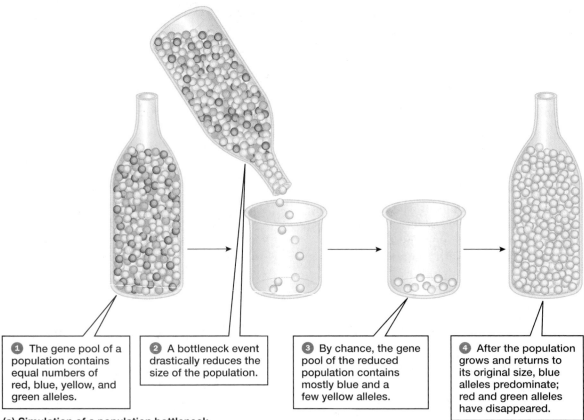

① The gene pool of a population contains equal numbers of red, blue, yellow, and green alleles.

② A bottleneck event drastically reduces the size of the population.

③ By chance, the gene pool of the reduced population contains mostly blue and a few yellow alleles.

④ After the population grows and returns to its original size, blue alleles predominate; red and green alleles have disappeared.

(a) Simulation of a population bottleneck

(b) Elephant seals

◀ **FIGURE 16-6 Population bottlenecks reduce variation (a)** A population bottleneck may drastically reduce genotypic and phenotypic variation because the few organisms that survive may all carry similar sets of alleles. **(b)** The northern elephant seal passed through a population bottleneck in the recent past. As a result, the population's genetic diversity is extremely low.

THINK CRITICALLY If a population grows large again after a bottleneck, genetic diversity will eventually increase. Why?

IN GREATER DEPTH | The Hardy–Weinberg Principle

The Hardy–Weinberg principle states that allele frequencies will remain constant over time in the gene pool of a large population in which there is random mating but no mutation, no gene flow, and no natural selection. In addition, Hardy and Weinberg showed that if allele frequencies do not change in an equilibrium population, the proportion of individuals with a particular genotype will also remain constant.

To better understand the relationship between allele frequencies and the occurrence of genotypes, picture an equilibrium population whose members carry a gene that has two alleles, A_1 and A_2. Note that each individual in this population must carry one of three possible diploid genotypes (combinations of alleles): A_1A_1, A_1A_2, or A_2A_2.

Suppose that in our population's gene pool, the frequency of allele A_1 is p, and the frequency of allele A_2 is q. Hardy and Weinberg demonstrated that the proportions of the different genotypes in the population can be calculated as:

Proportion of individuals with genotype $A_1A_1 = p^2$

Proportion of individuals with genotype $A_1A_2 = 2pq$

Proportion of individuals with genotype $A_2A_2 = q^2$

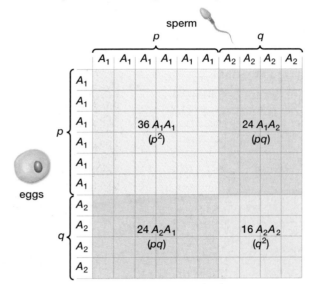

For example, if, in our population's gene pool, 60% of the alleles of a gene are A_1 and 40% are A_2 (that is, $p = 0.6$ and $q = 0.4$), then genotype proportions would be:

Proportion of individuals with genotype
$A_1A_1 = 36\%$
(because $p^2 = 0.6 \times 0.6 = 0.36$)

Proportion of individuals with genotype
$A_1A_2 = 48\%$
(because $2pq = 2 \times 0.6 \times 0.4 = 0.48$)

Proportion of individuals with genotype
$A_2A_2 = 16\%$
(because $q^2 = 0.4 \times 0.4 = 0.16$)

◀ **FIGURE E16-1 The relationship between allele and genotype frequencies** This Punnett square shows the expected genotypes of 100 zygotes formed by random mating in a population in which 60% of sperm and eggs carry allele A_1 ($p = 0.6$) and 40% carry A_2 ($q = 0.4$). As you can see, out of every 100 zygotes, the genotype of 36 will be A_1A_1 (36%, $p^2 = 0.36$), 48 will be A_1A_2 (48%, $2pq = 0.48$), and 16 will be A_2A_2 (16%, $q^2 = 0.16$).

This result is shown in graphical form in **FIGURE E16-1**. Because every member of the population must possess one of the three genotypes, the three proportions must always add up to one. For this reason, the expression that relates allele frequency to genotype proportions can be written as:

$$p^2 + 2pq + q^2 = 1$$

where the three terms on the left side of the equation represent the three genotypes.

---

of the future population in the new location will be quite unlike that of the larger population from which it sprang. Consider, for example, the Amish inhabitants of Lancaster County, Pennsylvania, who are descended from only 200 or so eighteenth-century immigrants. Among today's Lancaster County Amish, a genetic disorder known as Ellis–van Creveld syndrome is far more common than it is among the general population (**FIG. 16-7**). The prevalence of the syndrome among the Amish stems from a single immigrant couple who carried the Ellis–van Creveld allele. Because the founder population was so small, this single occurrence meant that the allele was carried by a comparatively high proportion of the population (1 or 2 carriers out of 200 versus about 1 in 1,000 in the general population). This high initial allele frequency, a result of the founder effect, combined with subsequent genetic drift, has led to extraordinarily high levels of Ellis–van Creveld syndrome among this Amish group.

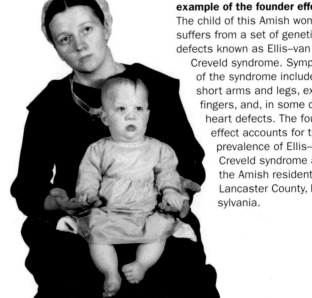

◀ **FIGURE 16-7 A human example of the founder effect** The child of this Amish woman suffers from a set of genetic defects known as Ellis–van Creveld syndrome. Symptoms of the syndrome include short arms and legs, extra fingers, and, in some cases, heart defects. The founder effect accounts for the prevalence of Ellis–van Creveld syndrome among the Amish residents of Lancaster County, Pennsylvania.

CASE STUDY \ CONTINUED
Evolution of a Menace

The mutant alleles that confer antibiotic resistance on members of a bacterial population can move to other populations by gene flow and increase by genetic drift. For example, MRSA living on a person's skin might be transferred to her roommate's skin via a shared towel or article of clothing. After joining the population of nonresistant bacteria on the roommate's skin, the newly arrived MRSA may reproduce and may even transfer resistance alleles directly to local bacteria through a process known as conjugation (see Chapter 20). The resistance alleles, introduced to the bacterial population on the roommate's skin by gene flow, may subsequently experience genetic drift that increases their frequency. But what will happen if antibiotics are introduced to the new environment (that is, the roommate's body)? Find out in Section 16.3.

▲ **FIGURE 16-8 Nonrandom mating among snow geese** Snow geese, which have either white plumage or blue-gray plumage, are most likely to mate with other birds of the same color.

Mating Within a Population Is Almost Never Random

The effects of *nonrandom mating* can play a significant role in evolution, because organisms seldom mate strictly randomly. For example, many organisms have limited mobility and tend to remain near their place of birth, hatching, or germination. In such species, most of the offspring of a given parent live in the same area; thus, when they reproduce, there is a good chance that they will be related to their reproductive partners. Such sexual reproduction between relatives is called *inbreeding*.

Because relatives are genetically similar, inbreeding tends to increase the number of individuals that inherit the same alleles from both parents and are therefore homozygous for many genes. This increase in homozygotes can have harmful effects, such as increased occurrence of genetic diseases or defects. Many gene pools include harmful recessive alleles that persist in the population because their negative effects are masked in heterozygous carriers (which have only a single copy of the harmful allele). Inbreeding, however, increases the odds of producing homozygous offspring with two copies of the harmful allele.

In animals, nonrandom mating can also arise if individuals have preferences or biases that influence their choice of mates. The snow goose is a case in point. Individuals of this species come in two "color phases"; some snow geese are white, and others are blue-gray (**FIG. 16-8**). Although both white and blue-gray geese belong to the same species, mate choice is not random with respect to color. The birds exhibit a strong tendency to mate with a partner of the same color. This preference for mates that are similar is known as *assortative mating*.

Neither inbreeding nor assortative mating by themselves will alter allele frequencies in a population. Nonetheless, they can have large effects on the distribution of different genotypes, and thus on the distribution of phenotypes, in the population.

All Genotypes Are Not Equally Beneficial

In a hypothetical equilibrium population, individuals of all genotypes survive and reproduce equally well; no genotype has any advantage over the others. This condition, however, is probably met only rarely, if ever, in real populations. Even though many alleles are neutral, in the sense that organisms possessing any of several alleles are equally likely to survive and reproduce, some alleles confer an advantage on their possessor. Any time an allele provides, in Alfred Russel Wallace's words, "some little superiority," the individuals who carry it are favored by **natural selection,** the process in which individuals with traits that help them survive and reproduce leave more offspring than do individuals that lack those traits. We examine the impact of natural selection in greater depth in the next section.

TABLE 16-1 summarizes the different causes of evolution.

CHECK YOUR LEARNING
Can you ...
- describe how mutation, gene flow, genetic drift, nonrandom mating, and natural selection affect evolution?

TABLE 16-1	Causes of Evolution
Process	**Consequence**
Mutation	Creates new alleles; increases variability
Gene flow	Increases similarity of different populations
Genetic drift	Causes random change of allele frequencies; can eliminate alleles
Nonrandom mating	Changes genotype frequencies, but not allele frequencies
Natural and sexual selection	Increases frequency of favored alleles; produces adaptations

Earth WATCH   The Perils of Shrinking Gene Pools

Many of Earth's species are in danger. According to the International Union for Conservation of Nature, more than 20,000 species of plants and animals are currently threatened with extinction. For most of these endangered species, the main threat is habitat destruction. When a species' habitat shrinks, its population size almost invariably follows suit. Many people, organizations, and governments are concerned about the plight of endangered species and are working to protect them and their habitats. Unfortunately, a population that has already become small enough to warrant endangered status is likely to undergo evolutionary changes that increase its chances of going extinct.

One problem is that, in small populations, mating choices are limited and a high proportion of matings may be between close relatives. This inbreeding increases the odds that offspring will be homozygous for harmful recessive alleles. These less-fit individuals may die before reproducing, further reducing the size of the population.

The greatest threat to small populations, however, stems from their inevitable loss of genetic diversity (FIG. E16-2). From our discussion of population bottlenecks, it is apparent that, when populations shrink to very small sizes, many of the alleles that were present in the original population will not be represented in the gene pool of the remnant population. Furthermore, we have seen that genetic drift in small populations will cause many of the surviving alleles to subsequently disappear permanently from the population (see Fig. 16-5c). Because genetic drift is a random process, many of the lost alleles will be advantageous ones that were previously favored by natural selection. Inevitably, the number of different alleles in the population grows ever smaller. Even if the size of an endangered population eventually begins to grow, it may take hundreds of generations to restore the lost genetic diversity.

Why does it matter if a population's genetic diversity is low? Low diversity creates two main risks. First, the fitness of the population as a whole is reduced by the loss of advantageous alleles that underlie adaptive traits. A less-fit population is unlikely to thrive. Second, a genetically impoverished population lacks the variation that will allow it to adapt when environmental conditions change. When the environment changes, as it inevitably will, a genetically uniform species is less likely to contain individuals well suited to survive and reproduce under the new conditions. A species unable to adapt to changing conditions is at very high risk of extinction.

▲ **FIGURE E16-2 Shrunken gene pools** The Ethiopian wolf is among the critically endangered species known to have extremely low genetic diversity.

What can be done to preserve the genetic diversity of endangered species? The best solution, of course, is to preserve plenty of diverse types of habitat so that species never become endangered in the first place. The human population, however, has grown so large and has thus appropriated so large a share of Earth's resources that this solution is impossible in many places. For many species, the only solution is to ensure that areas of preserved habitat are large enough to hold populations of sufficient size to contain most of a threatened species' total genetic diversity.

THINK CRITICALLY In many cases, circumstances prevent preservation of a large, unbroken area of a threatened species' habitat. Sometimes, however, several smaller areas can be protected. In such cases, conservation biologists stress that the small areas must be linked by corridors of the appropriate habitat. Can you explain why?

16.3 HOW DOES NATURAL SELECTION WORK?

Unlike the other causes of evolution that we have discussed, natural selection leads to the evolution of traits that enhance an organism's ability to survive and reproduce. Studying the adaptive evolution that results from natural selection has been a major focus of evolutionary biology.

Natural Selection Stems from Unequal Reproduction

The British economist Herbert Spencer, writing in 1864, coined the phrase "survival of the fittest" to summarize the process that Darwin had named natural selection. But this formulation is not quite accurate: Natural selection favors traits that increase their possessors' survival only to the extent that improved survival leads to improved

Cancer and Darwinian Medicine

In recent years, an increasing number of physicians and biomedical researchers have come to the realization that medical practice should be informed by an appreciation of how evolution affects human health and disease. Consider, for example, cancer. Although a cancerous tumor is an abnormal, harmful growth of tissue, it is also an evolving population of cells. The population of cells that forms the tumor inhabits a particular habitat within a body, and the cells in the population tend to have a variety of different genotypes. This variety arises because cancer cells typically contain multiple mutations. When the cells replicate, which they do far more frequently than do normal cells, they are likely to acquire additional mutations. The cells in a tumor essentially compete with one another to survive and reproduce. The cells that are best adapted to the local environment leave more offspring, and over time the tumor evolves—its genetic makeup changes.

If you understand how natural selection works, you can probably predict what happens when a tumor is treated with a chemotherapy drug. The drug changes the tumor's environment such that many cancer cells cannot survive. Often, however, some of the cells have mutant alleles that allow them to resist the drug. These resistant cells survive the attack, and they or their descendants may move to new environments in different parts of the body. Eventually, the resistant cells proliferate, and the cancer patient now has multiple tumors, all of which are resistant to treatment by chemotherapy. This evolutionary scenario helps explain why a post-chemotherapy cancer patient sometimes seems to be free of cancer (because all but a few tumor cells have been killed, and the few remaining ones are undetectable), only to later suffer an even more threatening recurrence (because natural selection has favored the evolution of drug-resistant cancer cell populations that eventually spread and grow).

Some cancer researchers believe that a better understanding of the details of tumor evolution could help improve cancer treatments. For example, if we knew which cancer cell genotypes were favored in the environment created by a particular drug, doctors could first administer that drug. Then, when the tumor population had been reduced to a small group of resistant cells, a second drug, known to make the environment inhospitable to the surviving cell type, could be introduced at just the right moment to reduce the cells' fitness before they have a chance to move to new environments in the body. As one cancer researcher put it, physicians must "play chess, not whack-a-mole" with evolving tumors. As researchers with an evolutionary perspective learn more about precisely how different types of cancer evolve, physicians will become increasingly able to play chess with tumors and win.

EVALUATE THIS A team of physicians treated four patients with breast cancer. Each patient received a course of chemotherapy (the same combination of drugs for each patient). In addition, researchers sequenced the genotype of cells in healthy skin tissue and in each patient's tumor at intervals during the treatment period. At the end of the course of treatment, all four patients were declared free of detectable cancer. But within 18 months, three of the patients had suffered a relapse. The researchers sequenced samples from the new tumors. If you compared the genetic information from the four patients, what kind of difference might you expect to find between the cured patient and the ones who suffered relapses? How might the theory of evolution by natural selection explain the difference?

reproduction. A trait that improves survival may, for example, increase the likelihood that an individual survives long enough to reproduce, or might increase an organism's life span and, therefore, its number of opportunities to reproduce. But ultimately, it is reproductive success that determines the future of an individual's alleles and the prevalence in the next generation of the traits associated with those alleles. Thus, the main driver of natural selection is differences in reproduction: Individuals bearing certain alleles leave more offspring (who inherit those alleles) than do other individuals with different alleles. In the terminology of evolutionary biology, individuals with greater lifetime reproductive success are said to have greater **fitness** than do individuals with lower reproductive success. (For an example of how the concept of fitness might aid medical practice, see "Health Watch: Cancer and Darwinian Medicine.")

Natural Selection Acts on Phenotypes

Although we have defined evolution as changes in the genetic composition of a population, it is important to recognize that natural selection does not act directly on the genotypes of individual organisms. Rather, natural selection acts on phenotypes, the structures and behaviors displayed by the members of a population. This selection of phenotypes, however, inevitably affects the genotypes present in a population, because phenotypes and genotypes are closely tied. For example, we know that a pea plant's height is strongly influenced by the plant's alleles of certain genes. If a population of pea plants were to encounter environmental conditions that favored taller plants, then taller plants would leave more offspring. These offspring would carry the alleles that contributed to their parents' height. Thus, if natural selection favors a particular phenotype, it will necessarily also favor the underlying genotype.

Some Phenotypes Reproduce More Successfully Than Others

As we have seen, natural selection simply means that some phenotypes reproduce more successfully than others do. This simple process is such a powerful agent of change because only the fittest phenotypes pass traits to subsequent generations. But what makes a phenotype fit? Successful phenotypes are those that have the best **adaptations**—characteristics that help an individual survive and reproduce in a particular environment.

An Environment Has Nonliving and Living Components

Individual organisms must cope with an environment that includes both nonliving physical factors and the other living organisms with which the individual interacts. The nonliving component of the environment includes such factors as climate, availability of water, and availability of nutrients. These nonliving factors play a large role in determining the traits that help an organism to survive and reproduce. However, adaptations also arise because of interactions with the living component of the environment, namely, other organisms. A simple example illustrates this concept.

Consider a buffalo grass plant growing in a small patch of soil in the eastern Wyoming plains. The plant's roots must be able to take up enough water and minerals for growth and reproduction, and to that extent, it must be adapted to its nonliving environment. But even in the dry prairies of Wyoming, this requirement is relatively trivial, provided that the plant is alone and protected in its square yard of soil. In reality, however, many other plants—other buffalo grass plants as well as other grasses, sagebrush bushes, and annual wildflowers—also sprout in that same patch of soil. If our buffalo grass is to survive, it must compete with the other plants for resources. Its long, deep roots and efficient methods of mineral uptake have evolved not only because the plains are dry but also because the buffalo grass must share the dry prairies with other plants. Further, buffalo grass must also coexist with animals that wish to eat it, such as the cattle and other plant-eating animals that graze the

prairie. So over time, tougher, harder-to-eat buffalo grass plants survived better and reproduced more on the prairie than did less-tough buffalo grass plants. As a result, buffalo grass leaves are quite tough; embedded silica compounds reinforce them.

Competition Acts As an Agent of Selection

As the example of buffalo grass shows, one of the major sources of natural selection is **competition** with other organisms for scarce resources. Competition for resources is most intense among members of the same species because, as Darwin wrote in *On the Origin of Species*, "they frequent the same districts, require the same food, and are exposed to the same dangers." In other words, no two competing organisms have such similar requirements for survival as do two members of the same species. Although different species may also compete for the same resources, they generally do so to a lesser extent than do individuals within a species.

Both Predators and Prey Act As Agents of Selection

When two species interact extensively, each exerts strong selection on the other. When one evolves a new feature or modifies an old one, the other typically evolves new adaptations in response. This process in which species mutually affect one another's evolution is called **coevolution.** Perhaps the most familiar form of coevolution is found in predator–prey relationships.

Predation describes any interaction in which one organism consumes another. In some instances, coevolution between predators (those that do the consuming) and prey (those that are consumed) is a sort of biological arms race, with each side evolving new adaptations in response to escalations by the other. Darwin used the example of wolves and deer: Wolf predation selects against slow or careless deer, thus leaving faster, more-alert deer to reproduce and pass on these traits. The resulting alert, swift deer select in turn against slow, clumsy wolves, because such predators cannot acquire enough food.

CASE STUDY CONTINUED

Evolution of a Menace

Antibiotic resistance evolves by natural selection. To see how, imagine a hospital patient with an infected wound. A doctor decides to treat the infection with an intravenous drip of penicillin. As the antibiotic courses through the patient's blood vessels, millions of bacteria die before they can reproduce. A few bacteria, however, carry a rare allele that codes for an enzyme that destroys penicillin. The bacteria carrying this rare allele are able to survive and reproduce, and their offspring inherit the penicillin-destroying allele. After a few generations, the frequency of the penicillin-destroying allele in the bacteria has soared to nearly 100%, and the frequency of the normal allele has declined to near zero. As a result of natural selection imposed by the antibiotic's killing power, the population of bacteria within the patient's body has evolved. The gene pool of the population has changed, and natural selection, in the form of bacterial destruction by penicillin, has caused the change.

Buffalo grass

Antibiotic Resistance Illustrates Key Points About Natural Selection

The example of antibiotic resistance highlights some important features of natural selection.

- **Natural selection does not cause genetic changes in individuals.** Alleles for antibiotic resistance arise spontaneously in some bacteria, long before the bacteria encounter an antibiotic. Antibiotics do not cause resistance to appear; their presence merely favors the survival of bacteria with antibiotic-destroying alleles over that of bacteria without such alleles.
- **Natural selection acts on individuals, but it is populations that are changed by evolution.** The agent of natural selection—in this example, antibiotics—acts on individual bacteria. As a result, some individuals reproduce and some do not. However, it is the population as a whole that evolves as its allele frequencies change.
- **Evolution by natural selection is not progressive; it does not make organisms "better."** The traits favored by natural selection change as the environment changes. Resistant bacteria are favored only when antibiotics are present. At a later time, when the environment no longer contains antibiotics, resistant bacteria may be at a disadvantage relative to other bacteria.

Sexual Selection Favors Traits That Help an Organism Mate

In many animal species, males have conspicuous features such as bright colors, long feathers or fins, or elaborate antlers. Males may also exhibit elaborate courtship behaviors. Although these extravagant features and behaviors typically play a role in mating, they also seem to be at odds with efficient survival and reproduction. Exaggerated ornaments and displays may help males gain access to females, but they may also make the males more conspicuous and thus vulnerable to predators. Darwin was intrigued by this apparent contradiction. He coined the term **sexual selection** to describe the special kind of selection that acts on traits that help an animal acquire a mate.

Darwin recognized that sexual selection could be driven either by sexual contests among males or by female preference for particular male phenotypes. Male–male competition for access to females can favor the evolution of features that provide an

▲ **FIGURE 16-10 The peacock's showy tail has evolved through sexual selection** The ancestors of today's peahens were apparently picky when deciding on a male with which to mate, favoring males with longer and more colorful tails.

advantage in fights or ritual displays of aggression (**FIG. 16-9**). In animal species in which females actively choose their mates, females often seem to prefer males with the most elaborate ornaments or most extravagant displays (**FIG. 16-10**). Why?

One hypothesis is that male structures, colors, and displays that do not enhance survival might instead provide a female with an outward sign of a male's condition. Only a vigorous, energetic male can survive when burdened with conspicuous coloration or a large tail that might make him more vulnerable to predators. Conversely, males that are sick or under parasitic attack are dull and frumpy compared with healthy males. A female that chooses the brightest, most ornamented male is also choosing the healthiest male. By doing so, she gains fitness if, for example, the healthiest male provides superior parental care to offspring or if he carries alleles for disease resistance that will be inherited by offspring and help ensure their survival. Females thus gain a reproductive advantage by choosing the most highly ornamented

◀ **FIGURE 16-9 Competition between males favors the evolution, through sexual selection, of structures for ritual combat** Two male bighorn sheep spar during the fall mating season. In many species, the losers of such contests are unlikely to mate, while winners enjoy tremendous reproductive success.

THINK CRITICALLY If we studied a population of bighorn sheep and were able to identify the father and mother of each lamb born, would you predict that the difference in number of offspring between the most reproductively successful adult and the least successful adult would be greater for males or for females?

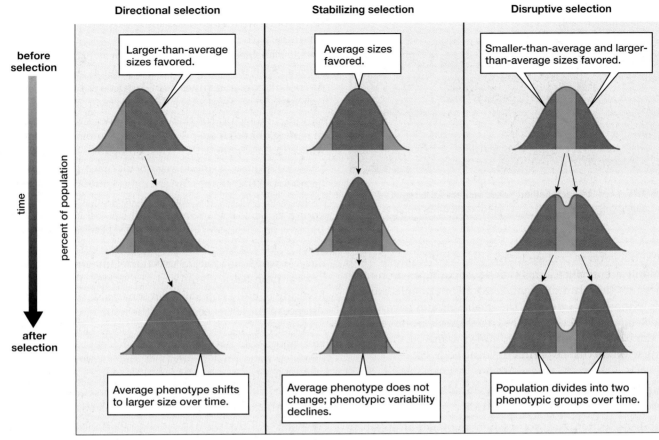

Directional selection

before selection

Larger-than-average sizes favored.

Average phenotype shifts to larger size over time.

Stabilizing selection

Average sizes favored.

Average phenotype does not change; phenotypic variability declines.

Disruptive selection

Smaller-than-average and larger-than-average sizes favored.

Population divides into two phenotypic groups over time.

time

after selection

percent of population

trait, such as size

▲ **FIGURE 16-11 Three ways that selection affects a population over time** A graphical illustration of three ways natural and/or sexual selection, acting on a normal distribution of phenotypes, can affect a population over time. In all graphs, the blue areas represent individuals that are selected against— that is, the individuals that do not reproduce as successfully as do the individuals in the purple range.

THINK CRITICALLY When selection is directional, is there any limit to how extreme the trait under selection will become? Why or why not?

males, and the traits (including the exaggerated ornament) of these flashy males will be passed to subsequent generations.

Selection Can Influence Populations in Three Ways

Natural selection and sexual selection can lead to various patterns of evolutionary change. Evolutionary biologists group these patterns into three categories (**FIG. 16-11**):

- **Directional selection** favors individuals with an extreme value of a trait and selects against both average individuals and individuals at the opposite extreme. For example, directional selection might favor small size and select against both average and large individuals in a population.
- **Stabilizing selection** favors individuals with the average value of a trait (for example, intermediate body size) and selects against individuals with extreme values.
- **Disruptive selection** favors individuals at both extremes of a trait (for example, both large and small body sizes) and selects against individuals with intermediate values.

Directional Selection Shifts Character Traits in a Specific Direction

If environmental conditions change in a consistent way, a species may respond by evolving in a consistent direction. For example, during past ice age periods in which Earth's climate cooled considerably, many mammal species evolved thicker fur. The evolution of antibiotic resistance in bacteria is another example of directional selection: When antibiotics are present in a bacterial species' environment, individuals with greater resistance reproduce more prolifically than do individuals with less resistance.

Stabilizing Selection Acts Against Individuals Who Deviate Too Far from the Average

Directional selection can't go on forever. What happens once a species is well adapted to a particular environment? If the environment is unchanging, most new variations that appear will be harmful. Under these conditions, we expect species to be subject to stabilizing selection, which favors the survival

▲ **FIGURE 16-12 Black-bellied seedcrackers** As a result of disruptive selection, each black-bellied seedcracker has either a large beak (left) or a small beak (right).

and reproduction of average individuals. Stabilizing selection commonly occurs when a trait is under opposing environmental pressures from two different sources. For example, among lizards of the genus *Aristelliger*, the smallest lizards have a hard time defending territories, but the largest lizards are more likely to be eaten by owls. As a result, *Aristelliger* lizards are under stabilizing selection that favors intermediate body size.

Disruptive Selection Adapts Individuals Within a Population to Different Habitats

Disruptive selection may occur when a population inhabits an area with more than one type of useful resource. In this situation, the most adaptive characteristics may be different for each type of resource. For example, the food source of the black-bellied seedcracker (**FIG. 16-12**), a small, seed-eating bird found in the forests of Africa, includes both hard seeds and soft seeds. Cracking hard seeds requires a large, stout beak, but a smaller, pointier beak is a more efficient tool for processing soft seeds. Consequently, black-bellied seedcrackers have beaks in one of two sizes. A bird may have a large beak or small beak, but very few birds have a medium-sized beak; individuals with intermediate-sized beaks have a lower survival rate than individuals with either large or small

beaks. Disruptive selection in black-bellied seedcrackers thus favors birds with large beaks and birds with small beaks, but not those with medium-sized beaks.

Black-bellied seedcrackers represent an example of *balanced polymorphism,* in which two or more phenotypes are maintained in a population. In many cases of balanced polymorphism, multiple phenotypes persist because each is favored by a separate environmental factor. For example, consider two different forms of hemoglobin that are present in some human populations in Africa. In these populations, the hemoglobin molecules of people who are homozygous for a particular allele produce defective hemoglobin that clumps up into long chains, which distort and weaken red blood cells. This distortion causes a serious illness known as sickle-cell anemia, which can kill its victims. Before the advent of modern medicine, people homozygous for the sickle-cell allele were unlikely to survive long enough to reproduce. So why hasn't natural selection eliminated the allele?

Far from being eliminated, the sickle-cell allele is present in nearly half the population in some areas of Africa. The persistence of the allele seems to be the result of counterbalancing selection that favors heterozygous carriers of the allele. Heterozygotes, who have one allele for defective hemoglobin and one allele for normal hemoglobin, suffer from mild anemia but also exhibit increased resistance to malaria, a deadly disease affecting red blood cells that is widespread in equatorial Africa. In areas of Africa with high risk of malaria infection, heterozygotes must have survived and reproduced more successfully than either type of homozygote. As a result, both the normal hemoglobin allele and the sickle-cell allele have been preserved.

CHECK YOUR LEARNING

Can you ...

- describe why selection of phenotypes can affect the evolution of genotypes?
- explain how competition and predation influence evolution?
- explain how sexual selection works and describe examples of its outcome?
- compare and contrast directional selection, stabilizing selection, and disruptive selection?

CASE STUDY \ REVISITED
Evolution of a Menace

Overuse of antibiotics has accelerated the evolution of antibiotic resistance. Each year, U.S. physicians write more than 100 million prescriptions for antibiotics; the Centers for Disease Control and Prevention estimates that about half of these prescriptions are unnecessary.

Although medical use and misuse of antibiotics are the most important sources of natural selection for antibiotic resistance, antibiotics also pervade the environment outside our bodies. More than 29 million pounds of antibiotics are fed to farm animals in the United States each year. In addition, Earth's soils and water are laced with antibiotics that enter the environment through human and livestock wastes, and from the antibacterial soaps and cleansers that are now routinely used in many households and workplaces. As a result of this massive alteration of the environment, resistant

bacteria are now found not only in hospitals and the bodies of sick people but also in our food, water, and soil. Susceptible bacteria are under constant attack, and resistant strains have little competition. In our fight against disease, we have rashly overlooked some basic principles of evolutionary biology and are now paying a heavy price.

THINK CRITICALLY Microbiologists have discovered that alleles associated with antibiotic resistance are present in bacteria that live in soil, even in environments that are comparatively free of antibiotic pollution from human activities. Why are such alleles present (albeit at low levels) in bacterial populations? Conversely, if resistance alleles are beneficial, why are they rare in natural populations of bacteria?

CHAPTER REVIEW

(MB) Go to **MasteringBiology** for practice quizzes, activities, eText, videos, current events, and more.

Answers to Think Critically, Evaluate This, Multiple Choice, and Fill-in-the-Blank questions can be found in the Answers section at the back of the book.

Summary of Key Concepts

16.1 How Are Populations, Genes, and Evolution Related?

Evolution is change in the frequencies of alleles in a population's gene pool. Allele frequencies in a population will remain constant over generations only if the following conditions are met: (1) There is no mutation, (2) there is no gene flow, (3) the population is very large, (4) all mating is random, and (5) all genotypes reproduce equally well (that is, there is no natural selection).

16.2 What Causes Evolution?

Evolutionary change is caused by mutation, gene flow, small population size, nonrandom mating, and natural selection.

Mutations are random, undirected changes in DNA composition. Although most mutations are neutral or harmful to the organism, some prove advantageous in certain environments. Mutations are rare and do not by themselves change allele frequencies very much, but they provide the raw material for evolution by other processes.

Gene flow is the movement of alleles between different populations of a species. Gene flow tends to reduce differences in the genetic composition of different populations.

If a population is small, chance events may reduce the survival and reproduction of a disproportionate number of individuals that bear a particular allele, thereby greatly changing the allele's frequency in the population; this is genetic drift.

Nonrandom mating, such as assortative mating and inbreeding, can change the distribution of genotypes in a population, in particular by increasing the proportion of homozygotes.

The survival and reproduction of organisms are influenced by their phenotypes. Because phenotype depends at least partly on genotype, natural selection tends to favor the persistence of certain alleles at the expense of others.

16.3 How Does Natural Selection Work?

Natural selection is driven by differences in reproductive success among different genotypes. Natural selection stems from the interactions of organisms with both the living and nonliving parts of their environments. When two species interact intensively, both of them may evolve in response. Phenotypes that help organisms mate can evolve by sexual selection.

Key Terms

adaptation *293*
allele frequency *283*
coevolution *293*
competition *293*
directional selection *295*
disruptive selection *295*
equilibrium
   population *284*
fitness *292*
founder effect *288*
gene flow *284*
gene pool *283*
genetic drift *286*
Hardy–Weinberg
   principle *284*
mutation *284*
natural selection *290*
population *282*
population bottleneck *287*
predation *293*
sexual selection *294*
stabilizing selection *295*

Thinking Through the Concepts

Multiple Choice

1. The alleles responsible for antibiotic resistance in bacteria
   a. arise in response to the presence of antibiotics.
   b. are identical to the alleles responsible for pesticide resistance in insects.
   c. are present in bacterial populations that have never been exposed to antibiotics.
   d. are formed by interactions between antibiotic molecules and bacterial DNA.

2. Stabilizing selection on a trait tends to
   a. make the trait more extreme.
   b. reduce variability in the trait.
   c. decrease the frequency of alleles associated with the trait.
   d. result in elaborate male ornaments.

3. An adaptation is
   a. any trait that arises from a mutation.
   b. a trait that increases the reproductive success of its bearer.
   c. any trait that changes during the lifetime of an organism.
   d. a trait that arises due to gene flow or genetic drift.

4. Which of the following statements about mutations is False?
   a. Mutations at a given chromosomal site are rare.
   b. The genomes of most people contain some mutant alleles that are present in neither parent.
   c. Mutations are the ultimate source of genetic variability.
   d. Beneficial mutations are more likely to occur when an organism's needs change.

5. Genetic drift occurs
   a. when different phenotypes have different reproductive success.
   b. in small populations but not in large populations.
   c. when only a random subset of a population reproduces.
   d. in mammals but not in bacteria.

Fill-in-the-Blank

1. The _____ provides a simple mathematical model for a non-evolving population, also called a(n) _____ population, in which _____ frequencies do not change over time. Are such populations likely to be found in nature? _____

2. Different versions of the same gene are called _____ . These versions arise as a result of changes in the sequence of _____ that form the gene. These changes are known as _____ . An individual with two identical copies of a given gene is described as being _____ for that gene, while an individual with two different versions of that gene is described as _____ .

3. An organism's _____ refers to the specific alleles found within its chromosomes, while the traits that these alleles produce are called its _____ . Which of these does natural selection act on? _____

4. A random form of evolution is called _____ . This form of evolution occurs more rapidly and has a greater effect in populations that are _____ . Two important causes of this form of evolution are _____ and _____ . Which of these would apply to a population started by a breeding pair that was stranded on an island? _____

5. Competition is most intense between members of a(n) _____ . Predators and their prey act as agents of _____ on one another, resulting in a form of evolution called _____ . This results in the evolution of characteristics called _____ that help both predators and their prey survive and reproduce.

6. The evolutionary fitness of an organism is measured by its success at _____ . The fitness of an organism can change if its _____ changes.

Review Questions

1. What is a gene pool? How would you determine the allele frequencies in a gene pool?

2. Define *equilibrium population*. Outline the conditions that must be met for a population to stay in genetic equilibrium.

3. How does population size affect the likelihood of changes in allele frequencies by chance alone? Can significant changes in allele frequencies (that is, evolution) occur as a result of genetic drift?

4. If you measured the allele frequencies of a gene and found large differences from those predicted by the Hardy–Weinberg principle, would that prove that natural selection is occurring in the population you are studying? Review the conditions that lead to an equilibrium population, and explain your answer.

5. People like to say that "you can't prove a negative." Study the experiment in Figure 16-3 again, and comment on what it demonstrates.

6. Describe the three ways in which natural selection can affect a population over time. Which way(s) is (are) most likely to occur in stable environments, and which way(s) might occur in rapidly changing environments?

7. What is sexual selection? How is sexual selection similar to and different from other forms of natural selection?

Applying the Concepts

1. In North America, the average height of adult humans has been increasing steadily for decades. Is directional selection occurring? What data would justify your answer?

2. By the 1940s, the whooping crane population had been reduced to only 15 individuals. Thanks to conservation measures, its numbers are now increasing. What special evolutionary problems do whooping cranes face now that they have passed through a population bottleneck?

17

THE ORIGIN OF SPECIES

Paedophryne amauensis, a minuscule frog unknown to science until 2013, is one of a number of previously undiscovered species recently found in the forests of New Guinea.

CASE STUDY

Discovering Diversity

AS DARKNESS FELL ON A WARM, HUMID EVENING in a rain forest on the island of New Guinea, a team of scientists carefully scrutinized the forest floor. They were seeking the source of a high-pitched call that sounded like an insect. But the scientists' careful search detected no animal that might have produced the sound. Finally, the frustrated searchers scooped up a bit of leaf litter from a spot that seemed to be the source of the calls, put it in a plastic bag, and headed back to the lab. There, concealed among the decaying leaves in the bag, they found what they had earlier missed: a tiny frog, smaller than a shirt button. The

frog was smaller, in fact, than any other known animal with a backbone. And what's more, the frog was of a type previously unknown to science: It was a newly discovered species. The new frog species, which its discoverers dubbed *Paedophryne amauensis*, is one of many species recently discovered in New Guinea, including birds, mammals, butterflies, and flowering plants.

New Guinea is not the only location to yield interesting new species. One particularly surprising find was in the Annamite Mountains of Vietnam, where the saola, a hoofed, horned antelope, was discovered in the early 1990s. The discovery of a new species of large mammal at that late date was a complete shock—after centuries of human exploration and exploitation of nearly every corner of the world, scientists had been certain that no large mammal species could have escaped detection. More recent surprises include the discovery, reported in 2013, of the olinguito, a nocturnal relative of the raccoon that inhabits high-altitude cloud forests in the Andes Mountains of South America.

Our discussion of newly discovered species leads us to some important questions: What do we mean when we say that some organisms constitute a species? And how do such species originate?

AT A GLANCE

17.1 WHAT IS A SPECIES?

Although Darwin brilliantly explained how evolution shapes complex organisms, his ideas did not fully explain life's diversity. In particular, the process of natural selection cannot by itself explain how living things came to be divided into groups, with each group distinctly different from all other groups. When we look at big cats, we don't see a continuous array of different tiger phenotypes that gradually grades into the phenotype of a lion. We see lions and tigers as separate, distinct types with no overlap. Each distinct type is known as a species.

In everyday life, most of us make unthinking use of an informal, nonscientific conception of species. We perceive sparrows as clearly different from eagles, which are obviously different from ducks. But we sometimes run into trouble when we try to make finer distinctions. It is not easy, for example, to distinguish among different species of sparrows, especially if we don't have a precise idea of what constitutes a species. How, then, do scientists make these finer distinctions?

Each Species Evolves Independently

Today, biologists define a **species** as a group of populations that evolves independently. Each species follows a separate evolutionary path. This definition, however, does not clearly state the standard by which such evolutionary independence is judged. The most widely used standard defines species as "groups of actually or potentially interbreeding natural populations, which are reproductively isolated from other such groups." This definition, known as the *biological species concept*, is based on the observation that **reproductive isolation** (inability to successfully breed outside the group) ensures evolutionary independence.

The biological species concept has two major limitations. First, because the definition is based on patterns of sexual reproduction, it does not help us determine species boundaries among asexually reproducing organisms. Second, it is not always practical or even possible to directly observe whether members of two different groups interbreed. Thus, a biologist who wishes to determine if a group of organisms is a separate species must often make the determination without knowing for sure if group members breed with organisms outside the group.

Despite the limitations of the biological species concept, most biologists accept it for identifying species of sexually

HAVE YOU EVER WONDERED...

One way to determine the number of species on Earth might be to simply count them. You could comb the scientific literature to find all the species that scientists have discovered and named, and then tally up the total number. One attempt to do just that, the Catalogue of Life project, has compiled an online searchable database that listed 1,612,941 species as of 2015. But even the Catalogue of Life can't tell you how many species are on Earth.

Why doesn't counting work? Because most of the planet's species remain undiscovered. Relatively few scientists are engaged in the search for new species, and nearly all undiscovered species are small and inconspicuous, or live in poorly explored habitats such as the floor of the ocean or the topmost branches of tropical rain forests. So no one knows the actual number of species on Earth. But biologists agree that the number must be much higher than the number of named species. A commonly held view is that the actual number is close to 8.7 million, the number estimated by a recent analysis that used sophisticated statistical methods to extrapolate past trends in species discovery.

How Many Species Inhabit the Planet?

reproducing organisms. However, alternative definitions are required by scientists who study bacteria and other organisms that mainly reproduce asexually. Even some biologists who study sexually reproducing organisms prefer species definitions that do not depend on a property (reproductive isolation) that can be difficult to measure. Several such alternatives to the biological species concept have been proposed. (One alternative, the phylogenetic species concept, is described in Chapter 19.)

Appearance Can Be Misleading

Biologists have found that differences in appearance do not always mean that two populations belong to different species. For example, a Northwestern garter snake may be brown, black, gray, green, or some shade in between, and it may be striped or unstriped (FIG. 17-1). If it is striped, the stripes may be broad or narrow, and they could be any of a variety of colors. Despite their diversity of appearance, though, all Northwestern garter snakes are members of the same species.

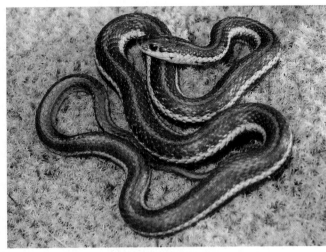

(a) A green-striped Northwestern garter snake

(b) A red-striped Northwestern garter snake

▲ FIGURE 17-1 **Members of a species may differ in appearance (a)** This green-striped Northwestern garter snake and **(b)** this red-striped Northwestern garter snake are members of the same species.

THINK CRITICALLY Southern Wisconsin is home to several populations of squirrels with black fur. These squirrels are hypothesized to be members of the squirrel species *Sciurus carolinensis*, which usually has gray fur. How could you determine if the squirrels with black fur are in fact of the same species as the squirrels with gray fur?

Conversely, some organisms with very similar appearances belong to different species. For example, the cordilleran flycatcher and the Pacific-slope flycatcher are so similar that even experienced birdwatchers cannot tell them apart (FIG. 17-2). Likewise, there are virtually no visible differences between the Asian mosquito species *Anopheles dirus* and *Anopheles harrisoni*. This similarity in appearance, however, disguises a crucial difference: *A. dirus* spreads malaria from person to person, but *A. harrisoni* generally does not.

CHECK YOUR LEARNING

Can you ...

- describe how biologists define *species* and explain why it is difficult to develop a criterion for distinguishing species?
- describe the biological species concept and discuss its limitations?
- list some reasons why it can be hard to tell different species apart?

(a) Cordilleran flycatcher

(b) Pacific-slope flycatcher

▲ FIGURE 17-2 **Members of different species may be similar in appearance** The **(a)** cordilleran flycatcher and **(b)** Pacific-slope flycatcher are different species.

TABLE 17-1	Mechanisms of Reproductive Isolation
Premating Isolating Mechanisms	Factors that prevent organisms of two species from mating
Geographic isolation	The species do not interbreed because a physical barrier separates them.
Ecological isolation	The species do not interbreed even if they are within the same area because they occupy different habitats.
Temporal isolation	The species do not interbreed because they breed at different times.
Behavioral isolation	The species do not interbreed because they have different courtship and mating rituals.
Mechanical incompatibility	The species do not interbreed because their reproductive structures are incompatible.
Postmating Isolating Mechanisms	Factors that prevent organisms of two species from producing vigorous, fertile offspring after mating
Gametic incompatibility	Sperm from one species cannot fertilize eggs of another species.
Hybrid inviability	Hybrid offspring fail to survive.
Hybrid infertility	Hybrid offspring are sterile or have low fertility.

CASE STUDY \ CONTINUED
Discovering Diversity

The tiny frog *Paedophryne amauensis* was discovered by researchers purposefully searching for the source of an unusual sound, but sometimes similarity between two species can lead scientists to accidental discoveries. A previously unknown species of wild cat was inadvertently discovered by researchers who were sequencing the DNA of a house cat–sized South American wild cat called the tigrina. The DNA analysis revealed that many alleles found in the supposed tigrinas living in northeastern Brazil are not shared with other tigrinas. This finding suggested that the northeastern cats do not interbreed with other tigrinas and are therefore a different species. What prevents the two species from interbreeding?

17.2 HOW IS REPRODUCTIVE ISOLATION BETWEEN SPECIES MAINTAINED?

The traits that prevent interbreeding and maintain reproductive isolation are called **isolating mechanisms**

(**TABLE 17-1**). Isolating mechanisms provide a clear benefit to individuals. An individual that breeds with a member of another species will probably produce no offspring (or offspring that are unfit or sterile), thereby wasting its reproductive effort and failing to contribute to future generations. Thus, natural selection favors traits that prevent reproduction across species boundaries.

Premating Isolating Mechanisms Prevent Mating Between Species

Reproductive isolation can be maintained by a variety of mechanisms, but those that prevent mating are especially effective. The mechanisms that prevent mating between species are collectively called **premating isolating mechanisms.**

Members of Different Species May Not Meet

Members of different species cannot mate if they never get near one another. **Geographic isolation** prevents interbreeding between populations that do not come into contact because they live in different, physically separated places (**FIG. 17-3**). However, we cannot determine if geographically separated populations are actually distinct species. Should the physical barrier separating the two populations disappear, the reunited

(a) Kaibab squirrel

(b) Abert's squirrel

◀ **FIGURE 17-3 Geographic isolation** To determine if these two squirrels are members of different species, we must know if they are "actually or potentially interbreeding." Unfortunately, it is hard to tell, because **(a)** the Kaibab squirrel lives only on the north rim of the Grand Canyon and **(b)** the Abert's squirrel lives on the south rim. The two populations are geographically isolated but still quite similar. Have they diverged enough since their separation to become reproductively isolated? Because they remain geographically isolated, we cannot say for sure.

populations might interbreed freely and not be separate species after all. Geographic isolation, therefore, is usually not considered to be a mechanism that maintains reproductive isolation between species. Instead, it is a mechanism that allows new species to form. If populations cannot interbreed after geographic barriers have been eliminated, then other premating isolating mechanisms must have developed.

Different Species May Occupy Different Habitats

Two populations that use different resources may spend time in different habitats within the same general area and thus exhibit **ecological isolation.** White-crowned sparrows and white-throated sparrows, for example, have extensively overlapping geographic ranges. The white-throated sparrow, however, frequents dense thickets, whereas the white-crowned sparrow inhabits fields and meadows, seldom penetrating far into dense growth. The two species may coexist within a few hundred yards of one another and yet seldom meet during the breeding season. A more dramatic example is provided by the more than 300 species of fig wasp (FIG. 17-4). In most cases, fig wasps of a given species breed in (and pollinate) the fruits of one particular species of fig, and each fig species hosts only one or two species of pollinating wasp. Thus, fig wasps of different species only rarely encounter one another during breeding, and pollen from one fig species is not ordinarily carried to flowers of a different species.

Different Species May Breed at Different Times

Even if two species occupy similar habitats, they cannot mate if they have different breeding seasons, a phenomenon called **temporal isolation** (time-based isolation). For example, the spring field cricket and the fall field cricket both can be found in many areas of North America, but as their names suggest, the former species breeds in spring and the latter in autumn. As a result, the two species do not interbreed.

In plants, the reproductive structures of different species may mature at different times. For example, Bishop pines and

▲ **FIGURE 17-4 Ecological isolation** This female fig wasp's eggs were fertilized by mating that took place within a fig. She will find another fig of the same species, enter it through a pore, lay eggs, and die. Her offspring will hatch, develop, and mate within the fig. Because each species of fig wasp reproduces only in its own particular fig species, each wasp species is reproductively isolated.

Monterey pines grow together near Monterey on the California coast (FIG. 17-5), but the two species release their sperm-containing pollen (and have eggs ready to be fertilized) at different times: The Monterey pine releases pollen in early spring, the Bishop pine in summer. For this reason, the two species do not interbreed under natural conditions.

Different Species May Have Different Courtship Signals

Among animals, elaborate courtship colors and behaviors can prevent mating with members of other species. Signals and behaviors that differ from species to species create **behavioral isolation.** For example, the extravagant plumes and arresting pose of a courting male Raggiana bird of paradise are conspicuous indicators of his species, and there is little chance that females of another species will mate with him by mistake (FIG. 17-6). Among frogs, males are often impressively indiscriminate, jumping on every female in sight regardless of the species. Females, however, approach only male frogs that utter the call appropriate to their species. If they do find themselves in an unwanted embrace, they give the "release call," which causes the male to let go. As a result, few **hybrids**—offspring of parents of different species—are produced.

(a) Bishop pine   **(b) Monterey pine**

◀ **FIGURE 17-5 Temporal isolation (a)** Bishop pines and **(b)** Monterey pines coexist in nature. In the laboratory they produce fertile hybrids. In the wild, however, they do not interbreed, because they release pollen at different times of the year.

▲ **FIGURE 17-6 Behavioral isolation** The mate-attraction display of a male Raggiana bird of paradise includes distinctive posture, movements, plumage, and vocalizations that do not resemble those of other bird of paradise species.

Differing Sexual Organs May Foil Mating Attempts

If a male and a female of different species attempt to mate, the attempt may be disrupted by physical differences between the species that are collectively known as **mechanical incompatibilities.** Among animal species with internal fertilization (in which the sperm is deposited inside the female's reproductive tract), the male's and female's sexual organs simply may not fit together. Incompatible body shapes may also make copulation between species impossible. For example, snails of species whose shells have left-handed spirals may be unable to successfully copulate with closely related snails whose shells have right-handed spirals, because the shell mismatch is accompanied by a body orientation mismatch that prevents the genitals of the two species from lining up properly during attempted copulations. Among plants, flowers with different structures may attract different pollinators, thereby preventing pollen transfer between species.

Postmating Isolating Mechanisms Limit Hybrid Offspring

When premating isolating mechanisms fail or have not yet evolved, members of different species may mate. However, **postmating isolating mechanisms** may prevent the formation of vigorous, fertile hybrid offspring, with the result that the two species remain separate, with little or no gene flow between them.

One Species' Sperm May Fail to Fertilize Another Species' Eggs

Even if a male inseminates a female of a different species, his sperm may not be able to fertilize her eggs, an isolating mechanism called **gametic incompatibility.** For example, in animals with internal fertilization, fluids in the female reproductive tract may weaken or kill sperm of other species.

Gametic incompatibility may be an especially important isolating mechanism in species that reproduce by scattering gametes in the water or in the air, such as marine invertebrate animals and wind-pollinated plants. For example, sea urchin sperm cells contain a protein that allows them to bind to eggs. The structure of the protein differs among species so that sperm of one sea urchin species cannot bind to the eggs of another species. In abalones (a type of mollusk), eggs are surrounded by a membrane that can be penetrated only by sperm containing a particular enzyme. Each abalone species has a distinctive version of the enzyme, so hybrids are rare, even though several species of abalones coexist in the same waters and spawn during the same period. Among plants, a similar chemical incompatibility may prevent the germination of pollen from one species that lands on the stigma (pollen-catching structure) of the flower of another species.

Hybrid Offspring May Fail to Survive or Reproduce

If cross-species fertilization does occur, the resulting hybrid may be unable to survive, a situation called **hybrid inviability.** The genetic instructions directing development of the two species may be so different that hybrids abort early in development. For example, captive leopard frogs can be induced to mate with wood frogs, and the matings generally yield fertilized eggs. The resulting embryos, however, inevitably fail to survive more than a few days.

In other animal species, a hybrid might survive to adulthood but fail to reproduce because it exhibits ineffective breeding behavior. Hybrids between certain species of lovebirds, for example, have great difficulty building nests. Members of each parental species inherit a particular behavior for carrying nest material; one species tucks the material under its rump feathers and the other carries it in its beak. Hybrids, however, use a nonfunctional mixture of the two behaviors. They repeatedly attempt to tuck nest material under their feathers, but are unable to do so because they don't release the material from their beaks. Hybrids with such ineffective nest-building behavior probably could not reproduce in the wild.

Hybrid Offspring May Be Infertile

Most animal hybrids, such as the mule (a cross between a horse and a donkey) and the liger (a zoo-based cross between a lion and a tiger), are sterile (**FIG. 17-7**). This **hybrid infertility** prevents hybrids from passing on their genetic material to offspring, thus blocking gene flow between the two parent species. A common reason for hybrid infertility is the failure of chromosomes to pair properly during meiosis, so that eggs and sperm fail to develop.

CHECK YOUR LEARNING

Can you ...

- describe the main types of premating and postmating reproductive isolating mechanisms?
- provide examples of each type of mechanism?

▲ **FIGURE 17-7 Hybrid infertility** This liger, the hybrid offspring of a lion and a tiger, is sterile. The gene pools of its parent species remain separate.

17.3 HOW DO NEW SPECIES FORM?

Despite his exhaustive exploration of the process of natural selection, Charles Darwin did not propose a complete mechanism of **speciation,** the process by which new species form. Today, however, biologists recognize that speciation depends on two processes: isolation and genetic divergence.

- **Isolation of populations:** If individuals (or their gametes) move freely between two populations, interbreeding and the resulting gene flow will cause changes in one population to become widespread in the other as well. Thus, two populations cannot grow increasingly different unless something happens to block interbreeding between them. Speciation depends on isolation.
- **Genetic divergence of populations:** It is not sufficient for two populations simply to be isolated. They will become separate species only if, during the period of isolation, they evolve sufficiently large differences. The differences must be large enough that, if the isolated populations were reunited, they could no longer interbreed and produce vigorous, fertile offspring. That is, speciation is complete only if divergence results in evolution of an isolating mechanism. Such differences can arise by chance (genetic drift), especially if at least one of the isolated populations is small (see Chapter 16). Large genetic differences can also arise through natural selection if the isolated populations experience different environmental conditions.

Speciation always requires isolation followed by divergence, but these steps can take place in several different ways. Evolutionary biologists group the different pathways to speciation into two broad categories: **allopatric speciation,** in which two populations are geographically separated from one another, and **sympatric speciation,** in which two populations share the same geographic area. (To learn more about how scientists study the outcome of speciation, see "How Do We Know That? Seeking the Secrets of the Sea" on page 306.)

1 Part of a mainland population reaches an isolated island.

many generations pass

2 Over many generations, the isolated populations begin to diverge due to genetic drift and natural selection.

many generations pass

3 Divergence may eventually become sufficient to cause reproductive isolation.

▲ **FIGURE 17-8 Allopatric isolation and divergence** In allopatric speciation, some event causes a population to be divided by an impassable geographic barrier. One way the division can occur is by colonization of an isolated island. The two now-separated populations may diverge genetically. If the genetic differences between the two populations become large enough to prevent interbreeding, then the two populations constitute separate species.

THINK CRITICALLY Make a list of events or processes that could cause geographic subdivision of a population. Do you think items on your list are sufficient to account for formation of the millions of species that have inhabited Earth?

Geographic Separation of a Population Can Lead to Allopatric Speciation

New species can arise by allopatric speciation when an impassible barrier physically separates different parts of a population.

Organisms May Colonize Isolated Habitats

A small population can become isolated if it moves to a new location (**FIG. 17-8**). For example, some members of a

HOW DO WE KNOW THAT?

Seeking the Secrets of the Sea

In addition to studying how new species arise, biologists also investigate the outcome of millennia of speciation: life's current diversity of species. However, even after several centuries of scientific exploration, much of this diversity remains poorly understood. One ambitious effort to increase our understanding of life's diversity is the recently concluded Census of Marine Life.

The Census of Marine Life was a huge collaborative effort to systematically explore the least explored part of Earth: its oceans. The census aimed to "assess and explain the diversity, distribution, and abundance of marine life." The project lasted 10 years, concluding in 2010, and involved 2,700 scientists from more than 80 countries (**FIG. E17-1**). The scientists conducted 540 expeditions at a cost of $650 million. The expeditions spanned the globe from the tropics to polar regions, explored coastal waters and the open ocean, and examined life from the water's surface down to its deepest depths. The census scientists studied a huge range of organisms—from microbes to whales—counting them, tracking their movements (**FIG. E17-2**), mapping their locations,

▲ **FIGURE E17-1 Investigating marine biodiversity** A researcher with the Census of Marine Life uses a light box to examine organisms in a shallow bay.

analyzing their DNA, and bringing many back to the lab for further study.

The efforts of the census scientists yielded a massive amount of information. The researchers found living organisms in every habitat that they explored, including ocean depths that lack oxygen. They documented previously unknown animal migration routes and compiled millions of records of organisms' locations and abundance that can serve as a baseline for tracking the effects of human activities on marine organisms. The census also identified "hotspots" in the ocean where life is especially abundant and discovered more than 6,000 new species. Overall, the Census of Marine Life dramatically increased our knowledge of life in the oceans and demonstrated the value of large-scale, coordinated scientific investigation of biodiversity.

THINK CRITICALLY How might conservation scientists use the map shown in Fig. E17-2 to help choose the best location for a proposed marine reserve?

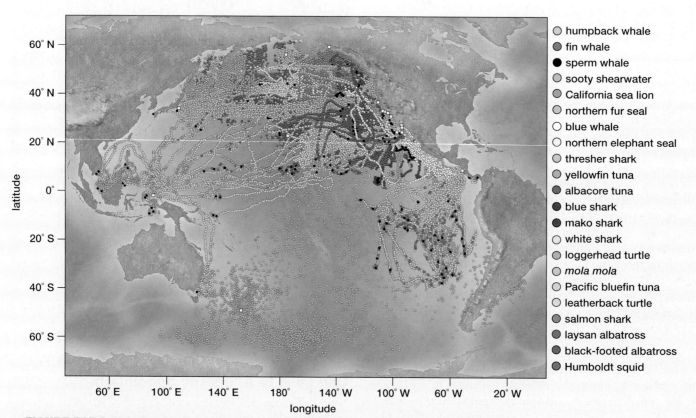

- ○ humpback whale
- ● fin whale
- ● sperm whale
- ○ sooty shearwater
- ○ California sea lion
- ○ northern fur seal
- ○ blue whale
- ○ northern elephant seal
- ○ thresher shark
- ○ yellowfin tuna
- ● albacore tuna
- ● blue shark
- ● mako shark
- ○ white shark
- ○ loggerhead turtle
- ○ *mola mola*
- ○ Pacific bluefin tuna
- ○ leatherback turtle
- ● salmon shark
- ● laysan albatross
- ● black-footed albatross
- ● Humboldt squid

▲ **FIGURE E17-2 Tracking species** To understand the movements and geographic distribution of species, Census researchers attached tracking tags to 22 species of top predators in the Pacific Ocean. The tags beamed signals to satellites, which relayed them to the scientists' computers. The resulting data were plotted on maps, revealing a wealth of information about where and when the animals move.

Block et al. 2011. *Nature* 475:86-90.

population of land-dwelling organisms might colonize an oceanic island. The colonists might be birds, flying insects, fungal spores, or wind-borne seeds blown by a storm. More earthbound organisms might reach the island on a drifting

CASE STUDY CONTINUED
Discovering Diversity

It is not surprising that the forests of New Guinea are home to a variety of distinctive species like the miniature frog *Paedophryne amauensis*. New Guinea is, after all, an island. It is likely that, in the past, populations colonized the island and became genetically isolated from mainland populations, thereby initiating the process of speciation. But what about non-island species, such as the saola, the olinguito (**FIG. 17-9**), and the other unique species of the Annamite and Andes Mountains? How might populations inhabiting mainland forests in Vietnam or Ecuador have become isolated from other populations?

(a) Saola

(b) Olinguito

▲ **FIGURE 17-9 Hidden mammals** Discoveries of previously unknown mammal species are uncommon, but remote forests can conceal species like the recently discovered **(a)** saola and **(b)** olinguito.

raft of vegetation torn from the mainland coast. Whatever the means, we know that such colonization occurs regularly, given the presence of living things on even the most remote islands.

Isolation by colonization is not limited to islands. For example, different coral reefs may be separated by miles of open ocean, so any reef-dwelling sponges, fishes, or algae that were carried by ocean currents to a distant reef would be effectively isolated from their original populations. Any habitat that is bounded by a large expanse of a very different habitat can isolate arriving colonists.

Geological and Climate Changes May Divide Populations

Isolation can also result from landscape changes that divide a population. For example, rising sea levels might transform a coastal hilltop into an island, isolating the residents. New rock from a volcanic eruption can divide a previously continuous sea or lake, splitting populations. A river that changes course can also divide populations, as can a newly formed mountain range. Climate shifts, such as those that happened in past ice ages, can change the distribution of vegetation and strand portions of populations in isolated patches of suitable habitat.

Throughout the history of Earth, many populations have been divided by continental drift. Earth's continents float on molten rock and slowly move about the surface of the planet. On a number of occasions during Earth's long history, continental landmasses have broken into pieces that subsequently moved apart (see Fig. 18-11). Each of these breakups must have split a multitude of populations.

Natural Selection and Genetic Drift May Cause Isolated Populations to Diverge

If two populations become geographically isolated for any reason, there will be no gene flow between them. If the environments of the locations differ, then natural selection may favor different traits in the different locations, and the populations may accumulate genetic differences. Alternatively, genetic differences may arise if one or more of the separated populations is small enough that substantial genetic drift occurs, which may be especially likely in the aftermath of a founder event in which a few individuals become isolated from the main body of the species. In either case, genetic differences between the separated populations may eventually become large enough to make interbreeding impossible. At that point, the two populations will have become separate species. Most evolutionary biologists believe that geographic isolation followed by allopatric speciation has been the most common source of new species, especially among animals.

Genetic Isolation Without Geographic Separation Can Lead to Sympatric Speciation

Genetic isolation—limited gene flow—is required for speciation, but populations can become genetically isolated without geographic separation. Thus, new species can arise by sympatric speciation.

Ecological Isolation Can Reduce Gene Flow

If a geographic area contains two distinct types of habitats (each with distinct food sources, places to raise young, and so on), different members of a single species may begin to specialize in one habitat or the other. If conditions are right, natural selection in the two different habitats may lead to the evolution of different traits in the two groups. Eventually, these differences may become large enough to prevent members of the two groups from interbreeding, and the formerly single species will have split into two species.

Such a split seems to be taking place right before biologists' eyes, so to speak, in the case of the fruit fly *Rhagoletis pomonella*. *Rhagoletis* is a parasite of the American hawthorn tree. This fly lays its eggs in the hawthorn's fruit; when the maggots hatch, they eat the fruit. About 150 years ago, scientists noticed that *Rhagoletis* had begun to infest apple trees, which were introduced into North America from Europe. Today, it appears that *Rhagoletis* is splitting into two species—one that breeds on apples, and one that breeds on hawthorns (**FIG. 17-10**). The two groups have evolved substantial genetic differences, some of which—such as those that affect the time of year at which adult flies emerge and begin to mate—are important for survival on a particular host plant.

The two kinds of flies will become two species only if they maintain reproductive separation. Apple trees and hawthorns typically grow in the same areas, and flies, after all, can fly. So why don't apple flies and hawthorn flies interbreed and cancel out any genetic differences between them? First, female flies usually lay their eggs in the same type of fruit in which they developed. Males also tend to prefer the same type of fruit in which they developed. Therefore, apple-liking males will encounter and mate with apple-liking females. Second, apples mature 2 to 3 weeks later than do hawthorn fruits, and the two types of flies emerge with timing appropriate for their chosen host fruit. Thus, the two varieties of flies have very little chance of meeting. Although some interbreeding between the two types of flies occurs, they seem to be well on their way to speciation. Will they make it? Entomologist Guy Bush suggests, "Check back with me in a few thousand years."

Mutations Can Lead to Genetic Isolation

In some instances, new species can arise nearly instantaneously as a result of mutations that change the number of chromosomes in an organism's cells. The acquisition of multiple copies of each chromosome is known as **polyploidy** and has been a frequent cause of sympatric speciation. In general, polyploid individuals cannot mate successfully with normal diploid individuals. Thus, a polyploid mutant is genetically isolated from its parent species. If, however, it somehow reproduces and leaves offspring, its descendants may form a new, reproductively isolated species.

Polyploid plants are more likely than polyploid animals to be able to reproduce, so speciation by polyploidy is more common in plants than in animals. Unlike animals, many plants can either self-fertilize or reproduce asexually, or both. Therefore, a polyploid plant is much more likely than a

① Part of a fly population that lives only on hawthorn trees moves to an apple tree.

many generations pass

② If flies living on the apple tree rarely encounter flies living on the hawthorn tree, the populations may diverge over many generations.

▲ **FIGURE 17-10 Sympatric isolation and divergence** In sympatric speciation, some event blocks gene flow between two parts of a population that remains in a single geographic area. One way in which this genetic isolation can occur is if a portion of a population begins to use a previously unexploited resource, such as when some members of an insect population shift to a new host plant species (as has occurred in the fruit fly species *Rhagoletis pomonella*). The two now-isolated populations may diverge genetically. If the genetic differences between the two populations become large enough to prevent interbreeding, then the two populations constitute separate species.

THINK CRITICALLY How might future scientists test whether *R. pomonella* has become two species?

polyploid animal to become the founding member of a new, polyploid species.

Under Some Conditions, Many New Species May Arise

In the same way that the history of your family can be represented by a family tree, the history of life can be represented by an *evolutionary tree*. The base of the evolutionary tree of life represents Earth's earliest organisms, and each of the endmost branches represents one of today's living species. Each fork in the branches represents a speciation event, when one species split into two. Hypotheses and discoveries about the evolutionary

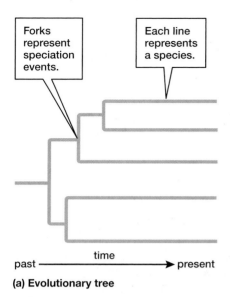

Forks represent speciation events.

Each line represents a species.

In an adaptive radiation, multiple speciation events may occur rapidly enough that biologists cannot be certain of their order.

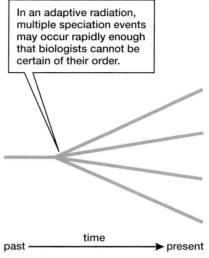

past ⟶ time ⟶ present

(a) Evolutionary tree

past ⟶ time ⟶ present

(b) Evolutionary tree representing adaptive radiation

◀ **FIGURE 17-11 Interpreting evolutionary trees** Evolutionary history is often represented by **(a)** an evolutionary tree, a graph in which the horizontal axis plots time. In **(b)**, an evolutionary tree representing an adaptive radiation, many lines may branch from a single point. This pattern reflects biologists' uncertainty about the order in which the multiple speciation events of the radiation took place. With more research, it may be possible to replace the "starburst" pattern with a more informative tree.

relationships among species are often communicated by depictions of a portion of life's evolutionary tree (FIG. 17-11a).

In some cases, many new species have arisen in a relatively short time (FIG. 17-11b). This process, called **adaptive radiation,** can occur when populations of one species invade a variety of new habitats and evolve in response to the differing environmental pressures in those habitats.

Adaptive radiation has occurred many times and in many groups of organisms, typically when species encounter a wide variety of unoccupied habitats. For example, episodes of adaptive radiation took place when some wayward finches

colonized the Galápagos Islands, when a population of cichlid fish reached isolated Lake Malawi in Africa, and when an ancestral silversword plant species arrived at the Hawaiian Islands (FIG. 17-12). These events gave rise to adaptive radiations of 13 species of Darwin's finches in the Galápagos, more than 300 species of cichlids in Lake Malawi, and 30 species of silversword plants in Hawaii. In these examples, the invading species faced no competitors except other members of their own species, and all the available habitats were rapidly exploited by new species that evolved from the original invaders.

(a) Ahinahina

(b) Waialeale dubautia

(c) Kupaoa

(d) Na'ena'e 'ula

CHECK YOUR LEARNING

Can you ...

- describe the two general steps that are required for a new species to arise?
- explain the difference between allopatric and sympatric speciation, and describe each process?
- explain adaptive radiation and describe the process by which it might arise?
- interpret an evolutionary tree diagram?

◀ **FIGURE 17-12 Adaptive radiation** About 30 species of silversword plants inhabit the Hawaiian Islands. These species are found nowhere else, and all of them descended from a single ancestral population within a few million years. This adaptive radiation has led to a collection of closely related species of diverse form and appearance, with an array of adaptations for exploiting the many different habitats in Hawaii, from warm, moist rain forests to cool, barren volcanic mountaintops.

THINK CRITICALLY Why do you suppose there are so many *endemic* species—that is, species found nowhere else—on islands? Why have the overwhelming majority of recent extinctions occurred on islands?

CASE STUDY \ CONTINUED
Discovering Diversity

One possible explanation for the distinctive collection of species found in the Annamite and Andes Mountains lies in the geological history of these regions. During the ice ages that have occurred repeatedly during the past million years or so, the area covered by tropical forests must have shrunk dramatically. Organisms that depended on the forests for survival would have been restricted to any remaining "islands" of forest, isolated from their fellows in other, distant patches of forest. As we have learned, this kind of isolation can set the stage for allopatric speciation and may have created the conditions that gave rise to the saola, the olinguito, and other unique denizens of tropical forests. Once a new species arises, can we expect it to persist indefinitely?

17.4 WHAT CAUSES EXTINCTION?

The ultimate fate of any species is **extinction,** the death of all of its members. In fact, at least 99.9% of all the species that have ever existed are now extinct. The natural course of evolution, as revealed by fossils, is continual turnover of species as new ones arise and old ones become extinct.

The immediate cause of extinction is probably always environmental change, in either the nonliving or the living parts of the environment. Environmental changes that can lead to extinction include habitat destruction and increased competition among species. In the face of such changes, species with small geographic ranges or highly specialized adaptations are especially susceptible to extinction.

Localized Distribution Makes Species Vulnerable

Species vary widely in their range of distribution and, hence, in their vulnerability to extinction. Some species, such as herring gulls, white-tailed deer, and humans, inhabit entire continents or even the whole Earth; others, such as the Devil's Hole pupfish, which is found in only one spring-fed water hole in the Nevada desert, have extremely limited ranges. Obviously, if a species inhabits only a very small area, any disturbance of that area could easily result in extinction. If Devil's Hole dries up due to a drought or well drilling nearby, its pupfish will immediately vanish. Conversely, wide-ranging species will not succumb to local environmental catastrophes.

Specialization Increases the Risk of Extinction

Another factor that may make a species vulnerable to extinction is extreme specialization. Each species evolves adaptations that help it survive and reproduce in its environment. In some cases, these adaptations include specializations that favor survival in a particular and limited set of environmental conditions. The Karner blue butterfly, for example, feeds only on the blue lupine plant (**FIG. 17-13**). The butterfly is therefore found only where the plant thrives. But the blue lupine

▲ **FIGURE 17-13 Extreme specialization places species at risk** The Karner blue butterfly feeds exclusively on the blue lupine, found in dry forests and clearings in the northeastern United States. Such behavioral specialization renders the butterfly extremely vulnerable to any environmental change that may exterminate its single host plant species.

THINK CRITICALLY If specialization puts a species at risk for extinction, how could this hazardous trait have evolved?

has become quite rare because farms and development have largely replaced its habitat of sandy, open woods and clearings in northeast North America. If the lupine disappears, the Karner blue butterfly will surely become extinct along with it. (see "Earth Watch: Why Preserve Biodiversity?")

Interactions with Other Species May Drive a Species to Extinction

Interactions such as predation and competition serve as agents of natural selection (see Chapter 16). In some cases, these same interactions can lead to extinction rather than to adaptation.

Predation is especially likely to contribute to extinction when species encounter predators to which they had not previously been exposed. For example, when the predatory brown tree snake was accidentally introduced to the Pacific island of Guam in the 1940s, bird populations began to decline. No predatory snakes had been previously present on Guam, so the island's birds had evolved no defenses against them. Within a few decades, almost all of Guam's native birds had disappeared, including two species that were found only on Guam, and are therefore now extinct.

Increased competition can also contribute to extinction. A possible example of extinction through competition began about 3 million years ago, when the isthmus of Panama rose

Earth WATCH

Why Preserve Biodiversity?

Today, most extinctions occur in the tropics, where the vast majority of species live. The main cause of these extinctions is environmental change, especially habitat destruction. Unfortunately, tropical habitats are being rapidly destroyed and disrupted by human activities (**FIG. E17-3**). For example, a recent United Nation report based on analysis of satellite photos and other data estimated that worldwide tropical rainforest cover decreased by about 40,000 square miles per year between 1980 and 2010, though the rate of loss has recently slowed in some countries (**FIG. E17-4**). Most of the lost forest was destroyed by logging or clearing land for agriculture. Similarly, a worldwide survey of coral reefs revealed that about 20% of Earth's reef area has already been destroyed and an additional 20% is severely damaged, again, mostly as the result of human influences such as pollution.

The rapid destruction of habitats in the tropics is causing many species to go extinct, as their homes disappear. Recent ecological research suggests that the current rate of extinction is extremely high, perhaps higher than ever before in the history of life on Earth. Does it matter? Is there any reason for us to try to slow the loss of biodiversity?

One reason to protect Earth's biodiversity is that our ecological self-interest may be at stake. For example, Earth's species form communities, highly complex webs of interdependent life-forms whose interactions sustain one another. These communities play a crucial role in processes that purify the air we breathe and the water we drink, build the rich topsoil in which we grow our crops, provide the bounty of food that we harvest from the oceans, and decompose and detoxify our waste. We depend entirely on these "ecosystem services." When our activities cause species to disappear from communities, we take a big risk. If we remove too many species, or remove some especially crucial species, we may disrupt the finely tuned processes of the community and undermine its ability to sustain us.

▲ **FIGURE E17-3 Biodiversity threatened** Destruction of tropical rain forests by indiscriminate logging threatens Earth's greatest storehouse of biological diversity.

CONSIDER THIS The precarious state of rare species around the world poses profound ethical dilemmas. For example, in many cases, the habitat destruction that endangers some species also helps people by making space for farmland, housing, and workplaces needed by a growing human population. How can we reconcile the conflict between valid human needs and the needs of endangered species? Furthermore, it is becoming clear that, even with the best of intentions, we cannot save all of the species currently threatened with extinction. The resources available to preserve and manage protected habitats are limited, and we must make choices that will cause some species to survive and others to perish. Who should decide which species will live and which will die? On what criteria should such decisions be based?

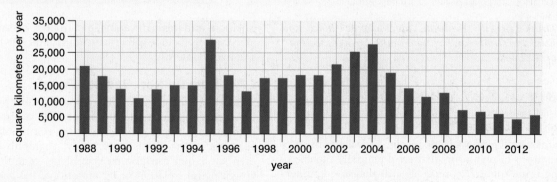

▲ **FIGURE E17-4 Yearly deforestation in the Brazilian Amazon** Earth and its forests are vast, which makes accurate measurement of deforestation difficult. One of the best tools for this task is satellite imagery; photos from space can be used to compare forest cover from year to year. Brazil's National Institute for Space Research has used this tool to track the country's loss of rain forest. The resulting data, displayed in this graph, show that, even as forest destruction has accelerated worldwide, the rate of deforestation in the Brazilian Amazon has slowed. Nonetheless, even the slower rate of destruction is still substantial; an area in Brazil larger than the state of Delaware was cleared in 2013.

above sea level and formed a land bridge between North America and South America. After the previously separated continents were connected, the mammal species that had evolved in isolation on each continent were able to mix. Ultimately, the North American species that moved south diversified and underwent an adaptive radiation that displaced the vast majority of the South American species, many of which went extinct. Although the reasons for the extinctions are not completely understood, it is likely that competition played a role; the species arriving from North America could exploit resources more efficiently than could their South American counterparts.

Habitat Change and Destruction Are the Leading Causes of Extinction

Habitat change, both contemporary and prehistoric, is the single greatest cause of extinctions. Present-day habitat

destruction due to human activities is proceeding at a rapid pace. Many biologists believe that we are presently in the midst of the fastest-paced and most widespread episode of species extinction in the history of life. Loss of tropical forests is especially devastating to species diversity. As many as half the species presently on Earth may be lost during the next 50 years as the tropical forests that contain them are cut for timber or to clear land for cattle and crops. (In Chapter 18, we will discuss extinctions due to prehistoric habitat change.)

CHECK YOUR LEARNING

Can you ...

- describe the main causes of extinction?
- describe some examples of living species that are at risk of extinction?

Discovering Diversity

Ironically, recent discoveries of previously unknown species come at a time when the remote forests that host many of them are in danger of disappearing. Economic development has brought logging and mining to ever more remote regions, and forests in New Guinea (home of *Paedophryne amauensis*), Vietnam (home of the saola), and many other developing nations are being cleared at an unprecedented rate. As a result, newly discovered species are often very rare. For example, despite intensive searching by biologists, there have been only two

verified observations of a live saola in the past two decades: a photo taken by an unattended wildlife camera in 1999 and a second photo in 2013.

THINK CRITICALLY Given that genetic isolation is the first step in speciation, could human activities that reduce many species to small, isolated populations actually increase biodiversity by creating conditions that lead to the formation of new species? Why or why not?

CHAPTER REVIEW

Go to **MasteringBiology** for practice quizzes, activities, eText, videos, current events, and more.

Answers to Think Critically, Evaluate This, Multiple Choice, and Fill-in-the-Blank questions can be found in the Answers section at the back of the book.

Summary of Key Concepts

17.1 What Is a Species?

According to the biological species concept, a species consists of all the populations of organisms that are potentially capable of interbreeding under natural conditions and that are reproductively isolated from other populations.

17.2 How Is Reproductive Isolation Between Species Maintained?

Reproductive isolation between species may be maintained by one or more of several mechanisms, collectively known as premating isolating mechanisms and postmating isolating mechanisms. Premating isolating mechanisms include geographic isolation, ecological isolation, temporal isolation, behavioral

isolation, and mechanical incompatibility. Postmating isolating mechanisms include gametic incompatibility, hybrid inviability, and hybrid infertility.

17.3 How Do New Species Form?

Speciation, the formation of new species, takes place when gene flow between two populations is reduced or eliminated and the populations diverge genetically. Most commonly, speciation is allopatric—gene flow is restricted by geographic isolation. However, speciation can also be sympatric—gene flow is restricted by ecological isolation or by mutations that cause polyploidy. Whether genetic isolation initially arises allopatrically or sympatrically, speciation is completed by subsequent genetic divergence of the separated populations through genetic drift or natural selection.

17.4 What Causes Extinction?

Factors that cause extinctions include competition among species and habitat destruction. Localized distribution and extreme specialization increase a species' vulnerability to extinction.

Key Terms

adaptive radiation *309*
allopatric speciation *305*
behavioral isolation *303*
ecological isolation *302*
extinction *310*
gametic incompatibility *304*
geographic isolation *302*
hybrid *303*
hybrid infertility *304*
hybrid inviability *304*
isolating mechanism *302*
mechanical incompatibility *304*

polyploidy *308*
postmating isolating
mechanism *304*
premating isolating
mechanism *302*
reproductive
isolation *300*
speciation *305*
species *300*
sympatric
speciation *305*
temporal isolation *303*

Thinking Through the Concepts

Multiple Choice

1. The biological species concept is difficult or impossible to apply to
   a. asexually reproducing organisms.
   b. large organisms.
   c. rapidly evolving organisms.
   d. plants.

2. Which of the following does *not* describe a premating isolating mechanism?
   a. the courtship display of a bird of paradise
   b. the sterility of the offspring of a horse and a donkey
   c. the difference between the flowering periods of the Monterey pine and the Bishop pine
   d. the tendency of each species of fig wasp to breed only in the fruits of a particular species of fig

3. All instances of speciation require
   a. genetic isolation and divergence.
   b. genetic drift.
   c. geographic subdivision of a population.
   d. adaptive radiation.

4. Analysis of *Rhagoletis* fly populations in North America provides evidence that if some members of an insect population shift to a new host species,
   a. sympatric speciation may eventually result.
   b. adaptive radiation is likely.
   c. little or no subsequent genetic divergence is likely.
   d. sympatric speciation is probably impossible in insects.

5. In the initial phase of allopatric speciation, gene flow between parts of a population is inhibited by
   a. competition.
   b. hybrid inviability.
   c. geographical isolation.
   d. ecological isolation.

Fill-in-the-Blank

1. A species is a group of _____ that evolves _____. The biological species concept identifies species on the basis of their _____. The biological species concept cannot be applied to species that reproduce _____.

2. Fill in the following with the appropriate isolating mechanism: Occurs when members of two populations have different courtship behaviors: _____; occurs when hybrid offspring fail to survive to reproduce: _____; occurs when members of two populations have different breeding seasons: _____; occurs when sperm from one species fails to fertilize the eggs of another species: _____; occurs when the sexual organs of two species are incompatible: _____.

3. Formation of a new species occurs when two populations of an existing species first become _____ and then _____. The process in which geographic separation of parts of a population leads to the formation of new species is called _____. Isolated populations may diverge through the action of _____ or _____.

4. The process by which many new species arise in a relatively short period of time is known as _____. This process often occurs when a species arrives in a previously unoccupied _____.

5. A species may be at higher risk of extinction if its geographic range includes a(n) _____ area, or if its food or habitat requirements are _____. The leading direct cause of extinction is _____.

Review Questions

1. Define the following terms: *species, speciation, allopatric speciation,* and *sympatric speciation.* Explain how allopatric and sympatric speciation might work, and give a hypothetical example of each.

2. Many of the oak tree species in central and eastern North America hybridize (interbreed). Are they "true species"?

3. Review the material on the possibility of sympatric speciation in *Rhagoletis* flies. What types of genotypic, phenotypic, or behavioral data would convince you that the two forms have become separate species?

4. A drug called colchicine prevents cell division after the chromosomes have doubled at the start of meiosis. Describe how you would use colchicine to produce a new polyploid plant species.

5. What are the two major types of reproductive isolating mechanisms? Give examples of each type, and describe how they work.

Applying the Concepts

1. It is difficult to perform experiments that test hypotheses about how new species form. But what if people lived for a really long time? Design an experiment lasting 100,000 years to test whether allopatric separation leads to speciation. What would your study organism be? Why? What would you measure, how often would you measure it, and what would you expect to find if the allopatric speciation hypothesis is correct?

Ancient DNA Has Stories to Tell

The people of Tibet are adapted to life at high altitude, thanks in part to past interbreeding between humans and a now-extinct species.

THE PEOPLE OF TIBET LIVE AND WORK at altitudes higher than 13,000 feet, where there is much less oxygen in the air than at lower elevations. If you are not a Tibetan and you tried to live at that altitude, you would probably get sick, or at least tire easily and have a hard time catching your breath. How do the Tibetans manage? Their bodies have a number of adaptations to life at high altitude, including a special variant of a gene known as *EPAS1*. The Tibetan's version of *EPAS1*, which is not found in other human populations, improves their bodies' ability to function efficiently in low-oxygen conditions.

How did Tibetans come to have their special version of the gene? Did it originate as a lucky mutation in the early human inhabitants of Tibet? Apparently not. Researchers recently discovered that the gene variant has a surprising history: Its appearance in Tibetans is the result of past interbreeding with members of a hominin species, known as the Denisovans, that has been extinct for tens of thousands of years. (The word *hominin* describes the group that includes humans and the extinct species that are our closest relatives.)

How did researchers discover this fascinating bit of evolutionary history? Until recently, its discovery would have been impossible. But researchers have learned how to extract and sequence DNA from the ancient remains of extinct organisms. And when they examined the Denisovan genome, they found that it contained the same variant of *EPAS1* that today is found only in Tibetans.

In the past, our knowledge of life's history came only from fossils and, more recently, the DNA of living organisms. But the fossil record can be spotty, and modern DNA provides only indirect inferences about past organisms, rather than direct evidence. These tools have nonetheless provided a tremendous amount of information about the past, but access to ancient DNA opens a new and exciting window to the later chapters of life's history, including the history of humans.

Aside from the surprising source of the Tibetans' adaptation, what else have we learned from ancient DNA? What other parts of life's story have been pieced together from DNA clues?

AT A GLANCE

18.1 HOW DID LIFE BEGIN?

Before Darwin, most people thought that all species were simultaneously created by God a few thousand years ago. Further, until the nineteenth century most people thought that new members of existing species sprang up all the time, through **spontaneous generation** from both nonliving matter and other, unrelated forms of life. Microorganisms were thought to arise spontaneously from broth, maggots from meat, and mice from mixtures of sweaty shirts and wheat.

In 1668, the Italian physician Francesco Redi disproved the maggots-from-meat hypothesis simply by keeping flies (whose eggs hatch into maggots) away from uncontaminated meat (see "How Do We Know That? Controlled Experiments Provide Reliable Data" in Chapter 1). In the mid-1800s, Louis Pasteur in France and John Tyndall in England disproved the broth-to-microorganism idea by showing that microorganisms did not appear in sterile broth unless the broth was first exposed to existing microorganisms in the surrounding environment (**FIG. 18-1**). Although Pasteur and Tyndall's work effectively demolished the notion of spontaneous generation, it did not address the question of how life on Earth originated in the first place. Or, as the biochemist Stanley Miller put it, "Pasteur never proved it didn't happen once; he only showed that it doesn't happen all the time."

The First Living Things Arose from Nonliving Ones

Modern scientific ideas about the origin of life began to emerge in the 1920s, when Alexander Oparin in Russia and J. B. S. Haldane in England noted that today's oxygen-rich atmosphere would not have permitted the spontaneous formation of the complex organic molecules necessary for life. Oxygen reacts readily with other molecules, disrupting chemical bonds. Thus, an oxygen-rich environment tends to keep molecules simple.

Oparin and Haldane speculated that the atmosphere of the young Earth must have contained very little oxygen and that, under such atmospheric conditions, complex organic molecules could have arisen through ordinary chemical reactions. Some kinds of molecules could persist in the lifeless environment of early Earth better than others and would therefore become more common over time. This chemical version of the "survival of the fittest" is called *prebiotic* (meaning "before life") evolution. In the scenario envisioned by Oparin and Haldane, prebiotic chemical evolution gave rise to progressively more complex molecules and eventually to living organisms.

Organic Molecules Can Form Spontaneously Under Prebiotic Conditions

Inspired by the ideas of Oparin and Haldane, Stanley Miller and Harold Urey set out in 1953 to simulate prebiotic evolution

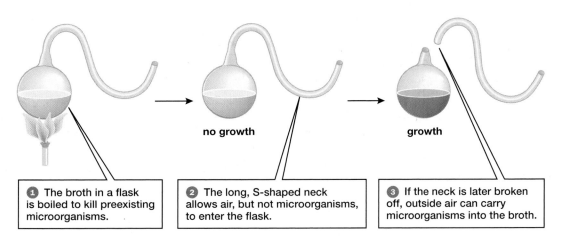

no growth   **growth**

① The broth in a flask is boiled to kill preexisting microorganisms.

② The long, S-shaped neck allows air, but not microorganisms, to enter the flask.

③ If the neck is later broken off, outside air can carry microorganisms into the broth.

▲ **FIGURE 18-1 Spontaneous generation refuted** Louis Pasteur's experiment disproved the spontaneous generation of microorganisms in broth.

in the laboratory. They knew that, on the basis of the chemical composition of the rocks that formed early in Earth's history, geochemists had concluded that the early atmosphere probably contained virtually no oxygen gas, but did contain methane (CH_4), ammonia (NH_3), hydrogen (H_2), and water vapor (H_2O). Miller and Urey simulated the oxygen-free atmosphere of early Earth by mixing these components in a flask. Electrical sparks mimicked the intense energy of early Earth's lightning storms. In this experimental microcosm, the researchers found that simple organic molecules appeared after just a few days (**FIG. 18-2**). The experiment showed that small molecules likely present in the early atmosphere can combine to form larger organic molecules if electrical energy is present. (Recall from Chapter 6 that reactions that synthesize biological molecules from smaller ones are endergonic—they consume energy.) Similar experiments by Miller and others produced amino acids, peptides, nucleotides, adenosine triphosphate (ATP), and other molecules characteristic of living things.

In recent years, new evidence has convinced most geochemists that the actual composition of Earth's early atmosphere probably differed from the mixture of gases used in the pioneering Miller–Urey experiment. However, more recent experiments with simulated atmospheres that more closely resembled the probable atmosphere of early Earth have also yielded organic molecules. In addition, these experiments have shown that electricity is not the only suitable energy source. Other energy sources that were available on early Earth, such as heat or ultraviolet (UV) light, can also drive the formation of organic molecules in experimental simulations of prebiotic conditions. Thus, even though we may never know exactly what the earliest atmosphere was like, we can be confident that organic molecules formed on early Earth.

Additional organic molecules probably arrived from space when meteorites and comets crashed into Earth's surface. Analysis of present-day meteorites recovered from impact craters on Earth has revealed that some meteorites contain relatively high concentrations of amino acids and other simple organic molecules. Laboratory experiments suggest that these molecules could have formed in interstellar space before plummeting to Earth.

Organic Molecules Can Accumulate Under Prebiotic Conditions

Prebiotic synthesis was neither very efficient nor very fast. Nonetheless, large quantities of organic molecules eventually accumulated. Today, most organic molecules have a short life because they are either digested by living organisms or they react with atmospheric oxygen. Early Earth, however, lacked both life and free oxygen, so organic molecules would not have been exposed to these threats.

Still, prebiotic molecules could have been broken down by other chemical reactions or by the sun's high-energy UV radiation. Although UV light can provide energy for the formation of organic molecules, it can also break them apart. However, laboratory researchers have identified conditions under which molecules likely to have been present on prebiotic Earth are stable and can persist and even join together to form more complex molecules. Where on early Earth might such conditions have been found? Possibilities include the waters of mineral-rich hot springs, sheltered spots beneath rock ledges at the sea's edge, pores in the rock columns that form at hydrothermal vents on the ocean floor, and tiny crevices between ice crystals.

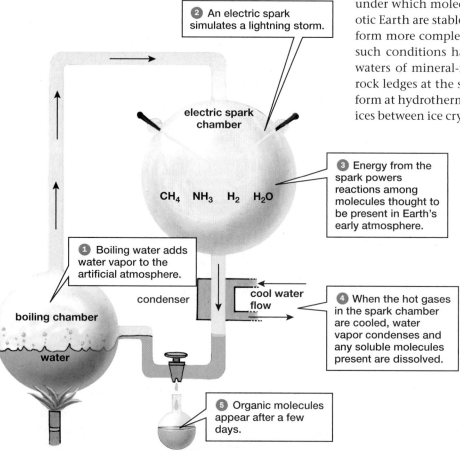

2 An electric spark simulates a lightning storm.

electric spark chamber

3 Energy from the spark powers reactions among molecules thought to be present in Earth's early atmosphere.

CH_4 NH_3 H_2 H_2O

1 Boiling water adds water vapor to the artificial atmosphere.

condenser

cool water flow

boiling chamber

water

4 When the hot gases in the spark chamber are cooled, water vapor condenses and any soluble molecules present are dissolved.

5 Organic molecules appear after a few days.

◀ **FIGURE 18-2 The experimental apparatus of Stanley Miller and Harold Urey** Life's very earliest stages left no fossils, so evolutionary scientists pursued a strategy of re-creating in the laboratory the conditions that may have prevailed on early Earth.

THINK CRITICALLY How would the experiment's result change if oxygen (O_2) were included in the spark chamber?

Clay May Have Catalyzed the Formation of Larger Organic Molecules

In the next stage of prebiotic evolution, wherever its location, simple molecules must have combined to form larger molecules. The chemical reactions that formed the larger molecules required that the reacting molecules be packed closely together. Scientists have proposed several processes by which the required high concentrations might have been achieved on early Earth. One possibility is that small molecules accumulated on the surfaces of clay particles, which often have a small electrical charge that attracts dissolved molecules with the opposite charge. Clustered on such a clay particle, small molecules would have been sufficiently close together to allow chemical reactions between them. Researchers have demonstrated the plausibility of this scenario with experiments in which adding clay to solutions of dissolved small organic molecules catalyzed the formation of larger, more complex molecules, including RNA. Such molecules might have gone on to become the building blocks of the first living organisms.

RNA May Have Been the First Self-Reproducing Molecule

Although all modern organisms use DNA to encode and store genetic information, it is unlikely that DNA was the earliest informational molecule. DNA can reproduce itself only with the help of large, complex protein enzymes, but the instructions for building these enzymes are encoded in DNA itself. For this reason, the origin of DNA's role as life's information storage molecule poses a "chicken and egg" puzzle: DNA requires proteins, but those proteins require DNA. It is thus difficult to construct a plausible scenario for the origin of self-replicating DNA unless we assume that the current DNA-based system of information storage evolved from an earlier system.

RNA Can Act As a Catalyst

A prime candidate for the first self-replicating informational molecule is RNA. In the 1980s, Thomas Cech and Sidney Altman, working with the single-celled organism *Tetrahymena thermophila,* discovered a cellular reaction that was catalyzed not by a protein, but by a small RNA molecule. Because this special RNA molecule performed a function previously thought to be performed only by protein enzymes, Cech and Altman gave their catalytic RNA molecule the name **ribozyme** (FIG. 18-3).

In the years since the discovery of ribozymes, researchers have found dozens of naturally occurring ones that catalyze a variety of different reactions, including cutting other RNA molecules and splicing RNA fragments together. Ribozymes are also found in ribosomes, where they catalyze the attachment of amino acid molecules to growing proteins. In addition, researchers have been able to synthesize various ribozymes in the laboratory, including some that can catalyze the replication of small RNA molecules. The most effective replication ribozyme so far synthesized can copy RNA sequences up to 206 nucleotides long.

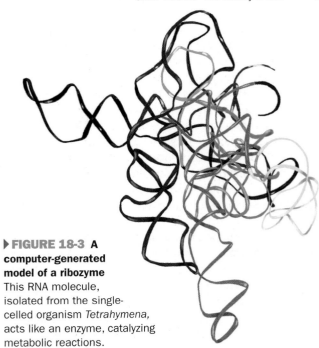

▶ **FIGURE 18-3 A computer-generated model of a ribozyme** This RNA molecule, isolated from the single-celled organism *Tetrahymena,* acts like an enzyme, catalyzing metabolic reactions.

Earth May Once Have Been an RNA World

The discovery that RNA molecules can act as catalysts for diverse reactions, including RNA replication, provides support for the hypothesis that life arose in an "RNA world." According to this view, the current era of DNA-based life was preceded by one in which RNA served as both the information-carrying genetic molecule and the catalyst for its own replication. This RNA world may have emerged after hundreds of millions of years of prebiotic chemical synthesis, during which RNA nucleotides would have been among the molecules synthesized. After reaching a sufficiently high concentration, perhaps on clay particles, the nucleotides probably bonded together to form short RNA chains.

Let's suppose that, purely by chance, one of these RNA chains was a ribozyme that could catalyze the production of copies of itself. This first self-reproducing ribozyme probably wasn't very good at its job and likely produced copies with lots of errors. These mistakes were the first mutations. Like modern mutations, most undoubtedly ruined the catalytic abilities of the "daughter molecules," but a few may have been improvements. Such improvements set the stage for natural selection among RNA molecules, as variant ribozymes with increased speed and accuracy of replication copied themselves more rapidly than did less efficient RNA molecules, and thereby became increasingly common. Molecular evolution in the RNA world proceeded until, by some still-unknown chain of events, RNA gradually receded into its present role as an intermediary between DNA and protein enzymes.

Membrane-like Vesicles May Have Enclosed Ribozymes

Self-replicating molecules on their own do not constitute life; in all living cells such molecules are contained within some kind of enclosing membrane. The precursors of the earliest biological membranes may have been simple structures that formed spontaneously from purely mechanical processes. For example, chemists have shown that if water containing proteins and lipids is agitated to simulate waves beating against ancient shores, the proteins and lipids combine to form hollow structures called *vesicles*. These hollow balls resemble living cells in several respects. They have a well-defined outer boundary that separates their internal contents from the external solution. If the composition of the vesicle is right, a "membrane" forms that is remarkably similar in appearance to a real cell membrane. Under certain conditions, vesicles can absorb material from the external solution, grow, and even divide.

If a vesicle happened to surround the right ribozymes, it would form something resembling a living cell. We could call it a **protocell,** structurally similar to a cell but not alive. In the protocell, ribozymes and any other enclosed molecules would have been protected from degradation by free-roaming reactive molecules. Nucleotides and other small molecules might have diffused across the membrane and been used to synthesize new ribozymes and other complex molecules. After sufficient growth, the vesicle may have divided, with a few copies of the ribozymes becoming incorporated into each daughter vesicle. If this process occurred, the evolution of the first cells would be nearly complete.

Was there a particular moment when a nonliving protocell gave rise to a living organism? Probably not. Like most evolutionary transitions, the change from protocell to living cell was a gradual process, with no sharp boundary between one state and the next.

But Did All This Really Happen?

The scenario just described, although plausible and consistent with many research findings, is by no means certain. One of the most striking aspects of origin-of-life research is a great diversity of assumptions, experiments, and contradictory hypotheses. Researchers disagree about whether life arose in quiet terrestrial pools, at the sea's edge, in hot deep-sea vents, or in polar ice. A few researchers even argue that life arrived on Earth from space. Can we draw any firm conclusions from the research conducted so far? No, but we can make a few reasonable deductions.

First, the experiments of Miller and others show that amino acids, nucleotides, and other organic molecules, along with simple membrane-like structures, are likely to have formed in abundance on early Earth. Second, chemical evolution had long periods of time and huge areas of the Earth available to it. Given sufficient time and a sufficiently large pool of reactant molecules, even extremely rare events can occur many times. And given the vast expanses of time and space available, each small step on the path from primordial soup to living cell had ample opportunity to take place.

No particular account of life's origin can be tested definitively. The origin of life left no record, and researchers exploring this mystery can proceed only by developing a hypothetical scenario and then conducting laboratory investigations to determine if the scenario's steps are chemically and biologically plausible.

CHECK YOUR LEARNING

Can you ...
- describe a likely scenario for the origin of life?
- describe, for each step in the scenario, some evidence that suggests the step is plausible?

18.2 WHAT WERE THE EARLIEST ORGANISMS LIKE?

When Earth first formed about 4.55 billion years ago, it was quite hot (**FIG. 18-4**). A multitude of meteorites

▼ FIGURE 18-4 **Early Earth** In the immediate aftermath of Earth's formation 4.5 billion years ago, the planet was characterized by intense heat, abundant volcanic activity, and repeated meteorite strikes.

smashed into the forming planet, and the kinetic energy of these extraterrestrial rocks was converted into heat on impact. Still more heat was released by the decay of radioactive atoms. Earth melted, and heavier elements such as iron and nickel sank to the center of the planet, where they remain molten even today. Nonetheless, geological evidence suggests that Earth had cooled enough for water to exist in liquid form by 4.3 billion years ago. Once liquid water was available, the prebiotic evolution that ultimately led to the first living organisms could begin.

The oldest fossil organisms found so far are in rocks that are about 3.4 billion years old. (Their age was determined using radiometric dating techniques; see "How Do We Know That? Discovering the Age of a Fossil" on page 321.) Chemical traces in older rocks have led some paleontologists to believe that life is even older, perhaps as old as 3.9 billion years.

The immense span of time in which life's origin and early history took place is known as the Precambrian. This name is among those assigned by geologists and paleontologists, who have devised a hierarchical naming system of eras, periods, and epochs to delineate geological time (**TABLE 18-1**).

The First Organisms Were Anaerobic Prokaryotes

The first cells to arise in Earth's oceans were **prokaryotes,** cells whose genetic material was not contained within a nucleus. These cells probably obtained nutrients and energy by absorbing organic molecules from their environment. There was no oxygen gas in the atmosphere, so the cells must have metabolized the organic molecules anaerobically. (You may recall from Chapter 8 that anaerobic metabolism yields only small amounts of energy.)

Thus, the earliest cells were primitive anaerobic bacteria. As these bacteria multiplied, they must have eventually used up the organic molecules produced by prebiotic chemical reactions. Simpler molecules, such as carbon dioxide and water, would still have been very abundant, as was energy in the form of sunlight. What was lacking, then, was not materials or energy but energetic molecules—molecules in which energy is stored in chemical bonds.

Some Organisms Evolved the Ability to Capture the Sun's Energy

Eventually, some cells evolved the ability to use the energy of sunlight to drive the synthesis of complex, high-energy molecules from simpler molecules; in other words, photosynthesis appeared. Photosynthesis requires a source of hydrogen, and the very earliest photosynthetic bacteria probably used hydrogen sulfide gas dissolved in water for this purpose (as purple photosynthetic bacteria do today). Eventually, however, Earth's supply of hydrogen sulfide

(which is produced mainly by volcanoes) must have run low. The shortage of hydrogen sulfide set the stage for the evolution of photosynthetic bacteria that were able to use the planet's most abundant source of hydrogen—water (H_2O).

Water-based photosynthesis converts water and carbon dioxide to energy-containing molecules of sugar, releasing oxygen as a by-product. The emergence of this new method for capturing energy introduced significant amounts of free oxygen into the atmosphere for the first time. At first, the newly liberated oxygen was quickly consumed by reactions with other molecules in the atmosphere and in Earth's crust. One especially common reactive atom in the crust was iron, and much of the new oxygen combined with iron atoms to form huge deposits of iron oxide (rust). As a result, iron oxide is abundant in rocks formed during this period.

After most of the accessible iron had turned to rust, the concentration of oxygen gas in the atmosphere began to increase. Chemical analysis of rocks suggests that significant amounts of oxygen first appeared in the atmosphere about 2.4 billion years ago, produced by bacteria that were probably very similar to modern photosynthetic bacteria.

Aerobic Metabolism Arose in Response to Dangers Posed by Oxygen

Oxygen is potentially very dangerous to living things, because it can react with organic molecules, breaking them down. Many of today's anaerobic bacteria perish when exposed to oxygen, which is for them a deadly poison. The accumulation of oxygen in the atmosphere of early Earth probably exterminated many organisms and fostered the evolution of cellular mechanisms for detoxifying oxygen. This crisis for evolving life also provided the environmental pressure for the next great advance: the ability to use oxygen in metabolism. This ability not only provides a defense against the chemical action of oxygen, but actually channels oxygen's destructive power through aerobic respiration to generate useful energy for the cell (see Chapter 8). Because the amount of energy available to a cell is vastly increased when oxygen is used to metabolize food molecules, aerobic cells had a significant selective advantage.

Some Organisms Acquired Membrane-Enclosed Organelles

Hordes of bacteria would offer a rich food supply to any organism that could eat them. Paleobiologists speculate that, once this potential prey population appeared, predation would have evolved quickly. These early predators were probably prokaryotes that were larger than typical bacteria. In addition, they must have lost the rigid cell wall that surrounds most bacterial cells, so that their flexible plasma membrane was in contact with the surrounding environment. Thus, the predatory cells were able to envelop smaller bacteria in

TABLE 18-1	The History of Life on Earth			
Era	**Period**	**Epoch**	**Millions of Years Ago**	**Major Events**
Cenozoic	Quaternary	Holocene	0.01–present	Evolution of genus *Homo*
		Pleistocene	2.6–0.01	
	Neogene	Pliocene	5.3–2.6	First grasslands and kelp forests, earliest hominins
		Miocene	23–5.3	
	Paleogene	Oligocene	34–23	Widespread flourishing of birds, mammals, insects, and flowering plants
		Eocene	56–34	
		Paleocene	66–56	
Mesozoic	Cretaceous		145–66	Flowering plants appear and become dominant Mass extinction of marine and terrestrial life, including dinosaurs
	Jurassic		201–145	Dominance of dinosaurs and conifers First birds
	Triassic		252–201	First mammals and dinosaurs Forests of gymnosperms and tree ferns
Paleozoic	Permian		299–252	Massive marine extinctions, including trilobites Flourishing of reptiles and the decline of amphibians
	Carboniferous		359–299	Forests of tree ferns and club mosses Dominance of amphibians and insects First reptiles and conifers
	Devonian		419–359	Fishes and trilobites flourish First amphibians, insects, seeds, and pollen
	Silurian		444–419	Many fishes, trilobites, and mollusks First vascular plants
	Ordovician		485–444	Dominance of arthropods and mollusks in the ocean Invasion of land by plants and arthropods First fungi
	Cambrian		541–485	Marine algae flourish Origin of most marine invertebrate phyla First fishes
Precambrian			630	First animals (soft-bodied marine invertebrates)
			1,200	First multicellular organisms
			1,700	First eukaryotes
			2,400	Accumulation of free oxygen in the atmosphere
			3,500	Origin of photosynthesis (in cyanobacteria)
			3,900–3,500	First living cells (prokaryotes)
			4,000–3,900	Appearance of the first rocks on Earth
			4,550	Origin of the solar system and Earth

HOW DO WE KNOW THAT?

Discovering the Age of a Fossil

Until the twentieth century, geologists could date rock layers and their accompanying fossils only in a relative way: Fossils found in deeper layers of rock were generally older than those found in shallower layers. But a few decades after the discovery of radioactivity in 1896, it became possible to determine absolute dates, with reasonable accuracy. The nuclei of radioactive elements spontaneously break down, or decay, into other elements. For example, carbon-14 (usually written ^{14}C) decays by emitting an electron to become nitrogen-14 (^{14}N). Each radioactive element decays at a rate that is independent of temperature, pressure, or the chemical compound of which the element is a part. The time it takes for half of a radioactive element's nuclei to decay at this characteristic rate is called its *half-life*. The half-life of ^{14}C, for example, is 5,730 years.

How are radioactive elements used in determining the age of rocks? If we know the rate of decay and measure the proportion of decayed nuclei to undecayed nuclei, we can estimate how much time has passed since these radioactive elements became trapped in the rock. This process is called *radiometric dating*.

A particularly straightforward radiometric dating technique measures the decay of potassium-40 (^{40}K), which has a half-life of about 1.25 billion years, into argon-40 (^{40}Ar) gas. Potassium-40 is commonly found in volcanic rocks such as granite and basalt. Suppose that a volcano erupts with a massive lava flow, covering the countryside. All the ^{40}Ar, being a gas, will bubble out of the molten lava, so when the lava first cools and solidifies into rock, it will not contain any ^{40}Ar (**FIG. E18-1**). Over time, however, any ^{40}K present in the hardened lava will decay into ^{40}Ar, with half of the ^{40}K decaying every 1.25 billion years. This ^{40}Ar gas will be trapped in the rock. A geologist could take a sample of the rock and measure the ratio of ^{40}K to ^{40}Ar to determine the rock's age. For example, if the analysis finds equal amounts of the two

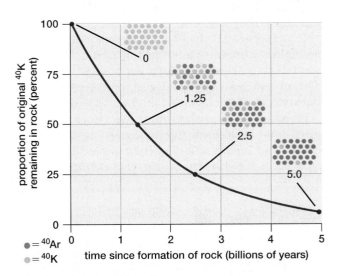

$\bullet = {}^{40}Ar$
$\circ = {}^{40}K$

▲ **FIGURE E18-1** The relationship between time and the decay of radioactive ^{40}K to ^{40}Ar

elements, the geologist will conclude that the lava hardened 1.25 billion years ago. Such age estimates are quite reliable. If a fossil is found beneath a lava flow dated at, say, 500 million years, then we know that the fossil is at least that old.

THINK CRITICALLY Uranium-235, with a half-life of 713 million years, decays to lead-207. If you analyze a rock and find that it contains uranium-235 and lead-207 in a ratio of 1:1, how old is the rock (assuming that decay of uranium-235 is the only source of lead-207)?

---

an infolded pouch of membrane and in this fashion engulf whole bacteria as prey.

These early predators were probably capable of neither photosynthesis nor aerobic metabolism. Although they could ingest smaller bacteria, they metabolized them inefficiently. By about 1.7 billion years ago, however, one predator probably gave rise to the first eukaryotic cell. Eukaryotic cells differ from prokaryotic cells in that they have an elaborate system of internal membranes, many of which enclose organelles such as a nucleus that contains the cell's genetic material. Organisms composed of one or more eukaryotic cells are known as **eukaryotes.**

The Internal Membranes of Eukaryotes May Have Arisen Through Infolding of the Plasma Membrane

The internal membranes of eukaryotic cells may have originally arisen through inward folding of the cell

membrane of a single-celled predator. If, as in most of today's bacteria, the DNA of the eukaryotes' ancestor was attached to the inside of its cell membrane, an infolding of the membrane near the point of DNA attachment may have pinched off and become the precursor of the cell nucleus.

In addition to the nucleus, other key eukaryotic structures include the organelles used for energy metabolism: mitochondria (in all eukaryotes) and chloroplasts (in plants and algae). How did these organelles evolve?

Mitochondria and Chloroplasts May Have Arisen from Engulfed Bacteria

The **endosymbiont hypothesis** proposes that early eukaryotic cells acquired the precursors of mitochondria and chloroplasts by engulfing certain types of bacteria

(**FIG. 18-5**). These cells and the bacteria trapped inside them (*endo* means "within") gradually entered into a *symbiotic* relationship, a close association between different types of organisms over an extended time. How might this have happened?

Let's suppose that an anaerobic predatory cell captured an aerobic bacterium for food, as it often did, but for some reason failed to digest this particular prey ❶. The aerobic bacterium remained alive and well, protected from other predatory cells. In fact, it was better off than ever, because the cytoplasm of its predator-host was chock-full of half-digested food molecules, the remnants of anaerobic metabolism. The aerobe absorbed these molecules and used oxygen to metabolize them, thereby gaining enormous

amounts of energy. So abundant were the aerobes' food resources, and so bountiful their energy production, that the aerobes must have leaked energy, probably as ATP or similar molecules, back into their host's cytoplasm. The anaerobic predatory cell with its symbiotic bacteria could now metabolize food aerobically, gaining a great advantage over other anaerobic cells and leaving a greater number of offspring. Eventually, the endosymbiotic bacterium lost its ability to live independently of its host, and the mitochondrion was born ❷.

One of these successful new cellular partnerships managed a second feat: It captured a photosynthetic bacterium and again failed to digest its prey ❸. The bacterium flourished in its new host and gradually evolved into the first chloroplast ❹. Other eukaryotic organelles may have also originated through endosymbiosis. Cilia, flagella, centrioles, and microtubules may all have evolved from a symbiosis between a spirilla-like bacterium (a form of bacterium with an elongated corkscrew shape) and an early eukaryotic cell.

Evidence for the Endosymbiont Hypothesis Is Strong

Evidence that supports the endosymbiont hypothesis includes the many distinctive biochemical features shared by eukaryotic organelles and living bacteria. In addition, mitochondria and chloroplasts each contain their own minute supply of DNA, which many researchers interpret as remnants of the DNA originally contained within the engulfed bacteria.

Another kind of support comes from *living intermediates,* organisms alive today that are similar to hypothetical ancestors and thus help show that a proposed evolutionary pathway is plausible. For example, the amoeba *Pelomyxa palustris* lacks mitochondria but hosts a permanent population of aerobic bacteria that carry out much the same role. A variety of other protists also harbor symbiotic bacteria inside their cells (**FIG. 18-6**), as do many insect species. These examples of modern cells that host bacterial endosymbionts suggest that similar symbiotic associations could have occurred almost 2 billion years ago and led to the first eukaryotic cells.

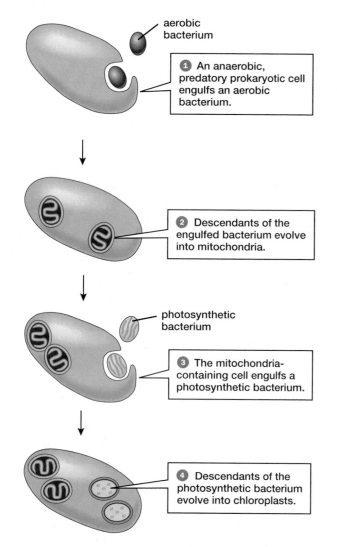

aerobic bacterium

❶ An anaerobic, predatory prokaryotic cell engulfs an aerobic bacterium.

❷ Descendants of the engulfed bacterium evolve into mitochondria.

photosynthetic bacterium

❸ The mitochondria-containing cell engulfs a photosynthetic bacterium.

❹ Descendants of the photosynthetic bacterium evolve into chloroplasts.

▲ **FIGURE 18-5 The probable origin of mitochondria and chloroplasts in eukaryotic cells**

THINK CRITICALLY Scientists have identified a free-living bacterium believed to be descended from the endosymbiont that gave rise to mitochondria. Would you expect the DNA sequence of this modern bacterium to be most similar to the sequence of DNA from a plant chloroplast, an animal cell nucleus, or a plant mitochondrion?

◀ **FIGURE 18-6 Symbiosis within a modern cell** The ancestors of the chloroplasts in today's plant cells may have been similar to the green, photosynthetic bacteria living symbiotically within the cytoplasm of the protist *Paulinella chromatophora,* pictured here.

Can you ...

- describe scenarios for the major evolutionary events and innovations that occurred during the period in which all organisms were single celled, including the origins of photosynthesis, atmospheric oxygen, aerobic respiration, and eukaryotic organelles?
- state the order in which these events occurred, and list evidence that supports these scenarios?

18.3 WHAT WERE THE EARLIEST MULTICELLULAR ORGANISMS LIKE?

Once predation had evolved, increased size became an advantage. In the marine environments to which life was restricted, a larger cell could easily engulf a smaller cell and would also be difficult for other predatory cells to ingest. But enormous single cells have problems. The larger a cell becomes, the less surface membrane is available per unit volume of cytoplasm (see Fig. 5-13). Thus, as a cell grows larger, the process of diffusion across its plasma membrane becomes progressively less able to accommodate the oxygen and nutrients that must move into the cell and the waste products that must move out. One way for an organism to overcome this limit on cell size is to be multicellular; that is, it can consist of many small cells packaged into a larger, unified body.

Some Algae Became Multicellular

The oldest fossils of multicellular organisms are about 1.2 billion years old. They consist of impressions of multicellular algae that arose from single-celled eukaryotic organisms containing chloroplasts. Multicellularity would have provided at least two advantages for these organisms. First, large, many-celled algae would have been difficult for single-celled predators to engulf. Second, specialization of cells would have provided the potential for staying in one place in the brightly lit waters of the shoreline, as rootlike structures burrowed in sand or clutched onto rocks, while leaflike structures floated above in the sunlight. The green, brown, and red algae lining our shores today are the descendants of these early multicellular algae.

Animal Diversity Arose in the Precambrian

The oldest known unequivocal traces of animals include fossil embryos found in Precambrian deposits that are 630 million years old. Fossils of apparently adult animal bodies first appear in rocks laid down between 610 million and 541 million years ago. Some of these ancient invertebrate animals (animals lacking a backbone) are quite different in appearance from any animals that appear in later fossil layers and may represent types of animals that left no descendants.

Other fossils in these rock layers, however, appear to be ancestors of today's animals. Ancestral sponges and jellyfish appear in the oldest layers, followed later by ancestors of worms, mollusks, and arthropods.

The full range of modern invertebrate animals, however, does not appear in the fossil record until the Cambrian period, marking the beginning of the Paleozoic era, about 541 million years ago. (The phrase "fossil record" is a shorthand reference to the entire collection of all fossil evidence that has been found to date.) These Cambrian fossils reveal an adaptive radiation (see Chapter 17) that had already yielded a diverse array of complex body plans. Almost all of the major groups of animals on Earth today were already present in the early Cambrian. The seemingly sudden appearance of so many different kinds of animals suggests that these groups actually arose earlier, but their early evolutionary history is not preserved in the fossil record.

Predation Favored the Evolution of Improved Mobility and Senses

The early diversification of animals was probably driven in part by the emergence of predatory lifestyles. For example, coevolution of predator and prey favored animals that were more mobile than their predecessors. Mobile predators gained an advantage from an ability to travel over wide areas in search of suitable prey; mobile prey benefited if they were able to make a speedy escape. The evolution of efficient movement was often associated with the evolution of greater sensory capabilities and more complex nervous systems. Senses for detecting touch, chemicals, and light became highly developed, along with nervous systems capable of handling the sensory information and directing appropriate behaviors.

By the Silurian period (444 million to 419 million years ago), life in Earth's seas included an array of anatomically complex animals, including armored trilobites, shelled ammonites, and the chambered nautilus (FIG. 18-7). The nautilus survives today in almost unchanged form in deep Pacific waters.

Skeletons Improved Mobility and Protection

In many Paleozoic animal species, mobility was enhanced in part by the origin of hard external body coverings known as **exoskeletons.** Exoskeletons improved mobility by providing hard surfaces to which muscles attached. These attachments made it possible for animals to use their muscles to move appendages used to swim through the water or crawl over the seafloor. Exoskeletons also provided support for animals' bodies and protection from predators.

About 530 million years ago, one group of animals—the fishes—developed a new form of body support and muscle attachment: an internal skeleton. These early fishes were inconspicuous members of the ocean community, but by 400 million years ago, fishes were a diverse and prominent

(a) Silurian scene

(b) Trilobite

(c) Ammonite

(d) *Nautilus*

▲ **FIGURE 18-7 Diversity of ocean life during the Silurian period (a)** An artist's rendition of life characteristic of the oceans during the Silurian period, 444 million to 419 million years ago. Among the most common fossils from that time are **(b)** the trilobites and their predators, the nautiloids, and **(c)** the ammonites. **(d)** This living *Nautilus* is very similar in structure to the Silurian nautiloids, showing that a successful body plan may exist virtually unchanged for hundreds of millions of years.

group. By and large, fishes proved to be faster than the invertebrates, with more acute senses and larger brains. Eventually, they became the dominant predators of the open seas.

CHECK YOUR LEARNING

Can you ...

- describe fossil evidence of the earliest multicellular organisms and the earliest animals?
- describe the advantages that fostered the origin of multicellularity?
- describe the adaptations associated with the later increase in animal diversity?

18.4 HOW DID LIFE INVADE THE LAND?

A compelling subplot in the long tale of life's history is the story of life's invasion of land. In moving to solid ground after more than 3 billion years of a strictly watery existence, organisms had many obstacles to overcome. Life in the sea provides buoyant support against gravity, but on land an organism must bear its weight against the crushing force of gravity. The sea provides ready access to life-sustaining water, but adequate water may not be easily available to a terrestrial organism. Sea-dwelling plants and animals can reproduce by means of mobile gametes that swim or drift to each other through the water. The sperm and eggs of land-dwellers, however, must be protected from drying out.

Despite the obstacles to life on land, the vast empty spaces of the Paleozoic landmass represented a tremendous evolutionary opportunity. The potential rewards of terrestrial life were especially great for plants. Water strongly absorbs light, so even in the clearest water, photosynthesis is at best possible only within a few hundred meters of the surface, and usually only at much shallower depths. Out of the water, the dazzling brightness of the sun permits rapid photosynthesis. Furthermore, terrestrial soils are rich storehouses of nutrients, whereas seawater tends to be low in nutrients, particularly nitrogen and phosphorus. Finally, the Paleozoic sea swarmed with plant-eating animals, but the land was devoid of animal life. Thus, the plants that first colonized the land would have had ample sunlight, abundant nutrients, and no predators.

Some Plants Became Adapted to Life on Dry Land

In moist soils at the water's edge, a few small green algae began to grow, taking advantage of the sunlight and nutrients. These algae didn't have large bodies to support them against the force of gravity, and, living right in the film of water on the soil, they could easily obtain water. About 475 million years ago, some of these algae gave rise to the first multicellular land plants. Initially simple and low-growing, land plants eventually evolved solutions to two of the main difficulties of plant life on land: obtaining and conserving water and staying upright despite gravity and winds. New adaptations that helped obtain and conserve water included water-resistant coatings on aboveground parts that reduced water loss by evaporation; rootlike structures that delved into the soil to absorb water and minerals; and specialized tissues (called vascular tissues) that contained tubes to conduct water from roots to leaves. Extra-thick walls surrounding certain cells enabled stems to stand erect, and the rootlike structures helped anchor erect plant bodies firmly to the soil.

Early Land Plants Retained Swimming Sperm and Required Water to Reproduce

Reproduction out of water presented challenges. Plants produce sperm and eggs, as animals do, and these gametes must meet to produce the next generation. The first land plants had swimming sperm, presumably much like those of today's mosses and ferns. Consequently, the earliest plants were restricted to swamps and marshes or to areas with abundant rainfall, where the ground would occasionally be covered with water. Here, the sperm and eggs could be released into the water, and sperm could swim to reach an egg. Later plants with swimming sperm prospered during periods in which the climate was warm and moist. For example, the Carbon-iferous period (359 million to 299 million years ago) was characterized by vast forests of giant tree ferns, club mosses, and horsetails (**FIG. 18-8**).

Seed Plants Encased Sperm in Pollen Grains

Meanwhile, some plants inhabiting drier regions had evolved a means of reproduction that no longer depended on water. The eggs of these plants were retained on the parent plant, and the sperm were encased in drought-resistant pollen grains that were carried by the wind from plant to plant. When the pollen grains landed near an egg, they released sperm cells directly into living tissue, eliminating the need for a surface film of water. The fertilized egg remained on the parent plant, where it developed inside a seed, which provided protection and nutrients for the embryo.

The earliest seed-bearing plants appeared in the late Devonian period (375 million years ago) and produced their seeds along branches, without any specialized structures to hold them. By the middle of the Carboniferous period, however, a new kind of seed-bearing plant had arisen. These plants, called **conifers,** protected their developing seeds inside cones. Conifers, which are wind-pollinated and do not depend on water for reproduction, flourished and spread during the Permian period (299 to 252 million years ago), when mountains rose, swamps drained, and the climate became much drier. The conifers' good fortune, however, was not shared by the tree ferns and giant club mosses, which, with their swimming sperm, largely went extinct.

Flowering Plants Enticed Animals to Carry Pollen

About 140 million years ago, during the Cretaceous period, the flowering plants appeared, having evolved from a group of conifer-like plants. Many flowering plants are pollinated by animals, especially insects, and this mode of pollination seems to have conferred an evolutionary advantage. Flower pollination by animals can be far more efficient than pollination by wind. Wind-pollinated plants must produce an enormous amount of pollen because the vast majority of pollen grains fail to reach their target. Today, flowering plants dominate the land, except in cold northern regions, where conifers still prevail. In some cases, flowering plants have re-evolved wind pollination, most likely in response to a past or ongoing reduction in the availability of animal pollinators.

Some Animals Became Adapted to Life on Dry Land

After land plants evolved, providing potential food sources for other organisms, animals emerged from the sea. The

◀ **FIGURE 18-8 The swamp forest of the Carboniferous period** Many of the treelike plants in this artist's reconstruction are extinct relatives of today's club mosses and horsetails.

THINK CRITICALLY Why are today's ferns, horsetails, and club mosses so small in comparison to their giant ancestors?

earliest evidence of land animals comes from fossils that are about 430 million years old. The first animals to move onto land were **arthropods** (the group that today includes insects, spiders, scorpions, centipedes, and crabs). Why arthropods? The answer seems to be that they already possessed certain structures that, purely by chance, were suited to life on land. Foremost among these structures was an exoskeleton, such as the shell of a lobster or crab. Exoskeletons are both waterproof and strong enough to support a small animal against the force of gravity.

For millions of years, arthropods had the land and its plants to themselves, and for tens of millions of years more, they were the dominant land animals. Dragonflies with a wingspan of 28 inches (70 centimeters) flew among the Carboniferous tree ferns, while millipedes 6.5 feet (2 meters) long munched their way across the swampy forest floor. Eventually, however, the arthropods' splendid isolation came to an end.

Amphibians Evolved from Lobefin Fishes

About 400 million years ago, a group of Devonian fishes called the lobefins appeared, probably in fresh water. **Lobefins** had two important features that would later enable their descendants to colonize land: (1) stout, fleshy fins with which they crawled about on the bottoms of shallow, quiet waters, and (2) an outpouching of the digestive tract that could be filled with air, like a primitive lung. One group of lobefins inhabited very shallow ponds and streams, which shrank during droughts and often became oxygen poor. By taking air into their lungs, these lobefins could still obtain oxygen. Some began to use their fins to crawl from pond to pond in search of prey or water, as some fish do today (**FIG. 18-9**).

The benefits of feeding on land and moving from pool to pool favored the evolution of a group of animals

▲ **FIGURE 18-9 A fish that walks on land** Some modern fishes, such as this mudskipper, walk on land. As did the ancient lobefin fishes that gave rise to amphibians, mudskippers use their strong pectoral fins to move across dry areas in their swampy habitats.

THINK CRITICALLY Does the mudskipper's ability to walk on land constitute evidence that lobefin fishes were the ancestors of amphibians?

that could stay out of water for longer periods and that could move about more effectively on land. With improvements in lungs and legs, **amphibians** evolved from lobefins, first appearing in the fossil record about 370 million years ago. To an amphibian, the Carboniferous swamp forests were a kind of paradise: no predators to speak of, abundant prey, and a warm, moist climate. As had the insects and millipedes, some amphibians evolved gigantic size, including salamanders more than 10 feet (3 meters) long.

Despite their success, early amphibians were not fully adapted to life on land. Their lungs were simple sacs without very much surface area, so they had to obtain some of their oxygen through their skin. Therefore, their skin had to be kept moist, a requirement that restricted them to swampy habitats. Further, amphibian sperm and eggs could not survive in dry surroundings and had to be deposited in water. So, although amphibians could move about on land, they could not stray too far from the water's edge. Along with the tree ferns and club mosses, amphibians declined when the climate turned dry at the beginning of the Permian period about 299 million years ago.

Reptiles Evolved from Amphibians

As the conifers were evolving on the fringes of the swamp forests, a group of amphibians was also evolving adaptations to drier conditions. These amphibians ultimately gave rise to the **reptiles,** which had three major adaptations to life on land. First, reptiles evolved shelled, waterproof eggs that enclosed a supply of food and water for the developing embryo. Thus, reptiles could lay eggs on land and avoid the dangerous swamps full of fish and amphibian predators. Second, ancestral reptiles evolved scaly, water-resistant skin that reduced the loss of body water to the dry air. Finally, reptiles evolved improved lungs that were able to provide the entire oxygen supply of an active animal. As the climate dried during the Permian period, reptiles became the dominant land vertebrates, relegating amphibians to the swampy backwaters where most remain today.

A few tens of millions of years later, the climate became wetter again. This period saw the evolution of some very large reptiles, in particular, the dinosaurs (**FIG. 18-10**). The variety of dinosaur forms was enormous—large and small, fleet-footed and ponderous, predators and plant-eaters. Dinosaurs were among the most successful animals ever, if we consider persistence as a measure of success. They flourished for more than 100 million years, until about 66 million years ago, when the last dinosaurs went extinct. No one is certain why they died out, but the aftereffects of a gigantic meteorite's impact with Earth seem to have been the final blow (as discussed in Section 18.5).

Even during the age of dinosaurs, many reptiles remained quite small. One major difficulty faced by small reptiles is maintaining a high body temperature. A warm

◀ **FIGURE 18-10 A reconstruction of a Cretaceous forest** By the Cretaceous period, flowering plants dominated terrestrial vegetation. Dinosaurs, such as the predatory pack of 6-foot-long *Velociraptors* shown here, were the preeminent land animals. Although small by dinosaur standards, *Velociraptors* were formidable predators with great running speed, sharp teeth, and deadly, sickle-like claws on their hind feet.

body is advantageous for an active animal, because warmer nerves and muscles work more efficiently. But a warm body loses heat to the environment unless the air is also warm. Heat loss is an especially big problem for small animals, which have a larger surface area per unit of volume than do larger animals. Many species of small reptiles have slow metabolisms and cope with the heat loss problem by confining activity to times when the air is sufficiently warm. One group of reptiles, however, followed a different evolutionary pathway. Members of this group, the birds, evolved insulation, in the form of feathers. (Birds were formerly placed in their own taxonomic group, separate from reptiles. For more information on why birds are now understood to be a type of reptile, see "In Greater Depth: Phylogenetic Trees" in Chapter 19.)

In ancestral birds, feathers, which are modified scales, helped retain body heat. Consequently, these animals could be active in cool habitats and during the night, when their scaly relatives became sluggish. Later, some ancestral birds evolved longer, stronger feathers on their forelimbs, perhaps under selection for better ability to glide from trees or to jump after insect prey. Ultimately, feathers evolved into structures capable of supporting powered flight. Fully developed, flight-capable feathers are present in 150-million-year-old fossils, so the earlier insulating structures that eventually developed into flight feathers must have been present well before that time.

Reptiles Gave Rise to Mammals

Unlike the egg-laying reptiles, **mammals** evolved live birth and the ability to feed their young with secretions of the mammary (milk-producing) glands. Ancestral mammals

CASE STUDY   CONTINUED

Ancient DNA Has Stories to Tell

Can ancient DNA reveal the secrets of dinosaur evolutionary history? Sadly, no. DNA decays far too quickly to be present in fossils as old as dinosaur fossils are. But all is not lost; the paleontologist Mary Schweitzer and her colleagues have discovered, in some exceptionally well preserved dinosaur fossils, what appear to be preserved soft tissues, such as blood, bone marrow, and skin. These discoveries were initially met with great skepticism that soft tissues could be preserved for so long, but as additional evidence has accumulated, an increasing number of paleontologists have accepted the discoveries. Researchers have extracted proteins such as hemoglobin, keratin, and collagen from the fossil tissue, and the amino acid sequences of these proteins may reveal previously unknown information about the evolution of dinosaurs. Nonetheless, evolution's historians must, for the most part, rely on more traditional methods. What have such methods revealed about the dinosaurs' successors as Earth's dominant large animals, the mammals?

also developed hair, which provided insulation. Because soft tissues like the uterus and mammary glands do not generally fossilize, we may never know when these structures first appeared or what their intermediate forms looked like. Hair, however, is occasionally preserved in fossils. The oldest known hair was fossilized about 160 million years ago, so mammals have presumably had hair for at least that long.

The earliest mammals arose more than 200 million years ago. Early mammals thus coexisted with the dinosaurs. They were mostly small creatures. The largest known mammal from the dinosaur era was about the size of a modern raccoon, but most early mammal species were far smaller. When the dinosaurs went extinct, however, mammals colonized newly empty habitats, prospered, and diversified into the array of forms that we see today.

CHECK YOUR LEARNING

Can you ...

- describe the transitions and innovations associated with the origin and evolution of the major groups of land plants and vertebrates?
- describe the advantages gained by the first plants and animals to colonize land?

CASE STUDY \ **CONTINUED**

Ancient DNA Has Stories to Tell

Although it may never be possible to recover DNA from dinosaurs, ancient DNA of more recent vintage can help us understand more about the physiology and behavior of extinct animals. For example, researchers have extracted DNA from 43,000-year-old wooly mammoths that were preserved in the permafrost of Siberia. (Cold climates are especially favorable for preserving ancient DNA.) The investigators were able to sequence some of the DNA, including the genes that produced hemoglobin (a protein that transports oxygen in the blood). The researchers then inserted the mammoth hemoglobin genes into bacteria. The bacteria produced hemoglobin molecules just like those that circulated in the mammoth's blood when it was alive.

Unlike hemoglobin from modern elephants, mammoth hemoglobin releases oxygen readily not only at core body temperature, but also at temperatures near freezing. Thus, though a modern elephant must keep its legs warm in order to provide oxygen to its leg muscles, a mammoth's legs could get very cold and still function, an adaptation that helped the animals survive in ice-age Siberia.

In the end, mammoths became extinct, as do all species, eventually. What was responsible for history's largest waves of extinction?

18.5 WHAT ROLE HAS EXTINCTION PLAYED IN THE HISTORY OF LIFE?

If there is a lesson in the great tale of life's history, it is that nothing lasts forever. The story of life can be read as a long series of evolutionary dynasties, with each new dominant group rising, ruling the land or the seas for a time, and, inevitably, falling into decline and extinction. Dinosaurs are the most famous of these fallen dynasties, but the list of extinct groups known only from fossils is impressively long. Despite the inevitability of extinction, however, the overall trend has been for species to arise at a faster rate than they disappear, so the number of species on Earth has tended to increase over time.

HAVE YOU EVER
WONDERED...

Scientists have cloned a number of animal species, including mice, dogs, cats, horses, and cows. Could the technology of cloning be used to bring back extinct species? In principle, yes, provided that perfectly preserved DNA of the extinct species is available. Such DNA could be transferred to an egg from a closely related, living species, and the egg implanted in a surrogate mother of that species.

If Extinct Species Can Be Revived by Cloning?

For example, researchers have suggested that it might be possible to clone a woolly mammoth, using an elephant surrogate mother and DNA extracted from 20,000-year-old mammoths found frozen beneath the Siberian tundra. Most scientists, however, believe that any DNA recovered from a fossil mammoth would be far too degraded for use in cloning, and synthesizing an entire mammoth genome (its sequence is now almost fully known) is beyond the capabilities of current technology. The odds of success might be greater for another proposed project, which would use DNA from a preserved museum specimen to revive the Tasmanian tiger, an Australian mammal that has been extinct for only 70 years. If cloning recently extinct species proves to be possible, do you think it would be a good idea?

Evolutionary History Has Been Marked by Periodic Mass Extinctions

Over much of life's history, the origin and disappearance of species have proceeded in a steady, relentless manner. This slow and steady turnover of species, however, has been interrupted by episodes of **mass extinction.** These mass extinctions are characterized by the relatively sudden disappearance of a wide variety of species over a large part of Earth. The most dramatic episode of all, which occurred 252 million years ago, at the end of the Permian period, wiped out more than 90% of the world's species in only 60,000 years. Life came perilously close to disappearing altogether.

Climate Change Contributed to Mass Extinctions

Mass extinctions have had a profound impact on the course of life's history. What could have caused such dramatic changes in the fortunes of so many species? Many evolutionary biologists believe that changes in climate must have played an important role. When the climate changes, as it has done many times over the course of Earth's history, organisms that are adapted for survival in one climate may be unable to survive in a drastically different climate. In particular, at times when warm climates gave way to drier, colder climates with more variable temperatures, species may have gone extinct after failing to adapt to the harsh new conditions.

One cause of climate change is the shifting positions of continents. These movements are sometimes called *continental drift.* Continental drift is caused by **plate tectonics,** in which the Earth's surface, including the continents and the seafloor, is divided into plates that rest atop a viscous but fluid layer

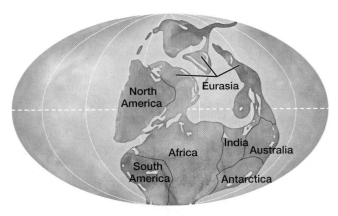

(a) 340 million years ago

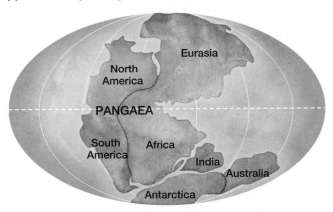

(b) 225 million years ago

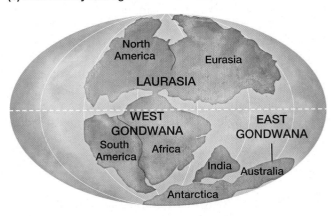

(c) 135 million years ago

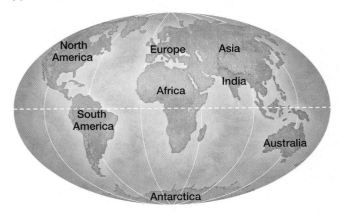

(d) Present

and move slowly about. As the plates wander, their positions may change in latitude (FIG. 18-11). For example, 340 million years ago, much of North America was located at or near the equator, an area characterized by consistently warm and wet tropical weather. But as time passed, plate tectonics carried the continent up into temperate and arctic regions. As a result, the once tropical climate was replaced by a regime of seasonal changes, cooler temperatures, and less rainfall. Plate tectonics continues today; the Atlantic Ocean, for example, widens by a few centimeters each year.

Catastrophic Events May Have Caused the Biggest Mass Extinctions

Geological data indicate that most mass extinction events coincided with periods of climatic change. But more sudden events may also have played a role. For example, catastrophic geological events, such as massive volcanic eruptions, could rapidly kill many organisms. Geologists have found evidence of an immense eruption that began just prior to the end of the Permian, and many suspect that the volcanic activity may have been a cause of the subsequent mass extinction.

The search for the causes of mass extinctions took a fascinating turn in the early 1980s, when Luis and Walter Alvarez proposed that the extinction event of 66 million years ago, which wiped out the dinosaurs and many other species, was caused by the impact of a huge meteorite. The Alvarezes' idea was met with great skepticism when it was first introduced, but geological research since that time has generated a great deal of evidence that a massive impact did indeed occur 66 million years ago. In fact, researchers have identified the Chicxulub crater, a 100-mile-wide crater buried beneath the Yucatan Peninsula of Mexico, as the impact site of a giant meteorite—6 miles (10 kilometers) in diameter—that collided with Earth around the time that dinosaurs disappeared.

Could this immense meteorite strike have caused the mass extinction that coincided with it? No one knows for sure, but scientists suggest that such a massive impact would have thrown so much debris into the atmosphere that the entire planet would have been plunged into darkness for a period of years. With little light reaching the planet, temperatures would have dropped precipitously and the photosynthetic capture of energy (on which all life ultimately depends) would have declined drastically. The worldwide "impact winter" would have spelled doom for the dinosaurs and a host of other species.

◀ FIGURE 18-11 **Continental drift from plate tectonics** The continents are passengers on plates moving on Earth's surface as a result of plate tectonics. **(a)** About 340 million years ago, much of what is now North America was positioned at the equator. **(b)** All the plates eventually fused together into one gigantic landmass, which geologists call Pangaea. **(c)** Gradually Pangaea broke up into Laurasia and Gondwanaland, which itself eventually broke up into West and East Gondwana. **(d)** Further plate motion eventually resulted in the current positions of the modern-day continents.

18.6 HOW DID HUMANS EVOLVE?

Scientists are intensely interested in the origin and evolution of humans. The outline of human evolution that we present in this section represents an interpretation that is widely shared among paleontologists. However, fossil evidence of human evolution is comparatively scarce and therefore open to a variety of interpretations. Thus, some paleontologists would disagree with some aspects of the scenario we present.

Humans Inherited Some Early Primate Adaptations for Life in Trees

Humans belong to the mammal group known as **primates,** which also includes lemurs, monkeys, and apes. The oldest primate fossils are 55 million years old, but because primate fossils are relatively rare compared with those of many other animals, the first primates probably arose considerably earlier but left no fossil record. Early primates were adapted for life in the trees, and many modern primates retain the tree-dwelling lifestyle of their ancestors (**FIG. 18-12**). The common heritage of humans and other primates is reflected in a set of physical characteristics that was present in the earliest primates and that persists in many modern primates, including humans.

(a) Tarsier

(b) Lemur

(c) Macaque

Binocular Vision Provided Early Primates with Accurate Depth Perception

One of the earliest primate adaptations seems to have been large, forward-facing eyes (see Fig. 18-12). Jumping from branch to branch is risky business unless an animal can accurately judge where the next branch is located. Accurate depth perception was made possible by binocular vision, provided by forward-facing eyes with overlapping fields of view. Another key adaptation was color vision. We cannot, of course, tell if a fossil animal had color vision, but since modern primates have excellent color vision, it seems reasonable to assume that earlier primates did as well. Many primates feed on fruit, and color vision helps to distinguish ripe fruit from green leaves and unripe fruit.

Early Primates Had Grasping Hands

Early primates had long, grasping fingers that could wrap around and hold onto tree limbs. This adaptation to tree dwelling was the basis for later evolution of human hands that could perform both a *precision grip* (used for delicate maneuvers such as picking up small objects and sewing) and a *power grip* (used for powerful actions, such as thrusting with a spear or swinging a hammer).

A Large Brain Facilitated Hand–Eye Coordination and Complex Social Interactions

Primates have brains that are larger, relative to their body size, than the brains of almost all other animals. No one really knows for certain which environmental factors favored the evolution of large brains. It seems reasonable, however, that controlling and coordinating rapid locomotion through trees, the dexterous movements of the hands in manipulating objects, and binocular, color vision would be facilitated by increased brain power. Most primates also have fairly complex social systems, which require relatively high intelligence. If sociality promoted increased survival and reproduction, the benefits to individuals of successful social interaction might have favored the evolution of a larger brain.

The Oldest Hominin Fossils Are from Africa

Based on analysis of human mutation rates and DNA sequences from modern chimpanzees, gorillas, and humans, researchers have estimated that the **hominin** line (humans and their fossil relatives) diverged from the ape lineage at least 7 million years ago. The fossil record is in accord with this estimate, as the oldest hominin fossil so far found is between 6 and

◀ **FIGURE 18-12 Representative primates** The **(a)** tarsier, **(b)** lemur, and **(c)** lion-tail macaque monkey all have a relatively flat face, with forward-looking eyes providing binocular vision. All also have color vision and grasping hands. These features, retained from the earliest primates, are shared by humans.

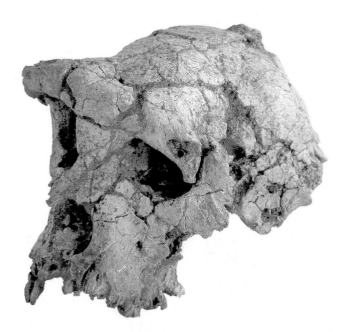

▲ **FIGURE 18-13 The earliest hominin** This nearly complete skull of *Sahelanthropus tchadensis,* which is more than 6 million years old, is the oldest hominin fossil yet found.

7 million years old (**FIG. 18-13**). This fossil species, *Sahelanthropus tchadensis,* was unearthed in the African country Chad and is clearly a hominin because it shares several anatomical features with later members of the group. But because this oldest known member of our family also exhibits other features that are more characteristic of apes, it may represent a point on our family tree that is close to the split between apes and hominins.

Two other hominin species—*Orrorin tugenensis* and *Ardipithecus ramidus*—are known from African fossils appearing in rocks that are between 4 million and 6 million years old. Most of our knowledge of these species is based on fossil finds that include only small portions of skeletons. But one specimen, a fairly complete 4.4-million-year-old *Ardipithecus* skeleton, has revealed some intriguing features of this early hominin. The structure of its legs, feet, hands, and pelvis suggests that *Ardipithecus* could walk upright, though it probably also climbed trees in its forest habitat. Its canine teeth were small, like those of modern humans and unlike the large, fang-like canines of today's apes.

A more extensive record of early hominin evolution begins about 4 million years ago. That date marks the beginning of the fossil record of the genus *Australopithecus* (**FIG. 18-14**), a group of African hominin species with brains larger than those of their forebears but still much smaller than those of modern humans.

Early Hominins Could Stand and Walk Upright

It is possible that even the earliest hominins walked upright. The discoverers of *Sahelanthropus* and *Orrorin* argue that the leg and foot bones of these earliest hominins have characteristics that indicate bipedal locomotion, but this conclusion will remain speculative until more complete skeletons of these species are found. However, the *Ardipithecus* skeleton shows that hominins capable of upright posture had arisen by

4.4 million years ago, and the earliest australopithecines (as the various species of *Australopithecus* and the related genus *Paranthropus* are collectively known) had knee joints that allowed them to straighten their legs fully, permitting efficient bipedal (upright, two-legged) locomotion. Footprints almost 4 million years old, discovered in Tanzania by anthropologist Mary Leakey, show that the earliest australopithecines could, and at least sometimes did, walk upright.

The reasons for the evolution of bipedal locomotion among the early hominins remain poorly understood. Perhaps hominins that could stand upright gained an advantage in gathering or carrying food. Whatever its cause, the early evolution of upright posture was extremely important in the evolutionary history of hominins, because if the hands were no longer needed for walking, they were free to serve other functions. Later hominins were thus able to carry weapons, manipulate tools, and eventually achieve the cultural revolutions produced by modern *Homo sapiens.*

Several Species of *Australopithecus* Emerged in Africa

The oldest australopithecine species, represented by fossilized teeth, skull fragments, and arm bones, was unearthed near an ancient lake bed in Kenya from sediments that were dated as being between 3.9 million and 4.1 million years old. The species was named *Australopithecus anamensis* by its discoverers. The second most ancient australopithecine, called *Australopithecus afarensis,* was discovered in the Afar region of Ethiopia. Fossil remains of this species as old as 3.9 million years have been unearthed. The *A. afarensis* line apparently gave rise to at least two distinct forms: smaller species such as *A. africanus* and *A. sediba,* and larger, more robust species such as *Paranthropus robustus* and *P. boisei* that had bigger molar teeth and heavier jaws, suggesting that their diet included hard foods such as nuts. All of the australopithecine species had gone extinct by 1.2 million years ago. Before disappearing, however, one of these species gave rise to a new branch of the hominin family tree, the genus *Homo* (see Fig. 18-14).

The Genus *Homo* Diverged from the Australopithecines 2.5 Million Years Ago

Hominins that are sufficiently similar to modern humans to be placed in the genus *Homo* first appear in African fossils that are about 2.5 million years old. Among the earliest African *Homo* fossils are *H. habilis* (see Fig. 18-14), a species whose body and brain were larger than those of the australopithecines but that retained the apelike long arms and short legs of their australopithecine ancestors. In contrast, the skeletal anatomy of *H. ergaster,* a species whose fossils first appear 2 million years ago, has limb proportions more like those of modern humans. This species is believed by many paleontologists to be on the evolutionary branch that led ultimately to our own species, *H. sapiens.* In this view, *H. ergaster* was the common ancestor of two distinct branches of hominins. The first branch led to *H. erectus,* which was the first hominin species to leave Africa. The second branch from *H. ergaster* ultimately led to

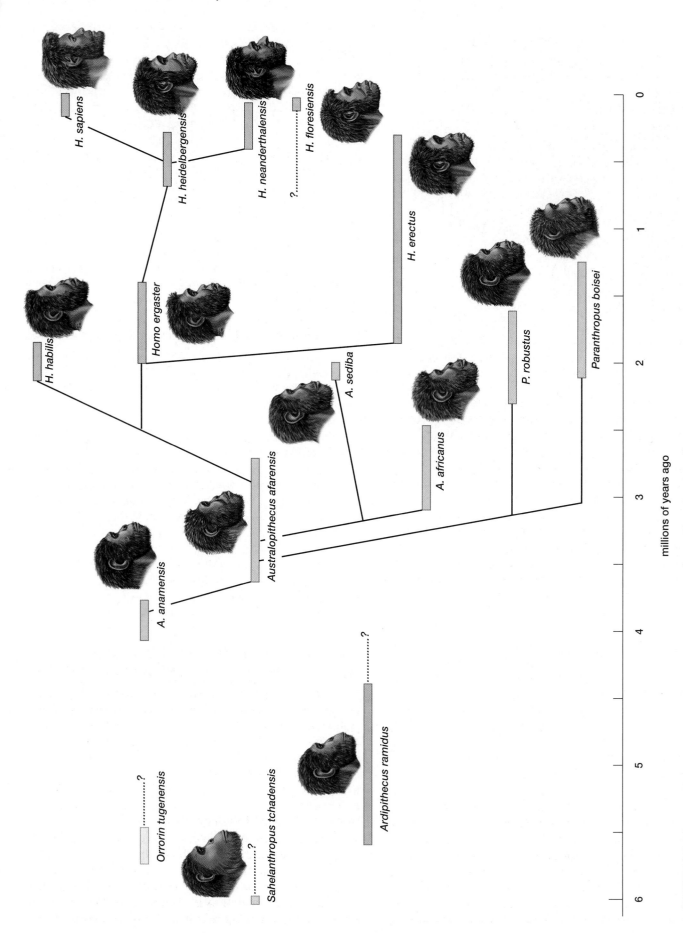

▲ **FIGURE 18-14 A possible evolutionary tree for humans** This hypothetical family tree shows facial reconstructions of representative specimens. Although many paleontologists consider this to be the most likely human family tree, there are several alternative interpretations of the known hominin fossils. Fossils of the earliest hominins are scarce and fragmentary, so the relationship of these species to later hominins remains unknown.

H. heidelbergensis, some of which migrated to Europe and gave rise to the Neanderthals, *H. neanderthalensis.* Meanwhile, back in Africa, another branch split off from the *H. heidelbergensis* lineage. This branch became *H. sapiens*—modern humans.

The Evolution of *Homo* Was Accompanied by Advances in Tool Technology

Hominin evolution is closely tied to the development of tools, a hallmark of hominin behavior. The oldest tools discovered so far were found in 2.5-million-year-old East African rocks, concurrent with the early emergence of the genus *Homo.* Early *Homo,* whose molar teeth (the rearmost teeth in the jaw) were much smaller than those of the australopithecines, might first have used stone tools to break and crush tough foods that were hard to chew. Hominins constructed their earliest tools by striking one rock with another to chip off fragments. During the next several hundred thousand years, toolmaking techniques in Africa gradually became more advanced. By 1.7 million years ago, tools had become more sophisticated. Flakes were chipped symmetrically from both sides of a rock to form double-edged tools ranging from hand axes, used for cutting and chopping, to points, probably used on spears (FIG. 18-15a, b). *Homo ergaster* and other bearers of these weapons presumably ate meat, probably acquired both from hunting and from scavenging for the remains of prey killed by other predators. Double-edged tools were carried to Europe at least 600,000 years ago by migrating populations of *H. heidelbergensis,* and the Neanderthal descendants of these emigrants took stone tool construction to new heights of skill and delicacy (FIG. 18-15C).

Neanderthals Had Large Brains and Excellent Tools

Neanderthals first appeared in the European fossil record about 150,000 years ago. By about 70,000 years ago, they had spread throughout Europe and western Asia. By 30,000 years ago, however, the species was extinct.

Contrary to the popular image of a hulking "caveman," Neanderthals were quite similar to modern humans in many ways. They walked fully erect, were dexterous enough to manufacture finely crafted stone tools, and had brains that, on average, were slightly larger than those of modern humans. Many European Neanderthal fossils show heavy brow ridges and a broad, flat skull, but others, particularly from areas around the eastern shores of the Mediterranean Sea, are somewhat more physically similar to *H. sapiens.*

Despite the physical and technological similarities between *H. neanderthalensis* and *H. sapiens,* there is no solid archaeological evidence that Neanderthals ever developed an advanced culture that included such characteristically human endeavors as art, music, and rituals. Some anthropologists argue that, because their skeletal anatomy shows that they were physically capable of making the sounds required for speech, Neanderthals might have acquired language. This interpretation of Neanderthal anatomy, however, is not unanimously accepted. In general, the available evidence of the Neanderthal way of life is limited and open to different interpretations,

(a) *Homo habilis*

(b) *Homo ergaster*

(c) *Homo neanderthalensis*

▲ **FIGURE 18-15 Representative hominin tools (a)** *Homo habilis* produced only fairly crude chopping tools called hand axes, usually unchipped on one end to hold in the hand. **(b)** *Homo ergaster* manufactured much finer tools. The tools were typically sharp all the way around the stone; at least some of these blades were probably tied to spears rather than held in the hand. **(c)** Neanderthal tools were works of art, with extremely sharp edges made by flaking off tiny bits of stone. In comparing these weapons, note the progressive increase in the number of flakes taken off the blades and the corresponding decrease in flake size. Smaller, more numerous flakes produce a sharper blade and suggest more insight into toolmaking and finer control of hand movements.

and anthropologists are engaged in a sometimes heated debate about how advanced Neanderthal culture became.

Neanderthals and *Homo sapiens* May Have Interbred

Our understanding of the evolutionary relationship between *H. sapiens* and *H. neanderthalensis* has improved dramatically in recent years, thanks to evidence from ancient DNA. Researchers have sequenced the entire Neanderthal genome from DNA that was extracted from 38,000-year-old bones found in a cave in Croatia. Based on comparison of the

Neanderthal genome to whole-genome sequences of modern humans, researchers have deduced that the evolutionary branch leading to Neanderthals diverged from the ancestral human line at least 400,000 years ago, thousands of years before the emergence of modern *H. sapiens*. However, the sequence comparison also revealed that up to 4% of a modern non–African human's DNA is similar to distinctively Neanderthal sequences. This finding suggests that our ancestors interbred with Neanderthals, probably about 60,000 years ago. Because modern Africans do not carry the Neanderthal sequences but all other people do, interbreeding with Neanderthals must have occurred after *H. sapiens* had left Africa but before modern humans spread around the world.

Two Other *Homo* Species Survived Until Relatively Recently

Scientists' ability to obtain DNA from ancient bones has also revealed a previously unknown hominin. The new hominin's existence was discovered by sequencing DNA extracted from a single finger bone found in deposits laid down between 30,000 and 50,000 years ago in Denisova Cave in Siberia. Analysis of the sequence showed that the bone came from a hominin that is evolutionarily distinct from both *H. neanderthalensis* and *H. sapiens*. Though the Denisovan hominin is so far known only by its DNA, one bone, and two teeth, paleoanthropologists suspect that it's only a matter of time until skeletons turn up.

Skeletons did turn up beneath the floor of a cave on the Indonesian island of Flores, where researchers discovered 18,000-year-old bones that they at first believed to be the fossil skeleton of a human child. Closer examination of the skeleton, however, revealed that it belonged to a fully grown adult, but one that was no more than 3 feet tall. The researchers gave this creature the nickname "Hobbit." Unlike today's small humans, such as pygmies or pituitary dwarves, Hobbit had a very small brain, smaller even than the brain of a typical chimpanzee (**FIG. 18-16**). Furthermore, the shapes and arrangements of the bones in Hobbit's wrist, shoulder, and other parts of its skeleton were unlike those of anatomically modern humans. On the basis of these findings, researchers concluded that Hobbit is not simply a small *H. sapiens* but represents a different species, now named *Homo floresiensis*. Thus, it appears that modern humans at one time shared the planet (or at least some parts of it) not only with Neanderthals, but also with Denisovans and *H. floresiensis*.

Modern Humans Emerged Less Than 200,000 Years Ago

The fossil record shows that anatomically modern humans appeared in Africa at least 160,000 years ago and possibly as long as 195,000 years ago. The location of these fossils suggests that *Homo sapiens* originated in Africa, but most of our knowledge about our own early history comes from European and Middle Eastern fossils of *H. sapiens,* collectively known as Cro-Magnons (after the district in France in which their remains were first discovered). Cro-Magnons appeared about

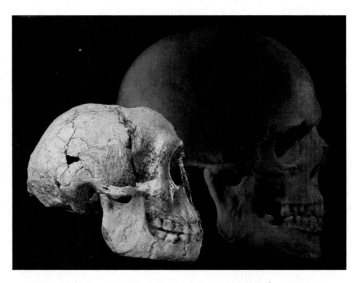

▲ **FIGURE 18-16 The hominin known as "Hobbit"** The skull of *Homo floresiensis*, a recently discovered diminutive human relative, is dwarfed by the skull of a modern *Homo sapiens*.

90,000 years ago. They had domed heads, smooth brows, and prominent chins (just like us). Their tools were precision instruments similar to the stone tools that were still used in a few cultures as recently as the 1960s.

Behaviorally, Cro-Magnons seem to have been similar to, but more sophisticated than, Neanderthals. Artifacts from 30,000-year-old Cro-Magnon archaeological sites include elegant bone flutes, graceful carved ivory sculptures, and evidence of elaborate burial ceremonies (**FIG. 18-17**).

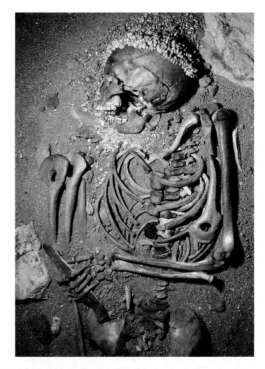

▲ **FIGURE 18-17 Paleolithic burial** This 24,000-year-old grave shows evidence that Cro-Magnon people ritualistically buried their dead. The body was covered with a dye known as red ocher and then buried with a headdress made of snail shells and a flint tool in its hand.

ingly clear that the genus *Homo* made repeated long-distance emigrations. What is less clear is how all this wandering is related to the origin of modern *H. sapiens*. According to the "African replacement" hypothesis (the basis of the scenario outlined earlier), *H. sapiens* emerged in Africa and dispersed less than 120,000 years ago, spreading into the Near East, Europe, and Asia and replacing all other hominins. But some paleoanthropologists believe that populations of *H. sapiens* evolved simultaneously in many regions from the already widespread populations of *H. erectus*. According to this "multiregional origin" hypothesis, continued migrations and interbreeding among *H. erectus* populations in different regions of the world maintained them as a single species as they gradually evolved into *H. sapiens* (**FIG. 18-19b**). Although an increasing number of studies of modern human DNA support the African replacement model, both hypotheses are consistent with the fossil record. Therefore, the question remains unsettled.

▲ **FIGURE 18-18 The sophistication of Cro-Magnon people** Cave paintings by Cro-Magnons have been remarkably preserved by the relatively constant underground conditions of the Chauvet-Pont-d'Arc cave in France.

Perhaps the most remarkable accomplishment of Cro-Magnons is the magnificent art left in caves in places such as Altamira in Spain and Lascaux and Chauvet in France (**FIG. 18-18**). The oldest cave paintings so far found are more than 30,000 years old, and even these make use of sophisticated artistic techniques. No one knows exactly why these paintings were made, but they attest to minds as capable as our own.

Cro-Magnons and Neanderthals Lived Side by Side

Cro-Magnons coexisted with Neanderthals in Europe and the Middle East for perhaps as many as 50,000 years before the Neanderthals disappeared. The genetic analyses described earlier show that Cro-Magnons interbred with Neanderthals, so some researchers hypothesize that Neanderthals were essentially absorbed into the human genetic mainstream. Other scientists disagree, noting that the DNA evidence reveals only relatively limited interbreeding, and suggest that the later-arriving Cro-Magnons simply overran and displaced the less-well-adapted Neanderthals.

Several Waves of Hominins Emigrated from Africa

The human family tree is rooted in Africa, but hominins found their way out of Africa on numerous occasions. For example, *H. erectus* reached tropical Asia almost 2 million years ago and apparently thrived there, eventually spreading across the continent (**FIG. 18-19a**). Similarly, *H. heidelbergensis* made it to Europe at least 780,000 years ago. It is increas-

CASE STUDY   CONTINUED
Ancient DNA Has Stories to Tell

We might be able to more easily distinguish between the African replacement and multiregional origin hypotheses (and to answer a host of other unanswered questions about the origin and early evolution of *H. sapiens*) if we had access to DNA sequences from the earliest representatives of our genus. Is it possible that researchers will one day extract useable DNA from, say, early *H. erectus*? Perhaps, but the odds of success are not great. *H. erectus* fossils are up to 1.8 million years old, but the oldest ancient genome so far obtained is from a 700,000-year-old fossil horse. What's more, the fossil horse was found in northern Canada, where the cold climate is excellent for preserving DNA. *H. erectus*, however, inhabited warmer regions where DNA degrades more quickly. Nonetheless, some evolutionary biologists hold out hope that useable DNA might be found in early *H. erectus* bones that fossilized in an environment conducive to preservation, perhaps in a deep cave or in oxygen-depleted underwater sediments. Could DNA survive for more than a million years under the right conditions? Current evidence says no, but then it wasn't too long ago that recovering Neanderthal DNA seemed like an impossible dream.

The Evolutionary Origin of Large Brains May Be Related to Meat Consumption and Cooking

The main physical features that distinguish us from our closest relatives, the apes, are our upright posture and large, highly developed brains. As described earlier, upright posture arose very early in hominin evolution, and hominins walked

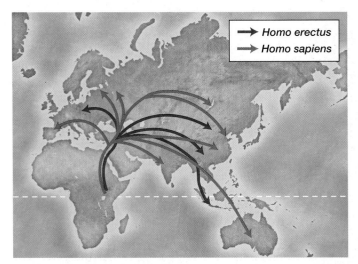

(a) African replacement hypothesis

(b) Multiregional hypothesis

▶ **FIGURE 18-19 Competing hypotheses for the evolution of
Homo sapiens (a)** The "African replacement" hypothesis suggests
that *H. sapiens* evolved in Africa and then migrated throughout the
Near East, Europe, and Asia, displacing the other hominin species
that were present in those regions. **(b)** The "multiregional" hypothesis suggests that populations of *H. sapiens* evolved in many regions
simultaneously from the already widespread populations of *H. erectus*.

THINK CRITICALLY Paleontologists recently discovered fossil
hominins with features characteristic of modern humans in
160,000-year-old sediments in Africa. Which hypothesis does this
new evidence support?

upright for several million years before large-brained *Homo*
species arose. What circumstances might have caused the
evolution of increased brain size? Many explanations have
been proposed, but little direct evidence is available; hypotheses about the evolutionary origins of large brains are necessarily speculative.

One proposed explanation for the origin of large brains
suggests that they evolved in response to increasingly
complex social interactions. In particular, fossil evidence
suggests that, beginning about 2 million years ago, hominin
social life began to include a new type of activity—the cooperative hunting of large game. The resulting access to significant amounts of meat must have fostered a need to develop
methods for distributing this valuable, limited resource
among group members. Some anthropologists hypothesize
that the individuals best able to manage this social interaction would have been more successful at gaining a large
share of meat and using their share advantageously. Perhaps
this social management was best accomplished by individuals with larger, more powerful brains, and natural selection
therefore favored such individuals. Observations of chimpanzee societies have shown that the distribution of group-
hunted meat often involves intricate social interactions in
which meat is used to form alliances, repay favors, gain access to sexual partners, placate rivals, and so on. Perhaps the
mental skill required to plan, assess, and remember such interactions was the driving force behind the evolution of our
large, clever brains.

Whatever the nature of the advantages that favored
individuals with larger brains, such brains could not
have evolved without some mechanism to provide the
large amount of energy necessary to grow and maintain
a large volume of brain tissue. Some researchers speculate
that cooking was the breakthrough that freed up the required extra energy. Cooked food is more digestible than
raw food and requires far less chewing, so cooked food
provides more nutrients with less effort expended. Thus,
cooking by early hominins might have removed the limit
that had previously restricted brain size. However, larger
brains first arose in *H. erectus* at least 2 million years ago,
and the earliest direct archaeological evidence of controlled fires is only 1 million years old. Proponents of the
cooking hypothesis suggest that cooking actually did arise
2 million years ago, and the lack of evidence of cooking
fires that old is simply due to the incompleteness of the
hominin fossil record.

Sophisticated Culture Arose
Relatively Recently

Even after the evolution of comparatively large brains in species such as *H. erectus,* more than a million years passed before the origin of modern humans and their extremely large
brains. And even after the first appearance of modern *H. sapiens,* more than 100,000 years passed before the appearance of
any archaeological evidence of the distinctively human characteristics that were made possible by a large brain: language,
abstract thought, and advanced culture. The evolutionary
origin of these human traits is another unresolved question, in part because direct evidence of the transition to advanced culture may never be found. Early humans capable of
language and symbolic thought would not necessarily have
created artifacts that indicated these capabilities. We can uncover some clues by studying our ape relatives, which possess
less-complex versions of many human behaviors and mental

processes. Their behavior might resemble that of ancestral hominins. Nonetheless, the late, seemingly rapid origin of advanced human culture remains a puzzle.

Biological Evolution Continues in Humans

Until recently, most evolutionary biologists agreed that the evolution of human bodies by natural selection had slowed or halted after we began to live in advanced societies. Today, however, our ever-growing ability to rapidly sequence DNA has made it possible for researchers to analyze sequences of a large and growing number of human genomes, and these analyses have led to a surprising conclusion: People have evolved rapidly since the advent of music, art, language, and the other hallmarks of advanced culture, and we continue to evolve today. Many of our genes show the telltale signs of evolution by natural selection in recent millennia. In many cases, the exact functions of these genes remain unknown, but researchers have determined the functions of some recent evolutionary changes. For example, the alleles required to digest milk as adults have arisen and become fixed in some populations within the past 7,000 years.

Culture Also Evolves

Human evolution in recent millennia has also included a great deal of cultural evolution, the evolution of information and behaviors that are transmitted from generation to generation by learning. Our recent evolutionary success, for example, was engendered not so much by new physical adaptations as by a series of cultural and technological revolutions. The first such revolution was the development of tools, which began with the early hominins. Tools increased the efficiency with which food and shelter could be acquired and thus increased the number of individuals that could survive within a given ecosystem. About 10,000 years ago, human culture underwent a second revolution as people discovered how to grow crops and domesticate animals. This agricultural revolution dramatically increased the amount of food that could be extracted from the environment, and the human population surged, increasing from about 5 million at the dawn of agriculture to around 750 million by 1750. The subsequent Industrial Revolution gave rise to the modern economy and its attendant improvements in public health. Longer lives and lower infant mortality led to truly explosive population growth, and today Earth's population is more than 7 billion and still growing.

Human cultural evolution and the accompanying increases in human population have had profound effects on the continuing biological evolution of other life-forms. Our agile hands and minds have transformed much of Earth's terrestrial and aquatic habitats. Humans have become the most powerful agent of natural selection. In the words of the late evolutionary biologist Stephen Jay Gould, "We have become, by the power of a glorious evolutionary accident called intelligence, the stewards of life's continuity on Earth. We did not ask for this role, but we cannot renounce it. We may not be suited for it, but here we are."

CHECK YOUR LEARNING

Can you ...

- describe the evolutionary history of humans and the factors that may have fostered humans' distinctive adaptations?
- name and describe some characteristics of the hominin species that played key roles in humans' evolutionary history?
- describe the key features of the most recent phase of human evolution?

CASE STUDY \ REVISITED
Ancient DNA Has Stories to Tell

The unexpected discovery that humans interbred with Neanderthals was a triumph for the experts who developed the techniques for extracting, isolating, and sequencing ancient DNA. But perhaps the most stunning revelation made possible by ancient DNA was the discovery of the Denisovans, a hominin species whose existence would still be unknown if not for analysis of its ancient DNA. The fossil fragments from which the DNA was extracted were too few, too small, and too nondescript to have even been recognized as belonging to a previously unknown species. A newfound ability to identify new extinct species on the basis of DNA alone raises the intriguing possibility of future discovery of other previously unsuspected species, hominin and otherwise.

Like Neanderthals, Denisovans left a genetic trace in modern humans. One example is the Denisovan gene variant that helps Tibetans live at high altitude. Additionally, the people native to New Guinea and other Pacific islands carry a substantial number of Denisovan sequences. Almost 5% of the genome of these people is of Denisovan origin. This finding suggests that Denisovans interbred with the ancestors of Pacific Islanders, either in mainland Asia before the islands were first colonized by people, or later, if Denisovans were somehow able to get to multiple islands.

CONSIDER THIS Knowledge of the genomes of ancient hominins might help us better understand not only the evolutionary history of hominins, but also the traits that differ between us and our relatives—the traits that make us human. Our understanding of the functions of different genes is growing rapidly, so detailed comparisons of our genes to those present in, say, Neanderthals and Denisovans are very revealing. What do you think are the functions most likely to be related to the genetic differences between us and our hominin relatives?

CHAPTER REVIEW

MB Go to **MasteringBiology** for practice quizzes, activities, eText, videos, current events, and more.

Answers to Think Critically, Evaluate This, Multiple Choice, and Fill-in-the-Blank questions can be found in the Answers section at the back of the book.

Summary of Key Concepts

18.1 How Did Life Begin?

Before life arose, energy from lightning, ultraviolet light, and heat formed organic molecules from water and the components of primordial Earth's atmosphere. The organic molecules formed probably included nucleic acids, amino acids, and lipids. By chance, some molecules of RNA may have had enzymatic properties, catalyzing the assembly of copies of themselves from nucleotides in Earth's waters. Protein-lipid vesicles enclosing these ribozymes may have formed protocells, the forerunners of life.

18.2 What Were the Earliest Organisms Like?

The oldest fossils, about 3.4 billion years old, are of prokaryotic cells that fed by absorbing organic molecules from their environment. Because there was no free oxygen in the atmosphere, their energy metabolism must have been anaerobic. As the cells multiplied, they depleted the organic molecules that had been formed by prebiotic synthesis. Some cells developed the ability to synthesize their own food molecules by using simple inorganic molecules and the energy of sunlight. These earliest photosynthetic cells were probably ancestors of today's cyanobacteria.

Photosynthesis releases oxygen as a by-product, and by about 2.4 billion years ago significant amounts of free oxygen had accumulated in the atmosphere. Aerobic metabolism, which generates more cellular energy than does anaerobic metabolism, probably arose about this time.

Eukaryotic cells had evolved by about 1.7 billion years ago. The first eukaryotic cells probably arose as symbiotic associations between predatory prokaryotic cells and other bacteria. Mitochondria may have evolved from aerobic bacteria engulfed by predatory cells. Chloroplasts may have evolved from photosynthetic bacteria by a similar process.

18.3 What Were the Earliest Multicellular Organisms Like?

Multicellular organisms evolved from eukaryotic cells and first appeared in the seas about 1.2 billion years ago. Multicellularity offers several advantages, including greater size. In plants, increased size offered some protection from predation. Specialization of cells allowed plants to anchor themselves in the nutrient-rich, well-lit waters near the shore. For animals, multicellularity allowed more efficient predation and more effective escape from predators. These in turn provided environmental pressures for faster locomotion, improved senses, and greater intelligence. Diverse animal forms appear in the fossil record beginning about 600 million years ago; fish were the predominant marine animals by about 400 million years ago.

18.4 How Did Life Invade the Land?

The first land organisms were probably algae. The first multicellular land plants appeared about 475 million years ago. Life on land required special adaptations for support of the body, reproduction, and the acquisition, distribution, and retention of water, but the land also offered abundant sunlight and freedom from aquatic herbivores. Soon after land plants evolved, arthropods invaded the land.

The earliest land vertebrates evolved from lobefin fishes, which had leglike fins and a primitive lung. A group of lobefins evolved into the amphibians about 370 million years ago. Reptiles evolved from amphibians, with several further adaptations for life on land. One reptile group, the birds, evolved feathers that provided insulation and facilitated flight. Mammals, whose bodies are insulated by hair, descended from a reptile group.

18.5 What Role Has Extinction Played in the History of Life?

The history of life has been characterized by constant turnover of species as some go extinct and are replaced by new ones. Mass extinctions, in which large numbers of species disappear within a relatively short time, have occurred periodically. Mass extinctions were probably caused by some combination of climate change and catastrophic events, such as volcanic eruptions and meteorite impacts.

18.6 How Did Humans Evolve?

One group of mammals evolved into the tree-dwelling primates. Some primates descended from the trees, and these were the ancestors of apes and humans. The oldest known hominin fossils are between 6 million and 7 million years old and were found in Africa. The australopithecines arose in Africa about 4 million years ago. These hominins walked erect, had larger brains than did their forebears, and fashioned primitive tools. One group of australopithecines gave rise to a line of hominins in the genus *Homo*. *Homo* arose in Africa, but populations of several *Homo* species migrated from Africa and spread to other geographic areas. In the last of these migrations, *Homo sapiens*, characterized by a large brain and advanced tool technology, dispersed from Africa to Asia and Europe.

Key Terms

amphibian   *326*
arthropod   *326*
conifer   *325*
endosymbiont
    hypothesis   *321*
eukaryote   *321*
exoskeleton   *323*
hominin   *330*
lobefin   *326*
mammal   *327*

mass extinction   *328*
plate tectonics   *329*
primate   *330*
prokaryote   *319*
protocell   *318*
reptile   *326*
ribozyme   *317*
spontaneous
    generation   *315*

Thinking Through the Concepts

Multiple Choice

1. Almost all of the oxygen gas in today's atmosphere is present as a result of
   a. outgassing from volcanoes.
   b. aerobic respiration.
   c. endosymbiosis.
   d. photosynthesis.

2. Extinction
   a. generally does not occur except during unpredictable mass extinctions.
   b. ordinarily occurs at a relatively slow but steady rate.
   c. has eliminated species at a faster rate than they have been formed.
   d. has not played a major role in the history of life.

3. In the endosymbiotic origin of the mitochondrion, the host cell benefited from the engulfed cell's ability to _____; in the endosymbiotic origin of the chloroplast, the host cell benefited from the engulfed cell's ability to _____.
   a. detoxify waste material; produce toxins that deterred predators
   b. use aerobic respiration to produce ATP; use photosynthesis to produce sugar
   c. use photosynthesis to generate oxygen; provide a shield against UV radiation
   d. move rapidly using flagella; form a cell nucleus

4. Which of the following does *not* list evolutionary events in chronological order (oldest to most recent)?
   a. first algae, first animals, first land plants
   b. first mammals, first reptiles, first hominins
   c. first amphibians, first dinosaurs, first flowering plants
   d. first photosynthesis, first exoskeletons, first hair

5. Which of the following lists hominin traits in the order in which they evolved?
   a. upright posture, large brain, symbolic culture
   b. large brain, upright posture, symbolic culture
   c. large brain, stone tools, symbolic culture
   d. upright posture, symbolic culture, stone tools

Fill-in-the-Blank

1. Because there was no oxygen in the earliest atmosphere, the first cells must have derived energy by _____ metabolism of organic molecules. Oxygen was introduced into the atmosphere when some microbes developed the ability to _____ and released oxygen as a by-product. Oxygen was _____ to many of the earliest cells, but some cells evolved the ability to use oxygen in _____ respiration, which provided far more _____.

2. The molecule _____ became a candidate for the first self-replicating information-carrying molecule when Thomas Cech and Sidney Altman discovered that some of these molecules can act as _____, which they called _____.

3. Complex cells that contain a nucleus and other organelles are called _____ cells. A compelling explanation for the origin of these complex cells is the _____ hypothesis. One observation that supports this hypothesis is that mitochondria have their own _____.

4. The sperm of early land plants had to reach the egg by _____, limiting them to _____ environments. An important adaptation of plants to dry land was the evolution of _____, which enclosed sperm in a drought-resistant coat.

5. Early plants that protected their seeds within cones are called _____. These relied on _____ to carry their pollen. Later, some plants evolved _____, which attracted animals, particularly _____ that carried their pollen. Animal pollination is much more _____ than wind pollination.

6. The first animals to live on land were _____ because their external skeletons, also called _____, supported the animals' weight, while protecting their bodies from _____.

7. Amphibians gave rise to _____, which had three important adaptations to life on dry land: shelled, waterproof _____; scaly, water-resistant _____; and more efficient _____.

Review Questions

1. What is the evidence that life might have originated from nonliving matter on early Earth?

2. How did the origin of photosynthesis affect subsequent evolution of life on Earth?

3. Explain the endosymbiont hypothesis for the origin of chloroplasts and mitochondria.

4. Name two advantages of multicellularity for plants and two for animals.

5. What advantages and disadvantages would terrestrial existence have had for the first plants to invade the land? For the first land animals?

6. Outline the major adaptations that emerged during the evolution of vertebrates from fish to amphibians to reptiles to birds and mammals. Explain how these adaptations increased the fitness of the various groups for life on land.

7. Outline the evolution of humans from early primates. Include in your discussion such features as binocular vision, grasping hands, bipedal locomotion, toolmaking, and brain expansion.

Applying the Concepts

1. Extinctions have occurred throughout the history of life on Earth. Why should we care if humans are causing a mass extinction event now?

2. In biological terms, what do you think was the most significant event in the history of life? Explain your answer.

SYSTEMATICS: SEEKING ORDER AMID DIVERSITY

Origin of a Killer

ONE OF THE WORLD'S most frightening diseases is also one of its most mysterious. Acquired immune deficiency syndrome (AIDS) appeared seemingly out of nowhere, and when it was first recognized in the early 1980s, no one knew what caused it or where it came from. Scientists raced to solve the mystery and, within a few years, identified the infectious agent that causes AIDS: human immunodeficiency virus (HIV). Once HIV had been identified, researchers turned their attention to the question of its origin.

Finding the origin of HIV required an evolutionary approach. To ask "Where did HIV come from?" is really to ask "What kind of virus was the ancestor of HIV?" To answer this question, researchers began by identifying the closest relatives of HIV; when a biologist concludes that two viruses are closely related, it means that they share a recent common ancestor from which both evolved. Thus, comparing HIV with its closest relatives allowed researchers to infer the characteristics of their common ancestor.

The researchers who explored the ancestry of HIV discovered that its closest relatives are found not among other viruses that infect humans, but among those that infect monkeys and apes. In fact, the latest research on HIV's evolutionary history has concluded that the closest relative of HIV-1 (the type of HIV that is most responsible for the worldwide AIDS

Biologists studying the evolutionary history of type 1 human immunodeficiency virus (HIV-1) discovered that the virus, which causes AIDS, probably originated in chimpanzees.

epidemic) is a virus strain that infects a chimpanzee subspecies that inhabits a limited range in West Africa. Therefore, the ancestor of the virus that we now know as HIV-1 did not evolve from a preexisting human virus. Instead, a chimpanzee virus must have acquired mutations that allowed it to infect humans and cause a deadly disease.

In this chapter, we examine *systematics,* the branch of biology that helped solve the mystery of HIV's origin. What methods did the researchers use to discover the relatives of HIV? What other kinds of questions can be answered with these methods?

AT A GLANCE

19.1 HOW ARE ORGANISMS NAMED AND CLASSIFIED?

To study and discuss organisms, biologists must name them. The branch of biology that is concerned with naming and classifying organisms is known as **taxonomy.** (A *taxon*—plural, *taxa*—is a named species or a named group of species). The basis of modern taxonomy was established by the Swedish naturalist Carl von Linné (1707–1778), who called himself Carolus Linnaeus, a Latinized version of his name. One of Linnaeus's most enduring achievements was the introduction of the two-part scientific name.

Each Species Has a Unique, Two-Part Name

The **scientific name** of an organism is a two-part Latin name that designates its genus and species. A **genus** is a group that includes a number of very closely related species; each **species** within a genus includes populations of organisms that can potentially interbreed under natural conditions. For example, the genus *Sialia* (bluebirds) includes three species: the eastern bluebird (*Sialia sialis*), the western bluebird (*Sialia mexicana*), and the mountain bluebird (*Sialia currucoides*) (FIG. 19-1). Although the three species are similar, bluebirds normally breed only with members of their own species.

In a scientific name, the genus name is presented first, followed by the species name. By convention, scientific names are always underlined or *italicized*. The first letter of the genus name is always capitalized, and the first letter of the species name is always lowercase. The species name is never used alone but is always paired with its genus name.

Each two-part scientific name is unique, so referring to an organism by its scientific name rules out any chance of ambiguity or confusion. For example, the bird *Gavia immer* is commonly known as the common loon in North America, as the northern diver in Great Britain, and by still other names in non–English-speaking countries. But the Latin scientific name *Gavia immer* is recognized by biologists worldwide, overcoming language barriers and allowing precise communication.

Modern Classification Emphasizes Patterns of Evolutionary Descent

In addition to naming species, biologists also classify them. Prior to the 1859 publication of Darwin's *On the Origin of Species,* classification served mainly to facilitate the study and discussion of organisms, much as a library's online catalog facilitates our ability to find a book. But after Darwin demonstrated that all organisms are linked by common ancestry, biologists began to recognize that classification ought to reflect and describe the pattern of evolutionary relatedness among organisms. Today, the process of classification focuses almost exclusively on reconstructing **phylogeny,** or evolutionary history. The science of reconstructing phylogeny is known as **systematics.** Systematists communicate their hypotheses about phylogeny by constructing evolutionary trees (see Fig. 17-11).

Systematists Identify Features That Reveal Evolutionary Relationships

As systematists seek to reconstruct the tree of life, they must do so without much direct knowledge of evolutionary history. Because systematists can't see into the past, they must infer it as best they can on the basis of similarities among living organisms. Not all similarities are useful for constructing phylogenetic trees, however. Some observed similarities stem from convergent

(a) Eastern bluebird

(b) Western bluebird

(c) Mountain bluebird

◀ **FIGURE 19-1 Three species of bluebird** Despite their obvious similarity, these three species of bluebird—**(a)** the eastern bluebird (*Sialia sialis*), **(b)** the western bluebird (*Sialia mexicana*), and **(c)** the mountain bluebird (*Sialia currucoides*)—evolve independently because they do not interbreed.

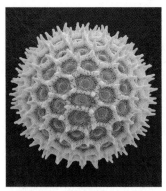

Morning glory pollen   **Bitter melon pollen**

▲ **FIGURE 19-2 Microscopic structures may be used to classify organisms** The shape and surface features of pollen grains are among the finely detailed structures that can be useful in classification. Such structures can reveal similarities and differences between species that are not apparent in larger and more easily visible structures.

evolution (see Chapter 15) in organisms that are not closely related, and such similarities are not useful for inferring evolutionary history. Instead, systematists use similarities that exist because two kinds of organisms both inherited a characteristic from a common ancestor. In the search for these informative similarities, biologists look at many kinds of characteristics.

Historically, the most important and useful distinguishing characteristics have been anatomical. Systematists look carefully at similarities in both external body structure and internal structures, such as skeletons and muscles. For example, homologous structures such as the finger bones of dolphins, bats, seals, and humans provide evidence of a common ancestor (see Fig. 15-8). To detect relationships among more closely related species, biologists may use microscopes to discern finer details, such as the external structure of the pollen grains of a flowering plant (**FIG. 19-2**).

Modern Systematics Relies on Molecular Similarities to Reconstruct Phylogeny

Recent advances in the techniques of molecular genetics have revolutionized studies of evolutionary relationships by allowing scientists to determine genetic similarities among organisms. Today's systematists rely mainly on the nucleotide sequences of DNA (that is, organisms' genotypes) to investigate relatedness among different types of organisms.

The logic underlying such molecular systematics is straightforward. It is based on the observation that when a single species divides into two species, the gene pool of each resulting species begins to accumulate mutations. The particular mutations present in each species' gene pool, however, will differ because the species are now evolving independently, with no gene flow between them. As time passes, more and more genetic differences accumulate. So a systematist who has obtained DNA sequences from representatives of both species can compare the two species' nucleotide sequences. Fewer differences indicate more closely related organisms

(species with a relatively recent common ancestor). The process by which systematists use genetic (and anatomical) similarities to reconstruct evolutionary history is discussed in "In Greater Depth: Phylogenetic Trees" on page 344.

In some cases, similarity of DNA sequences will be reflected in the structure of chromosomes. For example, both the DNA sequences and the chromosomes of chimpanzees and humans are extremely similar, showing that these two species shared a common ancestor in the not too distant past (**FIG. 19-3**).

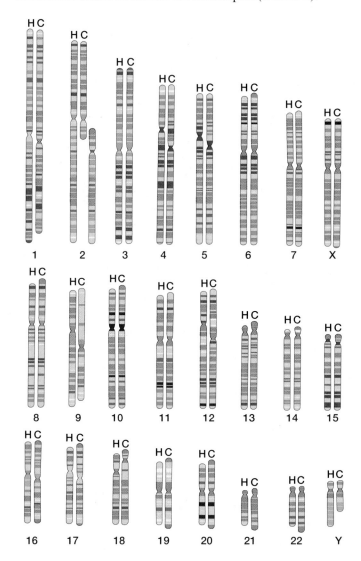

▲ **FIGURE 19-3 Human and chimpanzee chromosomes are similar** Chromosomes from different species can be compared by means of banding patterns that are revealed by staining. The comparison illustrated here, between human chromosomes (left member of each pair; H) and chimpanzee chromosomes (C), reveals that the two species are genetically very similar. The numbering system shown is that used for human chromosomes; note that human chromosome 2 corresponds to a combination of two chimp chromosomes. Data from Yunis, J. J., et al. 1980. *Science* 208:1145–1148.

THINK CRITICALLY Analysis of human chromosome 2 revealed that it contains both a functional centromere and the remnants of a second one. What does this finding suggest about the evolutionary origin of chromosome 2?

Origin of a Killer

Analysis of nucleotide sequences was absolutely required to construct the phylogeny of viruses that revealed the ancestor of HIV. The closely related viruses included in the phylogeny are all but indistinguishable on the basis of appearance and structure; the differences between them are revealed only by their nucleotide sequences. Thus, scientific sleuthing about the origin of HIV would have been impossible before the modern era of routine DNA sequencing.

DNA analysis has also shown that HIV is a member of a group called the lentiviruses. How do taxa get their names? Can the names tell us anything about the evolutionary histories of taxa?

Systematists Name Groups of Related Species

Although systematists communicate their findings about phylogeny mainly by presenting evolutionary trees, they also name groups of species. In keeping with their emphasis on reconstructing evolutionary history, systematists give formal names only to groups that include all the organisms descended from a common ancestor. Such groups are known as **clades.** If you examine an evolutionary tree, you will see that clades can be arranged in a hierarchy, with smaller clades nested within larger ones (**FIG. 19-4**).

Ranks Add Information to the Clade Name

When systematists name a clade, the name itself does not convey much information about the clade. For example, the name does not reveal much about the clade's size or breadth. Is the clade a relatively large, broadly inclusive one, such as the one that

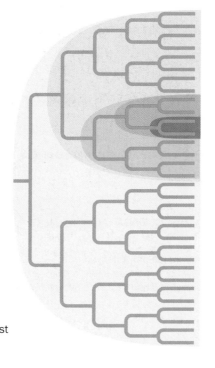

▶ **FIGURE 19-4**
Clades form a nested hierarchy Any group that includes all the descendants of a common ancestor is a clade. Some of the clades represented on this evolutionary tree are shaded in different colors. Note that smaller clades nest within larger clades.

includes all mammals? Or is it a smaller, narrower clade, perhaps one that includes only the three species of zebra? The clade's size and breadth would be obvious if we could view its evolutionary tree, but if we don't have access to the tree, we will need clues other than the clade's name in order to understand its scope.

One possible way to signal the relative size and inclusiveness of named clades is to place them into categories called *taxonomic ranks*. This approach has a long history; Linnaeus placed each species into a series of ranked categories on the basis of its resemblance to other species. The Linnaean classification system eventually came to include eight major ranks: *domain, kingdom, phylum, class, order, family, genus,* and *species*. These ranks form a nested hierarchy in which each level includes all of the other levels below it; each domain contains a number of kingdoms; each kingdom contains a number of phyla; each phylum includes a number of classes; each class includes a number of orders; and so on. As we move down the hierarchy, smaller and smaller groups are included. Thus, if you knew that a certain named clade was a phylum, and a second named clade was a genus within the phylum, you would understand that the second clade was a subset of the first one.

Use of Taxonomic Ranks Is Declining

Although the use of taxonomic ranks has a long history, today's systematists have de-emphasized the Linnaean ranking system. Historically assigned ranks may misrepresent evolutionary history as it is currently understood, and implementing a scientifically sound revision of the ranking system would present some difficult technical challenges. These days, many systematists do not assign taxonomic ranks to the clades they name and instead concentrate on using data to construct accurate evolutionary trees, rather than on subjective evaluations of whether a given clade should be called a kingdom, a phylum, a class, an order, or a family. As a result, use of Linnaean taxonomic ranks is declining.

In the chapters of this text that describe the diversity of life (Chapters 20–25), our use of taxonomic ranks will vary according to the practices of the biologists who study the different organisms we discuss. In most chapters, we will follow the emerging convention of avoiding ranks, instead using the term "taxonomic group" (as a synonym for clade) to describe a collection of related species. In some chapters, however, we will make selective use of a few Linnaean ranks. For example, we will follow the tradition of using "kingdom" to refer to the three clades that contain, respectively, all animals, all plants, and all fungi. Similarly, in the two chapters about animals, we will designate certain clades as phyla, in keeping with tradition in animal systematics. And we will refer to the three broadest, most inclusive of life's clades as **domains.**

CHECK YOUR LEARNING

Can you ...
- explain why scientific names are necessary?
- describe the type of similarities that systematists use to reconstruct phylogeny?
- describe the system of Linnaean taxonomic ranks?

IN GREATER DEPTH | Phylogenetic Trees

Systematists strive to develop a system of classification that reflects the phylogeny of organisms. Thus, the main job of systematists is to reconstruct phylogeny. Reconstructing the evolutionary history of all of Earth's organisms is, of course, a huge task, so each systematist typically chooses to work on some particular portion of the history.

Phylogenies Are Illustrated by Trees

The result of a phylogenetic reconstruction is usually represented by a diagram. These diagrams can take a number of different forms, but all of them show the sequence of branching events in which ancestral species split to give rise to descendant species. For this reason, diagrams of phylogeny are generally tree-like (though the tree may be oriented in any of a variety of directions; in this text we usually orient the tree horizontally, with the branch tips on the right.)

These trees can present the phylogeny of any specified set of taxa. Thus, phylogenetic trees can show evolutionary history at different scales. Systematists might reconstruct, for example, a tree of 10 species in a particular genus of clams, or a tree of 25 clades of animals, or a tree of the three domains of life.

Phylogenies Are Based on Shared Derived Characteristics

After selecting the taxa to include, a systematist is ready to begin building a tree. Most systematists use the *cladistic* approach to reconstruct phylogenetic trees. Under the cladistic approach, relationships among taxa are revealed by the occurrence of similarities known as *synapomorphies*. A synapomorphy is a trait that is similar in two or more taxa because these taxa inherited a "derived" version of the trait that had changed from its original state in a common ancestor. For example, the presence of feathers is a synapomorphy that links all living birds and distinguishes them from other vertebrates. The common ancestor of birds and crocodiles (their closest living relatives) had scales, which evolved into feathers—the derived state—in the lineage leading to birds but not in the lineage leading to crocodiles. The formation of synapomorphies is illustrated in **FIGURE E19-1**.

In the imaginary scenario illustrated in Figure E19-1, we can easily identify synapomorphies because we know the ancestral state of the trait (the DNA sequence CGT AGA TAC) and the subsequent changes that took place (T replacing A in the sixth position and C replacing G in the second position). In real life, however, systematists would not have direct knowledge of the ancestor, which lived in the distant past and whose identity is unknown. Without

this direct knowledge, a systematist observing a similarity between two taxa is faced with a challenge. Is the observed similarity a synapomorphy, or does it have some other cause, such as convergent evolution? The cladistic approach provides methods for identifying synapomorphies, but it remains possible to mistakenly use a shared trait that is not in fact a synapomorphy. To guard against such errors, systematists use numerous traits to build a tree, thereby minimizing the influence of any single trait.

Choosing Among Alternative Phylogenetic Hypotheses

In the last phase of the tree-building process, the systematist compares different possible trees. For example, three taxa can be arranged in three different branching patterns (**FIG. E19-2**). In Figure E19-2, each branching pattern represents a different hypothesis about the evolutionary history of sharks, frogs, and rodents. Which hypothesis is most likely to represent the true history of the three taxa? The one in which the taxa on adjacent branches share synapomorphies. For example, imagine that a systematist identified a number of synapomorphies that are shared by the taxa frogs and rodents but are not found in sharks, but has found no synapomorphies that link sharks and frogs or sharks and rodents.

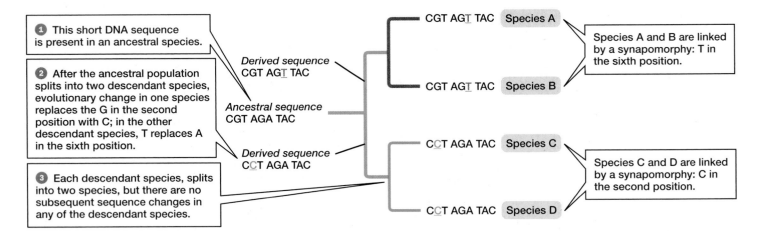

1 This short DNA sequence is present in an ancestral species.

2 After the ancestral population splits into two descendant species, evolutionary change in one species replaces the G in the second position with C; in the other descendant species, T replaces A in the sixth position.

3 Each descendant species, splits into two species, but there are no subsequent sequence changes in any of the descendant species.

Derived sequence CGT AGT TAC

Ancestral sequence CGT AGA TAC

Derived sequence CCT AGA TAC

CGT AGT TAC   Species A
CGT AGT TAC   Species B

Species A and B are linked by a synapomorphy: T in the sixth position.

CCT AGA TAC   Species C
CCT AGA TAC   Species D

Species C and D are linked by a synapomorphy: C in the second position.

▲ **FIGURE E19-1 Related taxa are linked by shared derived traits (synapomorphies)** A derived trait is one that has been modified from the ancestral version of the trait. When two or more taxa share a derived trait, the shared trait is said to be a synapomorphy. The hypothetical scenario illustrated here shows how synapomorphies arise.

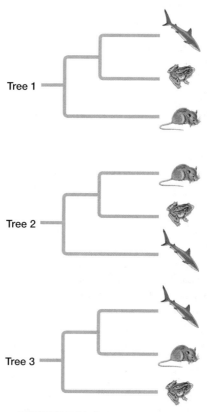

▲ FIGURE E19-2 **The three possible trees for three taxa**

In this case, tree 2 in Figure E19-2 is the best-supported hypothesis.

With large numbers of taxa, the number of possible trees grows dramatically. Similarly, a large number of traits also complicates the problem of identifying the tree best supported by the data. Fortunately, systematists have developed sophisticated computer programs to help cope with these complications.

Only Monophyletic Groups Are Named

Under the cladistic approach, phylogenetic trees play a key role in classification. Each named group should contain only organisms that are more closely related to one another than to any organisms outside the group. So, for example, the members of the clade Canidae (which includes dogs, wolves, foxes, and coyotes) are more closely related to each other than to any member of any other clade. Another way to state this principle is to say that each designated group should contain *all* of the living descendants of a common ancestor. In the terminology of cladistic systematics, such groups are said to be *monophyletic* (FIG. E19-3a). (Note that "monophyletic group" is therefore synonymous with "clade.")

Some names, especially names that predate the cladistic approach, designate groups that contain some, but not all, of the descendants of a common ancestor. Such groups are *paraphyletic* (FIG. E19-3b). For example, the taxon historically known as the reptiles—snakes, lizards, turtles, and crocodilians—is paraphyletic. To see why, examine the tree in Figure E19-3b. Find the branch that represents the common ancestor of crocodilians, snakes, lizards, and turtles (it is at the base of the tree). Then examine the tree again and make a list of all of the descendants of that common ancestor. Your list, if you performed this mental exercise correctly, includes the birds. That is, birds are part of the monophyletic group that includes all living descendants of the common ancestor that gave rise to crocodilians, snakes, lizards, and turtles. Therefore, the reptiles (Reptilia) constitute a monophyletic clade only if birds are included in the group. If we omit the birds, the taxon Reptilia is paraphyletic and, according to cladistic principles, is not a valid group name. Nonetheless, you will probably continue to encounter the word "reptiles" used in its older, technically incorrect sense, because so many people are accustomed to using it that way.

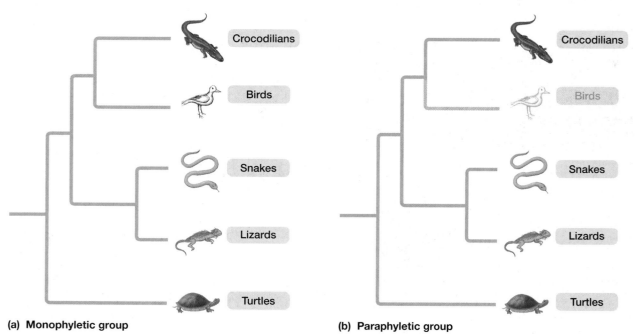

(a) Monophyletic group

(b) Paraphyletic group

▲ FIGURE E19-3 **Reptiles are a monophyletic group only if birds are included** Only groups that contain all of the descendants of a common ancestor are considered to be monophyletic groups. **(a)** The group consisting of turtles, lizards, snakes, birds, and crocodilians is monophyletic, because it includes all descendants of the group's common ancestor. **(b)** The group traditionally known as the reptiles is paraphyletic, because it excludes birds, which are descended from the group's common ancestor.

19.2 WHAT ARE THE DOMAINS OF LIFE?

If we picture the common ancestor of all living things as the trunk at the very base of the tree of life, we might ask: Which clades arose from the earliest branching of the trunk?

By the 1970s, most systematists had concluded from the evidence then available that early splits in the tree of life divided all species into five kingdoms. The five-kingdom system placed all prokaryotic organisms into a single kingdom and divided the eukaryotes into four kingdoms. Among the eukaryotes, the five-kingdom system recognized three kingdoms of multicellular organisms (plants, animals, and fungi) and placed all of the remaining, mostly single-celled, eukaryotes in a single kingdom.

As new data accumulated and understanding of phylogeny grew, however, scientific assessment of life's fundamental categories was gradually revised. A key element of this revision stemmed from the pioneering work of microbiologist Carl Woese, who showed that biologists had overlooked a key event in the early history of life, one that demanded a new and more evolutionarily accurate classification.

Woese and other biologists interested in the evolutionary history of microorganisms studied the biochemistry of prokaryotic organisms. The researchers, by studying nucleotide sequences of the genes that encode the RNA that is found in organisms' ribosomes, discovered that prokaryotes fall into two large groups, each with its own distinctive version of ribosomal RNA. Woese dubbed these two groups the Bacteria and the Archaea (FIG. 19-5). The very large number of differences between the ribosomal RNA sequences of Bacteria and Archaea indicate that their common ancestor may have lived more than 3 billion years ago.

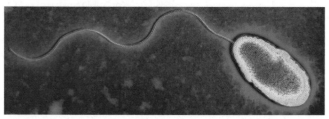

(a) A bacterium

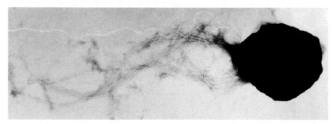

(b) An archaean

▲ FIGURE 19-5 **Two domains of prokaryotic organisms** Although similar in appearance, **(a)** *Pseudomonas aeruginosa,* in the domain Bacteria, and **(b)** *Methanococcus jannaschii,* in the domain Archaea, are less closely related than a mushroom and an elephant.

THINK CRITICALLY Given that bacteria and archaea are only very distantly related, why are they often similar in appearance?

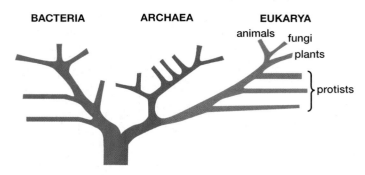

▲ FIGURE 19-6 **The tree of life** The three domains of life represent the three main "branches" on the tree of life. The term "protist" refers to the many eukaryotes that are not plants, animals, or fungi.

Despite superficial similarities in their appearance under the microscope, Bacteria and Archaea differ at the most fundamental molecular level; for example, they differ dramatically in the chemical composition of their cell walls and in the structure of their RNA polymerase molecules.

Bacteria and Archaea are no more closely related to one another than either one is to any eukaryote. The tree of life split into three parts very early in the history of life, long before the appearance of plants, animals, and fungi. This early split is reflected in a modern classification scheme that divides life into three domains: **Bacteria, Archaea,** and **Eukarya** (FIG. 19-6). Eukarya includes all organisms with eukaryotic cells, including plants, fungi, animals, and an array of clades of mostly single-celled organisms collectively known as *protists.* FIGURE 19-7 shows the evolutionary relationships among some members of the domain Eukarya.

CHECK YOUR LEARNING

Can you ...

• name and briefly describe the three domains of life?
• explain how scientists discovered that prokaryotes fall into two domains?

19.3 WHY DO CLASSIFICATIONS CHANGE?

As the emergence of the three-domain system shows, the hypotheses of evolutionary relationships on which classification is based are subject to revision as new data become available. Even the largest, most inclusive clades—which represent the earliest branchings of the tree of life—must sometimes be rearranged. Such changes at the top levels of classification occur only rarely, but at the other end of the classification hierarchy, among species designations, revisions are more frequent.

Species Designations Change When New Information Is Discovered

As researchers uncover new information, systematists regularly propose changes in species-level classifications. For

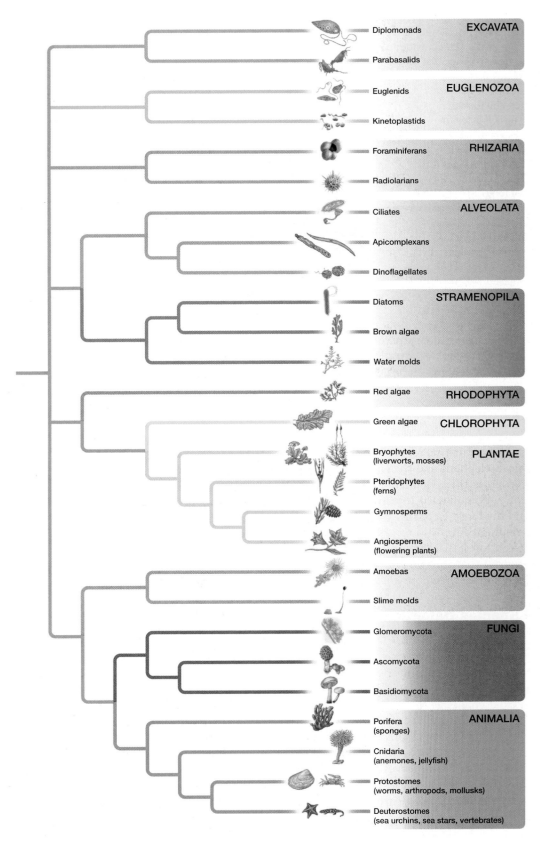

▲ **FIGURE 19-7 An evolutionary tree of eukaryotes** Some of the major evolutionary lineages within the domain Eukarya are shown.

HAVE YOU EVER
WONDERED...

When People Started Wearing Clothes?

Clothing decays pretty rapidly, so archeological discoveries can't answer the question of when our naked ancestors first started wearing clothes. But you might be surprised to learn that systematists can provide an answer. You might be even more surprised to learn that the answer comes from systematists who study lice. Most species of these parasitic insects live on only one host species, and human lice are no exception. The lice that can infest people include head lice that can survive only on the scalp, and body lice that can survive only in clothing (they leave the clothes temporarily to feed on the body, but can't thrive there for long). Systematists comparing the DNA of a number of louse species found that human head lice and body lice are close relatives. And, using a method known as the molecular clock, they estimated that the split between the two occurred between 80,000 and 170,000 years ago, most likely at the earlier end of that range. Because the common ancestor of the two kinds of lice could not have moved from the head to the body until there were clothes to live in, people must have been wearing clothes at least 80,000 years ago, probably even earlier. This timing suggests that we were already clothed when we left Africa and began to spread across Earth.

Body louse

example, until recently, systematists recognized two species of elephant, the African elephant and the Indian elephant. Now, however, we recognize three elephant species; the former African elephant is now divided into two species, the savanna elephant and the forest elephant. Why the change? Genetic analysis of elephants in Africa revealed that there is little gene flow between forest-dwelling and savanna-dwelling elephants. It turns out that the two groups of elephants are no more genetically similar than lions are to tigers.

The Biological Species Definition Can Be Difficult or Impossible to Apply

In some cases, systematists find themselves unable to say with certainty where one species ends and another begins. As discussed earlier (see Chapter 17), asexually reproducing organisms pose a particular challenge to systematists, because the criterion of interbreeding (the basis of the biological species definition that we have used in this text) cannot be used to distinguish among species. The irrelevance of this criterion in studies of asexual organisms leaves plenty of room for investigators to disagree about which asexual populations constitute a species, especially when comparing groups with similar phenotypes.

The difficulty of applying the biological species definition to asexual organisms applies to a significant portion of Earth's organisms. Most bacteria, archaea, and protists, for example, reproduce asexually most of the time. Some systematists argue that we need a more universally applicable definition of species, one that won't exclude asexual organisms and that doesn't depend on the criterion of reproductive isolation.

A number of alternative species definitions have been proposed, but none has been sufficiently compelling to displace completely the biological species definition. One alternative definition, however, has been gaining adherents in recent years. The *phylogenetic species concept* defines a species as the smallest possible group whose members descended from a common ancestor and share defining characteristics that distinguish them from other groups. In other words, if we draw an evolutionary tree that describes the pattern of ancestry among a collection of organisms, each distinctive branch on the tree constitutes a separate species. As you might suspect, rigorous application of the phylogenetic species concept would vastly increase the number of different species recognized by systematists.

Proponents and opponents of the phylogenetic species concept are currently engaged in a vigorous debate about the merits of this definition of species. Perhaps one day the phylogenetic species concept will replace the biological species concept as the "textbook definition" of species. In the meantime, classifications will continue to be debated and revised as systematists learn more about evolutionary relationships, particularly with the application of techniques used in molecular genetics.

CHECK YOUR LEARNING

Can you ...

- explain why phylogenetic classifications sometimes change?
- describe the limitations of the biological species concept, and explain how it differs from the phylogenetic species concept?

19.4 HOW MANY SPECIES EXIST?

The challenge of reconstructing the evolutionary history of Earth's species is complicated by the fact that most species remain undiscovered and undescribed. Scientists do not know exactly how many species share our world. Each year, around 15,000 new species are named, most of them insects, many from tropical rain forests. The total number of named species is currently about 1.6 million. However, scientists agree that many more species exist. A recent, well-received study estimated that 8.7 million species are present on Earth, and other studies have produced even higher estimates.

The number and variety of Earth's species constitute its **biodiversity.** Of all the species that have been identified thus far, about 5% are prokaryotes and protists. An additional 20% or so are plants and fungi, and the rest are animals. This distribution has little to do with the actual diversity of these organisms and a lot to do with the size of the organisms, how easy they are to classify, how accessible they are, and the number of scientists studying them. Historically, systematists have chiefly focused on large or conspicuous organisms in temperate regions, but biodiversity is greatest among small, inconspicuous organisms in the Tropics. In addition to the overlooked species on land and in shallow waters, an entire "continent" of species lies largely unexplored on the deep-sea floor. From the relatively limited samples available, scientists estimate that hundreds of thousands of unknown species may reside there.

Although about 5,000 species of prokaryotes have been described and named, most prokaryotic diversity remains undiscovered. Consider a study by Norwegian scientists, who analyzed DNA to count the number of different bacterial species present in a small sample of forest soil. To distinguish among species, the researchers arbitrarily defined bacterial DNA as coming from separate species if it differed by at least 30% from that of any other bacterial DNA in the sample. Using this criterion, they reported more than 4,000 species of bacteria in their soil sample and an equal number of species in a sample of shallow marine sediment.

Our ignorance of the full extent of life's diversity adds a new dimension to the tragedy of the destruction of tropical rain forests. Although these forests cover only about 6% of Earth's land area, they are believed to be home to two-thirds of the world's existing species, most of which have never been studied or named. Because these forests are being destroyed so rapidly, Earth is losing many species that we will never even know existed! Consider this: In 1990, a new species of primate, the black-faced lion tamarin, was discovered in a small patch of dense rain forest on an island just off the east coast

of Brazil (**FIG. 19-8**). Had the patch of forest been cut before this squirrel-sized monkey was discovered, its existence would have remained undocumented. At current rates of deforestation, most of the tropical rain forests, with their undescribed wealth of life, will be gone by the end of the century.

CHECK YOUR LEARNING

Can you …
- explain why the number of described species is much lower than the actual number?
- explain why it is difficult to accurately estimate how many species are on Earth?

▶ **FIGURE 19-8**
The black-faced lion tamarin Researchers estimate that no more than 400 individuals remain in the wild; captive breeding may be the black-faced lion tamarin's only hope for survival.

CASE STUDY \ **REVISITED**

Origin of a Killer

What evidence has persuaded evolutionary biologists that HIV originated in apes and monkeys? To understand the evolutionary thinking behind this conclusion, examine the evolutionary tree shown in **FIGURE 19-9**. This tree illustrates the phylogeny of HIV and its close relatives, the simian immunodeficiency viruses (SIVs), as revealed by a comparison of RNA sequences among different viruses.

Notice the positions on the tree of the four human viruses (two strains of HIV-1 and two of HIV-2; a strain is a genetically distinct subgroup of a particular type of virus). The branch leading to strain 1 of HIV-1 is directly adjacent to the branch leading to strain 1 of chimpanzee SIV. These adjacent branches indicate that strain 1 of HIV-1 is more closely related to a chimpanzee virus than to strain 2 of HIV-1. Similarly, strain 1 of HIV-2 is more closely related to pig-tailed macaque SIV than to strain 2 of HIV-2.

▶ **FIGURE 19-9 Evolutionary analysis helps reveal the origin of HIV** In this phylogeny of some immunodeficiency viruses, the viruses with human hosts do not cluster together. This lack of congruence between the evolutionary histories of the viruses and their host species suggests that the viruses must have jumped between host species.

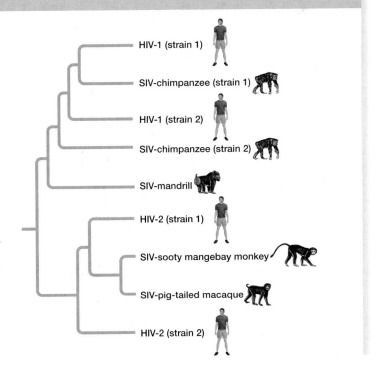

HIV-1 (strain 1)
SIV-chimpanzee (strain 1)
HIV-1 (strain 2)
SIV-chimpanzee (strain 2)
SIV-mandrill
HIV-2 (strain 1)
SIV-sooty mangebay monkey
SIV-pig-tailed macaque
HIV-2 (strain 2)

Both HIV-1 and HIV-2 are more closely related to ape or monkey viruses than to one another.

The only way for the evolutionary history shown in the tree to have emerged is if viruses jumped between host species. If HIV had evolved strictly within human hosts, the human viruses would be each other's closest relatives. Because the human viruses do not cluster together on the phylogenetic tree, we can infer that cross-species infection occurred, probably on multiple occasions. The most likely means of transmission is human consumption of monkeys (HIV-2) and chimpanzees (HIV-1). Recently, a strain of

SIV that is especially closely related to HIV-1 was found in members of a population of chimpanzees that inhabit forests in the southeastern corner of the Central African country of Cameroon. It is likely that viruses from this chimpanzee population made the chimpanzee-to-human jump that started the HIV epidemic.

CONSIDER THIS Can understanding the evolutionary origin of HIV help researchers devise better ways to treat AIDS and control its spread? More generally, how can evolutionary thinking help advance medical research?

CHAPTER REVIEW

 Go to **MasteringBiology** for practice quizzes, activities, eText, videos, current events, and more.

Summary of Key Concepts

Answers to Think Critically, Evaluate This, Multiple Choice, and Fill-in-the-Blank questions can be found in the Answers section at the back of the book.

19.1 How Are Organisms Named and Classified?

The scientific name of an organism is composed of its genus name and species name. Systematists use anatomical and molecular similarities among organisms to reconstruct the evolutionary relationships among species, and depict the results of their reconstructions in tree diagrams. On the basis of these evolutionary trees, systematists name clades (groups that include the species descended from a common ancestor). Clades nest within larger clades to form a nested hierarchy of categories. In Linnaean classification, the different clades in a hierarchy are assigned taxonomic ranks. The eight major ranks, in order of decreasing inclusiveness, are domain, kingdom, phylum, class, order, family, genus, and species.

19.2 What Are the Domains of Life?

The three domains of life, each representing one of three main branches of the tree of life, are Bacteria, Archaea, and Eukarya. Plants, fungi, and animals are among the clades within the domain Eukarya.

19.3 Why Do Classifications Change?

Classifications are subject to revision as new information is discovered. Species boundaries may be hard to define, particularly in the case of asexually reproducing species. However, systematics is essential for precise communication and contributes to our understanding of the evolutionary history of life.

19.4 How Many Species Exist?

Although only about 1.6 million species have been named, the total number of species is 8.7 million or higher. New species are being identified at the rate of 15,000 annually, mostly in tropical rain forests.

Key Terms

Archaea   *346*
Bacteria   *346*
biodiversity   *348*
clade   *343*
domain   *343*
Eukarya   *346*
genus   *341*
phylogeny   *341*
scientific name   *341*
species   *341*
systematics   *341*
taxonomy   *341*

Thinking Through the Concepts

Multiple Choice

1. To say that species A is more closely related to species B than to species C is to say that A and B
a. have a more ancient common ancestor than do A and C.
b. have a more recent common ancestor than do A and C.
c. are more similar in appearance than are A and C.
d. live in the same area, but C does not live there.

2. To be informative for reconstructing the phylogeny of a group of taxa, a characteristic must be
a. present in all the taxa due to its presence in a common ancestor.
b. present in all the taxa due to convergent evolution.
c. an anatomical feature.
d. a DNA sequence.

3. Which of the following illustrates the correct way to present the scientific name of humans?
a. *Homo Sapiens*   c. *homo sapiens*
b. Homo sapiens   d. *Homo sapiens*

4. In modern systematics, classifications are expected to
a. group organisms that share similar appearance.
b. include Linnaean taxonomic ranks.
c. reflect evolutionary history.
d. never change once established.

5. Which of the following includes all the domains that consist of prokaryotic organisms?
a. Bacteria and Fungi   c. Bacteria and Archaea
b. Archaea and Eukarya   d. Bacteria and Eukarya

Fill-in-the-Blank

1. The science of naming and classifying organisms is called _____. The related science of reconstructing and depicting evolutionary history is called _____. A group consisting of all organisms descended from a particular common ancestor is a(n) _____.

2. A scientific name consists of a(n) _____ name followed by a(n) _____ name. Both parts of a scientific name are in _____ (a language). The first letter of the first word in a scientific name is always _____, and both parts of the name are printed in _____ letters.

3. In Linnaean classification, the eight major taxonomic ranks, in descending order of inclusiveness, are _____, _____, _____, _____, _____, _____, _____, and _____. The three domains of life are _____, _____, and _____.

4. Systematists determine the evolutionary relationships among species mainly on the basis of similarities in _____ and _____.

5. The biological species definition is difficult to apply to organisms that _____. An alternative species definition, known as the _____, does not have this limitation.

6. The number of named species is about_____, but the actual number of species on Earth is estimated to be about _____ or higher.

Review Questions

1. What contributions did Linnaeus and Darwin make to modern taxonomy?

2. What features would you study to determine whether a dolphin is more closely related to a fish or to a bear?

3. What techniques might you use to determine whether the extinct cave bear is more closely related to a grizzly bear or to a black bear?

4. Only a small fraction of the total number of species on Earth has been scientifically described. Why?

5. In England, "daddy longlegs" refers to a long-legged fly, but the same name refers to a spider-like animal in the United States. How do scientists attempt to avoid such confusion?

6. Why are species designations of asexually reproducing organisms more likely to differ among different systematists than are the species designations of sexually reproducing organisms?

Applying the Concepts

1. The pressures created by human population growth and economic expansion place storehouses of biological diversity such as the Tropics in peril. The seriousness of the situation is clear when we consider that probably only 1 out of every 20 tropical species is known to science. What arguments can you make for preserving biological diversity in poor and developing countries, such as those in many areas of the Tropics? Does such preservation require that these countries sacrifice economic development? Suggest some solutions to the conflict between the growing demand for resources and the importance of conserving biodiversity.

2. During major floods, only the topmost branches of submerged trees may be visible above the water. If you were asked to sketch the branches below the surface of the water solely on the basis of the positions of the exposed tips, you would be attempting a reconstruction somewhat similar to the "family tree" by which systematists link various organisms according to their common ancestors (analogous to branching points). What sources of error do both exercises share? What advantages do modern systematists have?

3. Consider the following list of groups: (1) protists, (2) fungi, (3) seedless plants (ferns, mosses, and liverworts), (4) prokaryotes (bacteria and archaea), and (5) animals. Using Figures 19-6, 19-7, 22-4, 23-5, and 24-1 for reference, identify the monophyletic groups on the list.

20

THE DIVERSITY OF PROKARYOTES AND VIRUSES

Unwelcome Dinner Guests

ONE EVENING during his time as a student at the University of Michigan, Andrew Lekas ate a burrito at a favorite restaurant near campus. A few hours later, he began to feel sick, with symptoms that included vomiting, diarrhea, headache, and weakness. The illness ultimately became severe enough to require a hospital stay and more than a week in bed. What sickened Lekas? The lettuce in his burrito was contaminated.

As bad as Lekas's experience was, it could have been much worse. Other victims of foodborne illness have suffered more serious consequences. For example, Stephanie Smith, a former dance instructor from Minnesota, is paralyzed from the waist down as a result of the severe illness (hemolytic uremic syndrome) she developed after eating a tainted hamburger.

Regrettably, the ill effects of consuming contaminated food are all too common. The Centers for Disease Control and Prevention estimates that U.S. residents experience an astonishing 48 million cases of foodborne illness each year. Some of these cases are severe. For example, in 2011, at least 30 deaths were caused by a single source of cantaloupes, and in Germany, sprouts from a single supplier caused more than 850 cases of hemolytic uremic syndrome and 53 deaths. Overall, consumption of contaminated food results in about 125,000 hospitalizations and 3,000 deaths in the United States each year.

Hamburgers should be cooked thoroughly to eliminate dangerous bacteria.

What is it that contaminates food and causes so much illness? Bacteria. The nutrients in the food you consume during meals and snacks can also provide sustenance for a wide variety of disease-causing bacteria. Some of these invisible diners may accompany your lunch to your digestive tract and take up residence there, causing unpleasant symptoms or, in many cases, serious illness.

Devising effective strategies for protecting our food supply against bacterial contamination depends in part on how well we understand the biology of bacteria. What do scientists know about bacteria and their fellow prokaryotes, the archaea?

AT A GLANCE

20.1 WHICH ORGANISMS ARE MEMBERS OF THE DOMAINS ARCHAEA AND BACTERIA?

Earth's first organisms were prokaryotes, single-celled organisms that lacked organelles such as the nucleus, chloroplasts, and mitochondria. (See Chapter 4 for a comparison of prokaryotic and eukaryotic cells.) For the first 1.5 billion years or more of life's history, all life was prokaryotic. Even today, prokaryotes are extraordinarily abundant. A drop of seawater contains hundreds of thousands of prokaryotic organisms, and a spoonful of soil contains billions. The average human body is home to trillions of prokaryotes, which live on the skin, in the mouth, and in the stomach and intestines. In terms of abundance, prokaryotes are Earth's predominant form of life.

Prokaryotes are usually very small, ranging from about 0.2 to 10 micrometers in diameter. In comparison, the diameters of eukaryotic cells range from about 10 to 100 micrometers. About 250,000 average-sized prokaryotes could congregate on the period at the end of this sentence, though a few species are larger. The largest known bacterium (*Thiomargarita namibiensis*) can be up to 700 micrometers in diameter, as big as the tip of a ballpoint pen and visible to the naked eye.

The cell walls produced by prokaryotic cells give characteristic shapes to different types of prokaryotes. The most common shapes are spherical, rod-shaped, and corkscrew-shaped (**FIG. 20-1**).

Bacteria and Archaea Are Fundamentally Different

Two of life's three domains, **Bacteria** and **Archaea,** consist entirely of prokaryotes. Bacteria and archaea are superficially similar in appearance under the microscope, but they have striking structural and biochemical differences that reveal the ancient evolutionary separation between the two groups (**TABLE 20-1**). For example, the cell walls of bacteria are strengthened by molecules of *peptidoglycan,* a polysaccharide that also incorporates some amino acids. Peptidoglycan is unique to bacteria, and the cell walls of archaea do not contain it. Bacteria and archaea also differ in the structure and composition of their plasma membranes, ribosomes, and enzymes involved in RNA synthesis, as well as in the mechanics of basic processes such as transcribing the instructions encoded in DNA and synthesizing proteins.

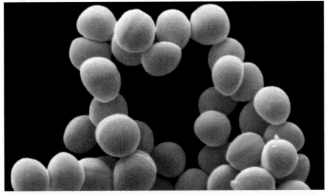

(a) Spherical

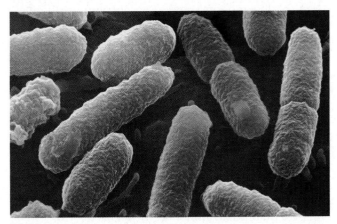

(b) Rod-shaped

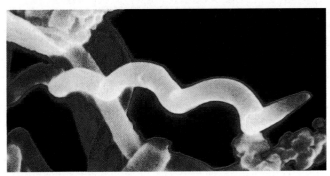

(c) Corkscrew-shaped

▲ **FIGURE 20-1** **Three common prokaryote shapes (a)** Spherical bacteria of the genus *Staphylococcus,* **(b)** rod-shaped bacteria of the genus *Escherichia,* and **(c)** corkscrew-shaped bacteria of the genus *Borrelia.*

TABLE 20.1	Differences Between Organisms in the Domains Archaea and Bacteria	
	Archaea	**Bacteria**
Peptidoglycan in cell wall	Absent	Present
Membrane lipid structure	Branched hydrocarbons	Unbranched hydrocarbons
Histone proteins associated with DNA	Present	Absent
Introns	Present in some genes	Absent
RNA polymerase	Several types	One type
Amino acid that initiates protein synthesis	Methionine	Formylmethionine

(For a refresher on the functions of histones, introns, and RNA polymerase, see Chapter 13.)

Classification Within the Prokaryotic Domains Is Based on DNA Sequences

The sharp differences between archaea and bacteria make distinguishing the two domains a straightforward matter, but classification within each domain is challenging. Historically, the main challenge arose because prokaryotes are very small and structurally simple; they do not exhibit the huge array of anatomical differences that can be used to infer the evolutionary history of plants, animals, and other eukaryotes. Consequently, prokaryotes have historically been classified on the basis of such features as shape, means of locomotion, pigments, nutrient requirements, the appearance of colonies (clusters of individuals that descended from a single cell), and staining properties. For example, the Gram stain, a staining technique, distinguishes two types of cell wall construction in bacteria. Based on the results of the stain, many bacteria can be classified as *gram-positive* or *gram-negative*.

In recent years, DNA sequence comparisons have revealed that many of the observable similarities and differences used in traditional prokaryotic classification do not accurately reflect evolutionary history. Today, prokaryote classification is based almost entirely on DNA sequence data. Systematists have used such data to determine the extent to which the traditional groupings within Bacteria and Archaea represent *clades* (groups of species united by descent from a common ancestor) and to identify additional clades. DNA sequence data have also allowed systematists to include in their classifications species that could not previously be confidently classified, because they could not be cultured in the lab and therefore could not be readily observed. Prokaryote classifications now even include some species that are so far known only from DNA sequences found during exploratory censuses in which all the DNA present in a microbial community is sampled and sequenced.

At present, systematists recognize about 30 major named groups within Bacteria and five within Archaea. Examples of some of the larger clades in Bacteria include Cyanobacteria, which includes many ecologically important photosynthetic species; Firmicutes, which includes the species that convert milk to yogurt and the species that causes tetanus; and Proteobacteria, which includes many of the species that cause food poisoning. The largest clades in Archaea are Crenarchaeota, which includes some species that live in super-hot environments, and Euryarchaeota, which includes some species that live in our intestines and produce methane (the main component of natural gas).

Determining the Evolutionary History of Prokaryotes Is Difficult

You may have noticed that, unlike the chapters in this text devoted to eukaryotic organisms, this chapter does not present an evolutionary tree for prokaryotes. No tree is shown because there is very little agreement among systematists with regard to the evolutionary branching patterns within Bacteria and Archaea. Although DNA comparisons have revealed clades within the two domains, they have not yielded a consensus view of the evolutionary relationships between those clades. Why haven't the usual methods of systematics yielded a widely accepted tree? One likely reason is that in prokaryotes, processes such as conjugation (described later in this chapter) make it possible for genes to move from one species to another. If a prokaryote species acquired some of its genes from other species, then different parts of its genome will have different evolutionary histories. Past (and ongoing) genetic mixing among even distantly related prokaryotes has resulted in a very complex evolutionary history that is extremely difficult to reconstruct.

CHECK YOUR LEARNING
Can you ...
- describe some differences between bacteria and archaea?
- describe the typical sizes and shapes of prokaryotes?

20.2 HOW DO PROKARYOTES SURVIVE AND REPRODUCE?

The abundance of prokaryotes is due in large part to adaptations that allow members of the two prokaryotic domains to inhabit and exploit a wide range of environments. In this section, we discuss some of the traits that help prokaryotes survive and thrive.

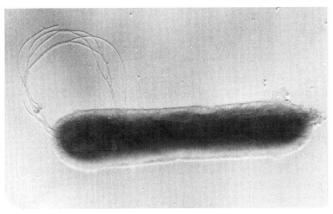

(a) A flagellated archaean

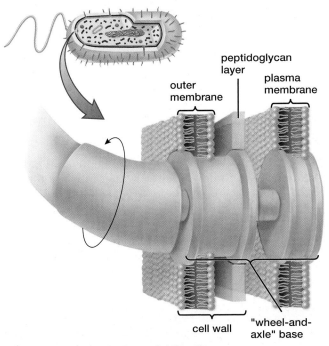

peptidoglycan
layer

outer
membrane

plasma
membrane

cell wall

"wheel-and-
axle" base

(b) The structure of the bacterial flagellum

▲ **FIGURE 20-2 The prokaryote flagellum (a)** A flagellated
archaean of the genus *Aquifex* uses its flagella to move toward
favorable environments. **(b)** In bacteria, a unique "wheel-and-axle"
arrangement anchors the flagellum within the cell wall and plasma
membrane, enabling the flagellum to rotate rapidly. This diagram
represents the flagellum of a gram-negative bacterium; the flagella
of gram-positive bacteria lack the outermost "wheels."

Some Prokaryotes Are Motile

Many bacteria and archaea adhere to a surface or drift passively in liquid surroundings, but some are motile—they can
move about. Many of these motile prokaryotes have **flagella**
(singular, flagellum), hair-like extensions that can rotate rapidly to propel the organism through its liquid environment.
Prokaryotic flagella may appear singly at one end of a cell,
in pairs (one at each end of the cell), as a tuft at one end of
the cell (**FIG. 20-2a**), or scattered over the entire cell surface.
The use of flagella to move allows prokaryotes to disperse into
new habitats, migrate toward nutrients, and leave unfavorable environments.

The structure of prokaryote flagella is different from the
structure of eukaryotic flagella (see Chapter 4 for a description of
the eukaryotic flagellum). In bacterial flagella, a unique wheel-
like structure embedded in the bacterial membrane and cell wall
allows the flagellum to rotate (**FIG. 20-2b**). Archaeal flagella
are thinner than bacterial flagella and are constructed of different proteins. The structure of the archaeal flagellum, however,
is not yet as well understood as that of the bacterial flagellum.

Many Bacteria Form Protective Films on Surfaces

Many prokaryotes continually synthesize signaling molecules and secrete them to the surrounding environment.
If a large number of prokaryotes gathers in one place, the
signaling molecules become concentrated enough that the
molecules begin to move across the plasma membranes of
neighboring cells and combine with receptors inside the
cells. The activated receptors trigger cellular processes that
would not otherwise be activated. Thus, the behavior of
prokaryotes may change when population density grows sufficiently high, a process known as *quorum sensing.*

Among the most common changes induced by quorum
sensing is formation of biofilms. In a **biofilm,** one or more
species of prokaryote aggregate to form a community that is
typically surrounded by sticky protective slime. The slime,
composed of polysaccharide or protein, is secreted by the
prokaryotes and both protects them and helps them adhere
to surfaces. One familiar biofilm is dental plaque, which is
formed by the bacteria that inhabit the mouth (**FIG. 20-3**).

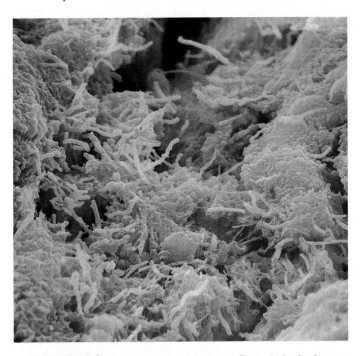

▲ **FIGURE 20-3 The cause of tooth decay** Bacteria in the human
mouth form a slimy biofilm that helps them cling to tooth enamel
and protects them from threats in the environment. In this micrograph, individual bacteria (colored green and yellow) are visible,
embedded in the brown biofilm. The bacteria-laden biofilm can
cause tooth decay.

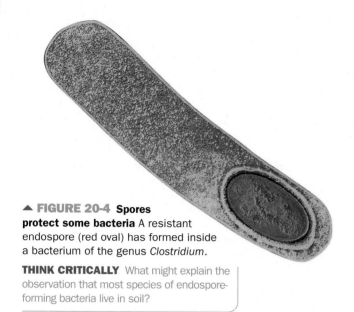

▲ **FIGURE 20-4 Spores protect some bacteria** A resistant endospore (red oval) has formed inside a bacterium of the genus *Clostridium*.

THINK CRITICALLY What might explain the observation that most species of endospore-forming bacteria live in soil?

A few of the bacteria that commonly cause foodborne illness form endospores. For example, *Bacillus cereus,* a bacterial species that includes strains that cause vomiting or diarrhea in people unlucky enough to consume them, forms endospores that are widespread in soil and dust. If some spores find their way into warm, moist food, they can develop and give rise to a thriving population of bacteria. *B. cereus* spores are somewhat resistant to heat, but fortunately for us, they can be destroyed by thorough cooking.

Although it is unsurprising that prokaryotes might thrive amidst the energy-rich substances that make up human food, our foodstuffs are hardly the only environment in which prokaryotes flourish. What are some of the more extreme environments in which bacteria and archaea can be found?

The protection afforded by biofilms helps defend the embedded prokaryotes against a variety of attacks, including those launched by antibiotics and disinfectants. As a result, biofilms can be very difficult to eradicate. Many infections of the human body take the form of biofilms, including those responsible for tooth decay, gum disease, and ear infections. Biofilms also cause many of the hospital-acquired infections that affect 725,000 Americans each year and kill 75,000 of them. Such biofilms can form in wounds and surgical incisions, as well as on implanted medical devices such as catheters, pacemakers, and artificial hips and knees.

Protective Endospores Allow Some Bacteria to Withstand Adverse Conditions

When environmental conditions become inhospitable, many rod-shaped bacteria form protective structures called **endospores.** An endospore, which forms inside a bacterium, consists of the bacterium's genetic material and a few enzymes encased within a thick protective coat (**FIG. 20-4**). After an endospore forms, the bacterial cell that contains it breaks open, and the spore is released to the environment. Metabolic activity ceases until the spore encounters favorable conditions, at which time metabolism resumes and the spore develops into an active bacterium.

Endospores are resistant even to extreme environmental conditions. Some can withstand boiling for an hour or more. Endospores are also able to survive for extraordinarily long periods. In the most astonishing example of such longevity, scientists discovered endospores that had been sealed inside rock for 250 million years. After being carefully extracted from their rocky tomb, the spores were incubated in test tubes. Amazingly, live bacteria developed from the ancient spores, which were older than the oldest dinosaur fossils.

Prokaryotes Are Specialized for Specific Habitats

Prokaryotes occupy virtually every habitat, including those where extreme conditions keep out other forms of life. For example, some bacteria thrive in near-boiling environments, such as the hot springs of Yellowstone National Park (**FIG. 20-5**). Many archaea live in even hotter environments, including deep-sea vents, where superheated water is spewed through cracks in Earth's crust at temperatures

▲ **FIGURE 20-5 Some prokaryotes thrive in extreme conditions** Hot springs harbor bacteria and archaea that are both heat and mineral tolerant. Several prokaryote species paint these hot springs in Yellowstone National Park with vivid colors; each species is confined to a specific area determined by temperature range.

THINK CRITICALLY Some of the enzymes that have important uses in molecular biology procedures are extracted from prokaryotes that live in hot springs. Can you guess why?

HAVE YOU EVER WONDERED ...

Unpleasant breath odors are caused mainly by prokaryotes that live in the mouth. The warm, moist human mouth cavity hosts a diverse microbial community that includes more than 2,000 prokaryote species. Many of these species acquire energy and nutrients by breaking down mucus, food particles, and dead cells. The by-products of this breakdown can include foul-smelling gases, some of which are also emitted by feces or decaying bodies.

What Causes Bad Breath?

The highest concentration of bad-breath prokaryotes is found at the base of the tongue. This location may be especially hospitable to microbes owing to the accumulation of mucus that drains down into the back of the throat from the nose. So, if you gargle with antiseptic mouthwash to control bad breath, thrust your tongue forward so the mouthwash can reach the base of your tongue.

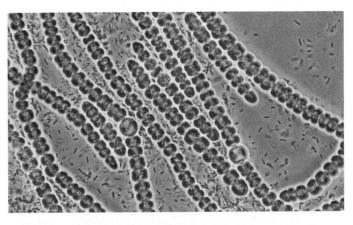

▲ **FIGURE 20-6 Photosynthetic prokaryotes** Photosynthetic cyanobacteria make up the filaments shown in this micrograph.

of up to 230°F (110°C). Prokaryotes can also survive at the extremely high pressures found on the sea floor and deep beneath Earth's surface, and in very cold environments, such as in Antarctic sea ice.

Extreme chemical conditions do not prevent colonization by prokaryotes, either. Thriving colonies of bacteria and archaea live in the Dead Sea, for example, where a salt concentration seven times that of the oceans precludes all other life, and in waters that are as acidic as vinegar or as alkaline as household ammonia. Given their ability to survive such extreme environments, it is not surprising that prokaryote communities also reside in a full range of more moderate habitats, including in and on the human body.

No single species of prokaryote, however, is as versatile as these examples may suggest. In fact, most prokaryotes are specialists. One species of archaea that inhabits deep-sea vents, for example, grows optimally at 223°F (106°C) and stops growing altogether at temperatures below 194°F (90°C). A bacteria species that lives deep underground has never been found less than 1.2 miles beneath the Earth's surface. Bacteria that live on the human body are also specialized; different species colonize the skin, the mouth, the respiratory tract, the large intestine, and the urogenital tract. (To learn more about how the bacteria in our bodies might affect our health, see "Health Watch: Is Your Body's Ecosystem Healthy?" on page 358.)

Prokaryotes Have Diverse Metabolisms

Prokaryotes are able to colonize diverse habitats in part because they have evolved diverse methods of acquiring energy and nutrients from the environment. For example, unlike eukaryotes, many prokaryotes are **anaerobes;** their metabolisms do not require oxygen. Their ability to inhabit oxygen-free environments allows prokaryotes to exploit habitats in which eukaryotes could not survive. Some anaerobes, such as

many of the archaea found in hot springs and the bacterium that causes tetanus, are actually poisoned by oxygen. Others are opportunists, engaging in anaerobic respiration when oxygen is lacking and switching to aerobic respiration (a more efficient process; see Chapter 8) when oxygen becomes available. Many prokaryotes are strictly aerobic and require oxygen at all times.

Whether aerobic or anaerobic, different prokaryote species acquire energy from an amazing array of sources. Some species of bacteria use photosynthesis to capture energy directly from sunlight (FIG. 20-6). Like green plants, photosynthetic bacteria possess chlorophyll. Most species produce oxygen as a by-product of photosynthesis; the extremely abundant marine photosynthetic bacterium *Prochlorococcus* produces about 20% of all the oxygen gas in the atmosphere. However, not all photosynthetic bacteria produce oxygen. Some, known as the sulfur bacteria, use hydrogen sulfide (H_2S) instead of water (H_2O) in photosynthesis, releasing sulfur instead of oxygen. No photosynthetic archaea are known.

Nonphotosynthetic prokaryotes extract energy from a wide assortment of substances. Prokaryotes subsist not only on the sugars, carbohydrates, fats, and proteins that we usually think of as foods, but also on compounds that are inedible or even poisonous to humans, including petroleum, methane, and solvents such as benzene and toluene. Some prokaryotes can even metabolize inorganic molecules, including hydrogen, sulfur, ammonia, and iron. This ability is one of the reasons that prokaryotes can live in habitats in which sunlight is absent and organic molecules scarce. For example, one bacterial species that lives deep underground metabolizes hydrogen that is produced when radiation from radioactive uranium breaks apart water molecules.

Prokaryotes Reproduce by Fission

Most prokaryotes reproduce asexually by **prokaryotic fission** (also called binary fission), a form of cell division that is much simpler than mitotic cell division (see

Health WATCH

Is Your Body's Ecosystem Healthy?

Microbiologists like to say, "You are born 100% human, but you die 90% microbial." A human fetus in the womb is more or less sterile, but as it passes down the birth canal to be born, it picks up some of the bacteria that live there, and more are transferred from the first hands to touch the new baby. From that point on, microbes accumulate steadily, acquired from the environment, from food, and from other humans. By the time a child is three years old, its body is home to an entire ecosystem of microorganisms. This "microbiome" includes hundreds of different species, living in huge numbers in the nose and mouth, on the scalp, in the urogenital tract, in the gut, and on almost every skin surface. All told, the microbiome of a typical person includes about 100 trillion microbial cells, about 10 microbial cells for every human one.

It is becoming increasingly apparent to physicians and researchers that the inhabitants of the microbiome are not mere hitchhikers, but instead play an important role in human health. Although some of the benefits provided by our microbial partners are well understood, such as the production by gut bacteria of nutrients essential to humans, most of the details about how the microbiome contributes to our health are more mysterious. However, much evidence suggests that its contribution is important. For example, researchers have shown that in infants with necrotizing enterocolitis, an often-fatal bowel disease, the composition of the community of microbes in the gut is different from that present in healthy babies. Similarly aberrant digestive tract microbiomes have been found in adults with bowel disorders, and even in people with disorders not directly related to digestion, such as diabetes and autoimmune diseases. Such findings have led to the hypothesis that, because a person's microbiome, like a coral reef or a tropical rain forest, is a diverse ecosystem characterized by complex interactions among species, disruptions of this ecosystem compromise the microbial ecosystem's ability to perform functions essential to human health. However, scientists are not yet able to say for sure if the disrupted microbiomes of diseased individuals are a cause or a consequence of disease.

Confirmation of the microbiome's possible role in maintaining health depends on improved knowledge of its composition and characteristics. For that reason, a major ongoing research effort is directed at identifying and characterizing all of the microbial species that compose the microbiome, and at identifying exactly how microbiomes differ among different people. The job is challenging, because the traditional way of identifying a prokaryotic species—by growing it in a culture dish in the lab—is not effective for the many species that cannot be easily cultured. Scientists are now focused on the alternative approach of taking a sample of a whole microbiome and sequencing all of the DNA present in it. Some of the initial results of this approach are astonishing. For example, samples taken from 18 different locations on the skin and in the airways, mouths, vaginas, and digestive tracts of 242 people yielded a total of about 10,000 microbial species and 8 million different genes, about 350 times as many genes as are present in the human genome.

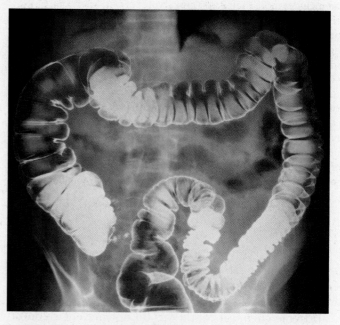

▲ **FIGURE E20-1 Transplanted feces could cure diseased bowels** By restoring a healthy microbial ecosystem in the recipient, fecal transplants can treat infected bowels and perhaps diabetes, obesity, and other disorders as well.

Further study of these microbial genes may reveal their functions and how they contribute to their human hosts.

In the meantime, some physicians are pushing ahead with treatments designed to restore ecological health to the microbiomes of sick people. For example, a few doctors have begun using fecal transplants to treat patients with severe bowel infections (**FIG. E20-1**). In this treatment, a small amount of feces from a healthy donor is transplanted into the bowel of the sick person, with the hope that the transplanted microbial community will become established and spread, displacing the harmful microbes responsible for the infection. The physicians using this treatment report very high success rates, much higher than those typical of treatment with antibiotics, which have been shown to devastate the gut microbiome. These reports have encouraged funding agencies in some countries to overcome the yuck factor and fund large-scale clinical trials of the fecal transplant treatment.

EVALUATE THIS As part of a study on the relationship between inflammatory bowel disease and the microbiome, researchers analyzed samples from the digestive tracts of 20 sets of identical twins in which one twin had the disease and the other did not. The researchers were able to determine the microbial diversity in each sample. Do you expect that samples from the healthy group will differ from samples from the ill group? If so, in what way? Why? Why did the researchers use twins in this study?

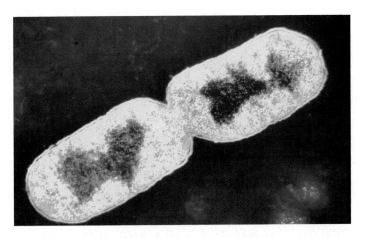

▲ **FIGURE 20-7** **Reproduction in prokaryotes** Prokaryotic cells reproduce by prokaryotic fission. In this color-enhanced electron micrograph, an *Escherichia coli*, a normal inhabitant of the human intestine, is dividing.

THINK CRITICALLY What is the main advantage of prokaryotic fission, compared to sexual reproduction?

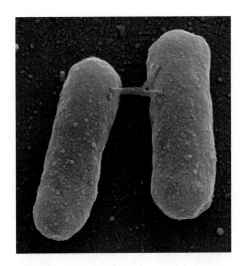

▲ **FIGURE 20-8** **Conjugation: prokaryotic "mating"** During conjugation, one prokaryote acts as a donor, transferring DNA to the recipient. In this micrograph, two *Escherichia coli* are connected by a sex pilus. The sex pilus will retract, drawing the recipient bacterium to the donor bacterium.

Chapter 9 for a detailed description of prokaryotic fission). Prokaryotic fission produces genetically identical copies of the original cell (**FIG. 20-7**). Under ideal conditions, some prokaryotic cells can divide about once every 20 minutes, potentially giving rise to sextillions (10^{21}) of offspring in a single day.

Rapid reproduction allows bacterial populations to evolve quickly. Recall that many mutations, the source of genetic variability, are the result of mistakes in DNA replication during cell division. Thus, the rapid, repeated cell division of prokaryotes provides ample opportunity for new mutations to arise and also allows mutations that enhance survival to spread quickly.

Prokaryotes May Exchange Genetic Material Without Reproducing

Although prokaryotic reproduction is generally asexual and does not involve genetic recombination, some bacteria and archaea nonetheless exchange genetic material. In these species, DNA is transferred from a donor to a recipient in a process called **conjugation.** The plasma membranes of two conjugating prokaryotes fuse temporarily to form a cytoplasmic bridge across which DNA travels. In bacteria, donor cells may use specialized extensions called *sex pili* that attach to a recipient cell, drawing it closer to allow conjugation (**FIG. 20-8**). Conjugation produces new genetic combinations that may allow the resulting bacteria to survive under a greater variety of conditions. In some cases, genetic material may be exchanged even between individuals of different species. Much of the DNA transferred during conjugation is contained within a structure called a **plasmid,** a small, circular DNA molecule that is separate from the single prokaryote chromosome.

CHECK YOUR LEARNING

Can you ...

- describe the range of environments inhabited by prokaryotes and the variety of methods by which they acquire energy?
- describe adaptations that help protect prokaryotes from environmental threats?
- explain how prokaryotes reproduce and exchange genetic material?

20.3 HOW DO PROKARYOTES AFFECT HUMANS AND OTHER ORGANISMS?

Although they are largely invisible to us, prokaryotes play a crucial role in life on Earth. Plants and animals (including humans) are utterly dependent on prokaryotes. Prokaryotes help plants and animals obtain vital nutrients and help break down and recycle wastes and dead organisms. We could not survive without prokaryotes, but their impact on us is not always beneficial. Some of humanity's most deadly diseases stem from microbes.

Prokaryotes Play Important Roles in Animal Nutrition

Many eukaryotic organisms depend on close associations with prokaryotes. For example, most animals that eat leaves—including cattle, rabbits, koalas, and deer—can't themselves digest cellulose, the principal component of plant cell walls. Instead, these animals depend on certain bacteria that have the ability to break down cellulose. These bacteria live in the animals' digestive tracts, where they

(a) Nodules on roots

(b) Nitrogen-fixing bacteria within nodules

▲ **FIGURE 20-9 Nitrogen-fixing bacteria in root nodules (a)** Special chambers called nodules on the roots of a legume provide a protected environment for nitrogen-fixing bacteria. **(b)** This scanning electron micrograph shows the nitrogen-fixing bacteria inside cells within the nodules.

THINK CRITICALLY If all of Earth's nitrogen-fixing prokaryotes were to die suddenly, what would happen to the concentration of nitrogen gas in the atmosphere?

liberate nutrients from plant tissue that the animals cannot digest themselves. Without such bacteria, leaf-eating animals could not survive.

Prokaryotes also have important effects on human nutrition. Many foods, including cheese, yogurt, and sauerkraut, are produced by the action of bacteria. Prokaryotes also inhabit your intestines. There they feed on undigested food, and some synthesize nutrients such as vitamin K and vitamin B_{12}, which the human body absorbs. In fact, good health and nutrition are highly dependent on the prokaryotes in our digestive system; we can't thrive without them.

Prokaryotes Capture the Nitrogen Needed by Plants

Humans could not live without plants, and plants are entirely dependent on bacteria. In particular, plants are unable to capture nitrogen from that element's most abundant reservoir: the atmosphere. Plants need nitrogen to grow. To acquire it, they depend on **nitrogen-fixing bacteria,** which live both in soil and in specialized nodules, which are small, rounded lumps on the roots of legumes (a group of plants that includes alfalfa, soybeans, lupines, and clover; **FIG. 20-9**). The nitrogen-fixing bacteria capture nitrogen gas (N_2) from air trapped in the soil and combine it with hydrogen to produce ammonium (NH_4^+), a nitrogen-containing nutrient that plants can use directly.

Prokaryotes Are Nature's Recyclers

Prokaryotes play a crucial role in recycling waste. Many prokaryotes obtain energy by breaking down complex organic molecules (molecules that contain carbon and hydrogen).

Such prokaryotes find a plentiful source of organic molecules in the waste products and dead bodies of plants and animals. By consuming and thereby decomposing these wastes, prokaryotes prevent wastes from accumulating in the environment. In addition, decomposition by prokaryotes releases the nutrients contained in wastes. Once released, the nutrients become available for reuse by other living organisms.

Prokaryotes perform their recycling service wherever organic matter is found. They are important decomposers in lakes and rivers, in the oceans, and in the soil and groundwater of forests, grasslands, deserts, and other terrestrial environments. The recycling of nutrients by prokaryotes and other decomposers provides the raw materials needed for continued life on Earth.

Prokaryotes Can Clean Up Pollution

Many of the pollutants that are produced as by-products of human activity are organic compounds. As such, these pollutants can potentially serve as food for archaea and bacteria. Nearly anything that human beings can synthesize—including detergents, many toxic pesticides, and harmful industrial chemicals such as benzene and toluene—can be broken down by some prokaryote.

Even oil can be broken down by prokaryotes. Soon after the tanker *Exxon Valdez* spilled 11 million gallons of crude oil into Prince William Sound, Alaska in 1989, researchers sprayed oil-soaked beaches with a fertilizer that encouraged the growth of natural populations of oil-eating bacteria. Within days, the oil deposits on these beaches were noticeably reduced in comparison with unsprayed areas. However, oil-eating bacteria are having a much slower effect on the 200 million gallons of oil released into the Gulf of Mexico during the *Deep Water Horizon* well blowout in 2010. The oil from the blowout was released deep underwater where the temperature is cold and prokaryote metabolism is slow. In addition, it is not practical to encourage bacterial growth by fertilizing the vast area over which the oil has spread.

The practice of manipulating conditions to stimulate breakdown of pollutants by living organisms is known as **bioremediation.** Improved methods of bioremediation could dramatically increase our ability to clean up toxic waste sites and polluted groundwater. A great deal of current research is therefore devoted to identifying prokaryote species that are especially effective for bioremediation and discovering practical methods for manipulating these organisms to improve their usefulness.

Some Bacteria Pose a Threat to Human Health

Despite the benefits some bacteria provide, the feeding habits of certain bacteria threaten our health and well-being. These **pathogenic** (disease-producing) bacteria synthesize toxic substances that cause disease symptoms. So far, no pathogenic archaea have been identified.

Some Anaerobic Bacteria Produce Dangerous Poisons

Some bacteria produce toxins that attack the nervous system. One such toxin is produced by *Clostridium tetani,* the bacterium that causes tetanus, a sometimes fatal disease whose symptoms include painful, uncontrolled contraction of muscles throughout the body. *C. tetani* are anaerobic bacteria that survive as spores until introduced into a favorable, oxygen-free environment. A deep puncture wound may allow tetanus bacteria to penetrate a human body and reach a place where they will be protected from contact with oxygen. As they multiply, the bacteria release their toxin into the body's bloodstream.

Another anaerobic *Clostridium* species that produces a dangerous neurotoxin is *Clostridium botulinum*. *C. botulinum* occurs naturally in soil, but may also thrive in a sealed container of canned food that has been incompletely sterilized. Foodborne *C. botulinum* is dangerous because botulinum toxin is among the most toxic substances known; a single gram is enough to kill 15 million people.

Humans Battle Bacterial Diseases Old and New

Bacterial diseases have had a significant impact on human history. Perhaps the most infamous example is bubonic plague, or "Black Death," which killed 100 million people during the mid-fourteenth century. In many parts of the world, one-third or more of the population died. Plague is caused by the highly infectious bacterium *Yersinia pestis,* which is spread by fleas that feed on infected rats and then move to human hosts. Although bubonic plague has not reemerged as a large-scale epidemic, about 2,000 to 3,000 people worldwide are diagnosed with the disease each year.

Some bacterial pathogens seem to emerge suddenly. Lyme disease, for example, was unknown until 1975 but is now diagnosed in about 30,000 people per year in the United States. This disease, named after the town of Old Lyme, Connecticut, where it was first described, is caused by the corkscrew-shaped bacterium *Borrelia burgdorferi* (see Fig. 20-1c). The bacterium is carried by deer ticks, which transmit it to the humans they bite. At first, the symptoms—chills, fever, and body aches—resemble those of flu. If untreated, weeks or months later the victim may experience rashes, bouts of arthritis, and in some cases abnormalities of the heart and nervous system.

Perhaps the most frustrating pathogens are those that come back to haunt us long after we believed that we had them under control. Two sexually transmitted bacterial diseases, gonorrhea and syphilis, have reached epidemic proportions around the globe. Cholera, a water-transmitted bacterial disease that flourishes when raw sewage contaminates drinking water or fishing areas, is under control in developed countries but remains a major killer in poorer parts of the world.

CHECK YOUR LEARNING
Can you ...
- explain how prokaryotes affect animal and plant nutrition?
- explain prokaryotes' role in nutrient recycling?
- describe how prokaryotes help clean up pollution?
- describe some of the pathogenic bacteria that threaten human health?

CASE STUDY   CONTINUED
Unwelcome Dinner Guests

Many of the bacteria responsible for foodborne illnesses do their damage by producing toxins. For example, different populations of the bacterial species *Escherichia coli* may differ genetically, and some genetic differences can transform this normally harmless inhabitant of the human digestive system into a toxin-producing pathogen. If one of these toxic strains, such as the ones designated O157:H7 and O104:H4, finds its way into a human digestive system, the bacteria attach firmly to the wall of the intestine and begin to release a toxin called shiga. Shiga toxin causes intestinal bleeding that results in painful cramping and bloody diarrhea. The toxin can damage other organs as well; victims of O157:H7 and O104:H4 often develop hemolytic uremic syndrome, a dangerous condition characterized by kidney failure and loss of red blood cells.

Bacteria are responsible for most foodborne illnesses. But besides bacteria, which other infectious agents can wreak havoc on the human body?

20.4 WHAT ARE VIRUSES, VIROIDS, AND PRIONS?

Although this chapter is devoted primarily to an overview of the prokaryotic domains, we also discuss here *viruses, viroids,* and *prions,* which are not organisms but have significant effects on organisms.

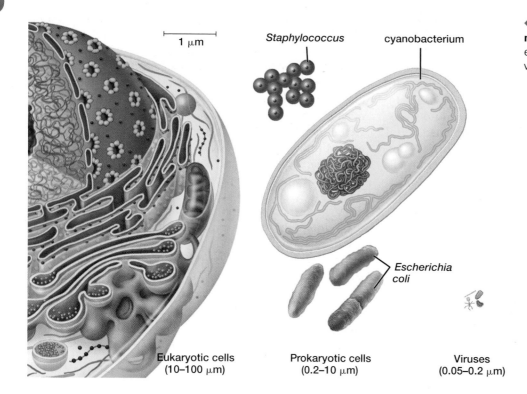

1 μm

Staphylococcus

cyanobacterium

Escherichia coli

Eukaryotic cells
(10–100 μm)

Prokaryotic cells
(0.2–10 μm)

Viruses
(0.05–0.2 μm)

◀ **FIGURE 20-10** **The sizes of microorganisms** The relative sizes of eukaryotic cells, prokaryotic cells, and viruses (1 μm = 1/1,000 millimeter).

Viruses Are Nonliving Particles

Although **viruses** are generally found in close association with living organisms, most biologists do not consider viruses to be alive because they lack many of the traits that characterize life. For example, they are not cells—nor are they composed of cells. Further, they cannot, on their own, accomplish the basic tasks that living cells perform. Viruses have no ribosomes on which to make proteins, no cytoplasm, no ability to synthesize organic molecules, and no capacity to extract and use the energy stored in such molecules. They possess no membranes of their own and cannot grow or reproduce on their own. The simplicity of viruses seems to place them outside the realm of living things. They can, however, evolve.

A Virus Consists of a Molecule of DNA or RNA Surrounded by a Protein Coat

Viruses are tiny; most are much smaller than even the smallest prokaryotic cell (**FIG. 20-10**). Virus particles are so small (0.05–0.2 micrometer in diameter) that they can be seen only under the enormous magnification of an electron microscope. Under such magnification, one can see that viruses assume a great variety of shapes (**FIG. 20-11**).

Viruses consist of two major parts: a molecule of hereditary material and a coat of protein surrounding the molecule. Depending on the type of virus, the hereditary molecule may be either DNA or RNA and may be single-stranded or

▶ **FIGURE 20-11** **Viruses come in a variety of shapes** Viral shape is determined by the nature of the virus's protein coat.

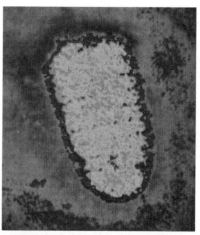

(a) Rabies virus

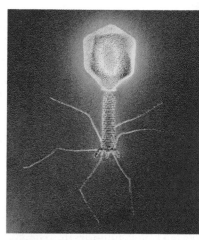

(b) Bacteriophage

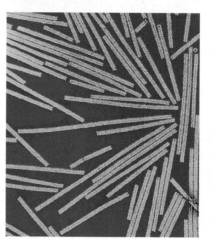

(c) Tobacco mosaic viruses

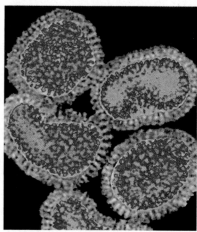

(d) Influenza viruses

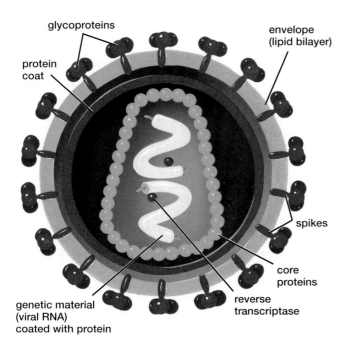

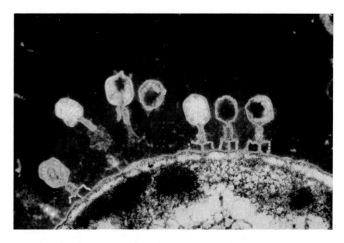

▲ **FIGURE 20-12 Viral structure and replication** A cross-section of HIV, the virus that causes AIDS. This virus is among those that have an outer envelope formed from the host cell's plasma membrane. Spikes made of glycoprotein (protein and carbohydrate) project from the envelope and help the virus attach to its host cell. Inside, a protein coat surrounds genetic material and molecules of reverse transcriptase, an enzyme that catalyzes the transcription of DNA from the viral RNA template after the virus enters a host cell.

THINK CRITICALLY Why are viruses unable to replicate outside of a host cell?

▲ **FIGURE 20-13 Some viruses infect bacteria** In this electron micrograph, bacteriophages are seen attacking a bacterium. They have injected their genetic material inside, leaving their protein coats clinging to the bacterial cell wall.

THINK CRITICALLY Biotechnologists often use viruses to transfer genes from the cells of one species to the cells of another. Which properties of viruses make them useful for this purpose?

double-stranded, linear or circular. The protein coat may be surrounded by an envelope formed from the plasma membrane of the host cell (**FIG. 20-12**).

Viruses Require a Host to Reproduce

A virus can reproduce only inside a **host** cell—the cell that the virus infects. Viral reproduction begins when a virus penetrates a host cell. After the virus enters the host cell, the viral genetic material takes command. The hijacked host cell then uses the instructions encoded in the viral genes to produce the components of new viruses. The pieces are rapidly assembled, and an army of new viruses bursts forth to invade and conquer neighboring cells (see "In Greater Depth: Virus Replication" on page 364).

Viruses Are Host Specific

Each type of virus is specialized to attack a specific host cell. As far as we know, no organism in any of life's three domains is immune to all viruses. The viruses that infect archaea or bacteria are often called **bacteriophages,** sometimes shortened to *phages* (**FIG. 20-13**). Phages may soon become important in treating diseases caused by bacteria because many disease-causing bacteria have become increasingly resistant

to antibiotics. Treatments based on phages could also take advantage of the viruses' specificity, attacking only the targeted bacteria and not the many other harmless or beneficial bacteria in the body.

In multicellular organisms such as plants and animals, different viruses specialize in attacking particular cell types. Viruses responsible for the common cold, for example, attack the membranes of the respiratory tract, and rabies viruses attack nerve cells. One type of herpes virus specializes in the mucous membranes of the mouth and lips, causing cold sores; a second type produces similar sores on or near the genitals. Herpes viruses take up permanent residence in the body, erupting periodically (typically during times of stress) as infectious sores. The devastating disease AIDS (acquired immune deficiency syndrome), which cripples the body's immune system, is caused by a virus that attacks a specific type of white blood cell that controls the body's immune response. Viruses also cause some types of cancer, such as T-cell leukemia (a cancer of the white blood cells), liver cancer, and cervical cancer.

Viral Infections Are Difficult to Treat

Because viruses depend on the cellular machinery of their hosts, the illnesses they cause are difficult to treat. The antibiotics that are often effective against bacterial infections are useless against viruses, and antiviral agents may destroy host cells as well as viruses. Despite the difficulty of attacking viruses as they "hide" within cells, a number of antiviral drugs have been developed. Many of these drugs destroy or block the function of enzymes that the targeted virus requires for replication.

IN GREATER DEPTH | Virus Replication

Viruses multiply, or replicate, using their own genetic material, which—depending on the virus—consists of single-stranded or double-stranded RNA or DNA. This material serves as a blueprint for the viral proteins and genetic material required to make new viruses. Viral enzymes may participate in replication as well, but the overall process depends on the biochemical machinery of the host cell. Viruses cannot replicate outside of living cells.

The process of viral replication varies considerably among the different types of viruses, but most modes of replication are variations of a general sequence of events:

- **Penetration** To replicate, a virus must enter a host cell. Some viruses are engulfed by a host cell (endocytosis)

after binding to receptors on the cell's plasma membrane that stimulate endocytosis. Other viruses are coated with an envelope that can fuse with the host's membrane. The viral genetic material is then released into the cytoplasm.

- **Synthesis** Viruses redirect the host cell's protein synthesis machinery to produce many copies of the viral proteins, and the viral genetic material is replicated many times. Transcription of the viral genome to messenger RNA uses nucleotides from the host cell, and protein synthesis uses the host cell's ribosomes, transfer RNAs, and amino acids.
- **Assembly** The viral genetic material and enzymes are surrounded by their protein coat.

- **Release** Viruses emerge from the host cell by "budding" from the plasma membrane or by bursting the cell.

FIGURE E20-2 shows the life cycle of the human immunodeficiency virus (HIV), the *retrovirus* that causes AIDS. Retroviruses are so named because a key step in their replication uses single-stranded RNA as a template to make double-stranded DNA, a process that reverses the normal DNA-to-RNA pathway. Retroviruses accomplish this reverse transcription by using a viral enzyme called *reverse transcriptase*.

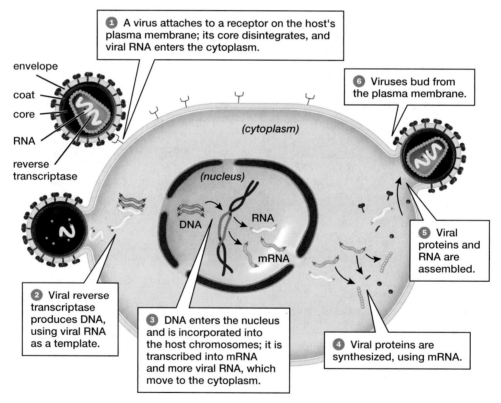

① A virus attaches to a receptor on the host's plasma membrane; its core disintegrates, and viral RNA enters the cytoplasm.

envelope
coat
core
RNA
reverse transcriptase

⑥ Viruses bud from the plasma membrane.

(cytoplasm)

(nucleus)

DNA   RNA

mRNA

⑤ Viral proteins and RNA are assembled.

② Viral reverse transcriptase produces DNA, using viral RNA as a template.

③ DNA enters the nucleus and is incorporated into the host chromosomes; it is transcribed into mRNA and more viral RNA, which move to the cytoplasm.

④ Viral proteins are synthesized, using mRNA.

▲ **FIGURE E20-2 How viruses replicate** The retrovirus HIV invades a white blood cell.

Unfortunately, the benefits of most antiviral drugs are limited because many viruses quickly evolve resistance to the drugs. Mutation rates can be very high in viruses, in part because many viruses lack mechanisms for correcting errors that occur during replication of genetic material. Thus, when a population of viruses is under attack by an antiviral drug, a mutation will often arise that confers resistance to the drug. The resistant viruses prosper and replicate in great numbers, eventually spreading to new human hosts. Ultimately, resistant viruses predominate, and a formerly helpful antiviral drug is rendered ineffective.

CASE STUDY CONTINUED
Unwelcome Dinner Guests

A few foodborne illnesses are caused by viruses. For example, the virus that causes hepatitis A is often transmitted in food, usually when the food has been handled by an infected person who has been lax about hand washing. Some people infected with hepatitis A do not exhibit symptoms, but many experience flu-like symptoms accompanied by jaundice (yellowish skin). Although additional complications can arise, most victims recover within a few months. Hepatitis A is one of the few foodborne diseases for which a vaccination exists.

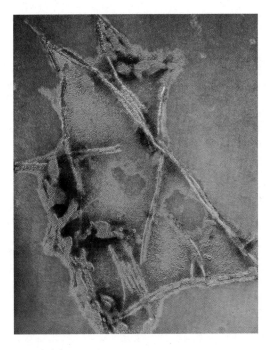

▲ **FIGURE 20-14 Prions: puzzling proteins** A section from the brain of a cow infected with bovine spongiform encephalopathy contains fibrous clusters of prion proteins.

Some Plant Diseases Are Caused by Infectious Agents Even Simpler Than Viruses

Viroids are infectious particles that lack a protein coat and consist of nothing more than short, circular strands of RNA. Despite their simplicity, viroids are able to enter the nucleus of a host cell and direct the synthesis of new viroids. About a dozen crop diseases, including cucumber pale fruit disease, avocado sunblotch, and potato spindle tuber disease, are caused by viroids. No viroid is known to infect animals.

Some Protein Molecules Are Infectious

Simple infectious agents known as *prions* attack mammalian nervous systems. In the 1950s, physicians studying the Fore, a primitive tribe in New Guinea, were puzzled to observe numerous cases of a fatal degenerative disease of the nervous system, which the Fore called kuru. The symptoms of kuru—loss of coordination, dementia, and ultimately death—were similar to those of the rare but more widespread Creutzfeldt-Jakob disease in humans and of scrapie and bovine spongiform encephalopathy diseases in domestic livestock (see "Case Study: Puzzling Proteins" in Chapter 3). Each of these diseases typically results in brain tissue that is spongy—riddled with holes. The researchers in New Guinea eventually determined that kuru was transmitted by ritual cannibalism; members of the Fore tribe honored their dead by consuming their brains. This practice has since stopped, and kuru has virtually disappeared. Clearly, kuru was caused by an infectious agent transmitted by infected brain tissue—but what was that agent?

In 1982, neurologist Stanley Prusiner published evidence that scrapie (and, by extension, kuru, Creutzfeldt-Jakob disease, and a number of other, similar afflictions) is caused by an infectious agent that consists only of protein. This idea seemed preposterous at the time, because most scientists believed that infectious agents must contain genetic material such as DNA or RNA to replicate. But Prusiner and his colleagues were able to isolate the infectious agent from scrapie-infected hamsters and demonstrate that it contained no nucleic acids. The researchers called these infectious protein particles **prions** (FIG. 20-14).

How can a protein replicate itself and be infectious? Research over the decades since Prusiner's discovery has shown that prions are misfolded versions of a common protein called PrP. PrP is found in the membranes of neurons and is required for normal neuron function. Sometimes, copies of the PrP molecule become folded into the wrong shape and are thus transformed into infectious prions. If misfolded prion protein molecules are introduced into a healthy mammal, they can induce other, normal copies of the PrP molecule to become transformed into prions, which in turn induce still other conversions of normal PrP to the prion version. Eventually, this chain reaction leads to a concentration of prions high enough to cause nerve cell damage and degeneration. Why would a slight alteration to a normally benign protein turn it into a dangerous cell killer? No one knows.

CHECK YOUR LEARNING

Can you ...
- describe the structure and characteristics of viruses, viroids, and prions?
- describe the effects they can have on host organisms?

Unwelcome Dinner Guests

How do harmful bacteria get into our food? Many foodborne illnesses result from consumption of contaminated beef. The intestinal tracts of about a third of the cattle in the United States carry bacteria that are harmful to humans, and these bacteria can be transmitted to humans when a meatpacker accidentally grinds some gut contents into hamburger. Similarly, chicken feces may splash onto eggs, setting the stage for harmful bacteria to enter the eggs through tiny cracks or when the consumer breaks the egg and its contents contact the shell. Produce such as lettuce, spinach, tomatoes, and melons can also become contaminated if farm fields are exposed to animal feces, which can be deposited by deer or wandering domestic animals or carried from nearby ranches and feedlots in dust or runoff. The warm, moist environments in which sprouts are grown provide excellent growing conditions for any harmful bacteria that may have been present on the seeds from which the sprouts were produced.

How can you protect yourself from the bacteria that share our food supply? It's easy: Clean, cook, and chill. Cleaning helps prevent the spread of pathogens. Wash your hands before preparing food, and wash all utensils and cutting boards after preparing each item. Thorough cooking is the best way to ensure that any bacteria present in food are killed. Meats, in particular, must be thoroughly cooked; food safety experts recommend using a meat thermometer to ensure that the thickest

part of cooked pork or ground beef has reached 160°F. The safe temperature for cuts of beef, veal, or lamb is 145°F; for all poultry, 165°F. The color of cooked meat can be an unreliable indicator of safety, but when a meat thermometer is unavailable, try to avoid eating meat that is still pink inside, especially ground beef. Fish should be cooked until it is opaque and flakes easily with a fork; cook eggs until both white and yolk are firm. Finally, keep stored food cold. Pathogens multiply most rapidly at temperatures between 40° and 140°F. So get your groceries home from the store and into the refrigerator or freezer as quickly as possible. Don't leave cooked leftovers unrefrigerated for more than 2 hours. Thaw frozen foods in the refrigerator or the microwave, not at room temperature. A little bit of attention to food safety can save you from unwelcome guests in your food.

CONSIDER THIS Consumer groups contend that we can improve food safety by giving government agencies additional funding and greater authority to inspect food processing plants and order recalls of contaminated food. Opponents of such steps argue that we need not empower government agencies because the best protection against food contamination is informed consumers, who will stop buying products from companies that have produced unsafe foods. Would you support or oppose additional government oversight of food safety?

CHAPTER REVIEW

Go to **MasteringBiology** for practice quizzes, activities, eText, videos, current events, and more.

Answers to Think Critically, Evaluate This, Multiple Choice, and Fill-in-the-Blank questions can be found in the Answers section at the back of the book.

Summary of Key Concepts

20.1 Which Organisms Are Members of the Domains Archaea and Bacteria?

Archaea and bacteria are unicellular and prokaryotic. Although archaea and bacteria are morphologically similar, they are not closely related and differ in several fundamental features, including cell wall composition, ribosomal RNA sequence, and membrane lipid structure. A cell wall determines the characteristic shapes of prokaryotes: spherical, rod-shaped, or corkscrew-shaped.

20.2 How Do Prokaryotes Survive and Reproduce?

Certain types of bacteria can move about using flagella; others form spores that disperse widely and withstand inhospitable environmental conditions. Bacteria and archaea have colonized nearly every habitat on Earth, including hot, acidic, very salty, and anaerobic environments.

Prokaryotes obtain energy in a variety of ways. Some rely on photosynthesis; others break down inorganic or organic

molecules to obtain energy. Many are anaerobic, able to obtain energy when oxygen is not available. Prokaryotes reproduce by prokaryotic fission and may exchange genetic material by conjugation, in which DNA is transferred from a donor to a recipient.

20.3 How Do Prokaryotes Affect Humans and Other Organisms?

Some bacteria are pathogenic, causing disorders such as pneumonia, tetanus, botulism, and the sexually transmitted infections gonorrhea and syphilis. Most prokaryotes, however, are harmless to humans and play important roles in natural ecosystems. Some live in the digestive tracts of animals that eat leaves, where the prokaryotes break down cellulose. Nitrogen-fixing bacteria enrich the soil and aid in plant growth. Many other bacteria live off the dead bodies and wastes of other organisms, liberating nutrients for reuse.

20.4 What Are Viruses, Viroids, and Prions?

Viruses are parasites consisting of a protein coat that surrounds genetic material. They are noncellular and unable to move, grow, or reproduce outside a living cell. They invade cells of a specific host and use the host cell's energy, enzymes, and ribosomes to produce more virus particles, which are liberated when the cell ruptures.

Many viruses are pathogenic to humans, including those causing colds and flu, herpes, AIDS, and certain forms of cancer.

Viroids are short strands of RNA that can invade a host cell's nucleus and direct the synthesis of new viroids. To date, viroids are known to cause only certain diseases of plants.

Prions have been implicated in diseases of the nervous system, such as kuru, Creutzfeldt-Jakob disease, scrapie, and bovine spongiform encephalopathy. Prions lack genetic material and are composed solely of mutated prion protein.

Key Terms

anaerobe   *357*
Archaea   *353*
Bacteria   *353*
bacteriophage   *363*
biofilm   *355*
bioremediation   *361*
conjugation   *359*
endospore   *356*
flagellum (plural, flagella)   *355*

host   *363*
nitrogen-fixing bacterium   *360*
pathogenic   *361*
plasmid   *359*
prion   *365*
prokaryotic fission   *357*
viroid   *365*
virus   *362*

Thinking Through the Concepts

Multiple Choice

1. The name of the process by which DNA is transferred from one prokaryote to another via a cytoplasmic bridge is
   a. conjugation.
   b. prokaryotic fission.
   c. meiosis.
   d. bioremediation.

2. A community of prokaryotes surrounded by slime and adhering to a surface is called a(n)
   a. plasmid.
   b. flagellum.
   c. endospore.
   d. biofilm.

3. Which of the following statements about archaea is *not* true?
   a. Some archaea can survive and thrive at temperatures above the boiling point of water.
   b. The cell walls of archaea contain peptidoglycan.
   c. Archaea are not closely related to bacteria.
   d. Archaea are not known to cause any disease of plants or animals.

4. Viruses
   a. are usually photosynthetic.
   b. consist of a single cell.
   c. consist of DNA or RNA and a protein coat.
   d. consist of ATP and a lipid coat.

5. Applying fertilizer near an oil spill to increase the population of oil-consuming bacteria is an example of
   a. bioremediation.
   b. conjugation.
   c. genetic engineering.
   d. lateral gene transfer.

Fill-in-the-Blank

1. _____ have peptidoglycan in their _____, but _____ do not.

2. Prokaryotic cells are _____ (larger/smaller) than eukaryotic cells. The most common shapes of prokaryotes are _____, _____, and _____.

3. Many prokaryotes use _____ to move about. Some prokaryotes secrete slime that protects them when they aggregate in communities called _____. Other prokaryotes can survive long periods and extreme conditions by producing protective structures called _____.

4. _____ bacteria inhabit environments that lack oxygen. _____ bacteria capture energy from sunlight.

5. Prokaryotes reproduce by _____ and may sometimes exchange genetic material through the process of _____.

6. The plant nutrient ammonium is produced by _____ bacteria in the soil and in nodules. Prokaryotes that live in the digestive tracts of cows and rabbits break down _____ in the leaves that those mammals eat.

7. Cholera, gonorrhea, and pneumonia are some of the diseases caused by pathogenic _____. Harmful strains of *E. coli* can be transmitted to humans by consumption of _____, _____, or _____.

8. A virus consists of a molecule of _____ or _____ surrounded by a(n) _____ coat. A virus cannot reproduce unless it enters a(n) _____ cell. A virus that infects bacteria is known as a(n) _____.

Review Questions

1. Describe some of the ways in which prokaryotes obtain energy and nutrients.

2. What are nitrogen-fixing bacteria, and what role do they play in ecosystems?

3. Describe some of the extreme environments in which prokaryotes are found. What parts of the human body are inhabited by prokaryotes?

4. What is an endospore? What is its function?

5. What is conjugation? What role do plasmids play in conjugation?

6. Why are prokaryotes especially useful in bioremediation?

7. Describe the structure of a typical virus. How do viruses replicate?

8. Describe some examples of how prokaryotes are helpful to humans and some examples of how they are harmful to humans.

9. How do archaea and bacteria differ? How do prokaryotes and viruses differ?

Applying the Concepts

1. In some developing countries, antibiotics can be purchased without a prescription. Why do you think this is done? What biological consequences would you predict?

2. Before the discovery of prions, many (perhaps most) biologists would have agreed with the statement "It is a fact that no infectious organism or particle can exist that lacks nucleic acid (such as DNA or RNA)." What lessons do prions teach us about nature, science, and scientific inquiry? (You may wish to review Chapter 1 to help answer this question.)

THE DIVERSITY OF PROTISTS

The photosynthetic protist *Caulerpa taxifolia* is an unwanted invader in temperate seas.

Green Monster

IN CALIFORNIA, IT IS A CRIME to possess, transport, or sell *Caulerpa*. Is *Caulerpa* an illicit drug or some kind of weapon? No—it is a small green seaweed. Why, then, do California's lawmakers want to ban it from their state?

The story of *Caulerpa*'s rise to public enemy status begins in the early 1980s at the Wilhelma Zoo in Stuttgart, Germany. There, the keepers of a saltwater aquarium found that the tropical seaweed *Caulerpa taxifolia* was an attractive companion and background for the tropical fish on display. Even better, years of captive breeding at the zoo had yielded a strain of the seaweed that was well suited to life in an aquarium. The new strain was particularly hardy and could survive in waters considerably cooler than the tropical waters in which wild *Caulerpa* is found. The zoo staff was happy to send cuttings to other institutions that wished to use it in aquarium displays.

One institution that received a cutting was the Oceanographic Museum of Monaco, located on the shore of the Mediterranean Sea. In 1984, a visiting marine biologist discovered a small patch of *Caulerpa* growing in the waters just below the museum. Presumably, some-

one cleaning an aquarium had dumped the water into the sea and thereby inadvertently introduced *Caulerpa* to the Mediterranean.

By 1989, the *Caulerpa* patch had grown to cover a few acres. It formed a continuous mat that seemed to exclude most of the other organisms that normally inhabit the Mediterranean Sea floor. The local herbivores, such as sea urchins and fish, did not feed on *Caulerpa*.

It soon became apparent that *Caulerpa* spreads rapidly, is not controlled by predation, and displaces native species. By the mid-1990s, biologists were alarmed to find *Caulerpa* all along the Mediterranean coast from Spain to Italy. Today, this invasive species grows in extensive beds throughout the Mediterranean and covers an ever-expanding area on the seafloor.

Despite the threat it poses to ecosystems, *Caulerpa* is a fascinating organism. Green algae such as *Caulerpa* are part of a diverse group of organisms informally known as protists. What other kinds of organisms are protists? How and where do protists live?

AT A GLANCE

21.1 What Are Protists?

21.2 What Are the Major Groups of Protists?

21.1 WHAT ARE PROTISTS?

Two of life's domains, Bacteria and Archaea, contain only prokaryotes. The third domain, Eukarya, includes all eukaryotic organisms. The most conspicuous Eukarya are plants, fungi, and animals (covered in Chapters 22 through 25). The remaining eukaryotes constitute a diverse array of organisms collectively known as **protists.** Protists do not form a clade—a group consisting of all the descendants of a particular common ancestor—so systematists do not use the term "protist" as a formal group name. Instead, protist is a term of convenience that refers to any eukaryote that is not a plant, animal, or fungus.

Most protists are single celled and are invisible to us as we go about our daily lives. If we could somehow shrink to their microscopic scale, we would be impressed with their beautiful forms, their varied lifestyles, their astonishingly diverse modes of reproduction, and the structural and physiological complexity that is possible within the limits of a single cell.

Protists Use Diverse Modes of Reproduction

Most protists reproduce asexually; an individual divides by mitotic cell division to yield two individuals that are genetically identical to the parent cell (FIG. 21-1a). Many protists, however, are also capable of sexual reproduction, in which two individuals contribute genetic material to an offspring that is genetically different from either parent. Nonreproductive processes that combine the genetic material of different individuals are also common among protists (FIG. 21-1b).

In many protist species that are capable of sexual reproduction, most reproduction is nonetheless asexual. Sexual reproduction occurs only infrequently, at a particular time of year or under certain circumstances, such as a crowded environment or a shortage of food. The details of sexual reproduction and the resulting life cycles vary tremendously among different types of protists.

Protists Use Diverse Modes of Nutrition

Three major modes of nutrition are represented among protists. Protists may ingest their food, absorb nutrients from their surroundings, or capture solar energy by photosynthesis. Protists that ingest their food are predators. Predatory single-celled protists may have flexible cell membranes that can change shape to surround and engulf food such as bacteria or another protist. Protists that feed in this manner typically use finger-like extensions called **pseudopods** to engulf prey. Other predatory protists create tiny currents that sweep food particles into mouth-like openings in the cell. Whatever

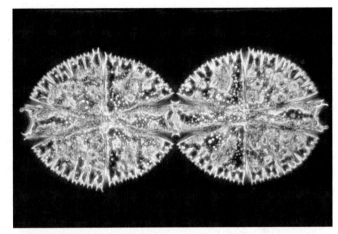

(a) Reproducing by cell division

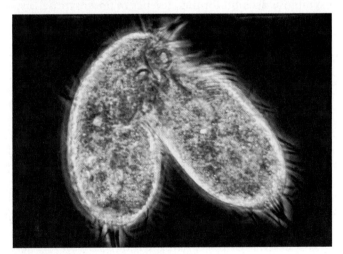

(b) Exchanging genetic material

▲ **FIGURE 21-1 Protistan reproduction and gene exchange**
(a) *Micrasterias*, a green alga, reproduces asexually by cell division.
(b) Two *Euplotes* ciliates exchange genetic material.

THINK CRITICALLY What do biologists mean when they say that sex and reproduction are uncoupled in most protists?

the means by which food is ingested, once it is inside the protist cell, it is typically packaged into a membrane-bound **food vacuole** for digestion.

Protists that absorb nutrients directly from the surrounding environment may be free living or may live inside the bodies of other organisms. The free-living types live in soil and other environments that contain dead organic matter, where they act as decomposers. Most absorptive feeders, however, live inside other organisms. In most cases, these protists are parasites whose feeding activity harms the host species.

Photosynthetic protists are abundant in oceans, lakes, and ponds. Most float suspended in the water, but some live inside the tissues of other organisms, such as corals or clams. These associations appear to be mutually beneficial; some of the solar energy captured by the photosynthetic protists is used by the host organism, which provides shelter and protection for the protists.

Protist photosynthesis takes place in chloroplasts. Chloroplasts are the descendants of ancient photosynthetic bacteria that took up residence inside a larger cell in a process known as *endosymbiosis* (see Chapter 18). In addition to the original instance of endosymbiosis that created the first protist chloroplast, there have been several later occurrences of *secondary endosymbiosis* in which a nonphotosynthetic protist engulfed a photosynthetic, chloroplast-containing protist. Ultimately, most components of the engulfed species disappeared, leaving only a chloroplast surrounded by four membranes. Two of these membranes are from the original, bacteria-derived chloroplast; one is from the engulfed protist; and one is from the food vacuole that originally contained the engulfed protist. Multiple occurrences of secondary endosymbiosis account for the presence of photosynthetic species in a number of different, unrelated protist groups.

Protists Affect Humans and Other Organisms

Protists have important impacts, both positive and negative, on human lives. The primary positive impact actually benefits all living things and stems from the ecological role of photosynthetic marine protists. Just as plants do on land, photosynthetic protists in the oceans capture solar energy and make it available to the other organisms in the ecosystem. Thus, the marine ecosystems on which humans depend for food in turn depend on protists. Further, in the process of using photosynthesis to capture energy, the protists release oxygen gas that helps replenish the oxygen removed from the atmosphere by respiration (recall from Chapter 8 that cellular respiration consumes oxygen).

On the negative side of the ledger, many human diseases are caused by parasitic protists. The diseases caused by protists include some of humanity's most prevalent ailments and some of its deadliest afflictions. Protists also cause a number of plant diseases, some of which damage crops that are important to humans.

CHECK YOUR LEARNING

Can you ...
- define *protist* and describe the various ways in which protists acquire nutrients and reproduce?
- describe a scenario for the evolutionary origin of protist chloroplasts?
- describe the major effects of protists on people and other organisms?

21.2 WHAT ARE THE MAJOR GROUPS OF PROTISTS?

Taking advantage of the advent of fast, inexpensive DNA sequencing, systematists have used genetic comparisons to gain a better understanding of protist clades and the evolutionary relationships among them. Some of these clades are listed in **TABLE 21–1**, along with the key characteristics of their members. An evolutionary tree that includes the major protist clades is depicted in Figure 19-7.

Past classifications of protists grouped species according to their mode of nutrition, but the old categories are now obsolete because they do not accurately reflect our current understanding of phylogeny. Nonetheless, biologists still use terminology that refers to groups of protists that share particular characteristics but are not necessarily related. For example, photosynthetic protists are collectively known as **algae** (singular, alga), and single-celled, nonphotosynthetic protists are collectively known as **protozoa** (singular, protozoan).

In the following sections, we explore a sample of protist diversity.

Excavates Lack Mitochondria

Excavates are named for a feeding groove that gives the appearance of having been "excavated" from the surface of the cell. Excavates are anaerobes (can live and grow without oxygen). They lack mitochondria, but they do possess other organelles that are probably evolutionarily derived from mitochondria. It is therefore likely that the excavates' ancestors did possess mitochondria, but these organelles were lost early in the evolutionary history of the group. The two largest groups of excavates are the diplomonads and the parabasalids.

Diplomonads Have Two Nuclei

The single cells of **diplomonads** have two nuclei and move about by means of multiple flagella. A parasitic diplomonad, *Giardia* (**FIG. 21-2**), poses a health problem worldwide. *Cysts*

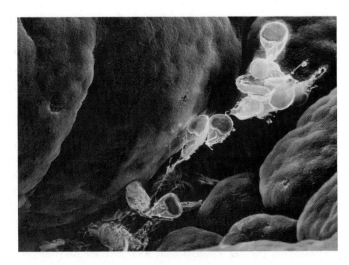

▲ **FIGURE 21-2 *Giardia:* The curse of campers** A diplomonad (genus *Giardia*) that may infect water—causing gastrointestinal disorders for the people who drink it—is shown here in a human small intestine.

TABLE 21-1 | The Major Groups of Protists

Group	Subgroup	Locomotion	Nutrition	Representative Features	Representative Genus
Excavates	Diplomonads	Swim with flagella	Heterotrophic (i.e., consume other organisms)	Lack mitochondria; inhabit soil or water, or may be parasitic; unicellular	*Giardia* (intestinal parasite of mammals)
	Parabasalids	Swim with flagella	Heterotrophic	Lack mitochondria; parasites or mutualistic symbionts; unicellular	*Trichomonas* (causes the sexually transmitted infection trichomoniasis)
Euglenozoans	Euglenids	Swim with one flagellum	Photosynthetic	Have an eyespot; all freshwater	*Euglena* (common pond-dweller)
	Kinetoplastids	Swim with flagella	Heterotrophic	Inhabit soil or water, or may be parasitic; unicellular	*Trypanosoma* (causes African sleeping sickness)
Stramenopiles (chromists)	Water molds	Swim with flagella (gametes)	Heterotrophic	Filamentous	*Plasmopara* (causes downy mildew)
	Diatoms	Glide along surfaces	Photosynthetic	Have silica shells; most marine; unicellular	*Navicula* (glides toward light)
	Brown algae	Nonmotile	Photosynthetic	Seaweeds of temperate oceans; multicellular	*Macrocystis* (forms kelp forests)
Alveolates	Dinoflagellates	Swim with two flagella	Photosynthetic	Many bioluminescent; often have cellulose walls; unicellular	*Gonyaulax* (causes red tide)
	Apicomplexans	Nonmotile	Heterotrophic	All parasitic; form infectious spores; unicellular	*Plasmodium* (causes malaria)
	Ciliates	Swim with cilia	Heterotrophic	Include the most complex single cells; unicellular	*Paramecium* (fast-moving pond-dweller)
Rhizarians	Foraminiferans	Extend thin pseudopods	Heterotrophic	Have calcium carbonate shells; unicellular	*Globigerina* (shells cover large areas of ocean floor)
	Radiolarians	Extend thin pseudopods	Heterotrophic	Have silica shells; unicellular	*Actinomma* (found in oceans worldwide)
Amoebozoans	Amoebas	Extend thick pseudopods	Heterotrophic	Have no shells; unicellular	*Amoeba* (common pond-dweller)
	Acellular slime molds	Slug-like mass oozes over surfaces	Heterotrophic	Form multinucleate plasmodium	*Physarum* (forms a large, bright orange mass)
	Cellular slime molds	Amoeboid cells extend pseudopods; slug-like mass crawls over surfaces	Heterotrophic	Form pseudoplasmodium with individual amoeboid cells	*Dictyostelium* (often used in laboratory studies)
Red algae		Nonmotile	Photosynthetic	Some deposit calcium carbonate; mostly marine; most multicellular	*Porphyra* (used to make sushi wrappers)
Chlorophyte algae		Swim with flagella (some species)	Photosynthetic	Close relatives of clade that includes land plants; unicellular and multicellular	*Ulva* (sea lettuce)

(tough structures that enclose the organism during one phase of its life cycle) of this diplomonad are released in the feces of infected humans, dogs, or other animals; a single gram of feces may contain 300 million cysts. The excreted cysts may enter streams and lakes, municipal water supplies, and even swimming pools and hot tubs. If a mammal drinks infected water, the cysts develop into the adult form in the small intestine. In humans, infections can cause severe diarrhea, dehydration, nausea, vomiting, and cramps. Fortunately, these infections can be cured with drugs, and deaths from *Giardia* infections are uncommon.

Parabasalids Include Mutualists

Parabasalids are anaerobic, flagellated protists named for the presence in their cells of a distinctive structure called the parabasal body, which consists of densely packed Golgi vesicles (see Chapter 4). All known parabasalids live inside animals. For example, this group includes several species that inhabit the digestive systems of some species of wood-eating termites (**FIG. 21-3**). The termites cannot themselves digest the cellulose in wood, but the parabasalids can. Thus, the insect and the protist are in a mutually beneficial relationship. The termite delivers food to the parabasalids in its gut; as the parabasalids digest the food, some of the nutrients released become available for use by the termite.

Some parabasalids are parasitic. For example. the parabasalid *Trichomonas vaginalis* infects the mucous layers of the urinary and reproductive tracts in people, causing the sexually transmitted disease trichomoniasis. Trichomoniasis affects about 3.7 million people in the United States each year.

Euglenozoans Have Distinctive Mitochondria

In most **euglenozoans,** the folds of the inner membrane of the cell's mitochondria have a distinctive shape that appears, under the microscope, as a stack of disks. Two major groups of euglenozoans are the euglenids and the kinetoplastids.

▲ **FIGURE 21-3 Parabasalids inhabit termite digestive tracts** The parabasalid *Trichonympha* lives in termite guts, where it digests cellulose in the woody plant material that termites consume.

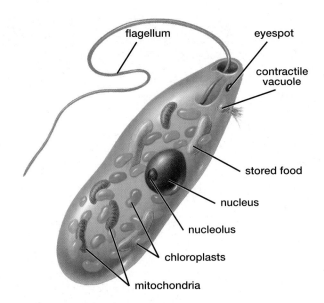

▲ **FIGURE 21-4** *Euglena,* **a representative euglenid** *Euglena's* elaborate single cell is packed with green chloroplasts, which will disappear if the protist is kept in darkness.

Euglenids Lack a Rigid Covering and Swim by Means of Flagella

Euglenids are single-celled protists that live mostly in fresh water and are named after the group's best-known representative, *Euglena,* a complex single cell that moves about by whipping its flagellum through water (**FIG. 21-4**). Euglenids lack a rigid outer covering, so some can move by wriggling as well as by whipping their flagella. Many euglenids are photosynthetic, but some species instead absorb or engulf food. Some euglenids possess simple light-sensing organelles consisting of a photoreceptor, called an *eyespot,* and an adjacent patch of pigment. The pigment shades the photoreceptor only when light strikes from certain directions, enabling the organism to determine the direction of the light source. Using information from the eyespot, the flagellum propels the protist toward light levels appropriate for photosynthesis.

Some Kinetoplastids Cause Human Diseases

The DNA in the mitochondria of **kinetoplastids** is arranged in complex assemblies called kinetoplasts, in which many copies of the circular mitochondrial genome are interlinked to form distinctive disk-shaped structures. Most kinetoplastids possess at least one flagellum, which may propel the organism, sense the environment, or ensnare food. Some kinetoplastids are free living, inhabiting soil and water; others live inside other organisms in a relationship that may be mutually beneficial or parasitic. A dangerous parasitic kinetoplastid in the genus *Trypanosoma* is responsible for African sleeping sickness, a potentially fatal disease (**FIG. 21-5**). Like many parasites, this organism has a complex life cycle, part of which is spent in the tsetse fly. While feeding on the blood of a mammal, an infected fly can transmit saliva containing the trypanosome to the mammal. The parasite then develops in the new host (which may be a human) and enters

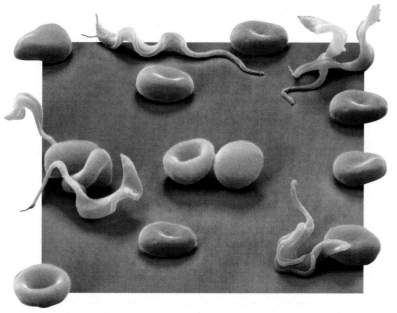

▲ **FIGURE 21-5 A disease-causing kinetoplastid** This photomicrograph shows human blood that is heavily infested with the corkscrew-shaped parasitic kinetoplastid *Trypanosoma*, which causes African sleeping sickness.

the bloodstream. The trypanosome may then be ingested by another tsetse fly that bites the host, thus beginning a new cycle of infection. For information on another disease-causing kinetoplastid, see "Health Watch: Neglected Protist Infections" on page 375.

Stramenopiles Have Distinctive Flagella

All **stramenopiles** have fine, hair-like projections on their flagella (though in many stramenopiles, flagella are present only at certain stages of the life cycle). Despite their shared evolutionary history, stramenopiles display a wide range of forms. Some are photosynthetic and some are not; most are single celled, but some are multicellular. Three major stramenopile groups are the water molds, the diatoms, and the brown algae.

Water Molds Have Had Important Impacts on Humans

The **water molds** form a small group of protists, many of which form long filaments that aggregate into cottony tufts. These tufts are superficially similar to structures produced by some fungi, but this resemblance is due to convergent evolution, not shared ancestry. Many water molds are decomposers that live

in water and damp soil. Some species have profound economic impacts on humans. For example, a water mold causes a disease of grapes known as *downy mildew*. Its inadvertent introduction into France from the United States in the late 1870s nearly destroyed the French wine industry. A water mold is also responsible for *late blight,* a disease of potatoes. When this protist was accidentally introduced into Ireland in about 1845, it destroyed nearly the entire potato crop, causing a devastating famine during which as many as 1 million people in Ireland starved and many more emigrated to the United States to escape the famine.

Diatoms Are Encased Within Glassy Walls

The **diatoms,** photosynthetic stramenopiles found in both fresh and salt water, produce protective cell walls that contain *silica* (glass) (**FIG. 21-6**). Accumulations of diatoms' glassy shells over millions of years have produced fossil deposits of "diatomaceous earth" that may be hundreds of meters thick. This slightly abrasive substance is widely used in products such as toothpaste and metal polish.

Diatoms form part of the **phytoplankton,** the single-celled photosynthesizers that float passively in the upper layers of Earth's lakes and oceans. Phytoplankton play an immensely important ecological role. Marine phytoplankton account for about 50% of all photosynthetic activity on Earth, absorbing carbon dioxide, recharging the atmosphere with oxygen, and supporting the complex web of aquatic life.

▶ **FIGURE 21-6 Some representative diatoms** This photomicrograph illustrates the intricate, microscopic beauty and variety of the glassy walls of diatoms.

(a) *Fucus*

(b) Kelp forest

▲ **FIGURE 21-7 Brown algae are multicellular protists (a)** *Fucus*, a genus found near shores, is shown here exposed at low tide. Notice the gas-filled floats, which provide buoyancy in water. **(b)** The giant kelp *Macrocystis* forms underwater forests off southern California.

Brown Algae Are Multicellular

Though most photosynthetic protists, such as diatoms, are single celled, some form multicellular structures that are commonly known as seaweeds. Although some seaweeds seem to resemble plants, they lack many of the distinctive features of plants. For example, no seaweed has true roots.

The stramenopiles include one group of seaweeds, the brown algae, which are named for the brownish-yellow pigments that (in combination with green chlorophyll) increase the seaweed's light-gathering ability. Almost all brown algae are marine. The group includes the dominant seaweed species that dwell along rocky shores in the temperate (cooler) oceans of the world, including the eastern and western coasts of the United States. Brown algae live in habitats ranging from nearshore waters, where they cling to rocks that are exposed at low tide, to far offshore. Several species use gas-filled floats to support their bodies (**FIG. 21-7a**). Some of the giant kelp found along the Pacific coast reach heights of 175 feet (53 meters) and may grow more than 6 inches (15 centimeters) in a single day. With their dense growth and towering height, kelp form undersea forests that provide food, shelter, and breeding areas for a large variety of marine animals (**FIG. 21-7b**).

Brown algae also contribute to familiar consumer goods. Sodium alginate, extracted from brown algae such as giant kelp, forms a gel that is used to thicken and stabilize ice cream, paint, shaving cream, and many other products.

Alveolates Include Parasites, Predators, and Phytoplankton

The **alveolates** are single-celled organisms that have distinctive, small cavities beneath the surface of their cells. Some alveolates are photosynthetic, some are parasitic, and some are predatory. The major alveolate groups are the dinoflagellates, apicomplexans, and ciliates.

Dinoflagellates Swim by Means of Two Whip-Like Flagella

Though most **dinoflagellates** are photosynthetic, there are also some nonphotosynthetic species. Dinoflagellates (from the Latin for "whirling whips") are named for the two whip-like flagella that propel them (**FIG. 21-8**). One flagellum encircles the cell, and the second projects behind it. Some dinoflagellates are enclosed only by a plasma membrane; others have cellulose walls that resemble armor plates. Although some species live in fresh water, dinoflagellates are especially abundant in the ocean, where they are an important component of the phytoplankton and a food source for larger organisms. Many dinoflagellates are bioluminescent, producing a brilliant blue-green light when disturbed by motion in the water.

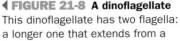

◀ **FIGURE 21-8 A dinoflagellate** This dinoflagellate has two flagella: a longer one that extends from a slot, and a shorter one that lies in a groove that encircles the cell.

Neglected Protist Infections

A number of protist species can live in or on the human body, and some of these species are parasites that cause infectious diseases. The best known and most feared of these diseases is malaria, which infects millions each year around the globe and kills hundreds of thousands of those infected. As a result, a great deal of money and effort are devoted to understanding and combating the *Plasmodium* species that cause the disease. Other protist infections, however, receive much less attention from researchers, physicians, and public health officials, even though they harm many people. Two of these under-the-radar diseases—Chagas disease and toxoplasmosis—have been included on the Centers for Disease Control's list of "neglected parasitic infections" because they affect large numbers of people in the United States.

Chagas disease is caused by the kinetoplastid *Trypanosoma cruzi,* which is transmitted from person to person mainly by blood-sucking insects known as triatomine bugs. However, it can also be transmitted from mother to child or by blood transfusions (blood donations in the United States are now screened for *T. cruzi*). Infections initially cause few or no symptoms, so the infection often goes undiagnosed. But years after infection, many victims develop heart diseases including abnormal rhythms and heart failure than can lead to sudden death. Because the bugs that transmit the disease are most common in Mexico, Central America, and South America, most Chagas disease cases occur in those regions. However, an estimated 300,000 victims of the disease live in the United States; many of them probably became infected elsewhere before moving to the U.S. Chagas infections can be cured with drugs during the weeks immediately following infection, but the disease is much more difficult to treat once it reaches the chronic, heart-damaging stage.

Toxoplasmosis is caused by the apicomplexan parasite *Toxoplasma gondii.* The parasite's primary host is cats, which shed *Toxoplasma* in their feces, from which they are transmitted to rodents that are later consumed by cats, completing the parasite's life cycle. However, humans can become infected if they come into contact with cat feces (for example, while gardening or cleaning a litter box) or consume contaminated food. Sources of contaminated food include unwashed produce and under-cooked meat from animals that had themselves ingested *Toxoplasma.* Initial infection usually results in, at worst, mild symptoms such as fever, but infections acquired during pregnancy are very dangerous to the baby, which may develop severe illness leading to epilepsy, blindness, and developmental disorders. In addition, even infected adults who are initially asymptomatic may be at risk, as the infection persists for life and can ultimately lead to reduced immune function and increased susceptibility to other infectious diseases.

One of *Toxoplasma*'s most fascinating adaptations is its ability to affect the behavior of infected mice and rats. The parasite invades the rodent's brain and induces neurological changes that make the animal less fearful and more reckless. That is, *Toxoplasma* changes rodent psychology in a way that makes the infected animal more likely to be eaten by a cat, thereby allowing the parasite to complete its life cycle and reproduce. Could *Toxoplasma* be affecting the behavior of the many millions of people it currently infects? Very preliminary evidence suggests that the parasite does indeed influence aspects of personality and may even trigger severe mental illness in susceptible individuals.

EVALUATE THIS Imagine that you are a physician consulting with a patient who has recently become pregnant for the first time. Should you recommend blood tests for Chagas diseases or toxoplasmosis? For each disease, why or why not? Would the patient's background and habits affect your recommendation? Also, what precautions, if any, would you recommend to help your patient avoid contracting these diseases?

Warm water that is rich in nutrients may bring on a dinoflagellate population explosion. Dinoflagellates can become so numerous that the water is dyed by the color of their bodies (**FIG. 21-9**). The water may turn yellow, pink, orange, or brown, but the most common result is a reddish tint, so dinoflagellate blooms are commonly called "red tides." During red tides, fish often die by the thousands, suffocated by clogged gills or by the oxygen depletion that results from the decay of billions of dinoflagellates. But dinoflagellate explosions can benefit oysters, mussels, and clams, which have a feast, filtering millions of the protists from the water and consuming them. In the process, however, the mollusks' bodies accumulate concentrations of a nerve toxin produced by the dinoflagellates. Dolphins, seals, sea otters, and humans who eat affected mollusks may be stricken with potentially lethal paralytic shellfish poisoning.

◀ **FIGURE 21-9**
A red tide The explosive reproductive rate of certain dinoflagellates under the right environmental conditions can produce concentrations so great that their microscopic bodies dye the seawater.

CASE STUDY \ **CONTINUED**
Green Monster

Just as the invasive *Caulerpa* seaweed often spreads uncontrollably when introduced to environments free of its normal predators and parasites, populations of toxin-producing dinoflagellates may grow explosively when released into new waters. Red tides have become increasingly common in recent years. One reason for this increased incidence is that dinoflagellate species that can cause red tides have been inadvertently spread around the world by humans. The dinoflagellates travel mainly in seawater that is pumped into the ballast tanks of cargo ships and then discharged at distant ports.

Sometimes, a protist released into a new environment has a damaging impact not because it overwhelms an ecosystem but because it directly causes a disease. What are some examples of such an introduced disease among the alveolates?

Apicomplexans Are Parasitic and Have No Means of Locomotion

All **apicomplexans** are parasitic, living inside the bodies and sometimes inside the individual cells of their hosts. They form infectious spores—resistant structures transmitted from one host to another through food, water, or the bite of an infected insect. As adults, apicomplexans have no means of locomotion. Many have complex life cycles, a common feature of parasites. A well-known example is the malaria parasite *Plasmodium* (**FIG. 21-10**). Parts of its life cycle are spent in the body of a female *Anopheles* mosquito. The mosquito is not harmed by the presence of *Plasmodium* and may eventually bite a human and pass the protist to the unfortunate victim. The protist develops in the victim's liver, then enters the blood, where it reproduces rapidly in red blood cells. When the blood cells rupture, they release large quantities of spores, which cause the recurrent fever of malaria. Uninfected mosquitoes may acquire the parasite by feeding on the blood of a malaria victim, spreading the parasite when they bite another person.

Plasmodium species infect many kinds of animals. *Plasmodium relictum,* for example, infects birds and causes avian malaria. It was introduced to Hawaii, which had previously been free of avian malaria, in the blood of exotic bird species intentionally released on the islands. Native Hawaiian birds, which had

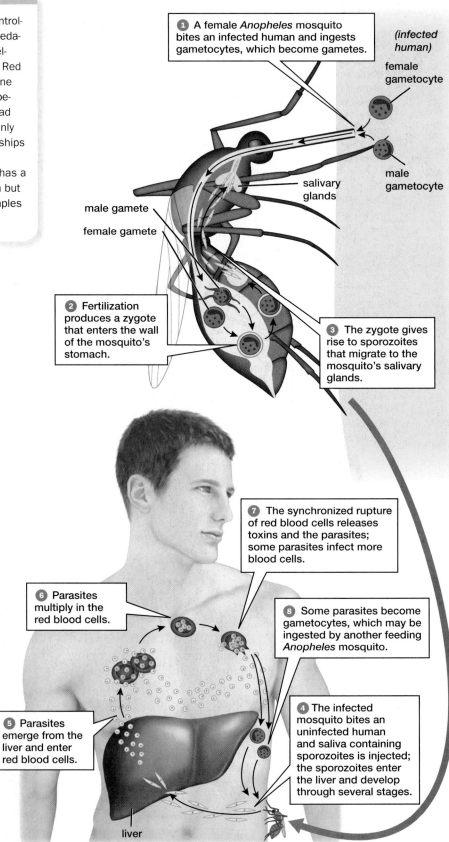

① A female *Anopheles* mosquito bites an infected human and ingests gametocytes, which become gametes.

(infected human)

female gametocyte

male gametocyte

salivary glands

male gamete

female gamete

② Fertilization produces a zygote that enters the wall of the mosquito's stomach.

③ The zygote gives rise to sporozoites that migrate to the mosquito's salivary glands.

⑦ The synchronized rupture of red blood cells releases toxins and the parasites; some parasites infect more blood cells.

⑥ Parasites multiply in the red blood cells.

⑧ Some parasites become gametocytes, which may be ingested by another feeding *Anopheles* mosquito.

⑤ Parasites emerge from the liver and enter red blood cells.

④ The infected mosquito bites an uninfected human and saliva containing sporozoites is injected; the sporozoites enter the liver and develop through several stages.

liver

▸ **FIGURE 21-10 The life cycle of the malaria parasite**

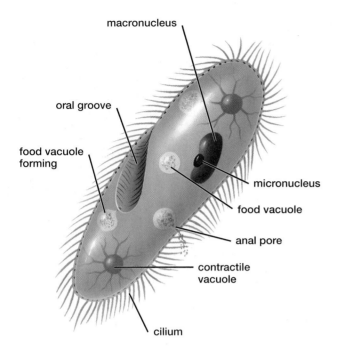

macronucleus

oral groove

food vacuole forming

micronucleus

food vacuole

anal pore

contractile vacuole

cilium

▲ **FIGURE 21-11 The complexity of ciliates** The ciliate *Paramecium* illustrates some important ciliate organelles. The oral groove acts as a mouth, food vacuoles—miniature digestive systems—form at its apex, and waste is expelled by exocytosis through an anal pore. Contractile vacuoles regulate water balance.

not evolved immune defenses against malaria, nonetheless mostly remained uninfected at first. In the 1920s, however, the mosquito species that transmits avian malaria was accidentally introduced to the island. Malaria then spread rapidly through native bird populations and was a major contributor to the extinction of many native species.

Ciliates Are the Most Complex of the Alveolates

Ciliates, which inhabit fresh and salt water, represent the peak of unicellular complexity. They possess many specialized organelles, including cilia, the short hairlike outgrowths for which they are named. In the well-known freshwater genus *Paramecium,* rows of cilia cover the organism's entire body surface (**FIG. 21-11**). The coordinated beating of the cilia propels the cell through the water at a rate of 1 millimeter (about 10 body lengths) per second—a protistan speed record. Although only a single cell, *Paramecium* responds to its environment as if it had a well-developed nervous system. Confronted with a noxious chemical or a physical barrier, the cell immediately backs up by reversing the beating of its cilia and then proceeds in a new direction. Some ciliates, such as *Didinium,* are accomplished predators (**FIG. 21-12**).

Rhizarians Have Thin Pseudopods

Protists in a number of different groups possess flexible plasma membranes that they can extend in any direction to form finger-like projections called pseudopods, which they

▲ **FIGURE 21-12 A microscopic predator** In this scanning electron micrograph, the predatory ciliate *Didinium* attacks a *Paramecium*. Notice that the cilia of *Didinium* are confined to two bands, whereas *Paramecium* has cilia over its entire body. Ultimately, the predator will engulf and consume its prey. This microscopic drama could take place on a pinpoint with room to spare.

use in locomotion and for engulfing food. The pseudopods of **rhizarians** are thin and thread-like. In many species in this group, the pseudopods extend through hard shells. Rhizarians include the foraminiferans and the radiolarians.

Foraminiferans Have Chalky Shells

The **foraminiferans** are primarily marine protists that produce beautiful shells. Their shells are constructed mostly of calcium carbonate (chalk; **FIG. 21-13**). These elaborate shells are pierced by myriad openings through which pseudopods extend. The chalky shells of dead foraminiferans,

▲ **FIGURE 21-13 The chalky shell of a foraminiferan** In a living foraminiferan, thin pseudopods would extend out through the openings in the shell, to sense the environment and capture food.

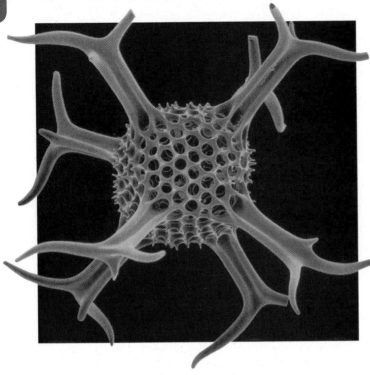

▲ **FIGURE 21-14 A radiolarian** Only the delicate, glassy shell is shown, so the pseudopods present in the living organism are not evident.

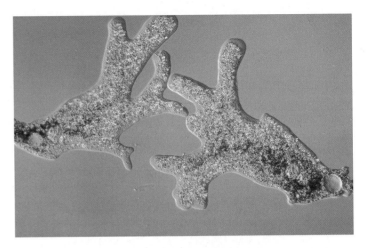

▲ **FIGURE 21-15 Amoebas** Amoebas are active predators that move through water to engulf food with thick, blunt pseudopods.

sinking to the ocean bottom and accumulating over millions of years, have resulted in immense deposits of limestone, such as those that form the famous White Cliffs of Dover, England.

Radiolarians Have Glassy Shells

Like foraminiferans, **radiolarians** have thin pseudopods that extend through hard shells. The shells of radiolarians, however, are made of glass-like silica (**FIG. 21-14**). The beauty of these microscopic glassy shells has long impressed architects, artists, and scientists. The prominent nineteenth-century biologist Ernst Haeckel wrote that "every morning I am newly amazed at the inexhaustible richness of these tiny and delicate structures," and said that his early study of radiolarians inspired him to pursue a career in science.

Amoebozoans Have Pseudopods and No Shells

Amoebozoans move by extending finger-shaped pseudopods, which may also be used for feeding. Amoebozoans generally do not have shells. The major groups of amoebozoans are the amoebas and the slime molds.

Amoebas Have Thick Pseudopods

Amoebas are common in freshwater lakes and ponds (**FIG. 21-15**). Many amoebas are predators that stalk and engulf prey, but some species are parasites. One parasitic form causes amoebic dysentery, a disease that is prevalent in warm climates. The dysentery-causing amoeba multiplies in the intestinal wall, triggering severe diarrhea.

Slime Molds Are Decomposers That Inhabit the Forest Floor

The physical form of *slime molds* seems to blur the boundary between a colony of separate individuals and a single, multicellular individual. The life cycle of the slime mold consists of two phases: a mobile feeding stage and a stationary reproductive stage called a *fruiting body*. There are two main types of slime molds: acellular and cellular.

Acellular Slime Molds Form a Multinucleate Mass of Cytoplasm Called a Plasmodium The **acellular slime molds**, also known as *plasmodial* slime molds, consist of a mass of cytoplasm that may spread thinly over an area of several square yards. Although the mass contains thousands of diploid nuclei, the nuclei are not confined in separate cells surrounded by plasma membranes. This structure, called a **plasmodium,** explains why these protists are described as "acellular" (without cells). The plasmodium oozes through decaying leaves and rotting logs, engulfing food such as bacteria and particles of organic material. The mass may be bright yellow or orange (**FIG. 21-16a**). Dry conditions or starvation stimulate the plasmodium to form fruiting bodies, on which haploid spores are produced (**FIG. 21-16b**). The spores are dispersed and germinate to produce mobile haploid cells. Two such cells may meet and fuse, forming a diploid zygote that gives rise to a new plasmodium.

Cellular Slime Molds Live as Independent Cells but Aggregate into a Pseudoplasmodium When Food Is Scarce The **cellular slime molds,** also known as *social amoebas,* live in soil as independent haploid cells that move and feed by producing pseudopods. In the best-studied genus, *Dictyostelium,* individual cells release a chemical signal when food becomes scarce. This signal attracts nearby cells into a dense aggregation that forms a slug-like mass called a **pseudoplasmodium** ("false plasmodium") because, unlike a true plasmodium, it consists of individual cells (**FIG. 21-17**).

(a) Plasmodium

(b) Fruiting bodies

▲ FIGURE 21-16 **An acellular slime mold (a)** A plasmodium oozes over a stone on the damp forest floor. **(b)** When food becomes scarce, the mass differentiates into fruiting bodies in which spores are formed.

A pseudoplasmodium can be viewed as a colony of individuals because the cells that compose it are not all genetically identical. In some ways, however, a pseudoplasmodium is more like a multicellular organism, because its cells differentiate into different cell types, with different types serving different functions. A pseudoplasmodium moves about in slug-like fashion, migrating toward an aboveground spot suitable for spore dispersal, where its cells differentiate to convert the structure to a fruiting body. Haploid spores formed within the fruiting body are dispersed by wind and germinate directly into new single-celled individuals. In some circumstances, two independent cells may fuse to form a diploid zygote, which develops into a larger cyst that ultimately releases haploid spores.

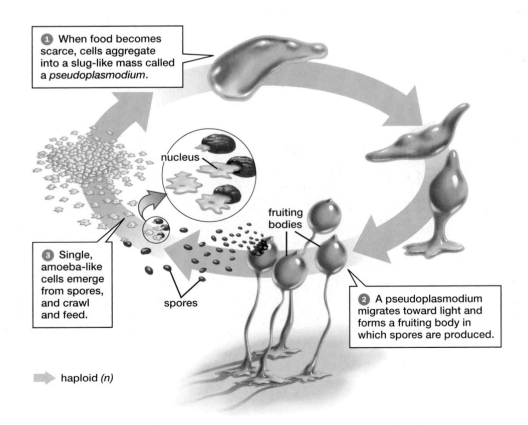

❶ When food becomes scarce, cells aggregate into a slug-like mass called a *pseudoplasmodium*.

nucleus

❸ Single, amoeba-like cells emerge from spores, and crawl and feed.

spores

fruiting bodies

❷ A pseudoplasmodium migrates toward light and forms a fruiting body in which spores are produced.

➡ haploid *(n)*

▲ FIGURE 21-17 **The life cycle of a cellular slime mold**

▲ **FIGURE 21-18 Red algae** Red coralline algae from the Mediterranean Sea. Coralline algae, which deposit calcium carbonate within their bodies, contribute to coral reefs in tropical waters.

Red Algae Contain Red Photosynthetic Pigments

The red algae are multicellular, photosynthetic seaweeds (**FIG. 21-18**). These protists range in color from bright red to nearly black; they derive their hue from red pigments that mask their green chlorophyll. Red algae are found almost exclusively in marine environments. They dominate in deep, clear tropical waters, where their red pigments absorb the deeply penetrating blue-green light and transfer this light energy to chlorophyll, where it is used in photosynthesis. The solar energy that red algae capture helps support non-photosynthetic organisms in marine ecosystems.

Red algae contain gelatinous substances with commercial uses. One of these substances is carrageenan, a combination of polysaccharides extracted from the cell walls of various species of red algae. Carrageenan melts at a relatively low temperature and, after cooling, forms a gel that remains stable at room temperature. These properties have proved useful to food processors, and carrageenan is widely used as a thickener and stabilizer in commercially produced foods including ice cream, yogurt, chocolate milk, soymilk, jellies, soups, salad dressings, and lunchmeats.

Chlorophytes Are Green Algae

The **chlorophytes** are a clade of green algae that includes both multicellular and unicellular species. Most chlorophyte algae live in freshwater ponds and lakes, but some live in the oceans. Some, such as *Oedogonium,* form thin filaments from long chains of cells (**FIG. 21-19a**). Other chlorophyte species form colonies containing clusters of cells that are somewhat interdependent and represent a structure intermediate between unicellular and multicellular forms. These colonies range from a few cells to a few thousand cells, as in *Volvox.* Most chlorophytes are small, but some marine species are large. For example, *Ulva,* or sea lettuce, is similar in size to the leaves of its namesake (**FIG. 21-19b**).

(a) Oedogonium

(b) Ulva

(c) Growing green algae for biofuel

▲ **FIGURE 21-19 Chlorophytes (a)** *Oedogonium* is a filamentous green alga composed of strands only one cell thick. **(b)** *Ulva* is a multicellular green alga that assumes a leaf-like shape. **(c)** Biofuels produced from algae grown at a facility like the one pictured may one day fill a significant portion of our energy needs, if technical obstacles can be overcome.

Some chlorophyte species are currently under intensive cultivation by companies that hope to use them for commercial production of biofuels (**FIG. 21-19c**). Fuels based on chlorophyte algae could in principle replace dwindling fossil fuels with a renewable fuel whose production and use release less carbon dioxide into the atmosphere. However, efforts to develop an efficient, economically viable process for converting algae to fuel have not yet been successful. Some scientists have argued that bioengineering holds the key to success and have begun research aimed at engineering chlorophyte

genomes to produce a novel organism capable of producing fuel efficiently under industrial conditions.

CHECK YOUR LEARNING

Can you ...

- list the major protist taxonomic groups and the key characteristics of each group?
- describe some examples of how members of each group affect humans?

CASE STUDY ⟩ **REVISITED**

Green Monster

Caulerpa taxifolia, the invasive seaweed that threatens to over-run the Mediterranean, is a chlorophyte. This species and other members of its genus have very unusual bodies. Outwardly, they appear plantlike, with rootlike structures that attach to the sea-floor and with other structures that look like stems and leaves, rising to a height of several inches. Despite its seeming similarity to a plant, however, a *Caulerpa* body consists of a single, extremely large cell. The entire body is surrounded by a single, continuous cell membrane. The interior consists of cytoplasm that contains many nuclei but is not subdivided. That a single cell can take such a complex shape is extraordinary.

A potential problem with *Caulerpa*'s single-celled organization might arise when its body is damaged, perhaps by wave action or when a predator takes a bite out of it. When the cell membrane is breached, all of the organism's cytoplasm could potentially leak out, an event that would be fatal. But *Caulerpa* has evolved a defense against this potential calamity. Shortly after the cell membrane breaks, it is quickly filled with a "wound plug" that closes the gap. After the plug is established, the cell begins to grow and regenerates any lost portion.

This ability to regenerate is a key component of the ability of *Caulerpa taxifolia* to spread rapidly in new environments. If part of a *Caulerpa* breaks off and drifts to a new location, the fragment can regenerate a whole new body. The regenerated individual becomes the founder of a new, quickly growing colony—and these quickly growing colonies might appear anywhere in the world. Authorities in many countries worry that the aquarium strain of *Caulerpa* could invade their coastal waters, unwittingly transported by ships from the Mediterranean or released by care-less aquarists. In fact, invasive *Caulerpa* is no longer restricted to the Mediterranean. It now thrives in at least 13 locations in Australia. It also appeared at two locations in California, but authorities there were able to eradicate it after seven years of intensive, expensive effort. Australia has not been so fortunate, and *Caulerpa taxifolia* continues to spread there.

CONSIDER THIS Is it important to stop the spread of *Caulerpa*? Governments invest substantial resources to combat introduced species and prevent their populations from increasing and dispersing. Is this a wise use of funds? Can you think of some arguments against spending time and money for this purpose?

CHAPTER REVIEW

 Go to **MasteringBiology** for practice quizzes, activities, eText, videos, current events, and more.

Answers to Think Critically, Evaluate This, Multiple Choice, and Fill-in-the-Blank questions can be found in the Answers section at the back of the book.

Summary of Key Concepts

21.1 What Are Protists?

"Protist" is a term of convenience that refers to any eukaryote that is not a plant, animal, or fungus. Most protists are single, highly complex eukaryotic cells, but some form colonies, and some, such as seaweeds, are multicellular. Protists exhibit diverse modes of

nutrition, reproduction, and locomotion. Photosynthetic protists form much of the phytoplankton, which plays a key ecological role. Some protists cause human diseases; others are crop pests.

21.2 What Are the Major Groups of Protists?

Protist groups include excavates (diplomonads and parabasalids), euglenozoans (euglenids and kinetoplastids), stramenopiles (water molds, diatoms, and brown algae), alveolates (dinoflag-ellates, apicomplexans, and ciliates), rhizarians (foraminiferans and radiolarians), amoebozoans (amoebas and slime molds), red algae, and chlorophyte algae.

Key Terms

acellular slime mold   *378*
alga (plural, algae)   *370*
alveolate   *374*
amoeba   *378*
amoebozoan   *378*
apicomplexan   *376*
cellular slime mold   *378*
chlorophyte   *380*
ciliate   *377*
diatom   *373*
dinoflagellate   *374*
diplomonad   *370*
euglenid   *372*
euglenozoan   *372*
excavate   *370*

food vacuole   *369*
foraminiferan   *377*
kinetoplastid   *372*
parabasalid   *372*
phytoplankton   *373*
plasmodium   *378*
protist   *369*
protozoan (plural,
   protozoa)   *370*
pseudoplasmodium   *378*
pseudopod   *369*
radiolarian   *378*
rhizarian   *377*
stramenopile   *373*
water mold   *373*

Thinking Through the Concepts

Multiple Choice

1. Which of the following statements about protists is False?
   a. Some protists are photosynthetic.
   b. All protists are eukaryotes.
   c. Although protists are diverse, they form a single clade.
   d. Protists include both unicellular and multicellular species.

2. The harmful protist blooms known as "red tides" are produced by
   a. apicomplexans.
   b. dinoflagellates.
   c. red algae.
   d. foraminiferans.

3. The organism that causes malaria belongs to which of the following groups?
   a. alveolates
   b. slime molds
   c. ciliates
   d. apicomplexans

4. The finger-like extensions of the cell membrane that some protists use for feeding or locomotion are called
   a. pseudopods.
   b. cilia.
   c. flagella.
   d. food vacuoles.

5. A person who develops severe diarrhea after drinking untreated water on a camping trip is likely to have been infected by
   a. *Plasmodium,* an apicomplexan.
   b. *Ulva,* a chlorophyte.
   c. *Paramecium,* a ciliate.
   d. *Giardia,* a diplomonad.

Fill-in-the-Blank

1. Protists that absorb nutrients from their surroundings may act as _____ of dead organic matter or as harmful _____ of larger living organisms.

2. Photosynthetic protists are collectively known as _____; nonphotosynthetic, single-celled protists are collectively known as _____.

3. Protist chloroplasts surrounded by four-layer membranes arose evolutionarily through _____, in which an ancestral nonphotosynthetic protist engulfed but did not digest a(n) _____.

4. The disease-causing protist that causes malaria is a member of the _____ group, and the protist that causes sleeping sickness is a member of the _____ group.

5. The plant diseases downy mildew and late blight are caused by protists in the _____ group. Slime molds are members of the _____ group.

6. Protists that make up a large proportion of Earth's phytoplankton include _____ and _____. The protist group containing the species most likely to one day be cultivated for biofuel production is _____.

Review Questions

1. List the major differences between prokaryotes and protists.

2. What is secondary endosymbiosis?

3. What is the importance of dinoflagellates in marine ecosystems? What can happen to marine ecosystems when certain dinoflagellate species reproduce rapidly?

4. What is the major ecological role played by single-celled algae?

5. Which protist group consists entirely of parasitic forms?

6. Which protist groups include seaweeds?

7. Which protist groups include species that use pseudopods?

Applying the Concepts

1. The internal structure of many protists is much more complex than that of cells of multicellular organisms. Does this mean that the protist is engaged in more complex activities than the multicellular organism is? If not, why are protistan cells more complicated?

2. What are some important benefits and services provided by protists to Earth's other organisms?

The huge, foul-smelling flower of the stinking corpse lily is a treat for visitors to Asian rain forests.

22

THE DIVERSITY OF PLANTS

Queen of the Parasites

THE FLOWER OF THE STINKING CORPSE LILY makes a strong impression. For one thing, it's huge; a single flower may be 3 feet across. It also has a rather strange appearance, consisting largely of fleshy lobes that are almost fungus-like. But as its name implies, the thing that makes a stinking corpse lily almost impossible to ignore is its aroma, which has been described as "a penetrating smell more repulsive than any buffalo carcass in an advanced stage of decomposition."

Unlike most plants, a stinking corpse lily has no visible leaves, roots, or stems. In fact, it is a parasite, and its body is completely embedded in the tissue of its host, a vine in the grape family. Because it has no leaves, the stinking corpse lily cannot produce any food of its own, but instead draws all of its nutrition from its host. The parasite becomes visible outside the body of its host only when one of its cabbage-shaped flower buds pushes through the surface of the host's stem and its gigantic, stinking flower opens for a week or so before shriveling and falling off. If a male and a female flower happen to be open simultaneously and close together, the female flower may be fertilized and produce seeds. A seed that is dispersed in the droppings of an animal that has consumed it, and that happens to land on a stem of the host species, may germinate and penetrate a new host.

When you think of plants, you might first think of their most obvious feature: green leaves that capture solar energy by photosynthesis. It may seem odd, then, that this chapter about plants begins with a peculiar plant that does not photosynthesize. Oddities such as the stinking corpse lily, however, serve as reminders that evolution does not always follow a predictable pathway, and that even an adaptation as seemingly valuable as the ability to live on sunlight can be lost. What other interesting characteristics have appeared over the evolutionary history of plants?

383

AT A GLANCE

22.1 WHAT ARE THE KEY FEATURES OF PLANTS?

What distinguishes plants from other organisms? Most plants exhibit three characteristic traits: photosynthesis, multicellular embryos, and alternation of generations, as explained below. Each of these traits also occurs in some other kinds of organisms, but only plants combine all three.

Plants Are Photosynthetic

Perhaps the most noticeable feature of nearly all plants is their green color. The color comes from the presence of chlorophyll in many plant tissues. Chlorophyll plays a crucial role in photosynthesis, the process by which plants use energy from sunlight to convert water and carbon dioxide to sugar (see Chapter 7). Chlorophyll and photosynthesis, however, are not unique to plants; they are also present in many types of protists and prokaryotes.

Plants Have Multicellular, Dependent Embryos

Plants are distinguished from other photosynthetic organisms by their characteristic embryos. A plant embryo is multicellular and is attached to and dependent on its parent. As it grows and develops, the embryo receives nutrients from the tissues of the parent plant. Multicellular, dependent embryos are not found among photosynthetic protists.

Plants Have Alternating Multicellular Haploid and Diploid Generations

Plant reproduction is characterized by a type of life cycle called **alternation of generations** (FIG. 22-1). In organisms with alternation of generations, separate multicellular diploid and haploid generations alternate with one another. (Recall that a diploid organism has paired chromosomes; a haploid organism has unpaired chromosomes.) In the diploid (2n) generation, the body consists of diploid cells and is known as the **sporophyte.** The multicellular embryo is part of the diploid sporophyte generation. Certain cells of sporophytes undergo meiosis to produce haploid (n) reproductive cells called *spores*. The haploid spores develop into multicellular, haploid bodies called **gametophytes.**

A gametophyte ultimately produces male and female haploid gametes (sperm and eggs) by mitosis. Gametes, like spores, are reproductive cells but, unlike spores, an individual gamete by itself cannot develop into a new individual. Instead, two gametes of opposite sexes must meet and fuse to form a new diploid individual. In plants, gametes produced by gametophytes fuse to form a diploid zygote (a fertilized egg), which develops into a diploid embryo. The embryo develops into a mature sporophyte, and the cycle begins again.

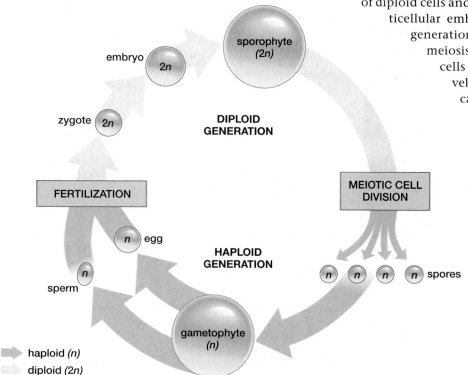

haploid (n)

diploid (2n)

▲ **FIGURE 22-1 Alternation of generations in plants** As shown in this generalized depiction of a plant life cycle, a diploid sporophyte generation produces haploid spores through meiotic cell division. The spores develop into a haploid gametophyte generation that produces haploid gametes by mitotic cell division. The fusion of these gametes results in a diploid zygote that develops into the sporophyte plant.

22.2 HOW HAVE PLANTS EVOLVED?

Plants form a clade within a larger clade that also includes several groups of green algae collectively known as *charophytes*. (A clade is a group consisting of all the descendants of a particular common ancestor.) The charophyte algae include plants' closest living relatives, the stoneworts (FIG. 22-2). The close evolutionary link between stoneworts and plants was revealed by DNA comparisons and is reflected in additional similarities between plants and charophytes. For example, plants and charophytes both store food as starch and have cell walls made of cellulose, and both use the same types of chlorophyll in photosynthesis (*chlorophylls a* and *b*).

The Ancestors of Plants Lived in Water

The ancestors of plants were photosynthetic protists, perhaps similar to stoneworts. Like modern stoneworts, the protists that gave rise to plants presumably lacked true roots, stems, leaves, and complex reproductive structures such as flowers or cones, features that appeared only later in the evolutionary history of plants. In addition, the ancestors of plants were confined to watery habitats.

For these ancestors of plants, life in water had many advantages. For example, in water, a body is bathed in a nutrient-rich solution, is supported by buoyancy, and is not likely to dry out. In addition, life in water facilitates reproduction, because gametes and zygotes can be carried by water currents or propelled by flagella.

Early Plants Invaded Land

Despite the benefits of aquatic environments, early plants invaded habitats on land. Today, most plants live on land. The move to land brought its own advantages, including access to sunlight unimpeded by water that might block its rays, access to nutrients contained in surface rocks, and freedom from predators. However, the move to land also imposed some challenges; plants could no longer rely on watery surroundings to provide support, moisture, nutrients, and transportation for gametes and zygotes. As a result, life on land has favored the evolution in plants of traits that help meet these environmental challenges.

Plant Bodies Evolved to Resist Gravity and Drying

Some of the key adaptations to life on land arose early in plant evolution and are now found in virtually all land plants (FIG. 22-3). These early adaptations include:

• Roots or rootlike structures that anchor the plant and absorb water and nutrients from the soil.
• A waxy **cuticle** that covers the surfaces of leaves and stems and that limits the evaporation of water.
• Pores called **stomata** (singular, stoma) in the leaves and stems that open to allow gas exchange but close when water is scarce, reducing the amount of water lost to evaporation.

Other key adaptations occurred somewhat later in the transition to terrestrial life and are now widespread but not universal among plants (most nonvascular plants, described later, lack these traits):

• Conducting tissues called **xylem** and **phloem** that transport water and dissolved substances. Xylem conducts water and minerals upward from the roots; phloem

▲ **FIGURE 22-2 *Chara*, a stonewort** The green algae known as stoneworts are plants' closest living relatives.

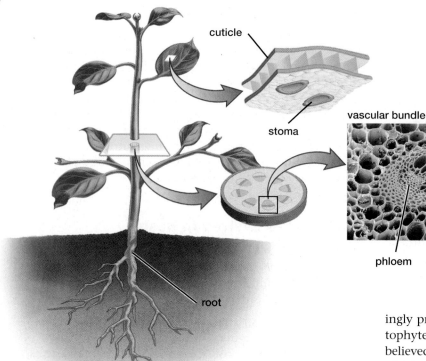

◀ FIGURE 22-3 Early adaptations for life on land Adaptations for life on land include roots that anchor the plant, a waxy surface cuticle that reduces evaporation, stomata that can be closed to conserve water, and (in vascular plants) lignin-impregnated xylem and phloem tissues that transport water and nutrients and help support the plant body.

conducts the products of photosynthesis to different parts of the plant body.

- The stiffening substance **lignin,** a rigid polymer that impregnates the cells of the conducting tissues and supports the plant body against the force of gravity.

Plants Evolved Sex Cells That Disperse Without Water and Protection for Their Embryos

The most widespread groups of plants, collectively known as seed plants, are characterized by sex cells that do not rely on water for dispersal and by especially well-protected and well-provisioned embryos. The key adaptations of these plant groups are pollen, seeds, and, in the flowering plants, flowers and fruits.

Early seed plants gained an advantage over their competitors by producing dry, microscopic pollen grains that allowed wind, instead of water, to carry the male gametes. Early seed plants also produced seeds, which provided protection and nourishment for developing embryos and the potential for more effective dispersal. Later came the evolution of flowers, which enticed animal pollinators that were able to deliver pollen more precisely than did wind. Fruits also attracted animals, which consumed the fruit and dispersed its seeds in their feces.

More Recently Evolved Plants Have Smaller Gametophytes

The evolutionary history of plants has been marked by a tendency for the sporophyte generation to become increas-

ingly prominent and for the longevity and size of the gametophyte generation to shrink. Thus, the earliest plants are believed to have been similar to today's nonvascular plants, which have a sporophyte that is smaller than the gametophyte and remains attached to it. In contrast, plants that originated somewhat later, such as ferns and the other seedless vascular plants, feature a life cycle in which the sporophyte is dominant, and the gametophyte is a much smaller, independent plant. Finally, in the most recently evolved group of plants, the seed plants, gametophytes are microscopic and barely recognizable as an alternate generation. These tiny gametophytes, however, still produce the eggs and sperm that unite to form the zygote that develops into the diploid sporophyte.

CHECK YOUR LEARNING

Can you ...

- describe the probable ancestor of plants?
- identify the closest living relatives of plants and explain their similarities to and differences from plants?
- describe the adaptations that equip plants for life on land?

CASE STUDY \ **CONTINUED**

Queen of the Parasites

The stinking corpse lily, with its huge, 3-foot-wide flowers, apparently evolved from an ancestor with tiny flowers. A recent analysis of DNA sequences revealed that the plant group most closely related to the group that includes the stinking corpse lily is the spurges, plants with mostly tiny flowers. The analysis also showed that the common ancestor of spurges and corpse lilies probably had flowers that were about 1/80th the size of modern stinking corpse lily flowers.

In stark contrast to the stinking corpse lily and its relatives, many plants have no flowers at all. What are these flowerless plants like?

22.3 WHAT ARE THE MAJOR GROUPS OF PLANTS?

Two major groups of land plants arose from ancient algal ancestors (**FIG. 22-4** and **TABLE 22-1**). Members of one group, the **nonvascular plants** (also called *bryophytes*), require a moist environment to reproduce and thus straddle the boundary between aquatic and terrestrial life. The other group, the **vascular plants** (also called *tracheophytes*), has colonized drier habitats.

Nonvascular Plants Lack Conducting Structures

Nonvascular plants retain some characteristics of their algal ancestors. Their gametes are dispersed by water, and they lack true roots, leaves, and stems. They do possess rootlike anchoring structures called *rhizoids* that bring water and nutrients into the plant body, but nonvascular plants lack well-developed structures for conducting water and nutrients. They must instead rely on slow diffusion or poorly developed conducting tissues to distribute water and other nutrients. As a result, their body size is limited. Size is also limited by the absence of the stiffening agent lignin in their bodies. Without lignin, nonvascular plants cannot grow upward very far. Most nonvascular plants are less than 1 inch (2.5 centimeters) tall.

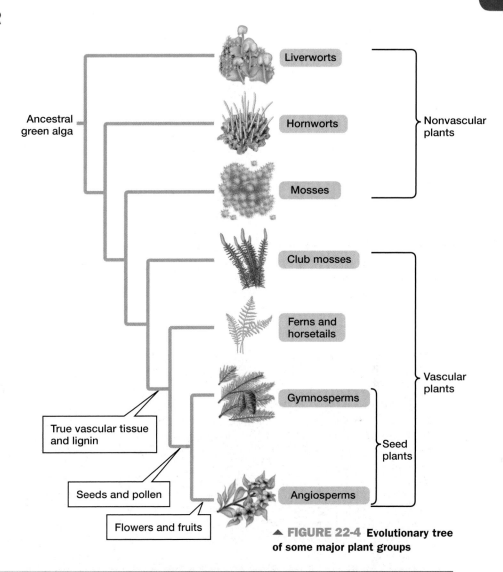

▲ **FIGURE 22-4 Evolutionary tree of some major plant groups**

TABLE 22-1	Features of the Major Plant Groups					
Group	**Subgroup**	**Relationship of Sporophyte and Gametophyte**	**Transfer of Reproductive Cells**	**Early Embryonic Development**	**Dispersal**	**Water and Nutrient Transport Structures**
Nonvascular plants	Liverworts Hornworts Mosses	The gametophyte is dominant—the sporophyte develops from a zygote retained on a gametophyte	Motile sperm swim to a stationary egg retained on a gametophyte	Occurs within the archegonium of a gametophyte	Haploid spores are carried by wind	Absent
Vascular plants	Club mosses Horsetails and ferns	The sporophyte is dominant—it develops from a zygote retained on a gametophyte	Motile sperm swim to a stationary egg retained on a gametophyte	Occurs within the archegonium of a gametophyte	Haploid spores are carried by wind	Present
	Gymnosperms	The sporophyte is dominant—the microscopic gametophyte develops within a sporophyte	Wind-dispersed pollen carries sperm to a stationary egg in a cone	Occurs within a protective seed containing a food supply	Seeds containing a diploid sporophyte embryo are dispersed by wind or animals	Present
	Angiosperms	The sporophyte is dominant—the microscopic gametophyte develops within a sporophyte	Pollen, dispersed by wind or animals, carries sperm to a stationary egg within a flower	Occurs within a protective seed containing a food supply; the seed is encased within fruit	Fruit, carrying seeds, is dispersed by animals, wind, or water	Present

Nonvascular Plants Include the Liverworts, Hornworts, and Mosses

The nonvascular plants include three groups: liverworts, hornworts, and mosses. Liverworts and hornworts are named for their shapes. The gametophytes of certain liverwort species have a lobed form reminiscent of the shape of a liver (**FIG. 22-5a**). Hornwort sporophytes generally have a spiky, somewhat horn-like shape (**FIG. 22-5b**). Liverworts and hornworts are most abundant in areas where moisture is plentiful, such as in moist forests and near the banks of streams and ponds.

Mosses are the most diverse and abundant of the nonvascular plants (**FIG. 22-5c**). Like liverworts and hornworts, mosses are most likely to be found in moist habitats. Some mosses, however, have a waterproof covering that reduces water loss. Many of these mosses are also able to survive the loss of much of the water in their bodies; they dehydrate and become dormant during dry periods but absorb water and resume growth when moisture returns. Such mosses can survive in deserts, on bare rock, and in far northern and southern latitudes where humidity is low and liquid water is scarce for much of the year.

Mosses of the genus *Sphagnum* are especially widespread, living in moist habitats in northern regions around the world. In many of these wet northern habitats, *Sphagnum* is the most abundant plant, forming extensive mats (**FIG. 22-5d**). Because decomposition is slow in cold climates and because *Sphagnum* contains compounds that inhibit bacterial growth, dead *Sphagnum* may decay very slowly. As a result, partially decayed moss tissue may accumulate in deposits that can, over thousands of years, become hundreds of feet thick. These deposits are known as peat. Peat has long been harvested for use as fuel, a practice that continues today in Ireland, Finland, Russia, and other northern countries. Now, however, peat is more often harvested for use in horticulture. Dried peat can absorb many times its own weight in water, making it useful as a soil conditioner and as a packing material for transporting live plants.

(a) Liverwort

(b) Hornwort

(c) Moss

(d) *Sphagnum* bog

▲ **FIGURE 22-5 Nonvascular plants** These plants are less than a half-inch (about 1 centimeter) in height. **(a)** Liverworts grow in moist, shaded areas. The palmlike structures on the female plants shown here hold eggs. Male plants produce sperm that swim through a film of water to reach and fertilize the eggs. **(b)** The hornlike sporophytes of hornworts grow upward from the gametophyte body. **(c)** Moss plants, showing the stalks that carry spore-bearing capsules. **(d)** Mats of *Sphagnum* moss cover moist bogs in northern regions.

THINK CRITICALLY Why are all nonvascular plants short?

The Reproductive Structures of Nonvascular Plants Are Protected

Nonvascular plants require moisture to reproduce, but they have evolved some traits that facilitate reproduction on land. For example, the reproductive structures of nonvascular plants are enclosed, which prevents the gametes from drying out (FIG. 22-6). There are two types of reproductive structures in which gametes are produced by mitotic cell division: **archegonia** (singular, archegonium), in which eggs develop, and **antheridia** (singular, antheridium), where sperm are formed ❶. In some nonvascular plant species, both archegonia and antheridia are located on the same plant; in other species, each individual plant is either male or female.

In all nonvascular plants, the sperm must swim to the egg through a film of water ❷, so nonvascular plants that live in drier areas can reproduce only when it rains. After fertilization, the zygote is retained in the archegonium, where the embryo grows and matures into a small diploid sporophyte that remains attached to the parent gametophyte plant ❸. At maturity, the sporophyte produces reproductive capsules. Within each capsule, haploid spores are produced by meiotic cell division ❹. When the capsule is opened, spores are released and dispersed by the wind ❺. If a spore lands in a suitable environment, it may develop into another haploid gametophyte plant ❻.

Vascular Plants Have Conducting Cells That Also Provide Support

Vascular plants are distinguished by the presence of xylem and phloem, specialized tissues consisting of tube-shaped conducting cells (see Fig. 22-3). These cells are impregnated

▼ **FIGURE 22-6 Life cycle of a moss** The photo shows moss plants; the short, leafy green plants are haploid gametophytes; the reddish brown stalks are diploid sporophytes.

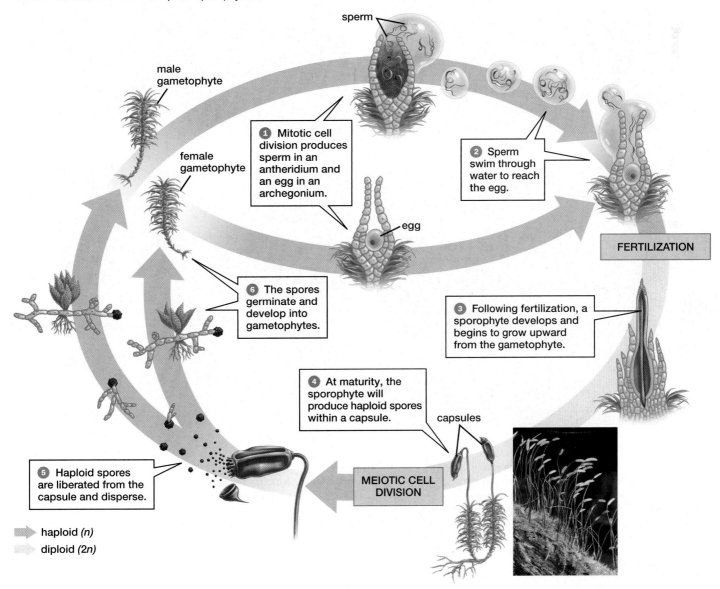

sperm

male gametophyte

female gametophyte

❶ Mitotic cell division produces sperm in an antheridium and an egg in an archegonium.

❷ Sperm swim through water to reach the egg.

egg

FERTILIZATION

❻ The spores germinate and develop into gametophytes.

❸ Following fertilization, a sporophyte develops and begins to grow upward from the gametophyte.

❹ At maturity, the sporophyte will produce haploid spores within a capsule.

capsules

❺ Haploid spores are liberated from the capsule and disperse.

MEIOTIC CELL DIVISION

⟹ haploid (n)

⟹ diploid (2n)

with the stiffening substance lignin and serve both support- ive and conducting functions. They allow vascular plants to grow taller than nonvascular plants, both because of the extra support provided by lignin and because the conducting cells allow water and nutrients absorbed by the roots to move to the upper portions of the plant. Another difference between vascular plants and nonvascular plants is that in vascular plants, the diploid sporophyte is the larger, more conspicuous generation; in nonvascular plants, the haploid gametophyte is more evident. The vascular plants can be divided into two groups: the seedless vascular plants and the seed plants.

The Seedless Vascular Plants Include the Club Mosses, Horsetails, and Ferns

Like nonvascular plants, seedless vascular plants (**FIG. 22-7**) have swimming sperm and require water for reproduction. As their name implies, they propagate by spores rather than seeds. Present-day seedless vascular plants—the club mosses, horsetails, and ferns—are much smaller than their ancestors, which dominated the landscape in the Carboniferous period (359 million to 299 million years ago; see Fig. 18-8).

(a) Club moss

(b) Horsetail

(c) Fern

(d) Tree fern

▲ FIGURE 22-7 **Some seedless vascular plants** Seedless vascular plants are found in moist woodland habitats. **(a)** The club mosses (sometimes called ground pines) grow in temperate forests. **(b)** The giant horsetail extends long, narrow branches in a series of rosettes at regular intervals along the stem. Its leaves are insignificant scales. At right is a cone-shaped spore-forming structure. **(c)** The leaves of this deer fern are emerging from coiled structures called fiddleheads. **(d)** Although most fern species are small, some, such as this tree fern, retain the large size that was common among ferns of the Carboniferous period.

THINK CRITICALLY In each of these photos, is the pictured structure a sporophyte or a gametophyte?

Club Mosses and Horsetails Are Seedless Plants with Tiny, Scalelike Leaves

The club mosses, which despite their common name are not actually mosses, are now limited to representatives a few inches in height (**FIG. 22-7a**). Their leaves are small and scale-like, resembling the leaflike structures of mosses. Club mosses of the genus *Lycopodium,* commonly known as ground pine, form a beautiful ground cover in some temperate coniferous and deciduous forests.

Modern horsetails belong to a single genus, *Equisetum,* that contains only 15 species, most less than 3 feet tall (**FIG. 22-7b**). The bushy branches of some species lend them the common name horsetails; the leaves are reduced to tiny scales on the branches. They are also called scouring rushes because all species of *Equisetum* have large amounts of silica (glass) in their outer layer of cells, giving them an abrasive texture. Early European settlers of North America used horsetails to scour pots and floors.

Ferns Are Broad-Leaved and Diverse

The ferns, with 12,000 species, are the most diverse of the seedless vascular plants (**FIG. 22-7c**). In the tropics, tree ferns still reach heights reminiscent of their ancestors from the Carboniferous period (**FIG. 22-7d**). Ferns are the only seedless vascular plants that have broad leaves.

In fern reproduction (**FIG. 22-8**), gametes are produced by mitotic cell division in archegonia and antheridia on the tiny fern gametophyte ❶. Sperm are released into water and swim to reach an egg in an archegonium ❷. If fertilization occurs, the resulting zygote develops into a sporophyte plant, which grows upward from its parent, the gametophyte ❸. On a mature sporophyte fern plant, which is much larger than the gametophyte, haploid spores are produced in structures called *sporangia* that form on special leaves of the sporophyte ❹. The sporangia open to release the spores, which are dispersed by the wind ❺. If a spore lands in a spot with suitable conditions, it germinates and develops into a gametophyte plant ❻.

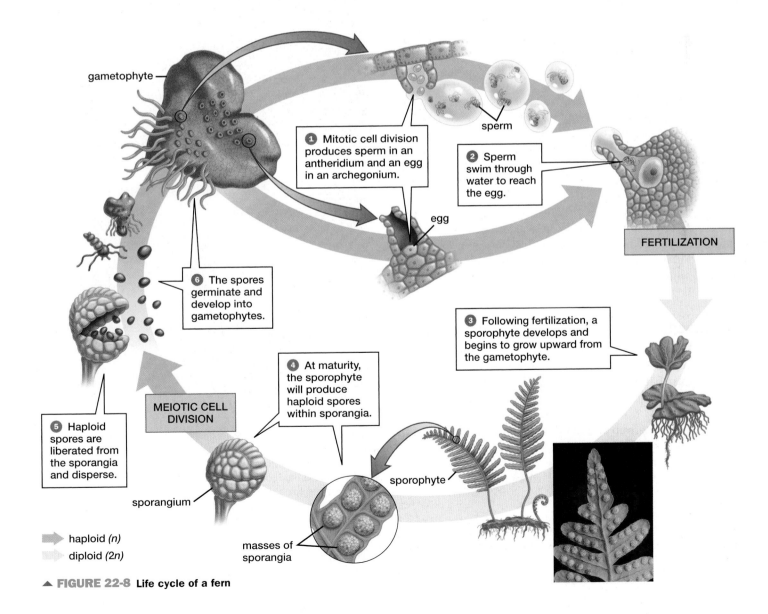

gametophyte

❶ Mitotic cell division produces sperm in an antheridium and an egg in an archegonium.

sperm

❷ Sperm swim through water to reach the egg.

egg

FERTILIZATION

❻ The spores germinate and develop into gametophytes.

❸ Following fertilization, a sporophyte develops and begins to grow upward from the gametophyte.

❹ At maturity, the sporophyte will produce haploid spores within sporangia.

MEIOTIC CELL DIVISION

❺ Haploid spores are liberated from the sporangia and disperse.

sporangium

sporophyte

masses of sporangia

haploid (n)

diploid (2n)

▲ **FIGURE 22-8 Life cycle of a fern**

The windborne spores of ferns make them especially effective at colonizing locations that lack abundant plant life. For example, just 2 years after a massive volcanic eruption that destroyed most life on the island of Krakatoa in 1883, visitors reported that ferns blanketed the previously denuded landscape. Similarly, fern abundance increased dramatically following the catastrophic asteroid impact that caused the extinction of dinosaurs and many other species about 66 million years ago. Spores of fossil ferns are extremely abundant in 66-million-year-old rocks at many locations around the world; these "spore spikes" are interpreted as evidence that massive fires followed the asteroid impact, burning up most vegetation and creating an opening for widespread colonization by ferns.

The Seed Plants Are Aided by Two Important Adaptations: Pollen and Seeds

The seed plants are distinguished from nonvascular plants and seedless vascular plants by their production of pollen and seeds. In seed plants, gametophytes (which produce the sex cells) are tiny. The female gametophyte is a small group of haploid cells that remains within the larger sporophyte and produces the egg. The male gametophyte is the **pollen** grain. Pollen grains are dispersed by wind or by animal pollinators such as bees. In this way, sperm move through the air to fertilize egg cells. This airborne transport means that the distribution of seed plants is not limited by the need for water through which sperm can swim to the egg.

Analogous to the eggs of birds and reptiles, **seeds** consist of an embryonic sporophyte plant, a supply of food for the embryo, and a protective outer coat (**FIG. 22-9**). The *seed coat* maintains the embryo in a state of dormancy until conditions are suitable for growth. The stored food helps sustain the emerging plant until it develops roots and leaves and can make its own food by photosynthesis.

Seed plants are grouped into two general types: gymnosperms, which lack flowers, and angiosperms, the flowering plants.

Gymnosperms Are Nonflowering Seed Plants

Gymnosperms evolved earlier than the flowering plants. Early gymnosperms coexisted with the forests of seedless vascular plants that prevailed during the Carboniferous period. During the subsequent Permian period (299 million to 252 million years ago), however, gymnosperms became the predominant plant group and remained so until the rise of the flowering plants more than 100 million years later. Most of these early gymnosperms are now extinct. Today, only four groups of gymnosperms survive: ginkgoes, cycads, gnetophytes, and conifers.

Only One Ginkgo Species Survives

Ginkgoes have a long evolutionary history. They were widespread during the Jurassic period, which began 201 million years ago. Today, however, they are represented by the

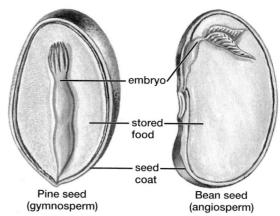

(a) Seeds

(b) Dandelion

(c) Coconut

▲ **FIGURE 22-9 Seeds (a)** Seeds from a gymnosperm (left) and an angiosperm (right). Both consist of an embryonic plant and stored food confined within a seed coat. **(b)** The tiny seeds of the dandelion are dispersed by the wind, held aloft by parachute-like tufts that are part of the fruit. **(c)** The massive, armored seeds (protected inside the fruit) of the coconut palm can survive prolonged immersion in seawater as they traverse oceans.

THINK CRITICALLY Can you think of some adaptations that help protect seeds from destruction by animal consumption?

single species *Ginkgo biloba,* the maidenhair tree (FIG. 22-10a). Ginkgo trees are either male or female; female trees bear foul-smelling, fleshy seeds the size of cherries. Because they are more resistant to pollution than are most other trees, ginkgoes (usually the male trees) have been extensively planted in U.S. cities.

Cycads Are Restricted to Warm Climates

Like ginkgoes, cycads were diverse and abundant in the Jurassic period but have since dwindled. Today approximately 160 species survive, most of which dwell in tropical or subtropical climates. Cycads have large, finely divided leaves and bear a superficial resemblance to palms or large ferns (FIG. 22-10b). Most cycads are about 3 feet (1 meter) in height, although some species can reach 65 feet (20 meters).

The tissues of cycads contain potent toxins. Despite the presence of these toxins, people in some parts of the world use cycad seeds, stems, and roots for food. Careful preparation and processing removes the toxins before the plants are consumed. Nonetheless, cycad toxins are the suspected cause of neurological problems that occur in societies, such as the Chamorro people of the Mariana Islands, that use cycads for food. Cycad toxins can also harm grazing livestock.

About half of all cycad species are classified as threatened or endangered. The main threats to cycads are habitat destruction, competition from introduced species, and harvesting for the horticultural trade. A large specimen of a rare cycad highly prized by collectors can sell for thousands of dollars. Because cycads grow slowly, recovery of endangered populations is uncertain.

Gnetophytes Include the Odd *Welwitschia*

The gnetophytes include about 70 species of shrubs, vines, and small trees. Leaves of gnetophyte species in the genus *Ephedra* contain alkaloid compounds that act in humans as stimulants and appetite suppressants. For this reason, *Ephedra* was once widely used as an energy booster and weight-loss aid. However, following reports of sudden deaths of *Ephedra* users and publication of several studies linking *Ephedra* consumption to increased risk of heart problems, the U.S. Food and Drug Administration banned the sale of products containing this gnetophyte.

(a) Gingko

(b) Cycad

(c) Gnetophyte

(d) Conifer

▲ **FIGURE 22-10 Gymnosperms (a)** The ginkgo, or maidenhair tree, is widely cultivated as a shade or ornamental tree. **(b)** Common in the age of dinosaurs, cycads are now limited to about 160 species. **(c)** The leaves of the gnetophyte *Welwitschia* can live 1,000 years. **(d)** The needle-shaped leaves of conifers are protected by a waxy surface layer.

The gnetophyte *Welwitschia mirabilis* is among Earth's most distinctive plants (FIG. 22-10c). Found only in the extremely dry deserts of southwest Africa, *Welwitschia* has a deep taproot that can extend as far as 100 feet (30 meters) down into the soil. Above the surface, the plant has a fibrous stem. Only two leaves ever grow from the stem. The leaves are never shed and remain on the plant for its entire life, which can be very long. A typical life span is about 1,000 years, and the oldest *Welwitschia* are more than 2,000 years old. The strap-like leaves continue to grow for that entire period, spreading over the ground. The older portions of the leaves, whipped by the wind for centuries, may shred or split, giving the plant its characteristic gnarled and tattered appearance.

Conifers Are Adapted to Cool Climates

Though other gymnosperm groups like gingkoes and cycads are drastically reduced from their former prominence, the **conifers** still dominate large areas of our planet. Conifers, whose 500 species include pines, firs, spruce, hemlocks, and cypresses, are most abundant in the far north and at high elevations, places where winters are long and conditions are dry (because water in the soil remains frozen and unavailable during winter).

Conifers are adapted to these dry, cold conditions in three ways. First, most conifers retain green leaves throughout the year, enabling these plants to continue photosynthesizing and growing slowly during times when most other plants become dormant. For this reason, conifers are often called evergreens. Second, conifer leaves are thin needles covered with a thick, waterproof surface that minimizes evaporation (FIG. 22-10d). Finally, conifers produce an "antifreeze" in their sap that enables them to continue transporting nutrients in below-freezing temperatures and also gives them their fragrant piney scent.

Reproduction is similar in all conifers, so let's examine the reproductive cycle of a pine tree (FIG. 22-11). The tree itself is the diploid sporophyte, and it produces both male and female cones ❶. Male cones are relatively small (typically about ¾ inch long), delicate structures consisting of scales in which pollen (the male gametophyte) develops. Each female cone consists of a series of woody scales arranged in a spiral around a central axis. At the base of each scale are two **ovules** (unfertilized seeds), within which diploid spore-forming cells arise.

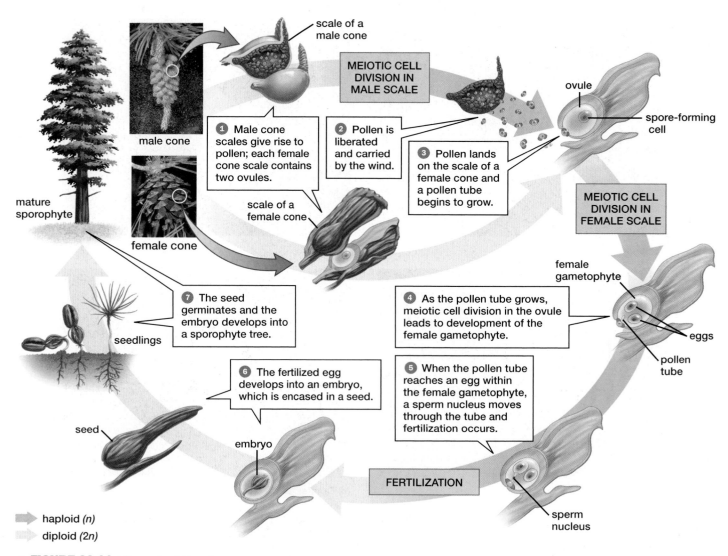

scale of a male cone

MEIOTIC CELL DIVISION IN MALE SCALE

ovule

spore-forming cell

male cone

❶ Male cone scales give rise to pollen; each female cone scale contains two ovules.

❷ Pollen is liberated and carried by the wind.

❸ Pollen lands on the scale of a female cone and a pollen tube begins to grow.

scale of a female cone

MEIOTIC CELL DIVISION IN FEMALE SCALE

female cone

mature sporophyte

female gametophyte

❼ The seed germinates and the embryo develops into a sporophyte tree.

❹ As the pollen tube grows, meiotic cell division in the ovule leads to development of the female gametophyte.

eggs

pollen tube

seedlings

❻ The fertilized egg develops into an embryo, which is encased in a seed.

❺ When the pollen tube reaches an egg within the female gametophyte, a sperm nucleus moves through the tube and fertilization occurs.

seed

embryo

FERTILIZATION

➡ haploid *(n)*

➡ diploid *(2n)*

sperm nucleus

▲ **FIGURE 22-11 Life cycle of the pine**

Male cones release pollen during the reproductive season and then disintegrate ❷. The amount of pollen released is immense; inevitably, some pollen grains land by chance on female cone scales ❸. The pollen grain then sends out a pollen tube that slowly burrows into an ovule. As the pollen tube grows, the diploid spore-forming cell in the ovule undergoes meiosis to produce haploid spores, one of which gives rise to a haploid female gametophyte, within which egg cells develop ❹. After nearly 14 months, the pollen tube finally reaches the egg cell and releases the sperm that fertilize it ❺. The resulting zygote becomes enclosed in a seed as it develops into an embryo—a tiny embryonic sporophyte plant ❻. The seed is liberated when the cone matures and its scales separate. If it lands in a suitable patch of soil, it may germinate and grow into a sporophyte tree ❼.

Angiosperms Are Flowering Seed Plants

Flowering plants, or **angiosperms,** have been Earth's predominant plants for more than 100 million years. The group is incredibly diverse, with more than 230,000 species. Angiosperms range in size from the diminutive duckweed (**FIG. 22-12a**) to the towering eucalyptus tree (**FIG. 22-12b**). From desert cactus to tropical orchids to grasses to parasitic stinking corpse lilies, angiosperms dominate the plant kingdom. Their enormous success is due in part to three major adaptations: flowers, fruits, and broad leaves.

(a) Duckweed

(c) Grass

(b) Eucalyptus

(d) Butterfly weed

◀ **FIGURE 22-12 Angiosperms (a)** The smallest angiosperm is the duckweed, found floating on ponds. These specimens are about 1/8 inch (3 millimeters) in diameter. **(b)** The largest angiosperms are eucalyptus trees, which can reach 325 feet (100 meters) in height. **(c)** Grasses (and many trees) have inconspicuous flowers and rely on wind for pollination. More conspicuous flowers, such as those on **(d)** this butterfly weed and on a eucalyptus tree (**b,** inset), entice insects and other animals that carry pollen between individual plants.

THINK CRITICALLY What are the advantages and disadvantages of wind pollination? What are the advantages and disadvantages of pollination by animals? Why do both types of pollination persist among the angiosperms?

Flowers Attract Pollinators

Flowers, the structures in which both male and female gametes are formed, probably evolved when gymnosperm ancestors formed an association with animals (most likely insects) that carried their pollen from plant to plant. According to this scenario, the relationship between these ancient gymnosperms and their animal pollinators was so advantageous that natural selection favored the evolution of showy flowers that advertised the presence of pollen to insects and other animals (**FIGS. 22-12b, d**). The animals benefited by eating some of the protein-rich pollen, whereas the plant benefited from the animals' unwitting transportation of pollen from plant to plant. With this animal assistance, many flowering plants no longer needed to produce prodigious quantities of pollen and send it flying on the fickle winds to ensure fertilization. But there are nonetheless many wind-pollinated angiosperms (**FIG. 22-12c**). These probably evolved from animal-pollinated ancestors when environmental changes resulted in the decline or extinction of the affected species' pollinators.

In the angiosperm life cycle, flowers develop on the dominant sporophyte plant (**FIG. 22-13**). In the flower, female gametophytes develop from ovules within a structure called the *ovary*; male gametophytes (pollen) are formed inside a structure called the *anther* ❶. During the reproductive season, pollen is released from the anthers and carried away on the wind or by animal pollinators ❷. If a pollen grain lands on a *stigma,* a sticky pollen-catching structure of the flower, a pollen tube begins to grow from the pollen grain ❸. The tube bores through the stigma and extends toward the female gametophyte, within which an egg cell has developed. Fertilization occurs when the pollen tube reaches the egg cell and releases sperm cells ❹. The resulting zygote develops into an embryo enclosed in a seed formed from the ovule ❺. After it is dispersed, the seed may germinate and give rise to a sporophyte plant ❻.

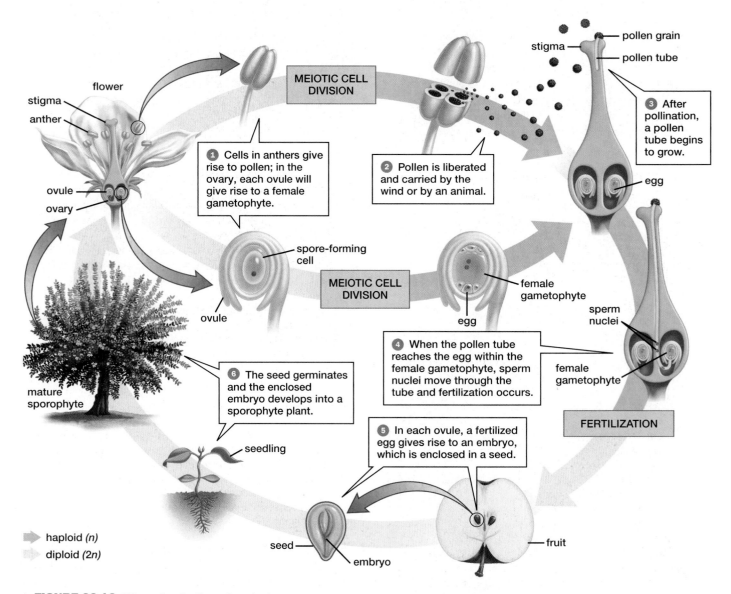

▲ **FIGURE 22-13 Life cycle of a flowering plant**

Fruits Encourage Seed Dispersal

The ovary surrounding the seeds of an angiosperm matures into a **fruit,** the second adaptation that has contributed to the success of angiosperms. Just as flowers encourage animals to transport pollen, so do many fruits entice animals to disperse seeds. If an animal eats a fruit, many of the enclosed seeds may pass through the animal's digestive tract unharmed, perhaps to fall at a suitable location for germination. Not all fruits, however, depend on edibility for dispersal. Dog owners are well aware, for example, that some fruits (called burrs) disperse by clinging to animal fur. Other fruits, such as those of maples, form wings that carry the seed through the air. The variety of dispersal mechanisms made possible by fruits has helped the angiosperms invade nearly all terrestrial habitats.

Broad Leaves Capture More Sunlight

The third feature that gives angiosperms an advantage in warmer, wetter climates is broad leaves. Broad leaves provide an advantage by collecting more sunlight for photosynthesis than the slender needles of conifers can. However, in regions with seasonal variations in growing conditions, many trees and shrubs drop their leaves during periods when water is in short supply, because being leafless reduces evaporative water loss. In temperate climates, such periods occur during the fall and winter. In the tropics and subtropics, species that inhabit areas where periods of drought are common may drop their leaves to conserve water during the dry season.

Broad leaves have costs as well as benefits. In particular, broad, tender leaves are much more appealing to herbivores than are the tough, waxy needles of conifers. As a result, angiosperms have evolved a range of defenses against mammalian and insect herbivores. These adaptations include physical defenses such as thorns, spines, and tough leaves. The evolutionary struggle for survival has also led to a host of chemical defenses—compounds that make plant tissue poisonous or distasteful to potential predators. Many of the compounds responsible for chemical defense have properties that humans have exploited for medicinal and culinary uses. Medicines such as aspirin and codeine, stimulants such as nicotine and caffeine, and spices such as mustard and pepper are all derived from angiosperm plants.

CHECK YOUR LEARNING

Can you ...

- explain how vascular plants and nonvascular plants differ?
- describe the major plant taxonomic groups and representative members of each group?
- describe the key steps in the life cycles of mosses, ferns, gymnosperms, and flowering plants?

CASE STUDY   CONTINUED
Queen of the Parasites

Why does the flower of a stinking corpse lily smell like rotting meat? Though the smell is utterly revolting to humans, it is attractive to blowflies and other insects that normally feed on and lay their eggs in decaying flesh. When such insects visit a male stinking corpse lily, they may carry away pollen that can fertilize a nearby female flower.

In many angiosperm species, flowers contain nectar that provides food for animal pollinators. But no such nectar reward awaits a fly that enters the flower of a stinking corpse lily. Instead, a fly attracted by the flower's stench searches in vain for putrefying meat, its movement guided toward the flower's cache of sticky pollen by grooves and hairs inside the flower. Eventually, the fly departs, coated in pollen. In essence, the plant tricks the fly into providing a service for no reward. Thus, the stinking corpse lily is a master exploiter: It takes advantage of both the host vines that provide its food and the flies that facilitate its reproduction.

The stinking corpse lily harms species it interacts with, but many plants benefit other species. How do other species benefit?

HAVE YOU EVER WONDERED ...

Which Plants Provide Us with the Most Food?

Although thousands of plant species have edible parts, people exploit only a small proportion of them for food. In fact, the vast majority of the plant-derived food consumed by humans comes from only 20 species. The fruits (grains) of just three grass species—corn (maize), wheat, and rice—provide about half the calories consumed by people worldwide. The average person in the United States consumes about 200 pounds of these grains each year. In terms of annual production, the big three are followed, in order, by soybeans, barley, sorghum, millet, and peanuts. Looking further down the list, the world's most abundantly produced foods that are *not* grains or legumes are potatoes and cassava (a root that is a staple in parts of Africa and South America).

22.4 HOW DO PLANTS AFFECT OTHER ORGANISMS?

As plants survive, grow, and reproduce, they alter and influence Earth's landscape and atmosphere in ways that are tremendously beneficial to the rest of the planet's inhabitants, including humans. Humans also reap additional benefits by actively exploiting plants.

Plants Play a Crucial Ecological Role

The complex ecosystems that host terrestrial life could not be maintained without the help of plants. Plants make vital contributions to the food, air, soil, and water that sustain life on land.

Green Lifesaver

Many of the drugs that physicians use to treat diseases contain substances that were originally discovered in plants. In most cases, once a pharmaceutically useful plant component has been identified and isolated, researchers devise a method to synthesize the drug without using any actual plants. An important exception, at least until recently, is the antimalarial drug artemisinin. Artemisinin is an extremely important medicine, a key component of the best available treatment for the millions of people infected with *Plasmodium*, the protist parasite that causes malaria. The parasite has evolved resistance to most of the other drugs used to treat the disease.

Artemisinin is found in the sweet wormwood plant, *Artemisia annua* (FIG. E22-1). Its effectiveness as an antimalarial was first discovered by Youyou Tu and her colleagues, who tested hundreds of different herbs used in traditional Chinese medicine before discovering that an extract of wormwood is an effective treatment for malaria. The researchers eventually identified artemisinin as the molecule responsible for wormwood's antimalarial capability.

Until recently, the only way to produce artemisinin was to extract it from wormwood plants. Now, however, researchers working on the cutting edge of synthetic biology have discovered how to insert *Artemisia* genes into yeast cells, and how to prompt the yeasts to excrete artemisinin while housed in industrial scale fermentation vats. The synthetic drug can be produced in weeks, rather than the 14–18 months it takes

▲ **FIGURE E22-1 Sweet wormwood, ready for harvest**

to grow and process wormwood plants. This change is expected to greatly increase the availability of artemisnin, though at the expense of the mostly poor farmers who have been growing the wormwood formerly needed to make the drug.

Although the discovery of a high-tech way to ramp up artemisinin production is good news for malaria victims, it comes at a time when some *Plasmodium* parasites are developing resistance to the drug. Will the next great antimalarial come from a plant? The big pharmaceutical companies have largely abandoned their efforts to systematically screen Earth's plant diversity for new drugs, but researchers working in the African country of Mali might be on to something. In early trials, they have found that an extract of the poppy species *Argemone mexicana* is quite effective against malaria.

THINK CRITICALLY In the initial trials with *Argemone mexicana* in Mali, researchers observed patients treated with different herbal medicines by a traditional healer. They performed a blood test on each patient to determine which ones had malaria and tracked each patient to determine if and how quickly he or she recovered. If you were in charge of the next follow-up study, how would you design it? Would you use the traditional healer's preparations of the plant? Would you have the healer determine dosage and frequency of treatment?

Plants Capture Energy That Other Organisms Use

Plants provide food, directly or indirectly, for all of the animals, fungi, and nonphotosynthetic microbes on land. Plants use photosynthesis to capture solar energy, and they convert part of the captured energy into leaves, shoots, seeds, and fruits that are eaten by other organisms. Many of these consumers of plant tissue are themselves eaten by still other organisms. Plants are the main providers of energy and nutrients to terrestrial ecosystems, and life on land depends on plants' ability to manufacture food from sunlight.

Plants Help Maintain the Atmosphere

In addition to providing food, plants produce oxygen gas as a by-product of photosynthesis. By doing so, they continually replenish oxygen in the atmosphere. Without plants' contribution, atmospheric oxygen would be rapidly depleted by the oxygen-consuming respiration of Earth's multitude of organisms. Plant photosynthesis also removes carbon dioxide from the atmosphere, converting it to compounds such as starch and cellulose, which are stored in plant bodies. Without plants, atmospheric carbon dioxide would soar to levels that would be fatal for almost all organisms.

Plants Build and Protect Soil

Plants also help create and maintain soil. When a plant dies, its stems, leaves, and roots become food for fungi, prokaryotes, and other decomposers. Decomposition breaks the plant tissue into tiny particles of organic matter that become part of the soil. Organic matter improves the ability of soil to hold water and nutrients, thereby making the soil more fertile and better able to support the growth of living plants. The roots of plants help stabilize soil and reduce erosion by wind and water.

Plants Help Keep Ecosystems Moist

Plants take up water from the soil and retain some of it in their tissues. By doing so, plants slow the rate at which water escapes from terrestrial ecosystems and increase the amount of water available to meet the needs of the ecosystems' inhabitants. By reducing the amount of water runoff, plants also reduce the chances of destructive flooding. Thus, floods can be more frequent in areas in which forests, grasslands, or marshes have been destroyed by human activities.

Plants Provide Humans with Necessities and Luxuries

It would be difficult to exaggerate the degree to which people depend on plants. Neither our explosive population growth nor our rapid technological advance would have been possible without plants.

Plants Provide Shelter, Fuel, and Medicine

Plants provide wood that is used to construct housing for much of Earth's human population. In addition, wood has historically been the main fuel for warming dwellings and cooking, and it remains so in many parts of the world. Coal, another important fuel, is composed of the remains of ancient plants that have been transformed by geological processes.

Plants have also supplied many of the medicines on which modern health care depends. Important drugs that were originally found in and extracted from plants include the painkillers aspirin, codeine, and morphine, the heart medication digoxin, the cancer treatments Taxol® and vinblastine, and many more (see "Health Watch: Green Lifesaver").

In addition to harvesting useful material from wild plants, humans have domesticated a host of useful plant species. Through generations of selective breeding, people have modified the seeds, stems, roots, flowers, and fruits of favored plant species to provide themselves with food and fiber. It is difficult to imagine life without corn, rice, potatoes, apples, tomatoes, cooking oil, cotton, and the myriad other staples that domestic plants provide.

Plants Provide Pleasure

Though we appreciate the practical value of wheat and wood, our most emotionally powerful connections with plants are purely sensual. We delight in the beauty and fragrance of flowers and present them to others as symbols of our most sublime and inexpressible emotions. Many of us spend hours of leisure time tending gardens and lawns, for no reward other than the pleasure and satisfaction we derive from observing the fruits of our labor. In our homes, we reserve space for our houseplant companions. We line our streets with trees and seek refuge from the stress of daily life in parks with abundant plant life. Clearly, plants help fill our emotional needs.

CHECK YOUR LEARNING
Can you …
- describe some of the effects that plants have on other organisms, including humans?

CASE STUDY   REVISITED

Queen of the Parasites

The 17 or so parasitic plant species of the genus *Rafflesia*, which includes the stinking corpse lily, are found in the moist forests of Southeast Asia, a habitat that is disappearing rapidly as forests are cleared for agriculture and development. The geographic range of the stinking corpse lily is limited to the dwindling forests of the Malaysian peninsula and the Indonesian islands of Borneo and Sumatra; the species is rare and endangered. The government of Indonesia has established parks and reserves in an effort to help protect the stinking corpse lily, but—as is often the case in developing countries—a forest that is protected on paper may still be vulnerable in reality.

Perhaps the best hope for the continued survival of the largest *Rafflesia* is the growing realization among the rural residents of Sumatra and Borneo that the spectacular, putrid-smelling flowers of the stinking corpse lily might lure interested tourists to their countries. Under an innovative conservation program that seeks to take advantage of this potential for ecotourism, people who live in the vicinity of the stinking corpse lily can become caretakers of the plants. These assigned caretakers watch over the plants and, in return, may charge a small fee to curious visitors. Local inhabitants have been given an economic incentive to protect this rare parasitic plant.

THINK CRITICALLY Perhaps surprisingly, a parasitic lifestyle is not terribly rare among plants. More than 4,400 plant species are parasites, and systematists estimate that parasitism has evolved at least 12 different times over the evolutionary history of plants. Given the obvious benefits of photosynthesis, why has parasitism (which is often accompanied by loss of photosynthetic capability) evolved repeatedly in photosynthetic plants?

CHAPTER REVIEW

(MB)   Go to **MasteringBiology** for practice quizzes, activities, eText, videos, current events, and more.

Answers to Think Critically, Evaluate This, Multiple Choice, and Fill-in-the-Blank questions can be found in the Answers section at the back of the book.

Summary of Key Concepts

22.1 What Are the Key Features of Plants?

Plants are multicellular organisms that exhibit alternation of generations, in which a haploid gametophyte generation alternates with a diploid sporophyte generation. Most plants are photosynthetic. Unlike their green algae relatives, plants have multicellular, dependent embryos.

22.2 How Have Plants Evolved?

Photosynthetic protists, probably aquatic green algae, gave rise to the first plants. Ancestral plants were probably similar to modern multicellular algae such as stoneworts, which are plants' closest living relatives.

Early plants invaded terrestrial habitats, and modern plants exhibit a number of key adaptations for terrestrial existence: rootlike structures for anchorage and for absorption of water and nutrients; a waxy cuticle that slows the loss of water through evaporation; stomata that can open, allowing gas exchange, or close, preventing water loss; the conducting tissues xylem and phloem that transport water and nutrients throughout the plant; and a stiffening substance, called lignin, that impregnates the conducting cells and helps support the plant body.

Plant reproductive structures suitable for life on land include a smaller male gametophyte (pollen) that allows wind to replace water in carrying sperm to eggs; seeds that nourish, protect, and help disperse developing embryos; flowers that attract animals, which carry pollen more precisely and efficiently than wind; and fruits that entice animals to disperse seeds.

There has been a general evolutionary trend toward a reduction in size of the haploid gametophyte, which is dominant in nonvascular plants but microscopic in seed plants.

22.3 What Are the Major Groups of Plants?

Two major groups of plants, nonvascular plants and vascular plants, arose from their ancient algal ancestors. Nonvascular plants, including the hornworts, liverworts, and mosses, are small, simple land plants that lack conducting cells and mostly live in moist habitats. Nonvascular plant reproduction requires water through which the sperm swim to the egg.

In vascular plants, a system of conducting cells—stiffened by lignin—conducts water and nutrients absorbed by the roots into the upper portions of the plant and supports the plant body. Thanks to this support system, seedless vascular plants, including the club mosses, horsetails, and ferns, can grow larger than nonvascular plants. The sperm of seedless vascular plants must swim to the egg for sexual reproduction to occur.

Gymnosperms and angiosperms are vascular plants with two major additional adaptive features for life on dry land: pollen and seeds. Gymnosperms, which include ginkgoes, cycads, gnetophytes, and conifers, were the first fully terrestrial plants to evolve.

Angiosperms, the flowering plants, dominate much of the land today. In addition to pollen and seeds, angiosperms also produce flowers and fruits.

22.4 How Do Plants Affect Other Organisms?

Plants play a key ecological role, capturing energy through photosynthesis for use by inhabitants of terrestrial ecosystems, replenishing atmospheric oxygen, sequestering carbon dioxide, creating and stabilizing soils, and slowing the loss of water from ecosystems. Plants are also used by humans to provide food, fuel, building materials, medicines, and aesthetic pleasure.

Key Terms

alternation of
   generations   *384*
angiosperm   *395*
antheridium (plural,
   antheridia)   *389*
archegonium (plural,
   archegonia)   *389*
conifer   *394*
cuticle   *385*
flower   *396*
fruit   *397*
gametophyte   *384*

gymnosperm   *392*
lignin   *386*
nonvascular plant   *387*
ovule   *394*
phloem   *385*
pollen   *392*
seed   *392*
sporophyte   *384*
stoma (plural,
   stomata)   *385*
vascular plant   *387*
xylem   *385*

Thinking Through the Concepts

Multiple Choice

1. In an alternation of generations life cycle, spores develop into _____ that produce gametes that fuse to give rise to _____.
   a. haploid gametophytes; diploid sporophytes
   b. diploid gametophytes; haploid sporophytes
   c. haploid sporophytes; diploid gametophytes
   d. diploid sporophytes; haploid gametophytes

2. Which of the following are *not* nonvascular plants?
   a. mosses
   b. liverworts
   c. ferns
   d. hornworts

3. Which of the following structures is present in angiosperms but not in gymnosperms?
   a. cones
   b. fruits
   c. seeds
   d. xylem

4. In which of the following is the gametophyte stage larger and more prominent than the sporophyte stage?
   a. nonvascular plants
   b. seed plants
   c. angiosperms
   d. gymnosperms

5. Which of the following is *not* an adaption that helps plants resist gravity and/or dry conditions on land?
   a. xylem
   b. lignin
   c. cuticle
   d. photosynthesis

Fill-in-the-Blank

1. Scientists hypothesize that the ancestors of plants were _____. There are two major types of plants; those that lack conducting cells are called _____ and those with conducting cells are called _____. All plants produce multicellular _____ and exhibit a complex life cycle called _____.

2. Plant adaptations to life on land include a(n) _____, which reduces evaporation of water, and _____, which open to allow gas exchange but close when _____ is scarce. In addition, the bodies of vascular plants gain increased support from _____ and _____ impregnated with the polymer _____; these structures also help _____ and _____ move within the plant body.

3. Seedless vascular plants must reproduce when conditions are wet because their sperm must _____. Two adaptations that allow seed plants to reproduce more efficiently on dry land are _____ and _____. The seed plants fall into two major categories: the nonflowering _____ and the flowering _____. Flowers were favored by natural selection because they _____. Fruits were favored by natural selection because they _____.

4. Three groups of nonvascular plants are _____, _____, and _____. Three groups of seedless vascular plants are _____, _____, and _____. Today, the most diverse group of plants is the _____.

Review Questions

1. What is meant by "alternation of generations"? What two generations are involved? How does each reproduce?

2. Explain the evolutionary changes in plant reproduction that adapted plants to increasingly dry environments.

3. Describe evolutionary trends in the life cycles of plants. Emphasize the relative sizes of the gametophyte and sporophyte.

4. From which algal group did green plants probably arise? Explain the evidence that supports this hypothesis.

5. List the structural adaptations necessary for the invasion of dry land by plants. Which of these adaptations are possessed by nonvascular plants? By ferns? By gymnosperms and angiosperms?

6. The number of species of flowering plants is greater than the number of species in the rest of the plant kingdom combined. What feature(s) are responsible for the enormous success of angiosperms? Explain why.

7. List the adaptations of gymnosperms that have helped them become the dominant trees in dry, cold climates.

8. What is a pollen grain? What role has it played in helping plants colonize dry land?

9. The majority of all plants are seed plants. What is the advantage of a seed? How do plants that lack seeds meet the needs served by seeds?

Applying the Concepts

1. Prior to the development of synthetic drugs, more than 80% of all medicines were of plant origin. Even today, indigenous tribes in remote Amazonian rain forests can provide a plant product to treat virtually any ailment. Herbal medicine is also widely and successfully practiced in China. Most of these drugs are unknown to the Western world. But the forests from which much of this plant material is obtained are being converted to agriculture. We are in danger of losing many of these potential drugs before they can be evaluated by Western medicine. What steps can you suggest to preserve these natural resources while also allowing nations to direct their own economic development?

2. Only a few hundred of the more than 200,000 species of plants have been domesticated for human use. One example is the almond. The domestic almond is nutritious and harmless, but its wild precursor can cause cyanide poisoning. The oak makes potentially nutritious seeds (acorns) that contain very bitter-tasting tannins. If we could breed the tannin out of acorns, they might become a delicacy. Why do you suppose we have failed to domesticate oaks?

23

THE DIVERSITY OF FUNGI

These honey mushrooms are part of the visible portion of the largest organism on Earth.

CASE STUDY

Humongous Fungus

WHAT IS THE LARGEST organism on Earth? A reasonable guess might be the world's largest animal, the blue whale, which can be 100 feet long and weigh 400,000 pounds. But the blue whale is dwarfed by the General Sherman tree, a giant sequoia that is 275 feet high and whose weight is estimated at 6,200 *tons*. Even these two behemoths, however, can't match the real record-holder, the fungus *Armillaria solidipes,* also known as the honey mushroom.

The largest known *Armillaria* is a specimen in Oregon that spreads over almost 2,400 acres and weighs more than 7,500 tons. Despite its huge size, no one has actually seen the monster fungus, because it is largely underground. Its only above-ground parts are brown mushrooms that sprout occasionally from its gigantic body. Just beneath the surface, however, the fungus spreads through the soil by means of long, string-like structures that extend until they encounter the tree roots on which *Armillaria* subsists.

How can researchers be sure that the Oregon fungus is truly one single individual and not many intertwined individuals? The strongest evidence is genetic. Researchers gathered *Armillaria* tissue samples from throughout the area thought to be inhabited by a single individual and compared DNA extracted from the samples. All were genetically identical, demonstrating that they came from the same individual.

The lives of fungi typically take place outside of our view, but they play a fascinating role in human affairs. How do fungi affect us and other organisms? How and where do they live, grow, and reproduce?

AT A GLANCE

23.1 WHAT ARE THE KEY FEATURES OF FUNGI?

When you think of a fungus, you probably picture a mushroom. Most fungi, however, do not produce mushrooms. And even in those that do, the mushrooms are just temporary reproductive structures. The main body is typically concealed beneath the soil or inside a piece of decaying wood. So, to fully appreciate fungi, we must look beyond the conspicuous structures we encounter on the forest floor, at the edges of our lawns, or on top of a pizza. A closer look at fungi reveals a group of eukaryotic, mostly multicellular organisms that play a key role in the web of life and whose lifestyle differs in fascinating ways from that of plants or animals.

Fungal Bodies Consist of Slender Threads

The body of almost every fungus is a **mycelium** (plural, mycelia; **FIG. 23-1a**), which is an interwoven mass of one-cell-thick, threadlike filaments called **hyphae** (singular, hypha; **FIGS. 23-1b, c**). In some species, hyphae consist of single elongated cells with numerous nuclei; in other species, hyphae are subdivided—by partitions called **septa** (singular,

septum)—into many cells, each containing one or more nuclei. Pores in the septa allow cytoplasm to stream between cells, distributing nutrients. Like plant cells, fungal cells are surrounded by cell walls. Unlike plant cells, however, fungal cell walls are strengthened by *chitin,* the same substance found in the hard outer surface (exoskeleton) of insects, crabs, and their relatives.

Most fungi cannot move. They compensate for this lack of mobility with hyphae that can grow rapidly in any direction within a suitable environment. In this way, the fungal mycelium can quickly spread into aging bread or cheese, beneath the bark of decaying logs, or into the soil. Periodically, the hyphae differentiate into reproductive structures that project above the surface. These structures, including mushrooms, puffballs, and the powdery molds on spoiled food, represent only a fraction of the complete fungal body, but are typically the only part of the fungus that we can easily see.

Fungi Obtain Their Nutrients from Other Organisms

Like animals, fungi survive by breaking down nutrients stored in the bodies or wastes of other organisms. Some fungi digest

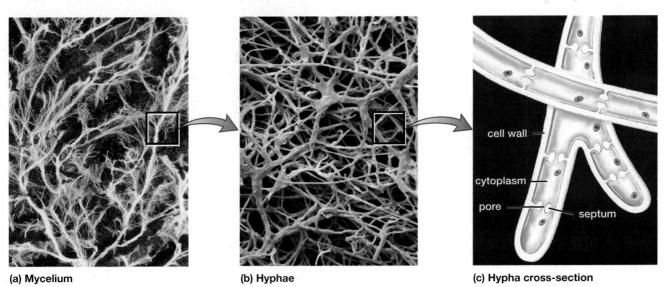

(a) Mycelium   **(b) Hyphae**   **(c) Hypha cross-section**

▲ **FIGURE 23-1 The filamentous body of a fungus (a)** A fungal mycelium spreads over decaying vegetation. The mycelium is composed of **(b)** a tangle of microscopic hyphae, only one cell thick, portrayed in cross-section **(c)** to show their internal organization.

THINK CRITICALLY Which features of a fungus's body structure are adaptations related to its method of acquiring nutrients?

▲ **FIGURE 23-2** **Nemesis of nematodes** The fungus *Arthrobotrys,* also known as the nematode (roundworm) strangler, traps its prey in a noose-like modified hypha. When a nematode wanders into the noose, its presence stimulates the noose cells to swell with water. In a fraction of a second, the noose constricts, trapping the worm. Fungal hyphae then penetrate and feast on their prey.

the bodies of dead organisms. Others are parasitic, feeding on living organisms and causing disease. Some live in close, mutually beneficial relationships with other organisms that provide food. There are even a few predatory fungi, which attack tiny worms in soil (**FIG. 23-2**).

Unlike most animals, fungi do not ingest food. Instead, they secrete enzymes that digest complex molecules outside their bodies, breaking down the molecules into smaller subunits that can be absorbed. Fungal hyphae can penetrate deeply into a source of nutrients, and because the hyphae are only one cell thick, each cell in a fungal body is in position to absorb nutrients directly from the surrounding environment.

Fungi Can Reproduce Both Asexually and Sexually

Fungi develop from **spores**—haploid cells that can give rise to a new individual. Fungal spores are tiny and extraordinarily mobile, even though most—but not all—lack a means of self-propulsion. They are distributed far and wide as hitchhikers on the outside of animal bodies, as passengers inside the digestive systems of animals that have eaten them, or as airborne drifters, cast aloft by chance or shot into the atmosphere by elaborate reproductive structures (**FIG. 23-3**). Spores are often produced in great numbers; a single giant puffball may contain 5 trillion spores.

In general, fungi are capable of both asexual and sexual reproduction (**FIG. 23-4**). For the most part, fungi reproduce asexually under stable conditions, but reproduce sexually under conditions of environmental change or stress.

Asexual Reproduction Produces Haploid Spores by Mitosis

The mycelia and spores of fungi are haploid. A haploid mycelium produces haploid asexual spores by mitosis. If

(a) Earthstar

(b) *Pilobolus*

▲ **FIGURE 23-3** **Some fungi can eject spores (a)** A ripe earthstar mushroom, struck by a drop of water, releases a cloud of spores that will be dispersed by air currents. **(b)** The delicate, translucent reproductive structures of *Pilobolus,* which inhabits horse manure, literally blow their tops when ripe, dispersing the black, spore-containing caps up to 3 feet away. Spores that adhere to grass remain there until consumed by a grazing herbivore, perhaps a horse. Later (likely some distance away), the horse will deposit a fresh pile of manure containing *Pilobolus* spores that have passed unharmed through its digestive tract.

an asexual spore is deposited in a favorable location, it will begin mitotic divisions and develop into a new mycelium. This simple reproductive cycle results in the rapid production of a genetically identical clone of the original mycelium.

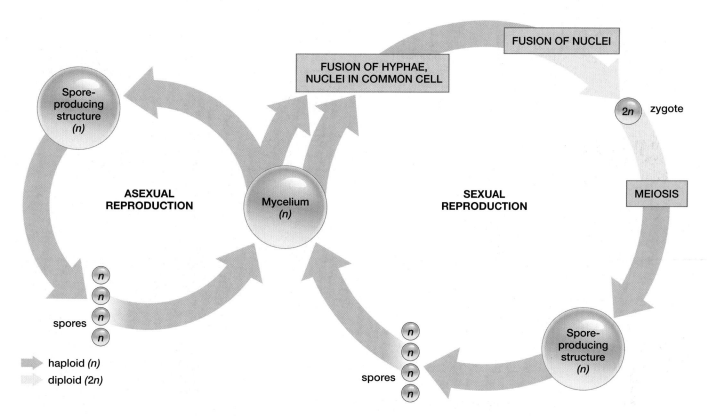

▲ **FIGURE 23-4 Generalized life cycle of fungi** In asexual reproduction, haploid hyphae in a mycelium give rise to structures that produce haploid spores by mitotic cell division. In sexual reproduction, haploid hyphae of different, compatible mating types fuse, resulting in cells that contain nuclei from both parents. These nuclei subsequently fuse, generating a diploid zygote that undergoes meiosis to yield haploid spores.

Sexual Reproduction Produces Haploid Spores by Meiosis

Diploid structures form only during a brief period of the sexual portion of the fungal life cycle. Sexual reproduction begins when a hypha of one mycelium comes into contact with a hypha from a second mycelium that is of a different, but compatible, mating type. (The different mating types of fungi are comparable to the different sexes of animals, except that in fungi there are often more than two mating types.) If conditions are suitable, the two hyphae may fuse, so that nuclei from the two different hyphae share a common cell. The merger of hyphae is followed by fusion of haploid nuclei, one from each of the two mating types, to form a diploid zygote. The zygote then undergoes meiosis to form haploid sexual spores. These spores are dispersed, germinate, and divide by mitosis to form new haploid mycelia. Unlike the cloned offspring produced by asexual spores, these sexually produced fungal bodies are genetically distinct from either parent.

CHECK YOUR LEARNING
Can you ...
- describe the structure of a typical fungus?
- explain how fungi obtain energy and nutrients and how they reproduce?

23.2 WHAT ARE THE MAJOR GROUPS OF FUNGI?

Nearly 125,000 species of fungi have been described, but this number represents only a fraction of the true diversity of these organisms. Many new species are discovered and described each year, and mycologists (scientists who study fungi) estimate that the number of undiscovered species of fungi is at least 1.5 million. Fungi are classified into six main taxonomic groups: Chytridiomycota (chytrids), Neocallimastigomycota (rumen fungi), Blastocladiomycota (blastoclades), Glomeromycota (glomeromycetes), Basidiomycota (basidiomycetes), and Ascomycota (ascomycetes). Some fungi, however, are not members of any of these six groups. Most of these unclassified fungal species were historically placed in the taxonomic group Zygomycota (zygomycetes), but recent analysis of DNA sequences has revealed that zygomycetes do not constitute a clade and so cannot form a named taxonomic group. (A clade is a group consisting of all the descendants of a particular common ancestor.) Systematists are gathering additional data on the evolutionary history of the species previously classified as zygomycetes, with the goal of classifying them into new named groups.

Mass sequencing of DNA extracted from soil and water samples has also revealed a host of new fungal species, most of them so far known only from their DNA sequences. Many of

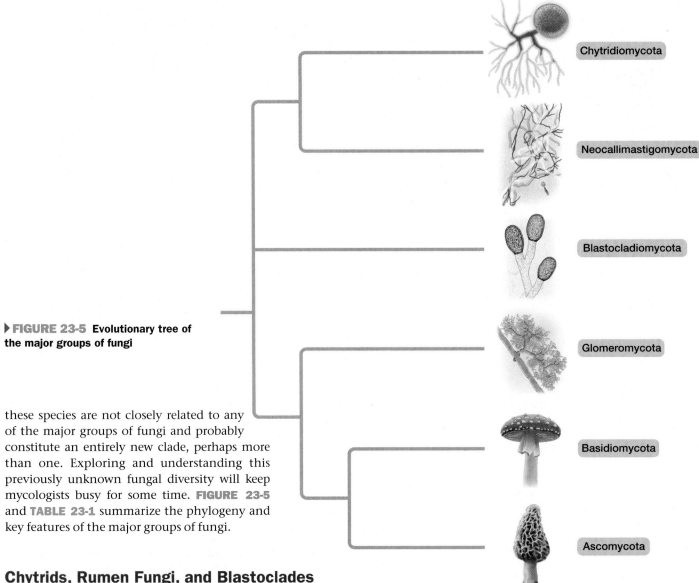

FIGURE 23-5 Evolutionary tree of the major groups of fungi

these species are not closely related to any of the major groups of fungi and probably constitute an entirely new clade, perhaps more than one. Exploring and understanding this previously unknown fungal diversity will keep mycologists busy for some time. **FIGURE 23-5** and **TABLE 23-1** summarize the phylogeny and key features of the major groups of fungi.

Chytrids, Rumen Fungi, and Blastoclades Produce Swimming Spores

The members of three taxonomic groups of fungi—the chytrids, rumen fungi, and blastoclades—are distinguished by their swimming spores, which require water for dispersal. Many members of these groups live in water, and even those that live on land require a film of water for reproduction. The spores propel themselves through the water by means of one or more flagella.

Chytrids Are Mostly Aquatic

Most **chytrids** live in fresh water, but a few species are marine. Chytrid spores have a single flagellum on one end. The oldest known fossil fungi are chytrids that are found in rocks more than 600 million years old. Ancestral fungi may well have been similar to today's aquatic and marine chytrids, so fungi probably originated in a watery environment before colonizing land.

Most chytrid species feed on dead aquatic plants or other debris in watery environments, but some species are parasites of plants or animals. One such parasitic chytrid is a major cause of the current worldwide die-off of frogs, which threatens many species and has already caused the extinction of several. (For more on the decline of frogs, see "Earth Watch: Frogs in Peril" in Chapter 25.)

Rumen Fungi Live in Animal Digestive Tracts

The **rumen fungi** are anaerobic (they do not require oxygen) and reside mainly in the digestive tracts of planteating animals such as cows, sheep, kangaroos, elephants, and iguanas. These animals are not able to digest cellulose (a major component of plant tissue) themselves but instead rely on symbiotic organisms that inhabit their guts. The rumen fungi are among these organisms; they produce enzymes that digest cellulose, and the resulting breakdown product nourishes both the fungi and their animal hosts. The spores of most rumen fungi have multiple flagella, which may form a tuft at one end of the spore.

TABLE 23-1	The Major Taxonomic Groups of Fungi			
Common Name (Latin Name)	Reproductive Structures	Cellular Characteristics	Economic and Health Impacts	Representative Genera
Chytrids (Chytridiomycota)	Form haploid or diploid flagellated spores	Septa are absent	Contribute to the decline of frog populations	*Batrachochytrium* (frog pathogen)
Rumen fungi (Neocallimastigomycota)	Form haploid or diploid flagellated spores	Septa are absent	Help enable cattle, horses, sheep to subsist on plants	*Neocallimastix* (lives in herbivore digestive systems)
Blastoclades (Blastocladiomycota)	Form haploid or diploid flagellated spores	Septa are absent	Cause brown spot disease of corn, crown wart disease of alfalfa	*Allomyces* (aquatic decomposer)
Glomeromycetes (Glomeromycota)	Form haploid asexual spores, often in clusters	Septa are absent	Form mycorrhizae (mutualistic, symbiotic associations with plant roots)	*Glomus* (widespread mycorrhizal partner)
Basidiomycetes (Basidiomycota)	Sexual reproduction involves formation of haploid basidiospores on club-shaped basidia	Septa are present	Cause smuts and rusts on crops; include some edible mushrooms	*Amanita* (poisonous mushroom); *Polyporus* (shelf fungus)
Ascomycetes (Ascomycota)	Form haploid sexual ascospores in saclike ascus	Septa are present	Cause molds on fruit; can damage textiles; cause Dutch elm disease and chestnut blight; include yeasts and morels	*Saccharomyces* (yeast); *Ophiostoma* (causes Dutch elm disease)
"Zygomycetes" (not a formally designated taxonomic group)	Form diploid sexual zygospores	Septa are absent	Cause soft fruit rot and black bread mold	*Rhizopus* (causes black bread mold); *Pilobolus* (dung fungus)

Blastoclades Have a Nuclear Cap

Blastoclades (FIG. 23-6) are distinguished by some characteristic features, such as a distinctive structure called the *nuclear cap* that is found near the nucleus of blastoclade spores. The nuclear cap consists of ribosomes. Blastoclades live in fresh water or in soil, and some are parasites of plants or aquatic invertebrates such as water fleas or mosquito larvae. Their spores have a single flagellum.

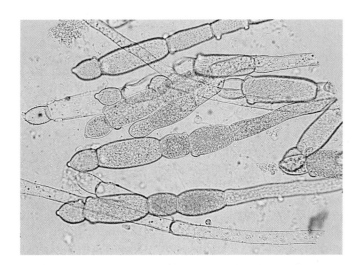

▲ **FIGURE 23-6 Blastoclade filaments** These filaments of the blastoclade fungus *Allomyces* are in the midst of sexual reproduction. The orange structures visible on many of the filaments will release male gametes; the clear swollen structures will release female gametes. Blastoclade gametes are flagellated, and these swimming reproductive structures aid dispersal of members of this mostly aquatic group.

▲ **FIGURE 23-7 Glomeromycete in a plant cell** Glomeromycete hyphae penetrate the cells of plants with which the fungus forms mutually beneficial associations. In this photo, a "trunk" hypha (red arrow) has branched off from the larger hypha at the bottom and divided into many smaller "fine branch" hyphae inside of a plant root cell.

Glomeromycetes Associate with Plant Roots

Almost all **glomeromycetes** live in intimate contact with the roots of plants. In fact, the hyphae of glomeromycetes actually penetrate the cells of the roots and form microscopic branching structures inside the cells (FIG. 23-7). This invasion of the plant's cells does not appear to harm the plant. On the contrary, glomeromycetes provide benefits to the plants they inhabit. This type of beneficial association between fungi and plant roots is known as a *mycorrhiza* and is described in more detail later in this chapter.

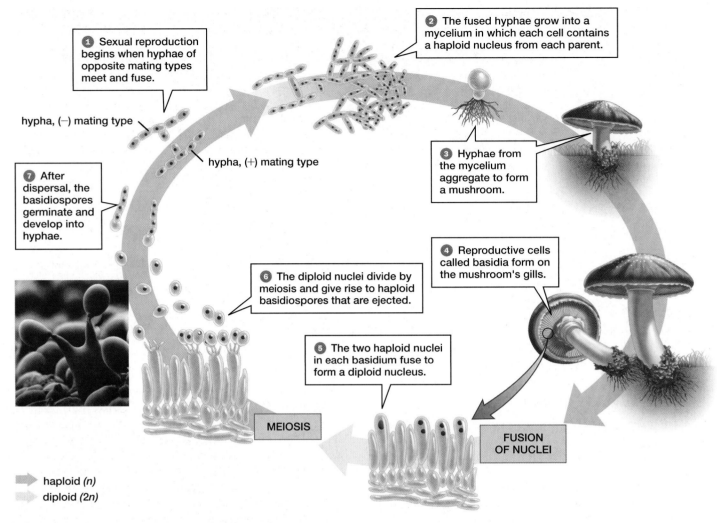

hypha, (−) mating type

hypha, (+) mating type

1 Sexual reproduction begins when hyphae of opposite mating types meet and fuse.

2 The fused hyphae grow into a mycelium in which each cell contains a haploid nucleus from each parent.

3 Hyphae from the mycelium aggregate to form a mushroom.

7 After dispersal, the basidiospores germinate and develop into hyphae.

4 Reproductive cells called basidia form on the mushroom's gills.

6 The diploid nuclei divide by meiosis and give rise to haploid basidiospores that are ejected.

5 The two haploid nuclei in each basidium fuse to form a diploid nucleus.

MEIOSIS

FUSION OF NUCLEI

➡ haploid *(n)*
➡ diploid *(2n)*

▲ **FIGURE 23-8 The life cycle of a typical basidiomycete** The photo shows two basidiospores attached to a basidium.

Glomeromycete reproduction is not fully understood; sexual reproduction by a member of the group is yet to be observed. During asexual reproduction, glomeromycetes produce clusters of spores by mitotic cell division. The spores form at the tips of hyphae that typically remain outside the host plant cell. When the spores germinate, hyphae grow into the surrounding soil, but the new fungus survives only if its germinating hyphae reach a plant root.

Basidiomycetes Produce Club-Shaped Reproductive Cells

Basidiomycetes typically reproduce sexually (**FIG. 23-8**). Hyphae of different mating types (designated "+" and "−") fuse **1** to form hyphae in which each cell contains two nuclei, one from each parent **2**. These hyphae grow into an underground mycelium that, in response to appropriate environmental conditions, gives rise to an aboveground fruiting body that consists of densely aggregated hyphae **3**. Some

of the hyphae in the fruiting body develop into club-shaped reproductive cells called **basidia** (singular, basidium), which contain two haploid nuclei **4**. In each basidium, the two nuclei fuse to yield a diploid nucleus **5**. The diploid nucleus divides by meiosis and gives rise to four haploid *basidiospores* **6**. If a basidiospore falls on fertile ground, it may germinate and form haploid hyphae **7**.

Basidiomycete fruiting bodies are familiar to most of us as mushrooms, puffballs, shelf fungi, and stinkhorns (**FIG. 23-9**). In many basidiomycetes, basidia are produced in leaflike gills on the undersides of mushrooms. In puffballs, the basidia are enclosed within the fruiting body. Basidiospores are released by the billions through openings in the tops of puffballs or from the gills of mushrooms and are dispersed by wind and water. In many cases, spores give rise to hyphae that grow outward from the original spore in a roughly circular pattern as the older hyphae in the center die. The subterranean body periodically sends up numerous mushrooms, which emerge in a ring-like pattern called a fairy ring (**FIG. 23-10**).

(a) Puffball

(b) Shelf fungus

(c) Stinkhorn

▲ **FIGURE 23-9 Diverse basidiomycetes (a)** The giant puffball *Lycoperdon giganteum* may produce up to 5 trillion spores. **(b)** Shelf fungi, some the size of dessert plates, are conspicuous on trees. **(c)** The spores of stinkhorns are carried on the outside of a slimy cap that smells terrible to humans, but appeals to flies. The flies lay their eggs on the stinkhorn and inadvertently disperse the spores that stick to their bodies.

CASE STUDY \ CONTINUED

Humongous Fungus

Because the underground bodies of basidiomycetes such as *Armillaria* grow at a relatively steady rate, the age of a mycelium can be estimated by measuring the area over which its above-ground reproductive structures spread. On the basis of such measurements, it is apparent that basidiomycetes can live for hundreds of years. Some are even older than that. For example, the researchers who discovered the gigantic *Armillaria* in Oregon estimate that it took at least 2,400 years to grow to its current size. Do the life cycles and habitats of other fungal groups permit them to grow as large and old as some basidiomycetes?

▲ **FIGURE 23-10 A mushroom fairy ring** Mushrooms emerge in a fairy ring from an underground fungal mycelium, growing outward from a central point where a single spore germinated, perhaps centuries ago.

Ascomycetes Form Spores in a Saclike Case

The **ascomycetes,** or **sac fungi,** reproduce both asexually and sexually (**FIG. 23-11**). In asexual reproduction, spores are produced at the tips of specialized hyphae and, after dispersal, develop into new hyphae ❶. During sexual reproduction, spores are produced by a more complex sequence of events that begins when hyphae of different mating types (+ and −) come into contact ❷. The two hyphae form reproductive structures that become linked by a connecting bridge. Haploid nuclei move across the bridge from the (−) reproductive structure to the (+) one, so that the (+) structure contains multiple nuclei from both parents ❸. The (+) structures that now contain the pooled nuclei develop into hyphae that are incorporated into a fruiting body ❹. At the tips of some of these hyphae, a saclike case called an **ascus** (plural, asci) forms ❺. At this stage, each ascus contains two haploid nuclei. These nuclei fuse to yield a single diploid nucleus ❻, which then divides by meiosis to yield four haploid nuclei ❼. These four nuclei divide by mitosis and develop into eight haploid spores known as *ascospores* ❽. Eventually, the ascus ruptures, liberating its ascospores. If the spores land in an appropriate location, they may germinate and develop into hyphae ❾.

Some ascomycetes live in decaying forest vegetation and form either beautiful cup-shaped reproductive structures (**FIG. 23-12a**) or corrugated, mushroom-like fruiting bodies called *morels* (**FIG. 23-12b**). The ascomycetes also include the species that produces penicillin, the first antibiotic, as well as many of the colorful molds that attack stored food and destroy fruit and grain crops and other plants. Ascomycetes may also harm animals (for an example, see "Earth Watch: Killer in the Caves" on page 414). The unicellular fungi known as yeasts are also ascomycetes. (When yeasts reproduce sexually, their single cell becomes the ascus.)

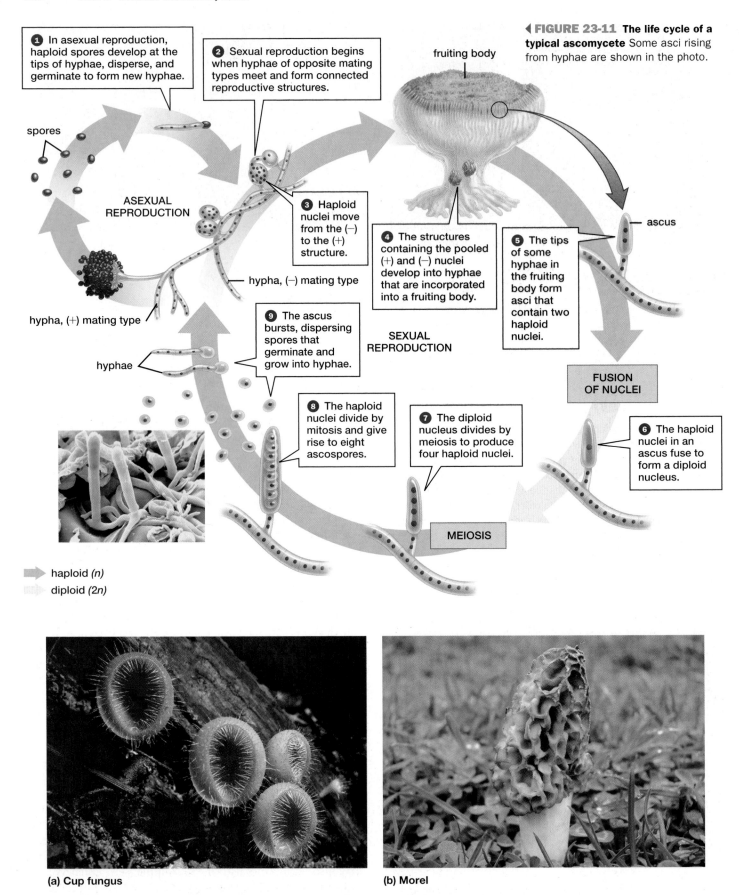

❶ In asexual reproduction, haploid spores develop at the tips of hyphae, disperse, and germinate to form new hyphae.

❷ Sexual reproduction begins when hyphae of opposite mating types meet and form connected reproductive structures.

◀ **FIGURE 23-11 The life cycle of a typical ascomycete** Some asci rising from hyphae are shown in the photo.

fruiting body

spores

ASEXUAL REPRODUCTION

❸ Haploid nuclei move from the (−) to the (+) structure.

hypha, (−) mating type

hypha, (+) mating type

ascus

❹ The structures containing the pooled (+) and (−) nuclei develop into hyphae that are incorporated into a fruiting body.

❺ The tips of some hyphae in the fruiting body form asci that contain two haploid nuclei.

❾ The ascus bursts, dispersing spores that germinate and grow into hyphae.

SEXUAL REPRODUCTION

hyphae

FUSION OF NUCLEI

❽ The haploid nuclei divide by mitosis and give rise to eight ascospores.

❼ The diploid nucleus divides by meiosis to produce four haploid nuclei.

❻ The haploid nuclei in an ascus fuse to form a diploid nucleus.

MEIOSIS

➡ haploid (n)
➡ diploid (2n)

(a) Cup fungus

(b) Morel

▲ **FIGURE 23-12 Diverse ascomycetes (a)** The cup-shaped fruiting body of the scarlet cup fungus. **(b)** The morel, an edible delicacy. (Consult an expert before sampling any wild fungus—some are deadly!)

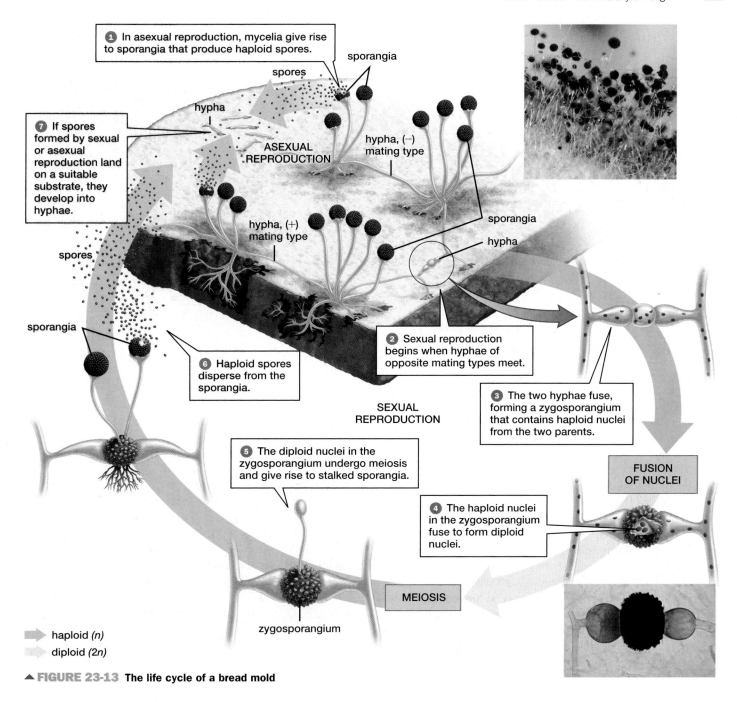

1 In asexual reproduction, mycelia give rise to sporangia that produce haploid spores.

sporangia

spores

hypha

7 If spores formed by sexual or asexual reproduction land on a suitable substrate, they develop into hyphae.

ASEXUAL REPRODUCTION

hypha, (−) mating type

spores

hypha, (+) mating type

sporangia

hypha

sporangia

6 Haploid spores disperse from the sporangia.

2 Sexual reproduction begins when hyphae of opposite mating types meet.

SEXUAL REPRODUCTION

3 The two hyphae fuse, forming a zygosporangium that contains haploid nuclei from the two parents.

5 The diploid nuclei in the zygosporangium undergo meiosis and give rise to stalked sporangia.

FUSION OF NUCLEI

4 The haploid nuclei in the zygosporangium fuse to form diploid nuclei.

MEIOSIS

zygosporangium

haploid (n)

diploid (2n)

▲ **FIGURE 23-13** **The life cycle of a bread mold**

Bread Molds Are Among the Fungi That Can Reproduce by Forming Diploid Spores

Many of the species formerly assigned to the zygomycetes live in soil or on decaying plant or animal material. These species include those belonging to the genus *Rhizopus,* which cause the familiar annoyances of soft fruit rot and black bread mold. In the bread mold, reproduction may be asexual or sexual (FIG. 23-13). Asexual reproduction is initiated by the formation of haploid spores in black spore cases called **sporangia** (singular, sporangium) **1**. These spores disperse through the air and, if they land on a suitable substrate (such as a piece of bread), germinate to form new haploid hyphae.

If two hyphae of different mating types (+ and −) come into contact, sexual reproduction may ensue **2**. The two hyphae fuse to form a *zygosporangium* that contains multiple haploid nuclei from the two parents **3**. As the zygosporangium develops, it becomes tough and resistant, and it can remain dormant for long periods until environmental conditions are favorable for growth. Inside the zygosporangium, the haploid nuclei fuse to produce diploid nuclei **4**. When conditions are favorable, the diploid nuclei undergo meiosis and give rise to stalked sporangia **5**. The sporangia produce haploid spores that disperse **6**, germinate, and develop into new haploid hyphae **7**.

CHECK YOUR LEARNING
Can you ...
- describe the six main taxonomic groups of fungi and explain why some fungal species are not assigned to any of these groups?
- describe the life cycles of a typical basidiomycete, ascomycete, and bread mold?

23.3 HOW DO FUNGI INTERACT WITH OTHER SPECIES?

Many fungi live in direct contact with another species for a prolonged period. Such intimate, long-term relationships are known as *symbiotic* relationships. In many cases, the fungal member of a symbiotic relationship is parasitic and harms its host. But some symbiotic relationships are mutually beneficial.

Lichens Are Formed by Fungi That Live with Photosynthetic Algae or Bacteria

Lichens are symbiotic associations between fungi and single-celled green algae or cyanobacteria (**FIG. 23-14a**). Lichens are sometimes described as fungi that have learned to garden, because the fungal member of the partnership "tends" the photosynthetic algal or bacterial partner by providing shelter and protection from harsh conditions. In this protected environment, the photosynthetic members of the partnership use sunlight to manufacture simple sugars, producing food for themselves but also some excess food that is consumed by the fungus. In fact, the fungus often consumes the lion's share of the photosynthetic product (up to 90% in some species), leading some researchers to conclude that the symbiotic relationship in lichens is really much more one-sided than it is usually portrayed.

Thousands of different fungal species (mostly ascomycetes) form lichens (**FIGS. 23-14b, c**), combining with one of a much smaller number of algal or bacterial species. Together, these organisms form a unit so tough and self-sufficient that lichens are among the first living things to colonize newly formed volcanic islands, because many lichens can grow on bare rock. Brightly colored lichens also invade other inhospitable habitats ranging from deserts to the Arctic. Understandably, lichens in extreme environments grow very slowly; arctic colonies, for example, may expand as slowly as

▶ **FIGURE 23-14 The lichen: a symbiotic partnership (a)** Most lichens have a layered structure bounded on the top and bottom by an outer layer formed from fungal hyphae. The fungal hyphae emerge from the lower layer, forming attachments that anchor the lichen to a surface, such as a rock or a tree. An algal layer in which the alga and fungus grow in close association lies beneath the upper layer of hyphae. **(b)** A colorful encrusting lichen, growing on dry rock, illustrates the tough independence of this symbiotic combination of fungus and algae. Pigments produced by the fungal partner are responsible for the bright orange color. **(c)** A leafy lichen grows on a rock.

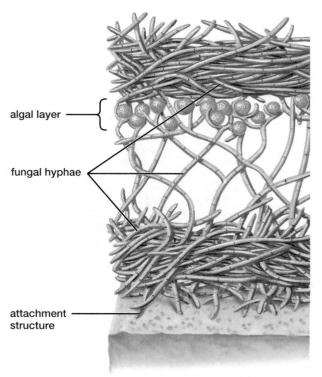

algal layer

fungal hyphae

attachment structure

(a) Lichen structure

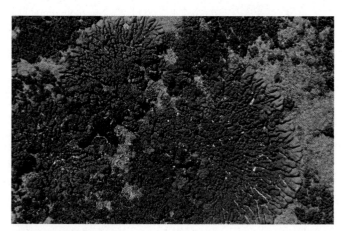

(b) Encrusting lichen

(c) Leafy lichen

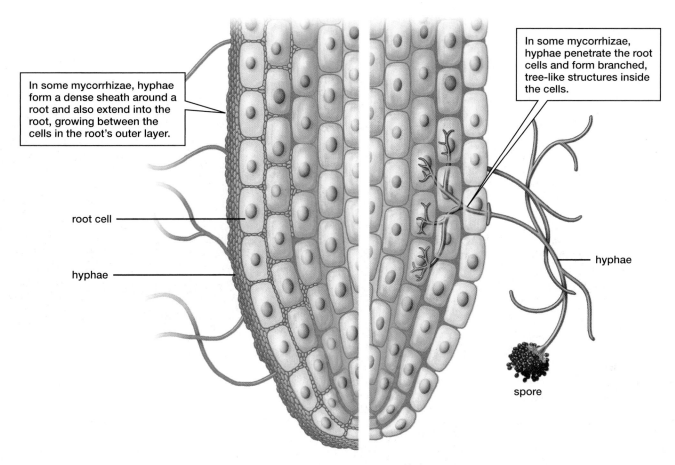

In some mycorrhizae, hyphae form a dense sheath around a root and also extend into the root, growing between the cells in the root's outer layer.

In some mycorrhizae, hyphae penetrate the root cells and form branched, tree-like structures inside the cells.

root cell

hyphae

hyphae

spore

▲ **FIGURE 23-15 Mycorrhizae enhance plant growth** Mycorrhizal associations between fungi and plant roots fall into two general types, shown in the left and right parts of this illustration (see also Fig. 23-7).

THINK CRITICALLY Fossil evidence suggests an important link between mycorrhizae and the successful invasion of land by plants. Why might mycorrhizae have been important in the colonization of terrestrial habitats by plants?

1 to 2 inches per 1,000 years. Despite their slow growth, lichens can persist for long periods of time; some arctic lichens are more than 4,000 years old.

Mycorrhizae Are Associations Between Plant Roots and Fungi

Mycorrhizae (singular, mycorrhiza) are important symbiotic associations between fungi and plant roots. More than 5,000 species of mycorrhizal fungi grow in intimate association with vascular plant roots, including those of most tree species. The hyphae of mycorrhizal fungi surround and invade roots (**FIG. 23-15**).

The association between plants and mycorrhizal fungi benefits both the fungi and their plant partners. The mycorrhizal fungi receive energy-rich sugar molecules that are produced photosynthetically by plants and passed from their roots to the fungi. In return, the fungi absorb mineral nutrients from the soil, passing some of them directly into the root cells. Mycorrhizal fungi also absorb water and pass it to the plant—an advantage for plants in dry, sandy soils.

The partnership between mycorrhizae and plants makes a crucial contribution to the health of Earth's plants. Plants without mycorrhizal fungi tend to be smaller and less vigorous than plants with mycorrhizal partners. Thus, the presence of mycorrhizae increases the overall productivity of Earth's plant communities, enhancing their ability to support the animals and other organisms that depend on them.

Endophytes Are Fungi That Live Inside Plant Stems and Leaves

The intimate association between fungi and plants is not limited to root mycorrhizae. Fungi have also been found living inside the aboveground tissues of virtually every plant species that has been tested for their presence. Some of these *endophytes* (organisms that live inside plants) are parasites that cause plant diseases, but many, perhaps most, are beneficial to the host plant. The best-studied examples of beneficial fungal endophytes are the ascomycete species that live inside the leaf cells of many species of grass. These fungi produce substances that are distasteful or toxic to insects

Earth WATCH   Killer in the Caves

The American chestnut and American elm have vanished from North American landscapes as a result of fungal diseases imported from other continents, and bats are now threatened with the same fate. An ascomycete species that arrived in the United States from Europe causes white-nose syndrome, an infectious disease that has killed an estimated 7 million bats of 11 species since its discovery in 2006. The disease is named for the white fungal growth that appears on the bare skin, especially the nose, of affected bats (FIG. E23-1).

The bat species that are most vulnerable to white-nose syndrome are those that hibernate during the winter. The fungus that causes the disease is adapted to cold temperatures; infections occur mainly during the winter and spread through groups of bats hibernating close together in caves and mines. White-nose syndrome is so far confined to bats in the eastern half of North America, but it has spread rapidly in recent years.

The disease has greatly reduced populations of the affected species. This abrupt decline of bats has serious implications for ecosystem health. All of the species that have been devastated by white-nose syndrome are insect-eaters that play an important role in regulating insect populations, helping ensure that plant communities are not overwhelmed by plant-eating insects. A recent study estimated that insect consumption in eastern North America has declined by 2,300 tons per year as a result of the onset of white-nose syndrome. This ecological loss has economic effects as well; insect consumption by bats saves U.S. farmers billions of dollars by reducing the amount of pesticide they must apply. Unfortunately, it is far from certain that we will continue to enjoy the ecological benefits of bats; there is so far no treatment or cure for white-nose syndrome.

▲ **FIGURE E23-1** **White nose syndrome** The fungus that causes the disease infects the skin of bats' faces, wings, and ears, breaking down skin and connective tissues. In addition, irritation from the infection causes bats to become active when they should be hibernating, which can lead to starvation.

THINK CRITICALLY The ascomycete that causes white-nose syndrome can remain viable on a cave floor for at least 5 years without infecting a bat host. As a result, even a cave in which all resident bats have become infected and died contains infectious spores that can potentially be carried to other caves. Given this circumstance, what steps could governments take to slow the spread of white-nose syndrome to previously unaffected areas?

---

and grazing mammals and thus help protect the grass plants from those predators.

The antipredator protection provided by fungal endophytes is sufficiently effective that agricultural scientists are working hard to discover a way to grow grasses free of endophytes and therefore more palatable to grazing animals. Horses, cows, and other agriculturally important grazers tend to avoid eating grasses that contain endophytes. When the only available food is endophyte-containing grass, animals that eat it experience poor health and slow growth.

Some Fungi Are Important Decomposers

Some fungi, acting as mycorrhizae and endophytes, play a major role in the growth and preservation of plant tissue. Other fungi, however, play a similarly major role in its destruction by acting as decomposers. Many fungal species can digest lignin or cellulose, the molecules that make up wood; some species can digest both molecules. Thus, when a tree or other woody plant dies, fungi can completely decompose its remains.

Fungal decomposition is not limited to wood; fungi digest dead organisms of all kinds. The fungi that are *saprophytes* (feeding on dead organisms) return the dead tissues' component substances to the ecosystems from which they came. The extracellular digestive activities of saprophytic fungi liberate nutrients that can be used by plants. If fungi and prokaryotes were suddenly to disappear, the consequences would be disastrous. Nutrients would remain locked in the bodies of dead plants and animals, the recycling of nutrients would grind to a halt, soil fertility would rapidly decline, and waste and organic debris would accumulate. In short, ecosystems would collapse.

CHECK YOUR LEARNING
Can you ...
- describe and explain the significance of some symbiotic associations involving fungi, including lichens, mycorrhizae, and endophytes?
- explain how fungi help recycle nutrients?

23.4 HOW DO FUNGI AFFECT HUMANS?

Most people give little thought to fungi, but they affect our lives in more ways than you might imagine.

Fungi Attack Plants That Are Important to People

Fungi cause the majority of plant diseases, and some of the plants that they infect are important to humans. For example, fungal pathogens have a devastating effect on the world's food supply. Especially damaging are the basidiomycete plant pests descriptively called *rusts* and *smuts,* which cause billions of dollars' worth of damage to crops each year (FIG. 23-16). For example, Ug99, an especially virulent strain of the wheat disease called black stem rust, is currently a major threat to the world's supply of wheat. (Wheat is one of the world's most important crops in terms of total calories contributed.) None of the wheat varieties that feed much of the world's

▲ FIGURE 23-17 **Pest-killing fungus** Fungi such as *Cordyceps* are used by farmers to control insect pests. The white spikes erupting from the body of this moth are *Cordyceps* fruiting bodies.

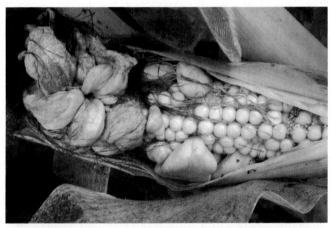

(a) Corn smut

(b) Black stem rust

▲ FIGURE 23-16 **Smuts and rusts (a)** Corn smut is a basidiomycete pathogen that destroys millions of dollars' worth of corn each year. Even a pest like corn smut has its admirers, though. In Mexico this fungus is known as *huitlacoche* and is considered a great delicacy. **(b)** Another basidiomycete, black stem rust, currently threatens millions of acres of wheat in Africa, Central Asia, and the Middle East.

population is resistant to Ug99, whose spores can travel long distances by wind. As a result, wheat crops have been decimated over a large and growing area of Africa, Central Asia, and the Middle East. If Ug99 spreads to the massive wheat crops of China and India, serious food shortages might occur.

Fungal diseases also affect the appearance of our landscape. The American elm and the American chestnut—two tree species that were once prominent in many of America's parks, yards, and forests—were destroyed on a massive scale by the ascomycetes that cause Dutch elm disease and chestnut blight. Today, few people can recall the graceful forms of large elms and chestnuts, which are now almost entirely absent from the landscape. There may be hope for a return of the chestnut, however. Researchers have engineered trees that are genetically modified to contain a gene that confers resistance to chestnut blight and have demonstrated that the modified trees pass the resistance gene to their offspring. The researchers are seeking regulatory approval to plant the transgenic trees in the wild.

The fungal impact on agriculture and forestry is not entirely negative. Fungal parasites that attack insects and other arthropod pests can be an important ally in pest control (FIG. 23-17). Farmers who wish to reduce their dependence on toxic and expensive chemical pesticides are increasingly turning to biological methods of pest control, including the application of "fungal pesticides." Fungal pathogens are currently used to control termites, rice weevils, tent caterpillars, aphids, citrus mites, and other pests.

CASE STUDY \ **CONTINUED**

Humongous Fungus

The *Armillaria* fungus species that grew to massive size in Oregon harms trees in the forests it inhabits. As the fungus feeds on roots, it causes "root rot" that weakens or kills trees. This root rot provides aboveground evidence of *Armillaria*'s existence; the giant Oregon specimen was first identified by examining aerial photos to find forested areas with many dead trees. Can people, as well plants, be victims of fungal attack?

Fungi Cause Human Diseases

The fungi include parasitic species that attack humans directly. Some of the most familiar fungal diseases are those caused by ascomycetes such as the yeast *Candida albicans,* which causes vaginal infections, and those that attack the skin, resulting in athlete's foot, jock itch, and ringworm. These diseases, though unpleasant, are not life threatening and can usually be treated with antifungal ointments.

Fungi can infect the lungs if victims inhale spores of disease-causing fungal species such as the ascomycetes that cause valley fever and histoplasmosis. In the United States, valley fever is found mainly in the southwest, whereas histoplasmosis is more prevalent in central and eastern regions. Like other fungal infections, these diseases can, if promptly diagnosed, be controlled with antifungal drugs. If untreated, however, they can develop into serious, systemic infections.

Serious disease can also result from inhaling spores of *Cryptococcus gattii,* a basidiomycete that is normally found in the tropics but has in recent years appeared in the Pacific Northwest. In all environments, the fungus has proved to be dangerous; the death rate of those infected ranges from 13% to 33%, depending on location. Fortunately, infections are so far relatively rare: about 30 cases per year in the United States, for example, compared to more than 20,000 annually for valley fever. However, the area affected by *C. gattii* has grown steadily, so the incidence of infections is likely to increase.

In addition to causing human diseases, fungi can also help combat them. Biologists have discovered that certain fungi attack and kill the mosquito species that transmit malaria (**FIG. 23-18**). Plans are under way to enlist these fungi in the fight against malaria, one of the world's deadliest diseases.

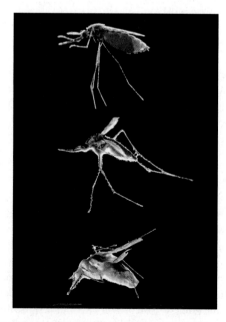

▲ **FIGURE 23-18 Disease-fighting fungus** A healthy malaria-carrying mosquito (top) infected by *Beauveria* is transformed to a fungus-encrusted corpse in less than 2 weeks.

Fungi Can Produce Toxins

In addition to their role as agents of infectious disease, some fungi produce toxins that are dangerous to humans. Of particular concern are toxins produced by fungi growing on grains and other foodstuffs that have been stored in moist conditions. For example, molds of the genus *Aspergillus* produce highly toxic, carcinogenic compounds known as aflatoxins. Some foods, such as peanuts, seem especially susceptible to attack by *Aspergillus.* Since aflatoxins were discovered in the 1960s, food growers and processors have developed methods for reducing the growth of *Aspergillus* in stored crops, so aflatoxins have been largely eliminated from the nation's peanut butter supply.

One infamous toxin-producing fungus is the ascomycete *Claviceps purpurea,* which infects rye plants and causes a disease known as ergot. This fungus produces several toxins, which can affect humans if infected rye is ground into flour and consumed. This happened frequently in northern Europe in the Middle Ages, with devastating effects. At that time, ergot poisoning was typically fatal, and victims experienced terrible symptoms before dying. One ergot toxin constricts blood vessels and reduces blood flow. The effect can be so extreme that gangrene develops and limbs shrivel and fall off. Other ergot toxins cause symptoms that include a burning sensation, vomiting, convulsive twitching, and vivid hallucinations. Today, new agricultural techniques have effectively eliminated ergot poisoning, but the hallucinogenic drug LSD, which is derived from a component of the ergot toxins, remains as a legacy of this disease.

Some of the deadliest poisons known to humankind are found in mushrooms. Especially noted for their poisons are certain species in the genus *Amanita,* which have suggestive common names such as death cap and destroying angel (**FIG. 23-19**). These names are apt, because even a single bite of one of these mushrooms can be lethal. Damage from *Amanita* toxins is most severe in the liver, where the toxins tend to accumulate. Often, a victim of *Amanita* poisoning

▲ **FIGURE 23-19 The destroying angel** Mushrooms produced by the basidiomycete *Amanita virosa* can be deadly.

can be saved only by undergoing a liver transplant. Each year, a number of small children, inexperienced collectors, and unlucky guests at gourmet dinners make unexpected trips to the hospital after eating poisonous wild mushrooms. So if you decide to collect some wild mushrooms to eat, be careful. Protect your health by inviting an expert to join your mushroom-hunting expeditions.

Many Antibiotics Are Derived from Fungi

Fungi also have positive impacts on human health. The modern era of lifesaving antibiotic medicines began with the discovery of penicillin, which is produced by an ascomycete mold (**FIG. 23-20**; also see Fig. 1-12). Penicillin is still used, along with other fungi-derived antibiotics such as oleandomycin and cephalosporin, to combat bacterial diseases. Other important drugs are also derived from fungi, including cyclosporin, which is used to suppress the immune response after an organ transplant so that the body is less likely to reject the transplanted organs.

Fungi Make Important Contributions to Gastronomy

Fungi make important contributions to the human diet. We consume some fungi directly, including wild and cultivated basidiomycete mushrooms and ascomycetes such as morels and the rare and prized truffle. The role of fungi in cuisine also has less-visible manifestations. For example, some of the

▲ **FIGURE 23-20** *Penicillium* *Penicillium* growing on an orange. Reproductive structures, which coat the fruit's surface, are visible, while hyphae, beneath, draw nourishment from inside. The antibiotic penicillin was first isolated from this fungus.

THINK CRITICALLY Why do some fungi produce antibiotic chemicals? How and where would you search for new antibiotics produced by fungi?

HAVE YOU EVER WONDERED ...

Why Truffles Are So Expensive?

Although many fungi are prized as food, none is as avidly sought as the truffle. The finest Italian truffles may sell for as much as $3,400 per pound, and unusually large specimens can fetch spectacular prices (**FIG. 23-21**). A 3.3-pound Italian white truffle once sold at auction for $330,000!

Why such high prices? Truffles develop underground, and it takes some work to find one. In fact, humans can't do it alone and need help from other species. Some animals, especially pigs, are attracted to the aroma of a mature truffle. If a pig follows the smell to a truffle, it will dig the fungus up and devour it. That's why, traditionally, truffle collectors have used muzzled pigs to hunt their quarry.

Today, trained dogs are the most common assistants to truffle hunters. Dogs are necessary even on the farms where much of today's truffle crop is laboriously grown. The difficulty of cultivating and harvesting truffles accounts in part for their high price. And no one has figured out how to cultivate the prized white truffle. The only way to acquire one is to follow the nose of a truffle-hunting dog or pig.

▲ **FIGURE 23-21** **The truffle** Truffles are the underground, spore-containing structures of an ascomycete that forms a mycorrhizal association with the roots of oak trees. Especially fine specimens are often sold at auction.

world's most famous cheeses, including Roquefort, Camembert, Stilton, and Gorgonzola, gain their distinctive flavors from ascomycete molds that grow on them as they ripen. Perhaps the most important and pervasive fungal contributors to our food supply, however, are the single-celled ascomycetes (and a few species of basidiomycetes) known as yeasts.

Wine and Beer Are Made Using Yeasts

The discovery that yeasts could be harnessed to enliven our culinary experience is surely a key event in human history. Among the many foods and beverages that depend on yeasts for their production are bread, wine, and beer, which have been consumed so widely for so long that it is difficult to imagine a world without them. All derive their special qualities from fermentation by yeasts. Fermentation occurs when yeasts extract energy from sugar and, as by-products of the metabolic process, emit carbon dioxide and ethyl alcohol.

As yeasts consume the fruit sugars in grape juice, the sugars are converted to alcohol, and wine is the result. Eventually, the increasing concentration of alcohol kills the yeasts, ending fermentation. If the yeasts in fermenting wine die before they have consumed all the available grape sugar, the wine will be sweet; if the yeasts have exhausted the supply of sugar, the wine will be dry.

Beer is brewed from grain (usually barley), but yeasts cannot effectively consume the carbohydrates in grain. For the yeasts to do their work, the barley grains must have sprouted (recall that grains are actually seeds). Germination converts the grains' carbohydrates to sugar, so the sprouted barley provides an excellent food source for the yeasts. As with wine, fermentation converts sugars to alcohol, but beer brewers capture the carbon dioxide by-product as well, giving the beer its characteristic bubbly carbonation.

Yeasts Make Bread Rise

In bread making, carbon dioxide is the crucial fermentation product. The yeasts added to bread dough do produce alcohol

as well as carbon dioxide, but the alcohol evaporates during baking. In contrast, the carbon dioxide is trapped in the dough, where it forms the bubbles that give bread its light, airy texture (and saves us from a life of eating sandwiches made with crackers).

So the next time you're enjoying a slice of French bread with Camembert cheese and a nice glass of chardonnay, or a slice of pizza and a cold bottle of your favorite brew, you might want to quietly give thanks to the yeasts. Our diets would certainly be a lot duller without the help we get from fungal assistants.

CHECK YOUR LEARNING

Can you ...

- explain how fungi affect agriculture?
- describe examples of how fungi affect human health?
- describe the role of fungi in the production of cheese, wine, beer, and bread?

Humongous Fungus

Why do *Armillaria* fungi grow so large? Their size is due in part to their ability to form structures called rhizomorphs, which consist of hyphae bundled together inside a protective rind. The rhizomorphs can extend long distances through nutrient-poor areas to reach new sources of food. The *Armillaria* fungus can thus grow beyond the boundaries of a particular food-rich area.

Another factor that may contribute to the gigantic size of the Oregon *Armillaria* is the climate in which it was found. In the dry climate of eastern Oregon, fungal fruiting bodies form only rarely, so the colossal *Armillaria* rarely produces spores. In the absence of spores that might grow into new individuals, the existing individual faces little competition for resources and is free to grow and fill an increasingly large area.

The discovery of the Oregon specimen is merely the latest chapter in a long-running, good-natured "fungus war" that began

with the discovery of the first humongous fungus, a 37-acre *Armillaria gallica* growing in Michigan. Since that initial landmark discovery, research groups in Michigan, Oregon, and elsewhere have engaged in a friendly competition to find the largest fungus. Will the current record ever be topped? Stay tuned.

CONSIDER THIS Because the entire Oregon *Armillaria* grew from a single spore, all of its cells are genetically identical. However, it is unlikely that any substances are transported through the entire 2,400-acre mycelium. And there is no continuous skin or bark or membrane that covers the entire mycelium and separates it from the environment as a unit. Is the fungus's genetic unity sufficient evidence for it to be considered a single individual, or do you need additional evidence? Do you think that the claim of "world's largest organism" is valid?

CHAPTER REVIEW

 Go to **MasteringBiology** for practice quizzes, activities, eText, videos, current events, and more.

Answers to Think Critically, Evaluate This, Multiple Choice, and Fill-in-the-Blank questions can be found in the Answers section at the back of the book.

Summary of Key Concepts

23.1 What Are the Key Features of Fungi?

Fungal bodies generally consist of filamentous hyphae, which are either multicellular or multinucleated and form large, intertwined networks called *mycelia*. Fungal nuclei are generally haploid. A cell wall of chitin surrounds fungal cells. All fungi feed by secreting digestive enzymes outside their bodies and absorbing the liberated nutrients.

Fungal reproduction is varied and complex. Asexual reproduction can occur through mitotic production of haploid spores. Sexual reproduction occurs when compatible haploid nuclei fuse to form a diploid zygote, which undergoes meiosis to form haploid spores. Both asexual and sexual spores produce haploid mycelia through mitosis.

23.2 What Are the Major Groups of Fungi?

The major taxonomic groups of fungi are Chytridiomycota (chytrids), Neocallimastigomycota (rumen fungi), Blastocladiomycota (blastoclades), Glomeromycota (glomeromycetes), Basidiomycota (basidiomycetes), and Ascomycota (ascomycetes).

23.3 How Do Fungi Interact with Other Species?

A lichen is a symbiotic association between a fungus and algae or cyanobacteria. The fungal partner provides shelter for the algae or cyanobacteria, which convert sunlight to energy-rich molecules that nourish the fungus. Mycorrhizae are associations between fungi and the roots of most vascular plants. The fungus derives photosynthetic nutrients from the plant roots and, in return, carries water and nutrients into the root from the surrounding soil. Endophytes are fungi that grow inside the leaves or stems of plants and that may help protect the plants that harbor them. Saprophytic fungi are extremely important decomposers in ecosystems. Their filamentous bodies penetrate soil and decaying organic material, liberating nutrients through extracellular digestion.

23.4 How Do Fungi Affect Humans?

Many plant diseases are caused by parasitic fungi. Some parasitic fungi can help control insect crop pests. Others can cause human diseases, including ringworm, athlete's foot, and common vaginal infections. Some fungi produce toxins that can harm humans. Nonetheless, fungi add variety to the human food supply, and fermentation by fungi helps make wine, beer, and bread. Fungi are also the source of many antibiotics.

Key Terms

ascomycete *409*	mycorrhiza (plural,
ascus (plural, asci) *409*	mycorrhizae) *413*
basidiomycete *408*	rumen fungus *406*
basidium (plural, basidia) *408*	sac fungus *409*
blastoclade *407*	septum (plural,
chytrid *406*	septa) *403*
glomeromycete *407*	sporangium (plural,
hypha (plural, hyphae) *403*	sporangia) *411*
lichen *412*	spore *404*
mycelium (plural, mycelia) *403*	

Thinking Through the Concepts

Multiple Choice

1. Fungi have cell walls composed primarily of
   a. lignin.
   c. chitin.
   b. cellulose.
   d. glucose.

2. Which of the following diseases is *not* caused by a fungus?
   a. valley fever
   c. histoplasmosis
   b. Dutch elm disease
   d. malaria

3. A symbiotic association of plant roots and fungi is known as a
   a. lichen.
   c. sporangium.
   b. mycorrhiza.
   d. chytrid.

4. The alcohol in beer and wine is a by-product of _____ by _____.
   a. infection; *Candida albicans*
   b. respiration; rumen fungi
   c. fermentation; yeast
   d. decomposition; saprophytes

5. _____ is an ecologically important function of fungi.
   a. Photosynthesis
   c. Alcohol production
   b. Decomposition
   d. Antibiotic production

Fill-in-the-Blank

1. The portions of a fungus that are visible to the naked eye are often structures specialized for _____. These structures release tiny _____, which are dispersed to produce new fungi.

2. The fungal body is a(n) _____ and is composed of microscopic threads called _____ that may be subdivided into many cells by _____. The cell walls of fungi are strengthened by _____.

3. In fungi, asexual spores are produced by _____ cell division and have _____ set(s) of chromosomes. Sexual spores are produced by _____ cell division in a(n) _____ and have _____ set(s) of chromosomes.

4. Fill in the blanks in the following sentences with the common names of fungal taxonomic groups. Almost all _____ live in intimate association with plant roots. Flagellated, swimming spores are produced by _____. Mushrooms and puffballs are reproductive structures of _____.

5. _____ are symbiotic associations of fungi and green algae. _____ are symbiotic, mutually beneficial associations of fungi and plant roots. Some fungi are _____ that live inside the aboveground tissues of plants.

6. Fungi are the only decomposers that can digest _____. The fungi that cause breads to rise and wines to ferment are _____. Human ailments caused by fungi include _____ and _____.

Review Questions

1. Describe the structure of the fungal body. How do fungal cells differ from most plant and animal cells?

2. What portion of the fungal body is represented by mushrooms, puffballs, and similar structures? Why are these structures elevated above the ground?

3. What two plant diseases, caused by parasitic fungi, have had an enormous impact on forests in the United States? In which taxonomic group are these fungi found?

4. List some fungi that attack crops. To which taxonomic group do they belong?

5. Describe asexual reproduction in fungi.

6. List the major taxonomic groups of fungi, describe some key features of each group, and give an example of a fungus in each group.

7. Describe how a fairy ring of mushrooms is produced. Why is its diameter related to its age?

8. Describe two symbiotic relationships between a fungus and another organism. In each case, explain how each partner in these associations is affected.

Applying the Concepts

1. Dutch elm disease in the United States is caused by an *exotic*—that is, an organism (in this case, a fungus) introduced from another part of the world. What damage has this introduction done? What other fungal pests fall into this category? Why are parasitic fungi particularly likely to be transported out of their natural habitat? What can governments do to limit this importation?

2. What ecological consequences would occur if humans, using a new and deadly fungicide, destroyed all fungi on Earth?

24 ANIMAL DIVERSITY I: INVERTEBRATES

The medicinal leech, long a symbol of premodern medical ignorance, is now part of modern medicine's toolkit.

CASE STUDY

Physicians' Assistants

MODERN MEDICINE is a high-tech enterprise, dependent on multimillion-dollar machines, sophisticated medical devices, and drugs developed via cutting-edge chemistry and biotechnology. But despite the prevalence of advanced technology in medicine, physicians also get low-tech assistance from an unlikely source: *invertebrates* (animals without a backbone). Consider, for example, the medicinal leech. For more than 2,000 years, healers have enlisted these parasitic segmented worms for treatment of a wide range of illnesses. For much of human medical history, treatment with leeches was based on the hope that the creatures would suck out the "tainted" blood that was believed to be the primary cause of disease. As the actual causes of disease were discovered, however, medical use of leeches declined. By the beginning of the twentieth century, leeches no longer had a place in the toolkit of modern medicine and had become a symbol of the ignorance of an earlier age. Today, however, medicinal leeches are making a surprising comeback.

Currently, leeches are used to treat a surgical complication known as venous insufficiency. This complication is especially common in reconstructive surgery, such as surgery to reattach a severed finger or repair a disfigured face. In such cases, surgeons are often unable to reconnect all of the veins that would normally carry blood away from tissues. Eventually, new veins will grow, but in the meantime blood may accumulate in the repaired tissue. Unless the excess blood is removed, it will coagulate, causing clots that can deprive the tissue of the oxygen and nutrients it needs to live. Fortunately, leeches can help. Applied to the affected area, the leeches get right to work, making small, painless incisions and sucking blood into their stomachs. To aid them in their blood-removal task, the leeches' saliva contains a mixture of chemicals that dilate blood vessels and prevent blood from clotting. Although the chemical brew in the saliva is an adaptation that helps leeches consume blood more efficiently, it also helps the patient by promoting blood flow in the damaged tissue. In this way, leeches provide a painless, effective treatment for venous insufficiency.

Although relatively few invertebrate animals have medical uses, invertebrates account for a large share of Earth's known biodiversity. What have scientists learned about these diverse and abundant organisms?

AT A GLANCE

24.1 What Are the Key Features of Animals?

24.2 Which Anatomical Features Mark Branch Points on the Animal Evolutionary Tree?

24.3 What Are the Major Animal Phyla?

24.1 WHAT ARE THE KEY FEATURES OF ANIMALS?

It is difficult to devise a concise definition of the term "animal." No single feature uniquely defines animals, so the group is defined by a list of characteristics. None of these characteristics is unique to animals, but together they distinguish animals from members of other taxonomic groups:

- Animals are eukaryotes.
- Animals are multicellular.
- Animal cells lack a cell wall.
- Animals obtain their energy by consuming other organisms.
- Animals typically reproduce sexually.
- Animals are motile (able to move about) during some stage of their lives.
- Most animals are able to respond rapidly to external stimuli.

CHECK YOUR LEARNING

Can you ...

- list the characteristics that collectively distinguish animals from other kinds of organisms?

24.2 WHICH ANATOMICAL FEATURES MARK BRANCH POINTS ON THE ANIMAL EVOLUTIONARY TREE?

By the Cambrian period, which began 541 million years ago, most of the animal phyla that currently populate Earth were already present. Unfortunately, the Precambrian fossil record is very sparse and does not reveal much about the early evolutionary history of animals. Therefore, systematists have looked to anatomy, embryological development, and DNA sequences for clues about animal history. These investigations have shown that certain features mark major branching points on the animal evolutionary tree and represent milestones in the evolution of the different body plans of modern animals (FIG. 24-1). In the following sections, we will explore these evolutionary milestones and their legacies in the bodies of modern animals.

Lack of Tissues Separates Sponges from All Other Animals

One of the earliest major innovations in animal evolution was the appearance of **tissues**—groups of similar cells integrated into a functional unit, such as muscle tissue or nerve tissue. Today, the bodies of almost all animals include tissues; the only animals that have retained the ancestral lack of tissues are the sponges. In sponges, individual cells often have specialized functions, but they act more or less independently and are not organized into tissues. This unique feature of sponges suggests that the split between sponges and the evolutionary branch leading to all other animal phyla must have occurred very early in the history of animals.

Animals with Tissues Exhibit Either Radial or Bilateral Symmetry

The first appearance of tissues coincided with the first appearance of body symmetry; all animals with true tissues also have symmetrical bodies. An animal is said to be symmetrical if it can be bisected along at least one plane such that the resulting halves are mirror images of one another.

The symmetrical, tissue-bearing animals can be divided into two groups, one containing animals that exhibit **radial symmetry** (FIG. 24-2a) and one with animals that exhibit **bilateral symmetry** (FIG. 24-2b). In radial symmetry, any plane through a central axis divides the object into roughly equal halves. In contrast, a bilaterally symmetrical animal can be divided into roughly mirror-image halves only along one particular plane through the central axis.

The difference between radially and bilaterally symmetrical animals reflects another major branching point in the animal evolutionary tree. This split separated the ancestors of the radially symmetrical cnidarians (sea jellies, corals, and anemones) and ctenophores (comb jellies) from the ancestors of the remaining animal phyla, all of which are bilaterally symmetrical.

Radially Symmetrical Animals Have Two Embryonic Tissue Layers; Bilaterally Symmetrical Animals Have Three

The distinction between radial and bilateral symmetry in animals is closely tied to a corresponding difference in the number of tissue layers, called *germ layers,* that arise during embryonic development. Embryos of animals with radial symmetry have two germ layers: an inner layer of **endoderm** (which gives rise to the tissues that line the gut cavity) and an outer layer of **ectoderm** (which gives rise to the tissues that cover the outside of the body.) Embryos of bilaterally symmetrical animals add a third germ layer, **mesoderm,** which lies between the endoderm and the ectoderm.

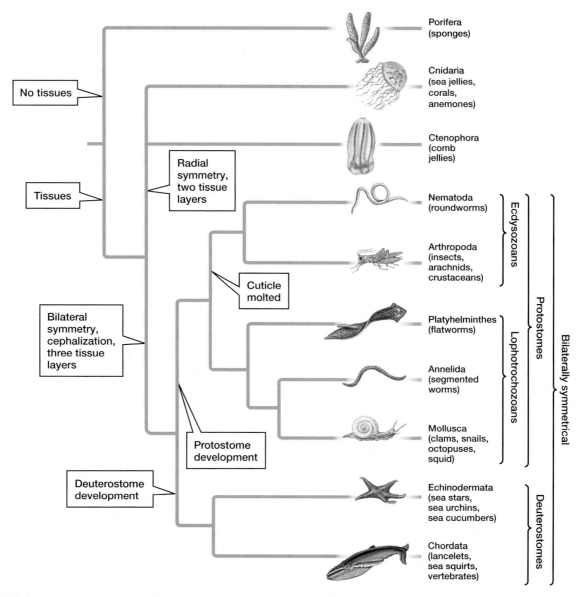

▲ FIGURE 24-1 **An evolutionary tree of some major animal phyla** Based on Dunn et al. 2008, Nature 452: 745–749.

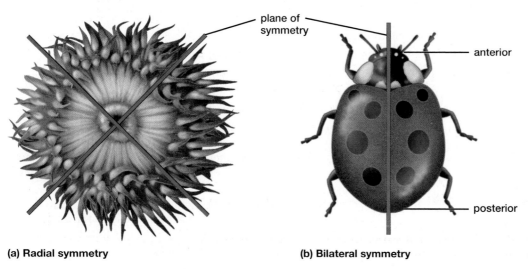

(a) Radial symmetry

(b) Bilateral symmetry

◀ FIGURE 24-2 **Body symmetry and cephalization** (a) Animals with radial symmetry, such as this sea anemone, lack a well-defined head. Any plane that passes through the central axis divides the body into mirror-image halves. (b) Animals with bilateral symmetry, such as this beetle, have an anterior head end and a posterior tail end. The body can be split into two mirror-image halves only along a particular plane that runs down the midline.

In bilaterally symmetrical animals, endoderm differentiates to form the tissues that line respiratory surfaces, the gut, and most hollow organs. (An **organ** is a discrete structure, such as a heart or stomach, in which two or more tissue types work together to perform functions.) Ectoderm forms nerve tissue and the tissues on the outer surface of the body. Mesoderm forms muscle and, when present, the circulatory and skeletal systems.

The connection between symmetry type and number of germ layers helps us make sense of the potentially puzzling case of the echinoderms (sea stars, sea urchins, and sea cucumbers). Adult echinoderms are radially symmetrical, yet our evolutionary tree places them squarely within the bilaterally symmetrical group. Why? Echinoderms have three germ layers, as well as several other characteristics (some described later) that unite them with the bilaterally symmetrical animals. So the immediate ancestors of echinoderms must have been bilaterally symmetrical, and the group subsequently evolved radial symmetry (a case of convergent evolution). To this day, larval echinoderms retain bilateral symmetry.

Bilaterally Symmetrical Animals Have Heads

Radially symmetrical animals tend either to be *sessile* (fixed to one spot, like sea anemones) or to drift around on currents (like sea jellies). Because such animals do not actively propel themselves in a particular direction, all parts of their bodies are more or less equally likely to encounter food. In contrast, most bilaterally symmetrical animals are *motile* (move under their own power), and resources such as food are most likely to be encountered by the part of the animal that is closest to the direction of movement. The evolution of bilateral symmetry was therefore accompanied by **cephalization,** the concentration of sensory organs and a brain in a defined head region. Cephalization produces an *anterior* (head) end, where sensory cells, sensory organs, clusters of nerve cells, and organs for ingesting food are concentrated. The other end of a cephalized animal is designated *posterior* and may feature a tail (see Fig. 24-2b).

Most Bilateral Animals Have Body Cavities

The members of many bilateral animal phyla have a fluid-filled cavity between the digestive tube (or gut, where food is digested and absorbed) and the outer body wall. In an animal with a body cavity, the gut and body wall are separated by a fluid-filled space, creating a "tube-within-a-tube" body plan. No radially symmetrical animal has a body cavity, so it is likely that this feature arose sometime after the split between radially and bilaterally symmetrical animals.

A fluid-filled body cavity can serve a variety of functions. In many invertebrate animals it acts as a kind of skeleton, providing support for the body and a framework against which muscles can act. In other animals, internal organs are suspended within the fluid-filled cavity, which serves as a protective buffer between the organs and the outside world.

Body Cavity Structure Varies Among Phyla

The most widespread type of body cavity is a **coelom,** a fluid-filled cavity that is completely lined with a thin layer of tissue that develops from mesoderm (**FIG. 24-3a**). Phyla whose members have a coelom are called *coelomates*. The annelids (segmented worms), arthropods (insects, spiders, crustaceans), mollusks (clams and snails), echinoderms, and chordates (which include humans) are coelomates.

Some animals have a body cavity that is *not* completely surrounded by mesoderm-derived tissue. This type of cavity is known as a **pseudocoelom.** Phyla whose members have a pseudocoelom are collectively known as *pseudocoelomates* (**FIG. 24-3b**). The roundworms (nematodes) are the largest pseudocoelomate group.

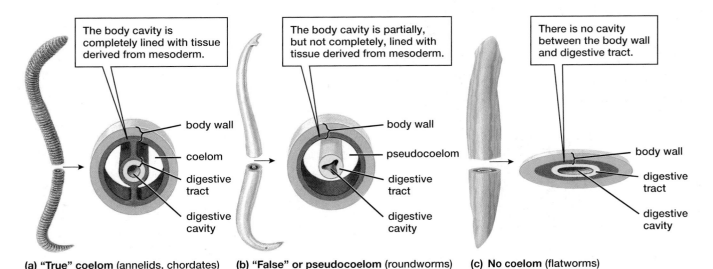

The body cavity is completely lined with tissue derived from mesoderm.

body wall
coelom
digestive tract
digestive cavity

(a) "True" coelom (annelids, chordates)

The body cavity is partially, but not completely, lined with tissue derived from mesoderm.

body wall
pseudocoelom
digestive tract
digestive cavity

(b) "False" or pseudocoelom (roundworms)

There is no cavity between the body wall and digestive tract.

body wall
digestive tract
digestive cavity

(c) No coelom (flatworms)

▲ **FIGURE 24-3 Body cavities (a)** Annelids have a true coelom. **(b)** Roundworms are pseudocoelomates. **(c)** Flatworms have no cavity between the body wall and digestive tract. (Tissues shown in blue are derived from ectoderm, those in red from mesoderm, and those in yellow from endoderm.)

Some phyla of bilateral animals have no body cavity and are known as *acoelomates*. For example, flatworms have no cavity between their gut and body wall; instead, the space is filled with solid tissue (**FIG. 24-3c**).

Bilateral Organisms Develop in One of Two Ways

Among the bilateral animal phyla, embryological development follows a variety of pathways. These diverse developmental pathways, however, can be grouped into two categories, known as **protostome** development and **deuterostome** development (**FIG. 24-4**). In protostome development, the coelom (when present) forms within the space between the body wall and the digestive cavity. In deuterostome development, the coelom forms as an outgrowth of the digestive cavity. The two types of development also differ in the pattern of cell division immediately after fertilization and in the method by which the mouth and anus are formed. Protostomes and deuterostomes represent distinct evolutionary branches within the bilateral animals. Annelids, arthropods, flatworms, roundworms, and mollusks exhibit protostome development; echinoderms and chordates are deuterostomes.

Protostomes Include Two Distinct Evolutionary Lines

The protostome animal phyla fall into two groups, which correspond to two different lineages that diverged early in the evolutionary history of protostomes. One group, the *ecdysozoans,* includes phyla such as the arthropods and roundworms, whose members have bodies covered by an outer layer that is periodically shed. The other group is known as the *lophotrochozoans*

(a) Lophophore

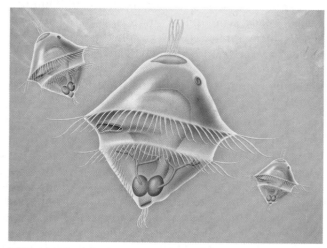

(b) Trochophore larva

▲ **FIGURE 24-5 Lophotrochozoan characteristics** Members of the lophotrochozoan phyla, which include flatworms, segmented worms, and mollusks, exhibit either **(a)** a feeding structure known as a lophophore or **(b)** a distinctive swimming larval form called a trochophore.

and includes phyla whose members have a special feeding structure called a lophophore (**FIG. 24-5a**), as well as phyla whose members pass through a particular type of developmental stage called a trochophore larva (**FIG. 24-5b**). The flatworms, annelids, and mollusks are lophotrochozoan phyla.

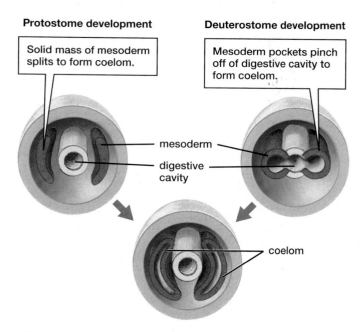

Protostome development

Solid mass of mesoderm splits to form coelom.

Deuterostome development

Mesoderm pockets pinch off of digestive cavity to form coelom.

mesoderm

digestive cavity

coelom

▲ **FIGURE 24-4 Formation of the coelom in protostomes and deuterostomes** The difference depicted here is one of several between protostome and deuterostome embryonic development. (Tissues shown in blue are derived from ectoderm, those in red from mesoderm, and those in yellow from endoderm.)

CHECK YOUR LEARNING

Can you ...

- describe how the body organization of sponges differs from that of all other animals?
- describe the different types of body symmetry, embryonic tissue layer arrangements, body cavities, and embryonic development present among tissue-bearing animals?
- give examples of animal groups with each type of body symmetry, tissue layer arrangement, body cavity, and embryonic development?
- name and describe the two main lineages within the protostome animals?

24.3 WHAT ARE THE MAJOR ANIMAL PHYLA?

For convenience, biologists often place animals in one of two major categories: **vertebrates,** those with a backbone (or vertebral column), and **invertebrates,** those lacking a backbone. The vertebrates—fish, amphibians, reptiles, and mammals (see Chapter 25)—are perhaps the most conspicuous animals from a human point of view, but less than 3% of all known animal species are vertebrates. The vast majority of animals are invertebrates. Biologists recognize about 32 phyla of animals; some key invertebrate phyla are summarized in TABLE 24-1.

Sponges Are Simple, Sessile Animals

Sponges (Porifera) are found in most aquatic environments. Most of Earth's 8,500 known sponge species inhabit ocean waters, but some live in freshwater habitats such as lakes and rivers. Adult sponges live attached to rocks or other underwater surfaces. They are generally sessile, though researchers have demonstrated that some species are able to move about (very slowly—a few millimeters per day). Sponges come in a variety of shapes and sizes. Some species have a well-defined shape, but others grow free-form over underwater rocks (FIG. 24-6). The largest sponges can grow to more than 3 feet (1 meter) in height.

Most sponges are **hermaphroditic**—they possess both male and female sexual organs. Sponges usually reproduce sexually, a process that begins when sperm are released into the water. Sperm may enter the body of another sponge and

be transported to eggs, which are retained in the sponge's body. Fertilized eggs develop inside the adult into active larvae that escape through the openings in the sponge body. Water currents disperse the larvae to new areas, where they settle and develop into adult sponges.

Sponges Lack Tissues

Sponge bodies do not contain tissues; in some ways, a sponge resembles a colony of single-celled organisms. The colony-like properties of sponges were revealed in an experiment performed by embryologist H. V. Wilson in 1907. Wilson mashed a sponge through a piece of silk, thereby breaking it apart into single cells and cell clusters. He then placed these tiny bits of sponge into seawater and waited for 3 weeks. By the end of the experiment, the cells had reaggregated into a functional sponge, demonstrating that individual sponge cells had been able to survive and move about independently.

A sponge's body is perforated by numerous tiny pores, through which water enters, and by fewer, large openings, through which water is expelled (FIG. 24-7). Within the sponge, water travels through canals. As water passes through the sponge, oxygen is extracted, microorganisms are filtered out and taken into individual cells where they are digested, and wastes are released.

Sponge Cells Are Specialized for Different Functions

Sponges have three major cell types, each with a specialized role (see Fig. 24-7). Flattened *epithelial cells* cover the animal's

(a) Encrusting sponge

(b) Tubular sponge

(c) Vase-shaped sponge

▲ **FIGURE 24-6 The diversity of sponges** Sponges come in a wide variety of sizes, shapes, and colors. Some, such as **(a)** this encrusting sponge, grow in a free-form pattern over undersea rocks. Others may be **(b)** tubular or **(c)** vase shaped.

THINK CRITICALLY Sponges are often described as the most "primitive" of animals. How can such a primitive organism have become so diverse and abundant?

TABLE 24-1 Comparison of the Major Animal Phyla

Common Name (Phylum)		Sponges (Porifera)	Sea Jellies, Corals, Anemones (Cnidaria)	Flatworms (Platyhelminthes)
Body plan	Level of organization	Cellular—lack tissues and organs	Tissue—lack organs	Organ system
	Germ layers	Absent	Two	Three
	Symmetry	Absent	Radial	Bilateral
	Cephalization	Absent	Absent	Present
	Body cavity	Absent	Absent	Absent
	Segmentation	Absent	Absent	Absent
Internal systems	Digestive system	Intracellular	Gastrovascular cavity; some intracellular	Gastrovascular cavity
	Circulatory system	Absent	Absent	Absent
	Respiratory system	Absent	Absent	Absent
	Excretory system (fluid regulation)	Absent	Absent	Canals with ciliated cells
	Nervous system	Absent	Nerve net	Head ganglia with longitudinal nerve cords
	Reproduction	Sexual; asexual (budding)	Sexual; asexual (budding)	Sexual (some hermaphroditic); asexual (body splits)
	Support	Endoskeleton of spicules	Hydrostatic skeleton	Hydrostatic skeleton
	Number of known species	8,500	10,000	10,000

outer body surfaces. Some epithelial cells are modified into *pore cells,* which surround the pores in the body wall, controlling their size and thereby regulating the entry of water. *Collar cells* bear flagella that extend into the inner cavity. The beating flagella maintain a flow of water through the sponge. The collars that surround the flagella act as fine sieves, filtering out microorganisms that are then ingested by the cell. Some of the food is passed to the *amoeboid cells.* These cells roam freely between the epithelial and collar cells, digesting and distributing nutrients, producing reproductive cells, and secreting small skeletal projections called *spicules.* Spicules, which may be composed of calcium carbonate (chalk), silica (glass), or protein, form an internal skeleton that provides support for the sponge's body (see Fig. 24-7).

Cnidarians Are Well-Armed Predators

The *cnidarians* (Cnidaria) include sea jellies (also known as jellyfish), corals, sea anemones, and hydrozoans (**FIG. 24-8**). The roughly 10,000 known species of cnidarians are confined to watery habitats; most are marine. Most species are small, from a few millimeters to a few inches in diameter, but the largest sea jellies can be 8 feet across and have tentacles 150 feet long. All cnidarians are predators.

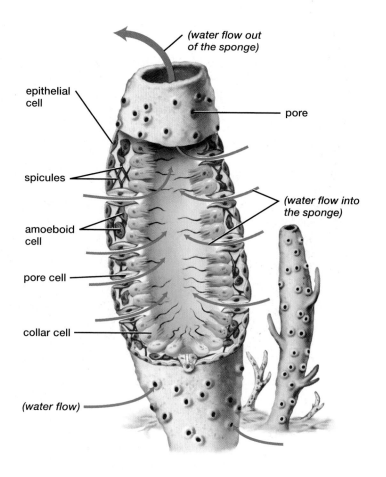

(water flow out of the sponge)

epithelial cell

pore

spicules

(water flow into the sponge)

amoeboid cell

pore cell

collar cell

(water flow)

▶ **FIGURE 24-7 The body plan of sponges** Water enters through numerous tiny pores in the sponge body and exits through a larger opening. Microscopic food particles are filtered from the water.

Segmented Worms (Annelida)	Clams, Snails, Octopuses, Squid (Mollusca)	Insects, Arachnids, Crustaceans (Arthropoda)	Roundworms (Nematoda)	Sea Stars, Sea Urchins, Sea Cucumbers (Echinodermata)
Organ system	Organ system	Organ system	Organ system	Organ system
Three	Three	Three	Three	Three
Bilateral	Bilateral	Bilateral	Bilateral	Bilateral larvae, radial adults
Present	Present	Present	Present	Absent
Coelom	Coelom	Coelom	Pseudocoel	Coelom
Present	Absent	Present	Absent	Absent
Separate mouth and anus	Separate mouth and anus	Separate mouth and anus	Separate mouth and anus	Separate mouth and anus (usually)
Closed	Open	Open	Absent	Absent
Absent	Gills, lungs	Tracheae, gills, or lungs	Absent	Tube feet, skin gills
Nephridia	Nephridia	Excretory glands resembling nephridia	Excretory gland	Absent
Head ganglia with paired ventral cords; ganglia in each segment	Well-developed brain in some cephalopods; several paired ganglia, most in the head; nerve network in the body wall	Head ganglia with paired ventral nerve cords; ganglia in segments, some fused	Head ganglia with dorsal and ventral nerve cords	Head ganglia absent; nerve ring and radial nerves; nerve network in the skin
Sexual (some hermaphroditic)	Sexual (some hermaphroditic)	Usually sexual	Sexual (some hermaphroditic)	Sexual (some hermaphroditic); asexual by regeneration (rare)
Hydrostatic skeleton	Hydrostatic skeleton	Exoskeleton	Hydrostatic skeleton	Endoskeleton of plates beneath outer skin
13,000	50,000	1,000,000	12,000	6,800

(a) Anemone

(b) Sea jelly

◀ FIGURE 24-8 Cnidarian diversity
(a) A red-spotted anemone spreads its tentacles to capture prey. **(b)** A sea jelly drifts in the ocean, its tentacles hanging down. **(c)** A close-up of coral reveals the extended tentacles of the polyps. **(d)** The sea wasp is a type of sea jelly whose stinging cells contain one of the most toxic of all known venoms. The venom quickly kills prey, such as the shrimp shown here, that brush against a sea wasp's tentacles.

(c) Coral

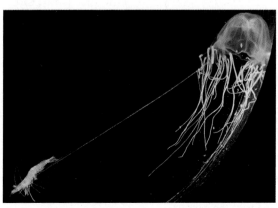

(d) Sea wasp

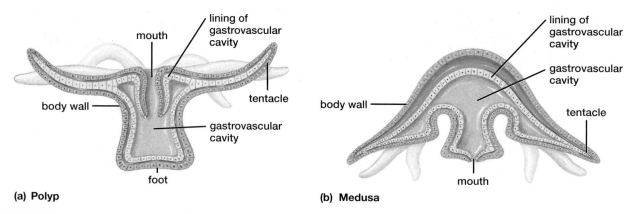

▲ **FIGURE 24-9** **Polyp and medusa (a)** The polyp form is seen in sea anemones and the individual polyps within a coral. **(b)** The medusa form, seen in the sea jelly, resembles an inverted polyp. (Tissues shown in blue are derived from ectoderm, those in yellow from endoderm.)

Cnidarians Have Tissues and Two Body Types

The cells of cnidarians are organized into distinct tissues, including contractile tissue that acts like muscle. Cnidarian nerve cells are organized into tissue called a *nerve net*, which branches through the body and stimulates the contractile tissue, allowing the body to move and capture prey. However, most cnidarians lack organs and have no brain.

Cnidarians come in a variety of forms, all of which are variations on two basic body plans: the *polyp* (**FIG. 24-9a**) and the *medusa* (**FIG. 24-9b**). The tubular polyp is adapted to a life spent quietly attached to rocks; it has tentacles that reach upward to grasp and immobilize prey. The medusa floats in the water and is carried by currents, its bell-shaped body trailing tentacles like multiple fishing lines.

Many cnidarian life cycles include both polyp and medusa stages, though some species live only as polyps and others only as medusae. Both polyps and medusae develop from just two germ layers—the interior endoderm and the exterior ectoderm; between those layers is a jelly-like substance. Polyps and medusae are radially symmetrical, with body parts arranged in a circle around the mouth and digestive cavity (see Fig. 24-2a).

Reproduction varies considerably among different types of cnidarians, but one pattern is fairly common in species with both polyp and medusa stages. In such species, polyps typically reproduce by asexual **budding** in which the polyp produces miniature versions of itself that drop off and assume an independent existence. Under certain conditions, however, budding may instead give rise to medusae. After a medusa grows to maturity, it may release gametes (sperm or eggs) into the water. If a sperm and egg meet, they may fuse to form a zygote that develops into a free-swimming, ciliated larva. The larva may eventually settle on a hard surface, where it develops into a polyp.

Cnidarians Have Stinging Cells

Cnidarian tentacles are armed with *cnidocytes,* cells containing structures that, when touched, explosively inject poisonous or sticky filaments into prey (**FIG. 24-10**). These stinging cells, found only in cnidarians, are used to capture prey. Cnidarians do not actively hunt. Instead, they wait for their victims to blunder, by chance, into the grasp of their enveloping tentacles. Stung and firmly grasped, the prey is forced through an expandable mouth into a digestive sac, the *gastrovascular cavity* (see Fig. 24-9). Digestive enzymes secreted into this cavity break down some of the food, and further digestion occurs within the cells lining the cavity. Because the gastrovascular cavity has only a single opening, undigested material is expelled through the mouth when digestion is completed. Although this two-way traffic prevents continuous feeding, it is adequate to support the low energy demands of these animals.

The venom of some cnidarians can cause painful stings in humans; the stings of a few sea jelly species can even be life threatening. The most deadly of these species is the sea wasp, *Chironex fleckeri* (see Fig. 24-8d), which is found in the waters off northern Australia and Southeast Asia and can grow to about 12 inches (30 centimeters) in diameter. The amount of venom in a single sea wasp could kill up to 60 people, and the victim of a serious sting may die within minutes.

Many Corals Secrete Hard Skeletons

One group of cnidarians, the corals, is of particular ecological importance (see Fig. 24-8c). In many coral species, polyps form colonies, and each member of the colony secretes a hard skeleton of calcium carbonate. The skeletons persist long after the organisms die, serving as a base to which other individuals attach themselves. The cycle continues until, after thousands of years, massive coral reefs are formed. Many of these reefs are threatened by climate change, as we describe in "Earth Watch: When Reefs Get Too Warm" on page 432.

Coral reefs are found in both cold and warm oceans. Cold-water reefs form in deep waters and, though widely distributed, have only recently attracted the attention of researchers and are not yet well studied. The more familiar

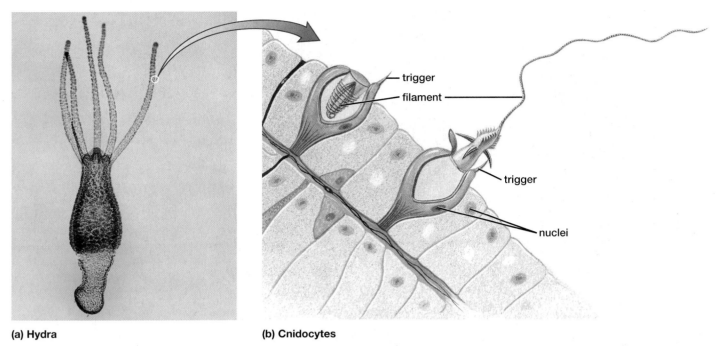

(a) Hydra

(b) Cnidocytes

▲ **FIGURE 24-10 Cnidarian weaponry: the cnidocyte (a)** Cnidarian tentacles, such as those of the hydra, contain cnidocyte cells. **(b)** At the slightest touch to its trigger, a structure within a cnidocyte cell violently expels a poisoned filament, which may penetrate prey.

warm-water coral reefs are restricted to clear, shallow waters in the tropics. Here, coral reefs form undersea habitats that are the basis of an ecosystem of stunning diversity and unparalleled beauty.

Comb Jellies Use Cilia to Move

The roughly 150 known species of radially symmetrical *comb jellies* (Ctenophora) are superficially similar in appearance to some cnidarians, but form a distinct evolutionary lineage. Most comb jellies are less than an inch (2.5 cm) in diameter, but a few species can grow to more than 3 feet (1 m) across. Comb jellies move by means of cilia, which are arranged in eight rows known as combs. Although most comb jellies are colorless and transparent or translucent, light scattered by the beating cilia of the combs can appear as eight ever-changing rainbows of color on the comb jelly's body (**FIG. 24-11**).

All comb jellies are carnivores. Most species inhabit coastal or oceanic waters and eat tiny invertebrate animals (including, in some cases, other, smaller comb jellies), which are captured with sticky tentacles. Almost all comb jellies are hermaphroditic. Each individual releases both sperm and eggs to the surrounding water. Fertilized eggs float freely in the water until they hatch into larvae that gradually develop into adult comb jellies.

▲ **FIGURE 24-11 A comb jelly** This comb jelly's body is unpigmented, but rows of cilia refract light to produce iridescent colors.

Flatworms May Be Parasitic or Free Living

The *flatworms* (Platyhelminthes) are aptly named; they have a flat, ribbon-like shape. They are bilaterally symmetrical (see Fig. 24-2b). Many of the approximately 10,000 described flatworm species are parasites (**FIG. 24-12a**). (**Parasites** are organisms that live in or on the body of another organism, called a *host*, which is harmed as a result of the relationship.) Nonparasitic, free-living flatworms inhabit freshwater, marine, and moist terrestrial habitats. They tend to be small and inconspicuous (**FIG. 24-12b**), but some are brightly colored, spectacularly patterned residents of tropical coral reefs (**FIG. 24-12c**).

All flatworm species can reproduce sexually; most are hermaphroditic. This trait allows a flatworm to reproduce through self-fertilization, a great advantage to a parasitic worm that may be the only one of its kind present in its host. Some flatworms can also reproduce asexually. For example, free-living species may reproduce by cinching themselves around the middle until they separate into two halves, each of which regenerates its missing parts.

Flatworms Have Organs but Lack Respiratory and Circulatory Systems

Unlike cnidarians, flatworms have organs. For example, most free-living flatworms have sense organs, including eyespots (see Fig. 24-12b) that detect light and dark, and cells that respond to

(a) Fluke

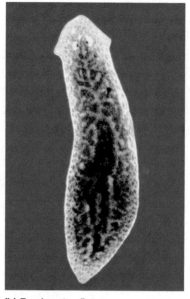
(b) Freshwater flatworm

(c) Marine flatworm

▲ **FIGURE 24-12 Flatworm diversity (a)** This fluke is an example of a parasitic flatworm. **(b)** Eyespots are clearly visible in the head of this free-living, freshwater flatworm. **(c)** Many of the flatworms that inhabit tropical coral reefs are brightly colored.

chemical and tactile stimuli. To process information, a flatworm has clusters of nerve cells called **ganglia** (singular, ganglion) in its head, forming a simple brain. Paired neural structures called **nerve cords** extend down the body and conduct nervous signals to and from the ganglia.

Flatworms lack respiratory and circulatory systems. Gases are exchanged by direct diffusion between body cells and the environment. This mode of respiration is possible because the small size and flat shape of flatworm bodies ensure that no body cell is very far from the surrounding environment. In the absence of a circulatory system, nutrients must move directly from the digestive tract to body cells. The digestive cavity has a branching structure that reaches all parts of the body and allows digested nutrients to diffuse into nearby cells. As in cnidarians and comb jellies, the digestive cavity has only one opening to the environment, so wastes pass out through the mouth.

Some Flatworms Are Harmful to Humans

Some parasitic flatworms can infect humans. For example, tapeworms can infect people who eat undercooked beef, pork, or fish that has been infected by the worms. Tapeworm larvae form encapsulated resting structures, called *cysts*, in the muscles of the cows, pigs, or fish. The cysts hatch in the human digestive tract, where the young tapeworms attach themselves to the lining of the intestine. There they may grow to a length of more than 20 feet (7 meters), absorbing digested nutrients directly through their outer surface and eventually releasing packets of eggs that are shed in the host's feces. If a pig, cow, or fish eats food contaminated with infected human feces, the eggs hatch in the animal's digestive tract, releasing larvae that

burrow into its muscles and form cysts, thereby continuing the infective cycle (**FIG. 24-13**).

Another group of parasitic flatworms includes the *flukes* (see Fig. 24-12a). Of these, the most devastating are liver flukes (common in Asia) and blood flukes, such as those of the genus *Schistosoma,* which cause the disease schistosomiasis. Like most parasites, flukes have a complex life cycle that includes an intermediate host (a snail, in the case of *Schistosoma*). Prevalent in Africa and parts of South America, schistosomiasis affects an estimated 200 million people worldwide. Its symptoms include diarrhea, anemia, and sometimes brain damage.

Annelids Are Segmented Worms

The bodies of *annelids* (Annelida) are divided into a series of similar repeating segments. Externally, this **segmentation** appears as a series of ring-like depressions on the surface. Internally, most of the segments contain identical copies of nerves, excretory structures, and muscles.

Sexual reproduction is common among annelids. Some species are hermaphroditic; others have separate sexes. Fertilization may be external or internal. External fertilization, in which sperm and eggs are released into the surrounding environment, is found mainly in species that live in water. In internal fertilization, two individuals copulate and sperm are transferred directly from one to the other. In hermaphroditic species, sperm transfer may be mutual, with each partner both donating and receiving sperm. In addition, some annelids can reproduce asexually, typically by a process in which the body breaks into two pieces, each of which regenerates the missing part.

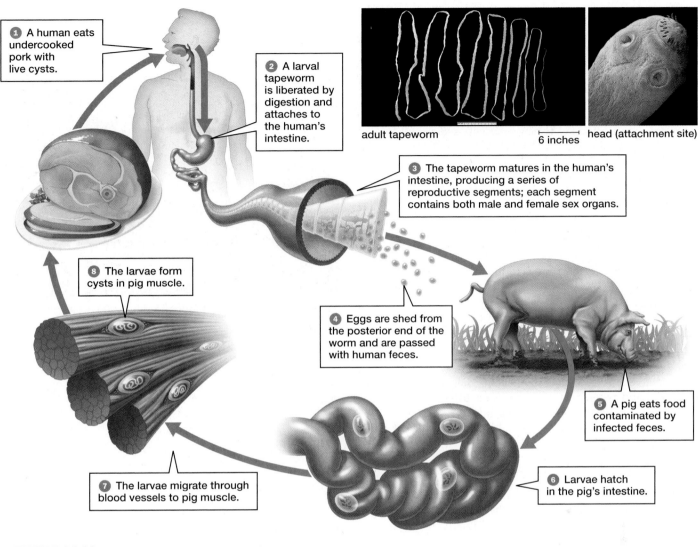

1 A human eats undercooked pork with live cysts.

2 A larval tapeworm is liberated by digestion and attaches to the human's intestine.

3 The tapeworm matures in the human's intestine, producing a series of reproductive segments; each segment contains both male and female sex organs.

4 Eggs are shed from the posterior end of the worm and are passed with human feces.

5 A pig eats food contaminated by infected feces.

6 Larvae hatch in the pig's intestine.

7 The larvae migrate through blood vessels to pig muscle.

8 The larvae form cysts in pig muscle.

adult tapeworm | 6 inches | head (attachment site)

▲ **FIGURE 24-13** **The life cycle of the human pork tapeworm**

THINK CRITICALLY What advantages favored the evolution of tapeworms' long, flat shape?

Annelids Are Coelomates and Have Organ Systems

Annelids have a fluid-filled true coelom between the body wall and the digestive tract (see Fig. 24-3a). The incompressible fluid in the coelom of many annelids is confined by the partitions between the segments and serves as a **hydrostatic skeleton,** a supportive framework against which muscles can act. A hydrostatic skeleton allows earthworms to burrow through soil.

Annelids have well-developed organ systems (groups of organs that act in a coordinated manner). For example, annelids have a **closed circulatory system** that distributes gases and nutrients throughout the body. In closed circulatory systems (including yours), blood remains confined to the heart and blood vessels. In the earthworm, for example, blood with oxygen-carrying hemoglobin is pumped through well-developed vessels by five pairs of "hearts" (**FIG. 24-14**).

These hearts are actually short segments of specialized blood vessels that contract rhythmically. The blood is filtered and wastes are removed by excretory organs called *nephridia* (singular, nephridium) and excreted to the environment through small pores. Nephridia resemble the individual tubules of the vertebrate kidney. The annelid nervous system consists of a simple brain in the head and a series of paired segmental ganglia joined by a pair of ventral nerve cords that pass along the length of the body. The annelid digestive system includes a tubular gut that runs from the mouth to the anus. This kind of digestive tract, with two openings and a one-way digestive path, is much more efficient than the single-opening digestive systems of cnidarians and flatworms. Digestion in annelids occurs in a series of compartments in the digestive tract, each specialized for a different phase of food processing.

Earth WATCH

When Reefs Get Too Warm

The coral reefs that form in shallow tropical oceans are among Earth's most diverse, beautiful, and economically important ecosystems. Unfortunately, coral reef ecosystems are declining worldwide, threatened by pollution, destructive fishing practices, sedimentation, and over-exploitation. One the most widespread threats to coral reefs is Earth's warming climate, which causes *bleaching*.

The cnidarians that build coral reefs do so with the help of photosynthetic protists called dinoflagellates. Dinoflagellates live inside the cells of the coral animals, providing the animals with nutritious products of photosynthesis in return for shelter and protection. This symbiotic relationship benefits both the dinoflagellates and the coral animals. However, when the water temperature gets too warm, the coral animals may eject their dinoflagellate partners. Water temperature as little as 1°C higher than the normal annual highs can result in dinoflagellate ejection. Because the dinoflagellates are responsible for the bright colors of many corals, reefs from which dinoflagellates have been expelled have a white, bleached appearance (**FIG. E24-1**).

Scientists do not yet understand exactly why warm temperature cause coral bleaching; one hypothesis is that warm temperatures stress the dinoflagellates, causing them to release molecules that are harmful to corals, which eject the dinoflagellates in self-defense. In any case, bleaching is bad for corals; if they remain without dinoflagellates for too long, they die. Even if the water temperature decreases in time for bleached corals to again take up dinoflagellates, the corals may be weakened and susceptible to diseases that increase mortality.

During warm years, coral bleaching can be widespread, affecting huge areas spread across Earth. As the planet

▲ **FIGURE E24-1** **Bleached coral**

warms due to human-caused climate change, mass coral bleaching events are expected to become more frequent (**FIG. E24-2**). The widespread coral deaths that will likely result are bad news for the thousands of reef-dwelling species (and millions of people) that directly or indirectly depend on healthy corals for their survival.

> **THINK CRITICALLY** Another threat to coral reefs is sediment that washes into the sea from farm field runoff and other sources of soil erosion. Why might such sediment be harmful to corals? In your answer, consider what you have learned about the symbiotic relationship between dinoflagellates and corals.

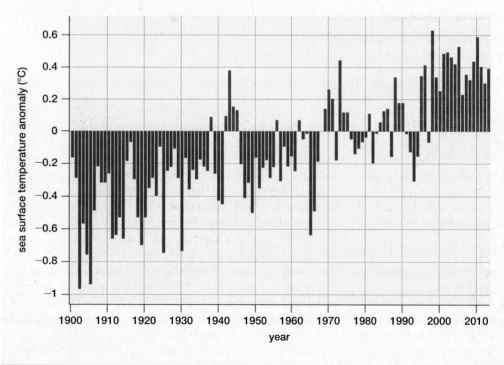

◀ **FIGURE E24-2** **Sea surface temperatures at the Great Barrier Reef, 1910–2013** Temperatures warm enough to cause coral bleaching have become increasingly frequent, as illustrated by these measurements from the waters around Earth's largest reef system, the Great Barrier Reef of Australia. The vertical axis represents "temperature anomaly," which measures the amount by which the average temperature in a given year differs from the long-term average.

▶ FIGURE 24-14 **An annelid, the earthworm** This diagram shows an enlargement of segments, many of which are repeating similar units separated by partitions.

THINK CRITICALLY What advantage does a digestive system with two openings have relative to digestive systems with only a single opening (like that of the flatworms)?

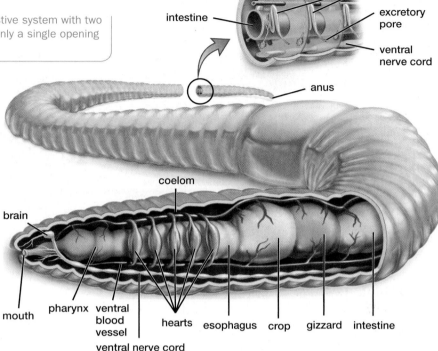

Annelids Include Oligochaetes, Polychaetes, and Leeches

The 13,000 known species of annelids fall into three main subgroups: the *oligochaetes,* the *polychaetes,* and the *leeches.* The oligochaetes include the familiar earthworm and its relatives. Charles Darwin devoted substantial time to the study of earthworms and was especially impressed by their role in improving soil fertility. More than a million earthworms may live in an acre of land, tunneling through soil, consuming and excreting soil particles and organic matter. These actions help ensure that air and water can move easily through the soil and that organic matter is continually mixed into it, creating conditions that are favorable for plant growth. The impact of earthworms, however, can also be negative. In some areas of North America, non-native, invasive earthworms disrupt the normal structure of forest soils, harming native forest plants.

Polychaetes live primarily in the ocean. In some polychaetes, most body segments have paired fleshy paddles that are used in locomotion. Other polychaetes live in tubes from which they project feathery gills that both exchange gases and filter the water for microscopic food (**FIGS. 24-15a, b**).

Leeches (**FIG. 24-15c**) live in freshwater or moist terrestrial habitats. Many leeches prey on smaller invertebrates; some suck the blood of larger animals.

(a) Polychaete gills

(b) Deep-sea polychaete

(c) Leech

▲ FIGURE 24-15 **Diverse annelids (a)** The brightly colored gills of a polychaete annelid. The rest of the worm's body is hidden inside a tube embedded in the coral that is visible in the background. **(b)** This polychaete lives near deep-sea vents where the water temperature may reach 175°F (80°C). **(c)** This leech, a freshwater annelid, shows numerous segments. The sucker encircles its mouth, allowing it to attach to its prey.

THINK CRITICALLY Why does pouring salt on a leech harm it?

Physicians' Assistants

The medicinal leech has been formally approved as a "medical device" by the U.S. Food and Drug Administration (FDA), mainly on the basis of its effectiveness as an aid to post-surgical healing. However, leeches may have additional medical uses. Especially promising is their potential for treatment of pain. Leech saliva contains pain-killing chemicals that help leeches evade detection when they puncture a victim's skin, so physicians hope that leeches can help reduce patients' pain. Several clinical trials have found that leeches provide substantial relief to people with osteoarthritis, a painful disorder of the joints. Anecdotal evidence suggests that leeches may help with other kinds of pain as well.

In addition to leeches, other invertebrate animals have medical potential, including insects and roundworms. You'll learn more about the biology of these animals later in this section.

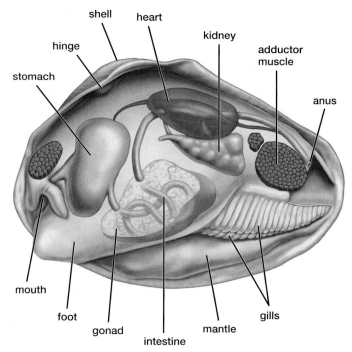

▲ **FIGURE 24-16 A bivalve mollusk** The body plan of a clam, showing the mantle, foot, gills, shell, and other features that are seen in most (but not all) mollusk species. The clam uses its adductor muscles to close its shell.

Most Mollusks Have Shells

If you have ever enjoyed a bowl of clam chowder, a plate of oysters on the half shell, or some sautéed scallops, you are indebted to *mollusks* (Mollusca). Mollusks are very diverse; in terms of number of known species, the 50,000 mollusks are second (albeit a distant second) only to the arthropods. These diverse species exhibit a range of lifestyles, from sessile forms such as mussels that spend their adult lives in one spot, filtering microorganisms from the water, to active, voracious predators of the ocean depths, such as the giant squid. Mollusks, with the exception of some snails and slugs, are aquatic. Although most mollusks are protected by hard shells of calcium carbonate, some lack shells.

The circulatory systems of most mollusks include a feature not seen in annelids: the **hemocoel,** or blood cavity. Blood empties into the hemocoel, where it bathes the internal organs directly. This arrangement is known as an **open circulatory system.** Mollusks also have a *mantle,* an extension of the body wall that forms a chamber for the gills and, in shelled species, secretes the shell (**FIG. 24-16**). The mollusk nervous system, like that of annelids, consists of ganglia connected by nerves, but in many mollusks more of the ganglia are concentrated in the brain. Reproduction is sexual; some species have separate sexes, and others are hermaphroditic.

Among the many subgroups of mollusks, we will discuss three in more detail: gastropods, bivalves, and cephalopods.

Gastropods Are One-Footed Crawlers

Snails and slugs—collectively known as *gastropods*—crawl on a muscular *foot,* and many are protected by shells that vary widely in form and color (**FIG. 24-17a**). Not all gastropods are shelled, however. Sea slugs, for example, lack shells, but their brilliant colors warn potential predators that they are poisonous or foul tasting (**FIG. 24-17b**).

(a) Snail

(b) Sea slug

▲ **FIGURE 24-17 The diversity of gastropod mollusks (a)** A Florida tree snail displays a brightly striped shell and eyes at the tip of stalks that retract instantly if touched. **(b)** The brilliant colors of many sea slugs warn potential predators that they are distasteful.

(a) Scallop

(b) Mussels

◀ **FIGURE 24-18 The diversity of bivalve mollusks (a)** This scallop parts its hinged shells to allow the intake of water from which food will be filtered. The blue spots visible along the mantle just inside the upper and lower shells are simple eyes. **(b)** Mussels attach to rocks in dense aggregations exposed at low tide. White barnacles (which are arthropods) are attached to the mussel shells and surrounding rock.

Gastropods feed with a *radula,* a flexible ribbon of tissue studded with spines that is used to scrape algae from rocks or to grasp larger plants or prey. For respiration, most gastropods use gills. In shelled species, gills are enclosed in a cavity beneath the shell. Gases can also diffuse readily through the skin of most gastropods. The few gastropod species that live in terrestrial habitats (including garden snails and slugs) use a simple lung for breathing.

Bivalves Are Filter Feeders

Included among the *bivalves* are scallops, oysters, mussels, and clams (**FIG. 24-18**). Bivalves possess two shells connected by a flexible hinge. A strong muscle clamps the shells closed in response to danger; this muscle is what you are served when you order scallops in a restaurant.

Clams use a muscular foot for burrowing in sand or mud. In mussels, which live attached to rocks, the foot is smaller and helps secrete threads that anchor the animal to the rocks. Scallops lack a foot and move by a sort of jet propulsion achieved by flapping their shells together.

Bivalves are filter-feeders, using their gills as both respiratory and feeding structures. Water circulates over the gills, which are covered with a thin layer of mucus that traps microscopic food particles. Beating cilia on the gills sweep food to the mouth.

Cephalopods Are Marine Predators

The *cephalopods* include octopuses, nautiluses, cuttlefish, and squids (**FIG. 24-19**). The largest invertebrates, the giant squid and the colossal squid, belong to this group (see "How Do We Know That? The Search for a Sea Monster" on page 436). All cephalopods are carnivores, and all are marine. In these mollusks, the foot has evolved into tentacles with well-developed sensory abilities for detecting prey. Prey are grasped by suction disks on the tentacles and may be immobilized by a paralyzing venom in the saliva before being torn apart by beaklike jaws.

Cephalopods move rapidly by jet propulsion, which is generated by the forceful expulsion of water from the mantle

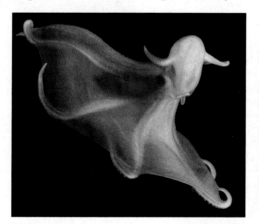

(a) Octopus

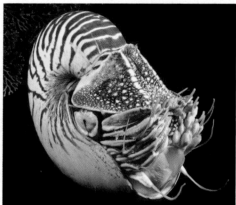

(b) Nautilus

▶ **FIGURE 24-19 The diversity of cephalopod mollusks (a)** In emergencies, an octopus can jet backward by vigorously contracting its mantle. Octopuses and squid can emit clouds of dark purple ink to confuse pursuing predators. **(b)** The chambered nautilus secretes a shell with internal, gas-filled chambers that provide buoyancy. Note the well-developed eyes and the tentacles used to capture prey. **(c)** A squid can move by contracting its mantle to generate jet propulsion, which pushes the animal backward through the water.

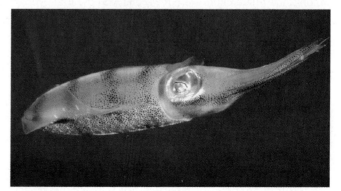

(c) Squid

HOW DO WE KNOW THAT?

The Search for a Sea Monster

The giant squid, one of the world's largest invertebrate animals, is also one of the most mysterious. The squid reaches lengths of 40 feet (13 meters) or more, with huge eyes as large as a human head. The squid's ten tentacles are covered with powerful suckers. The suckers contain sharp, clawlike hooks used to grasp prey, which is then pulled toward the squid's mouth, where a heavily muscled beak tears the food. Clearly, the giant squid is one of the most imposing organisms on Earth. However, we know almost nothing about its habits and lifestyle because almost all observations of giant squid are of dead animals washed ashore or hauled up in fishing nets.

Scientific interest in observing live giant squids in their natural, deep-water habitat resulted in a number of expeditions over the years, but none managed to observe or photograph a live giant squid until 2004. That year, a research team led by Tsunemi Kubodera placed a camera on a long, baited fishing line off the coast of Japan. Long hours of dragging the line through the water at a depth of 3,000 feet were eventually rewarded with a few photos of a giant squid that attacked the bait. Later, a squid attacking the bait was lured to the surface and photographed there.

In 2012, Kubodera undertook an expedition with two other scientists, Steve O'Shea and Edith Widder, to see if their combined efforts might yield better observations. Each researcher had a different hypothesis about how giant squids find prey, so each tried a different approach to attracting the animals. O'Shea attempted to attract giant squids with a mixture of chemicals extracted from dead ones. He believed that the chemicals would draw squids into the circle of bright light around his illuminated submarine. O'Shea's excursions attracted many squids, but none of them were giant squids.

Widder had a different idea. An expert on bioluminescence (the light produced naturally by many marine organisms), she hypothesized that giant squids eat small predators that in turn eat sea jellies, so giant squids would be attracted to the flashing bioluminescent alarm signal that certain deep-water sea jellies produce when under attack. Widder further hypothesized that giant squids avoid the bright lights and loud noises typical of submarines. So she devised a lure that emitted pulses of blue light similar to those of jellyfish and tethered this "e-jelly" beside a camera rig that illuminated its surroundings with red light, which is not detectable by most deep-sea animals. The whole contraption (**FIG. E24-3**) was tethered to a buoy and suspended 2,300 feet below the surface. After many hours of waiting, the researchers were rewarded with the first-ever video of a live giant squid, its tentacles spread as its mouth moved toward the camera.

Widder's e-jelly camera eventually produced video of several additional giant squid approaches. But, in the end, the most spectacular images were captured by Kubodera's method. Like O'Shea, he used a submarine and bait. But his bait was a smaller squid attached to the sub, illuminated only by dim red lights, and he allowed his sub to drift quietly. After many long excursions with no results, a giant squid finally approached the submarine. Kubodera eagerly observed the animal. After a few minutes, he took a big gamble by turning on some bright spotlights. To his great relief, the giant squid did not immediately flee, and the elated researcher captured

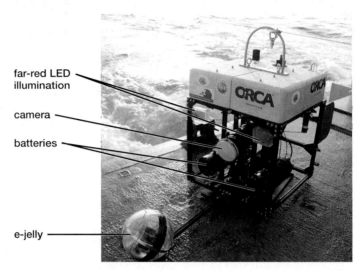

far-red LED illumination

camera

batteries

e-jelly

▲ **FIGURE E24-3** **Search tools** This "e-jelly" lure and camera rig were used to capture the first-ever live video of a giant squid.

▲ **FIGURE E24-4** **Giant squid** An image from the high-definition video of a giant squid encountered by researchers in deep Pacific waters. This squid is about ten feet long, though it would have been more than twice that long had it not lost its two longest tentacles.

high-definition color video of one of Earth's strangest and most magnificent creatures (**FIG. E24-4**).

THINK CRITICALLY The giant squids that were attracted to the e-jelly pictured in Figure E24-3 seemed to be attacking the camera rig beside the lure, rather than the lure itself. How does this observation support Edith Widder's hypothesis about giant squid foraging?

cavity. Octopuses may also travel along the seafloor by using their tentacles like multiple undulating legs. The rapid movements and active lifestyles of cephalopods are made possible in part by their closed circulatory systems. Cephalopods are the only mollusks with a closed circulatory system, which transports oxygen and nutrients more efficiently than does an open circulatory system.

Cephalopods have highly developed brains and sensory systems. The cephalopod eye rivals our own in complexity. The cephalopod brain, especially that of the octopus, is exceptionally large and complex. It endows the octopus with highly developed capabilities to learn and remember. In the laboratory, octopuses can rapidly learn to solve a maze, associate symbols with food, or open a screw-cap jar to obtain food. In the wild, some octopuses use tools. For example, veined octopuses clean mud from buried coconut shells, stack them up for transport, and turn them into a protective shelter.

Arthropods Are the Most Diverse and Abundant Animals

In terms of both number of individuals and number of species, no other animal phylum comes close to the *arthropods* (Arthropoda), which include insects, arachnids, myriapods, and crustaceans. About 1 million arthropod species have been discovered, and scientists estimate that millions more remain undescribed.

Arthropods Have Appendages and an External Skeleton

All arthropods have paired, jointed appendages and an **exoskeleton,** an external skeleton that encloses the arthropod body like a suit of armor. Secreted by the *epidermis* (the outer layer of skin), the exoskeleton is composed chiefly of protein and a polysaccharide called *chitin* (see Fig. 3-11). The exoskeleton provides rigid attachment sites for muscles, but also becomes thin and flexible at joints, thereby allowing the appendages to move. This combination of rigidity and flexibility makes possible the flight of the bumblebee and the intricate, delicate manipulations of the spider as it weaves its

▲ **FIGURE 24-21 The exoskeleton must be molted periodically** A cicada emerges from its outgrown exoskeleton (left).

web (**FIG. 24-20**). The exoskeleton also provides protection against predators. In addition, it contributed enormously to the arthropod invasion of land by providing a watertight covering for delicate, moist tissues such as those used for gas exchange.

Despite its benefits, the arthropod exoskeleton also poses some problems. For example, because it cannot expand as the animal grows, the exoskeleton must periodically be shed, or **molted,** and replaced with a larger one (**FIG. 24-21**). Molting uses energy and leaves the animal temporarily vulnerable to predators until the new skeleton hardens.

Arthropods Have Specialized Segments and Adaptations for Active Lifestyles

Arthropods are segmented, but their segments tend to be few and specialized for different functions such as sensing the environment, feeding, and movement (**FIG. 24-22**). For example, in insects, sensory and feeding structures are concentrated on the front segment, known as the *head,* and

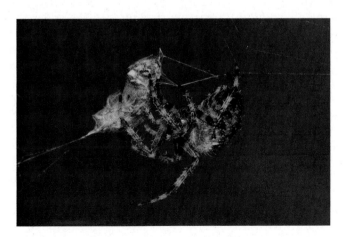

▲ **FIGURE 24-20 The exoskeleton allows precise movements** A garden orb spider begins to wrap a captured wasp in silk. Such dexterous manipulations are made possible by the exoskeleton and jointed appendages characteristic of arthropods.

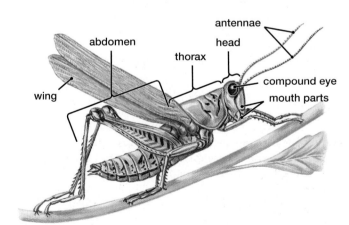

▲ **FIGURE 24-22 Segments are fused and specialized in insects** Insects, such as this grasshopper, show fusion and specialization of body segments into a distinct head, thorax, and abdomen. Segments are visible on the abdomen.

digestive structures are largely confined to the *abdomen,* which is the rear segment. Between the head and the abdomen is the *thorax,* the segment to which structures used in locomotion, such as wings and legs, are attached.

Efficient gas exchange is required to supply adequate oxygen to the muscles, which allows the rapid flying, swimming, or running displayed by many arthropods. In aquatic arthropods such as crustaceans, gas exchange is accomplished by gills. In terrestrial arthropods, gas exchange is performed either by lungs (in arachnids) or by *tracheae* (singular, trachea), a network of narrow, branching respiratory tubes that open to the surrounding environment and that penetrate all parts of the body. Most arthropods have open circulatory systems, like those of mollusks, in which blood directly bathes the organs in a hemocoel.

Most arthropods possess well-developed sensory and nervous systems. Arthropod sensory systems often include **compound eyes,** which have multiple light detectors (**FIG. 24-23**), and acute chemical and tactile senses. The arthropod nervous system consists of a brain composed of fused ganglia and a series of additional ganglia along the length of the body that are linked by a ventral nerve cord. This well-developed nervous system, combined with sophisticated sensory abilities, has permitted the evolution of complex behaviors in many arthropods.

Insects Are the Only Flying Invertebrates

The number of known *insect* species is about 850,000, roughly three times the total number of known species in all other groups of animals combined (**FIG. 24-24**). Insects have a single pair of antennae and three pairs of legs, usually supplemented by two pairs of wings. Insects' capacity for flight distinguishes them from all other invertebrates and has contributed to their enormous success. As anyone who has pursued a fly can testify, flight helps in escaping from predators. It also allows insects to find widely dispersed food. Swarms of locusts, for example, can travel 200 miles a day in search of food; researchers tracked one swarm on a journey that totaled almost 3,000 miles. Flight requires rapid and efficient gas exchange, which insects accomplish by means of tracheae.

All insects undergo **metamorphosis,** a change during development from a juvenile body form to an adult body form. In most insect species, metamorphosis is *complete.* In complete metamorphosis, the immature stage, called a **larva,** is worm shaped (for example, the maggot of a blowfly or the caterpillar of a moth or butterfly). The larva hatches from an egg, grows by eating voraciously and shedding its exoskeleton several times, and then forms a nonfeeding stage called a **pupa** (plural, pupae). Encased in an outer covering, the pupa body undergoes a drastic modification, emerging in its adult form. The adults mate and lay eggs, continuing the cycle. Metamorphosis may include a change in diet as well as in shape, thereby eliminating competition for food between adults and juveniles. Some insects, such as grasshoppers and crickets, undergo a more gradual metamorphosis (called

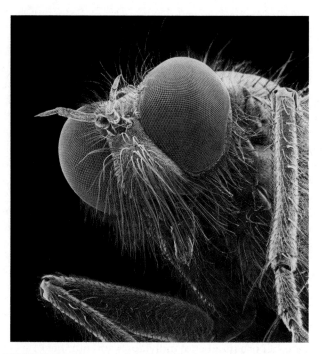

▲ **FIGURE 24-23 Arthropods possess compound eyes** This scanning electron micrograph shows the compound eye of a horse fly. Compound eyes consist of an array of similar light-gathering and sensing elements whose orientation gives the arthropod a wide field of view. Insects have reasonably good image-forming ability and good color discrimination.

incomplete metamorphosis), hatching as young that bear some resemblance to the adult, then gradually acquiring more adult features as they grow and molt.

Biologists classify the amazing diversity of insects into several dozen groups. We will describe three of the largest ones here.

Butterflies and Moths The butterflies and moths make up what is perhaps the most conspicuous and best-studied group of insects. The brightly colored, often iridescent wing patterns of many butterfly and moth species arise from pigments and light-refracting structures in the scales that cover the wings of all members of this group. (The scales are the powdery substance that rubs off onto your hand when you handle a butterfly or moth.) Butterflies fly mainly during the day and moths fly at night, though there are exceptions to this general rule, such as the hummingbird-like hawk moths that are often seen feeding by day in flower gardens.

The evolution of butterflies and moths has been closely tied to the evolution of flowering plants. Butterflies and moths feed almost exclusively on flowering plants, both as caterpillars and as adults. Many species of flowering plants depend in turn on butterflies and moths for pollination.

Bees, Ants, and Wasps The bees, ants, and wasps are known to many people by their painful stings. Many species in this group are equipped with a stinger that extends from

▶**FIGURE 24-24 The diversity of insects**
(a) Aphids suck sugar-rich juice from plants. **(b)** The bullet ant can inflict an extremely painful sting. **(c)** A June beetle displays its two pairs of wings as it comes in for a landing. The outer wings protect the abdomen and the inner wings, which are relatively thin and fragile.
(d) Caterpillars are larval forms of moths or butterflies. This caterpillar larva of a silk moth can produce clicking sounds with its mouthparts. The sounds may serve to warn predators that the caterpillar is distasteful.

THINK CRITICALLY How might insects' wings have contributed to their diversity and abundance?

(a) Aphid

(b) Ant

(c) Beetle flying

(d) Moth larva

the abdomen and can be used to inject venom into the victim. The venom may be extremely toxic, but fortunately for human stinging victims, each insect carries only a tiny amount. Nonetheless, the amount is often sufficient to cause considerable pain. Only females have stingers, which are used by many stinging species to help defend a nest from attack by a potential predator. Defense, however, is not the only use for stingers. Many wasps, for example, act as parasites of other insects when reproducing and may sting a host insect to paralyze it. The parasitic wasp then lays an egg inside the paralyzed body of the host, typically the caterpillar of a moth or butterfly, which becomes food for the wasp larva after it hatches.

The social behavior of some ant and bee species is extraordinarily intricate. They may form huge colonies with complex organization in which individuals specialize in particular tasks such as foraging, defense, reproduction, or rearing larvae. The organization and division of labor in these insect societies require sophisticated communication and learning. Social insects accomplish remarkable tasks. For example, honeybees manufacture and store food (honey), and some ant species "farm" by cultivating fungi in underground chambers or "milking" aphids by inducing them to secrete a nutritious liquid.

Beetles Roughly one-third of all known insect species are beetles. Beetles exhibit a huge variety of shapes, sizes, and lifestyles. All beetles, however, have hard, protective coverings over their wings. Many destructive agricultural pests are beetles, such as the Colorado potato beetle, the grain weevil, and the Japanese beetle. However, others, such as lady beetles, are predators that help control insect pests.

One of this group's most impressive adaptations is found in the bombardier beetle. This species defends itself against ants and other enemies by emitting a toxic spray from a nozzle-like structure at the end of its abdomen. The beetle is able to precisely aim the spray, which emerges with explosive force and at temperatures higher than 200°F (93°C). To avoid harming itself, the beetle produces this hot, toxic brew only when needed, by combining two otherwise harmless substances.

Most Arachnids Are Carnivores

The *arachnids* include spiders, mites, ticks, and scorpions (**FIG. 24-25**). All arachnids have eight walking legs, and most are carnivorous. Many subsist on a liquid diet of blood or predigested prey. For example, spiders, the most numerous arachnids, first immobilize their prey with a paralyzing venom. They then inject digestive enzymes into the helpless victim (typically an insect) and suck in the resulting soup. Arachnids breathe using tracheae, lungs, or both.

In contrast to the compound eyes of insects and crustaceans, arachnids have simple eyes, each with a single lens. Most spiders have eight eyes placed in such a way as to give them a panoramic view of predators and prey. The eyes are sensitive to movement, and in some spider species—especially those that hunt actively and have no webs—the eyes are thought to form images. Most spider perception,

(b) Scorpion

(a) Spider

(c) Ticks

▲ **FIGURE 24-25 The diversity of arachnids (a)** The tarantula is one of the largest spiders but is relatively harmless. **(b)** Scorpions, found in warm climates such as the deserts of the southwestern United States, paralyze their prey with venom from a stinger at the tip of the abdomen. A few species can harm humans. **(c)** In a tick that has not yet fed (left), the exoskeleton is flexible and folded. This allows the animal to become grotesquely bloated while feeding on blood (right).

however, is not through their eyes but through sensory hairs found over much of the body. Some of a spider's hairs are touch sensitive and help the animal perceive prey, mates, and surroundings. Other hairs are sensitive to chemicals and function as organs of smell and taste. Hairs also respond to vibrations in the air, ground, or web, allowing spiders to detect nearby movement by predators, prey, or other spiders.

Among the distinctive features of spiders is their production of protein threads known as silk. Spiders manufacture silk in special glands in their abdomens and use it to perform a variety of functions, such as building webs that capture prey, wrapping up and immobilizing captured prey (see Fig. 24-20), constructing protective shelters for themselves, making cocoons to surround their eggs, and making "draglines" that connect a spider to a web or other surface and support its weight if it drops from its perch. Spider silk is an amazingly light, strong, and elastic fiber. It can be stronger than a steel wire of the same size, yet is as elastic as rubber. Human engineers have long sought to develop a fiber with this combination of strength and elasticity. Despite careful study of the structure of spider silk, no comparable human-made substance has been successfully manufactured.

Myriapods Have Many Legs

The *myriapods* include the centipedes and millipedes, whose most prominent feature is an abundance of legs (**FIG. 24-26**).

Most millipede species have between 100 and 300 legs; the species with the largest number of legs can have up to 750.

HAVE YOU EVER WONDERED ...

Why Spiders Don't Stick to Their Own Webs?

If you have ever accidentally walked into a spider web, then you know that the strands of the web can be very sticky indeed. And the small insects one sees trapped in spider webs seem to be quite firmly snared. So how does the spider that built the web scuttle so easily across it without getting stuck? One key factor is that the web is not entirely sticky. The web's scaffolding is made from non-sticky strands, with sticky ones restricted to the capture area. So a moving spider can often simply stay on the non-sticky strands and avoid the sticky ones. But the spider's distinctive claws also help. These claws, in conjunction with specialized hairs on the tips of its legs, allow a spider leg to grip a single web strand and then release it. A spider moving in this fashion, grasping strands with only the very tips of its legs, will have only a tiny surface area in contact with the web at any one time. Thus, even if the spider grips a sticky strand with one of its legs, it can easily pull loose, just as you can if you step on a wad of chewing gum. In contrast, an unsuspecting fly that crashes into the web will contact a number of strands with many parts of its body simultaneously and be stuck fast.

(a) Centipede

(b) Millipede

▲ **FIGURE 24-26** **The diversity of myriapods (a)** Centipedes and **(b)** millipedes are common nocturnal arthropods. Each segment of a centipede's body holds one pair of legs, while each millipede segment has two pairs.

Centipedes are not quite as leggy; most have around 70 legs. Both centipedes and millipedes have one pair of antennae. The legs and antennae of centipedes are longer and more delicate than those of millipedes. Most myriapods have very simple eyes that detect light and dark but do not form images. In some species, the number of eyes can be high—up to 200— but other species lack eyes altogether. Myriapods respire by means of tracheae.

Myriapods inhabit terrestrial environments exclusively, living mostly in soil or leaf litter or under logs and rocks. Centipedes are generally carnivorous, capturing prey (mostly other

arthropods) with their frontmost legs, which are modified into sharp claws that inject venom into prey. Bites from large centipedes can be painful to humans. In contrast, most millipedes are not predators but instead feed on decaying vegetation and other debris. When attacked, many millipedes defend themselves by secreting a foul-smelling, distasteful liquid.

Most Crustaceans Are Aquatic

The *crustaceans,* including crabs, crayfish, lobsters, shrimp, and barnacles, are the only arthropods that live primarily in the water (**FIG. 24-27**). Crustaceans range in size from microscopic

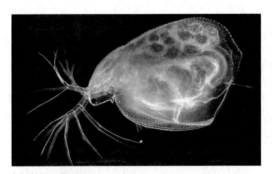

(a) Water flea

(b) Sowbug

(c) Hermit crab

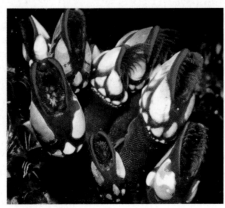

(d) Barnacles

◀ **FIGURE 24-27** **The diversity of crustaceans (a)** The microscopic water flea is common in freshwater ponds. Notice the eggs developing within the body. **(b)** The sowbug, found in dark, moist places such as under rocks, leaves, and decaying logs, is one of the few crustaceans to invade the land successfully. **(c)** This hermit crab protects its soft abdomen by inhabiting an abandoned snail shell. **(d)** The gooseneck barnacle uses a tough, flexible stalk to anchor itself to rocks, boats, or even animals such as whales. Other types of barnacles attach with shells that resemble miniature volcanoes (see Fig. 24-18b). Early naturalists thought barnacles were mollusks until they observed barnacles' jointed legs (seen here extending into the water).

species that live in the spaces between grains of sand to the largest of all arthropods, the Japanese spider crab, whose legs span nearly 12 feet (4 meters). Crustaceans have two pairs of sensory antennae, but the rest of their appendages are highly variable in form and number, depending on the habitat and lifestyle of the species. Most crustaceans have compound eyes similar to those of insects, and nearly all respire using gills.

Crustaceans are an important food source for larger animals. For example, small crustaceans called krill are abundant in the Southern Ocean and are the main food of whales, seals, seabirds, and other animals. People eat a lot of crustaceans, too. Shrimp, for example, are the most consumed seafood in the United States by far. Today, most of the shrimp we eat are farm raised, mainly in coastal areas of Asia and South America. Unfortunately, widespread shrimp farming has had adverse ecological consequences, mainly because large areas of ecologically important mangrove forests have been cleared for shrimp farming.

Roundworms Are Abundant and Mostly Tiny

Although you may be blissfully unaware of their presence, *roundworms* (Nematoda) are nearly everywhere. Roundworms, also called *nematodes,* have colonized nearly every habitat on Earth, and they play an important role in breaking down organic matter. They are extremely numerous; a single rotting apple may contain 100,000 roundworms. Billions thrive in each acre of topsoil. In addition, almost every plant and animal species hosts several parasitic nematode species.

In addition to being abundant and ubiquitous, roundworms are diverse. Although only about 12,000 roundworm species have been named, there may be as many as 500,000. Most, such as the one shown in **FIGURE 24-28**, are microscopic, but some parasitic forms reach a meter in length.

Roundworms Are Pseudocoelomates with a Simplified Body Plan

Roundworms have a rather simple body plan, featuring a tubular gut and a fluid-filled pseudocoelom that surrounds

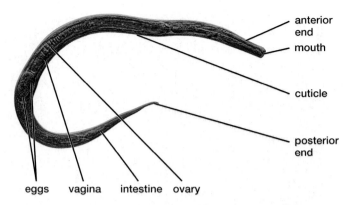

▲ **FIGURE 24-28** **A freshwater nematode** In this micrograph, eggs can be seen inside a female freshwater nematode.

the organs and forms a hydrostatic skeleton (see Fig. 24-3b). A tough, flexible, nonliving cuticle encloses and protects the thin, elongated body and is periodically molted. Sensory organs in the roundworm head transmit information to a simple brain composed of a nerve ring.

Nematodes lack circulatory and respiratory systems. Because most nematodes are extremely thin and have low energy requirements, diffusion suffices for gas exchange and distribution of nutrients. Most nematodes reproduce sexually, and the sexes are separate.

A Few Roundworm Species Are Harmful to Humans

Fifty roundworm species are known to infect humans. Most such worms are relatively harmless, but there are important exceptions. For example, hookworm larvae (found in soil in some tropical regions) can bore into human feet, enter the bloodstream, and travel to the intestine, where they cause continuous bleeding. Another dangerous roundworm parasite, *Trichinella,* causes the disease trichinosis. *Trichinella* worms can infect people who eat undercooked infected pork, which can contain up to 15,000 larval cysts per gram (**FIG. 24-29a**). The cysts hatch in the human digestive tract and invade blood vessels and muscles, causing bleeding and muscle damage.

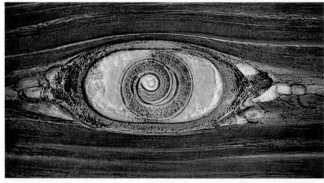

(a) *Trichinella*

(b) Heartworms

▲ **FIGURE 24-29** **Some parasitic nematodes (a)** Encysted larva of the *Trichinella* worm in the muscle tissue of a pig, where it may live for up to 20 years. **(b)** Adult heartworms in the heart of a dog. The juveniles are released into the bloodstream, where they may be ingested by mosquitoes and passed to another dog by the bite of an infected mosquito.

Parasitic roundworms can also endanger domestic animals. Dogs, for example, are susceptible to heartworm (**FIG. 24-29b**), which is transmitted by mosquitoes. In the southern United States, and increasingly in other parts of the country, heartworm poses a threat to the health of unprotected pets.

Echinoderms Have a Calcium Carbonate Skeleton

Echinoderms (Echinodermata) are found only in marine environments, and their common names tend to evoke their saltwater habitats: sand dollars, sea urchins, sea stars (or starfish), sea cucumbers, and sea lilies (**FIG. 24-30**). The name "echinoderm" ("hedgehog skin") stems from the bumps or spines that extend from the skin of most echinoderms. These spines are especially well developed in sea urchins and much reduced in sea stars and sea cucumbers. Echinoderm bumps and spines are actually extensions of an **endoskeleton** (internal skeleton) composed of plates of calcium carbonate that lie beneath the outer skin.

Echinoderms Are Bilaterally Symmetrical as Larvae and Radially Symmetrical as Adults

Echinoderms exhibit deuterostome development and are linked by common ancestry with the other deuterostome phyla, including the chordates (described in Chapter 25). Deuterostomes form a group of branches on the larger evolutionary tree of bilaterally symmetrical animals, but in echinoderms, only embryos and free-swimming larvae are bilaterally symmetrical. An adult echinoderm, in contrast, is radially symmetrical and lacks a head. This absence of cephalization is consistent with the sluggish lifestyle of echinoderms. Most echinoderms move very slowly as they feed on algae or small particles sifted from sand or water. Some echinoderms, however, are predators. Sea stars, for example, slowly pursue even slower-moving prey, such as snails or clams.

Echinoderms Have a Water-Vascular System

Echinoderms move on numerous tiny *tube feet,* delicate cylindrical projections that extend from the lower surface of the body and terminate in a suction cup. Tube feet are part of a unique echinoderm feature, the *water-vascular system,* which functions in locomotion, respiration, and food capture (**FIG. 24-31**). Seawater enters through an opening (the *sieve plate*) on the animal's upper surface and is conducted through a circular central canal, from which branch a number of radial canals. These canals conduct water to the tube feet, each of which is controlled by a muscular squeeze bulb known as an *ampulla.* Contraction of the bulb forces water into the tube foot, causing it to extend. The

(a) Sea cucumber

(b) Sea urchin

(c) Sea star

▲ **FIGURE 24-30 The diversity of echinoderms (a)** A sea cucumber feeds on debris in the sand. **(b)** A sea urchin's spines are actually projections of the internal skeleton. **(c)** Most sea stars have five arms, but some species have 20 or more.

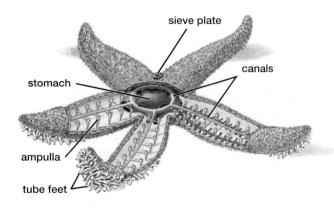

(a) Sea star body plan

(b) Sea star consuming a mussel

▲ **FIGURE 24-31 The water-vascular system of echinoderms (a)** Changing pressure inside the seawater-filled water-vascular system extends or retracts the tube feet. **(b)** The sea star often feeds on mollusks such as this mussel. A feeding sea star attaches numerous tube feet to the mussel's shells, exerting a relentless pull that slightly separates the shells. Then, the sea star turns the delicate tissue of its stomach inside out, extending it through its centrally located ventral mouth. The stomach can fit through an opening in the bivalve shells that is as narrow as 1 millimeter wide. Once pushed between the shells, the stomach tissue secretes digestive enzymes that weaken the mollusk, causing it to open further. Partially digested food is transported to the upper portion of the stomach, where digestion is completed.

suction cup may be pressed against the seafloor or a food object, to which it adheres tightly until its internal pressure is released.

Some Echinoderm Organ Systems Are Simplified

Echinoderms have a relatively simple nervous system with no distinct brain. Movements are loosely coordinated by a system consisting of a nerve ring that encircles the esophagus, radial nerves to the rest of the body, and a nerve network through the epidermis. In sea stars, simple receptors for light and chemicals are concentrated on the arm tips, and sensory cells are scattered over the skin. In some brittle star species, light receptors are associated with tiny lenses, smaller than the width of a human hair, that gather light and focus it on receptors. These "microlenses" are formed from crystals of calcite (calcium carbonate) and their optical quality is excellent, far superior to that of any human-created lens of comparable size. Researchers hypothesize that each of the thousands of lenses on a brittle star forms a tiny image and that the animal integrates the resulting information to detect changes in its surroundings, such as the approach of a predator.

Echinoderms lack a circulatory system, although movement of the fluid in their well-developed coelom serves this function. Gas exchange occurs through the tube feet and, in some forms, through numerous tiny "skin gills" that project through the epidermis. Most species have separate sexes and reproduce by shedding sperm and eggs into the water, where fertilization occurs.

Many echinoderms are able to regenerate lost body parts, and these regenerative powers are especially potent in sea stars. In fact, a single arm of a sea star is capable of developing into a whole animal, provided that part of the central body is attached to it. Before this ability was widely appreciated, mussel fishermen often tried to rid mussel beds of predatory sea stars by hacking them into pieces and throwing the pieces back. Needless to say, the strategy backfired.

Some Chordates Are Invertebrates

The *chordates* (Chordata) include the vertebrate animals and also a few groups of invertebrates, such as the sea squirts and the lancelets. We will discuss these invertebrate chordates and their vertebrate relatives in Chapter 25.

CHECK YOUR LEARNING

Can you ...

- describe the basic body plans of sponges, cnidarians, comb jellies, flatworms, annelids, mollusks, arthropods, roundworms, and echinoderms?
- list some member organisms in each of these groups, and describe the organisms' nervous and sensory systems and methods of circulation, gas exchange, digestion, and reproduction?
- give examples of the effects invertebrate animals have on humans?

Physicians' Assistants

Another invertebrate animal that has been recruited for medical duty is the blowfly or, more precisely, blowfly larvae, commonly

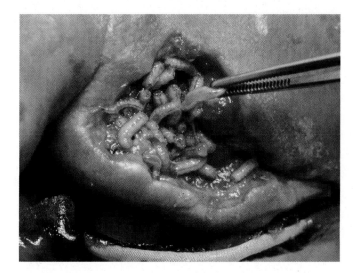

▲ **FIGURE 24-32 Blowfly maggots can clean wounds**

known as maggots (**FIG. 24-32**). Blowfly maggots have proved to be effective at ridding wounds and ulcers of dead and dying tissue. If such tissue is not removed, it can interfere with healing or lead to infection. Traditionally, dead tissue in wounds is removed by a physician wielding a scalpel, but maggots offer an increasingly common alternative treatment. In this treatment, a bandage containing day-old, sterile maggots is applied to the wound. The maggots consume dead or dying tissue, secreting digestive enzymes that do not harm healthy skin or bone. After a few days, the maggots have grown to the size of rice kernels and are removed. The treatment is repeated until the wound is clean.

CONSIDER THIS Medical treatment with invertebrate animals is typically much less expensive than drugs, surgery, and other options. Nonetheless, very little research funding is directed to promising but challenging medical uses of animals, such as treatment of autoimmune diseases with parasitic roundworms. Would it be a good idea to increase research on such treatments, or are they too risky?

CHAPTER REVIEW

Go to **MasteringBiology** for practice quizzes, activities, eText, videos, current events, and more.

Answers to Think Critically, Evaluate This, Multiple Choice, and Fill-in-the-Blank questions can be found in the Answers section at the back of the book.

Summary of Key Concepts

24.1 What Are the Key Features of Animals?

Animals are eukaryotic, multicellular, sexually reproducing organisms that acquire energy by consuming other organisms. Most animals can perceive and react rapidly to environmental stimuli and are motile at some stage in their lives. Their cells lack cell walls.

24.2 Which Anatomical Features Mark Branch Points on the Animal Evolutionary Tree?

The earliest animals had no tissues, a feature retained by modern sponges. All other modern animals have tissues. Animals with tissues can be divided into radially symmetrical and bilaterally symmetrical groups. During embryonic development, radially symmetrical animals have two germ layers; bilaterally symmetrical animals have three. Bilaterally symmetrical animals also tend to have sense organs and clusters of neurons concentrated in the head, a process called cephalization. Bilateral phyla can be divided into two main groups, one of which undergoes protostome development, the other of which undergoes deuterostome development. Protostome phyla can in turn be divided into ecdysozoans and lophotrochozoans. Some phyla of bilaterally

symmetrical animals lack body cavities, but most have either pseudocoeloms or true coeloms.

24.3 What Are the Major Animal Phyla?

The bodies of sponges (Porifera) come in a variety of shapes and are generally sessile. Sponges lack tissues but have three different types of specialized cells. Digestion occurs exclusively within the individual cells.

The sea jellies, corals, anemones, and hydrozoans (Cnidaria) have tissues. A simple network of nerve cells directs the activity of contractile cells, allowing loosely coordinated movements. Digestion is extracellular, occurring in a central gastrovascular cavity with a single opening. Cnidarians exhibit radial symmetry, an adaptation to both the free-floating lifestyle of the medusa and the sedentary existence of the polyp.

Comb jellies (Ctenophora) are superficially similar to sea jellies but form a separate taxonomic group. Comb jellies are small, radially symmetrical carnivores that move using eight rows of cilia.

Flatworms (Platyhelminthes) have a distinct head with sensory organs and a simple brain. A system of canals that form a network throughout the body aids in excretion. They lack a body cavity.

The segmented worms (Annelida) are the most complex worms, with a well-developed closed circulatory system and excretory organs. The segmented worms have a compartmentalized digestive system, which processes food in a sequence. Annelids also have a true coelom, a fluid-filled space between the body wall and the internal organs.

The clams, snails, octopuses, and squid (Mollusca) lack a skeleton; some forms protect the soft, moist, muscular body with a single shell (many gastropods and a few cephalopods) or a pair of hinged shells (the bivalves). Most mollusks live in aquatic or moist terrestrial environments and have an open circulatory system, where blood directly bathes internal organs in a hemocoel. The octopus has the most complex brain and the best-developed learning capacity of any invertebrate.

Arthropods (Arthropoda), the insects, arachnids, millipedes and centipedes, and crustaceans, are the most diverse and abundant animals on Earth. Jointed appendages and well-developed nervous systems make possible complex, finely coordinated behavior. The exoskeleton (which conserves water and provides support) and specialized respiratory structures (which remain moist and protected) enable the insects and arachnids to inhabit dry land. The diversification of insects has been enhanced by their ability to fly. Crustaceans, which include the largest arthropods, are restricted to moist, usually aquatic habitats and respire using gills.

The pseudocoelomate roundworms (Nematoda) possess a separate mouth and anus and a cuticle layer that is molted.

The sea stars, sea urchins, and sea cucumbers (Echinodermata) are an exclusively marine group. Echinoderm larvae are bilaterally symmetrical; however, the adults show radial symmetry. This, in addition to a primitive nervous system that lacks a definite brain, adapts them to a relatively sedentary existence. Echinoderm bodies are supported by an endoskeleton that sends projections through the skin. The water-vascular system, which functions in locomotion, feeding, and respiration, is a unique echinoderm feature.

The chordates (Chordata) include two invertebrate groups, the lancelets and sea squirts, as well as the vertebrates.

Key Terms

bilateral symmetry *421*	larva (plural, larvae) *438*
budding *428*	mesoderm *421*
cephalization *423*	metamorphosis *438*
closed circulatory system *431*	molt *437*
coelom *423*	nerve cord *430*
compound eye *438*	open circulatory
deuterostome *424*	system *434*
ectoderm *421*	organ *423*
endoderm *421*	parasite *429*
endoskeleton *443*	protostome *424*
exoskeleton *437*	pseudocoelom *423*
ganglion (plural, ganglia) *430*	pupa (plural, pupae) *438*
hemocoel *434*	radial symmetry *421*
hermaphroditic *425*	segmentation *430*
hydrostatic skeleton *431*	tissue *421*
invertebrate *425*	vertebrate *425*

Thinking Through the Concepts

Multiple Choice

1. Which of the following groups contains radially symmetrical organisms?
   a. arthropods
   b. cnidarians
   c. mollusks
   d. roundworms

2. To acquire nutrients, sponges
   a. filter microorganisms from water brought in through pores.
   b. use stinging cells to capture small prey.
   c. parasitize larger animals.
   d. suck nutrients from prey after injecting enzymes that dissolve prey tissues.

3. A coelom is
   a. a body cavity that is partially lined with tissue derived from mesoderm.
   b. a body cavity that is completely lined with tissue derived from mesoderm.
   c. found in sponges and cnidarians.
   d. never found in bilaterally symmetrical organisms.

4. Which of the following is *not* an advantage of arthropod exoskeletons?
   a. It provides protection from predators.
   b. It provides attachments sites for muscles.
   c. It improves sensory perception.
   d. It protects against water loss on land.

5. Which of the following groups does *not* include organisms in which food enters and digestive wastes exit through the same opening?
   a. annelids
   b. flatworms
   c. comb jellies
   d. cnidarians

Fill-in-the-Blank

1. Animals obtain energy by _____; they generally reproduce _____, and their cells lack _____.

2. Bilaterally symmetrical animals have _____ embryonic tissue layers, known as _____, _____, and _____. Radially symmetrical animals have _____ tissue layers; they lack the _____ layer.

3. Animals that have an anterior and posterior end are said to be _____. The anterior end of such animals often contains structures used for _____ and _____. Coelomate animals have a body _____ that is completely lined with tissue derived from _____.

4. Lophotrochozoans and ecdysozoans (molting animals) are two large clades of animals that exhibit _____ development. The other major type of development found in bilateral animals is known as _____ development, and is present in animals in the clades known as _____ and _____.

5. Animals that lack a backbone are described as _____; those that possess a backbone are _____. The vast majority of all animals fall into which of these two groups? _____ The only animals that lack tissues are _____, whose bodies resemble a colony of _____. Sea anemones and corals are _____. Earthworms and leeches are _____.

6. Three major groups within the mollusks are the two-shelled clams and scallops called _____; the foot-crawling snails and slugs called _____; and the tentacled squid

and octopuses called _____. Members of the largest animal phylum are called _____. Three important groups within this phylum are the six-legged, often flying _____; the eight-legged spiders and mites called _____; and the mostly aquatic _____.

7. Phyla that contain animals with segmented bodies include _____ and _____. In a(n) _____ system, blood is confined to blood vessels. In a(n) _____ system, blood bathes internal organs within a cavity called the _____.

8. For each of the following distinctive structures, name the animal group in which it is found:
   mantle: _____
   cnidocyte: _____
   water-vascular system: _____
   paired, jointed appendages: _____

Review Questions

1. List the characteristics that, taken together, distinguish animals from other kinds of organisms.

2. List the distinguishing characteristics of each phylum discussed in this chapter, and give an example of a member of each phylum.

3. Briefly describe each of the following adaptations, and explain its adaptive significance: bilateral symmetry, radial symmetry, cephalization, closed circulatory system, coelom, segmentation.

4. Describe and compare the respiratory systems of the four major arthropod groups.

5. Describe the advantages and disadvantages of the arthropod exoskeleton.

6. In which of the three major mollusk groups is each of the following characteristics found?
   a. two hinged shells
   b. a radula
   c. tentacles
   d. some sessile members
   e. the best-developed brains
   f. numerous eyes

7. Give three functions of the water-vascular system of echinoderms.

8. List some invertebrate animals that can harm humans and, for each animal that you list, name its taxonomic group.

Applying the Concepts

1. Insects are the largest group of animals on Earth. Insect diversity is greatest in the tropics, where habitat destruction and species extinction are occurring at an alarming rate. What biological, economic, and ethical arguments can you advance to persuade people and governments to preserve this biological diversity?

2. Discuss and defend the attributes you would use to define biological success among animals. Are humans a biological success by these standards? Why?

Would you be shocked to learn that *Tyrannosaurus* still walked the Earth? The discovery of modern coelacanth fishes was no less surprising.

Fish Story

ON DECEMBER 22, 1938,
Marjorie Courtenay-Latimer received a phone call that would lead to one of the most spectacular discoveries in biological history. The call was from a local fisherman whom Courtenay-Latimer, the curator of a small museum in South Africa, had asked to collect some fish specimens for the museum. His boat had returned from its most recent voyage and was waiting at the town dock. Dutifully, Courtenay-Latimer went to the boat and began sorting through the fish that were strewn across the deck. Later, she wrote, "I noticed a blue fin sticking up from beneath the pile. I uncovered the specimen, and, behold, there appeared the most beautiful fish I had ever seen." In addition to its beauty, the fish had some odd features, including fins that were stumpy and lobed, unlike the fins of any other living species.

Courtenay-Latimer did not recognize the strange fish, but she knew it was unusual. She tried to find a place to refrigerate it, but in her small town she was unable to find a cold storage facility willing to store a fish. In the end, she was able to save only the skin. Undaunted, she made some drawings of the fish and used them to attempt an identification. To her amazement, the creature did not resemble any species known to inhabit the waters off South Africa, but did seem similar to members of a family of fishes known as coelacanths. The only problem was that coelacanths were known only from fossils. As far as anyone knew, coelacanths had been extinct for 80 million years!

Perplexed, Courtenay-Latimer sent her drawings to J. L. B. Smith, a fish expert at Rhodes University. Smith was astounded when he saw the sketch, later writing that "a bomb seemed to burst in my brain." Although bitterly disappointed that the specimen's bones and internal organs had been lost, Smith arranged to view the preserved skin. Ultimately, he confirmed the astonishing news that coelacanths still swam in Earth's waters.

Although coelacanths remained hidden from science until the twentieth century, they share a phylum with frogs, dogs, snakes, and many other familiar animals, including humans. What do we know about these animals?

AT A GLANCE

25.1 What Are the Key Features of Chordates?

25.2 Which Animals Are Chordates?

25.3 What Are the Major Groups of Vertebrates?

25.1 WHAT ARE THE KEY FEATURES OF CHORDATES?

Humans are members of a taxonomic group known as the chordates (Chordata). Chordates (**FIG. 25-1**) include not only bony animals like birds and apes, but also the barrel-shaped tunicates and small fishlike creatures called lancelets. What characteristics do we share with these animals?

All Chordates Share Four Distinctive Structures

All chordates have deuterostome development (which is also characteristic of echinoderms; see Chapter 24) and are further united by four features they all possess at some stage of their lives: a dorsal nerve chord, a notochord, pharyngeal gill slits, and a post-anal tail.

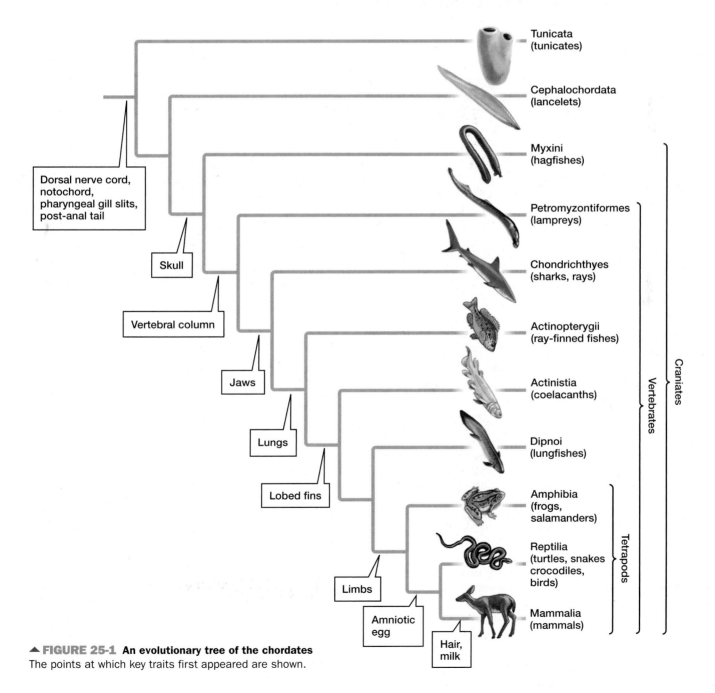

▲ **FIGURE 25-1 An evolutionary tree of the chordates**
The points at which key traits first appeared are shown.

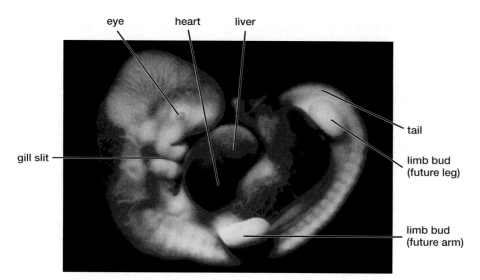

eye   heart   liver

gill slit

tail

limb bud
(future leg)

limb bud
(future arm)

◀ **FIGURE 25-2 Chordate features in the human embryo** This 5-week-old human embryo is about 1 centimeter long and clearly shows a tail and external gill slits (more properly called grooves, since they do not penetrate the body wall). Although the tail will disappear completely, the gill grooves contribute to the formation of the lower jaw.

- The **nerve cord** of chordates lies above the digestive tract, running lengthwise along the dorsal (upper) portion of the body. In contrast, the nerve cords of other animals lie in a ventral position, below the digestive tract (see Fig. 24-14). A chordate's nerve cord is hollow—its center is filled with fluid, unlike the nerve cords of other animals, which have solid nerve tissue throughout. During embryonic development in chordates, the nerve cord develops a thickening at its anterior end that becomes the brain.

- The **notochord** is a stiff rod that extends along the length of the body, between the digestive tract and the nerve cord. It provides support for the body and an attachment site for muscles. In many chordates, the notochord is present only during early stages of development and disappears as a skeleton develops.

- **Pharyngeal gill slits** are located in the pharynx (the cavity behind the mouth). In some chordates the slits form functional openings for gills (organs for gas exchange in water); in others they appear only as grooves during an early stage of development.

- The **post-anal tail** is a posterior extension of the chordate body that extends past the anus and contains muscle tissue and the rearmost portion of the nerve cord. Other animals lack this kind of tail. Most adult chordates have tails, but some species lose them during development.

This list of distinctive chordate structures may seem puzzling because, although humans are chordates, at first glance we seem to lack every feature except the nerve cord. But evolutionary relationships are sometimes seen most clearly during early stages of development, and it is during our embryonic phase that we develop, and subsequently lose, our notochord, our gill slits, and our tail (**FIG. 25-2**).

CHECK YOUR LEARNING

Can you ...

- describe the features that distinguish chordates from other animals?

25.2 WHICH ANIMALS ARE CHORDATES?

Chordates include three clades (groups that include all the descendants of a common ancestor): the tunicates, the lancelets, and the craniates.

Tunicates Are Marine Invertebrates

The tunicates (Tunicata) are a group of about 2,300 species of marine invertebrate chordates. Tunicates are small, with lengths ranging from a few millimeters to 1 foot (30 centimeters). The group includes immobile, filter-feeding, vase-shaped animals known as sea squirts (**FIG. 25-3**). Much of a sea squirt's body is occupied by its pharynx, which is like a basket perforated by gill slits and lined with mucus. Water enters the sea squirt's body through an *incurrent siphon,* passes into the pharynx at its top, moves through the gill slits, and exits the body through an *excurrent siphon.* Food particles are trapped in the basket's mucous lining and then moved to the digestive tract.

Adult sea squirts are sessile—they live firmly attached to a surface. Sea squirt larvae, however, swim actively and possess the four chordate features (see Fig. 25-3, left). Some other types of tunicates remain mobile throughout their lives. For example, barrel-shaped tunicates known as salps live in the open ocean and move by contracting an encircling band of muscle, which forces a jet of water out of the back of the animal, propelling it forward.

Most tunicates are *hermaphroditic* (each individual possesses both male and female sex organs). They may reproduce asexually or sexually. In asexual reproduction, miniature versions of an adult grow from its body and then drop off. In sexual reproduction, sperm are broadcast into the surrounding water and fertilize eggs that (depending on species) are either released into the water or retained inside the tunicate's body. In species that retain eggs inside the body, swimming sperm must enter the body to fertilize the eggs, and the resulting larvae must swim out.

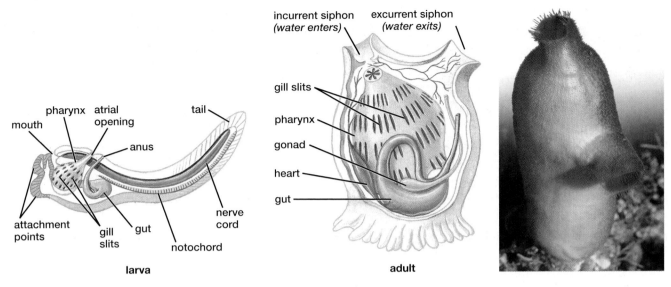

▲ **FIGURE 25-3 Sea squirt** The larva (left) of a sea squirt (a type of tunicate) exhibits all the diagnostic features of chordates. The adult sea squirt (middle) has lost its tail and notochord and has assumed a sedentary life (right).

Lancelets Live Mostly Buried in Sand

The 30 or so species of lancelets (Cephalochordata) form another group of invertebrate chordates. Lancelets are small (2 inches, or about 5 centimeters, long), fishlike animals that retain all four chordate features as adults (**FIG. 25-4**). An adult lancelet spends most of its time half-buried in the sandy sea bottom, with only the anterior end of its body exposed. The motion of cilia in the pharynx draws seawater into the lancelet's mouth. As the water passes through the pharyngeal gill slits, a film of mucus filters tiny food particles from the water. The captured food particles are transported to the lancelet's digestive tract.

Lancelets have separate sexes and always reproduce sexually. At particular times of year, most of the males and females in an area simultaneously release gametes (eggs and sperm) into the surrounding water. Fertilized eggs develop into microscopic larvae, which swim slowly and drift about during several weeks of continuing growth and development before dropping to the seabed and completing their transformation to the adult form.

Craniates Have a Skull

The **craniates** include all chordates that have a skull that encloses the brain. The skull may be composed of bone or cartilage, a tissue that resembles bone but is less brittle and more flexible. The earliest known craniates, whose fossils were found in 530-million-year-old rocks, resembled lancelets but had skulls and eyes. However, the mouths of the earliest craniates lacked jaws.

▼ **FIGURE 25-4 Lancelet** A lancelet, a fishlike invertebrate chordate. The adult organism exhibits all the chordate features.

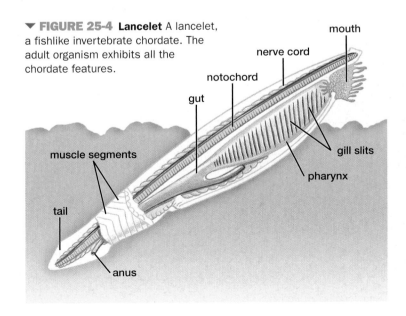

TABLE 25-1	Comparison of Craniate Groups			
Group	**Fertilization**	**Respiration**	**Heart Chambers**	**Body Temperature Regulation**
Hagfishes (Myxini)	External	Gills	Two	Ectothermic
Lampreys (Petromyzontiformes)	External	Gills	Two	Ectothermic
Cartilaginous fishes (Chondrichthyes)	Internal	Gills	Two	Ectothermic
Ray-finned fishes (Actinopterygii)	External[1]	Gills	Two	Ectothermic
Coelacanths (Actinistia)	Internal	Gills	Two	Ectothermic
Lungfishes (Dipnoi)	External	Gills and lungs	Two	Ectothermic
Amphibians (Amphibia)	External or internal[2]	Skin, gills, and lungs	Three	Ectothermic
Reptiles (Reptilia)	Internal	Lungs	Three[3]	Ectothermic[4]
Mammals (Mammalia)	Internal	Lungs	Four	Endothermic

[1]A relatively small number of ray-finned fish have internal fertilization.
[2]External in most frogs and toads; internal in caecilians and most salamanders.
[3]Except for birds and crocodilians, which have four chambers.
[4]Except for birds, which are endothermic.

Today, craniates include two subgroups: the hagfishes and the **vertebrates,** which are animals in which the embryonic notochord is replaced during development by a backbone, or **vertebral column,** composed of bone or cartilage. **TABLE 25-1** summarizes some characteristics of the craniate groups described in the rest of this chapter.

Hagfishes Are Slimy Residents of the Ocean Floor

As did ancestral craniates, hagfishes (Myxini) lack jaws. Instead, they use a tongue-like, tooth-bearing structure to grind and tear food. A hagfish body is stiffened by a notochord, but its skeleton is limited to a few small cartilaginous elements, one of which forms a rudimentary skull. Because hagfishes lack skeletal elements that surround the nerve cord to form a vertebral column, most systematists do not consider them to be vertebrates, although they are the vertebrates' closest relatives.

The 75 or so species of hagfishes are exclusively marine (**FIG. 25-5**). They respire using gills, have a two-chambered heart, and are ectothermic—that is, their body temperature depends on the temperature of their external environment. (Gills, two-chambered hearts, and ectothermy are also found in all vertebrate fishes.) Hagfishes live near the ocean floor, often burrowing in the mud, and feed primarily on worms. They will, however, eagerly attack dead and dying fish, using their teeth to burrow into a fish's body and consume its soft internal organs.

Hagfishes secrete slime as a defense against predators. When attacked by a predatory fish such as a shark, a hagfish quickly secretes a massive quantity of slime, which fills the mouth and gills of the would-be attacker, causing it to flee to avoid suffocation. A hagfish removes slime from its own body by twisting its body into a knot, which it slides forward over its head, scraping off the slime. Hagfish

▲ **FIGURE 25-5 Hagfishes** Hagfishes live in communal burrows in mud, feeding on worms.

slime contains a mixture of mucus and protein threads that are extremely long, highly elastic, and very strong. Researchers are currently investigating the structure of the protein threads in hope of developing useful materials, such as fabrics that mimic the slime's strength and elasticity.

Vertebrates Have a Backbone

The bony or cartilaginous vertebral column of a vertebrate supports its body, provides attachment sites for muscles, and protects the delicate nerve cord and brain. It is also part of a living internal skeleton that can grow and repair itself.

The early history of vertebrates was characterized by an array of strange, now-extinct jawless fishes, many of which were protected by bony armor plates. About 425 million years ago, jawless fishes gave rise to a group of fish that possessed an important new structure: jaws. Jaws allowed fish to grasp, tear, or crush their food, permitting them to exploit a much wider range of food sources than could jawless fish. Today, most (but not all) vertebrates have jaws.

Vertebrates have other adaptations that have contributed to their successful invasion of most habitats. One such adaptation is paired appendages. These first appeared as fins in fish and served as stabilizers for swimming. Over millions of years, some fins were modified by natural selection into legs that allowed animals to crawl onto dry land, and later into wings that allowed some to take to the air. Another adaptation that has contributed to the success of vertebrates is an increase in the size and complexity of their brains and sensory structures, which allow vertebrates to perceive their environment in detail and to respond to it in a great variety of ways.

CHECK YOUR LEARNING

Can you ...

- name and describe the chordates that are not craniates?
- name and describe the craniates that are not vertebrates?
- describe the key adaptations of vertebrates?

CASE STUDY CONTINUED

Fish Story

In the years since Courtenay-Latimer's discovery that coelacanths are not extinct, scientists have had the opportunity to investigate the creature's anatomy. The coelacanth body has some unusual features. For example, adult coelacanths retain a notochord, the body-stiffening rod that most other vertebrates lose during embryonic development. In addition, a coelacanth's brain is very small relative to its body size. The brain of a 90-pound (40-kilogram) coelacanth weighs only 1 or 2 grams (less than a tenth of an ounce). The tiny brain occupies less than 2% of the space in the cranial cavity; the rest is filled with fat.

Their anatomical oddities aside, coelacanths are vertebrates. What features distinguish them from other types of vertebrates? More generally, how do major groups of vertebrates differ?

25.3 WHAT ARE THE MAJOR GROUPS OF VERTEBRATES?

Vertebrates include lampreys, cartilaginous fishes, ray-finned fishes, coelacanths, lungfishes, amphibians, reptiles, and mammals.

Some Lampreys Parasitize Fish

Like hagfishes, the roughly 50 species of lampreys (Petromyzontiformes) are jawless. A lamprey is recognizable by the large, rounded sucker that surrounds its mouth and by the single nostril on the top of its head. The nerve cord of a lamprey is protected by segments of cartilage, so lampreys are considered to be true vertebrates. They live in both fresh and salt water, but the marine forms must return to fresh water to spawn. Lampreys migrate up shallow streams to spawn; eggs are deposited and fertilized in depressions that groups of lampreys excavate in the streambed. The adults die a short time after spawning. After the young hatch, they spend several years in the stream as larvae, eating algae, before maturing and moving downstream to their adult habitat in an ocean, lake, or river.

Adult lampreys of some species are parasitic. A parasitic lamprey uses its tooth-lined mouth to attach itself to a larger fish (**FIG. 25-6**). Using rasping teeth on its tongue, the lamprey excavates a hole in the host's body wall, through which it sucks blood and body fluids. Beginning in the 1920s, parasitic lampreys spread into the Great Lakes. There, in the absence of effective predators, they have multiplied prodigiously and greatly reduced commercial fish populations. Vigorous measures to control the lamprey population have allowed some recovery of fish populations in the Great Lakes.

Cartilaginous Fishes Are Marine Predators

The cartilaginous fishes (Chondrichthyes) include about 1,200 marine species, among them the sharks, skates, and rays

▲ **FIGURE 25-6 Lampreys** Some adult lampreys are parasitic, and use sucker-like mouths lined with rasping teeth to attach to fish.

(a) Shark

(b) Ray

▲ **FIGURE 25-7 Cartilaginous fishes (a)** A shark displaying several rows of teeth. As the frontmost teeth are lost, they are replaced by the new ones behind them. **(b)** The tropical blue-spotted stingray swims by graceful undulations of lateral extensions of its body. Both sharks and rays lack a swim bladder and tend to sink toward the bottom when they stop swimming.

(**FIG. 25-7**). Unlike hagfishes and lampreys (but like all other vertebrates), cartilaginous fishes have jaws. They are graceful predators whose skeleton is formed entirely of cartilage. Their bodies are protected by leathery skin roughened by tiny scales. Although some must swim to circulate water through their gills, most can pump water across their gills. In contrast to the external fertilization that characterizes reproduction in almost all other fish, cartilaginous fish have internal fertilization, in which a male deposits sperm directly into a female's reproductive tract. Some cartilaginous fishes are very large. A whale shark, for example, can grow to more than 45 feet (14 meters) in length, and a manta ray may be more than 20 feet (6 meters) wide.

Although some sharks feed by filtering plankton (tiny animals and protists) from the water, most are predators of larger prey such as other fishes, marine mammals, sea turtles, crabs, or squid. Many sharks attack their prey with strong jaws that contain several rows of razor-sharp teeth; the back rows move forward as the front teeth are lost to age and use.

Skates and rays are mostly bottom dwellers with flattened bodies, wing-shaped fins, and thin tails. Rays are generally larger than skates, but the most notable difference between the two groups is that rays give birth to live young, whereas skates lay eggs. Most skates and rays eat invertebrates. Some ray species defend themselves with a spine near their tail that can inflict dangerous wounds, and others produce a powerful electrical shock that can stun their prey.

Ray-Finned Fishes Are the Most Diverse Vertebrates

The vertebrate diversity crown belongs to the ray-finned fishes (Actinopterygii). About 32,000 species have been identified, and scientists estimate that perhaps twice this number exist, with many undiscovered species inhabiting deep waters and remote areas. Ray-finned fishes are found in nearly every watery habitat, both freshwater and marine.

Ray-finned fishes are distinguished by the structure of their fins, which consist of webs of skin supported by bony spines. In addition, ray-finned fishes have skeletons made of bone, a trait they share with the lobe-finned fishes and limbed vertebrates discussed later in this chapter. The skin of ray-finned fishes is covered with interlocking scales that provide protection while allowing for flexibility. Most ray-finned fishes have a swim bladder, a sort of internal balloon that allows a fish to float effortlessly at any level in the water. The swim bladder evolved from lungs, which were present (along with gills) in the ancestors of modern ray-finned fishes.

The ray-finned fishes include not only a large number of species but also a huge variety of different forms and lifestyles (**FIG. 25-8**). These range from snakelike eels to flattened flounders; from sluggish bottom feeders to speedy,

FIGURE 25-8 The diversity of ray-finned fishes Ray-finned fishes have colonized nearly every aquatic habitat. **(a)** This female deep-sea anglerfish attracts prey with a living lure that projects just above her mouth. The fish is ghostly white; at the 6,000-foot (1,800-meter) depth where anglers live, no light penetrates and thus colors are unnecessary. Male deep-sea anglerfish are extremely small and remain permanently attached to the female, always available to fertilize her eggs. Two parasitic males can be seen attached to this female. **(b)** This tropical green moray eel lives in rocky crevices. The small fish (a banded cleaner goby) on its lower jaw eats parasites that cling to the moray's skin. **(c)** A sea horse may anchor itself with its prehensile tail (adapted for grasping) while feeding on small crustaceans.

THINK CRITICALLY With regard to water regulation (maintaining the proper amount of water in the body), how does the challenge faced by a freshwater fish differ from that faced by a saltwater fish?

(a) Anglerfish   (b) Moray eel   (c) Sea horse

streamlined predators; from brightly colored reef dwellers to translucent, luminescent deep-sea dwellers; from the massive 3,000-pound (1,350-kilogram) mola to the tiny stout infant-fish, which weighs in at about 0.00003 ounce (1 milligram).

Ray-finned fishes are an extremely important source of food for humans. Unfortunately, however, the growing human population's appetite for ray-finned fishes, combined with increasingly effective high-tech methods for finding and catching them, has had a devastating impact on fish populations. Populations of almost all economically important ray-finned fish species have declined drastically. If overfishing continues, fish stocks are likely to collapse.

Coelacanths and Lungfishes Have Lobed Fins

Although almost all fish with bony skeletons belong to the ray-finned group, some bony fishes are coelacanths (Actinistia) or lungfishes (Dipnoi). Coelacanths are described in this chapter's case study. The six species of lungfishes are found in freshwater habitats in Africa, South America, and Australia (**FIG. 25-9**). Lungfishes have both gills and lungs. They tend to live in stagnant waters that may be low in oxygen, and their lungs allow them to supplement their supply of oxygen by breathing air. Lungfishes of several species are able to survive even if the pools they inhabit dry up completely. These fish burrow into mud and seal themselves in mucus-lined chambers. There, they breathe through their lungs and their metabolic rate declines drastically. When the rains return and the pools refill, the lungfishes leave their burrows and resume their underwater way of life.

Lungfishes and coelacanths are sometimes called *lobefins* because members of both groups have fleshy fins that contain rod-shaped bones surrounded by a thick layer of muscle.

This trait is indicative of the groups' shared ancestry, though the two lineages have been evolving separately for hundreds of millions of years.

In addition to the coelacanths and lungfishes, several other lineages of lobefins arose early in the evolutionary history of jawed fish. Members of one of these other lineages evolved modified fleshy fins that, in an emergency, could be used as legs, allowing the fish to drag itself from a drying puddle to a deeper pool. This lineage left descendants that survive today. These survivors are the **tetrapods** (from the Greek for "four feet"), which instead of fins have limbs that can support their weight on land. Tetrapods also have digits (fingers or toes) on the ends of their limbs. The tetrapods include amphibians, reptiles, and mammals.

▲ **FIGURE 25-9 Lungfishes are lobe-finned fish** Among the fishes, lungfishes are the group most closely related to land-dwelling vertebrates.

In recent years, advances in DNA sequencing technology have greatly increased the number of species whose genomes have been sequenced. In 2013, the African coelacanth joined the list. Comparing the coelacanth sequence with that of cartilaginous fishes revealed that the coelacanth genome has changed very slowly since the two groups diverged. Thus, coelacanth genes, like coelacanth bodies, are today much the same as they were in the group's heyday 300 million years ago.

Amphibians Live a Double Life

The first tetrapods to invade land were amphibians. Today, the 6,500 species of amphibians (Amphibia) straddle the boundary between aquatic and terrestrial existence (FIG. 25-10). The limbs of amphibians show varying degrees of adaptation to movement on land, from the belly-dragging crawl of salamanders to the long leaps of frogs. A three-chambered heart (in contrast to the two-chambered heart of fishes) circulates blood more efficiently, and lungs replace gills in most adult forms. Amphibian lungs, however, are relatively inefficient and must be supplemented by the skin, which serves as an additional respiratory organ. This respiratory function requires that the skin remain moist, a constraint that greatly restricts the range of amphibian habitats on land.

Many amphibians are also tied to moist habitats by their breeding behavior, which requires water. For example, as in most fishes, fertilization in frogs and toads is generally external and takes place in water, where the sperm can swim to the eggs. The eggs must remain moist, because they are protected only by a jelly-like coating that leaves them vulnerable to water loss by evaporation. Different amphibian species keep

(a) Tadpole

(b) Frog

(c) Salamander

(d) Caecilian

▲ **FIGURE 25-10** **"Amphibian" means "double life"** The double life of amphibians is illustrated by the bullfrog's transition from **(a)** a completely aquatic larval tadpole to **(b)** an adult leading a semi-terrestrial life. **(c)** The red salamander is restricted to moist habitats in the eastern United States. **(d)** Caecilians are legless, mostly burrowing amphibians.

THINK CRITICALLY What advantages might amphibians gain from their "double life"?

their eggs moist in different ways, but many species simply lay their eggs in water. In some amphibian species, fertilized eggs develop into aquatic larvae such as the tadpoles of some frogs and toads. These aquatic larvae undergo a dramatic transformation into semiterrestrial adults, a metamorphosis that gives the amphibians their name, which means "double life." Their double life and thin, permeable skin have made amphibians particularly vulnerable to pollutants and environmental degradation, as described in "Earth Watch: Frogs in Peril" on page 458.

Frogs and Toads Are Adapted for Jumping

The frogs and toads, with 5,700 species, are the most diverse group of amphibians. Adult frogs and toads move about by hopping and leaping, and their bodies are well adapted for this mode of locomotion, with hind legs that are long relative to their body size (much longer than their forelegs). The names "frog" and "toad" do not describe distinct evolutionary groups, but are instead used informally to distinguish two combinations of characteristics that are common among members of this branch of the amphibians. In general, frogs have smooth, moist skin, live in or near water, and have long hind limbs suitable for leaping; toads have bumpy, drier skin, live on land, and have shorter hind limbs suitable for hopping. Many frogs and toads (and other amphibians) contain toxic substances that make them distasteful to predators. In a few species, such as the golden poison dart frog of South America, the protective chemical is extremely toxic. The toxin from a single golden poison dart frog could kill several adult humans.

Most Salamanders Have Tails

Most salamanders have a lizard-like body: slender, with four legs of roughly equal size and a long tail. Some salamanders, however, have only very small legs; these species may have an eel-like appearance. Most of the roughly 580 species of salamanders live on land, often in moist, protected places, such as beneath rocks or logs on a forest floor. But members of some species are fully aquatic and spend their entire lives in the water. Even land-dwelling species generally move to ponds or streams to breed. In almost all salamander species, eggs hatch into aquatic larvae that use external gills to breathe. In some species, the larvae do not metamorphose, but instead retain the larval form throughout life.

Alone among vertebrates, salamanders can regenerate lost limbs. This ability has attracted the attention of researchers interested in regenerative medicine, which seeks treatments that would enable human bodies to repair or regenerate damaged tissues and organs. The researchers hope that learning how salamanders regenerate limbs will lead to effective treatments for humans.

Caecilians Are Limbless, Burrowing Amphibians

The caecilians form a small (175 species) group of legless amphibians that live in tropical regions. At first glance, a caecilian's appearance is reminiscent of an earthworm, though the larger species, which can be up to 5 feet (1.5 meters) long, might be mistaken for a snake. Most caecilians are burrowing animals that live underground, though a few species are aquatic. Caecilians eyes are very small and often covered by skin. As a result, caecilian vision is probably limited to detecting light.

Reptiles Are Adapted for Life on Land

The reptiles (Reptilia) include lizards, snakes, alligators, crocodiles, turtles, and birds (**FIG. 25-11**). Reptiles evolved from an amphibian ancestor about 250 million years ago.

(a) Alligator

(b) Snake

(c) Tortoise

▲ **FIGURE 25-11 The diversity of reptiles (other than birds) (a)** The outward appearance of the American alligator, found in swampy areas of the South, is almost identical to that of 150-million-year-old fossil alligators. **(b)** This scarlet king snake has a color pattern very similar to that of the poisonous coral snake, which potential predators avoid. This mimicry helps the harmless king snake elude predation. **(c)** The tortoises (a type of turtle) of the Galápagos Islands, Ecuador, may live to be more than 100 years old.

Earth WATCH

Frogs in Peril

During the past three decades, herpetologists (biologists who study reptiles and amphibians) from around the world have documented an alarming decline in amphibian populations. Thousands of species of frogs, toads, and salamanders are dramatically decreasing in number, and many have gone extinct.

This is a worldwide phenomenon; population crashes have been reported from every part of the globe. Of nearly 100 species of harlequin frog known from Central and South America, only 10 can still be found. In South Africa, the only remaining population of Rose's ghost frog has shrunk dramatically and the species is now critically endangered. The southern corroboree frog of Australia was once abundant, but there are now fewer than 50 of them in the wild (FIG. E25-1). In the United States, a recent study showed that populations of virtually all frog and toad species are shrinking (FIG. E25-2). Endangered species are disappearing most quickly, but more common species are declining as well. The declines are occurring even in protected areas, such as national parks.

The causes of the worldwide decline in amphibian diversity are not fully understood, but researchers have discovered that frogs and toads in many places are succumbing to infection by a pathogenic fungus. The fungus has been found in the skin of dead and dying amphibians of hundreds of different species at locations on every continent (except Antarctica, which lacks amphibians). Presence of the fungus has coincided with frog and toad die-offs, and most herpetologists agree that the fungus is causing the deaths.

It seems unlikely, however, that the fungus alone is responsible for the worldwide decline of amphibians. Many

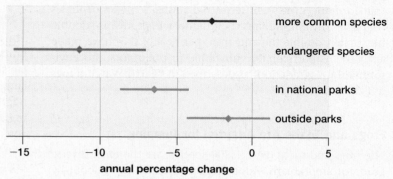

▲ **FIGURE E25-2 Shrinking populations** This graph shows estimated annual percentage population change of amphibians in the United States between 2001 and 2011, based on repeated counts at many locations across the country. The top part of the graph compares endangered species to more common species, and the bottom part compares populations in national parks to populations outside of parks. The horizontal lines passing through each data point show the margin of error for the estimated rates of change.

herpetologists believe that the fungal epidemic would not have arisen if the frogs and toads had not first been weakened by other stressors. What are the other possible causes of stress? All of the most likely causes stem from human modification of the biosphere—the portion of Earth that sustains life.

Habitat destruction, especially the draining of wetlands that are hospitable to amphibian life, is one major cause of the decline. Amphibians are also vulnerable to toxic substances in the environment because amphibian bodies are protected only by a thin, permeable skin that pollutants can easily penetrate. For example, researchers found that frogs exposed to trace amounts of atrazine, a widely used herbicide that washes from farm fields into streams and lakes and is found in virtually all fresh water in the United States, suffer severe damage to their reproductive tissues.

Many scientists believe that the troubles of amphibians signal an overall deterioration of Earth's ability to support life. According to this line of reasoning, the highly sensitive amphibians are providing an early warning of environmental degradation that will eventually affect more resistant organisms as well.

▲ **FIGURE E25-1 Amphibians in danger** The corroboree frog is rapidly declining in its native Australia.

THINK CRITICALLY Consider the graph shown in Figure E25-2. Imagine that the population of one particular endangered species has declined at the same rate each year, and that the rate of decline was equal to the estimated average rate for endangered species. If the initial population of this species included 1,000 individuals, what would its population size be 10 years later? Draw a graph showing how the species' population changed over 10 years, and then extend the graph to show its projected population after 50 years. What assumptions underlie your projection?

▲ **FIGURE 25-12** **The amniotic egg** A crocodile struggles free of its egg. The amniotic egg encapsulates the developing embryo in a fluid-filled membrane (the amnion), ensuring that development occurs in a watery environment, even if the egg is far from water.

Reptiles Haves Scales and Shelled Eggs

Most reptiles live on land. A number of adaptations make reptiles' life on land possible, three of which are especially notable: (1) Reptiles evolved a tough, scaly skin that reduces water loss and protects the body. (2) Reptiles evolved internal fertilization, in which the male deposits sperm within the female's body, eliminating the need to breed in water. (3) Reptiles evolved a shelled **amniotic egg.** The shell prevents the egg from drying out on land, and an internal membrane, the **amnion,** encloses the embryo in the watery environment that all developing animals require (**FIG. 25-12**).

In addition to these features, reptiles have more efficient lungs than do amphibians and do not use their skin as a respiratory organ. Reptile circulatory systems include a three-chambered or (in birds, alligators, and crocodiles) four-chambered heart that segregates oxygenated and deoxygenated blood more effectively than do amphibian hearts.

Lizards and Snakes Share a Common Evolutionary Heritage

Lizards and snakes together form a distinct lineage containing about 9,400 species. The common ancestor of snakes and lizards had limbs, which are retained by most lizards but have been lost in snakes. The limbed ancestry of snakes is revealed by remnants of hind limb bones found in some snake species.

Most lizards are small predators that eat insects or other small invertebrates, but a few lizard species are quite large. The Komodo dragon, for example, can reach 10 feet (3 meters) in length and weigh more than 200 pounds (90 kilograms). These giant lizards live in Indonesia and have powerful jaws and inch-long teeth that enable them to prey on large animals including deer, goats, and pigs. The Komodo dragon, however, does not rely on its teeth alone to kill its prey. It also produces a potent venom that flows from

a gland in its jaw into the wound of a bitten victim. If an animal bitten by a Komodo dragon is not immediately killed, the venom ensures that it will likely die soon after the attack. The lizard simply waits patiently until its wounded, poisoned prey dies.

Most snakes are active, predatory carnivores and have a variety of adaptations that help them acquire food. For example, many snakes have special sense organs that help track prey by detecting small temperature differences between a prey's body and its surroundings. Some snake species immobilize prey with venom that is delivered through hollow teeth. Snakes also have a distinctive jaw joint that allows the jaws to distend so that the snake can swallow prey much larger than its head. A snake's ribs are not attached to a breastbone (which snakes lack), so the ribs are easily pushed outward to accommodate passage of a large prey item down the body. Following one of its infrequent but large meals, a snake's body gears up to digest the food. The snake's heart, liver, kidneys, and intestine grow rapidly, almost doubling in size, and its metabolic rate increases dramatically, as if it were a sprinting racehorse rather than its motionless self. When digestion is complete, the snake's organs and metabolism return to their pre-meal state.

Alligators and Crocodiles Are Adapted for Life in Water

Crocodilians, as the 25 species of alligators and crocodiles are collectively known, are found in coastal and inland waters of the warmer regions of Earth. They are well adapted to an aquatic lifestyle, with eyes and nostrils located high on their heads so that they are able to remain submerged for long periods with only the uppermost portion of the head above the water's surface. Crocodilians have strong jaws and conical teeth that they use to crush and kill the fish, birds, mammals, turtles, and amphibians that they eat.

Parental care is extensive in crocodilians, which bury their eggs in mud nests. Parents guard the nest until the young hatch and then carry their newly hatched young in their mouths, moving them to safety in the water. Young crocodilians may remain with their mother for several years.

Turtles Have Protective Shells

The 325 species of turtles occupy a variety of habitats, including deserts, streams and ponds, and the ocean. These diverse habitats have fostered a variety of adaptations, but all turtles are protected by a hard, boxlike shell that is fused to the vertebrae, ribs, and collarbone. Turtles have no teeth but have instead evolved a horny beak. The beak is used to eat a variety of foods; some turtles are carnivores, some are herbivores, and some are scavengers. The largest turtle, the leatherback, is an ocean dweller that can grow to more than 6 feet (2 meters) in length and feeds largely on sea jellies. Leatherbacks and other marine turtles must return to land to breed and often undertake extraordinary long-distance migrations to reach the beaches on which they bury their eggs.

(a) Hummingbird

(b) Toucan

(c) Ostrich

▲ **FIGURE 25-13** **The diversity of birds (a)** The delicate hummingbird beats its wings about 60 times per second and weighs about 0.15 ounce (4 grams). **(b)** Toucans are fruit-eaters that inhabit the forests of Central and South America. **(c)** The ostrich, the largest of all birds, weighs more than 300 pounds (135 kilograms); its eggs weigh more than 3 pounds (1,500 grams).

THINK CRITICALLY Although the ancestor of all birds could fly, some bird species—such as the ostrich—cannot. Why do you suppose flightlessness has evolved repeatedly among birds?

Birds Are Feathered Reptiles

One very distinctive group of reptiles is the birds (**FIG. 25-13**). Although the 10,300 species of birds have traditionally been classified as a group separate from reptiles, biologists have shown that birds are really a subset of the reptiles (see Chapter 19 for a more complete explanation). The first birds appear in the fossil record roughly 150 million years ago. Modern birds are distinguished from other reptiles by feathers, which are essentially a highly specialized version of reptilian scales. Birds retain scales on their legs—evidence of the ancestry they share with the rest of the reptiles.

Bird anatomy and physiology are dominated by adaptations that help the animals fly. In particular, most birds are exceptionally light for their size. Lightweight bones reduce the weight of the bird skeleton, and many bones present in other reptiles have been lost in the course of evolution. Bird reproductive organs shrink considerably during nonbreeding periods, and female birds possess only a single ovary, further minimizing weight. Feathers serve as lightweight extensions of the wings and the tail, providing the lift and control required for flight; feathers also provide lightweight protection and insulation for the body.

Birds are also able to maintain body temperatures high enough to allow their muscles and metabolic processes to operate at peak efficiency, regardless of the temperature of the external environment. This physiological ability to maintain an internal temperature that is usually higher than that of the surrounding environment is characteristic of both birds and mammals, which are therefore sometimes described as warm blooded or endothermic. In contrast, the body temperature of ectothermic (cold-blooded) animals—invertebrates, fish, amphibians, and reptiles other than birds—varies with the temperature of their environment.

Endothermic animals such as birds have a high metabolic rate, which requires efficient oxygenation of tissues. Therefore, birds possess circulatory and respiratory adaptations that help meet the need for efficiency. A bird's four-chambered heart prevents mixing of oxygenated and deoxygenated blood. The respiratory system of birds is supplemented by air sacs that provide a continuous supply of oxygenated air to the lungs, even while the bird exhales.

Mammals Provide Milk to Their Offspring

One branch of the tetrapod evolutionary tree gave rise to a group that evolved hair and diverged to form the mammals (Mammalia). Mammals are named for the milk-producing **mammary glands** used by all female mammals to feed their young. The mammals first appeared approximately 250 million years ago but did not diversify and become prominent until after the dinosaurs went extinct roughly 66 million years ago. In most mammals, hair protects and insulates the warm body. Like birds and crocodilians, mammals have four-chambered hearts that increase the amount of oxygen delivered to the tissues. The 4,900 species of mammals include three main lineages: monotremes, marsupials, and placental mammals.

Monotremes Are Egg-Laying Mammals

Unlike other mammals, **monotremes** lay eggs rather than giving birth to live young. This group includes only five species: the platypus and four species of spiny anteaters, also known as echidnas (**FIG. 25-14**). Monotremes are found only in New Guinea (the four echidna species) and Australia (the platypus and one of the echidna species).

Echidnas are terrestrial and eat insects or earthworms that they dig out of the ground. Platypuses forage for food in the water, diving below the surface to capture small

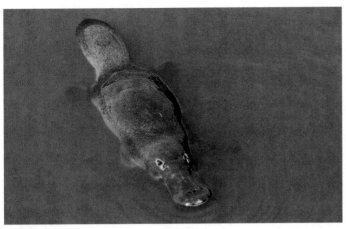

(a) Platypus

(b) Spiny anteater

▲ **FIGURE 25-14 Monotremes (a)** Monotremes, such as this platypus, lay leathery eggs resembling those of reptiles. Platypuses live in burrows that they dig in the banks of rivers, lakes, or streams. **(b)** The short limbs and heavy claws of spiny anteaters (also known as echidnas) help them unearth insects and earthworms to eat. The stiff spines that cover a spiny anteater's body are modified hairs.

vertebrate and invertebrate animals. Platypus bodies are well adapted for this aquatic lifestyle, with a streamlined shape, webbed feet, a broad tail, and a fleshy bill.

Monotreme eggs have leathery shells and are incubated for 10 to 12 days by the mother. Echidnas have a special pouch for incubating eggs, but a female platypus incubates her eggs by holding them between her tail and belly. Newly hatched monotremes are tiny and helpless and feed on milk secreted by the mother. Monotremes, however, lack nipples. Milk from the mammary glands oozes through ducts on the mother's abdomen and soaks the hair around the ducts. The young then suck the milk from the hair.

Marsupial Diversity Reaches Its Peak in Australia

In all mammals except the monotremes, embryos develop in the uterus, a hollow, muscular organ in the female reproductive tract. The lining of the uterus combines with membranes derived from the embryo to form the **placenta**, a structure that allows gases, nutrients, and wastes to be exchanged between the circulatory systems of the mother and embryo.

In **marsupials** (**FIG. 25-15**), embryos develop in the uterus for only a short period. Marsupial young are born at a very immature stage of development. Immediately after birth, a marsupial crawls to a nipple, firmly grasps it, and, nourished by milk, completes its development. In most but

(a) Wallaby

(b) Wombat

(c) Tasmanian devil

▲ **FIGURE 25-15 Marsupials (a)** Marsupials, such as the wallaby, give birth to extremely immature young who develop within the mother's protective pouch. **(b)** The wombat is a burrowing marsupial whose pouch opens toward the rear of its body to prevent dirt and debris from entering the pouch during tunnel digging. One of the wombat's predators is **(c)** the Tasmanian devil.

not all marsupial species, this postbirth development takes place in a protective pouch.

The majority of the 330 species of marsupials are found in Australia, where marsupials such as kangaroos have come to be seen as emblematic of the island continent. Kangaroos are the largest and most conspicuous of Australia's marsupials; the largest species, the red kangaroo, may be 7 feet tall (about 2 meters) and can make 30-foot (9-meter) leaps when moving at top speed. Though kangaroos are perhaps the most familiar marsupials, the group encompasses species with a range of sizes, shapes, and lifestyles, including koalas, wombats, and the Tasmanian devil. Only one marsupial species, the Virginia opossum, is native to North America.

The Tasmanian devil, a carnivorous predator the size of a small dog, is among the marsupial species at risk of extinction. Tasmanian devil populations were decimated by hunting until the species was protected by law and began to recover in the 1940s. Today, however, the recovery is threatened by a new form of cancer that appeared suddenly around 1996. Unlike most cancers, the one that afflicts Tasmanian devils is transmissible—it spreads from animal to animal. Tumors grow on the faces of affected animals. Tasmanian devils often bite each other on the face during fighting or sex, so tumor cells may enter a bite wound. The resulting facial tumors usually kill infected animals within a few months. Researchers estimate that the population of Tasmanian devils has decreased by 60% to 80% since the cancer epidemic began.

Placental Mammals Inhabit Land, Air, and Sea

Most mammal species are **placental** mammals (FIG. 25-16), so named because their placentas are far more complex than those of marsupials. Compared to marsupials, placental mammals retain their young in the uterus for a much longer period, so that offspring are more developed at birth.

The largest groups of placental mammals, in terms of number of species, are the rodents and the bats. Rodents account for almost 40% of all mammal species. Most rodent species are rats or mice, but the group also includes squirrels, hamsters, guinea pigs, porcupines, beavers, woodchucks, chipmunks, and voles. The largest rodent, the capybara, is found in South America and can weigh up to 110 pounds (50 kilograms).

About 20% of mammal species are bats, the only mammals to have evolved wings and powered flight. Bats are

(a) Capybara

(b) Bat

(c) Whale

(d) Cheetah

(e) Orangutan

◀ **FIGURE 25-16 The diversity of placental mammals (a)** The South American capybara is the world's largest rodent. It stands two feet tall and can weigh well over 100 pounds. **(b)** A bat, the only type of mammal capable of true flight, navigates at night by using a kind of sonar. Large ears help the animal detect echoes as its high-pitched cries bounce off nearby objects. **(c)** A humpback whale may migrate 15,000 or more miles each year. **(d)** Mammals are named after the mammary glands with which females nurse their young, as illustrated by this mother cheetah. **(e)** Orangutans are gentle, intelligent apes that occupy swamp forests in limited areas of the Tropics but are endangered by hunting and habitat destruction.

nocturnal and spend the daylight hours roosting in caves, rock crevices, or trees. Most bat species have evolved adaptations for feeding on a particular kind of food. Some bats eat fruit; others feed on nectar from night-blooming flowers. Most bats are predators, including species that hunt frogs, fish, or even other bats. A few species (vampire bats) subsist entirely on blood that they lap up from incisions they make in the skin of sleeping mammals or birds. Most predatory bats, however, feed on flying insects, which they detect by echolocation. To echolocate, a bat emits short pulses of high-pitched sound (too high for humans to hear). The sounds bounce off objects in the surrounding environment to produce echoes, which the bat hears and uses to identify and locate insect prey.

Although the majority of placental mammal species are rodents or bats, the other placental mammals are diverse in form and include many species that loom large in the human imagination. For example, many people are fascinated by the sometimes human-like social behavior of our closest relatives, the chimpanzees, gorillas, and other great apes. Some of us are awed by the grace and power of large carnivores such as lions, cheetahs, tigers, and wolves. And others are fascinated by the 70 species of whales, placental mammals that evolved from terrestrial ancestors and recolonized the ocean. The largest whale species, the blue whale, can grow to more than 100 feet long (more than 30 meters) and is the largest animal known to have existed in the history of Earth.

CHECK YOUR LEARNING

Can you ...
- describe the key features of lampreys, cartilaginous fishes, ray-finned fishes, coelacanths, lungfishes, amphibians, reptiles, and mammals?
- name and describe the main subgroups included within each of these groups?

CASE STUDY **REVISITED**

Fish Story

After Marjorie Courtenay-Latimer's discovery of the coelacanth, J. L. B. Smith dedicated himself to searching for more coelacanth specimens in the waters off South Africa. He didn't find one until 1952, when fishermen from the Comoro Islands, having seen leaflets that offered a reward for a coelacanth, contacted Smith with the news that they had one in their possession. Smith immediately booked a flight to the Comoros, and reportedly wept for joy upon holding the 88-pound coelacanth awaiting him.

In the years since Smith's trip, about 200 additional coelacanths have been caught by fishermen, mostly in waters around the Comoros but also around nearby Madagascar and off the coasts of Mozambique and South Africa. Scientists thought that the fish's range was restricted to this relatively small area in the western Indian Ocean, and it was a surprise when a few specimens were discovered in Indonesia, more than 6,000 miles away.

DNA tests showed that these Indonesian coelacanths were members of a second species.

The known populations of coelacanths are small, consisting of a few hundred individuals, and appear to be declining. Part of this decline is due to fishing, though coelacanths are mostly caught accidentally by fishermen searching for more commercially desirable species. Conservation efforts in South Africa and the Comoros thus focus largely on introducing fishing methods that will reduce the chances of accidentally snaring a coelacanth.

CONSIDER THIS Is it worth spending money to try to protect coelacanths from being accidentally killed by fishermen, or should scarce resources be instead devoted to preserving more ecologically important species and habitats?

CHAPTER REVIEW

 Go to **MasteringBiology** for practice quizzes, activities, eText, videos, current events, and more.

Answers to Think Critically, Evaluate This, Multiple Choice, and Fill-in-the-Blank questions can be found in the Answers section at the back of the book.

Summary of Key Concepts

25.1 What Are the Key Features of Chordates?

At some stage in their development, all chordates possess a notochord; a dorsal, hollow nerve cord; pharyngeal gill slits; and a post-anal tail.

25.2 Which Animals Are Chordates?

The chordates include three taxonomic groups: the tunicates, the lancelets, and the craniates. Tunicates are invertebrate filter-feeders that include the sessile sea squirts and the motile salps.

The lancelets are also invertebrate filter-feeders and live partially buried in sandy seafloors. Craniates include all animals with skulls: the hagfishes (jawless, eel-shaped craniates that lack a backbone) and the vertebrates.

25.3 What Are the Major Groups of Vertebrates?

Lampreys are jawless vertebrates; the best-known lamprey species are parasites of fish. Cartilaginous fishes have skeletons made entirely of cartilage and bodies protected by leathery skin. They breathe with gills and reproduce using internal fertilization. Ray-finned fishes have bony skeletons and fins consisting of webs of skin supported by bony spines. Their skin in protected by interlocking scales, and they breathe with gills.

Coelacanths and lungfishes are collectively known as lobe-fins because of their fleshy, bone-containing fins. Coelacanths breathe with gills; lungfishes have both gills and lungs and can survive out of the water during the dry season.

Most amphibians have simple lungs for breathing air. Most are confined to relatively damp terrestrial habitats because of their need to keep their skin moist for respiration, their need for water to facilitate external fertilization, and their aquatic larvae.

Reptiles—with their well-developed lungs, dry skin covered with relatively waterproof scales, internal fertilization, and an amniotic egg with its own water supply—are well adapted to the driest terrestrial habitats. One group of reptiles, the birds, has additional adaptations, such as an elevated body temperature, that allow the muscles to respond rapidly regardless of the temperature of the environment. The bird body is adapted for flight, with feathers, lightweight bones, and efficient circulatory and respiratory systems.

Mammals have insulating hair and nourish their young with milk. Except for monotreme mammals, they give birth to live young.

Key Terms

amnion *459*	pharyngeal gill slit *450*
amniotic egg *459*	placenta *461*
craniate *451*	placental *462*
mammary gland *460*	post-anal tail *450*
marsupial *461*	tetrapod *455*
monotreme *460*	vertebral column *452*
nerve cord *450*	vertebrate *452*
notochord *450*	

Thinking Through the Concepts

Multiple Choice

1. The two groups of mammals with the largest number of species are
   a. marsupials and monotremes.
   b. carnivores and whales.
   c. bats and rodents.
   d. apes and lampreys.

2. More than half of all vertebrate species are
   a. mammals.
   b. ray-finned fishes.
   c. amphibians.
   d. cartilaginous fishes.

3. Which of the following is *not* a characteristic common to all chordates?
   a. a notochord
   b. a dorsal, hollow nerve cord
   c. pharyngeal gill slits
   d. a vertebral column

4. Adult frogs and toads obtain oxygen through
   a. gills only.
   b. gills and lungs.
   c. lungs and skin.
   d. lungs only.

5. Hagfishes
   a. are predators of live fishes.
   b. feed on worms and dead and dying fishes.
   c. feed on algae attached to rocks.
   d. are filter-feeders.

Fill-in-the-Blank

1. In chordates, the nerve cord is _____ and runs along the _____ side of the body. During at least one stage of a chordate's life, it has a tail that extends past its _____ and its body is stiffened by a(n) _____ that runs along its length.

2. Animals that are chordates but not vertebrates include _____, _____, and _____. Craniates are animals that have a(n) _____. Animals that are craniates but not vertebrates include _____.

3. Both gills and lungs are present in adult _____. Sharks and rays have internal skeletons composed of _____. The vertebrate group with the largest number of species is _____. Lampreys have teeth but lack _____.

4. Among tetrapod groups, hair is found in _____; the skin is a respiratory organ in _____; shelled, amniotic eggs are found in _____; aquatic larvae with gills are found in _____.

5. The only mammals that lay eggs are _____. The only vertebrates that regenerate lost limbs are _____. The only mammals with powered flight are _____.

Review Questions

1. Briefly describe each of the following adaptations, and explain the adaptive significance of each: vertebral column, jaws, limbs, amniotic egg, feathers, placenta.

2. List the vertebrate groups that have each of the following.
   a. a skeleton of cartilage
   b. a two-chambered heart
   c. amniotic egg
   d. endothermy
   e. a four-chambered heart
   f. a placenta
   g. lungs supplemented by air sacs

3. List four distinguishing features of chordates.

4. Describe the ways in which amphibians are adapted to life on land. In what ways are amphibians still restricted to a watery or moist environment?

5. List the adaptations that distinguish reptiles from amphibians and help reptiles adapt to life in dry terrestrial environments.

6. List the adaptations of birds that contribute to their ability to fly.

7. How do mammals differ from birds, and what adaptations do they share?

Applying the Concepts

1. Are hagfishes vertebrates or invertebrates? On which characteristics did you base your answer? Is it important to be able to place them in one category or the other? Why?

2. Is the decline of amphibian populations of concern to humans? Why is it important to understand the causes of the decline?

UNIT 4
Behavior and Ecology

Human observers are captivated by the bright colors and ethereal beauty of coral reefs, which are among the most diverse, productive, and fragile of Earth's ecosystems.

"When we try to pick out anything by itself, we find it hitched to everything else in the universe." —J O H N M U I R, *in My First Summer in the Sierra (1911)*

26

ANIMAL BEHAVIOR

Sex and Symmetry

WHAT MAKES A MAN SEXY? According to a growing body of research, it's his symmetry. Female sexual preference for symmetrical males was first documented in insects. For example, biologist Randy Thornhill found that symmetry accurately predicts the mating success of male Japanese scorpionflies (see the inset photo). In Thornhill's experiments and observations, the most successful males were those whose left and right wings were equal or nearly equal in length. Males with one wing longer than the other were less likely to copulate; the greater the difference between the two wings, the lower the likelihood of mating success.

Thornhill's work with scorpionflies led him to wonder if the effects of male symmetry also extend to humans. To test the hypothesis that female humans find symmetrical males more attractive, Thornhill and colleagues began by measuring symmetry in some young adult males. Each man's degree of symmetry was assessed by measurements of his ear length and the width of his foot, ankle, hand, wrist, elbow, and ear. From these measurements, the researchers derived an index that

Both this male scorpionfly and this male human are exceptionally attractive to females of their species. The secret to their sex appeal may be that both have highly symmetrical bodies.

summarized the degree to which the size of these features differed between the right and left sides of the body.

The researchers next gathered a panel of heterosexual female observers who were unaware of the nature of the study and showed them photos of the faces of the measured males. As predicted by the researchers' hypothesis, men judged by the panel to be most attractive were also the most symmetrical. Apparently, a man's attractiveness to women is correlated with his body symmetry.

Why might females prefer symmetrical males? Consider this question as you read about animal behavior.

26.1 HOW DOES BEHAVIOR ARISE?

Behavior is any observable activity of a living animal. For example, a moth flies toward a bright light, a honeybee flies toward a cup of sugar water, and a housefly flies toward a piece of rotting meat. Bluebirds sing, wolves howl, and frogs croak. Mountain sheep butt heads in ritual combat, chimpanzees groom one another, ants attack a termite that approaches an anthill. Humans dance, play sports, and wage wars. Even the most casual observer encounters many fascinating examples of animal behavior each day.

An animal's behavior is influenced by its genes and by its environment. All behavior develops out of an interaction between the two.

Genes Influence Behavior

Several lines of evidence indicate that variation in behavior is influenced by variation in genes. The evidence includes observation of innate behaviors, behavioral experiments, and genetic analysis.

Innate Behaviors Can Be Performed Without Prior Experience

One indication that genes influence the development of behavior comes from behaviors that are performed by newborn animals and that therefore appear to be inherited. Such **innate** behaviors occur in reasonably complete form the first time an animal encounters a particular stimulus. For instance, a gull chick pecks at its parent's bill very soon after hatching, which stimulates the parent to feed it. Innate behavior also occurs in the common cuckoo, a bird species in which females lay eggs in the nests of other bird species, to be raised by the unwitting adoptive parents. Soon after a cuckoo egg hatches, the cuckoo chick performs the innate behavior of shoving the nest owner's eggs (or baby birds) out of the nest (**FIG. 26-1**).

(a) A cuckoo chick ejects an egg

(b) A foster parent feeds a cuckoo

▲ **FIGURE 26-1 Innate behavior (a)** The cuckoo chick, just hours after it hatches and before its eyes have opened, evicts the eggs of its foster parents from the nest. **(b)** The parents, responding to the stimulus of the cuckoo chick's wide-gaping mouth, feed the chick, unaware that it is not related to them.

THINK CRITICALLY The cuckoo chick benefits from its innate behavior, but the foster parent harms itself with its innate response to the cuckoo chick's begging. Why hasn't natural selection eliminated the foster parent's disadvantageous innate behavior?

The gull and the cuckoo benefit from their behaviors. The young gull acquires nutrition, and the young cuckoo eliminates its competitors for food. These examples illustrate a more general point about behavior: Much of it is adaptive and therefore may have evolved by natural selection.

Experiments Show That Behavior Can Be Inherited

Although the widespread occurrence of innate behaviors provides circumstantial evidence that behaviors can be inherited and are therefore influenced by genes, stronger evidence comes from experiments. One such experiment began when researchers noticed that fruit fly larvae exhibit two phenotypes with respect to feeding behavior. Some larvae are "rovers" and move about continually to search for food. Others are "sitters" and remain in more or less one place and eat whatever is there. When the researchers crossed adult rovers and sitters, all the offspring were rovers. When these first generation rovers bred with one another, however, the resulting generation contained both rovers and sitters in a ratio of roughly 3:1. As you may recall from your study of inheritance in Chapter 11, the pattern observed in this experiment is the one expected for a trait controlled by a single gene with two alleles, one of which is dominant.

Another cross-breeding experiment, this time with blackcap warblers, showed that these birds have a genetically influenced tendency to migrate in a particular direction. Blackcap warblers breed in Europe and migrate to Africa, but populations from different areas travel by different routes. Blackcaps from western Europe travel in a southwesterly direction to reach Africa, whereas birds from eastern Europe travel to the southeast. If birds from the two populations are crossbred in captivity, however, the hybrid offspring try to migrate due south, which is intermediate between the directions of the two parents. This result suggests that genes influence migratory direction.

Geneticists Can Identify Particular Genes That Influence Behavior

Sometimes researchers can pinpoint the particular genes that influence behavior. To determine which genes affect a behavior, an investigator may select a candidate gene and then examine the effects of mutations that inactivate the gene. In some species, researchers may be able to engineer "knockout" animals in which the candidate gene is disabled. For example, mice in which the *V1aR* gene has been knocked out exhibit increased levels of risky behaviors, such as lingering in brightly lit, open areas (which normal mice avoid, presumably because risk of predation is higher in such places). *V1aR* codes for a protein that is a receptor for the hormone arginine vasopressin (AVP). When AVP binds to receptors in the brain, it influences behavior, and mice lacking the receptor fail to respond appropriately to dangerous circumstances.

Before researchers can knock out candidate genes, they must have some idea of which genes are likely to affect the behavior of interest. To identify likely candidates, researchers often use techniques that identify chromosomal locations at which genetic variation is correlated with variation in a behavior. They also use techniques that reveal which genes are expressed in particular tissues (especially the brain) when a particular behavior occurs. These methods often reveal that complex behaviors are influenced by many genes. For example, researchers have identified more than 800 different genes whose expression in a zebra finch brain changes each time the bird sings.

The Environment Influences Behavior

The development and expression of behavior can be influenced by variation in an animal's environment, including both the animal's physical environment and its experiences.

Behaviors Are Influenced by the Physical Environment

The environment in which an animal develops can affect its adult behavior. Consider, for example, the zebrafish. This species is found in diverse habitats, including fast-flowing streams in which the water contains ample oxygen and stagnant ponds in which oxygen levels are very low. In an experiment, genetically similar zebrafish were reared from eggs in two different conditions: oxygen-rich or oxygen-poor water. When the fish grew up, the experimenters measured their response to **aggression,** or antagonistic behavior, from another fish in both an oxygen-rich and an oxygen-poor environment. In the high-oxygen environment, the behavior of fish that were reared in a high-oxygen environment was much more aggressive. In the low-oxygen environment, however, the most aggressive fish were those reared in a low-oxygen environment. It appears that the environment in which a fish develops causes the fish to develop the ability to behave appropriately in the environment that it is most likely to encounter as an adult.

An animal's behavior can also be influenced by the living part of its environment, especially its interactions with other animals. For example, in prairie voles (a small rodent), pups raised by a single mother receive less licking and grooming than do pups raised by a mother and father together. In adulthood, the mating and parental behavior of voles varies, depending on their experience as pups. Those cared for by two parents mate sooner and provide more attentive parental care than do voles raised by a single parent.

Behaviors Are Influenced by the Experiential Environment

The capacity to make relatively permanent changes in behavior on the basis of experience is called **learning.** This deceptively simple definition encompasses a vast array of phenomena. A toad learns to avoid distasteful insects; a baby shrew learns which adult is its mother; a human learns to speak a language; a sparrow learns to use the stars for

(a) Prairie dog alarm call

(b) Habituated prairie dogs

◀ **FIGURE 26-2 Habituation (a)** When a prairie dog detects the approach of a potential predator, it may produce a loud alarm call. However, if harmless intruders approach repeatedly, as when **(b)** human hikers routinely pass by, prairie dogs learn to stop responding with an alarm call.

navigation. Each of the many examples of animal learning represents the outcome of a unique evolutionary history, so learning is as diverse as animals themselves. Nonetheless, it can be useful to categorize types of learning, keeping in mind that the categories are only rough guides and that many examples of learning will not fit neatly into any category.

Habituation Is a Decline in Response to a Repeated Stimulus A common form of simple learning is **habituation,** defined as a decline in response to a repeated stimulus. The ability to habituate prevents an animal from wasting its energy and attention on irrelevant stimuli. For example, a sea anemone will retract its tentacles when first touched, but gradually stops retracting if touched repeatedly. Prairie dogs, which are ground-dwelling rodents, utter loud alarm calls and race to their burrows when potential predators approach their colony (**FIG. 26-2**). But prairie dogs habituate to repeated approaches by animals that prove to be nonthreatening, such as people on a well-used hiking path that passes near the colony, and stop responding with alarm calls.

The ability to habituate is generally adaptive. If a prairie dog ran to its burrow every time a harmless human passed by, the animal would waste a great deal of time and energy that could otherwise be spent on beneficial activities such as acquiring food. Habituation can also fine-tune an organism's innate responses to environmental stimuli. For example, newborn chicks instinctively crouch when any object moves over their heads, but birds that are a few weeks old crouch down when a hawk flies over but ignore harmless birds such as geese. The birds have habituated to things that soar by harmlessly and frequently, such as leaves, songbirds, and geese. Predators are much less common, and the novel shape of a hawk continues to elicit instinctive crouching. Thus, learning modified the innate response, making it more advantageous.

Researchers may intentionally induce habituation in animal subjects so they can be studied. For example, most of

what we know about the social behavior of primates such as chimpanzees, gorillas, and baboons comes from studies of wild populations in which animals have been laboriously habituated to human presence. Only then can researchers approach closely enough to observe behavior (**FIG. 26-3**).

Imprinting Is Rapid Learning by Young Animals Learning often occurs within limits that help increase the chances that only the appropriate behavior is acquired. Such constraints on learning are perhaps most strikingly illustrated by **imprinting,** a form of learning in which an animal's nervous system is rigidly programmed to learn a certain thing only during a certain period of development. The information learned during this *sensitive period* is incorporated into behaviors that are not easily altered by later experience.

Imprinting is best known in birds such as geese, ducks, and chickens. These birds learn to follow the animal or object that they most frequently encounter during an early sensitive period. In nature, a mother bird is likely to be nearby during

▲ **FIGURE 26-3 Chimpanzees habituate to researchers** Scientific investigation of animal behavior often relies on animals that have habituated to human presence.

▲ **FIGURE 26-4 Konrad Lorenz and imprinting** Konrad Lorenz, a pioneer in the scientific study of animal behavior, is followed by goslings that imprinted on him shortly after they hatched. They follow him as they would their mother.

the sensitive period, so her offspring imprint on her. In the laboratory, however, these birds may imprint on a toy train or other moving object (**FIG. 26-4**).

Imprinting is a concern for conservationists who aim to preserve endangered species by rearing animals in captivity for release into the wild. Such programs often go to great pains to ensure that captive-reared animals do not imprint on their human caretakers (**FIG. 26-5a**), so that released animals will be attracted to others of their species and not to people. However, conservationists can also take advantage of imprinting to help ensure that captive-reared animals develop behaviors necessary for survival (**FIG. 26-5b**).

Conditioning Is a Learned Association Between a Stimulus and a Response Behaviors generally occur in response to a particular stimulus, and many animals can learn to associate a behavior with a different stimulus. For example, in a classic experiment conducted by Ivan Pavlov, dogs that normally salivated in response to the sight of food were trained to salivate in response to hearing a ringing bell. This kind of learning, in which an animal learns a new association between a stimulus and an innate response, is known as **classical conditioning.** For example, lemon damselfish perform innate predator avoidance behavior—hiding and scanning their surroundings—in response to a chemical alarm signal released by other damselfish. After experimenters repeatedly exposed young damselfish to the alarm signal mixed with the scents of several unfamiliar fish species, the damselfish performed the alarm response when exposed only to

the scent of any of the previously unfamiliar species. As the damselfish experiment suggests, learning by classical conditioning can result in adaptive behavior, such as avoiding novel predators.

A more complex form of learning is **trial-and-error learning,** in which animals learn through experience to associate a behavior with a positive or negative outcome. Many animals are faced with naturally occurring rewards and punishments and learn to modify their behavior in response to them. For example, a hungry toad that captures a bee quickly learns to avoid future encounters with bees (**FIG. 26-6**). After only one experience with a stung tongue, a toad ignores bees and even other insects that resemble them.

Trial-and-error learning is sometimes known as **operant conditioning,** especially when the learning results from training in a laboratory setting. For example, an animal may learn to perform a behavior (such as pushing a lever or

(a) Feeding a condor chick

(b) Leading a flock of cranes

▲ **FIGURE 26-5 Imprinting and endangered species (a)** To prevent young California condors from imprinting on humans, caretakers use a condor puppet to feed them. When the captive-reared birds are released to the wild, they will be drawn to other condors rather than to people. **(b)** Early in life, captive-reared whooping cranes are exposed to and imprint on ultralight aircraft. The young cranes later follow an ultralight to learn the route for their southward migration.

1 A naive toad is presented with a bee.

2 While trying to eat the bee, the toad is stung painfully on the tongue.

3 Presented with a harmless robber fly, which resembles a bee, the toad cringes.

4 The toad is presented with a dragonfly.

5 The toad immediately eats the dragonfly, demonstrating that the learned aversion is specific to bees.

▲ **FIGURE 26-6 Trial-and-error learning in a toad**

pecking a button) to receive a reward or to avoid punishment. This technique is most closely associated with the psychologist B. F. Skinner, who designed the "Skinner box," in which an animal is isolated and allowed to train itself. The box might contain a lever that, when pressed, ejects a food pellet. If the animal accidentally bumps the lever, a food reward appears. After a few such occurrences, the animal learns the connection between pressing the lever and receiving food and begins to press the lever repeatedly.

Operant conditioning has been used to train animals to perform tasks far more complex than pressing a lever. For example, Gambian giant pouched rats have been trained to sniff out buried land mines and, in return for a banana or peanut reward, scratch the ground vigorously when they find a mine (**FIG. 26-7**). Unexploded land mines pose a major threat to the safety of millions of people in countries around the world; more than 100 million mines remain buried where they were planted during past wars and forgotten. The rats are very good at detecting mines and are too light to detonate the ones they find. Gambian giant pouched rats can also detect the scent of the bacteria that cause tuberculosis (TB) and have been trained to distinguish TB-infected mucus from uninfected mucus. This ability may soon become the basis of a cheap and effective diagnostic test for TB.

In Social Learning, Animals Learn from Other Animals In **social learning**, animals learn behaviors by watching or listening to others of their species. By observing and copying the behavior of others, animals may learn which foods to eat, where to find food or breeding sites, or how to avoid predators. Songbirds of many species acquire their songs by copying the songs of other birds. In animals that use tools, information about how to use them is usually gained by watching other animals use them. For example, some dolphins in Shark Bay, Australia, use sponges to help protect their snouts as they

▲ **FIGURE 26-7 A Gambian giant pouched rat at work, detecting land mines**

dig for prey in the stony seabed (**FIG. 26-8a**). Young dolphins learn this behavior by watching their mother perform it. Similarly, chimpanzees may use stones to crack open nuts or sticks to fish termites out of their mounds (**FIG. 26-8b**). These skills are transmitted from one chimpanzee to another by social learning.

Insight Is Problem Solving Without Trial and Error   In certain situations, animals seem able to solve problems suddenly, without the benefit of prior experience. This kind of sudden problem solving is sometimes called **insight learning,** because it seems at least superficially similar to the process by which humans mentally manipulate concepts to arrive at a solution. We cannot, of course, know for sure if non-human animals experience similar mental states when they solve problems.

One species that seems to be good at solving problems is the New Caledonian crow. These birds not only use tools, but also manufacture them, shaping twigs and leaves into

▲ **FIGURE 26-9   Insight learning** New Caledonian crows learn to solve fairly complex problems without prior training. Here, a crow obtains a reward by selecting the object that is most effective for raising the water level in a tube.

(a) Dolphin wearing a sponge tool

(b) Chimpanzees probing for termites

▲ **FIGURE 26-8   Social learning** Animals may learn helpful behaviors from other animals. By observing others, **(a)** bottlenose dolphins learn to use a sponge plucked from the seafloor as a snout protector, and **(b)** chimpanzees learn to use a twig to extract termites from a mound.

hooks that the birds use to extract insects from their hiding places. In a lab experiment, a New Caledonian crow was presented with a straight piece of wire and a bucket of meat that had been placed down a well, so the bird could not reach it. The crow quickly used its beak to bend the wire into a hook, which the bird used to lift the bucket up so it could eat the meat. New Caledonian crows also readily solved, on the first try, a multistep puzzle that required them to use a short stick to reach a long stick that was in turn used to reach a food reward. In experiments that tested crows' ability to gain access to food rewards floating on water at the bottom of a clear, vertical tube, researchers found that crows quickly reacted by dropping objects into the tube so that the water level rose, bringing the food to within the birds' reach (**FIG. 26-9**). What's more, the crows dropped solid objects that would sink rather than hollow objects that would float.

CHECK YOUR LEARNING
Can you ...
- describe evidence that genes influence behavior?
- describe evidence that the physical and experiential environments influence behavior?
- explain habituation, imprinting, classical conditioning, trial-and-error learning, social learning, and insight learning?

26.2 HOW DO ANIMALS COMPETE FOR RESOURCES?

Resources such as food, space, and mates are scarce relative to the reproductive potential of populations, leading to a contest to survive and reproduce. The resulting competition underlies many of the most frequent types of interactions between animals.

▲ **FIGURE 26-10** **Combat** Fighting can be dangerous for combatants, such as these male elephants.

Aggressive Behavior Helps Secure Resources

One of the most obvious manifestations of competition for resources is aggression between members of the same species. Aggressive behavior includes physical combat between rivals (FIG. 26-10). A fight, however, can injure its participants; even the victorious animal might not survive to pass on its genes. As a result, natural selection has favored the evolution of displays or rituals for resolving conflicts. Aggressive displays allow competitors to assess each other and determine a winner on the basis of size, strength, and motivation, rather than on the basis of wounds inflicted. Thanks to these displays, most aggressive encounters end without physical damage to the participants. We discuss signals of aggression in more detail in Section 26.5.

Dominance Hierarchies Help Manage Aggressive Interactions

Even when they do not cause injuries, aggressive interactions use a lot of energy and can disrupt other important tasks, such as finding food, watching for predators, or raising young. Thus, there are advantages to resolving conflicts with minimal aggression. When animals live in social groups in which individuals interact repeatedly, they may form a **dominance hierarchy,** in which each animal establishes a rank that determines its access to resources. Although aggressive encounters occur frequently while the dominance hierarchy is being established, once each animal learns its place in the hierarchy, disputes are infrequent, and the dominant individuals obtain the most access to the resources needed for reproduction, including food, space, and mates. For example, domestic chickens, after some initial squabbling, sort themselves into a reasonably stable "pecking order." Thereafter, all birds in the group defer to the dominant bird, all but the dominant bird give way to the second most dominant, and so on. In wolf packs, one member of each sex is the dominant, or "alpha," individual to whom all others of that sex are subordinate.

Animals May Defend Territories That Contain Resources

In many animal species, competition for resources takes the form of **territoriality,** the defense of an area where important resources are located. As with dominance hierarchies, territoriality tends to reduce aggression, because once a territory is established through aggressive interactions, relative peace prevails as boundaries are recognized and respected. One reason for this stability is that an animal is highly motivated to defend its territory and will often repel even larger, stronger animals that attempt to invade it.

Territories are most commonly defended by individual males, but may be defended by individual females, mated pairs, families, or larger social groups The defended area may include places to mate, raise young, feed, or store food. Territories are as diverse as the animals defending them. For example, a territory can be a tree where a woodpecker stores acorns (FIG. 26-11), a small depression in a lake floor used as a nesting site by a cichlid fish, a hole in the sand that is home to a crab, or a mouse carcass defended by a pair of burying beetles whose offspring are developing inside it.

Territoriality Occurs When Benefits Outweigh Costs

Acquiring and defending a territory require a costly investment of time and energy, but making the investment can increase an animal's reproductive success enough to offset the cost. For example, American redstarts are migratory songbirds that defend territories both on their breeding grounds in northern forests and in their wintering areas in the tropics. Birds that defend winter territories in lush, moist forests have a larger number of offspring than do birds whose winter territories are in less productive habitats. Thus, in redstarts, resources acquired via territorial behavior in the nonbreeding season have an important effect on breeding success months

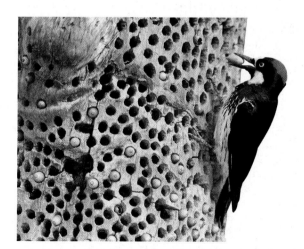

▲ **FIGURE 26-11** **A feeding territory** Acorn woodpeckers live in communal groups that excavate acorn-sized holes in dead trees and stuff the holes with green acorns for dining during the lean winter months. The group defends the trees vigorously against other groups of acorn woodpeckers and against acorn-eating birds of other species, such as jays.

later, probably because the birds with good winter territories are in better condition and can migrate north more quickly to acquire the best breeding territories and get an earlier start on reproduction.

The costs and benefits of defending a territory may change as conditions change, so many animals are territorial only at certain times or only under certain conditions. For example, nonbreeding pied wagtail birds defend feeding territories on days when the invertebrates they feed on are relatively scarce, but abandon territoriality on days when prey animals are especially abundant. When the breeding season arrives, male wagtails defend territories continuously, regardless of the abundance of food. If all goes well for a male, a female will join him on the territory. Many animal species defend territories only while breeding, when it is critical to monopolize the resources required to produce offspring.

CHECK YOUR LEARNING

Can you ...

- describe the function of aggressive behavior?
- explain why natural selection has favored the evolution of aggressive signals and displays, dominance hierarchies, and territoriality?

26.3 HOW DO ANIMALS BEHAVE WHEN THEY MATE?

Sexual selection (see Chapter 16) has fostered the evolution of behaviors that help animals compete for access to mates and that help them choose suitable mates. In most cases, males experience strong selection for traits that help them compete, and females experience selection for traits that help them make advantageous choices.

Males May Fight to Mate

The aggressive behaviors that we described in Section 26.2 often occur in the context of competition among males for opportunities to mate. For example, in the wasp *Sycoscapter australis*, males often must fight in order to mate. In this species, eggs develop within a fig, and newly hatched wasps emerge in the hollow space inside the fruit. Males hatch first and move about looking for emerging females to mate with. But when males encounter one another, they battle fiercely. The fights are intense; up to a quarter of combatants suffer fatal wounds. Winners earn the chance to mate.

In many species, males compete not for direct access to females, but for control of a territory that will attract females. In such species, males who successfully defend territories have the greatest chance of mating. Females usually prefer high-quality territories, which might have features such as large size, abundant food, and secure nesting areas. For example, male side-blotched lizards that defend territories containing many rocks are more successful in attracting mates than are males that defend territories with few rocks. Rockier territories provide more vantage points for detecting

▲ **FIGURE 26-12 Nuptial gifts** A male dance fly copulates with a female that has accepted his gift of a dead insect.

predators and a greater range of microclimates, which is important for "cold-blooded" animals like lizards that regulate body temperature by moving between warmer and cooler spots. Females that select males with the best territories increase their own reproductive success. When experimenters moved rocks from one male's territory to another, females strongly preferred to settle on the improved territories, showing that their preference was based on the territory rather than on the male defending it.

Males May Provide Gifts to Mates

In some species, females are induced to mate by males that provide resources more directly, in the form of a meal. For example, female dance flies mate only with males that bring them a dead insect to eat (**FIG. 26-12**). Copulation occurs while the female eats the insect. If the insect is too small, copulation does not last long enough for sperm to be transferred, so males that present larger insects have greater reproductive success. Similar *nuptial gifts* are presented by males of a number of spider and insect species and may consist of nutritious secretions rather than prey items. Perhaps the ultimate in nutritious gifts is presented by the male red-backed spider. Following copulation, a male throws his own body into the much larger female's jaws, and she usually obliges by eating him. The reproductive cost of this sacrifice is probably small, as even the males that are not eaten after mating are very unlikely to survive long enough to mate with a second female. The tiny males are preyed upon by many predators, and among those that do not sacrifice themselves during mating, almost all are killed and eaten before they can reach a second female's web.

Competition Between Males Continues After Copulation

Among animal species, it is common for both males and females to copulate with multiple partners. When females have multiple partners, males may evolve behaviors that increase

▲ **FIGURE 26-13 Mate guarding** After mating, a male may continue to guard his mate to prevent her from mating with other males. A male swamp milkweed beetle guards his mate by riding around on her back.

the odds that a male's sperm, and not those of a competitor, fertilize the female's eggs. For example, a male may continue to guard a female after copulation to prevent her from mating with other males. In many species of birds, lizards, fish, and insects, males spend days or weeks closely following previous sexual partners, fighting off the advances of other males. In some cases, the male even rides around on the female (**FIG. 26-13**).

If a male copulates with a female and does not subsequently guard her, his reproductive success may be usurped by competitors. For example, in the dunnock, a small songbird in which a female may copulate with multiple males, a male ready to copulate first pecks the female's genital opening. In response, the female ejects any sperm from recent prior copulations, increasing the odds that the current male's sperm will be the ones that fertilize her eggs. Other males take an even more direct approach. Before a male black-winged damselfly copulates with a female that has landed

in his territory, he uses the tip of his penis to scoop out any sperm that a previous sexual partner may have left in the female's sperm storage organ (**FIG. 26-14**).

Reproductive competition among males may continue even after conception. Lions live in social groups called *prides* that contain a few males that maintain exclusive reproductive access to a larger number of females. As ruling a pride is the only way for a male to reproduce, there is intense competition for the role; roaming males without a pride may attempt to overthrow a pride's current males and oust them from the group. If a takeover attempt is successful, the new males are likely to kill any cubs present in the pride. By eliminating the offspring of other males, the new males ensure that the females will reproduce again soon, this time investing their reproductive effort in the offspring of the new males. This kind of *infanticide* is widespread among animals.

Multiple Mating Behaviors May Coexist

In some species, a relatively small number of males can monopolize a large proportion of the opportunities for reproduction. For example, a few males in a population may be especially proficient at producing signals that females find attractive, or scarcity of resources may allow only a few males to secure a territory.

When some males are excluded from the optimal way of reproducing, alternative mating behaviors may arise. For example, in bluegill fish, two types of males, dubbed "parentals" and "cuckolders," coexist and exhibit distinctively different mating behavior (**FIG. 26-15**). Parental males are large, colorful, and territorial; they build and defend nests scooped out of the streambed. Females visit the nests and choose a territorial male to spawn with, laying eggs that mix with the male's sperm. Cuckolder males are very small "sneakers" that hide in vegetation near a nest and wait for an opportunity to deposit some sperm in the nest of a spawning pair without being noticed by the territory holder. When cuckolder males grow a bit larger, they may become female mimics, sporting drab colors that resemble those of females and behaving as a spawning female might. This subterfuge may allow a female-mimic male to slip between a spawning pair and add his sperm to the mix.

(a) Damselflies mating

(b) Damselfly penis

◀ **FIGURE 26-14 Sperm competition (a)** During copulation, a male damselfly (the upper fly in the photo) grasps the female behind her head with claspers on his rearmost segment. The male's penis is located on his underside just below his wings and transfers sperm to the genital opening on the female's rearmost segment, which she swings forward and upward toward the male's body. Before copulating, however, the male uses his **(b)** spiny penis to scrape out any sperm that a competing male may have deposited in the female's sperm storage organ.

(a) Territorial male

(b) Sneaker males lurk near a mating pair

(c) A female-mimic male approaches a mating pair

▲ **FIGURE 26-15 Alternative mating strategies** Different male mating behaviors coexist in populations of bluegills. **(a)** Some males are territorial, defending nests that attract females. **(b)** Other males are sneakers that do not defend territories but may be able to opportunistically fertilize eggs. **(c)** Female-mimic males look and act like females and may not be recognized as competitors by territorial males.

CHECK YOUR LEARNING

Can you ...

- describe some types of behavior that have evolved as a result of competition for mates?
- give specific examples of these types of behavior?
- explain why multiple mating behaviors may coexist in a population?

26.4 HOW DO ANIMALS COMMUNICATE?

Animals frequently broadcast information. If this information evokes a response from other individuals, and if that response tends to benefit the sender and the receiver, then a communication channel can form. **Communication** is the production of a signal by one organism that causes another organism to change its behavior. These changes in behavior, on average, benefit both the signaler and receiver. However, the overall benefit may be reduced if the communication channel is exploited by other animals that harm the communicators.

Although animals of different species may communicate, as when a cat hisses at a dog, most animals communicate primarily with members of their own species. The ways in which animals communicate are astonishingly diverse and use all of the senses. In the following sections, we will look at communication by visual displays, sound, chemicals, and touch.

Visual Communication Is Most Effective over Short Distances

Animals with well-developed eyes use visual signals to communicate. Visual communication may involve passive signals in which the size, shape, or color of the signal conveys important information. For example, when female mandrills become sexually receptive, they develop a large, brightly colored swelling on their buttocks (**FIG. 26-16**). Alternatively, visual signals can be active, consisting of specific movements or postures that convey a message. For example, a wolf signals aggression by lowering its head, raising its hackles (fur on its neck and back), and exposing its fangs (**FIG. 26-17a**).

Like all forms of communication, visual signals have both advantages and disadvantages. On the positive side, visual signals communicate instantaneously, and active visual signals can be rapidly changed to convey a variety of messages in a short period. For example, an initially aggressive wolf that encounters a more dominant individual can quickly shift to a submissive display, crouching with its rump lowered and its tail tucked (**FIG. 26-17b**). Visual communication is quiet and unlikely to alert distant predators, although the signaler does make itself conspicuous to those nearby. On the negative side, visual signals are generally ineffective in dense vegetation or in darkness and are limited to close-range communication.

Communication by Sound Is Effective over Longer Distances

Acoustic signals (messages broadcast by sound) overcome many of the shortcomings of visual displays. Like visual

▲ **FIGURE 26-16 A passive visual signal** The female mandrill's colorfully swollen buttocks serve as a passive visual signal that she is fertile and ready to mate.

(a) Aggressive display

(b) Submissive display

▲ **FIGURE 26-17 Active visual signals (a)** A wolf communicates aggressive intent by facing its opponent, lowering its head, erecting the fur along its backbone, and exposing its fangs. These signals can vary in intensity, communicating different levels of aggression. **(b)** A wolf communicates submission by lowering its rump and tucking its tail.

displays, acoustic signals reach receivers almost instantaneously. But unlike visual signals, acoustic signals can be transmitted through darkness, dense forests, and murky water. Acoustic signals can also be effective over longer distances than visual signals. For example, the low, rumbling calls of an African elephant can be heard by elephants several miles away, and the songs of humpback whales are audible for hundreds of miles. Even the small kangaroo rat produces a sound (by striking the desert floor with its hind feet) that is audible 150 feet (45 meters) away. The advantages of long-distance transmission, however, are offset by an important disadvantage: Predators and other unwanted receivers can also detect an acoustic signal from a distance and can use that signal to locate the signaler.

Like visual displays, acoustic signals can be varied to convey rapidly changing messages. An individual can convey different messages by varying the pattern, volume, or pitch of a sound. A wolf's vocal repertoire, for example, includes a variety of barks, howls, whimpers, and growls.

Some animals communicate with vibrations akin to the stimuli that humans perceive as sounds. For example, male water striders vibrate their legs, sending species-specific patterns of vibrations through the water to be detected by other water striders (**FIG. 26-18**). Caterpillars of some moth species communicate with other caterpillars by scraping or drumming on a leaf with a specialized structure, thereby sending vibrations through the surrounding vegetation.

Chemical Messages Persist Longer but Are Hard to Vary

Chemical substances that are produced by animals and that influence the behavior of other members of the species are called **pheromones.** Unlike visual or sound signals that may attract predators, pheromones are typically not detectable by other species. In addition, a pheromone can act as a kind of signpost, persisting over time and conveying a message long after the signaling animal has departed.

Pheromones can travel fairly far in air or water and so are effective for long-range communication. However, chemical communication requires animals to synthesize a different substance for each message. As a result, chemical signaling systems communicate fewer and simpler messages than do sight- or sound-based systems. In addition, pheromone signals cannot easily convey rapidly changing messages.

Humans have harnessed the power of pheromones to combat insect pests. The sex attractant pheromones of some agricultural pests, such as the Japanese beetle and the gypsy moth, have been successfully synthesized. These synthetic pheromones can be used to disrupt mating or to lure these insects into traps. Controlling pests with pheromones has major environmental advantages over conventional pesticides, which kill beneficial as well as harmful insects and foster the evolution of pesticide-resistant insects. In contrast,

▲ **FIGURE 26-18 Communication by vibration** The light-footed water strider relies on the surface tension of water to support its weight. By vibrating its legs, the water strider sends signals that radiate out over the surface of the water. These vibrations advertise the strider's species and sex to others nearby.

each pheromone is specific to a single species and does not promote the spread of resistance because insects resistant to the attraction of their own pheromones do not reproduce successfully.

CASE STUDY CONTINUED
Sex and Symmetry

Does symmetry have a scent? In one study, researchers measured the body symmetry of 80 men and then issued a clean T-shirt to each one. Each subject wore his shirt to bed for two consecutive nights. A panel of 82 women sniffed the shirts and rated their scents for "pleasantness" and "sexiness." Which shirts had the sexiest, most pleasant scents? The ones worn by the most symmetrical men. The researchers concluded that women can identify symmetrical men by their scent.

What other kinds of messages do animals send with pheromones and other signals? Find out just ahead, in Section 26.5.

Communication by Touch Requires Close Proximity

Communication by physical contact is, for obvious reasons, limited to signaling between animals that are very close together. As a result, it is most common in species that spend a great deal of time in social groups. However, even members of nonsocial species may come into close contact with other individuals during courtship or combat. So communication by touch may also occur in those contexts (**FIG. 26-19**).

Communication Channels May Be Exploited

Even though sending or receiving a signal is in general beneficial, communicators are sometimes harmed by organisms that exploit a communication channel. The exploiter may be an illegitimate receiver that intercepts a signal. For example,

▲ **FIGURE 26-19 Communication by touch** Touch is important in sexual communication. These land snails engage in courtship behavior that will culminate in mating.

▲ **FIGURE 26-20 Illegitimate receivers** A tungara frog's loud calls may be intercepted by a predatory fringe-lipped bat.

parasitic flies that lay their eggs in the bodies of field crickets find their hosts by moving toward the chirps that male crickets produce. Similarly, predatory fringe-lipped bats find tungara frogs to eat by homing in on the loud calls of male frogs (**FIG. 26-20**).

The exploiter of a communication channel may also be an illegitimate signaler. For example, as male fireflies in the genus *Photinus* fly about, they emit flashes in a distinctive pattern that identifies them as members of their species. If a receptive female on the ground sees the flashes, she may respond with flashes of her own, and the male flies down to mate with her. However, fireflies of the genus *Photuris* have evolved the ability to imitate the female *Photinus* flashing pattern. If a male *Photinus* approaches the deceptive signal, the larger *Photuris* firefly kills and eats him.

CHECK YOUR LEARNING
Can you ...
- compare the advantages and disadvantages of visual, acoustic, and chemical signals?
- describe examples of visual, acoustic, chemical, and tactile (touch) signals?

26.5 WHAT DO ANIMALS COMMUNICATE ABOUT?

The information shared by communicating helps animals manage a range of interactions with other animals. Animals may communicate to help resolve conflicts. They may communicate about sex, about food, or about predators. Members of a social group may communicate to help coordinate activities.

(a) A male baboon

(b) Sarcastic fringeheads

▲ FIGURE 26-21 **Aggressive displays (a)** Threat display of the male baboon. Despite the potentially lethal fangs so prominently displayed, aggressive encounters between baboons rarely cause injury. **(b)** The aggressive display of many male fish, such as these sarcastic fringeheads, includes elevating the fins and flaring the gill covers, thus making the body appear larger.

Animals Communicate to Manage Aggression

As you learned in our earlier discussion of aggression, the costliness of direct combat has fostered the evolution of signals that communicate aggression. During aggressive displays, animals may exhibit weapons, such as claws and fangs (FIG. 26-21a), and often make themselves appear larger (FIG. 26-21b). Competitors often stand upright and erect their fur, feathers, ears, or fins. These visual signals are typically accompanied by vocal signals such as growls, croaks, roars, or chirps. Fighting tends to be a last resort when displays fail to resolve a dispute.

In addition to aggressive visual and vocal displays, many animal species engage in ritualized combat. Deadly weapons may clash harmlessly (FIG. 26-22) or may be displayed without contacting the opponent. The ritual thus allows contestants to assess the strength and the motivation of their rivals, and the loser retreats. Ritual contests often involve communication by touch, as when two male sandperch fish lock mouths in ritualized wrestling, or two male zebras begin a contest over access to mates by placing their heads on one another's shoulders.

Aggressive visual and vocal displays are commonly deployed in territorial defense. For example, a male roe deer warns intruders away from its territory by displaying its antlers and a male anole lizard does so by displaying the colorful

dewlap that extends from his throat. A male sea lion defends a strip of beach by swimming up and down in front of it, calling continuously. A male cricket uses a special structure on its wings to produce chirps that advertise its ownership of its burrow. The often beautiful and elaborate vocalizations

▲ FIGURE 26-22 **Displays of strength** The oversized claws of fiddler crabs, which could severely injure another animal, grasp harmlessly. Eventually one crab, sensing greater vigor in his opponent, retreats unharmed.

of songbirds are also used in territory defense. For example, the husky trill of a male seaside sparrow warns other males to steer clear of his territory.

An animal that has a territory but cannot always be present may use pheromones to scent-mark the boundaries. For example, male sac-winged bats use secretions from their chin glands to mark the perimeter of the small territory that each male defends within a communal roost site that is occupied only during daytime. Scent-marking is also useful for territories that are too large for continuous monitoring of boundaries. A solitary tiger, for example, uses pheromones to mark its presence in a territory that may be as large as 385 square miles (1000 square kilometers).

Mating Signals Encode Sex, Species, and Individual Quality

Before animals can successfully mate, they must identify one another as members of the same species, as members of the opposite sex, and as being sexually receptive. In many species, finding an appropriate potential partner is only the first step. Often the male must demonstrate his quality before the female will accept him as a mate. The need to fulfill all of these requirements has resulted in the evolution of a diverse and fascinating array of courtship signals.

Animals often use sounds to advertise their sex and species. Male grasshoppers and crickets advertise their sex and species with chirps, and a male fruit fly does so with a buzz he produces by vibrating one wing. Acoustic signals may also be used by potential mates to compare rival suitors. During mating season, male hammer-headed bats gather together and produce a chorus of loud, honking calls. The males have a very large head that helps the vocalizations resonate (**FIG. 26-23**). Females fly among the gathered males, listening to their calls before choosing a male to mate with.

Many species use visual displays for courting. Male fence lizards, for example, bob their heads in a species-specific

▲ **FIGURE 26-23 Sexual vocalizations** The huge head of a male hammer-headed bat acts as a resonator that amplifies that male's loud courtship calls.

rhythm, and females prefer the rhythm of their own species. The elaborate construction projects of the male gardener bowerbird and the scarlet throat of the male frigatebird serve as flashy advertisements of sex, species, and male quality (**FIG. 26-24**).

Pheromones can also play an important role in reproductive behavior. A sexually receptive female silk moth, for example, sits quietly and releases a chemical message that can be detected by males up to 3 miles (5 kilometers) away. The exquisitely sensitive and selective receptors on the antennae of the male silk moth respond to just a few molecules of the substance, allowing him to travel upwind along a concentration gradient to find the female (**FIG. 26-25a**). Water is an excellent medium for dispersing chemical signals, and fish commonly use a combination of pheromones and elaborate courtship movements to ensure the synchronous release of gametes. Mammals, with their highly developed sense of smell, often rely on pheromones released by the female during her fertile periods to attract males (**FIG. 26-25b**).

(a) A bowerbird bower

(b) A male frigatebird

◀ **FIGURE 26-24 Sexual displays** **(a)** During courtship, a male gardener bowerbird builds a bower out of twigs and decorates it with colorful items that he gathers. **(b)** A male frigatebird inflates his scarlet throat pouch to attract passing females.

THINK CRITICALLY The male bowerbird provides no protection, food, or other resources to his mate or offspring. Why, then, do females carefully compare the bowers of different males before choosing a mate?

(a) Antennae detect pheromones

(b) Noses detect pheromones

▲ **FIGURE 26-25 Sexual pheromones (a)** Male moths find females not by sight but by following airborne pheromones released by females. These odors are sensed by receptors on the male's huge antennae, whose enormous surface area maximizes the chances of detecting the female scent. **(b)** When dogs meet, they typically sniff each other near the base of the tail. Scent glands there broadcast information about the bearer's sex and interest in mating.

THINK CRITICALLY Female dogs use a pheromone to signal readiness to mate, but female mandrills (see Fig. 26-16) signal mating readiness with a visual signal. What differences would you predict between the two species' methods of searching for food?

CASE STUDY \ **CONTINUED**
Sex and Symmetry

If females are attracted to symmetrical males, we might expect females to assess the symmetry of male visual signals that function in mate attraction. For example, a male house finch has a patch of bright red feathers on the crown of his head, and researchers have shown that males whose crown patches are brightly colored are more likely to attract a mate than are males with dull patches. But males with patches that are both colorful *and* highly symmetrical have the highest mating success of all.

Animals Warn One Another About Predators

Animals may communicate about threats, especially threats from predators. When aphids are attacked by predators, they secrete an alarm pheromone that causes receivers to stop feeding and move away from the signal. If a Belding's ground squirrel spots an approaching predator, it produces an alarm call that sends nearby squirrels scrambling for safety. In some cases, varying alarm signals can convey different messages. For example, vervet monkeys produce different calls in response to threats from each of their major predators: snakes, leopards, and eagles. The response of listening monkeys to each of these calls is appropriate to the particular predator. The "bark" that warns of a leopard or other four-legged carnivore causes monkeys on the ground to take to trees and those in trees to climb higher. The "rraup" call, which advertises the presence of an eagle or other hunting bird, causes monkeys on the ground to look upward and take cover, whereas monkeys already in trees drop to the shelter of lower, denser branches. The "chutter" call that indicates the presence of a snake causes the monkeys to stand up and search the ground for the predator.

Animals Share Information About Food

Animals that live in groups may share information about food. For example, foraging termites that discover food lay a trail of pheromones from the food to the nest, and other termites follow the trail (**FIG. 26-26**). Honeybees share

▲ **FIGURE 26-26 Communication about food** A trail of pheromones, secreted by termites from their own colony, orients foraging termites toward a source of food.

If the dance is performed on a vertical wall inside the hive, the angle (from vertical) of the waggle run represents the angle between the sun and the food source.

If the dance is performed on a horizontal surface outside, the waggle run is aimed at the food source.

The rate of circling communicates the distance to the food source.

▲ **FIGURE 26-27 Bee language: the waggle dance** A forager, returning from a rich source of nectar, performs a waggle dance that communicates the distance and direction of the food source as other foragers crowd around her, touching her with their antennae. The bee moves in a straight line while shaking her abdomen back and forth ("waggling") and buzzing her wings. She repeats this dance over and over in the same location, circling back in alternating directions.

information about the location of food by means of a **waggle dance** (FIG. 26-27) that is usually performed in the darkness inside a hive. Other bees crowd around the dancer, and a message is transmitted by vibrations from the dancer's moving wings and body.

Communication Aids Social Bonding

Social bonds between animals may be established or reinforced through communication by touch. For example, the return of a familiar individual to its group is facilitated by touch, as when an elephant puts its trunk in a new arrival's

▲ **FIGURE 26-28 Grooming** An adult olive baboon grooms a juvenile. Grooming both reinforces social relationships and removes debris and parasites from the fur.

mouth, two reunited coyotes touch noses, or an ant touches a returning colony member with its antennae. Social signaling through touch is especially apparent in primates, which have many gestures—including kissing, nuzzling, patting, petting, and grooming—that facilitate social cohesion (FIG. 26-28).

CHECK YOUR LEARNING

Can you ...

- describe how signals function in aggression, courtship, and mating?
- describe how animals share information about predators and food?

26.6 WHY DO ANIMALS PLAY?

Many animals play. Pygmy hippopotamuses push one another, shake and toss their heads, splash in the water, and pirouette on their hind legs. Otters delight in elaborate acrobatics. Bottlenose dolphins balance fish on their snouts, throw objects, and carry them in their mouths while swimming. Baby vampire bats chase, wrestle, and slap each other with their wings. Even octopuses have been seen playing a game: pushing objects away from themselves and into a current, then waiting for the objects to drift back, only to push them back into the current to start the cycle over again.

Although play behavior seems easy to recognize, it is challenging to formulate a precise definition of play. A widely used definition describes **play** as behavior that seems to lack any immediate function, and that often includes modified versions of behaviors used in other contexts.

(a) Chimpanzees

(b) Polar bears

(c) Red foxes

◀ **FIGURE 26-29 Young animals at play**

Animals Play Alone or with Other Animals

Play can be solitary, as when a single animal manipulates an object, such as a cat with a ball of yarn, or when a macaque monkey makes and plays with a snowball. Play can also be social. Often, young animals of the same species play together, but parents may join them. Social play typically includes chasing, fleeing, wrestling, kicking, and gentle biting (**FIG. 26-29**).

Young animals play more frequently than do adults. Play typically borrows movements from other behaviors (attacking, fleeing, stalking, and so on) and uses considerable energy. Also, play is potentially dangerous. Many young humans and other animals are injured, and some are killed, during play. In addition, play can distract an animal from the presence of danger while making it conspicuous to predators. So why do animals play?

Play Aids Behavioral Development

It is likely that play has survival value and that natural selection has favored those individuals who engage in playful activities. One of the best explanations for the survival value of play is the *practice hypothesis*. It suggests that play allows young animals to gain experience in behaviors that they will use as adults. By performing these acts repeatedly in play, an animal practices skills that will later be important in hunting, fleeing, and social interactions.

Play is most intense early in life when the brain develops and crucial neural connections form. Species with large brains tend to be more playful than species with small brains. Because larger brains are generally linked to greater learning ability, this relationship supports the hypothesis that adult skills are learned during juvenile play. Watch children roughhousing or playing tag, and you will see how play might foster strength and coordination and develop skills that might have helped our hunting ancestors survive.

CHECK YOUR LEARNING
Can you ...
- describe the characteristics of animal play and some examples of it?
- describe a hypothesis about the function of play?

26.7 WHAT KINDS OF SOCIETIES DO ANIMALS FORM?

Sociality, the tendency to associate with others and form groups, is a widespread feature of animal life. Most animals interact at least a little with other members of their species. Many spend the bulk of their lives in the company of others, and a few species have developed complex, highly structured societies.

Group Living Has Advantages and Disadvantages

Living in a group has both costs and benefits, and a species will not evolve social behavior unless the benefits of doing so outweigh the costs. Benefits to social animals include the following:

- Increased abilities to detect, repel, and confuse predators.
- Increased hunting efficiency or increased ability to spot localized food resources.
- Advantages resulting from the potential for division of labor within the group.
- Increased likelihood of finding mates.

On the negative side, social animals may encounter:

- Increased competition for limited resources.
- Increased risk of infection from contagious diseases.
- Increased risk that offspring will be killed by other members of the group.
- Increased risk of being spotted by predators.

Sociality Varies Among Species

The degree to which animals of the same species cooperate varies from one species to the next. Some types of animals, such as the mountain lion, are basically solitary; interactions between adults consist of brief aggressive encounters and mating. Other animals form loose social groups, such as schools of fish, flocks of birds, and herds of musk oxen (**FIG. 26-30**). Still others form more structured social groups that may include more complex relationships among members. For example, members of a baboon troop often form alliances; if a baboon is threatened by another in the troop, the other members of its alliance may come to its defense. Similarly, male bottlenose dolphins often form alliances of two or three animals that cooperate to monopolize and

▲ **FIGURE 26-31 Cooperation in a complex society** Naked mole rats are highly social rodents. The individuals in a colony belong to different castes.

defend reproductive females. Different dolphin alliances may band together to form super-alliances that engage in contests with other super-alliances to "steal" females.

In some species, social behavior includes division of labor. Consider the spider *Anelosimus studiosus*. These spiders live in groups of up to 50 that build large communal webs that capture larger prey than a solitary spider could trap on its own. Group members cooperate to rear all of the group's offspring. In this spider society, different members perform different roles. Some watch over eggs and regurgitate food to feed young spiders. Others maintain the web, capture prey, and defend the colony.

Division of labor is more extreme in the rigidly structured societies of many bees, ants, and termites, and of naked mole rats (**FIG. 26-31**). In these societies, individuals are born into one of several castes. For example, in naked mole rats, which live in large underground colonies in southern Africa, the colony is dominated by the *queen*, the only reproducing female, to whom all other members are subordinate *workers*. Some workers clean the tunnels and gather food. Others dig new tunnels or defend the colony against predators and members of other colonies.

A honeybee hive also contains a single queen. Her main function is to produce eggs (up to 1,000 per day). Male bees, called *drones*, serve merely as mates for the queen and die as soon as their mating duties are complete. The hive is run by sterile female workers. The workers' tasks include carrying food to the queen and to developing larvae, producing hexagonal cells of wax in which the queen deposits her eggs, cleaning the hive and removing the dead, protecting the hive against intruders, and foraging for food by gathering pollen and nectar for the hive.

Reciprocity or Relatedness May Foster the Evolution of Cooperation

In the animal societies we have described, individuals sometimes behave in ways that help others but seem to place the helper at an immediate disadvantage. A baboon helps defend

▲ **FIGURE 26-30 Cooperation in loosely organized social groups** A herd of musk oxen functions as a unit when threatened by predators such as wolves. Males form a circle, horns pointed outward, around the females and young.

another, putting itself at risk of injury. A social spider spends time and energy capturing insects for other spiders to eat. Worker bees spend their lives doing hard labor but do not reproduce. There are many other examples: Vampire bats may regurgitate part of a blood meal to feed another bat that did not find food; ground squirrels may risk their own safety to warn the rest of their group about an approaching predator; young, mature Florida scrub jays, instead of breeding, may remain at their parents' nest and help them raise subsequent broods.

At first glance, such cooperative behavior appears to be **altruism**—behavior that decreases the reproductive success of one individual to benefit another. But true altruism presents an evolutionary puzzle. If individuals perform self-sacrificing deeds that reduce their survival and reproduction, why aren't the alleles that contribute to this behavior eliminated from the gene pool?

One possibility is that cooperative behaviors that reduce an individual's fitness over the short term actually increase it over the longer term. For example, helping another individual now may pay benefits later when the favor is returned. Such *reciprocity* is most likely to be the basis of cooperation in groups that remain together for an extended period of time and in which individuals recognize and remember one another. Cooperation may also be favored when other members of the group are close relatives of the cooperating individual. Because close relatives share alleles, the altruistic individual may promote the survival of its own alleles through behaviors that maximize the survival of its close relatives. This concept is called **kin selection.** Kin selection is believed to underlie cooperation in many animal societies, perhaps including those of social insects like honeybees, whose distinctive system of sex determination produces female workers that are very genetically similar to one another.

CHECK YOUR LEARNING

Can you …

- list the advantages and disadvantages of living in a group and describe the different degrees of sociality that occur among mammals?
- describe some hypotheses for the evolution of cooperation in animal societies?

26.8 CAN BIOLOGY EXPLAIN HUMAN BEHAVIOR?

The behaviors of humans, like those of all other animals, have an evolutionary history. Thus, the methods and concepts that help us understand and explain the behavior of other animals can help us understand and explain human behavior as well.

The Behavior of Newborn Infants Has a Large Innate Component

Because newborn infants have not had time to learn, we can assume that much of their behavior is innate. The rhythmic

▲ **FIGURE 26-32 A human instinct** Thumb sucking is a difficult habit to discourage in young children because sucking on appropriately sized objects is an instinctive, food-seeking behavior. This fetus sucks its thumb at about 4 months of development.

movement of an infant's head in search of its mother's breast is an innate behavior that is expressed in the first days after birth. Sucking, which can be observed even in a human fetus, is also innate (**FIG. 26-32**). Other behaviors seen in newborns include grasping with the hands and feet and making walking movements when the body is held upright and supported.

Another example is smiling, which can occur soon after birth. Initially, smiling can be induced by almost any object looming over the newborn. This initial innate response, however, is soon modified by experience. Infants up to 2 months old will smile in response to a stimulus consisting of two eye-sized spots, which at that stage of development is a more potent stimulus for smiling than is an accurate representation of a human face. But as the child's development continues, learning and further development of the nervous system interact to limit the response to more correct representations of a face.

Newborns prefer their mothers' voices to other female voices. Even in their first 3 days of life, infants can be conditioned to produce certain rhythms of sucking when their mother's voice is used as reinforcement (**FIG. 26-33**). The infant's ability to learn his or her mother's voice and respond positively to it within days of birth has strong parallels to imprinting and may help initiate bonding with the mother.

Young Humans Acquire Language Easily

One of the most important insights from studies of animal learning is that animals tend to have an inborn tendency for specific types of learning that are important to their species' mode of life. In humans, one such inborn tendency is for the acquisition of language. Young children are able to acquire language rapidly and nearly effortlessly; they typically acquire a vocabulary of 28,000 words before the age of 8. Research suggests that we are born with a brain that is already primed for this early facility with language. For example, a human fetus

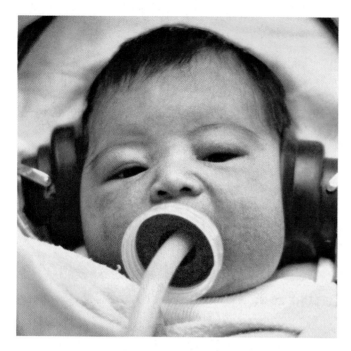

▲ **FIGURE 26-33 Newborns prefer their mother's voice** Using a nipple connected to a computer that plays audio tapes, researcher William Fifer demonstrated that newborns can be conditioned to suck at specific rates in order to listen to their own mothers' voices through headphones. For example, if the infant sucks faster than normal, her mother's voice is played; if she sucks more slowly, another woman's voice is played. Researchers found that infants easily learned and were willing to work hard at this task just to listen to their own mothers' voices, presumably because they had become used to her voice in the womb.

begins responding to sounds during the third trimester of pregnancy, and researchers have demonstrated that infants are able to distinguish among consonant sounds by 6 weeks after birth. In one experiment, infants sucked on a pacifier that contained a force transducer to record the sucking rate. The infants were conditioned to suck at a higher rate in response to playback of adult voices making various consonant sounds. When one sound (such as "ba") was presented repeatedly, the infants became habituated and decreased their sucking rate. But when a new sound (such as "pa") was presented, sucking rate increased, revealing that the infants perceived the new sound as different.

Behaviors Shared by Diverse Cultures May Be Innate

Another way to study the innate bases of human behavior is to compare simple acts performed by people from diverse cultures. This comparative approach has revealed several gestures that seem to form a universal, and therefore probably innate, human signaling system. Such gestures include facial expressions for pleasure, rage, and disdain, and greeting movements such as an upraised hand or the "eye flash" (in which the eyes are widely opened and the eyebrows rapidly elevated). The evolution of the neural pathways underlying these gestures presumably depended on the advantages that

accrued to both senders and receivers from sharing information about the emotional state and intentions of the sender. A species-wide method of communication was perhaps especially important before the advent of language and later remained useful during encounters between people who shared no common language.

Certain complex social behaviors are widespread among diverse cultures. For example, the incest taboo (avoidance of mating with close relatives) seems to be universal across human cultures (and even across many species of non-human primates). It seems unlikely, however, that a shared belief could be encoded in our genes. Some biologists have suggested that the taboo is instead a cultural expression of an evolved, adaptive behavior. According to this hypothesis, close contact among family members early in life suppresses sexual desire, and this response arose because of the negative consequences of inbreeding (such as a higher incidence of genetic diseases). The hypothesis does not require us to assume an innate social belief, but rather proposes that we inherit a learning program that causes us to undergo a kind of imprinting early in life.

Humans May Respond to Pheromones

Although the main channels of human communication are through the eyes and ears, humans also seem to respond to chemical messages. In one experiment, for example, researchers asked nine female volunteers to wear cotton gauze in their armpits for 8 hours each day during their menstrual cycles. The gauze was then disinfected with alcohol and swabbed above the upper lips of another set of 20 female subjects (who reported that they could detect no odors other than alcohol on the gauze). The subjects were exposed to the gauze in this way each day for 2 months, with half the group sniffing armpit secretions from women in the early (preovulation) part of the menstrual cycle, while the other half was exposed to armpit secretions from later in the cycle (postovulation). Women exposed to early-cycle secretions had shorter-than-usual menstrual cycles, and women exposed to late-cycle secretions

HAVE YOU EVER WONDERED ...

Which Is the World's Loudest Animal?

Sounds louder than about 120 decibels are painful to human ears, but some animals make sounds that are far louder than that. The songs of blue whales, for example, can reach 188 decibels, louder than the noise you would hear if you stood next to a jet airplane at takeoff. But the loudest known sound from an animal is produced by a much smaller organism, the pistol shrimp. The pistol shrimp doesn't sing, but it does use a specially modified claw to shoot a high-speed jet of water that stuns its prey. An air bubble forms behind the jet, and when the bubble collapses under the surrounding water pressure, the implosion produces a sound that can reach 200 decibels, much louder than a gun firing right beside your ear.

had delayed menstruation. It appears that women release different pheromones, with different effects on receivers, at different points in the menstrual cycle.

Other research suggests that people may also be able to detect chemical indicators of fear or stress. For example, researchers performed an experiment in which they collected sweat samples from 144 people. Half the people were sampled during a first-time skydive with a 1-minute freefall; the other half (the controls) were sampled during a bout of exercise. Test subjects then smelled one of the two types of samples while undergoing brain imaging. A brain region called the amygdala, which is associated with strong emotions such as fear and rage, was active in subjects who smelled the "stress sweat" from the skydivers, but not in the subjects who smelled the exercise sweat. A chemical present in the sweat of emotionally stressed people apparently triggered a similar emotion, or at least a similar brain response, in people exposed to the chemical.

Although the experiments described above offer strong evidence for the existence of human pheromones, little else is known about chemical communication in humans. The actual molecules that caused the effects documented by these and other experiments remain unknown. Receptors for chemical messages have not yet been found in humans, and we don't know if "menstrual pheromones" and "stress pheromones" are examples of an important communication system or merely isolated cases of a vestigial ability. Despite the hopeful advertisements for "sex attraction pheromones" on late-night television, chemical communication in humans is a scientific mystery awaiting a solution.

Biological Investigation of Human Behavior Is Controversial

The study of evolved, inherited human behavior is controversial, especially among nonscientists, because it challenges the long-held belief that environment is the most important determinant of human behavior. As discussed earlier in this chapter, we now recognize that all behavior has some genetic basis and that complex behavior in non-human animals typically combines elements of both innate and learned behaviors. Thus, it seems certain that our own behavior is influenced by both our evolutionary history and our cultural heritage. The debate over the relative importance of heredity and environment in determining human behavior continues and is unlikely ever to be fully resolved. The scientific study of human behavior will always be hampered because we can neither view ourselves with detached objectivity nor experiment with people as if they were laboratory rats. Despite these limitations, there is much to be learned about the interaction of learning and innate tendencies in humans.

CHECK YOUR LEARNING

Can you ...

- describe evidence that some human behaviors and learning predispositions are innate?
- describe evidence that human behavior is influenced by pheromones?
- explain why biological investigation of human behavior is controversial?

CASE STUDY \ **REVISITED**

Sex and Symmetry

In the experiment described at the beginning of this chapter, women found males with the most symmetrical bodies to have the most attractive faces. But how did the women know which males were most symmetrical? After all, the researchers' measurement of male symmetry was based on small differences in the sizes of body parts that the female judges did not even see during the test.

Perhaps male body symmetry is reflected in facial symmetry, and females prefer symmetrical faces. To test this hypothesis, a group of researchers used computers to alter photos of male faces, either increasing or decreasing their symmetry (**FIG. 26-34**). Then heterosexual female observers rated each face for attractiveness. The observers had a strong preference for more symmetrical faces.

Why would females prefer to mate with symmetrical males? The most likely explanation is that symmetry indicates good physical condition. Disruptions of normal embryological development can cause bodies to be asymmetrical, so a highly symmetrical body indicates healthy, normal development. Females that mate with individuals whose health and vitality are announced by their symmetrical bodies are likely to have offspring that are similarly healthy and vital.

▲ **FIGURE 26-34 Faces of varying symmetry** Researchers used sophisticated software to modify facial symmetry. The face at left is unaltered; the one on the right has been modified to be more symmetrical.

CONSIDER THIS Is our perception of human beauty determined by cultural standards, or is it part of our biological makeup, the product of our evolutionary heritage? What evidence would persuade you that beauty is a biological phenomenon? That it is a cultural one?

CHAPTER REVIEW

Go to **MasteringBiology** for practice quizzes, activities, eText, videos, current events, and more.

Answers to Think Critically, Evaluate This, Multiple Choice, and Fill-in-the-Blank questions can be found in the Answers section at the back of the book.

Summary of Key Concepts

26.1 How Does Behavior Arise?

All animal behavior is influenced by both genetic and environmental factors. Biologists distinguish between innate behaviors, whose development is not highly dependent on external factors, and learned behaviors, which require more extensive experiences with environmental stimuli in order to develop. Innate behaviors can be performed properly the first time an animal encounters the appropriate stimulus, whereas learned behavior changes in response to the animal's social and physical environment.

26.2 How Do Animals Compete for Resources?

Although many competitive interactions are resolved through aggression, serious injuries are rare. Most aggressive encounters are settled by displays that communicate the motivation, size, and strength of the combatants. Some species establish dominance hierarchies that minimize aggression and determine access to resources. On the basis of initial aggressive encounters, each animal acquires a status in which it defers to more dominant individuals and, in turn, dominates subordinates. Territoriality, a behavior in which animals defend areas where important resources are located, also minimizes aggressive encounters.

26.3 How Do Animals Behave When They Mate?

Sexual selection has favored traits that help males gain access to mates and traits that help females choose beneficial mates. Males may fight to obtain mates, offer food gifts to females to induce them to mate, or defend territories that attract females. Females prefer to mate with males that win contests, offer high-quality gifts, or defend high-quality territories. Males may also compete by guarding their mates or by killing the offspring of other males. Multiple alternative mating behaviors may coexist in a species.

26.4 How Do Animals Communicate?

Communication allows animals of the same species to interact effectively in their quest for mates, food, shelter, and other resources. Animals communicate through visual signals, sound, chemicals (pheromones), and touch. Visual communication is quiet and can convey rapidly changing information. Visual signals are active (body movements) or passive (body shape and color). Sound communication can also convey rapidly changing information, and it is effective when vision is impossible. Pheromones can be detected after the sender has departed, conveying simple messages over time. Animals that associate in close proximity may communicate by touch. On average, signalers and receivers benefit from communication, but communication channels are sometimes exploited by animals that harm the communicators.

26.5 What Do Animals Communicate About?

Animals communicate to manage aggressive interactions and reduce the occurrence of potentially dangerous fighting.

Mating signals help animals recognize the species, sex, sexual receptivity, and quality of potential mates. Alarm signals warn other animals of predators. Members of a social group may use signals to share information about the location of food. Physical contact reinforces social bonds and is a part of many premating rituals.

26.6 Why Do Animals Play?

Animals of many species engage in seemingly wasteful (and sometimes dangerous) play behavior. Play behavior in young animals has been favored by natural selection, probably because it provides opportunities to practice and perfect behaviors that will later be crucial for survival and reproduction.

26.7 What Kinds of Societies Do Animals Form?

Social living has both advantages and disadvantages, and species vary in the degree to which their members cooperate. Some species form cooperative societies. The most rigid and highly organized are those of the social insects such as the honeybee, in which the members fill rigidly defined roles throughout life.

26.8 Can Biology Explain Human Behavior?

Researchers are increasingly investigating the extent to which human behavior is influenced by evolved, genetically inherited factors. This emerging field is controversial. Because we cannot freely experiment on humans, and because learning plays a major role in nearly all human behavior, investigators must rely on studies of newborn infants and comparative cultural studies. Evidence is mounting that our genetic heritage plays a role in personality, intelligence, simple universal gestures, our responses to certain stimuli, and our tendency to learn specific things such as language at particular stages of development.

Key Terms

aggression   *468*
altruism   *485*
behavior   *467*
classical conditioning   *470*
communication   *476*
dominance hierarchy   *473*
habituation   *469*
imprinting   *469*
innate   *467*
insight learning   *472*

kin selection   *485*
learning   *468*
operant conditioning   *470*
pheromone   *477*
play   *482*
social learning   *471*
territoriality   *473*
trial-and-error
     learning   *470*
waggle dance   *482*

Thinking Through the Concepts

Multiple Choice

1. When a male lizard defends a certain area, he is exhibiting
   a. insight learning.
   b. kin selection.
   c. territoriality.
   d. altruism.

2. The benefits to an individual of living in a social group include
   a. increased protection against predators.
   b. decreased competition for food.
   c. less competition for mates.
   d. lower incidence of infectious disease.

3. Very early in their lives, young chicks crouch in fear when a leaf flutters overhead. Later they learn to disregard it. This kind of learning is called
   a. imprinting.
   b. habituation.
   c. operant conditioning.
   d. insight learning.

4. Young sea turtles head for the ocean immediately after they hatch. This behavior is most likely
   a. innate.
   b. learned through trial and error.
   c. classically conditioned.
   d. the result of habituation.

5. When animals engage in _____, they often perform displays that make them look as large and dangerous as possible.
   a. courtship
   b. altruism
   c. kin selection
   d. aggression

Fill-in-the-Blank

1. In general, animal behaviors arise from an interaction between the animal's _____ and its _____. Some behaviors are performed correctly the first time an animal encounters the proper _____. Such behaviors are described as _____.

2. Play is almost certainly an adaptive behavior because it uses considerable _____, and it can distract the animal from watching for _____. The most likely explanation for why animals play is the "_____ hypothesis," which states that play teaches the young animal _____ that will be useful as a(n) _____. This hypothesis is supported by the observation that animals that have larger _____ are more likely to play.

3. One of the simplest forms of learning is _____, defined as a decline in response to a(n) _____, harmless stimulus. A different type of learning in which an animal's nervous system is rigidly programmed to learn a certain behavior during a certain period in its life is called _____. The time frame during which such learning occurs is called the _____.

4. Animals often deal with competition for resources through _____ behavior. Such conflicts are often resolved through _____, which allow the competing animals to assess each other without _____

each other. Animals that resolve conflicts this way often display their _____ and make their bodies appear _____.

5. The defense of an area where important resources are located is called _____. Examples of important resources that may be defended include places to _____, _____, _____, and _____. Such resources are most commonly defended by which sex? _____ Are these spaces usually defended against members of the same species or against members of different species? _____

6. After each form of communication, list a major advantage in the first blank and a major disadvantage in the second blank. Pheromones: _____; _____. Visual displays: _____; _____.

Review Questions

1. Explain why animals play. Include the features of play in your answer.

2. List four senses through which animals communicate, and give one example of each form of communication. For each sense listed, present both advantages and disadvantages of that form of communication.

3. A bird will ignore a squirrel in its territory, but will act aggressively toward a member of its own species. Explain why.

4. Why are most aggressive encounters among members of the same species relatively harmless?

5. Discuss the advantages and disadvantages of group living.

6. What kinds of evidence might indicate that a particular human behavior is innate?

Applying the Concepts

1. Male mosquitoes orient toward the high-pitched whine of the female, and female mosquitoes, the only sex that sucks blood, are attracted to the warmth, humidity, and carbon dioxide exuded by their prey. Using this information, design a mosquito trap or killer that exploits a mosquito's innate behaviors. Then design one for moths.

2. Describe and give an example of a dominance hierarchy. What role does it play in social behavior? Give a human parallel, and describe its role in human society. Are the two roles similar? Why or why not? Repeat this exercise for territorial behavior in humans and in another animal.

3. You are the manager of an airport. Planes are being endangered by large numbers of flying birds, which can be sucked into and disable the engines. Without harming the birds, what might you do to discourage them from nesting and flying near the airport and its planes?

POPULATION GROWTH AND REGULATION

Northern elephant seals were hunted almost to extinction in the 1800s. Today, they are a tourist attraction on the central California coast.

CASE STUDY

The Return of the Elephant Seals

In the eighteenth and early nineteenth centuries, oil for lamps and for lubricating machine parts was often harvested from whales. In the mid-1800s, as hunting depleted whale populations in the eastern Pacific near California and Mexico, whalers turned to northern elephant seals as an alternative source of oil. A large bull, which might be 15 feet long and weigh over 5,000 pounds, could produce more than 100 gallons of oil.

Although no one knows for certain, there were probably a couple of hundred thousand elephant seals in the mid-1800s. Each year, in the early winter, thousands of bulls emerged from the Pacific onto the beaches of islands off the coast of southern California and Baja California, Mexico. A month or two later, smaller but much more abundant females would haul out on the same beaches. Some beaches were so packed with seals that hunters had to wait for the seals to get out of the way before they could land their boats.

Being such easy targets, elephant seals were slaughtered by the tens of thousands. Within about 20 years, elephant seals had all but disappeared. Between 1884 and 1892, not a single elephant seal was sighted anywhere. Because elephant seals had become so rare, museum expeditions searched year after year, looking for specimens for their collections. In 1892, a Smithsonian expedition found 9 seals on Isla de Guadalupe, about 150 miles off the coast of Baja California (and killed 7 as specimens). Modern molecular research suggests that as few as 10 to 20 seals remained alive in the late 1800s. These few seals bred on isolated beaches, and the population slowly increased. Still, an intensive search of Isla de Guadalupe in 1922 found only 264 seals. The Mexican government then banned killing or capturing elephant seals and even put soldiers on Isla de Guadalupe to enforce the ban. The United States banned elephant seal hunting soon thereafter.

Today, there are about 200,000 elephant seals, breeding on both islands and mainland beaches, including popular viewing locations such as Año Nuevo and San Simeon in central California. In this chapter, we will study the growth of populations, from bacteria to seals to people. How fast can populations grow? How long can populations keep increasing? Since Earth is not blanketed with elephant seals or any other species of plant or animal, there must be natural conditions that slow down and stop population growth; what are some of those limiting factors?

27.1 WHAT IS A POPULATION AND HOW DOES POPULATION SIZE CHANGE?

A **population** consists of all the members of a particular species that live in a specific area and can potentially breed with one another (see Fig. 1-10). Each population forms part of a **community,** defined as a group of interacting populations of multiple species living in the same area. A community and the nonliving components of the area (for example, soil, air, and water) form an **ecosystem** (from the Greek word *oikos*, a place to live). A natural ecosystem can be as small as a pond or as large as an ocean; it can be a field, a forest, or an island. All of the ecosystems in the entire habitable surface of Earth comprise the **biosphere. Ecology** is the study of the interrelationships of organisms with one another and with their nonliving environment. We begin our exploration of ecology with an overview of populations.

Changes in Population Size Result from Natural Increase and Net Migration

In natural ecosystems, some populations remain fairly stable in size over time, some undergo yearly cycles, and still others change, often dramatically, in response to complex environmental variables. For example, if members of a species invade new territory, such as an island, their population may grow rapidly for a while, but then the population usually either stabilizes or plummets.

Population size changes through births, deaths, and net migration. The **natural increase** of a population is the difference between births and deaths. Although it may sound strange, natural increase can be negative (a decrease if deaths exceed births. The net migration of a population is the difference between **immigration** (migration into the population) and **emigration** (migration out). A population grows when the sum of natural increase and net migration is positive and declines when this sum is negative. The equation for the change in population size within a given time span is:

change in = natural increase + net migration
population size (births − deaths) (immigration − emigration)

In the wild, most migration is emigration of young animals out of a population. For example, the term "lone wolf" comes from young wolves that emigrate from their home packs and immigrate into new areas where they may join an existing pack or (if they find a mate) start a pack of their own. As the members of a population reproduce, emigration may help to keep the original population at a size that is relatively stable and consistent with the resources available to support it. Although migration is significant in some natural populations, for simplicity, we will ignore migration and use only birth and death rates in our calculations of population growth.

CASE STUDY CONTINUED
The Return of the Elephant Seals

In 1960, about 14,000 elephant seals used the beaches of Isla de Guadalupe, over 90% of the total population at that time. As Guadalupe became crowded, the seals looked for new beaches. Because many seals migrated to new territory in California, immigration was an important factor in elephant seal population growth on California beaches and islands during the mid and late 1900s. But why did the Mexican beaches become crowded? And why are California beaches becoming crowded with elephant seals today?

Birth Rate, Death Rate, and the Initial Population Size Affect the Growth of Populations

Growing populations add individuals in proportion to the population's size, much like a bank account accumulates compound interest: Large bank accounts earn more money from interest than small ones do, even if the interest rate is the same. The **growth rate (r)** of a population is the change in population size per individual per unit time (we will use 1 year in our examples). A population's growth rate equals its **birth rate (b),** the births per individual per unit time, minus its **death rate (d),** the deaths per individual per unit time:

$$b \quad - \quad d \quad = \quad r$$
(birth rate) (death rate) (growth rate)

If the birth rate exceeds the death rate, the population growth rate will be positive and population size will increase. If the death rate exceeds the birth rate, the growth rate will be negative and population size will decrease. Birth and death rates are expressed as percentages or decimal fractions; for example, if half the population dies each year, then the death rate is 50%, or 0.5.

Let's calculate the growth rate of a herd of 100 adult white-tail deer, with 50 males and 50 females. Adult female white-tails typically give birth to twins each year, so the population produces 100 fawns (50 does × 2 fawns per doe = 100 fawns). The birth rate, therefore, is 1.0 births per individual per year.

We'll simplify the biology of white-tail deer and assume that female fawns mature and produce their own fawns the very next year. Predators and hunting kill many fawns and some adults, perhaps 70 each year, for a death rate of 0.7. The annual growth rate per individual, therefore, is 0.3:

$$\underset{\text{(birth rate)}}{1.0} \quad - \quad \underset{\text{(death rate)}}{0.7} \quad = \quad \underset{\text{(growth rate)}}{0.3}$$

To calculate population growth per unit time (G), we multiply the growth rate (r) by the population size (N) at the beginning of the time interval:

$$\underset{\substack{\text{(population growth} \\ \text{per unit time)}}}{G} \quad = \quad \underset{\text{(growth rate)}}{r} \quad \times \quad \underset{\text{(population size)}}{N}$$

The growth of the deer herd during the first year (G) is 0.3 × 100 = 30. At a constant growth rate (r), G and N both increase with each successive time interval. In our example, the herd begins its second year with 130 deer (the new value of N). If r remains the same, then the herd adds 39 deer (the new value of G) during this second year (0.3 × 130 = 39), for a total population of 169. In the third year, the herd adds 51 deer (0.3 × 169 = 51), and so on.

Real populations, of course, never follow this equation exactly. Rather, the equation only provides estimates of population sizes over time. Further, environmental conditions never remain the same for very long, so the birth and death rates of real populations fluctuate from year to year.

A Constant Positive Growth Rate Results in Exponential Growth

As this example shows, a constant positive growth rate adds ever-increasing numbers to a population during each succeeding time period; this is called **exponential growth.** If the size of an exponentially growing population is graphed against time, a characteristic shape called a **J-curve** will be produced (**FIG. 27-1**). Note that exponential growth does not require high growth rates. As you can see in Figure 27-1, exponential growth occurs at any constant positive rate of growth.

Differences in growth rates between populations or during changing environmental conditions can occur because of differences in birth rates, death rates, or both. Our deer herd, for example, might change from a growth rate of 0.3 to a growth rate of 0.1 either by a decreased birth rate (say from 1.0 to 0.8 per individual per year, perhaps because the does don't get enough food in a drought year) or by an increased death rate (from 0.7 to 0.9, if the number of predators increases). Of course, many different combinations of changes in birth and death rates could also reduce the growth rate to 0.1 per individual per year.

The Biotic Potential Is the Maximum Rate at Which a Population Can Grow

How many offspring an organism can produce is an inherited trait. Although the average number of offspring an individual produces each year varies from millions (for an oyster) to one or fewer (for people, porcupines, or elephants), a healthy organism of any species has the potential to replace itself many times during its reproductive lifetime. The **biotic potential** is the maximum rate at which a particular population can increase. Estimations of biotic potential for each species assume ideal conditions (unlimited resources, no predators, no diseases) that allow a maximum birth rate and a minimum death rate.

Factors that influence biotic potential include the following:

- The age of first reproduction
- The frequency of reproduction
- The average number of offspring produced each time the organism reproduces
- The organism's reproductive life span
- The death rate under ideal conditions

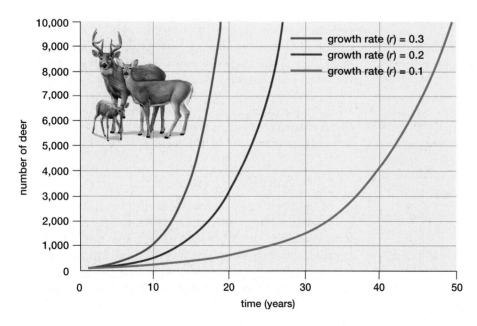

◀ **FIGURE 27-1 Exponential growth** The graph shows the growth of three hypothetical populations of deer, each starting with 100 adults with the same birth rate (1.0 per individual per year), but differing in death rate (0.7 to 0.9 per individual per year), producing three different growth rates: 0.3, 0.2, and 0.1 per individual per year. Any positive rate of growth produces an exponential (J-shaped) growth curve, but higher growth rates lead to higher populations much more quickly.

In modern times, Michelle Duggar of Arkansas probably comes close to demonstrating the realistic biotic potential of humans. Between 1988 and 2009, she gave birth to 19 children (including two sets of twins), all of whom remained healthy as of 2015. Duggar also had two miscarriages. Had all of her pregnancies resulted in live births, she would have borne about one child a year, for a little over 20 years. To see the population growth that this biotic potential can produce, let's start with one woman bearing one child each year from age 20 to age 40. Let's assume that half her children are female and that they all continue to bear children at this same reproductive rate, and then factor in reasonable estimates for death rates at various ages. In 100 years, this one woman could have 2.5 million descendants, and in 200 years, she could have almost 3 trillion!

How Many Children One Woman Can Bear?

A species with a low biotic potential, such as elephants (females may reproduce only once every 5 years), may take many years to reach the same population size that a species with a very high biotic potential, such as houseflies (which reproduce every few weeks), could reach in just a few months. Regardless of a species' biotic potential, however, unchecked exponential growth, even at very low rates, would eventually result in a staggering population size.

CHECK YOUR LEARNING

Can you ...

- explain how migration and natural increase cause population sizes to change?
- describe how population growth is calculated and how the growth rate interacts with population size to determine how populations grow per unit time?
- define *biotic potential* and list the factors that influence it?

CASE STUDY \ **CONTINUED**
The Return of the Elephant Seals

Female elephant seals reach sexual maturity at 3 to 4 years of age and produce one pup each year. Elephant seals breed harem-style, with a single large bull typically monopolizing and inseminating 30 to 100 females. Although males reach sexual maturity at 5 to 7 years of age, few are powerful enough to acquire a harem until 7 to 10 years of age; some are never able to compete. Further, although the maximum life span is about 20 years for females and 15 for males, only about 1 in 5 survives beyond 5 years of age. These factors combine to produce a biotic potential of about 12% per year. But elephant seal populations virtually never reproduce at their biotic potential. Why not?

27.2 HOW IS POPULATION GROWTH REGULATED?

In 1859, Charles Darwin wrote, "There is no exception to the rule that every organic being naturally increases at so high a rate that, if not destroyed, the Earth would soon be covered by the progeny of a single pair." In other words, a population cannot continue to grow indefinitely. In the following sections we discuss how population size results from the interaction between biotic potential and **environmental resistance,** which consists of all the factors that limit population growth, imposed by both the living and nonliving parts of the environment. Environmental resistance includes interactions among organisms, such as predation and competition for limited resources, and events such as freezing weather, storms, fires, floods, and droughts.

Exponential Growth in Natural Populations Is Always Temporary

Environmental resistance ensures that no natural population experiences exponential growth for very long.

Exponential Growth Occurs in Populations with Boom-and-Bust Cycles

Exponential growth occurs in populations that undergo regular cycles in which rapid population growth is followed by a sudden, massive die-off. These **boom-and-bust cycles** occur in a variety of organisms. Many short-lived, rapidly reproducing species—from photosynthetic microorganisms to insects—have seasonal population cycles that are linked to changes in rainfall, temperature, or nutrient availability, as is shown for a population of photosynthetic bacteria in **FIGURE 27-2**. Boom-and-bust cycles in these and other aquatic microorganisms can impact human health, as described in "Earth Watch: Boom-and-Bust Cycles Can Be Bad News" on page 494.

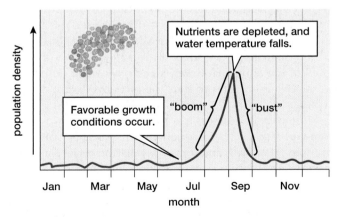

▲ **FIGURE 27-2 A boom-and-bust population cycle in photosynthetic bacteria** The density of a hypothetical population of photosynthetic bacteria in a lake. These microorganisms persist at a low level until early July. Then conditions become favorable for growth, and exponential growth occurs until early September, when the population plummets.

Earth WATCH

Boom-and-Bust Cycles Can Be Bad News

Although documented since Biblical times, during the past few decades, population explosions of sometimes toxic, single-celled photosynthetic protists and bacteria have occurred with increasing frequency throughout the world. Collectively called *harmful algal blooms (HABs)*, these population booms of toxic microorganisms kill fish and sicken people. They also cause major economic losses to the shellfish industry because clams, mussels, and scallops feed on these organisms and concentrate the poisons in their bodies, posing a hazard to human consumers. Some of the most damaging poisons are neurotoxins produced by dinoflagellates (protists), such as *Karenia brevis* (**FIG. E27-1**), which can reach densities of 20 million per liter of water. This and other protist species can cause red tides (see Fig. 21-9) that result in massive fish kills, usually in late summer.

Many of the bacteria and protists that produce harmful algal blooms are common residents of lakes and coastal waters. What causes these species to "bloom and boom"? Although the reasons are complex and vary with the species, warm water temperatures and adequate nutrients, such as phosphorus and nitrogen, are always required. Runoff of these nutrients from human agricultural activities has increased the frequency and intensity of HABs throughout the world. Global climate change may also contribute to the problem, because warmer waters foster more rapid growth of

◀ **FIGURE E27-1 One cause of harmful algal blooms** The dinoflagellate *Karenia brevis* (seen in this artificially colored SEM) causes harmful algal blooms in coastal waters of Florida and in the Gulf of Mexico. These blooms are sometimes called red tides, although the actual colors vary.

these protists and extend their growing season. The booming populations "bust" when the enormous populations of cells deplete the local water of nutrients and falling water temperatures in the autumn and winter further decrease their reproductive rate.

THINK CRITICALLY There are three major ways to minimize the effects of HABs: prevent their occurrence, disrupt or destroy ongoing blooms, and (to reduce human impacts) educate people about actions to take in case of a bloom. Find out more about each of these remedies. Which would be most preferable in the long run? Which are applicable right now? What are their costs and likely benefits?

In temperate climates, insect populations grow rapidly during the spring and summer, then crash with the freezing weather of winter. Female houseflies, for example, lay about 120 eggs at a time. The eggs hatch, and the resulting flies mature within 2 weeks. Thus, in many climates, seven generations can occur during spring and summer. If there were no environmental resistance, the seventh generation would "boom" to about 6 trillion flies—all descended from a single pregnant female. The actual population, of course, would be less than this, because many would be eaten by birds and bats or die from bacterial infections or other causes, but the fly population in early fall can still be very large. The "bust" part of the cycle occurs because hard frosts in autumn kill virtually all of the adults. However, a relatively small number of eggs survive the winter to start the cycle over again the following spring.

More complex factors can produce roughly 3- to 4-year boom-and-bust cycles for small rodents such as voles and lemmings (**FIG. 27-3**) and longer population cycles for hares, muskrats, and grouse. Lemming populations, for example, may grow until they overgraze their fragile arctic tundra ecosystem. Lack of food, increasing populations of predators, and social stress caused by crowding contribute to a sudden high mortality. Many more deaths occur as waves of lemmings emigrate from regions of high population density. During these mass migrations, lemmings are easy targets for predators. Many others drown as they encounter bodies of water and attempt to swim across. The reduced lemming

population eventually contributes to a decline in predator numbers, as well as a recovery of the plants on which the lemmings feed. These responses, in turn, set the stage for the next round of explosive growth in the lemming population.

Exponential Growth May Occur Temporarily if Environmental Resistance Is Reduced

In populations that do not experience boom-and-bust cycles, exponential growth may nevertheless occur under

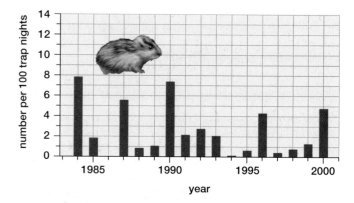

▲ **FIGURE 27-3 Boom-and-bust population cycles in a lemming population in the Canadian Arctic** This population follows a roughly 3- to 4-year cycle of boom and bust. The data are based on live trapping.

THINK CRITICALLY What factors might make these population data somewhat erratic?

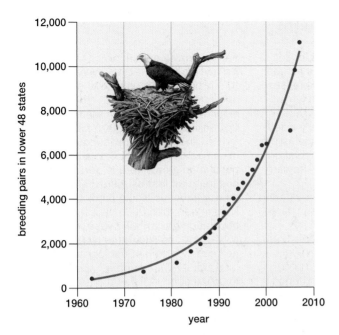

▲ **FIGURE 27-4 Exponential growth of the bald eagle popula-tion in the lower 48 United States** Hunting and the pesticide DDT reduced the bald eagle population in the lower 48 states to fewer than 500 nesting pairs in the early 1960s. Protection from hunting and banning DDT allowed the eagle population to grow exponen-tially for several decades, to more than 11,000 pairs by 2007. The smooth line is an exponential J-curve fit to the data points. Data from the U.S. Fish and Wildlife Service and the Center for Biological Diversity.

Protection from hunting and reduced DDT in the environ-ment allowed bald eagle populations to grow exponentially between the early 1960s and 2007, to about 11,000 nesting pairs (**FIG. 27-4**).

Exponential growth may also occur when individuals invade a new habitat with favorable conditions and little competition. **Invasive species** (see Chapter 28) are organ-isms with a high biotic potential that have been introduced (deliberately or accidentally) into ecosystems where they did not evolve and where they encounter little environmental resistance. Invasive species often show explosive population growth. For example, people introduced a few thousand cane toads into Australia in the 1930s to control beetles that were destroying sugar cane crops. In their new environment, cane toads, with their poisonous skin, suffered very little preda-tion. Cane toads also have a high biotic potential; females lay 7,000 to 35,000 eggs at a time. Spreading outward from their release point, cane toads now inhabit nearly 400,000 square miles in Australia and are migrating rapidly into new habi-tats, currently at a rate of about 30 miles a year. The cane toad population is now estimated at well over 200 million and growing. But even cane toads cannot increase their popula-tion forever. Why not?

Environmental Resistance Limits Population Growth Through Density-Dependent and Density-Independent Mechanisms

In all populations, environmental resistance eventually stops exponential growth.

Logistic Growth Occurs When New Populations Stabilize as a Result of Environmental Resistance

The maximum population size that can be sustained indefi-nitely without damage to an ecosystem is called the ecosys-tem's **carrying capacity (*K*).** Nutrients, energy, and space are the primary determinants of carrying capacity in many species. **Logistic population growth** is characteristic of populations that increase up to their environment's carrying capacity and then stabilize (**FIG. 27-5**). The curve that results when logistic growth is graphed is called an **S-curve,** after

special circumstances—for example, if food supply or habitat is increased, if predation is reduced, or if other environmental conditions improve. For example, before 1940, bald eagles in the United States were hunted, as were some of their favored prey species. Hunting, disturbing nests, or harassing eagles was prohibited in 1940, but shortly after World War II, the in-secticide DDT became widely used. An unintended side effect of DDT was to cause eagles to lay eggs with thin shells; the parents often crushed them during incubation (see "Health Watch: Biological Magnification of Toxic Substances" in Chapter 29 for more information). By the early 1960s there were only a few hundred nesting pairs of bald eagles in the lower 48 states. DDT was banned in the United States in 1972.

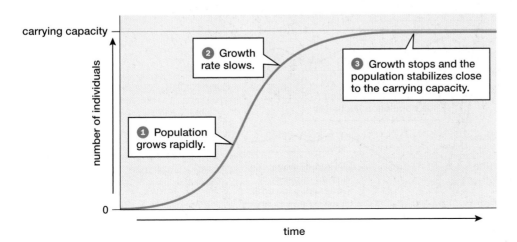

◀ **FIGURE 27-5 Logistic population growth** During logistic growth, a popula-tion begins growing exponentially, but the growth rate slows as the popula-tion encounters increasing density-dependent environmental resistance. Population growth finally ceases at or near carrying capacity (*K*). The result is a curve shaped like a "lazy S."

IN GREATER DEPTH | Logistic Population Growth

The equation for logistic population growth is:

$$G = rN\frac{(K - N)}{K}$$

This equation starts with the formula for exponential growth, $G = r \times N$, and multiplies the right-hand side by a factor, $(K - N)/K$, that models density-dependent environmental resistance. This new factor increasingly limits growth as the population size approaches the carrying capacity of its environment.

To see why, let's start with the value $(K - N)$. When we subtract the current population (N) from the carrying capacity (K), we get the number of individuals that can be added to the current population before it reaches carrying capacity. Now, if we divide this number by K, we get a fraction that expresses how far the population is from carrying capacity. If the fraction is close to 1.0, then the population is very small relative to the carrying capacity; as the fraction approaches 0, the population is getting closer and closer to its carrying capacity.

For example, consider a population of 10 foxes introduced onto an island with a carrying capacity of 1,000 foxes. The initial population ($N = 10$) is small, and the factor $(K - N)/K = (1,000 - 10)/1,000 = 990/1000$, or 0.99. The logistic growth equation therefore becomes $G = r \times N \times 0.99$, which is virtually identical to the exponential growth equation, $G = r \times N$. However, as N increases over time, $K - N$ becomes smaller and smaller, and the population growth curve begins to level off. When there are 900 foxes, $(1,000 - 900)/1,000 = 0.1$, so the logistic growth equation becomes $G = r \times N \times 0.1$, and the population growth rate is now only one-tenth as fast as the exponential growth equation would predict. When N equals K, then $(K - N) = 0$, and population growth will cease: $G = r \times N \times 0 = 0$, as illustrated by the final horizontal portion of the logistic curve (see Fig. 27-5). Therefore, when our fox population reaches the island's carrying capacity of 1,000, it stabilizes at about this number.

---

its general shape. "In Greater Depth: Logistic Population Growth" explains why logistic growth results in a population that stabilizes at carrying capacity.

An increase in population size above carrying capacity may be sustained for a short time. However, a population above its carrying capacity is living at the expense of resources that cannot regenerate as fast as they are being used up. A small overshoot above K is likely to be followed by a decrease in population, to a level below the original carrying capacity, until the resources recover and the original carrying capacity is restored. Repeated episodes of small overshoots and undershoots result in a population size that fluctuates around the carrying capacity (the green line in **FIG. 27-6**).

If a population far exceeds the carrying capacity of its environment, the consequences are more severe, because in this situation, the excess demands placed on the ecosystem are likely to destroy essential resources that may be unable to recover. This can permanently and severely reduce K, causing the population to decline to a fraction of its former size and then stabilize (the blue line in Fig. 27-6) or to disappear entirely (the red line in Fig. 27-6). For example, when reindeer were introduced onto St. Paul, an island off the coast of Alaska that lacks large predators, the reindeer population increased rapidly, seriously overgrazing the lichens that were their principal food source. Starvation then caused the reindeer population to plummet, as shown in **FIGURE 27-7**.

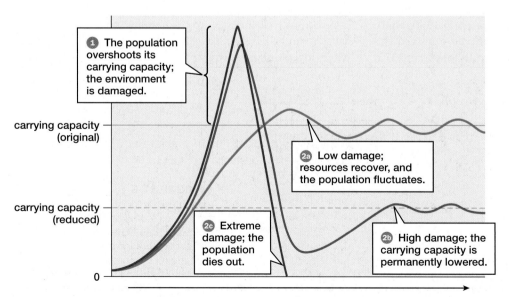

◀ **FIGURE 27-6 Consequences of exceeding carrying capacity** Populations can overshoot carrying capacity (K), but only for a limited time. Three possible results are illustrated.

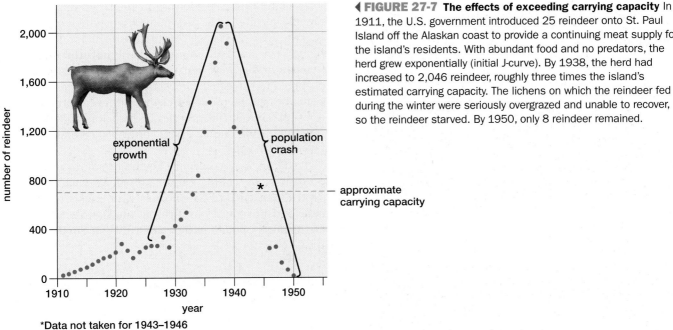

1911, the U.S. government introduced 25 reindeer onto St. Paul Island off the Alaskan coast to provide a continuing meat supply for the island's residents. With abundant food and no predators, the herd grew exponentially (initial J-curve). By 1938, the herd had increased to 2,046 reindeer, roughly three times the island's estimated carrying capacity. The lichens on which the reindeer fed during the winter were seriously overgrazed and unable to recover, so the reindeer starved. By 1950, only 8 reindeer remained.

Logistic population growth can occur in nature when a species moves into a new habitat, as ecologist Joseph Connell demonstrated for barnacle populations colonizing bare rock along a rocky ocean shoreline (**FIG. 27-8**). Initially, new settlers may find ideal conditions that allow their population to grow exponentially. As population density increases, however, individuals increasingly compete with one another, particularly for space, energy, and nutrients. Laboratory experiments using fruit flies have shown that competition for resources can control population size by reducing both the birth rate and the average life span. During logistic population growth, as environmental resistance increases, population growth slows and eventually stops at approximately the carrying capacity of the environment. In nature,

conditions are never completely stable, so carrying capacity and population size usually fluctuate somewhat from year to year (the green and blue lines in Fig. 27-6).

Two forms of environmental resistance restrict population growth. **Density-independent** factors limit population size regardless of the population density (the number of individuals per unit area). **Density-dependent** factors, in contrast, increase in effectiveness as the population density increases.

Density-Independent Factors Limit Populations Regardless of Density

The most important natural density-independent factors are weather (short-term atmospheric conditions such as temperature, rainfall, and wind) and climate (long-term weather patterns), which are responsible for most boom-and-bust population cycles. For example, in temperate climates, populations of many insects and annual plants are limited in size by the number of individuals that can be produced before the first hard freeze. Such populations are controlled by the climate because they typically do not reach carrying capacity before winter sets in. Hurricanes, floods, droughts, and fire can also have profound effects on local populations, regardless of density.

Density-Dependent Factors Become More Effective As Population Density Increases

For long-lived species, the most important elements of environmental resistance are density-dependent factors, such as consumer-prey interactions (including predation and parasitism) and competition, which limit population growth more strongly as population density increases.

Predators Exert Density-Dependent Controls on Populations

Predators are organisms that eat other organisms, called their **prey.** We will use a broad definition of predators, to

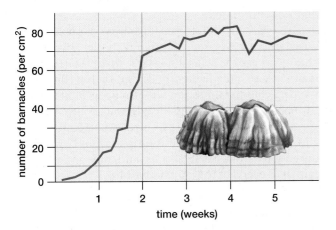

▲ **FIGURE 27-8 A logistic curve in nature** Barnacles are crustaceans whose larvae are carried in ocean currents to rocky seashores where they settle, attach permanently to rock, and grow into the shelled adult form. On a bare rock, the number of settling larvae and newly metamorphosed juveniles produces a logistic growth curve as competition for space limits their population density.

(a) Predators often kill weakened prey

(b) Predator populations often increase when prey are abundant

▲ **FIGURE 27-9 Predators help control prey populations (a)** A pack of grey wolves has brought down an elk that may have been weakened by age or parasites. **(b)** The snowy owl produces more chicks when prey (such as lemmings) are abundant.

include both *carnivores* (animals that eat other animals) and *herbivores* (animals that eat plants). Often, prey are killed directly and eaten (**FIG. 27-9a**), but not always. When deer browse on the buds of bushes and young trees, for example, or when gypsy moth larvae feed on the leaves of oaks, the plants are harmed but usually not killed.

Predators often eat a variety of prey, depending on which is most abundant and easiest to find. Coyotes might eat more field mice when the mouse population is high, but switch to eating more ground squirrels as the mouse population declines. In this way, predators often exert density-dependent population control over more than one prey population.

Predator populations often grow as their prey becomes more abundant, which make them even more effective as control agents. For predators such as the arctic fox and snowy owl, which rely heavily on lemmings for food, the number of offspring they can produce is determined by the abundance of prey. Snowy owls (**FIG. 27-9b**) hatch up to 12 chicks when lemmings are abundant, but may not reproduce at all in years when the lemming population has crashed.

In some cases, an increase in predators might cause a dramatic decline in the prey population, which in turn may result in an eventual decline in the predator population. This pattern can result in out-of-phase **population cycles** of predators and prey. In natural ecosystems, both predators and prey are subjected to a variety of other influences, so clear-cut examples of such cycles in nature are rare. However, out-of-phase population cycles of predators and their prey have been demonstrated under controlled laboratory conditions (**FIG. 27-10**).

Predators may contribute to the overall health of prey populations by culling those individuals that are poorly adapted, weakened by age or disease, or unable to find adequate food and shelter. In this way, predation may maintain healthy prey populations near a density that can be sustained by the resources of the ecosystem.

Parasites Spread More Rapidly in Dense Populations A **parasite** is an organism that lives in or on a larger organism, called its **host,** feeding on the host and harming it. Although some kill their hosts, many parasites benefit by having their host remain alive. Parasites include tapeworms that live in the intestines of mammals, lice that cling to a host's skin or hair, and disease-causing microorganisms. Parasites are density-dependent factors: Because most parasites cannot easily travel long distances, they spread more readily among hosts with dense populations. For example, plant diseases often spread rapidly through densely planted crops, and childhood diseases spread swiftly through schools and day-care centers. Even when parasites do not kill their hosts directly, they influence population size by weakening their hosts and making them more susceptible to death from other causes, such as harsh weather or predators. Organisms weakened by parasites are also less likely to reproduce. Parasites, like predators, often contribute to the death of less-fit individuals, producing a balance in which the host population is regulated but not eliminated.

This balance can be destroyed if parasites or predators are introduced into regions where local prey species have had no opportunity to evolve defenses against them. For example, the smallpox virus, inadvertently carried by traveling Europeans during colonial times, caused heavy loss of life among native inhabitants of North America, Hawaii, South America, and Australia. The chestnut blight fungus, introduced from Asia, has almost eliminated chestnut trees from U.S. forests. Introduced predators, such as rats and mongooses, have exterminated several of Hawaii's native birds.

Competition for Resources Helps Control Populations The resources that usually determine carrying capacity—space, energy, and nutrients—may be inadequate to support all the organisms that need them. **Competition,** interactions among individuals attempting to use the same finite resources, limits population size in a density-dependent manner. There are two major forms of competition:

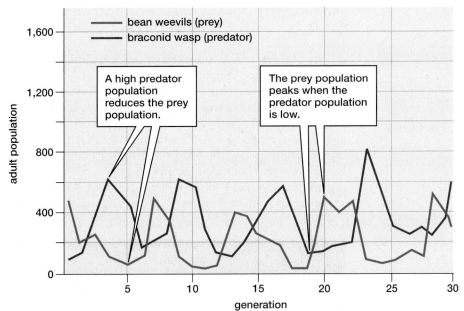

FIGURE 27-10 Experimental predator–prey cycles Tiny braconid wasps lay their eggs on bean weevil larvae, which provide food for the newly hatched wasp larvae. A large weevil population ensures a high survival rate for wasp offspring, increasing the predator population. Then, under intense predation, the weevil population plummets, reducing the food available to the next generation of wasps, whose population declines as a result. Reduced predation then allows the weevil population to increase rapidly, and so on.

Graph labels: bean weevils (prey); braconid wasp (predator). "A high predator population reduces the prey population." "The prey population peaks when the predator population is low."

interspecific competition (competition among individuals of different species) and **intraspecific competition** (competition among individuals of the same species). Because the needs of members of the same species for water and nutrients, shelter, breeding sites, and other resources are almost identical, intraspecific competition is an extremely important density-dependent mechanism of population control.

When population densities increase and competition becomes intense, some types of animals, for example, lemmings and locusts, react by emigrating. Emigrating swarms of locusts periodically plague parts of Africa, consuming all vegetation in their path (**FIG. 27-11**).

Density-Independent and Density-Dependent Factors Interact

The size of a population at any given time is the result of complex interactions between density-independent and density-dependent forms of environmental resistance. For example, a stand of pines weakened by drought (a density-independent factor) may more readily fall victim to the pine bark beetle (density-dependent). Likewise, a caribou weakened by hunger (density-dependent) and attacked by parasites (density-dependent) is more likely to be killed by an exceptionally cold winter (density-independent). Human activities increasingly impose both density-independent and density-dependent limitations on natural populations. Examples include bulldozing grasslands and their prairie dog towns to build shopping malls and housing tracts or felling rain forests to replace them with croplands: A high percentage of the prairie dogs or rain-forest animals will be exterminated, regardless of the population density before the land was cleared, so the initial effect of these human actions is density-independent. However, some of the animals may be able to move to nearby undisturbed habitat, increasing the population density in these areas. The resulting increased competition for food, nesting places, and other essentials will exert density-dependent controls on these now-larger populations. In some cases, the increased population density could result in overexploitation of crucial resources such as prey animals or nutritious plants, causing population crashes (see Figs. 27-6 and 27-7).

FIGURE 27-11 Emigration In response to overcrowding and lack of food, locusts emigrate in swarms, devouring nearly all vegetation (and even feeding on each other) as they go.

THINK CRITICALLY What benefits does mass emigration give to animals such as locusts or lemmings? Are there parallels to human emigrations?

CHECK YOUR LEARNING

Can you ...

• describe exponential growth, the conditions under which it occurs, and what shape of curve it produces?
• explain the stages of the logistic growth curve and describe the possible outcomes when populations overshoot carrying capacity?
• describe the two major forms of environmental resistance and provide examples of each?

CASE STUDY   CONTINUED

The Return of the Elephant Seals

Elephant seal populations experience both density-dependent and density-independent environmental resistance. Density-dependent factors include predation by great white sharks and orcas (probably a minor factor), potential food scarcity (unknown impact), and deaths during reproduction, which is probably the major density-dependent factor. The dominant males chase challengers away from their harems, stressing the females and sometimes crushing pups beneath tons of charging bull; the denser the seal population on the beach, the more pups are lost. El Niño weather years often bring strong storms to California, causing density-independent environmental resistance. Storms erode the beaches; high tides and waves sometimes submerge beaches and wash away pups. In severe El Niño years, as many as half the pups in a colony may be lost. Overall, however, elephant seals have secure breeding sites, abundant food, and few predators, and their populations are stable or increasing. Does this mean that their future is secure? Find out in "Case Study Revisited" at the end of the chapter.

27.3 HOW DO LIFE HISTORY STRATEGIES DIFFER AMONG SPECIES?

As you learned in Chapter 15, natural selection favors traits that promote the production of offspring that, in turn, survive and reproduce. All organisms have limited resources to invest in their offspring and in their own survival; no organism can produce endless numbers of offspring and provide extensive parental care for all of them. The resulting **life history** strategies—when to reproduce, how many offspring to produce at a time, and how much energy and resources to devote to each offspring—vary tremendously among species.

Many factors influence life history strategies. Important considerations are the stability of a species' environment, the mortality rate and the accompanying likelihood of multiple opportunities to reproduce before dying, and the likelihood that a population can reach, and remain at, the carrying capacity of its environment. Species occupying the two extremes of life history strategies are often called *r*-selected species and *K*-selected species, named after the *r* (rate of growth)

and *K* (carrying capacity) terms, respectively, in the logistic growth equation (see "In Greater Depth: Logistic Population Growth").

At one extreme, ***r*-selected species** typically live in a rapidly changing, unpredictable environment and are unlikely to reach carrying capacity before catastrophe intervenes: a hard freeze in temperate climates, months of drought in a desert, or (for parasites) the death of a host. These species have typically evolved characteristics that favor rapid reproduction (hence the term *r*-selected). They mature rapidly, have a short life span, produce a large number of small offspring, and provide little parental care (**FIG. 27-12a**). It is common for *r*-selected species to reproduce just a few times,

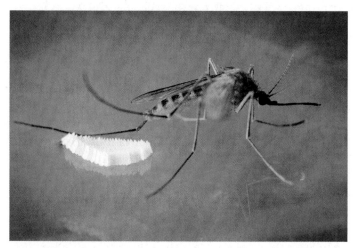

(a) *r*-selected: A female *Culex* mosquito lays a raft of 100 to 300 eggs every few days.

(b) *K*-selected: An African bush elephant and its calf.

▲ **FIGURE 27-12 Typical *r*-selected and *K*-selected species (a)** Most mosquitoes of the genus *Culex* have short life spans, produce numerous small eggs, and provide no parental care. Most offspring are eaten by predators in the larval stage. **(b)** African bush elephants have long life spans, produce single offspring after a 21-month-long pregnancy, and nurse their young for several years. Although a few infants fall prey to lions, elephants have no significant natural predators. Unless they fall victim to human poachers, elephants have a very high chance of surviving until old age.

sometimes only once, in their lives. The vast majority of the offspring die young. You will recognize these traits as typical of boom-and-bust species.

At the other extreme, **K-selected species** typically live in stable environments and often develop populations that persist near the carrying capacity of their environment for long periods of time (hence the term *K*-selected). *K*-selected species usually mature slowly, have a long life span, produce small numbers of fairly large offspring, and either provide significant parental care (as large mammals do) or package significant nutrients along with the embryo (as in acorns, walnuts, and other large-seeded trees). This parental investment allows a fairly high percentage of their offspring to live to maturity (**FIG. 27-12b**). Many *K*-selected species produce offspring one or more times a year for several, sometimes many, years.

Most species do not fit neatly into these two extreme categories, but have a combination of traits from the two extremes. For example, giant clams live mostly in stable tropical ocean environments, are slow to mature (5 to 10 years), and spawn many times in a typical life span of scores of years. However, they produce hundreds of millions of eggs at a time, have very few nutrients packaged in their eggs, and provide no parental care. As a result, perhaps only 1 egg in a billion successfully matures into an adult clam. Similarly, redwood trees live in a mild coastal climate, typically mature at 5 to 15 years of age, and may reproduce every year for 500 to 1,000 years, but the vast majority of their hundreds of thousands of tiny seeds never sprout. Both species frequently have populations that persist near the carrying capacity of their

environments for centuries, or even millennia. On the other hand, mice, which are considered to be strongly *r*-selected, invest significant parental care in their offspring, including about 3 weeks of pregnancy and another 3 weeks of nursing before the pups are weaned. This is not much compared to a human or an elephant, but much more than clams, insects, or dandelions. As these examples illustrate, there are many interacting factors that contribute to a species' life history. Many evolutionary ecologists suggest that a simple "*r*-selected" versus "*K*-selected" dichotomy does not do justice to crucial aspects of a species' evolution and interactions with living and nonliving components of its ecosystem.

A Species' Life History Predicts Survival Rates over Time

Species differ tremendously in their chances of dying at any given phase of their life cycle. **Survivorship tables** track groups of organisms (born at the same time) throughout their lives, recording how many survive in each succeeding year or other unit of time (**FIG. 27-13a**). When the proportion of the population that survives is plotted against the species' life span, the resulting **survivorship curves** are characteristic of the species in the environment where the data were collected (populations of the same species may show somewhat different survivorship curves if they live in different environments). Male and female survivorship may also differ. The three principal types of survivorship curves are shown in **FIGURE 27-13b**: late-loss, constant-loss, and early-loss, according to the part of the life span during which

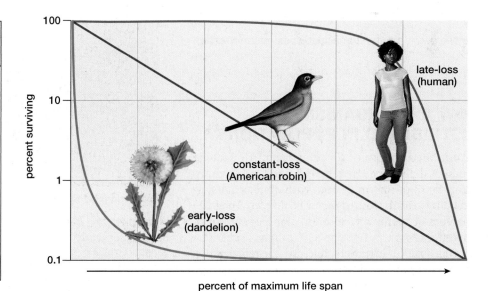

Age	Number of survivors
0	100,000
10	99,224
20	98,910
30	98,011
40	96,798
50	94,295
60	88,770
70	78,069
80	57,188
90	23,619
100	1,968
110	0

(a) A survivorship table

(b) Survivorship curves

▲ **FIGURE 27-13 Survivorship (a)** A survivorship table for the U.S. population in 2010, showing how many people are expected to remain alive at increasing ages, for each 100,000 people born. Plotting these data produce a curve similar to the blue curve in part (b). **(b)** Three types of survivorship curve are shown; note the logarithmic *y*-axis. Data in part (a) from Arias, E. 2014. "United States Life Tables, 2010." *National Vital Statistics Reports* 63(7): 1–63.

most deaths occur. The *y*-axis uses a logarithmic scale to compare the *proportion of the population* that survives at different times in the life span.

Late-loss populations have relatively low juvenile death rates, and most individuals survive to old age. The resulting survivorship curve is convex. Late-loss survivorship curves are characteristic of humans and other large and long-lived animals such as elephants and mountain sheep. These species produce relatively few offspring, which are protected and nourished by their parents during early life.

Constant-loss populations consist of individuals that have an equal chance of dying at any time during their life span, producing a straight line survivorship curve. This pattern is seen in some birds such as gulls and the American robin, in some species of turtles, and in laboratory populations of organisms that reproduce asexually, such as hydra and bacteria. *K*-selected species usually have late-loss or constant-loss survivorship curves.

Early-loss populations have a high death rate early in life, so the survivorship curve is concave. Although the death rate is very high among the young, those individuals that reach adulthood have a reasonable chance of surviving to old age. Early-loss survivorship curves are characteristic of organisms that produce large numbers of offspring but give them little or no care after they hatch or germinate. Most invertebrates (including giant clams), many fish and amphibians, and most plants (including redwoods) exhibit early-loss survivorship curves. Most *r*-selected species have early-loss survivorship curves.

CHECK YOUR LEARNING

Can you ...

- explain the important factors that determine a species' life history?
- describe and graph the three principal types of survivorship curves?

27.4 HOW ARE ORGANISMS DISTRIBUTED IN POPULATIONS?

Populations of different species show characteristic spacing of their members, determined by both behavior and environment. Spatial distributions may vary with time, changing with the breeding season, for example. Ecologists recognize three major types of spatial distribution: clumped, uniform, and random (**FIG. 27-14**).

Populations whose members live in clusters have a **clumped distribution.** Examples include family or social groupings, such as elephant herds, wolf packs, flocks of birds, and schools of fish (**FIG. 27-14a**). What are the advantages of clumping? Birds in flocks benefit from many eyes to spot food, such as a tree full of fruit. Schooling fish and flocks of birds may confuse predators with their sheer numbers. Predators, in turn, sometimes hunt in groups, cooperating to bring down large prey (see Fig. 27-9a). Some species, such as elephant seals, form temporary groups to mate and care for

(a) Clumped distribution

(b) Uniform distribution

(c) Random distribution

▲ **FIGURE 27-14 Spatial distribution in populations (a)** A school of fish may confuse predators with their numbers. **(b)** Colonies of gannets consist of a remarkably uniform pattern of nests, each just out of pecking range of its neighbors. **(c)** Dandelions and other plants with wind-borne seeds often have random population distributions because the seeds germinate wherever the wind deposits them.

their young. Other plant or animal populations cluster because resources are localized. In North American prairies, for example, stands of cottonwood trees cluster along the banks of streams, because streambanks are typically the only places with enough moisture in the soil for cottonwoods to grow.

Organisms with a **uniform distribution** maintain a relatively constant distance between individuals. Among plants, mature desert creosote bushes are often spaced very evenly. This spacing results from competition among their root systems, which occupy a roughly circular area around each plant. The roots efficiently absorb water and other nutrients from the desert soil, reducing the survival of nearby plants. A uniform distribution may also occur among animals, such as songbirds, that defend territories during their breeding seasons, thereby spacing themselves fairly evenly throughout suitable habitat. Sometimes a uniform distribution is simply a way to avoid constant harassment by neighbors. Some birds, including penguins, gulls, and gannets, nest in colonies; their nests are often evenly spaced, just out of pecking reach of one another (FIG. 27-14b).

Organisms with a **random distribution** are rare. Individuals of species with random distributions neither attract nor repel each other and do not form social groupings. The resources they need must be more or less equally available throughout the area they inhabit, and those resources must be fairly abundant, minimizing competition. The dandelions in your lawn are often randomly distributed, as they sprout from windblown seeds that land in random locations in well-fertilized soil (FIG. 27-14c). There are probably no vertebrate animal species that maintain a random distribution throughout the year; most interact socially, at least during the breeding season.

CHECK YOUR LEARNING
Can you ...
- describe the three types of spatial distribution and describe the typical characteristics of populations with each type of distribution?

27.5 HOW IS THE HUMAN POPULATION CHANGING?

Humans possess enormous brainpower and dexterous hands that can shape the environment to our demands. As our species was evolving, natural selection favored those with the ability and the drive to bear and nurture many offspring, which helped ensure that a few would survive. Ironically, this characteristic may now threaten us and the biosphere on which we depend.

The Human Population Has Grown Exponentially

Compare the graph of human population growth in FIG. 27-15a with the exponential growth curves in Figures 27-1 and 27-2:

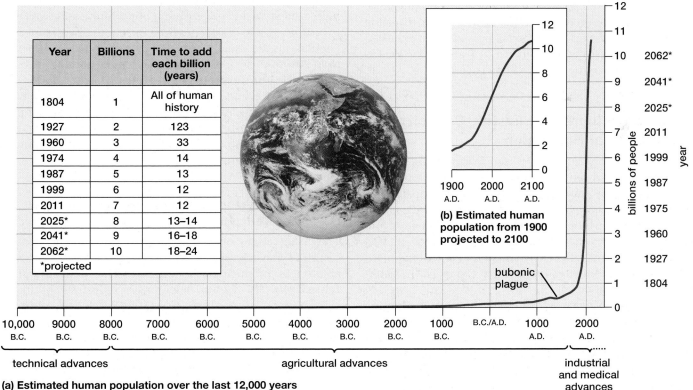

Year	Billions	Time to add each billion (years)
1804	1	All of human history
1927	2	123
1960	3	33
1974	4	14
1987	5	13
1999	6	12
2011	7	12
2025*	8	13–14
2041*	9	16–18
2062*	10	18–24
*projected		

(b) Estimated human population from 1900 projected to 2100

(a) Estimated human population over the last 12,000 years

▲ **FIGURE 27-15 Human population growth (a)** The human population from the Stone Age to the present has shown continued exponential growth as various advances overcame environmental resistance. Note the time intervals over which additional billions were added. **(b)** An expanded *x*-axis for the years 1900 to 2100 indicates that human population growth is slowing down. Note, however, that the population data after 2015 are projections. Actual population growth may not match these projections. Photo courtesy of NASA.

Each has the J-curve characteristic of exponential growth. Initially, the human population grew slowly. The population was probably about 1 million people in 10,000 B.C. (no one knows for sure, of course), and reached about 1 billion in the early 1800s—almost 12,000 years later. But the second billion was added in about 125 years, and less than 100 years after that, our population reached 7 billion (see the table in Fig. 27-15a). The human population has increased so fast that about 6% of all the people who have *ever* lived on Earth are alive today.

However, although the human population continues to grow rapidly, it may no longer be growing exponentially (**FIG. 27-15b**). According to the U.S. Census Bureau, the world's annual growth rate (rate of natural increase) declined from 2.2% in 1963 to 1.1% in 2015. Our numbers now grow by about 78 million each year, down from a peak of about 88 million in 1990. Are humans starting to enter the final bend of an S-shaped logistic growth curve (see Fig. 27-5) that will eventually lead to a stable population? Only time will tell. Nevertheless, by 2050, the human population is expected to exceed 9 billion, and still be growing, although more slowly than at present. How has humanity managed to sustain such prolonged, rapid population growth? Are human populations exempt from the effects of environmental resistance?

No, humans aren't exempt, but we have responded to environmental resistance by devising ways to overcome it. To accommodate our growing numbers, we have altered the face of the globe. Is there an ultimate limit to Earth's carrying capacity? Have we already reached or possibly even exceeded it? We explore these questions in "Earth Watch: Have We Exceeded Earth's Carrying Capacity?" on page 508.

People Have Increased Earth's Capacity to Support Our Population

Human population growth has been made possible by advances that circumvented various types of environmental resistance, thereby increasing Earth's carrying capacity for people. *Technical advances* by early humans included controlling fire, inventing tools and weapons, building shelters, and designing protective clothing. Tools and weapons allowed people to hunt more effectively and obtain additional high-quality food, while shelter and clothing expanded the habitable areas of the globe.

Domesticated crops and animals supplanted hunting and gathering in many parts of the world about 8,000 to 10,000 years ago. These *agricultural advances* provided people with a larger, more dependable food supply. An increased food supply resulted in a longer life span and more childbearing years. However, a high death rate from disease restrained population growth for thousands of years until major *industrial and medical advances* permitted a population explosion, beginning in the mid-eighteenth century. Improved sanitation and medical progress, including the discovery of antibiotics and vaccines, combined to dramatically decrease the death rate from infectious diseases.

World Population Growth Is Unevenly Distributed

As is true of any population, the population of an individual country is a function of natural increase and migration. In some countries, such as the United States, immigration is a significant cause of population growth. In most countries, however, natural increase is the principal cause of population changes. In the long run, natural increase depends on a population's *fertility rate:* the average number of children that each woman bears. If immigration and emigration are equal, a population will eventually stabilize if parents, on average, have just the number of children required to replace themselves; this is called **replacement level fertility (RLF).** Replacement level fertility is 2.1 children per woman (rather than exactly 2) because not all children survive to maturity. As we will see below, even at replacement level fertility, a population with a high proportion of children will continue to grow for many years, a phenomenon often called *momentum,* as this large number of children mature and have families of their own.

Today, countries are usually described as more or less developed. People in **more developed countries**—including Australia, New Zealand, Japan, and countries in North America and Europe—benefit from a relatively high standard of living, with access to modern technology and medical care, including readily available contraception. Average income is high, education and employment opportunities are available to both sexes, and death rates from infectious diseases are low. Populations are usually stable or declining. However, fewer than 20% of the world's people live in more developed countries. In the **less developed countries** of Central and South America, Africa, and much of Asia—home to more than 80% of humanity—the average person lacks these advantages. In addition, the populations of less developed countries are usually expanding, sometimes rapidly. How does development affect population growth?

Countries Progress from Less Developed to More Developed Through the Demographic Transition

The historical rate of population growth in more developed countries has changed over time in reasonably predictable stages, producing a pattern called the **demographic transition** (**FIG. 27-16**). Before major industrial and medical advances occurred, today's more developed countries were in the *pre-industrial stage*, with relatively small, stable, or slowly growing populations in which high birth rates were balanced by high death rates. This was followed by the *transitional stage*, in which food production increased and health care improved. These advances caused death rates to fall, while birth rates remained high, leading to an explosive rate of natural increase. During the *industrial stage*, birth rates fell as more people moved from small farms to cities (where children were less important as a source of labor), contraceptives became more readily available, and opportunities for women increased. More developed countries are now in the *post-industrial stage* of the demographic transition, and, in most cases, their populations are stable, or even decreasing, with low rates of both births and deaths.

In less developed countries, such as most countries in Central and South America, Asia (excluding China and Japan), and Africa, medical advances have decreased death rates and increased life span, but birth rates remain relatively high. Although China is also considered a less developed country, years ago, as its population approached 1 billion, the Chinese government recognized the negative impacts of continued population growth and instituted policies (some of them punitive and unpopular) that brought China's fertility rate below replacement level. As a result, China's population has almost stopped growing, and India is expected to become the world's most populous country by about 2025. China has since somewhat relaxed its "one-child" policies; it remains to be seen whether its birth rate will increase.

Most less developed countries are in the late-transitional or the industrial stage of the demographic transition. In many of these nations, adult children provide financial security for aging parents. Young children may also contribute significantly to family income by working on farms or in factories. Social factors drive population growth in countries where children confer prestige and where religious beliefs promote large families. Also, in less developed countries, many people who would like to limit their family size lack access to contraceptives. For example, in the West African nation of Nigeria, less than 20% of couples use modern contraceptive methods, and the average woman bears 5.2 children. Nigeria is suffering from soil erosion, water pollution, and the loss of forests and wildlife, suggesting that its carrying capacity is already compromised. With 43% of its 182 million people under the age of 15, continued population growth there is inevitable.

As in Nigeria, population growth is highest in the countries that can least afford it, as a result of positive feedback, in which past and current population growth tends to promote future population increases. As more people compete for the same limited resources, poverty continues. Poverty diverts children away from schools and into activities that help support their families. A lack of education and lack of access to contraceptives then contributes to continued high birth rates. As a result, the population increases, which tends to keep the people poor, so the cycle continues. Of the 7.2 billion people on Earth in 2015, about 6 billion resided in less developed countries. Although fertility rates in some less developed countries, such as Brazil, have declined because of social changes and increased access to contraceptives, most are still above RLF. Therefore, the human population will continue to grow, in less developed countries and the world as a whole, for many years.

The Age Structure of a Population Predicts Its Future Growth

Age structure diagrams show age groups on the vertical axis and the numbers or percentages of individuals in each age group on the horizontal axis, with males and females shown on opposite sides. These diagrams not only illustrate current age distributions, but also predict future population growth. Age structure diagrams all rise to a peak that reflects the maximum human life span, but the shape of the rest of the diagram reveals whether the population is expanding, stable, or shrinking. If adults in the reproductive age group (15 to 44 years) are having more children (the 0- to 14-year age group) than are needed to replace themselves, the population is above RLF and is expanding. Its age structure will be

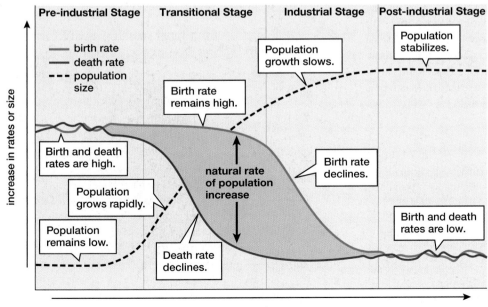

◀ **FIGURE 27-16 The demographic transition** A demographic transition typically begins with a relatively stable and small population with high birth and death rates. Death rates decline first, causing the population to increase. Then birth rates decline, causing the population to stabilize at a larger size with relatively low birth and death rates.

roughly triangular (**FIG. 27-17a**). If the adults of reproductive age have just the number of children needed to replace themselves, the population is at RLF. A population that has been at RLF for many years will have an age structure diagram with relatively straight sides (**FIG. 27-17b**). In shrinking populations, the reproducing adults have fewer children than are required to replace themselves, causing the age structure diagram to narrow at the base (**FIG. 27-17c**).

FIGURE 27-18 shows age structures for the populations of developed and developing countries for 2015, with projections for 2050. Even if rapidly growing countries were to achieve RLF immediately, their populations would continue to increase for decades. Why? When the number of children exceeds the number of reproducing adults, this creates momentum for future growth, as these children mature and enter their reproductive years. For example, when China reached RLF in the early 1990s, about 28% of its population was under age 15; in a stable human population, fewer than 20% are children. Because of this momentum, China has since grown by almost 200 million people. China will probably have a stable population soon, however, because now only 17% of its people are younger than 15. In contrast, children make up about 40% of the population of Africa, so its population will continue to increase rapidly.

Fertility in Some Nations Is Below Replacement Level

Although a reduced population will ultimately offer tremendous benefits for both the world's people and the biosphere that sustains them, current economic structures in many countries are based on growing populations. The difficult adjustments necessitated as populations decline—or even merely stabilize—motivate governments to adopt policies that encourage more childbearing and continued growth.

TABLE 27-1 provides growth rates for various world regions. In Europe, the population is shrinking slightly, by 0.1% per year, and the average fertility rate is 1.6—substantially below RLF—as many women delay or forgo having children. This situation raises concerns about the availability of future workers and taxpayers to support the resulting increase in the percentage of elderly people. As a result, several European countries are offering or considering incentives (such as large tax breaks) for couples to have children.

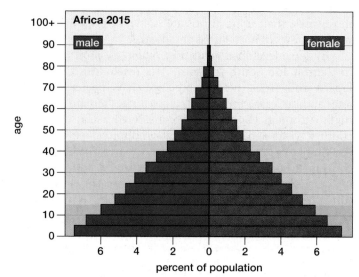

(a) Africa: A rapidly growing population

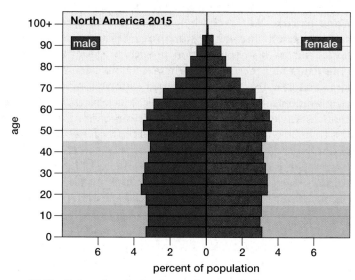

(b) North America: A slowly growing population

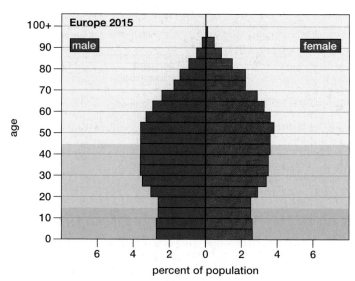

(c) Europe: A slowly declining population

▶ **FIGURE 27-17 Age structure diagrams (a)** The age structure for Africa illustrates a rapidly growing population that is projected to almost double by 2050. **(b)** North America represents a more slowly growing population that is still projected to add more than 100 million people by 2050 (about a 24% increase). **(c)** Europe's age structure is of a slowly shrinking population, projected to decline by more than 35 million (about 5%) by 2050. Background colors from bottom to top indicate age groups: prereproductive children (0 to 14 years), reproductive age adults (15 to 44 years), and postreproductive adults (45 to 100 years). Data from the U.S. Census Bureau, International Data Base.

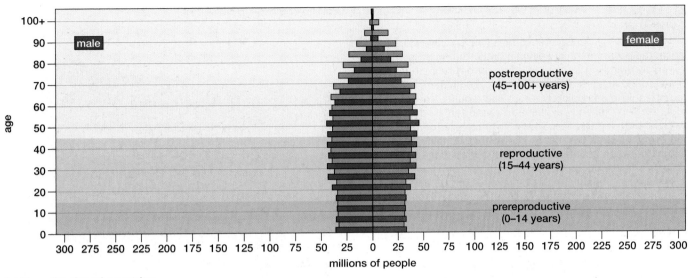

(a) More developed countries

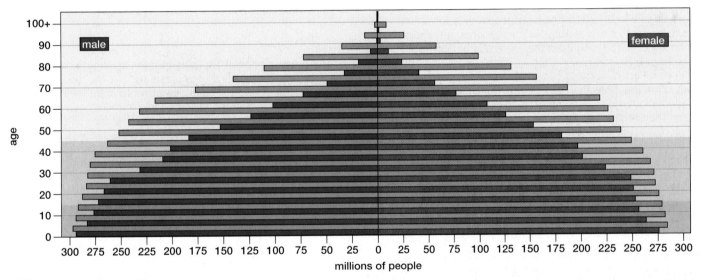

(b) Less developed countries

▲ **FIGURE 27-18 Age structure diagrams of more and less developed countries** Note that the predicted difference in the proportion of children compared to the proportion of adults in less developed countries is smaller in 2050 than in 2015, as these populations approach RLF. However, the large numbers of young people in less developed countries who will be entering their child-bearing years will cause continued population growth. Data from the U.S. Census Bureau, International Data Base.

THINK CRITICALLY How does a fertility rate above RLF produce positive feedback (in which a change creates a situation that amplifies itself) for population growth?

TABLE 27-1	Average Population Statistics by World Region: 2015	
Region	**Fertility Rate**	**Rate of Natural Increase (%)**
World	**2.4**	**1.1**
Less developed countries	**2.5**	**1.1**
Africa	4.3	2.3
Latin America/Caribbean	2.1	1.1
Asia*	2.5	1.4
China	1.6	0.4
More developed countries	**1.7**	**0.1**
Europe	1.6	−0.1
North America	2.0	0.5

*Excluding China.

Data from the U.S. Census Bureau International Data Base.

Earth WATCH

Have We Exceeded Earth's Carrying Capacity?

In Côte d'Ivoire, a country in western Africa, the government is waging a battle to protect some of its rapidly dwindling tropical rain forest from illegal hunters, farmers, and loggers. Officials destroy the shelters of the squatters, who immediately return and rebuild. One such squatter, Sep Djekoule, explained, "I have ten children and we must eat. The forest is where I can provide for my family, and everybody has that right." His words exemplify the conflict between population growth and wise management of Earth's finite resources.

How many people can Earth sustain? The Global Footprint Network, consisting of an international group of scientists and professionals from many fields, is attempting to assess humanity's *ecological footprint* (see Chapter 31). This project compares human demand for resources to Earth's capacity to supply these resources in a sustainable manner. "Sustainable" means that the resources can be renewed indefinitely and that the ability of the biosphere to supply them is not diminished over time. Is humanity living on the "interest" produced by our global endowment, or are we eating into the "principal"? The Global Footprint Network concluded that, in 2010 (the most recent year for which complete data are available), humanity consumed more than 150% of the resources that were sustainably available. In other words, to avoid damaging Earth's resources (thus reducing Earth's carrying capacity), our population in 2010 would require more than 1½ Earths. But by 2015, we had added about 400 million more people. Because people have used their technological prowess to overcome environmental resistance, our collective ecological footprint now dwarfs Earth's sustainable resource base, reducing Earth's future capacity to support us.

For example, the human population now uses almost 40% of Earth's productive land for crops and livestock. Despite this, the United Nations estimates that more than 800 million people are undernourished, including an estimated 25% of the population of sub-Saharan Africa. Erosion reduces the ability of land to support both crops and grazing livestock (FIG. E27-2). The quest for farmland drives people to clear-cut forests in places where the soil is poorly suited for agriculture. The demand for wood also causes large areas to be deforested annually, causing the runoff of much-needed fresh water, the erosion of topsoil, the pollution of rivers and oceans, and an overall reduction in the ability of the land and water to support not only future crops and live-stock, but also the fish and other wild animals that people harvest for food. Human consumption of food, wood, and, more recently, biofuels (crops that provide fuel) drives the destruction of tens of millions of acres of rain forest annually (see Chapter 31).

The United Nations estimates that about 60% of commercial ocean fish populations are being harvested at their maximum sustainable yield, and another 30% are being overfished. In parts of India, China, Africa, and the United States, underground water stores are being depleted to irrigate cropland far faster than they are being refilled by rain and snow. Because irrigated land supplies about 40% of human food crops, water shortages can rapidly lead to food shortages.

▲ **FIGURE E27-2 Overgrazing can lead to the loss of productive land** Human activities, including overgrazing, deforestation, and poor agricultural practices, reduce the productivity of the land.

Our present population, at its present level of technology, is clearly "overgrazing" the biosphere. As the 6 billion people in less developed countries strive to raise their standard of living, and the billion people in more developed countries continue to increase theirs, the damage to Earth's ecosystems accelerates. We all want to enjoy luxuries far beyond bare survival, but unfortunately, the resources currently demanded to support the high standard of living in developed countries are unattainable for most of Earth's inhabitants. For example, supporting the world population sustainably at the average standard of living in the United States would require about 4½ Earths. Technology can help us improve agricultural efficiency, conserve energy and water, reduce pollutants, and recycle far more of what we use. In the long run, however, it is extremely unlikely that technological innovation can compensate for continued population growth.

Inevitably, the human population will stop growing. Hope for the future lies in recognizing the signs of human overgrazing and responding by reducing our population before we cause further damage to the biosphere, diminishing its ability to support people and the other precious and irreplaceable forms of life on Earth.

THINK CRITICALLY Much of the easily visible ecological damage caused by humans occurs in less developed countries: overgrazed pastures, rain-forest devastation, and critically endangered species such as rhinos and orangutans. Some of this damage results from economic demand from more developed countries. Research the origins and ecological impacts of an imported product such as palm oil, gold, mahogany, or teak. Do developed countries export ecological damage when we import goods from less developed countries?

▶ **FIGURE 27-19 U.S. population growth** Since 1790, U.S. population growth has produced a J-shaped curve typical of exponential growth, with some slight slowing in recent decades. The U.S. Census Bureau predicts that the U.S. population will reach 334 million by 2020. Data from the U.S. Census Bureau.

THINK CRITICALLY At what stage of the S-curve is the U.S. population? What factors do you think might cause it to stabilize? Might the recent recession reduce population growth?

The U.S. Population Is Growing Rapidly

In 2015, the United States had a population of more than 321 million and a growth rate of about 0.8% per year (**FIG. 27-19**). The U.S. is one of the fastest-growing developed countries in the world. Continued immigration, which accounts for more than 30% of the population increase, will ensure growth for the indefinite future, unless the U.S. fertility rate (2.0 in 2015) drops sufficiently below RLF to compensate for the influx of people. Because the average U.S. resident has a large ecological footprint—about 2.5 times the global average—continued population growth in the U.S. has significant impact on both local and global environments.

CHECK YOUR LEARNING

Can you ...

- describe the advances that have allowed exponential growth of the human population?
- explain why rapid population growth continues today?
- explain the demographic transition?
- sketch age structure diagrams and describe how their shape predicts future changes in population size?

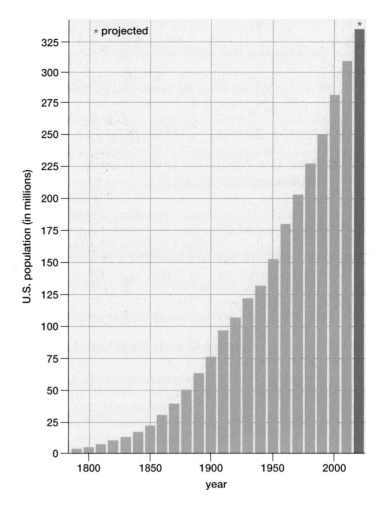

The Return of the Elephant Seals

The rapid recovery of elephant seal populations, from near-extinction in 1892 to about 200,000 animals today, is a triumph of wildlife conservation, with an invaluable role played by the exponential growth of a small population with abundant food, safe breeding sites, and few predators.

However, all modern elephant seals are the offspring of at most 20 ancestors in the 1890s. Therefore, all modern elephant seals may have descended from just one bull and a few females. Passing through such a "population bottleneck" (see Chapter 16) reduces the genetic diversity of future generations, because only the alleles that were present in this minuscule population were available to pass on to the entire present-day population. Indeed, when molecular biologists compared the genetic diversity of modern elephant seals with that of specimens collected before 1892, they found that today's elephant seals are almost genetically identical to one another. Much greater diversity was found

in pre-bottleneck specimens. It may take hundreds or even thousands of years for mutations to replenish allele diversity in elephant seals.

Does this matter? Right now, probably not very much, although elephant seals do have about 2 to 10 times more congenital abnormalities, such as cleft palate and defects in the heart or brain, than harbor seals and California sea lions do. A high incidence of congenital defects is fairly common in inbred populations with low genetic diversity.

CONSIDER THIS As a genetically homogeneous species, elephant seals have a diminished capacity to evolve in response to environmental changes that may occur in the future—the entire population will be affected, to about the same degree, by any adverse changes. Do you think that elephant seals, regardless of their abundance, should be considered potentially endangered for the foreseeable future? Why or why not?

CHAPTER REVIEW

(MB) Go to **MasteringBiology** for practice quizzes, activities, eText, videos, current events, and more.

Answers to Figure Caption questions, Multiple Choice questions, and Fill-in-the-Blank questions can be found in the Answers section at the back of the book.

Summary of Key Concepts

27.1 What Is a Population and How Does Population Size Change?

Populations change size through births and deaths, which result in natural increase, and through immigration and emigration, which produce net migration. Ignoring migration, the population growth rate (*r*) is its birth rate (*b*) minus its death rate (*d*). Population growth (*G*), the increase during a given time interval, equals the growth rate (*r*) multiplied by the population size (*N*). All organisms have the biotic potential to more than replace themselves over their lifetimes, resulting in population growth. A constant growth rate produces exponential growth.

27.2 How Is Population Growth Regulated?

Carrying capacity (*K*) is the maximum size at which a population may be sustained indefinitely by an ecosystem. *K* is determined by limited resources such as space, nutrients, and energy. Environmental resistance generally maintains populations at or below the carrying capacity. Above *K*, populations deplete their resource base, leading to (1) the population stabilizing near *K*; (2) reduction of *K* and a permanently reduced population; or (3) the population being eliminated from the area. Population growth is restrained by density-independent forms of environmental resistance (principally weather and climate) and density-dependent forms of resistance (competition, predation, and parasitism).

27.3 How Do Life History Strategies Differ Among Species?

r-selected species tend to live in unpredictable environments, mature early, produce large numbers of small offspring, and provide little or no parental care, whereas *K*-selected species tend to live in stable environments, mature slowly, and provide substantial parental care to small numbers of offspring. The life history strategies of most species fall between these two extremes. Late-loss survivorship curves are characteristic of *K*-selected species. Species with constant-loss curves have an equal chance of dying at any age. Early-loss curves are typical of *r*-selected species, or any species that produce numerous offspring, most of which die before reaching maturity.

27.4 How Are Organisms Distributed in Populations?

Clumped distribution may occur for social reasons or around limited resources. Uniform distribution is often the result of territorial spacing. Random distribution is rare, occurring when individuals do not interact socially and when resources are abundant and evenly distributed.

27.5 How Is the Human Population Changing?

The human population exhibited exponential growth for an unprecedented time, the result of a combination of high birth rates and technological, agricultural, industrial, and medical advances that have reduced the effects of environmental resistance and increased Earth's carrying capacity for humans. Age structure diagrams depict numbers of males and females in groups of increasing age. Expanding populations, mostly in less developed countries, have triangular age structures with a broad base. More developed countries typically have stable populations with relatively straight-sided age structures or shrinking populations with age structures that are narrow at the base. Most people live in less developed countries with growing populations and birth rates greater than replacement level fertility (RLF). Even if birth rates decline to RLF, momentum from earlier high birth rates and the resulting large number of children and reproductive adults ensure decades of continued population growth.

Key Terms

age structure diagram *505*
biosphere *491*
biotic potential *492*
birth rate (*b*) *491*
boom-and-bust cycle *493*
carrying capacity (*K*) *495*
clumped distribution *502*
community *491*
competition *498*
constant-loss population *502*
death rate (*d*) *491*
demographic transition *504*
density-dependent *497*
density-independent *497*
early-loss population *502*
ecology *491*
ecosystem *491*
emigration *491*
environmental resistance *493*
exponential growth *492*
growth rate (*r*) *491*
host *498*
immigration *491*
interspecific competition *499*
intraspecific competition *499*

invasive species *495*
J-curve *492*
K-selected species *501*
late-loss population *502*
less developed
   country *504*
life history *500*
logistic population
   growth *495*
more developed
   country *504*
natural increase *491*
parasite *498*
population *491*
population cycle *498*
predator *497*
prey *497*
r-selected species *500*
random distribution *503*
replacement level fertility
   (RLF) *504*
S-curve *495*
survivorship curve *501*
survivorship table *501*
uniform distribution *503*

Thinking Through the Concepts

Multiple Choice

1. Density-independent environmental resistance includes
a. predation.   c. parasitism.
b. floods.   d. competition.

2. Exponential growth often occurs when
a. a population nears carrying capacity.
b. a population exceeds carrying capacity.
c. organisms invade a new habitat.
d. the birth rate exceeds the death rate for a single generation.

3. Which of the following did *not* contribute to increasing Earth's carrying capacity for humans?
a. birth control   c. medical advances
b. agriculture   d. improved sanitation

4. A population is likely to increase in size if
a. there are more postreproductive adults than juveniles.
b. parasites increase in abundance.
c. its population pyramid is narrower at the base than in the middle.
d. its reproductive rate is greater than replacement level fertility.

5. Boom-and-bust populations
a. do not experience exponential growth.
b. are usually controlled by density-dependent environmental resistance.
c. may briefly exceed the carrying capacity of their environment.
d. level off at the carrying capacity of their environment.

Fill-in-the-Blank

1. Graphs that plot how the numbers of individuals born at the same time change over time are called _____. The specific type of curve that applies to a dandelion that releases 300 seeds, most of which never germinate, is called _____. The curve for humans is an example of _____.

2. The type of growth that occurs in a population that grows by a constant percentage per year is _____. Does this form of growth add the same number of individuals each year? _____ What shape of curve is generated if this type of growth is graphed? _____ Can this type of growth be sustained indefinitely? _____

3. The maximum population size that can be sustained indefinitely without damaging the environment is called the _____. A growth curve in which a population first grows exponentially and then levels off at (or below) this maximum sustainable size is called a(n) _____ curve, or a(n) _____ curve.

4. The type of spatial distribution likely to occur when resources are localized is _____. The type of spatial distribution that results when pairs of animals defend breeding territories is _____. The least common form of distribution is _____.

5. A population grows whenever the number of _____ plus _____ exceeds the number of _____ plus _____. The growth rate of a population increases when the frequency of reproduction _____.

Review Questions

1. Define *biotic potential*, list the factors that influence it, and explain why natural selection may favor a high biotic potential.

2. Write and describe the meaning of the equation for population growth using the variables *G*, *r*, and *N*.

3. Draw, name, and describe the properties of a growth curve of a population without environmental resistance.

4. Define *environmental resistance* and distinguish between density-independent and density-dependent forms of environmental resistance. Describe three examples of each.

5. What is logistic population growth? What is *K*?

6. Describe three different possible consequences of exceeding carrying capacity. Sketch these scenarios on a graph. Explain your answer.

7. Distinguish between populations showing concave and convex survivorship curves. What are their typical life history strategies?

8. Explain why environmental resistance has not prevented exponential human population growth since prehistoric times; provide examples. Can this continue? Explain why or why not.

9. Draw the general shape of age structure diagrams characteristic of (a) expanding, (b) stable, and (c) shrinking populations. Label all the axes. Explain why you can predict the next several decades of growth by the current age structure of populations.

10. Sketch and label the graph showing the general stages of the demographic transition and explain the changes that influence the growth and size of the population over time.

Applying the Concepts

1. Research a developing country (such as Nigeria, Afghanistan, or Uganda) with rapid population growth and find out what factors sustain that growth and why. Explain these factors and assess the likelihood that the fertility rate of this country will drop in the near future.

2. The U.S. Endangered Species Act (ESA) seeks to increase the population of threatened species to the point where they have a self-sustaining population that no longer needs special protection. Research two species that have been listed as endangered: one that has recovered to the extent that it has been "delisted" and one that has remained on the list for more than 20 years. Based on what you have learned in this chapter, what characteristics of the two species and their environment have significantly contributed to their differing fates?

Although smaller than the average housecat, the Channel Island fox plays a huge role in its ecolological community.

CASE STUDY

The Fox's Tale

The eight Channel Islands lie in the Pacific Ocean, scattered in an arc 11 to 70 miles off the southern California coast from Santa Barbara to Los Angeles. For thousands of years, the Channel Island fox, topping out at 4 to 6 pounds, was the largest terrestrial predator on the islands. The foxes ate everything from mice and large insects to fruit and seeds, and they flourished. They shared the islands with bald eagles, which are large enough to prey on the tiny foxes, but which fed primarily on the abundant fish in the surrounding ocean waters and posed no threat to the foxes. In the late 1900s, however, catastrophe struck the foxes, whose population crashed from more than 6,000 in the mid-1900s to just a few hundred by 2000. What happened?

Because the Channel Islands have never been connected to the mainland, their native animals and plants arrived by swimming, flying, rafting on mats of vegetation, or blowing in the wind. The islands became home to a unique collection of organisms adapted to the local conditions and containing some species found nowhere else on Earth, including the Channel Island fox. But starting in the 1800s, the tapestry of life on the islands began to unravel. Settlers in search of farmland cleared large areas of native vegetation and brought in cattle, pigs, and sheep. Some of the pigs and sheep escaped, colonizing the remaining wild areas of the islands, where they consumed or uprooted native plants. In the mid-1900s, most of the bald eagles were wiped out by the insecticide DDT (see Chapter 27). The cumulative effects of these changes devastated the islands' native flora and fauna, particularly the foxes, which almost became extinct. Some islands were down to their last 15 or 20 foxes, while populations of mice, skunks and non-native weeds increased.

These changes illustrate the interconnectedness of the species on the islands. Why would the addition of herbivores such as sheep and pigs, or the loss of fish-eating bald eagles, harm the foxes? How might the loss of foxes affect other island species? Is there any way of restoring connections that have been torn apart?

AT A GLANCE

28.1 WHY ARE COMMUNITY INTERACTIONS IMPORTANT?

An ecological **community** consists of all the populations of multiple species living and interacting with one another in a defined area. All community interactions involve access to resources. Interactions between species may be classified into three major categories (**TABLE 28-1**).

In **interspecific competition,** two or more species utilize the same resources (*interspecific* means "between species"). For example, acorn woodpeckers defend territories not only against other woodpeckers, but also against jays and squirrels, which may compete with them for acorns, and starlings, which compete with them for nesting holes. Plants of many different species often compete with one another for light, water, or nutrients such as soil nitrogen. Interspecific competition is detrimental to all of the species involved because it reduces their access to resources they need.

In **consumer–prey interactions,** one species (the consumer) uses another species (the prey) as a food source. Consumer–prey interactions include predation and parasitism. Obviously, predators and parasites benefit from their relationships with their prey, whereas the prey are harmed.

In **mutualism,** two species cooperate in ways that increase both species' access to resources. Therefore, mutualism may make it possible for species to thrive together where neither could survive alone.

Species interactions are typically sporadic, as when a lion chases down a zebra or a deer walks along nipping buds off bushes. In **symbiosis,** however, the relationships are prolonged and intimate, so that members of the two species are almost always found together. Unlike the common English meaning of symbiosis, in biology the relationship may be beneficial to both (mutualism) or beneficial to one and harmful to the other (parasitism).

TABLE 28-1	Interactions Between Species	
Type of Interaction	**Effect on Species A**	**Effect on Species B**
Interspecific competition between A and B	Harms	Harms
Consumer–prey interactions (A is the consumer, B is the prey)	Benefits	Harms
Mutualism between A and B	Benefits	Benefits

Community interactions exert strong evolutionary forces on the species involved. For example, by killing the prey that are easiest to catch, predators spare those individuals with better defenses against predation. These better-adapted individuals then produce offspring, and over time, their inherited characteristics increase within the prey population. This process, by which interacting species act as agents of natural selection on one another, is called **coevolution.**

CHECK YOUR LEARNING

Can you ...
- define a *community* and explain why community interactions are important?
- name the three major types of community interactions and describe their effects on the species involved?
- define *coevolution*?

28.2 HOW DOES THE ECOLOGICAL NICHE INFLUENCE COMPETITION?

Each species occupies a unique **ecological niche** that encompasses all aspects of its way of life. An important part of an ecological niche is the species' physical home, or habitat, including all of the environmental conditions necessary for its survival and reproduction. These can include nesting or denning sites, climate, the type of nutrients the species requires, its optimal temperature range, the amount of water it needs, the pH and salinity of the water or soil it may inhabit, and (for plants) the degree of sun or shade it can tolerate. An ecological niche also encompasses the entire "role" that a given species performs in an ecosystem, including what it eats (or whether it obtains energy from photosynthesis), which species prey upon it or parasitize it, and the other species with which it competes.

Although different species may share many aspects of their ecological niches, interspecific competition prevents any two species from occupying exactly the same niche.

Resource Partitioning Reduces the Overlap of Ecological Niches Among Coexisting Species

What happens if two species compete to precisely the same extent for every required resource—that is, if they attempt to occupy exactly the same niche? We might consider three possible outcomes: (1) the two species will coexist in the same

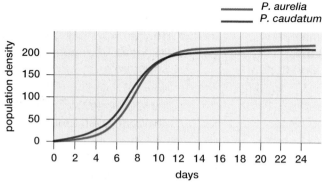

(a) Grown in separate flasks

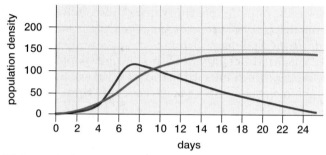

(b) Grown in the same flask

▲ **FIGURE 28-1 Interspecific competition in the lab (a)** Raised separately with a constant food supply, both *Paramecium aurelia* and *P. caudatum* show the S-curve typical of a population that initially grows rapidly and then stabilizes. **(b)** Raised together and forced to occupy the same niche, *P. aurelia* always eventually outcompetes *P. caudatum* and causes that population to die off. Data from Gause, G. F. 1934. *The Struggle for Existence.* Baltimore: Williams & Wilkins.

niche indefinitely; (2) one species will displace the other, which will go extinct (at least locally); or (3) one or both species will evolve slightly different niche requirements, thus avoiding complete competition.

In 1934, the Russian biologist G. F. Gause tested the first two possibilities. Gause grew two species of the protist *Paramecium* (*P. aurelia* and *P. caudatum*) in laboratory cultures. Grown separately, with bacteria as their only food, both species thrived (**FIG. 28-1a**). However, when Gause placed the two species together, they could not coexist: *P. aurelia* grew more rapidly and always eliminated *P. caudatum* (**FIG. 28-1b**). Gause's conclusion came to be known as the **competitive exclusion principle:** Two species with exactly the same niche cannot coexist indefinitely.

Can species coexist if their niches differ slightly? To find out, Gause repeated the experiment, but this time pairing *P. caudatum* with a different species, *P. bursaria*. *P. caudatum* feeds on bacteria suspended in the culture medium, but *P. bursaria* feeds mostly on bacteria that settle to the bottom of the cultures. These two species of *Paramecium* were able to coexist indefinitely because they preferred feeding in different places and thus occupied slightly different niches.

Gause didn't raise enough generations of *Paramecium* for significant evolution to occur, so he could not test the third possibility, the evolution of different niche requirements. In nature, however, there is plenty of time for competing species to evolve. Ecologist Robert MacArthur carefully observed five species of North American warbler under natural conditions. These birds all nest and hunt for insects in spruce trees. Although their niches overlap, MacArthur found that each species concentrates its search for food in different regions of the spruce trees, employs different hunting tactics, and nests at a slightly different time (**FIG. 28-2**). The five species of warblers have evolved behaviors that reduce the overlap of their niches, thereby reducing interspecific competition.

This phenomenon of dividing up resources, called **resource partitioning,** is often the outcome of the coevolution of different species with extensive (but not complete) niche overlap. A famous example of resource partitioning was discovered by Charles Darwin among related species of finches on the Galápagos Islands. Different finch species that share the same island evolved different bill sizes and shapes and different feeding behaviors that reduce competition among them (see Chapter 15).

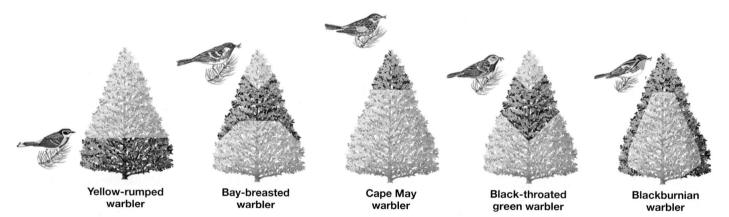

| Yellow-rumped warbler | Bay-breasted warbler | Cape May warbler | Black-throated green warbler | Blackburnian warbler |

▲ **FIGURE 28-2 Resource partitioning** Each of these five insect-eating species of North American warblers searches for food in a slightly different part of a spruce tree. This reduces niche overlap and competition. Adapted from MacArthur, R. H. 1958. Population ecology of some warblers of Northeastern coniferous forest. *Ecology* 39:599–619.

(a) Eurasian red squirrel

(b) Eastern gray squirrel

▲ **FIGURE 28-3 Competition limits population size and distribution (a)** Native red squirrels have been pushed out of most of England by competition from **(b)** gray squirrels imported from the United States. Gray squirrels may also have brought squirrelpox with them, which they resist better than red squirrels can.

Interspecific Competition Between Species May Limit the Population Size and Distribution of Each

Although natural selection often reduces niche overlap, there are still many species with similar niches; these species compete for limited resources. For example, lions and hyenas compete for prey in the African savanna, both by eating similar prey and by direct aggression toward the other species. In North America, similar competition occurs between wolves and coyotes and between bobcats, goshawks, and lynx.

Interspecific competition may restrict the population size and distribution of species, as is currently happening in Great Britain. The native squirrel in Britain and much of Europe is the Eurasian red squirrel, *Sciurus vulgaris* (**FIG. 28-3a**). In the late 1800s, rich landowners in Britain imported Eastern gray squirrels, *Sciurus carolinensis*, from the United States as "ornaments" for their manors (**FIG. 28-3b**). Gray squirrels, however, outcompete red squirrels in the deciduous forests of southern England. Gray squirrels are more efficient at eating acorns, often consuming them even before they are ripe, and raid red squirrels' seed caches. As a result, red squirrels are practically extinct in southern England. Red squirrels do much better in coniferous forests, probably because the large gray squirrels don't get enough nutrition from small conifer seeds, whereas the much smaller red squirrels can flourish on this diet. As a result of

these interactions, gray squirrels by the millions now dominate southern England, both species coexist in the mixed deciduous/conifer forests of northern England (although grays are probably still increasing), and red squirrels seem to be holding their own in the coniferous forests of Scotland.

Gray squirrels, of course, are not the only species that humans have moved across our planet. In "Earth Watch: Invasive Species Disrupt Community Interactions" on page 516, we describe what sometimes happens when people transport predatory or competing organisms into ecological communities whose members have not evolved to deal with them.

Competition Within a Species Is a Major Factor Controlling Population Size

Individuals of the same species have the same requirements for resources and thus occupy the same ecological niche. For this reason, **intraspecific competition** (*intra* means "within") is the most intense form of competition, because every member of the species competes for all of the same resources. Intraspecific competition exerts strong density-dependent environmental resistance, limiting population size (see Chapter 27). Intraspecific competition is one of the main factors driving evolution by natural selection, in which individuals that are better equipped to obtain scarce resources are more likely to reproduce successfully, passing their heritable traits to their offspring.

Earth
WATCH

Invasive Species Disrupt Community Interactions

Invasive species are non-native species that have been introduced into an ecosystem in which they did not evolve and that harm human health, the environment, or the local economy. Not all introduced species become invasive; ones that do typically reproduce rapidly, disperse widely, and thrive under a wide range of environmental conditions. Invasive species often encounter few competitors, predators, or parasites in their new environment. Their unchecked population growth may seriously damage the ecosystem as they outcompete or prey on native species.

English house sparrows were introduced into the United States on several occasions, starting in the 1850s, to control caterpillars feeding on shade trees. In 1890, European starlings were released into Central Park in New York City by a group attempting to introduce all the birds mentioned in the works of Shakespeare. Both bird species have spread throughout the continental United States. Their success has reduced populations of native songbirds, such as bluebirds and purple martins, with which they compete for nesting sites. Burmese pythons, probably released by pet owners, have become significant threats to natural communities in the Everglades of Florida (FIG. E28-1a). These huge predatory snakes have caused precipitous declines in populations of raccoons, opossums, deer, and rabbits.

Invasive plants also threaten natural communities. In the 1930s and 1940s, kudzu, a Japanese vine, was planted extensively in the southern United States to control soil erosion. Today, kudzu is a major pest, covering trees and engulfing abandoned buildings (FIG. E28-1b). Both water hyacinth and purple loosestrife were introduced into the United States as ornamental plants. Water hyacinth now clogs waterways in southern states, slowing boat traffic and displacing natural vegetation. Purple loosestrife aggressively invades wetlands, where it outcompetes native plants and reduces food and habitat for native animals.

Invasive species rank second only to habitat destruction in pushing endangered species toward extinction. Governments and land managers have sometimes tried to control invasive species by importing the species' natural predators or parasites (called *biocontrols*). However, biocontrols sometimes have unpredicted and even disastrous effects on native wildlife. The cane toad, for example, was introduced into Australia in the 1930s to control non-native beetles that threatened the sugarcane crop. Unfortunately, the toads have proven to be worse than the beetles; about 200 million toads now occupy northeastern Australia, outcompeting native frogs (FIG. E28-1c).

Despite the risks of biocontrols, there are often few realistic alternatives. Biologists now carefully screen proposed biocontrols to make sure they are likely to attack only the intended invasive species, and there have been successful introductions of biocontrols. For example, Eurasian beetles, released by the millions each year, are now among the most effective methods of controlling purple loosestrife in North America.

CONSIDER THIS Many invasive species were deliberately imported as pets or landscape plants and entered natural communities accidentally (escaped pets), deliberately (by owners who could no longer care for their exotic pets), or through natural spread (landscape plants). But probably the vast majority of non-native plants and animals never cause a major problem. Would you support laws to restrict the importation of non-native plants and animals? Why or why not?

(a) Burmese python

(b) Kudzu

(c) Cane toad

▲ FIGURE E28-1 **Invasive species (a)** Huge Burmese pythons introduced to the Florida Everglades can eat adult deer and even small alligators. **(b)** Kudzu, a Japanese vine imported to the southern United States, can rapidly cover entire trees and small buildings. **(c)** The cane toad (native to central and South America) was imported to Australia, where it outcompetes native toads and frogs and preys on many indigenous animals, such as this pygmy possum. Poisonous secretions on its skin protect the cane toad from predators.

CHECK YOUR LEARNING

Can you ...

- compare interspecific and intraspecific competition and explain which is more intense?
- explain how competitive exclusion leads to resource partitioning?

CASE STUDY **CONTINUED**

The Fox's Tale

Channel Island foxes do not encounter significant interspecific competition. Although foxes do compete with skunks for insects and mice, the competition seems to be very one-sided—the foxes suppress skunk populations, but skunks do not have much effect on fox populations. Bald and golden eagles, however, do seem to compete on the Channel Islands. When bald eagles were abundant, their presence apparently prevented golden eagles from settling on the islands. After bald eagles were eliminated by DDT, golden eagles colonized the islands. Around the same time, the fox population plummeted. Did the switch from bald to golden eagles cause the decline in foxes?

28.3 HOW DO CONSUMER–PREY INTERACTIONS SHAPE EVOLUTIONARY ADAPTATIONS?

Consumer–prey interactions are usually divided into two major categories: predation and parasitism. A **predator** is a free-living organism that eats other organisms. Although people generally think of predators as being *carnivores* (animals that eat other animals; **FIG. 28-4a**), we will include *herbivores* (animals that eat plants; **FIG. 28-4b**) in this category. An amoeba that eats bacteria or protists by phagocytosis is also a predator (**FIG. 28-4c**). A **parasite** usually lives in or on its prey (called its **host**) for a significant part of the parasite's life cycle, typically deriving all of its nourishment from its host. In these cases, the parasitism would be a special case of symbiosis. Tapeworms, fleas, and the many disease-causing protists, fungi, bacteria, and viruses are all parasites. There are very few parasitic vertebrates, but the lamprey (see Fig. 25-6), which attaches itself to a host fish and sucks its blood, is one example.

Not all organisms can be conveniently categorized as predators or parasites, but a few characteristics help to distinguish parasites from predators. Parasites are smaller and much more abundant than their hosts. Predators are often, but not always, larger than their prey, and they are generally less abundant than their prey, as explained in Chapter 29. Predators and parasites frequently differ in how much damage they do to their prey. Carnivorous predators, such as lions, eagles, and weasels, typically kill their prey. Herbivorous predators, such as deer and insects, usually do not immediately kill the plants they feed on; in fact, if herbivore numbers are low enough, they may never kill the plants. Parasites harm or weaken their host, but often do not immediately kill it. However, the weakened host may be more susceptible to predators, scarce food supplies, or harsh winter weather, so the parasite may hasten the death of the host.

Predators and Prey Coevolve Counteracting Adaptations

To survive and reproduce, predators must feed, and prey must avoid becoming food. Therefore, predator and prey are powerful forces of natural selection on one another, resulting in coevolution. Coevolution has produced the keen eyesight of the hawk and owl, which is countered by the earthy colors of their mouse and ground squirrel prey. Plants and their predators have also coevolved. For example, grasses embed tough silica substances in their blades that make them difficult to chew. In turn, natural selection has favored the evolution of long, hard teeth in grazing animals, such as horses.

(a) Eagle owl

(b) Pika

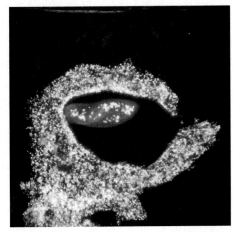
(c) Amoeba

▲ **FIGURE 28-4 Forms of predation (a)** An eagle owl feasts on a mouse. **(b)** A pika, whose preferred food is grass, is a small relative of the rabbit and lives in the Rocky Mountains. **(c)** Amoebas are microscopic protists that eat bacteria and smaller protists.

▲ **FIGURE 28-5 Coevolution between bats and moths** A long-eared bat uses a sophisticated echolocation system to hunt moths, which in turn have evolved specialized sound detectors and behaviors to avoid capture.

The adaptations of echolocating bats and their moth prey (**FIG. 28-5**) provide excellent examples of how both body structures and behaviors are molded by coevolution. Echolocating bats are usually nighttime hunters that emit pulses of sound, so high pitched that people can't hear them. The sounds bounce back from nearby objects. Bats use these echoes to produce a sonar "image" of their surroundings, allowing them to navigate around objects and detect prey.

As a result of natural selection imposed by echolocating bats, several species of moths have evolved ears that are particularly sensitive to the sounds produced by the bats. When they hear a bat, these moths take evasive action, flying erratically or dropping to the ground. Some species of bats, in turn, have evolved the ability to counter this defense by switching the frequency of their sound pulses away from the moth's sensitivity range.

Predators and Prey May Engage in Chemical Warfare

The evolution of counteracting defenses may give rise to a kind of "chemical warfare" between predators and prey. Many plants, including milkweeds, synthesize toxic and distasteful chemicals that deter predators. As plants evolved these defensive toxins, certain insects evolved increasingly efficient ways to detoxify or even use these substances. The result is that nearly every toxic plant is eaten by at least one type of insect. For example, monarch butterflies lay their eggs on milkweeds; after the eggs hatch, the monarch caterpillars eat nothing but milkweed (**FIG. 28-6**). The caterpillars not only tolerate the milkweed poison but also store it in their tissues as a defense against their own predators.

Other defensive chemicals include the clouds of ink that certain mollusks (including squid, octopuses, and some sea slugs) emit when a predator attacks. These chemical "smoke screens" confuse predators and mask the prey's escape. Another dramatic example of chemical defense is seen in the bombardier beetle. In response to the bite of an ant, the beetle releases secretions from defensive glands into a chamber in its abdomen. There, enzymes catalyze an explosive

chemical reaction that shoots a toxic, boiling-hot spray onto the attacker. Toxins can be used to attack as well as to defend. The venom of spiders and some snakes, such as rattlesnakes and cobras, both paralyzes their prey and deters predators.

Looks Can Be Deceiving for Both Predators and Prey

Have you ever heard the saying that the best hiding place may be in plain sight? Both predators and prey have evolved colors, patterns, and shapes that resemble their surroundings (**FIG. 28-7**). Such disguises, called **camouflage,** render plants and animals inconspicuous, even when they are in full view.

Some animals closely resemble generally inedible objects such as leaves, twigs, seaweed, thorns, or even bird droppings (**FIGS. 28-8a–c**). Camouflaged animals tend to remain motionless; a crawling bird dropping would ruin the disguise. Whereas many camouflaged animals resemble parts of plants, a few succulent desert plants look like small rocks, which hides them from animals seeking the water stored in the plants' bodies (**FIG. 28-8d**).

HAVE YOU EVER WONDERED ...

Why Rattlesnakes Rattle?

Rattlesnakes are venomous, feared predators. So why bother rattling? Actually, they don't rattle most of the time. Camouflaged, silent hunters, rattlers sneak up on mice, rats, and small birds and strike without warning. However, even though they have lethal venom, rattlers are nonetheless preyed upon by hawks, eagles, coyotes, badgers and a few other animals. If a rattler detects a potential predator, it will quietly slink away if it can. When confronted by a predator that it probably can't escape, the snake will coil up and rattle, giving its adversary time to have second thoughts. In places where they are frequently hunted by humans, rattlers don't seem to rattle as much as they used to. Because humans can kill from beyond a rattler's striking range, natural selection may be favoring the evolution of silent rattlers that have a better chance of going undetected.

▲ **FIGURE 28-6 Chemical warfare** A monarch caterpillar feeds on milkweed that contains a powerful toxin.

THINK CRITICALLY Why might the caterpillar be colored with conspicuous stripes?

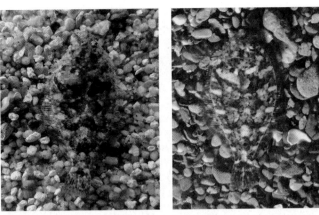

(a) Sand dab fish adjust camouflage to different backgrounds

(b) A camouflaged horned lizard

▲ **FIGURE 28-7 Camouflage by blending in (a)** Sand dabs are flat, bottom-dwelling ocean fish whose mottled colors closely resemble the sand on which they rest. Both their colors and patterns can be modified by nervous signals to better blend with their backgrounds. **(b)** This horned lizard helps protect itself from predation by snakes and hawks by resembling its surroundings of leaf litter.

THINK CRITICALLY How might predators evolve to detect camouflaged prey?

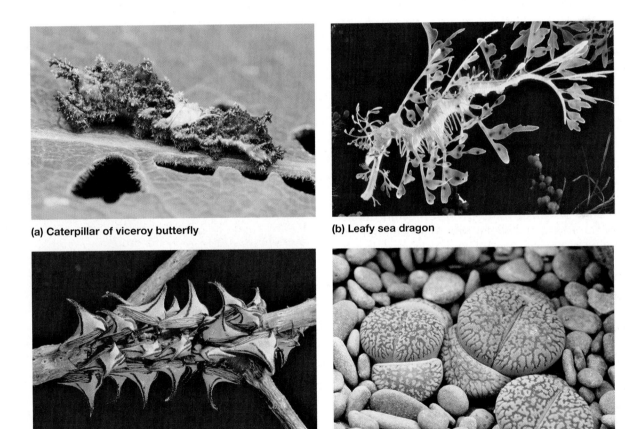

(a) Caterpillar of viceroy butterfly

(b) Leafy sea dragon

(c) Thorn treehoppers

(d) Living rock succulents

▲ **FIGURE 28-8 Camouflage by resembling specific objects (a)** A Viceroy butterfly caterpillar, whose color and shape resemble a bird dropping, sits motionless on a leaf. **(b)** The leafy sea dragon (a fish related to the seahorse) has evolved extensions of its body that mimic the algae in which it hides. **(c)** Thorn treehopper insects avoid detection by resembling thorns on a branch. **(d)** These South African succulents are appropriately called "living rocks."

THINK CRITICALLY How might such camouflage have evolved?

(a) A camouflaged snow leopard

(b) A camouflaged frogfish

▲ **FIGURE 28-9 Camouflage assists predators (a)** As it waits for prey, such as gazelles, wild sheep, and deer, the snow leopard in the mountains of Mongolia avoids detection with camouflage coloration. **(b)** Combining camouflage and aggressive mimicry, a frogfish waits in ambush, its lumpy, yellow body matching the sponge-encrusted rock on which it rests. Above its mouth dangles a tiny lure that closely resembles a small fish. The lure attracts small predators, which will suddenly find themselves to be prey instead.

Predators that ambush prey are also aided by camouflage. For example, the spotted snow leopard is almost invisible on a mountainside as it watches for prey (**FIG. 28-9a**). The frogfish, lurking motionless on the ocean floor awaiting smaller fish to swallow, closely resembles sponges and algae-covered rocks (**FIG. 28-9b**).

Some prey animals have evolved very different visual defenses: bright **warning coloration.** These animals may taste bad or make the predator sick (as monarch butterflies and their caterpillars do; see Fig. 28-6), inflict a venomous sting or bite (as bees and coral snakes do), or produce stinking chemicals when bothered (**FIG. 28-10**). The eye-catching colors seem to declare "Attack at your own risk!" An uneducated predator might attack, or even kill and eat, such a brightly colored prey once, but would avoid similar prey in the future, thus sparing others of the same species.

Mimicry refers to a situation in which members of one species have evolved to resemble another species. By sharing a similar warning-color pattern, several poisonous species may all benefit. Mimicry among different distasteful species is called *Müllerian mimicry*. For example, toxins from milkweed plants are stored in monarch butterfly caterpillars (see Fig. 28-6) and retained in the metamorphosed adult butterflies, which deters birds from preying on them. Similarly, the caterpillars of viceroy butterflies eat willow and poplar leaves, which contain bitter salicylic acid. Both the caterpillars and the adult viceroys store salicylic acid in their bodies. The wing patterns of monarch and viceroy butterflies are strikingly similar (**FIG. 28-11**). Birds that become ill from consuming one species are likely to avoid the other as well. Similarly, a toad that is stung while attempting to eat a honeybee is likely to avoid not only honeybees but other black and yellow striped insects (such as yellow jacket wasps) without ever tasting one.

Once warning coloration evolved, there arose a selective advantage for harmless animals to resemble venomous or distasteful ones, an adaptation called *Batesian mimicry*. For example, the harmless hoverfly avoids predation by resembling a bee (**FIG. 28-12a**), and the nonvenomous scarlet king snake is protected by brilliant warning coloration that closely resembles that of the highly venomous coral snake (**FIG. 28-12b**).

▲ **FIGURE 28-10 Warning coloration** The vivid stripe and the tail display behavior of the skunk advertise its ability to make any attacker miserable.

(a) Monarch

(b) Viceroy

▲ **FIGURE 28-11 Müllerian mimicry** Nearly identical warning coloration protects both **(a)** the distasteful monarch and **(b)** the equally distasteful viceroy butterfly. Their caterpillars, however, look very different from one another; see Figs. 28-6 and 28-8a.

(a) Bee (venomous)

Hoverfly (nonvenomous)

(b) Coral snake (venomous)

Scarlet king snake (nonvenomous)

▲ **FIGURE 28-12 Batesian mimicry (a)** A stinging bee (left) is mimicked by the stingless hoverfly (right). **(b)** The warning coloration of the venomous coral snake (left) is mimicked by the harmless scarlet king snake (right).

(a) False-eyed frog

(b) Peacock moth

(c) Swallowtail caterpillar

▲ **FIGURE 28-13 Startle coloration (a)** When threatened, the false-eyed frog of South America raises its rump, which resembles the eyes of a large predator. **(b)** The peacock moth from Trinidad is well camouflaged, but should a predator approach, it suddenly opens its wings to reveal spots resembling large eyes. **(c)** Predators of this caterpillar larva of the Eastern tiger swallowtail butterfly are deterred by its resemblance to a snake. Just behind the caterpillar's head is a bulge that resembles a snake's head, but—even more frightening—it bears two pairs of "eyes," not just one.

Some prey species use another form of mimicry: **startle coloration.** Several insects and even some vertebrates have evolved color patterns that closely resemble the eyes of a much larger, possibly dangerous animal (**FIG. 28-13**). If a predator gets close, the prey suddenly reveals its eyespots, startling the predator and sometimes allowing the prey to escape.

Some predators have evolved **aggressive mimicry,** in which they entice their prey to come close by resembling something attractive to the prey. For example, by using a rhythm of flashes that is unique to each species, female fireflies attract males to mate. But in one species, the females sometimes mimic the flashing pattern of a different species, attracting males that they kill and eat. The frogfish (see Fig. 28-9b) not only is camouflaged but also exhibits aggressive mimicry by dangling a wriggling lure that resembles a small fish just above its mouth. Small fish attracted by the lure are engulfed in a split second if they get too close.

CASE STUDY CONTINUED
The Fox's Tale

Predation played a major role in the decline of Channel Island foxes. The loss of bald eagles opened up the islands to invasion by golden eagles. Meanwhile, pigs that escaped from farms on the islands multiplied, bearing litters year-round and providing a steady supply of piglets for the golden eagles to eat. Although piglets were the main course, foxes were tasty snacks. For thousands of years, island foxes had no predators. As a result, unlike their mainland relatives, Channel Island foxes hunt during the day and spend a lot of time in the open—traits that made them easy targets for the eagles. As we will see, the loss of foxes reverberated throughout the islands' communities.

Parasites Coevolve with Their Hosts

Parasites can only survive and reproduce in suitable hosts; the hosts, in turn, are harmed by the parasites. Therefore, natural selection favors parasites that are better at invading hosts and hosts that can resist parasitic invasion, resulting in coevolution of parasites and hosts. For example, natural selection favored the evolution of the itch–scratch reflex because scratching an itch often dislodges blood-sucking parasites such as ticks and mosquitoes. These parasites, in turn, have evolved chemicals in their saliva that reduce the itch response, so that they are more likely to complete a blood meal before the host becomes aware of their attack and scratches them off.

Perhaps the best-studied example of parasite–host coevolution is the cellular warfare between infectious microorganisms and their mammalian hosts. Mammals defend themselves against these parasites by secreting antimicrobial molecules onto the skin and into the respiratory and digestive tracts; producing several types of white blood cells that eat and destroy microbes or secrete antibodies that help to kill microbes; and even killing their own body cells that have been infected by viruses (see Chapter 37). In response, microbes have evolved coatings that prevent the immune system from recognizing them; cluster together in *biofilms* that help to keep antimicrobial secretions from reaching them (see Chapter 20); and sometimes even invade and destroy parts of the immune system, as the human immunodeficiency virus (HIV) does.

Coevolution between parasite and host can result in a variety of outcomes: devastating disease; a truce of sorts, in which the host provides nutrients and housing for the parasite, but the parasite causes little harm; or even mutualism, in which both parasite and host derive some benefits, as we explore in "Health Watch: Parasitism, Coevolution, and Coexistence."

Parasitism, Coevolution, and Coexistence

A parasite inevitably weakens its host to some extent, at the very least by robbing nutrients from the host, thereby making it more vulnerable to predators or other parasites, less able to feed itself, or more susceptible to cold weather or drought. Sooner or later, when the host dies, the parasite must find a new host or perish. Until fairly recently, this observation led many biologists to conclude that natural selection will favor parasites that cause less harm to their hosts, postponing the day when the parasite must leave a dead host and find a new, live one. Recent research, however, has shown that this is not necessarily the case. If a parasite (or its offspring) can easily find a new host before its old host dies, then gobbling up the host and reproducing as fast as possible can be a winning evolutionary strategy.

Epidemic typhus is a good example. This type of typhus is caused by a bacterium that infects lice and is excreted in their feces (**FIG. E28-2**). When a person scratches an itchy louse bite, some louse feces may enter the scratch, infecting the person with typhus. Both louse and human often die within a few weeks, but by then the bacterium has moved on. How? Infected lice leave people with high fevers and seek out other, healthy people as new hosts. Before it dies, the sick louse may infect its new host with typhus. Uninfected lice may bite the newly infected person, ingesting the typhus bacterium and starting the cycle over again. Rapid bacterial reproduction at the expense of its host's health helps, rather than harms, the typhus bacterium.

Sometimes humans coevolve with their parasites, resulting in an uneasy truce. An example is *Helicobacter pylori*, a bacterium that colonizes the human stomach and often causes ulcers, sometimes even cancer. More than 3 billion people have *Helicobacter* in their stomachs. Strains of *Helicobacter* that have coevolved with particular human groups seem to cause little harm to their "home team" but may be deadly to outsiders. For example, Tumaco and Tuquerres, two villages in Colombia separated by only 125 miles, have a 25-fold difference in the incidence of stomach cancer, most of which is caused by *Helicobacter*. Researchers discovered that Tumaco, with a very low rate of stomach cancer, is mostly populated by the descendants of freed African slaves. Molecular evidence shows that these people also have *Helicobacter* strains that originated in Africa. In contrast, Tuquerres, which has an extremely high rate of stomach cancer, is populated mostly by native Americans. Their *Helicobacter* strains are European, probably an accidental legacy of conquistadors from Spain. The researchers hypothesize that coevolution of African people with African *Helicobacter* has reduced their cancer risk, while the nearby native Americans struggle to cope with introduced European *Helicobacter*.

▲ **FIGURE E28-2 A human body louse feeds on the blood of its host** Infected lice leave behind feces packed with the bacteria that cause epidemic typhus.

Humans and *Helicobacter* may even be coevolving into mutualism. In the human stomach, *Helicobacter* gets lots of nutrition. People infected with *Helicobacter* seem to benefit by having a lower incidence of acid reflux disease. They also have lower rates of asthma, allergies, and chronic inflammation, apparently because *Helicobacter* damps down their immune responses.

The incidence of asthma and allergies among children has steadily increased in recent years. Evidence is mounting for the "hygiene hypothesis": Being extremely sanitary might not be such a good idea, because people benefit from some of the microbes we pick up from our environment. Growing up in homes with dogs and cats—or even mice and cockroaches—decreases the risk of allergies and asthma. Perhaps the "10-second rule" is actually good for kids?

THINK CRITICALLY Malaria is a common, often lethal, disease of tropical countries, with about 200 million cases worldwide, resulting in more than 600,000 deaths. Malaria is carried by a few species of mosquitoes and transmitted to people when they are bitten by infected mosquitoes. In many parts of the Tropics, the dominant malaria-carrying mosquitoes feed mostly indoors, at night. Therefore, an effective tool for reducing malaria infections is the insecticide-treated bednet, which protects people while they sleep. Do you think that bednets will be a permanent solution to the transmission of malaria by mosquitoes?

28.4 HOW DO MUTUALISMS BENEFIT DIFFERENT SPECIES?

Mutualism is an interaction between species in which both benefit. Many, but not all, mutualistic relationships are symbiotic. You may have seen colored patches on rocks; they are probably lichens, a symbiotic mutualistic association between an alga and a fungus (FIG. 28-14a). The fungus provides support and protection while obtaining food from the brightly colored, photosynthetic alga. Mutualistic associations also occur in the digestive tracts of cattle and termites, where

(a) Lichen

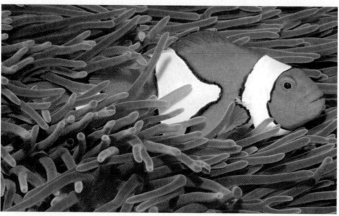

(b) Clownfish

▲ FIGURE 28-14 **Mutualism (a)** This brightly colored lichen growing on bare rock is a mutualistic relationship between an alga and a fungus. **(b)** The clownfish is coated with a protective coat of mucus, allowing it to snuggle unharmed among the stinging tentacles of an anemone.

protists and bacteria find food and shelter. The microorganisms break down cellulose, making its component sugar molecules available both to themselves and to the animals that harbor them. In human intestines, mutualistic bacteria synthesize vitamins, such as vitamin K, which we absorb and use.

In many other mutualisms, one partner does not live inside the other. For example, the clownfish of the southern Pacific Ocean shelters amidst the venomous tentacles of certain species of anemones (FIG. 28-14b). In this mutualistic symbiotic association, the anemone provides the clownfish with protection from predators, while the clownfish cleans the anemone and brings it bits of food.

Finally, many mutualistic relationships are not intimate and extended, so are not symbiotic. Consider the relationship between plants and the animals that pollinate them, including bees, moths, hummingbirds, and even some bats. The animals fertilize the plants by carrying plant sperm (found in pollen grains) from one plant to another and benefit by sipping nectar and sometimes eating pollen, so the relationship is mutualistic, but the partners actually spend very little time together.

28.5 HOW DO KEYSTONE SPECIES INFLUENCE COMMUNITY STRUCTURE?

In some communities, a particular species, called a **keystone species,** plays a major role in determining community structure—a role that is out of proportion to the species' abundance in the community. If the keystone species is removed, community interactions are significantly altered and the relative abundance of other species often changes dramatically.

In the African savanna, the bush elephant is a keystone species. By grazing on small trees and bushes (FIG. 28-15a), elephants prevent the encroachment of forests and help maintain the grassland community, along with its diverse population of grazing mammals and their predators.

Although elephants alter plant communities directly, large carnivores, such as wolves and cougars (FIG. 28-15b), may also have a major impact on plant communities. By keeping populations of deer and elk in check, wolves and cougars can help maintain the health of forests and stream banks that would otherwise be overgrazed. Because this vegetation provides food, nesting sites, and shelter for many species of smaller animals, the entire community structure depends on the presence of the predators.

Identifying a keystone species can be difficult. Many have been recognized only after their loss has had dramatic, unforeseen consequences. The intricate web of community interactions is illustrated by the effects of enormous fluctuations in populations of the northern sea otter (FIG. 28-15c)

(a) African elephant

(b) Cougar

(c) Northern sea otter

▲ **FIGURE 28-15 Keystone species (a)** The bush elephant is a keystone species on the African savanna. **(b)** The cougar, found in isolated habitats throughout North and South America, helps to control herbivores such as deer. **(c)** The northern sea otter rests in a kelp bed while cradling a sea urchin, one of its favorite foods.

along the coast of southwestern Alaska, including the Aleutian Islands, in the last 250 years. Between the mid-1700s and the early 1900s, fur hunters reduced the sea otter population, formerly numbering 200,000 to 300,000 animals, by about 99%. After otter hunting was banned in 1911, the population rebounded to about 100,000 animals. Kelp forests, sometimes described as the "rain forests of the ocean," flourished around islands in coastal waters where the otters were abundant. But starting in the mid-1990s, otter numbers plummeted again, with some populations declining by 90%. As a result, the numbers of sea urchins, a favored food of otters, skyrocketed. Sea urchins are a major predator of kelp, and they rapidly deforested the seabed, eliminating the diverse community of fish, mollusks, and crustaceans that the kelp forests once fed and sheltered. Clearly, sea otters are a keystone species near the Aleutian Islands. But what is killing them now? Killer whales, which had coexisted with sea otters in the past, have been increasingly observed dining on otters. Why? One hypothesis is that populations of seals and Steller sea lions—the whales' preferred prey—have declined drastically, forcing the whales to eat smaller prey, such as otters. Seal and sea lion populations have fallen, at least in part, because commercial fishing in the North Pacific has depleted their food supply.

CASE STUDY **CONTINUED**

The Fox's Tale

Foxes are keystone species on the Channel Islands. Predation by foxes controls deer mouse populations and probably limits the types and numbers of ground-nesting birds. Foxes control skunk populations by competing with them for food. Foxes also disperse the seeds of many island plants by eating fruits and expelling the undigested seeds in their feces, often a considerable distance from the parent plant. When fox populations plummeted while pig populations skyrocketed, community interactions changed dramatically, and ecological communities on many parts of the islands became barely recognizable shadows of their former selves. Can such disturbed communities ever recover?

CHECK YOUR LEARNING

Can you ...

- explain the concept of a keystone species?
- provide some examples of the effects that removing keystone species can have on their communities?

28.6 HOW DO SPECIES INTERACTIONS CHANGE COMMUNITY STRUCTURE OVER TIME?

In a mature terrestrial ecosystem, the populations that make up the community interact with one another and with their nonliving environment in intricate ways. But this tangled web of life did not spring fully formed from bare rock or naked soil; rather, it emerged in stages over a long period of time. These stages are called **succession:** a gradual change in a community and its nonliving environment in which groups of species replace one another in a reasonably predictable sequence. During succession, early-stage organisms modify the environment in ways that favor later organisms, while end-stage organisms suppress earlier organisms but tolerate one another's influences, producing a stable community. In general, succession produces a trend toward greater species diversity and species with longer life spans.

Succession begins with an ecological disturbance, an event that disrupts the ecosystem by altering its community, its *abiotic* (nonliving) environment, or both. The precise changes that occur during succession are as diverse as the environments in which succession occurs, but we can recognize certain general stages. Succession starts with a few hardy lichens or plants of species collectively known as **pioneers.** The pioneers alter the environment in ways that favor competing plants, which gradually displace the pioneers. If allowed to continue, succession progresses to a diverse and relatively stable **climax community.** Alternatively, recurring disturbances can maintain a community in an earlier, or **subclimax,** stage of succession. Our discussion of succession will focus on plant communities, which dominate the

(a) Kilauea, in Hawaii (primary succession)

(b) Yellowstone National Park, Wyoming (secondary succession)

▲ **FIGURE 28-16 Succession in progress (a)** Primary succession. (Left) The Hawaiian volcano Kilauea has erupted repeatedly since 1983, sending rivers of lava over the surrounding countryside. (Right) A pioneer fern takes root in a crack in hardened lava. **(b)** Secondary succession. (Left) In the summer of 1988, extensive fires swept through the forests of Yellowstone National Park in Wyoming. (Right) Trees and flowering plants are thriving in the sunlight, and wildlife populations are increasing as secondary succession occurs.

THINK CRITICALLY People have suppressed fires for decades. How might fire suppression affect forest ecosystems and succession?

landscape and provide both food and habitat for animals, fungi, and microorganisms.

There Are Two Major Forms of Succession: Primary and Secondary

During **primary succession,** a community gradually forms in a location where there are no remnants of a previous community and often no trace of life at all. The disturbance that sets the stage for primary succession may be a glacier scouring the landscape down to bare rock or a volcano producing a new island in the ocean or creating a layer of newly hardened lava on land (**FIG. 28-16a**). Building a community from scratch through primary succession typically requires thousands or even tens of thousands of years.

During **secondary succession,** a new community develops after an existing ecosystem is disturbed in a way that leaves significant remnants of the previous community behind, such as soil and seeds. For example, an abandoned farm typically has fertile soil and seeds from both crops and weeds. A clear-cut forest usually has soil, shrubs, and small trees remaining. Even a forest fire leaves behind the ingredients for secondary succession, such as residues of burnt trees that are high in plant nutrients. Fires also spare some trees and many healthy roots. Some plants produce seeds that can withstand fire or may even require it in order to sprout. The heat of a forest fire opens the cones of lodgepole pines, releasing their seeds. Thus, fires may promote rapid regeneration of forests and other communities (**FIG. 28-16b**).

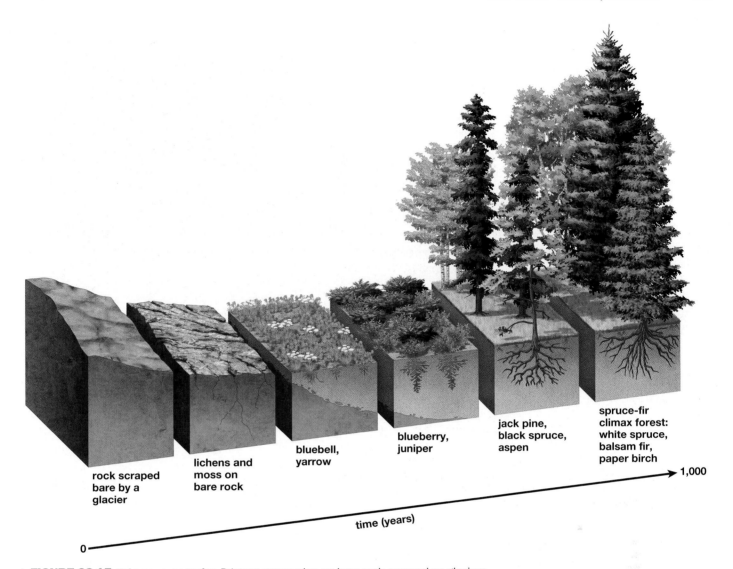

rock scraped
bare by a
glacier

lichens and
moss on
bare rock

bluebell,
yarrow

blueberry,
juniper

jack pine,
black spruce,
aspen

spruce-fir
climax forest:
white spruce,
balsam fir,
paper birch

time (years)   1,000

0

▲ **FIGURE 28-17 Primary succession** Primary succession on bare rock exposed as glaciers retreated from Isle Royale in Lake Superior in upper Michigan. Notice that the soil deepens over time, gradually burying the bedrock and allowing trees to take root.

Primary Succession May Begin on Bare Rock

FIGURE 28-17 illustrates primary succession on Isle Royale, Michigan, an island in northern Lake Superior that was scraped down to bare rock by glaciers that retreated roughly 10,000 years ago. Weathering includes cycles of freezing and thawing that cracks rocks and erodes their surface layers, producing small particles. Rainwater dissolves some of the minerals from rock particles, making them available to pioneer organisms.

Weathered rock provides an attachment site for lichens, which obtain energy through photosynthesis and acquire some of their mineral nutrients by dissolving rock with an acid that they secrete. As the lichens spread over the rock surface, drought-tolerant mosses begin growing in cracks. Fortified by nutrients liberated by the lichens, the mosses form a dense mat that traps dust, tiny rock particles and bits of organic debris. The mosses eventually cover and kill many of the lichens that made their growth possible.

The mat of mosses acts like a sponge, absorbing and trapping moisture. As mosses die and decompose, their bodies add nutrients to a thin layer of new soil. Within the moss mat, seeds of larger plants, such as bluebell and yarrow, germinate. As these plants die, their decomposing bodies further thicken the layer of soil.

As woody shrubs such as blueberry and juniper take advantage of the deeper soil, the mosses and remaining lichens may be shaded out and buried by decaying leaves and vegetation. Eventually, trees such as jack pine, black spruce, and aspen take root in the deeper crevices, and the sun-loving shrubs are shaded out. Within the forest, shade-tolerant seedlings of taller or faster-growing trees thrive, including balsam fir, paper birch, and white spruce. In time, these trees tower over and replace the earlier trees, which are intolerant of shade. After hundreds of years, a climax forest thrives on what was once bare rock.

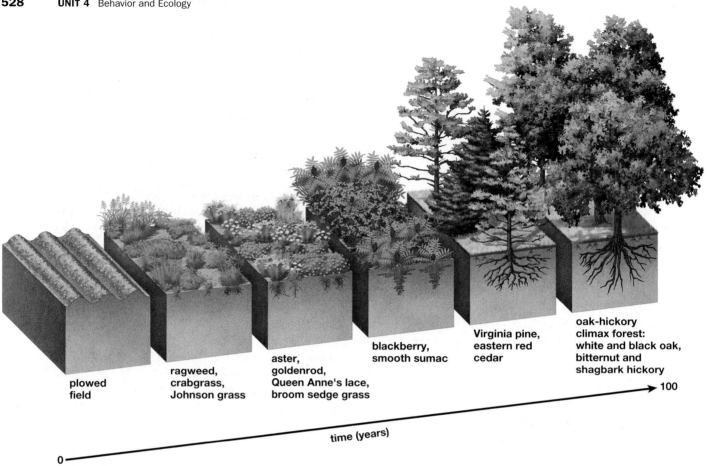

plowed
field

ragweed,
crabgrass,
Johnson grass

aster,
goldenrod,
Queen Anne's lace,
broom sedge grass

blackberry,
smooth sumac

Virginia pine,
eastern red
cedar

oak-hickory
climax forest:
white and black oak,
bitternut and
shagbark hickory

0

100

time (years)

▲ **FIGURE 28-18 Secondary succession** Secondary succession as it might occur on a plowed, abandoned farm field in North Carolina, in the southeastern United States. Notice that a thick layer of soil is present from the beginning, which greatly speeds up the process compared to primary succession.

An Abandoned Farm Undergoes Secondary Succession

FIGURE 28-18 illustrates secondary succession on an abandoned farm in the southeastern United States. The pioneer species are sun-loving, fast-growing plants such as ragweed, crabgrass, and Johnson grass. Such species generally produce large numbers of easily dispersed seeds that help them colonize open spaces. However, they don't compete well against longer-lived species that grow larger over the years and shade them out, so after a few years, plants such as asters, goldenrod, Queen Anne's lace, and perennial grasses move in, followed by woody shrubs such as blackberry and smooth sumac. Eventually, pine and cedar trees become established. About two decades after a field is abandoned, an evergreen forest dominated by pines takes over. However, the new forest alters conditions in ways that favor its successors. The shade of the pine forest inhibits the growth of its own seedlings while favoring the growth of hardwood trees, whose seedlings are shade tolerant and can grow beneath the pines. After about 70 years, slow-growing hardwoods, such as oak and hickory trees, begin to replace the aging pines. Roughly a century after the field was abandoned, the former farm is covered by relatively stable climax forest dominated by oak and hickory.

Succession Also Occurs in Ponds and Lakes

Succession in freshwater ponds and lakes occurs as soil and rock particles, washed in from the shore or carried in by streams entering the pond, settle to the bottom. This sediment gradually makes the pond more and more shallow. Eventually, moisture-loving plants take root along the shore. As more sediment fills the pond, the shoreline slowly moves toward the center of the pond. The soil of the original shoreline becomes drier, and plants from the surrounding land take over (**FIG. 28-19**). Eventually, no open water remains.

In forests, meadows are often produced by small lakes undergoing succession. As the lake fills in from the edges, grasses colonize the newly formed soil. As the lake shrinks and the area of meadow expands, trees encroach around the meadow's edges, eventually converting the meadow to forest.

Succession Culminates in a Climax Community

Succession ends with a reasonably stable climax community, which perpetuates itself if it is not disturbed by external forces (such as fire, parasites, invasive species, or human activities). The populations within a climax community have ecological niches that allow them to coexist without supplanting one another. In general, climax communities have more species and more types of community interactions than early stages of succession do. The plant species that dominate climax communities generally live longer and tend to be larger than pioneer species, particularly in climax forests.

Climax communities are determined by geological and climatic variables, including temperature, rainfall, elevation, latitude, type of rock (which determines the type of nutrients available in the soil), and exposure to sun and wind. These factors vary dramatically from one area to the next. For example, if you drive through Colorado or Wyoming, you will see a shortgrass prairie climax community on the eastern plains (in those rare areas where it has not been replaced by farms), pine-spruce forests in the mountains, tundra on the mountain summits, and sagebrush-dominated communities in the western valleys.

Natural events such as windstorms, avalanches, and fires started by lightning may destroy sections of climax communities, initiating secondary succession and producing a patchwork of various successional stages within an ecosystem. In many forests throughout the United States, rangers allow fires set by lightning to run their course, recognizing that fires are important for the maintenance of the ecosystem. Fires liberate nutrients and kill some (but usually not all)

of the trees. As a result, more nutrients and sunlight reach the forest floor, encouraging the growth of subclimax plants. The combination of climax and subclimax communities in different parts of the ecosystem provides habitats for a larger number of species of plants and animals than would either climax or subclimax vegetation alone.

Some Ecosystems Are Maintained in Subclimax Stages

Frequent disturbances can maintain some ecosystems in subclimax stages for very long periods of time. The tallgrass prairie that once covered northern Missouri and most of Illinois was a subclimax stage of an ecosystem whose climax community is deciduous forest. The subclimax prairie was maintained by periodic fires, some set by lightning and others deliberately set by Native Americans centuries ago to provide grazing land for bison. Where they have not been converted to farms, forest now encroaches on the grasslands, but people maintain some tallgrass prairie preserves with controlled burns.

Farms, gardens, and lawns are subclimax communities maintained by frequent, intentional disturbance. Grains are specialized grasses characteristic of the early stages of succession, and farmers spend a great deal of time, energy, and

▼ **FIGURE 28-19 Succession in a freshwater pond** This pond is rapidly being converted to dry land. Although there is a little open water in the center of the pond, water lilies dominate the pond, showing that it is only a few feet deep in most places. Shrubs and small trees are beginning to grow around the drier edges of the pond.

herbicides to prevent competitors, such as weeds, wildflowers, and woody shrubs, from taking over. The suburban lawn is a subclimax ecosystem maintained by regular mowing, which destroys woody colonizers, and the application of herbicides that selectively kill pioneer species such as crabgrass and dandelions.

CHECK YOUR LEARNING

Can you ...

- explain the process of succession and its general stages?
- define primary succession, secondary succession, subclimax ecosystem, and climax ecosystem?

CASE STUDY REVISITED

The Fox's Tale

Five of the Channel Islands are included in the Channel Islands National Park. Dismayed by scarred landscapes, invasive weeds replacing rare native plants, and especially by drastic declines in the island fox population, the National Park Service, in collaboration with other government and private organizations, set out to restore the islands' communities. Santa Cruz Island, the largest of the Channel Islands, has been a major focus of restoration efforts.

First, the cattle were removed. However, there were more than 20,000 feral sheep roaming the island. The sheep were rounded up or shot, and by the late 1990s, Santa Cruz was sheep-free. The feral pigs that plagued the island were shot by skilled hunters; the last pig was eliminated in 2007. Finally, golden eagles, which colonized the island and became major predators of foxes, were all trapped and relocated to the mainland, and a few bald eagles were successfully reintroduced. Conservation biologists hope that the bald eagles will keep golden eagles from returning.

Free of predators, the island fox population has rebounded—Santa Cruz alone is now home to about 1,300–1,500 foxes, and

the combined fox population on the islands is more than 6,000. Now that the pigs are gone, many native plants are making a comeback. Acorns, a favorite food of pigs, once again sprout into oak seedlings. Campaigns to control invasive weeds have helped the recovery of some of the Channel Islands' endangered plants, although several had already become extinct. The Channel Islands will probably always bear signs of past abuse, but future generations of visitors may see nearly intact, native climax communities once again.

THINK CRITICALLY Ecological succession sounds automatic—a farm is abandoned or a glacier retreats, and eventually the ecosystem regains its original communities. However, as in the Channel Islands, "letting nature take its course" doesn't always work, and ecosystems often need some help from restoration ecologists. If a disturbed ecosystem is simply left alone, what environmental conditions tend to favor natural succession, and what conditions may slow succession or even allow the ecosystem to remain permanently degraded?

CHAPTER REVIEW

(MB) Go to **MasteringBiology** for practice quizzes, activities, eText, videos, current events, and more.

Answers to Think Critically, Evaluate This, Multiple Choice, and Fill-in-the-Blank questions can be found in the Answers section at the back of the book.

Summary of Key Concepts

28.1 Why Are Community Interactions Important?

Ecological communities consist of all the interacting populations of different species within an ecosystem. Community interactions influence population size, and the interacting populations act on one another as agents of natural selection. Thus, community interactions shape the bodies and behaviors of the interacting populations.

28.2 How Does the Ecological Niche Influence Competition?

The ecological niche includes all aspects of a species' habitat and interactions with its living and nonliving environments. Each species occupies a unique ecological niche. Interspecific

competition occurs whenever the niches of two species within a community overlap. When two species with the same niche are forced (under laboratory conditions) to occupy the same ecological niche, one species always outcompetes the other. Species in natural communities have evolved in ways that avoid extensive niche overlap, with behavioral and physical adaptations that allow resource partitioning. Interspecific competition limits both the population size and the distribution of competing species. Intraspecific competition also limits population size because individuals of the same species occupy the same ecological niche and compete with one another for all of their needs.

28.3 How Do Consumer–Prey Interactions Shape Evolutionary Adaptations?

Predators and parasites are consumers of other organisms. In general, predators are free-living organisms that are less abundant than their prey, while parasites live in or on their hosts for a significant part of their life cycle and are much more abundant than their prey. Both types of consumers and their prey act as strong agents of natural selection on one another. Predators and prey have evolved a variety of toxic chemicals for attack and defense. Plants have evolved defenses ranging from poisons to overall toughness. These defenses, in turn, have selected for predators that can detoxify poisons and grind down tough tissues. Many prey animals have evolved protective colorations that render them either inconspicuous (camouflage) or startling (startle coloration) to their predators. Some prey are poisonous, distasteful, or venomous and exhibit warning coloration by which they are readily recognized and avoided by predators. Some harmless species have evolved to resemble distasteful organisms. Parasite–host coevolution includes the mammalian immune response against disease microbes and the many ways that these microbes have evolved to avoid detection or destruction by the immune response.

28.4 How Do Mutualisms Benefit Different Species?

Mutualism benefits two or more interacting species. Some mutualistic interactions are symbiotic, such as those of cows and their cellulose-digesting microorganisms. Other mutualistic interactions are more temporary, such as those that occur between plants and the animals that pollinate them.

28.5 How Do Keystone Species Influence Community Structure?

Keystone species have a greater influence on community structure than can be predicted by their numbers. If a keystone species is removed from a community, the structure of the community is significantly altered.

28.6 How Do Species Interactions Change Community Structure over Time?

Succession is a change in a community and its nonliving environment over time. During succession, plants alter the environment in ways that favor their competitors, thus producing a somewhat predictable progression of dominant species. Primary succession, which may take thousands of years, occurs where no remnant of a previous community exists, such as on bare rock. Secondary succession occurs much more rapidly because it builds on the remains of a disrupted community, such as an abandoned field or the remnants of a forest after a fire. Uninterrupted succession ends with a climax community, which tends to be self-perpetuating unless acted on by outside forces. Some ecosystems, including tallgrass prairie and farm fields, are maintained in relatively early, subclimax stages of succession by periodic disruptions.

Key Terms

aggressive mimicry   *522*
camouflage   *518*
climax community   *525*
coevolution   *513*
community   *513*
competitive exclusion
    principle   *514*
consumer–prey interaction   *513*
ecological niche   *513*
host   *517*
interspecific
    competition   *513*
intraspecific
    competition   *515*
keystone species   *524*
mimicry   *520*
mutualism   *513*
parasite   *517*
pioneer   *525*
predator   *517*
primary succession   *526*
resource partitioning   *514*
secondary succession   *526*
startle coloration   *522*
subclimax   *525*
succession   *525*
symbiosis   *513*
warning coloration   *520*

Thinking Through the Concepts

Multiple Choice

1. The orderly progress of communities, starting from bare rock with no soil or traces of a previous community, is called
   a. primary succession.
   b. secondary succession.
   c. subclimax community.
   d. climax community.

2. Which of the following statements is *not* true of an ecological niche?
   a. A niche includes the physical environment, such as climate and water availability.
   b. A niche includes predators and parasites.
   c. Niche overlap may help to drive evolution.
   d. Two species cannot have overlapping niches.

3. A keystone species
   a. is always a carnivore.
   b. has a minor influence on community structure.
   c. has a major influence on community structure out of proportion to its numbers.
   d. has a major influence on community structure only when present in large numbers.

4. Which of the following statements regarding invasive species is True?
   a. All introduced species are invasive.
   b. Almost all invasive species are plants.
   c. Invasive species have no predators or parasites in their new ecosystem.
   d. Invasive species multiply rapidly and may displace native species in a community.

5. Which of the following statements about predators and parasites is *not* true?
   a. Predators and parasites harm their prey.
   b. Predators live on or in their hosts for a significant part of their life cycle.
   c. Predators and parasites coevolve with their prey.
   d. Parasites may kill their prey.

Fill-in-the-Blank

1. Organisms that interact serve as agents of _____ on one another. This results in _____, which is the process by which species evolve adaptations to one another.

2. Predators may be meat-eaters, called _____, or plant-eaters, called _____. Both predators and their prey may blend into their surroundings by using _____.

3. The concept that no two species with identical niches can coexist indefinitely is called the _____ principle.

4. Fill in the types of coloration or mimicry: used by prey to signal that it is distasteful: _____; used by a moth with large eyespots on its wings:_____; mimicry of a poisonous animal by a nonpoisonous animal: _____; mimicry used by a predator to attract its prey: _____.

5. Fill in the appropriate type of community interaction: bacteria, living in the human gut, that synthesize vitamin K: _____; bacteria that cause illness: _____; a deer eating grass: _____; a bee pollinating a flower:_____.

6. A somewhat predictable change in community structure over time is _____. This process takes two forms. Which of these forms would start with bare rock? _____ Which would occur after a forest fire? _____ A relatively stable community that is the end product of this process is called a _____ community. A mowed lawn in suburbia is an example of a _____ community.

Review Questions

1. Define an *ecological community*, and describe the three major categories of community interactions, including the benefits and harms to the interacting species.

2. Explain how resource partitioning is a logical outcome of the competitive exclusion principle.

3. Describe examples of coevolution between consumers and their prey.

4. Define *succession*. Which type of succession would occur on a clear-cut forest (where all trees have been logged) and why?

5. Provide examples of two climax and two subclimax communities. How do they differ?

6. What is an invasive species? Why are they destructive? What general adaptations do invasive species possess?

7. What is a keystone species? How can a keystone species within a community be identified?

Applying the Concepts

1. An ecologist visiting an island finds two species of birds, one of which has a slightly larger bill than the other. Interpret this finding with respect to the ecological niche and the competitive exclusion principle, and explain both concepts.

2. Why is it difficult to study succession? Choose a specific location and describe how you would study succession there.

29

ENERGY FLOW AND NUTRIENT CYCLING IN ECOSYSTEMS

A brown bear intercepts a salmon on its journey upstream to spawn.

Dying Fish Feed an Ecosystem

THE SOCKEYE SALMON in Katmai National Park in Alaska have a remarkable life cycle. After hatching in shallow depressions in the gravel bed of a swiftly flowing stream, the juvenile salmon spend 1 to 3 years in fresh water, often in a nearby lake. Then the young fish begin the physiological changes that will prepare them for life in the ocean. They head downstream to estuaries, where fresh water and salt water mix. In the estuary, the salmon complete their transformation, then head out to sea.

In the ocean, the young salmon grow rapidly, feeding on small fish and crustaceans. A few years later, when they reach sexual maturity, an instinctive drive compels them to return to their home streams to spawn. If they are lucky enough to escape the jaws of brown bears and the talons of eagles on their journey, the salmon carry their precious payload of sperm and eggs upstream to renew the cycle of life. The salmon die soon after spawning.

The fishes' trip back to their birthplace is remarkable in another way: Nutrients almost always move downstream, washed from the land into the ocean. But salmon, filled with muscle and fat acquired from feeding in the ocean, bring ocean nutrients back to the land. They bring energy upstream, too. From bears to eagles to spruce trees, this upstream movement of nutrients and energy supports the extraordinarily rich terrestrial ecosystems of the southern coast of Alaska. Where do nutrients and energy originate? How do living organisms acquire nutrients and energy and transfer them among the members of biological communities?

AT A GLANCE

29.1 HOW DO NUTRIENTS AND ENERGY MOVE THROUGH ECOSYSTEMS?

All **ecosystems** consist of two components. The **biotic** components are the communities of living organisms in a given area. The **abiotic** components of an ecosystem consist of all the nonliving aspects of the environment, such as climate, light, temperature, water, and minerals in the soil. Interactions within biological communities and between communities and their abiotic environment determine the movement of energy and nutrients through ecosystems. Two basic principles underlie this movement: Nutrients cycle within and between ecosystems, whereas energy flows through ecosystems (**FIG. 29-1**).

Nutrients are atoms and molecules that organisms obtain from their environment. The same nutrient atoms have been sustaining life on Earth for about 3.5 billion years. Your body includes oxygen, carbon, hydrogen, and nitrogen atoms that were once part of a dinosaur or a woolly mammoth. Nutrients are transported around the planet and converted to different molecular forms, but they do not leave Earth.

Energy, in contrast, takes a one-way journey through ecosystems. Solar energy is captured by photosynthetic bacteria, algae, and plants and then flows from organism to organism. However, all chemical reactions are inefficient. Every time an organism synthesizes a protein or moves its body, some of the useful, concentrated energy found in the chemical bonds of biological molecules is converted to low-level heat given off to the environment (see Chapter 6). Therefore, life on Earth requires a continuous input of energy.

CHECK YOUR LEARNING

Can you ...

- explain why nutrients cycle within and between ecosystems, whereas energy flows through ecosystems?

29.2 HOW DOES ENERGY FLOW THROUGH ECOSYSTEMS?

Thermonuclear reactions in the sun transform a relatively small amount of matter into enormous quantities of energy. Much of this energy is in the form of electromagnetic radiation, including heat (infrared light), visible light, and ultraviolet light. However, the sun emits its energy equally in all directions and is 93 million miles away, so Earth intercepts only about 45 billionths of a percent of this energy. Still, that's almost as much energy per hour as humanity uses in a year. The atmosphere and its clouds absorb or reflect about half of this solar radiation. About half reaches Earth's surface as visible light. A small fraction of the light reaching the surface is used in photosynthesis.

Energy and Nutrients Enter Ecosystems Through Photosynthesis

Plants, algae, and photosynthetic bacteria acquire nutrients such as carbon, nitrogen, oxygen, and phosphorus from the abiotic portions of ecosystems. These photosynthetic organisms capture energy in sunlight and use it to join nutrient atoms into carbohydrates, proteins, nucleic acids, and the other biological molecules of their bodies, storing some of the sun's energy in the chemical bonds of these molecules. Thus, photosynthesis brings both energy and nutrients into ecosystems. Here, we will focus our attention on the flow of energy.

Energy Passes Through Ecosystems from One Trophic Level to the Next

In an ecosystem, each category of organisms through which energy passes is called a **trophic level** (literally, "feeding level"). Photosynthetic organisms form the first trophic level. These organisms are called **producers,** or **autotrophs** (Greek, meaning "self-feeders"), because they produce food for themselves using inorganic nutrients and solar energy. Through photosynthesis, producers directly or indirectly provide food for nearly all other forms of life. Organisms that cannot photosynthesize, called **consumers,** or **heterotrophs** ("other-feeders"), acquire energy and most of their nutrients prepackaged in the molecules that make up the bodies of the organisms that they eat.

There are several levels of consumers. **Primary consumers** feed directly on producers. These **herbivores** (from Latin words meaning "plant-eaters"), which include animals such as grasshoppers, mice, and zebras, form the second trophic level. **Carnivores** ("meat-eaters"), such as spiders, hawks, and salmon, are higher-level consumers. Carnivores act as **secondary consumers** when they prey on herbivores. Some carnivores at least occasionally eat other carnivores; when doing so, they occupy the fourth trophic level and are called **tertiary consumers.** In some instances, particularly in the oceans, there are even higher trophic levels.

▶ **FIGURE 29-1 Energy flow, nutrient cycling, and feeding relationships in ecosystems**
The energy of sunlight (yellow arrow) enters an ecosystem during photosynthesis by organisms called producers, which store some of the energy in the biological molecules of their bodies. This biologically available energy (red arrows) is then passed to nonphotosynthetic organisms called consumers. Energy in wastes and dead bodies supports detritivores and decomposers. Every organism loses some energy as heat (orange arrows), so useful energy gradually becomes unavailable to living organisms. Therefore, ecosystems require a continuous input of energy. In contrast, nutrients (purple arrows) are recycled.

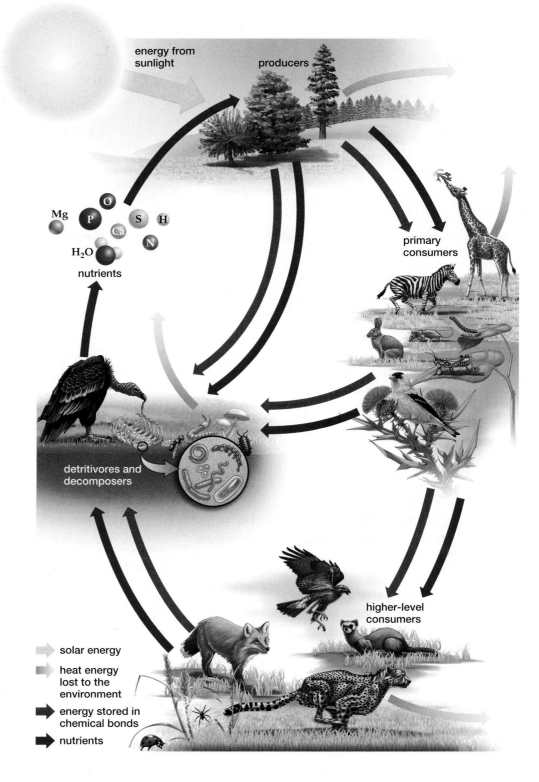

Net Primary Production Is a Measure of the Energy Stored in Producers

The amount of life that an ecosystem can support is determined by the amount of energy captured by its producers. The energy that photosynthetic organisms in a given area store in their bodies over a given period of time (for example, calories per square meter per year) is called **net primary production.** However, mass is much easier to determine than energy, and **biomass,** or dry biological material, is generally a good measure of the energy stored in organisms' bodies. Therefore, net primary production is usually given as grams of biomass per square meter per year.

The net primary production of an ecosystem is influenced by many factors, including the amount of sunlight reaching

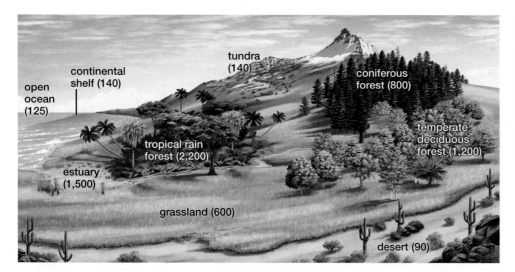

▶ **FIGURE 29-2 Net primary production in ecosystems** The average net primary production of some terrestrial and aquatic ecosystems is shown, measured in grams of biological material produced per square meter per year.

THINK CRITICALLY What factors contribute to the differences in primary production among ecosystems?

the producers, the availability of water and nutrients, and the temperature (**FIG. 29-2**). In deserts, for example, lack of water limits production. In the open ocean, light is a limiting factor in deep waters, and lack of nutrients limits production in most surface waters. In ecosystems where all resources are abundant, such as tropical rain forests, production is high.

An ecosystem's contribution to Earth's total production is determined both by the ecosystem's production per unit area and by the portion of Earth that the ecosystem covers. Although the oceans generally have low net primary production, they cover about 70% of Earth's surface, so they contribute about 25% of Earth's total production. This is about the same overall contribution as tropical rain forests, which have high net primary production, but which cover less than 5% of Earth's surface.

Food Chains and Food Webs Describe Feeding Relationships Within Communities

A **food chain** is a linear feeding relationship that includes a single species in each trophic level that is fed upon by a single species in the trophic level just above it (**FIG. 29-3**). Different ecosystems support radically different food chains. Plants are the dominant producers in land-based (terrestrial) ecosystems (**FIG. 29-3a**). Plants support plant-eating insects, reptiles, birds, and mammals, each of which may be preyed on by other animals. In contrast, microscopic photosynthetic protists and bacteria collectively called **phytoplankton** are the dominant producers in most aquatic food chains, such as those found in lakes and oceans (**FIG. 29-3b**). Phytoplankton support a diverse group of consumers called **zooplankton,** which consist mainly of protists and small shrimp-like crustaceans. These are eaten primarily by fish, which in turn are eaten by larger fish.

Animals in natural communities seldom fit neatly into simple food chains. A **food web** shows many interconnected food chains and more accurately describes the actual feeding relationships in a community (**FIG. 29-4**). Some animals, such as raccoons, bears, rats, and humans, are **omnivores** ("everything-eaters"); they act as primary, secondary, and occasionally tertiary consumers. A raccoon, for instance, is a primary consumer when it eats fruits and nuts, a secondary

consumer when it eats grasshoppers, and a tertiary consumer when it eats frogs or carnivorous fish. A carnivorous plant such as the Venus flytrap can tangle the food web further by acting as both a photosynthesizing producer and a spider-trapping tertiary consumer.

Detritivores and Decomposers Recycle Nutrients

Detritivores and decomposers are usually either left out of food webs or described as the final level because they feed on the wastes, dead bodies, or discarded body parts (such as fallen leaves) of all the other levels. **Detritivores** ("debris-eaters") are an army of mostly small and often unnoticed organisms, including nematode worms, earthworms, millipedes, dung beetles, and the larvae of some flies. A few large vertebrates such as vultures are also detritivores.

Decomposers are primarily fungi and bacteria. They feed mostly on the same material as detritivores, but they do not ingest chunks of organic matter, as detritivores do. Instead, they secrete digestive enzymes outside their bodies, where the enzymes break down organic material. The decomposers absorb some of the resulting nutrient molecules, but many of the nutrients remain in the environment.

Because they recycle nutrients, detritivores and decomposers are absolutely essential to life on Earth. They reduce the bodies and wastes of other organisms to simple molecules, such as carbon dioxide, ammonia, and minerals, that return to the atmosphere, soil, and water. Without detritivores and decomposers, ecosystems would gradually be buried by accumulated wastes and dead bodies, whose nutrients would be unavailable to other living organisms, including producers. If the producers died from a lack of nutrients, energy would cease to enter the ecosystem, and organisms at higher trophic levels, including people, would die as well.

Energy Transfer Between Trophic Levels Is Inefficient

A fundamental principle of thermodynamics is that energy use is never completely efficient (see Chapter 6). For example, as your car burns gasoline, only about 20% of the energy

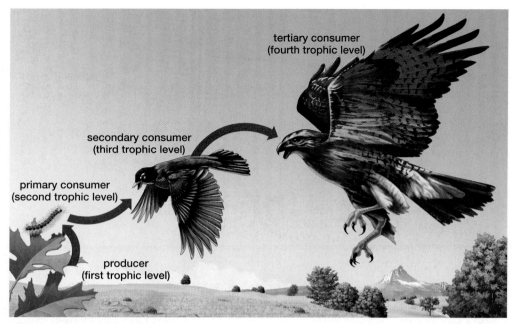

(a) A simple terrestrial food chain

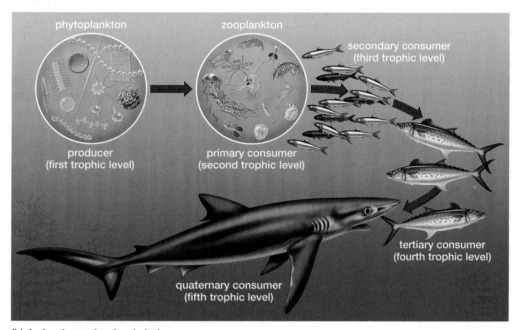

(b) A simple marine food chain

▲ FIGURE 29-3 Food chains on land and sea

is used to move the car; the other 80% is lost as heat. Inefficiency is also the rule in living systems: Waste heat is produced by all the biochemical reactions that keep cells alive. For example, splitting the chemical bonds of adenosine triphosphate (ATP) to power muscle contraction releases heat; that is why shivering or walking briskly on a cold day warms your body.

Most of the energy stored in the organisms at a given trophic level is not available to organisms in the next higher trophic level. When a grasshopper (a primary consumer) eats grass (a producer), only some of the solar energy trapped by the grass is available to the insect. The grass converted some

of the solar energy into the chemical bonds of cellulose, which a grasshopper cannot digest. In addition, although grasses don't feel warm to the touch, virtually every chemical reaction in the grass's leaves, stems, and roots gives off some energy as low-level heat, which thus isn't available to any other organism. Therefore, only a fraction of the energy captured by the producers of the first trophic level can be used by organisms in the second trophic level. If the grasshopper is eaten by a robin (the third trophic level), the bird will not obtain all the energy that the insect acquired from the plants. Some of the energy will have been used up to power hopping, flying, and eating. Some energy will be

▲ **FIGURE 29-4** **A simplified grassland food web** The animals pictured in the foreground include a vulture (a detritivore), a bull snake, a ground squirrel, a burrowing owl, a badger, a mouse, and a shrew (which looks like a small mouse but is carnivorous). In the middle distance you'll see a grouse, a meadowlark, a grasshopper, and a jackrabbit. In the distance, look for pronghorn antelope, a hawk, pheasants, a wolf, and bison.

found in the grasshopper's indigestible exoskeleton. Most of the energy will have been lost as heat. Similarly, most of the energy in a robin's body will be unavailable to a hawk that may consume it.

Although energy transfer between trophic levels varies significantly among communities, the average net transfer of energy from one trophic level to the next is roughly 10%. This means that, in general, the amount of energy stored in primary consumers is only about 10% of the energy stored in the bodies of producers. In turn, the bodies of secondary consumers contain roughly 10% of the energy stored in primary consumers. This inefficient energy transfer between trophic levels is called the "10% law." An **energy pyramid** illustrates the energy relationships between trophic levels—widest at the base and progressively narrowing in higher trophic levels (**FIG. 29-5**). A biomass pyramid for a given community typically has the same general shape as the community's energy pyramid.

The dominant organisms in a community are almost always photosynthesizers because they have the most energy available to them, as sunlight. The most abundant animals are herbivores. Carnivores are relatively scarce because there is far less energy available to support them.

Energy losses within and between trophic levels mean that long-lived animals at higher trophic levels often eat

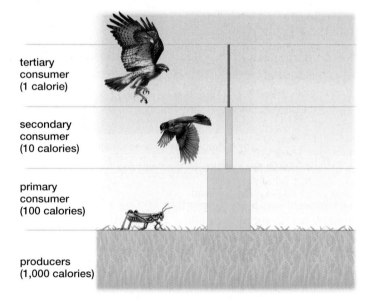

tertiary consumer (1 calorie)

secondary consumer (10 calories)

primary consumer (100 calories)

producers (1,000 calories)

▲ **FIGURE 29-5** **An energy pyramid for a grassland ecosystem** The width of each rectangle is proportional to the energy stored at that trophic level. Representative organisms for the first four trophic levels in a U.S. grassland ecosystem illustrated here are grass, a grasshopper, a robin, and a red-tailed hawk.

many times their body weight in organisms that occupy lower trophic levels—consider, for example, how much food you eat in a year, even if you stay the same weight. As long-lived animals continue to eat for months or years, they may consume, and often store, toxic substances that aren't easily broken down or excreted. Thus, toxic chemicals in plants may become concentrated in the bodies of high-level consumers. This **biological magnification** can lead to debilitating, even fatal, effects, as we explore in "Health Watch: Biological Magnification of Toxic Substances" on page 540.

CHECK YOUR LEARNING

Can you ...

- name the trophic levels in a community and give examples of organisms found in each trophic level?
- describe how energy flows through an ecosystem?
- explain why detritivores and decomposers are essential to ecosystem function?
- explain how the inefficiency of energy transfer between trophic levels determines the relative abundance of organisms in different trophic levels?

CASE STUDY CONTINUED

Dying Fish Feed an Ecosystem

When a sockeye salmon eats a smaller fish, the salmon is typically a tertiary consumer on the fourth trophic level, because the small fish was a secondary consumer that ate zooplankton, which were primary consumers that fed on photosynthetic phytoplankton. When an Alaskan brown bear eats a salmon, it is therefore feeding on the fifth trophic level. In a simplified food chain with only one type of organism at each trophic level, the 10% law means that a food chain containing a single 1,000-pound bear will also contain 10,000 pounds of salmon, 100,000 pounds of smaller fish, a million pounds of zooplankton, and 10 million pounds of phytoplankton.

All organisms contain not only energy, but also many types of nutrients. How are these nutrients recycled to future generations of living things?

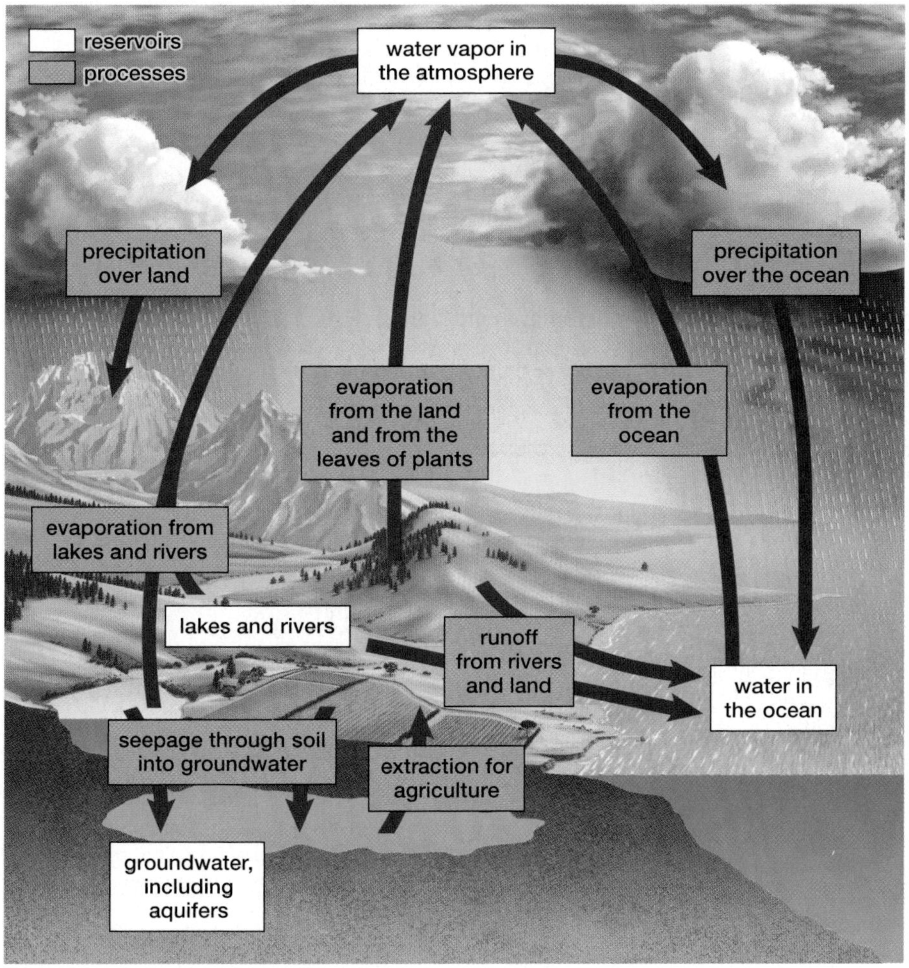

29.3 HOW DO NUTRIENTS CYCLE WITHIN AND AMONG ECOSYSTEMS?

As noted earlier, nutrients are elements and small molecules that form the chemical building blocks of life. Some, called **macronutrients,** are required by organisms in large quantities. These include water, carbon, hydrogen, oxygen, nitrogen, phosphorus, sulfur, and calcium. **Micronutrients,** including zinc, molybdenum, iron, selenium, and iodine, are required only in trace quantities. **Nutrient cycles,** also called *biogeochemical cycles,* describe the pathways that nutrients take as they move from their major sources in the abiotic parts of ecosystems, called **reservoirs,** through living communities and back again. In the following sections, we describe the cycles of water, carbon, nitrogen, and phosphorus.

The Hydrologic Cycle Has Its Major Reservoir in the Oceans

The **hydrologic cycle** (FIG. 29-6) is the pathway by which water travels from its major reservoir—the oceans—through the atmosphere, to smaller reservoirs in freshwater lakes, rivers, and groundwater, and then back again to the oceans. The oceans contain more than 97% of Earth's water. Another 2% of the total water is trapped in ice, leaving only 1% as liquid fresh water.

▲ **FIGURE 29-6 The hydrologic cycle**

Biological Magnification of Toxic Substances

The U.S. Food and Drug Administration advises that pregnant and breastfeeding women should avoid eating swordfish, shark, and king mackerel and limit eating albacore (white) tuna to no more than one can a week. But isn't eating fish supposed to be good for you? That depends on what fish you eat. These particular fish often contain a high concentration of mercury, which can damage the developing brains of fetuses and infants. Where does the mercury come from, and why do certain fish have so much of it?

Most mercury in the environment comes from two sources. Many small-scale gold mining operations, especially in less-developed countries, use mercury to separate gold from crude ore. In developed countries, including the United States, coal-fired power plants are the largest source of mercury contamination. Mercury, found in trace amounts in coal, is vaporized when the coal is burned and may be wafted thousands of miles on the winds. As a result, nowhere on Earth is free of mercury contamination.

Except in very localized areas, environmental concentrations of mercury are extremely low. Mercury becomes a health problem through *biological magnification*, the process by which toxic substances become concentrated in animals occupying high trophic levels. Most substances that undergo biological magnification share two properties. First, substances that biomagnify are stored in living tissue, particularly in fat. (Mercury is also stored in muscle.) Second, neither animals nor decomposers can readily break these substances down into harmless materials—they are not easily *biodegradable*. Mercury is an element, so it can never be broken down. When mercury becomes attached to certain organic functional groups, it becomes much more toxic. Although "organomercury" forms can be broken down, the process is very slow.

Biological magnification occurs because energy transfer between trophic levels is inefficient. By eating large quantities of producers that contain low levels of toxic chemicals and absorbing and storing the toxins in its body, an herbivore accumulates higher concentrations of the toxins. This sequence—eat, absorb, and store—continues up the trophic levels. Sharks, swordfish, and king mackerel are long-lived, predatory fish that feed at the top of long food chains, so they have ample opportunity to accumulate high concentrations of mercury. In ecological terms, the FDA's advice is for people to avoid eating at an even higher trophic level.

Organic compounds, including several pesticides, may also build up in animals that occupy high trophic levels. Biological magnification first came to public attention in the 1950s and 1960s, when wildlife biologists witnessed an alarming decline in populations of several fish-eating birds such as cormorants, ospreys, brown pelicans, and bald eagles. They were being poisoned by a pesticide called DDT. To control insects, many aquatic ecosystems had been sprayed with low amounts of DDT. However, these fish-eating birds contained DDT in concentrations a million times greater than the concentration in the water (**FIG. E29-1**). As a result, the birds laid thin-shelled eggs that broke under the

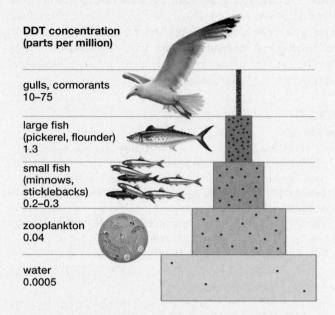

DDT concentration (parts per million)

gulls, cormorants
10–75

large fish (pickerel, flounder)
1.3

small fish (minnows, sticklebacks)
0.2–0.3

zooplankton
0.04

water
0.0005

▲ **FIGURE E29-1 Biological magnification of DDT** During the time when DDT was widely used as an insecticide, concentrations of DDT in a marsh on Long Island, New York, increased about a million-fold with increasing trophic levels, from extremely low levels in the water to toxic levels in predatory birds. Based on Woodwell, et al., *Science*, 1967.

parents' weight during incubation. Some other pesticides, and certain natural toxins produced by bacteria and algae, can also become concentrated as they move up the food chain.

Fortunately, there is good news. Populations of fish-eating birds have recovered significantly since DDT was banned in the United States in 1973. About 180 countries worldwide have agreed to ban or restrict the production and use of a dozen "persistent organic pollutants," including DDT and several other highly toxic pesticides. And in 2011, the U.S. Environmental Protection Agency issued new rules that require reductions in emissions of many toxic substances from power plants, including mercury, arsenic, nickel, and selenium, which should benefit both people and the other animals with which we share our planet.

EVALUATE THIS As you're having lunch with your friend Victoria, she confides in you that she's planning to become pregnant this year. You notice that Victoria is eating a tuna sandwich, and you remember reading that pregnant women should limit tuna intake because of mercury concerns. When you mention this to Victoria, she asks if there are fish that pose less risk. To answer Victoria's question, research the mercury levels typically found in catfish, salmon, light meat tuna and white meat tuna. Why might different species contain different levels of mercury?

The hydrologic cycle is driven by solar heat energy, which evaporates water from oceans, lakes, and streams. When water vapor condenses in the atmosphere, the water falls back to Earth as rain or snow. Because the oceans cover about 70% of Earth's surface, most evaporation occurs from them and most precipitation falls back onto them. Of the water that falls on land, some is absorbed by the roots of plants; much of this water is returned to the atmosphere by evaporation from plant leaves. Most of the rest of the water that falls on land evaporates from soil, lakes, and streams; a portion runs downhill in rivers back to the oceans; a minuscule fraction is stored in the bodies of living organisms; and some enters natural underground reservoirs called **aquifers.**

Aquifers are composed of sand, gravel, or water-permeable rock, such as sandstone, which are saturated with water. They are often tapped to supply water for household use and for irrigating crops. Water movement from the surface into aquifers is usually slow. In many areas of the world—including China, India, North Africa, California, and the Great Plains of the United States—water is being pumped out of aquifers faster than it is being replenished. If these aquifers are depleted, the resulting water shortages will force significant changes in agriculture. How do we know that aquifers are being depleted? There is the obvious way, of course, when wells go dry or have to be drilled deeper to reach water. But satellites can now survey underground water storage—and

much more—from space, as we explore in "How Do We Know That? Monitoring Earth's Health" on page 548.

The hydrologic cycle is crucial for terrestrial ecosystems because it provides the fresh water needed for land-based life. As you study the nutrient cycles that follow, keep in mind that nutrients in the soil must be dissolved in soil water to be taken up by the roots of plants or to be absorbed by bacteria. Plant leaves can take up carbon dioxide gas only after it has dissolved in a thin layer of water coating the cells inside the leaf. Without the hydrologic cycle, terrestrial organisms would rapidly disappear.

The Carbon Cycle Has Major Reservoirs in the Atmosphere and Oceans

Carbon atoms form the framework of all organic molecules. The **carbon cycle** (FIG. 29-7) is the pathway that carbon takes from its major short-term reservoirs in the atmosphere and oceans, through producers and into the bodies of consumers, detritivores, and decomposers, and then back to its reservoirs. Carbon enters a community when producers capture carbon dioxide (CO_2) during photosynthesis. On land, photosynthetic organisms acquire CO_2 from the atmosphere. Aquatic producers such as phytoplankton take up CO_2 dissolved in water.

Photosynthesizers "fix" carbon in biological molecules such as sugars and proteins (see Chapter 7). Producers return some of this carbon to the atmosphere or water as CO_2

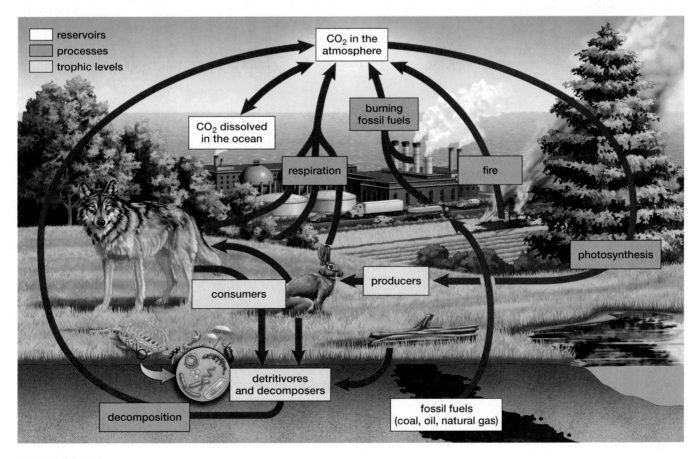

▲ **FIGURE 29-7 The carbon cycle**

generated by cellular respiration (see Chapter 8), but much of the carbon remains stored in the biological molecules of their bodies. When primary consumers eat the producers, they acquire this stored carbon. Primary consumers and the organisms in higher trophic levels release CO_2 during respiration, excrete carbon compounds in their feces, and store the rest of the carbon in their bodies. When organisms die, their bodies are broken down by detritivores and decomposers, whose cellular respiration returns CO_2 to the atmosphere and oceans. The complementary processes of uptake by photosynthesis and release by cellular respiration continually recycle carbon from the abiotic to the biotic portions of ecosystems and back again.

Some carbon, however, cycles much more slowly. Much of Earth's carbon is bound up in limestone, formed from calcium carbonate ($CaCO_3$) deposited on the ocean floor in the shells of prehistoric phytoplankton. The movement of carbon from this source to the atmosphere and back again takes millions of years. **Fossil fuels,** which include coal, oil, and natural

gas, are also long-term reservoirs for carbon. These substances were produced from the remains of prehistoric organisms buried deep underground and subjected to high temperature and pressure. In addition to carbon, the energy of prehistoric sunlight (originally captured by photosynthetic organisms) is trapped in these deposits, in the chemical bonds of carbon compounds. When humans burn fossil fuels to tap this stored energy, CO_2 is released into the atmosphere, with potentially serious consequences, as we describe in Section 29.4.

The Nitrogen Cycle Has Its Major Reservoir in the Atmosphere

Nitrogen is a crucial component of proteins, many vitamins, nucleotides (such as ATP), and nucleic acids (such as DNA). The **nitrogen cycle** (FIG. 29-8) is the pathway taken by nitrogen from its primary reservoir—nitrogen gas (N_2) in the atmosphere—to much smaller reservoirs of ammonia and

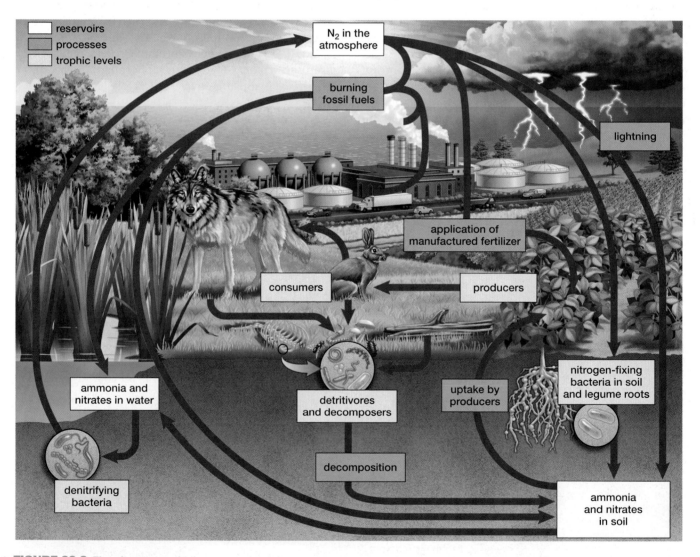

▲ **FIGURE 29-8 The nitrogen cycle**

THINK CRITICALLY What incentives cause humans to capture nitrogen from the air and pump it into the nitrogen cycle? What are some consequences of human augmentation of the nitrogen cycle?

nitrate in soil and water, through producers, consumers, detritivores and decomposers, and back to its reservoirs.

The atmosphere contains about 78% nitrogen gas, but plants and most other producers cannot use nitrogen in this form—they require either ammonia (NH_3) or nitrate (NO_3^-). A few types of bacteria that live in soil or water can convert N_2 into ammonia in a process called **nitrogen fixation.** Some nitrogen-fixing bacteria enter into a mutually beneficial relationship with certain plants, called *legumes,* in which the bacteria live in swellings on the plants' roots. Legumes such as alfalfa, soybeans, clover, and peas are extensively planted on farms, in part because they release excess ammonia produced by the bacteria, thus fertilizing the soil. Other bacteria in soil and water convert ammonia to nitrate. A small amount of nitrate is also produced during electrical storms, when the energy of lightning combines nitrogen and oxygen gases to form nitrogen oxide compounds. These nitrogen oxides fall to the ground dissolved in rain and are eventually converted to nitrate.

Producers absorb ammonia and nitrate and incorporate the nitrogen into biological molecules such as proteins and nucleic acids. The nitrogen passes through successively higher trophic levels as primary consumers eat the producers and are themselves eaten. At each trophic level, bodies and wastes are broken down by decomposers, which liberate ammonia back into the soil and water. The nitrogen cycle is completed by **denitrifying bacteria.** These residents of wet soil, swamps, and estuaries break down nitrate, releasing nitrogen gas back into the atmosphere (see Fig. 29-8).

CASE STUDY \ **CONTINUED**

Dying Fish Feed an Ecosystem

Nitrogen-fixing bacteria take up atmospheric nitrogen dissolved in ocean waters and produce ammonia, most of which is taken up by photosynthetic plankton. From the phytoplankton, nitrogen travels up the trophic levels, with some entering the bodies of salmon. As we will see in the Case Study Revisited, salmon bring nitrogen to the terrestrial ecosystems of the Alaskan coast.

People significantly manipulate the nitrogen cycle, both deliberately and unintentionally. As noted earlier, farmers plant legumes to fertilize their fields. Fertilizer factories combine N_2 from the atmosphere with hydrogen generated from natural gas, producing ammonia, which is then often converted to nitrate or urea (an organic nitrogen compound) for fertilizer. In addition, the heat produced by burning fossil fuels combines atmospheric N_2 and O_2, generating nitrogen oxides that form nitrates. These human activities now dominate the nitrogen cycle.

The Phosphorus Cycle Has Its Major Reservoir in Rock

Phosphorus is found in biological molecules such as nucleic acids and the phospholipids of cell membranes. It also forms a major component of vertebrate teeth and bones. The **phosphorus cycle** (FIG. 29-9) is the pathway taken by phosphorus from its primary reservoir in rocks to much smaller

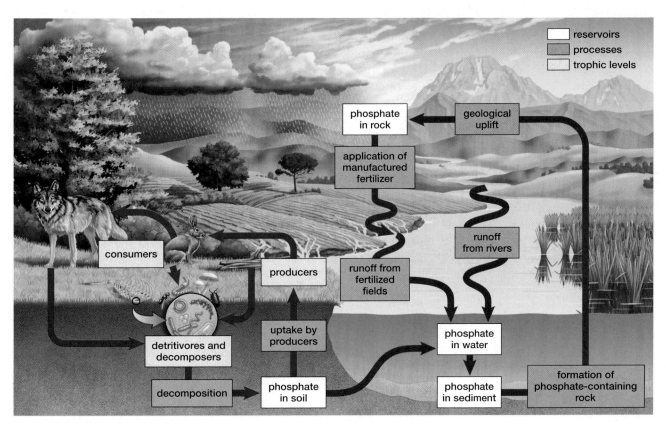

▲ **FIGURE 29-9 The phosphorus cycle**

reservoirs in soil and water, through living organisms, and then back to its reservoirs.

Throughout its cycle, almost all phosphorus is bound to oxygen, forming phosphate (PO_4^{3-}). There are no gaseous forms of phosphate, so there is no atmospheric reservoir in the phosphorus cycle. As phosphate-rich rocks are exposed by geological processes, some of the phosphate is dissolved by rain and flowing water, which carry it into soil, lakes, and the ocean, forming the smaller reservoirs of phosphorus that are available to ecological communities. Dissolved phosphate is absorbed by producers, which incorporate it into biological molecules. From producers, phosphate is passed through food webs; at each level, excess phosphate is excreted. Ultimately, detritivores and decomposers return the phosphate to the soil and water, where it may then be reabsorbed by producers or become bound to sediments and eventually re-formed into rock. Human production of phosphate fertilizers now dominates the phosphorus cycle: We extract about two to eight times more phosphate from rocks than was produced by natural processes in prehistoric times.

CHECK YOUR LEARNING

Can you ...

- explain why nutrients cycle within and among ecosystems?
- describe the hydrologic, nitrogen, carbon, and phosphorus cycles?

29.4 WHAT HAPPENS WHEN HUMANS DISRUPT NUTRIENT CYCLES?

Ancient peoples, with small populations and limited technology, had relatively little impact on nutrient cycles. However, as the human population grew and new technologies were developed, people began to significantly alter many nutrient cycles. Today, human use of fossil fuels and chemical fertilizers has disrupted the global nutrient cycles of nitrogen, phosphorus, sulfur, and carbon.

Overloading the Nitrogen and Phosphorus Cycles Damages Aquatic Ecosystems

Each year, fertilizers containing about 45 million tons of phosphate and 115 million tons of nitrogen (as ammonium, nitrate, and urea) are applied to farm fields to help satisfy the agricultural demands of a growing human population. When water washes over the land from rainfall or irrigation, it dissolves and carries away some of the nitrogen and phosphate. As the water drains into lakes, rivers, and ultimately the oceans, these nutrients can overstimulate the growth of phytoplankton. The resulting *harmful algal blooms* can turn clear water into an opaque green soup. As the phytoplankton die, their bodies sink into deeper water, where they provide a feast for decomposer bacteria. Cellular respiration by decomposers uses up most of the oxygen in the water, creating what is often called a dead zone. Deprived of oxygen, aquatic

▲ **FIGURE 29-10 Harmful algal blooms** Nutrients, especially nitrate and phosphate, wash off farmland in the Midwest and move down the Mississippi River, which flows from upper left to the center of the image, where it ends in the Mississippi Delta. When these nutrients enter the Gulf, they fertilize an explosive growth of algae, visible in this satellite photo as hazy green swirls near the coast.

invertebrates and fish either leave the area or die and decompose (making the problem worse).

Each summer, a huge dead zone occurs in the Gulf of Mexico off the coast of Louisiana. In spring, enormous quantities of nitrates and phosphates wash off fertilized farm fields into streams that flow into the Mississippi River. The Mississippi then empties the fertilizers into the Gulf of Mexico. In summer, when sunlight strengthens and the Gulf warms, the fertilizers create an algal bloom (**FIG. 29-10**), which soon produces a dead zone. Hurricanes and tropical storms break up the dead zone each autumn, but it reappears the following summer. The summer dead zone in the Gulf of Mexico now typically covers 5,000 to 8,000 square miles, an area about the size of Connecticut. Worldwide, dead zones are increasing in both size and number as agricultural activities intensify.

Overloading the Sulfur and Nitrogen Cycles Causes Acid Deposition

Natural processes put both nitrogen oxides and sulfur oxides into the atmosphere. Fires and lightning produce several types of nitrogen oxides, including nitrate; volcanoes, hot springs, and decomposers release sulfur dioxide (SO_2). However, combustion of fossil fuels now produces most of the nitrogen and sulfur oxides entering the atmosphere. When combined with water vapor in the atmosphere, nitrogen oxides and sulfur dioxide are converted into nitric acid and sulfuric acid. Days later and often hundreds of miles away, these acids fall to Earth in rain or snow. This **acid deposition** was first recognized in New Hampshire, where a sample of rain collected in 1963 had a pH of 3.7—about the same as orange juice, and 20 to 200 times more acidic than unpolluted rain, which usually has a pH between 5 and 6.

▲ **FIGURE 29-11 Acid deposition is corrosive** These identical building decorations in Brooklyn, New York, show the effects of acid deposition. On the left, the decoration has been restored to its original state; on the right, an unrestored decoration has eroded almost completely away.

▲ **FIGURE 29-12 Acid deposition can destroy forests** On the Camel's Hump in Vermont, virtually all of the older, mature trees are dead and bare. Since the 1990s, lower sulfur and nitrogen emissions from power plants have reduced acid rain in New England, and young trees are beginning to recolonize the forest.

Acid deposition damages forests, can render lakes lifeless, and even eats away at buildings and statues (FIG. 29-11). Acid deposition is most damaging to ecosystems where the soils have little buffering capacity to neutralize acids, such as much of New England, the mid-Atlantic states, the upper Midwest, Western mountains, and Florida. Upstate New York and New England are doubly vulnerable, because the prevailing westerly winds that sweep across North America carry sulfates and nitrates from coal-burning power plants in the Midwest directly over these states.

Acid deposition increases the exposure of organisms to toxic metals, such as aluminum, mercury, lead, and cadmium, which are far more soluble in acidified water than in water of neutral pH. Dissolved aluminum in the soil, for example, inhibits plant growth. When it runs off into lakes, it kills fish.

Plants growing in acidified soil often become weak and vulnerable to infection and damage by insects. Calcium and magnesium, which are essential nutrients for plants, are leached out of the soil by acid precipitation. Sugar maples in the Northeastern United States, often found in soils that are already low in calcium, have declined as a result. Acid rain directly damages the needles of conifers such as spruce and fir. About half of the red spruce and one-third of the sugar maples in the Green Mountains of Vermont have been killed over the past 40 years (FIG. 29-12).

Since 1990, government regulations have resulted in substantial reductions in emissions of both sulfur dioxide and nitrogen oxides from U.S. power plants—sulfur dioxide emissions are down about 40%, and nitrogen oxide levels have been reduced by more than 50%. Air quality has improved and rain has become less acidic, although large areas of the northeastern United States still receive rain with a pH below 5.0.

Damaged ecosystems recover slowly. If acid deposition is eliminated, eventually lakes in upstate New York and New England will return to their normal pH. Most aquatic life should then recover in 3 to 10 years, depending on the species, although the lakes may need to be restocked with fish. Forests usually take longer to recover because the life span of trees is so long and because soil chemistry changes slowly. Red spruce, however, is bouncing back more quickly than expected. In some locations, red spruce trees are growing faster than at any time in the past two centuries, although no one is sure why.

Interfering with the Carbon Cycle Is Changing Earth's Climate

The reservoir of carbon dioxide in the atmosphere not only provides producers with the starting material for photosynthesis, but also significantly affects Earth's climate. To understand why, let's begin with the fate of sunlight entering Earth's

HAVE YOU EVER
WONDERED ...

Each of us affects Earth through the choices we make. A *carbon footprint* is a measure of the impact that human activities have on climate based on the quantity of greenhouse gases they emit. Our personal carbon footprints give us a sense of our individual impacts. For example, each gallon of gasoline burned releases 19.6 pounds (8.9 kg) of CO_2 into the air. So, if your car gets 20 miles to the gallon, each mile that you drive will add about a pound of CO_2 to the atmosphere.

How Big Your Carbon Footprint Is?

The Web sites of the U.S. Environmental Protection Agency and several environmental organizations provide online carbon footprint calculators. Some calculators evaluate how your daily choices affect carbon emissions by asking questions such as: What kind of foods do you eat? How fuel efficient is your car? How well is your home insulated, and what temperature is your thermostat set to?

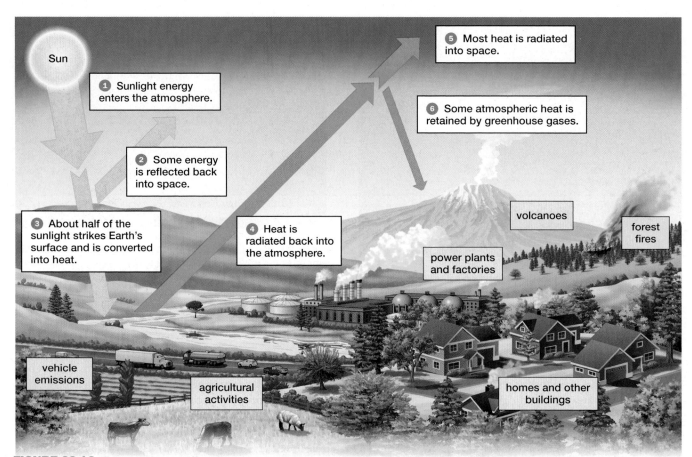

▲ **FIGURE 29-13 The greenhouse effect** Incoming sunlight warms Earth's surface and is radiated back to the atmosphere. Greenhouse gases, released by natural processes and substantially augmented by human activities (both shown in yellow rectangles), absorb increasing amounts of this heat, raising global temperatures.

atmosphere (**FIG. 29-13**). Some of the energy from sunlight ❶ is reflected back into space by the atmosphere and by Earth's surface, especially by areas covered with snow or ice ❷. About half of the sunlight strikes relatively dark surfaces (land, vegetation, and open water) and is converted into heat ❸ that is radiated into the atmosphere ❹. Most of this heat continues on into space ❺, but water vapor, CO_2 and several other **greenhouse gases** trap some of the heat in the atmosphere ❻. This natural process is called the **greenhouse effect,** and it keeps our atmosphere relatively warm. Without the natural greenhouse effect, the average temperature of Earth's surface would be far below freezing, and Earth would probably be lifeless.

The release of greenhouse gases by human activities increases the greenhouse effect. For Earth's temperature to remain constant, the total amount of energy entering and leaving Earth's atmosphere must be equal. When atmospheric concentrations of greenhouse gases increase, more heat is retained than is radiated into space, causing Earth to warm. Since burning fossil fuels began in earnest about 150 years ago during the Industrial Revolution, greenhouse gases, particularly CO_2, have increased. Other greenhouse gases have also increased, including methane (CH_4) and nitrous oxide (N_2O). At the present time, however, CO_2 contributes by far the largest share of the greenhouse effect caused by humans.

Burning Fossil Fuels Causes Climate Change

Human activities release about 35 to 40 billion tons of CO_2 into the atmosphere each year. Burning fossil fuels accounts for about 80% to 85% of this CO_2. A second source of added CO_2 is **deforestation,** which destroys tens of millions of forested acres annually and accounts for about 10% to 15% of humanity's CO_2 emissions. Deforestation is occurring principally in the Tropics, where rain forests are cleared for crops and cattle grazing. The carbon stored in the trees returns to the atmosphere when they are cut down and burned. A third, very minor source of CO_2 is volcanic activity. The U.S. Geological Survey estimates that less than 1% as much CO_2 enters the atmosphere from volcanoes as from human activities.

About half of the CO_2 released each year is absorbed by the oceans and terrestrial plants. The rest of the CO_2 remains in the atmosphere. Since 1850, the CO_2 content of the atmosphere has increased by over 40%—from 280 parts per million (ppm) to about 400 ppm in 2014—and is growing by about 2 ppm annually (**FIG. 29-14a**). Atmospheric CO_2 is now higher than at any time in the past 800,000 years.

A large and growing body of evidence indicates that human release of CO_2 and other greenhouse gases has amplified the natural greenhouse effect, altering the global climate. Air temperatures at Earth's surface, recorded at thousands of

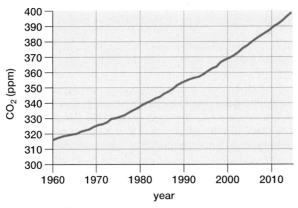

(a) Atmospheric CO₂

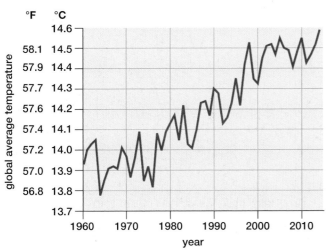

(b) Global surface temperature

▲ **FIGURE 29-14 Global temperature increases parallel atmospheric CO₂ increases (a)** Yearly average CO_2 concentrations in parts per million, measured 11,141 feet (3,396 meters) above sea level, on Mauna Loa, Hawaii. **(b)** Although global average temperatures fluctuate considerably from year to year, there is a clear upward trend over time. Data for both graphs from the National Oceanic and Atmospheric Administration.

THINK CRITICALLY If people stopped emitting CO_2 next year, do you think that global temperature would begin to decline immediately? Why or why not?

sites on land and sea, show that Earth has warmed by about 1.4°F (0.8°C) since the late 1800s, including an increase of 1°F (0.6°C) just since the 1970s (**FIG. 29-14b**). All but 1 of the 15 warmest years on record have occurred since 2000.

The widely reported "pause" in increasing temperatures since about 1998 (see Fig. 29-14b) is puzzling. Some researchers have hypothesized that the pause was caused by more heat being stored in the oceans rather than in the atmosphere during this time period. On the other hand, a study in mid-2015 suggested that warming has not actually slowed down—the apparent hiatus may be an artifact of outdated temperature recording methods, particularly in the oceans, and insufficient

data from the Arctic, which is warming more rapidly than the rest of the planet. If this study is correct, Earth's atmosphere is a little warmer than climate scientists had previously realized. Pauses do happen, though, such as the one that occurred from about 1960 through 1978, and then was followed by rapid warming between 1980 and 1998.

The overall impact of increased greenhouse gases is usually called **climate change,** reflecting the fact that greenhouse gases and the resulting global warming have many effects on our climate and Earth's ecosystems. Although a 1.4°F increase may not seem like much, our warming climate has already caused widespread changes. For example, glaciers are retreating worldwide: About 90% of the world's mountain glaciers are shrinking, and the trend seems to be accelerating (**FIG. 29-15**). Glacier National Park, Montana, named for its spectacular abundance of glaciers, had 150 glaciers in 1910; now, only 25 remain—and these are significantly smaller than they were in the recent past. During the past 30 years, the Arctic ice cap has become almost 50% thinner and 35% smaller in area. On the continent of Antarctica, although some ice sheets are shrinking while others are expanding, the total volume of ice has dramatically decreased in recent years

(a) Muir Glacier, 1941

(b) Muir Glacier, 2004

▲ **FIGURE 29-15 Glaciers are melting** Photos taken from the same vantage point in **(a)** 1941 and **(b)** 2004 document the retreat of the Muir Glacier in Glacier Bay National Park, Alaska.

HOW DO WE KNOW THAT?

Monitoring Earth's Health

Carbon dioxide concentrations in the atmosphere are increasing; Earth is getting warmer; oceans are acidifying; glaciers are retreating; Arctic sea ice is decreasing. You may wonder—how do we know all this?

Estimating some conditions on Earth is fairly straightforward. For example, atmospheric CO_2 is measured at hundreds of stations in dozens of countries, including Mauna Loa in Hawaii (see Fig. 29-14a). Estimates of CO_2 concentrations in the distant past are obtained by analyzing gas bubbles trapped in ancient Antarctic ice.

In some places on Earth, people began keeping accurate temperature records well over a century ago. Now, air temperatures are measured at about 1,500 locations, on both land and sea, each day. Sophisticated computational methods compensate for the uneven distribution of weather stations (more in England than in the Arctic or Sahara Desert) and produce global average temperatures. Ancient temperatures can be estimated by "natural proxies"—natural phenomena that vary with temperature and leave long-lasting records. For example, isotopes of oxygen in air trapped in bubbles inside ice vary with the air temperature at the time the bubble formed. Ice cores collected from glaciers in Antarctica or Greenland can therefore be used to estimate "paleotemperatures." Chemical measurements of corals and mollusk shells, and even some types of sediments and fossils, also provide estimates of paleotemperatures.

However, some measurements of Earth's environment wouldn't have been possible even 20 to 40 years ago. Many involve data collected by satellites. For example, measuring areas of forest is a simple, if tedious, matter of carefully examining satellite photos. Other measurements are much more sophisticated. Accurate estimates of Arctic sea ice started in 1979, with the launch of satellites that measure microwave

radiation emitted from Earth's surface. Ice emits more microwave radiation than liquid water does, so the satellites can easily distinguish the two. Satellite data show that the extent of Arctic sea ice has declined about 13% per decade since 1979 (**FIG. E29-2**). Many other features of Earth have distinctive "signature wavelengths" that satellites can detect, from sulfur dioxide emitted by power plants to chlorophyll in the oceans (**FIG. E29-3**).

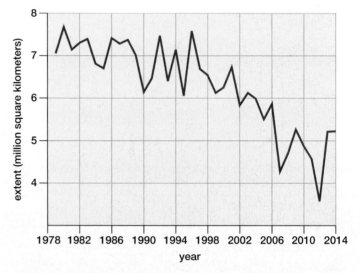

▲ **FIGURE E29-2 Changes in Arctic sea ice** Satellite measurements of Arctic sea ice began in 1979. By 2014, the area covered by ice at the end of the summer (September) had declined by more than a third.

(see "How Do We Know That? Monitoring Earth's Health"). The oceans are warming, which causes their water to expand and occupy more volume. This expansion, coupled with water flowing into the oceans from melting glaciers and ice sheets, causes sea levels to rise.

Continued Climate Change Will Disrupt Ecosystems and Endanger Many Species

What does the future hold? Climate scientists predict that a warming atmosphere will cause more severe storms, including stronger hurricanes; greater amounts of rain or snow in single storms (a phenomenon already observed in the northeastern United States during the past half-century); and more frequent, severe, and prolonged droughts. Increased CO_2 also makes the oceans more acidic, which disturbs many natural processes, including the ability of many marine

animals, such as snails and corals, to make their shells and skeletons.

Predictions of continued climate change are based on sophisticated computer models developed and run independently by climate scientists around the world. As the models continue to improve, they match past climate with ever-greater accuracy, providing increasing confidence in their predictions for the future. The models also provide evidence that natural causes, such as changes in the output of the sun, cannot account for the recent warming. The models match the data only when human greenhouse gas emissions are included in the calculations.

The Intergovernmental Panel on Climate Change (IPCC) is a consortium of hundreds of climate scientists and other experts from 130 nations who work together to address climate change. In their 2014 report, the IPCC predicted that even under the best-case scenario in which a

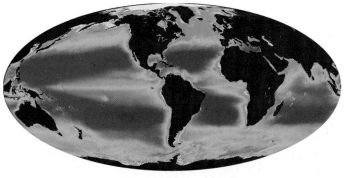

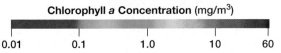

Chlorophyll *a* Concentration (mg/m³)

0.01   0.1   1.0   10   60

▲ **FIGURE E29-3** **Ocean chlorophyll** Satellite measurements of chlorophyll show which areas of the ocean have the greatest amount of phytoplankton. Purple/blue represent low chlorophyll concentrations, green/yellow intermediate amounts, and orange/red the highest concentrations.

Perhaps the most amazing measurements come from NASA's GRACE satellites—the Gravity Recovery and Climate Experiment. A satellite's orbiting speed is determined, in part, by the force of gravity exerted on it. Water and ice are heavy. Large volumes of ice on the land increase local gravity, tugging ever-so-slightly on the satellites, which then measure the extra gravitational pull. GRACE has found that land ice sheets in Antarctica and Greenland have declined dramatically over the past decade. Antarctica is losing about 150 billion tons of ice per year; Greenland is losing about 260 billion tons. GRACE can even measure water underground: the combination of prolonged drought and groundwater pumping for agriculture in California's Central Valley has greatly depleted the aquifers underlying the Valley (**FIG. E29-4**).

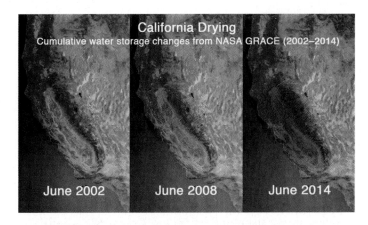

▲ **FIGURE E29-4** **Changes in gravity show depletion of water in California's aquifers** Underground aquifers in California's Central Valley are losing about 4 trillion gallons of water each year. The transition from green to red in these false-color images shows water lost between 2002 and 2014.

THINK CRITICALLY People tend to be much more attuned to what's happening right now and less aware of long-term trends. Every time there's a blast of cold weather in winter or hot weather in summer, opinion polls show lesser or greater concern about global warming. Climatologists, however, take a very long view and look for trends in climate data. Using a ruler, estimate trend lines for the data in Figures 29-14 and E29-2. What do the trend lines predict about the future of atmospheric CO_2 concentrations, global temperatures, and Arctic sea ice? If these trends persist, will the Arctic become ice-free in late summer? If so, in what year? When will CO_2 concentrations double from preindustrial levels and reach 560 parts per million? Is it reasonable to extrapolate straight (linear) trend lines into the future? Why or why not?

concerted worldwide effort is made to reduce greenhouse gas emissions, the average global temperature will rise by another 1.3°F (0.7°C) by the year 2100. Without major reductions in emissions, global temperatures might rise as much as 5.8°F (3.2°C). These changes in climate will be difficult to stop, let alone reverse, as we explore in "Earth Watch: Climate Intervention—A Solution to Climate Change?" on page 550.

Even if the more optimistic predictions are correct, the consequences for natural ecosystems will be profound. In 2011, scientists compiled the results of 53 studies that examined changes in the distribution of more than 1,000 species of terrestrial plants and animals. The species' ranges are moving toward the poles at an average rate of about 10.5 miles (17 kilometers) per decade—just what would be expected if they are moving in response

to a warming planet. As climate change continues, some plants and animals will find it easier to move than others will, either because they are intrinsically more mobile (such as some birds) or because they can move great distances while reproducing (such as some plants that produce lightweight, wind-borne seeds). Some species may not be able to move rapidly enough and will become rare or even go extinct.

Species on mountains or in the Arctic and Antarctic may have nowhere to go. For example, the loss of summer sea ice is bad news for polar bears and other marine mammals that rely on ice floes as nurseries for their young and as staging platforms for hunting fish or seals. As summer ice diminishes, both walrus and polar bear populations are moving onto land to give birth, putting the adults farther away from their prime hunting grounds. As walrus crowd together onto

Earth WATCH

Climate Intervention—A Solution to Climate Change?

What should be done about climate change? The obvious solution is to reduce our emissions of CO_2 and other greenhouse gases. But that isn't as simple as it sounds. Modern societies depend on the energy of fossil fuels and cannot just stop burning them overnight. As a result, some scientists and policymakers are pondering a temporary fix while we make the transition to carbon-neutral, renewable energy sources. Perhaps we can slow down climate change with *climate intervention*—altering fundamental characteristics of Earth to reduce, or counteract the warming effect of, greenhouse gases. Although climate intervention is a complex subject, there are two main approaches: shading the planet and removing CO_2 from the atmosphere.

Shading the Planet

Some molecules, such as a few sulfur compounds, can reflect sunlight back into space, cooling the planet. Whenever a massive volcano erupts, it spews millions of tons of sulfur dioxide miles high into the atmosphere. In 1991, when Mt. Pinatubo erupted in the Philippines (**FIG. E29-5**), it blasted about 20 million tons of sulfur dioxide as high as 22 miles skyward. For the next couple of years, global temperatures were slightly cooler. In the early 2000s, global emissions of sulfur from power plants, especially in China, rose by about 25%. At the same time, there appeared to be a pause in the upward trend of planetary temperatures. Although most of the pause was probably due to the oceans taking up more heat, the shading effect of added sulfur in the atmosphere may have contributed. Based on data such as these, some have suggested that governments could slow global warming by sending planes loaded with sulfur compounds high into the atmosphere, where they would release the sulfur. Other potential methods of shading the planet include spewing reflective metal particles from jet exhaust or modifying Earth's cloud cover.

Capturing and Storing CO_2

There are both engineering and biological approaches to capturing and storing carbon. For example, to reduce CO_2 in the atmosphere, air could be sucked through thousands of towers, planet-wide, in which CO_2 would be removed, concentrated, and then either injected underground or converted to solid forms (similar to limestone) that could be stored on land. Although technically feasible, this approach would be enormously expensive. However, the exhaust gases from power plants have a very high CO_2 concentration. It would be much less expensive to remove CO_2 at the emission source rather than from the atmosphere as a whole. In 2014, a coal-fired power plant in Canada became the first large-scale facility to capture and store carbon in this way.

A biological approach to removing CO_2 is to fertilize the oceans. In many parts of the open ocean, the nutrient that limits phytoplankton growth is iron, which is an essential part of enzymes involved in ATP production in mitochondria, chloroplasts, and bacteria. The proposal is to spread powdered iron on open ocean waters, triggering phytoplankton blooms. The phytoplankton would then take up CO_2 during

▲ **FIGURE E29-5 Sulfur emissions both cool and pollute** The 1991 eruption of Mt. Pinatubo in the Philippines injected millions of tons of sulfur dioxide miles into the atmosphere. The sulfur compounds reflected sunlight and cooled the planet for a few years.

photosynthesis and store some of the carbon in their bodies. When the phytoplankton die, they would sink, carrying the carbon to the ocean depths, where it should remain for many years.

Will Climate Intervention Work? Is It Worth the Risks?

Many people question the feasibility and desirability of climate intervention. As a panel of experts commissioned by the U.S. National Academy of Sciences put it, "Climate intervention is no substitute for reductions in carbon dioxide emissions." Why not? Atmospheric sulfur causes respiratory damage in people, decreases the ozone layer, and generates acid deposition that injures ecosystems. In addition, shading the planet would do nothing to reduce acidification of the oceans. Although small-scale tests have been promising, no one really knows if ocean fertilization would remove enough CO_2 to help very much, or what effects it might have on ocean ecosystems. Removing CO_2 from the atmosphere by chemical means would be extremely expensive. Nevertheless, the expert panel recommends studying both planetary shading and carbon storage: If climate change gets bad enough, people are likely to demand action—and action informed by careful research is far preferable to blundering about, with little understanding of the consequences.

CONSIDER THIS Investigate a climate intervention proposal in some depth. How much temperature change can be achieved, how fast, and at what cost? What are the likely health or environmental side effects? Compare these costs and benefits to "business as usual" emissions and to efforts to reduce carbon emissions.

small beaches (**FIG. 29-16**), instead of being spread out over sea ice, they sometimes endanger their own young. In 2009, for example, stampeding adults trampled 131 walrus calves to death. Complete loss of Arctic sea ice during the summer, which climate models predict could occur by mid-century, may cause the extinction of polar bears in the wild.

Some of the movement of species may have direct impacts on human health. Many diseases, especially those carried by mosquitoes and ticks, are currently restricted to tropical or subtropical parts of the planet. Like other animals, these disease vectors will probably spread poleward as a result of warming temperatures, bringing their diseases, such as malaria, dengue fever, yellow fever, and Rift Valley fever, with them. On the other hand, it may become so hot and dry in parts of the Tropics that mosquitoes and some other insects may have shortened life spans, thus reducing vector-borne diseases in these regions. At this time, no one can confidently predict the overall effects on human health.

CHECK YOUR LEARNING

Can you ...

- explain how human activities have disrupted nutrient cycles?
- describe how human interference with nutrient cycles causes acid deposition, damages aquatic ecosystems, enhances the greenhouse effect, and causes climate change?
- describe some of the evidence that Earth is warming and some of the impacts of climate change on Earth's ecosystems?

▲ **FIGURE 29-16 Walrus pack the beach at Point Lay, Alaska** Walrus are normally dispersed on floating sea ice in the summer. As the Arctic warms and sea ice disappears during the summer, they now congregate on a few beaches, a phenomenon first seen in 2007.

CASE STUDY REVISITED

Dying Fish Feed an Ecosystem

The sockeye salmon's return to an Alaskan stream is unforgettable (**FIG. 29-17**). Even after the fish have run the gauntlet of brown bears and bald eagles, hundreds remain, their brilliant red bodies writhing in water so shallow that it barely covers them. A female excavates a depression in the gravel where she releases her eggs; a male then showers them with sperm. But sperm and eggs aren't the only cargo the salmon carry upstream from the ocean. About 95% of a sockeye's body mass was acquired during its years of feeding in the ocean, so it carries enormous amounts of energy and nutrients upstream with it. Sunlight energy, originally captured by phytoplankton, then transferred to zooplankton and smaller fish, is now stored in the bodies of the salmon, where it becomes available to terrestrial predators and scavengers. During a summer gorging on salmon, a brown bear may put on as much as 400 pounds of fat, which serves as its vital energy supply while it hibernates during the long Alaskan winter. Minks also profit. Females nurse their young during the salmon runs, taking advantage of the virtually inexhaustible supply of half-eaten salmon carcasses left behind by the bears.

Of the nutrients that salmon transport from the ocean to the land, nitrogen is especially important. Nitrogen from ocean sources can be distinguished from terrestrial nitrogen by the

▲ **FIGURE 29-17 Spawning sockeye salmon**

ratio of two isotopes of nitrogen, ^{14}N and ^{15}N. Ecologists have found that 50% to 70% of the nitrogen near some streams in Alaska originated in the ocean and was brought upstream in the bodies of salmon. In some places, more than half of this salmon-derived nitrogen is carried some distance away from the streams by bears, which often eat only the choicest parts of the fish and leave the rest. Salmon-derived nitrogen is important to Sitka spruce, which can grow three times faster near salmon streams than near streams without salmon runs. It's also important to the next generation of salmon. Because salmon carcasses fertilize nearby lakes, stimulating the growth of phytoplankton and consequently the zooplankton upon which newly hatched salmon feed, the adult salmon indirectly feed their young.

CONSIDER THIS Dams, river pollution, and overfishing have depleted many salmon populations—some are listed as endangered or threatened under the Endangered Species Act. Some people argue that because these salmon are also raised commercially in fish farms, the depletion of wild populations doesn't really matter. Based on what you have learned in this chapter, do you think that wild populations of salmon should be protected and efforts made to increase their numbers?

CHAPTER REVIEW

MB   MB® Go to MasteringBiology for practice quizzes, activities, eText, videos, current events, and more.

Answers to Think Critically, Evaluate This, Multiple Choice, and Fill-in-the-Blank questions can be found in the Answers section at the back of the book.

Summary of Key Concepts

29.1 How Do Nutrients and Energy Move Through Ecosystems?

Ecosystems are sustained by a continuous input of energy from sunlight and the recycling of nutrients. Energy enters the biotic portion of ecosystems through photosynthesis and then flows through the ecosystem. Nutrients are obtained by organisms from their living and nonliving environments and are recycled within and among ecosystems.

29.2 How Does Energy Flow Through Ecosystems?

Photosynthetic organisms act as conduits of both energy and nutrients into biological communities. The energy of sunlight is captured by photosynthetic organisms (producers), the first trophic level in an ecosystem. Herbivores (plant-eaters, also called primary consumers) form the second trophic level. Carnivores (meat-eaters) are secondary consumers when they prey on herbivores and tertiary or higher-level consumers when they eat other carnivores. Omnivores occupy multiple trophic levels. Detritivores and decomposers feed on dead bodies and wastes. Decomposers (mostly bacteria and fungi) liberate nutrients as simple molecules that reenter nutrient cycles.

The higher the trophic level, the less energy is available to sustain it. In general, only about 10% of the energy captured by organisms at one trophic level is available to organisms in the next higher level.

29.3 How Do Nutrients Cycle Within and Among Ecosystems?

A nutrient cycle depicts the movement of a particular nutrient from its reservoir, usually in the abiotic portion of an ecosystem, through the biotic portion, and back to its reservoir.

In the hydrologic cycle, the major reservoir of water is the oceans. Solar energy evaporates water, which returns to Earth as precipitation. Water flows into lakes and underground aquifers and is carried by rivers to the oceans.

In the carbon cycle, the short-term reservoirs are CO_2 in the oceans and the atmosphere. Carbon enters producers via photosynthesis. From producers, carbon is passed through the food

web and released to the atmosphere as CO_2 during cellular respiration. Burning fossil fuels also releases CO_2 into the atmosphere.

In the nitrogen cycle, the major reservoir is N_2 in the atmosphere. N_2 is captured by nitrogen-fixing bacteria, which produce ammonia. Other bacteria convert ammonia to nitrate. Plants obtain nitrogen from nitrates and ammonia. Nitrogen passes from producers to consumers and is returned to the environment through excretion and the activities of detritivores and decomposers. N_2 is returned to the atmosphere by denitrifying bacteria.

In the phosphorus cycle, the principal reservoir consists of phosphate in rocks. Phosphate dissolves in water, is absorbed by photosynthetic organisms, and is passed through food webs. Some phosphate is excreted, and the rest is returned to the soil and water by decomposers. Some is carried to the oceans, where it may be deposited in marine sediments.

29.4 What Happens When Humans Disrupt Nutrient Cycles?

Human activities often produce and release more nutrients than nutrient cycles can efficiently process. Fertilizer use in agriculture has disrupted many aquatic ecosystems. By burning fossil fuels, humans have overloaded the natural cycles for sulfur, nitrogen, and carbon. In the atmosphere, sulfur dioxide and nitrogen oxide are converted to sulfuric acid and nitric acid, which fall to Earth as acid deposition, with harmful effects on lakes and forests.

Burning fossil fuels has substantially increased atmospheric carbon dioxide. Climate scientists have concluded that increased CO_2 causes increased global temperatures. Increased temperatures cause climate change that is manifested in more extreme weather, melting glaciers, thinning Arctic sea ice, warming and acidifying oceans, rising sea levels, and changing distributions and seasonal activities of wildlife.

Key Terms

abiotic *534*	carnivore *534*
acid deposition *544*	climate change *547*
aquifer *541*	consumer *534*
autotroph *534*	decomposer *536*
biological magnification *539*	deforestation *546*
biomass *535*	denitrifying bacteria *543*
biotic *534*	detritivore *536*
carbon cycle *541*	ecosystem *534*

Thinking Through the Concepts

Multiple Choice

1. Which trophic level(s) must be present to sustain any ecosystem?
   a. producers only
   b. producers and consumers
   c. producers, detritivores, and decomposers
   d. producers, consumers, detritivores, and decomposers

2. Which of the following is *not* a major reservoir in the carbon cycle?
   a. consumers
   b. the atmosphere
   c. the oceans
   d. fossil fuels

3. Denitrifying bacteria
   a. convert ammonia to nitrate.
   b. convert nitrate to ammonia.
   c. convert N_2 to ammonia.
   d. convert nitrate to N_2.

4. Net primary production per unit area is likely to be highest in which of the following ecosystems?
   a. grasslands
   b. deserts
   c. temperate deciduous forests
   d. tropical rain forests

5. The effect of CO_2 emissions from burning fossil fuels is likely to be
   a. nothing; photosynthesis by plants will remove all the added CO_2 from the atmosphere.
   b. nothing; the carbon cycle will simply run faster.
   c. planetary warming due to an increased greenhouse effect.
   d. planetary cooling due to a reduced greenhouse effect.

Fill-in-the-Blank

1. Nearly all life gets its energy from _____, which is captured by the process of _____. In contrast, _____ are constantly recycled during processes called _____.

2. Photosynthetic organisms are called either _____ or _____. The energy that these store and make available to other organisms is called _____.

3. Feeding levels within ecosystems are also called _____. An illustration of these levels with only one organism at each level is called a(n) _____. Feeding relationships are most accurately depicted as _____.

4. In general, only about _____ percent of the energy available in one trophic level is captured by the level above it.

5. Photosynthetic organisms make up the first trophic level. Organisms in higher trophic levels are collectively called _____ or _____. Photosynthetic organisms are consumed by organisms collectively called _____ or _____. Animals that feed on other animals are called _____ or _____. Organisms that feed on wastes and dead bodies are called _____ and _____.

6. During the nitrogen cycle, nitrogen gas is captured from its atmospheric reservoir by _____ in the soil, and is then returned to this reservoir by _____. The two forms of nitrogen that are used by plants are _____ and _____.

7. Two relatively short-term reservoirs for carbon are the _____ and _____. Carbon in these reservoirs is in the form of _____. Two long-term reservoirs for carbon are _____ and _____.

Review Questions

1. What makes the movement of energy through ecosystems fundamentally different from the movement of nutrients?

2. What is a producer? What trophic level does it occupy, and what is its importance in ecosystems?

3. Define *net primary production*. Would you predict higher productivity in a farm pond or an alpine lake? Explain your answer.

4. Name the first three trophic levels. Among the consumers, which are most abundant? Use the "10% law" to explain why you would predict that there will be a greater biomass of plants than herbivores in an ecosystem.

5. How do food chains and food webs differ? Which is the more accurate representation of feeding relationships in ecosystems?

6. Define *detritivore* and *decomposer* and explain their importance in ecosystems.

7. Trace the movement of carbon from one of its reservoirs through the biotic community and back to the reservoir. How have human activities altered the carbon cycle, and what are the implications for Earth's future climate?

8. Explain how nitrogen gets from the atmosphere into a plant's body.

9. Trace a pathway of a phosphorus molecule from a phosphate-rich rock into a carnivore. What makes the phosphorus cycle fundamentally different from the carbon and nitrogen cycles?

Applying the Concepts

1. Humans are omnivores who can feed on several trophic levels. Discuss how the inefficiency of energy transfer between trophic levels might apply to how many humans can be fed, with what environmental impacts, by people eating fundamentally different diets.

2. Discuss the contribution of human population growth to (a) acid precipitation and (b) global climate change.

30

EARTH'S DIVERSE ECOSYSTEMS

Many South American farmers use sustainable methods to grow cacao. The yellow pod contains seeds from which chocolate is made.

Food of the Gods

WHAT DO CHOCOLATE AND COFFEE have in common? Some would say they are both necessities of life. Every year, the average person in the United States eats about 11 pounds of chocolate and drinks about 340 cups of coffee. Scandinavians devour about twice as much chocolate and coffee as Americans do. Indeed, there must have been chocoholics in Sweden at least 300 years ago: Carolus Linnaeus, the Swedish scientist who invented scientific taxonomy, named the cacao tree *Theobroma cacao*—in Greek, "theobroma" means "food of the gods."

Cocoa is produced from the seeds of the cacao tree, originally from rain forests in South and Central America. Coffee "beans" are the seeds of two species of coffee plants, native to upland forests in the African country of Ethiopia. Both cacao and coffee are now widely cultivated in the Tropics, including South and Central America, Africa, and Southeast Asia. Worldwide, about 4.4 million tons of cacao seeds and 9 million tons of coffee beans are produced each year.

The original varieties of both cacao and coffee grew in the shade of taller trees—in fact, full sunlight killed the plants, especially seedlings. In Central and South America, cacao and coffee were traditionally cultivated as an understory plant in the rain forest. These plantations provided a multilevel, diverse habitat that supported monkeys, frogs, flowers, and about 200 species of birds. The forest vegetation absorbed water and protected the soil from erosion. The shade discouraged weed growth, and detritivores and decomposers recycled fallen leaves into plant nutrients.

In the 1960s and 1970s, however, new varieties of cacao and coffee plants were developed that thrived in full sun and yielded more seeds. As world demand for chocolate and coffee increases, more and more cacao and coffee plants are grown in full-sun plantations. What is lost when rain forests are cut down to make room for full-sun plantations? Or when we alter, or even destroy, natural communities in other ways, such as draining wetlands for agriculture or housing? To answer these questions, we must first understand the properties of the communities that make up life on Earth.

AT A GLANCE

30.1 WHAT DETERMINES THE DISTRIBUTION OF LIFE ON EARTH?

Both the type of living organisms—cacti or redwood trees, lobsters or jaguars—and their abundance vary enormously from place to place. This variability arises from an uneven distribution of four requirements for life on Earth: (1) energy, (2) nutrients, (3) liquid water, and (4) suitable temperatures.

Energy enters almost all ecosystems as sunlight, captured by plants and other photosynthetic organisms and transferred through food webs to consumers (see Chapter 29). Therefore, the distribution of life on Earth is largely determined by the requirements of the photosynthesizers. Although there are exceptions, we can make two major generalizations about how the location and abundance of the four requirements of life affect photosynthetic organisms.

First, in aquatic ecosystems, liquid water is almost always available, so sunlight energy, nutrients, and temperature are usually the factors that determine the distribution of life. Only the top 650 feet (200 meters) of a body of water, and usually much less, receives enough sunlight for photosynthesis. Even there, photosynthesis may be limited because surface waters usually contain low levels of many nutrients. Temperature may also be a limiting factor because many aquatic organisms, such as corals, thrive only within a fairly narrow temperature range.

Second, in terrestrial ecosystems, sunlight energy and most nutrients are relatively plentiful, so precipitation and temperature largely determine the distribution of life. Terrestrial plants require liquid water in the soil, at least during part of the year, because plants require water for their metabolic activities and to replace water that evaporates from their leaves. Soil moisture depends on precipitation and temperature. Generally speaking, the more precipitation, the more moisture the soil contains, but temperature also strongly influences soil moisture: High temperatures evaporate water from the soil, while prolonged freezing temperatures turn soil water to ice, making it unavailable to plants.

These requirements for life occur in specific patterns on our planet, resulting in characteristic communities of living organisms that extend over thousands, sometimes (in the oceans) even millions, of square miles. These large-scale communities are called **biomes,** often named after their principal types of vegetation, such as deciduous forests or grasslands.

The types of vegetation that dominate a terrestrial biome are determined by long-term patterns of temperature and precipitation. Measured over hours or days, temperature and precipitation are two of the principal components of weather. Weather patterns that prevail for years or centuries in a particular region make up its **climate.** Therefore, the climate in any particular region is a good predictor of its biome (**FIG. 30-1**). Before we begin our survey of terrestrial biomes, we will explore how the physical features of Earth and its movement in the solar system affect climate.

CHECK YOUR LEARNING

Can you ...

- name the four requirements for life on Earth?
- explain which of these requirements are most important in determining the distribution of life in aquatic and terrestrial ecosystems?

30.2 WHAT FACTORS INFLUENCE EARTH'S CLIMATE?

Weather and climate are driven by a great thermonuclear engine: the sun. Solar energy reaches Earth in a range of wavelengths—from short, high-energy ultraviolet (UV) rays, through visible light, to the long infrared wavelengths that we experience as heat (see Chapter 7). When sunlight enters the atmosphere, some is reflected back into space, but most solar energy is absorbed either by molecules in the atmosphere or by Earth's surface, thereby heating the planet. Fortunately, most UV radiation, which can damage biological molecules, including DNA, does not reach the surface. UV is absorbed by an **ozone layer** in the middle atmosphere, or stratosphere. During the twentieth century, humans produced a number of chemicals that began to deplete the ozone layer. In a remarkable case of international cooperation, almost all the nations of the world agreed to limit the production of ozone-depleting chemicals, as we explore in "Earth Watch: Plugging the Ozone Hole" on page 558.

The climate of different locations on Earth varies tremendously. These variations arise from the physical properties of our planet. Among the most important are Earth's curvature, its tilted axis, and the fact that it orbits the sun rather than staying in one place. As we will see, these factors cause uneven heating of Earth's surface. Uneven heating, in conjunction with Earth's rotation on its axis, generates air and ocean currents, which are in turn modified by the presence, location, and topography of the continents.

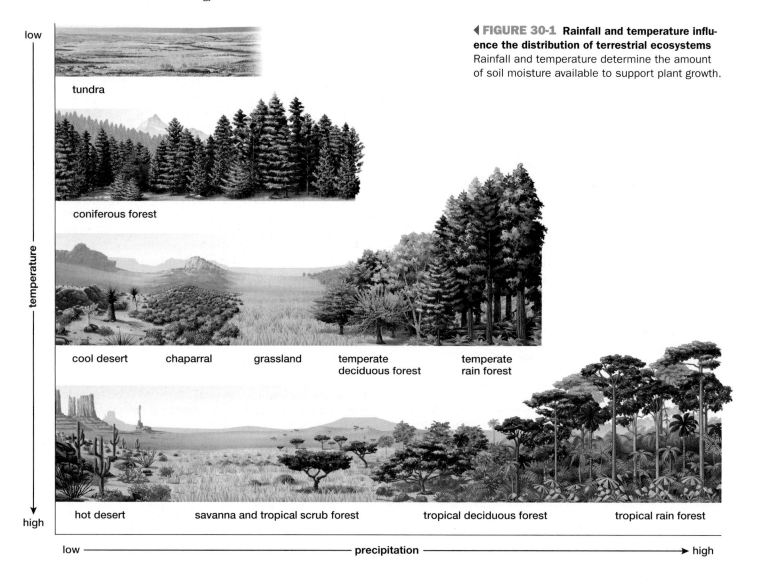

temperature
high

tundra

coniferous forest

cool desert   chaparral   grassland   temperate
deciduous forest   temperate
rain forest

hot desert   savanna and tropical scrub forest   tropical deciduous forest   tropical rain forest

low ———————————————— precipitation ———————————————————→ high

Earth's Curvature and Tilt on Its Axis Determine the Angle at Which Sunlight Strikes the Surface

Average yearly temperatures are determined by the amount of sunlight that reaches the surface in different regions, which in turn depends on latitude (**FIG. 30-2a**). Latitude is a measure of the distance north or south of the equator, expressed in degrees. The equator is defined as 0° latitude, and the poles are at 90° north and south latitudes. Sunlight hits the equator relatively directly (perpendicular to the surface) throughout the year. However, because Earth is a sphere, the farther away from the equator, the more slanted the sunlight, so a given amount of sunlight is spread out over a larger area. In addition, slanting sunlight at high latitudes must travel through more of Earth's atmosphere than vertical sunlight at the equator does, further reducing the amount of solar energy that reaches the surface. Earth is also tilted on its axis, about 23.5° relative to a line perpendicular to the plane of its orbit around the sun (**FIG. 30-2b**). During the course of a year, the tilted axis causes latitudes north and south of the equator to experience

significant changes in the angle and duration of sunlight, resulting in pronounced seasons. When Earth's position in its orbit causes the Northern Hemisphere to be tilted toward the sun, this hemisphere receives relatively direct sunlight and experiences summer (Fig. 30-2b, left). Simultaneously, the Southern Hemisphere is tilted away from the sun, receives slanted sunlight, and thus experiences winter. Six months later, conditions are reversed: It is summer in the Southern Hemisphere and winter in the Northern Hemisphere (Fig. 30-2b, right). Because sunlight hits the equator fairly directly throughout the year, the Tropics remain warm year-round.

Air Currents Produce Large-Scale Climatic Zones That Differ in Temperature and Precipitation

The angle at which sunlight strikes Earth's surface produces climatic zones with markedly different temperatures and precipitation (**FIG. 30-3**). Averaged over the course of a year, sunlight has the greatest warming effect at the equator ❶. The heat evaporates water from Earth's surface, especially from

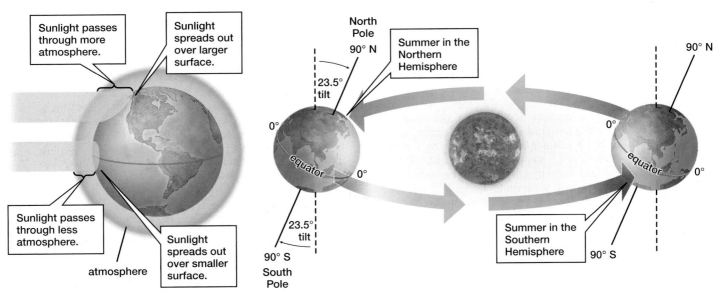

Sunlight passes through more atmosphere.

Sunlight spreads out over larger surface.

Sunlight passes through less atmosphere.

Sunlight spreads out over smaller surface.

atmosphere

(a) The intensity of sunlight hitting Earth's surface varies with latitude.

North Pole
90° N
23.5° tilt
0°
equator
0°

Summer in the Northern Hemisphere

23.5° tilt
90° S
South Pole

Summer in the Southern Hemisphere

90° N
0°
equator
0°
90° S

(b) The intensity of sunlight hitting different latitudes on Earth's surface varies with the seasons.

▲ **FIGURE 30-2 Earth's curvature and tilt cause temperature to vary with latitude and season of the year (a)** At the equator, sunlight passes through a minimum thickness of atmosphere and strikes Earth's surface nearly vertically year-round. Further toward the poles, sunlight passes through more atmosphere and hits Earth's surface at an angle that spreads a given amount of sunlight over a much larger land surface. Therefore, average temperatures are highest at the equator and lowest at the poles. **(b)** The tilt of Earth on its axis causes seasonal variations in how directly sunlight strikes different latitudes.

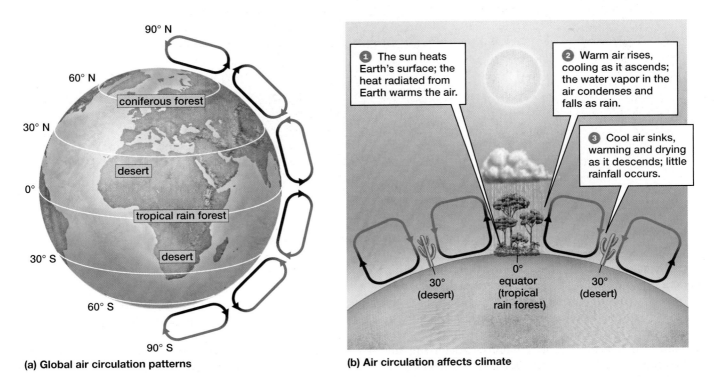

90° N
60° N
coniferous forest
30° N
desert
0°
tropical rain forest
30° S
desert
60° S
90° S

(a) Global air circulation patterns

① The sun heats Earth's surface; the heat radiated from Earth warms the air.

② Warm air rises, cooling as it ascends; the water vapor in the air condenses and falls as rain.

③ Cool air sinks, warming and drying as it descends; little rainfall occurs.

30° (desert)
equator (tropical rain forest)
0°
30° (desert)

(b) Air circulation affects climate

▲ **FIGURE 30-3 Air currents and climatic zones (a)** Warm air (red) rises at about 0° and 60° latitudes, and cool air (blue) falls at about 30° and 90° latitudes. **(b)** Air circulation patterns produce broad climatic zones.

Earth WATCH   Plugging the Ozone Hole

Ultraviolet light is so energetic that it can damage biological molecules. UV causes sunburn, premature aging of the skin, and skin cancer. Fortunately, more than 97% of UV radiation is filtered out by an ozone-enriched region of the stratosphere called the ozone layer, which begins about 6 miles (10 kilometers) above Earth's surface and extends up to about 30 miles (50 kilometers). UV light striking ozone and oxygen gas causes reactions that both break down and regenerate ozone. In the process, the UV radiation is converted to heat, and the overall level of ozone remains reasonably constant—or it did before humans intervened.

In 1985, British atmospheric scientists published the startling news that springtime levels of stratospheric ozone over Antarctica had declined by more than 30% since 1979. By the mid-1990s, the **ozone hole** over Antarctica had worsened, with springtime ozone only about 50% of its original levels (**FIG. E30-1**). Higher levels of UV light beneath the ozone hole reduce photosynthesis by phytoplankton—the producers in marine ecosystems, and the basis of food webs that support penguins, seals, and whales. Although ozone layer depletion is most severe over Antarctica, the ozone layer is somewhat reduced over most of the world. The ozone layer at mid-latitudes is about 5% less than its original level.

The ozone hole is caused primarily by human production and release of chlorofluorocarbons (CFCs). These chemicals were once widely used in the production of foam plastic, as coolants in refrigerators and air conditioners, as aerosol spray propellants, and as cleansers for electronic parts. CFCs are very stable and were considered safe. Their stability, however, proved to be a major problem because they remain chemically unchanged as they slowly rise into the stratosphere. In the stratosphere, UV light causes them to break down and release chlorine atoms, which in turn catalyze the breakdown of ozone.

Fortunately, major steps have been taken toward plugging the ozone hole. The 1987 Montreal Protocol set limits and established phase-out periods for several ozone-depleting chemicals. In a remarkable worldwide effort, 197 countries have signed the treaty. Since 2000, the ozone layer has begun to show signs of recovery. However, because CFCs persist in the atmosphere for many years, full recovery is not expected until 2050, possibly even later.

THINK CRITICALLY UV light damages some of the molecules involved in photosynthesis, not only in phytoplankton, but also in many terrestrial plants. Assume that the Montreal Protocol had not been implemented and that ozone-depleting chemicals had been released in ever-increasing amounts. In this situation, how would you expect ozone depletion to affect global climate change?

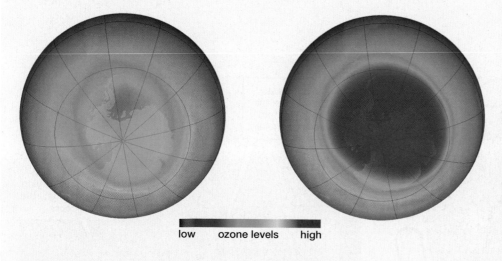

low   ozone levels   high

(a) Antarctic ozone hole, September 1979   **(b) Antarctic ozone hole, September 2014**

◀ **FIGURE E30-1** **The Antarctic ozone hole** Satellite images show ozone concentrations above the Antarctic in September 1979, before significant ozone depletion occurred, and in September 2014. Low concentrations of ozone are shown in blue and purple. There are fluctuations from year to year, but the area of the hole (about 9.5 million square miles; 25 million square kilometers) and its ozone concentration (about 50% of the concentration before depletion began) are currently both about the same as they were in the mid-1990s. Images courtesy of NASA.

the oceans, so equatorial air contains a lot of moisture. Warm air is less dense than cool air, so near the equator this warm, moist air rises. As it rises, the air cools, and water condenses out, falling as rain ❷. The direct rays of the sun and abundant rainfall at the equator create a warm, wet climate, where rain forests flourish.

After the moisture has fallen from the rising equatorial air, cooler, drier air remains. The continuing upward flow of air from the equator pushes this cool, dry air north and south. By about 30° N and 30° S latitudes, the air has cooled enough to sink. As it sinks, the air is warmed by heat radiated from Earth's surface. This warm, dry air produces very little rainfall ❸. As a result, the major deserts of the world are found near these latitudes. After reaching the desert surface, the warm air flows north and south, some moving back toward the equator and some moving toward the poles. As the air flows near the surface, its warmth evaporates water, so the air gradually becomes moist. This moderately warm, moist air rises at about 60° N and 60° S. The air cools as it rises, so water precipitates out as rain or (in winter) as snow. These climatic conditions favor the growth of coniferous and deciduous forests. Finally, the remaining dry air flows to the poles and sinks once more, producing very little precipitation. With highly slanted sunlight in summer and no sunlight at all in the winter, the poles are also extremely cold.

These patterns of air circulation predict that climate zones should occur in bands corresponding to latitude. The actual locations of Earth's biomes agree fairly well with this overall pattern, but the fit isn't perfect by any means (FIG. 30-4). The differences are largely caused by three factors: the rotation of Earth on its axis, the existence of the continents, and the presence of large mountain ranges on the continents.

Terrestrial Climates Are Affected by Prevailing Winds and Proximity to Oceans

The interaction between Earth's rotation and air flowing north and south from 30° N and 30° S latitudes determines average wind directions: east to west between the equator and 30° N and 30° S latitudes, and west to east north of 30° N and south of 30° S latitudes. Friction between winds and the ocean surface produces ocean currents. If there were no continents, then ocean currents would flow around the globe, east to west near the equator, and west to east north of 30° N and south of 30° S. However, the continents interrupt the currents, breaking them into roughly circular patterns called

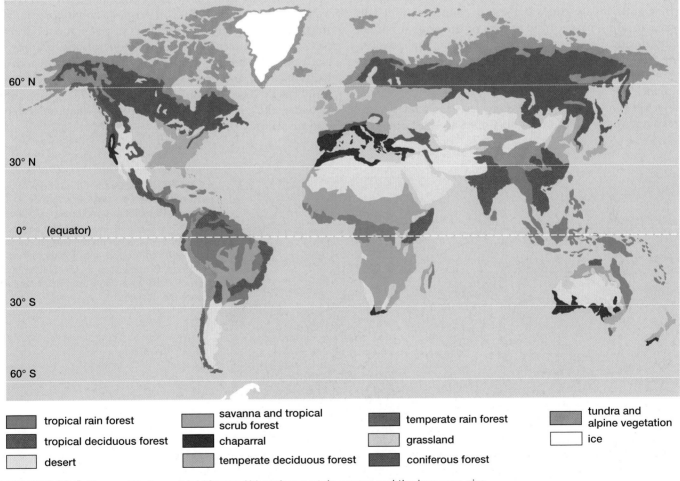

tropical rain forest	savanna and tropical scrub forest	temperate rain forest	tundra and alpine vegetation
tropical deciduous forest	chaparral	grassland	ice
desert	temperate deciduous forest	coniferous forest	

▲ **FIGURE 30-4 The world's terrestrial biomes** Although mountain ranges and the immense size of continents complicate the distribution of biomes, fairly consistent patterns remain. Tundra and coniferous forests are in the northernmost parts of the Northern Hemisphere, whereas the deserts of Mexico, the Sahara, Saudi Arabia, Southern Africa, and Australia are located around 30° N and 30° S latitudes. Tropical rain forests are found near the equator. Note that there is little land in the Southern Hemisphere between 45° S and the Antarctic continent; therefore, coniferous forest and tundra biomes are rare in the Southern Hemisphere.

gyres, which circulate clockwise in the Northern Hemisphere and counterclockwise in the Southern Hemisphere (**FIG. 30-5**).

Interactions between prevailing winds, ocean currents, and the sheer sizes of the continents profoundly affect terrestrial climates. Because water both heats and cools more slowly than land or air does, the interiors of continents have much more extreme temperatures than coastal regions at similar latitudes experience. For example, the average high temperatures in San Francisco, on the coast of California, range from 58°F (14°C) in winter to 71°F (22°C) in summer. Sacramento, only about 80 miles inland, has average high temperatures of 54°F (12°C) in winter and 92°F (33°C) in summer. In St. Louis, Missouri, about 1,700 miles east of San Francisco and 600 miles from the nearest ocean, high temperatures average about 38°F (3°C) in winter and 90°F (32°C) in summer.

The differential heating of land versus water causes the *monsoons* that bring summer rainfall to India and much of southeastern Asia. As the summer sun heats the land, hot air rises, pulling cooler, moist air inland off the ocean. The moist air rises up mountain slopes on the land, cooling still further until finally the water vapor condenses and falls as rain. In some parts of India, 80% of the annual rainfall occurs during the monsoons. Similar, but much weaker, monsoons bring rain to the southwestern United States, including parts of Arizona, New Mexico, Utah, Nevada, and Colorado.

Ocean gyres further modify some coastal climates. Some gyres carry warm water from the Tropics to coastal regions located relatively far from the equator. This creates warmer, moister climates than would be expected at these latitudes. For example, the Gulf Stream, carrying warm water from the Caribbean up the coast of North America and across the Atlantic (see Fig. 30-5), is responsible for the mild, infamously moist climate of the British Isles. Other ocean currents, such as the California current, move cold water from near the poles down toward the equator, causing cooler climates than would be expected in regions adjacent to these currents.

Finally, changes in water temperature in the central and eastern Pacific Ocean drive the *El Niño/Southern Oscillation,* a major shift in global climate that recurs every few years. Sporadically, surface waters in the equatorial Pacific warm and move eastward toward South America. Because the warm ocean water often appears off the coast of South America around Christmas, the local fishermen called it El Niño, "the (male) child," referring to Jesus. Evaporation from this pool of warm water causes increased winter precipitation in northern South America, Mexico, and the southern United States. In the western Pacific, cooler water replaces the warm water that has moved east. Less water evaporates from this cooler water, producing less rainfall in Australia and Indonesia. The shifting paths of these huge masses of water and air also affect weather patterns far from the equator; for example, El Niño typically brings warmer, drier winters to northern regions of North America. There are also years when the equatorial Pacific is cooler than usual; these conditions are called La Niña, "the female child." Its effects on climate are roughly opposite those of El Niño.

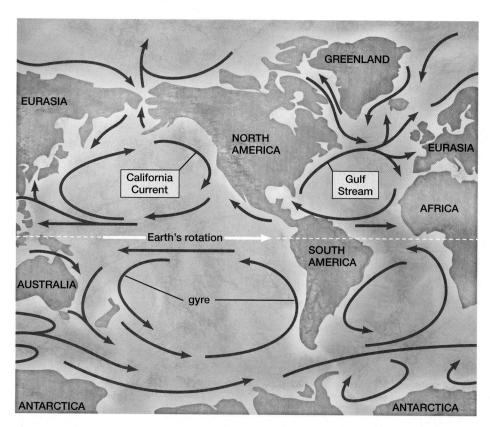

◀ **FIGURE 30-5 Ocean circulation patterns** Gyres flow clockwise in the Northern Hemisphere and counterclockwise in the Southern Hemisphere. Some ocean currents, such as the Gulf Stream, carry warm water from the Tropics toward the poles. Others, such as the California Current, carry cold water from polar regions toward the equator.

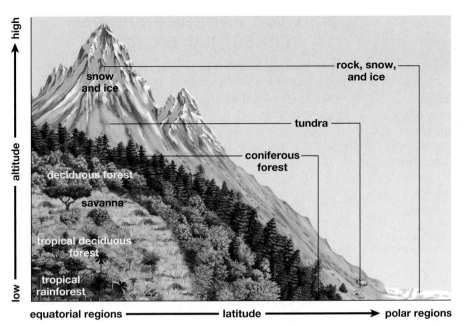

Climbing a mountain in the Northern Hemisphere is like heading north; in both cases, increasingly cool temperatures produce a similar series of biomes.

Mountains Complicate Climate Patterns

Variations in elevation within continents significantly affect climate. As elevation increases, air becomes thinner and cooler. The temperature drops approximately 3.5°F (2°C) for every 1,000 feet (305 meters) in elevation, so increasing elevation and increasing latitude have similar effects on terrestrial ecosystems (FIG. 30-6). Even near the equator, lofty mountains, such as Mount Kilimanjaro in Tanzania (19,341 feet) and Chimborazo in Ecuador (20,565 feet), may be snowcapped much of the year.

Mountains also modify patterns of precipitation. When water-laden air is forced to rise as it meets a mountain, it cools. Because cooling reduces the air's ability to hold water, the water condenses and falls as rain or snow on the windward side of the mountain. The air warms again as it travels down the far (lee) side of the mountain, so it absorbs water from the land, creating a local dry area called a **rain shadow** (FIG. 30-7). For example, the Sierra Nevada range of California wrings moisture from westerly winds blowing off the

Pacific Ocean. On the western side of the mountains, heavy winter snows provide moisture for forests of pine, fir, and massive sequoias. The Great Basin Desert, the Owens Valley, and the northern Mojave Desert, in the rain shadow on the east side of the Sierra Nevada, receive only 5 to 7 inches of rain a year and support mostly cacti and drought-resistant bushes.

CHECK YOUR LEARNING

Can you ...
- distinguish between weather and climate?
- explain how Earth's curvature, tilt on its axis, and orbit around the Sun affect climate?
- explain how temperature and precipitation interact to determine soil moisture and the distribution of terrestrial biomes?
- describe how winds, ocean currents, continents, and mountains affect climate and the distribution of terrestrial biomes?

◀ FIGURE 30-7 **Mountains create rain shadows**

CASE STUDY   **CONTINUED**

Food of the Gods

In recent years, a fungus called coffee rust has become a major threat to coffee production in Central America. The fungus infects the leaves of the coffee plant, preventing photosynthesis. If enough leaves are infected, the entire plant may die. Coffee rust is typically a major problem in full-sun plantations at warm temperatures. It grows slowly, if at all, at temperatures below 59°F (15°C). In the past, coffee rust was not a significant problem for farms in many Central American countries, where coffee is mostly grown in cool climates at altitudes above 4,000 feet. However, climate change has brought warmer temperatures and wetter weather, and the rust has proliferated, even in high-altitude farms.

Coffee rust has been known to occur in Africa for at least 150 years, generally without destroying the coffee plants. Can growing coffee in a more natural rain forest habitat help modern farmers to defeat coffee rust?

30.3 WHAT ARE THE PRINCIPAL TERRESTRIAL BIOMES?

In the following sections, we discuss the major terrestrial biomes, beginning at the equator and working our way poleward. We also discuss some of the impacts of human activities on these biomes.

Tropical Rain Forests

Near the equator, the average temperature is between 77° and 86°F (25° and 30°C), with little variation during the year. Rainfall ranges from 100 to 160 inches (250 to 400 centimeters) annually. These evenly warm, moist conditions create the most productive biome on Earth, the **tropical rain forest,** dominated by broadleaf evergreen trees (**FIG. 30-8**). Extensive rain forests are found in Central and South America, Africa, and Southeast Asia.

Rain forests have the highest **biodiversity,** or total number of species, of any biome on Earth. Although rain forests cover less than 5% of Earth's total land area,

◀ **FIGURE 30-8 The tropical rain-forest biome** Towering trees reach for the light in the dense tropical rain forest. Amid their branches dwells the most diverse assortment of life on Earth, including (from left to right) tree-dwelling orchids, red-eyed tree frogs, and fruit-eating toucans.

THINK CRITICALLY How does a biome with such poor soil support the highest plant productivity and the greatest animal diversity on Earth?

ecologists estimate that they contain half of the world's biodiversity. For example, in a 3-square-mile tract of rain forest in Peru (about 8 square kilometers), scientists counted more than 1,300 butterfly species and 600 bird species. For comparison, the entire continental United States is home to only about 600 butterfly species and 800 bird species.

Tropical rain forests typically have several layers of vegetation. The tallest trees may be more than 200 feet (60–70 meters) high, towering above the rest of the forest. Below these giants sits a fairly continuous canopy of treetops at about 90 to 120 feet (30 to 40 meters). Another layer of shorter trees typically stands below the canopy. Woody vines grow up the trees. Collectively, these plants capture most of the sunlight. Only about 2% of the sunlight reaches the forest floor, where the plants often have enormous, dark-green leaves, an adaptation that allows them to carry out photosynthesis in dim light.

Because of the lack of sunlight, edible plant material close to the ground is scarce, so most of the animals—including birds, monkeys, and insects—inhabit the trees. Competition for the nutrients that do reach the ground is intense among both plants and animals. For example, when a monkey defecates high up in the canopy, hundreds of dung beetles converge on the droppings within minutes after the waste hits the ground. Plants absorb nutrients almost as soon as soil decomposers release them from wastes or dead plants and animals. This rapid recycling means that almost all the nutrients in a rain forest are stored in the vegetation, leaving the soil relatively infertile.

Human Impacts Because of infertile soil and heavy rains, agriculture in rain forests is risky and often destructive. If the trees are cut and carried away for lumber, few nutrients remain to support crops. If the trees are burned, releasing nutrients into the soil, the heavy year-round rainfall quickly dissolves the nutrients and carries them away, leaving the soil depleted after only a few seasons of cultivation.

Nevertheless, rain forests are being felled for lumber or burned for ranching or farming at an alarming rate. Satellite images indicate that about 20 to 30 million acres of tropical rain forest are lost each year—the area of a football field every 1 to 1.5 seconds. About half of the world's rain forests have now been lost. In addition, like all forests, rain forests absorb carbon dioxide and release oxygen. About 10% of the CO_2 released into the atmosphere by human activities comes from cutting and burning tropical rain forests, intensifying the greenhouse effect and accelerating climate change. Fortunately, some areas have been set aside as protected preserves, and some reforestation efforts are underway.

Tropical Deciduous Forests

Slightly farther from the equator, annual rainfall is still high, but there are pronounced wet and dry seasons. In these

CASE STUDY **CONTINUED**

Food of the Gods

When coffee is grown in the shade of tall trees in a rain forest environment, coffee rust fungus, other diseases, and even some insect predators often cause relatively little damage, compared to full-sun plantations. Ecologists Ivette Perfecto and John Vandermeer have found that a second fungus, called the white halo, attacks the coffee rust fungus without harming the coffee plants. White halo is common in shady environments, but not in full-sun plantations. Birds, which are much more abundant in shady plantations, eat coffee berry borers, which are beetles whose larvae live in, and eat, coffee beans. Even the supposed advantage of full-sun plantations—higher yields—can be problematic. When grown in full sun with lots of fertilizer, coffee plants often produce so many berries that they stress themselves, becoming weaker and more susceptible to coffee rust. Full-sun cacao, too, is more susceptible to various fungal diseases, including witch's broom and frosty pod rot.

Although an ever-increasing demand for coffee and chocolate means that more rain forest is likely to be converted to plantations, shady plantations with a diverse assortment of forest trees can both produce these delicious foods and help to preserve many of the benefits of an intact rain forest for future generations. Can similar strategies that combine production and preservation be applied to other biomes?

areas, which include much of India as well as parts of Southeast Asia, South America, and Central America, **tropical deciduous forests** grow. During the dry season, the trees cannot get enough water from the soil to replace what evaporates from their leaves. Many shed their leaves during the dry season ("deciduous" literally means "falling off"), minimizing water loss.

Human Impacts Human activities impact tropical deciduous forests in much the same ways that they affect tropical rain forests. Logging, burning to clear land for agriculture, and cutting for firewood all contribute to deforestation of tropical deciduous forests. Fortunately, many tropical deciduous trees "stump-sprout" after logging. Therefore, if disturbances are not too severe and not too frequent, tropical deciduous forests often recover fairly quickly, with nearly the same species as were present before the disturbances occurred.

Tropical Scrub Forests and Savannas

Along the edges of the tropical deciduous forest, reduced rainfall produces the **tropical scrub forest** biome, dominated by deciduous trees that are shorter and more widely spaced than in tropical deciduous forests. Between the scattered trees, sunlight penetrates to ground level, which allows grass to grow. Still farther from the equator, the climate grows drier, and grasses become the dominant vegetation, with only scattered trees; this biome is the **savanna** (FIG. 30-9).

Rainfall in tropical scrub forests and savannas ranges from about 12 to 40 inches (30 to 100 centimeters) a year, almost all falling during a rainy season lasting 3 or 4 months. When the dry season arrives, rain might not fall for months, and the soil becomes hard, dry, and dusty. Grasses are well adapted to this type of climate, growing very rapidly during the rainy season and dying back to drought-resistant roots during the dry season. Only a few specialized trees, such as the thorny acacia or the water-storing baobab, can survive the dry seasons.

The African savanna supports the most diverse array of large mammals on Earth. These include herbivores such as antelope, wildebeest, rhinos, elephants, and giraffes and carnivores such as lions, leopards, hyenas, and wild dogs.

Human Impacts Africa's rapidly expanding human population threatens the wildlife of the savanna. The abundant grasses that make the savanna a suitable habitat for so much wildlife also make it suitable for grazing domestic cattle. Fences erected to contain cattle disrupt the migration of herds of wild herbivores as they search for food and water. In addition, black market sales, usually in Asia, of products derived from rare African animals may be the death knell for some species. Black market demand for rhino horns has already driven the black rhinoceros to the brink of extinction,

◀ **FIGURE 30-9 The African savanna** Giraffes feed on savanna trees and share this biome with (from left to right) lions, rare black rhinos, and zebras.

(a) Sahara dunes

(b) Utah desert

◀ **FIGURE 30-10** **The desert biome** **(a)** Under the most extreme conditions of heat and drought, deserts can be almost devoid of life, such as these sand dunes of the Sahara Desert in Africa. **(b)** Throughout much of Utah and Nevada, the Great Basin Desert presents a landscape of widely spaced shrubs, such as sagebrush and greasewood.

(a) Cactus

(b) Euphorb

▲ **FIGURE 30-11** **Environmental demands shape physical characteristics** Evolution in response to similar desert conditions has molded the bodies of **(a)** cacti and **(b)** euphorbs into nearly identical shapes, although they are not closely related to one another.

and poaching for ivory endangers African elephants; about 20,000 were killed in 2013, a number that far exceeds the elephants' reproductive rate.

Deserts

Even drought-resistant grasses need at least 10 to 20 inches (25 to 50 centimeters) of rain a year, depending on the temperature and the precipitation's seasonal distribution. Biomes where annual rainfall is 10 inches or less are called **deserts.** Although we tend to think of them as hot, deserts are defined by their lack of precipitation rather than by their temperatures. In the Gobi Desert of Asia, for example, although the summers are very hot, average temperatures are below freezing for half the year. Desert biomes are found on every continent, typically at around 30° N and 30° S latitudes, and in the rain shadows of mountain ranges.

Deserts vary in just how dry they are. At one extreme are the Atacama Desert in Chile and parts of the Sahara Desert in Africa, where it almost never rains and no vegetation grows (**FIG. 30-10a**). More commonly, deserts are characterized by widely spaced vegetation and large areas of bare ground (**FIG. 30-10b**).

Only highly specialized plants can grow in deserts. Although not closely related to each other, cacti (mostly in the Western Hemisphere) and euphorbs (mostly in the Eastern Hemisphere; **FIG. 30-11**) have shallow, spreading roots that rapidly absorb rainwater before it evaporates. Their thick stems store water when it is available. Spines protect the plants from herbivores that would otherwise eat the stems for both nutrition and water. Evaporation is minimized because the leaves, if any, are very small; typically, most photosynthesis occurs in the green, fleshy stem. A heavy wax coating on the stem further reduces water loss.

Some deserts have a very brief rainy season, in which a whole year's rain falls in just a few storms. Annual wildflowers take advantage of the brief period of moisture to sprout from seed, grow, flower, and produce seeds of their own in a month or two (**FIG. 30-12**).

◀ **FIGURE 30-12** **Desert wildflowers** After a relatively wet spring, this Arizona desert is carpeted with wildflowers. Through much of the year—and sometimes for several years—annual wildflower seeds lie dormant, waiting for adequate rains to fall.

Desert animals are also adapted to survive heat and drought. Few animals are active during the hot summer days. Many desert denizens take refuge from the heat in underground burrows that stay relatively cool and moist. In North American deserts, nocturnal (night-active) animals include jackrabbits, bats, kangaroo rats, and burrowing owls (**FIG. 30-13**). Reptiles such as snakes, turtles, and lizards adjust their activity cycles depending on the temperature. In summer, they may be active only around dawn and dusk. Kangaroo rats and many other small desert animals survive without ever drinking. They obtain water from their food and as a by-product of cellular respiration (see Chapter 8). Larger animals, such as desert bighorn sheep, depend on permanent water holes during the driest times of the year.

Human Impacts Desert ecosystems are fragile. Desert soil is stabilized by strands of bacteria that intertwine among sand grains. Driving motorized vehicles on the desert destroys this crucial bacterial network, causing soil erosion and reducing the nutrients available to the desert's slow-growing plants. Regeneration is extremely slow: In the Mojave Desert of California, tread marks left by tanks during World War II are still visible today. Desert soil may require hundreds of years to fully recover from heavy vehicle use.

(a) A kangaroo rat

(b) A burrowing owl

▲ **FIGURE 30-13 Desert dwellers (a)** Kangaroo rats and **(b)** burrowing owls spend the hottest part of the day in burrows, emerging at night to feed.

▲ **FIGURE 30-14 Desertification in the Sahel** A rapidly growing human population, coupled with drought and poor land use, has reduced the ability of many dry regions to support life.

Human activities also contribute to **desertification,** the process by which relatively dry regions are converted to desert as a result of drought coupled with misuse of the land. When people overharvest bushes and trees for firewood, graze too many livestock, and deplete both surface and groundwater to grow crops, the native vegetation becomes extremely vulnerable to drought. Loss of vegetation in turn allows the soil to erode, further decreasing the land's productivity. Desertification has severely impacted the Sahel region in Africa just south of the Sahara Desert (**FIG. 30-14**). In 2011, 11 African countries proposed building a "Great Green Wall" of trees and bushes, about 9 miles (15 kilometers) wide, clear across the continent, in an effort to revegetate the degraded environment and stop desertification in the Sahel. Senegal, on the Atlantic coast, has planted 50,000 acres of drought-resistant trees, including acacia trees that provide food for livestock and gum arabic that can be sold overseas as a food additive. In Niger, preventing overgrazing and planting grasses and bushes has reduced erosion and helped forests to grow back naturally.

Chaparral

Many coastal regions that border deserts, such as those in southern California and much of the Mediterranean, support the **chaparral** biome (**FIG. 30-15**). The annual rainfall is up to 30 inches (about 75 centimeters), nearly all of which falls during cool, wet winters. Summers are hot and dry. Chaparral plants consist mainly of drought-resistant shrubs and small trees. Their leaves are usually small and are often coated with tiny hairs or waxy layers that reduce evaporation during the dry summer months. Chaparral is adapted to fire. Many shrubs regrow from their roots after fires. Others have seeds that are stimulated to germinate by chemicals found in smoke.

Human Impacts People enjoy living in warm, dry climates adjacent to oceans, so development for housing is a major threat to chaparral biomes. In more rugged terrain, especially in

▲ **FIGURE 30-15 The chaparral biome** Limited to warm, dry coastal regions and maintained by fires caused by lightning, this biome is characterized by drought-resistant shrubs and small trees, such as these seen here in the foothills of the San Gabriel mountains in Southern California.

southern Europe, chaparral has been cleared for grazing, olive groves, and other types of agriculture.

Grasslands

In the centers of continents, such as North America and Eurasia, **grassland,** or *prairie,* biomes predominate (**FIG. 30-16**). Grassland biomes typically have hot summers and cold winters and receive 15 to 30 inches of rain annually. In general, these biomes have a continuous cover of grass and virtually no trees, except along rivers. In tallgrass prairie—in North America, originally found from Texas to southern Canada—grasses reach up to 6 feet in height. An acre of natural tallgrass prairie may support 200 to 400 different species of native plants. Areas further west, which receive less rainfall, support midgrass and shortgrass prairies. In these grasslands, prairie dogs and ground squirrels provide food for eagles, foxes, coyotes, and bobcats. Pronghorns browse in western grasslands, and bison survive in preserves.

Why do grasslands lack trees? Water and fire are the crucial factors in the competition between grasses and trees. The hot, dry summers and frequent droughts of the midgrass and shortgrass prairies can be tolerated by grass but are fatal to trees. In tallgrass prairies with more precipitation, forests are the climax ecosystems.

Historically, however, tree growth was suppressed by a combination of occasional severe drought and frequent fires caused by lightning or set by Native Americans. Although fire kills trees, the root systems of grasses survive.

Human Impacts Grasses growing and decomposing for thousands of years produced the most fertile soil in the world. In the early nineteenth century, North American grasslands supported an estimated 60 million bison. Today, the Midwestern U.S. grasslands have been largely converted to farm and range land, and cattle have replaced bison.

Prairie dog colonies and the eagles and ferrets that hunted them have become a rare sight as their habitat shrinks. In some regions, overgrazing has destroyed the native grasses, allowing woody sagebrush to flourish (**FIG. 30-17**). Undisturbed grasslands are now largely confined to protected areas. Tallgrass prairie is one of the most endangered ecosystems in the world. Only about 1% remains, in tiny remnants restored by planting native species and maintained by controlled burning.

▼ **FIGURE 30-16 Shortgrass prairie** Shortgrass prairie is characterized by low-growing grasses. In addition to many wildflowers, life in shortgrass prairies includes (from left to right) bison (in preserves), prairie dogs, and pronghorns.

▲ **FIGURE 30-17 Sagebrush desert or shortgrass prairie?** Biomes are influenced by human activities as well as by temperature, rainfall, and soil. The shortgrass prairie field on the right has been overgrazed by cattle, causing the grasses to be replaced by sagebrush.

Temperate Deciduous Forests

At their eastern edge, the North American grasslands merge into the **temperate deciduous forest** biome (**FIG. 30-18**). Temperate deciduous forests are also found in much of Europe and eastern Asia. More precipitation occurs in temperate deciduous forests than in grasslands (30 to 60 inches, or 75 to 150 centimeters). The soil retains enough moisture for trees to grow, shading out most grasses.

Winters in temperate deciduous forests often have long periods of below-freezing weather, when liquid water is not available. Deciduous trees drop their leaves in the fall and remain dormant through the winter, thus conserving their water. During the brief time in spring when the ground has thawed but emerging leaves on the trees have not yet blocked off the sunlight, abundant wildflowers grace the forest floor.

Decaying leaf litter on the forest floor provides food and suitable habitat for bacteria, earthworms, fungi, and small plants. A variety of vertebrates—including mice, shrews, squirrels, raccoons, deer, bears, and many species of birds—dwell in deciduous forests.

Human Impacts Large predatory mammals such as black bears, wolves, bobcats, and mountain lions were formerly abundant in the eastern United States, but hunting and habitat loss have severely reduced their numbers. Consequently, in many areas deer populations have skyrocketed due to a lack of predators. Clearing for lumber, agriculture, and housing dramatically reduced deciduous forests in the United States. Virgin (uncut) deciduous forests are almost nonexistent, but the last century has seen extensive regrowth of deciduous forests on abandoned farms and formerly logged land.

Temperate Rain Forests

On the Pacific coast of the United States and Canada, from northern California to southeast Alaska, lies a **temperate rain forest**

◀ **FIGURE 30-18 The temperate deciduous forest biome** Temperate deciduous forests of the eastern United States are inhabited by (from top to bottom) white-tailed deer and birds such as this blue jay; in spring, a profusion of woodland wildflowers (such as these hepaticas) blooms briefly before the trees produce leaves that shade the forest floor.

THINK CRITICALLY Both tropical deciduous and temperate deciduous forests are dominated by trees that drop their leaves for part of the year. Explain how dropping leaves is an effective adaptation in these two very different biomes.

▲ **FIGURE 30-19 The temperate rain-forest biome** The Hoh River temperate rain forest in Olympic National Park receives about 12 feet of rain annually. Ferns, mosses, and wildflowers grow in the pale green light of the forest floor. Denizens of this rain forest include (from top to bottom) ferns, such as this lady fern, elk, and flowering foxglove.

(FIG. 30-19). Temperate rain forests are also located along the southeastern coast of Australia, the southwestern coast of New Zealand, and parts of Chile and Argentina. In North America, these biomes typically receive more than 55 inches (140 centimeters) of rain annually—and as much as 12 feet a year in some areas. The nearby ocean keeps the temperature moderate.

Most of the trees in the temperate rain forest are huge conifers, such as spruce, Douglas fir, and hemlock, commonly 250 to 300 feet tall. The forest floor and tree trunks are typically covered with mosses and ferns. Fungi thrive in the moisture and enrich the soil. As in tropical rain forests, so little light reaches the forest floor that tree seedlings usually cannot become established. Whenever one of the forest giants falls, however, it opens up a patch of light, and new seedlings quickly sprout, often right atop the fallen log.

Human Impacts Tall, straight trees are extremely valuable for lumber, and consequently many temperate rain forests have been logged. In the mild, wet climate, the forests regrow quickly, providing a renewable supply of lumber. However, some animals, such as the spotted owl, dwell mainly in old-growth forests that are hundreds of years old. Fortunately, some pristine temperate rain forest is preserved in national parks, including Olympic in Washington and Glacier Bay in Alaska.

Northern Coniferous Forests

North of the grasslands and temperate forests stretches the **northern coniferous forest** (also called the *taiga*; FIG. 30-20). The northern coniferous forest, which is the largest terrestrial biome on Earth, stretches across Scandinavia, Siberia, central Alaska, Canada, and parts of the northern United States. Similar forests occur in many mountain ranges, including the Cascades, the Sierra Nevada, and the Rocky Mountains.

Conditions in the northern coniferous forest are much harsher than in temperate deciduous forests, with long, cold winters and short growing

◀ **FIGURE 30-20 The northern coniferous forest biome** The small needles and conical shape of conifers allow them to shed heavy snows. (Upper left) A Canada lynx catches a snowshoe hare. (Upper right) A great horned owl waits for nightfall, when it will begin to hunt.

seasons. About 16 to 40 inches (40 to 100 centimeters) of precipitation occur annually, much of it as snow. The conical shape and narrow, stiff needles of evergreen conifers allow them to shed snow efficiently. The waxy coating on the needles minimizes water loss during the long winters, when water remains frozen. By retaining their leaves during the winter, evergreen conifers conserve the energy that deciduous trees must expend to grow new leaves in the spring. Therefore, when spring arrives, conifers can begin photosynthesis immediately. Large mammals—including black bears, moose, deer, and wolves—still roam the northern coniferous forest, as do wolverines, lynxes, foxes, bobcats, and snowshoe hares. These forests also serve as breeding grounds for many migratory bird species.

Human Impacts Clear-cutting for papermaking and lumber has leveled huge expanses of northern coniferous forest in both Canada and the U.S. Pacific Northwest (**FIG. 30-21**). Demand is also increasing to extract natural gas and oil, often from unconventional sources such as oil sands. Nevertheless, much of Canada's coniferous forest remains intact. Encouragingly, the provincial governments of Ontario and Quebec have pledged to protect half of the publicly owned coniferous forest and to manage the remainder sustainably.

▲ FIGURE 30-21 **Clear-cutting** Coniferous forests are vulnerable to clear-cutting, as seen in this forest in Alberta, Canada. Clear-cutting is a relatively simple and inexpensive means of logging compared to selective harvesting of trees, but its environmental costs are high. Erosion diminishes the fertility of the soil, slowing new growth. Further, the dense stands of similarly aged trees that typically regrow are more susceptible to fires and parasites than a natural stand of trees of various ages would be.

Tundra

The biome furthest north is the arctic **tundra,** a vast treeless region bordering the Arctic Ocean (**FIG. 30-22**). Conditions in the tundra are severe. Winter temperatures are often −40°F (−55°C) or below, with howling winds. Precipitation averages 10 inches (25 centimeters) or less each year, making this region a freezing desert. Even during the summer, frosts are frequent, and the growing season may last only a few weeks. Similar climates and tundra vegetation are found at high elevations on mountains worldwide.

The cold climate of the arctic tundra results in **permafrost,** a permanently frozen layer of soil. Soil above the permafrost thaws each summer, often to a

◀ FIGURE 30-22 **The tundra biome** Life on the tundra is seen here in Denali National Park, Alaska, turning color in autumn. (From left to right) Perennial plants such as this frost-covered bearberry grow low to the ground, avoiding the chilling tundra wind. Tundra animals, such as arctic fox and caribou, can regulate blood flow in their legs, keeping them just warm enough to prevent frostbite while preserving precious body heat for the brain and other vital organs.

depth of 2 feet (60 centimeters) or more. When the summer thaws arrive, the underlying permafrost limits the ability of soil to absorb the water from melting snow and ice, and so the tundra becomes a marsh.

Trees do not grow in the tundra because of the extreme cold, the brief growing season, and the permafrost, which limits the depth of roots. Nevertheless, the ground is carpeted with small perennial flowers, dwarf willows, and large lichens called "reindeer moss," a favorite food of caribou. The summer marshes also provide superb mosquito habitat. These and other insects feed about 100 different species of birds, most of which migrate here to nest and raise their young during the brief summer feast. The tundra vegetation also supports arctic hares and lemmings (small rodents) that are eaten by wolves, owls, and arctic foxes.

Human Impacts The tundra is among the most fragile of all terrestrial biomes because of its short growing season. A willow 4 inches (10 centimeters) high may be 50 years old. Alpine tundra is easily damaged by off-road vehicles and even hikers. Fortunately for the inhabitants of the arctic tundra, the impacts of civilization are mostly localized around oil-drilling sites, pipelines, mines, and scattered military bases. The most significant threat to the tundra is climate change. Shrubs and trees are replacing tundra along its southern margins. Climate models suggest that more than a third of Earth's tundra may be lost by the end of this century.

CHECK YOUR LEARNING

Can you ...

- describe the principal terrestrial biomes and discuss how temperature and precipitation interact to determine their characteristic plant life?
- describe human impacts on terrestrial biomes?

30.4 WHAT ARE THE PRINCIPAL AQUATIC BIOMES?

Of the four requirements for life, aquatic ecosystems typically provide abundant water and appropriate temperatures. However, sunlight in aquatic ecosystems decreases with depth, as it is absorbed by water and blocked by suspended particles. In addition, nutrients in aquatic ecosystems tend to be concentrated in sediments on the bottom, so where nutrients are high, light levels tend to be low.

Freshwater Lakes

Freshwater lakes form when natural depressions fill with water from groundwater seepage, streams, and runoff from rain or melting snow. Large lakes in temperate climates have distinct zones of life (**FIG. 30-23**). Near the shore is the shallow **littoral zone,** which receives abundant sunlight and

▼ **FIGURE 30-23 Lake life zones** A typical large lake has three life zones: a nearshore littoral zone with rooted plants, an open-water limnetic zone, and a deep, dark profundal zone.

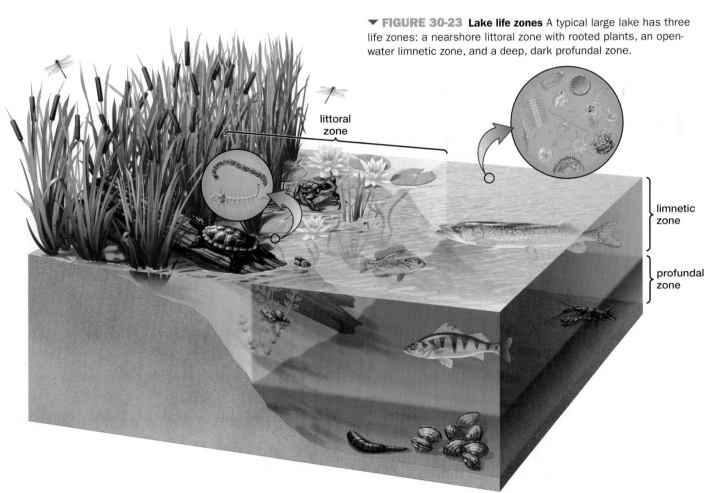

HAVE YOU EVER WONDERED ...

Remember *Jurassic Park* and *Jurassic World*? OK, maybe the idea of resurrecting *T. rex* and *Velociraptor* is a little far-fetched. But how about a Pleistocene Park? Russian scientist Sergey Zimov thinks that the present-day Siberian tundra, dominated by mosses and shrubs, is an artificial landscape created when prehistoric people wiped out most of Siberia's large herbivores, including mammoths, bison, and woolly rhinos, about 10,000 years ago. Zimov hypothesizes that grazing and trampling by these large herbivores destroyed mosses, bushes, and tree seedlings, but grasses thrived. When the megafauna were killed off, the whole ecosystem changed. Zimov wants to recreate the "mammoth steppe"—a vast grassland supporting herds of herbivores and the carnivores that prey on them. Of course, there aren't any mammoths or woolly rhinos anymore, but Zimov has introduced Yakutian horses, musk ox, European bison, and elk to a large protected area in Siberia called Pleistocene Park. Horses seem to be crucial to restoring the grasslands. Wherever there are enough horses, grasslands are returning. The "re-wilding" of Pleistocene Park is well under way.

If People Can Re-create Ancient Biomes?

Yakutian horses are an essential component of Pleistocene Park

dominate in the limnetic zone. Organisms that live in the profundal zone are nourished by organic matter that drifts down from the littoral and limnetic zones and by sediment washed in from the land. Inhabitants of the profundal zone include catfish, which mainly feed on the bottom, and detritivores and decomposers such as crayfish, aquatic worms, clams, leeches, and bacteria.

Freshwater Lakes Are Classified According to Their Nutrient Content

Freshwater lakes may be described as *oligotrophic* (Greek, "poorly fed"), *eutrophic* ("well fed"), or *mesotrophic* (between these two extremes, or "middle fed"). Here, we describe the characteristics of oligotrophic and eutrophic lakes.

Oligotrophic lakes contain few nutrients and support relatively little life. Many oligotrophic lakes were formed by glaciers that scraped depressions in bare rock and are now fed by mountain streams and snowmelt. Because there is little sediment or microscopic life to cloud the water, oligotrophic lakes are clear, and light penetrates deeply. Fish that require well-oxygenated water, such as trout, thrive in oligotrophic lakes.

Eutrophic lakes receive relatively large inputs of sediments, organic material, and inorganic nutrients (such as phosphates and nitrates) from their surroundings, allowing them to support dense plant communities (**FIG. 30-24**). They are murky from suspended sediment and dense phytoplankton populations, so the limnetic zone is shallow. The dead bodies of limnetic zone inhabitants sink into the profundal zone, where they feed decomposer organisms. The metabolic activities of these decomposers use up oxygen, so the

nutrients. Plants in the littoral zone include cattails, bullrushes, and water lilies, anchored in the bottom near the shore, and fully submerged plants that flourish in slightly deeper waters. Littoral waters are home to small organisms collectively called **plankton** (from a Greek word meaning "drifters"). Photosynthetic protists and bacteria are called **phytoplankton.** Nonphotosynthetic protists and tiny crustaceans that feed on phytoplankton make up the **zooplankton.**

A great diversity of animal life is also found in the littoral zone, although many of the animals, especially fish, spend time in more than one zone. Littoral vertebrates include frogs, aquatic snakes, turtles, and fish such as pike, bluegill, and perch; invertebrates include insect larvae, snails, flatworms, and crustaceans such as crayfish.

As the water increases in depth, plants are unable to anchor themselves to the bottom and still receive enough sunlight for photosynthesis. This open-water region is divided into an upper **limnetic zone,** in which enough light penetrates to support photosynthesis by phytoplankton, and a lower **profundal zone,** in which light is too weak for photosynthesis to occur (see Fig. 30-23). Plankton and fish

▲ **FIGURE 30-24 A eutrophic lake** Rich in dissolved nutrients carried from the land, eutrophic lakes support dense growths of algae, phytoplankton, and both floating and rooted plants.

profundal zone of eutrophic lakes is often very low in oxygen and supports little other life.

Gradually, as nutrient-rich sediment accumulates, oligotrophic lakes tend to become mesotrophic and then eutrophic, a process called eutrophication. Although large lakes may persist for millions of years, eutrophication may eventually cause lakes to undergo succession to dry land (see Chapter 29).

Human Impacts Nutrients carried into lakes from farms, feedlots, sewage, and even fertilized suburban lawns accelerate eutrophication. Overfertilized lakes sometimes experience massive algal blooms, followed by die-offs and decomposition that deplete the water of oxygen and kill most of the fish. Phosphate-free detergents, more effective sewage treatment, reduced fertilizer use, and proper location and operation of feedlots lessen the danger of eutrophication.

Streams and Rivers

Streams often originate in mountains, the *source region* shown in **FIGURE 30-25**, where runoff from rain and melting snow cascades over impervious rock. Little sediment reaches the streams, phytoplankton is sparse, and the water is clear and cold. Algae grow on rocks in the streambed, where insect larvae find food and shelter. Turbulence keeps mountain streams well oxygenated, providing a home for trout that feed on insect larvae and smaller fish.

At lower elevations, in the *transition zone,* small streams merge, forming wider, slower-moving streams and small rivers. The water warms slightly, and more sediment is carried in by tributaries, providing nutrients that allow aquatic plants, algae, and phytoplankton to proliferate. Fish such as bass, bluegills, and yellow perch (all of which require less oxygen than trout do) are found in such waterways.

As the land becomes lower and flatter, the river warms, widens, and slows, meandering back and forth. The water becomes murky with dense populations of phytoplankton. Decomposer bacteria deplete the oxygen in deeper water, but carp and catfish can still thrive despite the low oxygen levels. When precipitation or snowmelt is high, the river may flood the surrounding flat land, called a *floodplain,* depositing sediment over the adjoining terrestrial ecosystem.

Rivers drain into lakes or into other rivers that ultimately lead to an estuary, the area where a river meets the ocean (described below). Close to sea level, most rivers move slowly, depositing their sediment. In many cases, the sediment interrupts the river's flow, breaking it into small winding channels before it finally empties into the sea.

Human Impacts Rivers are sometimes channelized (deepened and straightened) to facilitate boat traffic, to prevent

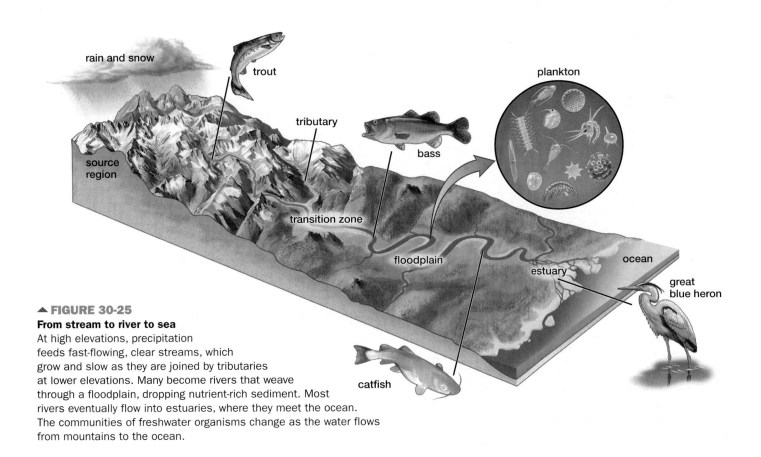

▲ **FIGURE 30-25**

From stream to river to sea

At high elevations, precipitation feeds fast-flowing, clear streams, which grow and slow as they are joined by tributaries at lower elevations. Many become rivers that weave through a floodplain, dropping nutrient-rich sediment. Most rivers eventually flow into estuaries, where they meet the ocean. The communities of freshwater organisms change as the water flows from mountains to the ocean.

flooding, and to allow farming along their banks. Channelization increases erosion of the riverbed and banks because water flows more rapidly in straightened rivers. In addition, where natural flooding has been prevented, floodplain soil no longer receives the nutrients formerly deposited by floodwaters.

In the United States, both Pacific and Atlantic salmon populations have been greatly reduced by hydroelectric dams, water diversion for agriculture, erosion from logging operations, and overfishing. On both coasts of the United States, federal, state, and local groups are working to restore clean, free-flowing rivers that support salmon and rich wildlife communities. Some dams in Washington State and Maine have been removed to allow salmon once again to migrate upstream to spawn, in some cases for the first time in more than 150 years.

Freshwater Wetlands

Freshwater **wetlands,** which include marshes, swamps, and bogs, are regions where the soil is covered or saturated with water. Wetlands support dense growths of algae and phytoplankton, as well as both floating and rooted plants, including cattails, marsh grasses, and water-tolerant trees, such as bald cypress. Wetlands provide breeding grounds, food, and shelter for a great variety of birds (cranes, grebes, herons, kingfishers, and ducks), mammals (beavers, muskrats, and otters), freshwater fish, and invertebrates such as crayfish and dragonflies.

Freshwater wetlands are among the most productive ecosystems in North America. Many occur around the margins of lakes or in the floodplains of rivers. Wetlands act as giant sponges, absorbing water and then gradually releasing it into rivers, making wetlands important safeguards against flooding and erosion. Wetlands also serve as natural water filters and purifiers. As water flows slowly through wetlands, suspended particles fall to the bottom. Wetland plants and phytoplankton absorb nutrients such as nitrates and phosphates that have washed from the land. Bacteria break down many organic pollutants, rendering them harmless.

Human Impacts About half of the freshwater wetlands in the United States (outside of Alaska) have been lost as a result of being drained and filled for agriculture, housing, and commercial uses. Destruction of wetlands makes nearby water more susceptible to pollutants, reduces wildlife habitat, and may increase the severity of floods.

Fortunately, local, state, and federal agencies have cooperated to protect existing wetlands and restore some that have been degraded. These actions have combined to slow wetland loss in the United States. A survey by the U.S. Fish and Wildlife Service found that, although individual types of freshwater wetlands have expanded or shrunk, their total area has remained fairly constant in recent years.

Marine Biomes

The oceans can be divided into life zones characterized by the amount of light they receive and their proximity to the shore (**FIG. 30-26**). The **photic zone** consists of relatively shallow waters (to a depth of about 650 feet, or 200 meters) where the light is strong enough to support photosynthesis. Below the photic zone lies the **aphotic zone,** which extends to the ocean floor, with a maximum depth of about 36,000 feet (11,000 meters) in the Marianas Trench in the Pacific Ocean. Light in the aphotic zone is inadequate for photosynthesis. Therefore, nearly all the energy to support life must be extracted from the excrement and bodies of organisms that sink down from the photic zone above.

Because their water levels rise and fall with the tides, oceans do not have a defined shoreline. Instead, the **intertidal zone,** where the land meets the ocean, is alternately covered and exposed by the tides. The **nearshore zone** extends out to sea from the low-tide line, with gradually increasing depth as the continental shelf slopes downward. The nearshore zone is usually considered to end, and the **open ocean** to begin, where the water is deep enough that wave action no longer affects the bottom, even during strong storms.

Shallow Water Marine Biomes

As in freshwater lakes, the major concentrations of life in the oceans are found in shallow waters where both nutrients and light are abundant. Such locations include estuaries, the intertidal zone, and kelp forests and coral reefs, which are mostly located in the nearshore zone.

Estuaries An **estuary** is an area of brackish water where fresh water from one or more rivers mixes with seawater (**FIG. 30-27a**). The waters of estuaries vary in salinity. High tides, for example, bring an influx of seawater, while heavy rains bring a pulse of fresh water down the river. Estuaries support enormous biological productivity and diversity. Many commercially important animal species, including shrimp, oysters, clams, crabs, and a variety of fish, spend part of their lives in estuaries. Dozens of species of birds, including ducks, swans, and shorebirds, feed and nest in estuaries.

Intertidal Zones In the intertidal zone, organisms must be adapted to survive both submerged in seawater and exposed to the air as the tides rise and fall. During heavy rains, organisms in tide pools and mudflats may also experience significant dilution of the seawater. On rocky shores, barnacles (shelled crustaceans) and mussels (mollusks) filter phytoplankton from the water at high tide and close their shells at low tide to resist drying. At high tide, sea stars pry open mussels to eat, sea urchins feast on algae coating the rocks, and anemones spread their tentacles to catch passing crustaceans and small fish (**FIG. 30-27b**). The intertidal zone of sandy shores and mudflats typically has less diversity but still

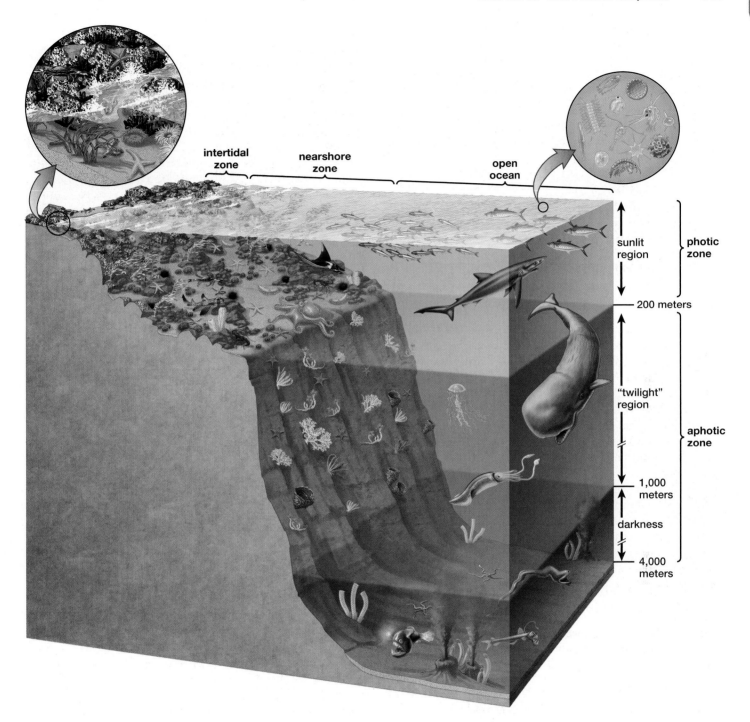

intertidal zone

nearshore zone

open ocean

sunlit region — photic zone

— 200 meters

"twilight" region

aphotic zone

1,000 meters

darkness

4,000 meters

▲ **FIGURE 30-26 Ocean life zones** Photosynthesis can occur only in the sunlit photic zone, which includes the intertidal zone, nearshore zone, and the upper waters of the open ocean. Approximate depths of various regions are shown, although these vary considerably depending on the clarity of the water; note that the depths are not drawn to scale. Nearly all the organisms that spend their lives in the aphotic zone rely on organic material that drifts down from the photic zone above.

contains life, including organisms such as sand crabs and burrowing worms.

Kelp Forests Kelp are enormous brown algae that can grow as tall as 160 feet (50 meters; see Chapter 21). Kelp often occur in dense stands called **kelp forests,** found throughout the world in cool waters of the nearshore zone (**FIG. 30-27c**). Kelp forests provide food and shelter for an amazing variety of animals, including annelid worms, sea anemones, sea urchins, snails, sea stars, lobsters, crabs, fish, seals, and otters.

(a) An estuary

(b) A tide pool

(c) An underwater kelp forest

(d) A tropical coral reef

▲ **FIGURE 30-27 Shallow water marine biomes (a)** Life flourishes in estuaries, where fresh river water mixes with seawater. Salt marsh grasses provide shelter for fish and invertebrates that are eaten by egrets (shown here) and many other birds. **(b)** Although pounded by waves and baked by the sun, tide pools in the intertidal zone harbor a brilliant diversity of invertebrates. **(c)** Kelp forests are home to a stunning array of invertebrates and fish, such as these bright orange Garibaldis. **(d)** Coral reefs provide habitat for many species of fish and invertebrates.

THINK CRITICALLY Why do estuaries and other coastal ecosystems have higher productivity than the open ocean?

Coral Reefs Corals are relatives of anemones and sea jellies. Some corals build skeletons of calcium carbonate. These skeletons accumulate over hundreds or thousands of years, building **coral reefs** (FIG. 30-27d). Coral reefs are most abundant in tropical waters, where temperatures typically range between 68° and 86°F (20° and 30°C). Large reefs are found in the Pacific and Indian Oceans, the Caribbean, and the Gulf of Mexico as far north as southern Florida. Coral reefs provide anchorage, shelter, and food for a diverse community of algae, fish, and invertebrates such as shrimp, sponges, and octopuses. The reefs are home to more than 90,000 known species, with possibly a million yet to be discovered.

Most reef-building corals harbor unicellular photosynthetic protists, called dinoflagellates, in their bodies. The relationship is mutually beneficial: The dinoflagellates benefit from high nutrient and carbon dioxide levels within the corals, and the dinoflagellates provide the corals with food produced by photosynthesis. Because their dinoflagellates require sunlight for photosynthesis, reef-building corals can thrive only within the photic zone, usually at depths of less than 130 feet (40 meters). Dinoflagellates give many corals their brilliant colors.

Human Impacts Human population growth is increasing the conflict between preserving coastal ecosystems as wildlife habitat and developing these areas for energy extraction, housing, harbors, and marinas. Estuaries are threatened by runoff from farming operations, which often provide a glut of nutrients from fertilizer and livestock excrement. This fosters excessive growth of algae and photosynthetic bacteria. When these organisms die, they provide nutrients that stimulate the growth of decomposers, whose metabolism depletes the water of oxygen, killing both fish and invertebrates.

Coral reefs face multiple threats. Anything that diminishes the water's clarity harms the coral's photosynthetic partners and hinders coral growth. Runoff from farming, agriculture, logging, and construction carries silt and excess nutrients. Mollusks, turtles, fish, crustaceans, and the corals themselves are often harvested from reefs faster than they can reproduce. Removing parrotfish and invertebrates often leads to an explosion of algae that smother the reefs.

Even though they require warm water, coral reefs are vulnerable to global warming caused by increased CO_2 in the atmosphere. When waters become too warm, corals expel their colorful photosynthetic dinoflagellates and appear to be bleached (see Chapter 31). The dinoflagellates return if the water cools, but when water temperatures remain too high for too long, the corals may starve. Increased CO_2 also slowly acidifies the oceans, reducing the ability of corals to build their skeletons of calcium carbonate.

There is some good news. Many countries now recognize the enormous benefits of coral reefs, including economic benefits from tourism, and are working to protect their reefs. Australia's Great Barrier Reef Marine Park and the Papahānaumokuākea Marine National Monument in the Hawaiian Islands protect enormous reef systems. Collectively, about 20,000 known species thrive in these two hotspots of biodiversity.

The Open Ocean

Beyond the coastal regions lie vast areas of the ocean in which the bottom is too deep to allow plants to anchor and still receive enough light to grow. Therefore, most life in the open ocean depends on photosynthesis by phytoplankton drifting in the photic zone. Phytoplankton are consumed by zooplankton, such as tiny shrimp-like crustaceans, which in turn are eaten by larger invertebrates, small fish, and even some marine mammals, such as humpback and blue whales (**FIG. 30-28**).

Even in the photic zone, the amount of life in the open ocean varies tremendously from place to place, largely due to differences in nutrient availability. Nutrients are provided by two major sources: runoff from the land and upwelling from

◀ **FIGURE 30-28 The open ocean** The open ocean supports fairly abundant life in the photic zone, including whales, such as these humpbacks feeding on krill (lower left). The animals in open ocean, including whales, krill, and sea jellies (lower middle), all ultimately depend on phytoplankton, the photic zone's photosynthetic producers (lower right).

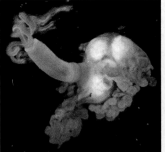

◀ **FIGURE 30-29 Denizens of the deep** The skeleton of a whale provides an undersea nutrient bonanza. A zombie worm (upper left) inserts its rootlike lower body into the bones of the decomposing whale carcass. Other denizens of the deep include viperfish (upper middle), whose huge jaws and sharp teeth allow it to grasp and swallow its prey whole, and nearly transparent squid with short tentacles below bulging eyes (upper right).

the ocean depths. **Upwelling** brings cold, nutrient-laden water from the ocean depths to the surface. Major areas of upwelling occur around Antarctica and along western coastlines, including those of California, Peru, and West Africa. Nutrient-rich waters that support a large phytoplankton community are greenish and relatively murky. In contrast, the blue clarity of many tropical waters is due to a lack of nutrients, which limits the concentration of phytoplankton in the water.

Human Impacts Two major threats to the open ocean are pollution and overfishing. For example, plastic refuse, blown off the land or deliberately dumped at sea, is often mistaken for food by sea turtles, gulls, porpoises, seals, and whales. Animals that consume this refuse may die from clogged digestive tracts. Oil from oil-tanker spills, runoff from improper disposal on land, and leakage from offshore oil wells contaminates the open ocean. Some components of oil cause lethal developmental defects in a variety of marine organisms.

Many fish populations are harvested unsustainably as a result of increased demand for fish and highly efficient fishing technologies (see Chapter 31). For example, the once abundant cod populations off eastern Canada collapsed in 1992,

prompting a fishing moratorium that still continues. In 2013, cod fishing was severely restricted off the coast of New England as well, with further restrictions enacted in 2014. Populations of Pacific bluefin tuna, haddock, mackerel, and many other fish have also declined dramatically because of overfishing. In 2014, the United Nations Food and Agriculture Organization estimated that 30% of marine fish stocks were being fished at unsustainable levels.

Efforts are now being made to prevent overfishing. Many countries have established quotas on fish whose populations are threatened. Fishing restrictions have succeeded in rebuilding the stocks of fish such as the Acadian redfish and Atlantic swordfish. Marine reserves, where fishing is prohibited, are increasingly being established throughout the world, causing substantial improvements in the diversity, number, and size of marine animals. Nearby areas also benefit because the reserves act as nurseries, helping to restore populations outside the reserves.

The Ocean Floor

Because the amount of light in the aphotic zone is inadequate for photosynthesis, most of the food on the ocean floor comes from the excrement and dead bodies that drift down from above. Nevertheless, life is found on the ocean floor in amazing quantity and variety, including worms, sea cucumbers, sea stars, mollusks, squid, and fish of bizarre shapes (**FIG. 30-29**). Little is known of the behavior and ecology of these exotic creatures, which almost never survive being brought to the surface.

Entire communities feed on the dead bodies of whales, each of which contains an average of 40 tons of food. When a whale carcass reaches the ocean floor, fish, crabs, worms, and snails swarm over it, extracting nutrients from its flesh and bones. Bone-eating zombie worms tunnel into the bone

◄ FIGURE 30-30 Hydrothermal vent communities Hydrothermal vents spew superheated water rich in minerals that provide both energy and nutrients to the vent community. Giant red tube worms may reach 9 feet (nearly 3 meters) in length and live up to 250 years (left). The foot of this snail is protected by scales coated with iron sulfide (right).

and absorb nutrients. Anaerobic bacteria complete the breakdown of bone, and the bacteria themselves provide food for clams, worms, mussels, and crustaceans.

Hydrothermal Vent Communities In 1977, geologists exploring the Galápagos Rift (an area of the Pacific floor where the plates that form Earth's crust are separating) found cracks in the seafloor, called hydrothermal ("hot water") vents. Hydrothermal vents emit superheated water containing sulfides and other minerals (**FIG. 30-30**). Surrounding the vents are **hydrothermal vent communities** of pink fish, blind white crabs, enormous mussels, white clams, sea anemones, giant tube worms, and a species of snail sporting iron-laden armor plates (Fig. 30-30, right). Hundreds of species have been found near vents, which have now been discovered in many deep-sea regions where separating tectonic plates allow material from Earth's interior to spew out.

In this unique, completely dark ecosystem, sulfur bacteria serve as the producers. Instead of photosynthesis, sulfur bacteria use **chemosynthesis** to manufacture organic molecules from carbon dioxide, harvesting energy from a source that is deadly to most other forms of life—hydrogen

sulfide, discharged from the vents. Many vent animals consume the sulfur bacteria directly; others, such as the giant tube worm, harbor chemosynthetic bacteria within their bodies and live off the by-products of bacterial metabolism. The tube worms derive their red color from a unique form of hemoglobin that transports hydrogen sulfide to its symbiotic bacteria.

The bacteria and archaea that inhabit the vent communities can survive at remarkably high temperatures; some can live at 248°F (106°C; the tremendous pressure in the deep ocean prevents water from boiling at temperatures well above its sea-level boiling point). Scientists are investigating how the enzymes and other proteins of these heat-loving microbes can continue to function at such high temperatures.

CHECK YOUR LEARNING

Can you …

- describe the principal freshwater and marine biomes?
- explain how water depth and proximity to the shore help to determine the nature and abundance of life in each?
- describe some effects humans have on aquatic biomes?

World demand for both coffee and chocolate will probably continue to soar. Can coffee and cacao farmers meet this demand, enjoy a reasonable income, and simultaneously help to preserve rain forests? That largely depends on how the coffee and cacao plants are grown.

Full-sun plantations, in which the original vegetation has been completely removed and replaced with monocultures of coffee or cacao, often have the highest coffee or cocoa production, but at the cost of lower quality, substantial inputs of fertilizers and pesticides, and little biodiversity. Shady plantations typically provide higher quality product and more biodiversity, but some shady plantations are better than others. For example, most coffee from Mexico is grown in shady plantations ("shade-grown coffee"), but often under only a sparse cover of a few species of trees, which does not provide a sufficiently diverse habitat for rain-forest birds and other species. In contrast, in rustic plantations, the canopy remains mostly intact, because only low-growing plants are removed from the rain forest to make way for coffee or cocoa plants. Rustic plantations that promote diversity and sustainable production can be certified by the Smithsonian Migratory Bird Center, the Rainforest Alliance, and similar organizations. To

achieve certification, plantations must include a diversity of tree species. Some rustic plantations provide a home for more than 150 different species of birds, which feast on fruit, or on insects that thrive in the trees and moist soil.

Are rustic plantations good for biodiversity but bad deals for farmers? Not if they're well managed. In many cases, the canopy trees serve as an additional source of food or income for the farmers, providing citrus fruit, bananas, guavas, and lumber. Higher quality cocoa and coffee, combined with certification by the Bird Center or Rainforest Alliance, usually bring farmers a higher price for their products and, consequently, higher incomes.

CONSIDER THIS Ecologists argue that it is both ecologically and economically advantageous to manage Earth's ecosystems as a long-term investment rather than for short-term profit, letting natural ecosystems thrive while simultaneously providing needed products for people. Research one sustainable natural product (for example, a forest product or a type of fish). Was the source of this product ever overexploited? What measures have been taken to ensure sustainability? What challenges might the future bring, and how might those challenges be overcome?

CHAPTER REVIEW

MB Go to **MasteringBiology** for practice quizzes, activities, eText, videos, current events, and more.

Answers to Think Critically, Evaluate This, Multiple Choice, and Fill-in-the-Blank questions can be found in the Answers section at the back of the book.

Summary of Key Concepts

30.1 What Determines the Distribution of Life on Earth?

The requirements for life on Earth include nutrients, energy, liquid water, and an appropriate temperature range. In aquatic ecosystems, liquid water is readily available; sunlight, nutrients, and temperature determine the distribution and abundance of life. On land, sunlight energy and nutrients are usually plentiful; the distribution of life is largely determined by soil moisture, which in turn is determined by precipitation and temperature. The requirements for life occur in specific patterns on Earth, resulting in characteristic large-scale communities called biomes.

30.2 What Factors Influence Earth's Climate?

Because of Earth's curvature, the sun's rays are nearly vertical and pass through the least amount of atmosphere at the equator; toward the poles, the rays are more slanted and must penetrate more atmosphere. Thus, the equator is uniformly warm, whereas higher latitudes have lower overall temperatures. Earth's tilt on its axis causes seasonal variations in climate at northern and southern latitudes as Earth orbits the sun. Rising warm air and sinking cool air in regular patterns from north to south produce

areas of low and high moisture. These patterns are modified by prevailing winds, the presence and topography of continents, and proximity to oceans.

30.3 What Are the Principal Terrestrial Biomes?

The tropical rain-forest biome, located near the equator, is warm and wet year-round. Tropical rain forests are dominated by broadleaf evergreen trees. Most nutrients are found in the vegetation. Most animals live in the trees. Rain forests have the highest productivity and biodiversity on Earth. Slightly farther from the equator, wet seasons alternate with dry seasons during which trees shed their leaves, producing tropical deciduous forests. Scrub forests and savannas receive less rain than tropical deciduous forests and have extended dry seasons. Savannas are characterized by widely spaced trees with grass growing beneath. Most deserts, which receive less than 10 inches of rain annually, are located around 30° N and 30° S latitudes, or in the rain shadows of mountain ranges. In deserts, plants are widely spaced and have adaptations to conserve water. Animals have both behavioral and physiological mechanisms to avoid excessive heat and to conserve water. Chaparral exists in desert-like conditions that are moderated by their proximity to a coastline, allowing drought-resistant bushes and small trees to thrive. Grasslands occur in the centers of continents. These biomes have a continuous grass cover and few trees. Relatively low precipitation, fires, and severe droughts prevent the growth of trees. Grasslands have the world's richest soils and have largely been converted to agriculture. Temperate deciduous forests, whose broadleaf trees drop their leaves in winter, dominate the eastern half of the United States and are also found in Europe and eastern Asia. Moderate precipitation and lack of severe droughts allow the growth of deciduous trees, which shade the forest floor, preventing the growth of grasses. Temperate rain forests, dominated by conifers, occur in coastal regions with both high rainfall and moderate temperatures. The northern coniferous forest nearly encircles Earth below the arctic region. It is dominated by conifers whose small, waxy needles are adapted to conserve water and take advantage of the short growing season. The tundra is a frozen desert where permafrost prevents the growth of trees and the bushes remain stunted. Tundra is found both in the Arctic and on mountain peaks.

30.4 What Are the Principal Aquatic Biomes?

Sunlight is strong enough for photosynthesis only in shallow waters. Nutrients are found in bottom sediments, washed in from surrounding land or provided by upwelling in nearshore ocean waters.

In freshwater lakes, the littoral zone receives both sunlight and nutrients and supports the most life. The limnetic zone is the well-lit region of open water where photosynthetic protists thrive. In the deep profundal zone of large lakes, light is inadequate for photosynthesis, and most energy is provided by detritus. Oligotrophic lakes are clear, are low in nutrients, and support sparse communities. Eutrophic lakes are rich in nutrients and support dense communities. During succession, lakes shift from an oligotrophic to a eutrophic condition.

Streams begin at a source region, often in mountains, where water is provided by rain and snow. Source water is generally clear, high in oxygen, and low in nutrients. In the transition zone, streams join to form rivers that carry sediment from land and support a larger community. On their way to lakes or oceans, rivers enter relatively flat floodplains, where they deposit nutrients, take a meandering path, and spill over the land during floods.

Most life in the oceans is found in shallow water, where sunlight can penetrate, and is concentrated near the continents, particularly in areas of upwelling, where nutrients are most plentiful. Estuaries are highly productive areas where rivers meet the ocean. The intertidal zone, alternately covered and exposed by tides, harbors organisms that can withstand waves and exposure to air. Kelp forests grow in cool, nutrient-rich coastal areas and provide food and shelter for many animals. Coral reefs, formed by the skeletons of corals, are primarily found in shallow water in warm tropical seas. Coral reefs support an extremely diverse ecosystem. In the open ocean, most life is found in the photic zone, where light supports photosynthesis by phytoplankton. In the aphotic zone, life is supported by nutrients that drift down from the photic zone. The deep ocean floor lies within the aphotic zone. Whale carcasses provide a nutrient bonanza that supports a succession of unique communities. Specialized hydrothermal vent communities, supported by chemosynthetic bacteria, thrive at great depths in geothermally heated water.

Key Terms

Thinking Through the Concepts

Multiple Choice

1. Which of the following does *not* characterize the tropical rain-forest biome?
   a. warm temperatures year-round
   b. abundant rainfall
   c. nutrient-rich soil
   d. high biodiversity

2. The biome that is mostly covered by grass and scattered trees, with warm temperatures year-round and pronounced wet and dry seasons, is the
   a. tropical deciduous forest.
   b. savanna.
   c. desert.
   d. tropical scrub forest.

3. The part of a freshwater lake that typically contains the most abundant plant and animal life is the
   a. profundal zone.
   b. aphotic zone.
   c. limnetic zone.
   d. littoral zone.

4. Which of the following is True about ocean ecosystems?
   a. Surface waters typically have abundant sunlight but low levels of nutrients.
   b. Because there isn't enough sunlight for photosynthesis, deep waters typically have no life.
   c. Surface waters in the open ocean have the highest productivity.
   d. Life near hydrothermal vents relies on photosynthesis for energy.

5. As the global climate warms, which of the following changes in the distribution of biomes is likely to occur?
   a. spread of coniferous forests to lower elevations on mountains
   b. spread of tundra to lower elevations on mountains
   c. spread of northern coniferous forests farther north
   d. spread of northern coniferous forests farther south

Fill-in-the-Blank

1. The tilt of Earth on its axis produces _____. Coastal climates are more moderate because of _____. A dry region on the side of a mountain range that faces away from the direction of prevailing winds is called a(n) _____.

2. Of the four major requirements of life, which are the most important in determining the nature and distribution of terrestrial biomes? _____, _____ Which are most important for aquatic ecosystems? _____, _____, _____

3. The most biologically diverse terrestrial ecosystems are _____. The most biologically diverse aquatic ecosystems are _____.

4. The shallow portion of a large freshwater lake is called the _____. Photosynthetic plankton are called _____; nonphotosynthetic plankton are called _____. The open-water portion of a lake is divided into two zones, the upper _____ and the lower _____. Lakes that are low in nutrients are described as _____. Lakes high in nutrients are described as _____. The most diverse freshwater ecosystems are _____.

5. The primary producers of the open ocean are mainly _____. Hydrothermal vent communities are supported by bacteria that obtain energy from the process of _____, using the compound _____ as an energy source. Water may be at temperatures above the surface boiling point near hydrothermal vents, but it does not boil because of the _____.

Review Questions

1. Explain how air currents contribute to the formation of rain forests and large deserts.

2. What are large, roughly circular ocean currents called? What effect do they have on climate, and where is that effect strongest?

3. Explain why traveling up a mountain in the Northern Hemisphere takes you through biomes similar to those you would encounter by traveling north for a long distance.

4. Where are the nutrients of the tropical rain-forest biome concentrated? Why is life in the tropical rain forest concentrated high above the ground?

5. List some adaptations of desert cactus plants and desert animals to heat and drought.

6. What is desertification?

7. How are trees of the northern coniferous forest adapted to a lack of water and a short growing season?

8. How do deciduous and coniferous biomes differ?

9. What environmental factor best explains why the natural biome is shortgrass prairie in eastern Colorado, tallgrass prairie in Illinois, and deciduous forest in Ohio?

10. Where is life in the oceans most abundant, and why?

11. Distinguish among the littoral, limnetic, and profundal zones of lakes in terms of their location and the communities they support.

12. Distinguish between oligotrophic and eutrophic lakes. Describe a natural scenario and a human-created scenario under which an oligotrophic lake might be converted to a eutrophic lake.

13. Compare the source, transition, and floodplain zones of streams and rivers.

14. Distinguish between the photic and aphotic zones of the ocean. How do organisms in the photic zone obtain nutrients? How are nutrients obtained in the aphotic zone?

Applying the Concepts

1. Fairbanks, Alaska, the plains of eastern Montana, and Tucson, Arizona, all have about the same annual precipitation. Explain why these locations contain very different vegetation.

2. Using Figures 30-3 and 30-4 as starting points, explain why terrestrial biomes are not evenly distributed in bands of latitude across Earth's surface. Explain how your proposed mechanisms apply to two specific locations.

31 CONSERVING EARTH'S BIODIVERSITY

Exterminating the wolves in Yellowstone National Park put an entire ecosystem in jeopardy.

The Wolves of Yellowstone

IN 1926, THE LAST TWO WOLVES in Yellowstone National Park were killed by park rangers. Why were wolves intentionally exterminated in a national park? Because, when Congress established the park in 1872, the Secretary of the Interior had been instructed to "provide against the wanton destruction of the fish and game." Wolf predation on elk was considered "wanton destruction," so all the wolves were killed. But it turned out that no wolves meant too many elk, which in turn devastated the park. As early as the 1930s, scientists noted that aspen, cottonwoods, and willows were being overgrazed. In fact, from the mid-1930s through 1997, at 87 study sites in northern Yellowstone, not a single aspen sapling survived.

In 1987, the U.S. Fish and Wildlife Service, charged with restoring endangered species, proposed transplanting wolves from Canada to Yellowstone. After considerable debate, 31 wolves were released in Yellowstone in 1995 and 1996. Since 2000, wolf populations in the park have ranged from about 100 to 170. Within a few years after wolves were reintroduced to the park, elk populations started to decline. Elk behavior changed, as well. They spent more time looking around and less time feeding. Aspen saplings suffered less browsing and began to grow again, providing increased habitat for many other species, including wildflowers and songbirds. Although it is very difficult to prove that wolf restoration is the cause of the decline in elk and the rebound in aspen, most ecologists studying the Yellowstone ecosystem have concluded that wolves indeed caused these profound changes.

The impact of wolves in Yellowstone is just one example of the importance of **biodiversity:** the sum total of Earth's tremendous variety of life, including genes, species, communities, and ecosystems. When wolves were exterminated from Yellowstone in 1926, there were still grizzly bears, coyotes, foxes, and mountain lions in the park, but all of these predators put together could not fill the role of the wolf in the Yellowstone ecosystem.

Is biodiversity truly important to ecosystem function, or are just a few species the key players? If biodiversity is essential, why is that so? How do human activities endanger biodiversity, and potentially the functioning of the ecosystems upon which all life on Earth depends?

AT A GLANCE

31.1 WHAT IS CONSERVATION BIOLOGY?

Conservation biology is the scientific discipline devoted to understanding and preserving Earth's biodiversity, including:

- *Genetic Diversity*   The long-term survival of a species depends on the variety of different alleles in its gene pool. Genetic diversity is crucial for a species to adapt to changing environments.
- *Species Diversity*   The variety and relative abundance of the different species that make up a community are important for the functioning of the community. Species diversity helps to buffer communities against disturbances such as drought, climate change, and even invasive species.
- *Ecosystem Diversity*   Ecosystem diversity includes the variety of communities and the nonliving environments on which the communities depend. Ecosystem diversity also includes the different types of ecosystems, both terrestrial and aquatic, found throughout the biosphere.

CHECK YOUR LEARNING

Can you ...

- describe the goals of conservation biology?
- explain the importance of the three levels of biodiversity that conservation biologists study and seek to protect?

31.2 WHY IS BIODIVERSITY IMPORTANT?

The vast majority of people in developed countries, such as most of the countries in North America and Europe, live in cities or suburbs. Why should preserving biodiversity be important to us? Many people would say that species and natural ecosystems are worth preserving for their own sake. But even if you disagree, a very practical reason for preserving biodiversity is simple self-interest: Ecosystems, and the biodiversity that sustains them, are essential for human well-being.

Ecosystem Services Are Practical Uses for Biodiversity

Ecosystem services are the benefits that people obtain from ecosystems. The *Millennium Ecosystem Assessment* groups ecosystem services into four interconnected categories: (1) provisioning services, (2) regulating services, (3) cultural services, and (4) supporting services. Here we provide a few examples of each type of ecosystem service.

Provisioning Services Are Products Directly Obtained from Ecosystems

Ecosystems provide many materials and sources of energy that are used by people. Provisioning services are often called "natural resources."

- *Food*   Most of our food comes from farms (agricultural ecosystems), but people throughout the world also eat wild-grown food. For example, the United Nations Food and Agriculture Organization estimates that about 20 pounds (9 kilograms) of wild fish and other seafood are caught per person per year, worldwide. In parts of Africa, Asia, and South America, wild animals provide an important source of protein for an often poorly nourished population.
- *Raw Materials*   Wood is used for construction, furniture, and paper worldwide. Natural ecosystems provide almost all of the fresh water used for agriculture, industry, and drinking.
- *Energy*   Hydroelectric power provides electricity in almost every country with enough rainfall and suitable dam sites. In less-developed countries, rural residents often rely on wood for heating and cooking.

Regulating Services Control Ecosystem Processes

Regulating services affect the quality or abundance of many of the products people obtain from ecosystems, including water, soil, and food. Regulating services also help to control weather and climate.

- *Water Purification*   Natural ecosystems, including forests, grasslands, and wetlands, purify water by removing sediments and pollutants.
- *Pollination*   Bees and other insects pollinate most plants, including important agricultural products such as coffee, cocoa, and many fruits.
- *Pest Control*   Animals as diverse as bats, frogs, birds, and wasps feed on insects that are often agricultural pests or carry human diseases such as malaria.
- *Erosion and Flood Control*   Vegetation helps to hold soil in place and prevent erosion. Plant roots also increase the

▲ **FIGURE 31-1 Loss of flood control services** Although triggered by heavy monsoon rains, the catastrophic flooding in northern India in 2014 was worsened by massive deforestation. Hundreds of people were killed.

soil's capacity to hold water, reducing both erosion and flooding. Deforestation is thought to have contributed to massive flooding triggered by heavy rains in India and Brazil in 2014 (**FIG. 31-1**).

- *Climate Regulation* By providing shade, reducing temperatures, and serving as windbreaks, plant communities affect local climates. Forests dramatically influence the water cycle, as water evaporating from leaves returns to the atmosphere. In the Amazon rain forest, one-third to one-half of the rain is water that evaporated from leaves.
- *Carbon Storage* Plants remove CO_2 from the atmosphere during photosynthesis. Some of this CO_2 is stored in the plants, especially in the trunks and roots of trees. Thus, forests slow down the increase in atmospheric CO_2 that causes climate change and acidifies the oceans. If forests are cut down or burned, they release CO_2 again. About 10% to 15% of the CO_2 produced by human activities results from deforestation.

Cultural Services Are Nonmaterial Benefits Desired by People

People use natural ecosystems to increase their enjoyment of life, in many cases returning home not with material products, but with pleasant memories and reduced stress.

- *Recreation* Many, perhaps most, people take pleasure in "returning to nature." In the United States, more than 430 million visitors flock to national parks and national forests each year. Hundreds of millions more go to wildlife refuges and state parks.
- *Tourism* Ecotourism, in which people travel to observe unique biological communities, is a rapidly growing recreational industry. Examples of ecotourism destinations include tropical coral reefs and rain forests, the Galápagos Islands, the African savanna, and even Antarctica (**FIG. 31-2**).
- *Mental and Physical Health* Scientific studies have found that being in, or often just looking at, natural environments hastens healing after surgery, reduces stress hormone levels, improves mental focus in children with ADHD, and boosts the immune system.

Supporting Services Are Crucial to Providing Other Ecosystem Services

Ecosystem services that either affect people only indirectly or take a very long time to affect human welfare are usually classed as supporting services.

- *Habitat* Ecosystems provide habitat for the organisms that supply provisioning and regulating services, including wild foods, wood, and pollinators.
- *Photosynthesis* Photosynthesis by plants, algae, and cyanobacteria provides all of the oxygen needed for life on Earth and is the first step in providing food for almost all living organisms, including people.
- *Genetic Resources* The wealth of genes found in wild plants is an often-overlooked ecosystem service that may help to protect our food supply. According to the UN Food and Agriculture Organization, most of our food is supplied

(a) Scuba diving in a coral reef in the Red Sea   **(b) Spotting penguins in Antarctica**

▲ **FIGURE 31-2 Ecotourism** Carefully managed ecotourism represents a sustainable use of natural ecosystems, generating revenue and providing an incentive to preserve wildlife habitat.

by only 12 crop plants, including rice, wheat, and corn. Researchers have identified genes in wild relatives of these domesticated plants that might be transferred into crops to increase their productivity or provide greater resistance to disease, drought, or salty soil.

- *Soil Formation*   It can take hundreds of years to build up a single inch of soil. For example, the rich soils of the Midwestern United States accumulated under natural grasslands over thousands of years. Farmers have converted these grasslands into one of the most productive agricultural regions in the world.
- *Nutrient Cycling*   As we described in Chapter 29, nutrients cycle within and between ecosystems, often moving from reservoirs that are not available to living organisms to chemical forms that organisms can use. For example, nitrogen-fixing bacteria in soil convert atmospheric nitrogen into ammonia and nitrate, which plants can then use in the synthesis of proteins and nucleic acids.

Ecological Economics Attempts to Measure the Monetary Value of Ecosystem Services

Historically, people have assumed that ecosystem services are free and unlimited. Therefore, the value of ecosystem services has seldom been taken into account when making decisions about land use, farming practices, power generation, and a host of other human activities. Fortunately, that is beginning to change. **Ecological economics** attempts to determine the monetary value of ecosystem services and to assess the trade-offs that occur when natural ecosystems are damaged to make way for human activities.

For example, a farmer planning to divert water from a wetland to irrigate a crop would traditionally weigh the benefit of increased crop production against the cost of the project's labor and materials. This analysis ignores the many services an intact wetland provides, such as neutralizing pollutants, controlling floods, and providing breeding grounds for fish, birds, and many other animals. If the loss of ecosystem services were factored into the cost–benefit analysis, the intact wetland might well be more valuable than the crop. In a market economy, however, the economic benefits from projects that damage ecosystems usually go to individuals, whereas the costs are borne by society as a whole. Thus, it is difficult to apply the principles of ecological economics except in projects designed and funded by government agencies.

New York City offers an excellent example of government planning to preserve ecosystem services. The city obtains most of its water from the Catskill Mountains, a 1600-square-mile watershed 120 miles away in upstate New York. In 1997, realizing that its water would be polluted by sewage and agricultural runoff as the Catskills were developed, city officials calculated that it would cost $6 billion to $8 billion to build a water filtration plant, plus an additional $250 million annually to run it. Recognizing that the same water purification service was provided by the ecosystems of the Catskill Mountains, city officials decided to invest about $1.5 billion in protecting them, purchasing large tracts of land and keeping them in a reasonably natural state

Earth's Ecosystem Services Have Enormous Monetary Value

In 2014, an international team of ecologists, economists, and geographers calculated that ecosystem services provide benefits to humanity worth between $125 trillion and $145 trillion per year, about twice the world's annual gross domestic product (an estimate of the market value of all goods and services produced everywhere in the world). However, humanity isn't taking very good care of our biosphere. For example, the *Millennium Ecosystem Assessment* concluded that 60% of Earth's ecosystem services were being degraded.

Biodiversity Supports Ecosystem Function

Several studies have concluded that areas with the highest biodiversity also tend to be areas providing the greatest ecosystem services. Why is biodiversity important to ecosystem function? More diverse communities tend to have higher productivity (see Chapter 29). Diverse communities are also better able to withstand disturbances, such as drought, severe winters, episodes of pollution, or delivery of excess nutrients, such as farm fertilizer runoff. When a community has a large number of different species, each with its own niche, resources are often used very efficiently, leaving few resources available for invasive species to gain a foothold.

One way in which biodiversity might protect ecosystems, sometimes called the "redundancy hypothesis," is that several species in a community may have functionally equivalent roles. For example, several species of bees in an ecosystem may pollinate flowers. If a few of these species are lost, the remaining ones may increase their population size and pollinate most or all of the flowers, as long as the ecosystem operates under typical conditions. If, however, the ecosystem is stressed—by drought, for example—some of the remaining bee species may not survive the stress, resulting in significantly less pollination and, hence, less plant reproduction.

The "rivet hypothesis" postulates that ecosystem function is analogous to an airplane wing, in which the loss of a couple of rivets may not be catastrophic, but the loss of rivets in strategic places causes the entire wing to fall apart. In an ecosystem, superficially similar species may have somewhat different positions in the web of ecosystem stability, and the loss of a few critical species may cause collapse. Returning to our bee example, some species of bees specialize in pollinating specific species of flowers. Eliminating one of these bee species may mean that some species of plants no longer reproduce. Any animals that specialize in feeding on those plants will die along with them. If just a few critical bee species disappear, then many plant and animal species will also die off.

Earth WATCH

Whales—The Biggest Keystones of All?

The oceans are a lot emptier than they used to be. No one knows for sure how many whales originally roamed the seas, but it is believed that commercial whaling reduced many whale populations by 90%. Even now, decades after almost all commercial whaling ceased, most whale populations are much less than half their pre-whaling size. Because whales swim at the top of the food chain, you would think that whale prey, from giant squid to shrimp-like krill, should be experiencing a population boom, right?

Actually, they're not. Populations of krill have not increased and may have even declined a little, despite the fact that the pre-whaling population of filter-feeding baleen whales probably ate something like 150 million tons of krill each year. How can that be? Studies by marine ecologists suggest an answer: The nutrients required by krill are carried from ocean depths to surface waters by whales.

Krill eat photosynthetic phytoplankton, which can only live in well-lit waters near the surface. Phytoplankton require nutrients such as iron and nitrogen, which tend to be scarce in surface waters. Why are nutrients scarce at the surface? Because most organisms and their feces are denser than seawater. For example, single-celled algae sink a few yards each day. Feces and dead animals sink much faster, some as much as a half-mile a day. As they sink, algae, dead animals, and feces all carry nutrients from the surface down to the depths. Ocean currents and winter storms bring some of these nutrients back up to the surface, but not all.

Enter the great whales. Many whales feed at substantial depths, from a few hundred feet to as much as a half mile below the surface. Whales, of course, return to the surface to breathe—and to poop (**FIG. E31-1**). Whales release huge plumes of buoyant feces that effectively bring nutrients from the depths back to the surface. Whale feces have about 10 million times as much iron as seawater does. Whale feces and urine also bring nitrogen up to the surface in the form of ammonia and urea. By bringing nutrients to sunlit surface water,

◀ **FIGURE E31-1**
Whale feces fertilize the oceans Whales, such as the sperm whale shown here, release huge plumes of semi-liquid feces that drift in surface waters, providing essential nutrients for photosynthetic algae.

the "whale pump" enhances the productivity of the oceans and may actually increase krill populations.

Fortunately, almost all whaling has ceased, and most whale populations are increasing, some by as much as 5% to 7% a year. As these giant keystones rise up, they will likely bring ocean ecosystems up with them.

THINK CRITICALLY How would you test the whale pump hypothesis? Assume that you can measure populations of whales and phytoplankton and the concentrations of iron, nitrogen, and other nutrients in ocean waters.

In some ecosystems, one or two "rivets," called **keystone species,** may be crucially important to ecosystem function. Think of the analogy that inspired the phrase: A keystone sits at the top of a stone arch and holds all the other pieces in place. Remove the keystone, and the whole arch collapses. Similarly, in a biological community, a keystone species is one whose role is much more important than would be predicted by the size of its population or by a superficial glance at its position in the food web. In the oceans, the great whales may be keystone species, as we explore in "Earth Watch: Whales—The Biggest Keystones of All?"

CHECK YOUR LEARNING

Can you ...

- describe the major categories of ecosystem services provided to humanity and provide examples of each?
- explain why biodiversity helps to maintain functioning ecosystems?

CASE STUDY CONTINUED
The Wolves of Yellowstone

In Yellowstone, the wolf is a top predator and keystone species that helps to determine populations not only of its direct prey, but also of other species in the ecosystem. In the early twentieth century, no one expected that exterminating wolves would wreak havoc on aspens and the many species that depend on them. Unfortunately, in many parts of the world, loss of biodiversity, including keystone predators, continues today. What are the principal causes?

31.3 IS EARTH'S BIODIVERSITY DIMINISHING?

No species lasts forever. Over the course of evolutionary time, species arise, flourish for various periods of time, and go extinct. If all species are fated to eventual extinction, why should we worry about modern extinctions? Because the rate of extinction during modern times has become extraordinarily high.

Extinction Is a Natural Process, but Rates Have Risen Dramatically in Recent Years

The fossil record indicates that, in the absence of cataclysmic events, extinctions occur naturally at a very low rate. This *background extinction rate* ranges from about 0.1 to 1 extinction per million species per year. However, the fossil record also provides evidence of five major **mass extinctions,** during which many species were eradicated in a relatively short period of time (see Chapter 18). The most recent mass extinction happened roughly 66 million years ago, abruptly ending the age of dinosaurs. Sudden changes in the environment, such as might be caused by enormous meteor impacts or rapid climate change, are the most likely reasons for these mass extinctions.

A recent study estimates that the modern extinction rate is about 1,000 times the background rate: 100 to 1,000 extinctions per million species per year. And species aren't being replaced; new ones appear at a far slower rate. As a result, many biologists are convinced that humans are causing a sixth mass extinction. Extinctions of birds and mammals are best documented, although these represent only about 0.1% of the world's species. Since the 1500s, we have lost about 1.7% of all mammal species and 1.6% to 2% of all bird species, an extinction rate probably more than 100 times the background rate.

Each year, a Red List of at-risk species is published by the International Union for Conservation of Nature (IUCN), the world's largest conservation network. Species are described as **vulnerable, endangered,** or **critically endangered,** depending on how likely they are to become extinct in the near future. Species that fall into any of these categories are described as **threatened.** In 2014, the Red List contained 22,176 threatened species, including 13% of all birds, 26% of mammals, and 41% of amphibians. The U.S. Fish and Wildlife Service lists more than 1,500 threatened and endangered species in the United States alone. Why are so many species in danger of extinction?

CHECK YOUR LEARNING
Can you ...
- define *mass extinction*?
- explain why biologists fear that a mass extinction is occurring as a result of human activities?

HAVE YOU EVER

WONDERED ...

What You Can Do to Prevent Extinctions?

You may feel helpless to prevent extinction, but in fact, you can do a lot. You can purchase coffee, chocolate, fish, and many other products that have been grown or harvested in a sustainable manner and certified by organizations such as the Marine Stewardship Council, Seafood Watch, the Forest Stewardship Council, or the Rainforest Alliance. You can contribute to organizations, such as the Nature Conservancy, the World Wildlife Fund, or Saving Species, that work directly to protect biodiversity. For example, Saving Species was the driving force behind preserving wildlife corridors that connect cloud forest habitat in Ecuador, home of the olinguito, a 2-pound relative of raccoons that was first discovered in the wild in 2013. The same habitat also supports 14 species of endangered hummingbirds and many other rare and endangered animals and plants.

Olinguito

31.4 WHAT ARE THE MAJOR THREATS TO BIODIVERSITY?

The 2014 edition of the *Living Planet Report,* a joint project of the World Wildlife Fund, the Zoological Society of London, the Global Footprint Network, and the Water Footprint Network, estimated that the total population of wild vertebrate animals on Earth is only about half of what it was 40 years earlier. This loss was primarily caused by habitat destruction and overexploitation by hunting and fishing, with smaller contributions from climate change, invasive species, pollution, and disease. Much of the loss of habitat is essentially irreversible, as people convert wildlands to farms, cities, and roads.

Humanity's Ecological Footprint Exceeds Earth's Resources

The human **ecological footprint** is an estimate of the area of Earth's surface required to produce the resources we use and absorb the wastes we generate. A complementary concept, **biocapacity,** is an estimate of the sustainable resources and waste-absorbing capacity actually available on Earth. Although related to the concept of carrying capacity (explained in Chapter 28), both footprint and biocapacity calculations are subject to change as new technologies influence the way people use resources. The calculations assume that humans can use the entire planet, without reserving any of it for the rest of life on Earth.

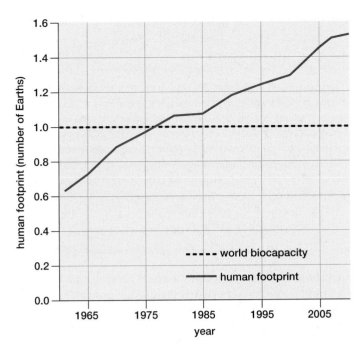

▲ **FIGURE 31-3 Human demand exceeds Earth's biocapacity** Humanity's ecological footprint from 1961 to 2010, expressed as a fraction of Earth's total sustainable biocapacity (dashed line at 1.0). In 1961, we were using a little more than half of Earth's biocapacity. In 2010, we would have needed about 1.5 Earths to support us, at current rates of consumption, in a sustainable manner. (Because of the time required to obtain and analyze the data, footprint calculations for 2010 were first published in 2014.) Data from the World Wildlife Fund, the Zoological Society of London, the Global Footprint Network, and the Water Footprint Network (2014), *The Living Planet Report.*

In 2010, the biocapacity available for each of the 6.7 billion people then living on Earth was 4.2 acres, but the average human footprint was 6.4 acres. In other words, we exceeded biocapacity by about 50%: In the long run, we would need about 1.5 Earths to support humanity at 2010 consumption and population levels (**FIG. 31-3**). Countries vary enormously in their ecological footprints, from about 12 to 24 acres per person for wealthy countries, such as most of Europe, Canada, Australia, New Zealand, and the United States, to as little as 1 to 2 acres per person for poor countries, such as most of those in Africa. Since these estimates were made, the human population has grown by about 400 million, while Earth's total biocapacity has not significantly increased.

Running such an ecological deficit is possible only on a temporary basis. Imagine a savings account that must support you for the rest of your life. If you preserve the capital and live on the interest, the account will last forever. But if you withdraw the capital to support an extravagant lifestyle or a growing family, you will soon run out of money. By degrading Earth's ecosystems, humanity is drawing down Earth's ecological capital. As the human population grows and highly populated countries such as India and China raise their living standards, the strain on Earth's resources will increase. By using up so much biocapacity, people inevitably reduce the resources available for the rest of life on Earth.

Many Human Activities Directly Threaten Biodiversity

Habitat destruction, overexploitation, invasive species, pollution, and global climate change pose the greatest dangers to biodiversity. Threatened species often face several of these perils simultaneously. For example, coral reefs, home to about one-third of marine fish species, suffer from a combination of overharvesting, pollution, ocean acidification, and global warming.

Habitat Destruction Is the Most Serious Threat to Biodiversity

Habitat loss imperils more than 85% of all endangered mammals, birds, and amphibians. The most serious threat is the loss of tropical rain forests, home to about half of Earth's plant and animal species. Satellite images indicate that about 30,000 to 45,000 square miles of rain forest are lost each year (the state of Kentucky is about 40,000 square miles), or the area of a football field every 1 to 1.5 seconds (**FIG. 31-4**). The primary cause of the destruction of tropical rain forests is converting the land to agriculture, to create both small subsistence farms and huge plantations and ranches that supply beef, soybeans, palm oil, sugarcane, and biofuels, mostly to developed countries (see "Earth Watch: Biofuels—Are Their Benefits Bogus?" in Chapter 7).

1975   **2012**

▲ **FIGURE 31-4 Habitat destruction** The loss of habitat due to human activities is the greatest single threat to biodiversity worldwide. These satellite images show a section of rain forest in Brazil in 1975 (left) and 2012 (right). More than half of the original rain forest has been cut down.

▲ **FIGURE 31-5 Habitat fragmentation** Fields isolate patches of forest in Paraguay.

THINK CRITICALLY Which types of species do you think are most likely to disappear from small patches of forest? What do you think would be the resulting effect on the ecosystem?

Even when a natural ecosystem is not destroyed, it may become broken into small pieces, separated by roads, farms, or housing developments (**FIG. 31-5**). This **habitat fragmentation** is a serious threat to wildlife. Some species of U.S. songbirds, such as the ovenbird and Acadian fly-catcher, need as much as 300 acres of continuous forest to find food, mates, and breeding sites; in smaller forest patches, reproductive success is much lower. Big cats are also threatened by habitat fragmentation. Starting in the 1970s, India has set aside 47 forest reserves intended to protect the endangered Bengal tiger. However, many of the reserves have become islands in a sea of development, forcing the tigers into isolated patches of woodland.

To be truly functional, a preserve must support a **minimum viable population (MVP),** the smallest isolated population that can persist in spite of natural events, including disease, fires, floods, and the loss of genetic diversity through inbreeding and genetic drift. The MVP for any species is influenced by many factors, including the quality of the environment, the species' average life span, its fertility, and the number of young that typically reach maturity. Most wildlife experts think that an MVP of Bengal tigers must include at least 50 females—more than are found in most of India's tiger reserves.

Many countries are working to preserve critical habitat. One of the largest protected habitats is the Papahānaumokuākea Marine National Monument in the Hawaiian Islands, designated in 2006. This national monument covers 84 million acres of the Pacific Ocean and is home to about 7,000 species of birds, fish, and marine mammals. Some species depend on

CASE STUDY \ **CONTINUED**

The Wolves of Yellowstone

As you learned in Chapter 29, large predators at the top of food webs, such as wolves and grizzly bears, are always relatively rare, because of energy losses between trophic levels. Even a national park the size of Yellowstone—almost 3,500 square miles—often cannot sustain a minimum viable population of such animals over long time periods, primarily because of the loss of genetic diversity and the dangers of epidemic diseases. How can an MVP be provided for these animals? We will return to this crucial concern in Section 31.5.

such huge reserves; for others, critical habitat may be a few patches of sandy beach. "Earth Watch: Saving Sea Turtles" on page 594 discusses an innovative sea turtle conservation program in Brazil, which not only preserves turtle nesting sites but also helps local communities to prosper.

Overexploitation Threatens Many Species

Overexploitation is the hunting or harvesting of natural populations at a rate that exceeds their ability to replenish their numbers. Overexploitation of many species has increased as a growing demand for wild animals and plants has been coupled with technological advances that have increased our efficiency at harvesting them. For example, overharvesting is the single greatest threat to marine life, causing dramatic declines of many species, including invertebrates such as abalone, oyster, and corals, and fish such as cod, many sharks, haddock, Pacific bluefin tuna, and mackerel. The UN Food and Agriculture Organization estimates that about 30% of global fish populations are overexploited, and another 60% are being fished to their maximum sustainable yield.

Both poverty and wealth can contribute to overexploitation, particularly of endangered species. Rapidly growing populations in less-developed countries increase the demand for animal products, as hunger and poverty drive people to harvest all that can be sold or eaten, legally or illegally, without regard to its rarity. Rich consumers often fuel the exploitation of endangered species by paying high prices for illegal products such as elephant-tusk ivory, rare orchids, and exotic birds. Although good data about black market activities are difficult to come by (for obvious reasons), the sale of endangered species, or products derived from them, is extremely lucrative, thought to total about $19 billion a year.

Invasive Species Displace Native Wildlife and Disrupt Community Interactions

Humans have transported a multitude of species around the world—everything from thistles to camels. In many cases, the introduced species cause no great harm. Sometimes, however, non-native species become invasive: They increase

(a) Blue rock hunter cichlid

(b) Nile perch

▲ **FIGURE 31-6 Invasive species endanger native wildlife (a)** Lake Victoria was home to hundreds of species of stunningly colored cichlid fish, such as the blue rock hunter cichlid pictured here. **(b)** The Nile perch, introduced into Lake Victoria for fishermen, has proven to be a disaster for native fish.

in number at the expense of native species, competing with them for food or habitat or preying on them directly (see Chapter 29). According to the Center for Invasive Species and Ecosystem Health, there are almost 2,900 invasive species in North America, mostly plants and insects. About half of all threatened U.S. species suffer from competition with or predation by invasive species.

Island ecosystems are particularly vulnerable to invasive species, because populations of native plants and animals are usually low, the native species are often found nowhere else in the world, and, if they can't compete with the invaders, the natives cannot easily move to new habitat. For example, the Hawaiian Islands have lost about 1,000 species of native plants and animals since their settlement by humans. Many of the losses have been caused by competition and predation by invasive species, beginning with pigs and rats brought by the original Polynesian settlers and accelerating in the nineteenth and twentieth centuries. Most of the native wildlife of Hawaii remains in danger: According to the U.S. Fish and Wildlife Service, more than 430 plant and animal species in Hawaii are endangered, by far the largest number in any state.

Lakes are also especially vulnerable to invasive species. For example, Lake Victoria in Africa was once home to about 400 to 500 different species of cichlid fish that were found nowhere else on Earth (**FIG. 31-6a**). Enormous predatory Nile perch (**FIG. 31-6b**) and much smaller plankton-feeding tilapia were introduced into Lake Victoria in the mid-1900s. The combination of predation by Nile perch, competition from tilapia, pollution, and algal blooms (brought on by nutrients from surrounding farms draining into the lake) has caused a mass extinction of cichlids; only about 200 species remain.

Pollution Is a Multifaceted Threat to Biodiversity

Pollution takes many forms, including synthetic chemicals such as plasticizers, flame retardants, and pesticides; toxic metals such as mercury, lead, and cadmium; and high levels of nutrients, usually from sewage or agricultural runoff.

Because synthetic chemicals are often lipid soluble, even small amounts in the environment may accumulate to toxic levels in the fatty tissue of animals (see Chapter 29). In the mid-twentieth century, for example, the insecticide DDT accumulated in many predatory birds, causing them to lay eggs with shells so thin that they cracked when the parents sat on them during incubation. Fortunately, DDT and 11 other persistent organic pollutants have been banned or heavily restricted by a treaty signed by about 180 countries. Disputes continue over possible environmental and human health effects of several other synthetic organic chemicals. Although the evidence is still controversial, bisphenol A, used in the manufacture of certain types of plastics, is suspected of causing reproductive and developmental abnormalities. A class of insecticides called neonicotinoids has been implicated in massive declines in honeybee populations.

Many heavy metals are naturally bound up in rocks and thus rendered harmless. However, mining, industrial processes, and burning fossil fuels release heavy metals into the environment. Even extremely low levels of certain heavy metals, such as mercury and lead, are toxic to virtually all organisms.

Finally, nutrients in excessive amounts become pollutants. For example, burning fossil fuels releases nitrogen and sulfur compounds, disrupting their natural biogeochemical cycles and causing acid precipitation that threatens forests and lakes (see Fig. 29-12). Fertilizer runoff from farms and lawns often enters nearby waters and may cause harmful algal blooms (see Fig. 29-10).

Global Climate Change Is an Emerging Threat to Biodiversity

The rapid pace of human-induced climate change challenges the ability of species to adapt. Scientists at the Convention on Biological Diversity, an international organization with more than 150 member countries, have concluded that warmer conditions have already contributed to some extinctions and are likely to cause many more. Although it is difficult to predict all the impacts of global climate change, they likely include the following:

- Deserts may become hotter and drier.
- Warmer conditions are forcing some species to retreat toward the poles or up mountains to stay within the climate zones in which they can survive and reproduce. Relatively immobile species, especially plants, may be unable to retreat fast enough to stay within a suitable temperature range, because they typically only "move" as far and as fast as wind or animals disperse their seeds.
- Cool habitat will probably disappear completely from mountaintops. Animals that live at high altitudes, such as pikas in the Rocky Mountains (FIG. 31-7a), face shrinking habitat as the mountains warm. Some local populations on isolated mountains have already vanished.
- Insect pests that were previously killed by frost or sustained freezes may spread and thrive. For example, in the northern and central Rocky Mountains, populations of pine bark beetles were formerly limited by sustained extremely cold weather in the winter. In the past 20 years, however, these beetles have reached epidemic levels, with over 4 million acres infested. The infestation in British Columbia, Canada, is much worse: over 40 million acres (FIG. 31-7b).
- Parasites and insect-borne diseases are spreading closer to the poles. From lungworms in musk ox in Arctic Canada to disease-carrying mosquitoes infecting people in Sweden with tularemia, a warmer climate allows many pathogens and their carriers both to move northward and often to reproduce faster.
- Coral reefs require warm water, but too much warming causes bleaching and coral death (FIG. 31-7c; see Chapter 24). Coral reefs have already suffered massive damage in the Seychelles Islands, American Samoa, Sri Lanka, the coasts of Tanzania and Kenya, and parts of the Australian Great Barrier Reef.

CHECK YOUR LEARNING

Can you ...

- explain the concepts of ecological footprint and biocapacity, and how they are interrelated?
- describe how habitat destruction, overexploitation, invasive species, pollution, and global climate change threaten biodiversity?

(a) A pika gathers plants for the winter

(b) Pine bark beetles are killing pine trees

(c) Bleached corals (white) are usually dead or dying

▲ **FIGURE 31-7 Global climate change threatens biodiversity**
(a) Pikas live at high altitudes in the Rocky Mountains; as the climate warms, suitable pika habitat may disappear off the tops of the mountains. **(b)** A forest infested with pine bark beetles often consists of a mosaic of uninfected trees (with green needles), newly killed trees (with reddish needles), and trees killed several years earlier (gray, without needles). **(c)** Corals usually contain photosynthetic algae that provide nourishment for the coral. When the water warms too much, corals eject their algae and become white; without the algae to help feed them, they often die.

31.5 WHY IS HABITAT PROTECTION NECESSARY TO PRESERVE BIODIVERSITY?

As we have seen, human activities pose many threats to biodiversity. Reversing some of these activities, for example, by reducing overexploitation and curbing greenhouse gas emissions, is crucial to preserving biodiversity. Without suitable natural habitat, however, many species cannot survive. Therefore, it is essential to set aside habitat in protected reserves and to connect small, fragmented reserves with wildlife corridors.

Core Reserves Preserve All Levels of Biodiversity

Core reserves are natural areas protected from most human uses except low-impact recreation. Ideally, a core reserve encompasses enough space to preserve ecosystems with all their biodiversity, withstanding storms, fires, and floods without losing species.

To establish effective core reserves, ecologists must estimate the smallest areas required to sustain MVPs of the species that require the most space. The sizes of these *minimum critical areas* vary significantly among species and also depend on the availability of food, water, and shelter. In general, large predators in arid environments need a larger minimum critical area than small herbivores in lush environments.

Wildlife Corridors Connect Habitats

One fact stands out in estimating minimum critical areas, especially for reserves that include large predators: In today's crowded world, an individual core reserve, even a large national park, is seldom large enough to maintain biodiversity by itself. **Wildlife corridors,** which are strips of protected land linking core reserves, allow animals to move relatively freely and safely between habitats that would otherwise be isolated. Corridors thereby increase the effective size of small reserves by connecting them. In India, for example, government and private groups are working to preserve forested corridors linking some tiger reserves. Researchers have found genetic evidence that tigers are traveling through the corridors to mate with other tigers in reserves as far as 230 miles away. In the increasingly fragmented Atlantic forest of Brazil, forested corridors connect reserves that are home to endangered black lion tamarins and golden lion tamarins (**FIG. 31-8**).

▶ **FIGURE 31-8 Wildlife corridors connect habitat** These continuous strips of forest winding through pastures provide vital corridors for the movement of jaguars, ocelots, and endangered black lion tamarins (inset) between larger patches of forest.

THINK CRITICALLY What would be the likely effect of isolated small reserves on the genetic diversity of endangered species? How would genetic diversity be affected by connecting the small reserves with wildlife corridors?

CHECK YOUR LEARNING
Can you ...
- describe some strategies that can preserve natural ecosystems and their associated biodiversity?
- define the terms *core reserve* and *wildlife corridor*, and explain the relationship between them?

CASE STUDY CONTINUED
The Wolves of Yellowstone

Wildlife corridors are often the only way to provide enough habitat for a minimum viable population of large predators such as wolves, grizzlies, and mountain lions. A coalition of conservation groups and scientists has proposed the Yellowstone to Yukon Conservation Initiative, which would provide corridors connecting habitat in the Rocky Mountains all the way from Yellowstone and Grand Teton National Parks in Wyoming to the Yukon Territory in northwestern Canada.

Wildlife corridors such as these must include private land interspersed between national parks and national forests. Can private landowners preserve wildlife habitat while still making a living and enjoying their land?

31.6 WHY IS SUSTAINABILITY ESSENTIAL FOR A HEALTHY FUTURE?

Natural ecosystems share certain features that allow them to persist and flourish. Important characteristics of sustainable ecosystems include diverse communities, relatively stable populations that remain within the carrying capacity of the environment, recycling and efficient use of raw materials, and reliance on renewable sources of energy. Environments

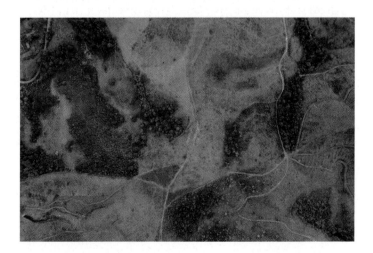

Earth WATCH   Saving Sea Turtles

Six of the seven species of sea turtles are threatened with extinction, endangered by their unusual life history, losses to predators, and human activities.

Most sea turtles don't begin to breed until they are 20 to 50 years old. When they reach reproductive age, the females must swim hundreds, even thousands, of miles to reach their nesting grounds, often on the same beaches where they hatched. The turtles drag themselves ashore, excavate a hole in the sand, deposit their eggs, and return to the sea (FIG. E31-2a). The eggs may be eaten by domestic dogs, foxes, wild pigs, raccoons, and a host of other predators. After about 2 months, baby turtles emerge from the surviving eggs and begin their difficult journey to adulthood. Seabirds and crabs attack them as they crawl to the ocean (FIG. E31-2b). Once there, the hatchlings may become a tasty morsel for fish.

As if these natural dangers weren't enough, female turtles and their eggs are easy prey for human poachers. Turtle meat and eggs are a delicacy in many cultures, turtle shells make beautiful jewelry, and turtle skin makes fancy leather. Turtles are also caught, both deliberately and accidentally, in fish lines and nets. The beautiful beaches attract tourists who may frighten nesting females. Finally, hatchlings find the sea by crawling toward the brightest area in sight—but the brightest place on the beach may now be not the moon reflecting off the ocean, but the lights of a resort.

These dangers plagued the five species of sea turtles that nest on the beaches of Brazil—until the late 1970s, when some students from a Brazilian university were vacationing on Rocas Atoll just off the coast. They watched a group of sea turtles come ashore to lay eggs, only to be slaughtered by the very fishermen who had been hired as the students' guides. Two of the students, José Albuquerque and Guy Marcovaldi, founded Projeto Tartarugas Marinhas (TAMAR for short, from the Portuguese tartarugas marinhas, or "sea turtles").

Albuquerque and Marcovaldi realized that for sea turtle conservation to succeed, fishermen and local villagers had to participate. Today, TAMAR has 22 bases on the Brazilian coast. Most of TAMAR's employees are former fishermen. Instead of hunting sea turtles, they free turtles caught in nets and patrol the beaches during nesting season. TAMAR biologists tag females and trace their travels. The fishermen

(a) A green turtle excavating a nest

(b) A turtle hatchling heads for the sea

▲ FIGURE E31-2 Endangered sea turtles (a) A female green turtle scoops sand with powerful flippers, creating a cavity where she will bury about 100 eggs. (b) After incubating in the sand for about 2 months, the eggs hatch. Here a hatchling heads for the sea, where (if it survives) it will spend 20 to 50 years before reaching sexual maturity.

that have been modified by human development often do not possess these characteristics. As a result, many human-modified ecosystems may not be sustainable in the long run. How can we meet our needs in ways that sustain the ecosystems on which we depend?

Sustainable Development Promotes Long-Term Ecological and Human Well-Being

In *Caring for the Earth: A Strategy for Sustainable Living,* the IUCN stated that **sustainable development** "meets the needs of the present without compromising the ability of future generations to meet their own needs" by "improving the quality of human life while living within the carrying capacity of supporting ecosystems." Therefore, sustainable development must minimize the use of nonrenewable resources and use renewable resources in a manner that allows them to be utilized year after year, far into the future. Here we will explore three specific issues of sustainability: the use of renewable resources, sustainable agriculture, and the preservation of reasonably natural ecosystems while still providing desired goods for people.

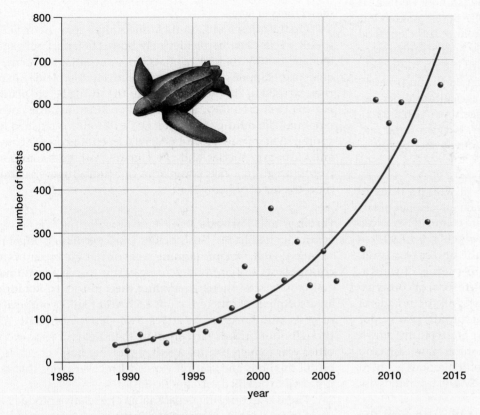

◀ **FIGURE E31-3** **Atlantic leather-back sea turtle populations in Florida are growing** Atlantic leatherback sea turtle nests on a group of beaches in Florida have been surveyed each year from 1989 to 2014. Because sea turtles are difficult to count at sea, nests are used as an indicator of population size. The population has been growing exponentially since the late 1980s (smooth curve).

fend off the (now rare) turtle poachers, identify nests in risky locations, and relocate the eggs to better beach sites or to a nearby hatchery. As of 2014, TAMAR had helped more than 15 million hatchlings reach the sea.

TAMAR has been successful because the project organizers have engaged local communities as partners in turtle protection. Money flows into the local economies as ecotourists come to see baby turtles, visit turtle museums, and buy souvenirs made by local residents. TAMAR also sponsors communal gardens, day-care centers, and environmental education activities. Recognizing that the economic benefits derived from preserving turtles far outweigh the money that can be made by hunting them, local residents eagerly participate in turtle conservation. As IUCN sea turtle specialist

Neca Marcovaldi put it, "Brazil's sea turtles are now worth more alive."

THINK CRITICALLY In 1970, Atlantic leatherback sea turtle populations were very low, and the U.S. Fish and Wildlife Service listed the species as endangered. Over the next several decades, steps were taken to protect turtle nests and prevent accidental killing of turtles at sea by fishing fleets. As a result, the population of leatherbacks nesting on beaches in Florida has been growing exponentially (**FIG. E31-3**). What factors probably contributed to this exponential growth? Can exponential growth continue? If not, what factors are likely to cause the population to stabilize?

Sustainable Development Relies on Renewable Resources

In principle, minerals such as aluminum, iron, and copper can be recycled by humans, just as nutrients are recycled in ecosystems, so that we never run out. Further, many minerals can substitute for each other in various uses, with enough ingenuity by inventors, engineers, and materials scientists. However, substitution cannot overcome absolute limits on supply. Some minerals, such as copper, are typically used in large quantities; certain rare minerals used in devices such as cell phones and computers are already in short supply. Other

minerals are highly toxic when dumped into the environment. For all of these reasons, the adage "reduce, reuse, and recycle" is very apt when applied to minerals.

Fossil fuels cannot be recycled. Further, burning fossil fuels releases carbon dioxide, which is the principal cause of global climate change, with profound effects on humans and natural ecosystems alike, as outlined in Section 31.4 and Chapter 30. Therefore, a concerted effort to switch from fossil fuels to renewable energy sources, such as solar, wind, and geothermal power, is an essential part of sustainable development.

Sustainable Agriculture Preserves Productivity with Reduced Impact on Natural Ecosystems

Agricultural land is by far the largest proportion of land appropriated by people for their own use, so how we farm is an important component of sustainable development. The goals of sustainable agriculture are to produce sufficient food to feed humankind, to ensure economic benefits to farmers, and to maintain ecosystem services so that both future generations of people and the rest of life on Earth can flourish. Many of the practices of sustainable agriculture differ significantly from those of traditional agriculture (TABLE 31-1). Farms differ in many of these practices, but two stand out: how fields are prepared for planting and whether synthetic fertilizers and pesticides are used.

Tillage There are two major methods of treating fields to grow crops: conventional tillage, in which all residues of last year's crop are removed and the fields are completely plowed each year, and conservation tillage, in which at least 30% of the previous crop's residue remains on the surface of the soil, with soil disturbance by plowing or other forms of cultivation. **No-till** farming is the most complete form of conservation tillage. No-till farming usually leaves all the residue of harvested crops in the fields as mulch for the next year's crops and/or grows a cover crop—plants that are typically grown during the fall or early spring and are not the main cash crop for the farmer. In the United States, no-till methods are used on about 35% of all croplands. Another 27% of farmlands use other types of conservation tillage.

No-till farming reduces soil erosion from both wind and rain and reduces fertilizer runoff. It also helps to improve soil structure and increase the amount of organic matter in the soil. The combination of crop mulch and increased organic matter also keeps the soil cooler and more moist during the hottest parts of the summer, helping crops to withstand drought. Because it requires less plowing, no-till farming also saves labor, wear and tear on tractors, and up to 14 gallons of diesel fuel per acre.

On the other hand, no-till farmers often spray herbicides to kill weeds. The herbicides may blow off farm fields and damage nearby natural habitats. In some situations, cover crops can suppress weeds fairly effectively. Herbicides may then be used to kill the cover crop in the spring before planting the cash crop (FIG. 31-9). In other situations, the cover crop may die during the winter or can be cut or crushed in spring, so herbicides are not needed. Depending on the cover and cash crop species and the climate, soil, and common weeds in a given locale, cover crops can reduce, or even eliminate, the need for herbicides.

Fertilizer and Pesticide Use The vast majority of farms use synthetic herbicides, insecticides, and fertilizers; organic farmers do not. Organic farming relies on natural predators to control pests and on soil microorganisms to degrade animal and crop wastes, thereby recycling their nutrients. Rotating an assortment of different crops each year reduces outbreaks of pests and diseases that attack a single type of plant. Although most organic farmers plow their fields at least every other year to help control weeds, a growing number employ no-till methods and control weeds with cover crops that do not need to be killed with herbicides.

There is an ongoing debate about the relative productivity of organic versus conventional farming. A recent analysis of over 100 studies found that organic farms average about 19% smaller yields compared to conventional farming, although for some crops the difference is much smaller. For

TABLE 31-1	Agricultural Practices Affect Sustainability	
	Unsustainable Agriculture	**Sustainable Agriculture**
Soil erosion	Allows soil to erode far faster than it can be replenished because the remains of crops are plowed under, leaving the soil exposed until new crops grow.	No-till agriculture greatly reduces soil erosion. Planting strips of trees as windbreaks reduces wind erosion.
Pest control	Uses large amounts of pesticides to control crop pests.	Trees and shrubs near fields provide habitat for insect-eating birds and predatory insects. Reducing insecticide use helps to protect birds and insect predators.
Fertilizer use	Uses large amounts of synthetic fertilizer.	No-till agriculture retains nutrient-rich soil. Animal wastes are used as fertilizer. Legumes that replenish soil nitrogen (such as soybeans and alfalfa) are alternated with crops that deplete soil nitrogen (such as corn and wheat).
Water quality	Allows runoff from bare soil to contaminate water with pesticides and fertilizers. Allows excessive amounts of animal wastes to drain from feedlots.	Animal wastes are used to fertilize fields. Plant cover left by no-till agriculture reduces nutrient runoff.
Irrigation	May excessively irrigate crops, often using groundwater pumped from aquifers at a rate faster than the water is replenished by precipitation.	Modern irrigation technology reduces evaporation and delivers water only when and where it is needed. No-till agriculture reduces evaporation.
Crop diversity	Relies on a small number of high-profit crops, which encourages outbreaks of insects or plant diseases and leads to reliance on large quantities of pesticides.	Alternating crops and planting a wider variety of crops reduce the likelihood of major outbreaks of insects and diseases.
Fossil fuel use	Uses large amounts of nonrenewable fossil fuels to run farm equipment, produce fertilizer, and apply fertilizers and pesticides.	No-till agriculture reduces the need for plowing and fertilizing.

(a) Cotton seedlings emerge in a no-till field in North Carolina   **(b) The same field one month later**

▲ **FIGURE 31-9 No-till agriculture (a)** A cover crop of wheat has been killed with an herbicide. Cotton seedlings thrive amid the dead wheat, which anchors soil and reduces evaporation. **(b)** Later in the season, the same field shows a healthy cotton crop mulched by the dead wheat.
Photos courtesy of Dr. George Naderman, Former Extension Soil Specialist (retired), College of Agriculture and Life Sciences, North Carolina State University, Raleigh, NC.

beans, peas, and lentils, organic and conventional farming produce the same yields. Organic food is usually sold at a higher price than nonorganic food, so the farmer may receive a higher income even if yields are lower. In addition, conventional farming has benefitted from decades of intensive research into crops that produce high yields under a high-fertilizer, and often high-pesticide, regime. Research into crop varieties that might produce higher yields in organic farming has barely begun; research into weed control in no-till fields without using herbicides is also in its infancy.

In the best-case scenario, farmers would grow a variety of crops, use agricultural practices that retain soil fertility, and use as little energy and as few potentially toxic chemicals as possible. Insect pests would be controlled by birds and predatory insects and by crop rotation, so that pests that specialize on particular crop plants would not find a feast laid out for them year after year. Fields would be relatively small, separated by strips of natural habitat for native plants and animals. Many projects, such as the University of California's Sustainable Agriculture Research and Education Program, support research and educate farmers about the advantages of sustainable agriculture and how to practice it.

Sustainable Development Balances Preservation of Natural Ecosystems with Providing Goods for People

With rare exceptions, most farms provide more wildlife habitat than cities do. To help keep rural land undeveloped, many states and counties offer conservation easements, whereby a landowner gives up the right to develop property, usually in return for some sort of tax credit. Conservation easements can be powerful tools for preserving natural habitat at low cost. As of 2014, more than 22 million acres of woods, farmland, and wildlife habitat in the United States have been preserved through conservation easements.

Although woods, fields, and farms typically provide more wildlife habitat, even cities and suburbs can provide homes for some types of wildlife. From apartment balconies to large suburban gardens, growing the right plants can provide nectar, pollen, berries, seeds, and nuts for birds, insects, and a variety of small mammals such as chipmunks and squirrels. The National Wildlife Federation offers free advice on how to make your home wildlife-friendly.

Providing for People and Wildlife: The Case of the Migrating Monarchs Monarch butterflies, probably the most recognizable butterfly in North America, provide a case study in preserving wildlife while providing for people's needs. Each fall, hundreds of millions of monarch butterflies in eastern North America migrate south to spend the winter in just a handful of forest groves in the mountains of central Mexico (**FIG. 31-10a**). Without these wintering sites, the entire monarch population east of the Rocky Mountains would vanish. Conditions in these stands of fir and pine trees are just right for the overwintering monarchs. A thick canopy of needles protects them from snow and rain. The groves are cool enough to slow down the butterflies' metabolism so they don't starve to death but are not so cold that they freeze.

However, the groves are owned not by the Mexican government but by the local people, most of whom are poor farmers. The stands of large trees that are essential to monarch survival have also traditionally been an important economic resource for the farmers, providing firewood and lumber. How can people provide for both butterflies and farmers? Several organizations are helping Mexican agricultural experts to train the farmers in sustainable agriculture. One of the most profitable "crops" in the reserve is, ironically, trees. The soil and climate provide ideal conditions

(a) Monarch butterflies at their wintering groves in Mexico

(b) Monarch butterfly feeding on milkweed

▲ **FIGURE 31-10 Preserving monarch butterflies (a)** In the winter, so many monarchs roost in a handful of forest groves in the mountains of central Mexico that their weight bends the branches of the trees. **(b)** Monarchs sip nectar from milkweed flowers and lay their eggs on the leaves, which are the only food eaten by monarch caterpillars.

for rapid tree growth. Some conifers mature in less than 20 years. Planting seedlings today allows some harvesting in only 5 years, for firewood and Christmas trees. After 15 years, the trees are large enough for commercial lumber. If, meanwhile, the trees are continually replanted, the cycle can continue indefinitely, and the old-growth groves can be left alone. Environmental and social organizations, in collaboration with the Mexican government, have planted more than 5 million trees in the past decade. The World Wildlife Fund has set up a $5,000,000 trust fund to help farmers find alternate sources of income rather than log their land.

The fund is also used to establish forested corridors connecting the groves where the monarchs overwinter. Meanwhile, ECOLIFE is building stoves that are 60% more fuel efficient for cooking and home heating than traditional open fires, which reduces deforestation, saves money, and reduces respiratory illness caused by indoor smoke from open fires.

Another source of income for the farmers is ecotourism. Tourists flock to the reserve each year to see the butterflies. If properly regulated, ecotourism can both preserve the forest and provide significant income opportunities for the local people, who serve as guides to the monarch groves and offer food, accommodations, and souvenirs for the tourists.

These efforts seem to have slowed, perhaps stopped, the loss of monarchs in recent years. Nevertheless, monarch populations are only about 20% as large as they were 15 or 20 years ago. One reason is the loss of summer habitat in the United States and Canada. Monarch caterpillars eat only milkweeds, which grow mainly in disturbed areas such as roadsides and pastures. Mowing roadsides, intensive farming, and especially the widespread use of herbicides have greatly reduced the milkweed population; one study estimates that milkweeds are about 60% less abundant in the Midwestern United States than they were 20 years ago. People in the United States and Canada can help by planting milkweeds (*Asclepias* species) in their gardens (**FIG. 31-10b**). Most have beautiful pink or orange flowers, with lots of nectar for both butterflies and hummingbirds. Several organizations provide seeds or seedlings and advice on planting. Environmental organizations and some states are planting milkweeds along roadsides and railroads. Iowa has restored 10,000 acres of roadsides to natural vegetation, including milkweeds. Such "milkweed highways" can provide vital habitat for monarchs both during the summer and during their autumn migration.

The Future of Earth Is in Your Hands

How should we manage our planet so that it provides a healthy, satisfying life for the current generation of people, while simultaneously retaining biodiversity and the resources needed for future generations? No one can give a simple, certain answer. However, three interrelated questions must be considered: (1) What should human lifestyles look like? (2) What technologies can support those lifestyles in a sustainable way? (3) How many people can Earth support and in what lifestyle?

Changes in Lifestyle and Use of Appropriate Technologies Are Essential

The billions of people on Earth will never all agree on exactly what is needed for a happy, fulfilling life. Nearly everyone would agree, however, that a minimal lifestyle should include adequate food and clothing, clean air and water, good health care and working conditions, educational and career opportunities, and access to natural environments. Most of Earth's people live in less-developed countries and lack at least some of these necessities.

Without a sustainable approach to development, there can be no long-term improvement in the quality of human life—in fact, it might even decline. We must make choices about which technologies are sustainable and how to make the transition from the realities of today to a hoped-for tomorrow.

Human Population Growth Is Unsustainable

The root causes of environmental degradation are simple: too many people using too many resources and generating too much waste. As the IUCN eloquently stated in *Who Will Care for the Earth?* ". . . the central issue [is] how to bring human populations into balance with the natural ecosystems that sustain them."

In the long run, that balance cannot be achieved if the human population continues to grow. Given the lifestyle to which the vast majority of people on Earth aspire, many are convinced that the balance cannot be maintained even with our current population, and yet we add 75 to 80 million people each year. No matter how simple our diets, how efficient our housing, how low-impact our farming techniques, or how much we reuse and recycle, continued population growth will eventually overwhelm our best efforts.

Let's return to our comparison of Earth's biocapacity and the human ecological footprint (**FIG. 31-11**). As you can see, the rapid increase in humanity's ecological footprint between 1961 and 2010 (red line) is roughly paralleled by our rapid population increase (blue line). The ecological footprint *per person* (green line), however, was only about 10% higher in 2010 than it was in 1961. In fact, the average person was using a bit less of Earth's biocapacity in 2010

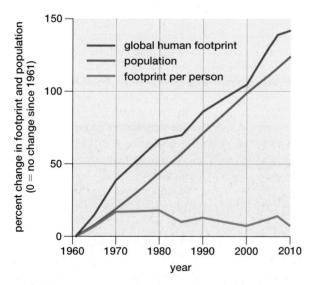

▲ **FIGURE 31-11 Human population growth threatens sustainability** Between 1961 and 2010, human population growth (blue line) increased at approximately the same rate as the global human ecological footprint (red line). The footprint per person (green line) has remained almost the same since 1970: The increase in our global footprint has resulted almost entirely from population growth. Data from the World Wildlife Fund, the Zoological Society of London, and the Global Footprint Network (2014), *The Living Planet Report.*

than in 1970. If the human population had not increased, the total human ecological footprint would still be well below Earth's biocapacity, but because there are so many more of us today, the total human footprint has climbed far above Earth's biocapacity. Eliminating, and probably reversing, population growth is essential if we wish to improve the quality of life for the 7 billion of us already here, provide the potential for a similar quality of life for our descendants, and save what is left of Earth's biodiversity for future generations.

The Choices Are Yours

It is all too easy to assume that "sustainable development" is solely the responsibility of industry, commercial farms, or governments, but we can all promote sustainable living by our individual actions. As Canadian educator and philosopher Marshall McLuhan noted 50 years ago, "There are no passengers on Spaceship Earth. We are all crew." Here are some ways to make a difference.

Conserve Energy

- *Heating and Cooling*   Don't heat your house over 68°F in winter or air condition it below 78°F in summer. Reduce the heating or cooling while you're away. When you purchase or remodel a home, consider energy-efficient features such as passive solar heating, good insulation, an attic fan, double-glazed windows (with "low-E" coating to reduce heat transfer), and tight weather stripping. Plant deciduous trees on the south side of your home for shade in summer (when the trees are covered with leaves) and sun in winter (after the leaves have dropped off). If possible, purchase renewable energy, usually wind or solar power, from your energy provider.
- *Lighting*   Use energy-efficient compact fluorescent or LED bulbs, which typically use about a quarter to a sixth as much energy as an incandescent bulb with the same light output and also last 8 to 30 times longer.
- *Hot Water*   Take shorter showers and switch to low-flow shower heads. Wash only full loads in your washing machine and dishwasher; use cold water to wash clothes; don't prewash your dishes. Insulate and turn down the temperature on your water heater.
- *Appliances*   When you choose a major appliance, look for the most energy-efficient models. Don't use your dryer in the summer—put up a clothesline.
- *Transportation*   Choose the most fuel-efficient car that meets your needs and use it efficiently by combining errands. Use public transit, carpool, walk, bicycle, or telecommute when possible.

Conserve Materials

- *Recycling*   Look into recycling options in your community and recycle everything that is accepted. Don't forget to recycle your cell phone when you replace it. Purchase recycled paper products. You can also buy decking and carpet made from recycled plastic bottles.

- *Reuse* Reuse anything possible, such as manila envelopes, file folders, and both sides of paper. Consider buying used furniture. Refill your water bottle. Reuse your grocery bags. Give away—rather than throw away—serviceable clothing, toys, and furniture. Make rags out of unusable old clothes and use them instead of disposable cleaning materials.
- *Conserve Water* If you live in a dry area, plant drought-resistant vegetation around your home to reduce water usage. A low-flow shower head not only saves energy, but also saves water.

Support Sustainable Practices

- *Food Choices* When possible, buy locally grown produce that does not require long-distance shipping. Look for coffee with the "Bird-Friendly™" or "Rainforest Alliance Certified" seal of approval, or other evidence of sustainable production such as the Starbucks "Coffee and Farmer Equity" program. Reduce your meat consumption. Search Internet sites such as the Seafood Watch of the Monterey Bay Aquarium to find out which fish at your local supermarket have been harvested sustainably.
- *Avoid Harmful Chemicals* Limit your use of harsh cleaners, insecticides, and herbicides that may contaminate water and soil.

Magnify Your Efforts

- *Support Organized Conservation Efforts* Join conservation groups and donate money to conservation efforts.

Sign up for e-mail alerts that educate you about environmental legislation and make it easy for you to contact your government representatives and express your views.
- *Volunteer* Volunteer for local campus and community projects that improve the environment.
- *Make Your Vote Count* Investigate candidates' stands and voting records on environmental issues and consider this information when choosing which candidate to support.
- *Educate* Through your words and actions, share your concern for sustainability with your family, friends, and community. Write letters to the editor of your school or local newspaper, to local businesses, and to elected officials. Talk with your campus administrators about reducing energy use on your campus. Recruit other concerned people and lobby for change.
- *Reduce Population Growth* Consider the consequences of human population growth when you plan your family. Adoption, for example, allows people to raise large families while simultaneously contributing to the welfare of humanity and the environment.

CHECK YOUR LEARNING
Can you ...
- describe the principles of sustainable development?
- explain how population, technology, and lifestyle choices interact to affect sustainability?

CASE STUDY \ **REVISITED**
The Wolves of Yellowstone

The simple trophic cascade from wolves to elk to aspen is far from the only impact of wolves on the Yellowstone ecosystem. Remember, real feeding relationships almost always form a web, not a chain (see Chapter 29). In the Yellowstone food web, wolves affect many trophic levels. For example, grizzly bears often scavenge on wolf-killed elk carcasses. Grizzlies also eat a lot of berries in the summer, and berry-producing bushes thrive without constant grazing by elk—again thanks to the wolves. Before wolf reintroduction, berries provided about 4% of the grizzlies' summer diet; now they provide about 10%.

Wolves also reduce coyote populations by chasing them away and sometimes killing them. Coyotes are major predators on pronghorn fawns, while wolves are not. In areas of the park with wolves, there are few coyotes, and four times as many pronghorn fawns survive as in areas without wolves. Wolves indirectly benefit birds, too. Crows and vultures scavenge on elk carcasses. Several types of songbirds are more abundant in aspen groves and willow thickets, many of which have grown only since wolves reduced elk populations.

According to the Green World Hypothesis, Earth is so green because top predators keep populations of herbivores in check, which in turn allows more plant growth. If this hypothesis is correct, then many ecosystems benefit from the presence of top predators. And they do. In the Pacific Ocean off the coast of

British Columbia, sea otters were hunted extensively for their fur. The decline in otters allowed an explosive increase in sea urchins, a favored prey of otters (**FIG. 31-12**). The urchins then virtually

▲ **FIGURE 31-12 A sea otter dines on sea urchins** Otters keep the urchin population in check and thereby promote the health of kelp forests.

eliminated kelp forests (see Chapter 28). When otter hunting was stopped and the otters returned, urchin populations declined drastically, and the kelp forests came back. In Venezuela, the construction of a hydroelectric dam isolated a number of hilltops, which became small islands in the resulting reservoir. Within just a few years, large predators vanished from the islands, herbivore numbers increased tremendously, and reproduction by canopy tree species almost ceased. As the research article describing these findings stated, there was "ecological meltdown in predator-free forest fragments."

CONSIDER THIS Before 1700, wolves roamed over almost all of North America; coyotes were found mostly west of the Mississippi. White-tailed deer lived throughout the eastern half of the continent. In the eastern United States, most of the land was covered with coniferous or deciduous forests. When Europeans arrived, they cleared most of the forests, planted crops, raised livestock, and exterminated the wolves. Based on what you have learned in this chapter and its Case Study, what do you think happened to the populations of coyotes and white-tailed deer between 1700 and now, in terms of their range and abundance, and to the types of vegetation in much of the eastern states (outside of cities, suburbs, and farms)? Why?

CHAPTER REVIEW

 Go to **MasteringBiology** for practice quizzes, activities, eText, videos, current events, and more.

Answers to Think Critically, Evaluate This, Multiple Choice, and Fill-in-the-Blank questions can be found in the Answers section at the back of the book.

Summary of Key Concepts

31.1 What Is Conservation Biology?

Conservation biology is the scientific discipline devoted to understanding and preserving Earth's biodiversity, including diversity at the genetic, species, and ecosystem levels.

31.2 Why Is Biodiversity Important?

Biodiversity provides provisioning, regulating, cultural, and supporting ecosystem services. Biodiversity is a source of goods, such as food, fuel, building materials, and medicines. Other ecosystem services include forming soil, purifying water, controlling floods, moderating climate, and providing genetic reserves and recreational opportunities. The emerging discipline of ecological economics attempts to measure the contribution of ecosystem goods and services to the economy and estimates the costs of losing them to unsustainable development.

31.3 Is Earth's Biodiversity Diminishing?

Natural communities have a low background extinction rate. Many biologists have concluded that human activities are currently causing a mass extinction, increasing extinction rates perhaps 1,000-fold.

31.4 What Are the Major Threats to Biodiversity?

The ecological footprint estimates the area of Earth required to support the human population at any given level of consumption and waste production. Biocapacity estimates the resources and waste-absorbing capacity actually available. The human footprint is already exceeding Earth's biocapacity, leaving less and less to support other forms of life. Major threats to biodiversity include habitat destruction and fragmentation, overexploitation, invasive species, pollution, and global climate change.

31.5 Why Is Habitat Protection Necessary to Preserve Biodiversity?

Conservation efforts include conserving wild ecosystems by establishing core reserves connected by wildlife corridors, which helps to preserve functional communities and self-sustaining wildlife populations.

31.6 Why Is Sustainability Essential for a Healthy Future?

Sustainable development meets present needs without compromising the future. Such development requires that people maintain biodiversity, recycle raw materials, and rely on renewable resources. A shift to sustainable farming is crucial for conserving soil and water, reducing pollution and energy use, and preserving biodiversity. Human population growth is unsustainable and is driving consumption of resources beyond nature's ability to replenish them.

Key Terms

biocapacity *588*
biodiversity *583*
conservation biology *584*
core reserve *593*
critically endangered species *588*
ecological economics *586*
ecological footprint *588*
ecosystem services *584*
endangered species *588*
habitat fragmentation *590*

keystone species *587*
mass extinction *588*
minimum viable population (MVP) *590*
no-till *596*
overexploitation *590*
sustainable development *594*
threatened species *588*
vulnerable species *588*
wildlife corridor *593*

Thinking Through the Concepts

Multiple Choice

1. A factor that increases humanity's ecological footprint is
   a. human population growth.
   b. technological innovations that improve the efficiency with which resources are used.

c. renewable energy sources.
d. sustainable agriculture.

2. A species that plays an essential role in an ecosystem, usually out of proportion to its population size, is called a
   a. predator.
   b. keystone species.
   c. rivet species.
   d. redundant species.

3. The modern rate of extinction is
   a. higher than the background rate.
   b. the same as the background rate.
   c. lower than the background rate.
   d. having little effect on biodiversity.

4. Biocapacity is
   a. the carrying capacity of an ecosystem.
   b. the area of Earth required to produce the resources used by humans.
   c. the number of species in an ecosystem.
   d. the amount of sustainable resources and waste-absorbing capacity of Earth.

5. Which of the following is *not* true of a population of large predators in a small reserve?
   a. The species may disappear from the reserve.
   b. The species will probably undergo a population explosion.
   c. The species will probably lose genetic diversity.
   d. The species may overeat its prey, causing a reduction in prey population.

Fill-in-the-Blank

1. Three levels of biodiversity are _____, _____, and _____. If the population of a species becomes too small, it is likely to have lost much of its _____ diversity.

2. Products or processes by which functioning ecosystems benefit humans are collectively called _____. Four important categories of these benefits are _____, _____, _____, and _____.

3. Many of the benefits that humans derive from functioning ecosystems, such as purifying water, have traditionally been considered to be free. The discipline of _____ tries to quantify the monetary value of these benefits.

4. The major threats to biodiversity include _____, _____, _____, _____, and _____. For most endangered species, _____ is probably the major threat.

5. The smallest population of a species that is likely to be able to survive in the long term is called the _____. When suitable habitat for a given species is split up into areas that are too small to support a large enough population, this is called _____. One way to maintain large enough populations is to set up core reserves of suitable habitat, connected by _____.

6. A Native American saying tells us that "We do not inherit the Earth from our ancestors, we borrow it from our children." If this principle guided our activities, we would practice _____ development.

Review Questions

1. What are the three different levels of biodiversity, and why is each one important?

2. What is ecological economics? Why is it important?

3. List the major categories of services that natural ecosystems provide, and give three examples of each.

4. What five specific threats to biodiversity are described in this chapter? Provide an example of each.

5. Why are efforts to protect monarch butterflies a good model for conservation and sustainable development?

Applying the Concepts

1. List some reasons that the ecological footprints of U.S. residents are very large. How could you reduce the size of your own footprint? How does the ecological footprint of U.S. residents extend into the Tropics?

2. Search for and describe some examples of habitat destruction, pollution, and invasive species in the region around your home or campus. Predict how each of these might affect specific local populations of native animals and plants.

APPENDIX I BIOLOGICAL VOCABULARY: COMMON ROOTS, PREFIXES, AND SUFFIXES

Biology has an extensive vocabulary often based on Greek or Latin rather than English words. Rather than memorizing every word as if it were part of a new, foreign language, you can figure out the meaning of many new terms if you learn a much smaller number of word roots, prefixes, and suffixes. We have provided common meanings in biology rather than literal translations from Greek or Latin. For each item in the list, the following information is given: meaning; part of word (prefix, suffix, or root); example from biology.

a–, an–: without, lack of (prefix); *abiotic*, without life

acro–: top, highest (prefix); *acrosome*, vesicle of enzymes at the tip of a sperm

ad–: to (prefix); *adhesion*, property of sticking to something else

allo–: other (prefix); *allopatric* (literally, "different fatherland"), restricted to different regions

amphi–: both, double, two (prefix); *amphibian*, a class of vertebrates that usually has two life stages (aquatic and terrestrial; e.g., a tadpole and an adult frog)

andro: man, male (root); *androgen*, a male hormone such as testosterone

ant–, anti–: against (prefix); *antibiotic* (literally "against life"), a substance that kills bacteria

antero–: front (prefix or root); *anterior*, toward the front of

aqu–, aqua–: water (prefix or root); *aquifer*, an underground water source, usually rock saturated with water

apic–: top, highest (prefix); *apical meristem*, the cluster of dividing cells at the tip of a plant shoot or root

arthr–: joint (prefix); *arthropod*, animals such as spiders, crabs, and insects, with exoskeletons that include jointed legs

–ase: enzyme (suffix); *protease*, an enzyme that digests protein

auto–: self (prefix); *autotrophic*, self-feeder (e.g., photosynthetic)

bi–: two (prefix); *bipedal*, having two legs

bio–: life (prefix or root); *biology*, the study of life

blast: bud, precursor (root); *blastula*, embryonic stage of development, a hollow ball of cells

bronch: windpipe (root); *bronchus*, a branch of the trachea (windpipe) leading to a lung

carcin, –o: cancer (root); *carcinogenesis*, the process of producing a cancer

cardi, –a–, –o–: heart (root); *cardiac*, referring to the heart

carn–, –i–, –o–: flesh (prefix or root); *carnivore*, an animal that eats other animals

centi–: one hundredth (prefix); *centimeter*, a unit of length, 1 one-hundredth of a meter

cephal–, –i–, –o–: head (prefix or root); *cephalization*, the tendency for the nervous system to be located principally in the head

chloro–: green (prefix or root); *chlorophyll*, the green, light-absorbing pigment in plants

chondr–: cartilage (prefix); *Chondrichthyes*, class of vertebrates including sharks and rays, with a skeleton made of cartilage

chrom–: color (prefix or root); *chromosome*, a threadlike strand of DNA and protein in the nucleus of a cell (*Chromosome* literally

means "colored body," because chromosomes absorb some of the colored dyes commonly used in microscopy.)

–cide: killer (suffix); *pesticide*, a chemical that kills "pests" (usually insects)

–clast: break down, broken (root or suffix); *osteoclast*, a cell that breaks down bone

co–: with or together with (prefix); *cohesion*, property of sticking together

coel–: hollow (prefix or root); *coelom*, the body cavity that separates the internal organs from the body wall

contra–: against (prefix); *contraception*, acting to prevent conception (pregnancy)

cortex: bark, outer layer (root); *cortex*, outer layer of kidney

crani–: skull (prefix or root); *cranium*, the skull

cuti–: skin (root); *cuticle*, the outermost covering of a leaf

–cyte, cyto–: cell (root or prefix); *cytokinin*, a plant hormone that promotes cell division

de–: from, out of, remove (prefix); *decomposer*, an organism that breaks down organic matter

dendr: treelike, branching (root); *dendrite*, highly branched input structures of nerve cells

derm: skin, layer (root); *ectoderm*, the outer embryonic germ layer of cells

deutero–: second (prefix); *deuterostome* (literally, "second opening"), an animal in which the coelom is derived from the gut

di–: two (prefix); *dicot*, an angiosperm with two cotyledons in the seed

diplo–: both, double, two (prefix or root); *diploid*, having paired homologous chromosomes

dys–: difficult, painful (prefix); *dysfunction*, an inability to function properly

eco–: house, household (prefix); *ecology*, the study of the relationships between organisms and their environment

ecto–: outside (prefix); *ectoderm*, the outermost tissue layer of animal embryos

–elle: little, small (suffix); *organelle* (literally, "little organ"), a subcellular structure that performs a specific function

end–, endo–, ento–: inside, inner (prefix); *endocrine*, pertaining to a gland that secretes hormones inside the body

epi–: outside, outer (prefix); *epidermis*, outermost layer of skin

equi–: equal (prefix); *equidistant*, the same distance

erythro–: red (prefix); *erythrocyte*, red blood cell

eu–: true, good (prefix); *eukaryotic*, pertaining to a cell with a true nucleus

ex–, exo–: out of (prefix); *exocrine*, pertaining to a gland that secretes a substance (e.g., sweat) outside of the body

extra–: outside of (prefix); *extracellular*, outside of a cell

–fer: to bear, to carry (suffix); *conifer*, a tree that bears cones

gastr–: stomach (prefix or root); *gastric*, pertaining to the stomach

–gen–: to produce (prefix, root, or suffix); *antigen*, a substance that causes the body to produce antibodies

glyc-, glyco-: sweet (prefix); *glycogen*, a starch-like molecule composed of many glucose molecules bonded together

gyn, –o: female (prefix or root); *gynecology*, the study of the female reproductive tract

haplo–: single (prefix); *haploid*, having a single copy of each type of chromosome

hem–, hemato–: blood (prefix or root); *hemoglobin*, the molecule in red blood cells that carries oxygen

hemi–: half (prefix); *hemisphere*, one of the halves of the cerebrum

herb–, herbi–: grass (prefix or root); *herbivore*, an animal that eats plants

hetero–: other (prefix); *heterotrophic*, an organism that feeds on other organisms

hom–, homo–, homeo–: same (prefix); *homeostasis*, to maintain constant internal conditions in the face of changing external conditions

hydro–: water (usually prefix); *hydrophilic*, being attracted to water

hyper–: above, greater than (prefix); *hyperosmotic*, having a greater osmotic strength (usually higher solute concentration)

hypo–: below, less than (prefix); *hypodermic*, below the skin

inter–: between (prefix); *interneuron*, a neuron that receives input from one (or more) neuron(s) and sends output to another neuron (or many neurons)

intra–: within (prefix); *intracellular*, pertaining to an event or substance that occurs within a cell

iso–: equal (prefix); *isotonic*, pertaining to a solution that has the same osmotic strength as another solution

–itis: inflammation (suffix); *hepatitis*, an inflammation (or infection) of the liver

kin–, kinet–: moving (prefix or root); *cytokinesis*, the movements of a cell that divide the cell in half during cell division

lac–, lact–: milk (prefix or root); *lactose*, the principal sugar in mammalian milk

leuc–, leuco–, leuk–, leuko–: white (prefix); *leukocyte*, a white blood cell

lip–: fat (prefix or root); *lipid*, the chemical category to which fats, oils, and steroids belong

–logy: study of (suffix); *biology*, the study of life

lyso–, –lysis: loosening, split apart (prefix, root, or suffix); *lysis*, to break open a cell

macro–: large (prefix); *macrophage*, a large white blood cell that destroys invading foreign cells

medulla: marrow, middle substance (root); *medulla*, inner layer of kidney

mega–: large (prefix); *megaspore*, a large, haploid (female) spore formed by meiotic cell division in plants

–mere: segment, body section (suffix); *sarcomere*, the functional unit of a vertebrate skeletal muscle cell

meso–: middle (prefix); *mesophyll*, middle layers of cells in a leaf

meta–: change, after (prefix); *metamorphosis*, to change body form (e.g., developing from a larva to an adult)

micro–: small (prefix); *microscope*, a device that allows one to see small objects

milli–: one-thousandth (prefix); *millimeter*, a unit of measurement of length; 1 one-thousandth of a meter

mito–: thread (prefix); *mitosis*, cell division (in which chromosomes appear as threadlike bodies)

mono–: single (prefix); *monocot*, a type of angiosperm with one cotyledon in the seed

morph–: shape, form (prefix or root); *polymorphic*, having multiple forms

multi–: many (prefix); *multicellular*, pertaining to a body composed of more than one cell

myo–: muscle (prefix); *myofibril*, protein strands in muscle cells

neo–: new (prefix); *neonatal*, relating to or affecting a newborn child

neph–: kidney (prefix or root); *nephron*, functional unit of mammalian kidney

neur–, neuro–: nerve (prefix or root); *neuron*, a nerve cell

neutr–: of neither gender or type (usually root); *neutron*, an uncharged subatomic particle found in the nucleus of an atom

non–: not (prefix); *nondisjunction*, the failure of chromosomes to distribute themselves properly during cell division

oligo–: few (prefix); *oligomer*, a molecule made up of a few subunits (see also *poly*–)

omni–: all (prefix); *omnivore*, an animal that eats both plants and animals

oo–, ov–, ovo–: egg (prefix); *oocyte*, one of the stages of egg development

opsi–: sight (prefix or root); *opsin*, protein part of light-absorbing pigment in eye

opso–: tasty food (prefix or root); *opsonization*, process whereby antibodies and/or complement render bacteria easier for white blood cells to engulf

–osis: a condition, disease (suffix); *atherosclerosis*, a disease in which the artery walls become thickened and hardened

oss–, osteo–: bone (prefix or root); *osteoporosis*; a disease in which the bones become spongy and weak

para–: alongside (prefix); *parathyroid*, referring to a gland located next to the thyroid gland

pater, patr–: father (usually root); *paternal*, from or relating to a father

path–, –i–, –o–: disease (prefix or root); *pathology*, the study of disease and diseased tissue

–pathy: disease (suffix); *neuropathy*, a disease of the nervous system

peri–: around (prefix); *pericycle*, the outermost layer of cells in the vascular cylinder of a plant root

phago–: eat (prefix or root); *phagocyte*, a cell (e.g., some types of white blood cell) that eats other cells

–phil, philo–: to love (prefix or suffix); *hydrophilic* (literally, "water loving"), pertaining to a water-soluble molecule

–phob, phobo–: to fear (prefix or suffix); *hydrophobic* (literally, "water fearing"), pertaining to a water-insoluble molecule

photo–: light (prefix); *photosynthesis*, the manufacture of organic molecules using the energy of sunlight

–phyll: leaf (root or suffix); *chlorophyll*, the green, light-absorbing pigment in a leaf

–phyte: plant (root or suffix); *gametophyte* (literally, "gamete plant"), the gamete-producing stage of a plant's life cycle

plasmo, –plasm: formed substance (prefix, root, or suffix); *cytoplasm*, the material inside a cell

ploid: chromosomes (root); *diploid*, having paired chromosomes

pneumo–: lung (root); *pneumonia*, a disease of the lungs

–pod: foot (root or suffix); *gastropod* (literally, "stomach-foot"), a class of mollusks, principally snails, that crawl on their ventral surfaces

poly–: many (prefix); *polysaccharide*, a carbohydrate polymer composed of many sugar subunits (see also *oligo–*)

post–, postero–: behind (prefix); *posterior*, pertaining to the hind part

pre–, pro–: before, in front of (prefix); *premating isolating mechanism*, a mechanism that prevents gene flow between species, acting to prevent mating (e.g., having different courtship rituals or different mating seasons)

prim–: first (prefix); *primary cell wall*, the first cell wall laid down between plant cells during cell division

pro–: before (prefix); *prokaryotic*, pertaining to a cell without (that evolved before the evolution of) a nucleus

proto–: first (prefix); *protocell*, a hypothetical evolutionary ancestor to the first cell

pseudo–: false (prefix); *pseudopod* (literally, "false foot"), the extension of the plasma membrane by which some cells, such as *Amoeba*, move and capture prey

quad–, quat–: four (prefix); *quaternary structure*, the "fourth level" of protein structure, in which multiple peptide chains form a complex three-dimensional structure

ren: kidney (root); *adrenal*, gland attached to the mammalian kidney

retro–: backward (prefix); *retrovirus*, a virus that uses RNA as its genetic material; this RNA must be copied "backward" to DNA during infection of a cell by the virus

sarco–: muscle (prefix); *sarcoplasmic reticulum*, a calcium-storing, modified endoplasmic reticulum found in muscle cells

scler–: hard, tough (prefix); *sclerenchyma*, a type of plant cell with a very thick, hard cell wall

semi–: one-half (prefix); *semiconservative replication*, the mechanism of DNA replication, in which one strand of the original DNA double helix becomes incorporated into the new DNA double helix

–some, soma–, somato–: body (prefix or suffix); *somatic nervous system*, part of the peripheral nervous system that controls the skeletal muscles that move the body

sperm, sperma–, spermato–: seed (usually root); *gymnosperm*, a type of plant producing a seed not enclosed within a fruit

–stasis, stat–: stationary, standing still (suffix or prefix); *homeostasis*, the physiological process of maintaining constant internal conditions despite a changing external environment

stoma, –to–: mouth, opening (prefix or root); *stoma*, the adjustable pore in the surface of a leaf that allows carbon dioxide to enter the leaf

sub–: under, below (prefix); *subcutaneous*, beneath the skin

sym–: same (prefix); *sympatric* (literally "same father"), found in the same region

tel–, telo–: end (prefix); *telophase*, the last stage of mitosis and meiosis

test–: witness (prefix or root); *testis*, male reproductive organ (derived from the custom in ancient Rome that only males had standing in the eyes of the law; *testimony* has the same derivation)

therm–: heat (prefix or root); *thermoregulation*, the process of regulating body temperature

trans–: across (prefix); *transgenic*, having genes from another organism (usually another species); the genes have been moved "across" species

tri–: three (prefix); *triploid*, having three copies of each homologous chromosome

–trop–, tropic: change, turn, move (suffix); *phototropism*, the process by which plants orient toward the light

troph: food, nourishment (root); *autotrophic*, self-feeder (e.g., photosynthetic)

ultra–: beyond (prefix); *ultraviolet*, light of wavelengths beyond the violet

uni–: one (prefix); *unicellular*, referring to an organism composed of a single cell

vita: life (root); *vitamin*, a molecule required in the diet to sustain life

–vor: eat (usually root); *herbivore*, an animal that eats plants

zoo–, zoa–: animal (usually root); *zoology*, the study of animals

APPENDIX II PERIODIC TABLE OF THE ELEMENTS

atomic number (number of protons)

element (chemical symbol)

atomic mass (total mass of protons + neutrons + electrons)

1	2	3	4	5	6	7	8	9	10	11	12	13	14	15	16	17	18
1 **H** 1.008																	2 **He** 4.003
3 **Li** 6.941	4 **Be** 9.012											5 **B** 10.81	6 **C** 12.01	7 **N** 14.01	8 **O** 16.00	9 **F** 19.00	10 **Ne** 20.18
11 **Na** 22.99	12 **Mg** 24.31											13 **Al** 26.98	14 **Si** 28.09	15 **P** 30.97	16 **S** 32.07	17 **Cl** 35.45	18 **Ar** 39.95
19 **K** 39.10	20 **Ca** 40.08	21 **Sc** 44.96	22 **Ti** 47.87	23 **V** 50.94	24 **Cr** 52.00	25 **Mn** 54.94	26 **Fe** 55.85	27 **Co** 58.93	28 **Ni** 58.69	29 **Cu** 63.55	30 **Zn** 65.39	31 **Ga** 69.72	32 **Ge** 72.61	33 **As** 74.92	34 **Se** 78.96	35 **Br** 79.90	36 **Kr** 83.80
37 **Rb** 85.47	38 **Sr** 87.62	39 **Y** 88.91	40 **Zr** 91.22	41 **Nb** 92.91	42 **Mo** 95.94	43 **Tc** (98)	44 **Ru** 101.1	45 **Rh** 102.9	46 **Pd** 106.4	47 **Ag** 107.9	48 **Cd** 112.4	49 **In** 114.8	50 **Sn** 118.7	51 **Sb** 121.8	52 **Te** 127.6	53 **I** 126.9	54 **Xe** 131.3
55 **Cs** 132.9	56 **Ba** 137.3	57 ***La** 138.9	72 **Hf** 178.5	73 **Ta** 180.9	74 **W** 183.8	75 **Re** 186.2	76 **Os** 190.2	77 **Ir** 192.2	78 **Pt** 195.1	79 **Au** 197.0	80 **Hg** 200.6	81 **Tl** 204.4	82 **Pb** 207.2	83 **Bi** 209.0	84 **Po** (209)	85 **At** (210)	86 **Rn** (222)
87 **Fr** (223)	88 **Ra** (226)	89 **†Ac** (227)	104 **Rf** (261)	105 **Db** (262)	106 **Sg** (263)	107 **Bh** (264)	108 **Hs** (265)	109 **Mt** (268)	110 **Ds** (281)	111 **Rg** (280)	112 ** ** ** (277)	113 ** ** **	114 ** ** ** (285)	115 ** ** **			

*Lanthanide series

58 **Ce** 140.1	59 **Pr** 140.9	60 **Nd** 144.2	61 **Pm** (145)	62 **Sm** 150.4	63 **Eu** 152.0	64 **Gd** 157.3	65 **Tb** 158.9	66 **Dy** 162.5	67 **Ho** 164.9	68 **Er** 167.3	69 **Tm** 168.9	70 **Yb** 173.0	71 **Lu** 175.0

†Actinide series

90 **Th** 232.0	91 **Pa** 231	92 **U** 238.0	93 **Np** (237)	94 **Pu** (244)	95 **Am** (243)	96 **Cm** (247)	97 **Bk** (247)	98 **Cf** (251)	99 **Es** (252)	100 **Fm** (257)	101 **Md** (258)	102 **No** (259)	103 **Lr** (262)

The periodic table of the elements was first devised by Russian chemist Dmitri Mendeleev. The atomic numbers of the elements (the numbers of protons in the nucleus) increase in normal reading order: left to right, top to bottom. The table is "periodic" because all the elements in a column possess similar chemical properties, and such similar elements therefore recur "periodically" in each row. For example, elements that usually form ions with a single positive charge—including H, Li, Na, K, and so on—occur as the first element in each row.

The gaps in the table are a consequence of the maximum numbers of electrons in the most reactive, usually outermost, electron shells of the atoms. It takes only two electrons to completely fill the first shell, so the first row of the table contains only two elements, H and He. It takes eight electrons to fill the second and third shells, so there are eight elements in the second and third rows. It takes 18 electrons to fill the fourth and fifth shells, so there are 18 elements in these rows.

The lanthanide series and actinide series of elements are usually placed below the main body of the table for convenience—it takes 32 electrons to completely fill the sixth and seventh shells, so the table would become extremely wide if all of these elements were included in the sixth and seventh rows.

The important elements found in living things are highlighted in color. The elements in pale red are the six most abundant elements in living things. The elements that form the five most abundant ions in living things are in purple. The trace elements important to life are shown as dark blue (more common) and lighter blue (less common). For radioactive elements, the atomic masses are given in parentheses, and represent the most common or the most stable isotope. Elements indicated as double asterisks have not yet been named.

APPENDIX III METRIC SYSTEM CONVERSIONS

To Convert Metric Units:	Multiply by:	To Get English Equivalent:
Length		
Centimeters (cm)	0.394	Inches (in)
Meters (m)	3.281	Feet (ft)
Meters (m)	1.094	Yards (yd)
Kilometers (km)	0.621	Miles (mi)
Area		
Square centimeters (cm²)	0.155	Square inches (in²)
Square meters (m²)	10.764	Square feet (ft²)
Square meters (m²)	1.196	Square yards (yd²)
Square kilometers (km²)	0.386	Square miles (mi²)
Hectare (ha) (10,000 m²)	2.471	Acres (a)
Volume		
Cubic centimeters (cm³)	0.0610	Cubic inches (in³)
Cubic meters (m³)	35.315	Cubic feet (ft³)
Cubic meters (m³)	1.308	Cubic yards (yd³)
Cubic kilometers (km³)	0.240	Cubic miles (mi³)
Liters (L)	1.057	Quarts (qt), U.S.
Liters (L)	0.264	Gallons (gal), U.S.
Mass		
Grams (g)	0.0353	Ounces (oz)
Kilograms (kg)	2.205	Pounds (lb)
Metric ton (tonne) (t)	1.102	Ton (tn), U.S.
Speed		
Meters/second (mps)	2.237	Miles/hour (mph)
Kilometers/hour (kmph)	0.621	Miles/hour (mph)

To Convert English Units:	Multiply by:	To Get Metric Equivalent:
Length		
Inches (in)	2.540	Centimeters (cm)
Feet (ft)	0.305	Meters (m)
Yards (yd)	0.914	Meters (m)
Miles (mi)	1.609	Kilometers (km)
Area		
Square inches (in²)	6.452	Square centimeters (cm²)
Square feet (ft²)	0.0929	Square meters (m²)
Square yards (yd²)	0.836	Square meters (m²)
Square miles (mi²)	2.590	Square kilometers (km²)
Acres (a)	0.405	Hectare (ha) (10,000 m²)
Volume		
Cubic inches (in³)	16.387	Cubic centimeters (cm³)
Cubic feet (ft³)	0.0283	Cubic meters (m³)
Cubic yards (yd³)	0.765	Cubic meters (m³)
Cubic miles (mi³)	4.168	Cubic kilometers (km³)
Quarts (qt), U.S.	0.946	Liters (L)
Gallons (gal), U.S.	3.785	Liters (L)
Mass		
Ounces (oz)	28.350	Grams (g)
Pounds (lb)	0.454	Kilograms (kg)
Ton (tn), U.S.	0.907	Metric ton (tonne) (t)
Speed		
Miles/hour (mph)	0.447	Meters/second (mps)
Miles/hour (mph)	1.609	Kilometers/hour (kmph)

Metric Prefixes

Prefix		Meaning	
giga-	G	$10^9 =$	1,000,000,000
mega-	M	$10^6 =$	1,000,000
kilo-	k	$10^3 =$	1,000
hecto-	h	$10^2 =$	100
deka-	da	$10^1 =$	10
		$10^0 =$	1
deci-	d	$10^{-1} =$	0.1
centi-	c	$10^{-2} =$	0.01
milli-	m	$10^{-3} =$	0.001
micro-	μ	$10^{-6} =$	0.000001

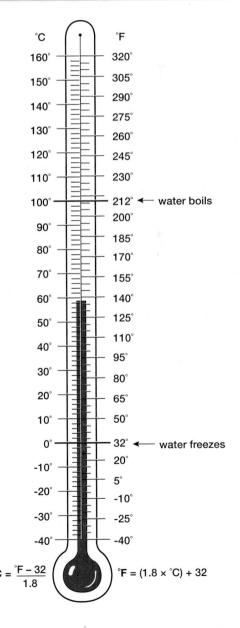

$$°C = \frac{°F - 32}{1.8} \qquad °F = (1.8 \times °C) + 32$$

APPENDIX IV CLASSIFICATION OF MAJOR GROUPS OF EUKARYOTIC ORGANISMS*

Kingdom†	Phylum or Class	Common Name
Excavata	Parabasalia	parabasalids
	Diplomonadida	diplomonads
Euglenozoa	Euglenida	euglenids
	Kinetoplastida	kinetoplastids
Stramenopila	Oomycota	water molds
	Phaeophyta	brown algae
	Bacillariophyta	diatoms
Alveolata	Apicomplexa	sporozoans
	Pyrrophyta	dinoflagellates
	Ciliophora	ciliates
Rhizaria	Foraminifera	foraminiferans
	Radiolaria	radiolarians
Amoebozoa	Tubulinea	amoebas
	Myxomycota	acellular slime molds
	Acrasiomycota	cellular slime molds
Rhodophyta		red algae
Chlorophyta		green algae
Plantae	Marchantiophyta	liverworts
	Anthocerotophyta	hornworts
	Bryophyta	mosses
	Lycopodiopsida	club mosses
	Polypodiopsida	ferns, horsetails
	Gymnospermae	cycads, ginkgos, gnetophytes, conifers
	Anthophyta	flowering plants
Fungi	Chytridiomycota	chytrids
	Neocallimastigomycota	rumen fungi
	Blastocladiomycota	blastoclades
	Glomeromycota	glomeromycetes
	Ascomycota	sac fungi
	Basidiomycota	club fungi
Animalia	Porifera	sponges
	Cnidaria	hydras, sea anemones, sea jellies, corals
	Ctenophora	comb jellies
	Platyhelminthes	flatworms
	Annelida	segmented worms
	Oligochaeta	earthworms
	Polychaeta	tube worms
	Hirudinea	leeches
	Mollusca	mollusks
	Gastropoda	snails
	Pelecypoda	mussels, clams
	Cephalopoda	squid, octopuses
	Nematoda	roundworms
	Arthropoda	arthropods
	Insecta	insects
	Arachnida	spiders, ticks
	Crustacea	crabs, lobsters
	Myriapoda	millipedes, centipedes
	Chordata	chordates
	Tunicata	tunicates
	Cephalochordata	lancelets
	Myxini	hagfishes
	Petromyzontiformes	lampreys
	Chondrichthyes	sharks, rays
	Actinopterygii	ray-finned fishes
	Actinistia	coelacanths
	Dipnoi	lungfishes
	Amphibia	amphibians (frogs, salamanders)
	Reptilia	reptiles (turtles, crocodiles, birds, snakes, lizards)
	Mammalia	mammals

*This table lists only those taxonomic categories described in the textbook.

†Although the major protist groups are not generally called "kingdoms," they are approximately the same taxonomic rank as the kingdoms Plantae, Fungi, and Animalia.

GLOSSARY

abiotic (ā-bī-ah´-tik): nonliving; the abiotic portion of an ecosystem includes soil, rock, water, and the atmosphere.

accessory pigment: a colored molecule, other than chlorophyll *a*, that absorbs light energy and passes it to chlorophyll *a*.

acellular slime mold: a type of organism that forms a multinucleate structure that crawls in amoeboid fashion and ingests decaying organic matter; also called *plasmodial slime mold*. Acellular slime molds are members of the protist clade Amoebozoa.

acid: a substance that releases hydrogen ions (H⁺) into solution; a solution with a pH less than 7.

acid deposition: the deposition of nitric or sulfuric acid, either in rain (acid rain) or in the form of dry particles, as a result of the production of nitrogen oxides or sulfur dioxide through burning, primarily of fossil fuels.

acidic: referring to a solution with an H⁺ concentration exceeding that of OH⁻; referring to a substance that releases H⁺.

activation energy: in a chemical reaction, the energy needed to force the electron shells of reactants together, prior to the formation of products.

active site: the region of an enzyme molecule that binds substrates and performs the catalytic function of the enzyme.

active transport: the movement of materials across a membrane through the use of cellular energy, normally against a concentration gradient.

adaptation: a trait that increases the ability of an individual to survive and reproduce compared to individuals without the trait.

adaptive radiation: the rise of many new species in a relatively short time; may occur when a single species invades different habitats and evolves in response to different environmental conditions in those habitats.

adenine (A): a nitrogenous base found in both DNA and RNA; abbreviated as A.

adenosine diphosphate (a-den´-ō-sēn-dī-fos´-fāt; ADP): a molecule composed of the sugar ribose, the base adenine, and two phosphate groups; a component of ATP.

adenosine triphosphate (a-den´-ō-sēn-trī ĭ fos´-fāt; ATP): a molecule composed of the sugar ribose, the base adenine, and three phosphate groups; the major energy carrier in cells. The last two phosphate groups are attached by "high-energy" bonds.

adhesion: the tendency of polar molecules (such as water) to adhere to polar surfaces (such as glass).

adhesive junctions: attachment structures that link cells to one another within tissues.

aerobic: using oxygen.

age structure diagram: a graph showing the distribution of males and females in a population according to age groups.

aggression: antagonistic behavior, normally among members of the same species, that often results from competition for resources.

aggressive mimicry (mim´ik-rē): the evolution of a predatory organism to resemble a harmless animal or a part of the environment, thus gaining access to prey.

albinism: a recessive hereditary condition caused by defective alleles of the genes that encode the enzymes required for the synthesis of melanin, the principal pigment in mammalian skin and hair; albinism results in white hair and pink skin.

alcoholic fermentation: a type of fermentation in which pyruvate is converted to ethanol (a type of alcohol) and carbon dioxide, using hydrogen ions and electrons from NADH; the primary function of alcoholic fermentation is to regenerate NAD⁺ so that glycolysis can continue under anaerobic conditions.

alga (al´-ga; pl., algae, al´-jē): any photosynthetic protist.

allele (al-ēl´): one of several alternative forms of a particular gene.

allele frequency: for any given gene, the relative proportion of each allele of that gene in a population.

allopatric speciation (al-ō-pat´-rik): the process by which new species arise following physical separation of parts of a population (geographical isolation).

allosteric regulation: the process by which enzyme action is enhanced or inhibited by small organic molecules that act as regulators by binding to the enzyme at a regulatory site distinct from the active site and altering the shape and/or function of the active site.

alternation of generations: a life cycle, typical of plants, in which a diploid sporophyte (spore-producing) generation alternates with a haploid gametophyte (gamete-producing) generation.

altruism: a behavior that benefits other individuals while reducing the fitness of the individual that performs the behavior.

alveolate (al-vē´-ō-lāt): a member of the Alveolata, a large protist clade. The alveolates, which are characterized by a system of sacs beneath the cell membrane, include ciliates, dinoflagellates, and apicomplexans.

amino acid: the individual subunit of which proteins are made, composed of a central carbon atom bonded to an amino group (–NH₂), a carboxyl group (–COOH), a hydrogen atom, and a variable group of atoms denoted by the letter *R*.

amniocentesis (am-nē-ō-sen-tē´-sis): a procedure for sampling the amniotic fluid surrounding a fetus: A sterile needle is inserted through the abdominal wall, uterus, and amniotic sac of a pregnant woman, and 10 to 20 milliliters of amniotic fluid are withdrawn. Various tests may be performed on the fluid and the fetal cells suspended in it to provide information on the developmental and genetic state of the fetus.

amnion (am´-nē-on): one of the embryonic membranes of reptiles (including birds) and mammals; encloses a fluid-filled cavity that envelops the embryo.

amniotic egg (am-nē-ōt´-ik): the egg of reptiles, including birds; contains a membrane, the amnion, that surrounds the embryo, enclosing it in a watery environment and allowing the egg to be laid on dry land.

amoeba: an amoebozoan protist that uses a characteristic streaming mode of locomotion by extending a cellular projection called a *pseudopod*. Also known as a *lobose amoeba*.

amoeboid cell: a protist or animal cell that moves by extending a cellular projection called a pseudopod.

amoebozoan: a member of the Amoebozoa, a protist clade. The amoebozoans, which generally lack shells and move by extending pseudopods, include the lobose amoebas and the slime molds.

amphibian: a member of the chordate clade Amphibia, which includes the frogs, toads, and salamanders, as well as the limbless caecilians.

anaerobe (an-ə-rōb): an organism that can live and grow in the absence of oxygen.

anaerobic: not using oxygen.

analogous structure: structures that have similar functions and superficially similar appearance but very different anatomies, such as the wings of insects and birds. The similarities are the result of similar environmental pressures rather than a common ancestry.

anaphase (an´-a-fāz): in mitosis, the stage in which the sister chromatids of each chromosome separate from one another and are moved to opposite poles of the cell; in meiosis I, the stage in which homologous chromosomes, consisting of two sister chromatids, are separated; in meiosis II, the stage in which the sister chromatids of each chromosome separate from one another and are moved to opposite poles of the cell.

angiosperm (an´-jē-ō-sperm): a flowering vascular plant.

antheridium (an-ther-id'-ē-um; pl., antheridia): a structure in which male sex cells are produced; found in nonvascular plants and certain seedless vascular plants.

anticodon: a sequence of three bases in transfer RNA that is complementary to the three bases of a codon of messenger RNA.

antidiuretic hormone (an-tē-dī-ūr-et'-ik; ADH): a hormone produced by the hypothalamus and released into the bloodstream by the posterior pituitary when blood volume is low; increases the permeability of the distal tubule and the collecting duct to water, allowing more water to be reabsorbed into the bloodstream.

antioxidant: any molecule that reacts with free radicals, neutralizing their ability to damage biological molecules. Vitamins C and E are examples of dietary antioxidants.

aphotic zone: the region of the ocean below 200 meters where sunlight does not penetrate.

apicomplexan (ā-pē-kom-pleks'-an): a member of the protist clade Apicomplexa, which includes mostly parasitic, single-celled eukaryotes such as *Plasmodium*, which causes malaria in humans. Apicomplexans are part of a larger group known as the alveolates.

aquaporin: a channel protein in the plasma membrane of a cell that is selectively permeable to water.

aquifer (ok'-wifer): an underground deposit of fresh water, often used as a source of water for irrigation.

archaea: prokaryotes that are members of the domain Archaea, one of the three domains of living organisms; only distantly related to members of the domain Bacteria.

Archaea: one of life's three domains; consists of prokaryotes that are only distantly related to members of the domain Bacteria.

archegonium (ar-ke-gō'-nē-um; pl., archegonia): a structure in which female sex cells are produced; found in nonvascular plants and certain seedless vascular plants.

arthropod: a member of the animal phylum Arthropoda, which includes the insects, spiders, ticks, mites, scorpions, crustaceans, millipedes, and centipedes.

artificial selection: a selective breeding procedure in which only those individuals with particular traits are chosen as breeders; used mainly to enhance desirable traits in domesticated plants and animals; may also be used in evolutionary biology experiments.

ascomycete: a member of the fungus clade Ascomycota, whose members form sexual spores in a saclike case known as an ascus.

ascus (as'-kus; pl., asci): a saclike case in which sexual spores are formed by members of the fungus clade Ascomycota.

asexual reproduction: reproduction that does not involve the fusion of haploid gametes.

atom: the smallest unit of an element that retains the properties of the element.

atomic mass: the total mass of all the protons, neutrons, and electrons within an atom.

atomic nucleus: the central part of an atom that contains protons and neutrons.

atomic number: the number of protons in the nuclei of all atoms of a particular element.

ATP synthase: a channel protein in the thylakoid membranes of chloroplasts and the inner membrane of mitochondria that uses the energy of H^+ ions moving through the channel down their concentration gradient to produce ATP from ADP and inorganic phosphate.

autosome (aw'-tō-sōm): a chromosome that occurs in homologous pairs in both males and females and that does not bear the genes determining sex.

autotroph (aw'-tō-trōf): literally, "self-feeder"; normally, a photosynthetic organism; a producer.

bacteria (sing., bacterium): prokaryotes that are members of the domain Bacteria, one of the three domains of living organisms; only distantly related to members of the domain Archaea.

Bacteria: one of life's three domains; consists of prokaryotes that are only distantly related to members of the domain Archaea.

bacteriophage (bak-tir'-ē-ō-fāj): a virus that specifically infects bacteria.

Barr body: a condensed, inactivated X chromosome in the cells of female mammals that have two X chromosomes.

basal body: a structure derived from a centriole that produces a cilium or flagellum and anchors this structure within the plasma membrane.

base: (1) a substance capable of combining with and neutralizing H^+ ions in a solution; a solution with a pH greater than 7; (2) one of the nitrogen-containing, single- or double-ringed structures that distinguishes one nucleotide from another. In DNA, the bases are adenine, guanine, cytosine, and thymine.

basic: referring to a solution with an H^+ concentration less than that of OH^-; referring to a substance that combines with H^+.

basidiomycete: a member of the fungus clade Basidiomycota, which includes species that produce sexual spores in club-shaped cells known as basidia.

basidiospore (ba-sid'-ē-ō-spor): a sexual spore formed by members of the fungus clade Basidiomycota.

basidium (pl., basidia): a diploid cell, typically club-shaped, formed by members of the fungus clade Basidiomycota; produces basidiospores by meiosis.

behavior: any observable activity of a living animal.

behavioral isolation: reproductive isolation that arises when species do not interbreed because they have different courtship and mating rituals.

bilateral symmetry: a body plan in which only a single plane through the central axis will divide the body into mirror-image halves.

binomial system: the method of naming organisms by genus and species, often called the scientific name, usually using Latin or Greek words or words derived from Latin or Greek.

biocapacity: an estimate of the sustainable resources and waste-absorbing capacity actually available on Earth. Biocapacity calculations are subject to change as new technologies change the way people use resources.

biodegradable: able to be broken down into harmless substances by decomposers.

biodiversity: the diversity of living organisms; measured as the variety of different species, the variety of different alleles in species' gene pools, or the variety of different communities and nonliving environments in an ecosystem or in the entire biosphere.

biofilm: a community of prokaryotes of one or more species, in which the prokaryotes secrete and are embedded in slime that adheres to a surface.

biogeochemical cycle: the pathways of a specific nutrient (such as carbon, nitrogen, phosphorus, or water) through the living and nonliving portions of an ecosystem; also called a *nutrient cycle*.

biological magnification: the increasing accumulation of a toxic substance in progressively higher trophic levels.

biological molecules: all molecules produced by living things.

biology: the study of all aspects of life and living things.

biomass: the total weight of all living material within a defined area.

biome (bī'-ōm): a terrestrial ecosystem that occupies an extensive geographical area and is characterized by a specific type of plant community; for example, deserts.

bioremediation: the use of organisms to remove or detoxify toxic substances in the environment.

biosphere (bī'-ō-sfēr): all life on Earth and the nonliving portions of Earth that support life.

biotechnology: any industrial or commercial use or alteration of organisms, cells, or biological molecules to achieve specific practical goals.

biotic (bī-ah'-tik): living.

biotic potential: the maximum rate at which a population is able to increase, assuming ideal conditions that allow a maximum birth rate and minimum death rate.

birth rate: the number of births per individual in a specified unit of time, such as a year.

blastoclade: a member of the fungus clade Blastocladiomycota, whose members have swimming spores with a single flagellum and ribosomes arranged to form a nuclear cap.

boom-and-bust cycle: a population cycle characterized by rapid exponential growth followed by a sudden massive die-off; seen in seasonal species, such as many insects living in temperate climates, and in some populations of small rodents, such as lemmings.

budding: asexual reproduction by the growth of a miniature copy, or bud, of the adult animal on the body of the parent. The bud breaks off to begin independent existence.

buffer: a compound that minimizes changes in pH by reversibly taking up or releasing H^+ ions.

bundle sheath cells: cells that surround the veins of plants; in C_4 (but not in C_3) plants, bundle sheath cells contain chloroplasts.

C_3 pathway: in photosynthesis, the cyclic series of reactions whereby carbon from carbon dioxide is fixed as phosphoglyceric acid, the simple sugar glyceraldehyde-3-phosphate is generated, and the carbon-capture molecule, RuBP, is regenerated. Also called the *Calvin cycle.*

C_3 plant: a plant that relies on the C_3 pathway to fix carbon.

C_4 pathway: the series of reactions in certain plants that fixes carbon dioxide into a four-carbon molecule, which is later broken down for use in the Calvin cycle of photosynthesis. This reduces wasteful photorespiration in hot, dry environments.

C_4 plant: a plant that relies on the C_4 pathway to fix carbon.

Calvin cycle: in photosynthesis, the cyclic series of reactions whereby carbon from carbon dioxide is fixed as phosphoglyceric acid, the simple sugar glyceraldehyde-3-phosphate is generated, and the carbon-capture molecule, RuBP, is regenerated. Also called the *C_3 pathway.*

cambium (kam′-bē-um; pl., cambia): a lateral meristem, parallel to the long axis of roots and stems, that causes secondary growth of woody plant stems and roots. See also *cork cambium; vascular cambium.*

camouflage (cam′-a-flaj): coloration and/or shape that renders an organism inconspicuous in its environment.

capillary action: the movement of water within narrow spaces resulting from its properties of adhesion and cohesion.

carbohydrate: a compound composed of carbon, hydrogen, and oxygen, with the approximate chemical formula $(CH_2O)_n$; includes sugars, starches, and cellulose.

carbon cycle: the biogeochemical cycle by which carbon moves from its reservoirs in the atmosphere and oceans through producers and into higher trophic levels, and then back to its reservoirs.

carbon fixation: the process by which carbon derived from carbon dioxide is captured in organic molecules during photosynthesis.

carnivore (kar′-neh-vor): literally, "meat-eater"; a predatory organism that feeds on herbivores or on other carnivores; a secondary (or higher) consumer.

carotenoid (ka-rot′-en-oid): a red, orange, or yellow pigment, found in chloroplasts, that serves as an accessory light-gathering pigment in thylakoid photosystems.

carrier: an individual who is heterozygous for a recessive condition; a carrier displays the dominant phenotype but can pass on the recessive allele to offspring.

carrier protein: a membrane protein that facilitates the diffusion of specific substances across the membrane. The molecule to be transported binds to the outer surface of the carrier protein; the protein then changes shape, allowing the molecule to move across the membrane.

carrying capacity (K): the maximum population size that an ecosystem can support for a long period of time without damaging the ecosystem; determined primarily by the availability of space, nutrients, water, and light.

catalyst (kat′-uh-list): a substance that speeds up a chemical reaction without itself being permanently changed in the process; a catalyst lowers the activation energy of a reaction.

cell: the smallest unit of life, consisting, at a minimum, of an outer membrane that encloses a watery medium containing organic molecules, including genetic material composed of DNA.

cell cycle: the sequence of events in the life of a cell, from one cell division to the next.

cell division: splitting of one cell into two; the process of cellular reproduction.

cell plate: in plant cell division, a series of vesicles that fuse to form the new plasma membranes and cell wall separating the daughter cells.

cell theory: the scientific theory stating that every living organism is made up of one or more cells; cells are the functional units of all organisms; and all cells arise from preexisting cells.

cell wall: a nonliving, protective, and supportive layer secreted outside the plasma membrane of fungi, plants, and most bacteria and protists.

cellular respiration: the oxygen-requiring reactions, occurring in mitochondria, that break down the end products of glycolysis into carbon dioxide and water while capturing large amounts of energy as ATP.

cellular slime mold: a type of organism consisting of individual amoeboid cells that can aggregate to form a slug-like mass, which in turn forms a fruiting body. Cellular slime molds are members of the protist clade Amoebozoa.

cellulose: an insoluble carbohydrate composed of glucose subunits; forms the cell wall of plants.

central vacuole (vak′-ū-ōl): a large, fluid-filled vacuole occupying most of the volume of many plant cells; performs several functions, including maintaining turgor pressure.

centriole (sen′-trē-ōl): in animal cells, a short, barrel-shaped ring consisting of nine microtubule triplets; a pair of centrioles is found near the nucleus and may play a role in the organization of the spindle; centrioles also give rise to the basal bodies at the base of each cilium and flagellum that give rise to the microtubules of cilia and flagella.

centromere (sen′-trō-mēr): the region of a replicated chromosome at which the sister chromatids are held together until they separate during cell division.

cephalization (sef-ul-ī-zā′-shun): concentration of sensory organs and nervous tissue in the anterior (head) portion of the body.

channel protein: a membrane protein that forms a channel or pore completely through the membrane and that is usually permeable to one or to a few water-soluble molecules, especially ions.

chaparral: a biome located in coastal regions, with very low annual rainfall; is characterized by shrubs and small trees.

checkpoint: a mechanism in the eukaryotic cell cycle by which protein complexes in the cell determine whether the cell has successfully completed a specific process that is essential to successful cell division, such as the accurate replication of chromosomes.

chemical bond: an attraction between two atoms or molecules that tends to hold them together. Types of bonds include covalent, ionic, and hydrogen.

chemical energy: a form of potential energy that is stored in molecules and may be released during chemical reactions.

chemical reaction: a process that forms and breaks chemical bonds that hold atoms together in molecules.

chemiosmosis (ke-mē-oz-mō′-sis): a process of ATP generation in chloroplasts and mitochondria. The movement of electrons down an electron transport system is used to pump hydrogen ions across a membrane, thereby building up a concentration gradient of hydrogen ions; the hydrogen ions diffuse back across the membrane through the pores of ATP-synthesizing enzymes; the energy of their movement down their concentration gradient drives ATP synthesis.

chemosynthesis (kē-mō-sin-the-sis): the process of oxidizing inorganic molecules, such as hydrogen sulfide, to obtain energy. Producers in hydrothermal vent communities, where light is absent, use chemosynthesis instead of photosynthesis.

chemosynthetic (kēm'-ō-sin-the-tik): capable of oxidizing inorganic molecules to obtain energy.

chiasma (kī-as'-muh; pl., chiasmata): a point at which a chromatid of one chromosome crosses with a chromatid of the homologous chromosome during prophase I of meiosis; the site of exchange of chromosomal material between chromosomes.

chitin (kī'-tin): a compound found in the cell walls of fungi and the exoskeletons of insects and some other arthropods; composed of chains of nitrogen-containing, modified glucose molecules.

chlorophyll (klor'-ō-fil): a pigment found in chloroplasts that captures light energy during photosynthesis; chlorophyll absorbs violet, blue, and red light but reflects green light.

chlorophyll *a* (klor'-ō-fil): the most abundant type of chlorophyll molecule in photosynthetic eukaryotic organisms and in cyanobacteria; chlorophyll *a* is found in the reaction centers of the photosystems.

chlorophyte (klor'-ō-fīt): a member of Chlorophyta, a protist clade. Chlorophytes are photosynthetic green algae found in marine, freshwater, and terrestrial environments.

chloroplast (klor'-ō-plast): the organelle in plants and plantlike protists that is the site of photosynthesis; is surrounded by a double membrane and contains an extensive internal membrane system that bears chlorophyll.

chorionic villus sampling (kōr-ē-on'-ik; CVS): a procedure for sampling cells from the chorionic villi produced by a fetus: A tube is inserted into the uterus of a pregnant woman, and a small sample of villi is suctioned off for genetic and biochemical analyses.

chromatid (krō'-ma-tid): one of the two identical strands of DNA and protein that forms a duplicated chromosome. The two sister chromatids of a duplicated chromosome are joined at the centromere.

chromatin (krō'-ma-tin): the complex of DNA and proteins that makes up eukaryotic chromosomes.

chromosome (krō'-mō-sōm): a DNA double helix and associated proteins that help to organize and regulate the use of the DNA.

chytrid: a member of the fungus clade Chytridomycota, which includes species with flagellated swimming spores.

ciliate (sil'-ē-et): a member of a protist group characterized by cilia and a complex unicellular structure. Ciliates are part of a larger group known as the alveolates.

cilium (sil'-ē-um; pl., cilia): a short, hair-like, motile projection from the surface of certain eukaryotic cells that contains microtubules in a 9 + 2 arrangement. The movement of cilia may propel cells through a fluid medium or move fluids over a stationary surface layer of cells.

citric acid cycle: a cyclic series of reactions, occurring in the matrix of mitochondria, in which the acetyl groups from the pyruvic acids produced by glycolysis are broken down to CO_2, accompanied by the formation of ATP and electron carriers; also called the Krebs cycle.

clade: a group that includes all the organisms descended from a common ancestor, but no other organisms; a monophyletic group.

class: in Linnaean classification, the taxonomic rank composed of related orders. Closely related classes form a phylum.

classical conditioning: a type of learning in which an animal learns to associate an innate behavior with a novel stimulus, as when a dog is trained to salivate in response to the sound of a bell.

climate: patterns of weather that prevail for long periods of time (from years to centuries) in a given region.

climate change: a long-lasting change in weather patterns, which may include significant changes in temperature, precipitation, the timing of seasons, and the frequency and severity of extreme weather events.

climax community: a diverse and relatively stable community that forms the endpoint of succession.

clone: offspring that are produced by mitosis and are, therefore, genetically identical to each other.

cloning: the process of producing many identical copies of a gene; also the production of many genetically identical copies of an organism.

closed circulatory system: a type of circulatory system, found in certain worms and vertebrates, in which the blood is always confined within the heart and vessels.

clumped distribution: the distribution characteristic of populations in which individuals are clustered into groups; the groups may be social or based on the need for a localized resource.

codominance: the relation between two alleles of a gene, such that both alleles are phenotypically expressed in heterozygous individuals.

codon: a sequence of three bases of messenger RNA that specifies a particular amino acid to be incorporated into a protein; certain codons also signal the beginning or end of protein synthesis.

coelom (sē'-lōm): in animals, a space or cavity, lined with tissue derived from mesoderm, that separates the body wall from the inner organs.

coenzyme: an organic molecule that is bound to certain enzymes and is required for the enzymes' proper functioning; typically, a nucleotide bound to a water-soluble vitamin.

coevolution: the evolution of adaptations in two species due to their extensive interactions with one another, such that each species acts as a major force of natural selection on the other.

cohesion: the tendency of the molecules of a substance to stick together.

communication: the act of producing a signal that causes a receiver, normally another animal of the same species, to change its behavior in a way that is, on average, beneficial to both signaler and receiver.

community: populations of different species that live in the same area and interact with one another.

competition: interaction among individuals who attempt to utilize a resource (for example, food or space) that is limited relative to the demand for that resource.

competitive exclusion principle: the concept that no two species can simultaneously and continuously occupy the same ecological niche.

competitive inhibition: the process by which two or more molecules that are somewhat similar in structure compete for the active site of an enzyme.

complementary base pair: in nucleic acids, bases that pair by hydrogen bonding. In DNA, adenine is complementary to thymine, and guanine is complementary to cytosine; in RNA, adenine is complementary to uracil, and guanine to cytosine.

compound eye: a type of eye, found in many arthropods, that is composed of numerous independent subunits called *ommatidia*. Each ommatidium contributes a piece of a mosaic-like image perceived by the animal.

concentration: the amount of solute (often in moles, a measurement that is proportional to the number of molecules) in a given volume of solvent.

concentration gradient: a difference in the concentration of a solute between different regions within a fluid or across a barrier such as a membrane.

conclusion: in the scientific method, a decision about the validity of a hypothesis, based on experiments or observations.

conifer (kon-eh-fer): a member of a group of nonflowering vascular plants whose members reproduce by means of seeds formed inside cones; retains its leaves throughout the year.

conjugation: in prokaryotes, the transfer of DNA from one cell to another via a temporary connection; in single-celled eukaryotes, the mutual exchange of genetic material between two temporarily joined cells.

connection protein: a protein in the plasma membrane of a cell that attaches to the cytoskeleton inside the cell, to other cells, or to the extracellular matrix.

conservation biology: the application of knowledge from ecology and other areas of biology to understand and conserve biodiversity.

constant-loss population: a population characterized by a relatively constant death rate; constant-loss populations have a roughly linear survivorship curve.

consumer: an organism that eats other organisms; a heterotroph.

consumer–prey interaction: an interaction between species where one species (the consumer) uses another (the prey) as a food source.

contractile vacuole: a fluid-filled vacuole in certain protists that takes up water from the cytoplasm, contracts, and expels the water outside the cell through a pore in the plasma membrane.

control: that portion of an experiment in which all possible variables are held constant; in contrast to the "experimental" portion, in which a particular variable is altered.

convergent evolution: the independent evolution of similar structures among unrelated organisms as a result of similar environmental pressures; see also *analogous structures*.

coral reef: an ecosystem created by animals (reef-building corals) and plants in warm tropical waters.

core reserve: a natural area protected from most human uses that encompasses enough space to preserve most of the biodiversity of the ecosystems in that area.

coupled reaction: a pair of reactions, one exergonic and one endergonic, that are linked together such that the energy produced by the exergonic reaction provides the energy needed to drive the endergonic reaction.

covalent bond (kō-vā′-lent): a chemical bond between atoms in which electrons are shared.

craniate: an animal that has a skull.

crassulacean acid metabolism (CAM): a biochemical pathway used by some plants in hot, dry climates, that increases the efficiency of carbon fixation during photosynthesis. Mesophyll cells capture carbon dioxide at night and use it to produce sugar during the day.

critically endangered species: a species that faces an extreme risk of extinction in the wild in the immediate future.

cross-fertilization: the union of sperm and egg from two individuals of the same species.

crossing over: the exchange of corresponding segments of the chromatids of two homologous chromosomes during meiosis I; occurs at chiasmata.

cuticle (kū′-ti-kul): a waxy or fatty coating on the surfaces of the aboveground epidermal cells of many land plants; aids in the retention of water.

cytokinesis (sī-tō-ki-nē′-sis): the division of the cytoplasm and organelles into two daughter cells during cell division; normally occurs during telophase of mitotic and meiotic cell division.

cytoplasm (sī′-tō-plaz-um): all of the material contained within the plasma membrane of a cell, exclusive of the nucleus.

cytosine (C): a nitrogenous base found in both DNA and RNA; abbreviated as C.

cytoskeleton: a network of protein fibers in the cytoplasm that gives shape to a cell, holds and moves organelles, and is typically involved in cell movement.

cytosol: the fluid portion of the cytoplasm; the substance within the plasma membrane exclusive of the nucleus and organelles.

daughter cell: one of the two cells formed by cell division.

death rate: the number of deaths per individual in a specified unit of time, such as a year.

decomposer: an organism, usually a fungus or bacterium, that digests organic material by secreting digestive enzymes into the environment, in the process liberating nutrients into the environment.

deductive reasoning: the process of generating hypotheses about the results of a specific experiment or the nature of a specific observation.

deforestation: the excessive cutting of forests. In recent years, deforestation has occurred primarily in rain forests in the Tropics, to clear space for agriculture.

dehydration synthesis: a chemical reaction in which two molecules are joined by a covalent bond with the simultaneous removal of a hydrogen from one molecule and a hydroxyl group from the other, forming water; the reverse of hydrolysis.

deletion mutation: a mutation in which one or more pairs of nucleotides are removed from a gene.

demographic transition: a change in population dynamic in which a fairly stable population with both high birth rates and high death rates experiences rapid growth as death rates decline and then returns to a stable (although much larger) population as birth rates decline.

demography: the study of the changes in human numbers over time, grouped by world regions, age, sex, educational levels, and other variables.

denature: to disrupt the secondary and/or tertiary structure of a protein while leaving its amino acid sequence intact. Denatured proteins can no longer perform their biological functions.

denatured: having the secondary and/or tertiary structure of a protein disrupted, while leaving the amino acid sequence unchanged. Denatured proteins can no longer perform their biological functions.

denitrifying bacteria (dē-nī′-treh-fī-ing): bacteria that break down nitrates, releasing nitrogen gas to the atmosphere.

density-dependent: referring to any factor, such as predation, that limits population size to an increasing extent as the population density increases.

density-independent: referring to any factor, such as floods or fires, that limits a population's size regardless of its density.

deoxyribonucleic acid (dē-ox-ē-rī-bō-noo-klā′-ik; DNA): a molecule composed of deoxyribose nucleotides; contains the genetic information of all living cells.

desert: a biome in which less than 10 inches (25 centimeters) of rain fall each year; characterized by cacti, succulents, and widely spaced, drought-resistant bushes.

desertification: the process by which relatively dry, drought-prone regions are converted to desert as a result of drought and overuse of the land, for example, by overgrazing or cutting of trees.

desmosome (dez′-mō-sōm): a strong cell-to-cell junction that attaches adjacent cells to one another.

detritivore (de-trī′-ti-vor): one of a diverse group of organisms, ranging from worms to vultures, that eat the wastes and dead remains of other organisms.

deuterostome (doo′-ter-ō-stōm): an animal with a mode of embryonic development in which the coelom is derived from outpocketings of the gut; characteristic of echinoderms and chordates.

diatom (dī′-uh-tom): a member of a protist group that includes photosynthetic forms with two-part glassy outer coverings; important photosynthetic organisms in fresh water and salt water. Diatoms are part of a larger group known as the stramenopiles.

differentiate: the process whereby a cell becomes specialized in structure and function.

differentiated cell: a mature cell specialized for a specific function; in plants, differentiated cells normally do not divide.

diffusion: the net movement of solute particles from a region of high solute concentration to a region of low solute concentration, driven by a concentration gradient; may occur within a fluid or across a barrier such as a membrane.

dinoflagellate (dī-nō-fla′-jel-et): a member of a protist group that includes photosynthetic forms in which two flagella project through armor-like plates; abundant in oceans; can reproduce rapidly, causing "red tides." Dinoflagellates are part of a larger group known as the alveolates.

diploid (dip′-loid): referring to a cell with pairs of homologous chromosomes.

diplomonad: a member of a protist group characterized by two nuclei and multiple flagella. Diplomonads, which include disease-causing parasites such as *Giardia*, are part of a larger group known as the excavates.

directional selection: a type of natural selection that favors one extreme of a range of phenotypes.

disaccharide (dī-sak′-uh-rīd): a carbohydrate formed by the covalent bonding of two monosaccharides.

disruptive selection: a type of natural selection that favors both extremes of a range of phenotypes.

dissolve: the process by which solvent molecules completely surround and disperse the individual atoms or molecules of another substance, the solute.

disturbance: any event that disrupts an ecosystem by altering its community, its abiotic structure, or both; disturbance precedes succession.

disulfide bond: the covalent bond formed between the sulfur atoms of two cysteines in a protein; typically causes the protein to fold by bringing otherwise distant parts of the protein close together.

DNA cloning: any of a variety of technologies that are used to produce multiple copies of a specific segment of DNA (usually a gene).

DNA helicase: an enzyme that helps unwind the DNA double helix during DNA replication.

DNA ligase: an enzyme that bonds the terminal sugar in one DNA strand to the terminal phosphate in a second DNA strand, creating a single strand with a continuous sugar-phosphate backbone.

DNA polymerase: an enzyme that bonds DNA nucleotides together into a continuous strand, using a preexisting DNA strand as a template.

DNA probe: a sequence of nucleotides that is complementary to the nucleotide sequence of a gene or other segment of DNA under study; used to locate the gene or DNA segment during gel electrophoresis or other methods of DNA analysis.

DNA profile: the pattern of short tandem repeats of specific DNA segments; using a standardized set of 13 short tandem repeats, DNA profiles identify individual people with great accuracy.

DNA replication: the copying of the double-stranded DNA molecule, producing two identical DNA double helices.

DNA sequencing: the process of determining the order of nucleotides in a DNA molecule.

domain: the broadest category for classifying organisms; organisms are classified into three domains: Bacteria, Archaea, and Eukarya.

dominance hierarchy: a social structure that arises when the animals in a social group establish individual ranks that determine access to resources; ranks are usually established through aggressive interactions.

dominant: an allele that can determine the phenotype of heterozygotes completely, such that they are indistinguishable from individuals homozygous for the allele; in the heterozygotes, the expression of the other (recessive) allele is completely masked.

double helix (hē′-liks): the shape of the two-stranded DNA molecule; similar to a ladder twisted lengthwise into a corkscrew shape.

Down syndrome: a genetic disorder caused by the presence of three copies of chromosome 21; common characteristics include learning disabilities, distinctively shaped eyelids, a small mouth, heart defects, and low resistance to infectious diseases; also called *trisomy 21*.

duplicated chromosome: a eukaryotic chromosome following DNA replication; consists of two sister chromatids joined at the centromeres.

early-loss population: a population characterized by a high birth rate, a high death rate among juveniles, and lower death rates among adults; early loss populations have a concave survivorship curve.

ecdysone (ek-dī′-sōn): a steroid hormone that triggers molting in insects and other arthropods.

ecological economics: the branch of economics that attempts to determine the monetary value of ecosystem services and to compare the monetary value of natural ecosystems with the monetary value of human activities that may reduce the services that natural ecosystems provide.

ecological footprint: the area of productive land needed to produce the resources used and absorb the wastes (including carbon dioxide) generated by an individual person, or by an average person of a specific part of the world (for example, an individual country), or of the entire world, using current technologies.

ecological isolation: reproductive isolation that arises when species do not interbreed because they occupy different habitats.

ecological niche (nitch): the role of a particular species within an ecosystem, including all aspects of its interaction with the living and nonliving environments.

ecology (ē-kol′-uh-jē): the study of the interrelationships of organisms with each other and with their nonliving environment.

ecosystem (ē′-kō-sis-tem): all the organisms and their nonliving environment within a defined area.

ecosystem services: the processes through which natural ecosystems and their living communities sustain and fulfill human life. Ecosystem services include purifying air and water, replenishing oxygen, pollinating plants, reducing flooding, providing wildlife habitat, and many more.

ectoderm (ek′-tō-derm): the outermost embryonic tissue layer, which gives rise to structures such as hair, the epidermis of the skin, and the nervous system.

electromagnetic spectrum: the range of all possible wavelengths of electromagnetic radiation, from wavelengths longer than radio waves, to microwaves, infrared, visible light, ultraviolet, X-rays, and gamma rays.

electron: a subatomic particle, found in an electron shell outside the nucleus of an atom, that bears a unit of negative charge and very little mass.

electron carrier: a molecule that can reversibly gain or lose electrons. Electron carriers generally accept high-energy electrons produced during an exergonic reaction and donate the electrons to acceptor molecules that use the energy to drive endergonic reactions.

electron shell: a region in an atom within which electrons orbit; each shell corresponds to a fixed energy level at a given distance from the nucleus.

electron transport chain (ETC): a series of electron carrier molecules, found in the thylakoid membranes of chloroplasts and the inner membrane of mitochondria, that extract energy from electrons and generate ATP or other energetic molecules.

element: a substance that cannot be broken down, or converted, to a simpler substance by ordinary chemical reactions.

emerging infectious disease: a previously unknown infectious disease (one caused by a microbe), or a previously known infectious disease whose frequency or severity has significantly increased in the past two decades.

emigration (em-uh-grā′-shun): migration of individuals out of an area.

endangered species: a species that faces a high risk of extinction in the wild in the near future.

endergonic (en-der-gon′-ik): pertaining to a chemical reaction that requires an input of energy to proceed; an "uphill" reaction.

endocytosis (en-dō-sī-tō′-sis): the process in which the plasma membrane engulfs extracellular material, forming membrane-bound sacs that enter the cytoplasm and thereby move material into the cell.

endoderm (en′-dō-derm): the innermost embryonic tissue layer, which gives rise to structures such as the lining of the digestive and respiratory tracts.

endomembrane system: internal membranes that create loosely connected compartments within the eukaryotic cell. It includes the nuclear envelope, the endoplasmic reticulum, vesicles, the Golgi apparatus, and lysosomes.

endoplasmic reticulum (en-dō-plaz′-mik re-tik′-ū-lum; ER): a system of membranous tubes and channels in eukaryotic cells; the site of most protein and lipid synthesis.

endoskeleton (en′-dō-skel′-uh-tun): a rigid internal skeleton with flexible joints that allow for movement.

endospore: a protective resting structure of some rod-shaped bacteria that withstands unfavorable environmental conditions.

endosymbiont hypothesis: the hypothesis that certain organelles, especially chloroplasts and mitochondria, arose as mutually beneficial associations between the ancestors of eukaryotic cells and captured bacteria that lived within the cytoplasm of the pre-eukaryotic cell.

energy: the capacity to do work.

energy-carrier molecule: high-energy molecules that are synthesized at the site of an exergonic reaction, where they capture one or two energized electrons and hydrogen ions. They include nicotinamide adenine dinucleotide (NADH) and flavin adenine dinucleotide ($FADH_2$).

energy pyramid: a graphical representation of the energy contained in succeeding trophic levels, with maximum energy at the base (primary producers) and steadily diminishing amounts at higher levels.

energy-requiring transport: the transfer of substances across a cell membrane using cellular energy; includes active transport, endocytosis, and exocytosis.

entropy (en′-trō-pē): a measure of the amount of randomness and disorder in a system.

environmental resistance: any factor that tends to counteract biotic potential, limiting population growth and the resulting population size.

enzyme (en′zīm): a biological catalyst, usually a protein, that speeds up the rate of specific biological reactions.

epidermal tissue: dermal tissue in plants that forms the epidermis, the outermost cell layer that covers leaves, young stems, and young roots.

epidermis (ep-uh-der′-mis): in animals, specialized stratified epithelial tissue that forms the outer layer of the skin; in plants, the outermost layer of cells of a leaf, young root, or young stem.

epigenetics: the study of the mechanisms by which cells and organisms change gene expression and function without changing the base sequence of their DNA; usually, epigenetic controls over DNA expression involve modification of DNA, modification of chromosomal proteins, or alteration of transcription or translation through the actions of noncoding RNA molecules.

equilibrium population: a population in which allele frequencies and the distribution of genotypes do not change from generation to generation.

estuary (es′-choō-ār-ē): a wetland formed where a river meets the ocean; the salinity is quite variable, but lower than in seawater and higher than in fresh water.

euglenid (ū′-gle-nid): a member of a protist group characterized by one or more whip-like flagella, which are used for locomotion, and by a photoreceptor, which detects light. Euglenids are photosynthetic and are part of a larger group known as euglenozoans.

euglenozoan: a member of the Euglenozoa, a protist clade. The euglenozoans, which are characterized by mitochondrial membranes that appear under the microscope to be shaped like a stack of disks, include the euglenids and the kinetoplastids.

Eukarya (ū-kar′-ē-a): one of life's three domains; consists of all eukaryotes (plants, animals, fungi, and protists).

eukaryote (ū-kar′-ē-ōt): an organism whose cells are eukaryotic; plants, animals, fungi, and protists are eukaryotes.

eukaryotic (ū-kar-ē-ot′-ik): referring to cells of organisms of the domain Eukarya (plants, animals, fungi, and protists). Eukaryotic cells have genetic material enclosed within a membrane-bound nucleus, and they contain other membrane-bound organelles.

eutrophic lake: a lake that receives sufficiently large inputs of sediments, organic material, and inorganic nutrients from its surroundings to support dense communities, especially of plants and phytoplankton; contains murky water with poor light penetration.

evolution: (1) the descent of modern organisms, with modification, from preexisting life-forms; (2) the theory that all organisms are related by common ancestry and have changed over time; (3) any change in the genetic makeup (the proportions of different genotypes) of a population from one generation to the next.

excavate: a member of the Excavata, a protist clade. The excavates, which generally lack mitochondria, include the diplomonads and the parabasalids.

exergonic (ex-er-gon′-ik): pertaining to a chemical reaction that releases energy (either as heat or in the form of increased entropy); a "downhill" reaction.

exocytosis (ex-ō-sī-tō′-sis): the process in which intracellular material is enclosed within a membrane-bound sac that moves to the plasma membrane and fuses with it, releasing the material outside the cell.

exon: a segment of DNA in a eukaryotic gene that codes for amino acids in a protein; see also *intron*.

exoskeleton (ex′-ō-skel′-uh-tun): a rigid external skeleton that supports the body, protects the internal organs, and has flexible joints that allow for movement.

experiment: in the scientific method, the use of carefully controlled observations or manipulations to test the predictions generated by a hypothesis.

exponential growth: a continuously accelerating increase in population size; this type of growth generates a curve shaped like the letter "J."

extinction: the death of all members of a species.

extracellular matrix (ECM): material secreted by and filling the spaces between cells. Animal cells secrete supporting and adhesive proteins embedded in a gel of polysaccharides linked by proteins; plant cells secrete a matrix of cellulose that forms cell walls.

facilitated diffusion: the diffusion of molecules across a membrane, assisted by protein pores or carriers embedded in the membrane.

family: in Linnaean classification, the taxonomic rank composed of related genera. Closely related families make up an order.

fat (molecular): a lipid composed of three saturated fatty acids covalently bonded to glycerol; fats are solid at room temperature.

fatty acid: an organic molecule composed of a long chain of carbon atoms, with a carboxylic acid (–COOH) group at one end; may be saturated (all single bonds between the carbon atoms) or unsaturated (one or more double bonds between the carbon atoms).

feedback inhibition: in enzyme-mediated chemical reactions, the condition in which the product of a reaction inhibits one or more of the enzymes involved in synthesizing the product.

fermentation: anaerobic reactions that convert the pyruvic acid produced by glycolysis into lactic acid or alcohol and CO_2, using hydrogen ions and electrons from NADH; the primary function of fermentation is to regenerate NAD^+ so that glycolysis can continue under anaerobic conditions.

fibrin (fī′-brin): a clotting protein formed in the blood in response to a wound; binds with other fibrin molecules and provides a matrix around which a blood clot forms.

fibrinogen (fī-brin′-ō-jen): the inactive form of the clotting protein fibrin. Fibrinogen is converted into fibrin by the enzyme thrombin, which is produced in response to injury.

first law of thermodynamics: the principle of physics that states that within any isolated system, energy can be neither created nor destroyed, but can be converted from one form to another; also called the law of conservation of energy.

fitness: the reproductive success of an organism, relative to the average reproductive success in the population.

flagellum (fla-jel′-um; pl., flagella): a long, hair-like, motile extension of the plasma membrane; in eukaryotic cells, it contains microtubules arranged in a 9 + 2 pattern. The movement of flagella propels some cells through fluids.

flavin adenine dinucleotide (FAD or FADH₂): an electron-carrier molecule produced in the mitochondrial matrix by the Krebs cycle; subsequently donates electrons to the electron transport chain.

flower: the reproductive structure of an angiosperm plant.

flower bud: a cluster of meristem cells (a bud) that forms a flower.

fluid: any substance whose molecules can freely flow past one another; "fluid" can describe liquids, cell membranes, and gases.

fluid mosaic model: a model of cell membrane structure; according to this model, membranes are composed of a double layer of phospholipids in which various proteins are embedded. The phospholipid bilayer is a somewhat fluid matrix that allows the movement of proteins within it.

food chain: a linear feeding relationship in a community, using a single representative from each of the trophic levels.

food vacuole: a membranous sac, within a single cell, in which food is enclosed. Digestive enzymes are released into the vacuole, where intracellular digestion occurs.

food web: a representation of the complex feeding relationships within a community, including many organisms at various trophic levels, with many of the consumers occupying more than one level simultaneously.

foraminiferan (for-am-i-nif'-er-un): a member of a protist group characterized by pseudopods and elaborate calcium carbonate shells. Foraminiferans are generally aquatic (largely marine) and are part of a larger group known as rhizarians.

fossil: the remains of a dead organism, normally preserved in rock. Fossils may be petrified bones or wood shells; eggs; feces; impressions of body forms, such as feathers, skin, or leaves; or markings made by organisms, such as footprints.

fossil fuel: a fuel, such as coal, oil, and natural gas, derived from the remains of ancient organisms.

founder effect: the result of an event in which an isolated population is founded by a small number of individuals; may result in genetic drift if allele frequencies in the founder population are by chance different from those of the parent population.

free nucleotide: a nucleotide that has not been joined with other nucleotides to form a DNA or RNA strand.

free radical: a molecule containing an atom with an unpaired electron, which makes it highly unstable and reactive with nearby molecules. By removing an electron from the molecule it attacks, it creates a new free radical and begins a chain reaction that can lead to the destruction of biological molecules crucial to life.

fruit: in flowering plants, the ripened ovary (plus, in some cases, other parts of the flower), which contains the seeds.

functional group: one of several groups of atoms commonly found in an organic molecule, including hydrogen, hydroxyl, amino, carboxyl, and phosphate groups, that determine the characteristics and chemical reactivity of the molecule.

gamete (gam'-ēt): a haploid sex cell, usually a sperm or an egg, formed in sexually reproducing organisms.

gametophyte (ga-mēt'-ō-fīt): the multicellular haploid stage in the life cycle of plants.

ganglion (gang'-lē-un; pl., ganglia): a cluster of neurons.

gametic incompatibility: a postmating reproductive isolating mechanism that arises when sperm from one species cannot fertilize eggs of another species.

gap junction: a type of cell-to-cell junction in animals in which channels connect the cytoplasm of adjacent cells.

gel electrophoresis: a technique in which molecules (such as DNA fragments) are placed in wells in a thin sheet of gelatinous material and exposed to an electric field; the molecules migrate through the gel at a rate determined by certain characteristics, most commonly size.

gene: the unit of heredity; a segment of DNA located at a particular place on a chromosome that usually encodes the information for the amino acid sequence of a protein and, hence, a particular trait.

gene flow: the movement of alleles from one population to another owing to the movement of individual organisms or their gametes.

gene linkage: the tendency for genes located on the same chromosome to be inherited together.

gene pool: the total of all alleles of all genes in a population; for a single gene, the total of all the alleles of that gene that occur in a population.

gene therapy: the attempt to cure a disease by inserting, deleting, or altering a patient's genes.

genetic code: the collection of codons of mRNA, each of which directs the incorporation of a particular amino acid into a protein during protein synthesis or causes protein synthesis to start or stop.

genetic drift: a change in the allele frequencies of a small population purely by chance.

genetic engineering: the modification of the genetic material of an organism, usually using recombinant DNA techniques.

genetic recombination: the generation of new combinations of alleles on homologous chromosomes due to the exchange of DNA during crossing over.

genetically modified organism (GMO): a plant or animal that contains DNA that has been modified or that has been obtained from another species.

genomic imprinting: a form of epigenetic control by which a given gene is expressed in an offspring only if the gene has been inherited from a specific parent; the copy of the gene that was inherited from the other parent is usually not expressed.

genotype (jēn'-ō-tīp): the genetic composition of an organism; the actual alleles of each gene carried by the organism.

genus (jē'-nus): in Linnaean classification, the taxonomic rank composed of related species. Closely related genera make up a family.

geographic isolation: reproductive isolation that arises when species do not interbreed because a physical barrier separates them.

glomeromycete: a member of the fungus clade Glomeromycota, which includes species that form mycorrhizal associations with plant roots and that form bush-shaped branching structures inside plant cells.

glucose: the most common monosaccharide, with the molecular formula $C_6H_{12}O_6$; most polysaccharides, including cellulose, starch, and glycogen, are made of glucose subunits covalently bonded together.

glycerol (glis'-er-ol): a three-carbon alcohol to which fatty acids are covalently bonded to make fats and oils.

glycogen (glī'-kō-jen): a highly branched polymer of glucose that is stored by animals in the muscles and liver and metabolized as a source of energy.

glycolysis (glī-kol'-i-sis): reactions, carried out in the cytoplasm, that break down glucose into two molecules of pyruvic acid, producing two ATP molecules; does not require oxygen but can proceed when oxygen is present.

glycoprotein: a protein to which a carbohydrate is attached.

Golgi apparatus (gōl-jē): a stack of membranous sacs, found in most eukaryotic cells, that is the site of processing and separation of membrane components and secretory materials.

gradient: a difference in concentration, pressure, or electrical charge between two regions.

grana: stacks of thylakoid membranes that form disks that surround the thylakoid space within the chloroplast.

grassland: a biome, located in the centers of continents, that primarily supports grasses; also called a *prairie*.

greenhouse effect: the process in which certain gases such as carbon dioxide and methane trap sunlight energy in a planet's atmosphere as heat; the glass in a greenhouse causes a similar warming effect. The result, global warming that causes climate change, is being enhanced by the production of these gases by humans.

greenhouse gas: a gas, such as carbon dioxide or methane, that traps sunlight energy in a planet's atmosphere as heat; a gas that participates in the greenhouse effect.

growth factor: small molecules, usually proteins or steroids, that bind to receptors on or in target cells and enhance their rate of cell division or differentiation.

growth rate: a measure of the change in population size per individual per unit of time.

guanine (G): a nitrogenous base found in both DNA and RNA; abbreviated as G.

gymnosperm (jim'-nō-sperm): a nonflowering seed plant, such as a conifer, gnetophyte, cycad, or gingko.

gyre (jīr): a roughly circular pattern of ocean currents, formed because continents interrupt the flow of the current; rotates clockwise in the Northern Hemisphere and counterclockwise in the Southern Hemisphere.

habitat fragmentation: the process by which human development and activities produce patches of wildlife habitat that may not be large enough to sustain minimum viable populations.

habituation (heh-bich-oo-ā′-shun): a type of simple learning characterized by a decline in response to a repeated stimulus.

haploid (hap′-loid): referring to a cell that has only one member of each pair of homologous chromosomes.

Hardy–Weinberg principle: a mathematical model proposing that, under certain conditions, the allele frequencies and genotype frequencies in a sexually reproducing population will remain constant over generations.

heat of fusion: the energy that must be removed from a compound to transform it from a liquid into a solid at its freezing temperature.

heat of vaporization: the energy that must be supplied to a compound to transform it from a liquid into a gas at its boiling temperature.

helix (hē′-liks): a coiled, spring-like secondary structure of a protein.

hemocoel (hē′-mō-sēl): a cavity within the bodies of certain invertebrates in which a fluid, called hemolymph, bathes tissues directly; part of an open circulatory system.

hemodialysis (hē-mō-dī-al′-luh-sis): a procedure that simulates kidney function in individuals with damaged or ineffective kidneys; blood is diverted from the body, artificially filtered, and returned to the body.

hemophilia: a recessive, sex-linked disease in which the blood fails to clot normally.

herbivore (erb′-i-vor): literally, "plant-eater"; an organism that feeds directly and exclusively on producers; a primary consumer.

hermaphroditic (her-maf′-ruh-dit′-ik): possessing both male and female sexual organs. Some hermaphroditic animals can fertilize themselves; others must exchange sex cells with a mate.

heterotroph (het′-er-ō-trōf′): literally, "other-feeder"; an organism that eats other organisms; a consumer.

heterozygous (het′-er-ō-zī′-gus): carrying two different alleles of a given gene; also called *hybrid*.

homeostasis (hōm-ē-ō-stā′-sis): the maintenance of the relatively constant internal environment that is required for the optimal functioning of cells.

hominin: a human or a prehistoric relative of humans; the oldest known hominin is *Sahelanthropus*, whose fossils are more than 6 million years old.

homologous chromosome: a chromosome that is similar in appearance and genetic information to another chromosome with which it pairs during meiosis; also called *homologue*.

homologous structures: structures that may differ in function but that have similar anatomy, presumably because the organisms that possess them have descended from common ancestors.

homologue (hō′-mō-log): a chromosome that is similar in appearance and genetic information to another chromosome with which it pairs during meiosis; also called *homologous chromosome*.

homozygous (hō-mō-zī′-gus): carrying two copies of the same allele of a given gene; also called *true-breeding*.

host: the prey organism on or in which a parasite lives; the host is harmed by the relationship.

Huntington disease: an incurable genetic disorder, caused by a dominant allele, that produces progressive brain deterioration, resulting in the loss of motor coordination, flailing movements, personality disturbances, and eventual death.

hybrid: an organism that is the offspring of parents differing in at least one genetically determined characteristic; also used to refer to the offspring of parents of different species.

hybrid infertility: a postmating reproductive isolating mechanism that arises when hybrid offspring (offspring of parents of two different species) are sterile or have low fertility.

hybrid inviability: a postmating reproductive isolating mechanism that arises when hybrid offspring (offspring of parents of two different species) fail to survive.

hydrogen bond: the weak attraction between a hydrogen atom that bears a partial positive charge (due to polar covalent bonding with another atom) and another atom (oxygen, nitrogen, or fluorine) that bears a partial negative charge; hydrogen bonds may form between atoms of a single molecule or of different molecules.

hydrologic cycle (hī-drō-loj′-ik): the biogeochemical cycle by which water travels from its major reservoir, the oceans, through the atmosphere to reservoirs in freshwater lakes, rivers, and groundwater, and back into the oceans. The hydrologic cycle is driven by solar energy. Nearly all water remains as water throughout the cycle (rather than being used in the synthesis of new molecules).

hydrolysis (hī-drol′-i-sis): the chemical reaction that breaks a covalent bond by means of the addition of hydrogen to the atom on one side of the original bond and a hydroxyl group to the atom on the other side; the reverse of dehydration synthesis.

hydrophilic (hī-drō′-fil′-ik): pertaining to molecules that dissolve readily in water, or to molecules that form hydrogen bonds with water; polar.

hydrophobic (hī-drō-fō′-bik): pertaining to molecules that do not dissolve in water or form hydrogen bonds with water; nonpolar.

hydrostatic skeleton (hī-drō-stat′-ik): in invertebrate animals, a body structure in which fluid-filled compartments provide support for the body and change shape when acted on by muscles, which alters the animal's body shape and position or causes the animal to move.

hydrothermal vent community: a community of unusual organisms, living in the deep ocean near hydrothermal vents, that depends on the chemosynthetic activities of sulfur bacteria.

hypertonic (hī-per-ton′-ik): referring to a solution that has a higher concentration of solute (and therefore a lower concentration of free water) than has the cytosol of a cell.

hypha (hī′-fuh; pl., hyphae): in fungi, a thread-like structure that consists of elongated cells, typically with many haploid nuclei; many hyphae make up the fungal body.

hypothesis (hī-poth′-eh-sis): a proposed explanation for a phenomenon based on available evidence that leads to a prediction that can be tested.

hypotonic (hī-pō-ton′-ik): referring to a solution that has a lower concentration of solute (and therefore a higher concentration of free water) than has the cytosol of a cell.

immigration (im-uh-grā′-shun): migration of individuals into an area.

immune system: a system of cells, including macrophages, B cells, and T cells, and molecules, such as antibodies and cytokines, that work together to combat microbial invasion of the body.

imprinting: a type of learning in which an animal acquires a particular type of information during a specific sensitive phase of development.

incomplete dominance: a pattern of inheritance in which the heterozygous phenotype is intermediate between the two homozygous phenotypes.

inductive reasoning: the process of creating a generalization based on many specific observations that support the generalization, coupled with an absence of observations that contradict it.

inheritance: the genetic transmission of characteristics from parent to offspring.

innate (in-āt′): inborn; instinctive; an innate behavior is performed correctly the first time it is attempted.

inorganic: describing any molecule that does not contain both carbon and hydrogen.

insertion: the site of attachment of a muscle to the relatively movable bone on one side of a joint.

insertion mutation: a mutation in which one or more pairs of nucleotides are inserted into a gene.

insight learning: a type of learning in which a problem is solved by understanding the relationships among the components of the problem rather than through trial and error.

intermediate filament: part of the cytoskeleton of eukaryotic cells that is composed of several types of proteins and probably functions mainly for support.

intermembrane space: the fluid-filled space between the inner and outer membranes of a mitochondrion.

interphase: the stage of the cell cycle between cell divisions in which chromosomes are duplicated and other cell functions occur, such as growth, movement, and acquisition of nutrients.

interspecific competition: competition among individuals of different species.

interstitial fluid: fluid that bathes the cells of the body; in mammals, interstitial fluid leaks from capillaries and is similar in composition to blood plasma, but lacking the large proteins found in plasma.

intertidal zone: an area of the ocean shore that is alternately covered by water during high tides and exposed to the air during low tides.

intraspecific competition: competition among individuals of the same species.

intrinsically disordered proteins: proteins or segments of proteins with no stable secondary or tertiary structure.

intron: a segment of DNA in a eukaryotic gene that does not code for amino acids in a protein; see also *exon*.

invasive species: organisms with a high biotic potential that are introduced (deliberately or accidentally) into ecosystems where they did not evolve and where they encounter little environmental resistance and tend to displace native species.

inversion: a mutation that occurs when a piece of DNA is cut out of a chromosome, turned around, and reinserted into the gap.

invertebrate (in-vert'-uh-bret): an animal that lacks a vertebral column.

ion (ī-on): a charged atom or molecule; an atom or molecule that either has an excess of electrons (and, hence, is negatively charged) or has lost electrons (and is positively charged).

ionic bond: a chemical bond formed by the electrical attraction between positively and negatively charged ions.

isolated system: in thermodynamics, a hypothetical space where neither energy nor matter can enter or leave.

isolating mechanism: a morphological, physiological, behavioral, or ecological difference that prevents members of two species from interbreeding.

isotonic (ī-sō-ton'-ik): referring to a solution that has the same concentration of solute (and therefore the same concentration of free water) as has the cytosol of a cell.

isotope (ī'-suh-tōp): one of several forms of a single element, the nuclei of which contain the same number of protons but different numbers of neutrons.

Jacob syndrome: a set of characteristics typical of human males possessing one X and two Y chromosomes (XYY); most XYY males are phenotypically normal, but XYY males tend to be taller than average and to have a slightly increased risk of learning disabilities.

J-curve: the J-shaped growth curve of an exponentially growing population in which increasing numbers of individuals join the population during each succeeding time period.

karyotype: a preparation showing the number, sizes, and shapes of all of the chromosomes within a cell.

K-selected species: species that typically live in a stable environment and often develop a population size that is close to the carrying capacity of that environment. *K*-selected species usually mature slowly, have a long life span, produce small numbers of fairly large offspring, and provide significant parental care or nutrients for the offspring, so that a large percentage of the offspring live to maturity.

kelp forest: a diverse ecosystem consisting of stands of tall brown algae and associated marine life. Kelp forests occur in oceans worldwide in nutrient-rich cool coastal waters.

keratin (ker'-uh-tin): a fibrous protein in hair, nails, and the epidermis of skin.

keystone species: a species whose influence on community structure is greater than its abundance would suggest.

kin selection: a type of natural selection that favors traits that enhance the survival or reproduction of an individual's relatives, even if the traits reduce the fitness of the individuals bearing them.

kinetic energy: the energy of movement; includes light, heat, mechanical movement, and electricity.

kinetochore (ki-net'-ō-kor): a protein structure that forms at the centromere regions of chromosomes; attaches the chromosomes to the spindle.

kinetoplastid: a member of a protist group characterized by distinctively structured mitochondria. Kinetoplastids are mostly flagellated and include parasitic forms such as *Trypanosoma*, which causes sleeping sickness. Kinetoplastids are part of a larger group known as euglenozoans.

kingdom: the second broadest taxonomic category, consisting of related phyla. Related kingdoms make up a domain.

Klinefelter syndrome: a set of characteristics typically found in individuals who have two X chromosomes and one Y chromosome; these individuals are phenotypically males but are usually unable to father children without the use of artificial reproductive technologies, such as in vitro fertilization. Men with Klinefelter syndrome may have several female-like traits, including broad hips and partial breast development.

Krebs cycle: a cyclic series of reactions, occurring in the matrix of mitochondria, in which the acetyl groups from the pyruvic acids produced by glycolysis are broken down to CO_2, accompanied by the formation of ATP and electron carriers; also called the citric acid cycle.

lactic acid fermentation: anaerobic reactions that convert the pyruvic acid produced by glycolysis into lactic acid, using hydrogen ions and electrons from NADH; the primary function of lactic acid fermentation is to regenerate NAD^+ so that glycolysis can continue under anaerobic conditions.

lactose (lak'-tōs): a disaccharide composed of glucose and galactose; found in mammalian milk.

lactose intolerance: the inability to digest lactose (milk sugar) because lactase, the enzyme that digests lactose, is not produced in sufficient amounts; symptoms include bloating, gas pains, and diarrhea.

lactose operon: in prokaryotes, the set of genes that encodes the proteins needed for lactose metabolism, including both the structural genes and a common promoter and operator that control transcription of the structural genes.

larva (lar'-vuh; pl., larvae): an immature form of an animal that subsequently undergoes metamorphosis into its adult form; includes the caterpillars of moths and butterflies, the maggots of flies, and the tadpoles of frogs and toads.

late-loss population: a population in which most individuals survive into adulthood; late-loss populations have a convex survivorship curve.

law of conservation of energy: the principle of physics that states that within any isolated system, energy can be neither created nor destroyed, but can be converted from one form to another; also called the first law of thermodynamics.

law of independent assortment: the independent inheritance of two or more traits, assuming that each trait is controlled by a single gene with no influence from gene(s) controlling the other trait; states that the alleles of each gene are distributed to the gametes independently of the alleles for other genes; this law is true only for genes located on different chromosomes or very far apart on a single chromosome.

law of segregation: the principle that each gamete receives only one of each parent's pair of alleles of each gene.

laws of thermodynamics: the physical laws that define the basic properties and behavior of energy.

learning: the process by which behavior is modified in response to experience.

legume (leg'-ūm): a member of a family of plants characterized by root swellings in which nitrogen-fixing bacteria are housed; includes peas, soybeans, lupines, alfalfa, and clover.

less developed country: a country that has not completed the demographic transition; usually has relatively low income and educational level for most of its population, with high birth rates and sometimes fairly high death rates, resulting in moderate to rapid population growth.

lichen (lī'-ken): a symbiotic association between an alga or cyanobacterium and a fungus, resulting in a composite organism.

life history: the characteristic survivorship and reproductive features of a species, particularly when and how often reproduction occurs, how many offspring are produced, how many resources are provided to each offspring, and what proportion of the offspring survive to maturity.

life table: a data table that groups organisms born at the same time and tracks them throughout their life span, recording how many continue to survive in each succeeding year (or other unit of time). Various parameters such as sex may be used in the groupings. Human life tables may include many other parameters (such as socioeconomic status) used by demographers.

light reactions: the first stage of photosynthesis, in which the energy of light is captured in ATP and NADPH; occurs in thylakoids of chloroplasts.

lignin: a hard material that is embedded in the cell walls of vascular plants and that provides support in terrestrial species; an early and important adaptation to terrestrial life.

limnetic zone: the part of a lake in which enough light penetrates to support photosynthesis.

linkage: the inheritance of certain genes as a group because they are parts of the same chromosome. Linked genes do not show independent assortment.

lipid (li'-pid): one of a number of organic molecules containing large nonpolar regions composed solely of carbon and hydrogen, which make lipids hydrophobic and insoluble in water; includes oils, fats, waxes, phospholipids, and steroids.

littoral zone: the part of a lake, usually close to the shore, in which the water is shallow and plants find abundant light, anchorage, and adequate nutrients.

lobefin: fish with fleshy fins that have well-developed bones and muscles. Lobefins include two living clades: the coelacanths and the lungfishes.

locus (pl., loci): the physical location of a gene on a chromosome.

logistic population growth: population growth characterized by an early exponential growth phase, followed by slower growth as the population approaches its carrying capacity, and finally reaching a stable population at the carrying capacity of the environment; this type of growth generates a curve shaped like a stretched-out letter "S."

lysosome (lī'-sō-sōm): a membrane-bound organelle containing intracellular digestive enzymes.

macronutrient: a nutrient required by an organism in relatively large quantities.

maltose (mal'-tōs): a disaccharide composed of two glucose molecules.

mammal: a member of the chordate clade Mammalia, which includes vertebrates with hair and mammary glands.

mammary gland (mam'-uh-rē): a milk-producing gland used by female mammals to nourish their young.

marsupial (mar-soo'-pē-ul): a member of the clade Marsupialia, which includes mammals whose young are born at an extremely immature stage and undergo further development in a pouch, where they remain attached to a mammary gland; kangaroos, opossums, and koalas are marsupials.

mass extinction: a relatively sudden extinction of many species, belonging to multiple major taxonomic groups, as a result of environmental change. The fossil record reveals five mass extinctions over geologic time.

mass number: the total number of protons and neutrons in the nucleus of an atom.

matrix: the fluid contained within the inner membrane of the mitochondrion.

mechanical incompatibility: a reproductive isolating mechanism that arises when differences in the reproductive structures of two species make the structures incompatible and prevent interbreeding.

meiosis (mī-ō'-sis): in eukaryotic organisms, a type of nuclear division in which a diploid nucleus divides twice to form four haploid nuclei.

meiosis I: the first division of meiosis, which separates the pairs of homologous chromosomes and sends one homologue from each pair into each of two daughter nuclei, which are therefore haploid.

meiosis II: the second division of meiosis, which separates the chromatids into independent chromosomes and parcels one chromosome into each of two daughter nuclei.

meiotic cell division: meiosis followed by cytokinesis.

mesoderm (mēz'-ō-derm): the middle embryonic tissue layer, lying between the endoderm and ectoderm, and normally the last to develop; gives rise to structures such as muscles, the skeleton, the circulatory system, and the kidneys.

mesophyll (mez'-ō-fil): loosely packed, usually photosynthetic cells located beneath the epidermis of a leaf.

messenger RNA (mRNA): a strand of RNA, complementary to the DNA of a gene, that conveys the genetic information in DNA to the ribosomes to be used during protein synthesis; sequences of three bases (codons) in mRNA that specify particular amino acids to be incorporated into a protein.

metabolic pathway: a sequence of chemical reactions within a cell in which the products of one reaction are the reactants for the next reaction.

metabolism: the sum of all chemical reactions that occur within a single cell or within all the cells of a multicellular organism.

metamorphosis (met-a-mor'-fō-sis): in animals with indirect development, a radical change in body form from larva to sexually mature adult, as seen in amphibians (e.g., tadpole to frog) and insects (e.g., caterpillar to butterfly).

metaphase (met'-a-fāz): in mitosis, the stage in which the chromosomes, attached to spindle fibers at kinetochores, are lined up along the equator of the cell; also the approximately comparable stages in meiosis I and meiosis II.

microfilament: part of the cytoskeleton of eukaryotic cells that is composed of the proteins actin and (in some cases) myosin; functions in the movement of cell organelles, locomotion by extension of the plasma membrane, and sometimes contraction of entire cells.

micronutrient: a nutrient required by an organism in relatively small quantities.

microRNA: small molecules of RNA that interfere with the translation of specific genes.

microtubule: a hollow, cylindrical strand, found in eukaryotic cells, that is composed of the protein tubulin; part of the cytoskeleton used in the movement of organelles, cell growth, and the construction of cilia and flagella.

mimicry (mim'-ik-rē): the situation in which a species has evolved to resemble something else, typically another type of organism.

minimum viable population (MVP): the smallest isolated population that can persist indefinitely and survive likely natural events such as fires and floods.

mitochondrion (mī-tō-kon'-drē-un; pl., mitochondria): an organelle, bounded by two membranes, that is the site of the reactions of aerobic metabolism.

mitosis (mī-tō'-sis): a type of nuclear division, used by eukaryotic cells, in which one copy of each chromosome (already duplicated during interphase before mitosis) moves into each of two daughter nuclei; the daughter nuclei are therefore genetically identical to each other.

mitotic cell division: mitosis followed by cytokinesis.

molecule (mol'-e-kūl): a particle composed of one or more atoms held together by chemical bonds; the smallest particle of a compound that displays all the properties of that compound.

molt: to shed an external body covering, such as an exoskeleton, skin, feathers, or fur.

monomer (mo'-nō-mer): a small organic molecule, several of which may be bonded together to form a chain called a polymer.

monosaccharide (mo-nō-sak'-uh-rīd): the basic molecular unit of all carbohydrates, normally composed of a chain of carbon atoms bonded to hydrogen and hydroxyl groups.

monotreme: a member of the clade Monotremata, which includes mammals that lay eggs; platypuses and spiny anteaters are monotremes.

more developed country: a country that has completed the demographic transition; usually has relatively high income and educational level for most of its population, with low birth and death rates that result in low or even negative population growth.

multicellular: many-celled; most members of the kingdoms Fungi, Plantae, and Animalia are multicellular, with intimate cooperation among cells.

multiple alleles: many alleles of a single gene, perhaps dozens or hundreds, as a result of mutations.

muscle tissue: tissue composed of one of three types of contractile cells (smooth, skeletal, or cardiac).

muscular dystrophy: an inherited disorder, almost exclusively found in males, in which defective dystrophin proteins cause the skeletal muscles to degenerate.

mutation: a change in the base sequence of DNA in a gene; often used to refer to a genetic change that is significant enough to alter the appearance or function of the organism.

mutualism (mū'-choo-ul-iz-um): a symbiotic relationship in which both participating species benefit.

mycelium (mī-sēl'-ē-um; pl., mycelia): the body of a fungus, consisting of a mass of hyphae.

mycorrhiza (mī-kō-rī'-zuh; pl., mycorrhizae): a symbiotic association between a fungus and the roots of a land plant that facilitates mineral extraction and absorption.

natural causality: the scientific principle that natural events occur as a result of preceding natural causes.

natural increase: the difference between births and deaths in a population. This number will be positive if the population is increasing and negative if it is decreasing.

natural laws: basic principles derived from the study of nature that have never been disproven by scientific inquiry. Natural laws include the laws of gravity, the behavior of light, and the way atoms interact with one another.

natural selection: unequal survival and reproduction of organisms due to heritable differences in their phenotypes, with the result that better adapted phenotypes become more common in the population.

nearshore zone: the region of coastal water that is relatively shallow but constantly submerged and that can support large plants or seaweeds; includes bays and coastal wetlands.

nerve cord: a major nervous pathway consisting of a cord of nervous tissue extending lengthwise through the body, paired in many invertebrates and unpaired in chordates.

net primary production: the energy stored in the autotrophs of an ecosystem over a given time period.

neutral mutation: a mutation that does not detectably change the function of the encoded protein.

neutron: a subatomic particle that is found in the nuclei of atoms, bears no charge, and has a mass approximately equal to that of a proton.

nicotinamide adenine dinucleotide (NAD$^+$ or NADH): an electron carrier molecule produced in the cytoplasmic fluid by glycolysis and in the mitochondrial matrix by the Krebs cycle; subsequently donates electrons to the electron transport chain.

nicotinamide adenine dinucleotide phosphate (NADP$^+$ or NADPH): an electron carrier molecule used in photosynthesis to transfer high-energy electrons from the light reactions to the Calvin cycle.

nitrogen cycle: the biogeochemical cycle by which nitrogen moves from its primary reservoir of nitrogen gas in the atmosphere via nitrogen-fixing bacteria to reservoirs in soil and water, through producers and into higher trophic levels, and then back to its reservoirs.

nitrogen fixation: the process that combines atmospheric nitrogen with hydrogen to form ammonia (NH_3).

nitrogen-fixing bacterium: a bacterium that possesses the ability to remove nitrogen (N_2) from the atmosphere and combine it with hydrogen to produce ammonia (NH_3).

no-till: a method of growing crops that leaves the remains of harvested crops in place, with the next year's crops being planted directly in the remains of last year's crops without significant disturbance of the soil.

noncompetitive inhibition: the process by which an inhibitory molecule binds to a site on an enzyme that is distinct from the active site. As a result, the enzyme's active site is distorted, making it less able to catalyze the reaction involving its normal substrate.

nondisjunction: an error in meiosis in which chromosomes fail to segregate properly into the daughter cells.

nonpolar covalent bond: a covalent bond with equal sharing of electrons.

nonvascular plant: a plant that lacks lignin and well-developed conducting vessels. Nonvascular plants include mosses, hornworts, and liverworts.

northern coniferous forest: a biome with long, cold winters and only a few months of warm weather; dominated by evergreen coniferous trees; also called *taiga*.

notochord (nōt'-ō-kord): a stiff, but somewhat flexible, supportive rod that extends along the head-to-tail axis and is found in all members of the phylum Chordata at some stage of development.

nuclear envelope: the double-membrane system surrounding the nucleus of eukaryotic cells; the outer membrane is typically continuous with the endoplasmic reticulum.

nuclear pore complex: an array of proteins that line pores in the nuclear membrane and control which substances enter and leave the nucleus.

nucleic acid (noo-klā'-ik): an organic molecule composed of nucleotide subunits; the two common types of nucleic acids are ribonucleic acid (RNA) and deoxyribonucleic acid (DNA).

nucleoid (noo-klē-oid): the location of the genetic material in prokaryotic cells; not membrane enclosed.

nucleolus (noo-klē'-ō-lus; pl., nucleoli): the region of the eukaryotic nucleus that is engaged in ribosome synthesis; consists of the genes encoding ribosomal RNA, newly synthesized ribosomal RNA, and ribosomal proteins.

nucleotide: a subunit of which nucleic acids are composed; a phosphate group bonded to a sugar (deoxyribose in DNA), which is in turn bonded to a nitrogen-containing base (adenine, guanine, cytosine, or thymine in DNA). Nucleotides are linked together, forming a strand of nucleic acid, by bonds between the phosphate of one nucleotide and the sugar of the next nucleotide.

nucleotide substitution mutation: a mutation in which a single base pair in DNA has been changed.

nucleus (atomic): the central region of an atom, consisting of protons and neutrons.

nucleus (cellular): the membrane-bound organelle of eukaryotic cells that contains the cell's genetic material.

nutrient: a substance acquired from the environment and needed for the survival, growth, and development of an organism.

nutrient cycle: the pathways of a specific nutrient (such as carbon, nitrogen, phosphorus, or water) through the living and nonliving portions of an ecosystem; also called a *biogeochemical cycle*.

observation: in the scientific method, the recognition of and a statement about a specific phenomenon, usually leading to the formulation of a question about the phenomenon.

oil: a lipid composed of three fatty acids, some of which are unsaturated, covalently bonded to a molecule of glycerol; oils are liquid at room temperature.

oligotrophic lake: a lake that is very low in nutrients and hence supports little phytoplankton, plant, and algal life; contains clear water with deep light penetration.

omnivore: an organism that consumes both plants and animals.

open circulatory system: a type of circulatory system found in some invertebrates, such as arthropods and most mollusks, that includes an open space (the hemocoel) in which blood directly bathes body tissues.

open ocean: that part of the ocean in which the water is so deep that wave action does not affect the bottom, even during strong storms.

operant conditioning: a laboratory training procedure in which an animal learns to perform a response (such as pressing a lever) through reward or punishment.

operator: a sequence of DNA nucleotides in a prokaryotic operon that binds regulatory proteins that control the ability of RNA polymerase to transcribe the structural genes of the operon.

operon (op'-er-on): in prokaryotes, a set of genes, often encoding the proteins needed for a complete metabolic pathway, including both the structural genes and a common promoter and operator that control transcription of the structural genes.

order: in Linnaean classification, the taxonomic rank composed of related families. Related orders make up a class.

organ: a structure (such as the liver, kidney, or skin) composed of two or more distinct tissue types that function together.

organ system: two or more organs that work together to perform a specific function; for example, the digestive system.

organelle (or-guh-nel'): a membrane-enclosed structure found inside a eukaryotic cell that performs a specific function.

organic: describing a molecule that contains both carbon and hydrogen.

organic molecule: a molecule that contains both carbon and hydrogen.

organism (or'-guh-niz-um): an individual living thing.

origin: the site of attachment of a muscle to the relatively stationary bone on one side of a joint.

osmosis (oz-mō'-sis): the diffusion of water across a differentially permeable membrane, normally down a concentration gradient of free water molecules. Water moves into the solution that has a lower concentration of free water from a solution that has a higher concentration of free water.

overexploitation: hunting or harvesting natural populations at a rate that exceeds those populations' ability to replenish their numbers.

ovule: a structure within the ovary of a flower, inside which the female gametophyte develops; after fertilization, it develops into the seed.

ozone hole: a region of severe ozone loss in the stratosphere caused by ozone-depleting chemicals; maximum ozone loss occurs from September to early October over Antarctica.

ozone layer: the ozone-enriched layer of the upper atmosphere (stratosphere) that filters out much of the sun's ultraviolet radiation.

parabasalid: a member of a protist group characterized by mutualistic or parasitic relationships with the animal species inside which they live. Parabasalids are part of a larger group known as the excavates.

parasite (par'-uh-sīt): an organism that lives in or on a larger organism (its host), harming the host but usually not killing it immediately.

passive transport: the movement of materials across a membrane down a gradient of concentration, pressure, or electrical charge without using cellular energy.

pathogenic (path'-ō-jen-ik): capable of producing disease; referring to an organism with such a capability (a pathogen).

pedigree: a diagram showing genetic relationships among a set of individuals, normally with respect to a specific genetic trait.

pelagic (puh-la'-jik): free-swimming or floating.

peptide (pep'-tīd): a chain composed of two or more amino acids linked together by peptide bonds.

peptide bond: the covalent bond between the nitrogen of the amino group of one amino acid and the carbon of the carboxyl group of a second amino acid, joining the two amino acids together in a peptide or protein.

periodic table: a chart first devised by Russian chemist Dmitri Mendeleev that includes all known elements and organizes them according to their atomic numbers in rows and their general chemical properties in columns (see Appendix II).

permafrost: a permanently frozen layer of soil, usually found in tundra of the Arctic or high mountains.

pH scale: a scale, with values from 0 to 14, used for measuring the relative acidity of a solution; at pH 7 a solution is neutral, pH 0 to 7 is acidic, and pH 7 to 14 is basic; each unit on the scale represents a tenfold change in H^+ concentration.

phagocytosis (fa-gō-sī-tō'-sis): a type of endocytosis in which extensions of a plasma membrane engulf extracellular particles, enclose them in a membrane-bound sac, and transport them into the interior of the cell.

pharyngeal gill slit (far-in'-jē-ul): one of a series of openings, located just posterior to the mouth, that connects the throat to the outside environment; present (as some stage of life) in all chordates.

phenotype (fēn'-ō-tīp): the physical characteristics of an organism; can be defined as outward appearance (such as flower color), as behavior, or in molecular terms (such as glycoproteins on red blood cells).

pheromone (fer'-uh-mōn): a chemical produced by an organism that alters the behavior or physiological state of another member of the same species.

phloem (flō-um): a conducting tissue of vascular plants that transports a concentrated solution of sugars (primarily sucrose) and other organic molecules up and down the plant.

phospholipid (fos-fō-li'-pid): a lipid consisting of glycerol bonded to two fatty acids and one phosphate group, which bears another group of atoms, typically charged and containing nitrogen. A double layer of phospholipids is a component of all cellular membranes.

phospholipid bilayer: a double layer of phospholipids that forms the basis of all cellular membranes. The phospholipid heads, which are hydrophilic, face the watery interstitial fluid, a watery external environment, or the cytosol; the tails, which are hydrophobic, are buried in the middle of the bilayer.

phosphorus cycle (fos'-for-us): the biogeochemical cycle by which phosphorus moves from its primary reservoir—phosphate-rich rock—to reservoirs of phosphate in soil and water, through producers and into higher trophic levels, and then back to its reservoirs.

photic zone: the region of an ocean where light is strong enough to support photosynthesis.

photon (fō'-ton): the smallest unit of light energy.

photorespiration: a series of reactions in plants in which O_2 replaces CO_2 during the Calvin cycle, preventing carbon fixation; this wasteful process dominates when C_3 plants are forced to close their stomata to prevent water loss.

photosynthesis: the complete series of chemical reactions in which the energy of light is used to synthesize high-energy organic molecules, usually carbohydrates, from low-energy inorganic molecules, usually carbon dioxide and water.

photosystem: in thylakoid membranes, a cluster of chlorophyll, accessory pigment molecules, proteins, and other molecules that collectively capture light energy, transfer some of the energy to electrons, and transfer the energetic electrons to an adjacent electron transport chain.

phylogeny (fī-lah'-jen-ē): the evolutionary history of a group of species.

phylum (fī'-lum): in Linnaean classification, the taxonomic rank composed of related classes. Related phyla make up a kingdom.

phytoplankton (fī'-tō-plank-ten): photosynthetic protists that are abundant in marine and freshwater environments.

pigment molecule: a light-absorbing, colored molecule, such as chlorophyll, carotenoid, or melanin molecules.

pilus (pl., pili): a hair-like protein structure that projects from the cell wall of many bacteria. Attachment pili help bacteria adhere to structures. Sex pili assist in the transfer of plasmids.

pinocytosis (pi-nō-sī-tō′-sis): the nonselective movement of extracellular fluids and their dissolved substances, enclosed within a vesicle formed from the plasma membrane, into a cell.

pioneer: an organism that is among the first to colonize an unoccupied habitat in the first stages of succession.

placenta (pluh-sen′-tuh): in mammals, a structure formed by a complex interweaving of the uterine lining and the embryonic membranes, especially the chorion; functions in gas, nutrient, and waste exchange between embryonic and maternal circulatory systems and also secretes the hormones estrogen and progesterone, which are essential to maintaining pregnancy.

placental (pluh-sen′-tul): referring to a mammal possessing a complex placenta (that is, species that are not marsupials or monotremes).

plankton: microscopic organisms that live in marine or freshwater environments; includes phytoplankton and zooplankton.

plasma membrane: the outer membrane of a cell, composed of a bilayer of phospholipids in which proteins are embedded.

plasmid (plaz′-mid): a small, circular piece of DNA located in the cytoplasm of many bacteria; usually does not carry genes required for the normal functioning of the bacterium, but may carry genes, such as those for antibiotic resistance, that assist bacterial survival in certain environments.

plasmodesma (plaz-mō-dez′-muh; pl., plasmodesmata): a cell-to-cell junction in plants that connects the cytosol of adjacent cells.

plasmodium (plaz-mō′-dē-um): a slug-like mass of cytoplasm containing thousands of nuclei that are not confined within individual cells.

plastid (plas′-tid): in plant cells, an organelle bounded by two membranes that may be involved in photosynthesis (chloroplasts), pigment storage, or food storage.

plate tectonics: the theory that Earth's crust is divided into irregular plates that are converging, diverging, or slipping by one another; these motions cause continental drift, the movement of continents over Earth's surface.

play: behavior that seems to lack any immediate function and that often includes modified versions of behaviors used in other contexts.

pleated sheet: a form of secondary structure exhibited by certain proteins, such as silk, in which many protein chains lie side by side, with hydrogen bonds holding adjacent chains together.

pleiotropy (plē′-ō-trō-pē): a situation in which a single gene influences more than one phenotypic characteristic.

polar covalent bond: a covalent bond with unequal sharing of electrons, such that one atom is relatively negative and the other is relatively positive.

pollen: the male gametophyte of a seed plant; also called a *pollen grain*.

polygenic inheritance: a pattern of inheritance in which the interactions of two or more functionally similar genes determine phenotype.

polymer (pah′-li-mer): a molecule composed of three or more (perhaps thousands) smaller subunits called monomers, which may be identical (for example, the glucose monomers of starch) or different (for example, the amino acids of a protein).

polymerase chain reaction (PCR): a method of producing virtually unlimited numbers of copies of a specific piece of DNA, starting with as little as one copy of the desired DNA.

polypeptide: a long chain of amino acids linked by peptide bonds. A protein consists of one or more polypeptides.

polyploid (pahl′-ē-ploid): having more than two copies of each homologous chromosome.

polysaccharide (pahl-ē-sak′-uh-rīd): a large carbohydrate molecule composed of branched or unbranched chains of repeating monosaccharide subunits, normally glucose or modified glucose molecules; includes starches, cellulose, and glycogen.

population: all the members of a particular species within a defined area, found in the same time and place and actually or potentially interbreeding.

population bottleneck: the result of an event that causes a population to become extremely small; may cause genetic drift that results in changed allele frequencies and loss of genetic variability.

population cycle: regularly recurring, cyclical changes in population size.

post-anal tail: a tail that extends beyond the anus and contains muscle tissue and the most posterior part of the nerve cord; found in all chordates at some stage of development.

postmating isolating mechanism: any structure, physiological function, or developmental abnormality that prevents organisms of two different species, once mating has occurred, from producing vigorous, fertile offspring.

potential energy: "stored" energy including chemical energy (stored in molecules), elastic energy (such as stored in a spring), or gravitational energy (stored in the elevated position of an object).

prairie: a biome, located in the centers of continents, that primarily supports grasses; also called *grassland*.

predation (pre-dā′-shun): the act of eating another living organism.

predator: an organism that eats other organisms.

prediction: in the scientific method, a statement describing an expected observation or the expected outcome of an experiment, assuming that a specific hypothesis is true.

premating isolating mechanism: any structure, physiological function, or behavior that prevents organisms of two different species from exchanging gametes.

prey: organisms that are eaten, and often killed, by another organism (a predator).

primary consumer: an organism that feeds on producers; an herbivore.

primary electron acceptor: a molecule in the reaction center of each photosystem that accepts an electron from one of the two reaction center chlorophyll *a* molecules and transfers the electron to an adjacent electron transport chain.

primary structure: the amino acid sequence of a protein.

primary succession: succession that occurs in an environment, such as bare rock, in which no trace of a previous community is present.

primate: a member of the mammalian clade Primates, characterized by the presence of an opposable thumb, forward-facing eyes, and a well-developed cerebral cortex; includes lemurs, monkeys, apes, and humans.

prion (prē′-on): a protein that, in mutated form, acts as an infectious agent that causes certain neurodegenerative diseases, including kuru and scrapie.

producer: a photosynthetic organism; an autotroph.

product: an atom or molecule that is formed from reactants in a chemical reaction.

profundal zone: the part of a lake in which light is insufficient to support photosynthesis.

prokaryote (prō-kar′-ē-ōt): an organism whose cells are prokaryotic (their genetic material is not enclosed in a membrane-bound nucleus and they lack other membrane-bound organelles); bacteria and archaea are prokaryotes.

prokaryotic (prō-kar-ē-ot′-ik): referring to cells of the domains Bacteria or Archaea. Prokaryotic cells have genetic material that is not enclosed in a membrane-bound nucleus; they also lack other membrane-bound organelles.

prokaryotic fission: the process by which a single bacterium divides in half, producing two identical offspring.

promoter: a specific sequence of DNA at the beginning of a gene, to which RNA polymerase binds and starts gene transcription.

prophase (prō′-fāz): the first stage of mitosis, in which the chromosomes first become visible in the light microscope as thickened, condensed threads and the spindle begins to form; as the spindle is completed, the nuclear envelope breaks apart, and the spindle microtubules invade the nuclear region and attach to the kinetochores of the chromosomes. Also, the first stage of meiosis: In meiosis I, the homologous chromosomes pair up, exchange parts at chiasmata, and attach to spindle microtubules; in meiosis II, the spindle re-forms and chromosomes attach to the microtubules.

protein: a polymer composed of amino acids joined by peptide bonds.

protist: a eukaryotic organism that is not a plant, animal, or fungus. The term encompasses a diverse array of organisms and does not represent a monophyletic group.

protocell: the hypothetical evolutionary precursor of living cells, consisting of a mixture of organic molecules within a membrane.

proton: a subatomic particle that is found in the nuclei of atoms; it bears a unit of positive charge and has a relatively large mass, roughly equal to the mass of a neutron.

protostome (prō′-tō-stōm): an animal with a mode of embryonic development in which the coelom is derived from splits in the mesoderm; characteristic of arthropods, annelids, and mollusks.

protozoan (prō-tuh-zō′-an; pl., protozoa): a nonphotosynthetic, single-celled protist.

pseudocoelom (soo′-dō-sēl′-ōm): in animals, a "false coelom," that is, a space or cavity, partially but not fully lined with tissue derived from mesoderm, that separates the body wall from the inner organs; found in roundworms.

pseudoplasmodium (soo′-dō-plaz-mō′-dē-um): an aggregation of individual amoeboid cells that form a slug-like mass.

pseudopod (soo′-dō-pod): an extension of the plasma membrane by which certain cells, such as amoebas, locomote and engulf prey.

Punnett square method: a method of predicting the genotypes and phenotypes of offspring in genetic crosses.

pupa (pl., pupae): a developmental stage in some insect species in which the organism stops moving and feeding and may be encased in a cocoon; occurs between the larval and the adult phases.

quaternary structure (kwat′-er-nuh-rē): the complex three-dimensional structure of a protein consisting of more than one peptide chain.

question: in the scientific method, a statement that identifies a particular aspect of an observation that a scientist wishes to explain.

***r*-selected species:** species that typically live in rapidly changing, unpredictable environments and usually do not have population sizes that approach carrying capacity. *r*-selected species usually mature rapidly, have a short life span, produce a large number of small offspring, and provide little parental care, so most offspring die before reaching maturity.

radial symmetry: a body plan in which any plane along a central axis will divide the body into approximately mirror-image halves. Cnidarians and many adult echinoderms exhibit radial symmetry.

radioactive: pertaining to an atom with an unstable nucleus that spontaneously breaks apart or decays, with the emission of radiation.

radiolarian (rā-dē-ō-lar′-ē-un): a member of a protist group characterized by pseudopods and typically elaborate silica shells. Radiolarians are largely aquatic (mostly marine) and are part of a larger group known as rhizarians.

rain shadow: a local dry area, usually located on the downwind side of a mountain range that blocks the prevailing moisture-bearing winds.

random distribution: the distribution characteristic of populations in which the probability of finding an individual is equal in all parts of an area.

reactant: an atom or molecule that is used up in a chemical reaction to form a product.

reaction center: two chlorophyll *a* molecules and a primary electron acceptor complexed with proteins and located near the center of each photosystem within the thylakoid membrane. Light energy is passed to one of the chlorophylls, which donates an energized electron to the primary electron acceptor, which then passes the electron to an adjacent electron transport chain.

receptor protein: a protein, located in a membrane or the cytosol of a cell, that binds to specific molecules (for example, a hormone or neurotransmitter), triggering a response in the cell, such as endocytosis, changes in metabolic rate, cell division, or electrical changes.

receptor-mediated endocytosis: the selective uptake of molecules from the interstitial fluid by binding to a receptor located at a coated pit on the plasma membrane and pinching off the coated pit into a vesicle that moves into the cytosol.

recessive: an allele that is expressed only in homozygotes and is completely masked in heterozygotes.

recognition protein: a protein or glycoprotein protruding from the outside surface of a plasma membrane that identifies a cell as belonging to a particular species, to a specific individual of that species, and in many cases to one specific organ within the individual.

recombinant DNA: DNA that has been altered by the addition of DNA from a different organism, typically from a different species.

recombination: the formation of new combinations of the different alleles of each gene on a chromosome; the result of crossing over.

regulatory gene: in prokaryotes, a gene encoding a protein that binds to the operator of one or more operons, controlling the ability of RNA polymerase to transcribe the structural genes of the operon.

replacement level fertility (RLF): the average number of offspring per female that is required to maintain a stable population.

repressor protein: in prokaryotes, a protein encoded by a regulatory gene, which binds to the operator of an operon and prevents RNA polymerase from transcribing the structural genes.

reproductive isolation: the failure of organisms of one population to breed successfully with members of another; may be due to premating or postmating isolating mechanisms.

reptile: a member of the chordate group that includes the snakes, lizards, turtles, alligators, birds, and crocodiles.

reservoir: the major source and storage site of a nutrient in an ecosystem, normally in the abiotic portion.

resource partitioning: the coexistence of two species with similar requirements, each occupying a smaller niche than either would if it were by itself; a means of minimizing the species' competitive interactions.

restriction enzyme: an enzyme, usually isolated from bacteria, that cuts double-stranded DNA at a specific nucleotide sequence; the nucleotide sequence that is cut differs for different restriction enzymes.

restriction fragment: a piece of DNA that has been isolated by cleaving a larger piece of DNA with restriction enzymes.

restriction fragment length polymorphism (RFLP): a difference in the length of DNA fragments that were produced by cutting samples of DNA from different individuals of the same species with the same set of restriction enzymes; fragment length differences occur because of differences in nucleotide sequences, and hence in the ability of restriction enzymes to cut the DNA, among individuals of the same species.

rhizarian: a member of Rhizaria, a protist clade. Rhizarians, which use thin pseudopods to move and capture prey and which often have hard shells, include the foraminiferans and the radiolarians.

ribonucleic acid (RNA) (rī-bō-noo-klā′-ik; RNA): a molecule composed of ribose nucleotides, each of which consists of a phosphate group, the sugar ribose, and one of the bases adenine, cytosine, guanine, or uracil; involved in converting the information in DNA into protein; also the genetic material of some viruses.

ribosomal RNA (rī-bō-sō′-mul; rRNA): a type of RNA that combines with proteins to form ribosomes.

ribosome (rī′-bō-sōm): a complex consisting of two subunits, each composed of ribosomal RNA and protein, found in the cytoplasm of cells or attached to the endoplasmic reticulum, that is the site of protein synthesis, during which the sequence of bases of messenger RNA is translated into the sequence of amino acids in a protein.

ribozyme: an RNA molecule that can catalyze certain chemical reactions, especially those involved in the synthesis and processing of RNA itself.

RNA polymerase: in RNA synthesis, an enzyme that catalyzes the bonding of free RNA nucleotides into a continuous strand, using RNA nucleotides that are complementary to those of the template strand of DNA.

rubisco: in the carbon fixation step of the Calvin cycle, the enzyme that catalyzes the reaction between ribulose bisphosphate (RuBP) and carbon dioxide, thereby fixing the carbon of carbon dioxide in an organic molecule; short for ribulose bisphosphate carboxylase.

rumen fungus: a member of the fungus clade Neocallimastigomycota, whose members have swimming spores with multiple flagella. Rumen fungi are anaerobic and most live in the digestive tracts of plant-eating animals.

S-curve: the S-shaped growth curve produced by logistic population growth, usually describing a population of organisms introduced into a new area; consists of an initial period of exponential growth, followed by a decreasing growth rate, and finally, relative stability around a growth rate of zero.

sac fungus: a member of the fungus clade Ascomycota, whose members form spores in a saclike case called an ascus.

sarcoplasmic reticulum (sark'-ō-plas'-mik re-tik'-ū-lum; SR): the specialized endoplasmic reticulum in muscle cells; forms interconnected hollow tubes. The sarcoplasmic reticulum stores calcium ions and releases them into the interior of the muscle cell, initiating contraction.

saturated: referring to a fatty acid with as many hydrogen atoms as possible bonded to the carbon backbone (therefore, a saturated fatty acid has no double bonds in its carbon backbone).

savanna: a biome that is dominated by grasses and supports scattered trees; typically has a rainy season during which most of the year's precipitation falls, followed by a dry season during which virtually no precipitation occurs.

science: the organized and systematic inquiry, through observation and experiment, into the origins, structure, and behavior of our living and nonliving surroundings.

scientific method: a rigorous procedure for making observations of specific phenomena and searching for the order underlying those phenomena.

scientific name: the two-part Latin name of a species; consists of the genus name followed by the species name.

scientific theory: an explanation of natural phenomena developed through extensive and reproducible observations; more general and reliable than a hypothesis.

scientific theory of evolution: the theory that modern organisms descended, with modification, from preexisting life-forms.

second law of thermodynamics: the principle of physics that states that any change in an isolated system causes the quantity of concentrated, useful energy to decrease and the amount of randomness and disorder (entropy) to increase.

secondary consumer: an organism that feeds on primary consumers; a type of carnivore.

secondary structure: a repeated, regular structure assumed by a protein chain, held together by hydrogen bonds; for example, a helix.

secondary succession: succession that occurs after an existing community is disturbed—for example, after a forest fire; secondary succession is much more rapid than primary succession.

seed: the reproductive structure of a seed plant, protected by a seed coat; contains an embryonic plant and a supply of food for it.

segmentation (seg-men-tā'-shun): an animal body plan in which the body is divided into repeated, typically similar units.

selectively permeable: the quality of a membrane that allows certain molecules or ions to move through it more readily than others.

self-fertilization: the union of sperm and egg from the same individual.

semiconservative replication: the process of replication of the DNA double helix; the two DNA strands separate, and each is used as a template for the synthesis of a complementary DNA strand. Consequently, each daughter double helix consists of one parental strand and one new strand.

septum (pl., septa): a partition that separates the fungal hypha into individual cells; pores in septa allow the transfer of materials between cells.

sex chromosome: either of the pair of chromosomes that usually determines the sex of an organism; for example, the X and Y chromosomes in mammals.

sex-linked: referring to a pattern of inheritance characteristic of genes located on one type of sex chromosome (for example, X) and not found on the other type (for example, Y); in mammals, in almost all cases, the gene controlling the trait is on the X chromosome, so this pattern is often called X-linked. In X-linked inheritance, females show the dominant trait unless they are homozygous recessive, whereas males express whichever allele, dominant or recessive, is found on their single X chromosome.

sexual reproduction: a form of reproduction in which genetic material from two parent organisms is combined in the offspring; usually, two haploid gametes fuse to form a diploid zygote.

sexual selection: a type of natural selection that acts on traits involved in finding and acquiring mates.

short tandem repeat (STR): a DNA sequence consisting of a short sequence of nucleotides (usually two to five nucleotides in length) repeated multiple times, with all of the repetitions side by side on a chromosome; variations in the number of repeats of a standardized set of 13 STRs produce DNA profiles used to identify people by their DNA.

sickle-cell anemia: a recessive disease caused by a single amino acid substitution in the hemoglobin molecule. Sickle-cell hemoglobin molecules tend to cluster together, distorting the shape of red blood cells and causing them to break and clog capillaries.

simple diffusion: the diffusion of water, dissolved gases, or lipid-soluble molecules through the phospholipid bilayer of a cellular membrane.

social learning: learning that is influenced by observation of, or interaction with, other animals, usually of the same species.

solute: a substance dissolved in a solvent.

solution: a solvent containing one or more dissolved substances (solutes).

solvent: a liquid capable of dissolving (uniformly dispersing) other substances in itself.

speciation: the process of species formation, in which a single species splits into two or more species.

species (spē'-sēs): the basic unit of taxonomic classification, consisting of a population or group of populations that evolves separately from other populations. In sexually reproducing organisms, a species can be defined as a population or group of populations whose members interbreed freely with one another under natural conditions but do not interbreed with members of other populations.

specific heat: the amount of energy required to raise the temperature of 1 gram of a substance by 1°C.

spermatophore: a package of sperm formed by the males of some invertebrate animals; the spermatophore can be inserted into the female reproductive tract, where it releases its sperm.

spindle: an array of microtubules that moves the chromosomes to opposite poles of a cell during mitotic and meiotic cell division.

spindle microtubule: one of the microtubules organized in a spindle shape that separate chromosomes during meiotic and meiotic cell division.

spontaneous generation: the proposal that living organisms can arise from nonliving matter.

sporangium (spor-an'-jē-um; pl., sporangia): a structure in which spores are produced.

spore: (1) in plants and fungi, a haploid cell capable of developing into an adult without fusing with another cell (without fertilization); (2) in bacteria and some other organisms, a stage of the life cycle that is resistant to extreme environmental conditions.

sporophyte (spor´-ō-fīt): the multicellular diploid stage in the life cycle of a plant; produces haploid, asexual spores through meiosis.

stabilizing selection: a type of natural selection that favors the average phenotype in a population.

starch: a polysaccharide that is composed of branched or unbranched chains of glucose molecules; used by plants as a carbohydrate-storage molecule.

start codon: the first AUG codon in a messenger RNA molecule.

startle coloration: a form of mimicry in which a color pattern (in many cases resembling large eyes) can be displayed suddenly by a prey organism when approached by a predator.

stem cell: an undifferentiated cell that is capable of dividing and giving rise to one or more distinct types of differentiated cell(s).

steroid: a lipid consisting of four fused carbon rings, with various functional groups attached.

stoma (stō´-muh; pl., stomata): an adjustable opening in the epidermis of a leaf or young stem, surrounded by a pair of guard cells, that regulates the diffusion of carbon dioxide and water into and out of the leaf or stem.

stop codon: a codon in messenger RNA that stops protein synthesis and causes the completed protein chain to be released from the ribosome.

stramenopile: a member of Stramenopila, a large protist clade. Stramenopiles, which are characterized by hair-like projections on their flagella, include the water molds, the diatoms, and the brown algae.

strand: a single polymer of nucleotides; DNA is composed of two strands wound about each other in a double helix; RNA is usually single-stranded.

stroma (strō´-muh): the semifluid material inside chloroplasts in which the thylakoids are located; the site of the reactions of the Calvin cycle.

structural gene: in the prokaryotic operon, the genes that encode enzymes or other cellular proteins.

subclimax: a community in which succession is stopped before the climax community is reached; it is maintained by regular disturbance—for example, a tallgrass prairie maintained by periodic fires.

substrate: the atoms or molecules that are the reactants for an enzyme-catalyzed chemical reaction.

succession (suk-seh´-shun): a structural change in a community and its nonliving environment over time. During succession, species replace one another in a somewhat predictable manner until a stable, self-sustaining climax community is reached.

sucrose: a disaccharide composed of glucose and fructose.

sugar: a simple carbohydrate molecule, either a monosaccharide or a disaccharide.

sugar-phosphate backbone: a chain of sugars and phosphates in DNA and RNA; the sugar of one nucleotide bonds to the phosphate of the next nucleotide in a DNA or RNA strand. The bases in DNA or RNA are attached to the sugars of the backbone.

surface tension: the property of a liquid to resist penetration by objects at its interface with the air, due to cohesion between molecules of the liquid.

survivorship curve: the curve that results when the number of individuals of each age in a population is graphed against their age, usually expressed as a percentage of their maximum life span.

survivorship table: a data table that groups organisms born at the same time and tracks them throughout their life span, recording how many continue to survive in each succeeding year (or other unit of time). Various parameters such as gender may be used in the groupings. Human life tables may include many other parameters (such as socioeconomic status) used by demographers.

sustainable development: human activities that meet current needs for a reasonable quality of life without exceeding nature's limits and without compromising the ability of future generations to meet their needs.

symbiosis: a relationship between species in which the two or more species share a close, long-term physical association.

sympatric speciation (sim-pat´-rik): the process by which new species arise in populations that are not physically divided; the genetic isolation required for sympatric speciation may be due to ecological isolation or chromosomal aberrations (such as polyploidy).

systematics: the branch of biology concerned with reconstructing phylogenies and with naming clades.

taiga (tī´-guh): a biome with long, cold winters and only a few months of warm weather; dominated by evergreen coniferous trees; also called *northern coniferous forest*.

taxonomy (tax-on´-uh-mē): the branch of biology concerned with naming and classifying organisms.

telomere (tēl´-e-mēr): the nucleotides at the end of a chromosome that protect the chromosome from damage during condensation, and prevent the end of one chromosome from attaching to the end of another chromosome.

telophase (tēl´-ō-fāz): in mitosis and both divisions of meiosis, the final stage, in which the spindle fibers usually disappear, nuclear envelopes re-form, and cytokinesis generally occurs. In mitosis and meiosis II, the chromosomes also relax from their condensed form.

temperate deciduous forest: a biome having cold winters and warm summers, with enough summer rainfall for trees to grow and shade out grasses; characterized by trees that drop their leaves in winter (deciduous trees), an adaptation that minimizes water loss when the soil is frozen.

temperate rain forest: a temperate biome with abundant liquid water year-round, dominated by conifers.

template strand: the strand of the DNA double helix from which RNA is transcribed.

temporal isolation: reproductive isolation that arises when species do not interbreed because they breed at different times.

territoriality: the defense of an area in which important resources are located.

tertiary consumer (ter´-shē-er-ē): a carnivore that feeds on other carnivores (secondary consumers).

tertiary structure (ter´-shē-er-ē): the complex three-dimensional structure of a single peptide chain; held in place by disulfide bonds between cysteines.

test cross: a breeding experiment in which an individual showing the dominant phenotype is mated with an individual that is homozygous recessive for the same gene. The ratio of offspring with dominant versus recessive phenotypes can be used to determine the genotype of the phenotypically dominant individual.

tetrapod: an organism descended from the first four-limbed vertebrate. Tetrapods include all extinct and living amphibians, reptiles (including birds), and mammals.

therapeutic cloning: the production of a clone for medical purposes. Typically, the nucleus from one of a patient's own cells would be inserted into an egg whose nucleus had been removed; the resulting cell would divide and produce embryonic stem cells that would be compatible with the patient's tissues and therefore would not be rejected by the patient's immune system.

threatened species: all species classified as critically endangered, endangered, or vulnerable.

thylakoid (thī´-luh-koid): a disk-shaped, membranous sac found in chloroplasts, the membranes of which contain the photosystems, electron transport chains, and ATP-synthesizing enzymes used in the light reactions of photosynthesis.

thymine (T): a nitrogenous base found only in DNA; abbreviated as T.

tight junction: a type of cell-to-cell junction in animals that prevents the movement of materials through the spaces between cells.

tissue: a group of (normally similar) cells that together carry out a specific function; a tissue may also include extracellular material produced by its cells.

trans fat: a type of fat, produced during the process of hydrogenating oils, that may increase the risk of heart disease. The fatty acids of trans fats include an unusual configuration of double bonds that is not normally found in fats of biological origin.

transcription: the synthesis of an RNA molecule from a DNA template.

transfect: to introduce foreign DNA into a host cell; usually includes mechanisms to regulate the expression of the DNA in the host cell.

transfer RNA (tRNA): a type of RNA that binds to a specific amino acid, carries it to a ribosome, and positions it for incorporation into the growing protein chain during protein synthesis. A set of three bases in tRNA (the anticodon) is complementary to the set of three bases in mRNA (the codon) that codes for that specific amino acid in the genetic code.

transformation: a method of acquiring new genes, whereby DNA from one bacterium (normally released after the death of the bacterium) becomes incorporated into the DNA of another, living bacterium.

transgenic: referring to an animal or a plant that contains DNA derived from another species, usually inserted into the organism through genetic engineering.

translation: the process whereby the sequence of bases of messenger RNA is converted into the sequence of amino acids of a protein.

translocation: a mutation that occurs when a piece of DNA is removed from one chromosome and attached to another chromosome.

transport protein: a protein that regulates the movement of water-soluble molecules through the plasma membrane.

trial-and-error learning: a type of learning in which behavior is modified in response to the positive or negative consequences of an action.

triglyceride (trī-glis′-er-īd): a lipid composed of three fatty acid molecules bonded to a single glycerol molecule.

trisomy 21: see *Down syndrome*.

trisomy X: a condition of females who have three X chromosomes instead of the normal two; most such women are phenotypically normal and are fertile.

trophic level: literally, "feeding level"; the categories of organisms in a community, and the position of an organism in a food chain, defined by the organism's source of energy; includes producers, primary consumers, secondary consumers, and so on.

tropical deciduous forest: a biome, warm all year-round, with pronounced wet and dry seasons; characterized by trees that shed their leaves during the dry season (deciduous trees), an adaptation that minimizes water loss.

tropical rain forest: a biome with evenly warm, evenly moist conditions year-round, dominated by broadleaf evergreen trees; the most diverse biome.

tropical scrub forest: a biome, warm all year-round, with pronounced wet and dry seasons (drier conditions than in tropical deciduous forests); characterized by short, deciduous, often thorn-bearing trees with grasses growing beneath them.

true-breeding: pertaining to an individual all of whose offspring produced through self-fertilization are identical to the parental type. True-breeding individuals are homozygous for a given trait.

tundra: a biome with severe weather conditions (extreme cold and wind, and little rainfall) that cannot support trees.

turgor pressure: pressure developed within a cell (especially the central vacuole of plant cells) as a result of osmotic water entry.

Turner syndrome: a set of characteristics typical of a woman with only one X chromosome; women with Turner syndrome are sterile, with a tendency to be very short and to lack typical female secondary sexual characteristics.

unicellular: single-celled; most members of the domains Bacteria and Archaea and the kingdom Protista are unicellular.

uniform distribution: the distribution characteristic of a population with a relatively regular spacing of individuals, commonly as a result of territorial behavior.

unsaturated: referring to a fatty acid with fewer than the maximum number of hydrogen atoms bonded to its carbon backbone (therefore, an unsaturated fatty acid has one or more double bonds in its carbon backbone).

upwelling: an upward flow that brings cold, nutrient-laden water from the ocean depths to the surface.

variable: a factor in a scientific experiment that is deliberately manipulated in order to test a hypothesis.

vascular plant (vas′-kū-lar): a plant that has conducting vessels for transporting liquids; also called a tracheophyte.

vertebral column (ver-tē′-brul): a column of serially arranged skeletal units (the vertebrae) that protect the nerve cord in vertebrates; the backbone.

vertebrate: an animal that has a vertebral column.

vesicle (ves′-i-kul): a small, temporary, membrane-bound sac within the cytoplasm.

vestigial structure (ves-tij′-ē-ul): a structure that is the evolutionary remnant of structure that performed a useful function in an ancestor, but is currently either useless or used in a different way.

viroid (vī′-roid): a particle of RNA that is capable of infecting a cell and of directing the production of more viroids; responsible for certain plant diseases.

virus (vī′-rus): a noncellular parasitic particle that consists of a protein coat surrounding genetic material; multiplies only within a cell of a living organism (the host).

vulnerable species: a species that is likely to become endangered unless conditions that threaten its survival improve.

waggle dance: a symbolic form of communication used by honeybee foragers to communicate the location of a food source to their hive mates.

warning coloration: bright coloration that warns predators that the potential prey is distasteful or even poisonous.

water mold: a member of a protist group that includes species with filamentous shapes that give them a superficially fungus-like appearance. Water molds, which include species that cause economically important plant diseases, are part of a larger group known as the stramenopiles.

wax: a lipid composed of fatty acids covalently bonded to long-chain alcohols.

weather: short-term fluctuations in temperature, humidity, cloud cover, wind, and precipitation in a region over periods of hours to days.

wetlands: a region (sometimes called a marsh, swamp, or bog) in which the soil is covered by, or saturated with, water for a significant part of the year.

wildlife corridor: a strip of protected land linking larger areas. Wildlife corridors allow animals to move freely and safely between habitats that would otherwise be isolated by human activities.

work: energy transferred to an object, usually causing the object to move.

working memory: the first phase of learning; short-term memory that is electrical or biochemical in nature.

X chromosome: the female sex chromosome in mammals and some insects.

xylem (zī-lum): a conducting tissue of vascular plants that transports water and minerals from root to shoot.

Y chromosome: the male sex chromosome in mammals and some insects.

zooplankton: nonphotosynthetic protists that are abundant in marine and freshwater environments.

zygomycete: a fungus species formerly placed in the now-defunct taxonomic group Zygomycota. Zygomycetes, which include the species that cause fruit rot and bread mold, do not constitute a true clade and are now distributed among several other taxonomic groups.

zygosporangium (zī′-gō-spor-an-jee-um): a tough, resistant reproductive structure produced by some fungi, such as bread molds; encloses diploid nuclei that undergo meiosis and give rise to haploid spores.

zygote (zī′-gōt): in sexual reproduction, a diploid cell (the fertilized egg) formed by the fusion of two haploid gametes.

Chapter 1

Think Critically

Figure Captions

Figure 1-3 The bun is made from wheat, a photosynthetic organism that can capture sunlight directly. The meat came from a cow, which feeds on photosynthetic organisms and fuels its body with energy stored from photosynthesis.

Figure 1-9 Of all the animals, plant-eaters have access to the largest amount of energy because they feed on plants, which capture the energy directly from sunlight. The huge size of some herbivores is also an adaptation for reaching leaves on tall trees.

Figure 1-10 Global climate change

How Do We Know That? Controlled Experiments Provide Reliable Data

Anything that could have gotten through the gauze might possibly have produced the maggots, or they could have come from eggs that were in the meat before it was placed in the jars. An experiment to support the hypothesis that flies produce maggots would be to use Redi's experimental setup with meat in each of two gauze-covered jars, but place flies in one jar and not the other.

Multiple Choice

1. b; **2.** c; **3.** a; **4.** d; **5** a

Fill-in-the-Blank

1. stimuli; energy, materials; complex, organized; evolve
2. atom; cell; tissues; population; community; ecosystem
3. scientific theory; hypothesis; scientific method
4. evolution; natural selection
5. deoxyribonucleic acid, DNA; genes

Chapter 2

Think Critically

Figure Captions

Figure 2-1 The mass number of H is 1; the mass number of He is 4.

Figure 2-2 Atoms with outer shells that are not full become stable by filling (or emptying) their outer shells.

Figure 2-3 Heat from the fire excites electrons into higher energy levels. When the electrons spontaneously revert to their original stable level, they give off light as well as heat.

Figure 2-9 No. The hydrophobic oil exerts no attraction for the ions in salt, so the salt would remain in solid crystals.

Figure 2-10 The water drop would spread out on clean glass because of adhesion to the glass and cohesion among the water molecules. On an oil-covered slide, the droplet would round up due to hydrophobic interactions.

Figure 2-11 There is more empty space between the molecules in ice than between molecules of liquid water, so ice is less dense and floats.

How Do We Know That? Radioactive Revelations

Fluid-filled space occupies a far larger portion of the brain of the Alzheimer's patient, indicating a significant loss of neurons.

Evaluate This

Health Watch: Free Radicals—Friends and Foes?

Eating dark chocolate by itself is unlikely to reverse high blood pressure. If Thomas added a 6-oz chocolate bar (about 1000 Calories) to his regular daily diet, he could gain 50 pounds over 6 months, which would likely worsen his high blood pressure. Thomas should maintain a healthy weight, gradually increase his exercise, and eat lots of fruits and vegetables. In addition, it would be okay for him to add a far smaller amount of dark chocolate daily.

Multiple Choice

1. d; **2.** a; **3.** d; **4.** b; **5.** d

Fill-in-the-Blank

1. protons, neutrons; electrons, electron shells
2. ion; positive; ionic
3. isotopes; elements; radioactive
4. inert; reactive; share
5. polar; hydrogen; cohesion

Chapter 3

Think Critically

Figure Captions

Figure 3-1 Hydrogen cyanide is a polar molecule, because the nitrogen atom exerts a much stronger pull on carbon's electrons than does the hydrogen atom.

Figure 3-8 During hydrolysis of sucrose, water is split; a hydrogen from water is added to the oxygen from glucose (that formerly linked the two subunits), and the remaining OH from water is added to the carbon (that formerly bonded to oxygen) of the fructose subunit.

Figure 3-14 Other amino acids with hydrophobic functional groups are glycine, alanine, valine, isoleucine, methionine, tryptophan, and proline.

Figure 3-16 Heat energy can break hydrogen bonds and disrupt a protein's three-dimensional structure. This prevents the protein from carrying out its usual function(s).

Figure 3-22 Triglycerides are broken apart by hydrolysis reactions.

Figure 3-27 As lipids, steroids are soluble in the phospholipid-based cell membranes and can cross them to act inside the cell.

Evaluate This

Health Watch: Fake Foods

The doctor should do the following:

- Emphasize the importance of losing weight and checking the calories in food. Sugar-free cakes and donuts will likely have at least as many calories as the regular type.
- Explain that a sweet tooth can (and should) be tamed by gradually cutting back on sweets.
- Suggest the patient get more exercise.
- Provide standard advice for pre-diabetes patients (this information is beyond the scope of this chapter but is available online on many reputable Web sites).

Health Watch: Cholesterol, *Trans* Fats, and Your Heart

The doctor should do the following:

- Do a blood test for high LDL cholesterol; this could cause partial artery blockage that would deprive the heart of adequate oxygen during exercise and cause chest pain.
- Recommend that the patient lose weight.
- Ask about the patient's exercise regimen (if any) and recommend that the patient try a variety of exercise classes to find one she enjoys.
- Ask if her diet includes a lot of saturated fat; if so, strongly recommend she switch to oil.
- Ask if she smokes, which contributes to heart disease; if so, encourage cessation.
- Suggest additional tests that are used to evaluate arterial health and explain how obesity damages the heart (use reputable Web sites to research this information).

Multiple Choice

1. c; **2.** c; **3.** a; **4.** d; **5** a

Fill-in-the-Blank

1. monomer, polymers; polysaccharides; hydrolysis; (any three) cellulose, starch, glycogen, chitin
2. hydrogen bond, hydrogen and disulfide bonds, hydrogen bond, peptide bond
3. dehydration, water; amino acids; primary; helix, pleated sheet; denatured
4. ribose sugar, base, phosphate; adenosine triphosphate (ATP); adenine, guanine, cytosine, thymine; deoxyribonucleic acid (DNA), ribonucleic acid (RNA); phosphate
5. oil, wax, fat, steroids, cholesterol, phospholipid

Chapter 4

Think Critically

Figure Captions

Figure 4-8 Fluid would not move upward because the beating of the flagella would direct fluid straight out from the cell membranes. Mucus and trapped particles would accumulate in the trachea.

Figure 4-10 Condensation allows the chromosomes to become organized and separated from one another so that a complete copy of genetic information can be distributed to each of the daughter cells that results from cell division.

Figure 4-14 The fundamentally similar composition of membranes allows them to merge with one another. This allows molecules in vesicles to be transferred from one membrane-enclosed structure (such as endoplasmic reticulum) to another (such as the Golgi apparatus). The material can be exported from the cell when the vesicle merges with the plasma membrane.

Figure 4-15 If lysosomal enzymes were active at a pH found in the cytoplasm, the enzymes would break down the membranes of the endoplasmic reticulum, Golgi, and the vesicles exposed to the enzymes.

How Do We Know That? The Search for the Cell

The light micrograph provides overall views of the structures in relationship to one another, and it also allows you to see the living cell, undistorted by preservation techniques. You can observe how the cilia beat and how they move the cell. In the SEM of the intact cell, you can observe its overall three-dimensional shape and the fact that it is completely covered with cilia. The TEM provides a more diagrammatic view of the mitochondria with a clearer image of the internal organization and folded cristae (see Fig. 4-17). The SEM of mitochondria shows that the internal membranes create confined spaces separate from the fluid inside this organelle.

Earth Watch: Would You Like Fries with Your Cultured Cow Cells?

Your graph projections will differ somewhat based on different trend estimation lines. However, China's per person meat consumption if current trends continue will still not exceed that of the United States in 2050, but it will be similar to that of the UK. The United States and UK will increase slightly; India will remain stable. China currently consumes the most total meat; in 1980, the United States exceeded China in total meat consumption.

Multiple Choice

1. a; **2.** b; **3.** b; **4.** d; **5.** b

Fill-in-the-Blank

1. phospholipids, proteins; phospholipid, protein
2. microfilaments, intermediate filaments, microtubules; microtubules; microtubules; microfilaments; intermediate filaments
3. ribosomes, endoplasmic reticulum, nucleolus, Golgi apparatus, cell wall, messenger RNA
4. rough endoplasmic reticulum; vesicles, Golgi apparatus; carbohydrate; plasma membrane
5. mitochondria, chloroplasts, extracellular matrix, nucleoid, cilia, cytoplasm

6. mitochondria, chloroplasts; double, ATP, size
7. cell walls; nucleoid; plasmids; sex pili

Chapter 5

Think Critically

Figure Captions

Figure 5-6 The distilled water made the solution hypotonic to the blood cells, causing enough water to flow in to burst their fragile cell membranes.

Figure 5-7 The rigid cell wall of plant cells counteracts the pressure exerted by water entering by osmosis, so although the cell would stiffen from internal water pressure, it would not burst.

Figure 5-8 No. Actively transporting water across a membrane against its concentration gradient would waste energy because the water would simply diffuse (osmose) back through the membrane.

Figure 5-12 Exocytosis uses cellular energy, whereas diffusion occurs passively. During exocytosis, materials are expelled without passing directly through the plasma membrane, allowing the cell to eliminate materials that are too large to pass through membranes.

How Do We Know That? The Discovery of Aquaporins

The control eggs would have shrunk very slightly. Because they swelled slightly in distilled water, it is clear that they are somewhat permeable to water, and they would have lost some water by osmosis in the hypertonic solution. The eggs with aquaporins inserted into their plasma membranes would have shrunken considerably and would be much smaller than the control eggs because water can flow either way through aquaporins, and it would have flowed out into the hypertonic environment by osmosis.

Health Watch: Membrane Fluidity, Phospholipids, and Fumbling Fingers

It is adaptive to feel enhanced pain when parts of the body are in danger of being damaged by the cold, because this will stimulate urgent behavior to warm up the affected body region.

Case Study Revisited: Vicious Venoms

Venom phospholipases are injected to injure and help immobilize the prey, whereas phospholipases in the digestive tract break down membrane phospholipids into nutrients that can be absorbed.

Multiple Choice

1. c; **2.** b; **3** b; **4** c; **5** a

Fill-in-the-Blank

1. phospholipids; receptor, recognition, enzymes, attachment, transport
2. selectively permeable; diffusion; osmosis; aquaporins; active transport
3. channel, carrier; simple, lipids
4. adhesive junctions, tight junctions, gap junctions; plasmodesmata
5. simple diffusion, simple diffusion, facilitated diffusion, facilitated diffusion
6. exocytosis; yes; vesicles

Chapter 6

Think Critically

Figure Captions

Figure 6-1 Yes, but this would be a boring roller coaster because each successive "hill" would need to be much lower than the previous one.

Figure 6-4 Glucose breakdown is exergonic; photosynthesis is endergonic.

Figure 6-7 Both parts of the coupled reaction occur with a loss of useable energy in the form of heat.

Figure 6-8 No, only exergonic reactions can occur spontaneously after the activation energy is surmounted.

Case Study Revisited: Energy Unleashed

Excess heat in the runner's body stimulates responses that reduce heat, an example of feedback inhibition. This same principle is at work when the end product of a metabolic process inhibits an enzyme involved in its production.

Evaluate This

Health Watch: Lack of an Enzyme Leads to Lactose Intolerance

He would suspect lactose intolerance, which would be even more likely if the child's parents were of East Asian, West African, or Native American origin. If tests confirmed his suspicions, he might suggest that the child reduce lactose-containing foods to a level she can tolerate and that the parents try giving her special formulations of dairy products that have lactase added to them, or lactase pills that can be taken prior to a meal.

Multiple Choice

1. b; **2.** a; **3.** d; **4.** b; **5.** d

Fill-in-the-Blank

1. created, destroyed; kinetic, potential
2. less; entropy
3. exergonic; endergonic; exergonic; endergonic
4. adenosine triphosphate; adenosine diphosphate, phosphate; energy
5. proteins; catalysts, activation energy; active site
6. inhibiting; competitive

Chapter 7
Think Critically
Figure Captions

Figure 7-5 If only red light hit the leaf, much less light energy would be gathered from chlorophyll *b* and essentially none from the carotenoids, so photosynthesis and its resulting O_2 production would decline significantly. No oxygen would be generated if only infrared light was present because there are no infrared-capturing pigments in chloroplasts. Shining green light on the leaf would substantially reduce O_2 production because neither chlorophyll *a* nor *b* would absorb it. Most green light is reflected from the leaf.

Figure E7-1 In cool, moist conditions, evaporation is not a problem, so C_3 plants can leave their stomata open, allowing adequate CO_2 to enter and O_2 to diffuse out. The C_3 pathway is more energy efficient than the C_4 pathway; C_4 uses one extra ATP per CO_2 molecule (to regenerate PEP). Thus, when CO_2 is abundant and photorespiration is not a problem, C_3 plants produce sugar at lower energy cost, and they outcompete C_4 plants.

Earth Watch: Biofuels—Are Their Benefits Bogs?

It is reasonable to assume that algal and/or cellulose biofuels will gradually (or rapidly) replace corn. Corn prices would be expected to drop and stabilize at a lower price than they are currently. Other scenarios are possible, however, and any scenario with a reasonable rationale is correct.

Multiple Choice

1. a; **2.** b; **3.** a; **4.** b; **5.** d

Fill-in-the-Blank

1. stomata, oxygen (O_2), carbon dioxide (CO_2); water loss (evaporation); chloroplasts, mesophyll
2. red, blue, violet; green; carotenoids; photosystems, thylakoid
3. reaction center; primary electron acceptor molecule; electron transport chain; hydrogen ions (H^+); chemiosmosis
4. water (H_2O), carbon dioxide (CO_2); Calvin cycle; carbon fixation
5. rubisco, oxygen (O_2); photorespiration; C_4 pathway, CAM (or crassulacean acid metabolism) pathway
6. ATP, NADH, Calvin; RuBp (or ribulose bisphosphate); G3P (glyceraldehyde-3-phosphate), glucose

Chapter 8
Think Critically
Figure Captions

Figure 8-3 Glycolysis produces a net of two NADH and two ATP molecules.

Figure 8-6 Without oxygen, electrons would be unable to continue entering the electron transport chain, and no further ATP would be produced.

Figure 8-11 NAD^+ would become unavailable for further glycolysis or cellular respiration.

Evaluate This

Health Watch: How Can You Get Fat by Eating Sugar?

Hypothesis: Colin is eating many more carbohydrates than he did previously to stay satisfied. Questions: What is he eating instead of fat? Has he decreased his average daily caloric intake? Has he increased his exercise? Recommendations: Eat a lot of fruits and vegetables; include some healthy fats from foods like avocados and nuts. Avoid refined carbohydrates, count calories, exercise regularly, and gradually increase daily exercise.

Case Study Revisited: Raising a King

Jeremy does not need to worry about a child inheriting his disorder, because mitochondrial mutations are only inherited from the mother.

Multiple Choice

1. d; **2** a; **3.** b; **4.** d; **5.** b

Fill-in-the-Blank

1. glycolysis, cellular respiration; cytosol, mitochondria; aerobic
2. anaerobic; glycolysis, two; fermentation, NAD^+
3. ethanol (alcohol), carbon dioxide; lactic acid; lactate fermentation
4. matrix, intermembrane space, concentration gradient; chemiosmosis; ATP synthase
5. Krebs; acetyl CoA; two; NADH, $FADH_2$ (order not important)

Chapter 9
Think Critically
Figure Captions

Figure 9-9 If the sister chromatids of one replicated chromosome failed to separate, then one daughter cell would not receive any copy of that chromosome, whereas the other daughter cell would receive both copies.

Figure 9-11 If the receptors were constantly "on," the cell and its daughters would divide rapidly all the time. See "Health Watch: Cancer—Running the Stop Signs at the Cell Cycle Checkpoints" for further information about the consequences.

Evaluate This

Health Watch: Cancer—Running the Stop Signs at the Cell Cycle Checkpoints

For a malignant tumor, the pathology lab would report mutated oncogenes, either causing excessive production of growth factors or making the cells of the tumor more likely divide when growth factors are present (such as mutations in cyclin genes), and mutated tumor suppressor genes (such as p53), rendering the cell likely to divide even if it has damaged DNA.

Multiple Choice

1. c; **2.** b; **3.** a; **4.** d; **5.** d

Fill-In-the-Blank

1. DNA (deoxyribonucleic acid)
2. prokaryotic fission
3. mitotic, differentiation; Stem
4. growth factors; Checkpoints; oncogenes, tumor suppressor genes
5. prophase, metaphase, anaphase, telophase; cytokinesis, telophase
6. kinetochores; polar

Chapter 10

Think Critically

Figure Captions

Figure 10-5 Draw the chromosomes and follow them through meiosis. If one pair of homologues failed to separate at anaphase I, one of the resulting daughter cells (and the gametes produced from it) would have both homologues and the other daughter cell (and the gametes produced from it) would not have any copies of that homologue. Assuming that these defective gametes fused with normal gametes, then the offspring would have either three copies or only one copy of that homologue.

Figure 10-6 Draw the chromosomes and follow them through meiosis, *including the chromosomes of each homologous pair that did not cross over at all*. Don't forget that the homologues are randomly separated during meiosis I. Therefore, some gametes would receive both of the incorrectly crossed-over chromosomes, but many would receive only one incorrectly crossed-over chromosome. When gametes were produced, some would contain a chromosome that was missing one of its own segments but contained a segment from another, nonhomologous chromosome. Assuming that these defective gametes fused with normal gametes, then the offspring would receive only one copy of some genes and three copies of other genes. For those genes that were removed from the incorrectly crossed-over chromosome, many offspring would receive those genes only from the other parent's gamete. For those genes that were added to the incorrectly crossed-over chromosome, many offspring would receive one copy of the genes added to the incorrect chromosome, a second copy of those same genes from a chromatid that never crossed over, and a third copy from the other parent's gamete.

How Do We Know That? The Evolution of Sexual Reproduction

In the graph of part (a), snails exposed to worms (worms only and worms with bacteria) mated more often than snails not exposed to worms did, indicating that parasitism caused an increase in sexual reproduction. In part (b), female snails exposed to worms mated with a greater number of different males than unexposed females did, which should increase the genetic variability of the females' offspring. Both experiments suggest that parasitism selects for sexual reproduction and increased genetic variability in mud snails.

Multiple Choice

1. b; **2.** d; **3.** a; **4.** a; **5.** b

Fill-in-the-Blank

1. four; gametes OR sperm and eggs
2. prophase, chiasmata; crossing over
3. shuffling of homologues, crossing over, union of gametes
4. alternation of generations; meiotic, mitotic
5. Turner; do not, cannot; Klinefelter; XXY

Chapter 11

Think Critically

Figure Captions

Figure 11-8 Half of the gametes produced by a *Pp* plant will have the *P* allele, and half will have the *p* allele. All of the gametes produced by a *pp* plant will have the *p* allele. Therefore, half of the offspring of a *Pp* × *pp* cross will be *Pp* (purple) and half will be *pp* (white), whereas all of the offspring of a *PP* × *pp* cross will be *Pp* (purple). See Figure 11-9.

Figure 11-11 A plant with wrinkled, green seeds has the genotype *ssyy*. A plant with smooth, yellow seeds could be *SSYY*, *SsYY*, *SSYy*, or *SsYy*. Set up four Punnett squares to see if the smooth yellow plant's genotype can be revealed by a test cross.

Figure 11-12 Chromosomes, not individual genes, assort independently during meiosis. Therefore, if the genes for seed color and seed shape were on the same chromosome, they would tend to be inherited together and would not assort independently.

Figure 11-13 A palomino has the genotype C_1C_2. The only way to ensure a palomino foal is to breed a cremello (C_2C_2) with a chestnut (C_1C_1).

Figure 11-25 One of Victoria and Albert's sons, Leopold, had hemophilia. To be male, Leopold must have inherited Albert's Y

chromosome. The X chromosome, not the Y chromosome, bears the gene for blood clotting, so Leopold must have inherited the hemophilia allele from his mother, Victoria.

Evaluate This

Health Watch: The Sickle-Cell Allele and Athletics

A sickling crisis is most likely to occur when blood oxygen concentrations drop. Strenuous exercise reduces blood oxygen, as does breathing "thin" air at high altitude. Playing football in Denver's mile-high altitude brings both risk factors into play. Even though Clark didn't experience any problems playing in Pittsburgh or the other NFL cities, playing in Denver would increase his risks.

Health Watch: Muscular Dystrophy

If the woman is a carrier, she is heterozygous for the defective dystrophin allele. Statistically, half of her eggs will receive the defective allele. Therefore, if her next child is a son, he has a 50% chance of having muscular dystrophy. If her next child is a daughter, she will not be affected (the father almost certainly does not have the defective allele, so daughters will receive at least one normal allele from him). Assuming that both of her parents were phenotypically normal, the woman inherited the defective allele from her mother, who would be heterozygous for the defective allele. Therefore, her sisters each have a 50% chance of being a carrier.

Multiple Choice

1. a; **2.** d; **3.** b; **4.** d; **5.** a

Fill-in-the-Blank

1. genotype, phenotype; heterozygous
2. independently; as a group; linked
3. XY, XX; sperm
4. sex linked
5. incomplete dominance; codominance; polygenic inheritance

Chapter 12

Think Critically

Figure Captions

Figure 12-6 It takes more energy to break apart a C–G base pair because these bases are held together by three hydrogen bonds, compared with the two hydrogen bonds that bind A to T.

Figure E12-6 DNA polymerase always moves in the 3′ to 5′ direction on a parental strand. Because the two strands of a DNA double helix are oriented in opposite directions, the 5′ direction on one strand leads toward the replication fork, and the 5′ direction on the other strand leads away from the fork. Therefore, DNA polymerase must move in opposite directions on the two strands.

How Do We Know That? DNA Is the Hereditary Molecule

If nucleic acid is the genetic material, then the protein coats of the offspring viruses should be encoded by the viral nucleic acid. Therefore, offspring viruses with normal RNA should have normal protein coats, and offspring viruses with HR RNA should have HR protein coats. That is what Fraenkel-Conrat found, confirming that nucleic acid (RNA), not protein, is the genetic material of TMV.

Multiple Choice

1. b; **2.** a; **3.** b; **4.** d; **5.** b

Fill-in-the-Blank

1. nucleotides; sugar (deoxyribose), phosphate, base (order not important)
2. phosphate, sugar (order not important); double helix
3. thymine, cytosine; complementary
4. semiconservative
5. DNA helicase; DNA polymerase; DNA ligase
6. mutations; nucleotide substitution mutation

Chapter 13

Think Critically

Figure Captions

Figure 13-3 RNA polymerase always travels in the 3′ to 5′ direction. Because the two DNA strands run in opposite directions, if the other DNA strand were the template strand, then RNA polymerase must travel in the opposite direction (that is, right to left in this illustration).

Figure 13-4 Cells produce far larger amounts of some proteins than others. Obvious examples include cells that produce antibodies or protein hormones, which they secrete into the bloodstream in vast quantities, affecting functioning throughout the body. If a cell needs to produce more of certain proteins, it will probably synthesize more mRNA that will be translated into that protein.

Figure 13-7 Grouped in codons, the original mRNA sequence visible here is CGA AUC UAG UAA. Changing all G to U would produce the sequence CUA AUC UAU UAA. The two changes are in the first codon (CGA to CUA) and the third codon (UAG to UAU). Refer to the genetic code (Table 13-3). First, CGA encodes arginine, while CUA encodes leucine, so the first G → U change would substitute leucine for arginine in the protein. Second, UAG is a stop codon, but UAU encodes tyrosine. Therefore, the second G → U change would add tyrosine to the protein instead of stopping translation. The final codon in the illustration, UAA, is a stop codon, so the new protein would end with tyrosine.

Health Watch: The Strange World of Epigenetics

Generally, methyl groups attached to DNA reduce transcription, and acetyl groups attached to histone proteins increase transcription. Therefore, you would expect that people with type 2 diabetes will probably have methylated DNA in the insulin gene and/or in its promoter. To increase insulin production, you could try to remove methyl groups on the insulin gene and/or its promoter or to add acetyl groups to histone proteins in the vicinity of the insulin gene.

Evaluate This

Health Watch: Androgen Insensitivity Syndrome

The gene for the androgen receptor is on the X chromosome. XY offspring inherit X chromosomes from their mother. Therefore, a girl with androgen insensitivity who has XY chromosomes must have inherited the mutated androgen receptor gene from her mother. Because her mother has XX chromosomes and can bear children, she must be heterozygous for androgen insensitivity; the father cannot have androgen insensitivity; otherwise, he would be phenotypically female. All future XX children will be phenotypically female and able to bear children, because they will inherit one X chromosome from their father, with a functional androgen receptor gene. Half of their XY children will have androgen insensitivity (because they inherited the X chromosome from their mother that has a mutated androgen receptor gene) and the other half will not have androgen insensitivity (because they inherited the X chromosome from their mother that has a functional androgen receptor gene). A Punnett square could be used to illustrate the probabilities.

Multiple Choice

1. b; **2.** a; **3.** a; **4.** c; **5.** d

Fill-in-the-Blank

1. transcription; translation; ribosome
2. messenger RNA, transfer RNA, ribosomal RNA (order not important); microRNA
3. three; codon; anticodon
4. RNA polymerase; template; promoter; termination signal
5. start, stop; transfer RNA; peptide
6. substitution; Insertion; Deletion

Chapter 14

Think Critically

Figure Captions

Figure 14-7 Each person normally has two copies of each STR gene, one on each of a pair of homologous chromosomes. A person may be homozygous (two copies of the same allele) or heterozygous (one copy of each of two alleles) for each STR. The bands on the gel represent individual alleles of an STR gene. Therefore, a single person can have one band (if homozygous) or two bands (if heterozygous). If a person is homozygous for an STR allele, then he has two copies of the same allele. The DNA from both (identical) alleles will run in the same place on the gel and, therefore, that (single) band will have twice as much DNA as each of the two bands of DNA from a heterozygote. The more DNA, the brighter the band.

Figure 14-9 The genetic material of bacteriophages is DNA. Each bacterial restriction enzyme cuts DNA at a specific nucleotide sequence. A given bacterium is likely to have evolved restriction enzymes that cut DNA at sequences found in bacteriophages but not in their own chromosome.

How Do We Know That? Prenatal Genetic Screening

Examine each STR in turn. TPOX: The child is homozygous for 10 repeats, so both the mother and father must have at least one allele with 10 repeats; either man could be the father. CSF: The child is heterozygous, with 6 and 8 repeats. If we assume that the mother contributed the 6-repeat allele, then the father must have contributed an 8-repeat allele. Again, either man could be the father. D5S: The child is heterozygous, with 9 and 12 repeats. The mother does not have a 12-repeat allele, so this allele must have come from the father. Man 1 does not have a 12-repeat allele, so he cannot be the father. Man 2 has a 12-repeat allele, so he could be the father. Continuing through the rest of the STRs in the same manner, we see that man 2 could be the father.

Earth Watch: What's Really in That Sushi?

Animals leave feces and hair (caught on thorns, for example) in their habitat. Barcoding hair (if the hair samples included bits of the follicles, which contain live cells with DNA) could reveal what species of animals are in the rain forest. Barcoding feces (which contain cells from both predators and their prey) could reveal what species of predators are in the forest, and what species of prey they eat.

Multiple Choice

1. a; **2.** c; **3.** c; **4.** a; **5.** b

Fill-in-the-Blank

1. Genetically modified organisms OR Transgenic organisms
2. Transformation; plasmids
3. polymerase chain reaction
4. short tandem repeats (STRs); length, or size, or number of repeats; DNA profile
5. electrophoresis OR gel electrophoresis; DNA probe, base pairing OR hydrogen bonds

Chapter 15

Think Critically

Figure Captions

Figure 15-6 No. Mutations, the ultimate source of the variation on which natural selection acts, occur in all organisms, including those that reproduce asexually.

Figure 15-7 No. Evolution can include changes in traits that are not revealed in morphology (the physical form of an organism), such as physiological systems and metabolic pathways. More generally, evolution in the sense of changes in a species' genetic makeup is inevitable; genetic evolution is not necessarily reflected in morphological change.

Figure 15-10 Analogous. Bird tails consist of feathers; dog tails do not (they consist of bone/muscle/skin). If the two structures were homologous, they would both consist of bone and muscle, or both consist of feathers.

How Do We Know That? Charles Darwin and the Mockingbirds

The species of mockingbird that descended from the birds that originally colonized the islands arose as birds gradually dispersed, moving from island to island away from the site of the original colonization.

Earth Watch: People Promote High-Speed Evolution

Natural selection favoring pesticide resistance is absent in the pesticide-free zones, so among the insects that live there, the frequency of alleles that confer resistance remains very low. If insects hatched in these areas later interbreed with resistant insects hatched in the pesticide-used zones, gene flow between the nonresistant and resistant subpopulations will result in a lower frequency of resistance alleles in the pesticide-used zones, thus slowing the evolution of resistance.

Multiple Choice

1. c; **2.** d; **3.** c; **4.** c; **5.** b

Fill-in-the-Blank

1. wing, arm; analogous, convergent; vestigial
2. common ancestor; amino acids, ATP
3. catastrophism; uniformitarianism; old
4. evolution; mutations, DNA
5. natural selection; artificial selection
6. many traits are inherited; Gregor Mendel

Chapter 16

Think Critically

Figure Captions

Figure 16-3 The surviving colonies would be in different places on each treated plate, and there would be a different number of surviving colonies on each plate (because the antibiotic-caused mutations would arise unpredictably, depending on which bacteria happened to interact with the antibiotic such that mutations were caused). Another possibility is that all of the colonies would survive (if the antibiotic always caused mutations in every colony).

Figure 16-4 For a locus with two alleles, one dominant and one recessive, there are two possible phenotypes. A mating between a heterozygote and a homozygote-recessive yields offspring with a 50:50 ratio of the two phenotypes.

	B	*b*
b	*Bb* (black)	*bb* (brown)
b	*Bb* (black)	*bb* (brown)

Figure 16-6 Mutations inevitably and continually add variability to a population, and after the population becomes larger, the counteracting, diversity-reducing effects of drift decrease. The net result is an increase in genetic diversity.

Figure 16-9 Greater for males. A female's reproductive success is limited by her maximum litter size, but a male's potential reproductive success is limited only by the number of available females. When, as in bighorn sheep, males battle for access to females, the most successful males can impregnate many females, while unsuccessful males may not fertilize any females at all. Thus, the difference between the most and least successful male can be very large. In contrast, even the most successful female can have only one litter of offspring per breeding season, which is not that many more offspring than a female who fails to reproduce.

Figure 16-11 There is always a limit to directional selection. As a trait becomes more extreme, eventually the cost of increasing it further outweighs the benefits (for example, the cost of obtaining extra food may outweigh the benefit of larger size), or it may be physically impossible for the trait to become more extreme (for example, a limb's length may be limited by the maximum length that a bone can attain without breaking under its own weight).

Earth Watch: The Perils of Shrinking Gene Pools

One way to counter loss of genetic diversity due to genetic drift in small populations is to foster gene flow between populations. Because genetic drift is a random process, for many genes, different alleles will be lost (and therefore different alleles will be present) in different small populations. Thus, gene flow can introduce new alleles to an isolated population and increase its genetic diversity. Such gene flow is promoted by habitat corridors, which allow individuals, spores, or seeds to move between habitat patches.

Case Study Revisited: Evolution of a Menace

Many species of bacteria and fungi that live in soil secrete antibiotic chemicals to help them compete for access to space and food. As a result of natural selection imposed by these poisons in their environment, some soil bacteria have evolved antibiotic resistance. Because the alleles that provide resistance may have negative side effects on the fitness of their possessors, they tend to be rare in environments that contain few antibiotic producers.

Evaluate This

Health Watch: Cancer and Darwinian Medicine

If you sequenced the genotypes in the original and new tumors of the relapsed patients, mutations that were present in both the original and new tumors could be the cause of the drug resistance. It is likely that these mutations would not be present in the original tumor of the patient who did not suffer a relapse.

Multiple Choice

1. c; **2.** b; **3.** b; **4.** d; **5.** c

Fill-in-the-Blank

1. Hardy–Weinberg principle, equilibrium, allele; no
2. alleles; nucleotides; mutations; homozygous, heterozygous
3. genotype, phenotype; phenotype
4. genetic drift; small; founder effect, population bottleneck; founder effect
5. species; natural selection, coevolution; adaptations
6. reproduction; environment

Chapter 17

Think Critically

Figure Captions

Figure 17-1 The key question is whether the gray-furred and black-furred squirrels interbreed freely. Tests would involve careful observation of the squirrels in areas where both types occur to check for mixed mating and perhaps genetic comparisons to determine the degree of gene flow between the two types. If hybrid matings are observed, it would be important to determine if hybrid offspring are viable and fertile.

Figure 17-8 Possibilities include continental drift; climate changes (especially glacial advances) that cause habitat fragmentation; formation of islands by volcanic activity or rising sea level; movements of organisms to existing islands (including "islands" of isolated habitats such as lakes, mountaintops, and deep-ocean vents); and formation of barriers to movement (e.g., new mountain ranges, deserts, rivers). These processes are indeed sufficiently common and widespread to account for a multitude of speciation events over the history of life.

Figure 17-10 The key question is whether the two populations inhabiting the two species of trees (apple and hawthorn) interbreed. Tests might involve careful observation of flies under natural conditions, lab experiments in which captive flies of the two types are provided with opportunities to interbreed, or genetic comparisons to determine the degree of gene flow between the two types of flies.

Figure 17-12 Small groups of organisms can colonize islands and become genetically isolated from the mainland population of their species. If such isolated island populations ultimately become separate species, the new species will (at least initially) be endemic to the island on which speciation occurred. Populations of species endemic to islands, especially small islands, are likely to be small. Species with small populations are at higher risk of extinction.

Figure 17-13 Natural selection cannot look forward and ensure that the only traits that evolve are those that ensure survival of the species as a whole. Instead, natural selection ensures only the preservation of traits that help individuals survive and reproduce more successfully than individuals lacking the trait. So if, in a particular species, highly specialized individuals survive and reproduce better than less-specialized individuals, the specialized phenotype will eventually dominate, even if it ultimately puts the species at greater risk of extinction.

How Do We Know That? Seeking the Secrets of the Sea

The map reveals areas in which multiple species are concentrated for extended time periods. It would be helpful to protect these "hotspots."

Case Study Revisited: Discovering Diversity

The negative effects of very small population size on species will act more quickly than any potential new speciation could occur. Very small populations are at high risk of extinction, in part due to the negative effects of inbreeding and loss of genetic diversity through genetic drift (see Chapter 16). Even if new species were to arise in isolated populations, populations of the new species would be small and at risk of extinction.

Multiple Choice

1. a; **2.** b; **3.** a; **4.** a; **5.** c

Fill-in-the-Blank

1. populations, independently; reproductive isolation; asexually
2. behavioral isolation, hybrid inviability, temporal isolation, gametic incompatibility, mechanical incompatibility
3. genetically isolated, diverge; allopatric speciation; genetic drift, natural selection
4. adaptive radiation; habitat (or environment)
5. small, specialized; habitat destruction

Chapter 18
Think Critically
Figure Captions

Figure 18-2 The presence of oxygen would prevent the accumulation of organic compounds by quickly oxidizing them or their precursors. All of the successful abiotic synthesis experiments used oxygen-free "atmospheres."

Figure 18-5 The bacterial sequence would be most similar to that of the plant mitochondrion, because (as the descendant of the immediate ancestor of the mitochondrion) the bacterium shares with the mitochondrion a more recent common ancestor than it does with the chloroplast or the nucleus.

Figure 18-8 Today's ferns, horsetails, and club mosses are small most likely because of competition with seed plants, which had not yet arisen during the period when ferns and club mosses reached large sizes. After seed plants arose, competition from them eventually eliminated other types of plants from many ecological niches, presumably including those niches that favored evolution of large size.

Figure 18-9 No. The mudskipper merely demonstrates the plausibility of a hypothetical intermediate step in the proposed scenario for the origin of land-dwelling tetrapods. But the existence of a modern example similar in form to the hypothetical intermediate form does not provide information about the actual identity of that intermediate form.

Figure 18-19 The African replacement hypothesis. These fossils are the oldest modern humans found so far, and their presence in Africa suggests that modern humans were present in Africa before they were present anywhere else, which, if true, would mean that they originated in Africa.

How Do We Know That? Discovering the Age of a Fossil

713 million years old. (1:1 ratio means that $\frac{1}{2}$ of the original uranium-235 is left, so it has reached its half-life.)

Multiple Choice

1. d; **2.** b; **3.** b; **4.** b; **5.** a

Fill-in-the-Blank

1. anaerobic; photosynthesize; poisonous (toxic, harmful), aerobic, energy
2. RNA (ribonucleic acid), enzymes (catalysts), ribozymes
3. eukaryotic; endosymbiotic; DNA (deoxyribonucleic acid)
4. swimming, moist (wet); pollen

5. conifers; wind; flowers, insects; efficient
6. arthropods, exoskeletons, drying
7. reptiles, eggs, skin, lungs

Chapter 19
Think Critically
Figure Captions

Figure 19-3 The finding suggests that chromosome 2 arose from the fusion of two separate chromosomes, each of which contained a centromere. The two ancestral chromosomes must have been present in the common ancestor of chimpanzees and humans, and chimpanzees retained the separate chromosomes.

Figure 19-5 Over their long evolutionary histories, both bacteria and archaea retained a single-celled, prokaryotic structure, the complexity of which is limited by a lack of organelles. This simple structure limits the variety of forms that such an organism can take and still survive. The limited array of possible options made the evolution of similar forms in the two domains likely.

Multiple Choice

1. b; **2.** a; **3.** d; **4.** c; **5.** c

Fill-in-the-Blank

1. taxonomy; systematics; clade
2. genus, species; Latin; capitalized, italic
3. domain, kingdom, phylum, class, order, family, genus, species; Archaea, Bacteria, Eukarya
4. anatomy, DNA sequence
5. reproduce asexually; phylogenetic species concept
6. 1.6 million, 8.7 million

Chapter 20
Think Critically
Figure Captions

Figure 20-4 Protective structures like endospores are most likely to evolve in environments in which protection is especially advantageous. Compared to other environments inhabited by bacteria, soils are especially vulnerable to drying out, which can be fatal to unprotected bacteria. Bacteria that could resist long dry periods would gain an evolutionary advantage.

Figure 20-5 Enzymes from bacteria that live in hot environments are active at high temperatures (temperatures that usually denature enzymes in organisms from more temperate environments). This ability to function at high temperatures makes the enzymes useful in test tube reactions (such as the polymerase chain reaction) that are run at high temperatures.

Figure 20-7 The main advantage is efficiency. In prokaryotic fission, every individual produces new individuals. In sexual reproduction, only some individuals (e.g., females) produce offspring. So the average individual of a species produces twice as many offspring by fission as it would by sexual reproduction.

Figure 20-9 The concentration of nitrogen gas would increase, because the major process for removing atmospheric nitrogen would end, while the processes that add nitrogen gas to the atmosphere would continue.

Figure 20-12 Viruses lack ribosomes and the rest of the "machinery" required to manufacture proteins.

Figure 20-13 Viruses replicate by integrating their genetic material into the host cell's genome. Thus, if biotechnologists can insert foreign genetic material into a virus, the virus will naturally tend to transfer the foreign genes to the cells they infect.

Evaluate This

Health Watch: Is Your Body's Ecosystem Healthy?

Given the increasing evidence that the composition of the microbiome is associated with the health of the digestive system, we might expect to find that the microbiomes of the healthy twins in the sample differ from those of the sick twins. Because a microbiome is an ecosystem, and healthier ecosystems typically have greater species diversity, the microbiomes of the healthy twins might well be more diverse than those of the sick twins. The researchers compared sets of twins to help rule out the possibility that the differences between the healthy and sick twins in a pair were due to genetic differences between them (identical twins are genetically identical).

Multiple Choice

1. a; **2.** d; **3.** b; **4.** c; **5.** a

Fill-in-the-Blank

1. Bacteria, cell walls, archaea
2. smaller; spherical, rod-shaped, corkscrew-shaped
3. flagella; biofilms; endospores
4. Anaerobic; Photosynthetic
5. prokaryotic fission, conjugation
6. nitrogen-fixing; cellulose
7. bacteria; meat, eggs, produce (order not important)
8. DNA, RNA (order not important), protein; host; bacteriophage

Chapter 21

Think Critically

Figure Caption

Figure 21-1 Sex is the process that combines the genomes of two different individuals. In plants and animals, this mixing of genomes occurs only during reproduction. But in many protists (and prokaryotes), genome mixing may occur through conjugation and other processes that take place independently of reproduction (which in many cases occurs by mitotic cell division).

Evaluate This

Health Watch: Neglected Protist Infections

Testing for Chagas disease would probably be most appropriate for women who have spent time in areas where triatomine bugs are present or whose parents have lived in such areas. The best protection from the disease for people who live in affected areas is to take steps to protect homes against triatomine bug incursions. Toxiplasmosis is not so clearly tied to a particular risk factor, so more widespread testing during pregnancy might be warranted. However, women with cats in their households may be at particular risk. To reduce risk of infection, you could advise pregnant women to avoid litter boxes and gardening, to wash produce thoroughly, and to consume only fully cooked meat.

Multiple Choice

1. c; **2** b; **3.** d; **4.** a; **5.** d

Fill-in-the-Blank

1. decomposers, parasites
2. algae; protozoa
3. secondary endosymbiosis, photosynthetic protist
4. apicomplexan (or alveolate); kinetoplastid (or euglenozoan)
5. water mold (or oomycete or stramenopile); amoebozoan
6. dinoflagellates, diatoms; chlorophytes

Chapter 22

Think Critically

Figure Captions

Figure 22-5 Bryophytes lack lignin (which provides stiffness and support) and xylem and phloem (conducting tissues that transport materials to distant parts of the body). Xylem, phloem, and stiff stems seem to be required to achieve heights greater than a few inches.

Figure 22-7 All of the pictured structures are sporophytes. In ferns, horsetails, and club mosses, the gametophyte is small and inconspicuous.

Figure 22-9 The most common adaptations are hard, protective shells and incorporation of toxic and/or distasteful chemicals.

Figure 22-12

Type of Pollination	Advantages	Disadvantages
Wind	Not dependent on presence of animals; no investment in nectar or showy flowers; pollen can disperse over large distances	Larger investment in pollen because most fails to reach an egg; higher chance of failure to fertilize any egg
Animal	Each pollen grain has much greater chance of reaching suitable egg	Depends on presence of animals; must invest in nectar and showy flowers

Both types of pollination persist in angiosperms because the cost–benefit balance, and therefore the most adaptive pollination system, differs depending on the ecological circumstances of a species.

Health Watch: Green Lifesaver

Although the initial study took advantage of the traditional healer's experience in devising effective methods of preparing and administering herbal remedies, the design of a follow-up study should probably include methods to standardize preparations and treatments so as to test the effects of different preparations and different dosages. It would also be a good idea to introduce control groups, double-blind testing procedures, and standardized methods for assessing patient condition.

Case Study Revisited: Queen of the Parasites

Photosynthesis is a very useful adaptation, but it comes with costs, such as the energy expenditure required to access and acquire the nutrients needed to produce the molecules and structures used in photosynthesis. In environments in which the necessary nutrients are scarce or competition for them is especially intense, a nonphotosynthetic plant might gain an advantage, provided it evolved an alternative means of acquiring energy, such as by stealing it from other photosynthetic plants.

Multiple Choice

1. a; **2.** c; **3.** b; **4.** a; **5.** d

Fill-in-the-Blank

1. green algae; nonvascular plants (bryophytes), vascular plants (tracheophytes); embryos, alternation of generations
2. cuticle, stomata, water; xylem and phloem (order not important), lignin, water, nutrients
3. swim to the egg; seeds, pollen; gymnosperms, angiosperms; attract pollinators; facilitate seed dispersal
4. hornworts, liverworts, mosses; club mosses, horsetails, ferns; angiosperms

Chapter 23

Think Critically

Figure Captions

Figure 23-1 Its filamentous shape helps the fungal body penetrate and extend into its food sources and also maximizes the ratio of surface area to interior volume (which maximizes the area available for absorbing nutrients). The extreme thinness of the filaments ensures that no cell is very far from the surface at which nutrients are absorbed.

Figure 23-15 Compared to plants that lack mycorrhizae, plants that have them are much more effective at absorbing water and nutrients from the soil. Therefore, when early plants were first spreading over Earth's land, individuals whose roots were associated with fungi would have gained an advantage over those that lacked such associations.

Figure 23-20 In nature, bacteria compete with fungi for access to food and living space. The antibiotic chemicals produced by fungi serve as a defense against competition from bacteria. New antibiotics are most likely to be found in environments in which fungi and bacteria coexist, such as soil. One might begin a search by testing extracts of various candidate fungi to see if they kill bacterial cultures of different types.

Earth Watch: Killer in the Caves

The animal most likely to carry white-nose syndrome from cave to cave is people. Spelunkers (people who explore caves) can carry fungal spores between caves on their boots and equipment. So one way to slow the spread of the disease would be to ban people from exploring caves (and in fact such bans are currently in place in many places).

Multiple Choice

1. c; **2.** d; **3.** b; **4.** c; **5.** b

Fill-in-the-Blank

1. reproduction; spores
2. mycelium, hyphae, septa; chitin
3. mitotic, one; meiotic, zygote, one
4. glomeromycetes; chytrids, rumen fungi, and blastoclades; basidiomycetes
5. Lichens; Mycorrhizae; endophytes
6. wood (cellulose and lignin); yeasts; athlete's foot, jock itch, vaginal infections, ringworm, histoplasmosis, valley fever (many possible answers)

Chapter 24

Think Critically

Figure Captions

Figure 24-6 Sponges are "primitive" only in the sense that their lineage arose early in the evolutionary history of animals and their body plan is comparatively simple. But early origin and simplicity do not determine effectiveness, and the sponge body plan and way of life are clearly suitable for excellent survival and reproduction in many habitats.

Figure 24-13 Parasitic tapeworms have no gut and absorb nutrients across their body surfaces. Their ribbon-like shape maximizes surface area for absorption and allows the worm body to extend through the greatest possible area of the host's body (to be in contact with as many nutrients as possible).

Figure 24-14 Two openings allow one-way travel of food through the gut, which allows continuous feeding. One-way movement allows more efficient digestion than two-way movement; digestive waste from which all nutrients have been extracted can be excreted quickly without the need for reverse travel back along the gut, and food can be processed more quickly.

Figure 24-15 Water travels easily through the moist epidermis of a leech. When a high concentration of salt is dissolved in the moisture on the outside of a leech's body, water moves rapidly out of the leech's body by osmosis, dehydrating and ultimately killing the animal.

Figure 24-24 The mobility provided by flight may have allowed ancestral insects to more easily exploit new food sources, habitats, and geographic areas. This ability to disperse would have promoted formation of new species and increased population sizes.

How Do We Know That? The Search for a Sea Monster

Widder hypothesized that a giant squid's preferred food is not jellyfish, but the small (compared to a giant squid) predators that eat jellyfish. The observation that a giant squid attacked an object near the e-jelly lure rather than the lure itself is consistent with the hypothesis's prediction that giant squids will attack animals near jellyfish, rather than jellyfish themselves.

Earth Watch: When Reefs Get Too Warm

Sediments can negatively affect reef-building corals in two ways. (1) Suspended sediments make the water murky so that less light passes through, reducing the light available to fuel photosynthesis in corals' dinoflagellate symbionts. (2) Sediments can settle directly on the corals, forcing them to spend a lot of energy keeping their surfaces clean, or even burying them so completely that they suffocate.

Multiple Choice

1. b; **2.** a; **3.** b; **4.** c; **5.** a

Fill-in-the-Blank

1. consuming other organisms; sexually; cell walls
2. three, endoderm, mesoderm, ectoderm; two, mesoderm
3. cephalized; sensing the environment, ingesting food; cavity, mesoderm
4. protostome; deuterostome, echinoderms, chordates
5. invertebrates, vertebrates; invertebrates; sponges, single-celled organisms; cnidarians; annelids
6. bivalves, gastropods, cephalopods; arthropods; insects, arachnids, crustaceans
7. annelids, arthropods (chordates is also correct); closed circulatory; open circulatory, hemocoel
8. mollusks, cnidarians, echinoderms, arthropods

Chapter 25

Think Critically

Figure Captions

Figure 25-8 A freshwater fish's body is immersed in a hypotonic solution, so water tends to continuously enter the body by osmosis. The physiological challenge is to get rid of all this excess water. For a saltwater fish, the challenge is reversed. The surrounding solution is hypertonic, so water tends to leave the body. The physiological challenge is to retain sufficient water.

Figure 25-10 One advantage is that adults and juveniles occupy different habitats and therefore do not compete with one another for resources (the niche occupied by an individual over its lifetime is broadened).

Figure 25-13 Flight is a very expensive trait (consumes a lot of energy, requires many special structures). In circumstances in which the benefits of flight are low, such as in habitats without predators, natural selection may favor individuals that forgo an investment in flight, and flightlessness can arise.

Earth Watch: Frogs in Peril

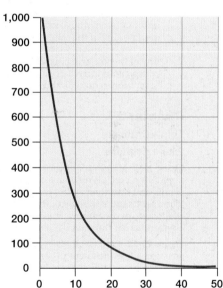

On the graph in the figure, the annual rate of decrease for endangered species appears to be about 12%. Assuming a decrease of 12% per year, the population after 10 years will be about 278. (Remember that each year's starting population differs from the starting population from the prior year.) A graph of population size extrapolated to 50 years shows that the population will shrink to nearly zero, assuming a constant rate of decrease (that is, exponential decrease).

Multiple Choice

1. c; **2.** b; **3.** d; **4.** c; **5.** b

Fill-in-the-Blank

1. hollow, dorsal; anus, notochord
2. tunicates, lancelets, hagfishes (order not important); skull; hagfishes
3. lungfishes; cartilage; ray-finned fishes; jaws
4. mammals, amphibians, reptiles, amphibians
5. monotremes; salamanders; bats

Chapter 26

Think Critically

Figure Captions

Figure 26-1 One possibility is that the variation necessary for selection has never arisen. (If no members of the species by chance gain the ability to discriminate between their own chicks and cuckoo chicks, then selection has no opportunity to favor the novel behavior.) Another possibility is that the cost of the behavior is relatively low. (If parasitism by cuckoos is rare, a parent that feeds any begging chick in its nest is, on average, much more likely to benefit than suffer.)

Figure 26-24 If females do indeed gain any benefits from their mates, those benefits would necessarily be genetic (as the male provides no material benefits). Male fitness may vary, and to the extent that this fitness can be passed to offspring, females would benefit by choosing the most fit males. If a male's fitness is reflected in his ability to build and decorate a bower, females would benefit by preferring to mate with males that build especially good bowers.

Figure 26-25 Canines forage mainly by smell; mandrills are mostly visual foragers. Modes of sexual signaling are affected not only by the nature of the information to be encoded, but also by the sensory biases and sensitivities of the species involved. Communication systems may evolve to take advantage of traits that originally evolved for other functions.

Multiple Choice

1. c; **2.** a; **3.** b; **4.** a; **5.** d

Fill-in-the-Blank

1. genes, environment; stimulus; innate
2. energy, predators; practice, skills; adult; brains
3. habituation, repeated; imprinting; sensitive period
4. aggressive; displays, injuring (wounding, damaging); weapons (such as fangs, claws), larger
5. territoriality (territorial behavior); mate, raise young, feed, store food; male; same species
6. For pheromones; any of the following advantages are correct: long lasting, use little energy to produce, are species specific, don't attract predators, may convey messages over long distances. Any of the following disadvantages are correct: cannot convey changing information, convey fewer different types of information. For visual displays, any of the following advantages are correct: instantaneous, can change rapidly, many different messages can be sent, quiet. Any of the following disadvantages are correct: make animals conspicuous to predators, generally ineffective in darkness or dense vegetation, require that recipient be in visual range (close by).

Chapter 27
Think Critically
Figure Captions

Figure 27-3 Many variables interact in complex ways to produce real population cycles. Weather, for example, affects the lemmings' food supply and thus their ability to survive and reproduce. Predation of lemmings is influenced by both the number of predators and the availability of other prey, which in turn is influenced by multiple environmental variables.

Figure 27-11 Emigration relieves population pressure in an overpopulated area, spreading the migrating animals into new habitats that may have more resources. Human emigration within and between countries is often driven by the desire or need for more resources, although social factors—such as wars and religious or racial persecution—also fuel human emigration. (This subjective question can lead to discussion of the extent to which overpopulation may drive human emigration.)

Figure 27-18 When fertility exceeds RLF, there are more children than parents. As the additional children mature and become parents themselves, this more numerous generation produces still more children, and so on, in a positive feedback cycle.

Figure 27-19 U.S. population growth resembles the rapidly rising "exponential" phase of the S-curve. Stabilization will require some combination of reduction in immigration rates and birth rates. An increase in death rates is less likely, but cannot be ruled out entirely. People often have fewer children during recessions, so the "great recession" of 2008 through 2014 probably reduced the U.S. birth rate temporarily.

Earth Watch: Boom-and-Bust Cycles Can Be Bad News

In the short term (while a bloom is occurring), educating people about the dangers of the bloom (for example, don't eat shellfish

from the affected area) and/or destroying the bloom, if possible, are the only realistic alternatives. In the long run, preventing blooms provides both human and ecological health benefits. Investigate the causes of and remedies for HABs, for example, at the mouth of the Mississippi River. What social and economic costs would be required to prevent blooms?

Earth Watch: Have We Exceeded Earth's Carrying Capacity?

Answers will vary, but they should consider the following questions: What are the ecological and economic impacts of producing the product in its native country? Of shipping the product overseas? What domestic products could substitute for the imported product? Would the domestic product have a smaller ecological impact than the imported product?

Multiple Choice

1. b; **2.** c; **3.** a; **4.** d; **5.** c

Fill-in-the Blank

1. survivorship curves; early-loss; late-loss
2. exponential; no; J-curve; no
3. carrying capacity; logistic, S-
4. clumped; uniform; random
5. births, immigrants, deaths, emigrants; increases

Chapter 28
Think Critically
Figure Captions

Figure 28-6 Monarch caterpillars store toxic chemicals from the milkweeds they eat, so they would likely kill or at least sicken birds or other predators that might eat them. However, the caterpillar would still be dead. The bold stripes of monarch caterpillars probably evolved as a warning to predators, advertising that they are toxic. When a predator has eaten and been sickened by one monarch caterpillar, it would probably not want to eat another. The evolution of bright colors would make it easier for predators to learn to avoid monarch caterpillars. This would enhance survival of the monarchs.

Figure 28-7 Many predators hunt by detecting odors, sounds, or even electric fields, in addition to, or instead of, sighting their prey visually.

Figure 28-8 Ancestors of these organisms that happened to slightly resemble their surroundings would be a bit less likely to be eaten, so they would be slightly more likely to reproduce and leave offspring. Chance mutations that increased their camouflage would enhance their survival and reproduction even more, eventually resulting in the modern, superb camouflage.

Figure 28-16 Fire has been a natural part of the forest environment for many thousands of years. Some forest plants might directly depend on fire for reproduction (for example, opening cones to release seeds) or might benefit from clearing out the trees, which lets sunlight hit the ground and reduces competition for water and nutrients. Fire might maintain some forests in permanent, or recurring, subclimax stages.

Health Watch: Parasitism, Coevolution, and Coexistence

Mosquitoes, even those of the same species, have some genetic variability, including having differences in their preferred time and place of feeding. For simplicity, let's assume that the bed nets are 100% effective, and that people are never bitten while sleeping beneath the net. A mosquito that flew about, especially inside houses, earlier in the day would have the opportunity to feed on people before they went to bed beneath a net. These "early" mosquitoes would reproduce more often than the rest of the mosquito population, passing their genes for early feeding on to their offspring. Eventually, one would expect that early mosquitoes would dominate the population, rendering bed nets ineffective.

Case Study Revisited: The Fox's Tale

If the ecosystem is severely degraded, for example by overgrazing that almost completely eliminates native vegetation, then it would be difficult for succession to restore the original community. This is especially true on islands, which may be separated by many miles from sources of native seeds. If invasive species are introduced simultaneously with overgrazing, then the bare soil would provide an opening for seeds of invasive plants to sprout and possibly take over permanently.

Multiple Choice

1. a; **2.** d; **3.** c; **4.** d; **5.** b

Fill-in-the- Blank

1. natural selection; coevolution
2. carnivores, herbivores; camouflage
3. competitive exclusion
4. warning coloration, startle coloration, Batesian mimicry, aggressive mimicry
5. mutualism (or symbiosis), parasitism, predation, mutualism
6. succession; primary succession; secondary succession; climax; subclimax

Chapter 29
Think Critically
Figure Captions

Figure 29-2 On land, high productivity is supported by optimal temperatures for plant growth, a long growing season, and plenty of moisture, such as is found in rain forests. Lack of water limits desert productivity. In aquatic ecosystems, high productivity is supported by an abundance of nutrients and adequate light, such as is found in estuaries. Lack of nutrients limits the productivity of the open ocean, even in well-lit surface waters.

Figure 29-8 Humanity's need to grow crops to feed our growing population has led to the fixing of nitrogen for fertilizer using industrial processes. Additionally, large-scale livestock feedlots generate enormous amounts of nitrogenous waste. Nitrogen oxides are also generated when fossil fuels are burned in power plants, vehicles, and factories, and when forests are burned. Consequences include the overfertilization of lakes and rivers and the creation of dead zones in coastal waters that receive excessive nutrient runoff from land. Another important consequence is acid deposition, in which nitrogen oxides formed by combustion produce nitric acid in the atmosphere; this acid is then deposited on land.

Figure 29-14 Global temperatures would not begin to decline immediately. CO_2 stays in the atmosphere for years, providing a long lag time before its contribution to the greenhouse effect would decline significantly. Further, there are other greenhouse gases, such as methane, produced by human activities. Finally, large amounts of CO_2 are stored in the oceans; no one really knows how much of this CO_2 might be emitted into the atmosphere.

How Do We Know That: Monitoring Earth's Health

It isn't a trivial task to "eyeball" a trend line (scientists use computer programs). Further, not all trends are linear: the yearly increase in atmospheric CO_2 concentrations, for example, has been getting larger over time. Nevertheless, if you draw a straight line through the data for Arctic sea ice and, say, the last 25 years for CO_2 concentrations, you will find that the data predict the Arctic to become ice-free sometime in the 2050s, and CO_2 concentrations to have doubled by about 2070—both probably within your lifetime. Actually, if major changes in human activities do not occur, these effects may happen sooner. For example, ice reflects more than half of the sunlight that hits it, whereas open water absorbs more than 90%. Therefore, as Arctic sea ice declines in area, more open water exists, which absorbs sunlight, heating the water and melting more ice, creating even more open water, which warms up even more, and so on, so the decline in sea ice may accelerate. On the other hand, if people use more renewable energy and less fossil fuel energy, then CO_2 levels may eventually stabilize or even drop, slowing down or reversing the loss of ice.

Evaluate This
Health Watch: Biological Magnification of Toxic Substances

Refer Victoria to the "food" tab on the U.S. Food and Drug Administration's Web site for information on the mercury content of seafoods. She will find that catfish and salmon have the least mercury; light meat tuna (mostly from skipjack) have intermediate amounts; and white meat tuna (from

albacore) have the most. Other websites can help to explain why. Assuming that the fish live in water of similar mercury content, there are two major factors that contribute to the amount of mercury in fish: (1) Their average trophic level: the lower the level, the less mercury. (2) Their age when they are eaten: on average, older fish have eaten more prey, and have had more chance to accumulate mercury. Most catfish in your supermarket are farmed fish, eaten at about 2 years old, that feed at a low trophic level (algae, aquatic plants, insects and crayfish). Salmon are mostly 2 to 4 years old when eaten (farmed salmon are eaten when younger, and have less mercury), and feed at a moderately low trophic level (mollusks, small fish, and crustaceans that feed on phytoplankton). Skipjack tuna are typically eaten when 2 to 3 years old, and feed at a somewhat higher trophic level than salmon do. Albacore tuna are eaten when 2 to 4 years old, and feed at a still higher trophic level. Albacore are also extremely active, and eat a great deal of food—as much as 25% of their body weight daily—which exposes them to a higher total dose of mercury.

Multiple Choice

1. c; **2.** a; **3.** d; **4.** d; **5.** c

Fill-in-the-Blank

1. sunlight, photosynthesis; nutrients, nutrient cycles
2. autotrophs, producers; net primary production
3. trophic levels; food chain; food webs
4. 10
5. heterotrophs, consumers; herbivores, primary consumers; carnivores, secondary consumers; detritivores, decomposers
6. nitrogen-fixing bacteria, denitrifying bacteria; ammonia, nitrate
7. atmosphere, oceans; CO_2 (carbon dioxide); limestone, fossil fuels

Chapter 30
Think Critically
Figure Captions

Figure 30-8 Nutrients are abundant in tropical rain forests, but they are not stored in the soil. The optimal temperature and moisture of tropical climates allow plants to make such efficient use of nutrients that nearly all nutrients are stored in plant bodies, and to a lesser extent, in the bodies of the animals they support. These growing conditions support a vast array of plants, and these, in turn, provide a wealth of habitats and food sources for diverse animals.

Figure 30-18 Tropical deciduous forests have a dry season in which the soil has little moisture. Temperate deciduous forests have freezing winter weather, in which soil moisture becomes frozen and unavailable to the trees. In both cases, dropping the leaves reduces the trees' water loss at a time of year when the water cannot be replaced by absorbing water from the soil.

Figure 30-27 Coastal ecosystems have an abundance of the two limiting factors for life in water: nutrients and light to support photosynthetic organisms. Both upwelling from ocean depths and runoff from the land can provide nutrients, depending on the location of the ecosystem. The shallow water in these areas allows adequate light to penetrate to support rooted plants and/or anchored algae, which in turn provide food and shelter for a wealth of marine life.

Earth Watch: Plugging the Ozone Hole

Ozone depletion would lead to higher UV radiation, which would reduce photosynthesis. Less photosynthesis means that less CO_2 would be removed from the atmosphere. More CO_2 remaining in the atmosphere would increase the greenhouse effect, leading to more global warming and other changes in climate. Increased CO_2 in the atmosphere would also cause more CO_2 to dissolve in the oceans, increasing ocean acidification.

Multiple Choice

1. c; **2.** b; **3.** d; **4.** a; **5.** c

Fill-in-the-Blank

1. seasons; ocean currents *or* the presence of the ocean; rain shadow
2. suitable temperatures, availability of liquid water; light, nutrients, suitable temperature
3. tropical rain forests; coral reefs

4. littoral zone; phytoplankton, zooplankton; limnetic zone, profundal zone; oligotrophic; eutrophic; wetlands

5. phytoplankton; chemosynthesis, hydrogen sulfide; high pressure

Chapter 31

Think Critically

Figure Captions

Figure 31-5 Any rare organism faces a greater chance of extinction in a small reserve. For example, large carnivores need a large prey population to support them (see discussion of energy pyramids in Chapter 29), which in turn needs a large area of suitable habitat. Large herbivores are likely to be rare for similar reasons; for example, it takes a very large area of vegetation to support a population of elephants. A chance event (storm, fire, disease) may well kill all of the members of such a population in a small reserve. The effect on the ecosystem will depend on which species disappear. The loss of large predators, as in Yellowstone, may cause a cascading effect in which herbivore populations skyrocket, to the detriment of populations of at least some plants and the animals that depend on them.

Figure 31-8 A small reserve will have small populations of many organisms, especially large, predatory animals. Small populations typically lose genetic diversity through genetic drift (see Chapter 16). As genetically diverse organisms migrate through corridors, they help to reduce genetic drift and maintain genetic diversity.

Earth Watch: Whales—The Biggest Keystones of All?

In the absence of whaling, whale populations would be expected to increase. If the whale pump hypothesis is correct, then iron, nitrogen, and perhaps other nutrients in surface waters should increase, which should increase populations of phytoplankton and perhaps krill.

Earth Watch: Saving Sea Turtles

With a low turtle population, food supplies were probably abundant. Listing leatherbacks as endangered and enforcing other protection measures meant that they were much less likely to be killed in fishing nets. When turtle nesting beaches in Florida were protected, people no longer disturbed the nests and eggs were no longer collected. Abundant food, less predation by fishing, and protected nesting sites greatly increased turtle survival and reproduction, allowing exponential population growth (see Chapter 27). Exponential growth cannot continue indefinitely. Density-dependent factors usually slow down population growth in long-lived species such as sea turtles. These factors might include increasing natural predation, for example, by gulls and raccoons learning that the beaches with large numbers of turtle nests are easy sources of food during turtle nesting season. As turtle populations increase, food supplies in the ocean might also become depleted.

Multiple Choice

1. a; **2.** b; **3.** a; **4.** d; **5.** b

Fill-in-the-Blank

1. genetic, species, ecosystem; genetic

2. ecosystem services; provisioning services, regulating services, cultural services, supporting services

3. ecological economics

4. habitat destruction, overexploitation, invasive species, pollution, global warming; habitat destruction

5. minimum viable population; habitat fragmentation; wildlife corridors

6. sustainable

CREDITS

Photo Credits

Unit Openers: Unit 1: Volker Steger/Christian Bardele/Science Source; Unit 2: Erik Lam/Shutterstock; Unit 3: Thomas Kitchin/Victoria Hurst/Getty Images; Unit 4: Rodger Klein/WaterFrame/AGE Fotostock; Unit 5: Erik Isakson/AGE Fotostock; Unit 6: Terry Audesirk

Chapter 1 opener: Misha Hussain/Reuters; 1-CO-inset: Frederick A. Murphy/CDC; 1-1: Mary Martin/Biophoto Associates/Science Source; 1-2: Melba Photo Agency/Alamy; 1-4: Rido/Shutterstock; 1-5: Terry Audesirk; 1-6a: Dr. Tony Brain/Science Source; 1-6b: Terry Audesirk; 1-6c: Eric Baccega/AGE Fotostock; 1-7: NASA; 1-8: Penny Tweedie/Alamy; 1-9: Jose Maria Farfagl/Wenn Photos/Newscom; 1-12: John Durham/Science Source; Have You Ever Wondered? box: Joel Sartore/Getty Images

Chapter 2 opener: Air Photo Service/ABACAUSA.COM/Newscom; 2-4b: Piotr Sosnowski; 2-8a: Basil/Shutterstock; 2-8b: Danita Delimont/Alamy; 2-8c: Terry Audesirk; 2-10: Renn Sminkey/Pearson Education, Inc.; 2-12: Vicky Kasala/SuperStock; E2-1a: National Institutes of Health; E2-2a–b: Science Source; E2-3: Hunor Kristo/Fotolia; E2-4-1-2: Terry Audesirk

Chapter 3 opener: Terry Audesirk; 3-9a: Jeremy Burgess/Science Source; 3-10a: Mike Norton/Shutterstock; 3-10b: Jeremy Burgess/Science Source; 3-10c: Biophoto Associates/Science Source; 3-11: Terry Audesirk; 3-12a: Tracy Morgan/DK Images; 3-12b: B G Smith/Shutterstock; 3-12c: Xiaodong Ye/iStock/Getty Images; 3-18: Fotokostic/Shutterstock; 3-23a: Sylvie Bouchard/Fotolia; 3-23b: Terry Audesirk; 3-24a-b: Terry Audesirk; 3-25: Tischenko Irina/Shutterstock; E3-1: Terry Audesirk; E3-3: GJLP/Science Source; E3-4-1: VersusStudio/Fotolia; E3-4-2: Valua Vital/Fotolia

Chapter 4 opener: Matt Dunham/AP Images; E4-1-a1: Cecil H. Fox/Science Source; E4-1-a2: The Project Gutenberg Literary Archive Foundation; E4-1b: The Print Collector/Alamy; E4-1c: Pacific Northwest National Laboratory; E4-2a: M. I. Walker/Science Source; E4-2b: SPL/Science Source; E4-2c: CNRI/Science Source; E4-2d: Science Source; UN-1: Stacie Kirsch/Electron Microscopy Sciences; 4-3b-c: Science Source; 4-3d: John Cardamone Jr./Biological Photo Service; 4-3e: Proceedings of the National Academy of Sciences; 4-6b: Steve Gschmeissner/Science Source; 4-7b: Jennifer Waters/Science Source; 4-8a: Don W. Fawcett/Science Source; 4-8b: Charles Daghlian/Science Source; 4-8c: David M. Phillips/Science Source; 4-9b: Dr. Elena Kiseleva/Science Source; 4-10: Melba/AGE Fotostock; E4-3: Toby Melville/Reuters; 4-11: Oscar L. Miller Jr.; 4-12b, bottom: Don W. Fawcett/Science Source; 4-12b, top: MedImage/Science Source; 4-13: Meredith Carlson/Biophoto Associates/Science Source; 4-16b: Walter Dawn/Science Source; 4-17: The Keith R. Porter Endowment for Cell Biology; 4-18: Robin Treadwell/Science Source; 4-19: Biophoto Associates/Science Source

Chapter 5 opener: Fred LaBounty/Alamy; E5-1: Ted Kinsman/Science Source; E5-3: El Pais Photos/Newscom; E5-4a: Maarten J. Chrispeels and Peter Agre. Elsevier; 5-6a: David M. Phillips/Science Source; 5-6b: Amar/Phanie/AGE Fotostock; 5-6c: David M. Phillips/Science Source; 5-7a-b: Nigel Cattlin/Science Source; 5-9b: Don W. Fawcett/Science Source; 5-10b-1-3: The Company of Biologists Ltd; 5-11b: Eric V. Grave/Science Source; 5-11c: SPL/Science Source; 5-12: Linda Hufnagel; 5-14a,c: Don W. Fawcett/Science Source; 5-14b: Claude and Goodenough. *The Journal of Cell Biology*; 5-14d: Biology Pics/Science Source; 5-15a: Justin Schwartz; 5-15b: American College of Physicians; 5-15b inset: Larry F Jernigan/Getty Images

Chapter 6 opener: Lucas Jackson/Reuters; 6-1: David Wall/Alamy; 6-3: Terry Audesirk; 6-5a: Terry Audesirk; Have You Ever Wondered? box right: Cathy Keifer/Shutterstock; Have You Ever Wondered? box left: Bioglow/REX/AP Images; E6-2: Terry Audesirk; 6-14: Terry Audesirk

Chapter 7 opener: Detlev van Ravensway/Science Source; 7-2a-b: Jeremy Burgess/Science Source; 7-3a: Haveseen/Shutterstock; 7-3c: John Durham/Science Source; 7-3d: Robin Treadwell/Science Source; 7-6: Terry Audesirk; E7-1 left: Terry Audesirk; E7-1 center: Martin Haas/Shutterstock; E7-1 right: Terry Audesirk; E7-2 left: Mathisa/Shutterstock; E7-2 center, right: Terry Audesirk; E7-4 inset: Eric Gevaert/Shutterstock; E7-4: Beawiharta/Reuters

Chapter 8 opener: University of Leicester/Rex Fe/AP Images; 8-CO-inset: The Print Collector/Glow Images; 8-4: CNRI/Science Source; 8-8: Mara Zemgaliete/Fotolia; E8-3 left: Tassii/Getty Images; E8-3 right: William Leaman/Alamy; 8-10: Walter Bieri/EPA/Newscom; 8-12: Terry Audesirk; 8-13: Rui Vieira/PA Wire/AP Images

Chapter 9 opener: Mitchell Layton/Getty Images; 9-3a: Hazel Appleton/Health Protection Agency Centre for Infections/Science Source; 9-3b: M. I. Walker/Science Source; 9-3c: Biophoto Associates/Science Source; 9-3d: Terry Audesirk; 9-4: Lee Hoon-Koo/AFP/Getty Images; 9-6: Andrew Syred/Science Source; 9-9a: Jennifer Waters/Science Source; 9-9b: Science Source; 9-9c-g: Jennifer Waters/Science Source; 9-9h: Michael Davidson/Molecular Expressions; E9-1: Du Cane Medical Imaging Ltd./Science Source

Chapter 10 opener: Mikael Buck Photography Limited; 10-1: Biophoto Associates/Science Source; E10-1a: Paddy Ryan; 10-13a: Science Source; 10-13b: Philidor/Fotolia

Chapter 11 opener: Ronald C. Modra/Getty Images; 11-2: Science Source; 11-14: Science Source; 11-15: Sarah Leen/National Geographic Stock; 11-16: David Hosking/Alamy; UN-1: Deanne Fitzmaurice/AP Photo; 11-19: Indigo Instruments; 11-21: Gerry Audesirk; 11-23a: Friedrich Stark/Alamy; 11-23b: Jvoisey/Getty Images; 11-24a-b: Omikron/Science Source; E11-1: Newscom; 11-25: Mary Evans/Science Source; E11-2: Yves Herman/Thomson Reuters (Markets) LLC; E11-3: Vetpathologist/Shutterstock; 11-26: David J. Phillip/AP Photo

Chapter 12 opener: Yuri Maselov/Alamy; 12-CO-inset: David Bagnall/Alamy; 12-4: Science Source; 12-5: National Cancer Institute; 12-6: Kenneth Eward/BioGrafx/Science Source; E12-6: Science Source; 12-11: Bruce Stotesburgy/Times Colonist

Chapter 13 opener: Martineau, Luke; 13-4: Oscar L.Miller, Jr.; 13-6b: Annual Reviews, Inc.; E13-1: AIS-DSD Support Group; E13-2: Randy Jirtle; E13-11a-c: Bo Hong et al. PNAS; 13-12: Gavni/Getty Images; 13-13: Thomas Lohnes/Newscom

Chapter 14 opener: P. Kevin Morley/Richmond Times-Dispatch/AP Images; 14-4: Janet Chapple/Granite Peak Publications; 14-7 left, right: Dr Margaret Kline/National Institute of Standards and Technology; 14-8: Mark Thiessen/National Geographic Image Collection/Alamy; E14-1: Andy Schlink; E14-2-1: Mark Stoeckle; E14-2-2: Daniel Prudek/Shutterstock; E14-2-3: Jan S./Shutterstock; E14-2-4: Studiotouch/Shutterstock; E14-2-5: Stubblefield Photography/Shutterstock; 14-11: ZEPHYR/SPL/Science Photo Library /Alamy; 14-12: Monsanto Company; 14-13: Tim Flach/Stone/Getty Images; 14-14: Stephen Lock/ZUMA Press/Newscom; E14-3: Lada/Science Source; E14-5: International Rice Research Institute; 14-17: Richmond Times Dispatch, Carl Lynn/AP Images

Chapter 15 opener: Tony Heald/Nature Picture Library; 15-4a: Scott Orr/Getty Images; 15-4b: Woudloper; 15-4c: Smithsonian Institution Libraries; How Do We Know That? box: SuperStock; E15-1-1: Ingo Schulz/AGE Fotostock; E15-1-2: Stefan Huwiler/imageBROKER/AGE Fotostock; E15-1-3: Therin Weise/Arco Images/AGE Fotostock; E15-1-4: Ross Hoddinott/Nature Picture Library; E15-2: Letz/SIPA/Newscom; 15-6: NHPA/Science Source; 15-10a: mite/Fotolia; 15-10b: miyamotokei/Fotolia; 15-11a-c: TT Nyhetsbyrån AB; 15-13a: John Pitcher/Getty Images; 15-13b: GK Hart/Vikki Hart/Getty Images; 15-14a-b: Professor Robin M. Tinghitella, University of Denver; E15-3: H. Reinhard/Arco Images/Alamy; 15-15: VEM/Science Source

Chapter 16 opener: Science Source; 16-6b: Fogstock/AGE Fotostock; 16-7: The Alan Mason Chesney Medical Archives; 16-8: Johann Schumacher/Getty Images; E16-2: M. Watson/Ardea.c/Mary Evans Picture Library Ltd/AGE Fotostock; UN-1: Danita Delimont/Alamy; 16-9: NHPA/SuperStock; 16-10: zjk/Fotolia; 16-12: Smith, Thomas B.

Chapter 17 opener: National News/ZUMAPRESS/Newscom; 17-1a-b: Jeremiah Easter/Tetrachromat Design; 17-2a: Rick and Nora Bowers/Alamy; 17-2b: Tim Zurowski/AGE Fotostock; 17-3a-b: Thomas Leeson; Pat Leeson/Science Source; 17-4: National Geographic Stock; 17-5a: Ken Owen/Channel Islands Restoration; 17-5b: Hey Paul/Public domain; 17-6: Phil Savoie/Nature Picture Library; 17-7: Boaz Rottem/Alamy; E17-1: Newscom; 17-9a: Newscom; 17-9b: Mark Gurney/Smithsonian/Getty Images; 17-12a: Goss Images/Alamy; 17-12b: A.C. Medeiros/Carr Botanical Consultation; 17-12c: Jack Jeffrey/Photo Resource Hawaii/Alamy; 17-12d: Gerald D. Carr/Carr Botanical Consultation; 17-13: Steve Apps/Alamy; E17-3: Crack Palinggi/Reuters

Chapter 18 opener: Keren Su/Getty Images; 18-4: Mark Garlick/Science Source; 18-6: Michael Plewka; 18-7a: Chase Studio, OMNH; 18-7b: Peter Halasz; 18-7c: Grauy/Getty Images; 18-7d: Douglas Faulkner/Science Source; 18-8: Richard Bizley/Nature Picture Library; 18-9: Terry Whittaker/Science Source; 18-10: Science Source; 18-12a: Tom McHugh/Science Source; 18-12b Kevin Schafer/Alamy; 18-12c: TUNS/Arco Images GmbH TUNS/Arco Images GmbH/Glow Images;

Text Credits

INDEX

Citations followed by *b* refer to material in boxes; citations followed by *f* refer to material in figures or illustrations; and citations followed by *t* refer to material in tables.